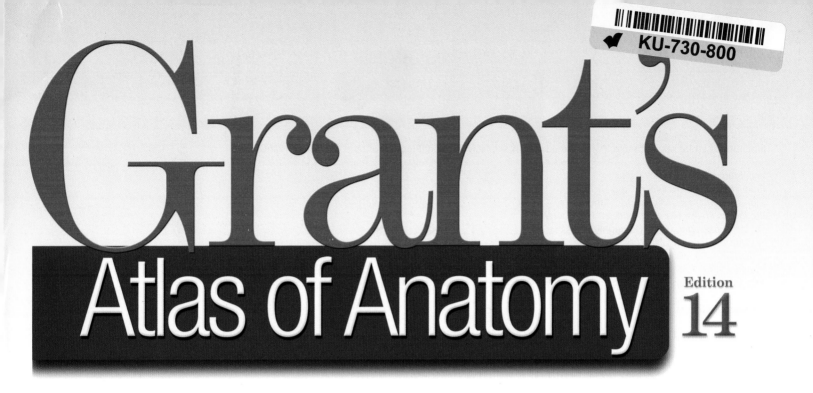

Grant's
Atlas of Anatomy

Edition 14

ANNE M.R. AGUR, BSc(OT), MSc, PhD

Professor, Division of Anatomy, Department of Surgery, Faculty of Medicine
Division of Physical Medicine and Rehabilitation, Department of Medicine
Department of Physical Therapy, Department of Occupational Science and Occupational Therapy
Division of Biomedical Communications, Institute of Medical Science
Rehabilitation Sciences Institute, Graduate Department of Dentistry
University of Toronto
Toronto, Ontario, Canada

ARTHUR F. DALLEY II, PhD, FAAA

Professor, Department of Cell and Developmental Biology
Adjunct Professor, Department of Orthopaedic Surgery
Vanderbilt University School of Medicine
Adjunct Professor of Anatomy
Belmont University School of Physical Therapy
Nashville, Tennessee

 Wolters Kluwer

Philadelphia • Baltimore • New York • London
Buenos Aires • Hong Kong • Sydney • Tokyo

Not authorised for sale in United States, Canada, Australia, New Zealand, Puerto Rico, and U.S. Virgin Islands.

Acquisitions Editor: Crystal Taylor
Product Development Editor: Greg Nicholl
Marketing Manager: Michael McMahon
Production Project Manager: Bridgett Dougherty
Design Coordinator: Holly McLaughlin
Art Director: Jennifer Clements
Artist/Illustrator: Nick Woolridge, Nicole Clough, Marissa Webber
Manufacturing Coordinator: Margie Orzech
Prepress Vendor: Absolute Service, Inc.

Fourteenth Edition

Library of Congress Cataloging-in-Publication Data

Names: Agur, A. M. R., author. | Dalley, Arthur F., II, author.
Title: Grant's atlas of anatomy / Anne M.R. Agur, Arthur F. Dalley II.
Other titles: Atlas of anatomy
Description: Fourteenth edition. | Philadelphia : Wolters Kluwer, [2017] |
   Includes bibliographical references and index.
Identifiers: LCCN 2015042750 | ISBN 9781469890685
Subjects: | MESH: Anatomy, Regional—Atlases.
Classification: LCC QM25 | NLM QS 17 | DDC 611.0022/2—dc23 LC record available at http://lccn.loc.gov/2015042750

RRS1511

Dr. John Charles Boileau Grant

1886–1973

by Dr. Carlton G. Smith, MD, PhD (1905–2003)
Professor Emeritus, Division of Anatomy,
  Department of Surgery
Faculty of Medicine, University of Toronto,
  Toronto, Ontario, Canada

Dr. J.C. Boileau Grant in his office, McMurrich Building, University of Toronto, 1946. Through his textbooks, Dr. Grant made an indelible impression on the teaching of anatomy throughout the world. (Courtesy of Dr. C. G. Smith.)

The life of Dr. J.C. Boileau Grant has been likened to the course of the seventh cranial nerve as it passes out of the skull: complicated but purposeful.[1] He was born in the parish of Lasswade in Edinburgh, Scotland, on February 6, 1886. Dr. Grant studied medicine at the University of Edinburgh from 1903 to 1908. Here, his skill as a dissector in the laboratory of the renowned anatomist, Dr. Daniel John Cunningham (1850–1909), earned him a number of awards.

Following graduation, Dr. Grant was appointed the resident house officer at the Infirmary in Whitehaven, Cumberland. From 1909 to 1911, Dr. Grant demonstrated anatomy in the University of Edinburgh, followed by 2 years at the University of Durham, at Newcastle-on-Tyne in England, in the laboratory of Professor Robert Howden, editor of *Gray's Anatomy*.

With the outbreak of World War I in 1914, Dr. Grant joined the Royal Army Medical Corps and served with distinction. He was mentioned in dispatches in September 1916, received the Military Cross in September 1917 for "conspicuous gallantry and devotion to duty during attack," and received a bar to the Military Cross in August 1918.[1]

In October 1919, released from the Royal Army, he accepted the position of Professor of Anatomy at the University of Manitoba in Winnipeg, Canada. With the frontline medical practitioner in mind, he endeavored to "bring up a generation of surgeons who knew exactly what they were doing once an operation had begun."[1] Devoted to research and learning, Dr. Grant took interest in other projects, such as performing anthropometric studies of Indian tribes in northern Manitoba during the 1920s. In Winnipeg, Dr. Grant met Catriona Christie, whom he married in 1922.

Dr. Grant was known for his reliance on logic, analysis, and deduction as opposed to rote memory. While at the University of Manitoba, Dr. Grant began writing *A Method of Anatomy, Descriptive and Deductive*, which was published in 1937.[2]

In 1930, Dr. Grant accepted the position of Chair of Anatomy at the University of Toronto. He stressed the value of a "clean" dissection, with the structures well defined. This required the delicate touch of a sharp scalpel, and students soon learned that a dull tool was anathema. Instructive dissections were made available in the Anatomy Museum, a means of student review on which Dr. Grant placed a high priority. Illustrations of these actual dissections are included in *Grant's Atlas of Anatomy*.

The first edition of the *Atlas*, published in 1943, was the first anatomical atlas to be published in North America.[3] *Grant's Dissector* preceded the *Atlas* in 1940.[4]

Dr. Grant remained at the University of Toronto until his retirement in 1956. At that time, he became Curator of the Anatomy Museum in the University. He also served as Visiting Professor of Anatomy at the University of California at Los Angeles, where he taught for 10 years.

Dr. Grant died in 1973 of cancer. Through his teaching method, still presented in the Grant's textbooks, Dr. Grant's life interest—human anatomy—lives on. In their eulogy, colleagues and friends Ross MacKenzie and J. S. Thompson said, "Dr. Grant's knowledge of anatomical fact was encyclopedic, and he enjoyed nothing better than sharing his knowledge with others, whether they were junior students or senior staff. While somewhat strict as a teacher, his quiet wit and boundless humanity never failed to impress. He was, in the very finest sense, a scholar and a gentleman."[1]

[1]Robinson C. *Canadian Medical Lives: J.C. Boileau Grant: Anatomist Extraordinary.* Ontario, Canada: Associated Medical Services Inc/Fithzenry & Whiteside, 1993.

[2]Grant JCB. *A Method of Anatomy: Descriptive and Deductive.* Baltimore, MD: Williams & Wilkins Co, 1937.

[3]Grant JCB. *Grant's Atlas of Anatomy.* Baltimore, MD: Williams & Wilkins Co, 1943.

[4]Grant JCB, Cates HA. *Grant's Dissector (A Handbook for Dissectors).* Baltimore, MD: Williams & Wilkins Co, 1940.

Reviewers

RADIOLOGIC FIGURE CONTRIBUTORS

Joel A. Vilensky, PhD
Professor, Department of Anatomy and Cell Biology
Indiana University School of Medicine
Fort Wayne, Indiana

Edward C. Weber, DO
The Imaging Center
Fort Wayne, Indiana

FACULTY REVIEWERS

Ernest Adeghate, MD, PhD, DSc
Professor and Chair
College of Medicine and Health Sciences
United Arab Emirates University
Al-Ain, United Arab Emirates

Jean-pol Beauthier, MD, PhD
Professor of Forensic Pathology
Université libre de Bruxelles
Brussels, Belgium

Jennifer A. Carr, PhD
Preceptor
Harvard University
Cambridge, Massachusetts

Donald J. Fletcher, PhD
Professor and Vice Chair
Department of Anatomy and Cell Biology
Brody School of Medicine, East Carolina University
Greenville, North Carolina

Douglas J. Gould, PhD
Professor and Vice Chair
Department of Biomedical Sciences
William Beaumont School of Medicine, Oakland University
Rochester, Michigan

Robert Hage, MD, PhD, DLO, MBA
Professor and Co-chair
School of Medicine, St. George's University
Grenada, West Indies

Jonathan Kalmey, PhD
Assistant Dean of Preclinical Education, Professor of Anatomy
Lake Erie College of Osteopathic Medicine
Erie, Pennsylvania

Randy J. Kulesza, PhD
Associate Professor
Lake Erie College of Osteopathic Medicine
Erie, Pennsylvania

Diana Rhodes, DVM, PhD
Professor of Anatomy and Chair
Department of Anatomy
Pacific Northwest University of Health Sciences
Yakima, Washington

Bruce Wainman, PhD
Associate Professor, Pathology and Molecular Medicine
Director, Education Program in Anatomy
McMaster University
Ontario, Canada

STUDENT REVIEWERS

Todd Christensen
University of Medicine and Health Sciences, St. Kitts

Margaret Connolly
Tufts University School of Medicine

Laura Deschamps
Philadelphia College of Osteopathic Medicine

Kyle Diamond
Charles E. Schmidt College of Medicine

Dustun Field
Trinity School of Medicine

Tripp Hines
James H. Quillen College of Medicine,
East Tennessee State University

Kimber Johnsen
University of Medicine and Health Sciences, St. Kitts

Nalin Lalwani
University of Medicine and Health Sciences, St. Kitts

Amy Leshner
St. George's, University of London

Garren Low
Keck School of Medicine of USC

Milcris N. Calderon Maduro
Ponce Health Sciences University School of Medicine

Katherine Morganti
Louisiana State University Health Sciences Center, Shreveport

Elizabeth Nelson
University of Utah School of Medicine

Fabian Nelson
Avalon University School of Medicine

Nina Nguyen
Université de Sherbrooke

Preface

This edition of *Grant's Atlas* has, like its predecessors, required intense research, market input, and creativity. It is not enough to rely on a solid reputation; with each new edition, we have adapted and changed many aspects of the *Atlas* while maintaining the commitment to pedagogical excellence and anatomical realism that has enriched its long history. Medical and health sciences education, and the role of anatomy instruction and application within it, continually evolve to reflect new teaching approaches and educational models. The health care system itself is changing, and the skills and knowledge that future health care practitioners must master are changing along with it. Finally, technologic advances in publishing, particularly in online resources and electronic media, have transformed the way students access content and the methods by which educators teach content. All of these developments have shaped the vision and directed the execution of this fourteenth edition of *Grant's Atlas*, as evidenced by the following key features.

Recolorization of the original carbon-dust *Grant's Atlas* images from high-resolution scans. The entire collection of carbon-dust illustrations were remastered and recolored for the fourteenth edition using a vibrant new palette. The stunning detail and contrast of the original Grant's art was maintained while adding a new level of luminosity of organs and especially transparency of tissues, enabling demonstrations of deeper relationships not possible with merely recolored grayscale illustrations, thereby enhancing the student learning experience. The student is able to visualize and appreciate clearly the newly revealed relationships between structures, enabling the formation of three-dimensional (3D) constructs for each region of the body. The recolorization, enabled by modern image processing, allows reproduction and viewing of the images—both in print and electronically—with unprecedented high resolution and fidelity, continuing their vital role informing future generations of medical and health care providers about the structure and function of the human body.

A unique feature of *Grant's Atlas* is that rather than providing an idealized view of human anatomy, the classic illustrations represent actual dissections that the student can directly compare with specimens in the lab. Because the original models used for these illustrations were real cadavers, the accuracy of these illustrations is unparalleled, offering students the best introduction to anatomy possible.

Schematic illustrations. Updated for the fourteenth edition with a modern uniform style and consistent color palette, the full-color schematic illustrations and orientation figures supplement the dissection figures to clarify anatomical concepts, show the relationships of structures, and give an overview of the body region being studied.

The illustrations conform to Dr. Grant's admonition to "keep it simple": Extraneous labels were deleted, and some labels were added to identify key structures and make the illustrations as useful as possible to students.

Legends with easy-to-find clinical applications. Admittedly, artwork is the focus of any atlas; however, the *Grant's* legends have long been considered a unique and valuable feature of the *Atlas*. The observations and comments that accompany the illustrations assist orientation and draw attention to salient points and significant structures that might otherwise escape notice. Their purpose is to interpret the illustrations without providing exhaustive description. Readability, clarity, and practicality were emphasized in the editing of this edition. Clinical comments, which deliver practical "pearls" that link anatomical features with their significance in health care practice, appear in blue text within the figure legends. New clinical comments based on current practices have been added in this edition, providing even more relevance for students searching for medical application of anatomical concepts.

Enhanced diagnostic imaging and surface anatomy. Because medical imaging has taken on increased importance in the diagnosis and treatment of injuries and illnesses, diagnostic images are used liberally throughout and at the end of each chapter. Over 100 clinically significant magnetic resonance images (MRIs), computed tomography (CT) scans, ultrasound scans, and corresponding orientation drawings are included, many of which are new to or updated for this edition. Labeled surface anatomy photographs which, like the illustrations, feature ethnic diversity continue to be an important feature in this new edition.

Updated and improved tables. Tables help students organize complex information in an easy-to-use format ideal for review and study. In addition to muscles, tables summarizing nerves, arteries, and other relevant structures are included. Tables are made more meaningful with illustrations strategically placed on the same page, demonstrating the structures and relationships described in the tables.

Logical organization and layout. The organization and layout of the *Atlas* have always been determined with ease of use as the goal. In this edition, to facilitate dissection, the body regions have been reordered in the same sequence as the more recent and current editions of *Grant's Dissector*. The order of plates within every chapter was scrutinized to ensure that it is logical and pedagogically effective.

We hope that you enjoy using this fourteenth edition of *Grant's Atlas* and that it becomes a trusted partner in your educational experience. We believe that this new edition safeguards the *Atlas's* historical strengths while enhancing its usefulness to today's students.

Anne M.R. Agur
Arthur F. Dalley II

Recoloring *Grant's Atlas*

The principal illustrations for *Grant's Atlas*, created in the 1940s and 1950s, use classic techniques of carbon dust or wash in pure grayscale. Although the detail of the grayscale carbon-dust illustrations was outstanding (see below figure on the left), the need for color was soon obvious. Early editions of the *Atlas* layered solid colors over parts of the grayscale artwork to highlight the presence and relationships of important structures such as veins, arteries, and nerves. This didactic approach and technology persisted throughout the first eight editions.

In the early 1990s, the *Atlas* was revised using a complex pre-digital technique where the original illustrations were photographed and printed on photographic paper. The prints were then colorized by hand with photo dyes, and the resulting colored prints were rephotographed for reproduction in print. Although this process resulted in a significant enrichment of the illustrations, the technique sometimes led to loss of detail and reduction of contrast. Over the next several editions, the color of the digital images were adjusted and enhanced (see below figure in the middle).

In the late 1990s, the University of Toronto assumed care of the original illustrations. The illustrations had been handled roughly over their long lives and were in many cases deteriorating due to their non-archival substrates. In 2008, an interdisciplinary team[1] of communications scholars, illustrators, and archivists applied for and received funding from the Social Sciences and Humanities Research Council of Canada to study the illustrations and to create a digital archive of the corpus. The team catalogued, documented, and scanned the artifacts at high resolution. The effort revealed a number of "lost" illustrations among the more than 1,000 images. Some of these images have been restored to the current edition.

Once the database of high-resolution images was compiled, the possibility arose to "remaster" and recolor the images for the next edition of *Grant's Atlas*. A system was set up to clean the images and create new layers of color.

- Almost all of the original illustrations contained hand-lettered labels and leader lines that had to be removed. This was accomplished by the careful use of digital cloning and retouching tools.
- The tonal range and contrast was adjusted to maximize clarity and dynamic range.
- A series of color layers were added over the cleaned scans, based on a carefully chosen palette. Most layers were set to the color transfer mode, which was chosen to assure that the grayscale balance of the underlying scans would not be altered.
- All of the recolored illustrations went through numerous rounds of revision with the authors to assure accuracy and reflect the pedagogic needs of the new edition.

This work was overseen by Nicholas Woolridge and carried out by two graduates of the Master of Science in Biomedical Communications (MScBMC) program: Nicole Clough and Marissa Webber. The retouching process was designed to preserve the detail, texture, and contrast of the original artwork (see below image on the right), allowing the illustrations to continue informing students about the structure and function of the human body for decades to come.

Nicholas Woolridge
Director, Master of Science
in Biomedical Communications Program
University of Toronto
September 2015

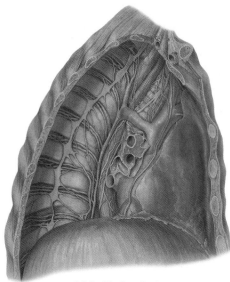

Original Carbon-Dust

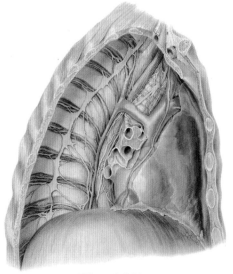

**Thirteenth Edition
with Added Color**

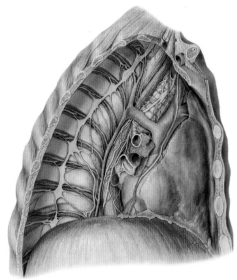

**Fourteenth Edition
with Enhanced Color and Detail**

[1]Led by Kim Sawchuk, from Concordia University, and included Nancy Marrelli, Nicholas Woolridge, Brian Sutherland, Nina Czegledy, Mél Hogan, Dave Mazierski, and Margot Mackay.

Acknowledgments

Starting with the first edition of *Grant's Atlas* published in 1943, many people have given generously of their talents and expertise and we acknowledge their participation with heartfelt gratitude. Most of the original carbon-dust halftones on which this book is based were created by Dorothy Foster Chubb, a pupil of Max Brödel and one of Canada's first professionally trained medical illustrators. She was later joined by Nancy Joy. Mrs. Chubb was mainly responsible for the artwork of the first two editions and the sixth edition; Professor Joy, for those in between. In subsequent editions, additional line and halftone illustrations by Elizabeth Blackstock, Elia Hopper Ross, and Marguerite Drummond were added. In recent editions, the artwork of Valerie Oxorn and the surface anatomy photography of Anne Rayner of Vanderbilt University Medical Center's Medical Art Group have augmented the modern look and feel of the atlas.

Much credit is also due to Charles E. Storton for his role in the preparation of the majority of the original dissections and preliminary photographic work. We also wish to acknowledge the work of Dr. James Anderson, a pupil of Dr. Grant, under whose stewardship the seventh and eighth editions were published.

The following individuals also provided invaluable contributions to previous editions of the atlas and are gratefully acknowledged: C.A. Armstrong, P.G. Ashmore, D. Baker, D.A. Barr, J.V. Basmajian, S. Bensley, D. Bilbey, J. Bottos, W. Boyd, J. Callagan, H.A. Cates, S.A. Crooks, M. Dickie, C. Duckwall, R. Duckwall, J.W.A. Duckworth, F.B. Fallis, J.B. Francis, J.S. Fraser, P. George, R.K. George, M.G. Gray, B.L. Guyatt, C.W. Hill, W.J. Horsey, B.S. Jaden, M.J. Lee, G.F. Lewis, I.B. MacDonald, D.L. MacIntosh, R.G. MacKenzie, S. Mader, K.O. McCuaig, D. Mazierski, W.R. Mitchell, K. Nancekivell, A.J.A. Noronha, S. O'Sullivan, V. Oxorn, W. Pallie, W.M. Paul, D. Rini, C. Sandone, C.H. Sawyer, A.I. Scott, J.S. Simpkins, J.S. Simpson, C.G. Smith, I.M. Thompson, J.S. Thompson, N.A. Watters, R.W. Wilson, B. Vallecoccia, and K. Yu.

FOURTEENTH EDITION

We are indebted to our students, colleagues, and former professors for their encouragement—especially Joel Vilensky, Sherry Downie, Ryan Splittgerber, Mitchell T. Hayes, Edward Weber, and Douglas J. Gould for their invaluable input.

We wish to thank Dr. Joel A. Vilensky and Dr. Edward C. Weber for their contribution of new images to update and enhance the imaging sections of this edition.

We extend our gratitude to Professors Nick Woolridge and David Mazerski who developed the carbon-dust recolorization process and along with Nicole Clough and Marissa Webber who recolorized all of the carbon-dust images. Their artistic skills and anatomical insights made substantial contributions to this edition. We would also like to acknowledge Jennifer Clements, Art Director at Wolters Kluwer, who managed the art program for this edition.

Special thanks go to everyone at Wolters Kluwer—especially Crystal Taylor, Senior Acquisitions Editor, and Greg Nicholl, Senior Product Development Editor. We also thank Bridgett Dougherty, Production Project Manager. All of your efforts and expertise are much appreciated.

We would like to thank the hundreds of instructors and students who have over the years communicated via the publisher and directly with the editor their suggestions for how this *Atlas* might be improved. Finally, we would like to acknowledge the reviewers who reviewed previous editions of the *Atlas* as well as the reviewers who reviewed the fourteenth edition and provided expert advice on the development of this edition.

Contents

List of Tables

Figure and Table Credits

CHAPTER 1
Back

Figures 1.3D&E, 1.4, and 1.17B. Modified from Moore KL, Dalley AF, Agur MR. *Clinically Oriented Anatomy*, 7th ed. Baltimore, MD: Lippincott Williams & Wilkins, 2014.

Figures 1.7A–D, 1.9A,B,D&E, 1.14B, 1.15C, 1.18A–C, 1.19A&B, 1.21A&B, 1.31A–E, 1.32A–D, 1.38C, 1.41A&C, 1.42A&B, 1.45B, 1.46A–E, 1.47, 1.48, and 1.49A&B. Modified from Moore KL, Agur MR, Dalley AF. *Essential Clinical Anatomy*, 5th ed. Baltimore, MD: Lippincott Williams & Wilkins, 2015.

Figure 1.8A&B. Courtesy of J. Heslin, University of Toronto, Ontario, Canada.

Figures 1.8C&D and 1.50C. Courtesy of D. Armstrong, University of Toronto, Ontario, Canada.

Figures 1.9C and 1.53A–D. Courtesy of D. Salonen, University of Toronto, Ontario, Canada.

Figure 1.43A–E. Modified from Tank PW, Gest TR. *Lippincott Williams & Wilkins Atlas of Anatomy*. Baltimore, MD: Lippincott Williams & Wilkins, 2009.

Figures 1.50A&B, 1.51A&B, and 1.52A&B. Courtesy of the Visible Human Project; National Library of Medicine; Visible Man 1805.

CHAPTER 2
Upper Limb

Figures 2.3A,B,D,&E, 2.5A&B, 2.7A–D, 2.19, 2.22B, 2.25B, 2.34F, 2.45C, 2.48B, 2.53D, 2.61A&B, 2.62, 2.70B, 2.72D, 2.73, 2.80, 2.81A&B, 2.86C&D, 2.87D, and Table 2.8. Modified from Moore KL, Agur MR, Dalley AF. *Essential Clinical Anatomy*, 5th ed. Baltimore, MD: Lippincott Williams & Wilkins, 2015.

Figures 2.4A–C, 2.6, 2.8A–D, 2.9A&B, 2.12A&B, 2.13A–C, 2.23B&C, 2.24A&B, 2.29B, 2.44B, 2.47B&D, and 2.67B. Modified from Moore KL, Dalley AF, Agur MR. *Clinically Oriented Anatomy*, 7th ed. Baltimore, MD: Lippincott Williams & Wilkins, 2014.

Figure 2.10. Modified from Tank PW, Gest TR. *Lippincott Williams & Wilkins Atlas of Anatomy*. Baltimore, MD: Lippincott Williams & Wilkins, 2009.

Figures 2.18A–D, 2.31A–D, 2.33D, 2.35A–D, 2.63A, 2.64A, 2.65A, 2.72A–C, and 2.83A&B. Modified from Clay JH, Pounds DM. *Basic Clinical Massage Therapy*. Baltimore, MD: Lippincott Williams & Wilkins, 2002.

Figures 2.24C and 2.90F. Courtesy of D. Armstrong, University of Toronto, Ontario, Canada.

Figures 2.48C, 2.55B, 2.96A–C, 2.97B–D, and 2.98A–C. Courtesy of D. Salonen, University of Toronto, Ontario, Canada.

Figures 2.48D and 2.99B. Courtesy of R. Leekam, University of Toronto and West End Diagnostic Imaging, Ontario, Canada.

Figure 2.54A&B (MRIs). Courtesy of J. Heslin, University of Toronto, Ontario, Canada.

Figure 2.90C&D. Courtesy of E. Becker, University of Toronto, Ontario, Canada.

CHAPTER 3
Thorax

Figures 3.7B, 3.14A&B, 3.15B, 3.19, 3.20, 3.27A–C, 3.28A,C,&D, 3.34A–F, 3.43C, 3.48A–C, 3.49A&D, 3.50A&C, 3.53A–C, 3.60C, 3.65A–C, 3.71A&B, 3.77E, and 3.78F&H. Modified from Moore KL, Agur MR, Dalley AF. *Essential Clinical Anatomy*, 5th ed. Baltimore, MD: Lippincott Williams & Wilkins, 2015.

Figures 3.14C, 3.15A, 3.28A, 3.51A&C–E, 3.52A&B, 3.54B, 3.55B, 3.56B&C, 3.57C, 3.58B, 3.70, and 3.72B. Modified from Moore KL, Dalley AF, Agur MR. *Clinically Oriented Anatomy*, 7th ed. Baltimore, MD: Lippincott Williams & Wilkins, 2014.

Figures 3.43B&E, 3.49C, and 3.57B. Courtesy of I. Verschuur, Joint Department of Medical Imaging, UHN/Mount Sinai Hospital, Toronto, Ontario, Canada.

Figure 3.50B&D. Courtesy of I. Morrow, University of Manitoba, Canada.

Figure 3.51B. Courtesy of Dr. J. Heslin, Toronto, Ontario, Canada.

Figure 3.52C. Feigenbaum H, Armstrong WF, Ryan T. *Feigenbaum's Echocardiography*, 5th ed. Philadelphia, PA: Lippincott Williams & Wilkins, 2005:116.

Figure 3.64B. Courtesy of Dr. E.L. Lansdown, University of Toronto, Ontario, Canada.

Figures 3.79A–E, 3.80A&B, and 3.81A&B. Courtesy of Dr. M.A. Haider, University of Toronto, Ontario, Canada.

CHAPTER 4
Abdomen

Figures 4.3, 4.5, 4.7A, 4.10D&E, 4.17A–E, 4.18, 4.20C, 4.22B, 4.24A&B, 4.27B, 4.31A–C, 4.32A, 4.33A&B, 4.35A, 4.44 (insets), 4.51B&C, 4.54A, 4.55, 4.66A, 4.72A, 4.76B, 4.79C, 4.80A–D, 4.81, 4.83, 4.85A&B, 4.89A,B,&D–F, and 4.93A–C (schematics on left). Modified from Moore KL, Agur MR, Dalley AF. *Essential Clinical Anatomy*, 5th ed. Baltimore, MD: Lippincott Williams & Wilkins, 2015.

Figure 4.7B. Lockhart RD, Hamilton GF, Fyfe FW. *Anatomy of the Human Body*. Philadelphia, PA: JB Lippincott, 1959.

Figure 4.9A–E. Modified from Clay JH, Pounds DM. *Basic Clinical Massage Therapy*, 2nd ed. Baltimore, MD: Lippincott Williams & Wilkins, 2008.

Figures 4.10A&B, 4.42C–E, 4.43B, 4.58B&C, 4.62A–H, 4.73A–E, and 4.85C. Modified from Moore KL, Dalley AF, Agur MR. *Clinically Oriented Anatomy*, 7th ed. Baltimore, MD: Lippincott Williams & Wilkins, 2014.

Figures 4.32C (photo) and 4.34A. Dudek RW, Louis TM. *High-Yield Gross Anatomy*, 4th ed. Baltimore, MD: Lippincott Williams & Wilkins, 2010.

Figures 4.34B, 4.36, 4.45B, and 4.61A&B. Courtesy of Dr. J. Heslin, Toronto, Ontario, Canada.

Figures 4.34C&D, 4.42B, 4.45A, and 4.72B. Courtesy of Dr. E.L. Lansdown, University of Toronto, Ontario, Canada.

Figure 4.42A. Courtesy of Dr. C.S. Ho, University of Toronto, Ontario, Canada.

Figure 4.47. Courtesy of Dr. K. Sniderman, University of Toronto, Ontario, Canada.

Figure 4.53B. Courtesy of A.M. Arenson, University of Toronto, Ontario, Canada.

Figure 4.66B (MRI). Courtesy of G.B. Haber, University of Toronto, Ontario, Canada.

Figure 4.66B (photo). Courtesy of Mission Hospital Regional Center, Mission Viejo, California.

Figure 4.73B (MRI). Courtesy of M. Asch, University of Toronto, Ontario, Canada.

Figures 4.91B&D, 4.92B&C, and 4.93A–C (MRIs). Courtesy of Dr. M.A. Haider, University of Toronto, Ontario, Canada.

CHAPTER 5
Pelvis and Perineum

Figures 5.3C, 5.4B&C, 5.11B, 5.12B, 5.16B–D, 5.18A–D, 5.19, 5.26B, 5.27A&B, 5.28B–D, 5.29A&B, 5.38A&B, 5.39B–D, 5.47B–E, 5.48A–F, 5.51B, 5.52B, and 5.54C. Modified from Moore KL, Agur MR, Dalley AF. *Essential Clinical Anatomy*, 5th ed. Baltimore, MD: Lippincott Williams & Wilkins, 2015.

Figure 5.7A&B. Snell R. *Clinical Anatomy by Regions*, 9th ed. Baltimore, MD: Lippincott Williams & Wilkins, 2012.

Figures 5.24A&B (MRIs), 5.30B, 5.43A, 5.57B&E–H, and 5.64A–D,F,&H. Courtesy of Dr. M.A. Haider, University of Toronto, Ontario, Canada.

Figure 5.24C. Modified from Bickley LS. *Bates' Guide to Physical Examination and History Taking*, 10th ed. Philadelphia, PA: Wolters Kluwer Health, 2009.

Figures 5.28A, 5.30E&F, 5.33A–C, 5.39A, 5.40, 5.41, and 5.59B. Modified from Moore KL, Dalley AF, Agur MR. *Clinically Oriented Anatomy*, 7th ed. Baltimore, MD: Lippincott Williams & Wilkins, 2014.

Figures 5.30C and 5.34A&B. Courtesy of A.M. Arenson, University of Toronto, Ontario, Canada.

Figure 5.35D. Reprinted with permission from Stuart GCE, Reid DF. Diagnostic studies. In: Copeland LJ. *Textbook of Gynecology*. Philadelphia, PA: WB Saunders, 1993.

Figures 5.43B and 5.57C. From the Visible Human Project; National Library of Medicine; Visible Woman Image Numbers 1870 and 1895.

CHAPTER 6
Lower Limb

Figures 6.2A&B, 6.9A&B, 6.12A, 6.13A, 6.15A&B, 6.17B, 6.19C, 6.29A&B, 6.30A, 6.32B&C, 6.38A, 6.45 (schematics), 6.48B&C, 6.53A, 6.58A&B, 6.61A&B, 6.63D, 6.65A&B, 6.66D, 6.67B, and 6.72A–C. Modified from Moore KL, Agur MR, Dalley AF. *Essential Clinical Anatomy*, 5th ed. Baltimore, MD: Lippincott Williams & Wilkins, 2015.

Figure 6.3A&C. Courtesy of P. Babyn, University of Toronto, Ontario, Canada.

Figures 6.7A–D, 6.12B, 6.13B, 6.24B&C, 6.33B, 6.59A&E, 6.67E, 6.68B, 6.71A&B, 6.74A, 6.75A, 6.76A, 6.77A, 6.80B&C, 6.81D, and 6.87A. Modified from Moore KL, Dalley AF, Agur MR. *Clinically Oriented Anatomy*, 7th ed. Baltimore, MD: Lippincott Williams & Wilkins, 2014.

Figure 6.8A&B. Based on Foerster O. The dermatomes in man. *Brain*. 1933;56(1):1–39.

Figure 6.8C&D. Based on Keegan JJ, Garrett FD. The segmental distribution of the cutaneous nerves in the limbs of man. *Anat Rec*. 1948;102:409–437.

Figure 6.14B. Courtesy of Dr. E.L. Lansdown, University of Toronto, Ontario, Canada.

Figures 6.22A–E&H, 6.29C–F, 6.30B–D, and 6.62C&D. Modified from Clay JH, Pounds DM. *Basic Clinical Massage Therapy*. Baltimore, MD: Lippincott Williams & Wilkins, 2002.

Figure 6.34A&B. Modified from Tank PW, Gest TR. *Lippincott Williams & Wilkins Atlas of Anatomy*. Baltimore, MD: Lippincott Williams & Wilkins, 2009.

Figure 6.39A. Courtesy of E. Becker, University of Toronto, Ontario, Canada.

Figures 6.39C, 6.56C&D, 6.92C–E (MRIs), and 6.94A–D (MRIs). Courtesy of Dr. D. Salonen, University of Toronto, Ontario, Canada.

Figure 6.49C. Courtesy of Dr. Robert Peroutka, Cockeysville, MD.

Figure 6.70A. Courtesy of Dr. D. K. Sniderman, University of Toronto, Ontario, Canada.

Figure 6.82B. Courtesy of E. Becker, University of Toronto, Ontario, Canada.

Figures 6.85B and 6.86B. Courtesy of Dr. W. Kucharczyk, University of Toronto, Ontario, Canada.

Figure 6.90E. Courtesy of Dr. P. Bobechko, University of Toronto, Ontario, Canada.

CHAPTER 7
Head

Figures 7.1B,E,&F, 7.76B, 7.103A–F, 7.107A–E (MRIs), 7.108A–F, and 7.109A–C. Courtesy of Dr. D. Armstrong, University of Toronto, Ontario, Canada.

Figures 7.3C, 7.6B, 7.17A&B, 7.19, 7.21B&C, 7.29, 7.31B, 7.44A, 7.45B, 7.60B, 7.63C, 7.64A&C, 7.68B, 7.70A&B, 7.71A&B, 7.72A (top), 7.82A&B, 7.84D, 7.98A&C, and Table 7.15. Modified from Moore KL, Dalley AF, Agur MR. *Clinically Oriented Anatomy*, 7th ed. Baltimore, MD: Lippincott Williams & Wilkins, 2014.

Figures 7.14A, 7.15A&B, 7.18A&B, 7.20B, 7.21A, 7.22A–D, 7.24B, 7.25A&B, 7.30B&C, 7.33B&C, 7.39B,C,&E, 7.42B–E, 7.43A&B, 7.44B, 7.45D, 7.46B, 7.48A&D, 7.51, 7.52A&B, 7.55B&C, 7.56A–C, 7.57A–D, 7.58A&B, 7.59A–C, 7.67A–C, 7.78A–C, 7.79D&E, 7.85A, 7.86A, 7.89B, 7.90C–E, 7.91A&B, and 7.92A–D. Modified from Moore KL, Agur MR, Dalley AF. *Essential Clinical Anatomy*, 5th ed. Baltimore, MD: Lippincott Williams & Wilkins, 2015.

Figure 7.34A–C. Courtesy of I. Verschuur, Joint Department of Medical Imaging, UHN/Mount Sinai Hospital, Toronto, Ontario, Canada.

Figures 7.35A&B, 7.38D, 7.94B&C, and 7.95B. Courtesy of Dr. W. Kucharczyk, University of Toronto, Ontario, Canada.

Figure 7.46A. Courtesy of J.R. Buncic, University of Toronto, Ontario, Canada.

Figure 7.53A–C. Modified from Clay JH, Pounds DM. *Basic Clinical Massage Therapy*. Baltimore, MD: Lippincott Williams & Wilkins, 2002.

Figure 7.56 (MRIs). Langland OE, Langlais RP, Preece JW. *Principles of Dental Imaging*, 2nd ed. Baltimore, MD: Lippincott Williams & Wilkins, 2002.

Figure 7.65D. Courtesy of M.J. Phatoah, University of Toronto, Ontario, Canada.

Figure 7.66E. Courtesy of Dr. B. Libgott, Division of Anatomy/Department of Surgery, University of Toronto, Ontario, Canada.

Figures 7.76C and 7.77B. Courtesy of E. Becker, University of Toronto, Ontario, Canada.

Figure 7.96A&B. Courtesy of the Visible Human Project; National Library of Medicine; Visible Man 1107 and 1168.

Figures 7.99–7.102, 7.104, 7.105B&C, and 7.106. Colorized from photographs provided courtesy of Dr. C.G. Smith, which appears in Smith CG. *Serial Dissections of the Human Brain*. Baltimore, MD: Urban & Schwarzenber, Inc and Toronto: Gage Publishing Ltd, 1981. (© Carlton G. Smith)

CHAPTER 8
Neck

Figures 8.2A–C, 8.3A, 8.5A&C–G, 8.6B&C, 8.8B, 8.12B, 8.15A–C, 8.17B, 8.19A, 8.36B–F&H–J, 8.37D, and 8.39. Modified from Moore KL, Agur MR, Dalley AF. *Essential Clinical Anatomy*, 5th ed. Baltimore, MD: Lippincott Williams & Wilkins, 2015.

Figures 8.4A&B, 8.8D&E, 8.23A, 8.28C, and 8.31C. Modified from Moore KL, Dalley AF, Agur MR. *Clinically Oriented Anatomy*, 7th ed. Baltimore, MD: Lippincott Williams & Wilkins, 2014.

Figure 8.5B. Courtesy of J. Heslin, University of Toronto, Ontario, Canada.

Figures 8.7B&C, 8.12A, and 8.24A&B. Modified from Clay JH, Pounds DM. *Basic Clinical Massage Therapy*. Baltimore, MD: Lippincott Williams & Wilkins, 2002.

Figure 8.15D. Courtesy of Dr. D. Armstrong, University of Toronto, Ontario, Canada.

Figures 8.28A and 8.43B. Modified from Tank PW, Gest TR. *Lippincott Williams & Wilkins Atlas of Anatomy*. Baltimore, MD: Lippincott Williams & Wilkins, 2009.

Figure 8.30B. From Liebgott B. *The Anatomical Basis of Dentistry*. Philadelphia, PA: Mosby, 1982.

Figure 8.37A. Rohen JW, Yokochi C, Lutjen-Drecoll E, et al. *Color Atlas of Anatomy*, 5th ed. Baltimore, MD: Lippincott Williams & Wilkins, 2002.

Figures 8.37C and 8.40A–C. Courtesy of Dr. D. Salonen, University of Toronto, Ontario, Canada.

Figure 8.42A. Courtesy of Dr. E. Becker, University of Toronto, Ontario, Canada.

Figure 8.43A. Siemens Medical Solutions USA, Inc.

CHAPTER 9
Cranial Nerves

Figures 9.3, 9.5A&B, 9.6A–C, 9.7, 9.8C&D, 9.10A, 9.11B, 9.13B–E, 9.14A, 9.15B&C, 9.16D, 9.17A, 9.18A,B,&D, 9.19A, 9.20B, and 9.21. Modified from Moore KL, Agur MR, Dalley AF. *Essential Clinical Anatomy*, 5th ed. Baltimore, MD: Lippincott Williams & Wilkins, 2015.

Figure 9.16C. Modified from Moore KL, Dalley AF, Agur MR. *Clinically Oriented Anatomy*, 7th ed. Baltimore, MD: Lippincott Williams & Wilkins, 2014.

Figures 9.23A–F and 9.24A–C. Courtesy of Dr. W. Kucharczyk, University of Toronto, Ontario, Canada.

CHAPTER 1

Back

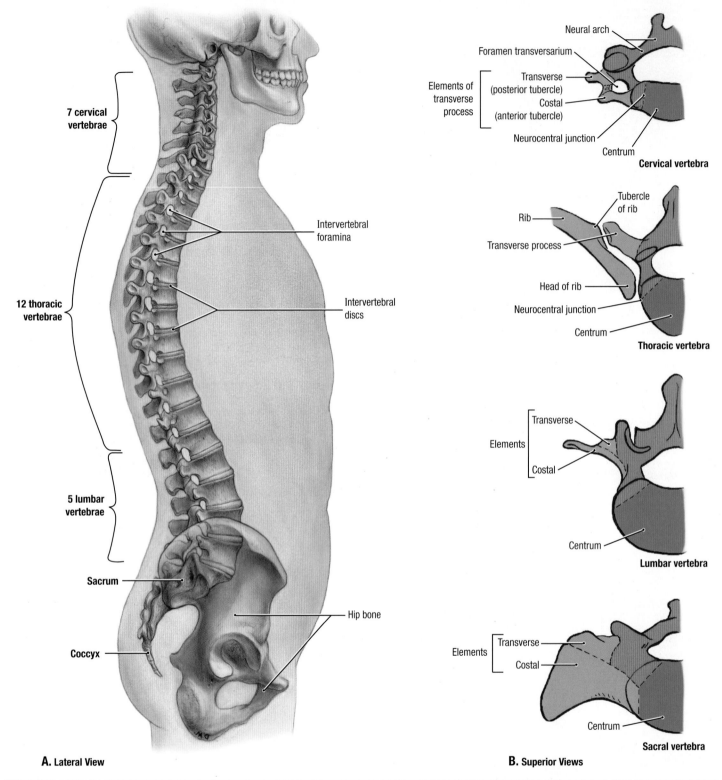

A. Lateral View

B. Superior Views

1.1 OVERVIEW OF VERTEBRAL COLUMN

A. Vertebral column showing articulation with skull and hip bone.
- The vertebral column usually consists of 24 separate (presacral) vertebrae, 5 fused vertebrae in the sacrum, and variably 4 fused or separated coccygeal vertebrae. Of the 24 separate vertebrae, 12 support the ribs (thoracic vertebrae), 7 are in the neck (cervical vertebrae, and 5 are in the lumbar region (lumbar vertebrae).

- The spinal nerves exit the vertebral (spinal) canal via the intervertebral (IV) foramina. There are 8 cervical, 12 thoracic, 5 lumbar, 5 sacral, and 1 to 2 coccygeal spinal nerves.

B. Homologous parts of vertebrae. A rib is a free costal element in the thoracic region; in the cervical and lumbar regions, it is represented by the anterior part of a transverse process, and in the sacrum, by the anterior part of the lateral mass.

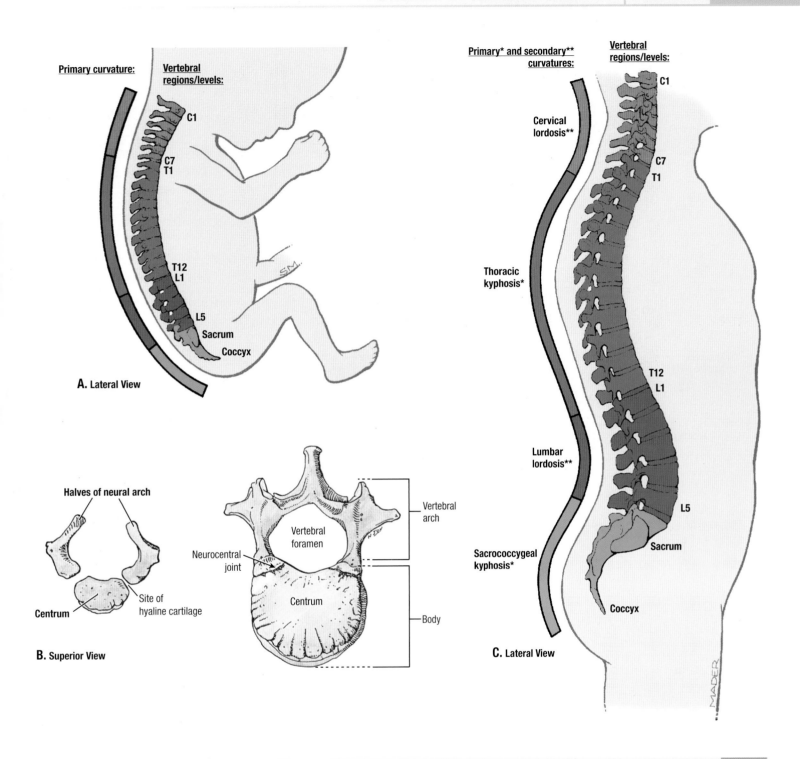

Primary curvature:

Vertebral regions/levels:

C1
C7
T1
T12
L1
L5
Sacrum
Coccyx

A. Lateral View

Halves of neural arch

Centrum

Site of hyaline cartilage

B. Superior View

Vertebral arch

Vertebral foramen

Neurocentral joint

Centrum

Body

Primary* and secondary curvatures:**

Vertebral regions/levels:

Cervical lordosis**

C1
C7
T1

Thoracic kyphosis*

Lumbar lordosis**

Sacrococcygeal kyphosis*

T12
L1
L5
Sacrum
Coccyx

C. Lateral View

CURVATURES OF VERTEBRAL COLUMN

1.2

A. Fetus. Note the C-shaped curvature of the fetal spine, which is concave anteriorly over its entire length. **B.** Development of the vertebrae. At birth, a vertebra consists of three bony parts (two halves of the neural arch and the centrum) united by hyaline cartilage. At age 2, the halves of each neural arch begin to fuse, proceeding from the lumbar to the cervical region; at approximately age 7, the arches begin to fuse to the centrum, proceeding from the cervical to lumbar regions. **C.** Adult. The four curvatures of the adult vertebral column include the cervical lordosis, which is convex anteriorly and lies between vertebrae C1 and T2; the thoracic kyphosis, which is concave anteriorly, between vertebrae T2 and T12; the lumbar lordosis, convex anteriorly and lying between T12 and the lumbosacral joint; and the sacrococcygeal kyphosis, concave anteriorly and spanning from the lumbosacral joint to the tip of the coccyx. The anteriorly concave thoracic kyphosis and sacrococcygeal kyphosis are primary curves, and the anteriorly convex cervical lordosis and lumbar lordosis are secondary curves that develop after birth. The cervical lordosis develops when the child begins to hold the head up, and the lumbar kyphosis develops when the child begins to walk.

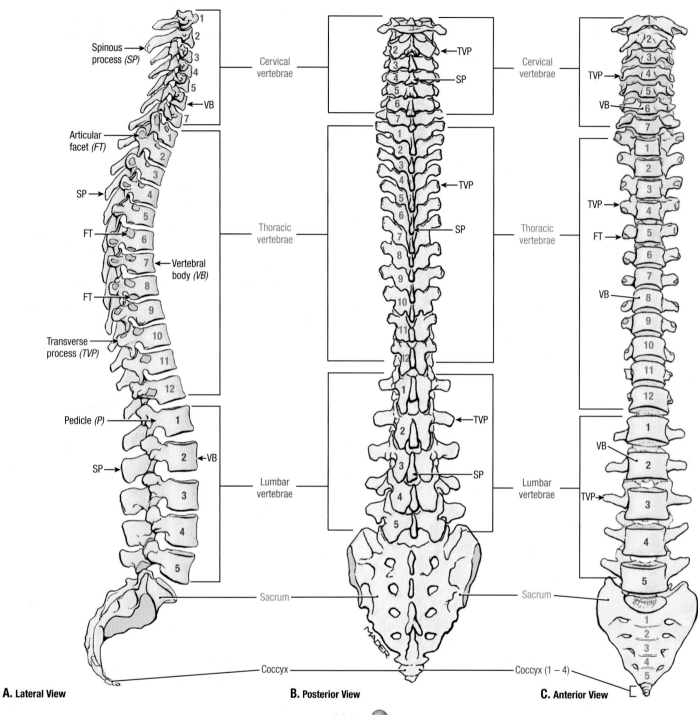

A. Lateral View

B. Posterior View

C. Anterior View

1.3 PARTS OF VERTEBRAL COLUMN

A. Lateral view. **B.** Posterior view. **C.** Anterior view. **D.** and **E.** Parts of a typical vertebra (e.g., the 2nd lumbar vertebra). *FT*, facet for articulation with the ribs; *L*, lamina; *P*, pedicle; *SP*, spinous process; *TVP*, transverse process; *VB*, vertebral body.

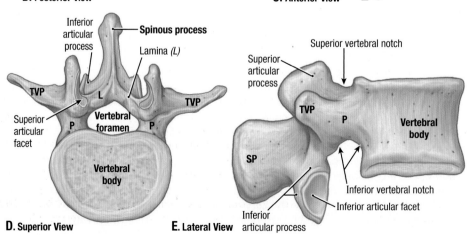

D. Superior View

E. Lateral View

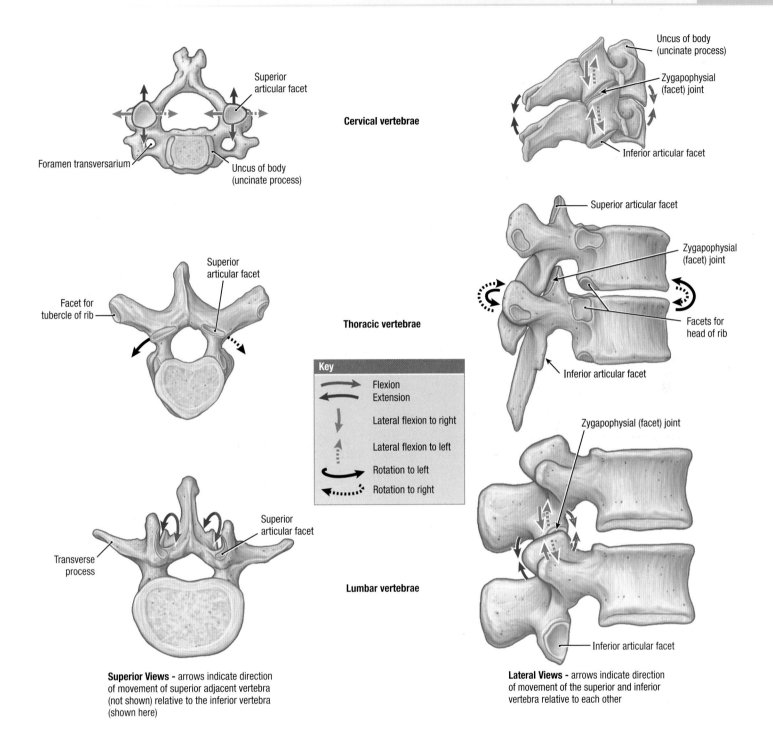

Cervical vertebrae

Superior articular facet

Foramen transversarium

Uncus of body (uncinate process)

Uncus of body (uncinate process)

Zygapophysial (facet) joint

Inferior articular facet

Thoracic vertebrae

Facet for tubercle of rib

Superior articular facet

Superior articular facet

Zygapophysial (facet) joint

Facets for head of rib

Inferior articular facet

Key

→ ←	Flexion Extension
↓	Lateral flexion to right
↑	Lateral flexion to left
↷	Rotation to left
↶	Rotation to right

Lumbar vertebrae

Transverse process

Superior articular facet

Zygapophysial (facet) joint

Inferior articular facet

Superior Views - arrows indicate direction of movement of superior adjacent vertebra (not shown) relative to the inferior vertebra (shown here)

Lateral Views - arrows indicate direction of movement of the superior and inferior vertebra relative to each other

VERTEBRAL FEATURES AND MOVEMENTS

1.4

- In the thoracic and lumbar regions, the articular processes/facets lie posterior to the vertebral bodies and in the cervical region posterolateral to the bodies. Superior articular facets in the cervical region face mainly superiorly, in the thoracic region, mainly posteriorly, and in the lumbar region, mainly medially. The change in direction is gradual from cervical to thoracic but abrupt from thoracic to lumbar.
- Although movements between adjacent vertebrae are relatively small, the summation of all the small movements produces a considerable range of movement of the vertebral column as a whole.

- Movements of the vertebral column are freer (have greater range of motion) in the cervical and lumbar regions than in the thoracic region. Lateral bending is freest in the cervical and lumbar regions; flexion is greatest in the cervical region; extension is most marked in the lumbar region, but the interlocking articular processes prevent rotation.
- The thoracic region is most stable because of the external support gained from the articulations of the ribs and costal cartilages with the sternum. The direction of the articular facets permits rotation, but flexion, extension, and lateral bending are severely restricted.

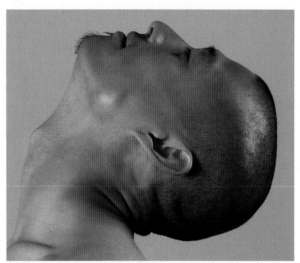

A. Lateral View

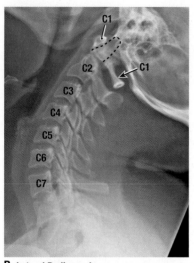

B. Lateral Radiograph

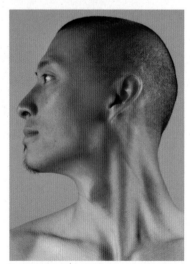

C. Lateral View

D. Lateral Radiograph

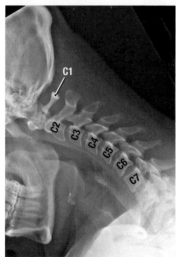

E. Anterior View

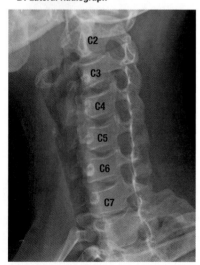

F. Oblique Radiograph

1.5 SURFACE ANATOMY WITH RADIOGRAPHIC CORRELATION OF SELECTED MOVEMENTS OF THE CERVICAL SPINE

A. Extension of the neck. **B.** Radiograph of the vertebded cervical spine. **C.** Flexion of the neck. **D.** Radiograph of the flexed cervical spine. **E.** Head rotated (turned) to left. **F.** Radiograph of cervical spine rotated to left.

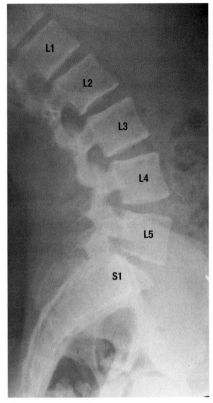

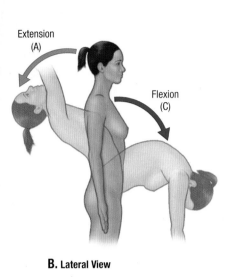

B. Lateral View

A. Lateral Radiograph, Lumbar Vertebrae Extended

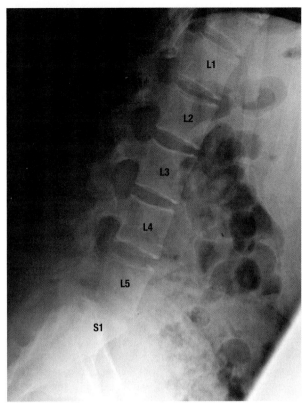

C. Lateral Radiograph, Lumbar Vertebrae Flexed

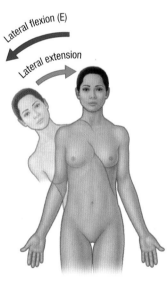

D. Anterior View

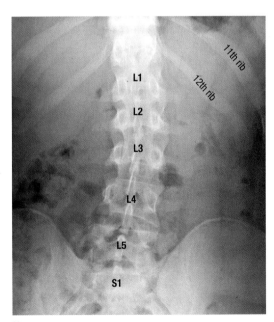

E. Anteroposterior Radiograph, Lumbar Vertebrae Laterally Flexed to Right

SURFACE ANATOMY WITH RADIOGRAPHIC CORRELATION OF SELECTED MOVEMENTS OF THE LUMBAR SPINE 1.6

A. Radiograph of the extended lumbar spine. **B.** Schematic illustration of flexion and extension of the trunk. **C.** Radiograph of the flexed lumbar spine. **D.** Schematic illustration of lateral (side) flexion of the trunk. **E.** Radiograph of the lumbar spine during lateral bending.

The range of movement of the vertebral column is limited by the thickness, elasticity, and compressibility of the IV discs; shape and orientation of the zygapophysial joints; tension of the joint capsules of the zygapophysial joints; resistance of the ligaments and back muscles; connection to thoracic (rib) cage and bulk of surrounding tissue.

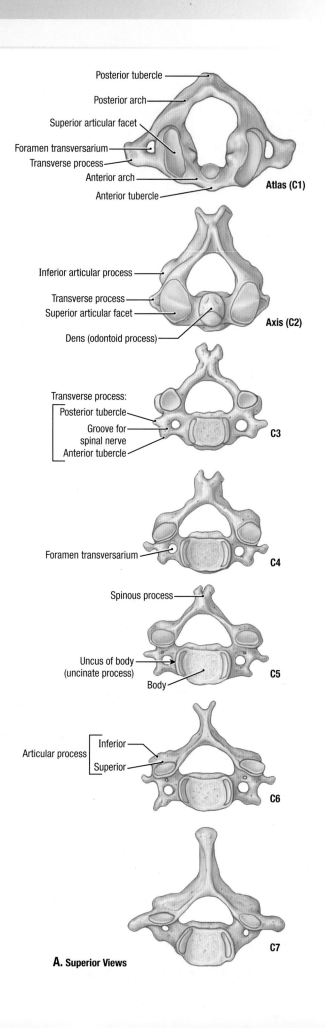

Posterior tubercle
Posterior arch
Superior articular facet
Foramen transversarium
Transverse process
Anterior arch
Anterior tubercle
Atlas (C1)

Inferior articular process
Transverse process
Superior articular facet
Dens (odontoid process)
Axis (C2)

Transverse process:
Posterior tubercle
Groove for spinal nerve
Anterior tubercle
C3

Foramen transversarium
C4

Spinous process
Uncus of body (uncinate process)
Body
C5

Articular process
Inferior
Superior
C6

C7

A. Superior Views

1.7   CERVICAL SPINE

A. Disarticulated cervical vertebrae. The bodies of the cervical vertebrae can be dislocated in neck injuries with less force than is required to fracture them. Because of the large vertebral canal in the cervical region, some dislocation can occur without damaging the spinal cord. When a cervical vertebra is severely dislocated, it injures the spinal cord. If the dislocation does not result in "facet jumping" with locking of the displaced articular processes, the cervical vertebrae may self-reduce ("slip back into place") so that a radiograph may not indicate that the cord has been injured. Magnetic resonance imaging (MRI) may reveal the resulting soft tissue damage.

Aging of the IV disc combined with the changing shape of the vertebrae results in an increase in compressive forces at the periphery of the vertebral bodies, where the disc attaches. In response, osteophytes (bony spurs) commonly develop around the margins of the vertebral body, especially along the outer attachment of the IV disc. Similarly, as altered mechanics place greater stresses on the zygapophysial joints, osteophytes develop along the attachments of the joint capsules, especially those of the superior articular process.

TABLE 1.1	**TYPICAL CERVICAL VERTEBRAE (C3–C7)**[a]
Part	*Distinctive Characteristics*
Body	Small and wider from side to side than anteroposteriorly; superior surface is concave with an uncus of body (uncinate process bilaterally); inferior surface is convex
Vertebral foramen	Large and triangular
Transverse processes	Foramina transversaria small or absent in vertebra C7; vertebral arteries and accompanying venous and sympathetic plexuses pass through foramina, except C7 foramina, which transmits only small accessory vertebral veins; anterior and posterior tubercles separated by groove for spinal nerve
Articular processes	Superior articular facets directed superoposteriorly; inferior articular facets directed infero-anteriorly; obliquely placed facets are most nearly horizontal in this region
Spinous process	Short (C3–C5) and bifid, only in Caucasians (C3–C5); process of C6 is long but that of C7 is longer; C7 is called "vertebra prominens"

[a]C1 and C2 vertebrae are atypical.

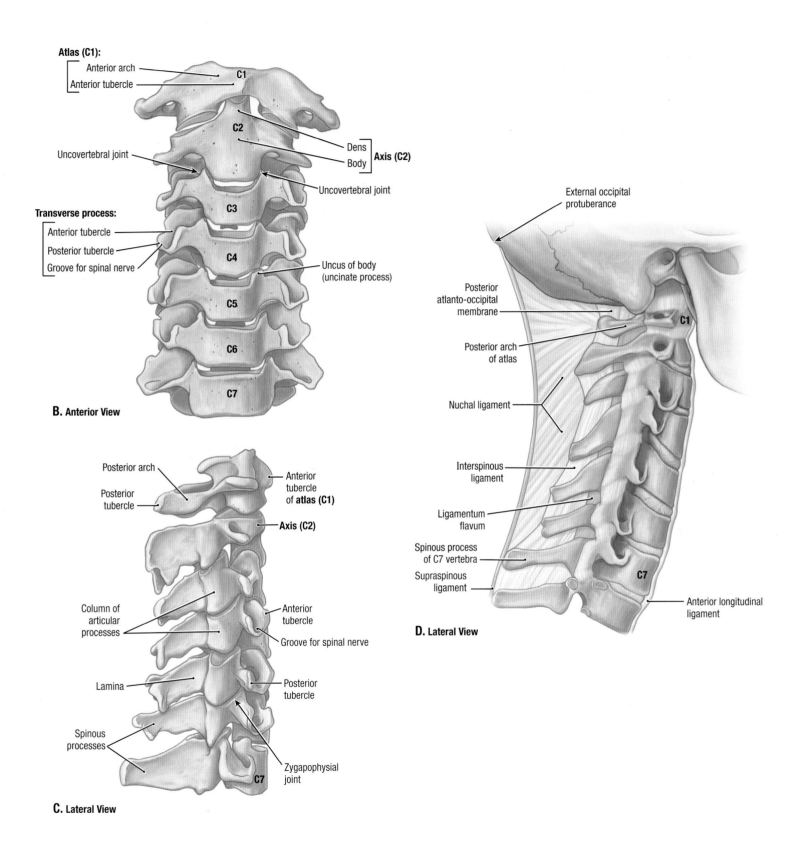

Atlas (C1):
Anterior arch
Anterior tubercle

C1

C2

Dens
Body
Axis (C2)

Uncovertebral joint

Uncovertebral joint

Transverse process:
Anterior tubercle
Posterior tubercle
Groove for spinal nerve

C3

C4

Uncus of body
(uncinate process)

C5

C6

C7

B. Anterior View

Posterior arch

Posterior tubercle

Anterior tubercle
of **atlas (C1)**

Axis (C2)

Column of articular processes

Anterior tubercle

Groove for spinal nerve

Lamina

Posterior tubercle

Spinous processes

Zygapophysial joint

C7

C. Lateral View

External occipital protuberance

Posterior atlanto-occipital membrane

Posterior arch of atlas

C1

Nuchal ligament

Interspinous ligament

Ligamentum flavum

Spinous process of C7 vertebra

Supraspinous ligament

C7

Anterior longitudinal ligament

D. Lateral View

B. and **C.** Articulated cervical vertebrae. **D.** Ligaments.

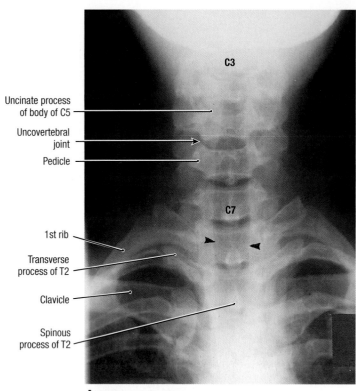

Uncinate process of body of C5

Uncovertebral joint

Pedicle

1st rib

Transverse process of T2

Clavicle

Spinous process of T2

C3

C7

A. Anteroposterior View

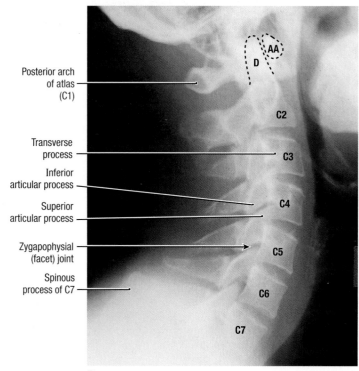

Posterior arch of atlas (C1)

Transverse process

Inferior articular process

Superior articular process

Zygapophysial (facet) joint

Spinous process of C7

AA

D

C2

C3

C4

C5

C6

C7

B. Lateral View

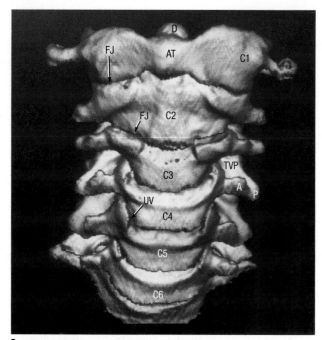

D

FJ

AT

C1

FJ

C2

TVP

C3

A P

UV

C4

C5

C6

C. Anterior View

Key			
A	Anterior tubercle of transverse process	PA	Posterior arch of C1
AA	Anterior arch of C1	PT	Posterior tubercle of C1
AT	Anterior tubercle of C1	SF	Superior articular facet of C1
C1–C7	Vertebrae	SP	Spinous process
D	Dens (odontoid) process of C2	T	Foramen transversarium
FJ	Zygapophysial (facet) joint	TVP	Transverse process
La	Lamina	UV	Uncovertebral joint
P	Posterior tubercle of transverse process	VC	Vertebral canal

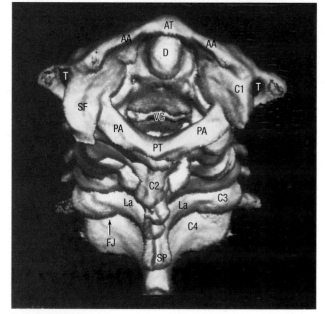

AT

AA

D

AA

T

C1

T

SF

VC

PA

PA

PT

C2

C3

La

La

C4

FJ

SP

D. Posterior View

1.8 IMAGING OF THE CERVICAL SPINE

A. and **B.** Radiographs. The *arrowheads* demarcate the margins of the *(black)* column of air in the trachea. **C.** and **D.** Three-dimensional reconstructed computed tomographic (CT) images.

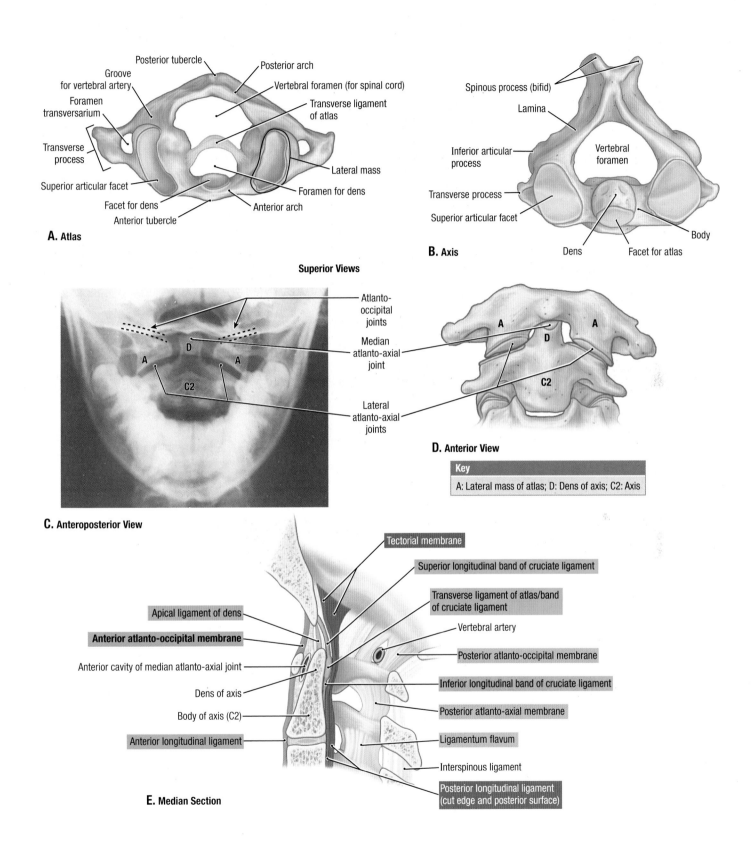

A. Atlas

- Posterior tubercle
- Groove for vertebral artery
- Foramen transversarium
- Transverse process
- Superior articular facet
- Facet for dens
- Anterior tubercle
- Posterior arch
- Vertebral foramen (for spinal cord)
- Transverse ligament of atlas
- Lateral mass
- Foramen for dens
- Anterior arch

B. Axis

- Spinous process (bifid)
- Lamina
- Inferior articular process
- Transverse process
- Superior articular facet
- Vertebral foramen
- Dens
- Facet for atlas
- Body

Superior Views

C. Anteroposterior View

- Atlanto-occipital joints
- Median atlanto-axial joint
- Lateral atlanto-axial joints

D. Anterior View

Key

A: Lateral mass of atlas; D: Dens of axis; C2: Axis

E. Median Section

- Tectorial membrane
- Superior longitudinal band of cruciate ligament
- Transverse ligament of atlas/band of cruciate ligament
- Vertebral artery
- Posterior atlanto-occipital membrane
- Inferior longitudinal band of cruciate ligament
- Posterior atlanto-axial membrane
- Ligamentum flavum
- Interspinous ligament
- Posterior longitudinal ligament (cut edge and posterior surface)
- Apical ligament of dens
- **Anterior atlanto-occipital membrane**
- Anterior cavity of median atlanto-axial joint
- Dens of axis
- Body of axis (C2)
- Anterior longitudinal ligament

ATLAS AND AXIS AND THE ATLANTO-AXIAL JOINT | 1.9

A. Atlas. **B.** Axis. **C.** Radiograph taken through the open mouth. **D.** Articulated atlas and axis. **E.** Median section with ligaments. The structures highlighted in the same color are continuous.

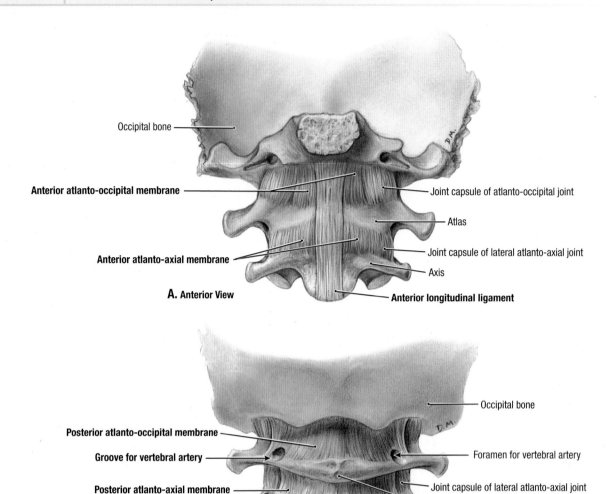

Occipital bone

Anterior atlanto-occipital membrane

Joint capsule of atlanto-occipital joint

Atlas

Joint capsule of lateral atlanto-axial joint

Anterior atlanto-axial membrane

Axis

A. Anterior View

Anterior longitudinal ligament

Posterior atlanto-occipital membrane

Occipital bone

Groove for vertebral artery

Foramen for vertebral artery

Posterior atlanto-axial membrane

Joint capsule of lateral atlanto-axial joint

Posterior tubercle of atlas

Spinous process of axis (bifid)

B. Posterior View

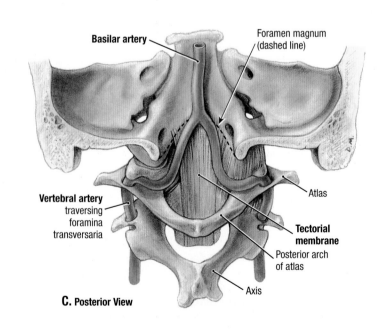

Basilar artery

Foramen magnum (dashed line)

Vertebral artery traversing foramina transversaria

Atlas

Tectorial membrane

Posterior arch of atlas

Axis

C. Posterior View

1.10 CRANIOVERTEBRAL JOINTS AND VERTEBRAL ARTERY

A. Anterior atlanto-axial and atlanto-occipital membranes. The anterior longitudinal ligament ascends to blend with, and form a central thickening in, the anterior atlanto-axial and atlanto-occipital membranes. **B.** Posterior atlanto-axial and atlanto-occipital membranes. Inferior to the axis (C2 vertebra), ligamenta flava occur in this position. **C.** Tectorial membrane and vertebral artery. The tectorial membrane is a superior continuation of the posterior longitudinal ligament superior to the body of the axis. After coursing through the foramina transversaria of vertebrae C6–C1, the vertebral arteries turn medially, grooving the superior aspect of the posterior arch of the atlas and piercing the posterior atlanto-occipital membrane **(B)**. The right and left vertebral arteries traverse the foramen magnum and merge intracranially, forming the basilar artery.

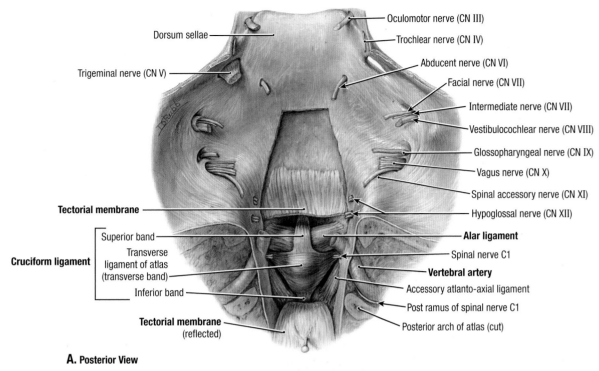

Dorsum sellae

Oculomotor nerve (CN III)

Trochlear nerve (CN IV)

Abducent nerve (CN VI)

Trigeminal nerve (CN V)

Facial nerve (CN VII)

Intermediate nerve (CN VII)

Vestibulocochlear nerve (CN VIII)

Glossopharyngeal nerve (CN IX)

Vagus nerve (CN X)

Spinal accessory nerve (CN XI)

Tectorial membrane

Hypoglossal nerve (CN XII)

Cruciform ligament
- Superior band
- Transverse ligament of atlas (transverse band)
- Inferior band

Alar ligament

Spinal nerve C1

Vertebral artery

Accessory atlanto-axial ligament

Post ramus of spinal nerve C1

Tectorial membrane (reflected)

Posterior arch of atlas (cut)

A. Posterior View

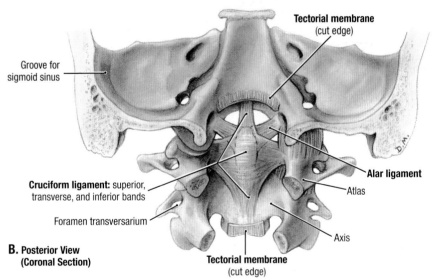

Tectorial membrane (cut edge)

Groove for sigmoid sinus

Alar ligament

Atlas

Cruciform ligament: superior, transverse, and inferior bands

Foramen transversarium

Axis

B. Posterior View (Coronal Section)

Tectorial membrane (cut edge)

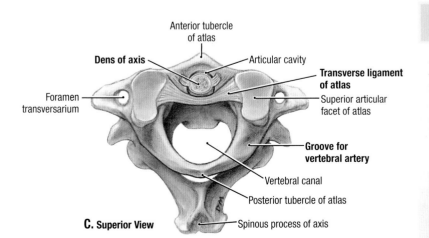

Anterior tubercle of atlas

Dens of axis

Articular cavity

Transverse ligament of atlas

Superior articular facet of atlas

Foramen transversarium

Groove for vertebral artery

Vertebral canal

Posterior tubercle of atlas

Spinous process of axis

C. Superior View

LIGAMENTS OF ATLANTO-OCCIPITAL AND ATLANTO-AXIAL JOINTS

1.11

A. Cranial nerves and dura mater of posterior cranial fossa with dura mater and tentorial membrane incised and removed to reveal the medial atlanto-axial joint. The alar ligaments serve as check ligaments for the rotary movements of the atlanto-axial joints. **B.** and **C.** Transverse ligament of the atlas. The transverse band of the cruciform ligament, forms the posterior wall of a socket that receives the dens of the axis, forming a pivot joint.

Fracture of atlas. The atlas is a bony ring, with two wedge-shaped lateral masses, connected by relatively thin anterior and posterior arches and the transverse ligament of the atlas (see Figs. 1.12A and C). Vertical forces (e.g., striking the head on bottom of pool) may force the lateral masses apart fracturing one or both of the anterior or posterior arches. If the force is sufficient, rupture of the transverse ligament of the atlas will also occur.

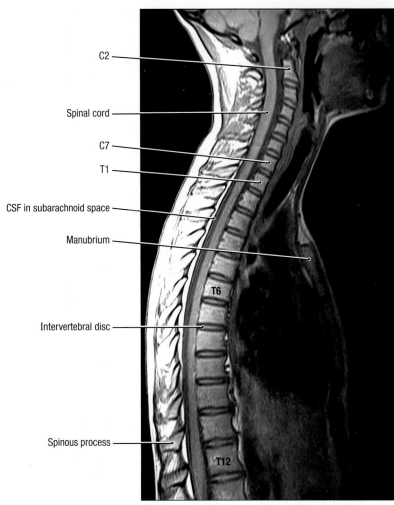

C2

Spinal cord

C7

T1

CSF in subarachnoid space

Manubrium

T6

Intervertebral disc

Spinous process

T12

A. Midsagittal MRI

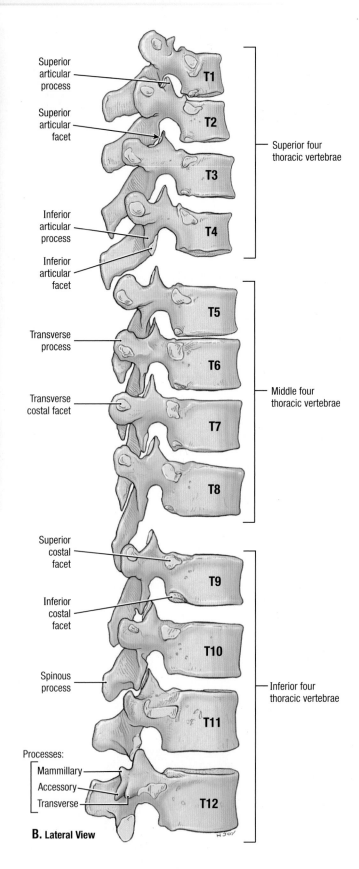

Superior articular process — T1

Superior articular facet — T2

T3

Inferior articular process — T4

Inferior articular facet

T5

Transverse process — T6

Transverse costal facet — T7

T8

Superior costal facet — T9

Inferior costal facet — T10

Spinous process — T11

Processes:
Mammillary
Accessory
Transverse — T12

B. Lateral View

Superior four thoracic vertebrae

Middle four thoracic vertebrae

Inferior four thoracic vertebrae

1.12   THORACIC VERTEBRAE

A. MRI of thoracic spine. **B.** Features.

Fracture of thoracic vertebrae. Although the characteristics of the superior aspect of vertebra T12 are distinctly thoracic, its inferior aspect has lumbar characteristics for articulation with vertebra L1. The abrupt transition allowing primarily rotational movements with vertebra T11 while disallowing rotational movements with vertebral L1 makes vertebra T12 especially susceptible to fracture.

TABLE 1.2	THORACIC VERTEBRAE
Part	*Distinctive Characteristics*
Body	Heart-shaped; has one or two costal facets for articulation with head of rib
Vertebral foramen	Circular and smaller than those of cervical and lumbar vertebrae
Transverse processes	Long and extend posterolaterally; length diminishes from T1 to T12; T1–T10 have transverse costal facets for articulation with a tubercle of ribs 1–10 (ribs 11 and 12 have no tubercle and do not articulate with a transverse process)
Articular processes	Superior articular facets directed posteriorly and slightly laterally; inferior articular facets directed anteriorly and slightly medially
Spinous process	Long and slopes postero-inferiorly; tip extends to level of vertebral body below

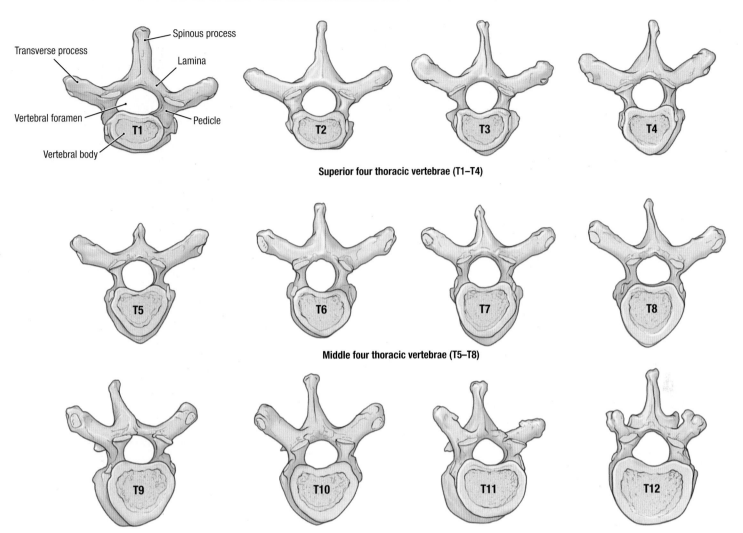

Superior four thoracic vertebrae (T1–T4)

Middle four thoracic vertebrae (T5–T8)

C. Superior Views

Inferior four thoracic vertebrae (T9–T12)

Labels (T1): Spinous process, Transverse process, Lamina, Vertebral foramen, Pedicle, Vertebral body

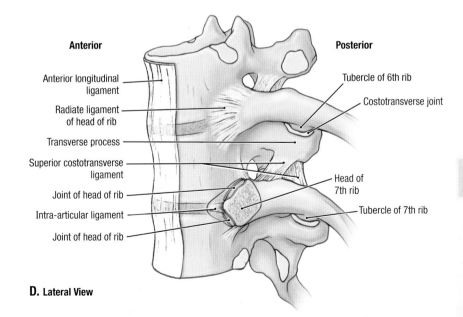

D. Lateral View

Anterior

Posterior

Anterior longitudinal ligament
Radiate ligament of head of rib
Transverse process
Superior costotransverse ligament
Joint of head of rib
Intra-articular ligament
Joint of head of rib

Tubercle of 6th rib
Costotransverse joint
Head of 7th rib
Tubercle of 7th rib

THORACIC VERTEBRAE (*continued*) | 1.12

C. Disarticulated thoracic vertebrae. The vertebral bodies increase in size as the vertebral column descends, each bearing an increasing amount of weight transferred by the vertebra above. **D.** Intra- and extra-articular ligaments of the costovertebral articulations. Typically, the head of each rib articulates with the bodies of two adjacent vertebrae and the IV disc between them, and the tubercle of the rib articulates with the transverse process of the inferior vertebra.

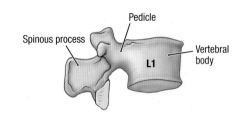

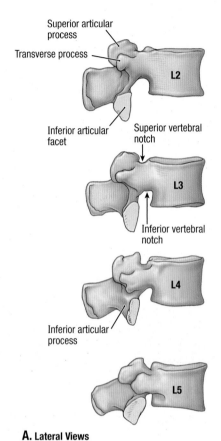

Pedicle

Spinous process

Vertebral body

L1

Superior articular process

Transverse process

L2

Inferior articular facet

Superior vertebral notch

L3

Inferior vertebral notch

L4

Inferior articular process

L5

A. Lateral Views

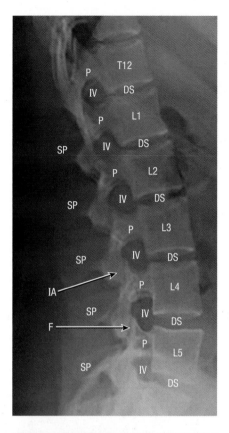

B. Lateral Radiograph

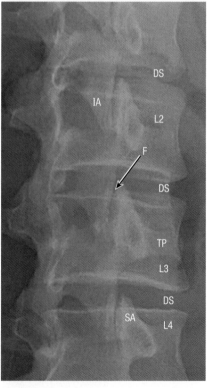

C. Oblique Radiograph

1.13 LUMBAR VERTEBRAE

A, D, and **E.** Features. **B, C,** and **F.** Radiographs. **G.** Laminectomy.

TABLE 1.3	**LUMBAR VERTEBRAE**
Part	*Distinctive Characteristics*
Body	Massive; kidney-shaped when viewed superiorly
Vertebral	Triangular; larger than in thoracic vertebrae and foramen smaller than in cervical vertebrae
Transverse	Long and slender; accessory process on posterior surface of base of each transverse process
Articular processes	Superior articular facets directed posteromedially (or medially); inferior articular facets directed anterolaterally (or laterally); mammillary process on posterior surface of each superior articular process
Spinous process	Short and sturdy; thick, broad, and rectangular

Key for B, C, and D			
F	Zygapophysial (facet) joint	P	Pedicle
DS	Intervertebral disc space	SA	Superior articular process
IA	Inferior articular process	SP	Spinous process
IV	Intervertebral foramen	T12–L5	Vertebral bodies
L	Lamina	TP	Transverse process

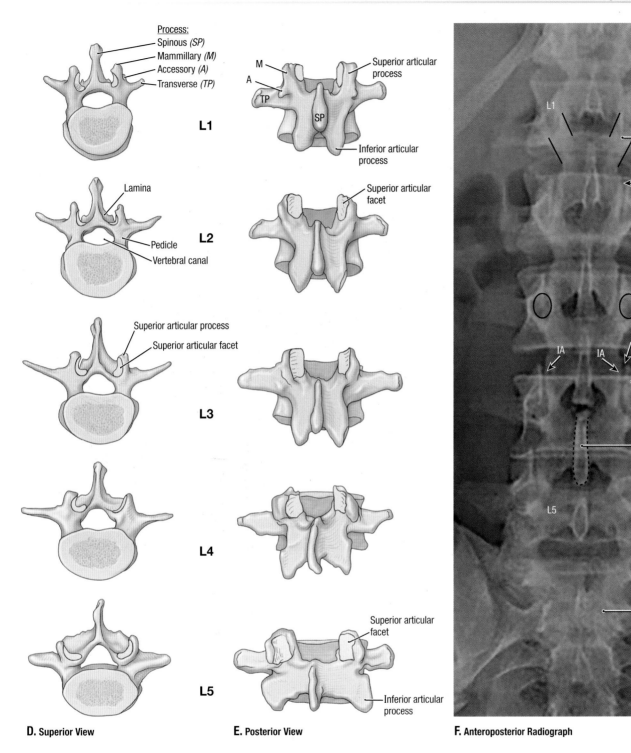

Process:
- Spinous *(SP)*
- Mammillary *(M)*
- Accessory *(A)*
- Transverse *(TP)*

L1

Lamina

Pedicle

Vertebral canal

L2

Superior articular process

Superior articular facet

L3

L4

L5

D. Superior View

M

A

TP

SP

Superior articular process

Inferior articular process

Superior articular facet

Superior articular facet

Inferior articular process

E. Posterior View

L1

L

F

P

F

IA   IA

SA

L4 SP

L5

Sacrum

F. Anteroposterior Radiograph

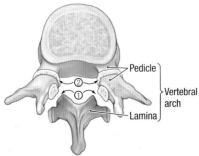

Pedicle

Vertebral arch

Lamina

G. Superior View, Sites of Laminectomy (1 and 2)

LUMBAR VERTEBRAE (*continued*) **1.13**

A **laminectomy** is the surgical excision of one or more spinous processes and their supporting laminae in a particular region of the vertebral column by transecting the interarticular part (Fig. 1.13G, *1*). The term is also commonly used to denote the removal of most of the vertebral arch by transecting the pedicles (Fig. 1.13G, *2*). Laminectomies provide access to the vertebral canal to relieve pressure on the spinal cord or nerve roots, commonly caused by a tumor or herniated IV disc.

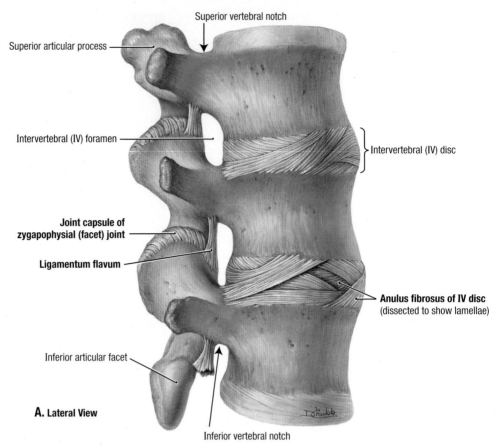

Superior vertebral notch

Superior articular process

Intervertebral (IV) foramen

Intervertebral (IV) disc

Joint capsule of zygapophysial (facet) joint

Ligamentum flavum

Anulus fibrosus of IV disc
(dissected to show lamellae)

Inferior articular facet

A. Lateral View

Inferior vertebral notch

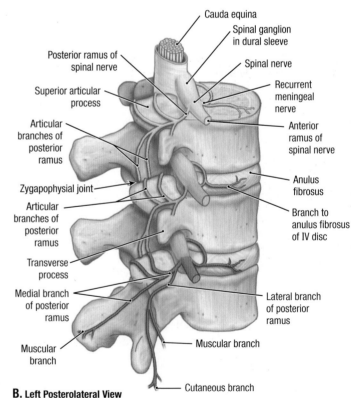

Cauda equina

Spinal ganglion in dural sleeve

Posterior ramus of spinal nerve

Spinal nerve

Superior articular process

Recurrent meningeal nerve

Articular branches of posterior ramus

Anterior ramus of spinal nerve

Zygapophysial joint

Anulus fibrosus

Articular branches of posterior ramus

Branch to anulus fibrosus of IV disc

Transverse process

Medial branch of posterior ramus

Lateral branch of posterior ramus

Muscular branch

Muscular branch

Cutaneous branch

B. Left Posterolateral View

1.14 STRUCTURE AND INNERVATION OF INTERVERTEBRAL DISCS AND ZYGAPOPHYSIAL JOINTS

A. Intervertebral discs and intervertebral foramen. Sections have been removed from the superficial layers of the anulus fibrosus of the inferior IV disc to show the change in direction of the fibers in the concentric layers of the anulus. Note that the IV discs form the inferior half of the anterior boundary of the IV foramen. **B.** Innervation of zygapophysial joints and the anulus fibrosus of IV discs.

When the **zygapophysial joints are injured** or develop osteophytes during aging (osteoarthritis), the related spinal nerves are affected. This causes pain along the distribution pattern of the dermatomes and spasm in the muscles derived from the associated myotomes. Denervation of lumbar zygapophysial joints is a procedure that may be used for treatment of back pain caused by disease of these joints. The denervation process is directed at the articular branches of two adjacent posterior rami of the spinal nerves because each joint receives innervation from both the nerve exiting that level and the superjacent nerve.

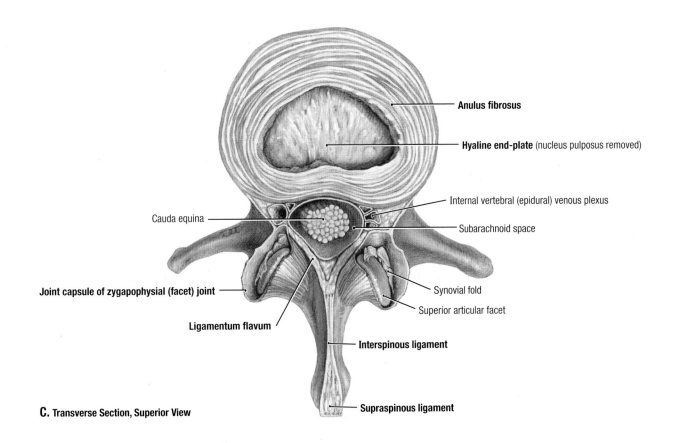

Anulus fibrosus

Hyaline end-plate (nucleus pulposus removed)

Internal vertebral (epidural) venous plexus

Cauda equina

Subarachnoid space

Joint capsule of zygapophysial (facet) joint

Synovial fold

Superior articular facet

Ligamentum flavum

Interspinous ligament

C. Transverse Section, Superior View

Supraspinous ligament

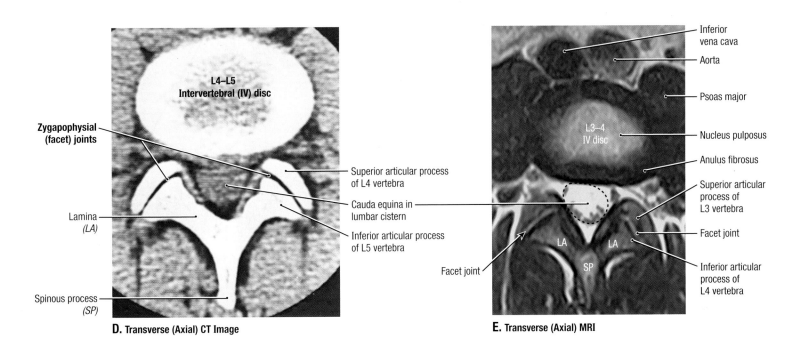

L4–L5 Intervertebral (IV) disc

Zygapophysial (facet) joints

Superior articular process of L4 vertebra

Cauda equina in lumbar cistern

Lamina (LA)

Inferior articular process of L5 vertebra

Spinous process (SP)

D. Transverse (Axial) CT Image

Inferior vena cava

Aorta

Psoas major

L3–4 IV disc

Nucleus pulposus

Anulus fibrosus

Superior articular process of L3 vertebra

Facet joint

LA   LA

SP

Inferior articular process of L4 vertebra

Facet joint

E. Transverse (Axial) MRI

STRUCTURE AND INNERVATION OF INTERVERTEBRAL DISCS AND ZYGAPOPHYSIAL JOINTS (*continued*)   1.14

C. Transverse section. The nucleus pulposus has been removed, and the cartilaginous epiphysial plate exposed. There are fewer rings of the anulus fibrosus posteriorly, and consequently, this portion of the anulus fibrosus is thinner. The ligamentum flavum, interspinous, and supraspinous ligaments are continuous. **D.** CT image of L4/L5 IV disc. **E.** MRI.

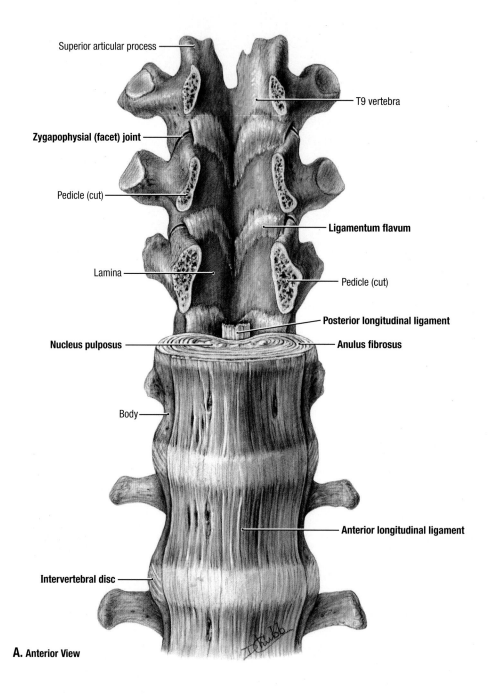

Superior articular process

T9 vertebra

Zygapophysial (facet) joint

Pedicle (cut)

Ligamentum flavum

Lamina

Pedicle (cut)

Posterior longitudinal ligament

Nucleus pulposus

Anulus fibrosus

Body

Anterior longitudinal ligament

Intervertebral disc

A. Anterior View

1.15   INTERVERTEBRAL DISCS: LIGAMENTS AND MOVEMENTS

A. Anterior longitudinal ligament and ligamenta flava. The pedicles of the superior vertebrae were sawed through to show the ligamenta flava.

- The anterior and posterior longitudinal ligaments are ligaments of the vertebral bodies; the ligamenta flava are ligaments of the vertebral arches.
- The anterior longitudinal ligament consists of broad, strong, fibrous bands that are attached to the IV discs and vertebral

bodies anteriorly and are perforated by the foramina for arteries and veins passing to and from the vertebral bodies.
- The ligamenta flava, composed of elastic fibers, extend between adjacent laminae and converge in the median plane. They extend laterally to blend with the joint capsule of the zygapophysial joints.

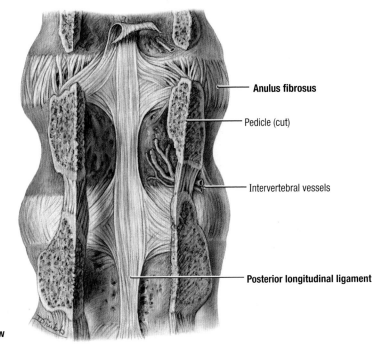

Anulus fibrosus

Pedicle (cut)

Intervertebral vessels

Posterior longitudinal ligament

B. Posterior View

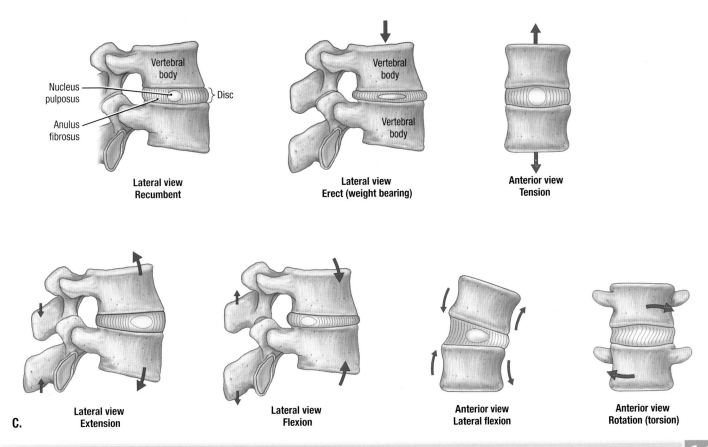

C.

INTERVERTEBRAL DISCS: LIGAMENTS AND MOVEMENTS (*continued*)

1.15

B. Posterior longitudinal ligament. The pedicles of vertebra T9–T11 were sawed through and the vertebral arch removed to show the posterior aspect of the vertebral bodies. The posterior longitudinal ligament is a narrow band passing from disc to disc, spanning the posterior surfaces of the vertebral bodies. **C.** IV disc during loading and movement. The movement or loading of the IV disc changes its shape and the position of the nucleus pulposus.

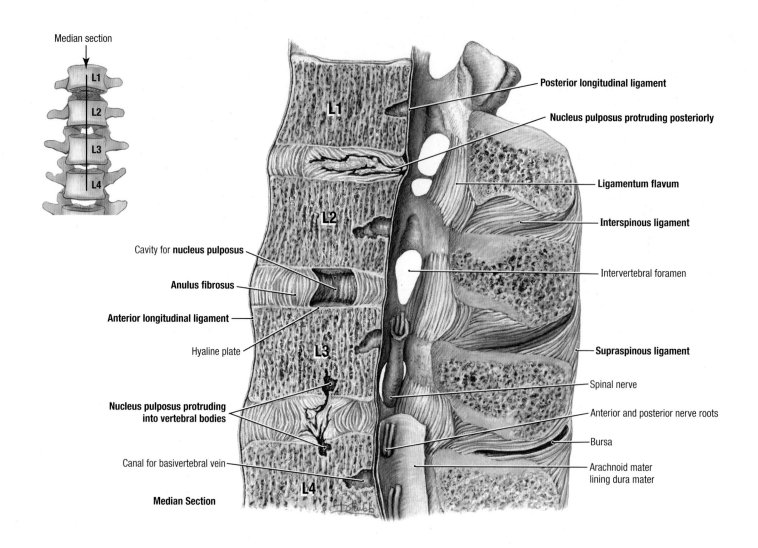

Median section

L1
L2
L3
L4

Posterior longitudinal ligament

Nucleus pulposus protruding posteriorly

Ligamentum flavum

Interspinous ligament

Intervertebral foramen

Supraspinous ligament

Spinal nerve

Anterior and posterior nerve roots

Bursa

Arachnoid mater lining dura mater

Cavity for **nucleus pulposus**

Anulus fibrosus

Anterior longitudinal ligament

Hyaline plate

Nucleus pulposus protruding into vertebral bodies

Canal for basivertebral vein

Median Section

L1
L2
L3
L4

1.16 LUMBAR REGION OF VERTEBRAL COLUMN

The nucleus pulposus of the normal disc between vertebrae L2 and L3 has been removed from the enclosing anulus fibrosus. The bursa between L3 and L4 spines is presumably the result of habitual hyperextension, which brings the lumbar spines into contact.

The nucleus pulposus of the disc between L1 and L2 has herniated posteriorly through the ianulus. **Herniation** or **protrusion of the gelatinous nucleus pulposus** into or through the anulus

fibrosus is a well-recognized cause of low back and lower limb pain. If degeneration of the posterior longitudinal ligament and wearing of the anulus fibrosus has occurred, the nucleus pulposus may herniate into the vertebral canal and compress the spinal cord or nerve roots of spinal nerves in the cauda equina. Herniations usually occur posterolaterally, where the anulus is relatively thin and does not receive support from the ligaments.

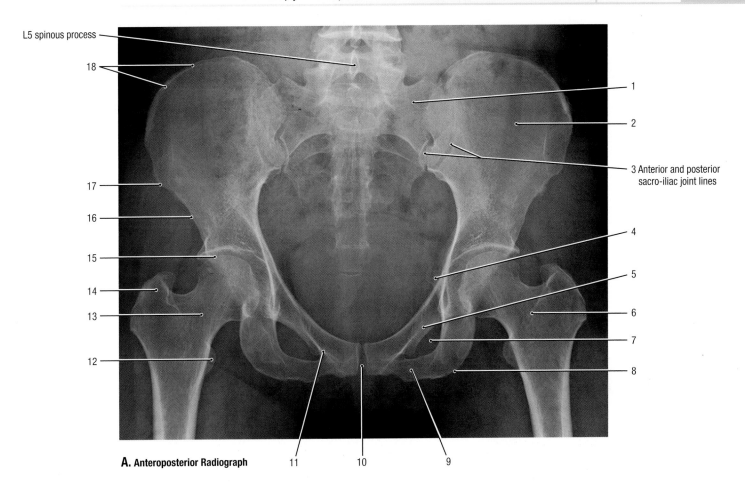

L5 spinous process

18

17

16

15

14

13

12

1

2

3 Anterior and posterior
sacro-iliac joint lines

4

5

6

7

8

A. Anteroposterior Radiograph   11   10   9

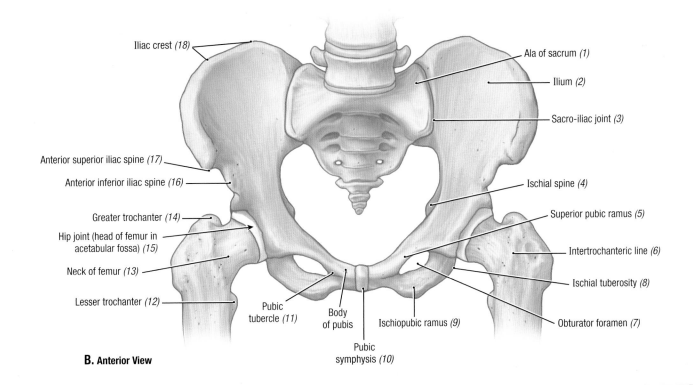

Iliac crest (18)

Anterior superior iliac spine (17)

Anterior inferior iliac spine (16)

Greater trochanter (14)

Hip joint (head of femur in
acetabular fossa) (15)

Neck of femur (13)

Lesser trochanter (12)

Pubic
tubercle (11)

Body
of pubis

Pubic
symphysis (10)

Ala of sacrum (1)

Ilium (2)

Sacro-iliac joint (3)

Ischial spine (4)

Superior pubic ramus (5)

Intertrochanteric line (6)

Ischial tuberosity (8)

Ischiopubic ramus (9)

Obturator foramen (7)

B. Anterior View

PELVIS   1.17

A. Radiograph of pelvis. **B.** Bony pelvis with articulated femora.

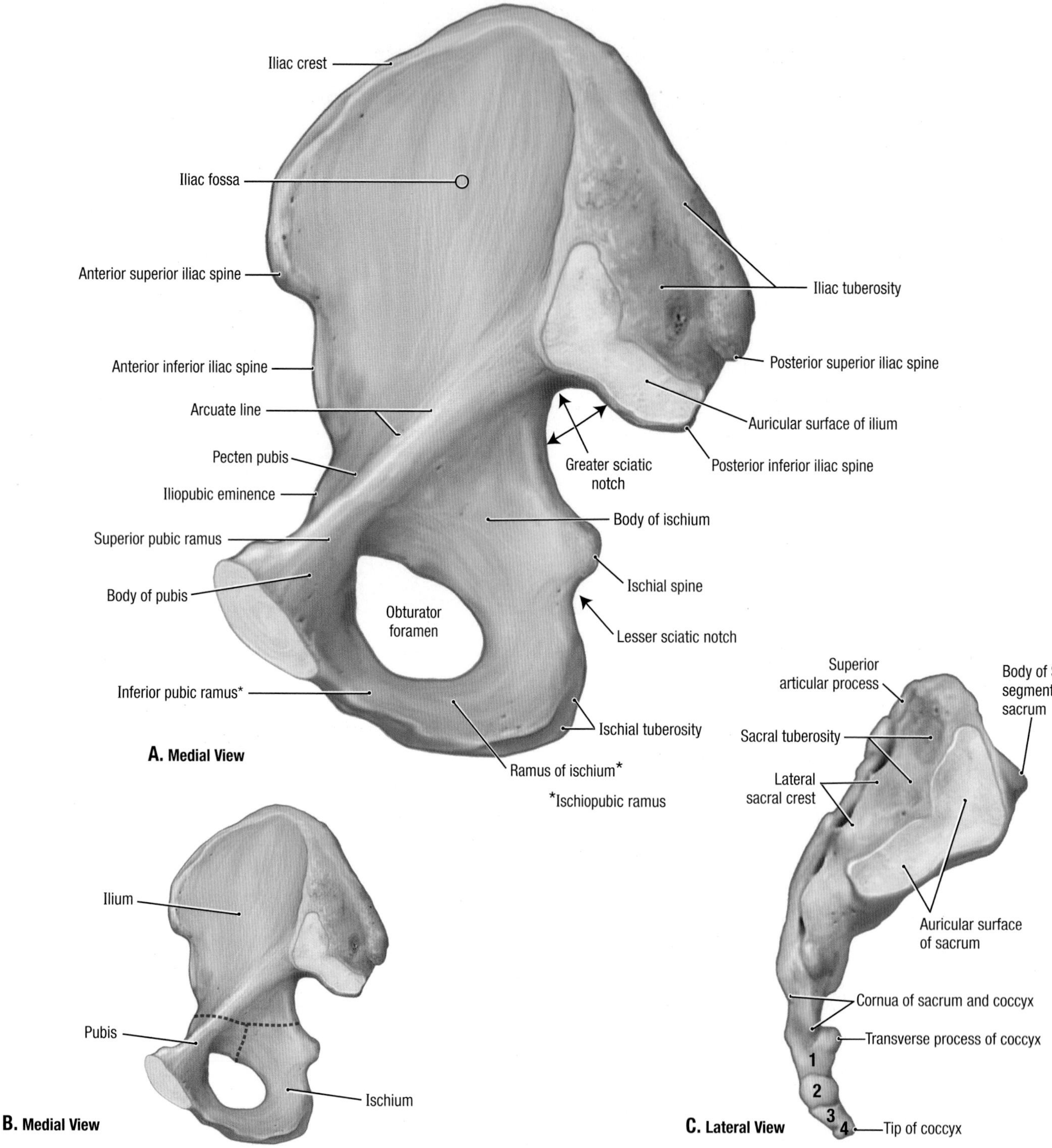

Iliac crest

Iliac fossa

Anterior superior iliac spine

Iliac tuberosity

Anterior inferior iliac spine

Posterior superior iliac spine

Arcuate line

Auricular surface of ilium

Pecten pubis

Greater sciatic notch

Posterior inferior iliac spine

Iliopubic eminence

Superior pubic ramus

Body of ischium

Body of pubis

Ischial spine

Obturator foramen

Lesser sciatic notch

Inferior pubic ramus*

Ischial tuberosity

A. Medial View

Ramus of ischium*

*Ischiopubic ramus

Ilium

Superior articular process

Body of S1 segment of sacrum

Sacral tuberosity

Lateral sacral crest

Auricular surface of sacrum

Pubis

Cornua of sacrum and coccyx

Transverse process of coccyx

Ischium

1
2
3
4   Tip of coccyx

B. Medial View

C. Lateral View

| **1.18** | HIP BONE, SACRUM, AND COCCYX |

A. Features of hip bone. **B.** Ilium, ischium, and pubis. **C.** Sacrum and coccyx.

- Each hip bone consists of three bones: ilium, ischium, and pubis.
- Anterosuperiorly, the auricular, ear-shaped surface of the sacrum articulates with the auricular surface of the ilium; the sacral and

iliac tuberosities are for the attachment of the posterior sacro-iliac and interosseous sacro-iliac ligaments.

- The five sacral vertebrae are fused to form the sacrum.

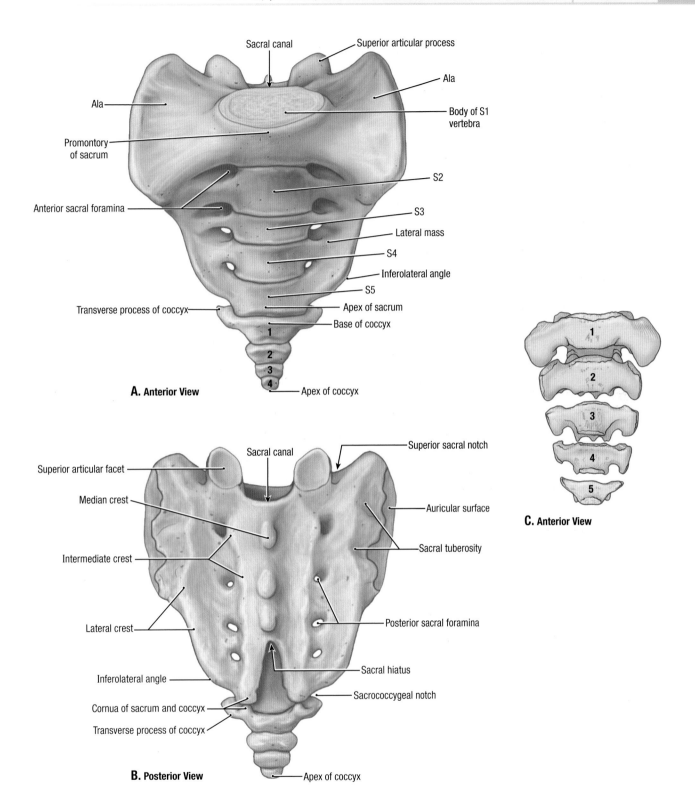

A. Anterior View

Sacral canal

Superior articular process

Ala

Ala

Body of S1 vertebra

Promontory of sacrum

S2

Anterior sacral foramina

S3

Lateral mass

S4

Inferolateral angle

S5

Transverse process of coccyx

Apex of sacrum

Base of coccyx

1
2
3
4

Apex of coccyx

B. Posterior View

Sacral canal

Superior sacral notch

Superior articular facet

Median crest

Auricular surface

Intermediate crest

Sacral tuberosity

Lateral crest

Posterior sacral foramina

Inferolateral angle

Sacral hiatus

Cornua of sacrum and coccyx

Sacrococcygeal notch

Transverse process of coccyx

Apex of coccyx

C. Anterior View

1
2
3
4
5

SACRUM AND COCCYX

1.19

A. Pelvic (anterior) surface. **B.** Dorsal (posterior) surface. **C.** Sacrum in youth.

- The bodies of the five sacral vertebrae are demarcated in the mature sacrum by four transverse lines ending laterally in four pairs of anterior sacral foramina **(A)**. The coccyx has four vertebrae

(segments)—the first having a pair of transverse processes and a pair of cornua (horns).

- The ossification and fusion of the sacral vertebrae may not be complete until age 35.

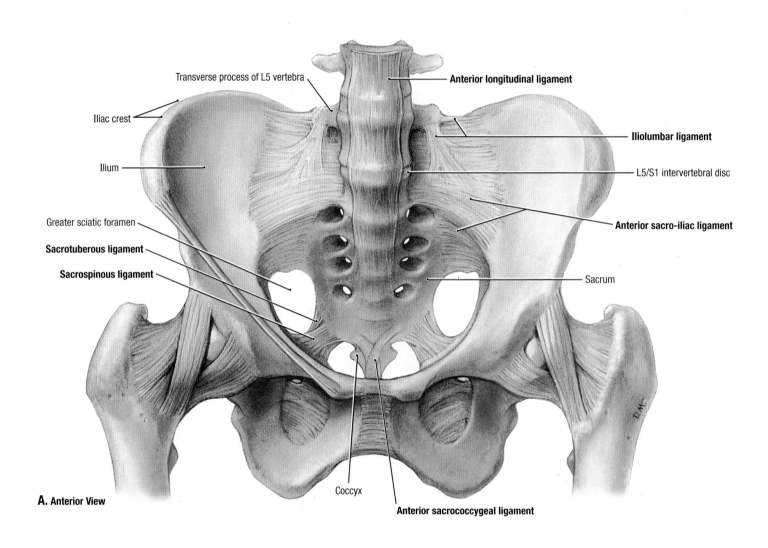

Transverse process of L5 vertebra

Iliac crest

Ilium

Greater sciatic foramen

Sacrotuberous ligament

Sacrospinous ligament

Anterior longitudinal ligament

Iliolumbar ligament

L5/S1 intervertebral disc

Anterior sacro-iliac ligament

Sacrum

Coccyx

Anterior sacrococcygeal ligament

A. Anterior View

1.20   LUMBAR AND PELVIC LIGAMENTS

The anterior sacro-iliac ligament is part of the fibrous capsule of the sacro-iliac joint anteriorly and spans between the lateral aspect of the sacrum and the ilium, anterior to the auricular surfaces.

During **pregnancy**, the pelvic joints and ligaments relax, and pelvic movements increase. The sacro-iliac interlocking mechanism is less effective because the relaxation permits greater rotation of the pelvis and contributes to the lordotic posture often assumed during pregnancy with the change in the center of gravity. Relaxation of the sacro-iliac joints and pubic symphysis permits as much as 10% to 15% increase in diameters (mostly transverse), facilitating passage of the fetus through the pelvic canal. The coccyx is also allowed to move posteriorly.

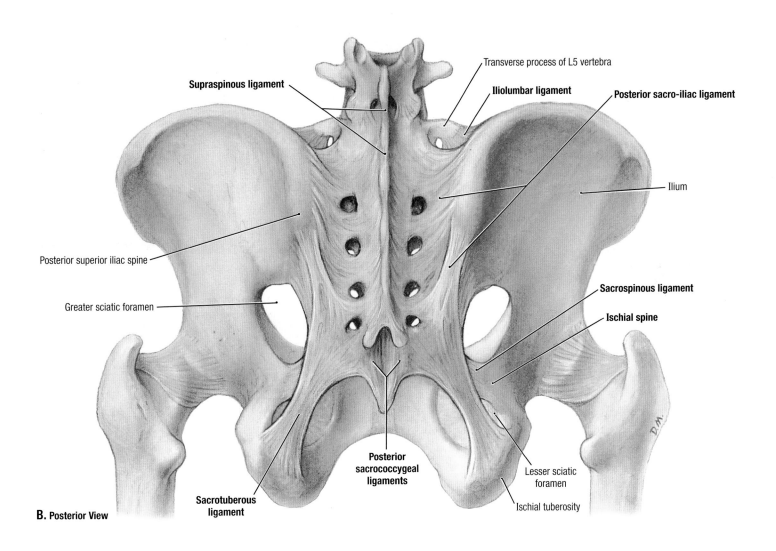

Transverse process of L5 vertebra

Supraspinous ligament

Iliolumbar ligament

Posterior sacro-iliac ligament

Ilium

Posterior superior iliac spine

Sacrospinous ligament

Greater sciatic foramen

Ischial spine

Posterior sacrococcygeal ligaments

Lesser sciatic foramen

Sacrotuberous ligament

Ischial tuberosity

B. Posterior View

LUMBAR AND PELVIC LIGAMENTS (*continued*)

`1.20`

- The sacrotuberous ligaments attach the sacrum, ilium, and coccyx to the ischial tuberosity; the sacrospinous ligaments unite the sacrum and coccyx to the ischial spine. The sacrotuberous and sacrospinous ligaments convert the sciatic notches of the hip bones into greater and lesser sciatic foramina.
- The fibers of the posterior sacro-iliac ligament vary in obliquity; the superior fibers are shorter and lie between the ilium and superior part of the sacrum; the longer, obliquely oriented inferior fibers span between the posterior superior iliac spine and the inferior part of the sacrum.
- The iliolumbar ligaments unite the ilia and transverse processes of L5.

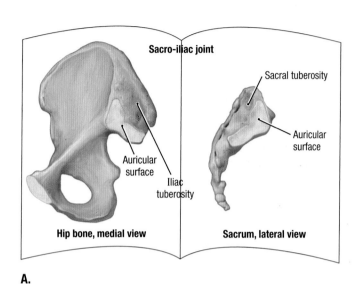

A.

Sacro-iliac joint

Sacral tuberosity

Auricular surface

Iliac tuberosity

Auricular surface

Hip bone, medial view

Sacrum, lateral view

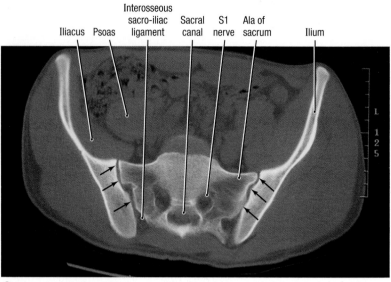

C. Transverse (Axial) CT Image

Iliacus Psoas Interosseous sacro-iliac ligament Sacral canal S1 nerve Ala of sacrum Ilium

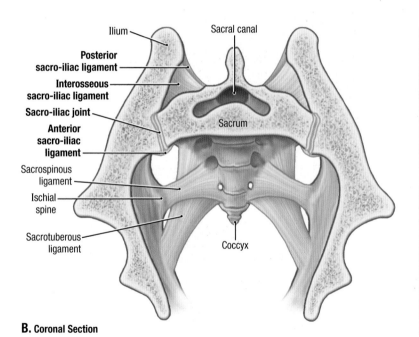

B. Coronal Section

Ilium

Sacral canal

Posterior sacro-iliac ligament

Interosseous sacro-iliac ligament

Sacro-iliac joint

Anterior sacro-iliac ligament

Sacrospinous ligament

Ischial spine

Sacrotuberous ligament

Sacrum

Coccyx

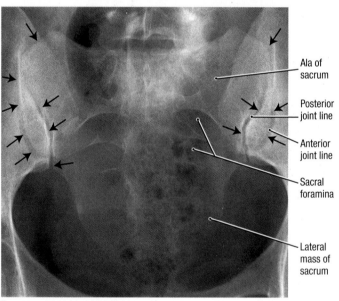

D. Anteroposterior Radiograph

Ala of sacrum

Posterior joint line

Anterior joint line

Sacral foramina

Lateral mass of sacrum

1.21   ARTICULAR SURFACES OF SACRO-ILIAC JOINT AND LIGAMENTS

A. Articular surfaces. Note the auricular surface *(blue)* of the sacrum and hip bone and the roughened areas superior and posterior to the auricular areas for the attachment of the interosseous sacro-iliac ligament. **B.** Sacro-iliac ligaments. The interosseous sacro-iliac ligament consists of short fibers connecting the sacral tuberosity to the iliac tuberosity. **C.** CT image. The sacro-iliac joint is indicated *(arrows)*. Note that the articular surfaces of the ilium and sacrum have irregular shapes that result in partial interlocking of the bones. **D.** Radiograph. Due to the oblique placement of the sacro-iliac joints, the anterior and posterior joint lines appear separately.

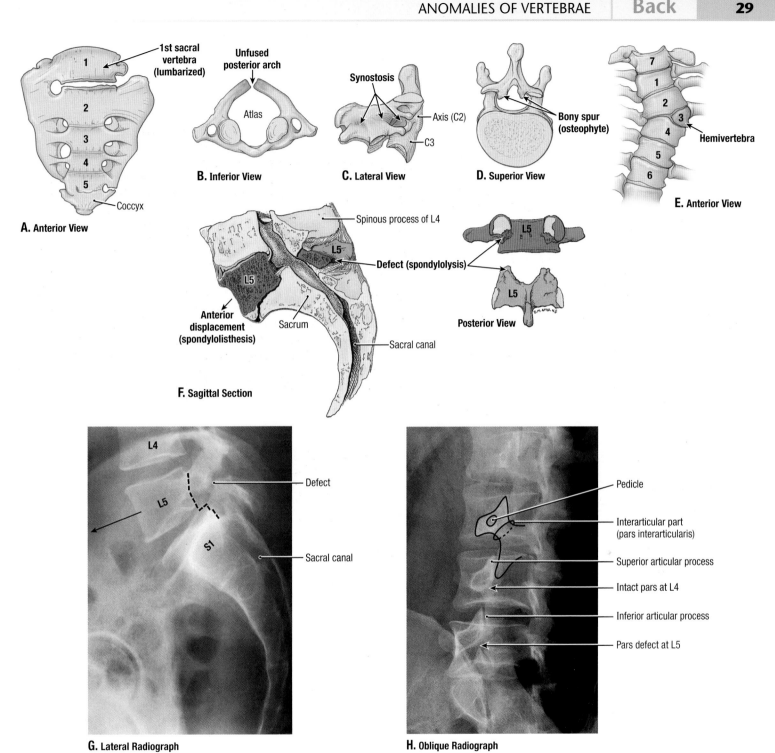

1st sacral vertebra (lumbarized)

1
2
3
4
5

Coccyx

A. Anterior View

Unfused posterior arch

Atlas

B. Inferior View

Synostosis

Axis (C2)

C3

C. Lateral View

Bony spur (osteophyte)

D. Superior View

7
1
2
3
4
5
6

Hemivertebra

E. Anterior View

Spinous process of L4

L5

Defect (spondylolysis)

L5

L5

L5

Posterior View

Anterior displacement (spondylolisthesis)

Sacrum

Sacral canal

F. Sagittal Section

L4

L5

Defect

S1

Sacral canal

G. Lateral Radiograph

Pedicle

Interarticular part (pars interarticularis)

Superior articular process

Intact pars at L4

Inferior articular process

Pars defect at L5

H. Oblique Radiograph

ANOMALIES OF VERTEBRAE AND SPONDYLOLYSIS AND SPONDYLOLISTHESIS | 1.22

A. Transitional lumbosacral vertebra. Here, the 1st sacral vertebra is partly free (lumbarized). Not uncommonly, the 5th lumbar vertebra may be partly fused to the sacrum (sacralized). **B.** Unfused posterior arch of the atlas. **C.** Synostosis (fusion) of vertebrae C2 (axis) and C3. **D.** Bony spurs. Sharp bony spurs may grow from the laminae inferiorly into the ligamenta flava. **E.** Hemivertebra. The entire right half of vertebra T3 and the corresponding rib are absent. The left lamina and the spine are fused with those of T4, and the left IV foramen is reduced in size. Observe the associated scoliosis (lateral curvature of the spine). **F.** Articulated and isolated spondylolytic L5 vertebra. The vertebra has an oblique defect (spondylolysis) through the interarticular part (pars interarticularis). Also, the vertebral body of L5 has slipped anteriorly (spondylolisthesis). **G.** and **H.** Radiographs. The posterior vertebral margins of L5 *(dotted line)* and the sacrum shows the anterior displacement of L5 *(arrow)* **(G)**. Note the superimposed outline of a dog: the nose is the transverse process, the eye is the pedicle, the neck is the interarticular part and the ear is the superior articular process **(H)**. The lucent *(dark)* cleft across the "neck" of the dog is the **spondylolysis**; the anterior displacement *(arrow)* is the **spondylolisthesis**.

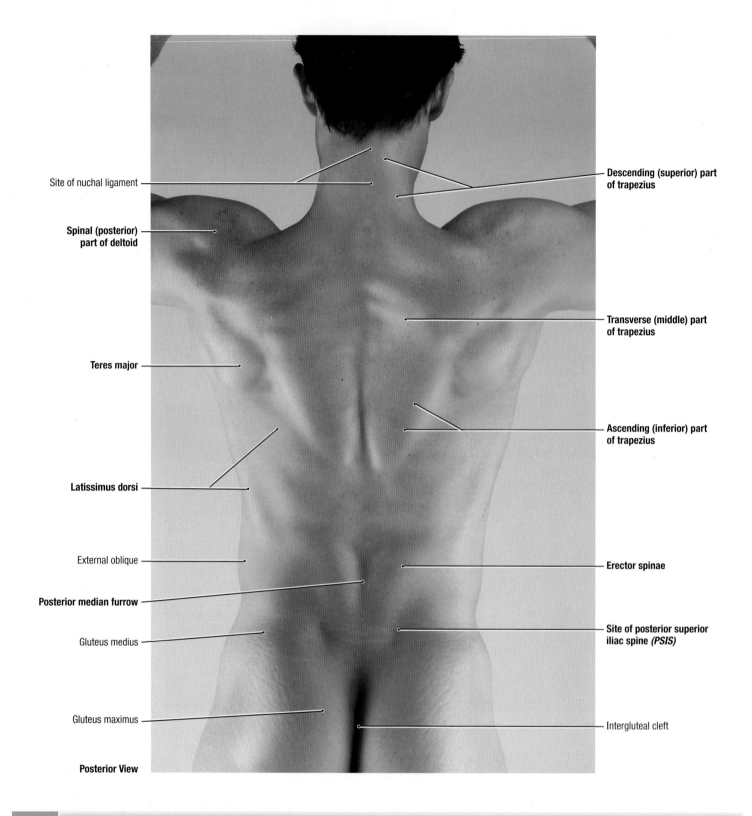

Site of nuchal ligament

Spinal (posterior) part of deltoid

Teres major

Latissimus dorsi

External oblique

Posterior median furrow

Gluteus medius

Gluteus maximus

Posterior View

Descending (superior) part of trapezius

Transverse (middle) part of trapezius

Ascending (inferior) part of trapezius

Erector spinae

Site of posterior superior iliac spine (PSIS)

Intergluteal cleft

1.23   SURFACE ANATOMY OF BACK

- The arms are abducted, so the scapulae have rotated superiorly on the thoracic wall.
- The latissimus dorsi and teres major muscles form the posterior axillary fold.
- The trapezius muscle has three parts: descending, transverse, and ascending.

- Note the deep median furrow that separates the longitudinal bulges formed by the contracted erector spinae group of muscles.
- Dimples (depressions) indicate the site of the posterior superior iliac spines, which usually lie at the level of the sacro-iliac joints.

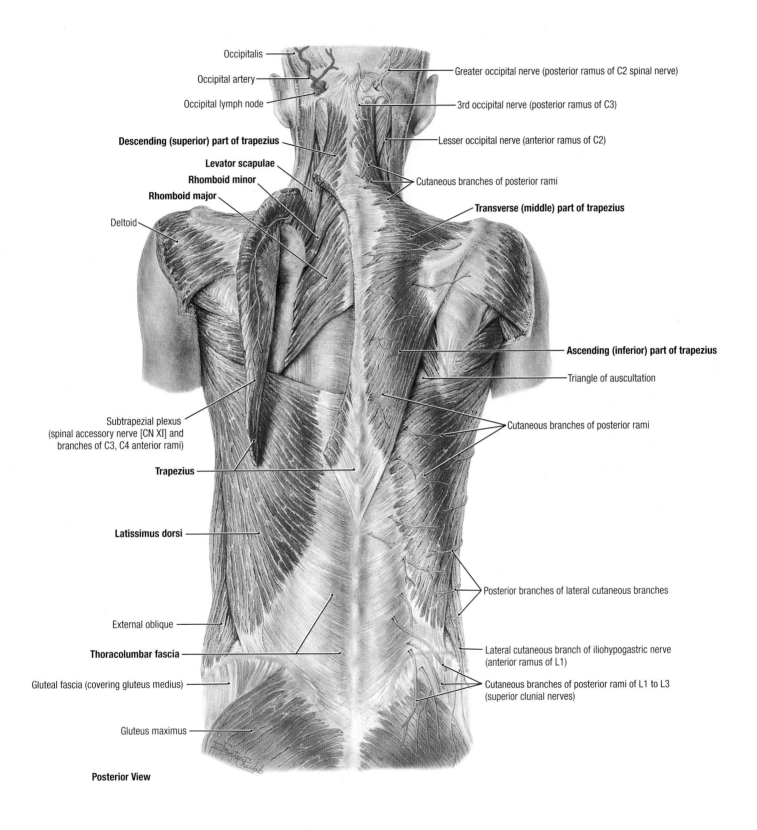

Occipitalis

Occipital artery

Occipital lymph node

Descending (superior) part of trapezius

Levator scapulae

Rhomboid minor

Rhomboid major

Deltoid

Subtrapezial plexus
(spinal accessory nerve [CN XI] and
branches of C3, C4 anterior rami)

Trapezius

Latissimus dorsi

External oblique

Thoracolumbar fascia

Gluteal fascia (covering gluteus medius)

Gluteus maximus

Posterior View

Greater occipital nerve (posterior ramus of C2 spinal nerve)

3rd occipital nerve (posterior ramus of C3)

Lesser occipital nerve (anterior ramus of C2)

Cutaneous branches of posterior rami

Transverse (middle) part of trapezius

Ascending (inferior) part of trapezius

Triangle of auscultation

Cutaneous branches of posterior rami

Posterior branches of lateral cutaneous branches

Lateral cutaneous branch of iliohypogastric nerve
(anterior ramus of L1)

Cutaneous branches of posterior rami of L1 to L3
(superior clunial nerves)

SUPERFICIAL MUSCLES OF BACK

1.24

The left trapezius muscle is reflected. Observe two layers: the trapezius and latissimus dorsi muscles, and the levator scapulae and rhomboids minor and major. These axio-appendicular muscles help attach the upper limb to the trunk.

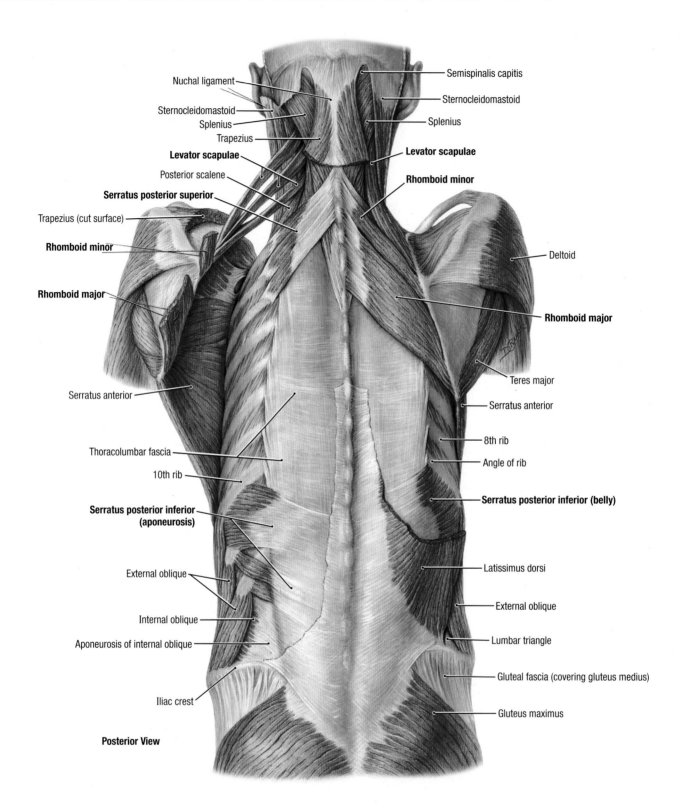

Nuchal ligament

Sternocleidomastoid

Splenius

Trapezius

Levator scapulae

Posterior scalene

Serratus posterior superior

Trapezius (cut surface)

Rhomboid minor

Rhomboid major

Serratus anterior

Thoracolumbar fascia

10th rib

Serratus posterior inferior (aponeurosis)

External oblique

Internal oblique

Aponeurosis of internal oblique

Iliac crest

Posterior View

Semispinalis capitis

Sternocleidomastoid

Splenius

Levator scapulae

Rhomboid minor

Deltoid

Rhomboid major

Teres major

Serratus anterior

8th rib

Angle of rib

Serratus posterior inferior (belly)

Latissimus dorsi

External oblique

Lumbar triangle

Gluteal fascia (covering gluteus medius)

Gluteus maximus

1.25 INTERMEDIATE MUSCLES OF BACK

The trapezius and latissimus dorsi muscles are largely cut away on both sides. The left rhomboid muscles have been reflected, allowing the vertebral border of the scapula to be raised from the thoracic wall. The serratus posterior superior and inferior form the intermediate layer of muscles, passing from the vertebral spines to the ribs; the two muscles slope in opposite directions and are accessory muscles of respiration. The thoracolumbar fascia extends laterally to the angles of the ribs, becoming thin superiorly and passing deep to the serratus posterior superior muscle. The fascia gives attachment to the latissimus dorsi and serratus posterior inferior muscles (see Fig. 1.30).

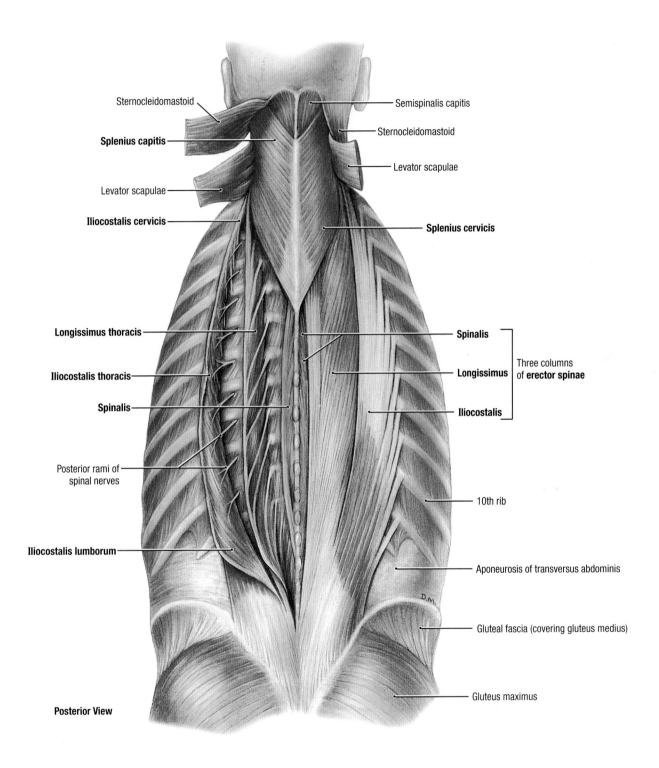

Sternocleidomastoid

Splenius capitis

Levator scapulae

Iliocostalis cervicis

Longissimus thoracis

Iliocostalis thoracis

Spinalis

Posterior rami of
spinal nerves

Iliocostalis lumborum

Posterior View

Semispinalis capitis

Sternocleidomastoid

Levator scapulae

Splenius cervicis

Spinalis

Longissimus

Iliocostalis

Three columns
of erector spinae

10th rib

Aponeurosis of transversus abdominis

Gluteal fascia (covering gluteus medius)

Gluteus maximus

DEEP MUSCLES OF BACK: SPLENIUS AND ERECTOR SPINAE

1.26

The right erector spinae muscles are *in situ*, lying between the spinous processes medially and the angles of the ribs laterally. The erector spinae are split into three longitudinal columns: iliocostalis laterally, longissimus in the middle, and spinalis medially. The left longissimus muscle is pulled laterally to show the insertion into the transverse processes and ribs; not shown here are its extensions to the neck and head, longissimus cervicis and capitis.

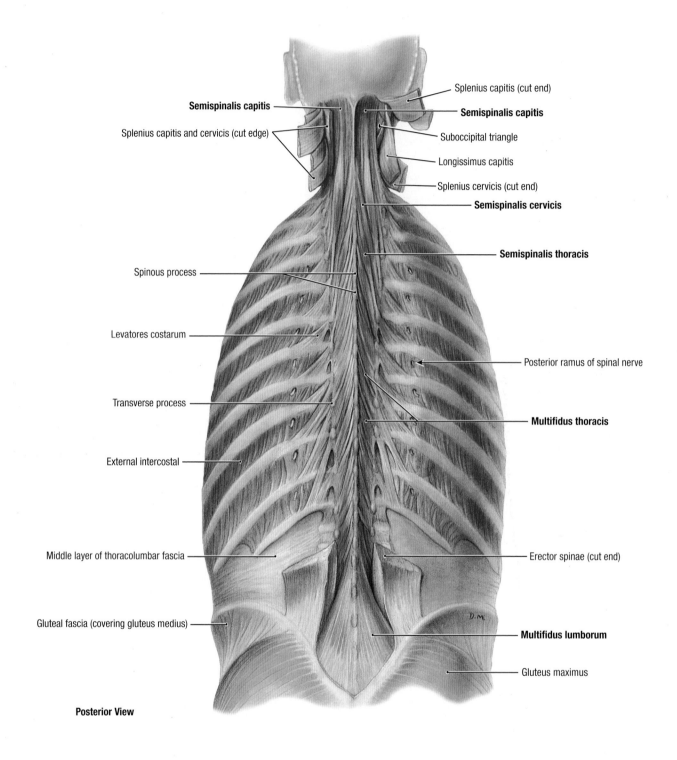

Semispinalis capitis

Splenius capitis and cervicis (cut edge)

Splenius capitis (cut end)

Semispinalis capitis

Suboccipital triangle

Longissimus capitis

Splenius cervicis (cut end)

Semispinalis cervicis

Semispinalis thoracis

Spinous process

Levatores costarum

Posterior ramus of spinal nerve

Transverse process

Multifidus thoracis

External intercostal

Middle layer of thoracolumbar fascia

Erector spinae (cut end)

Gluteal fascia (covering gluteus medius)

Multifidus lumborum

Gluteus maximus

Posterior View

1.27   DEEP MUSCLES OF BACK: SEMISPINALIS AND MULTIFIDUS

- The semispinalis, multifidus, and rotatores muscles constitute the transversospinalis group of deep muscles. In general, their bundles pass obliquely in a superomedial direction, from transverse processes to spinous processes in successively deeper layers. The bundles of semispinalis span approximately five interspaces, those of multifidus, approximately three, and those of rotatores, one or two.

- The semispinalis (thoracis, cervicis, and capitis) muscles span the lower thoracic region to the cranium.
- The multifidus muscle extends from the sacrum to the spinous process of the axis. In the lumbosacral region, it emerges from the aponeurosis of the erector spinae and extends from the sacrum, and mammillary processes of the lumbar vertebrae, to insert into spinous processes approximately three segments higher.

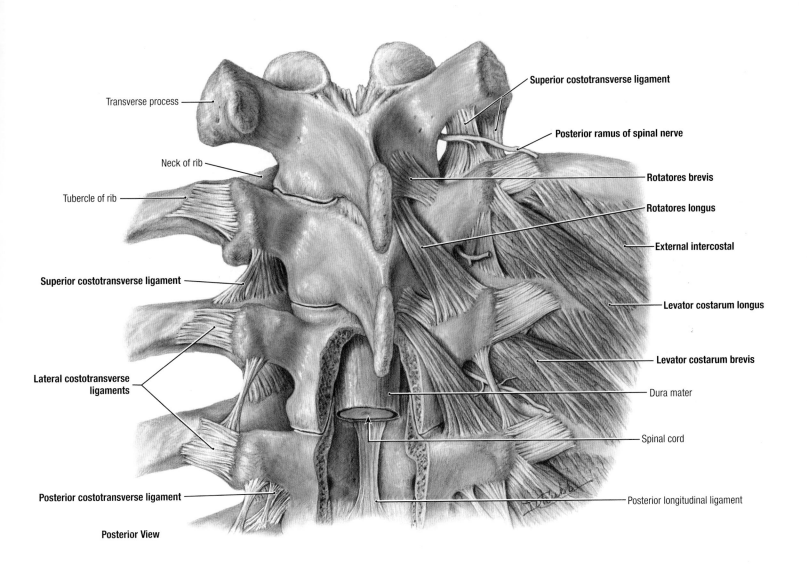

Transverse process

Superior costotransverse ligament

Neck of rib

Posterior ramus of spinal nerve

Tubercle of rib

Rotatores brevis

Rotatores longus

External intercostal

Superior costotransverse ligament

Levator costarum longus

Levator costarum brevis

Lateral costotransverse ligaments

Dura mater

Spinal cord

Posterior costotransverse ligament

Posterior longitudinal ligament

Posterior View

ROTATORES AND COSTOTRANSVERSE LIGAMENTS

1.28

- Of the three layers of transversospinalis muscles, the rotatores are the deepest and shortest. They pass from the root of one transverse process superomedially to the junction of the transverse process and lamina of the vertebra above. Rotatores longus span two vertebrae.
- The levatores costarum pass from the tip of one transverse process inferiorly to the rib below (brevis); some span two ribs (longus).

- The posterior ramus passes posterior to the superior costotransverse ligament.
- The lateral costotransverse ligament is strong and joins the tubercle of the rib to the tip of the transverse process. It forms the posterior aspect of the joint capsule of the costotransverse joint.

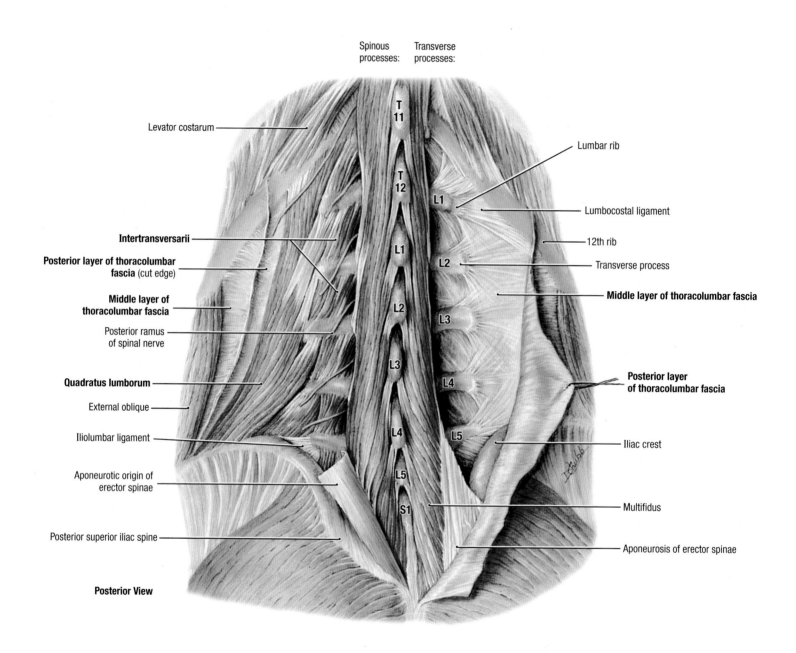

Spinous processes:

Transverse processes:

Levator costarum

Lumbar rib

T 11

T 12

L1

Lumbocostal ligament

Intertransversarii

12th rib

Posterior layer of thoracolumbar fascia (cut edge)

L1

L2

Transverse process

Middle layer of thoracolumbar fascia

Middle layer of thoracolumbar fascia

Posterior ramus of spinal nerve

L2

L3

Quadratus lumborum

L3

L4

Posterior layer of thoracolumbar fascia

External oblique

Iliolumbar ligament

L4

L5

Iliac crest

Aponeurotic origin of erector spinae

L5

S1

Multifidus

Posterior superior iliac spine

Aponeurosis of erector spinae

Posterior View

1.29 BACK: MULTIFIDUS, QUADRATUS LUMBORUM, AND THORACOLUMBAR FASCIA

After removal of right erector spinae at the L1 level, the middle layer of thoracolumbar fascia is seen to extend from the tip of each lumbar transverse process in a fan-shaped manner. A short lumbar rib is present at the level of L1.

After removal of the left posterior and middle layers of thoracolumbar fascia, the lateral border of the quadratus lumborum muscle is oblique, and the medial border is in continuity with the intertransversarii.

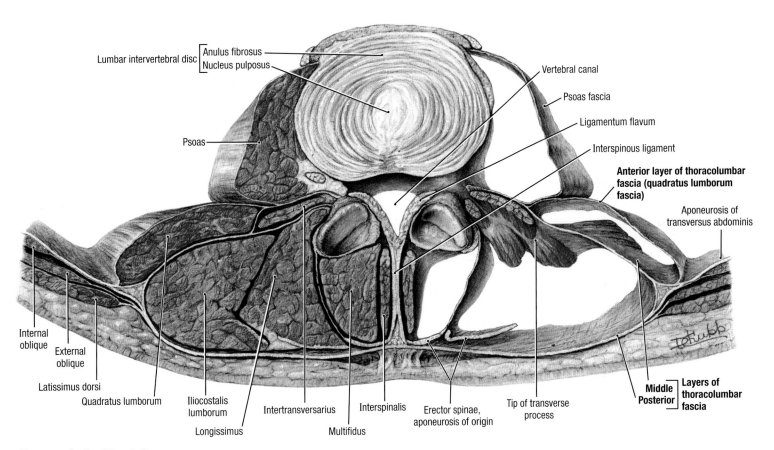

Lumbar intervertebral disc [Anulus fibrosus
Nucleus pulposus

Vertebral canal

Psoas fascia

Ligamentum flavum

Interspinous ligament

Anterior layer of thoracolumbar fascia (quadratus lumborum fascia)

Psoas

Aponeurosis of transversus abdominis

Internal oblique

External oblique

Latissimus dorsi

Quadratus lumborum

Iliocostalis lumborum

Longissimus

Intertransversarius

Interspinalis

Multifidus

Erector spinae, aponeurosis of origin

Tip of transverse process

Middle / Posterior] **Layers of thoracolumbar fascia**

Transverse Section (Dissected), Superior View

TRANSVERSE SECTION OF BACK MUSCLES AND THORACOLUMBAR FASCIA

1.30

- The left muscles are seen in their fascial sheaths or compartments; the right muscles have been removed from their sheaths.
- The aponeurosis of transversus abdominis and posterior aponeurosis of internal oblique muscles split into two strong sheets, the middle and posterior layers of thoracolumbar fascia. The anterior layer of thoracolumbar fascia is the deep fascia of the quadratus lumborum (quadratus lumborum fascia). The posterior layer of the thoracolumbar fascia provides proximal attachment for

the latissimus dorsi muscle and, at a higher level, the serratus posterior inferior muscle.

Back strain is a common back problem that usually results from extreme movements of the vertebral column, such as extension or rotation. Back strain refers to some stretching or microscopic tearing of muscle fibers and/or ligaments of the back. The muscles usually involved are those producing movements of the lumbar IV joints.

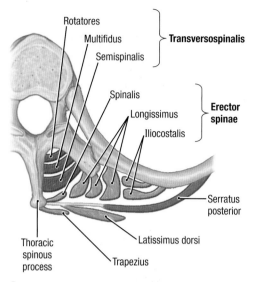

A. Transverse Section

Key for A: Back muscles

- Superficial extrinsic
- Intermediate extrinsic
- Erector spinae (intermediate intrinsic)
- Transversospinales (deep intrinsic)

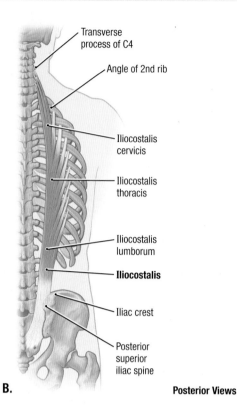

B.   **Posterior Views**   **C.**

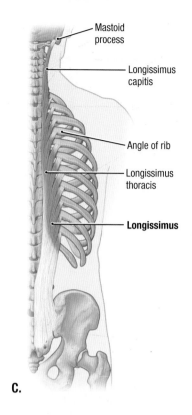

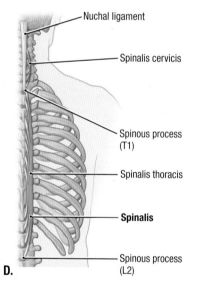

D.

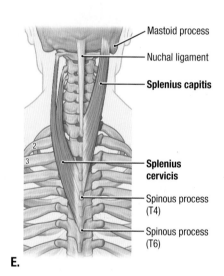

E.

1.31 SUPERFICIAL AND INTERMEDIATE LAYERS OF INTRINSIC BACK MUSCLES

A. Transverse section. The erector spinae consists of three columns and the transversospinalis consists of three layers. **B.** Iliocostalis. **C.** Longissimus. **D.** Spinalis. **E.** Splenius capitis and cervicis.

TABLE 1.4	**SUPERFICIAL AND INTERMEDIATE LAYERS OF INTRINSIC BACK MUSCLES**				
Muscles	*Caudal (Inferior) Attachment*	*Rostral (Superior) Attachment*	*Nerve Supply*	*Main Actions*	
Superficial layer Splenius	Nuchal ligament and spinous processes of C7–T6 vertebrae	*Splenius capitis:* fibers run superolaterally to mastoid process of temporal bone and lateral third of superior nuchal line of occipital bone *Splenius cervicis:* posterior tubercles of transverse processes of C1–C3/C4 vertebrae	Posterior rami of spinal nerves	*Acting unilaterally:* laterally flex neck and rotate head to side of active muscles *Acting bilaterally:* extend head and neck	
Intermediate layer Erector spinae	Arises by a broad tendon from posterior part of iliac crest, posterior surface of sacrum, sacral and inferior lumbar spinous processes, and supraspinous ligament	*Iliocostalis (lumborum, thoracis, and cervicis):* fibers run superiorly to angles of lower ribs and cervical transverse processes *Longissimus (thoracis, cervicis, and capitis):* fibers run superiorly to ribs between tubercles and angles to transverse processes in thoracic and cervical regions, and to mastoid process of temporal bone *Spinalis (thoracis, cervicis, and capitis):* fibers run superiorly to spinous processes in the upper thoracic region and to skull		*Acting unilaterally:* laterally bend vertebral column to side of active muscles *Acting bilaterally:* extend vertebral column and head; as back is flexed, control movement by gradually lengthening their fibers	

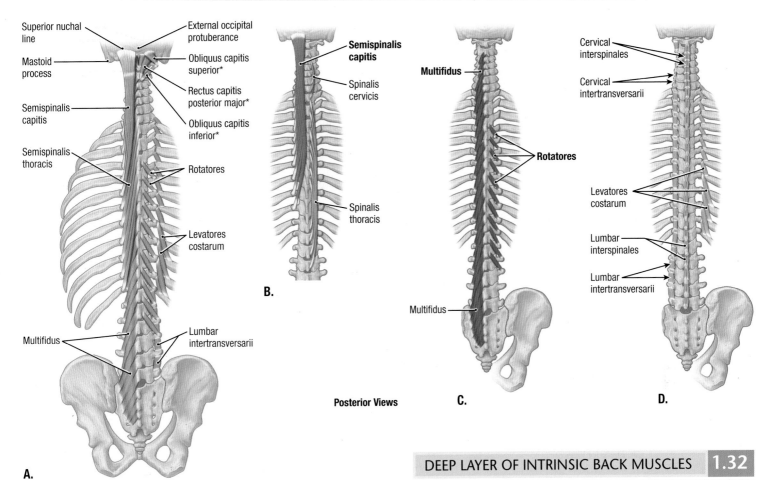

Superior nuchal line
External occipital protuberance
Mastoid process
Obliquus capitis superior*
Semispinalis capitis
Rectus capitis posterior major*
Obliquus capitis inferior*
Semispinalis thoracis
Rotatores
Levatores costarum
Multifidus
Lumbar intertransversarii

A.

Semispinalis capitis
Spinalis cervicis
Rotatores
Spinalis thoracis

B.

Multifidus
Rotatores
Multifidus

Posterior Views

C.

Cervical interspinales
Cervical intertransversarii
Levatores costarum
Lumbar interspinales
Lumbar intertransversarii

D.

DEEP LAYER OF INTRINSIC BACK MUSCLES 1.32

A. Overview. **B.** Semispinalis. **C.** Multifidus and rotatores. **D.** Interspinalis, intertransversarii, and levatores costarum.

TABLE 1.5	**DEEP LAYERS OF INTRINSIC BACK MUSCLES**			
Muscles	*Caudal (Inferior) Attachment*	*Rostral (Superior) Attachment*	*Nerve Supply[a]*	*Main Actions*
Deep layer Transversospinalis	*Semispinalis:* arises from thoracic and cervical transverse processes *Multifidus:* arises from sacrum and ilium, transverse processes of T1–L5, and articular processes of C4–C7 *Rotatores:* arise from transverse processes of vertebrae; best developed in thoracic region	*Semispinalis: thoracis, cervicis, and capitis:* fibers run superomedially and attach to occipital bone and spinous processes in thoracic and cervical regions, spanning four to six segments *Multifidus (lumborum, thoracis, and cervicis):* fibers pass superomedially to spinous processes, spanning two to four segments *Rotatores (thoracis and cervicis):* Pass superomedially and attach to junction of lamina and transverse process of vertebra of origin or into spinous process above their origin, spanning one to two segments	Posterior rami of spinal nerves	**Extension** *Semispinalis:* extends head and thoracic and cervical regions of vertebral column and rotates them contralaterally *Multifidus:* stabilizes vertebrae during local movement of vertebral column *Rotatores:* stabilize vertebrae and assist with local extension and rotary movements of vertebral column; may function as organ of proprioception
Minor deep layer Interspinales	Superior surfaces of spinous processes of cervical and lumbar vertebrae	Inferior surfaces of spinous processes of vertebrae superior to vertebrae of origin		Aid in extension and rotation of vertebral column
Intertransversarii	Transverse processes of cervical and lumbar vertebrae	Transverse processes of adjacent vertebrae	Posterior and anterior rami of spinal nerves	Aid in lateral flexion of vertebral column *Acting bilaterally:* stabilize vertebral column
Levatores costarum	**Medial attachment:** tips of transverse processes of C7 and T1–T11 vertebrae	**Lateral attachment:** pass inferolaterally and insert on rib between its tubercle and angle	Posterior rami of C8–T11 spinal nerves	Elevate ribs, assisting inspiration Assist with lateral flexion of vertebral column

[a]Most back muscles are innervated by posterior rami of spinal nerves, but a few are innervated by anterior and posterior rami.

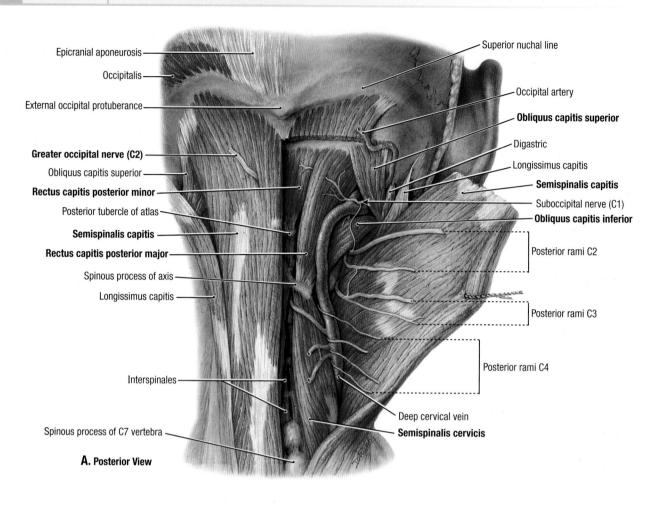

Epicranial aponeurosis

Occipitalis

External occipital protuberance

Greater occipital nerve (C2)

Obliquus capitis superior

Rectus capitis posterior minor

Posterior tubercle of atlas

Semispinalis capitis

Rectus capitis posterior major

Spinous process of axis

Longissimus capitis

Interspinales

Spinous process of C7 vertebra

Superior nuchal line

Occipital artery

Obliquus capitis superior

Digastric

Longissimus capitis

Semispinalis capitis

Suboccipital nerve (C1)

Obliquus capitis inferior

Posterior rami C2

Posterior rami C3

Posterior rami C4

Deep cervical vein

Semispinalis cervicis

A. Posterior View

1.33   SUBOCCIPITAL REGION I

A. Superficial dissection. The trapezius, sternocleidomastoid, and splenius muscles are removed. The right semispinalis capitis muscle is cut and reflected laterally. **B.** Transverse section at the level of the axis.

- The semispinalis capitis, the great extensor muscle of the head and neck, forms the posterior wall of the suboccipital region. It is pierced by the greater occipital nerve (posterior ramus of C2) and has free medial and lateral borders at this level.
- The greater occipital nerve, when followed caudally, leads to the inferior border of the obliquus capitis inferior muscle, around which it turns. Following the inferior border of the obliquus capitis inferior muscle medially from the nerve leads to the spinous process of the axis; followed laterally, this leads to the transverse process of the atlas.

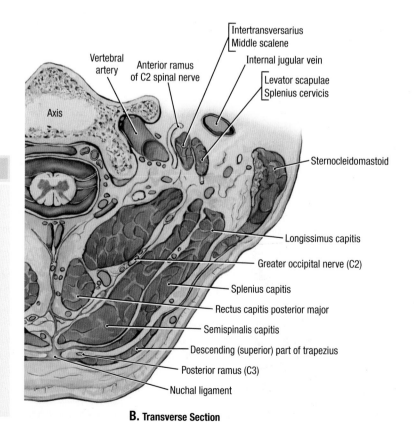

Vertebral artery

Anterior ramus of C2 spinal nerve

Intertransversarius
Middle scalene

Internal jugular vein

Levator scapulae
Splenius cervicis

Sternocleidomastoid

Axis

Longissimus capitis

Greater occipital nerve (C2)

Splenius capitis

Rectus capitis posterior major

Semispinalis capitis

Descending (superior) part of trapezius

Posterior ramus (C3)

Nuchal ligament

B. Transverse Section

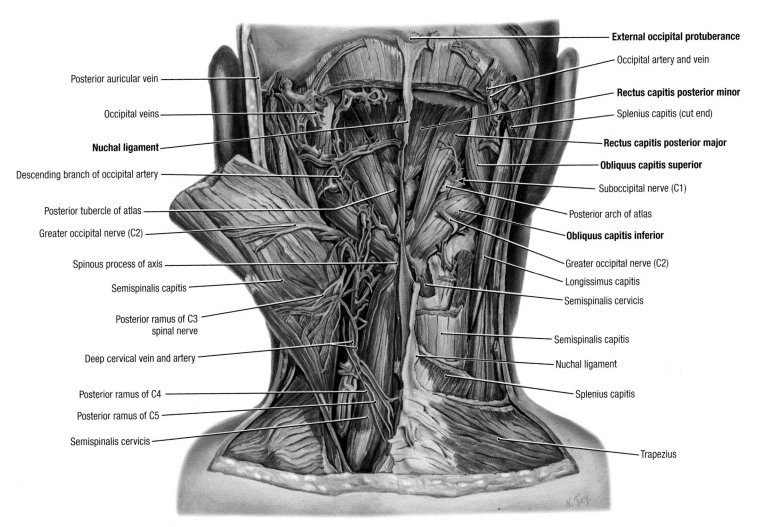

External occipital protuberance

Occipital artery and vein

Rectus capitis posterior minor

Splenius capitis (cut end)

Rectus capitis posterior major

Obliquus capitis superior

Suboccipital nerve (C1)

Posterior arch of atlas

Obliquus capitis inferior

Greater occipital nerve (C2)

Longissimus capitis

Semispinalis cervicis

Semispinalis capitis

Nuchal ligament

Splenius capitis

Trapezius

Posterior auricular vein

Occipital veins

Nuchal ligament

Descending branch of occipital artery

Posterior tubercle of atlas

Greater occipital nerve (C2)

Spinous process of axis

Semispinalis capitis

Posterior ramus of C3 spinal nerve

Deep cervical vein and artery

Posterior ramus of C4

Posterior ramus of C5

Semispinalis cervicis

A. Posterior View

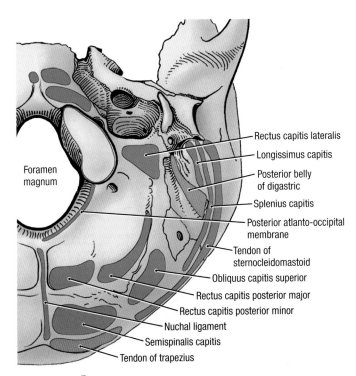

Foramen magnum

Rectus capitis lateralis

Longissimus capitis

Posterior belly of digastric

Splenius capitis

Posterior atlanto-occipital membrane

Tendon of sternocleidomastoid

Obliquus capitis superior

Rectus capitis posterior major

Rectus capitis posterior minor

Nuchal ligament

Semispinalis capitis

Tendon of trapezius

B. Inferior View

SUBOCCIPITAL REGION II

1.34

A. Deep dissection. The left semispinalis capitis is reflected and the right muscle is removed; neck is flexed. **B.** Muscle attachments on the inferior aspect of the cranium.

- The suboccipital region contains four pairs of structures: two straight muscles, the rectus capitis posterior major and minor; two oblique muscles, the obliquus capitis superior and obliquus capitis inferior; two nerves (posterior rami), C1 suboccipital (motor) and C2 greater occipital (sensory); and two arteries, the occipital and vertebral.
- The nuchal ligament, which represents the cervical part of the supraspinous ligament, is a median, thin, fibrous partition attached to the spinous processes of cervical vertebrae and the external occipital protuberance.

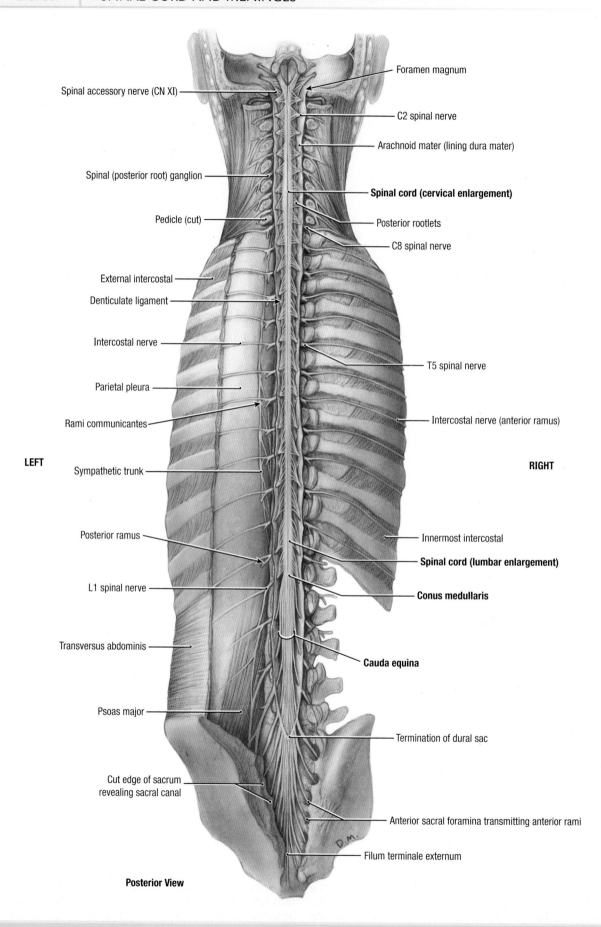

Spinal accessory nerve (CN XI)

Spinal (posterior root) ganglion

Pedicle (cut)

External intercostal

Denticulate ligament

Intercostal nerve

Parietal pleura

Rami communicantes

LEFT

Sympathetic trunk

Posterior ramus

L1 spinal nerve

Transversus abdominis

Psoas major

Cut edge of sacrum revealing sacral canal

Foramen magnum

C2 spinal nerve

Arachnoid mater (lining dura mater)

Spinal cord (cervical enlargement)

Posterior rootlets

C8 spinal nerve

T5 spinal nerve

Intercostal nerve (anterior ramus)

RIGHT

Innermost intercostal

Spinal cord (lumbar enlargement)

Conus medullaris

Cauda equina

Termination of dural sac

Anterior sacral foramina transmitting anterior rami

Filum terminale externum

Posterior View

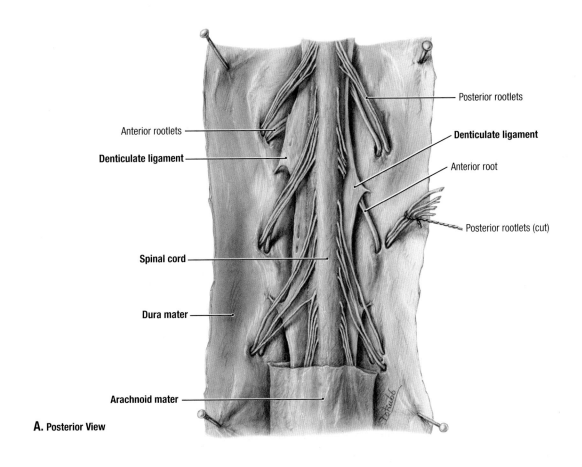

Posterior rootlets

Anterior rootlets

Denticulate ligament

Denticulate ligament

Anterior root

Posterior rootlets (cut)

Spinal cord

Dura mater

Arachnoid mater

A. Posterior View

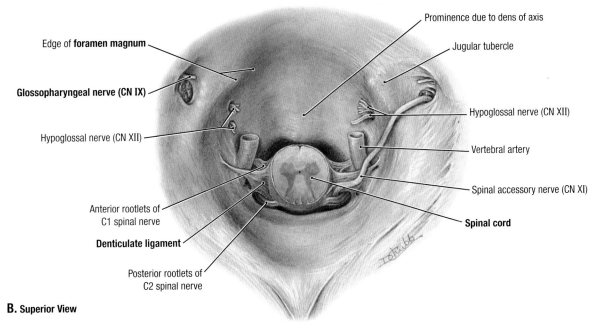

Prominence due to dens of axis

Edge of **foramen magnum**

Jugular tubercle

Glossopharyngeal nerve (CN IX)

Hypoglossal nerve (CN XII)

Hypoglossal nerve (CN XII)

Vertebral artery

Spinal accessory nerve (CN XI)

Anterior rootlets of
C1 spinal nerve

Spinal cord

Denticulate ligament

Posterior rootlets of
C2 spinal nerve

B. Superior View

SPINAL CORD AND MENINGES

1.36

A. Dural sac cut open. The denticulate ligament anchors the cord to the dural sac between successive nerve roots by means of strong, toothlike processes. The anterior nerve roots (rootlets) lie anterior to the denticulate ligament, and the posterior nerve roots (rootlets) lie posterior to the ligament. **B.** Structures of vertebral canal seen through foramen magnum. The spinal cord, vertebral arteries, spinal accessory nerve (CN XI), and most superior part of the denticulate ligament pass through the foramen magnum within the meninges.

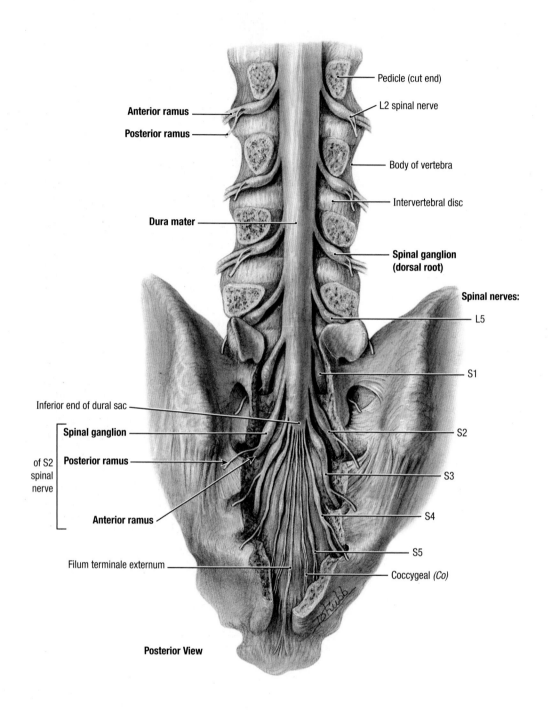

Pedicle (cut end)

L2 spinal nerve

Anterior ramus

Posterior ramus

Body of vertebra

Intervertebral disc

Dura mater

**Spinal ganglion
(dorsal root)**

Spinal nerves:

L5

S1

Inferior end of dural sac

Spinal ganglion

S2

of S2
spinal
nerve

Posterior ramus

S3

Anterior ramus

S4

S5

Filum terminale externum

Coccygeal *(Co)*

Posterior View

1.37   INFERIOR END OF DURAL SAC I

The posterior parts of the lumbar vertebrae and sacrum were re-moved, along with the fat and internal (epidural) venous plexus that occupy the epidural space. Note that the inferior limit of the dural sac is at the level of the posterior superior iliac spine (body of 2nd sacral vertebra); the dura continues as the filum terminale externum.

Epidural anesthesia (block). An anesthetic can be injected into the extradural space. The anesthetic has direct effect on the spinal nerve roots in the epidural space. The patient loses sensation infe-rior to the level of the block (see Fig. 1.38C).

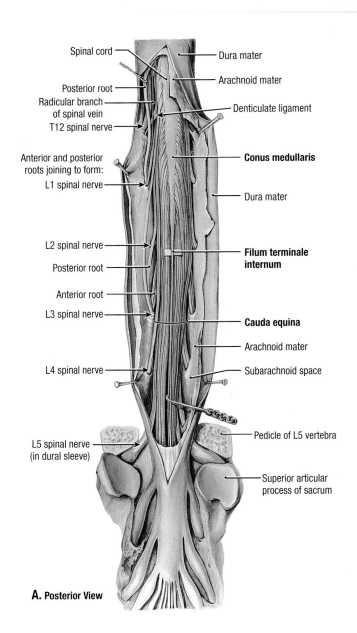

Spinal cord

Posterior root

Radicular branch
of spinal vein

T12 spinal nerve

Anterior and posterior
roots joining to form:
L1 spinal nerve

L2 spinal nerve

Posterior root

Anterior root

L3 spinal nerve

L4 spinal nerve

L5 spinal nerve
(in dural sleeve)

Dura mater

Arachnoid mater

Denticulate ligament

Conus medullaris

Dura mater

**Filum terminale
internum**

Cauda equina

Arachnoid mater

Subarachnoid space

Pedicle of L5 vertebra

Superior articular
process of sacrum

A. Posterior View

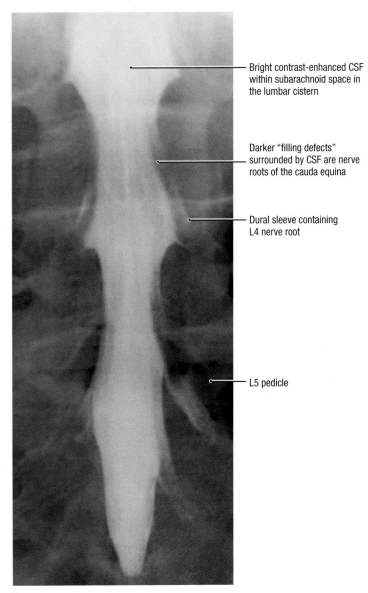

Bright contrast-enhanced CSF
within subarachnoid space in
the lumbar cistern

Darker "filling defects"
surrounded by CSF are nerve
roots of the cauda equina

Dural sleeve containing
L4 nerve root

L5 pedicle

B. Frontal Myelogram

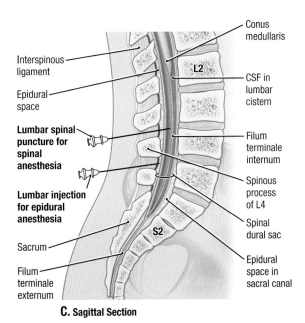

Conus
medullaris

Interspinous
ligament

Epidural
space

**Lumbar spinal
puncture for
spinal
anesthesia**

**Lumbar injection
for epidural
anesthesia**

Sacrum

Filum
terminale
externum

L2

CSF in
lumbar
cistern

Filum
terminale
internum

Spinous
process
of L4

Spinal
dural sac

Epidural
space in
sacral canal

S2

C. Sagittal Section

INFERIOR END OF DURAL SAC II   1.38

A. Inferior dural sac and lumbar cistern of subarachnoid space (opened).
B. Myelogram of the lumbar region of the vertebral column. Contrast medium
was injected into the subarachnoid space. **C.** Lumbar spinal puncture and epidural anesthesia.

• The conus medullaris continues as a glistening thread, the filum terminale internum, which descends with the nerve roots, constituting the cauda equina.
• In the adult, the spinal cord usually ends at the level of the disc between
vertebrae L1 and L2. Variations: 95% of cords end within the limits of the
bodies of L1 and L2, whereas 3% end posterior to the inferior half of T12,
and 2% posterior to L3.

To obtain a **sample of CSF from the lumbar cistern**, a lumbar puncture needle, fitted with a stylet, is inserted into the subarachnoid space. Flexion of the
vertebral column facilitates insertion of the needle by stretching the ligamenta
flava and spreading the laminae and spinous processes apart. The needle is
inserted in the midline between the spinous processes of the L3 and L4 (or
the L4 and L5) vertebrae. At these levels in adults, there is little danger of
damaging the spinal cord.

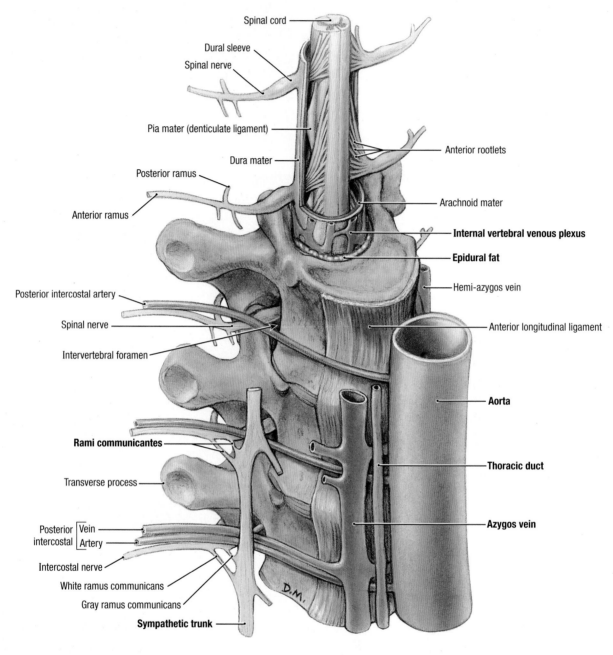

Spinal cord

Dural sleeve

Spinal nerve

Pia mater (denticulate ligament)

Dura mater

Posterior ramus

Anterior ramus

Posterior intercostal artery

Spinal nerve

Intervertebral foramen

Rami communicantes

Transverse process

Posterior Vein
intercostal Artery

Intercostal nerve

White ramus communicans

Gray ramus communicans

Sympathetic trunk

Anterior rootlets

Arachnoid mater

Internal vertebral venous plexus

Epidural fat

Hemi-azygos vein

Anterior longitudinal ligament

Aorta

Thoracic duct

Azygos vein

D.M.

Right Anterolateral View

1.39 SPINAL CORD AND PREVERTEBRAL STRUCTURES

The vertebrae have been removed superiorly to expose the spinal cord and meninges.

- The aorta descends to the left of the midline, with the thoracic duct and azygos vein to its right.
- Typically, the azygos vein is on the right side of the vertebral bodies, and the hemi-azygos vein is on the left.

- The thoracic sympathetic trunk and ganglia lie lateral to the thoracic vertebrae; the rami communicantes connect the sympathetic ganglia with the spinal nerve.
- A sleeve of dura mater surrounds the spinal nerves and blends with the sheath (epineurium) of the spinal nerve.
- The dura mater is separated from the walls of the vertebral canal by epidural fat and the internal vertebral venous plexus.

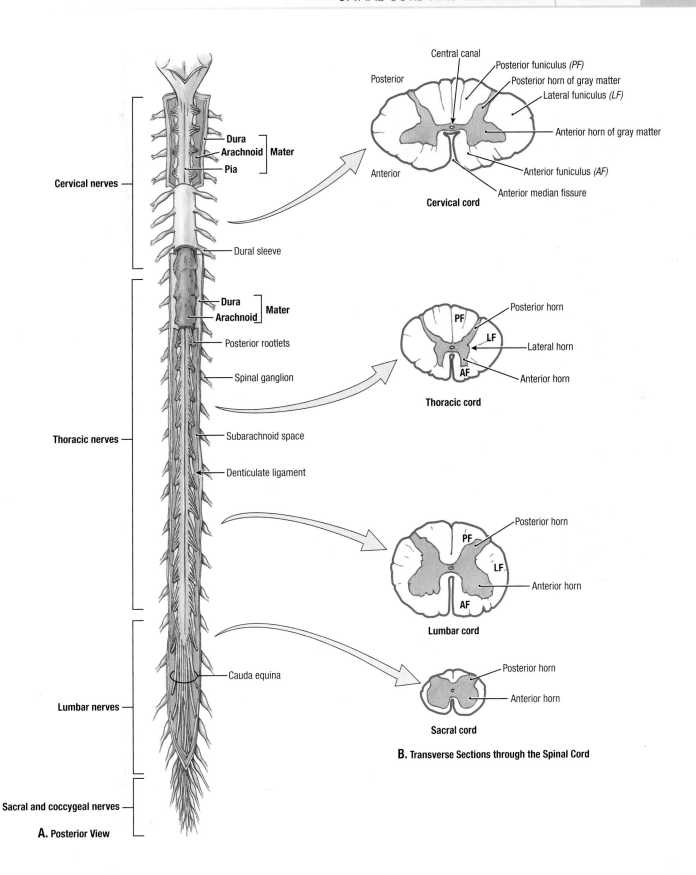

Central canal
Posterior funiculus (PF)
Posterior horn of gray matter
Lateral funiculus (LF)
Posterior
Anterior horn of gray matter
Anterior
Anterior funiculus (AF)
Anterior median fissure

Cervical cord

Cervical nerves

Dura
Arachnoid } **Mater**
Pia

Dural sleeve

Dura
Arachnoid } **Mater**

Posterior rootlets

Spinal ganglion

Thoracic nerves

Subarachnoid space

Denticulate ligament

Cauda equina

Lumbar nerves

Sacral and coccygeal nerves

A. Posterior View

Posterior horn
PF
LF
Lateral horn
AF
Anterior horn

Thoracic cord

Posterior horn
PF
LF
AF
Anterior horn

Lumbar cord

Posterior horn
Anterior horn

Sacral cord

B. Transverse Sections through the Spinal Cord

ISOLATED SPINAL CORD AND SPINAL NERVE ROOTS WITH COVERINGS AND REGIONAL SECTIONS 1.40

A. The spinal dural sac has been opened to reveal arachnoid and pia mater as well as spinal cord and posterior nerve roots. **B.** Cervical, thoracic, lumbar, and sacral spinal cord.

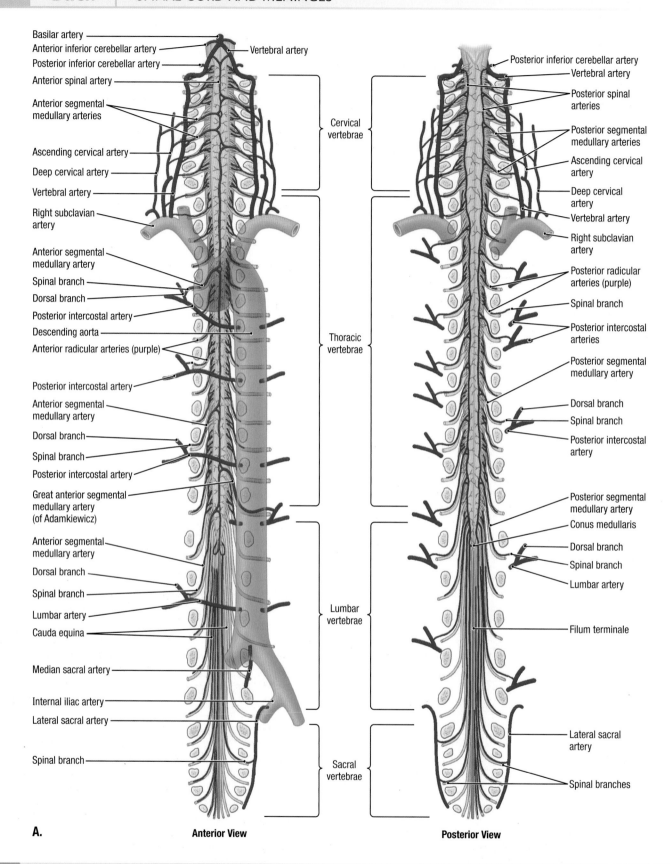

Basilar artery
Anterior inferior cerebellar artery
Posterior inferior cerebellar artery
Anterior spinal artery
Anterior segmental medullary arteries
Ascending cervical artery
Deep cervical artery
Vertebral artery
Right subclavian artery
Anterior segmental medullary artery
Spinal branch
Dorsal branch
Posterior intercostal artery
Descending aorta
Anterior radicular arteries (purple)
Posterior intercostal artery
Anterior segmental medullary artery
Dorsal branch
Spinal branch
Posterior intercostal artery
Great anterior segmental medullary artery (of Adamkiewicz)
Anterior segmental medullary artery
Dorsal branch
Spinal branch
Lumbar artery
Cauda equina
Median sacral artery
Internal iliac artery
Lateral sacral artery
Spinal branch

Vertebral artery

Cervical vertebrae

Thoracic vertebrae

Lumbar vertebrae

Sacral vertebrae

A.

Anterior View

Posterior inferior cerebellar artery
Vertebral artery
Posterior spinal arteries
Posterior segmental medullary arteries
Ascending cervical artery
Deep cervical artery
Vertebral artery
Right subclavian artery
Posterior radicular arteries (purple)
Spinal branch
Posterior intercostal arteries
Posterior segmental medullary artery
Dorsal branch
Spinal branch
Posterior intercostal artery
Posterior segmental medullary artery
Conus medullaris
Dorsal branch
Spinal branch
Lumbar artery
Filum terminale
Lateral sacral artery
Spinal branches

Posterior View

1.41 **BLOOD SUPPLY OF SPINAL CORD**

A. Arteries of spinal cord. The segmental reinforcements of blood supply from the segmental medullary arteries are important in supplying blood to the anterior and posterior spinal arteries.

Fractures, dislocations, and fracture-dislocations may interfere with the blood supply to the spinal cord from the spinal and medullary arteries.

POSTERIOR

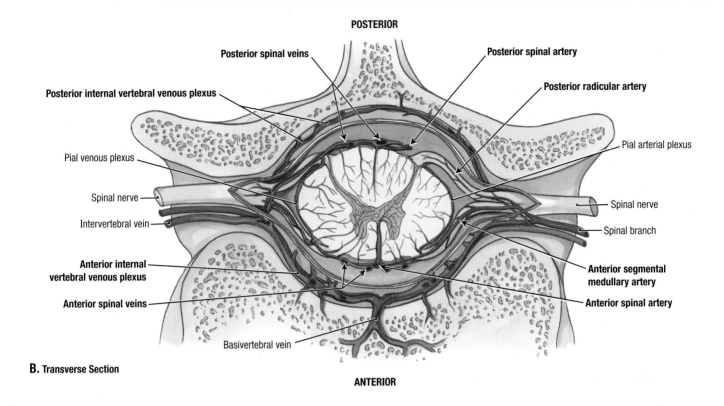

Posterior spinal veins

Posterior spinal artery

Posterior internal vertebral venous plexus

Posterior radicular artery

Pial arterial plexus

Pial venous plexus

Spinal nerve

Spinal nerve

Intervertebral vein

Spinal branch

Anterior internal vertebral venous plexus

Anterior segmental medullary artery

Anterior spinal veins

Anterior spinal artery

Basivertebral vein

B. Transverse Section

ANTERIOR

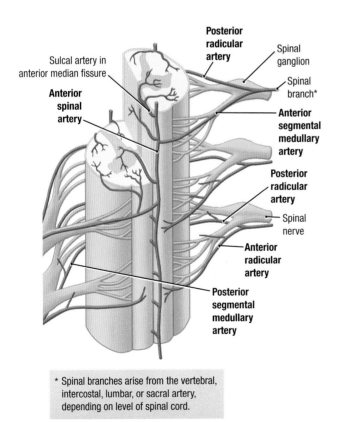

Sulcal artery in anterior median fissure

Posterior radicular artery

Spinal ganglion

Spinal branch*

Anterior spinal artery

Anterior segmental medullary artery

Posterior radicular artery

Spinal nerve

Anterior radicular artery

Posterior segmental medullary artery

* Spinal branches arise from the vertebral, intercostal, lumbar, or sacral artery, depending on level of spinal cord.

C. Anterolateral View

BLOOD SUPPLY OF SPINAL CORD (*continued*)   1.41

B. Arterial supply and venous drainage. **C.** Segmental medullary and radicular arteries.

- The spinal arteries run longitudinally from the brainstem to the conus medullaris of the spinal cord. By themselves, the anterior and posterior spinal arteries supply only the short superior part of the spinal cord.
- The anterior and posterior segmental medullary arteries enter the IV foramen to unite with the spinal arteries to supply blood to the spinal cord. The great anterior segmental medullary artery (Adamkiewicz artery) occurs on the left side in 65% of people. It reinforces the circulation to two thirds of the spinal cord.
- Posterior and anterior roots of the spinal nerves and their coverings are supplied by posterior and anterior radicular arteries, which run along the nerve roots. These vessels do not reach the posterior or anterior spinal arteries.
- The anterior and posterior spinal veins are arranged longitudinally; they communicate freely with each other and are drained by anterior and posterior medullary and radicular veins. The veins draining the spinal cord join the internal vertebral plexus in the epidural space.

Ischemia. Deficiency of blood supply (ischemia) of the spinal cord can lead to muscle weakness and paralysis. The spinal cord may also suffer circulatory impairment if the segmental medullary arteries, particularly the great anterior segmental medullary artery (of Adamkiewicz), are narrowed by obstructive arterial disease or aortic clamping during surgery.

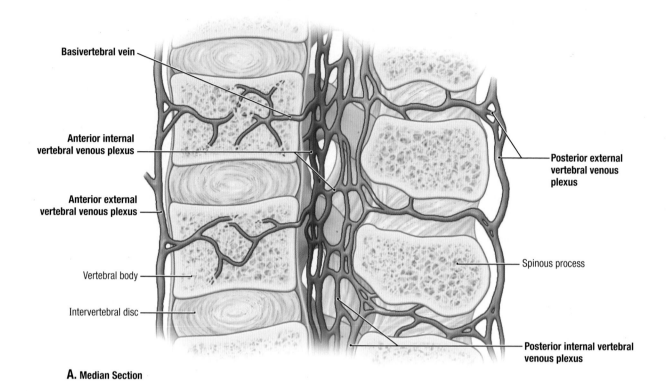

Basivertebral vein

Anterior internal vertebral venous plexus

Anterior external vertebral venous plexus

Vertebral body

Intervertebral disc

Posterior external vertebral venous plexus

Spinous process

Posterior internal vertebral venous plexus

A. Median Section

1.42   VERTEBRAL VENOUS PLEXUSES

A. Median section of lumbar spine. **B.** Superior view of lumbar vertebra with the vertebral body sectioned transversely.

- There are internal and external vertebral venous plexuses, communicating with each other and with both systemic veins and the portal system. **Infection and tumors can spread** from the areas drained by the systemic and portal veins to the vertebral venous system and lodge in the vertebrae, spinal cord, brain, or skull.
- The internal vertebral venous plexus, located in the vertebral canal, consists of a plexus of thin-walled, valveless veins that surround the dura mater. Cranially, the internal venous plexus communicates through the foramen magnum with the occipital and basilar sinuses; at each spinal segment, the plexus receives veins from the spinal cord and a basivertebral vein from the vertebral body. The plexus is drained by IV veins that pass through the intervertebral and sacral foramina to the vertebral, intercostal, lumbar, and lateral sacral veins.
- The anterior external vertebral venous plexus is formed by veins that course through the body of each vertebra. Veins that pass through the ligamenta flava form the posterior external vertebral venous plexus. In the cervical region, these plexuses communicate with the occipital and deep cervical veins. In the thoracic, lumbar, and pelvic regions, the azygos (or hemi-azygos), ascending lumbar, and lateral sacral veins, respectively, further link segment to segment.

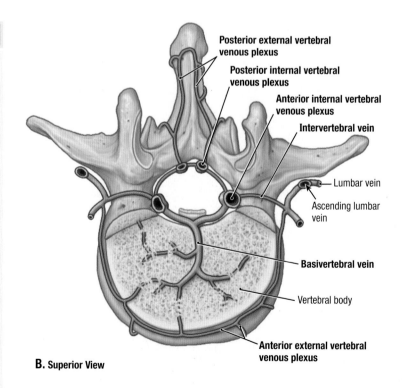

Posterior external vertebral venous plexus

Posterior internal vertebral venous plexus

Anterior internal vertebral venous plexus

Intervertebral vein

Lumbar vein

Ascending lumbar vein

Basivertebral vein

Vertebral body

Anterior external vertebral venous plexus

B. Superior View

Nerves carrying somatic and sympathetic nerve fibers to the body wall and limbs:

All dorsal rami and:

Cervical plexus (C1–C4)

Brachial plexus (C5–T1)

Intercostal nerves (T1–T11) and subcostal nerve (T12)

Lumbar plexus (L1–L4)

Sacral plexus (L4–S4)

Coccygeal plexus (S4–Co)

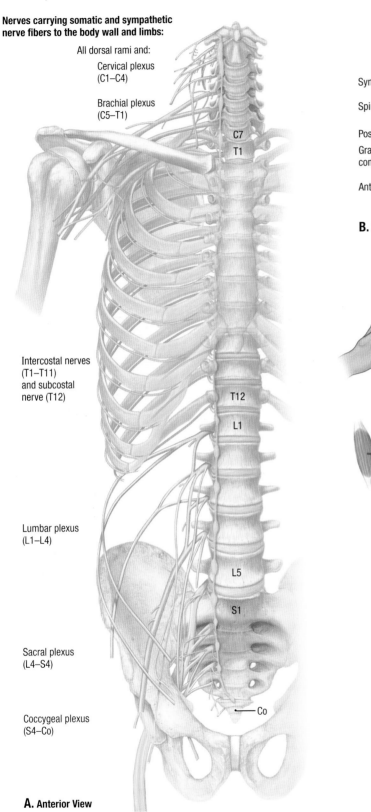

C7
T1
T12
L1
L5
S1
Co

A. Anterior View

Posterior root and rootlets

Sympathetic trunk

Spinal nerve

Posterior ramus

Gray ramus communicans

Anterior ramus

Anterior root and rootlets

Sympathetic ganglion

White ramus communicans

B. Parts of spinal nerves

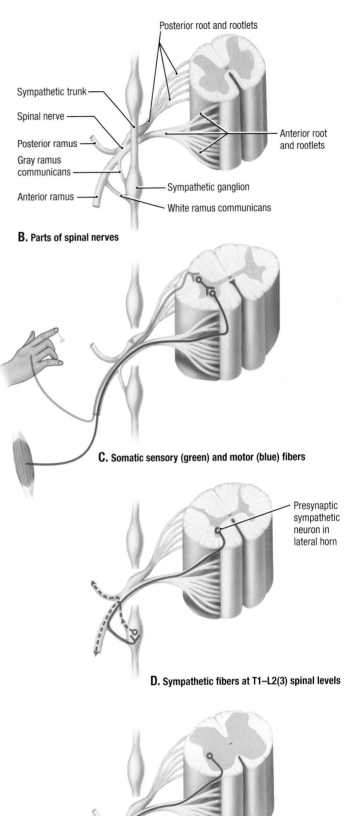

C. Somatic sensory (green) and motor (blue) fibers

Presynaptic sympathetic neuron in lateral horn

D. Sympathetic fibers at T1–L2(3) spinal levels

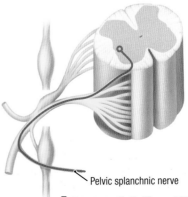

Pelvic splanchnic nerve

E. Parasympathetic fibers at S2–4 spinal cord levels

B–E. Anterolateral Views

OVERVIEW OF THE INNERVATION OF THE LIMBS AND BODY WALL

1.43

A. Overview. **B.** Parts of spinal nerve. **C.** Somatic sensory and motor fibers. **D.** Sympathetic fibers at T1–L2 levels. **E.** Parasympathetic fibers at S1–S4 levels coursing with pudendal nerve.

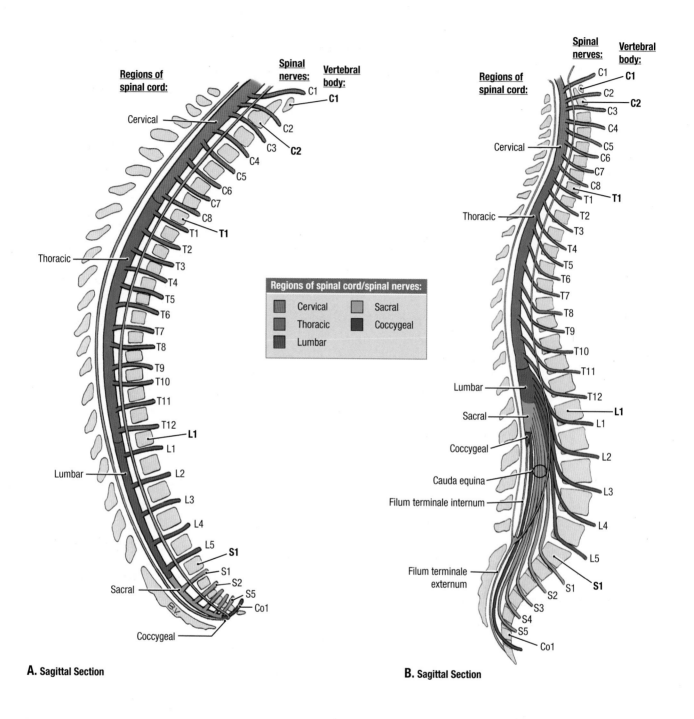

A. Sagittal Section

B. Sagittal Section

Regions of spinal cord/spinal nerves:
- Cervical
- Sacral
- Thoracic
- Coccygeal
- Lumbar

1.44 SPINAL CORD AND SPINAL NERVES

A. Spinal cord at 12 weeks gestation. **B.** Spinal cord of an adult.

- Early in development, the spinal cord and vertebral (spinal) canal are nearly equal in length. The canal grows longer, so spinal nerves have an increasingly longer course to reach the IV foramen at the correct level for their exit. The spinal cord of adults terminates between vertebral bodies L1–L2. The remaining spinal nerves, seeking their IV foramen of exit, form the cauda equina.

- All 31 pairs of spinal nerves—8 cervical (C), 12 thoracic (T), 5 lumbar (L), 5 sacral (S), and 1 coccygeal (Co)—arise from the spinal cord and exit through the IV foramina in the vertebral column.

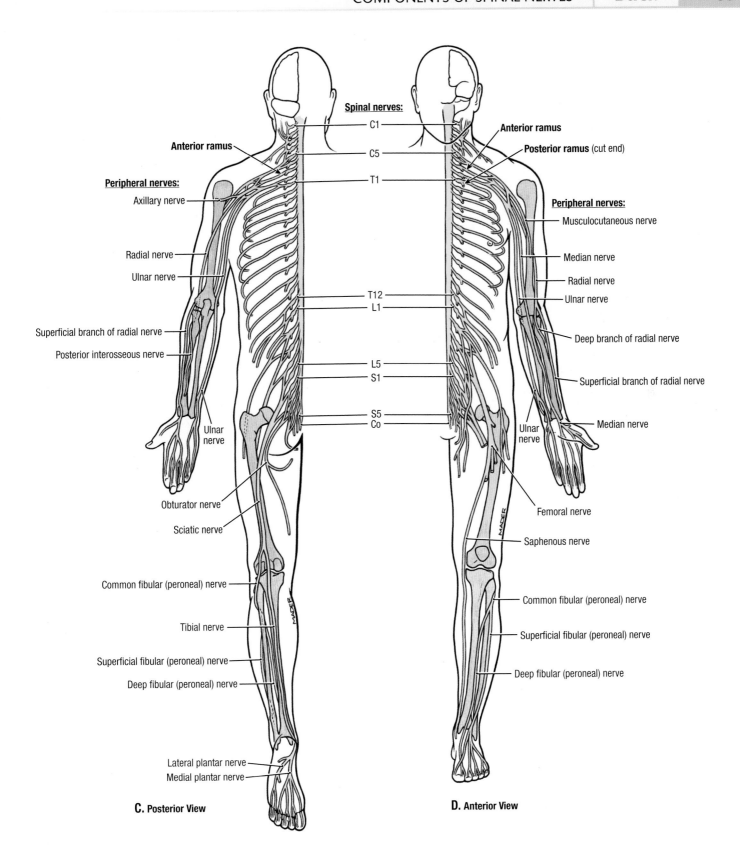

Spinal nerves:
C1
C5
T1
T12
L1
L5
S1
S5
Co

Anterior ramus

Anterior ramus
Posterior ramus (cut end)

Peripheral nerves:
Axillary nerve
Radial nerve
Ulnar nerve
Superficial branch of radial nerve
Posterior interosseous nerve
Ulnar nerve
Obturator nerve
Sciatic nerve
Common fibular (peroneal) nerve
Tibial nerve
Superficial fibular (peroneal) nerve
Deep fibular (peroneal) nerve
Lateral plantar nerve
Medial plantar nerve

Peripheral nerves:
Musculocutaneous nerve
Median nerve
Radial nerve
Ulnar nerve
Deep branch of radial nerve
Superficial branch of radial nerve
Median nerve
Ulnar nerve
Femoral nerve
Saphenous nerve
Common fibular (peroneal) nerve
Superficial fibular (peroneal) nerve
Deep fibular (peroneal) nerve

C. Posterior View

D. Anterior View

SPINAL CORD AND SPINAL NERVES (*continued*)

1.44

C. and **D.** Peripheral nerves.
- The anterior rami supply nerve fibers to the anterior and lateral regions of the trunk and upper and lower limbs.
- The posterior rami supply nerve fibers to synovial joints of the vertebral column, deep muscles of the back, and overlying skin.

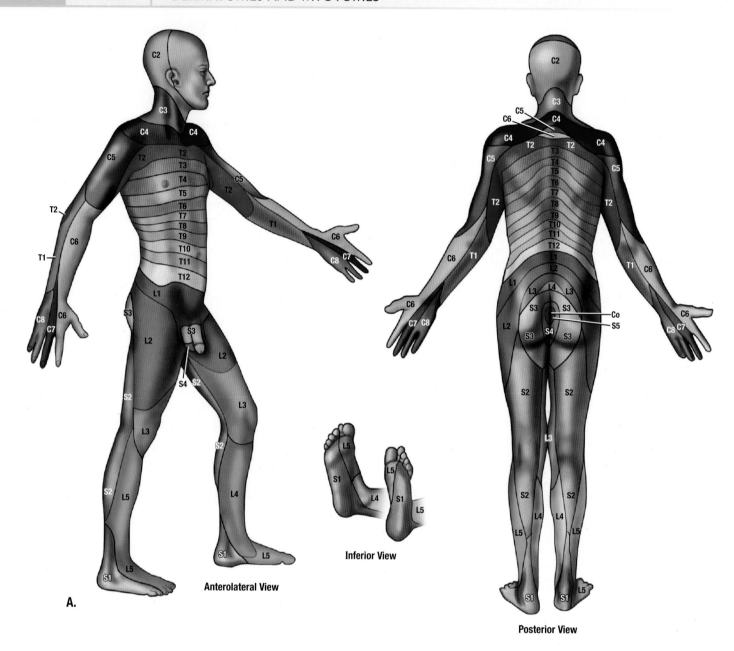

Anterolateral View

A.

Inferior View

Posterior View

1.45 DERMATOMES

A. Dermatome map. From clinical studies of lesions in the posterior roots or spinal nerves, dermatome maps have been devised that indicate the typical patterns of innervation of the skin by specific spinal nerves. (Based on Foerster O. The dermatomes in man. *Brain*. 1933;56:1.) **B.** Schematic illustration of a dermatome and myotome. The unilateral area of skin innervated by the general sensory fibers of a single spinal nerve is called a dermatome.

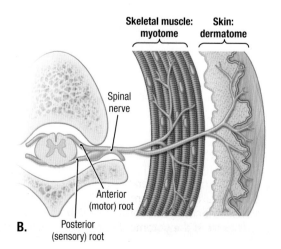

Skeletal muscle: myotome **Skin: dermatome**

Spinal nerve

Anterior (motor) root

Posterior (sensory) root

B.

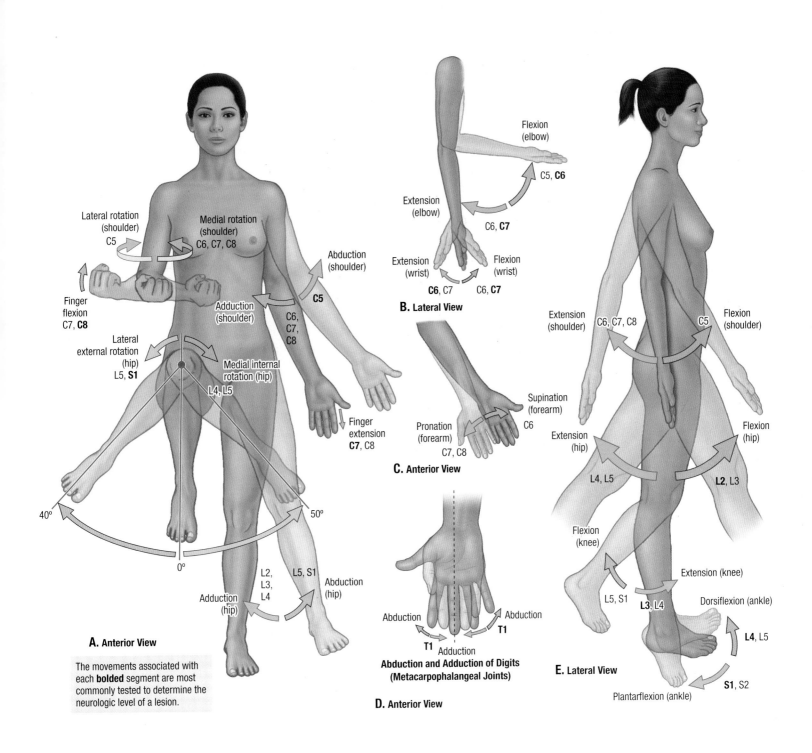

Lateral rotation
(shoulder)
C5

Medial rotation
(shoulder)
C6, C7, C8

Abduction
(shoulder)

Finger
flexion
C7, **C8**

Adduction
(shoulder)

C5

C6,
C7,
C8

Lateral
external rotation
(hip)
L5, **S1**

Medial internal
rotation (hip)
L4, L5

Finger
extension
C7, C8

40°

50°

0°

L2,
L3,
L4

L5, S1

Adduction
(hip)

Abduction
(hip)

A. Anterior View

The movements associated with
each **bolded** segment are most
commonly tested to determine the
neurologic level of a lesion.

Flexion
(elbow)
C5, **C6**

Extension
(elbow)
C6, **C7**

Extension
(wrist)
C6, C7

Flexion
(wrist)
C6, **C7**

B. Lateral View

Supination
(forearm)
C6

Pronation
(forearm)
C7, C8

C. Anterior View

Abduction

Abduction
T1

T1 Adduction

**Abduction and Adduction of Digits
(Metacarpophalangeal Joints)**

D. Anterior View

Extension
(shoulder)
C6, C7, C8

Flexion
(shoulder)
C5

Flexion
(hip)
L2, L3

Extension
(hip)
L4, L5

Flexion
(knee)
L5, S1

Extension (knee)
L3, L4

Dorsiflexion (ankle)
L4, L5

Plantarflexion (ankle)
S1, S2

E. Lateral View

MYOTOMES

1.46

Somatic motor (general somatic efferent) fibers transmit impulses
to skeletal (voluntary) muscles. The unilateral muscle mass receiving
innervation from the somatic motor fibers conveyed by a single spinal nerve is a myotome. Each skeletal muscle is innervated by the
somatic motor fibers of several spinal nerves; therefore, the muscle
myotome will consist of several segments. The muscle myotomes
have been grouped by joint movement to facilitate clinical testing.
The intrinsic muscles of the hand constitute a single myotome—T1.

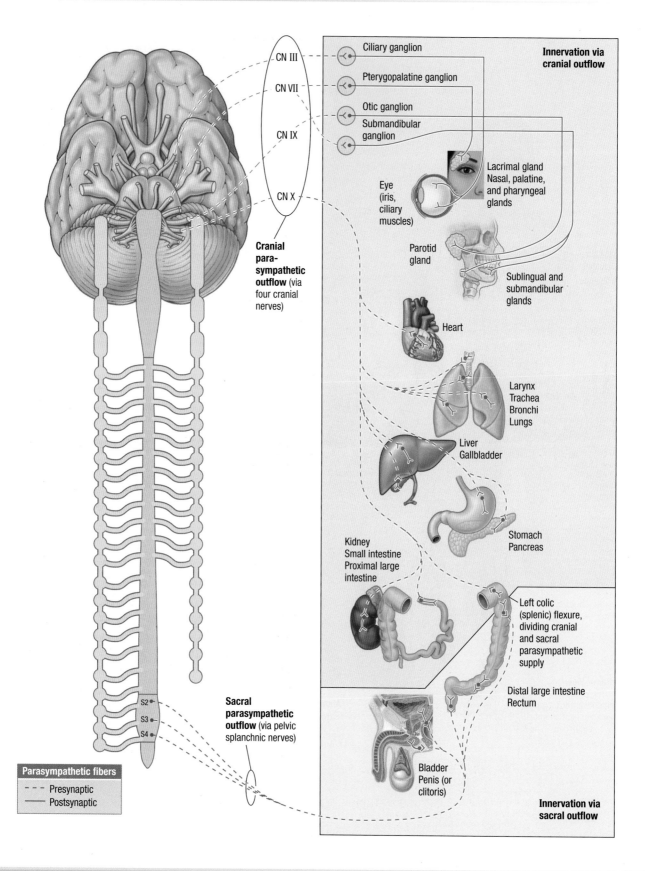

Ciliary ganglion

Pterygopalatine ganglion

Otic ganglion

Submandibular ganglion

CN III

CN VII

CN IX

CN X

Innervation via cranial outflow

Lacrimal gland
Nasal, palatine, and pharyngeal glands

Eye (iris, ciliary muscles)

Parotid gland

Sublingual and submandibular glands

Heart

Larynx
Trachea
Bronchi
Lungs

Liver
Gallbladder

Stomach
Pancreas

Kidney
Small intestine
Proximal large intestine

Left colic (splenic) flexure, dividing cranial and sacral parasympathetic supply

Distal large intestine
Rectum

Bladder
Penis (or clitoris)

Innervation via sacral outflow

Cranial para-sympathetic outflow (via four cranial nerves)

Sacral parasympathetic outflow (via pelvic splanchnic nerves)

S2
S3
S4

Parasympathetic fibers
- - - Presynaptic
—— Postsynaptic

1.47   DISTRIBUTION OF PARASYMPATHETIC NERVE FIBERS

The presynaptic nerve cell bodies of the parasympathetic system are located in two sites: the gray matter of the brainstem (cranial parasympathetic outflow) and in the gray matter of the sacral segments of the spinal cord (sacral parasympathetic outflow).

Parietal distribution

(via gray rami communicans)

Visceral distribution

(via splanchnic nerves and peri-arterial plexuses)

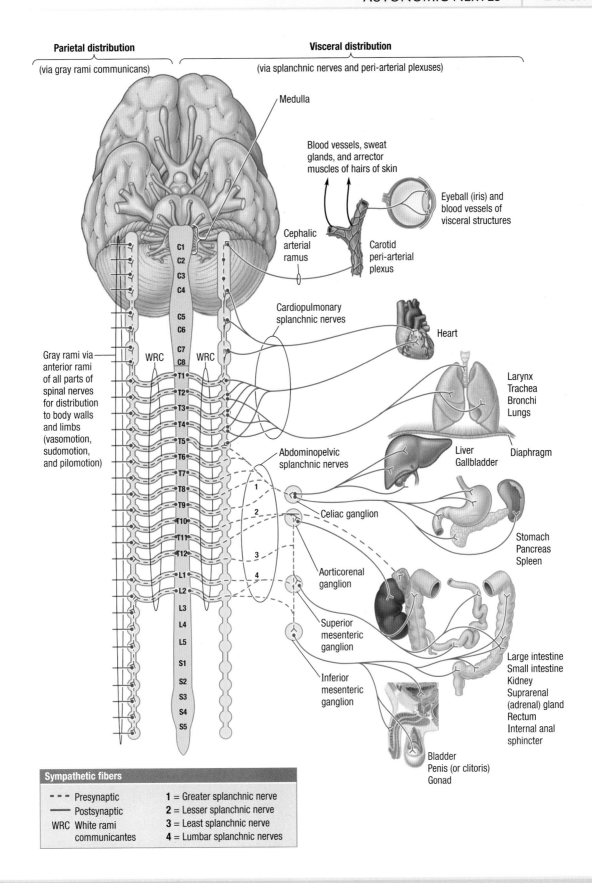

Medulla

Blood vessels, sweat glands, and arrector muscles of hairs of skin

Eyeball (iris) and blood vessels of visceral structures

Cephalic arterial ramus

Carotid peri-arterial plexus

Cardiopulmonary splanchnic nerves

Heart

Larynx
Trachea
Bronchi
Lungs

Gray rami via anterior rami of all parts of spinal nerves for distribution to body walls and limbs (vasomotion, sudomotion, and pilomotion)

WRC   WRC

Liver
Gallbladder

Diaphragm

Abdominopelvic splanchnic nerves

Celiac ganglion

Stomach
Pancreas
Spleen

Aorticorenal ganglion

Superior mesenteric ganglion

Inferior mesenteric ganglion

Large intestine
Small intestine
Kidney
Suprarenal (adrenal) gland
Rectum
Internal anal sphincter

Bladder
Penis (or clitoris)
Gonad

Sympathetic fibers

- - - Presynaptic   **1** = Greater splanchnic nerve
——— Postsynaptic   **2** = Lesser splanchnic nerve
WRC White rami   **3** = Least splanchnic nerve
       communicantes **4** = Lumbar splanchnic nerves

DISTRIBUTION OF SYMPATHETIC NERVE FIBERS

1.48

The cell bodies of presynaptic neurons of the sympathetic system are located in the intermediolateral cell column and extend between the first thoracic and the second lumbar segments of the spinal cord.

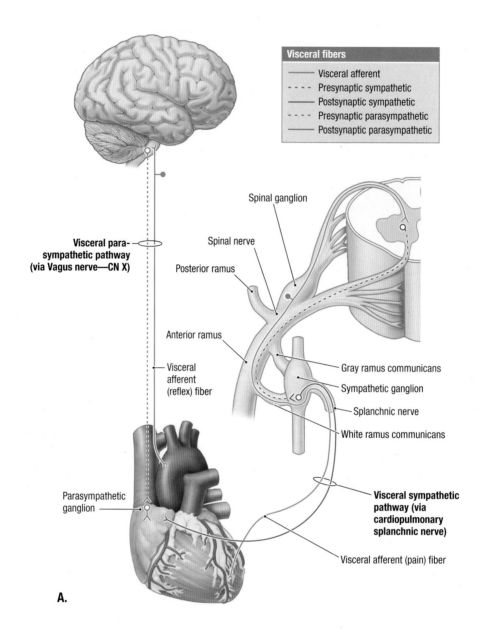

Visceral fibers
— Visceral afferent
---- Presynaptic sympathetic
— Postsynaptic sympathetic
---- Presynaptic parasympathetic
— Postsynaptic parasympathetic

Spinal ganglion

Spinal nerve

Posterior ramus

Anterior ramus

Visceral para-sympathetic pathway (via Vagus nerve—CN X)

Visceral afferent (reflex) fiber

Gray ramus communicans

Sympathetic ganglion

Splanchnic nerve

White ramus communicans

Parasympathetic ganglion

Visceral sympathetic pathway (via cardiopulmonary splanchnic nerve)

Visceral afferent (pain) fiber

A.

1.49   VISCERAL AFFERENT AND VISCERAL EFFERENT (MOTOR) INNERVATION

A. Schematic illustration. Visceral afferent fibers have important relationships to the central nervous system (CNS), both anatomically and functionally. We are usually unaware of the sensory input of these fibers, which provides information about the condition of the body's internal environment. This information is integrated in the CNS, often triggering visceral or somatic reflexes or both. Visceral reflexes regulate blood pressure and chemistry by altering such functions as heart and respiratory rates and vascular resistance. Visceral sensation that reaches a conscious level is generally categorized as pain that is usually poorly localized and may be perceived as hunger or nausea. However, adequate stimulation may elicit true pain. Most visceral/reflex (unconscious) sensation and some pain travel in visceral afferent fibers that accompany the parasympathetic fibers retrograde. Most visceral pain impulses (from the heart and most organs of the peritoneal cavity) travel along visceral afferent fibers accompanying sympathetic fibers.

Visceral efferent (motor) innervation. The efferent nerve fibers and ganglia of the ANS are organized into two systems or divisions.
1. **Sympathetic (thoracolumbar) division.** In general, the effects of sympathetic stimulation are catabolic (preparing the body for "flight or fight").
2. **Parasympathetic (craniosacral) division.** In general, the effects of parasympathetic stimulation are anabolic (promoting normal function and conserving energy).
Conduction of impulses from the CNS to the effector organ involves a series of two neurons in both sympathetic and parasympathetic systems. The cell body of the presynaptic (preganglionic) neuron (first neuron) is located in the gray matter of the CNS. Its fiber (axon) synapses on the cell body of a postsynaptic (postganglionic) neuron, the second neuron in the series. The cell bodies of such second neurons are located in autonomic ganglia outside the CNS, and the postsynaptic fibers terminate on the effector organ (smooth muscle, modified cardiac muscle, or glands).

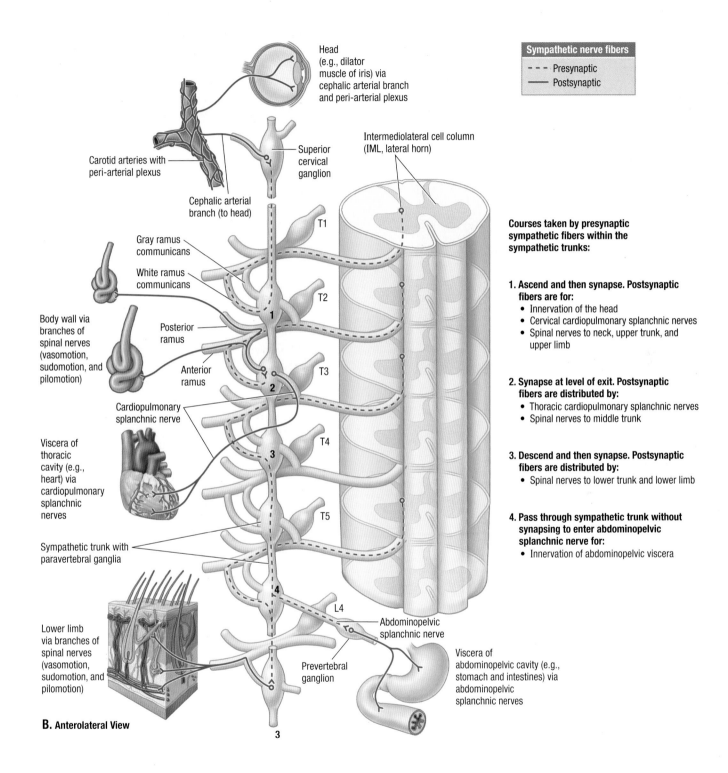

Head
(e.g., dilator
muscle of iris) via
cephalic arterial branch
and peri-arterial plexus

Sympathetic nerve fibers
- - - Presynaptic
——— Postsynaptic

Carotid arteries with
peri-arterial plexus

Superior
cervical
ganglion

Intermediolateral cell column
(IML, lateral horn)

Cephalic arterial
branch (to head)

T1

Gray ramus
communicans

White ramus
communicans

T2

Body wall via
branches of
spinal nerves
(vasomotion,
sudomotion, and
pilomotion)

Posterior
ramus

Anterior
ramus

T3

Cardiopulmonary
splanchnic nerve

Viscera of
thoracic
cavity (e.g.,
heart) via
cardiopulmonary
splanchnic
nerves

T4

Sympathetic trunk with
paravertebral ganglia

T5

Lower limb
via branches of
spinal nerves
(vasomotion,
sudomotion, and
pilomotion)

L4

Abdominopelvic
splanchnic nerve

Prevertebral
ganglion

Viscera of
abdominopelvic cavity (e.g.,
stomach and intestines) via
abdominopelvic
splanchnic nerves

B. Anterolateral View

**Courses taken by presynaptic
sympathetic fibers within the
sympathetic trunks:**

**1. Ascend and then synapse. Postsynaptic
fibers are for:**
- Innervation of the head
- Cervical cardiopulmonary splanchnic nerves
- Spinal nerves to neck, upper trunk, and
  upper limb

**2. Synapse at level of exit. Postsynaptic
fibers are distributed by:**
- Thoracic cardiopulmonary splanchnic nerves
- Spinal nerves to middle trunk

**3. Descend and then synapse. Postsynaptic
fibers are distributed by:**
- Spinal nerves to lower trunk and lower limb

**4. Pass through sympathetic trunk without
synapsing to enter abdominopelvic
splanchnic nerve for:**
- Innervation of abdominopelvic viscera

VISCERAL AFFERENT AND VISCERAL EFFERENT (MOTOR) INNERVATION (continued)   `1.49`

B. Courses taken by sympathetic motor fibers. Presynaptic fibers all follow the same course until they reach the sympathetic trunks. In the sympathetic trunks, they follow one of four possible courses. Fibers involved in providing sympathetic innervation to the body wall and limbs or viscera above the level of the diaphragm follow paths 1 to 3. They synapse in the paravertebral ganglia of the sympathetic trunks. Fibers involved in innervating abdominopelvic viscera follow path 4 to prevertebral ganglion via abdominopelvic splanchnic nerves. Postsynaptic fibers usually do not ascend or descend within the sympathetic trunks, exiting at the level of synapse.

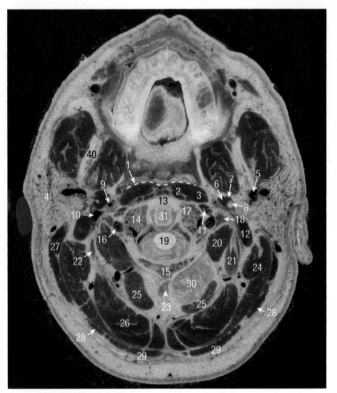

A. Inferior View

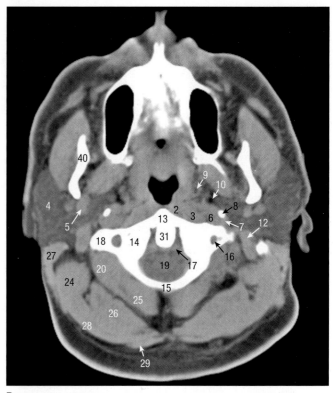

B. Transverse CT Image

Key	
1	Site of retropharyngeal space
2	Longus colli
3	Longus capitis
4	Parotid gland
5	Retromandibular vein
6	Stylopharyngeus
7	Styloglossus
8	Stylohyoid muscle and ligament/process
9	Internal carotid artery
10	Internal jugular vein
11	Rectus capitis lateralis
12	Posterior belly of digastric
13	Anterior arch of atlas (C1 vertebra)
14	Lateral mass of atlas (C1)
15	Posterior arch of atlas (C1)
16	Vertebral artery
17	Transverse ligament of atlas (C1)
18	Transverse process of atlas (C1)
19	Spinal cord
20	Rectus capitis posterior major
21	Obliquus capitis inferior
22	Obliquus capitis superior
23	Spinous process of atlas (C1)
24	Longissimus capitis
25	Rectus capitis posterior minor
26	Semispinalis capitis
27	Sternocleidomastoid
28	Splenius capitis
29	Trapezius
30	Fatty mass
31	Dens of axis (C2 vertebra)
32	Anterior tubercle of atlas (C1)
33	Inferior articular facet of atlas (C1)
34	Foramen magnum
35	Foramen transversarium
36	Posterior tubercle of atlas (C1)
37	Mastoid process
38	Occipital bone of skull
39	External occipital protuberance
40	Ramus of mandible

ANTERIOR

RIGHT — LEFT

POSTERIOR

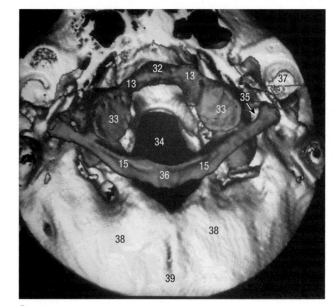

C. Reconstructed CT Image

1.50 IMAGING OF SUPERIOR NUCHAL REGION AT LEVEL OF ATLAS

A. Transverse section of specimen. **B.** Transverse CT image. **C.** Three-dimensional CT image of the base of the skull and atlas.

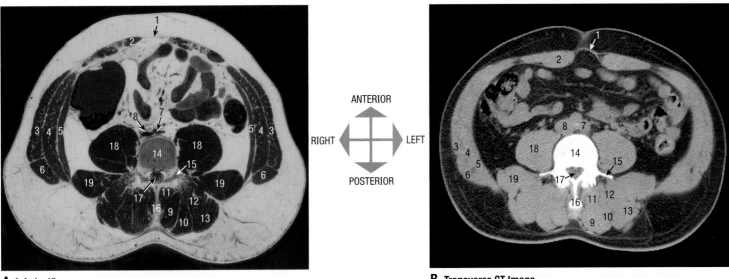

A. Inferior View

B. Transverse CT Image

ANTERIOR

RIGHT — LEFT

POSTERIOR

Key							
1	Linea alba	6	Latissimus dorsi	11	Multifidus	16	Spinous process
2	Rectus abdominis	7	Descending aorta	12	Rotatores	17	Cauda equina
3	External oblique	8	Inferior vena cava	13	Iliocostalis	18	Psoas major
4	Internal oblique	9	Spinalis	14	4th lumbar vertebra	19	Quadratus lumborum
5	Transversus abdominis	10	Longissimus	15	Transverse process		

IMAGING OF LUMBAR SPINE AT L4   1.51

A. Transverse section of specimen. **B.** Transverse CT image.

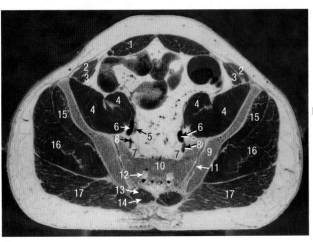

A. Inferior View

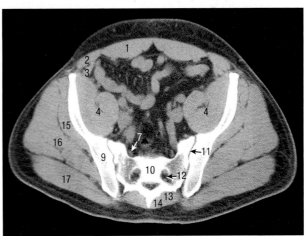

B. Transverse CT Image

ANTERIOR

RIGHT — LEFT

POSTERIOR

Key							
1	Rectus abdominis	6	Internal iliac vein	10	2nd sacral vertebra	14	Erector spinae
2	External oblique	7	Anterior rami	11	Sacro-iliac joint	15	Gluteus minimus
3	Internal oblique	8	Superior gluteal vessels	12	Sacral nerve root	16	Gluteus medius
4	Iliopsoas	9	Body of ilium	13	Multifidus	17	Gluteus maximus
5	Internal iliac artery						

IMAGING OF SACRO-ILIAC JOINT   1.52

A. Transverse section of specimen. **B.** Transverse CT image.

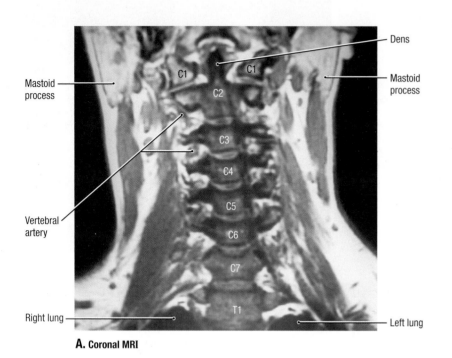

Mastoid process

Vertebral artery

Right lung

C1 C1

Dens

Mastoid process

C2

C3

C4

C5

C6

C7

T1

Left lung

A. Coronal MRI

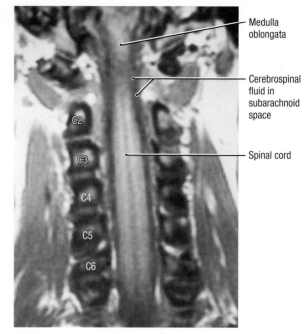

Medulla oblongata

Cerebrospinal fluid in subarachnoid space

Spinal cord

C2

C3

C4

C5

C6

B. Coronal MRI

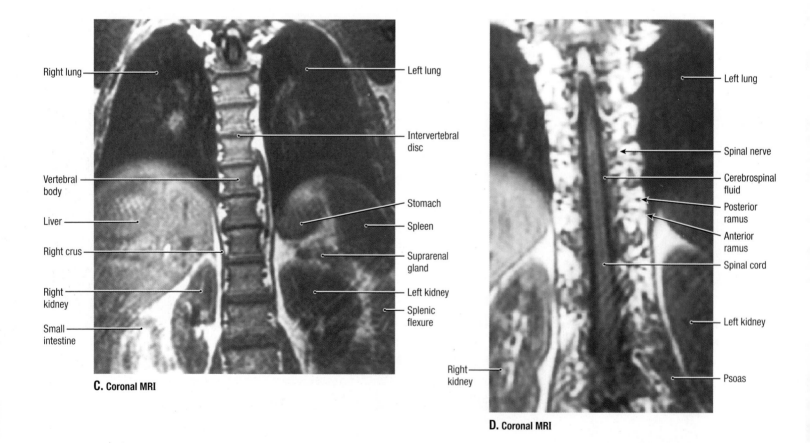

Right lung

Vertebral body

Liver

Right crus

Right kidney

Small intestine

Left lung

Intervertebral disc

Stomach

Spleen

Suprarenal gland

Left kidney

Splenic flexure

C. Coronal MRI

Right kidney

Left lung

Spinal nerve

Cerebrospinal fluid

Posterior ramus

Anterior ramus

Spinal cord

Left kidney

Psoas

D. Coronal MRI

1.53 CORONAL MRIs OF CERVICAL AND THORACIC SPINE

A. and **B.** Cervical spine. **C.** and **D.** Thoracic spine.

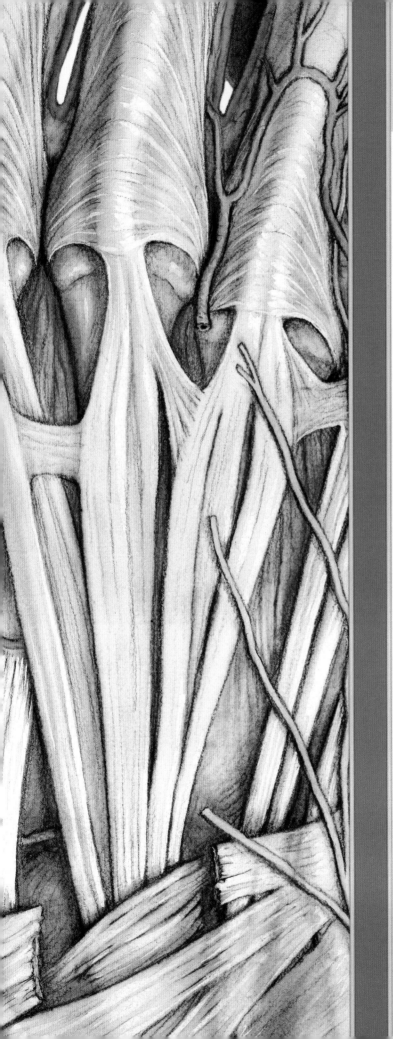

CHAPTER 2

Upper Limb

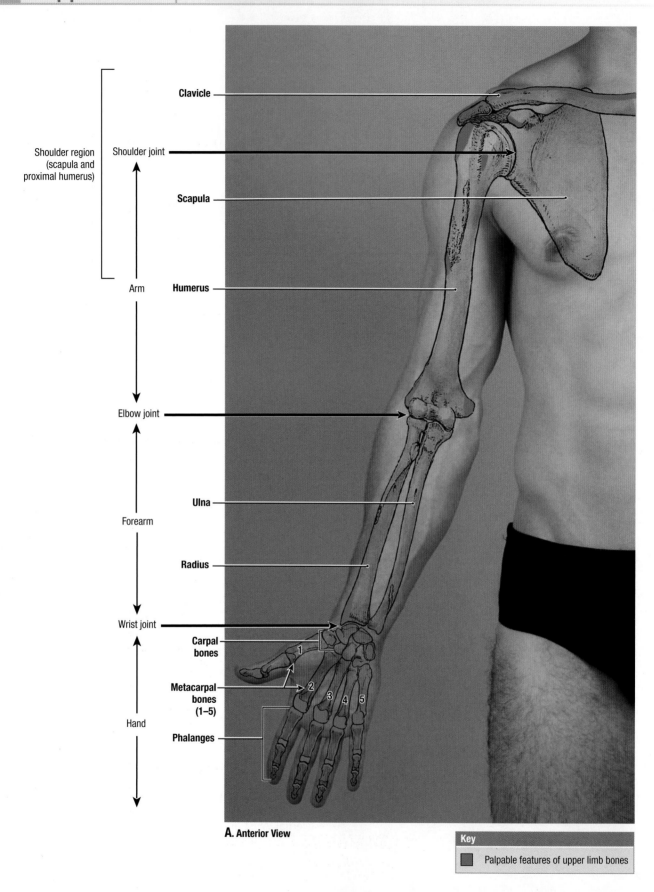

Clavicle

Shoulder region (scapula and proximal humerus)

Shoulder joint

Scapula

Arm

Humerus

Elbow joint

Ulna

Forearm

Radius

Wrist joint

Carpal bones

Metacarpal bones (1–5)

Hand

Phalanges

A. Anterior View

Key

Palpable features of upper limb bones

2.1   REGIONS, BONES, AND MAJOR JOINTS OF UPPER LIMB

Joints divide the upper limb into four main regions: the shoulder, arm, forearm, and hand.

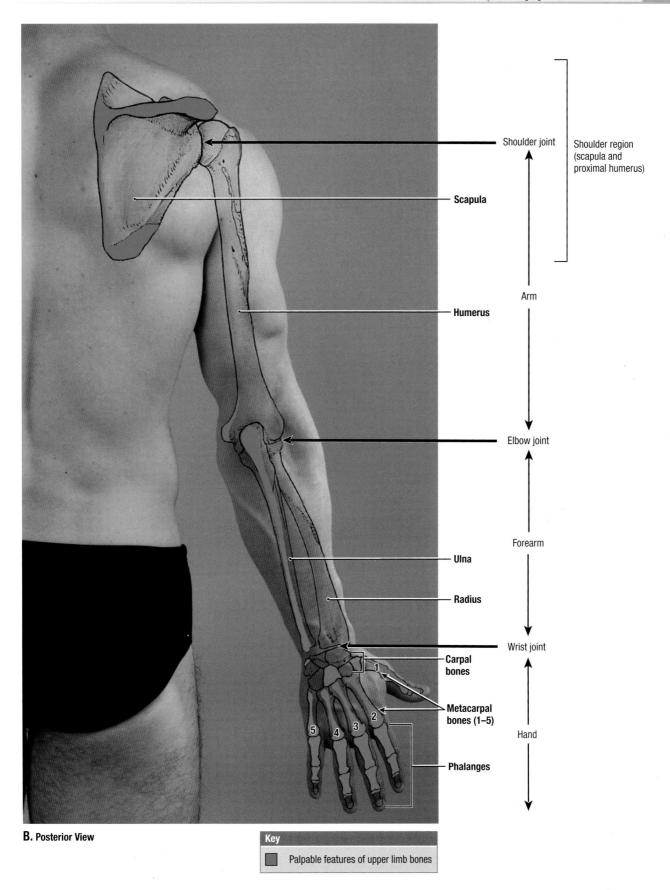

Shoulder joint

Shoulder region (scapula and proximal humerus)

Scapula

Arm

Humerus

Elbow joint

Forearm

Ulna

Radius

Wrist joint

Carpal bones

Metacarpal bones (1–5)

Hand

Phalanges

B. Posterior View

Key

Palpable features of upper limb bones

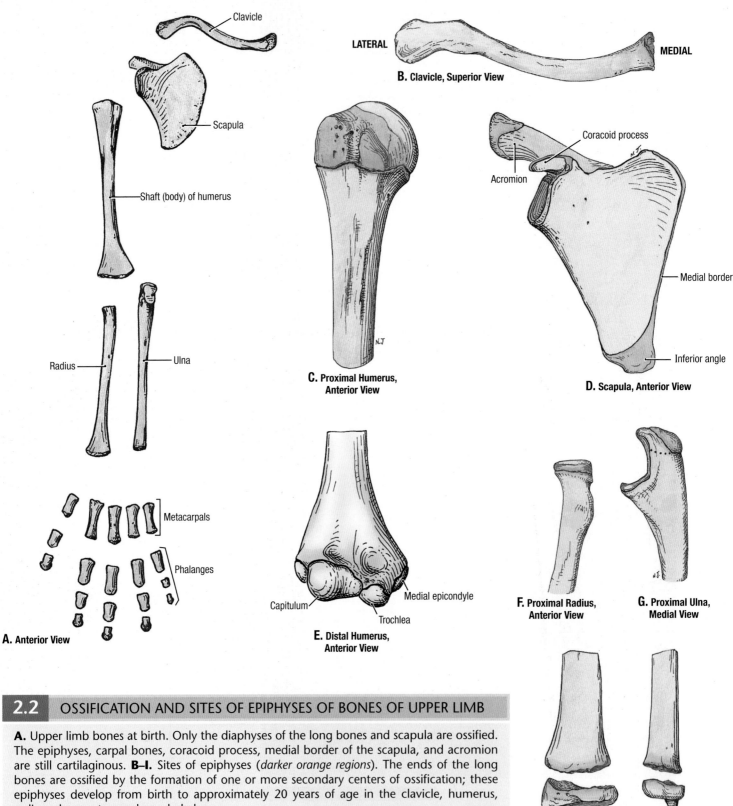

B. Clavicle, Superior View

LATERAL — MEDIAL

Clavicle

Scapula

Shaft (body) of humerus

Radius — Ulna

Metacarpals

Phalanges

A. Anterior View

C. Proximal Humerus, Anterior View

Coracoid process

Acromion

Medial border

Inferior angle

D. Scapula, Anterior View

Capitulum — Medial epicondyle

Trochlea

E. Distal Humerus, Anterior View

F. Proximal Radius, Anterior View

G. Proximal Ulna, Medial View

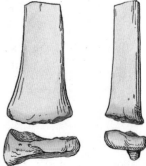

H. Distal Radius, Anterior View

I. Distal Ulna, Anterior View

2.2	OSSIFICATION AND SITES OF EPIPHYSES OF BONES OF UPPER LIMB

A. Upper limb bones at birth. Only the diaphyses of the long bones and scapula are ossified. The epiphyses, carpal bones, coracoid process, medial border of the scapula, and acromion are still cartilaginous. **B–I.** Sites of epiphyses (*darker orange regions*). The ends of the long bones are ossified by the formation of one or more secondary centers of ossification; these epiphyses develop from birth to approximately 20 years of age in the clavicle, humerus, radius, ulna, metacarpals, and phalanges.

Epiphyses. Without knowledge of bone growth and the appearance of bones in radiographic and other diagnostic images at various ages, a displaced epiphysial plate could be mistaken for a fracture, and separation of an epiphysis could be interpreted as a displaced piece of fractured bone. Knowledge of the patient's age and the location of epiphyses can prevent these errors.

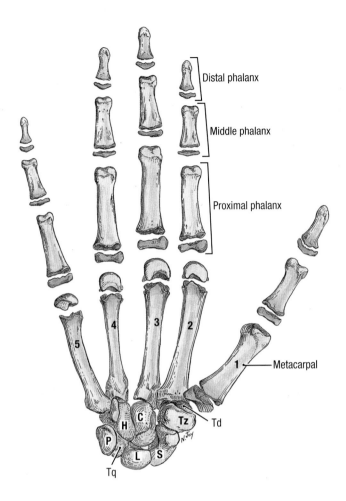

J. Anterior View (Right Hand)

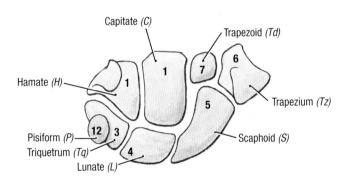

Numbers: approximate age of ossification of carpal bones in years

K. Anterior View

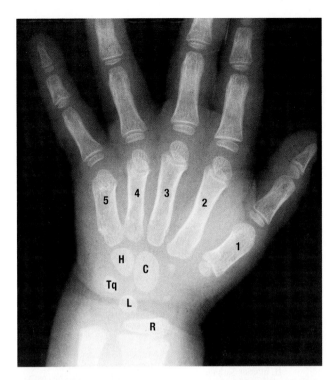

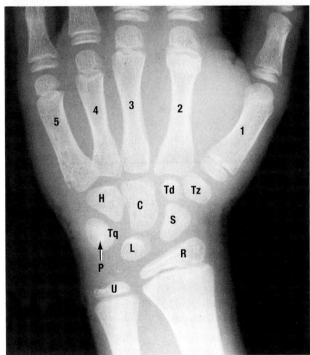

L. Anteroposterior View, Right Hand

Epiphyses in radiographs appear as radiolucent lines

OSSIFICATION AND SITES OF EPIPHYSES OF BONES OF UPPER LIMB (*continued*)

2.2

J. Ossification of bones of hand. Note the phalanges have a single proximal epiphysis and metacarpals 2, 3, 4, and 5 have single distal epiphyses. The 1st metacarpal behaves as a phalanx by having proximal epiphysis. Short-lived epiphyses may appear at the other ends of metacarpals 1 and/or 2. There are individual and gender differences in sequence and timing of ossification. **K.** Sequence of ossification of carpal bones. **L.** Radiographs of stages of ossification of wrist and hand. A 2½-year-old child *(top)*; the lunate is ossifying, and the distal radial epiphysis (*R*) is present (*C*, capitate; *H*, hamate; *L*, lunate; *Tq*, triquetrum). An 11-year-old child *(bottom)*. All carpal bones are ossified (*S*, scaphoid; *Td*, trapezoid; *Tz*, trapezium; *arrowhead*, pisiform), and the distal epiphysis of the ulna (*U*) has ossified.

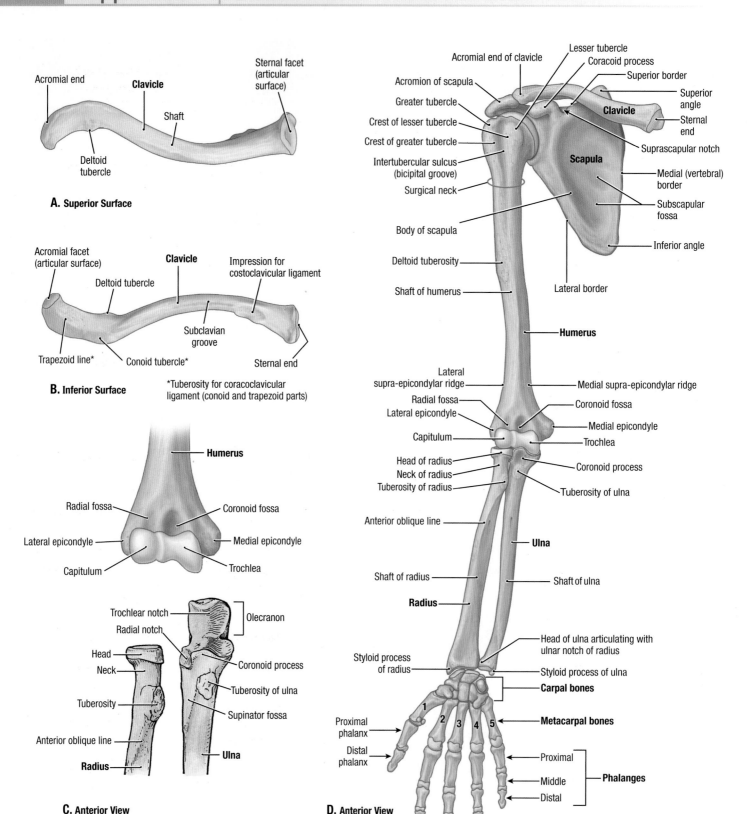

A. Superior Surface

Acromial end
Clavicle
Shaft
Sternal facet (articular surface)
Deltoid tubercle

B. Inferior Surface

Acromial facet (articular surface)
Clavicle
Deltoid tubercle
Impression for costoclavicular ligament
Subclavian groove
Trapezoid line*
Conoid tubercle*
Sternal end
*Tuberosity for coracoclavicular ligament (conoid and trapezoid parts)

C. Anterior View

Humerus
Radial fossa
Coronoid fossa
Lateral epicondyle
Medial epicondyle
Capitulum
Trochlea

Trochlear notch
Olecranon
Radial notch
Head
Neck
Coronoid process
Tuberosity
Tuberosity of ulna
Supinator fossa
Anterior oblique line
Radius
Ulna

D. Anterior View

Acromial end of clavicle
Lesser tubercle
Coracoid process
Acromion of scapula
Superior border
Greater tubercle
Superior angle
Crest of lesser tubercle
Clavicle
Sternal end
Crest of greater tubercle
Suprascapular notch
Intertubercular sulcus (bicipital groove)
Scapula
Medial (vertebral) border
Surgical neck
Subscapular fossa
Body of scapula
Deltoid tuberosity
Inferior angle
Shaft of humerus
Lateral border
Humerus
Lateral supra-epicondylar ridge
Medial supra-epicondylar ridge
Radial fossa
Coronoid fossa
Lateral epicondyle
Medial epicondyle
Capitulum
Trochlea
Head of radius
Neck of radius
Coronoid process
Tuberosity of radius
Tuberosity of ulna
Anterior oblique line
Ulna
Shaft of radius
Shaft of ulna
Radius
Head of ulna articulating with ulnar notch of radius
Styloid process of radius
Styloid process of ulna
Carpal bones
Proximal phalanx
Metacarpal bones
Distal phalanx
Proximal
Middle
Phalanges
Distal

2.3 FEATURES OF BONES OF UPPER LIMB

A. and **B.** Clavicle. **C.** Anterior aspect of disarticulated distal end of humerus and proximal end of radius and ulna. **D.** Anterior aspect of articulated upper limb.

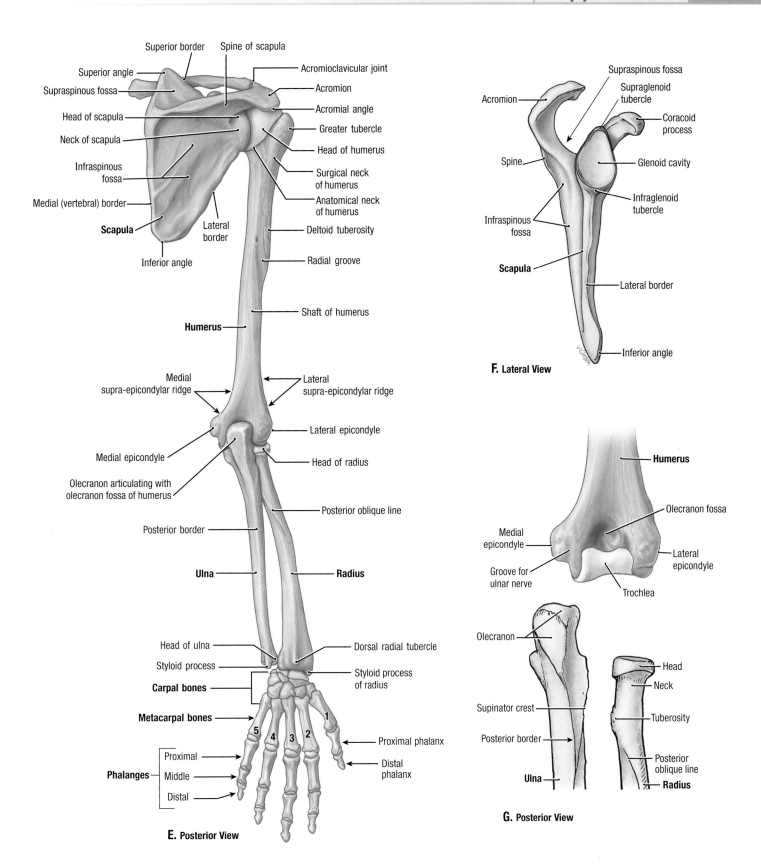

Superior border
Spine of scapula
Acromioclavicular joint
Superior angle
Acromion
Supraspinous fossa
Acromial angle
Head of scapula
Greater tubercle
Neck of scapula
Head of humerus
Surgical neck of humerus
Infraspinous fossa
Anatomical neck of humerus
Medial (vertebral) border
Scapula
Lateral border
Deltoid tuberosity
Inferior angle
Radial groove
Shaft of humerus
Humerus
Medial supra-epicondylar ridge
Lateral supra-epicondylar ridge
Lateral epicondyle
Medial epicondyle
Head of radius
Olecranon articulating with olecranon fossa of humerus
Posterior oblique line
Posterior border
Ulna
Radius
Head of ulna
Dorsal radial tubercle
Styloid process
Styloid process of radius
Carpal bones
Metacarpal bones
Proximal phalanx
Proximal
Distal phalanx
Phalanges — Middle
Distal

E. Posterior View

Supraspinous fossa
Supraglenoid tubercle
Acromion
Coracoid process
Spine
Glenoid cavity
Infraglenoid tubercle
Infraspinous fossa
Scapula
Lateral border
Inferior angle

F. Lateral View

Humerus
Olecranon fossa
Medial epicondyle
Lateral epicondyle
Groove for ulnar nerve
Trochlea
Olecranon
Head
Neck
Supinator crest
Tuberosity
Posterior border
Posterior oblique line
Ulna
Radius

G. Posterior View

FEATURES OF BONES OF UPPER LIMB (*continued*)

2.3

E. Posterior aspect of articulated upper limb bones. **F.** Lateral aspect of scapula. **G.** Posterior aspect of disarticulated distal end of humerus and proximal ends of radius and ulna.

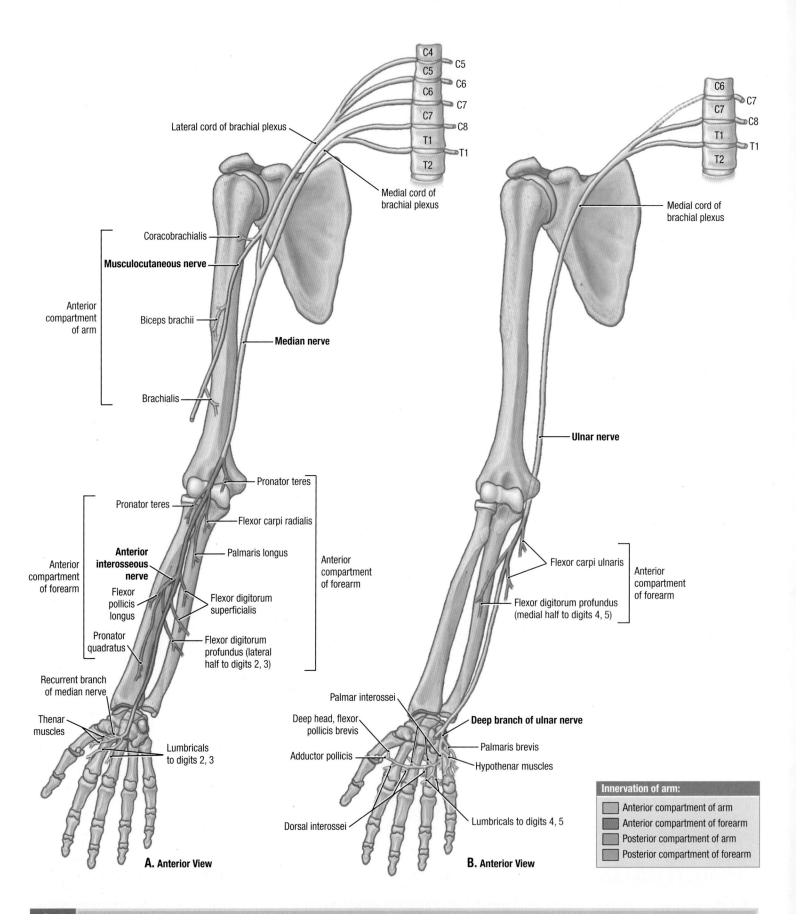

Lateral cord of brachial plexus

Medial cord of brachial plexus

Medial cord of brachial plexus

Coracobrachialis

Musculocutaneous nerve

Anterior compartment of arm

Biceps brachii

Median nerve

Brachialis

Ulnar nerve

Pronator teres

Pronator teres

Flexor carpi radialis

Palmaris longus

Anterior compartment of forearm

Anterior interosseous nerve

Flexor pollicis longus

Flexor digitorum superficialis

Anterior compartment of forearm

Pronator quadratus

Flexor digitorum profundus (lateral half to digits 2, 3)

Flexor carpi ulnaris

Anterior compartment of forearm

Flexor digitorum profundus (medial half to digits 4, 5)

Recurrent branch of median nerve

Thenar muscles

Lumbricals to digits 2, 3

Palmar interossei

Deep head, flexor pollicis brevis

Adductor pollicis

Deep branch of ulnar nerve

Palmaris brevis

Hypothenar muscles

Dorsal interossei

Lumbricals to digits 4, 5

A. Anterior View

B. Anterior View

Innervation of arm:

Anterior compartment of arm
Anterior compartment of forearm
Posterior compartment of arm
Posterior compartment of forearm

2.4   **OVERVIEW OF MOTOR INNERVATION OF UPPER LIMB**

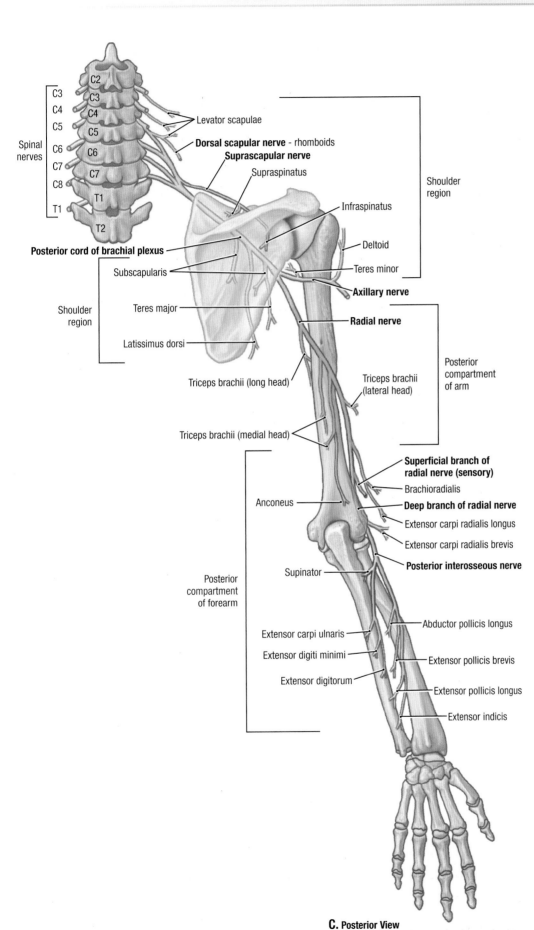

C2
C3
C3
C4
C4
C5
C5
C6
C6
C7
C7
C8
T1
T1
T2

Spinal nerves

Levator scapulae

Dorsal scapular nerve - rhomboids
Suprascapular nerve

Supraspinatus

Infraspinatus

Shoulder region

Posterior cord of brachial plexus

Deltoid

Teres minor

Subscapularis

Axillary nerve

Teres major

Radial nerve

Shoulder region

Latissimus dorsi

Triceps brachii (long head)

Triceps brachii (lateral head)

Posterior compartment of arm

Triceps brachii (medial head)

Superficial branch of radial nerve (sensory)

Brachioradialis

Anconeus

Deep branch of radial nerve

Extensor carpi radialis longus

Extensor carpi radialis brevis

Posterior interosseous nerve

Supinator

Posterior compartment of forearm

Abductor pollicis longus

Extensor carpi ulnaris

Extensor digiti minimi

Extensor pollicis brevis

Extensor digitorum

Extensor pollicis longus

Extensor indicis

C. Posterior View

OVERVIEW OF MOTOR INNERVATION OF UPPER LIMB (*continued*)   **2.4**

A. Musculocutaneous and median nerves. The musculocutaneous nerve innervates all the muscles of the anterior compartment of the arm. The median nerve innervates muscles of the anterior compartment of the forearm (with 1½ exceptions that are innervated by the ulnar nerve), the lumbricals to digits 2 and 3, and the intrinsic muscles of the thumb (thenar muscles). **B.** Ulnar nerve. The ulnar nerve innervates the flexor carpi ulnaris and ulnar half of the flexor digitorum profundus in the forearm, the hypothenar and interosseous muscles of the hand, the lumbricals to digits 3 and 4, and 1½ thenar muscles (adductor pollicis and the deep head of the flexor pollicis brevis). **C.** Radial nerve. The radial nerve innervates all muscles of the posterior compartments of the arm and forearm.

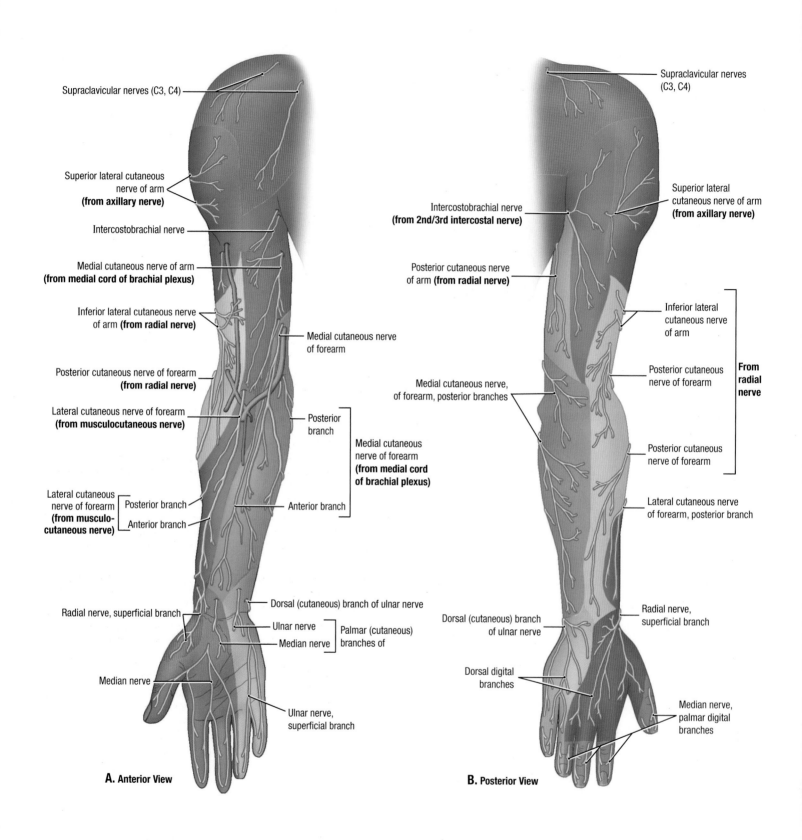

Supraclavicular nerves (C3, C4)

Superior lateral cutaneous nerve of arm **(from axillary nerve)**

Intercostobrachial nerve

Medial cutaneous nerve of arm **(from medial cord of brachial plexus)**

Inferior lateral cutaneous nerve of arm **(from radial nerve)**

Posterior cutaneous nerve of forearm **(from radial nerve)**

Lateral cutaneous nerve of forearm **(from musculocutaneous nerve)**

Lateral cutaneous nerve of forearm **(from musculo-cutaneous nerve)** — Posterior branch / Anterior branch

Radial nerve, superficial branch

Median nerve

Medial cutaneous nerve of forearm

Posterior branch

Medial cutaneous nerve of forearm **(from medial cord of brachial plexus)**

Anterior branch

Dorsal (cutaneous) branch of ulnar nerve

Ulnar nerve

Median nerve — Palmar (cutaneous) branches of

Ulnar nerve, superficial branch

A. Anterior View

Supraclavicular nerves (C3, C4)

Intercostobrachial nerve **(from 2nd/3rd intercostal nerve)**

Posterior cutaneous nerve of arm **(from radial nerve)**

Medial cutaneous nerve, of forearm, posterior branches

Superior lateral cutaneous nerve of arm **(from axillary nerve)**

Inferior lateral cutaneous nerve of arm

Posterior cutaneous nerve of forearm

Posterior cutaneous nerve of forearm

From radial nerve

Lateral cutaneous nerve of forearm, posterior branch

Dorsal (cutaneous) branch of ulnar nerve

Dorsal digital branches

Radial nerve, superficial branch

Median nerve, palmar digital branches

B. Posterior View

Summary of distribution of the peripheral (named) cutaneous nerves in upper limb. Most nerves are branches of nerve plexuses and therefore contain fibers from more than one spinal nerve.

TABLE 2.1 | **CUTANEOUS NERVES OF UPPER LIMB**

Nerve	Spinal Nerve Components	Source	Course/Distribution
Supraclavicular nerves	C3–C4	Cervical plexus	Pass anterior to clavicle, immediately deep to platysma, and supply the skin over the clavicle and superolateral aspect of the pectoralis major muscle
Superior lateral cutaneous nerve of arm	C5–C6	Axillary nerve (posterior cord of brachial plexus)	Emerges from posterior margin of deltoid to supply skin over lower part of this muscle and the lateral side of the midarm
Inferior lateral cutaneous nerve of arm		Radial nerve (posterior cord of brachial plexus)	Arises with the posterior cutaneous nerve of forearm; pierces lateral head of triceps brachii to supply skin over the inferolateral aspect of the arm
Posterior cutaneous nerve of arm	C5–C8		Arises in axilla and supplies skin on posterior surface of the arm to olecranon
Posterior cutaneous nerve of forearm			Arises with the inferior lateral cutaneous nerve of the arm; pierces lateral head of triceps brachii to supply skin over the posterior aspect of the arm
Superficial branch of radial nerve	C6–C7		Arises in cubital fossa; supplies lateral (radial) half of the dorsal aspect of hand and thumb, and proximal portion of the dorsal aspects of digits 2 and 3, and the lateral (radial) half of dorsal aspect of digit 4
Lateral cutaneous nerve of forearm		Musculocutaneous nerve (lateral cord of brachial plexus)	Arises between biceps brachii and brachialis muscle as continuation of musculocutaneous nerve distal to branch to brachialis; emerges in cubital fossa lateral to biceps tendon and median cubital vein; supplies skin along radial (lateral) border of forearm to base of thenar eminence
Median nerve	C6–C7 (via lateral root); C8–T1 (via medial root)	Lateral and medial cords of brachial plexus	Courses with brachial artery in arm and deep to flexor digitorum superficialis in forearm; distal to origin of palmar cutaneous branch, traverses carpal tunnel to supply skin of palmar aspect of radial 3½ digits and adjacent palm, plus distal dorsal aspects of same, including nail beds
Ulnar nerve	(C7), C8–T1	Medial cord of brachial plexus	Courses with brachial, superior ulnar collateral, and ulnar arteries; supplies skin of palmar and dorsal aspects of medial (ulnar) 1½ digits and palm and dorsum of hand proximal to those digits
Medial cutaneous nerve of forearm	C8–T1		Pierces deep fascia with basilic vein in midarm; divides into anterior and posterior branches supplying skin over anterior and medial surfaces of forearm to wrist
Medial cutaneous nerve of arm	C8–T2		Smallest and most medial branch of brachial plexus; communicates with intercostobrachial nerve and then descends medial to brachial artery and basilic vein to innervate skin of distal medial arm
Intercostobrachial nerve	T2	Lateral cutaneous branch of 2nd intercostal nerve	Arises distal to angle of 2nd rib; supplies skin of axilla and proximal medial arm

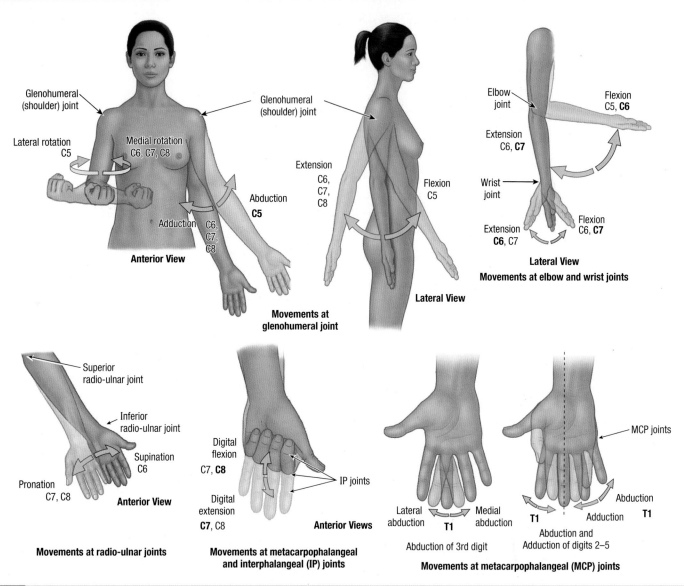

Glenohumeral (shoulder) joint

Glenohumeral (shoulder) joint

Lateral rotation C5

Medial rotation C6, C7, C8

Abduction **C5**

Adduction C6, C7, C8

Anterior View

Movements at glenohumeral joint

Extension C6, C7, C8

Flexion C5

Lateral View

Elbow joint

Flexion C5, **C6**

Extension C6, **C7**

Wrist joint

Extension **C6**, C7

Flexion C6, **C7**

Lateral View
Movements at elbow and wrist joints

Superior radio-ulnar joint

Inferior radio-ulnar joint

Supination C6

Pronation C7, C8

Anterior View

Movements at radio-ulnar joints

Digital flexion C7, **C8**

Digital extension **C7**, C8

IP joints

Anterior Views

Movements at metacarpophalangeal and interphalangeal (IP) joints

Lateral abduction

Medial abduction

T1

Abduction of 3rd digit

MCP joints

Abduction T1

Adduction

Abduction and Adduction of digits 2–5

Movements at metacarpophalangeal (MCP) joints

2.6 MYOTOMES AND MYOTATIC (DEEP TENDON STRETCH) REFLEXES

Myotomes. Somatic motor (general somatic efferent) fibers transmit impulses to skeletal (voluntary) muscles. The unilateral muscle mass receiving information from the somatic motor fibers conveyed by a single spinal nerve is a myotome. The movements associated with each bolded segment in Table 2.2 are most commonly tested to determine the neurologic level of a lesion. **Myotatic reflexes.** A myotatic reflex (deep tendon or stretch reflex) is an involuntary contraction of a muscle in response to sudden stretching. Myotatic reflexes are elicited by briskly tapping the tendon with a reflex hammer. Each tendon reflex is mediated by specific spinal nerves. Stretch reflexes control muscle tone.

TABLE 2.2	**CLINICAL MANIFESTATIONS OF NERVE ROOT COMPRESSION: UPPER LIMB**				
Herniated Disc Between	*Compressed Nerve Root*	*Dermatome Affected*	*Muscles Affected*	*Movement Weakness*	*Nerve and Myotatic Reflex Involved*
C4 and C5	C5	C5 Shoulder Lateral surface UL	Deltoid	Abduction of shoulder	Axillary nerve ↓ Biceps jerk
C5 and C6	C6	C6 Thumb	Biceps Brachialis Brachioradialis	Flexion of elbow Supination/pronation of forearm	Musculocutaneous nerve ↓ Biceps jerk ↓ Brachioradialis jerk
C6 and C7	C7	C7 Posterior surface UL Middle and index fingers	Triceps Wrist extensors	Extension of elbow Extension of wrist	↓ Triceps reflex

UL, upper limb.

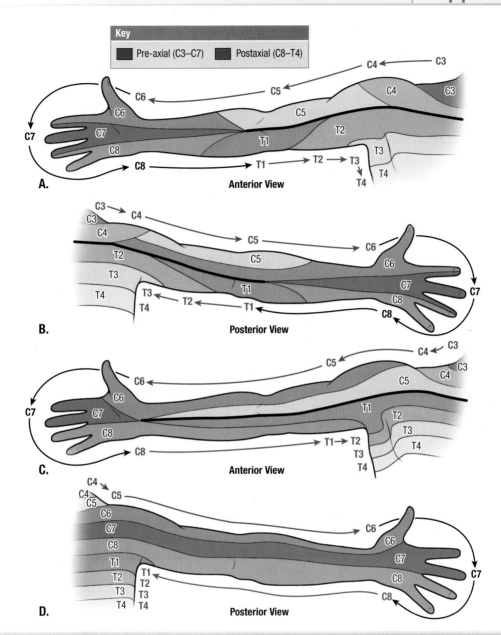

Key
■ Pre-axial (C3–C7) ■ Postaxial (C8–T4)

A. Anterior View

B. Posterior View

C. Anterior View

D. Posterior View

DERMATOMES OF UPPER LIMB

2.7

Two different dermatome maps are commonly used. **A.** and **B.** The dermatome pattern according to Foerster (1933) is preferred by many because of its correlation with clinical findings. In the Foerster schema, dermatomes C6–T1 are displaced from the trunk to limbs. **C.** and **D.** The dermatome pattern according to Keegan and Garrett (1948) is preferred by others for its correlation with development. Although depicted as distinct zones, adjacent dermatomes overlap considerably except along the axial line.

TABLE 2.3	**DERMATOMES OF UPPER LIMB**
Spinal Segment/Nerve(s)	*Description of Dermatome(s)*
C3, C4	Region at base of neck extending laterally over shoulder
C5	Lateral aspect of arm (i.e., superior aspect of abducted arm)
C6	Lateral forearm and thumb
C7	Middle and ring fingers (or middle three fingers) and center of posterior aspect of forearm
C8	Little finger, medial side of hand and forearm (i.e., inferior aspect of abducted arm)
T1	Medial aspect of forearm and inferior arm
T2	Medial aspect of superior arm and skin of axilla[a]

[a]Not indicated on the Keegan and Garrett map. However, pain experienced during a heart attack, considered to be mediated by T1 and T2, is commonly described as "radiating down the medial side of the left arm."

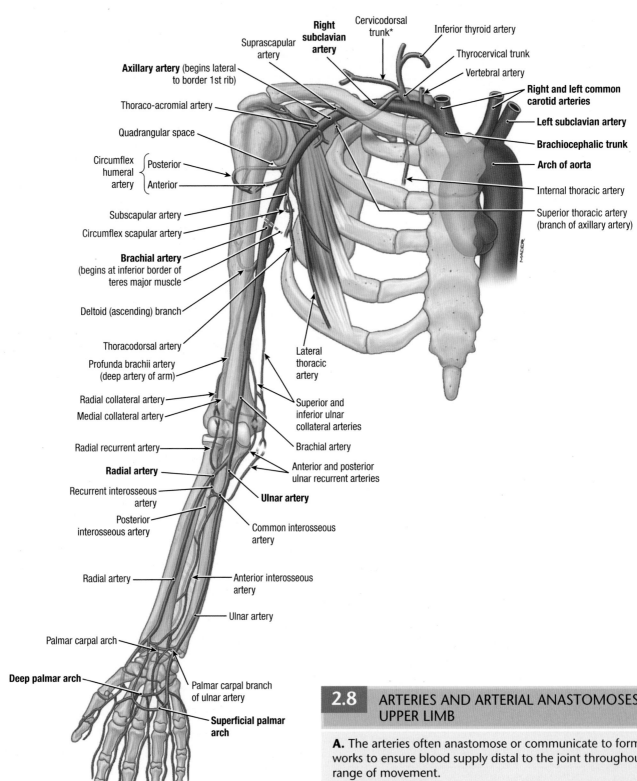

Right subclavian artery

Cervicodorsal trunk*

Inferior thyroid artery

Suprascapular artery

Thyrocervical trunk

Vertebral artery

Axillary artery (begins lateral to border 1st rib)

Right and left common carotid arteries

Thoraco-acromial artery

Left subclavian artery

Quadrangular space

Brachiocephalic trunk

Circumflex humeral artery { Posterior / Anterior }

Arch of aorta

Internal thoracic artery

Subscapular artery

Circumflex scapular artery

Superior thoracic artery (branch of axillary artery)

Brachial artery (begins at inferior border of teres major muscle)

Deltoid (ascending) branch

Thoracodorsal artery

Lateral thoracic artery

Profunda brachii artery (deep artery of arm)

Radial collateral artery

Medial collateral artery

Superior and inferior ulnar collateral arteries

Radial recurrent artery

Brachial artery

Radial artery

Anterior and posterior ulnar recurrent arteries

Recurrent interosseous artery

Ulnar artery

Posterior interosseous artery

Common interosseous artery

Radial artery

Anterior interosseous artery

Ulnar artery

Palmar carpal arch

Deep palmar arch

Palmar carpal branch of ulnar artery

Superficial palmar arch

A. Palmar View

2.8 ARTERIES AND ARTERIAL ANASTOMOSES OF UPPER LIMB

A. The arteries often anastomose or communicate to form networks to ensure blood supply distal to the joint throughout the range of movement.

Arterial occlusion. If a main channel is occluded, the smaller alternate channels can usually increase in size, providing a collateral circulation that ensures the blood supply to structures distal to the blockage. However, collateral pathways require time to develop; they are usually insufficient to compensate for sudden occlusions.

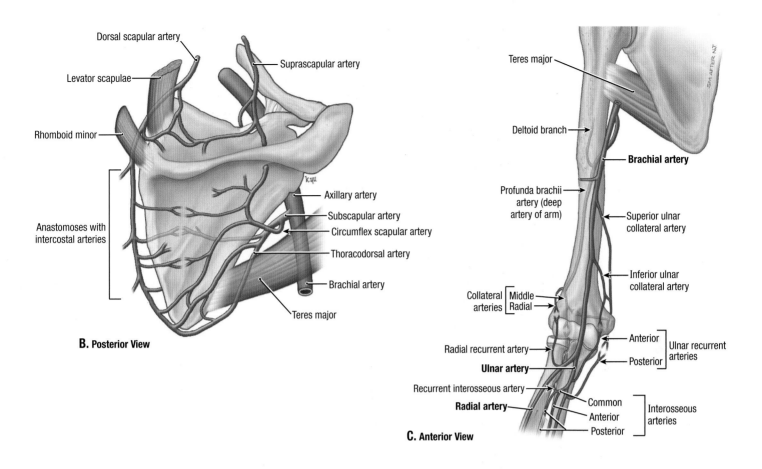

B. Posterior View

C. Anterior View

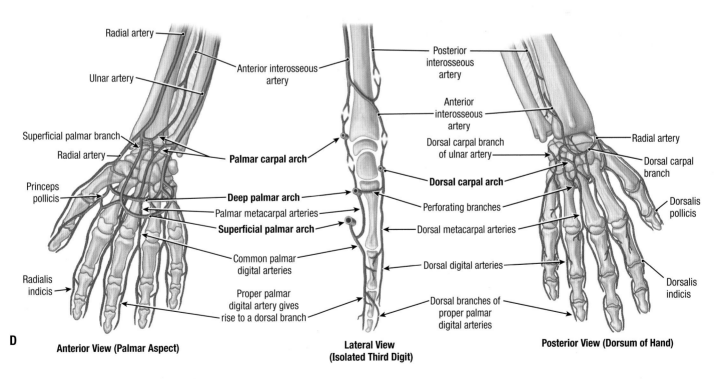

D

Anterior View (Palmar Aspect)

Lateral View (Isolated Third Digit)

Posterior View (Dorsum of Hand)

ARTERIES AND ARTERIAL ANASTOMOSES OF UPPER LIMB (*continued*)

2.8

B. Scapular anastomoses. **C.** Anastomoses of the elbow. **D.** Anastomoses of the hand. Joints receive blood from articular arteries that arise from vessels around joints.

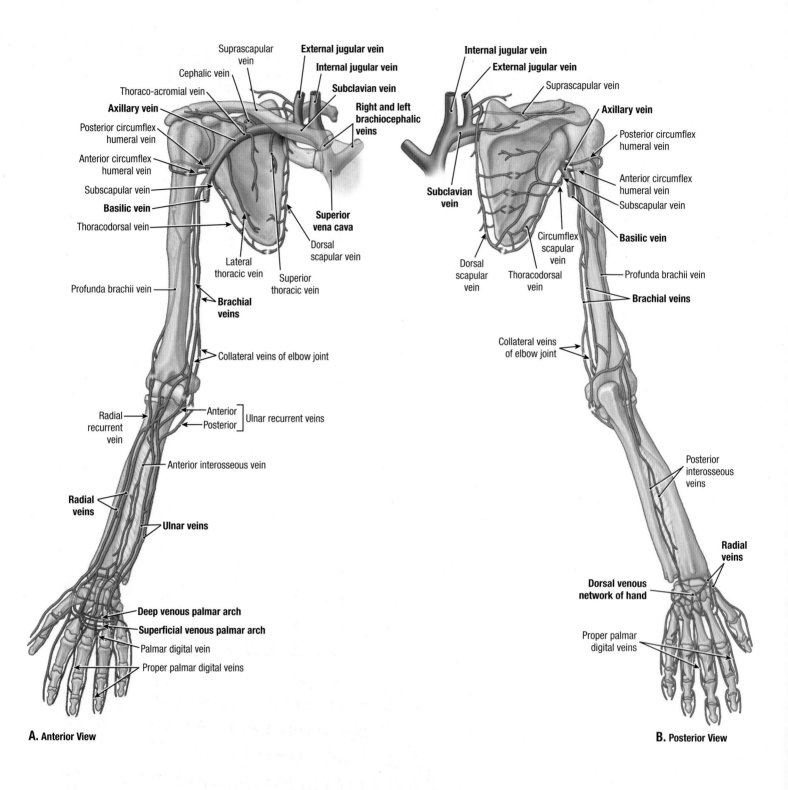

Suprascapular vein
Cephalic vein
Thoraco-acromial vein
Axillary vein
Posterior circumflex humeral vein
Anterior circumflex humeral vein
Subscapular vein
Basilic vein
Thoracodorsal vein
Profunda brachii vein
Brachial veins
Collateral veins of elbow joint
Radial recurrent vein
Anterior } Posterior } Ulnar recurrent veins
Anterior interosseous vein
Radial veins
Ulnar veins
Deep venous palmar arch
Superficial venous palmar arch
Palmar digital vein
Proper palmar digital veins

External jugular vein
Internal jugular vein
Subclavian vein
Right and left brachiocephalic veins
Superior vena cava
Dorsal scapular vein
Lateral thoracic vein
Superior thoracic vein

A. Anterior View

Internal jugular vein
External jugular vein
Suprascapular vein
Axillary vein
Posterior circumflex humeral vein
Anterior circumflex humeral vein
Subscapular vein
Basilic vein
Profunda brachii vein
Brachial veins
Subclavian vein
Dorsal scapular vein
Circumflex scapular vein
Thoracodorsal vein
Collateral veins of elbow joint
Posterior interosseous veins
Radial veins
Dorsal venous network of hand
Proper palmar digital veins

B. Posterior View

2.9 OVERVIEW OF DEEP VEINS OF UPPER LIMB

Deep veins lie internal to the deep fascia and occur as paired, continually interanastomosing accompanying veins (e.g., venae comitantes) surrounding and sharing the name of the artery they accompany.

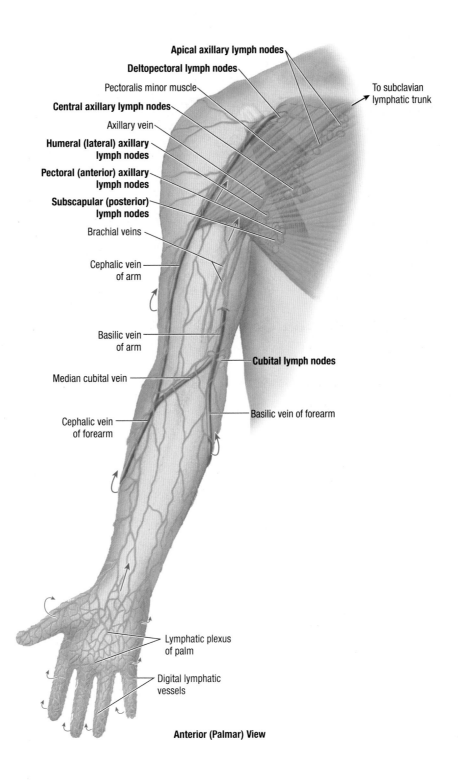

Apical axillary lymph nodes

Deltopectoral lymph nodes

Pectoralis minor muscle

Central axillary lymph nodes

Axillary vein

Humeral (lateral) axillary lymph nodes

Pectoral (anterior) axillary lymph nodes

Subscapular (posterior) lymph nodes

Brachial veins

Cephalic vein of arm

Basilic vein of arm

Median cubital vein

Cephalic vein of forearm

To subclavian lymphatic trunk

Cubital lymph nodes

Basilic vein of forearm

Lymphatic plexus of palm

Digital lymphatic vessels

Anterior (Palmar) View

SUPERFICIAL VENOUS AND LYMPHATIC DRAINAGE OF UPPER LIMB

2.10

Superficial lymphatic vessels arise from lymphatic plexuses in the digits, palm, and dorsum of the hand and ascend with the superficial veins of the upper limb. The superficial lymphatic vessels ascend through the forearm and arm, converging toward the cephalic and especially to the basilic vein to reach the axillary lymph nodes. Some lymph passes through the cubital nodes at the elbow and the deltopectoral (infraclavicular) nodes at the shoulder. Deep lymphatic vessels accompany the neurovascular bundles of the upper limb and end primarily in the humeral (lateral) and central axillary lymph nodes.

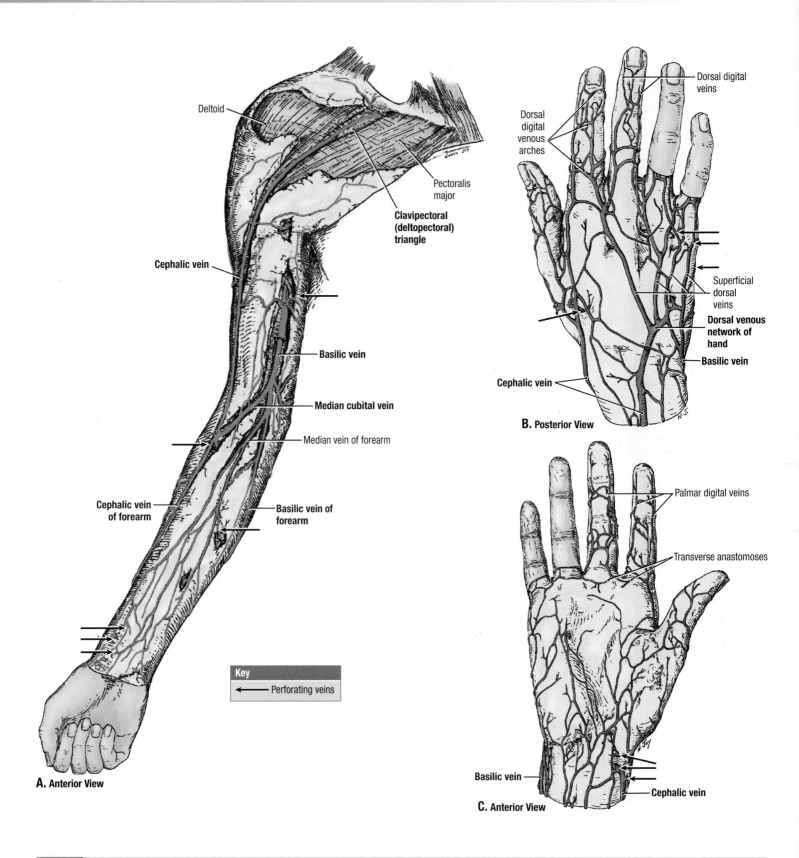

Deltoid

Pectoralis major

Clavipectoral (deltopectoral) triangle

Cephalic vein

Basilic vein

Median cubital vein

Median vein of forearm

Cephalic vein of forearm

Basilic vein of forearm

Key

← Perforating veins

A. Anterior View

Dorsal digital veins

Dorsal digital venous arches

Superficial dorsal veins

Dorsal venous network of hand

Basilic vein

Cephalic vein

B. Posterior View

Palmar digital veins

Transverse anastomoses

Basilic vein

Cephalic vein

C. Anterior View

2.11 | **SUPERFICIAL VENOUS DRAINAGE OF UPPER LIMB**

A. Forearm, arm, and pectoral region. **B.** Dorsal surface of hand. **C.** Palmar surface of hand. *Arrows* indicate where perforating veins penetrate the deep fascia. Blood is continuously shunted from these superficial veins in the subcutaneous tissue to deep veins via the perforating veins.

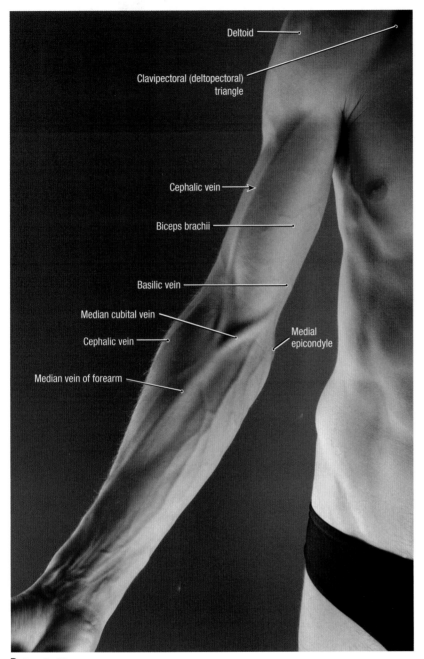

Deltoid

Clavipectoral (deltopectoral)
triangle

Cephalic vein

Biceps brachii

Basilic vein

Median cubital vein

Cephalic vein

Medial
epicondyle

Median vein of forearm

D. Anterior View

Superficial
dorsal veins

Cephalic
vein

Dorsal venous
network of hand

E. Posterior View

SUPERFICIAL VENOUS DRAINAGE OF UPPER LIMB (*continued*)   2.11

D. Surface anatomy of veins of forearm and arm. **E.** Surface anatomy of veins of the dorsal surface of hand.

Because of the prominence and accessibility of the superficial veins, they are commonly used for **venipuncture** (puncture of a vein to draw blood or inject a solution). By applying a tourniquet to the arm, the venous return is occluded, and the veins distend and usually are visible and/or palpable. Once a vein is punctured, the tourniquet is removed so that when the needle is removed the vein will not bleed extensively. The median cubital vein is commonly used for venipuncture. The veins forming the dorsal venous network of the hand and the cephalic and basilic veins arising from it are commonly used for long-term introduction of fluids (**intravenous feeding**). The cubital veins are also a site for the **introduction of cardiac catheters** to secure blood samples from the great vessels and chambers of the heart.

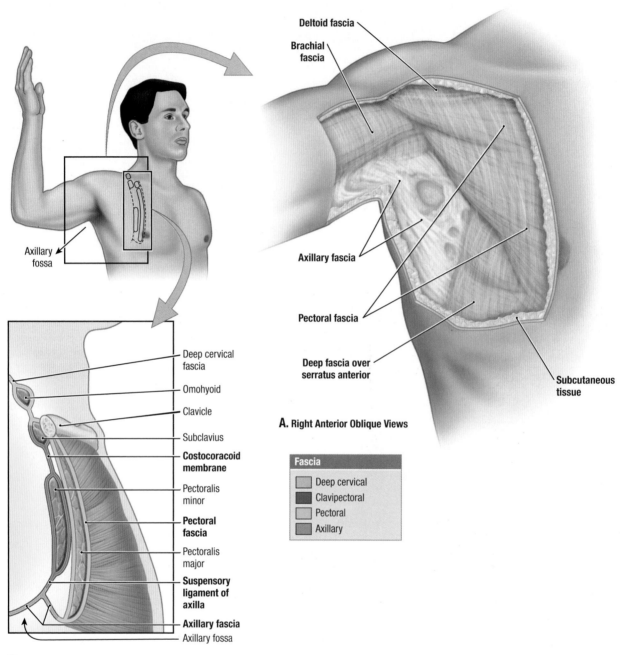

A. Right Anterior Oblique Views

Fascia	
	Deep cervical
	Clavipectoral
	Pectoral
	Axillary

B. Lateral View of Sagittal Section

2.12   DEEP FASCIA OF UPPER LIMB—AXILLARY AND CLAVIPECTORAL FASCIA

A. Axillary fascia. The axillary fascia forms the floor of the axillary fossa and is continuous with the pectoral fascia covering the pectoralis major muscle and the brachial fascia of the arm. **B.** Clavipectoral fascia. The clavipectoral fascia extends from the axillary fascia to enclose the pectoralis minor and subclavius muscles and then attaches to the clavicle. The part of the clavipectoral fascia superior to the pectoralis minor is the costocoracoid membrane, and the part of the clavipectoral fascia inferior to the pectoralis minor is the suspensory ligament of the axilla. The suspensory ligament of the axilla, an extension of the axillary fascia, supports the axillary fascia and pulls the axillary fascia and the skin inferior to it superiorly when the arm is abducted, forming the axillary fossa or "armpit."

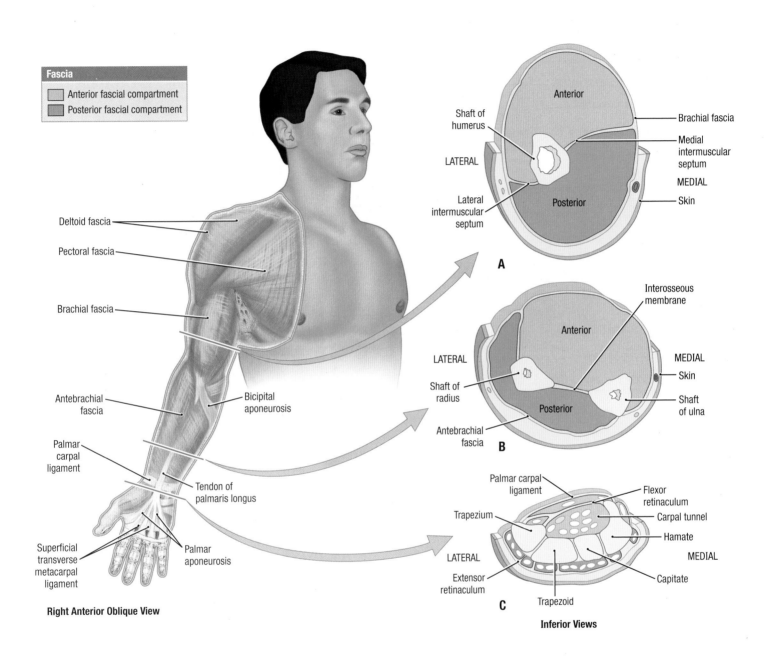

Fascia
- Anterior fascial compartment
- Posterior fascial compartment

Deltoid fascia

Pectoral fascia

Brachial fascia

Antebrachial fascia

Palmar carpal ligament

Bicipital aponeurosis

Tendon of palmaris longus

Superficial transverse metacarpal ligament

Palmar aponeurosis

Right Anterior Oblique View

A

Shaft of humerus

LATERAL

Lateral intermuscular septum

Anterior

Brachial fascia

Medial intermuscular septum

MEDIAL

Skin

Posterior

B

Interosseous membrane

LATERAL

Anterior

MEDIAL

Shaft of radius

Skin

Antebrachial fascia

Posterior

Shaft of ulna

C

Palmar carpal ligament

Trapezium

LATERAL

Extensor retinaculum

Trapezoid

Flexor retinaculum

Carpal tunnel

Hamate

MEDIAL

Capitate

Inferior Views

DEEP FASCIA OF UPPER LIMB, BRACHIAL AND ANTEBRACHIAL FASCIA

2.13

A. Brachial fascia. The brachial fascia is the deep fascia of the arm and is continuous superiorly with the pectoral and axillary layers of fascia. Medial and lateral intermuscular septa extend from the deep aspect of the brachial fascia to the humerus, dividing the arm into anterior and posterior musculofascial compartments. **B.** Antebrachial fascia. The antebrachial fascia surrounds the forearm and is continuous with the brachial fascia and deep fascia of the hand. The interosseous membrane separates the forearm into anterior and posterior musculofascial compartments. Distally, the fascia thickens to form the palmar carpal ligament, which is continuous with the flexor retinaculum and dorsally with the extensor expansion. The deep fascia of the hand is continuous with the antebrachial fascia, and on the palmar surface of the hand, it thickens to form the palmar aponeurosis. **C.** Flexor retinaculum (transverse carpal ligament). The flexor retinaculum extends between the medial and lateral carpal bones to form the carpal tunnel.

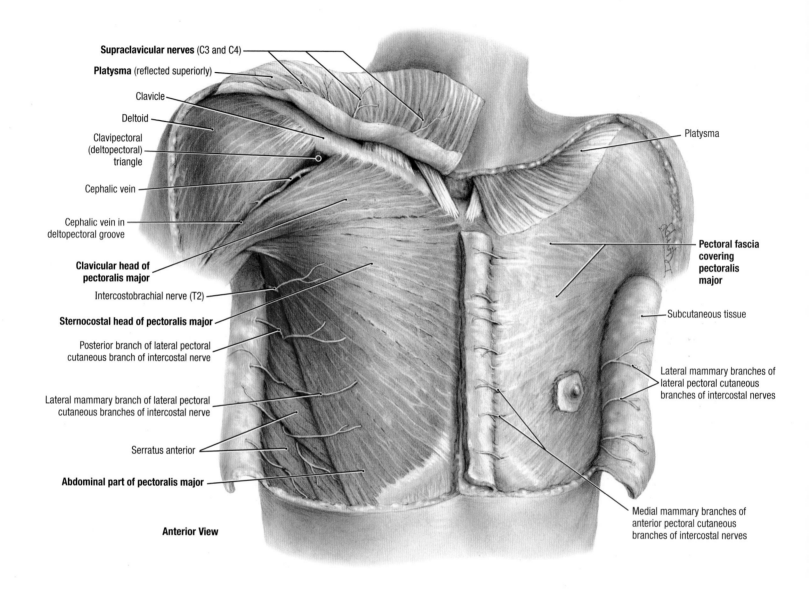

Supraclavicular nerves (C3 and C4)

Platysma (reflected superiorly)

Clavicle

Deltoid

Clavipectoral (deltopectoral) triangle

Cephalic vein

Cephalic vein in deltopectoral groove

Clavicular head of pectoralis major

Intercostobrachial nerve (T2)

Sternocostal head of pectoralis major

Posterior branch of lateral pectoral cutaneous branch of intercostal nerve

Lateral mammary branch of lateral pectoral cutaneous branches of intercostal nerve

Serratus anterior

Abdominal part of pectoralis major

Platysma

Pectoral fascia covering pectoralis major

Subcutaneous tissue

Lateral mammary branches of lateral pectoral cutaneous branches of intercostal nerves

Medial mammary branches of anterior pectoral cutaneous branches of intercostal nerves

Anterior View

2.14 | SUPERFICIAL DISSECTION, MALE PECTORAL REGION

- The platysma muscle, which usually descends to the 2nd or 3rd rib, is cut short on the right side and, together with the supraclavicular nerves, is reflected on the left side.
- The exposed intermuscular bony strip of the clavicle is subcutaneous and subplatysmal.
- The cephalic vein passes deeply to join the axillary vein in the clavipectoral (deltopectoral) triangle.

- The cutaneous innervation of the pectoral region is by the supraclavicular nerves (C3 and C4) and upper thoracic nerves (T2–T6); the brachial plexus (C5–T1) does not supply cutaneous branches to the pectoral region.

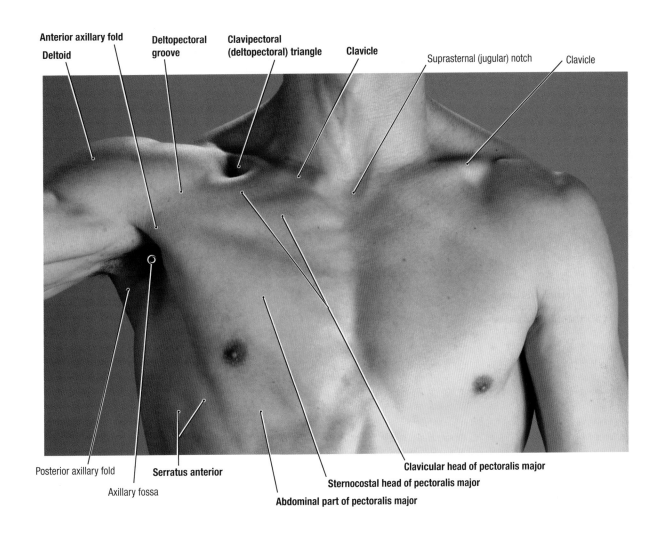

Anterior axillary fold

Deltoid

Deltopectoral groove

Clavipectoral (deltopectoral) triangle

Clavicle

Suprasternal (jugular) notch

Clavicle

Posterior axillary fold

Serratus anterior

Axillary fossa

Abdominal part of pectoralis major

Sternocostal head of pectoralis major

Clavicular head of pectoralis major

SURFACE ANATOMY, MALE PECTORAL REGION

2.15

The clavipectoral (deltopectoral) triangle is the depressed area just inferior to the lateral part of the clavicle, bounded by the clavicle superiorly, the deltoid laterally, and the clavicular head of pectoralis major medially. The clavipectoral triangle and the intermuscular deltopectoral groove extending from its inferior apex demarcate an "internervous plane" (plane not crossed by motor nerves) for an **anterior or deltopectoral surgical incision** to approach the axilla, shoulder joint, or proximal humerus.

When the arm is abducted and then adducted against resistance, the two heads of the pectoralis major are visible and palpable. As this muscle extends from the thoracic wall to the arm, it forms the anterior axillary fold. Digitations of the serratus anterior appear inferolateral to the pectoralis major. The coracoid process of the scapula is covered by the anterior part of deltoid; however, the tip of the process can be felt on deep palpation in the clavipectoral triangle. The deltoid forms the contour of the shoulder.

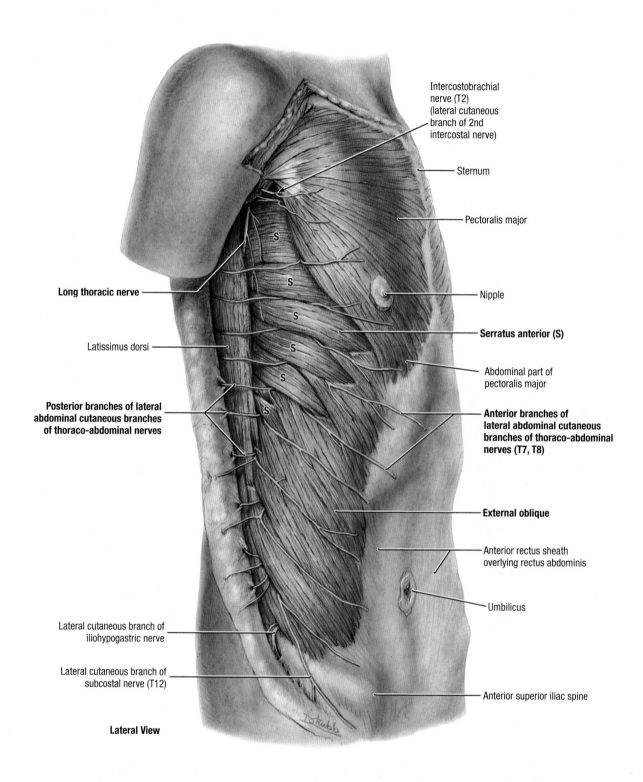

Intercostobrachial nerve (T2) (lateral cutaneous branch of 2nd intercostal nerve)

Sternum

Pectoralis major

Nipple

Serratus anterior (S)

Abdominal part of pectoralis major

Anterior branches of lateral abdominal cutaneous branches of thoraco-abdominal nerves (T7, T8)

External oblique

Anterior rectus sheath overlying rectus abdominis

Umbilicus

Anterior superior iliac spine

Long thoracic nerve

Latissimus dorsi

Posterior branches of lateral abdominal cutaneous branches of thoraco-abdominal nerves

Lateral cutaneous branch of iliohypogastric nerve

Lateral cutaneous branch of subcostal nerve (T12)

Lateral View

2.16 SUPERFICIAL DISSECTION OF TRUNK

- The slips of the serratus anterior interdigitate with the external oblique.
- The long thoracic nerve (nerve to serratus anterior) lies on the lateral (superficial) aspect of the serratus anterior; this nerve is vulnerable to damage from **stab wounds** and during surgery (e.g., radical mastectomy).

- The anterior and posterior branches of the lateral thoracic and abdominal cutaneous branches of intercostal and thoraco-abdominal nerves are dissected.

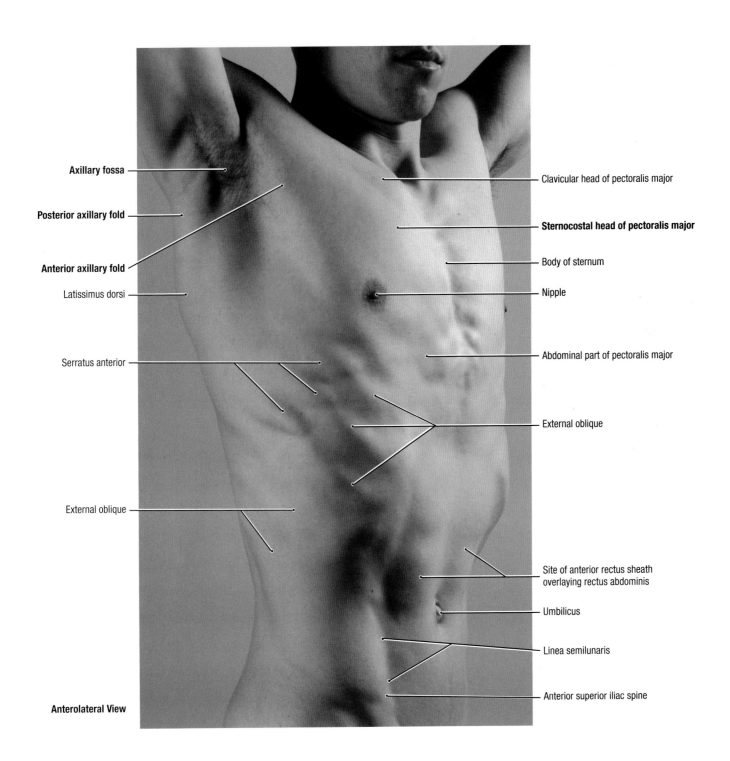

Axillary fossa

Posterior axillary fold

Anterior axillary fold

Latissimus dorsi

Serratus anterior

External oblique

Anterolateral View

Clavicular head of pectoralis major

Sternocostal head of pectoralis major

Body of sternum

Nipple

Abdominal part of pectoralis major

External oblique

Site of anterior rectus sheath overlaying rectus abdominis

Umbilicus

Linea semilunaris

Anterior superior iliac spine

SURFACE ANATOMY OF ANTEROLATERAL ASPECT OF TRUNK

2.17

When the arm is abducted and then adducted against resistance, the sternocostal part of the pectoralis major can be seen and palpated. If the anterior axillary fold bounding the axilla is grasped between the fingers and thumb, the inferior border of the sternocostal head of the pectoralis major can be felt. Several digitations of the serratus anterior are visible inferior to the anterior axillary fold. The posterior axillary fold is composed of skin and muscular tissue (latissimus dorsi and teres major) bounding the axilla posteriorly.

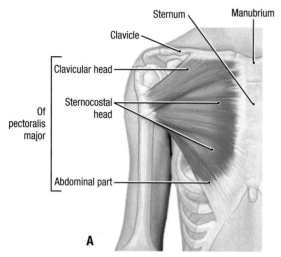

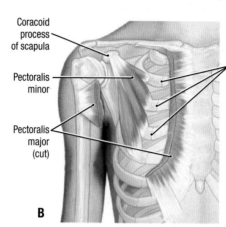

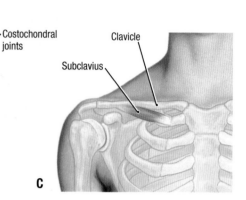

A

B

C

Anterior Views

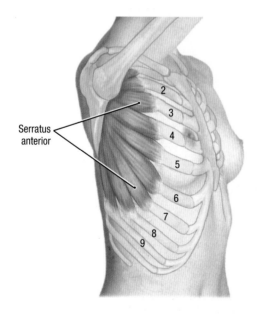

D. Right Anterolateral View

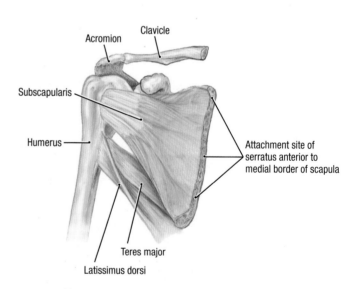

E. Anterior View

2.18   PECTORALIS MAJOR AND MINOR AND SERRATUS ANTERIOR

A. Pectoralis major. **B.** Pectoralis minor. **C.** Subclavius. **D.** and **E.** Serratus anterior and its scapular attachment.

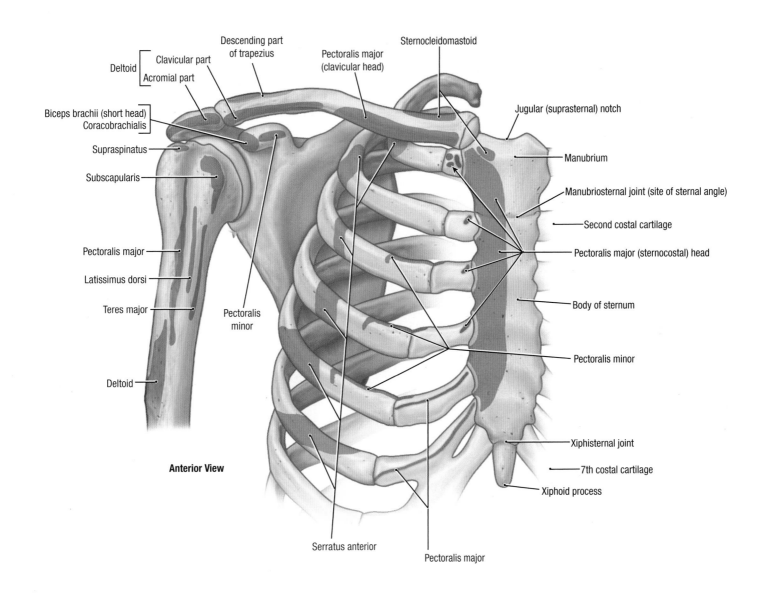

Descending part of trapezius

Deltoid [Clavicular part / Acromial part]

Biceps brachii (short head)
Coracobrachialis

Supraspinatus

Subscapularis

Pectoralis major

Latissimus dorsi

Teres major

Pectoralis minor

Deltoid

Anterior View

Serratus anterior

Pectoralis major (clavicular head)

Sternocleidomastoid

Jugular (suprasternal) notch

Manubrium

Manubriosternal joint (site of sternal angle)

Second costal cartilage

Pectoralis major (sternocostal) head

Body of sternum

Pectoralis minor

Xiphisternal joint

7th costal cartilage

Xiphoid process

Pectoralis major

ANTERIOR ATTACHMENTS OF ANTERIOR AND POSTERIOR AXIO-APPENDICULAR AND SCAPULOHUMERAL MUSCLES

2.19

TABLE 2.4   ANTERIOR AXIO-APPENDICULAR MUSCLES

Muscle	Proximal Attachment (red)	Distal Attachment (blue)	Innervation[a]	Main Actions
Pectoralis major	*Clavicular head:* anterior surface of medial half of clavicle *Sternocostal head:* anterior surface of sternum, superior six costal cartilages *Abdominal part:* aponeurosis of external oblique muscle	Crest of greater tubercle of intertubercular sulcus (lateral lip of bicipital groove)	Lateral and medial pectoral nerves; clavicular head (C5 and **C6**), sternocostal head (**C7**, **C8**, and T1)	Adducts and medially rotates humerus at shoulder joint; draws scapula anteriorly and inferiorly Acting alone: clavicular head flexes shoulder joint, and sternocostal head extends it from the flexed position
Pectoralis minor	3rd to 5th ribs near their costal cartilages	Medial border and superior surface of coracoid process of scapula	Medial pectoral nerve (C8 and T1)	Stabilizes scapula by drawing it inferiorly and anteriorly against thoracic wall
Subclavius	Junction of 1st rib and its costal cartilage	Inferior surface of middle third of clavicle	Nerve to subclavius (**C5** and C6)	Anchors and depresses clavicle at sternoclavicular joint
Serratus anterior	External surfaces of lateral parts of 1st to 8th–9th ribs	Anterior surface of medial border of scapula (see Fig. 2.18E)	Long thoracic nerve (C5, **C6**, and **C7**)	Protracts scapula and holds it against thoracic wall; rotates scapula

[a]Numbers indicate spinal cord segmental innervation (e.g., C5 and C6 indicate that nerves supplying the clavicular head of pectoralis major are derived from 5th and 6th cervical segments of spinal cord). Boldface numbers indicate the main segmental innervation. Damage to these segments or to motor nerve roots arising from them results in paralysis of the muscles concerned.

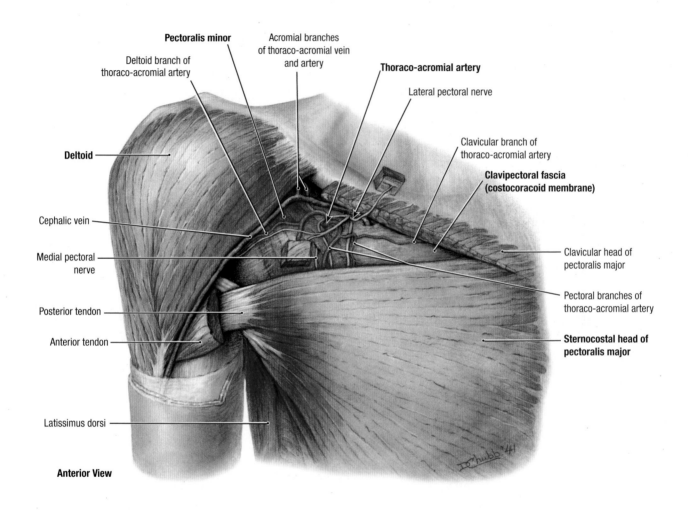

Pectoralis minor

Deltoid branch of
thoraco-acromial artery

Acromial branches
of thoraco-acromial vein
and artery

Thoraco-acromial artery

Lateral pectoral nerve

Clavicular branch of
thoraco-acromial artery

**Clavipectoral fascia
(costocoracoid membrane)**

Deltoid

Cephalic vein

Medial pectoral
nerve

Posterior tendon

Anterior tendon

Latissimus dorsi

Clavicular head of
pectoralis major

Pectoral branches of
thoraco-acromial artery

**Sternocostal head of
pectoralis major**

Anterior View

2.20   ANTERIOR WALL OF AXILLA AND CLAVIPECTORAL FASCIA

Anterior wall of axilla. The clavicular head of the pectoralis major is excised, except for two cubes of muscle that remain to identify the branches of the lateral pectoral nerve.

- The clavipectoral fascia superior to the pectoralis minor (costo-coracoid membrane) is pierced by the cephalic vein, the lateral pectoral nerve, and the thoraco-acromial vessels.

- The pectoralis minor and clavipectoral fascia are pierced by the medial pectoral nerve.
- Observe the insertion of the pectoralis major from deep to superficial: inferior part of the sternocostal head, superior part of the sternocostal head (posterior tendon), and clavicular head (anterior tendon).

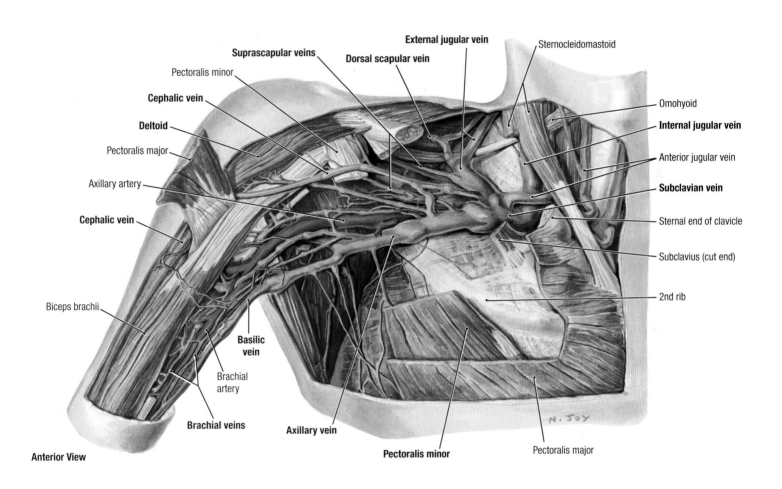

External jugular vein

Suprascapular veins

Dorsal scapular vein

Sternocleidomastoid

Pectoralis minor

Cephalic vein

Omohyoid

Deltoid

Internal jugular vein

Pectoralis major

Anterior jugular vein

Axillary artery

Subclavian vein

Cephalic vein

Sternal end of clavicle

Subclavius (cut end)

2nd rib

Biceps brachii

Basilic vein

Brachial artery

Brachial veins

Axillary vein

Pectoralis minor

Pectoralis major

N. JOY

Anterior View

VEINS OF AXILLA

2.21

- The basilic vein joins the brachial veins to become the axillary vein near the inferior border of teres major, the axillary vein becomes the subclavian vein at the lateral border of the 1st rib, and the subclavian joins the internal jugular to become the brachiocephalic vein posterior to the sternal end of the clavicle.

- Numerous valves, enlargements in the vein, are shown.
- The cephalic vein in this specimen bifurcates to end in the axillary and external jugular veins.

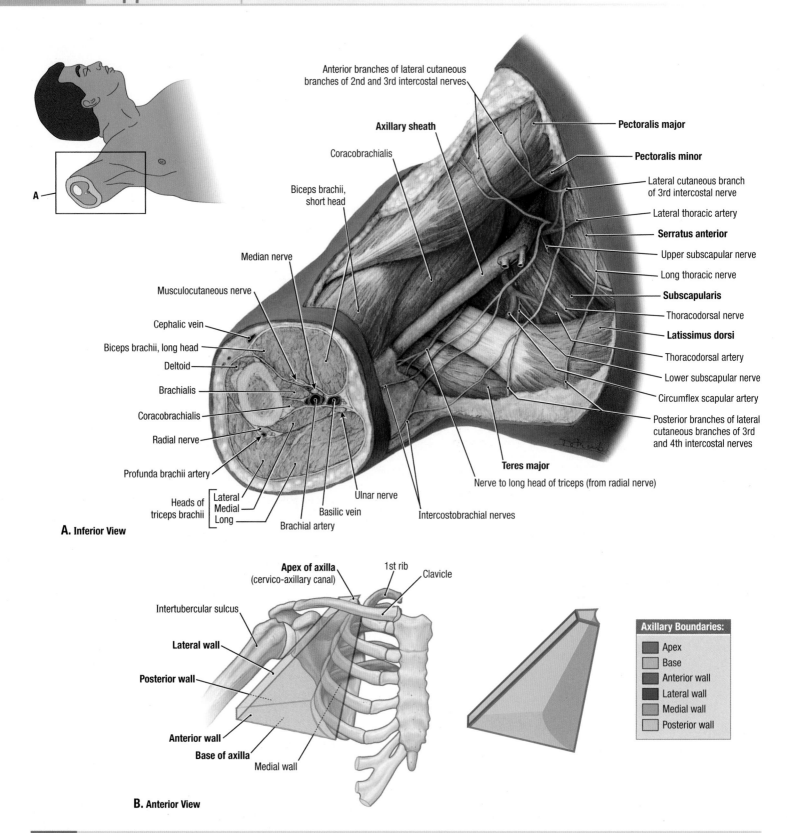

Anterior branches of lateral cutaneous branches of 2nd and 3rd intercostal nerves

Axillary sheath

Coracobrachialis

Biceps brachii, short head

Median nerve

Musculocutaneous nerve

Cephalic vein

Biceps brachii, long head

Deltoid

Brachialis

Coracobrachialis

Radial nerve

Profunda brachii artery

Heads of triceps brachii { Lateral, Medial, Long }

A. Inferior View

Pectoralis major

Pectoralis minor

Lateral cutaneous branch of 3rd intercostal nerve

Lateral thoracic artery

Serratus anterior

Upper subscapular nerve

Long thoracic nerve

Subscapularis

Thoracodorsal nerve

Latissimus dorsi

Thoracodorsal artery

Lower subscapular nerve

Circumflex scapular artery

Posterior branches of lateral cutaneous branches of 3rd and 4th intercostal nerves

Teres major

Nerve to long head of triceps (from radial nerve)

Intercostobrachial nerves

Basilic vein

Ulnar nerve

Brachial artery

Apex of axilla (cervico-axillary canal)

1st rib

Clavicle

Intertubercular sulcus

Lateral wall

Posterior wall

Anterior wall

Base of axilla

Medial wall

B. Anterior View

Axillary Boundaries:
- Apex
- Base
- Anterior wall
- Lateral wall
- Medial wall
- Posterior wall

2.22 WALLS AND CONTENTS OF THE AXILLA

A. Dissection. **B.** Location and walls of axilla.
- The walls of the axilla are anterior (formed by the pectoralis major, pectoralis minor, and subclavius muscles), posterior (formed by subscapularis, latissimus dorsi, and teres major muscles), medial (formed by the serratus anterior muscle), and lateral (formed by the intertubercular sulcus [bicipital groove] of the humerus [concealed by the biceps and coracobrachialis muscles]).
- The axillary sheath surrounds the nerves and vessels (neurovascular bundle) of the upper limb.

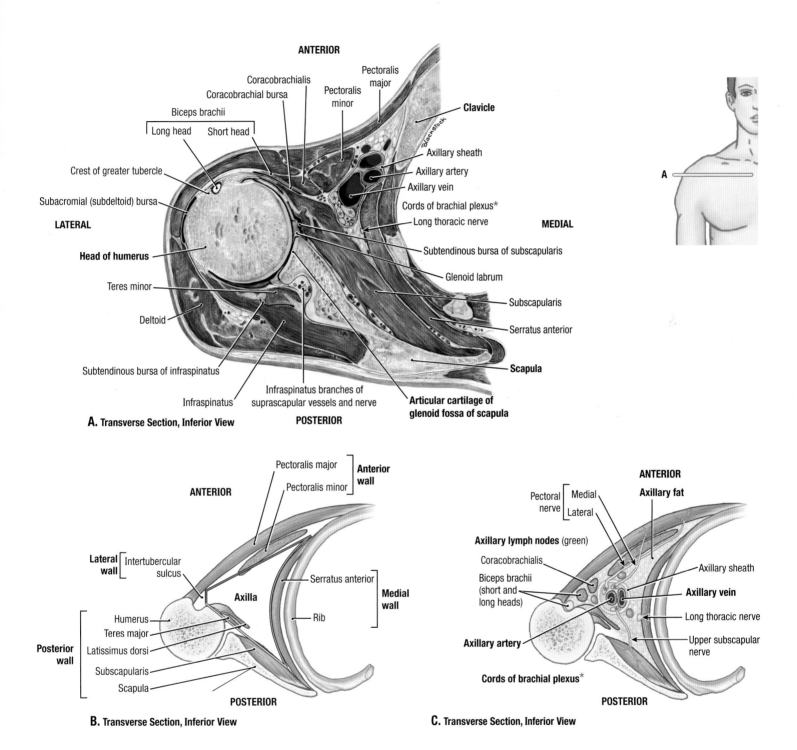

A. Transverse Section, Inferior View

ANTERIOR

Pectoralis major
Pectoralis minor
Coracobrachialis
Coracobrachial bursa
Biceps brachii
Long head   Short head
Clavicle
Axillary sheath
Crest of greater tubercle
Axillary artery
Axillary vein
Subacromial (subdeltoid) bursa
Cords of brachial plexus*
LATERAL
Long thoracic nerve
MEDIAL
Head of humerus
Subtendinous bursa of subscapularis
Teres minor
Glenoid labrum
Subscapularis
Deltoid
Serratus anterior
Subtendinous bursa of infraspinatus
Scapula
Infraspinatus branches of suprascapular vessels and nerve
Articular cartilage of glenoid fossa of scapula
Infraspinatus
POSTERIOR

A —

B. Transverse Section, Inferior View

Pectoralis major
Anterior wall
Pectoralis minor
ANTERIOR
Lateral wall { Intertubercular sulcus
Serratus anterior
Axilla
Medial wall
Humerus
Rib
Teres major
Latissimus dorsi
Posterior wall
Subscapularis
Scapula
POSTERIOR

C. Transverse Section, Inferior View

ANTERIOR
Pectoral nerve { Medial / Lateral
Axillary fat
Axillary lymph nodes (green)
Coracobrachialis
Axillary sheath
Biceps brachii (short and long heads)
Axillary vein
Long thoracic nerve
Upper subscapular nerve
Axillary artery
Cords of brachial plexus*
POSTERIOR

TRANSVERSE SECTIONS THROUGH SHOULDER JOINT AND AXILLA   **2.23**

A. Anatomical section. **B.** Walls of axilla. **C.** Walls and contents of axilla.
- The intertubercular sulcus (bicipital groove) containing the tendon of the long head of the biceps brachii muscle is directed anteriorly; the short head of the biceps muscle and the coracobrachialis and pectoralis minor muscles are sectioned just inferior to their attachments to the coracoid process.
- The small glenoid cavity is deepened by the glenoid labrum.

- Bursae include the subdeltoid (subacromial) bursa, between the deltoid and greater tubercle; the subtendinous bursa of subscapularis, between the subscapularis tendon and scapula; and coracobrachial bursa, between the coracobrachialis and subscapularis.
- The axillary sheath encloses the axillary artery and vein and the three cords of the brachial plexus to form a neurovascular bundle, surrounded by axillary fat.

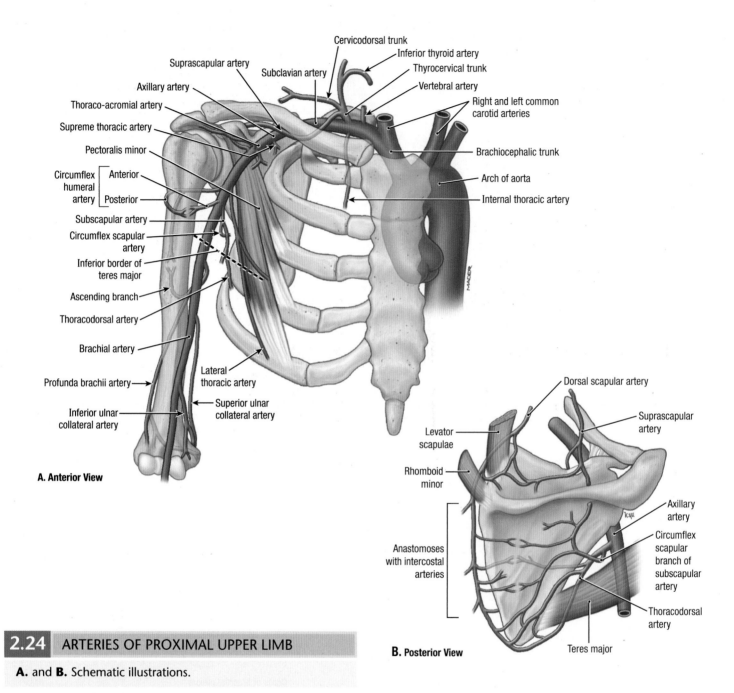

A. Anterior View

B. Posterior View

2.24 ARTERIES OF PROXIMAL UPPER LIMB

A. and **B.** Schematic illustrations.

TABLE 2.5	ARTERIES OF PROXIMAL UPPER LIMB (SHOULDER REGION AND ARM)	
Artery	*Origin*	*Course*
Internal thoracic	Subclavian artery	Descends, inclining anteromedially, posterior to sternal end of clavicle and 1st costal cartilage; enters thorax to descend in parasternal plane; gives rise to perforating branches, anterior intercostal, musculophrenic, and superior epigastric arteries
Thyrocervical trunk		Ascends as a short, wide trunk, often giving rise to the suprascapular artery and/or cervicodorsal trunk and terminating by bifurcating into the ascending cervical and inferior thyroid arteries
Suprascapular	Cervicodorsal trunk from thyrocervical trunk (or as direct branch of subclavian artery[a])	Passes inferolaterally over anterior scalene muscle and phrenic nerve, subclavian artery and brachial plexus running laterally posterior and parallel to clavicle; next passes over transverse scapular ligament to supraspinous fossa and then lateral to scapular spine (deep to acromion) to infraspinous fossa

[a]See Weiglein AH, Moriggl B, Schalk C, et al. Arteries in the posterior cervical triangle in man. *Clin Anat.* 2005;18:553–557.

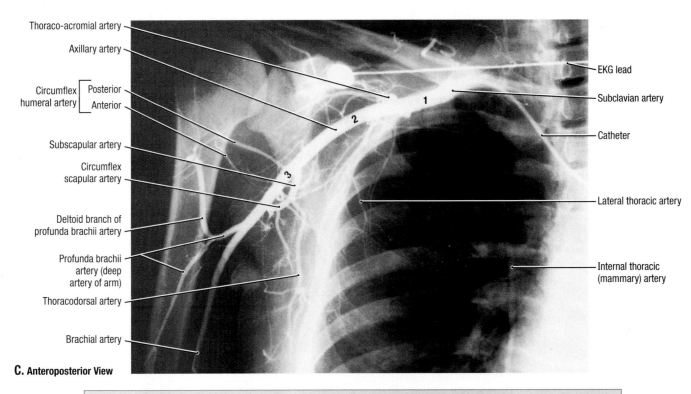

Thoraco-acromial artery

Axillary artery

Circumflex humeral artery — Posterior / Anterior

Subscapular artery

Circumflex scapular artery

Deltoid branch of profunda brachii artery

Profunda brachii artery (deep artery of arm)

Thoracodorsal artery

Brachial artery

EKG lead

Subclavian artery

Catheter

Lateral thoracic artery

Internal thoracic (mammary) artery

C. Anteroposterior View

1: First part of the axillary artery between lateral border of 1st rib and medial border of pectoralis minor.
2: Second part of the axillary artery posterior to pectoralis minor.
3: Third part of the axillary artery from lateral border of pectoralis minor to inferior border of teres major, where it becomes brachial artery.

ARTERIES OF PROXIMAL UPPER LIMB (continued)   2.24

C. Axillary arteriogram.

TABLE 2.5	ARTERIES OF PROXIMAL UPPER LIMB (SHOULDER REGION AND ARM) (continued)		
Artery	*Origin*		*Course*
Supreme thoracic	1st part (as only branch)	Axillary artery	Runs anteromedially along superior border of pectoralis minor; then passes between it and pectoralis major to thoracic wall; helps supply 1st and 2nd intercostal spaces and superior part of serratus anterior
Thoraco-acromial	2nd part (medial branch)		Curls around superomedial border of pectoralis minor, pierces costocoracoid membrane (clavipectoral fascia), and divides into four branches: pectoral, deltoid, acromial, and clavicular
Lateral thoracic	2nd part (lateral branch)		Descends along axillary border of pectoralis minor; follows it onto thoracic wall, supplying lateral aspect of breast
Circumflex humeral (anterior and posterior)	3rd part (sometimes via a common trunk)		Encircle surgical neck of humerus, anastomosing with each other laterally; larger posterior branch traverses quadrangular space
Subscapular	3rd part (largest branch)		Descends from level of inferior border of subscapularis along lateral border of scapula, dividing within 2–3 cm into terminal branches, the circumflex scapular and thoracodorsal arteries
Circumflex scapular	Subscapular artery		Curves around lateral border of scapula to enter infraspinous fossa, anastomosing with subscapular artery
Thoracodorsal	Near its origin		Continuation of subscapular artery; accompanies thoracodorsal nerve to enter latissimus dorsi
Profunda brachii (deep brachial) artery	Near middle of arm	Brachial artery	Accompanies radial nerve through radial groove of humerus, supplying posterior compartment of arm and participating in peri-articular arterial anastomosis around elbow joint
Superior ulnar collateral	Inferior to teres major		Accompanies ulnar nerve to posterior aspect of elbow; anastomoses with posterior ulnar recurrent artery
Inferior ulnar collateral	Superior to medial epicondyle of humerus		Passes anterior to medial epicondyle of humerus to anastomose with anterior ulnar collateral artery around elbow joint

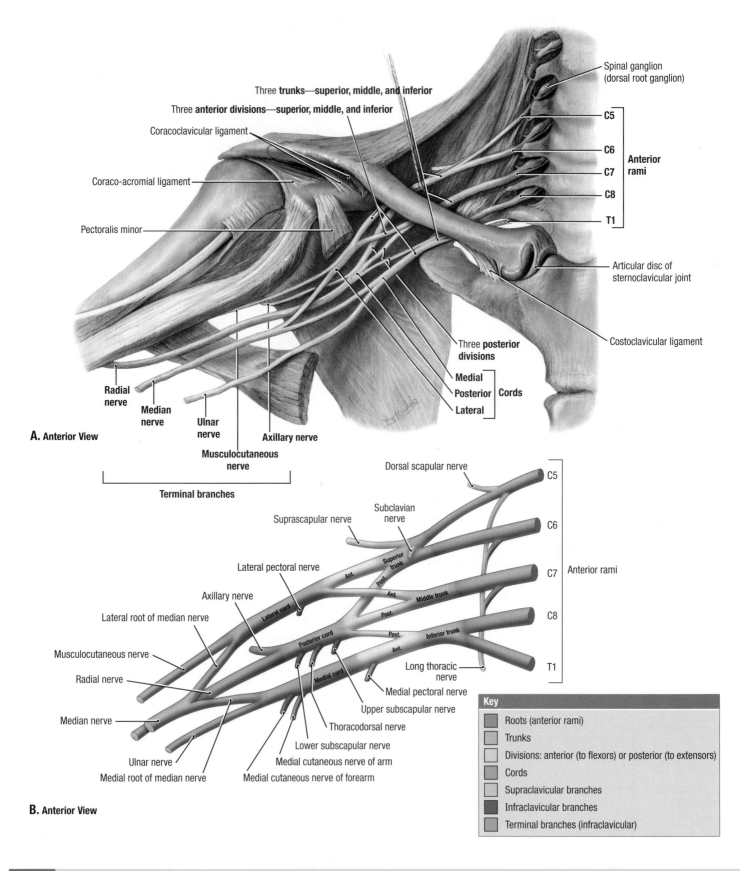

Three **trunks**—superior, middle, and inferior

Three **anterior divisions**—superior, middle, and inferior

Coracoclavicular ligament

Coraco-acromial ligament

Pectoralis minor

Spinal ganglion
(dorsal root ganglion)

C5
C6
C7   Anterior
C8   rami
T1

Articular disc of
sternoclavicular joint

Costoclavicular ligament

Three **posterior
divisions**

Medial
Posterior Cords
Lateral

**Radial
nerve**

**Median
nerve**

**Ulnar
nerve**

Axillary nerve

**Musculocutaneous
nerve**

A. Anterior View

Terminal branches

Dorsal scapular nerve

C5

Suprascapular nerve

Subclavian
nerve

C6

Lateral pectoral nerve

Superior
trunk

C7   Anterior rami

Axillary nerve

Ant.

Post.

Lateral root of median nerve

Ant.

Middle trunk

Lateral cord

Musculocutaneous nerve

Post.

C8

Posterior cord

Post.

Inferior trunk

Radial nerve

Ant.

Long thoracic
nerve

T1

Medial cord

Median nerve

Medial pectoral nerve

Upper subscapular nerve

Ulnar nerve

Thoracodorsal nerve

Lower subscapular nerve

Medial root of median nerve

Medial cutaneous nerve of arm

Medial cutaneous nerve of forearm

B. Anterior View

Key	
	Roots (anterior rami)
	Trunks
	Divisions: anterior (to flexors) or posterior (to extensors)
	Cords
	Supraclavicular branches
	Infraclavicular branches
	Terminal branches (infraclavicular)

2.25 BRACHIAL PLEXUS

A. Dissection. **B.** Schematic illustration.

TABLE 2.6	**BRACHIAL PLEXUS**		
Nerve	*Origin*	*Course*	*Distribution/Structure(s) Supplied*
Supraclavicular branches			
Dorsal scapular	Anterior ramus of C5 with a frequent contribution from C4	Pierces scalenus medius, descends on deep surface of rhomboids	Rhomboids and occasionally supplies levator scapulae
Long thoracic	Anterior rami of C5–C7	Descends posterior to C8 and T1 rami and passes distally on external surface of serratus anterior	Serratus anterior
Subclavian	Superior trunk receiving fibers from C5 and C6 and often C4	Descends posterior to clavicle and anterior to brachial plexus and subclavian artery	Subclavius and sternoclavicular joint
Suprascapular		Passes laterally across posterior triangle of neck, through suprascapular notch deep to superior transverse scapular ligament	Supraspinatus, infraspinatus, and glenohumeral (shoulder) joint
Infraclavicular branches			
Lateral pectoral	Lateral cord receiving fibers from C5–C7	Pierces clavipectoral fascia to reach deep surface of pectoral muscles	Primarily pectoralis major but sends a loop to medial pectoral nerve that innervates pectoralis minor
Musculocutaneous		Pierces coracobrachialis and descends between biceps brachii and brachialis	Coracobrachialis, biceps brachii, and brachialis; continues as lateral cutaneous nerve of forearm
Median	Lateral root of median nerve is a terminal branch of lateral cord (C6, C7); medial root of median nerve is a terminal branch of medial cord (C8, T1)	Lateral and medial roots merge to form median nerve lateral to axillary artery; crosses anterior to brachial artery to lie medial to artery in cubital fossa	Flexor muscles in forearm (except flexor carpi ulnaris, ulnar half of flexor digitorum profundus), 3½ thenar and lateral 2 lumbrical muscles in hand, and skin of palm and 3½ digits lateral to a line bisecting 4th digit and the dorsum of the distal halves of these digits
Medial pectoral	Medial cord receiving fibers from C8, T1	Passes between axillary artery and vein and enters deep surface of pectoralis minor	Pectoralis minor and part of pectoralis major
Medial cutaneous nerve of arm		Runs along the medial side of axillary vein and communicates with intercostobrachial nerve	Skin on medial side of arm
Medial cutaneous nerve of forearm		Runs between axillary artery and vein	Skin over medial side of forearm
Ulnar	Terminal branch of medial cord receiving fibers from C8, T1, and often C7	Passes down medial view of arm and runs posterior to medial epicondyle to enter forearm	Innervates 1½ flexor muscles in forearm (see Median nerve), 1½ thenar, 2 medial lumbricals, all interossei and adductor pollicis muscles in hand, and skin of hand medial to a line bisecting 4th digit (ring finger) anteriorly and posteriorly
Upper subscapular	Branch of posterior cord receiving fibers from C5	Passes posteriorly and enters subscapularis	Superior portion of subscapularis
Thoracodorsal	Branch of posterior cord receiving fibers from C6 to C8	Arises between upper and lower subscapular nerves and runs inferolaterally to latissimus dorsi	Latissimus dorsi
Lower subscapular	Branch of posterior cord receiving fibers from C6	Passes inferolaterally, deep to subscapular artery and vein, to subscapularis and teres major	Inferior portion of subscapularis and teres major
Axillary	Terminal branch of posterior cord receiving fibers from C5 and C6	Passes to posterior aspect of arm through quadrangular space[a] with posterior circumflex humeral artery and then winds around surgical neck of humerus; gives rise to lateral cutaneous nerve of arm	Teres minor and deltoid, glenohumeral (shoulder) joint, and skin of superolateral arm
Radial	Terminal branch of posterior cord receiving fibers from C5 to T1	Descends posterior to axillary artery; enters radial groove to pass between long and medial heads of triceps brachii	Triceps brachii, anconeus, brachioradialis, and extensor muscles of forearm; supplies skin on posterior and inferolateral aspect of arm and forearm and dorsum of hand lateral to axial line of digit 4

[a]Quadrangular space is bounded superiorly by subscapularis and teres minor, inferiorly by teres major, medially by long head of triceps brachii, and laterally by humerus.

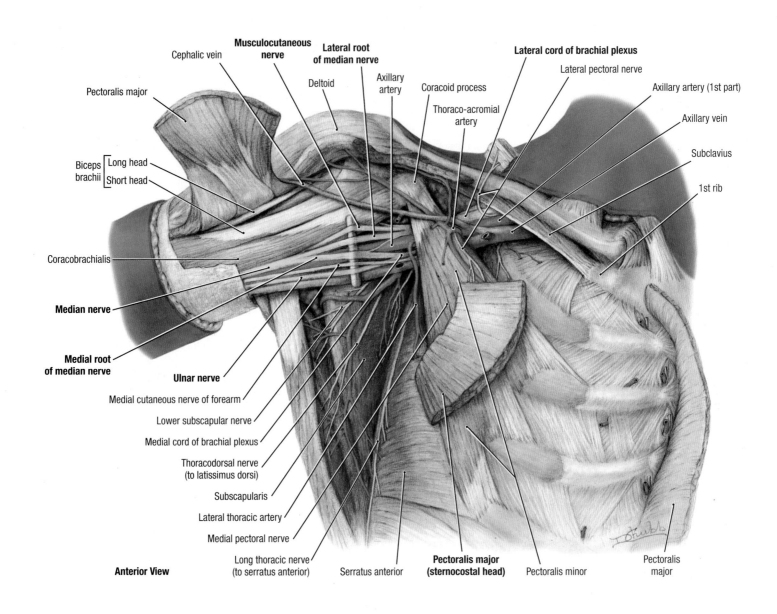

Cephalic vein

Musculocutaneous nerve

Lateral root of median nerve

Lateral cord of brachial plexus

Pectoralis major

Deltoid

Axillary artery

Coracoid process

Lateral pectoral nerve

Axillary artery (1st part)

Thoraco-acromial artery

Axillary vein

Subclavius

1st rib

Biceps brachii — Long head / Short head

Coracobrachialis

Median nerve

Medial root of median nerve

Ulnar nerve

Medial cutaneous nerve of forearm

Lower subscapular nerve

Medial cord of brachial plexus

Thoracodorsal nerve (to latissimus dorsi)

Subscapularis

Lateral thoracic artery

Medial pectoral nerve

Long thoracic nerve (to serratus anterior)

Anterior View

Serratus anterior

Pectoralis major (sternocostal head)

Pectoralis minor

Pectoralis major

2.26   STRUCTURES OF AXILLA: DEEP DISSECTION I

- The pectoralis major muscle is reflected, and the clavipectoral fascia is removed; the cube of muscle superior to the clavicle is cut from the clavicular head of the pectoralis major muscle.
- The subclavius and pectoralis minor are the two deep muscles of the anterior wall.
- The second part of the axillary artery passes posterior to the pectoralis minor muscle, a fingerbreadth from the tip of the coracoid process; the axillary vein lies anterior and then medial to the axillary artery.

- The median nerve, followed proximally, leads by its lateral root to the lateral cord and musculocutaneous nerve and by its medial root to the medial cord and ulnar nerve. These four nerves and the medial cutaneous nerve of the forearm are derived from the anterior divisions of the brachial plexus and are raised on a stick. The lateral root of the median nerve may occur as several strands.
- The musculocutaneous nerve enters the flexor compartment of the arm by piercing the coracobrachialis muscle.

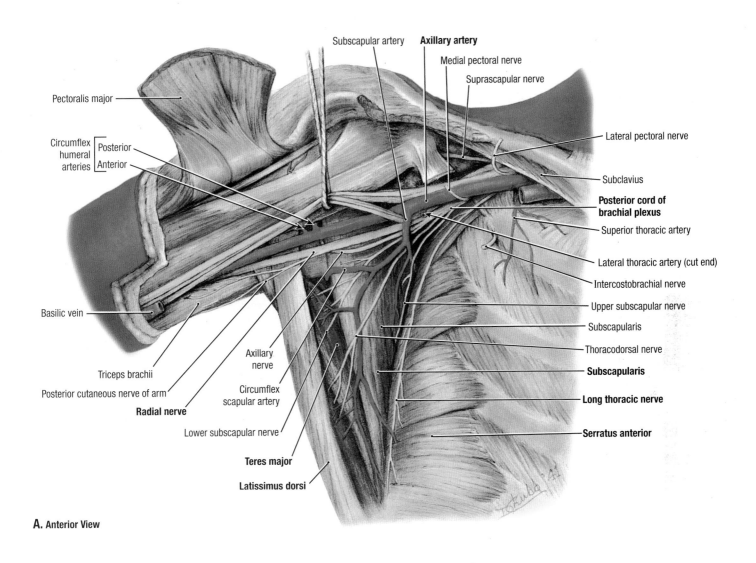

Subscapular artery
Axillary artery
Medial pectoral nerve
Suprascapular nerve

Pectoralis major

Lateral pectoral nerve

Circumflex humeral arteries [Posterior / Anterior]

Subclavius

Posterior cord of brachial plexus
Superior thoracic artery
Lateral thoracic artery (cut end)
Intercostobrachial nerve
Upper subscapular nerve
Subscapularis
Thoracodorsal nerve
Subscapularis
Long thoracic nerve
Serratus anterior

Basilic vein

Triceps brachii
Posterior cutaneous nerve of arm
Radial nerve

Axillary nerve
Circumflex scapular artery
Lower subscapular nerve

Teres major

Latissimus dorsi

A. Anterior View

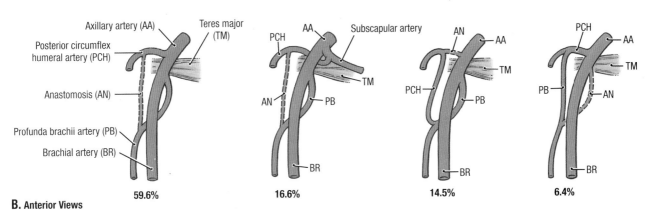

Axillary artery (AA)   Teres major (TM)
Posterior circumflex humeral artery (PCH)
Anastomosis (AN)
Profunda brachii artery (PB)
Brachial artery (BR)

PCH   AA   Subscapular artery
TM
AN   PB
BR

AN   AA
TM
PCH   PB
BR

PCH   AA
TM
PB   AN
BR

59.6%   16.6%   14.5%   6.4%

B. Anterior Views

POSTERIOR AND MEDIAL WALLS OF AXILLA: DEEP DISSECTION II

2.27

A. Dissection. The pectoralis minor muscle is excised, the lateral and medial cords of the brachial plexus are retracted, and the axillary vein is removed. **B.** Variations of the posterior circumflex humeral artery and profunda brachii artery. Percentages are based on 235 specimens dissected in Dr. Grant's laboratory.

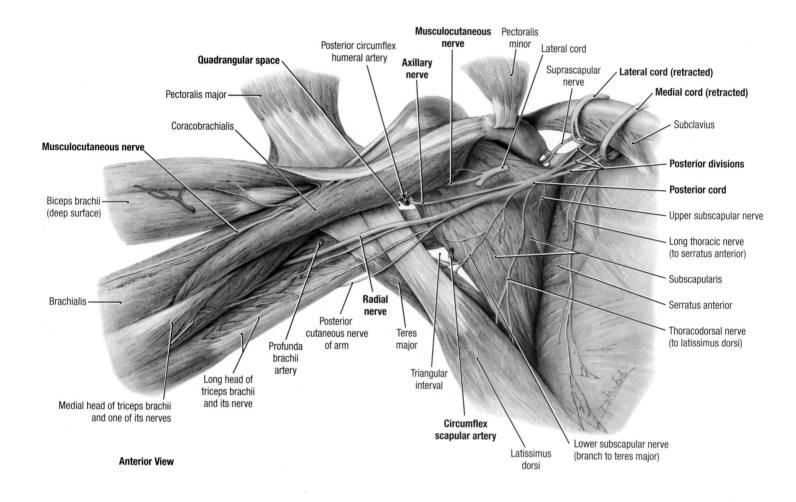

Anterior View

2.28 POSTERIOR WALL OF AXILLA, MUSCULOCUTANEOUS NERVE, AND POSTERIOR CORD: DEEP DISSECTION III

- The pectoralis major and minor muscles are reflected laterally; the lateral and medial cords of the brachial plexus are reflected superiorly; and the arteries, veins, and median and ulnar nerves are removed.
- Coracobrachialis arises with the short head of the biceps brachii muscle from the tip of the coracoid process and attaches halfway down the medial aspect of the humerus.
- The musculocutaneous nerve pierces the coracobrachialis muscle and supplies it, the biceps, and the brachialis before becoming the lateral cutaneous nerve of the forearm.
- The posterior cord of the plexus is formed by the union of the three posterior divisions; it supplies the three muscles of the posterior wall of the axilla and then bifurcates into the radial and axillary nerves.

- In the axilla, the radial nerve gives off the nerve to the long head of the triceps brachii muscle and a cutaneous branch; in this specimen, it also gives off a branch to the medial head of the triceps. It then enters the radial groove of the humerus with the profunda brachii (deep brachial) artery.
- The axillary nerve passes through the quadrangular space along with the posterior circumflex humeral artery. The borders of the quadrangular space are superiorly, the lateral border of the scapula; inferiorly, the teres major; laterally, the humerus (surgical neck); and medially, the long head of triceps brachii. The circumflex scapular artery traverses the triangular interval.

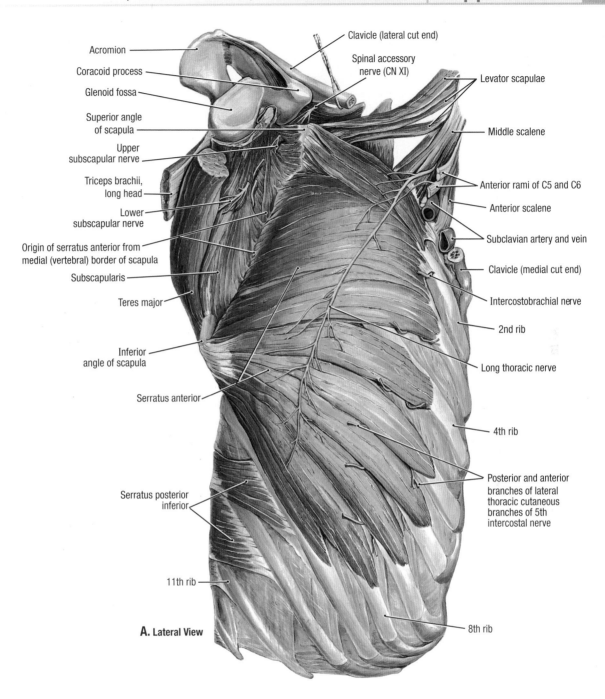

Acromion

Coracoid process

Glenoid fossa

Superior angle
of scapula

Upper
subscapular nerve

Triceps brachii,
long head

Lower
subscapular nerve

Origin of serratus anterior from
medial (vertebral) border of scapula

Subscapularis

Teres major

Inferior
angle of scapula

Serratus anterior

Serratus posterior
inferior

11th rib

Clavicle (lateral cut end)

Spinal accessory
nerve (CN XI)

Levator scapulae

Middle scalene

Anterior rami of C5 and C6

Anterior scalene

Subclavian artery and vein

Clavicle (medial cut end)

Intercostobrachial nerve

2nd rib

Long thoracic nerve

4th rib

Posterior and anterior
branches of lateral
thoracic cutaneous
branches of 5th
intercostal nerve

8th rib

A. Lateral View

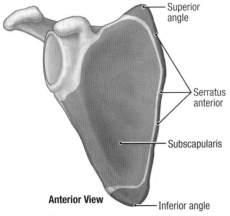

Superior
angle

Serratus
anterior

Subscapularis

Anterior View

Inferior angle

B. Sites of Muscle Attachment to Scapula

SERRATUS ANTERIOR AND SUBSCAPULARIS | 2.29

A. The serratus anterior muscle, which forms the medial wall of the axilla, has a fleshy belly extending from the superior 8 or 9 ribs in the midclavicular line to the medial border of the scapula (**B**).

Winged scapula. When the serratus anterior is paralyzed because of injury to the long thoracic nerve, the medial border of the scapula moves laterally and posteriorly, away from the thoracic wall. When the arm is abducted, the medial border and the inferior angle of the scapula pull away from the posterior thoracic wall, a deformation known as a winged scapula. In addition, the arm cannot be abducted above the horizontal position because the serratus anterior is unable to rotate the glenoid cavity superiorly.

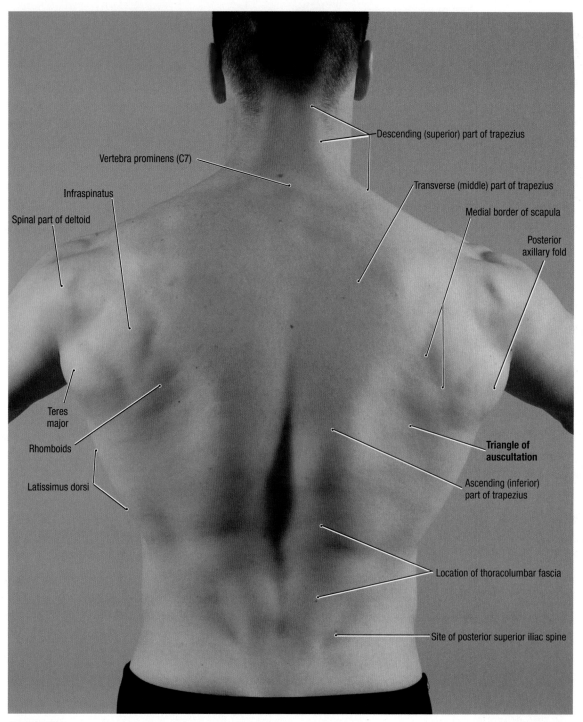

Descending (superior) part of trapezius

Vertebra prominens (C7)

Transverse (middle) part of trapezius

Infraspinatus

Medial border of scapula

Spinal part of deltoid

Posterior axillary fold

Teres major

Triangle of auscultation

Rhomboids

Ascending (inferior) part of trapezius

Latissimus dorsi

Location of thoracolumbar fascia

Site of posterior superior iliac spine

Posterior View

2.30 SURFACE ANATOMY OF SUPERFICIAL BACK

The superior border of the latissimus dorsi and a part of the rhomboid major are overlapped by the trapezius. The area formed by the superior border of latissimus dorsi, the medial border of the scapula, and the inferolateral border of the trapezius is called the **triangle of auscultation.** This gap in the thick back musculature is a good place to examine posterior segments of the lungs with a stethoscope. When the scapulae are drawn anteriorly by folding the arms across the thorax and the trunk is flexed, the auscultatory triangle enlarges. The teres major forms a raised oval area on the inferolateral third of the posterior aspect of the scapula when the arm is adducted against resistance. The posterior axillary fold is formed by the teres major and the tendon of the latissimus dorsi.

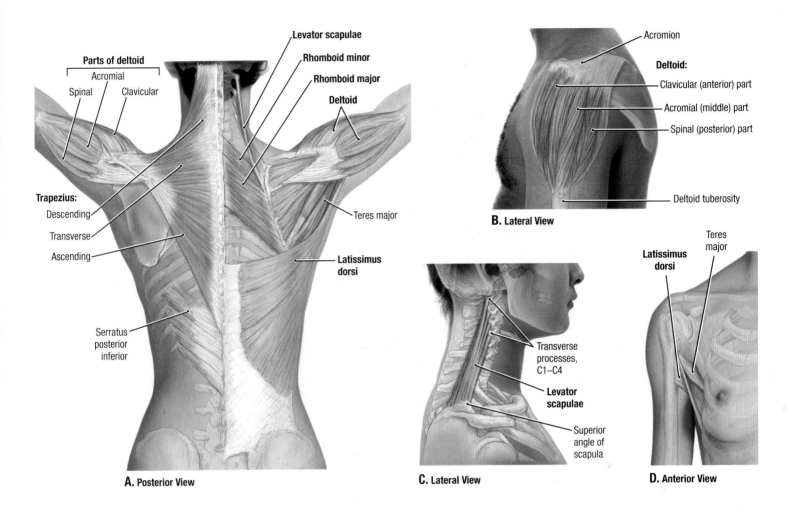

Parts of deltoid
Acromial
Spinal
Clavicular

Levator scapulae
Rhomboid minor
Rhomboid major
Deltoid

Trapezius:
Descending
Transverse
Ascending

Teres major

Latissimus dorsi

Serratus posterior inferior

A. Posterior View

Acromion
Deltoid:
Clavicular (anterior) part
Acromial (middle) part
Spinal (posterior) part
Deltoid tuberosity

B. Lateral View

Transverse processes, C1–C4
Levator scapulae
Superior angle of scapula

C. Lateral View

Teres major
Latissimus dorsi

D. Anterior View

SUPERFICIAL BACK AND DELTOID MUSCLES

2.31

A. Overview. **B.** Deltoid. **C.** Levator scapulae. **D.** Latissimus dorsi and teres major.

TABLE 2.7	**SUPERFICIAL BACK (POSTERIOR AXIO-APPENDICULAR) AND DELTOID MUSCLES**			
Muscle	*Proximal Attachment*	*Distal Attachment*	*Innervation*	*Main Actions*
Trapezius	Medial third of superior nuchal line; external occipital protuberance, nuchal ligament, and spinous processes of C7–T12 vertebrae	Lateral third of clavicle, acromion, and spine of scapula	Spinal accessory nerve (CN XI—motor) and cervical nerves (C3–C4—sensory)	Elevates, retracts, and rotates scapula; *descending part* elevates, *transverse part* retracts, and *ascending part* depresses scapula; descending and ascending part act together in superior rotation of scapula
Latissimus dorsi	Spinous processes of inferior six thoracic vertebrae, thoracolumbar fascia, iliac crest, and inferior three or four ribs	Intertubercular sulcus (bicipital groove) of humerus	Thoracodorsal nerve (**C6**, **C7**, C8)	Extends, adducts, and medially rotates shoulder joint; elevates body toward arms during climbing
Levator scapulae	Posterior tubercles of transverse processes of C1–C4 vertebrae	Superior part of medial border of scapula	Dorsal scapular (C5) and cervical (C3–C4) nerves	Elevates scapula and tilts its glenoid cavity inferiorly by rotating scapula
Rhomboid minor and major	*Minor:* inferior part of nuchal ligament and spinous processes of C7 and T1 vertebrae *Major:* spinous processes of T2–T5 vertebrae	Medial border of scapula from level of spine to inferior angle	Dorsal scapular nerve (C4–**C5**)	Retract scapula and rotate it to depress glenoid cavity; fix scapula to thoracic wall
Deltoid	Lateral third of clavicle (*clavicular part*), acromion (*acromial part*), and spine (*spinal part*) of scapula	Deltoid tuberosity of humerus	Axillary nerve (**C5**–C6)	*Clavicular (anterior) part:* flexes and medially rotates shoulder joint *Acromial (middle) part:* abducts shoulder joint *Spinal (posterior) part:* extends and laterally rotates shoulder joint

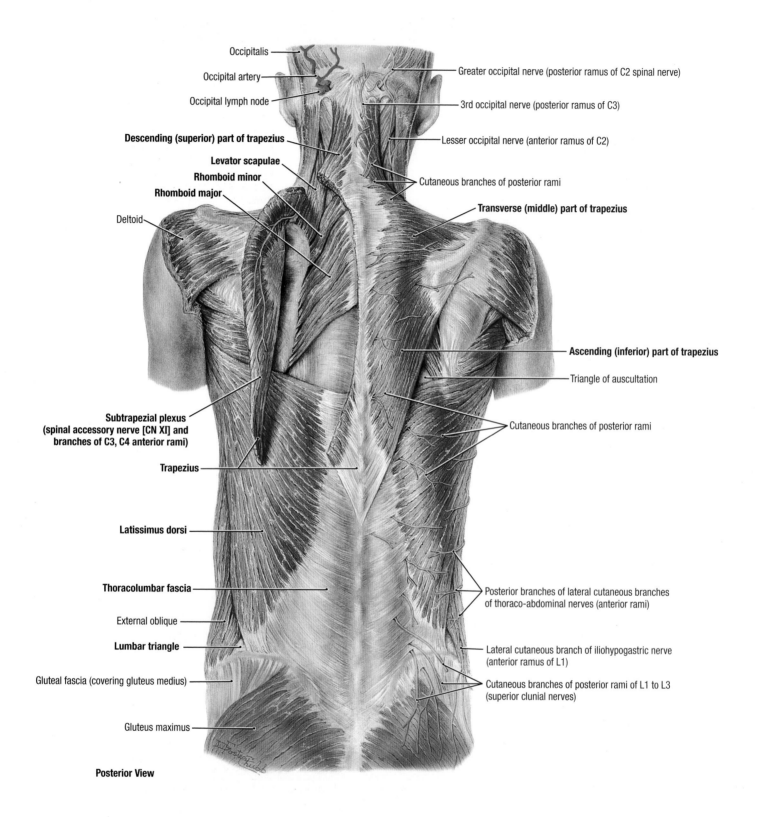

Occipitalis

Occipital artery

Occipital lymph node

Descending (superior) part of trapezius

Levator scapulae

Rhomboid minor

Rhomboid major

Deltoid

Subtrapezial plexus
(spinal accessory nerve [CN XI] and
branches of C3, C4 anterior rami)

Trapezius

Latissimus dorsi

Thoracolumbar fascia

External oblique

Lumbar triangle

Gluteal fascia (covering gluteus medius)

Gluteus maximus

Posterior View

Greater occipital nerve (posterior ramus of C2 spinal nerve)

3rd occipital nerve (posterior ramus of C3)

Lesser occipital nerve (anterior ramus of C2)

Cutaneous branches of posterior rami

Transverse (middle) part of trapezius

Ascending (inferior) part of trapezius

Triangle of auscultation

Cutaneous branches of posterior rami

Posterior branches of lateral cutaneous branches
of thoraco-abdominal nerves (anterior rami)

Lateral cutaneous branch of iliohypogastric nerve
(anterior ramus of L1)

Cutaneous branches of posterior rami of L1 to L3
(superior clunial nerves)

2.32 CUTANEOUS NERVES OF SUPERFICIAL BACK AND POSTERIOR AXIO-APPENDICULAR MUSCLES

The trapezius muscle is cut and reflected on the left side. A superficial or first muscle layer consists of the trapezius and latissimus dorsi muscles, and a second layer of the levator scapulae and rhomboids.

Cutaneous branches of posterior rami penetrate but do not supply the superficial back muscles.

TABLE 2.8 | **SCAPULAR MOVEMENTS**

Scapula moves on the thoracic wall at the conceptual "scapulothoracic joint." *Dotted lines*, starting position for each movement. *Boldface*, prime movers.

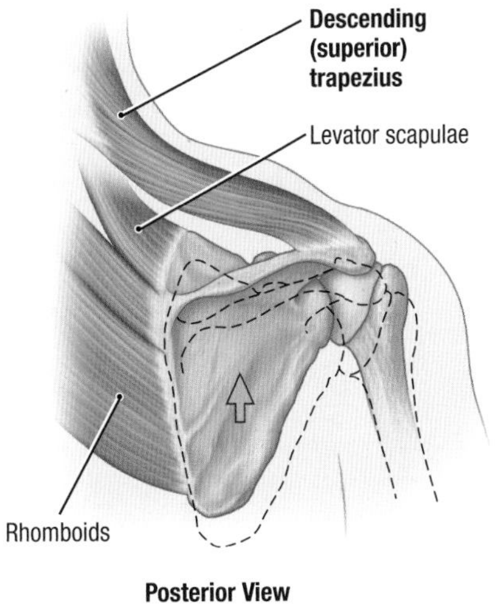

Descending (superior) trapezius
Levator scapulae
Rhomboids
Posterior View
A. Elevation

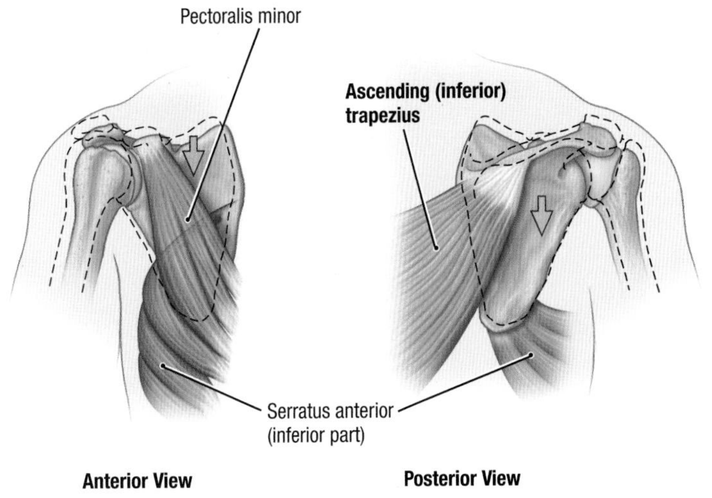

Pectoralis minor
Ascending (inferior) trapezius
Serratus anterior (inferior part)
Anterior View
Posterior View
B. Depression

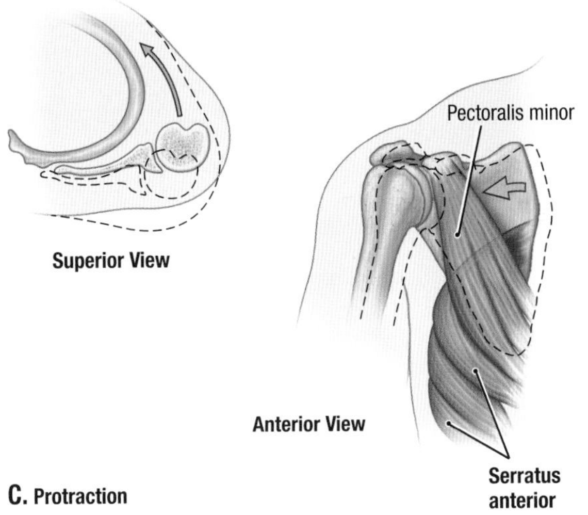

Superior View
C. Protraction

Pectoralis minor
Anterior View
Serratus anterior

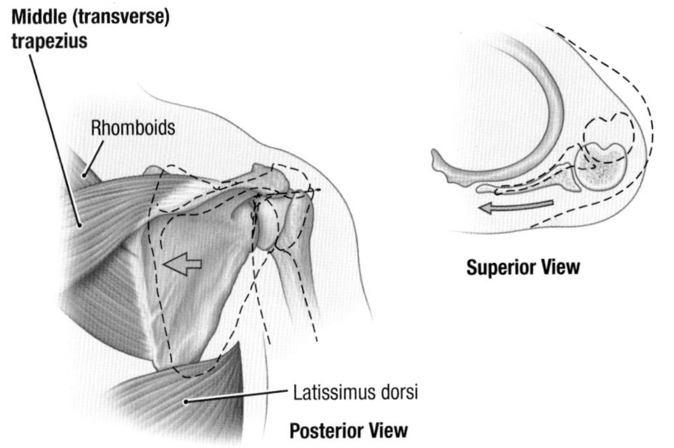

Middle (transverse) trapezius
Rhomboids
Latissimus dorsi
Posterior View
Superior View
D. Retraction

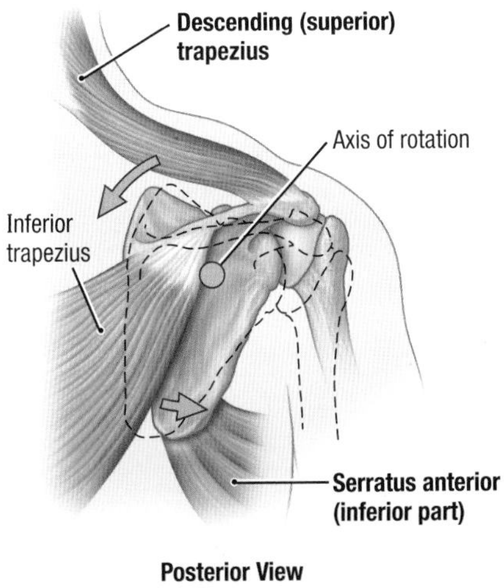

Descending (superior) trapezius
Axis of rotation
Inferior trapezius
Serratus anterior (inferior part)
Posterior View
E. Rotation Elevating Glenoid Cavity

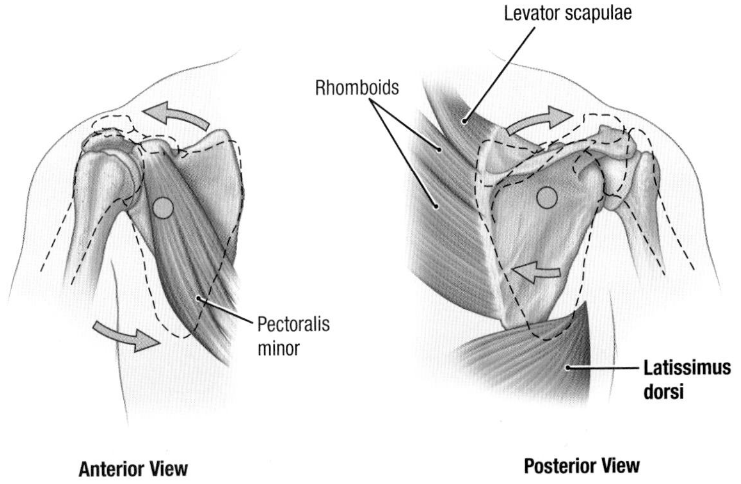

Pectoralis minor
Anterior View

Levator scapulae
Rhomboids
Latissimus dorsi
Posterior View
F. Rotation Depressing Glenoid Cavity

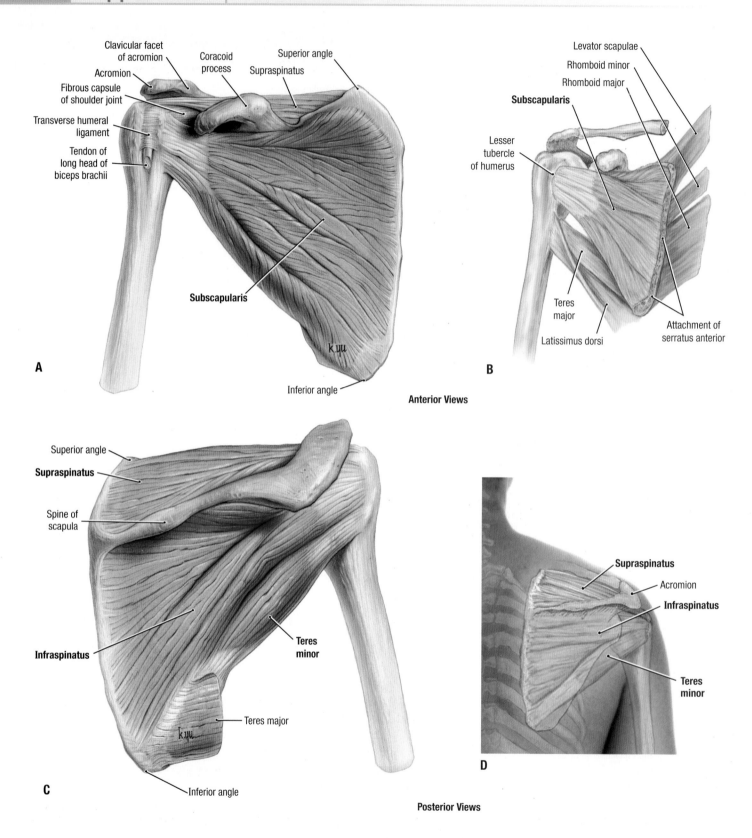

Anterior Views

Posterior Views

2.33 ROTATOR CUFF

A. and **B.** Subscapularis. **C.** and **D.** Supraspinatus, infraspinatus, and teres minor.

Four of the scapulohumeral muscles (supraspinatus, infraspinatus, teres minor, and subscapularis) are called rotator cuff muscles because they form a musculotendinous rotator cuff around the glenohumeral (shoulder) joint. All except the supraspinatus are rotators of the humerus.

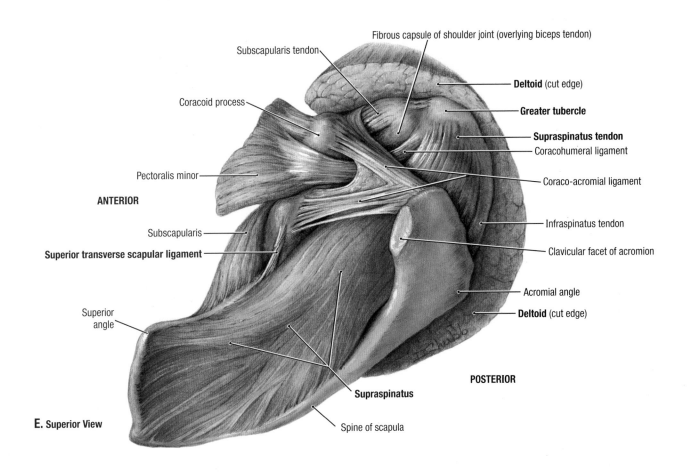

Fibrous capsule of shoulder joint (overlying biceps tendon)

Subscapularis tendon

Deltoid (cut edge)

Greater tubercle

Supraspinatus tendon

Coracohumeral ligament

Coracoid process

Coraco-acromial ligament

Pectoralis minor

ANTERIOR

Infraspinatus tendon

Subscapularis

Clavicular facet of acromion

Superior transverse scapular ligament

Acromial angle

Deltoid (cut edge)

Superior angle

POSTERIOR

Supraspinatus

E. Superior View

Spine of scapula

ROTATOR CUFF (continued)

2.33

E. Supraspinatus. The supraspinatus, also part of the rotator cuff, initiates and assists the deltoid in abducting the shoulder joint. The tendons of the rotator cuff muscles blend with and reinforce the joint capsule of the glenohumeral joint, protecting the joint and giving it stability.

Injury or disease may damage the rotator cuff, producing instability of the glenohumeral joint. **Rupture or tear of the supraspinatus tendon** is the most common injury of the rotator cuff. **Degenerative tendinitis of the rotator cuff** is common, especially in older people.

TABLE 2.9	**SCAPULOHUMERAL MUSCLES**			
Muscle	*Proximal Attachment*	*Distal Attachment*	*Innervation*	*Main Actions*
Supraspinatus (S)	Supraspinous fossa of scapula	Superior facet on greater tubercle of humerus	Suprascapular nerve (C4, **C5**, and C6)	Initiates abduction at shoulder joint and acts with rotator cuff muscles[a]
Infraspinatus (I)	Infraspinous fossa of scapula	Middle facet on greater tubercle of humerus	Suprascapular nerve (**C5** and C6)	Laterally rotates shoulder joint; helps to hold humeral head in glenoid cavity of scapula
Teres minor (T)	Superior part of lateral border of scapula	Inferior facet on greater tubercle of humerus	Axillary nerve (**C5** and C6)	
Subscapularis (S)	Subscapular fossa	Lesser tubercle of humerus	Upper and lower subscapular nerves (C5, **C6**, and C7)	Medially rotates shoulder joint and adducts it; helps to hold humeral head in glenoid cavity
Teres major[b]	Posterior surface of inferior angle of scapula	Crest of lesser tubercle (medial lip of bicipital groove) of humerus	Lower subscapular nerve (**C6** and C7)	Adducts and medially rotates shoulder joint

[a]Collectively, the supraspinatus, infraspinatus, teres minor, and subscapularis muscles are referred to as the rotator cuff muscles or "SITS" muscles. They function together during all movements of the shoulder joint to hold the head of the humerus in the glenoid cavity of scapula.
[b]Not a rotator cuff muscle.

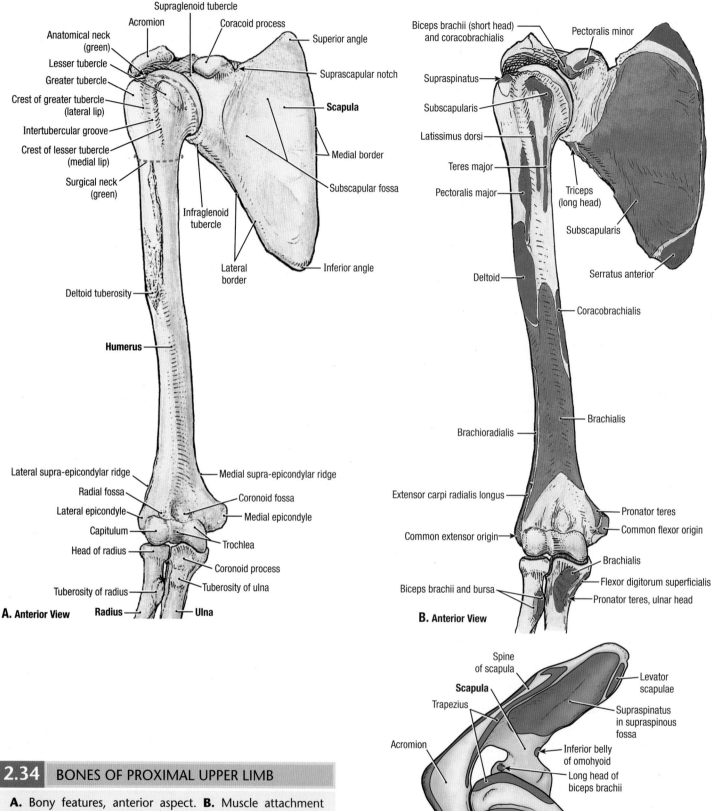

Supraglenoid tubercle
Acromion
Coracoid process
Superior angle
Anatomical neck (green)
Suprascapular notch
Lesser tubercle
Greater tubercle
Scapula
Crest of greater tubercle (lateral lip)
Intertubercular groove
Medial border
Crest of lesser tubercle (medial lip)
Subscapular fossa
Surgical neck (green)
Infraglenoid tubercle
Deltoid tuberosity
Humerus
Lateral border
Inferior angle
Lateral supra-epicondylar ridge
Medial supra-epicondylar ridge
Radial fossa
Coronoid fossa
Lateral epicondyle
Medial epicondyle
Capitulum
Trochlea
Head of radius
Coronoid process
Tuberosity of radius
Tuberosity of ulna
A. Anterior View **Radius** **Ulna**

Biceps brachii (short head) and coracobrachialis
Pectoralis minor
Supraspinatus
Subscapularis
Latissimus dorsi
Teres major
Triceps (long head)
Pectoralis major
Subscapularis
Deltoid
Serratus anterior
Coracobrachialis
Brachialis
Brachioradialis
Extensor carpi radialis longus
Pronator teres
Common flexor origin
Common extensor origin
Brachialis
Flexor digitorum superficialis
Biceps brachii and bursa
Pronator teres, ulnar head
B. Anterior View

Spine of scapula
Scapula
Levator scapulae
Trapezius
Supraspinatus in supraspinous fossa
Acromion
Inferior belly of omohyoid
Long head of biceps brachii
Clavicle
Pectoralis major
Sternocleidomastoid (SCM)
Deltoid
Coracobrachialis and short head of biceps brachii
Coracoid process
C. Superior View

2.34 BONES OF PROXIMAL UPPER LIMB

A. Bony features, anterior aspect. **B.** Muscle attachment sites, anterior aspect. **C.** Muscle attachment sites, clavicle and scapula. **Fractures of the clavicle** are common, often caused by indirect force transmitted from an outstretched hand through the bones of the forearm and arm to the shoulder during a fall. A fracture may also result from a fall directly on the shoulder. The weakest part of the clavicle is at the junction of its middle and lateral thirds.

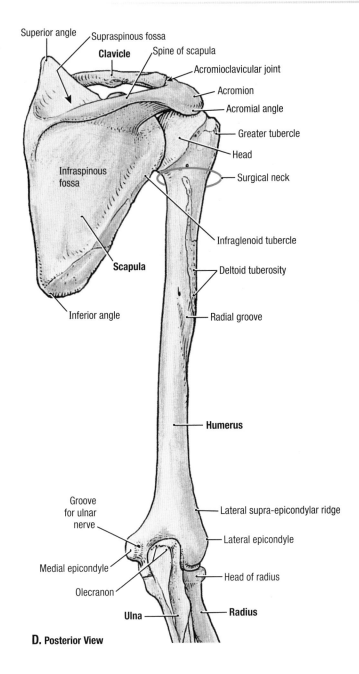

D. Posterior View

Labels: Superior angle, Suprasupinous fossa, **Clavicle**, Spine of scapula, Acromioclavicular joint, Acromion, Acromial angle, Greater tubercle, Head, Surgical neck, Infraspinous fossa, Infraglenoid tubercle, **Scapula**, Deltoid tuberosity, Inferior angle, Radial groove, **Humerus**, Groove for ulnar nerve, Lateral supra-epicondylar ridge, Lateral epicondyle, Medial epicondyle, Head of radius, Olecranon, **Ulna**, **Radius**

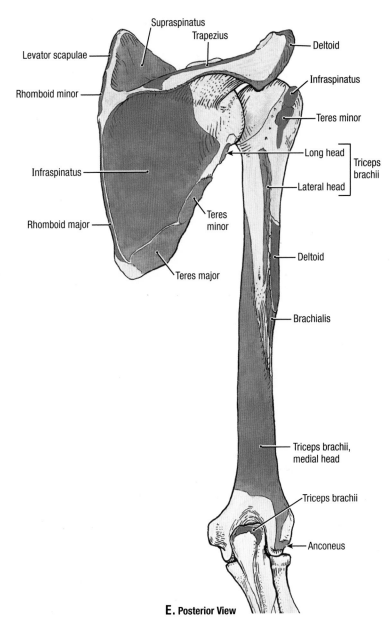

E. Posterior View

Labels: Supraspinatus, Trapezius, Deltoid, Levator scapulae, Infraspinatus, Rhomboid minor, Teres minor, Infraspinatus, Long head, Lateral head, Triceps brachii, Rhomboid major, Teres minor, Deltoid, Teres major, Brachialis, Triceps brachii, medial head, Triceps brachii, Anconeus

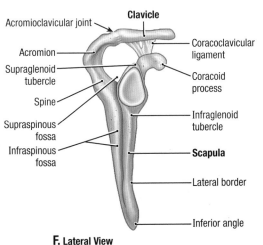

F. Lateral View

Labels: Acromioclavicular joint, **Clavicle**, Acromion, Coracoclavicular ligament, Supraglenoid tubercle, Coracoid process, Spine, Infraglenoid tubercle, Supraspinous fossa, Infraspinous fossa, **Scapula**, Lateral border, Inferior angle

BONES OF PROXIMAL UPPER LIMB (*continued*) | 2.34

D. Bony features, posterior aspect. **E.** Muscle attachment sites, posterior aspect. **F.** Lateral aspect of scapula.

Fractures of the surgical neck of the humerus are especially common in elderly people with **osteoporosis** (degeneration of bone). Even a low energy fall on the hand, with the force being transmitted up the forearm bones of the extended limb, may result in a fracture. **Transverse fractures of the shaft of humerus** frequently result from a direct blow to the arm. Fracture of the distal part of the humerus, near the supra-epicondylar ridges, is a **supra-epicondylar (supracondylar) fracture.**

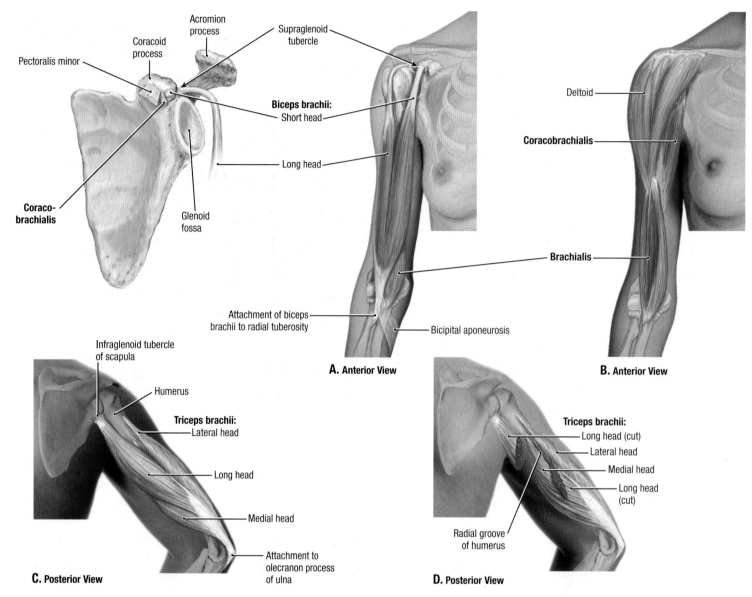

Pectoralis minor

Coracoid process

Acromion process

Supraglenoid tubercle

Biceps brachii:
Short head

Long head

Coraco-brachialis

Glenoid fossa

Attachment of biceps brachii to radial tuberosity

Bicipital aponeurosis

A. Anterior View

Deltoid

Coracobrachialis

Brachialis

B. Anterior View

Infraglenoid tubercle of scapula

Humerus

Triceps brachii:
Lateral head

Long head

Medial head

Attachment to olecranon process of ulna

C. Posterior View

Triceps brachii:
Long head (cut)

Lateral head

Medial head

Long head (cut)

Radial groove of humerus

D. Posterior View

2.35 ARM MUSCLES

TABLE 2.10	**ARM MUSCLES**			
Muscle	*Proximal Attachment*	*Distal Attachment*	*Innervation*	*Main Actions*
Biceps brachii	*Short head:* tip of coracoid process of scapula *Long head:* supraglenoid tubercle of scapula and glenoid labrum	Tuberosity of radius and fascia of forearm through bicipital aponeurosis	Musculocutaneous nerve (C5, **C6**, C7)	Supinates forearm and, when forearm is supine, flexes elbow joint; short head flexes shoulder joint; long head helps to stabilize shoulder joint during abduction.
Brachialis	Distal half of anterior surface of humerus	Coronoid process and tuberosity of ulna	Musculocutaneous nerve (C5–C7) and radial (C5–C7)	Flexes elbow joint in all positions
Coracobrachialis	Tip of coracoid process of scapula	Middle third of medial surface of humerus	Musculocutaneous nerve (C5, **C6**, C7)	Assists with flexion and adduction of shoulder joint
Triceps brachii	*Long head:* infraglenoid tubercle of scapula *Lateral head:* posterior surface of humerus, superior to radial groove *Medial head:* posterior surface of humerus, inferior to radial groove	Proximal end of olecranon of ulna and fascia of forearm	Radial nerve (C6, **C7**, **C8**)	Extends the elbow joint; long head steadies head of humerus when shoulder joint is abducted
Anconeus	Lateral epicondyle of humerus	Lateral surface of olecranon and superior part of posterior surface of ulna	Radial nerve (C7–T1)	Assists triceps in extending elbow joint; stabilizes elbow joint; abducts ulna during pronation

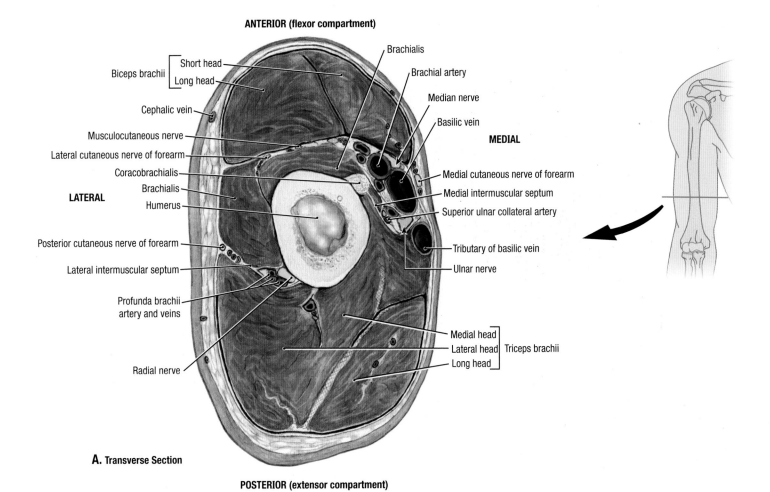

ANTERIOR (flexor compartment)

Biceps brachii { Short head | Long head

Brachialis

Brachial artery

Median nerve

Cephalic vein

Basilic vein

Musculocutaneous nerve

MEDIAL

Lateral cutaneous nerve of forearm

Coracobrachialis

Medial cutaneous nerve of forearm

LATERAL

Brachialis

Medial intermuscular septum

Humerus

Superior ulnar collateral artery

Posterior cutaneous nerve of forearm

Tributary of basilic vein

Lateral intermuscular septum

Ulnar nerve

Profunda brachii artery and veins

Medial head]
Lateral head } Triceps brachii
Long head]

Radial nerve

A. Transverse Section

POSTERIOR (extensor compartment)

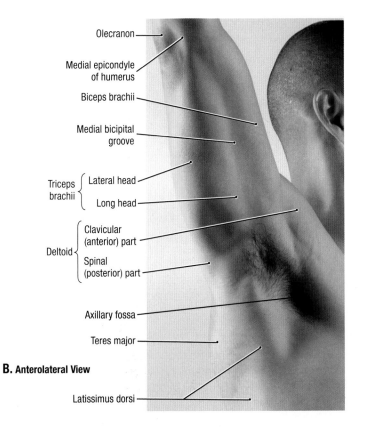

Olecranon

Medial epicondyle of humerus

Biceps brachii

Medial bicipital groove

Triceps brachii { Lateral head | Long head

Deltoid { Clavicular (anterior) part | Spinal (posterior) part

Axillary fossa

Teres major

B. Anterolateral View

Latissimus dorsi

ANTERIOR AND POSTERIOR COMPARTMENTS OF ARM `2.36`

A. Anatomical section. **B.** Surface anatomy.

- Three muscles, the biceps brachii, brachialis, and coracobrachialis, lie in the anterior compartment of the arm; the triceps brachii lies in the posterior compartment.
- The medial and lateral intermuscular septum separates these two muscle groups.
- The radial nerve and profunda brachii artery and veins serving the posterior compartment lie in contact with the radial groove of the humerus.
- The musculocutaneous nerve serving the anterior compartment lies in the plane between the biceps and the brachialis muscles.
- The median nerve crosses to the medial side of the brachial artery.
- The ulnar nerve passes posteriorly onto the medial side of the triceps muscle.

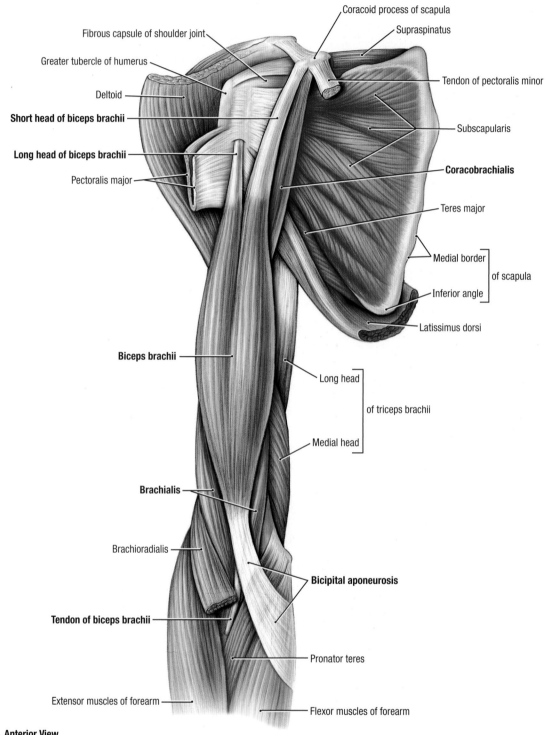

Coracoid process of scapula

Fibrous capsule of shoulder joint

Supraspinatus

Greater tubercle of humerus

Tendon of pectoralis minor

Deltoid

Short head of biceps brachii

Subscapularis

Long head of biceps brachii

Coracobrachialis

Pectoralis major

Teres major

Medial border

of scapula

Inferior angle

Latissimus dorsi

Biceps brachii

Long head

of triceps brachii

Medial head

Brachialis

Brachioradialis

Bicipital aponeurosis

Tendon of biceps brachii

Pronator teres

Extensor muscles of forearm

Flexor muscles of forearm

A. Anterior View

2.37 MUSCLES OF ANTERIOR ASPECT OF ARM I

- The biceps brachii has two heads: a long head and a short head.
- When the elbow joint is flexed approximately 90 degrees, the biceps is a flexor from the supinated position of the forearm but a very powerful supinator from the pronated position.

- A triangular membranous band, the bicipital aponeurosis, runs from the biceps tendon across the cubital fossa and merges with the antebrachial (deep) fascia covering the flexor muscles on the medial side of the forearm.

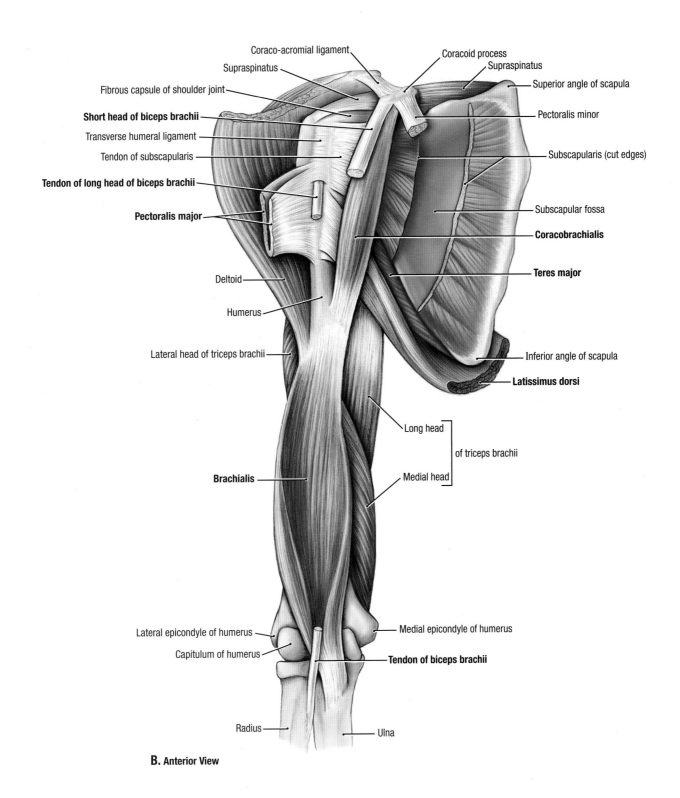

Coraco-acromial ligament

Coracoid process

Supraspinatus

Supraspinatus

Fibrous capsule of shoulder joint

Superior angle of scapula

Short head of biceps brachii

Pectoralis minor

Transverse humeral ligament

Tendon of subscapularis

Subscapularis (cut edges)

Tendon of long head of biceps brachii

Subscapular fossa

Pectoralis major

Coracobrachialis

Teres major

Deltoid

Humerus

Lateral head of triceps brachii

Inferior angle of scapula

Latissimus dorsi

Long head

of triceps brachii

Medial head

Brachialis

Lateral epicondyle of humerus

Medial epicondyle of humerus

Capitulum of humerus

Tendon of biceps brachii

Radius

Ulna

B. Anterior View

MUSCLES OF ANTERIOR ASPECT OF ARM II

- The brachialis, a flattened fusiform muscle, lies posterior (deep) to the biceps and produces the greatest amount of flexion force.
- The coracobrachialis, an elongated muscle in the superomedial part of the arm, is pierced by the musculocutaneous nerve. It helps flex and adduct the shoulder joint.

- **Rupture of the tendon of the long head of the biceps** usually results from wear and tear of an inflamed tendon **(biceps tendinitis)**. Normally, the tendon is torn from its attachment to the supraglenoid tubercle of the scapula. The detached muscle belly forms a ball near the center of the distal part of the anterior aspect of the arm.

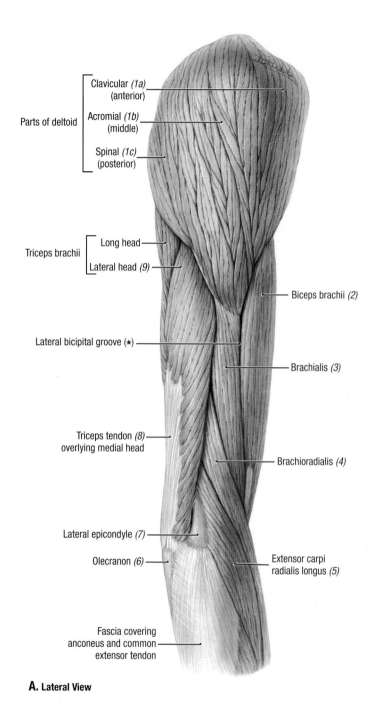

Parts of deltoid
- Clavicular *(1a)* (anterior)
- Acromial *(1b)* (middle)
- Spinal *(1c)* (posterior)

Triceps brachii
- Long head
- Lateral head *(9)*

Biceps brachii *(2)*

Lateral bicipital groove *(⋆)*

Brachialis *(3)*

Triceps tendon *(8)* overlying medial head

Brachioradialis *(4)*

Lateral epicondyle *(7)*

Olecranon *(6)*

Extensor carpi radialis longus *(5)*

Fascia covering anconeus and common extensor tendon

A. Lateral View

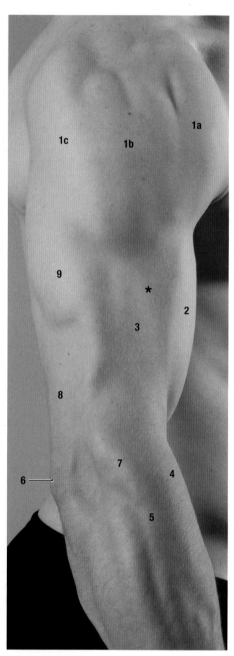

B. Lateral View

2.38 LATERAL ASPECT OF ARM

A. Dissection. Numbers in parentheses refer to structures **(B)**.
B. Surface anatomy.

Atrophy of the deltoid occurs when the axillary nerve (C5 and C6) is severely damaged (e.g., as might occur when the surgical neck of the humerus is fractured). As the deltoid atrophies, the rounded contour of the shoulder disappears. This gives the shoulder a flattened appearance and produces a slight hollow inferior to the acromion. A loss of sensation may occur over the lateral side of the proximal part of the arm, the area supplied by the superior lateral cutaneous nerve of the arm. To test the deltoid (or the function of the axillary nerve), the shoulder joint is abducted against resistance, starting from approximately 15 degrees. Supraspinatus initiates abduction at the shoulder joint.

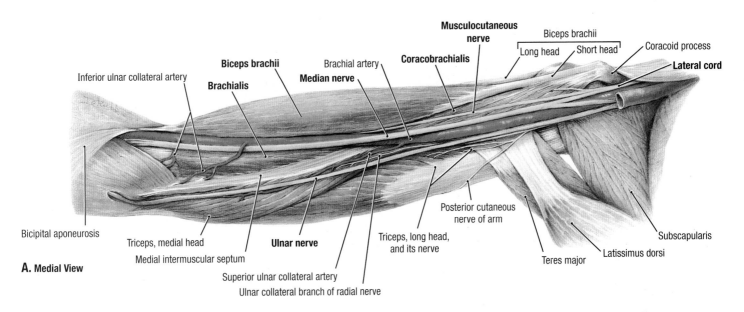

A. Medial View

Labels (clockwise):
- Inferior ulnar collateral artery
- Biceps brachii
- Brachialis
- Brachial artery
- Median nerve
- Coracobrachialis
- Musculocutaneous nerve
- Biceps brachii — Long head / Short head
- Coracoid process
- Lateral cord
- Subscapularis
- Latissimus dorsi
- Teres major
- Triceps, long head, and its nerve
- Posterior cutaneous nerve of arm
- Ulnar collateral branch of radial nerve
- Superior ulnar collateral artery
- Ulnar nerve
- Medial intermuscular septum
- Triceps, medial head
- Bicipital aponeurosis

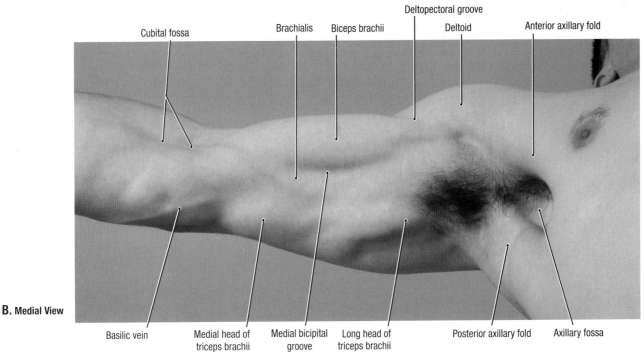

B. Medial View

Labels:
- Cubital fossa
- Brachialis
- Biceps brachii
- Deltopectoral groove
- Deltoid
- Anterior axillary fold
- Basilic vein
- Medial head of triceps brachii
- Medial bicipital groove
- Long head of triceps brachii
- Posterior axillary fold
- Axillary fossa

MEDIAL ASPECT OF ARM

2.39

A. Dissection. **B.** Surface anatomy.

- The axillary artery passes just inferior to the tip of the coracoid process and courses posterior to the coracobrachialis. At the inferior border of the teres major, the axillary artery changes names to become the brachial artery and continues distally on the anterior aspect of the brachialis.
- Although collateral pathways confer some protection against gradual temporary and partial occlusion, sudden complete **occlusion or laceration of the brachial artery** creates a surgical emergency because paralysis of muscles results from ischemia within a few hours.

- The median nerve lies adjacent to the axillary and brachial arteries and then crosses the artery from lateral to medial.
- Proximally, the ulnar nerve is adjacent to the medial side of the artery, passes posterior to the medial intermuscular septum, and descends on the medial head of triceps to pass posterior to the medial epicondyle; here, the ulnar nerve is palpable.
- The superior ulnar collateral artery and ulnar collateral branch of the radial nerve (to medial head of the triceps) accompany the ulnar nerve in the arm.

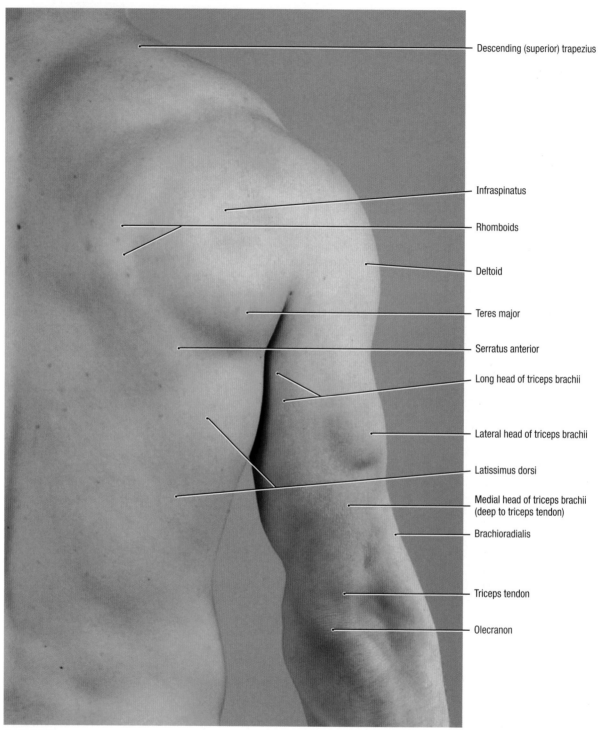

Descending (superior) trapezius

Infraspinatus

Rhomboids

Deltoid

Teres major

Serratus anterior

Long head of triceps brachii

Lateral head of triceps brachii

Latissimus dorsi

Medial head of triceps brachii
(deep to triceps tendon)

Brachioradialis

Triceps tendon

Olecranon

Posterior View

2.40 SURFACE ANATOMY OF SCAPULAR REGION AND POSTERIOR ASPECT OF ARM

The three heads of the triceps brachii form a bulge on the posterior aspect of the arm and are identifiable in a lean individual when the elbow joint is extended from the flexed position against resistance.

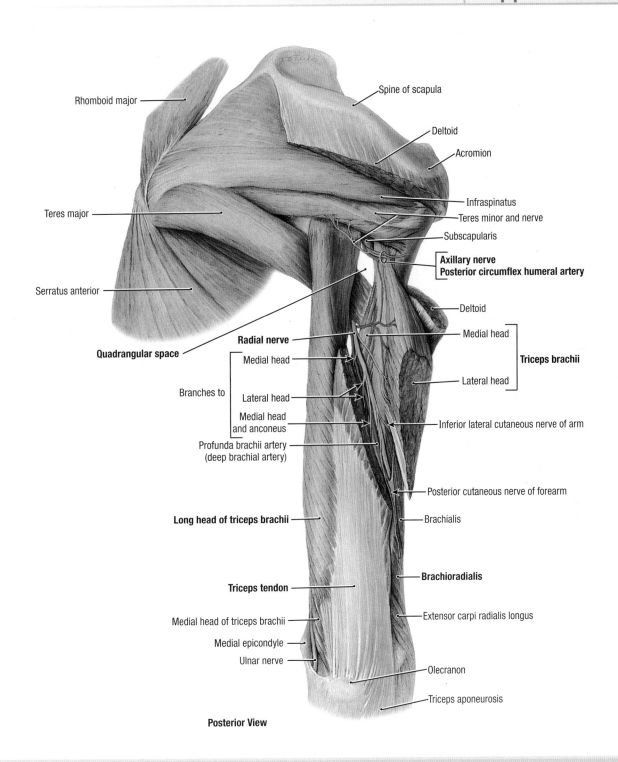

Rhomboid major

Spine of scapula

Deltoid

Acromion

Infraspinatus

Teres minor and nerve

Subscapularis

Axillary nerve
Posterior circumflex humeral artery

Teres major

Serratus anterior

Deltoid

Radial nerve

Quadrangular space

Medial head

Branches to

Lateral head

Medial head
and anconeus

Profunda brachii artery
(deep brachial artery)

Medial head

Triceps brachii

Lateral head

Inferior lateral cutaneous nerve of arm

Posterior cutaneous nerve of forearm

Long head of triceps brachii

Brachialis

Brachioradialis

Triceps tendon

Extensor carpi radialis longus

Medial head of triceps brachii

Medial epicondyle

Ulnar nerve

Olecranon

Triceps aponeurosis

Posterior View

TRICEPS BRACHII AND RELATED NERVES

2.41

- The lateral head is reflected laterally, and the medial head is attached to the deep surface of the triceps tendon, which attaches to the olecranon.
- The radial nerve and profunda brachii artery pass between the proximal attachments of the long and medial heads of the triceps brachii in the middle third of the arm, directly contacting the radial groove of the humerus.
- **Midarm fracture.** The middle third of the arm is a common site for fractures of the humerus, often with associated **radial nerve trauma**. When the radial nerve is injured in the radial groove, the triceps brachii muscle typically is only weakened because only the medial head is affected. However, the muscles in the posterior compartment of the forearm, supplied by more distal branches of the radial nerve, are paralyzed. The characteristic clinical sign of radial nerve injury is **wrist drop** (inability to extend the wrist joint and fingers at the metacarpophalangeal joints).
- The axillary nerve passes through the quadrangular space along with the posterior circumflex humeral artery.
- The ulnar nerve follows the medial border of the triceps then passes posterior to the medial epicondyle.

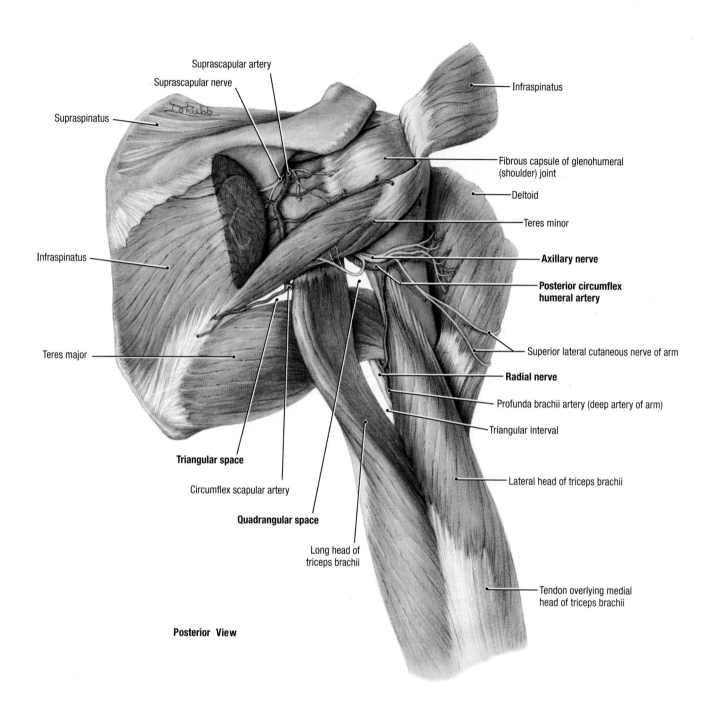

Suprascapular artery

Suprascapular nerve

Supraspinatus

Infraspinatus

Teres major

Triangular space

Circumflex scapular artery

Quadrangular space

Long head of triceps brachii

Infraspinatus

Fibrous capsule of glenohumeral (shoulder) joint

Deltoid

Teres minor

Axillary nerve

Posterior circumflex humeral artery

Superior lateral cutaneous nerve of arm

Radial nerve

Profunda brachii artery (deep artery of arm)

Triangular interval

Lateral head of triceps brachii

Tendon overlying medial head of triceps brachii

Posterior View

2.42 DORSAL SCAPULAR AND SUBDELTOID REGIONS

- The infraspinatus muscle, aided by the teres minor and spinal (posterior) fibers of the deltoid muscle, rotates the shoulder joint laterally.
- The long head of the triceps muscle passes between the teres minor and teres major and separates the quadrangular space from the triangular interval.
- Regarding the distribution of the suprascapular and axillary nerves, each comes from C5 and C6; each supplies two muscles— the suprascapular nerve innervates the supraspinatus and infraspinatus, and the axillary nerve innervates the teres minor and

deltoid muscles. Both nerves supply the shoulder joint, but only the axillary nerve has a cutaneous branch.
- **Axillary nerve injury** may occur when the glenohumeral (shoulder) joint dislocates because of its close relation to the inferior part of the joint capsule. Subglenoid displacement of the head of the humerus into the quadrangular space may damage the axillary nerve. Axillary nerve injury is indicated by paralysis of the deltoid and sensory loss over the lateral side of the proximal part of the arm.

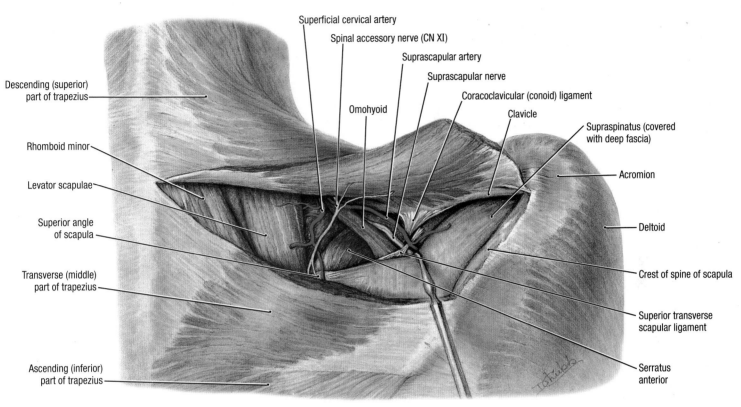

A. Posterior View

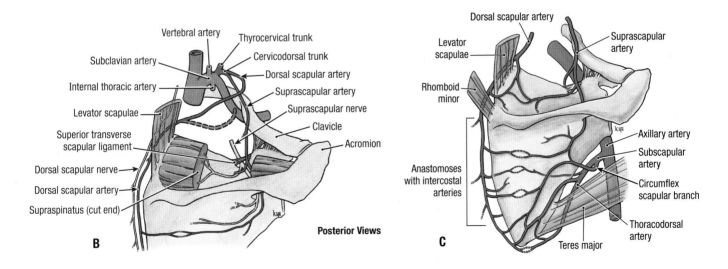

Posterior Views

B

C

SUPRASCAPULAR REGION

2.43

A. Dissection. At the level of the superior angle of the scapula, the transverse part of the trapezius muscle is reflected. **B.** Suprascapular and dorsal scapular arteries. **C.** Scapular anastomosis.

Several arteries join to form anastomoses on the anterior and posterior surfaces of the scapula. The importance of the collateral circulation made possible by these anastomoses becomes apparent when **ligation of a lacerated subclavian or axillary artery** is necessary or there is occlusion of these vessels. The direction of blood flow in the subscapular artery is then reversed, enabling blood to reach the third part of the axillary artery. In contrast to a sudden occlusion, slow occlusion of an artery often enables sufficient lateral circulation to develop, preventing **ischemia** (deficiency of blood).

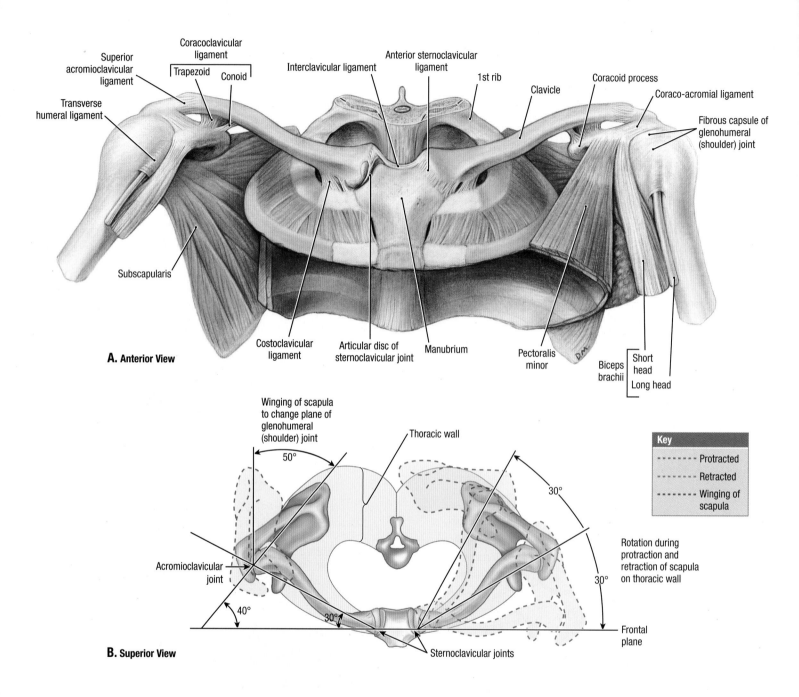

A. Anterior View

Coracoclavicular ligament — Trapezoid, Conoid

Superior acromioclavicular ligament

Transverse humeral ligament

Interclavicular ligament

Anterior sternoclavicular ligament

1st rib

Clavicle

Coracoid process

Coraco-acromial ligament

Fibrous capsule of glenohumeral (shoulder) joint

Subscapularis

Costoclavicular ligament

Articular disc of sternoclavicular joint

Manubrium

Pectoralis minor

Biceps brachii — Short head, Long head

B. Superior View

Winging of scapula to change plane of glenohumeral (shoulder) joint

50°

Thoracic wall

Acromioclavicular joint

40°

30°

Sternoclavicular joints

30°

30°

30°

Rotation during protraction and retraction of scapula on thoracic wall

Frontal plane

Key
- - - - - Protracted
- - - - - Retracted
- - - - - Winging of scapula

2.44   PECTORAL GIRDLE

A. Dissection. **B.** Clavicular movements at the sternoclavicular and acromioclavicular joints during rotation, protraction, and retraction of the scapula on the thoracic wall and winging of the scapula.

- The shoulder region includes the sternoclavicular, acromioclavicular, and shoulder (glenohumeral) joints; the mobility of the clavicle is essential to the movement of the upper limb.
- The sternoclavicular joint is the only joint connecting the upper limb (appendicular skeleton) to the trunk (axial skeleton).

The articular disc of the sternoclavicular joint divides the joint cavity into two parts and attaches superiorly to the clavicle and inferiorly to the first costal cartilage; the disc resists superior and medial displacement of the clavicle.

Paralysis of serratus anterior. Note that when the serratus anterior is paralyzed because of injury to the long thoracic nerve **(B)**, the medial border of the scapula moves laterally and posteriorly away from the thoracic wall, giving the scapula the appearance of a wing **(winged scapula)**. See Clinical Comment for Figure 2.29.

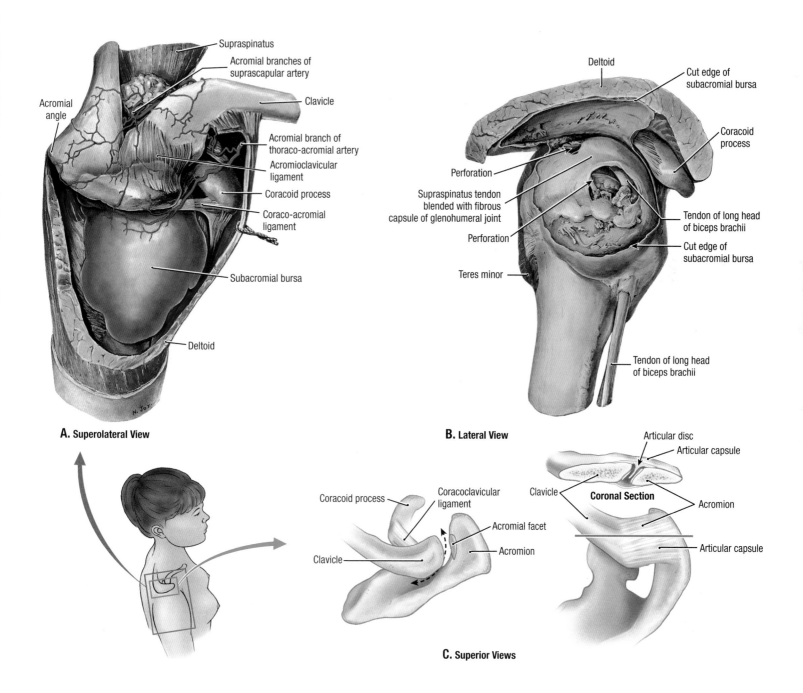

A. Superolateral View

- Supraspinatus
- Acromial branches of suprascapular artery
- Clavicle
- Acromial angle
- Acromial branch of thoraco-acromial artery
- Acromioclavicular ligament
- Coracoid process
- Coraco-acromial ligament
- Subacromial bursa
- Deltoid

B. Lateral View

- Deltoid
- Cut edge of subacromial bursa
- Coracoid process
- Perforation
- Supraspinatus tendon blended with fibrous capsule of glenohumeral joint
- Perforation
- Tendon of long head of biceps brachii
- Cut edge of subacromial bursa
- Teres minor
- Tendon of long head of biceps brachii

C. Superior Views

- Coracoid process
- Coracoclavicular ligament
- Clavicle
- Acromial facet
- Acromion
- Clavicle
- Articular disc
- Articular capsule
- Coronal Section
- Acromion
- Articular capsule

SUBACROMIAL BURSA AND ACROMIOCLAVICULAR JOINT

2.45

A. Subacromial bursa. The bursa has been injected with purple latex. **B.** Acromioclavicular joint. **C.** Attrition of supraspinatus tendon. As a result of wearing away of the supraspinatus tendon and underlying capsule, the subacromial bursa and shoulder joint come into communication. The intracapsular part of the tendon of the long head of biceps muscle becomes frayed, leaving it adherent to the intertubercular groove. Of 95 dissecting room subjects in Dr. Grant's lab, none of the 18 younger than 50 years of age had a perforation, but 4 of the 19 who were 50 to 60 years and 23 of the 57 older than 60 years had perforations. The perforation was bilateral in 11 subjects and unilateral in 14.

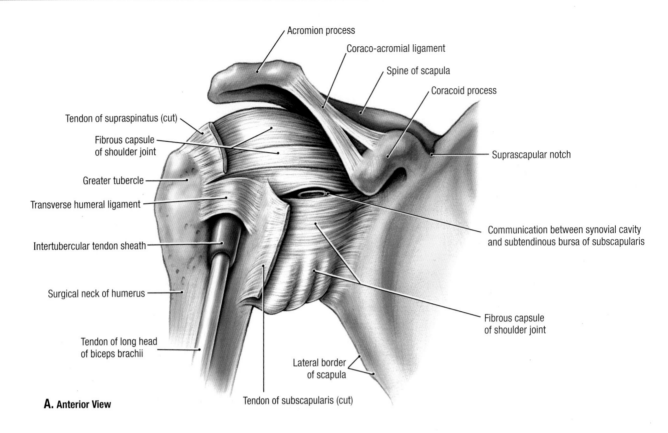

Acromion process

Coraco-acromial ligament

Spine of scapula

Coracoid process

Tendon of supraspinatus (cut)

Fibrous capsule of shoulder joint

Greater tubercle

Transverse humeral ligament

Intertubercular tendon sheath

Surgical neck of humerus

Tendon of long head of biceps brachii

Suprascapular notch

Communication between synovial cavity and subtendinous bursa of subscapularis

Fibrous capsule of shoulder joint

Lateral border of scapula

A. Anterior View

Tendon of subscapularis (cut)

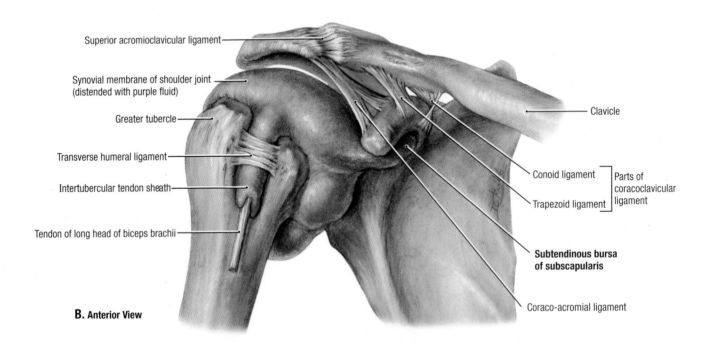

Superior acromioclavicular ligament

Synovial membrane of shoulder joint (distended with purple fluid)

Greater tubercle

Transverse humeral ligament

Intertubercular tendon sheath

Tendon of long head of biceps brachii

Clavicle

Conoid ligament | Parts of coracoclavicular ligament

Trapezoid ligament

Subtendinous bursa of subscapularis

Coraco-acromial ligament

B. Anterior View

2.46 LIGAMENTS AND ARTICULAR CAPSULE OF GLENOHUMERAL (SHOULDER) JOINT

A. Fibrous capsule.
- The loose fibrous capsule is attached to the margin of the glenoid cavity and to the anatomical neck of the humerus.
- The strong coracoclavicular ligament provides stability to the acromioclavicular joint and prevents the scapula from being driven medially and the acromion from being driven inferior to the clavicle.
- The coraco-acromial ligament prevents superior displacement of the head of the humerus.

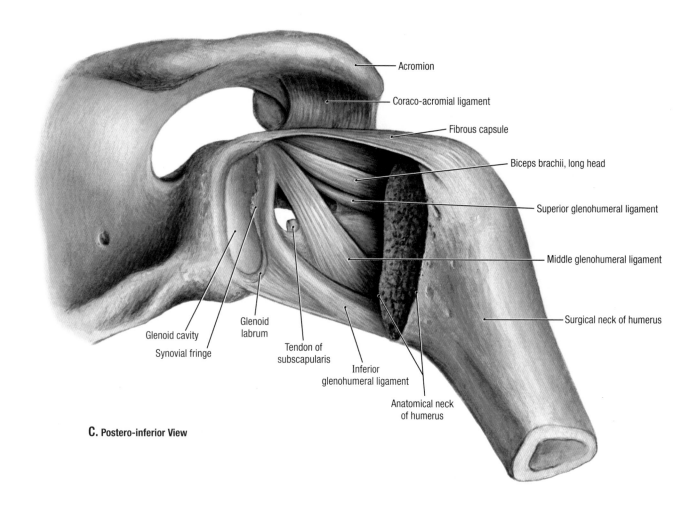

Acromion

Coraco-acromial ligament

Fibrous capsule

Biceps brachii, long head

Superior glenohumeral ligament

Middle glenohumeral ligament

Surgical neck of humerus

Glenoid cavity

Glenoid labrum

Synovial fringe

Tendon of subscapularis

Inferior glenohumeral ligament

Anatomical neck of humerus

C. Postero-inferior View

LIGAMENTS AND ARTICULAR CAPSULE OF GLENOHUMERAL (SHOULDER) JOINT (*continued*)   2.46

B. Synovial membrane of joint capsule. The synovial membrane lines the fibrous capsule and has two prolongations: (1) where it forms a synovial sheath for the tendon of the long head of the biceps muscle in its osseofibrous tunnel and (2) inferior to the coracoid process, where it forms a bursa between the subscapularis tendon and margin of the glenoid cavity—the subtendinous bursa of the subscapularis. **C.** Glenohumeral ligaments viewed from the interior of the shoulder joint.

- The joint is exposed from the posterior aspect by cutting away the thinner postero-inferior part of the capsule and sawing off the head of the humerus.
- The glenohumeral ligaments are visible from within the joint but are not easily seen externally.
- The glenohumeral ligaments and tendon of the long head of biceps brachii muscle converge on the supraglenoid tubercle.

- The slender superior glenohumeral ligament lies parallel to the tendon of the long head of biceps brachii. The middle ligament is free medially because the subtendinous bursa of subscapularis communicates with the joint cavity; usually, there is only a single site of communication. In this individual, there are openings on both sides of the ligament.

Because of its freedom of movement and instability, the glenohumeral joint is commonly dislocated by direct or indirect injury. Most **dislocations of the humeral head** occur in the downward (inferior) direction but are described clinically as anterior or (more rarely) posterior dislocations, indicating whether the humeral head has descended anterior or posterior to the infraglenoid tubercle and the long head of triceps. Anterior dislocation of the glenohumeral joint occurs most often in young adults, particularly athletes. It is usually caused by excessive extension and lateral rotation of the humerus.

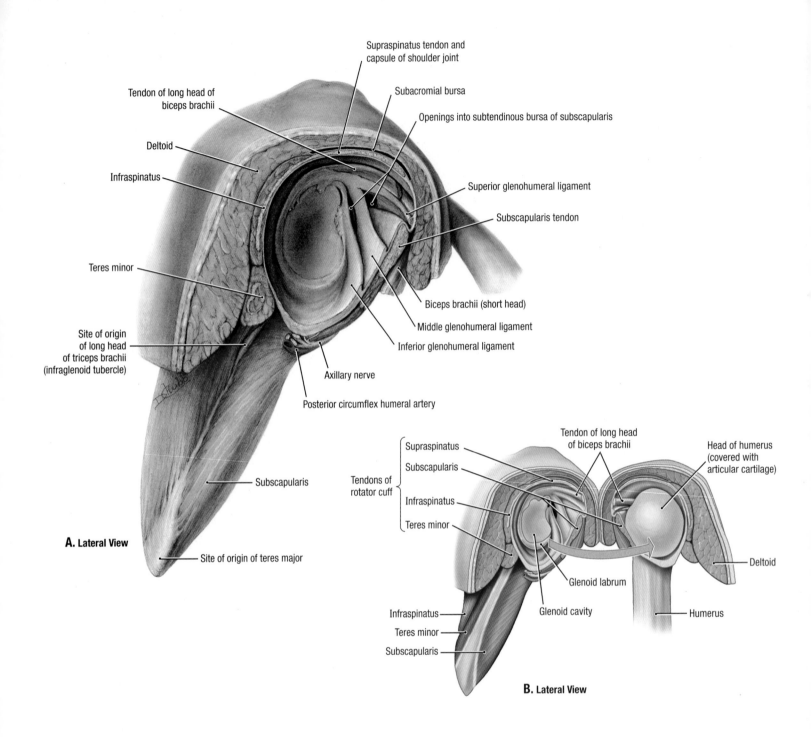

Tendon of long head of biceps brachii

Deltoid

Infraspinatus

Teres minor

Site of origin of long head of triceps brachii (infraglenoid tubercle)

Supraspinatus tendon and capsule of shoulder joint

Subacromial bursa

Openings into subtendinous bursa of subscapularis

Superior glenohumeral ligament

Subscapularis tendon

Biceps brachii (short head)

Middle glenohumeral ligament

Inferior glenohumeral ligament

Axillary nerve

Posterior circumflex humeral artery

Subscapularis

A. Lateral View

Site of origin of teres major

Supraspinatus

Subscapularis

Tendons of rotator cuff

Infraspinatus

Teres minor

Tendon of long head of biceps brachii

Head of humerus (covered with articular cartilage)

Glenoid labrum

Glenoid cavity

Deltoid

Humerus

Infraspinatus

Teres minor

Subscapularis

B. Lateral View

2.47 INTERIOR OF GLENOHUMERAL (SHOULDER) JOINT AND RELATIONSHIP OF ROTATOR CUFF

A. Dissection. **B.** Schematic illustration.
- The subacromial bursa is between the acromion and deltoid superiorly and the tendon of supraspinatus inferiorly.
- The four short rotator cuff muscles (supraspinatus, infraspinatus, teres minor, and subscapularis) cross the glenohumeral joint and blend with the capsule.
- The axillary nerve and posterior circumflex humeral artery are in contact with the capsule inferiorly and may be injured when the glenohumeral joint dislocates.

- Inflammation and calcification of the subacromial bursa result in pain, tenderness, and limitation of movement of the glenohumeral joint. This condition is also known as **calcific scapulohumeral bursitis**. Deposition of calcium in the supraspinatus tendon may irritate the overlying subacromial bursa, producing an inflammatory reaction, **subacromial bursitis**.

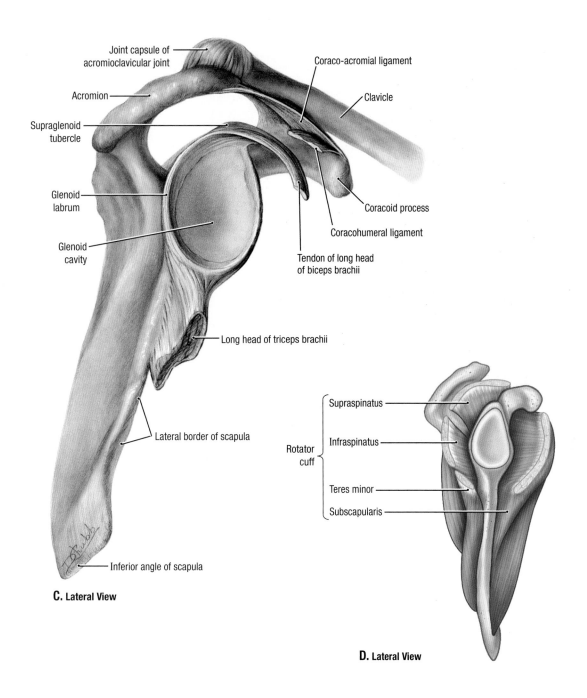

C. Lateral View

Joint capsule of acromioclavicular joint

Coraco-acromial ligament

Acromion

Clavicle

Supraglenoid tubercle

Glenoid labrum

Glenoid cavity

Coracoid process

Coracohumeral ligament

Tendon of long head of biceps brachii

Long head of triceps brachii

Lateral border of scapula

Inferior angle of scapula

Rotator cuff
- Supraspinatus
- Infraspinatus
- Teres minor
- Subscapularis

D. Lateral View

INTERIOR OF GLENOHUMERAL (SHOULDER) JOINT AND RELATIONSHIP OF ROTATOR CUFF (*continued*) | **2.47**

C. Dissection. **D.** Schematic illustration of the rotator cuff muscles and their relationship to the glenoid cavity.

- The coraco-acromial arch (coracoid process, coraco-acromial ligament, and acromion) prevents superior displacement of the head of the humerus.
- The long head of the triceps brachii muscle arises just inferior to the glenoid cavity; the long head of biceps just superior to it.
- The main function of the musculotendinous rotator cuff is to hold the large head of the humerus in the smaller and shallow glenoid cavity of the scapula, both during the relaxed state (by tonic contraction) and during active abduction.

Tearing of the fibrocartilaginous glenoid labrum commonly occurs in the athletes who throw (e.g., a baseball) and in those who have shoulder instability and subluxation (partial dislocation) of the glenohumeral joint. The tear often results from sudden contraction of the biceps or forceful subluxation of the humeral head over the glenoid labrum. Usually, a tear occurs in the anterosuperior part of the labrum.

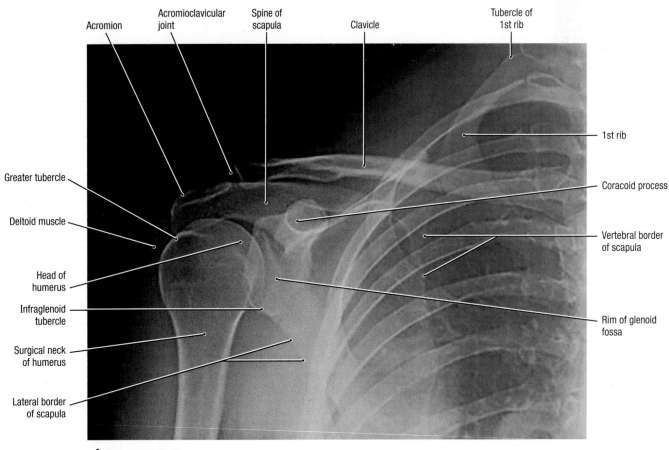

Acromion
Acromioclavicular joint
Spine of scapula
Clavicle
Tubercle of 1st rib

Greater tubercle

Deltoid muscle

Head of humerus

Infraglenoid tubercle

Surgical neck of humerus

Lateral border of scapula

1st rib

Coracoid process

Vertebral border of scapula

Rim of glenoid fossa

A. Anteroposterior View

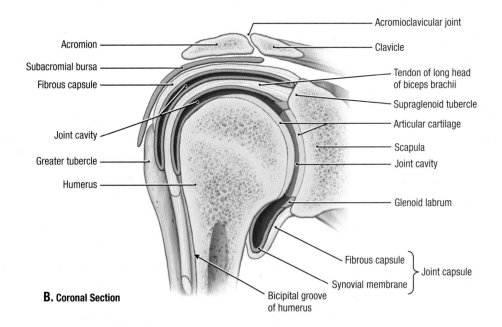

Acromion

Subacromial bursa

Fibrous capsule

Joint cavity

Greater tubercle

Humerus

Acromioclavicular joint

Clavicle

Tendon of long head of biceps brachii

Supraglenoid tubercle

Articular cartilage

Scapula

Joint cavity

Glenoid labrum

Fibrous capsule

Synovial membrane

Joint capsule

Bicipital groove of humerus

B. Coronal Section

2.48 IMAGING OF GLENOHUMERAL (SHOULDER) JOINT

A. Radiograph. **B.** Sectioned joint to show location of subacromial bursa and joint cavity.

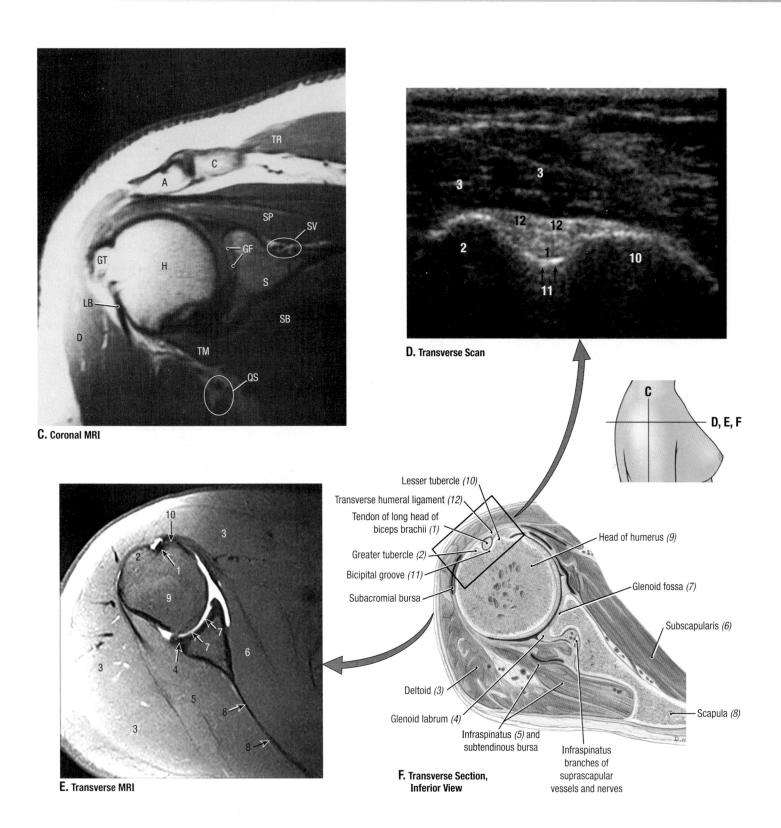

C. Coronal MRI

D. Transverse Scan

E. Transverse MRI

Lesser tubercle *(10)*
Transverse humeral ligament *(12)*
Tendon of long head of biceps brachii *(1)*
Greater tubercle *(2)*
Bicipital groove *(11)*
Subacromial bursa
Head of humerus *(9)*
Glenoid fossa *(7)*
Subscapularis *(6)*
Deltoid *(3)*
Glenoid labrum *(4)*
Infraspinatus *(5)* and subtendinous bursa
Infraspinatus branches of suprascapular vessels and nerves
Scapula *(8)*

F. Transverse Section, Inferior View

IMAGING OF GLENOHUMERAL (SHOULDER) JOINT (*continued*) | **2.48**

C. Coronal MRI. *A,* acromion; *C,* clavicle; *D,* deltoid; *GF,* glenoid cavity; *GT,* crest of greater tubercle; *H,* head of humerus; *LB,* long head of biceps brachii; *QS,* quadrangular space; *S,* scapula; *SB,* subscapularis; *SP,* supraspinatus; *SV,* suprascapular vessels and nerve; *TM,* teres minor; *TR,* trapezius. **D.** Transverse ultrasound scan of area indicated (**F**). **E.** Transverse MRI with contrast agent. **F.** Transverse section. Numbers (**F**) refer to structures labeled in **D** and **E**.

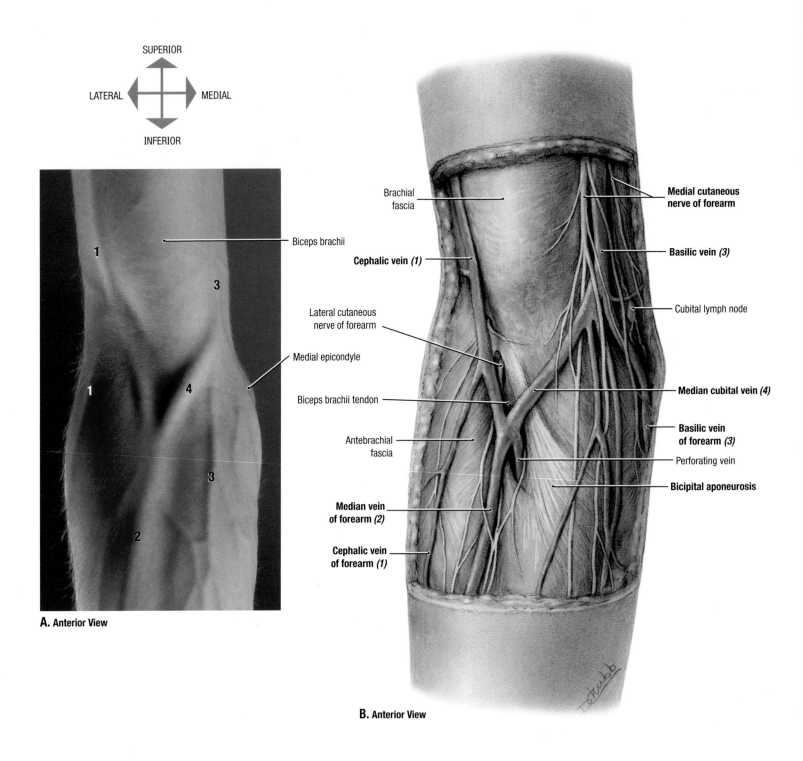

SUPERIOR

LATERAL — MEDIAL

INFERIOR

A. Anterior View

Biceps brachii

Lateral cutaneous
nerve of forearm

Medial epicondyle

Biceps brachii tendon

Antebrachial
fascia

**Median vein
of forearm (2)**

**Cephalic vein
of forearm (1)**

Brachial
fascia

Cephalic vein (1)

**Medial cutaneous
nerve of forearm**

Basilic vein (3)

Cubital lymph node

Median cubital vein (4)

**Basilic vein
of forearm (3)**

Perforating vein

Bicipital aponeurosis

B. Anterior View

2.49 CUBITAL FOSSA: SURFACE ANATOMY AND SUPERFICIAL DISSECTION

A. Surface anatomy. **B.** Cutaneous nerves and superficial veins. Numbers in parentheses refer to structures **(A)**.

- The cubital fossa is a triangular space (compartment) inferior to the elbow crease, roofed by deep fascia.
- In the forearm, the superficial veins (cephalic, median, basilic, and their connecting veins) make a variable, M-shaped pattern.
- The cephalic and basilic veins occupy the bicipital grooves, one on each side of the biceps brachii. In the lateral bicipital groove,

the lateral cutaneous nerve of the forearm appears just superior to the elbow crease; in the medial bicipital groove, the medial cutaneous nerve of the forearm becomes cutaneous at approximately the midpoint of the arm.

- The cubital fossa is the common site for **sampling and transfusion of blood and intravenous injections** because of the prominence and accessibility of veins. Usually, the median cubital vein or basilic vein is selected.

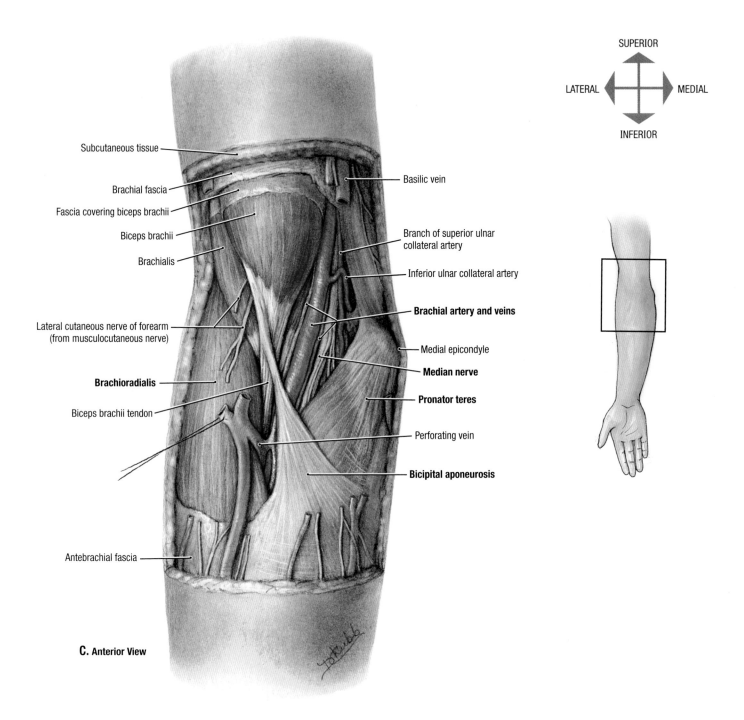

SUPERIOR

LATERAL ⟷ MEDIAL

INFERIOR

Subcutaneous tissue

Brachial fascia

Fascia covering biceps brachii

Biceps brachii

Brachialis

Lateral cutaneous nerve of forearm
(from musculocutaneous nerve)

Brachioradialis

Biceps brachii tendon

Antebrachial fascia

Basilic vein

Branch of superior ulnar
collateral artery

Inferior ulnar collateral artery

Brachial artery and veins

Medial epicondyle

Median nerve

Pronator teres

Perforating vein

Bicipital aponeurosis

C. Anterior View

CUBITAL FOSSA: DEEP DISSECTION I (*continued*)

2.49

C. Boundaries and contents of the cubital fossa.
- The cubital fossa is bound laterally by the brachioradialis and medially by the pronator teres and superiorly by a line joining the medial and lateral epicondyles.
- The three chief contents of the cubital fossa are the biceps brachii tendon, brachial artery, and median nerve.
- The biceps brachii tendon, on approaching its insertion, rotates through 90 degrees, and the bicipital aponeurosis extends medially from the proximal part of the tendon.

- A fracture of the distal part of the humerus, near the supra-epicondylar ridges, is called a **supra-epicondylar (supracondylar) fracture**. The distal bone fragment may be displaced anteriorly or posteriorly. Any of the nerves or branches of the brachial vessels related to the humerus may be injured by a displaced bone fragment.

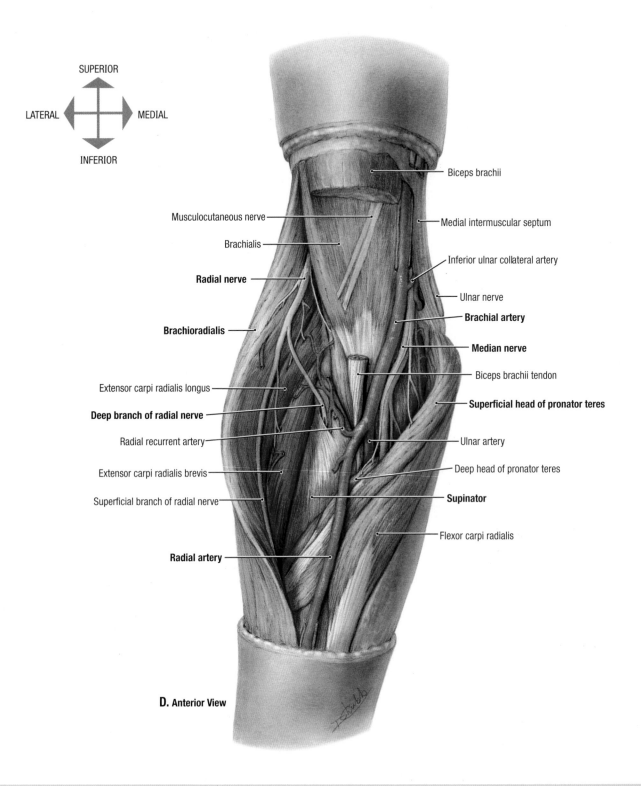

SUPERIOR

LATERAL ◀▶ MEDIAL

INFERIOR

Musculocutaneous nerve

Brachialis

Radial nerve

Brachioradialis

Extensor carpi radialis longus

Deep branch of radial nerve

Radial recurrent artery

Extensor carpi radialis brevis

Superficial branch of radial nerve

Radial artery

Biceps brachii

Medial intermuscular septum

Inferior ulnar collateral artery

Ulnar nerve

Brachial artery

Median nerve

Biceps brachii tendon

Superficial head of pronator teres

Ulnar artery

Deep head of pronator teres

Supinator

Flexor carpi radialis

D. Anterior View

2.49 CUBITAL FOSSA: DEEP DISSECTION II

D. Floor of the cubital fossa.
- Part of the biceps brachii muscle is excised, and the cubital fossa is opened widely, exposing the brachialis and supinator muscles in the floor of the fossa.
- The deep branch of the radial nerve pierces the supinator.
- The brachial artery lies between the biceps tendon and median nerve and divides into two branches, the ulnar and radial arteries.

- The median nerve supplies the flexor muscles. With the exception of the twig to the deep head of pronator teres, its motor branches arise from its medial side.
- The radial nerve supplies the extensor muscles. With the exception of the twig to brachioradialis, its motor branches arise from its lateral side. In this specimen, the radial nerve has been displaced laterally, so here its lateral branches appear to run medially.

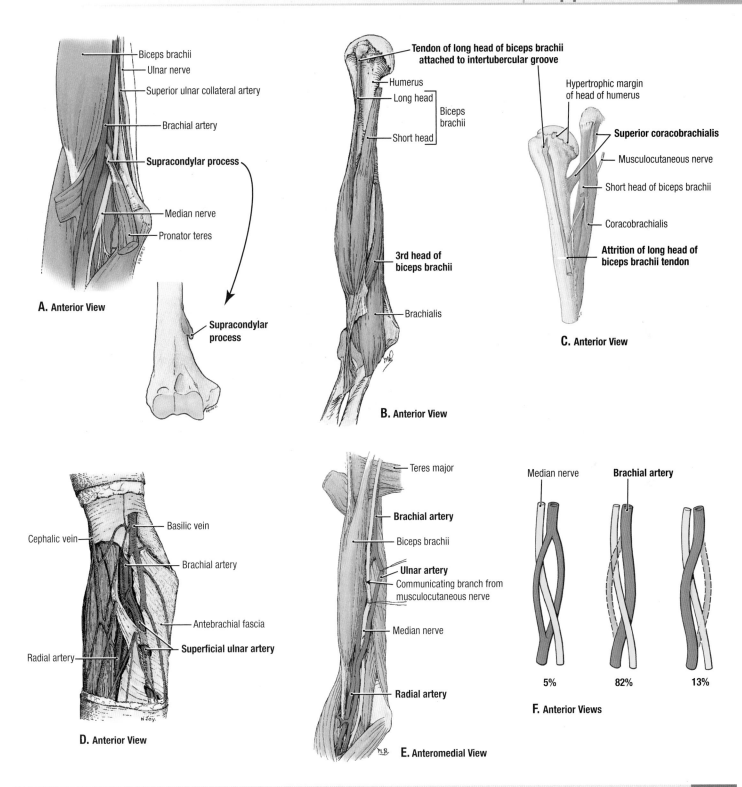

A. Anterior View

- Biceps brachii
- Ulnar nerve
- Superior ulnar collateral artery
- Brachial artery
- **Supracondylar process**
- Median nerve
- Pronator teres

Supracondylar process

B. Anterior View

- **Tendon of long head of biceps brachii attached to intertubercular groove**
- Humerus
- Long head
- Short head
- Biceps brachii
- 3rd head of biceps brachii
- Brachialis

C. Anterior View

- Hypertrophic margin of head of humerus
- **Superior coracobrachialis**
- Musculocutaneous nerve
- Short head of biceps brachii
- Coracobrachialis
- **Attrition of long head of biceps brachii tendon**

D. Anterior View

- Cephalic vein
- Basilic vein
- Brachial artery
- Antebrachial fascia
- **Superficial ulnar artery**
- Radial artery

E. Anteromedial View

- Teres major
- **Brachial artery**
- Biceps brachii
- **Ulnar artery**
- Communicating branch from musculocutaneous nerve
- Median nerve
- **Radial artery**

F. Anterior Views

- Median nerve
- **Brachial artery**
- 5%
- 82%
- 13%

ANOMALIES

2.50

A. Supracondylar process of humerus. A fibrous band, from which the pronator teres muscle arises, joins this supra-epicondylar process to the medial epicondyle. The median nerve, often accompanied by the brachial artery, passes through the foramen formed by this band. This may be a cause of nerve entrapment. **B.** Third head of biceps brachii. In this case, there is also attrition of the biceps tendon. **C.** Attrition of the tendon of the long head of biceps brachii and presence of a coracobrachialis. **D.** Superficial ulnar artery.

E. Anomalous division of brachial artery. In this case, the median nerve passes between the radial and ulnar arteries, which arise high in the arm. **F.** Relationship of median nerve and brachial artery. The variable relationship of these two structures can be explained developmentally. In a study of 307 limbs in Dr. Grant's lab, portions of both primitive brachial arteries persisted in 5%, the posterior in 82%, and the anterior in 13%.

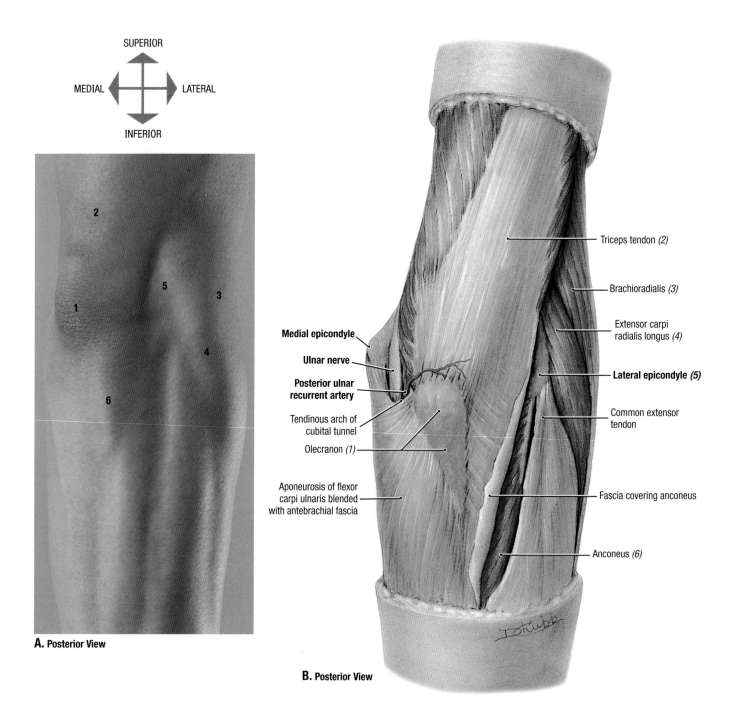

SUPERIOR

MEDIAL — LATERAL

INFERIOR

A. Posterior View

Triceps tendon *(2)*

Brachioradialis *(3)*

Extensor carpi radialis longus *(4)*

Medial epicondyle

Ulnar nerve

Lateral epicondyle *(5)*

Posterior ulnar recurrent artery

Common extensor tendon

Tendinous arch of cubital tunnel

Olecranon *(1)*

Aponeurosis of flexor carpi ulnaris blended with antebrachial fascia

Fascia covering anconeus

Anconeus *(6)*

B. Posterior View

2.51   POSTERIOR ASPECT OF ELBOW I

A. Surface anatomy. **B.** Superficial dissection. Numbers in parentheses refer to structures **(A)**.

- The triceps brachii is attached distally to the superior surface of the olecranon and, through the deep fascia covering the anconeus, into the lateral border of olecranon.

- The posterior surfaces of the medial epicondyle, lateral epicondyle, and olecranon are subcutaneous and palpable.

- The ulnar nerve, also palpable, runs subfascially posterior to the medial epicondyle; distal to this point, it disappears deep to the two heads of the flexor carpi ulnaris.

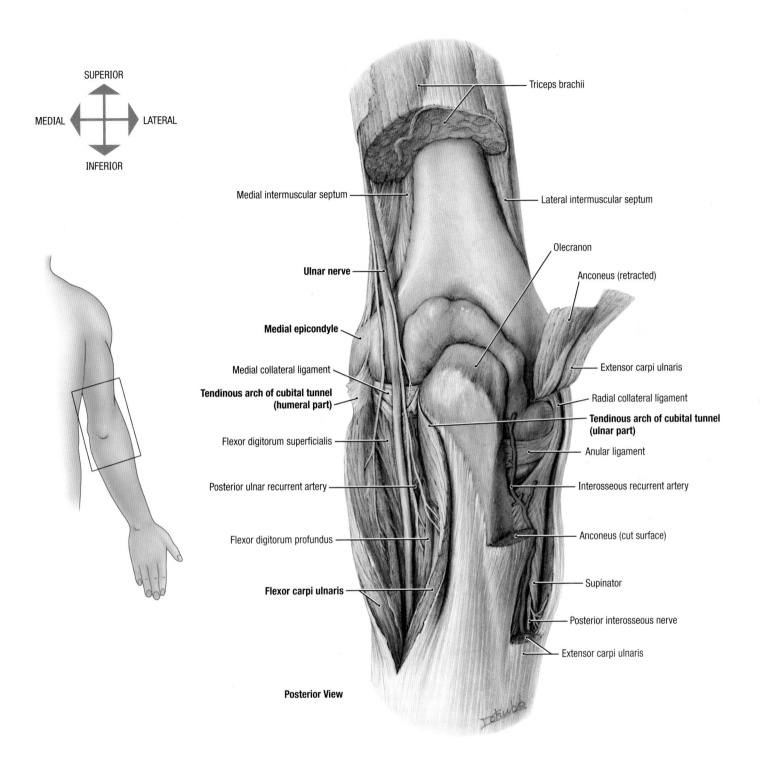

SUPERIOR

MEDIAL ←→ LATERAL

INFERIOR

Triceps brachii

Medial intermuscular septum

Lateral intermuscular septum

Olecranon

Ulnar nerve

Anconeus (retracted)

Medial epicondyle

Medial collateral ligament

Extensor carpi ulnaris

Tendinous arch of cubital tunnel (humeral part)

Radial collateral ligament

Tendinous arch of cubital tunnel (ulnar part)

Flexor digitorum superficialis

Anular ligament

Posterior ulnar recurrent artery

Interosseous recurrent artery

Flexor digitorum profundus

Anconeus (cut surface)

Supinator

Flexor carpi ulnaris

Posterior interosseous nerve

Extensor carpi ulnaris

Posterior View

POSTERIOR ASPECT OF ELBOW II

Deep dissection. The distal portion of the triceps brachii muscle was removed. Note that the ulnar nerve descends subfascially within the posterior compartment of the arm, passing posterior to the medial epicondyle in the groove for the ulnar nerve. Next, it passes posterior to the ulnar collateral ligament of the elbow joint and then between the flexor carpi ulnaris and flexor digitorum profundus muscles.

Ulnar nerve injury occurs most commonly where the nerve passes posterior to the medial epicondyle of the humerus. The injury results when the medial part of the elbow hits a hard surface, fracturing the medial epicondyle. The ulnar nerve may be compressed in the cubital tunnel, resulting in **cubital tunnel syndrome**. The cubital tunnel is formed by the tendinous arch joining the humeral and ulnar heads of attachment of the flexor carpi ulnaris muscle. Ulnar nerve injury can result in extensive motor and sensory loss to the hand.

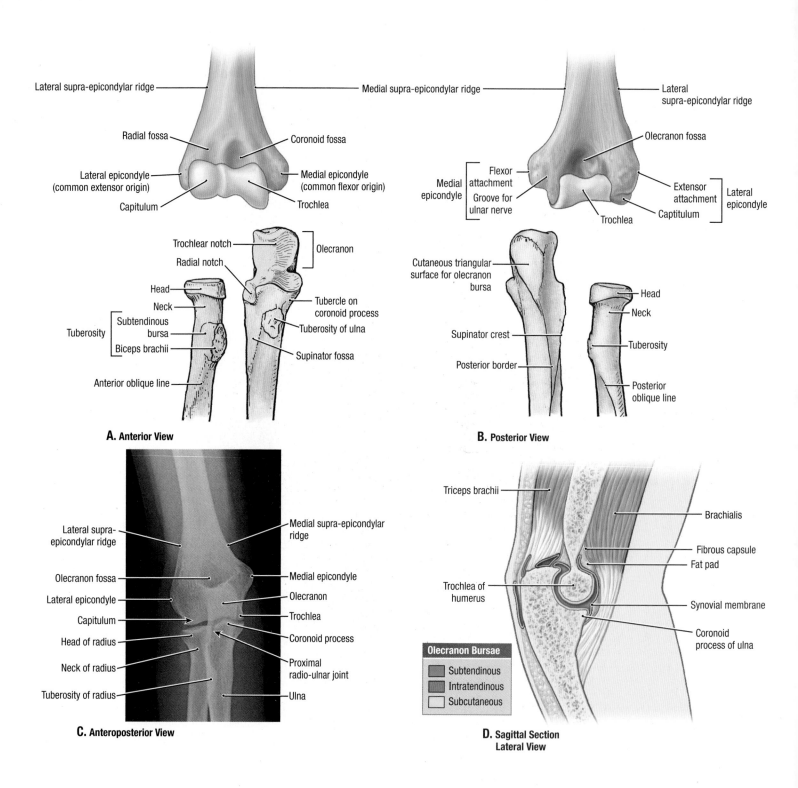

A. Anterior View

B. Posterior View

C. Anteroposterior View

Olecranon Bursae

- Subtendinous
- Intratendinous
- Subcutaneous

D. Sagittal Section Lateral View

2.53 BONES AND IMAGING OF ELBOW REGION

A. Anterior bony features. **B.** Posterior bony features. **C.** Radiograph of elbow joint. **D.** Section of humero-ulnar joint.

The subcutaneous olecranon bursa is exposed to injury during falls on the elbow and to infection from abrasions of the skin covering the olecranon. Repeated excessive pressure and friction produces a friction **subcutaneous olecranon bursitis** (e.g., "student's elbow").

Subtendinous olecranon bursitis results from excessive friction between the triceps tendon and the olecranon. For example, it may occur due to repeated flexion-extension of the forearm during certain assembly-line jobs. The pain is severe during flexion of the forearm because of pressure exerted on the inflamed subtendinous olecranon bursa by the triceps tendon.

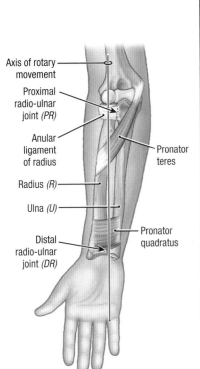

Axis of rotary movement
Proximal radio-ulnar joint (PR)
Anular ligament of radius
Radius (R)
Ulna (U)
Distal radio-ulnar joint (DR)
Pronator teres
Pronator quadratus

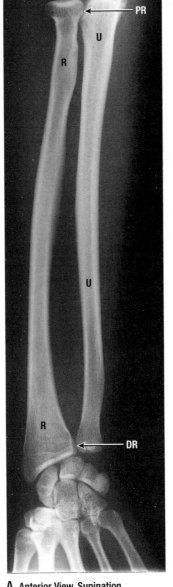

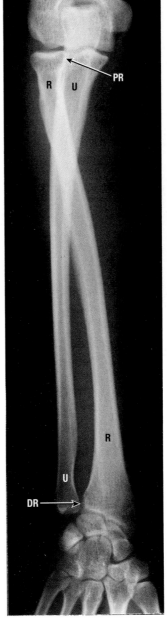

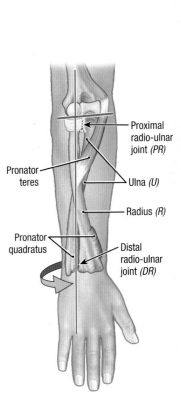

Proximal radio-ulnar joint (PR)
Pronator teres
Ulna (U)
Radius (R)
Pronator quadratus
Distal radio-ulnar joint (DR)

A. Anterior View, Supination   **B.** Anterior View, Pronation

SUPINATION AND PRONATION AT SUPERIOR, MIDDLE, AND INFERIOR RADIO-ULNAR JOINTS   2.54

A. Radiograph of forearm in supination. **B.** Radiograph of fore-arm in pronation. The radius crosses the ulna when the forearm is pronated. The superior and inferior radio-ulnar joints are synovial joints; the middle radio-ulnar joint is a syndesmosis (fibrous joint) in which the interosseous membrane connects the forearm bones.

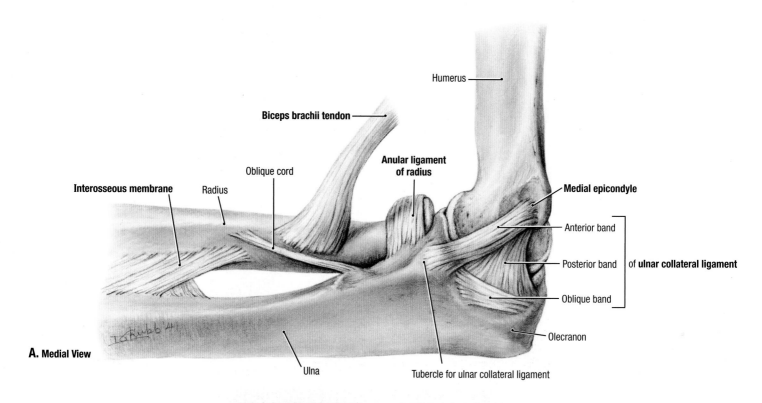

Humerus

Biceps brachii tendon

Anular ligament of radius

Oblique cord

Interosseous membrane   Radius

Medial epicondyle

Anterior band

Posterior band   of **ulnar collateral ligament**

Oblique band

Olecranon

A. Medial View

Ulna   Tubercle for ulnar collateral ligament

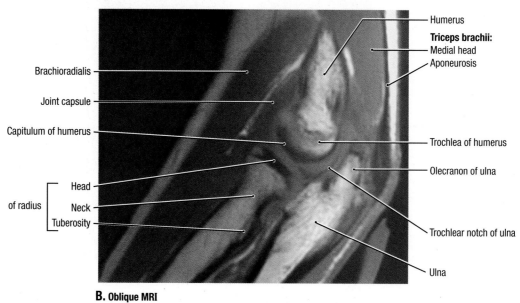

Humerus

Triceps brachii:
Medial head
Aponeurosis

Brachioradialis

Joint capsule

Capitulum of humerus

Trochlea of humerus

Olecranon of ulna

Head
of radius   Neck
Tuberosity

Trochlear notch of ulna

Ulna

Trochlear notch of ulna

B. Oblique MRI

2.55   MEDIAL ASPECT OF BONES AND LIGAMENTS OF ELBOW REGION

A. Ligaments. The anterior band of the ulnar (medial) collateral ligament is a strong, round cord that is taut when the elbow joint is extended. The posterior band is a weak fan that is taut in flexion of the joint. **B.** MRI of elbow joint.

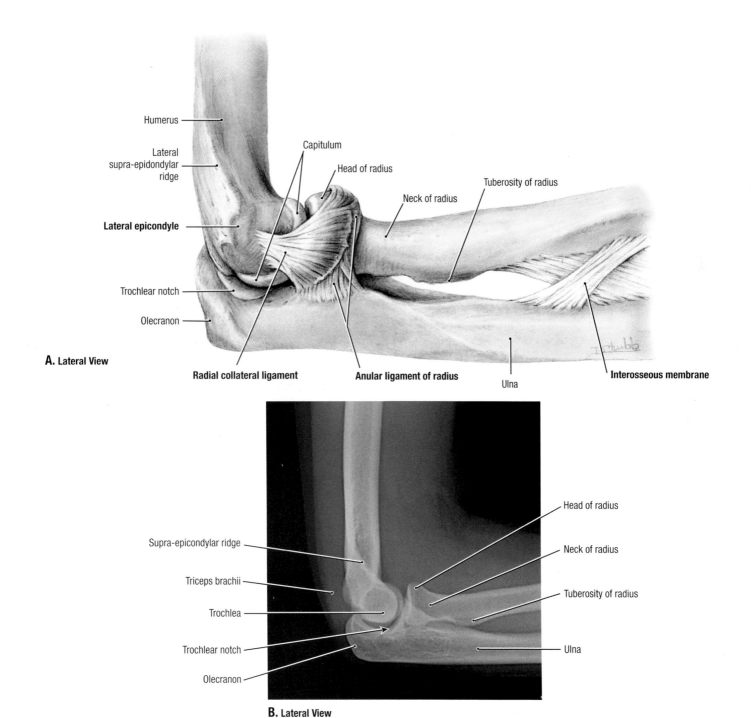

Humerus

Lateral supra-epicondylar ridge

Capitulum

Head of radius

Tuberosity of radius

Neck of radius

Lateral epicondyle

Trochlear notch

Olecranon

A. Lateral View

Radial collateral ligament

Anular ligament of radius

Ulna

Interosseous membrane

Supra-epicondylar ridge

Head of radius

Triceps brachii

Neck of radius

Trochlea

Tuberosity of radius

Trochlear notch

Ulna

Olecranon

B. Lateral View

LATERAL ASPECT OF BONES AND LIGAMENTS OF ELBOW REGION

2.56

A. Ligaments. The fan-shaped radial (lateral) collateral ligament is primarily attached to the anular ligament of the radius; superficial fibers of the lateral ligament blend with the fibrous capsule and continue onto the radius. **B.** Lateral radiograph.

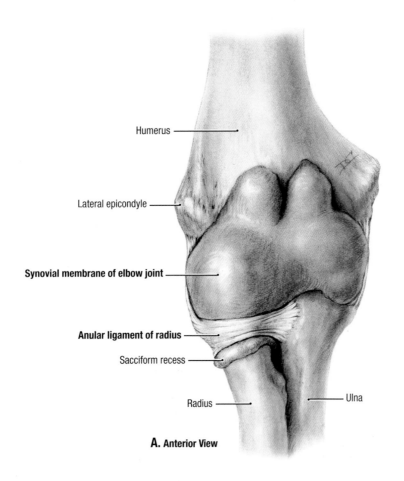

Humerus

Lateral epicondyle

Synovial membrane of elbow joint

Anular ligament of radius

Sacciform recess

Radius

Ulna

A. Anterior View

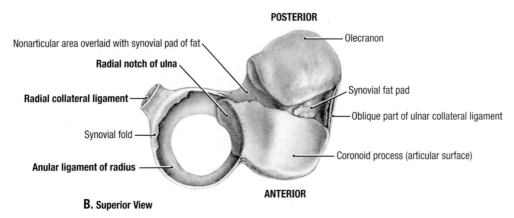

POSTERIOR

Olecranon

Nonarticular area overlaid with synovial pad of fat

Radial notch of ulna

Radial collateral ligament

Synovial fat pad

Oblique part of ulnar collateral ligament

Synovial fold

Anular ligament of radius

Coronoid process (articular surface)

ANTERIOR

B. Superior View

2.57 SYNOVIAL CAPSULE OF ELBOW JOINT AND ANULAR LIGAMENT

A. Synovial capsule of elbow and proximal radio-ulnar joints. The cavity of the elbow was injected with purple fluid (wax). The fibrous capsule was removed, and the synovial membrane remains. **B.** Anular ligament.

- The anular ligament secures the head of the radius to the radial notch of the ulna and with it forms a tapering columnar socket (i.e., wide superiorly, narrow inferiorly).
- The anular ligament is bound to the humerus by the radial collateral ligament of the elbow.

A common childhood injury is **subluxation and dislocation of the head of the radius** after traction on a pronated forearm (e.g., when lifting a child onto a bus). The sudden pulling of the upper limb tears or stretches the distal attachment of the less tapering anular ligament of a child. The radial head then moves distally, partially out of the anular ligament. The proximal part of the torn ligament may become trapped between the head of the radius and the capitulum of the humerus. The source of pain is the pinched anular ligament.

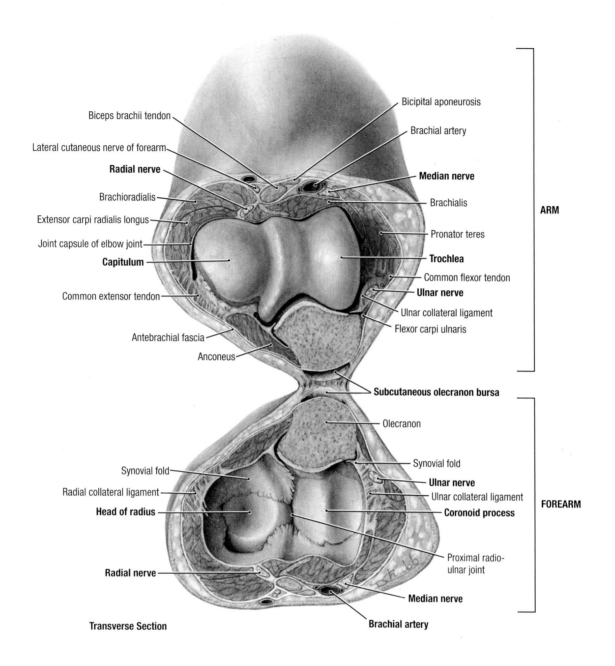

Biceps brachii tendon

Lateral cutaneous nerve of forearm

Radial nerve

Brachioradialis

Extensor carpi radialis longus

Joint capsule of elbow joint

Capitulum

Common extensor tendon

Antebrachial fascia

Anconeus

Bicipital aponeurosis

Brachial artery

Median nerve

Brachialis

Pronator teres

Trochlea

Common flexor tendon

Ulnar nerve

Ulnar collateral ligament

Flexor carpi ulnaris

Subcutaneous olecranon bursa

ARM

Synovial fold

Radial collateral ligament

Head of radius

Radial nerve

Olecranon

Synovial fold

Ulnar nerve

Ulnar collateral ligament

Coronoid process

Proximal radio-ulnar joint

Median nerve

Brachial artery

FOREARM

Transverse Section

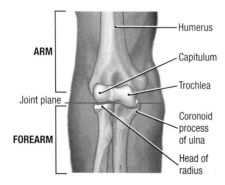

ARM — Humerus

Capitulum

Trochlea

Joint plane

Coronoid process of ulna

FOREARM — Head of radius

ARTICULAR SURFACES OF ELBOW JOINT | 2.58

The tissue surrounding the condyles of the humerus has been sectioned in a transverse plane, followed by disarticulation of the elbow joint, revealing the articular surfaces. Compare the forearm (inferior) component with Figure 2.57B.

- Synovial folds containing fat overlie the periphery of the head of the radius and the nonarticular indentations on the trochlear notch of the ulna.
- The radial nerve is in contact with the joint capsule, the ulnar nerve is in contact with the ulnar collateral ligament, and the median nerve is separated from the joint capsule by the brachialis muscle.

TABLE 2.11	ARTERIES OF FOREARM

Radial artery

Origin:
In cubital fossa, as smaller terminal branch of brachial artery

Course/Distribution:
Runs distally under brachioradialis, lateral to flexor carpi radialis, defining boundary between the flexor and extensor compartments and supplying the radial aspect of both. Gives rise to a superficial palmar branch near the radiocarpal joint; it then transverses the anatomical snuff box to pass between the heads of the first dorsal interosseous muscle joining the deep branch of the ulnar artery to form the deep palmar arch

Ulnar artery

Origin: In cubital fossa, as larger terminal branch of brachial artery

Course/Distribution: Passes distally between second and third layers of forearm flexor muscles, supplying ulnar aspect of flexor compartment; passes superficial to flexor retinaculum at wrist, continuing as the superficial palmar arch (with superficial branch of radial) after its deep palmar branch joins the deep palmar arch

Radial recurrent artery

Origin: In cubital fossa, as first (lateral) branch of radial artery

Course/Distribution: Courses proximally, superficial to supinator, passing between brachioradialis and brachialis to anastomose with radial collateral artery

Anterior and posterior ulnar recurrent arteries

Origin: In and immediately distal to cubital fossa, as first and second medial branches of ulnar artery

Course/Distribution:
Course proximally to anastomose with the inferior and superior ulnar collateral arteries, respectively, forming collateral pathways anterior and posterior to the medial epicondyle of the humerus

Common interosseous artery

Origin: Immediately distal to the cubital fossa, as first lateral branch of ulnar artery

Course/Distribution: Terminates almost immediately, dividing into anterior and posterior interosseous arteries

Anterior and posterior interosseous arteries

Origin: Distal to radial tubercle, as terminal branches of common interosseous

Course/Distribution: Pass to opposite sides of interosseous membrane; anterior artery runs on interosseous membrane; posterior artery runs between superficial and deep layers of extensor muscles as primary artery of compartment

Interosseous recurrent artery

Origin: Initial part of posterior interosseous artery

Course/Distribution:
Courses proximally between lateral epicondyle and olecranon, deep to anconeus, to anastomose with middle collateral artery

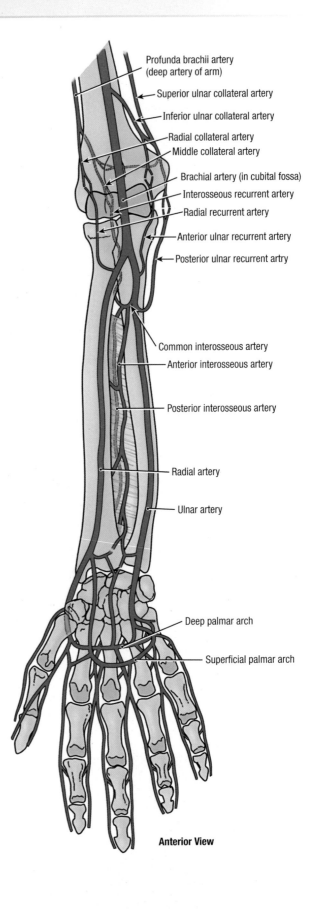

Profunda brachii artery
(deep artery of arm)
Superior ulnar collateral artery
Inferior ulnar collateral artery
Radial collateral artery
Middle collateral artery
Brachial artery (in cubital fossa)
Interosseous recurrent artery
Radial recurrent artery
Anterior ulnar recurrent artery
Posterior ulnar recurrent artry
Common interosseous artery
Anterior interosseous artery
Posterior interosseous artery
Radial artery
Ulnar artery
Deep palmar arch
Superficial palmar arch

Anterior View

2.59   ARTERIES OF FOREARM

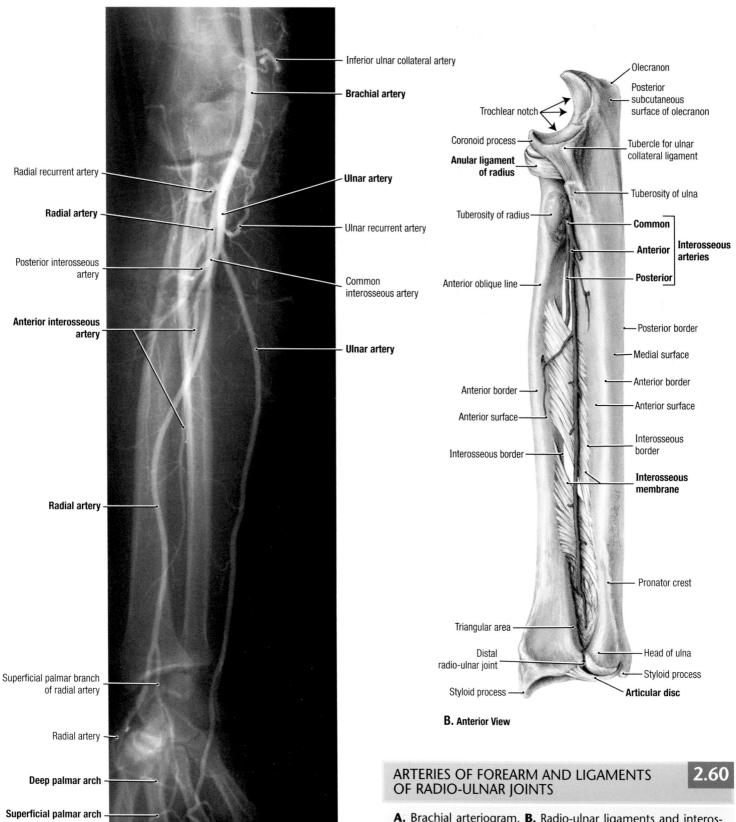

A. Anteroposterior View

Inferior ulnar collateral artery
Brachial artery
Radial recurrent artery
Radial artery
Ulnar artery
Ulnar recurrent artery
Posterior interosseous artery
Common interosseous artery
Anterior interosseous artery
Ulnar artery
Radial artery
Superficial palmar branch of radial artery
Radial artery
Deep palmar arch
Superficial palmar arch

Olecranon
Posterior subcutaneous surface of olecranon
Trochlear notch
Coronoid process
Tubercle for ulnar collateral ligament
Anular ligament of radius
Tuberosity of ulna
Tuberosity of radius
Common
Anterior **Interosseous arteries**
Posterior
Anterior oblique line
Posterior border
Medial surface
Anterior border
Anterior border
Anterior surface
Anterior surface
Interosseous border
Interosseous border
Interosseous membrane
Pronator crest
Triangular area
Head of ulna
Distal radio-ulnar joint
Styloid process
Styloid process
Articular disc

B. Anterior View

ARTERIES OF FOREARM AND LIGAMENTS OF RADIO-ULNAR JOINTS

2.60

A. Brachial arteriogram. **B.** Radio-ulnar ligaments and interosseous arteries. The ligament maintaining the proximal radio-ulnar joint is the anular ligament, that for the distal joint is the articular disc, and that for the middle joint is the interosseous membrane. The interosseous membrane is attached to the interosseous borders of the radius and ulna, but it also spreads onto their surfaces.

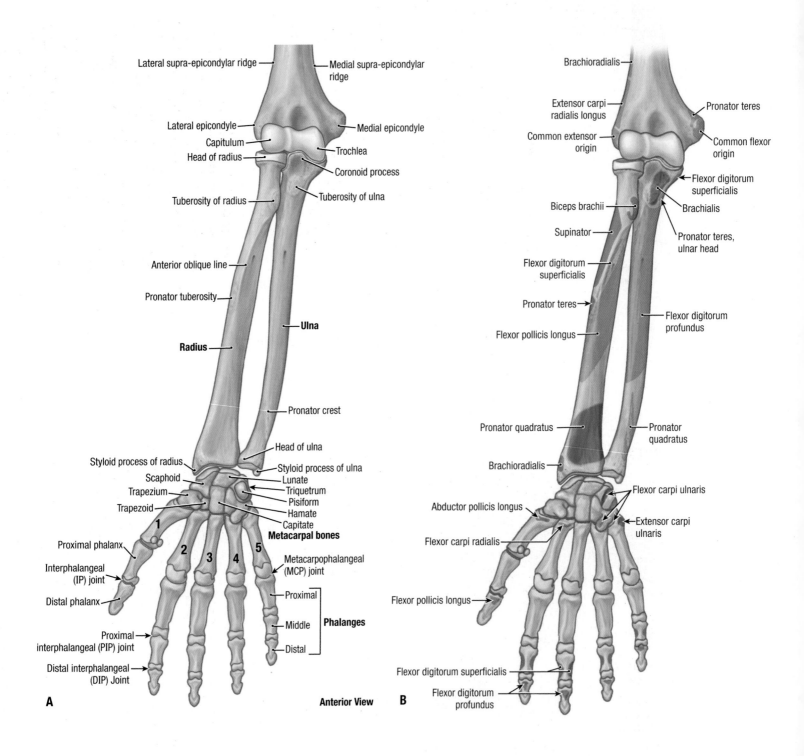

Lateral supra-epicondylar ridge
Medial supra-epicondylar ridge
Lateral epicondyle
Capitulum
Head of radius
Tuberosity of radius
Medial epicondyle
Trochlea
Coronoid process
Tuberosity of ulna
Anterior oblique line
Pronator tuberosity
Ulna
Radius
Pronator crest
Head of ulna
Styloid process of radius
Scaphoid
Trapezium
Trapezoid
Styloid process of ulna
Lunate
Triquetrum
Pisiform
Hamate
Capitate
Metacarpal bones
1 2 3 4 5
Proximal phalanx
Interphalangeal (IP) joint
Distal phalanx
Metacarpophalangeal (MCP) joint
Proximal
Middle
Distal
Phalanges
Proximal interphalangeal (PIP) joint
Distal interphalangeal (DIP) Joint
A
Anterior View

Brachioradialis
Extensor carpi radialis longus
Common extensor origin
Pronator teres
Common flexor origin
Flexor digitorum superficialis
Biceps brachii
Brachialis
Supinator
Pronator teres, ulnar head
Flexor digitorum superficialis
Pronator teres
Flexor pollicis longus
Flexor digitorum profundus
Pronator quadratus
Pronator quadratus
Brachioradialis
Abductor pollicis longus
Flexor carpi ulnaris
Extensor carpi ulnaris
Flexor carpi radialis
Flexor pollicis longus
Flexor digitorum superficialis
Flexor digitorum profundus
B

2.61 BONES OF FOREARM AND HAND AND ATTACHMENTS OF FOREARM MUSCLES

A. Bony features. **B.** Sites of muscle attachments.

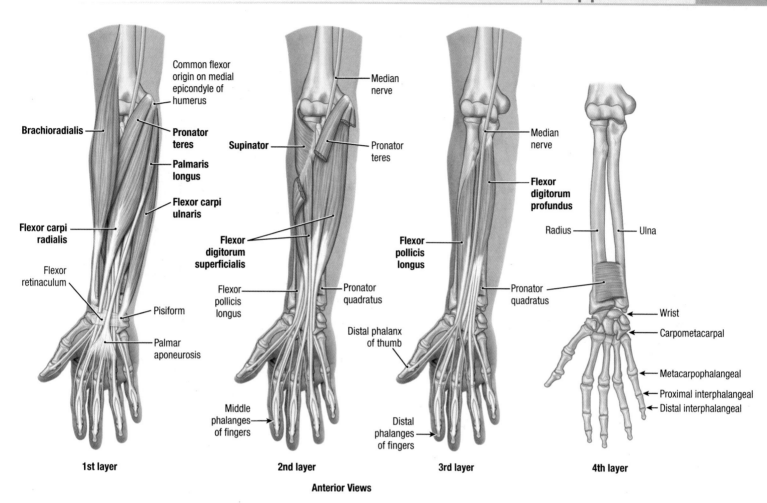

1st layer

Common flexor origin on medial epicondyle of humerus

Brachioradialis

Pronator teres

Palmaris longus

Flexor carpi ulnaris

Flexor carpi radialis

Flexor retinaculum

Pisiform

Palmar aponeurosis

2nd layer

Median nerve

Supinator

Pronator teres

Flexor digitorum superficialis

Flexor pollicis longus

Pronator quadratus

Middle phalanges of fingers

3rd layer

Median nerve

Flexor digitorum profundus

Flexor pollicis longus

Pronator quadratus

Distal phalanx of thumb

Distal phalanges of fingers

4th layer

Radius

Ulna

Wrist

Carpometacarpal

Metacarpophalangeal

Proximal interphalangeal

Distal interphalangeal

Anterior Views

MUSCLES OF ANTERIOR FOREARM

2.62

The muscles of the anterior aspect of the forearm are arranged in three layers.

TABLE 2.12	MUSCLES OF ANTERIOR FOREARM			
Muscle	*Proximal Attachment*	*Distal Attachment*	*Innervation*	*Main Actions*
Pronator teres	Medial epicondyle of humerus and coronoid process of ulna	Middle of lateral surface of radius (pronator tuberosity)	Median nerve (C6–**C7**)	Pronates forearm and flexes elbow joint
Flexor carpi radialis	Medial epicondyle of humerus	Base of 2nd and 3rd metacarpals		Flexes and abducts wrist joint
Palmaris longus		Distal half of flexor retinaculum and palmar aponeurosis	Median nerve (C7–**C8**)	Flexes wrist joint and tightens palmar aponeurosis
Flexor carpi ulnaris	*Humeral head:* medial epicondyle of humerus *Ulnar head:* olecranon and posterior border of ulna	Pisiform, hook of hamate, and 5th metacarpal	Ulnar nerve (C7–**C8**)	Flexes and adducts wrist joint
Flexor digitorum superficialis	*Humero-ulnar head:* medial epicondyle of humerus, ulnar collateral ligament, and coronoid process of ulna *Radial head:* superior half of anterior border of radius	Bodies of middle phalanges of medial four digits	Median nerve (C7, **C8**, and T1)	Flexes PIPs of medial four digits; acting more strongly, it flexes MCPs and wrist joint
Flexor digitorum profundus	Proximal three quarters of medial and anterior surfaces of ulna and interosseous membrane	Bases of distal phalanges of medial four digits	*Medial part:* ulnar nerve (**C8**–T1) *Lateral part:* median nerve (**C8**–T1)	Flexes DIPs of medial four digits; assists with flexion of wrist joint
Flexor pollicis longus	Anterior surface of radius and adjacent interosseous membrane	Base of distal phalanx of thumb	Anterior interosseous nerve from median (**C8**–T1)	Flexes IP joints of 1st digit (thumb) and assists flexion of wrist joint
Pronator quadratus	Distal fourth of anterior surface of ulna	Distal fourth of anterior surface of radius		Pronates forearm; deep fibers bind radius and ulna together

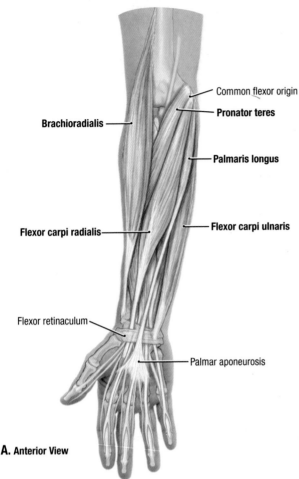

Common flexor origin

Pronator teres

Brachioradialis

Palmaris longus

Flexor carpi radialis

Flexor carpi ulnaris

Flexor retinaculum

Palmar aponeurosis

A. Anterior View

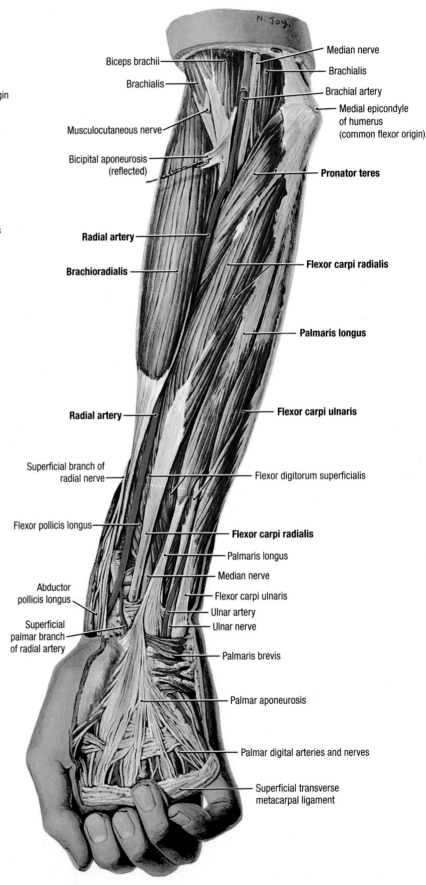

N. Joy

Biceps brachii

Brachialis

Musculocutaneous nerve

Bicipital aponeurosis (reflected)

Radial artery

Brachioradialis

Radial artery

Superficial branch of radial nerve

Flexor pollicis longus

Abductor pollicis longus

Superficial palmar branch of radial artery

Median nerve

Brachialis

Brachial artery

Medial epicondyle of humerus (common flexor origin)

Pronator teres

Flexor carpi radialis

Palmaris longus

Flexor carpi ulnaris

Flexor digitorum superficialis

Flexor carpi radialis

Palmaris longus

Median nerve

Flexor carpi ulnaris

Ulnar artery

Ulnar nerve

Palmaris brevis

Palmar aponeurosis

Palmar digital arteries and nerves

Superficial transverse metacarpal ligament

B. Anterior View

2.63 SUPERFICIAL MUSCLES OF FOREARM AND PALMAR APONEUROSIS

- At the elbow, the brachial artery lies between the biceps tendon and median nerve. It then bifurcates into the radial and ulnar arteries.
- At the wrist, the radial artery is lateral to the flexor carpi radialis tendon, and the ulnar artery is lateral to flexor carpi ulnaris tendon.
- In the forearm, the radial artery lies between the flexor and extensor compartments. The muscles lateral to the artery are supplied by the radial nerve, and those medial to it by the median and ulnar nerves; thus, no motor nerve crosses the radial artery.
- The brachioradialis muscle slightly overlaps the radial artery, which is otherwise superficial.
- The four superficial muscles all attach proximally to the medial epicondyle of the humerus (common flexor origin).
- The palmaris longus muscle, in this specimen, has an anomalous distal belly; this muscle usually has a small belly at the common flexor origin and a long tendon that is continued into the palm as the palmar aponeurosis. The palmaris longus is absent unilaterally or bilaterally in approximately 14% of limbs.

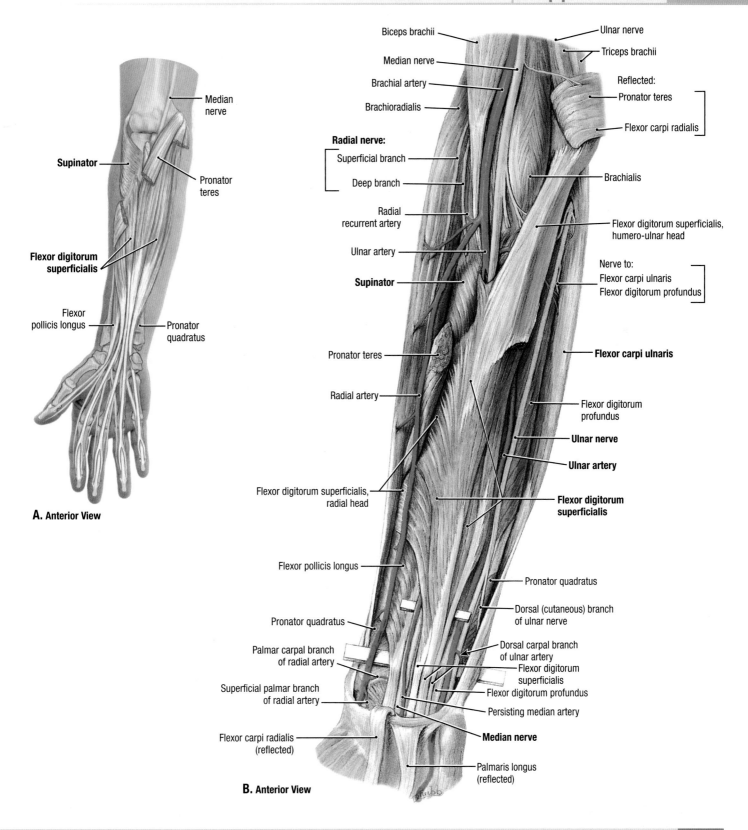

A. Anterior View

Median nerve

Supinator

Pronator teres

Flexor digitorum superficialis

Flexor pollicis longus

Pronator quadratus

B. Anterior View

Biceps brachii

Median nerve

Brachial artery

Brachioradialis

Radial nerve:

Superficial branch

Deep branch

Radial recurrent artery

Ulnar artery

Supinator

Pronator teres

Radial artery

Flexor digitorum superficialis, radial head

Flexor pollicis longus

Pronator quadratus

Palmar carpal branch of radial artery

Superficial palmar branch of radial artery

Flexor carpi radialis (reflected)

Ulnar nerve

Triceps brachii

Reflected:

Pronator teres

Flexor carpi radialis

Brachialis

Flexor digitorum superficialis, humero-ulnar head

Nerve to:
Flexor carpi ulnaris
Flexor digitorum profundus

Flexor carpi ulnaris

Flexor digitorum profundus

Ulnar nerve

Ulnar artery

Flexor digitorum superficialis

Pronator quadratus

Dorsal (cutaneous) branch of ulnar nerve

Dorsal carpal branch of ulnar artery

Flexor digitorum superficialis

Flexor digitorum profundus

Persisting median artery

Median nerve

Palmaris longus (reflected)

FLEXOR DIGITORUM SUPERFICIALIS AND RELATED STRUCTURES

2.64

- The flexor digitorum superficialis muscle is attached proximally to the humerus, ulna, and radius.
- The ulnar artery passes obliquely posterior to the flexor digitorum superficialis; at the medial border of the muscle, the ulnar artery joins the ulnar nerve.
- The median nerve descends vertically posterior to the flexor digitorum superficialis and appears distally at its lateral border.
- The median artery of this specimen is a variation resulting from persistence of an embryologic vessel that usually disappears.

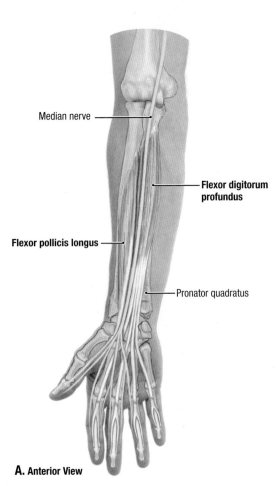

Median nerve

Flexor digitorum profundus

Flexor pollicis longus

Pronator quadratus

A. Anterior View

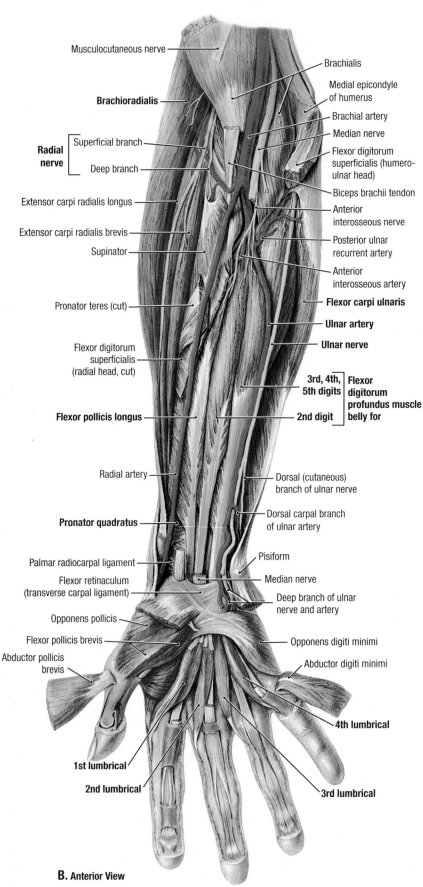

Musculocutaneous nerve

Brachioradialis

Radial nerve
- Superficial branch
- Deep branch

Extensor carpi radialis longus

Extensor carpi radialis brevis

Supinator

Pronator teres (cut)

Flexor digitorum superficialis (radial head, cut)

Flexor pollicis longus

Radial artery

Pronator quadratus

Palmar radiocarpal ligament

Flexor retinaculum (transverse carpal ligament)

Opponens pollicis

Flexor pollicis brevis

Abductor pollicis brevis

1st lumbrical

2nd lumbrical

Brachialis

Medial epicondyle of humerus

Brachial artery

Median nerve

Flexor digitorum superficialis (humero-ulnar head)

Biceps brachii tendon

Anterior interosseous nerve

Posterior ulnar recurrent artery

Anterior interosseous artery

Flexor carpi ulnaris

Ulnar artery

Ulnar nerve

3rd, 4th, 5th digits | **Flexor digitorum profundus muscle belly for**

2nd digit

Dorsal (cutaneous) branch of ulnar nerve

Dorsal carpal branch of ulnar artery

Pisiform

Median nerve

Deep branch of ulnar nerve and artery

Opponens digiti minimi

Abductor digiti minimi

4th lumbrical

3rd lumbrical

B. Anterior View

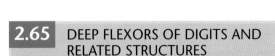

2.65 DEEP FLEXORS OF DIGITS AND RELATED STRUCTURES

- The ulnar nerve enters the forearm posterior to the medial epicondyle, then descends between the flexor digitorum profundus and flexor carpi ulnaris, and is joined by the ulnar artery. At the wrist, the ulnar nerve and artery pass anterior to the flexor retinaculum and lateral to the pisiform to enter the palm.
- At the elbow, the ulnar nerve supplies the flexor carpi ulnaris and the medial half of the flexor digitorum profundus muscles; proximal to the wrist, it gives off the dorsal (cutaneous) branch.
- The four lumbricals arise from the flexor digitorum profundus tendons.

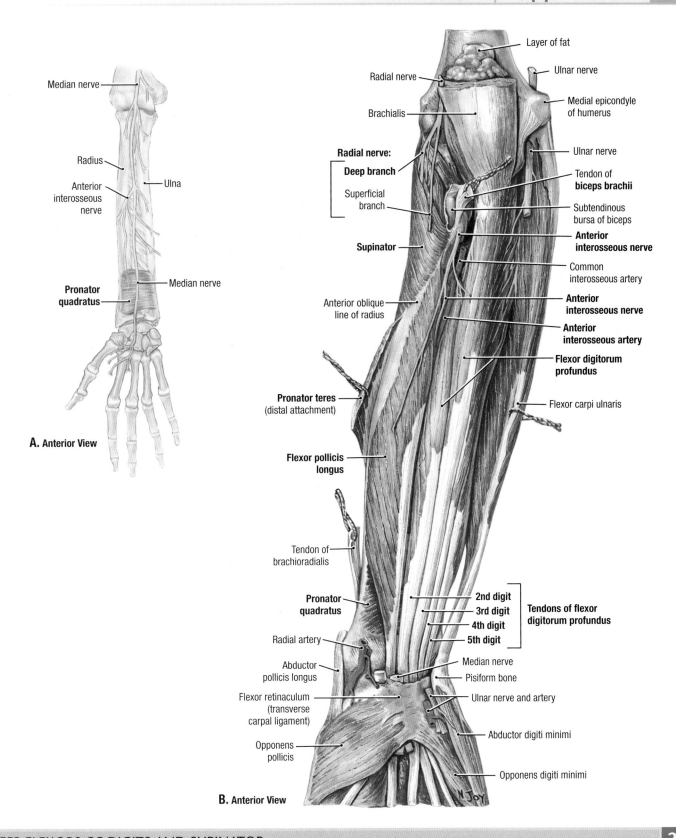

Median nerve

Radius

Anterior interosseous nerve

Ulna

Pronator quadratus

Median nerve

A. Anterior View

Layer of fat

Radial nerve

Ulnar nerve

Brachialis

Medial epicondyle of humerus

Radial nerve:

Deep branch

Ulnar nerve

Superficial branch

Tendon of **biceps brachii**

Subtendinous bursa of biceps

Anterior interosseous nerve

Supinator

Common interosseous artery

Anterior interosseous nerve

Anterior oblique line of radius

Anterior interosseous artery

Flexor digitorum profundus

Pronator teres (distal attachment)

Flexor carpi ulnaris

Flexor pollicis longus

Tendon of brachioradialis

Pronator quadratus

2nd digit

3rd digit

Tendons of flexor digitorum profundus

4th digit

5th digit

Radial artery

Abductor pollicis longus

Median nerve

Pisiform bone

Flexor retinaculum (transverse carpal ligament)

Ulnar nerve and artery

Abductor digiti minimi

Opponens pollicis

Opponens digiti minimi

B. Anterior View

DEEP FLEXORS OF DIGITS AND SUPINATOR | 2.66

- The anterior interosseous nerve and artery pass deeply between the flexor pollicis longus and flexor digitorum profundus muscles to lie on the interosseous membrane.
- The deep branch of the radial nerve pierces and innervates the supinator muscle.

Severance of the deep branch of the radial nerve results in an inability to extend the thumb and MCP joints of the other digits. Loss of sensation does not occur because the deep branch is entirely muscular and articular in distribution.

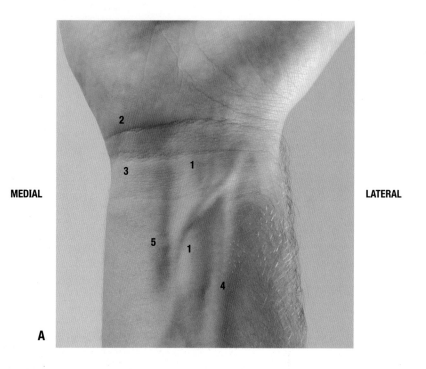

MEDIAL   LATERAL

A

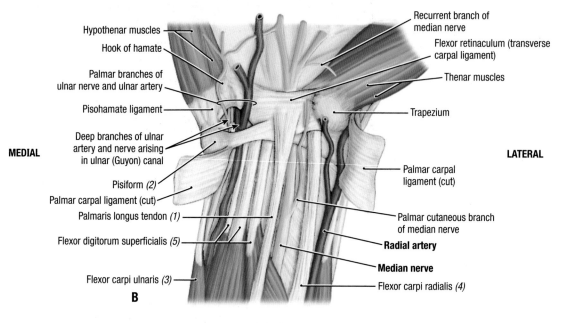

MEDIAL   LATERAL

Hypothenar muscles

Hook of hamate

Palmar branches of
ulnar nerve and ulnar artery

Pisohamate ligament

Deep branches of ulnar
artery and nerve arising
in ulnar (Guyon) canal

Pisiform (2)

Palmar carpal ligament (cut)

Palmaris longus tendon (1)

Flexor digitorum superficialis (5)

Flexor carpi ulnaris (3)

B

Recurrent branch of
median nerve

Flexor retinaculum (transverse
carpal ligament)

Thenar muscles

Trapezium

Palmar carpal
ligament (cut)

Palmar cutaneous branch
of median nerve

Radial artery

Median nerve

Flexor carpi radialis (4)

Anterior Views of Right Hand and Wrist

2.67 STRUCTURES OF ANTERIOR WRIST

A. Surface anatomy. **B.** Schematic illustration. **C.** Dissection.

- The distal skin incision follows the transverse skin crease at the wrist. The incision crosses the pisiform, to which the flexor carpi ulnaris muscle attaches, and the tubercle of the scaphoid, to which the tendon of flexor carpi radialis muscle is a guide.
- The palmaris longus tendon bisects the transverse skin crease; deep to the lateral margin of the tendon is the median nerve.
- Note the ulnar (Guyon) canal through which the ulnar vessels and nerve pass medial to the pisiform.

- The radial artery passes deep to the tendon of the abductor pollicis longus muscle.
- The flexor digitorum superficialis tendons to the 3rd and 4th digits become anterior to those of the 2nd and 5th digits.
- The recurrent branch of the median nerve to the thenar muscles lies within a circle whose center is 2.5 to 4 cm distal to the tubercle of the scaphoid.

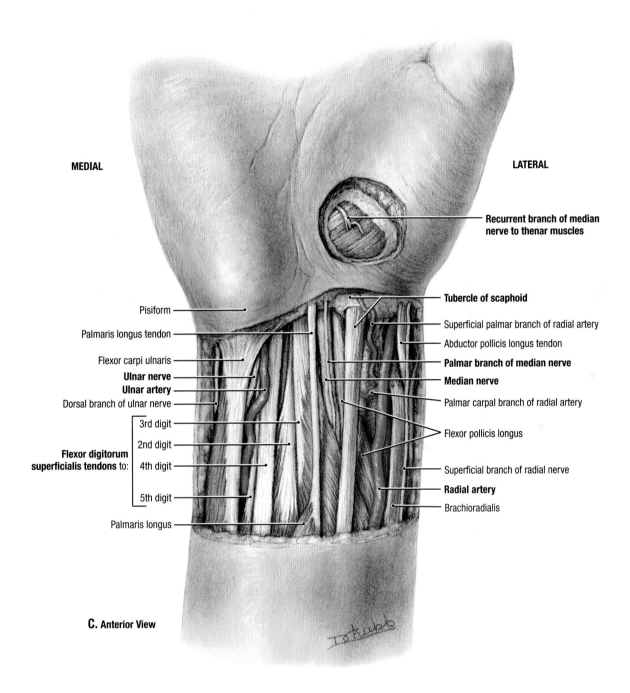

MEDIAL

LATERAL

Recurrent branch of median nerve to thenar muscles

Pisiform

Palmaris longus tendon

Flexor carpi ulnaris

Ulnar nerve

Ulnar artery

Dorsal branch of ulnar nerve

Tubercle of scaphoid

Superficial palmar branch of radial artery

Abductor pollicis longus tendon

Palmar branch of median nerve

Median nerve

Palmar carpal branch of radial artery

Flexor pollicis longus

Superficial branch of radial nerve

Radial artery

Brachioradialis

Flexor digitorum superficialis tendons to:
- 3rd digit
- 2nd digit
- 4th digit
- 5th digit

Palmaris longus

C. Anterior View

STRUCTURES OF ANTERIOR WRIST (*continued*)

2.67

Lesions of the median nerve usually occur in two places: the forearm and wrist. The most common site is where the nerve passes though the carpal tunnel. Lacerations of the wrist often cause median nerve injury because this nerve is relatively close to the surface. This results in paralysis of the thenar muscles and the first two lumbricals. Hence, opposition of the thumb is not possible and fine control movements of the 2nd and 3rd digits are impaired. Sensation is also lost over the thumb and adjacent two and a half digits.

Median nerve injury resulting from a perforating wound in the elbow region results in loss of flexion of the proximal and distal interphalangeal joints of the 2nd and 3rd digits. The ability to flex the metacarpophalangeal joints of these digits is also affected because digital branches of the median nerve supply the 1st and 2nd lumbricals. The palmar cutaneous branch of the median nerve does not traverse the carpal tunnel. It supplies the skin of the central palm, which remains sensitive in carpal tunnel syndrome.

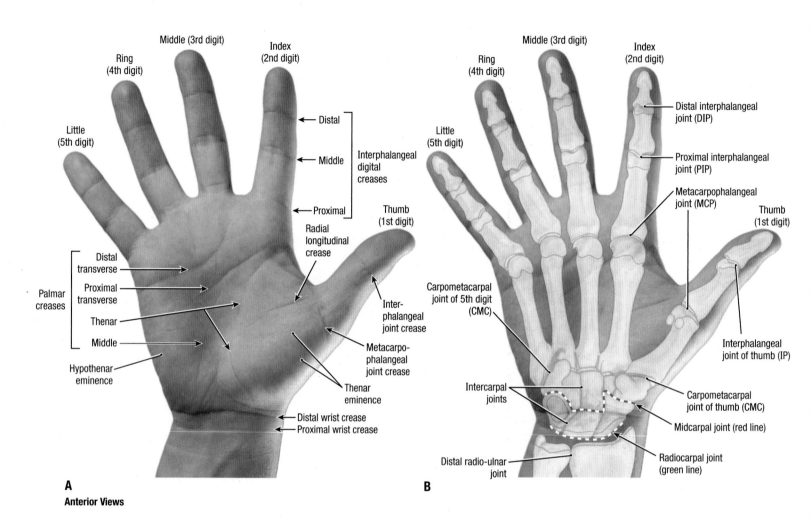

A Anterior Views

2.68   SURFACE ANATOMY OF HAND AND WRIST

A. Skin creases of wrist and hand. **B.** Surface projection of joints of wrist and hand. Note relationship of bones and joints to features of the hand.

The palmar skin presents several more or less constant *flexion creases* where the skin is firmly bound to the deep fascia:
- *Wrist creases*: **proximal, middle, distal**. The distal wrist crease indicates the proximal border of the flexor retinaculum.
- *Palmar creases*: **radial longitudinal crease** (the "life line" of palmistry), proximal and distal transverse palmar creases

- *Transverse digital flexion creases*: The **proximal digital crease** is located at the root of the digit, approximately 2 cm distal to the metacarpophalangeal joint. The proximal digital crease of the thumb crosses obliquely, proximal to the 1st metacarpophalangeal joint. The **middle digital crease** lies over the proximal interphalangeal joint, and the **distal digital crease** lies proximal to the distal interphalangeal joint. The thumb, having two phalanges, has only two flexion creases.

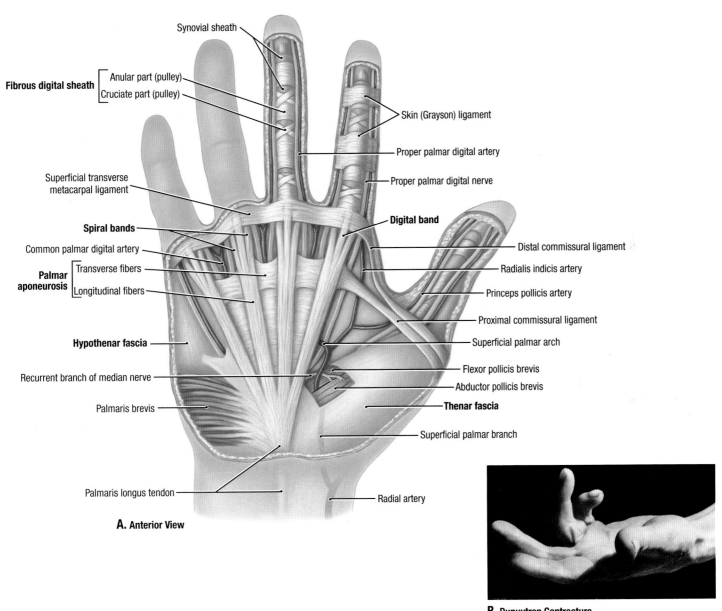

Fibrous digital sheath
- Anular part (pulley)
- Cruciate part (pulley)

Synovial sheath

Skin (Grayson) ligament

Proper palmar digital artery

Proper palmar digital nerve

Superficial transverse metacarpal ligament

Digital band

Spiral bands

Distal commissural ligament

Common palmar digital artery

Radialis indicis artery

Palmar aponeurosis
- Transverse fibers
- Longitudinal fibers

Princeps pollicis artery

Proximal commissural ligament

Hypothenar fascia

Superficial palmar arch

Recurrent branch of median nerve

Flexor pollicis brevis

Abductor pollicis brevis

Palmaris brevis

Thenar fascia

Superficial palmar branch

Palmaris longus tendon

Radial artery

A. Anterior View

B. Dupuytren Contracture

PALMAR (DEEP) FASCIA: PALMAR APONEUROSIS, THENAR AND HYPOTHENAR FASCIA

2.69

A. Anterior view. The palmar fascia is thin over the thenar and hypothenar eminences but thick centrally, where it forms the palmar aponeurosis, and in the digits, where it forms the fibrous digital sheaths. At the distal end (base) of the palmar aponeurosis, four bundles of digital and spiral bands continue to the bases and fibrous digital sheaths of digits 2 to 5.

B. Dupuytren contracture is a disease of the palmar fascia resulting in progressive shortening, thickening, and fibrosis of the palmar fascia and palmar aponeurosis. The fibrous degeneration of the longitudinal digital bands of the aponeurosis on the medial side of the hand pulls the 4th and 5th fingers into partial flexion at the metacarpophalangeal and proximal interphalangeal joints. The contracture is frequently bilateral. Treatment of Dupuytren contracture usually involves surgical excision of all fibrotic parts of the palmar fascia to free the fingers.

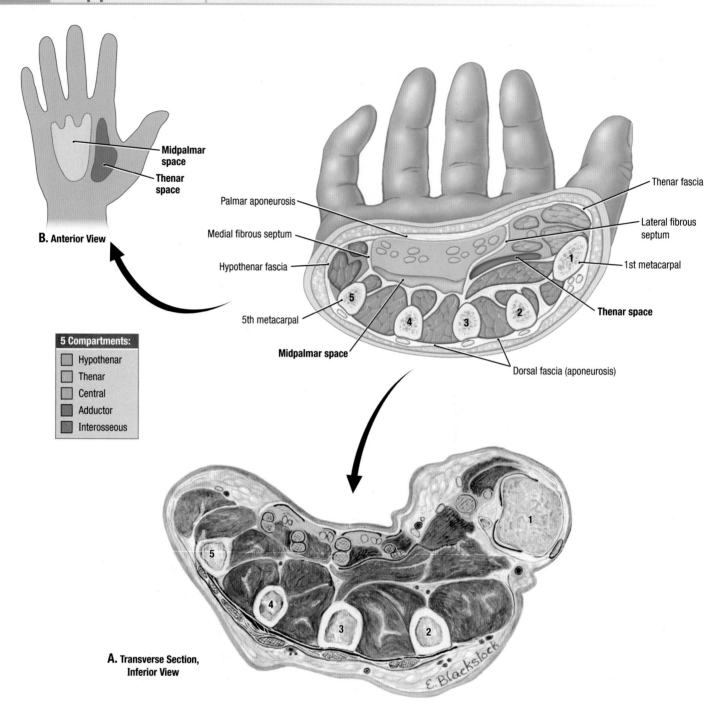

B. Anterior View

- Midpalmar space
- Thenar space

Palmar aponeurosis
Medial fibrous septum
Hypothenar fascia
5th metacarpal
Midpalmar space

Thenar fascia
Lateral fibrous septum
1st metacarpal
Thenar space
Dorsal fascia (aponeurosis)

5 Compartments:
- Hypothenar
- Thenar
- Central
- Adductor
- Interosseous

E. Blackstock

A. Transverse Section, Inferior View

2.70 SYNOVIAL CAPSULE OF ELBOW JOINT AND ANULAR LIGAMENT

A. Transverse section through the middle of the palm showing the fascial compartments for the musculotendinous structures of the hand. **B.** Potential fascial spaces of palm.

- The potential midpalmar space lies posterior to the central compartment, is bounded medially by the hypothenar compartment, and is related distally to the synovial sheath of the 3rd, 4th, and 5th digits.
- The potential thenar space lies posterior to the thenar compartment and is related distally to the synovial sheath of the index finger.
- The potential midpalmar and thenar spaces are separated by a septum that passes from the palmar aponeurosis to the 3rd metacarpal.

Because the palmar fascia is thick and strong, **swellings resulting from hand infections** usually appear on the dorsum of the hand where the fascia is thinner. The potential fascial spaces of the palm are important because they may become infected. The fascial spaces determine the extent and direction of the spread of pus formed in the infected areas. Depending on the site of infection, pus will accumulate in the thenar, hypothenar, or adductor compartments. Antibiotic therapy has made infections that spread beyond one of these fascial compartments rare, but an untreated infection can spread proximally through the carpal tunnel into the forearm anterior to the pronator quadratus and its fascia.

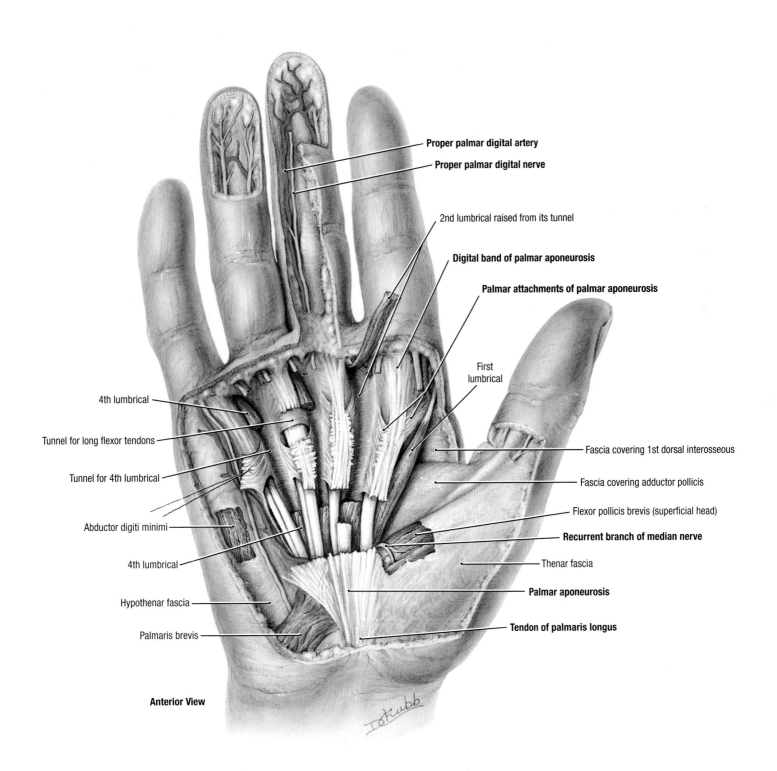

Proper palmar digital artery

Proper palmar digital nerve

2nd lumbrical raised from its tunnel

Digital band of palmar aponeurosis

Palmar attachments of palmar aponeurosis

First lumbrical

4th lumbrical

Tunnel for long flexor tendons

Tunnel for 4th lumbrical

Abductor digiti minimi

4th lumbrical

Hypothenar fascia

Palmaris brevis

Fascia covering 1st dorsal interosseous

Fascia covering adductor pollicis

Flexor pollicis brevis (superficial head)

Recurrent branch of median nerve

Thenar fascia

Palmar aponeurosis

Tendon of palmaris longus

Anterior View

PALMAR APONEUROSIS

2.71

- From the palmar aponeurosis, four longitudinal digital bands enter the fingers; the other fibers form extensive fibro-areolar septa that pass posteriorly to the palmar ligaments (see Fig. 2.78) and, more proximally, to the fascia covering the interossei. Thus, two sets of tunnels exist in the distal half of the palm: (1) tunnels for long flexor tendons and (2) tunnels for lumbricals, digital vessels, and digital nerves.
- In the dissected middle finger, note the absence of fat deep to the skin creases of the fingers.

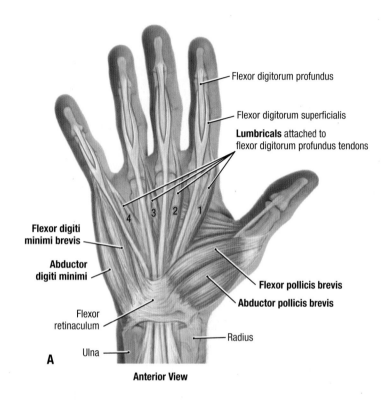

Flexor digitorum profundus

Flexor digitorum superficialis

Lumbricals attached to flexor digitorum profundus tendons

4 3 2 1

Flexor digiti minimi brevis

Abductor digiti minimi

Flexor retinaculum

Ulna

Radius

Flexor pollicis brevis

Abductor pollicis brevis

A **Anterior View**

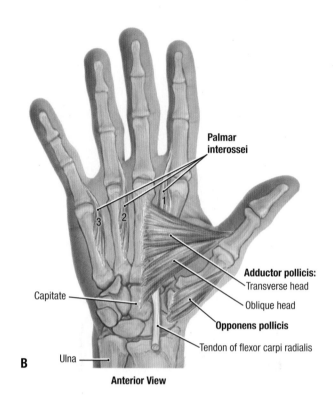

Palmar interossei

3 2 1

Capitate

Ulna

Adductor pollicis:
Transverse head

Oblique head

Opponens pollicis

Tendon of flexor carpi radialis

B **Anterior View**

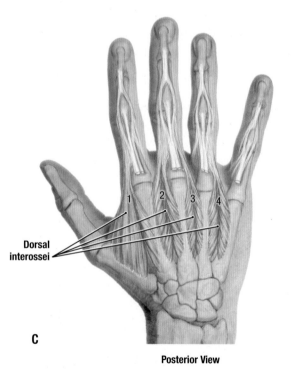

1 2 3 4

Dorsal interossei

C **Posterior View**

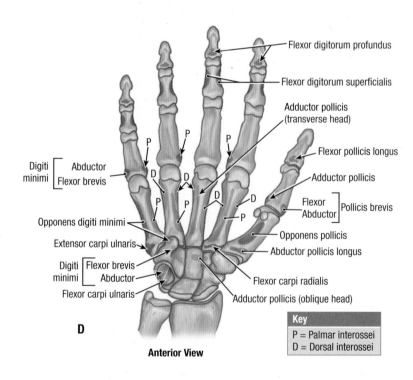

Flexor digitorum profundus

Flexor digitorum superficialis

Adductor pollicis (transverse head)

Flexor pollicis longus

Adductor pollicis

Flexor / Abductor } Pollicis brevis

Opponens pollicis

Abductor pollicis longus

Flexor carpi radialis

Adductor pollicis (oblique head)

Digiti minimi { Abductor / Flexor brevis

Opponens digiti minimi

Extensor carpi ulnaris

Digiti minimi { Flexor brevis / Abductor

Flexor carpi ulnaris

P P P

D D

D D

P P P

Key
P = Palmar interossei
D = Dorsal interossei

D **Anterior View**

2.72 MUSCULAR LAYERS OF PALM

A. Lumbricals. **B.** Adductor pollicis. **C.** Dorsal (*D*) and palmar (*P*) interossei. **D.** Bony attachments.

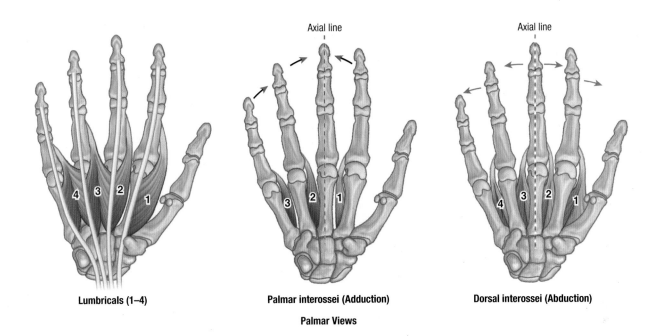

Lumbricals (1–4)

Palmar interossei (Adduction)

Dorsal interossei (Abduction)

Palmar Views

LUMBRICALS AND INTEROSSEI

2.73

The lumbricals and interossei are intrinsic muscles of the hand. The actions of the palmar (adduction) and dorsal (abduction) interossei are shown with *arrows*.

TABLE 2.13	MUSCLES OF HAND			
Muscle	*Proximal Attachment*	*Distal Attachment*	*Innervation*	*Main Actions*
Abductor pollicis brevis	Flexor retinaculum and tubercles of scaphoid and trapezium	Lateral side of base of proximal phalanx of thumb	Recurrent branch of median nerve (**C8** and T1)	Abducts thumb and helps oppose it
Flexor pollicis brevis	Flexor retinaculum (transverse carpal ligament) and tubercle of trapezium			Flexes thumb
Opponens pollicis		Lateral side of 1st metacarpal		Opposes thumb toward center of palm and rotates it medially
Adductor pollicis	*Oblique head:* bases of 2nd and 3rd metacarpals, capitate, and adjacent carpal bones *Transverse head:* anterior surface of shaft of 3rd metacarpal	Medial side of base of proximal phalanx of thumb	Deep branch of ulnar nerve (C8 and **T1**)	Adducts thumb toward lateral border of palm
Abductor digiti minimi	Pisiform	Medial side of base of proximal phalanx of digit 5		Abducts digit 5, assists in flexion of its PIP joint
Flexor digiti minimi brevis	Hook of hamate and flexor retinaculum (transverse carpal ligament)	Medial border of 5th metacarpal		Flexes PIP joint of digit 5
Opponens digiti minimi				Draws 5th metacarpal anteriorly and rotates it, bringing digit 5 into opposition with thumb
Lumbricals 1 and 2	Lateral two tendons of flexor digitorum profundus	Lateral sides of extensor expansions of digits 2–5	Median nerve (C8 and **T1**)	Flex MCP joints and extend IP joints of digits 2–5
Lumbricals 3 and 4	Medial three tendons of flexor digitorum profundus		Deep branch of ulnar nerve (C8 and **T1**)	
Dorsal interossei 1–4	Adjacent sides of two metacarpals	Extensor expansions and bases of proximal phalanges of digits 2–4		Abduct 2–4 MCP joints; act with lumbricals to flex MCP and extend IP joints
Palmar interossei 1–3	Palmar surfaces of 2nd, 4th, and 5th metacarpals	Extensor expansions of digits and bases of proximal phalanges of digits 2, 4, and 5		Adduct 2, 4, and 5 MCP joints; act with lumbricals to flex MCP and extend IP joints

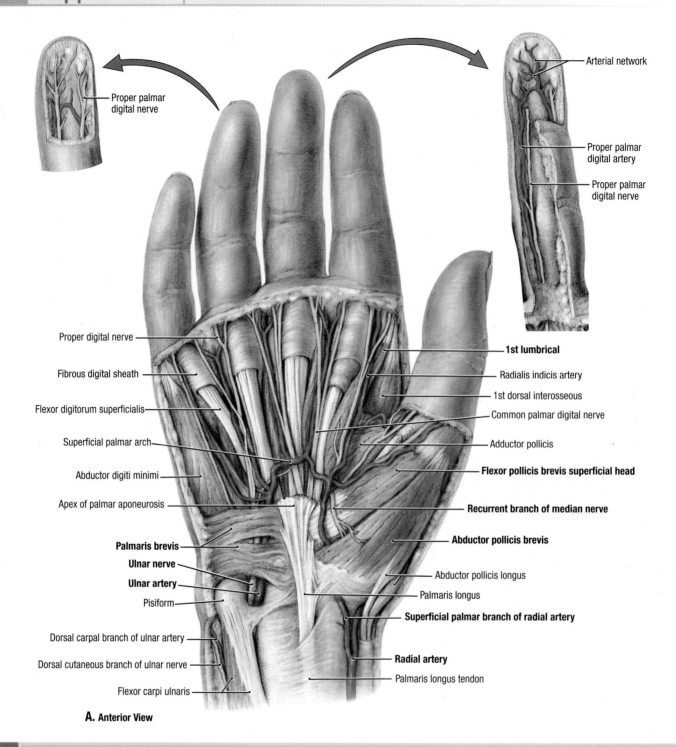

Proper palmar digital nerve

Arterial network

Proper palmar digital artery

Proper palmar digital nerve

Proper digital nerve

Fibrous digital sheath

Flexor digitorum superficialis

Superficial palmar arch

Abductor digiti minimi

Apex of palmar aponeurosis

Palmaris brevis

Ulnar nerve

Ulnar artery

Pisiform

Dorsal carpal branch of ulnar artery

Dorsal cutaneous branch of ulnar nerve

Flexor carpi ulnaris

1st lumbrical

Radialis indicis artery

1st dorsal interosseous

Common palmar digital nerve

Adductor pollicis

Flexor pollicis brevis superficial head

Recurrent branch of median nerve

Abductor pollicis brevis

Abductor pollicis longus

Palmaris longus

Superficial palmar branch of radial artery

Radial artery

Palmaris longus tendon

A. Anterior View

2.74 SUPERFICIAL DISSECTION OF PALM, ULNAR, AND MEDIAN NERVES

A. Superficial palmar arch and digital nerves and vessels.
- The skin, superficial fascia, palmar aponeurosis, and thenar and hypothenar fasciae have been removed.
- The superficial palmar arch is formed by the ulnar artery and completed by the superficial palmar branch of the radial artery.
- The four lumbricals lie posterior to the digital vessels and nerves. The lumbricals arise from the lateral sides of the flexor digitorum profundus tendons and are inserted into the lateral sides of the dorsal expansions of the corresponding digits. The medial two lumbricals are bipennate and also arise from the medial sides of adjacent flexor digitorum profundus tendons.

- In the digits, a proper palmar digital artery and nerve lie on each side of the fibrous digital sheath.
- Note the canal (Guyon) through which the ulnar vessels and nerve pass medial to the pisiform.

Laceration of palmar (arterial) arches. Bleeding is usually profuse when the palmar (arterial) arches are lacerated. It may not be sufficient to ligate (tie off) only one forearm artery when the arches are lacerated because these vessels usually have numerous communications in the forearm and hand and thus bleed from both ends.

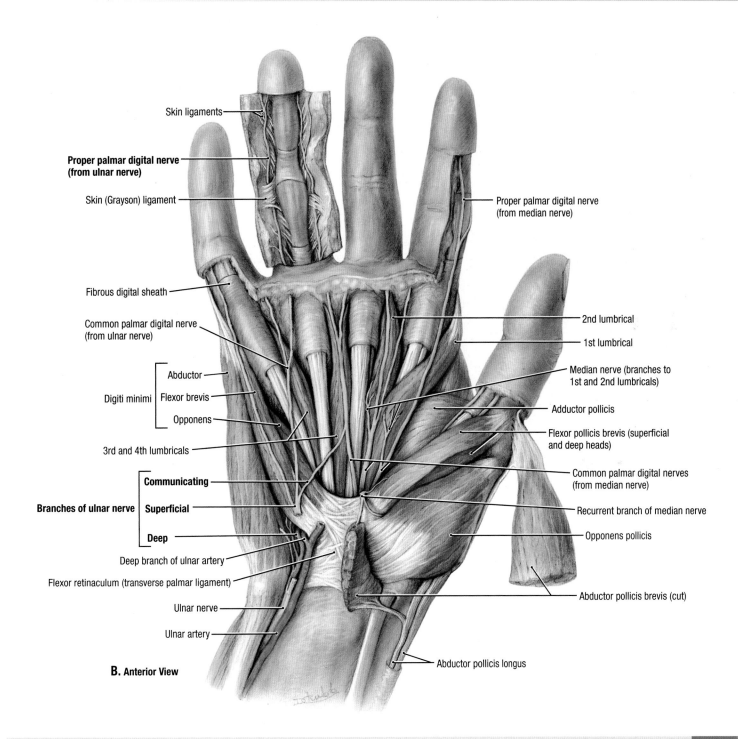

Skin ligaments

**Proper palmar digital nerve
(from ulnar nerve)**

Skin (Grayson) ligament

Fibrous digital sheath

Common palmar digital nerve
(from ulnar nerve)

Abductor
Digiti minimi — Flexor brevis
Opponens

3rd and 4th lumbricals

Communicating

Branches of ulnar nerve — **Superficial**

Deep

Deep branch of ulnar artery

Flexor retinaculum (transverse palmar ligament)

Ulnar nerve

Ulnar artery

Proper palmar digital nerve
(from median nerve)

2nd lumbrical

1st lumbrical

Median nerve (branches to
1st and 2nd lumbricals)

Adductor pollicis

Flexor pollicis brevis (superficial
and deep heads)

Common palmar digital nerves
(from median nerve)

Recurrent branch of median nerve

Opponens pollicis

Abductor pollicis brevis (cut)

Abductor pollicis longus

B. Anterior View

SUPERFICIAL DISSECTION OF PALM, ULNAR, AND MEDIAN NERVES (*continued*)   2.74

B. Ulnar and median nerves.

Carpal tunnel syndrome results from any lesion that significantly reduces the size of the carpal tunnel or, more commonly, increases the size of some of the structures (or their coverings) that pass through it (e.g., inflammation of the synovial sheaths). The median nerve is the most vulnerable structure in the carpal tunnel. The median nerve has two terminal sensory branches that supply the skin of the hand; hence, paresthesia (tingling), hypoesthesia (diminished sensation), or anesthesia (absence of tactile sensation) may occur in the lateral three and a half digits. However, recall that the palmar cutaneous branch of the median nerve arises proximal to and does not pass through the carpal tunnel; thus, sensation in the central palm remains unaffected. This nerve also has one terminal motor branch, the recurrent branch, which innervates the three thenar muscles. Wasting of the thenar eminence and progressive loss of coordination and strength in the thumb may occur. To relieve the compression, partial or complete surgical division of the flexor retinaculum, a procedure called **carpal tunnel release**, may be necessary. The incision is made toward the medial side of the wrist and flexor retinaculum to avoid possible injury to the recurrent branch of the median nerve. This procedure is also done laparoscopically.

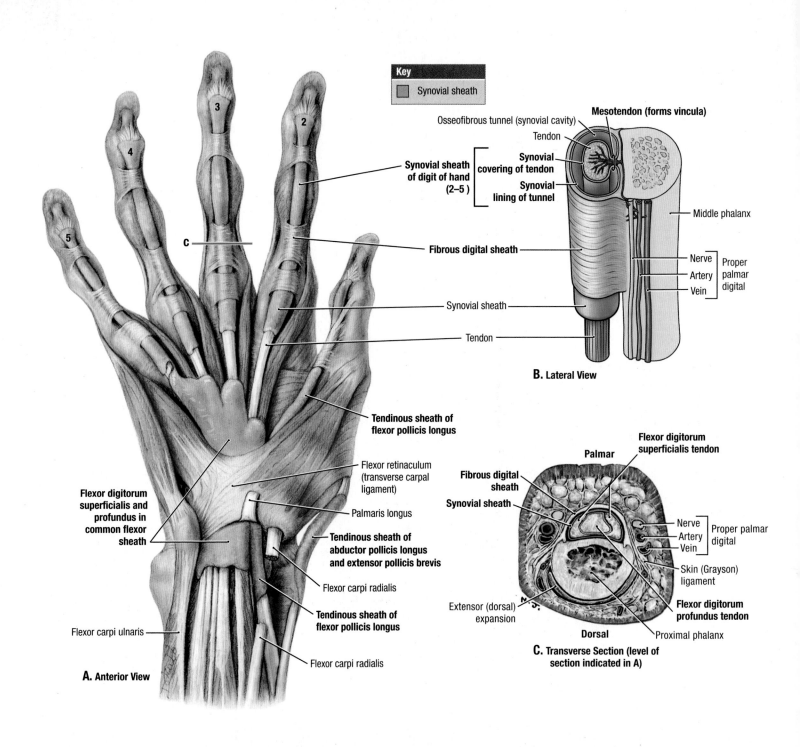

Key
Synovial sheath

Osseofibrous tunnel (synovial cavity)
Mesotendon (forms vincula)
Tendon

Synovial sheath of digit of hand (2–5)
Synovial covering of tendon
Synovial lining of tunnel

Middle phalanx

Fibrous digital sheath

Nerve
Artery
Vein
Proper palmar digital

Synovial sheath

Tendon

B. Lateral View

3
2
4
5
c

Tendinous sheath of flexor pollicis longus

Flexor digitorum superficialis and profundus in common flexor sheath

Flexor retinaculum (transverse carpal ligament)

Palmaris longus

Tendinous sheath of abductor pollicis longus and extensor pollicis brevis

Flexor carpi radialis

Tendinous sheath of flexor pollicis longus

Flexor carpi ulnaris

Flexor carpi radialis

A. Anterior View

Palmar
Flexor digitorum superficialis tendon
Fibrous digital sheath
Synovial sheath

Nerve
Artery
Vein
Proper palmar digital

Skin (Grayson) ligament

Flexor digitorum profundus tendon

Extensor (dorsal) expansion

Dorsal
Proximal phalanx

C. Transverse Section (level of section indicated in A)

2.75 SYNOVIAL SHEATHS OF PALM OF HAND

A. Tendinous (synovial) sheaths of long flexor tendons of the digits. **B.** Osseofibrous tunnel and tendinous (synovial) sheath. **C.** Transverse section through the proximal phalanx.

Injuries such as puncture of a finger by a rusty nail can cause **infection of the digital synovial sheaths.** When inflammation of the tendon and synovial sheath (**tenosynovitis**) occurs, the digit swells and movement becomes painful. Because the tendons of the 2nd to 4th digits nearly always have separate synovial sheaths, the infection usually is confined to the infected digits. If the infection is untreated, the proximal ends of these sheaths may rupture, allowing the infection to spread to the midpalmar space. Because the synovial sheath of the little finger is usually continuous with the common flexor sheath, tenosynovitis in this finger may spread to the common flexor sheath and through the palm and carpal tunnel to the anterior forearm. Likewise, tenosynovitis in the thumb may spread through the continuous tendinous sheath of flexor pollicis longus.

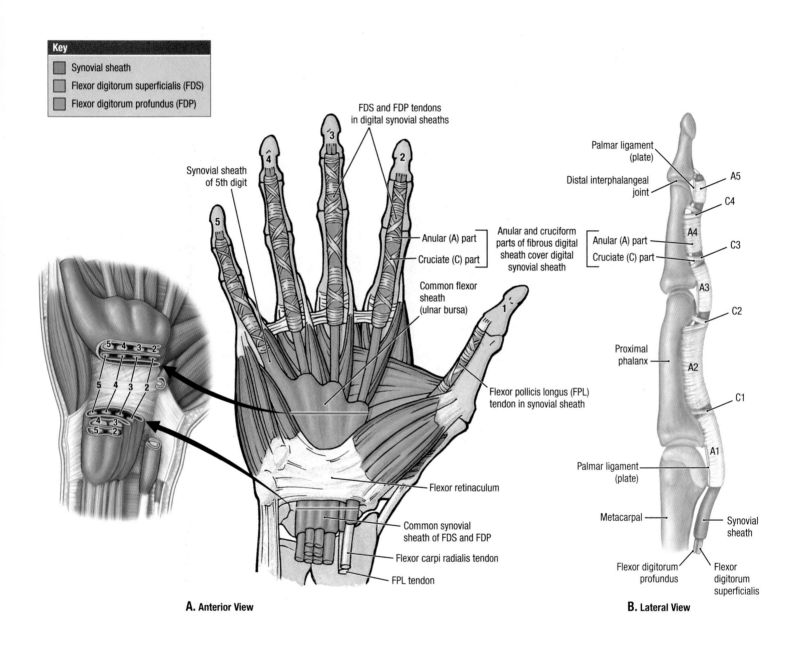

Key
- Synovial sheath
- Flexor digitorum superficialis (FDS)
- Flexor digitorum profundus (FDP)

FDS and FDP tendons in digital synovial sheaths

Synovial sheath of 5th digit

Anular (A) part

Cruciate (C) part

Anular and cruciform parts of fibrous digital sheath cover digital synovial sheath

Common flexor sheath (ulnar bursa)

Flexor pollicis longus (FPL) tendon in synovial sheath

Flexor retinaculum

Common synovial sheath of FDS and FDP

Flexor carpi radialis tendon

FPL tendon

A. Anterior View

Palmar ligament (plate)

Distal interphalangeal joint

A5

C4

Anular (A) part

Cruciate (C) part

A4

C3

A3

C2

Proximal phalanx

A2

C1

A1

Palmar ligament (plate)

Metacarpal

Synovial sheath

Flexor digitorum profundus

Flexor digitorum superficialis

B. Lateral View

FIBROUS DIGITAL SHEATHS

2.76

A. Fibrous digital and synovial sheaths. **B.** Anular and cruciate parts (pulleys) of the fibrous digital sheath.

Fibrous digital sheaths are the strong ligamentous tunnels containing the flexor tendons and their synovial sheaths. The sheaths extend from the heads of the metacarpals to the bases of the distal phalanges. These sheaths prevent the tendons from pulling away from the digits (bowstringing). The fibrous digital sheaths combine with the bones to form osseofibrous tunnels through which the tendons pass to reach the digits. The anular and cruciform (cruciate) parts, often referred to clinically as "pulleys," are thickened reinforcements of the fibrous digital sheaths.

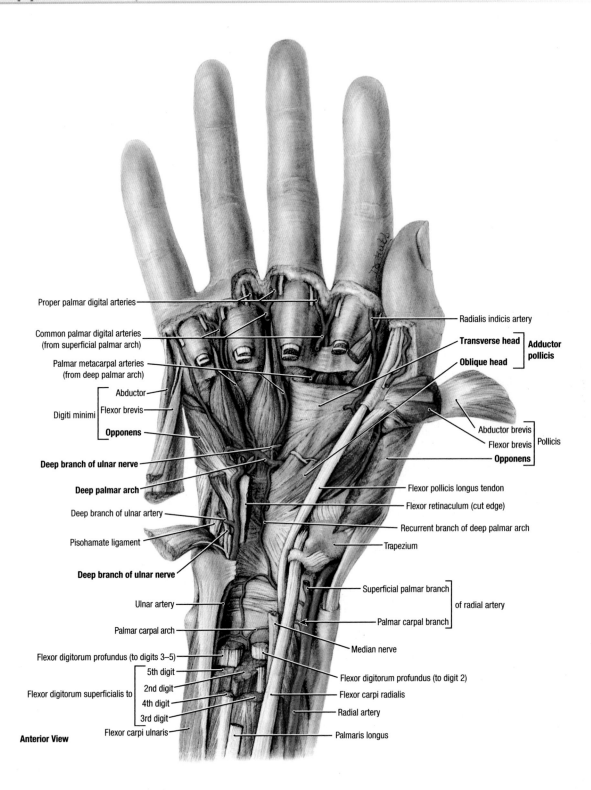

Proper palmar digital arteries

Common palmar digital arteries
(from superficial palmar arch)

Palmar metacarpal arteries
(from deep palmar arch)

Abductor

Flexor brevis

Digiti minimi

Opponens

Deep branch of ulnar nerve

Deep palmar arch

Deep branch of ulnar artery

Pisohamate ligament

Deep branch of ulnar nerve

Ulnar artery

Palmar carpal arch

Flexor digitorum profundus (to digits 3–5)

Flexor digitorum superficialis to

5th digit

2nd digit

4th digit

3rd digit

Flexor carpi ulnaris

Anterior View

Radialis indicis artery

Transverse head } **Adductor pollicis**

Oblique head

Abductor brevis }

Flexor brevis } Pollicis

Opponens

Flexor pollicis longus tendon

Flexor retinaculum (cut edge)

Recurrent branch of deep palmar arch

Trapezium

Superficial palmar branch }

} of radial artery

Palmar carpal branch }

Median nerve

Flexor digitorum profundus (to digit 2)

Flexor carpi radialis

Radial artery

Palmaris longus

2.77 DEEP DISSECTION OF PALM

- The deep branch of the ulnar artery joins the radial artery to form the deep palmar arch.
- The pisohamate ligament is often considered a continuation of the tendon of flexor carpi ulnaris, making the pisiform a sesamoid bone.

Compression of the ulnar nerve may occur at the wrist where it passes between the pisiform and the hook of hamate. The depression between these bones is converted by the pisohamate ligament into an osseofibrous ulnar canal. **Ulnar canal syndrome** is manifested by hypoesthesia in the medial one and one half digits and weakness of the intrinsic hand muscles. Clawing of the 4th and 5th digits may occur, but in contrast to proximal nerve injury, their ability to flex the wrist joint is unaffected.

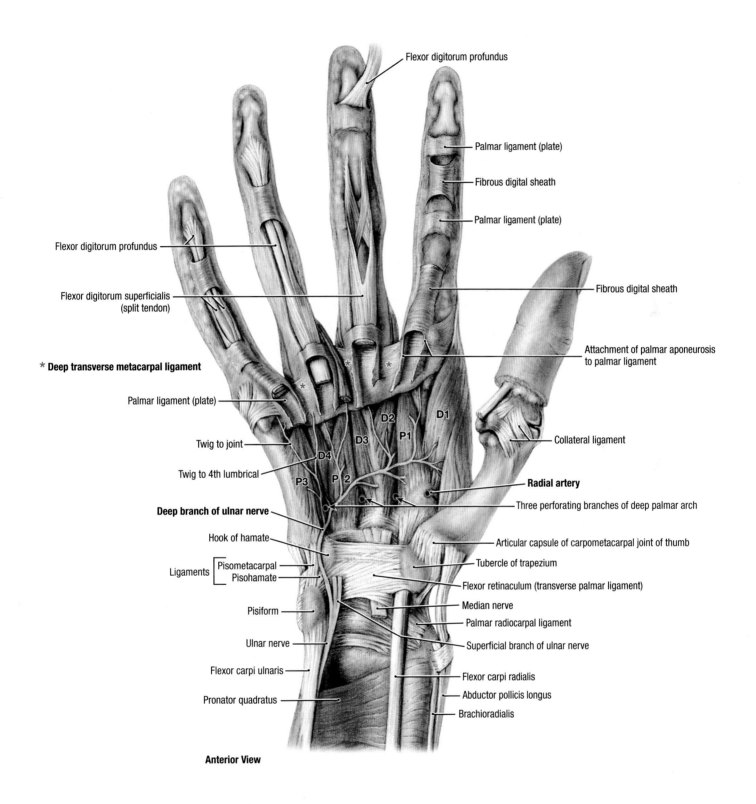

Flexor digitorum profundus

Palmar ligament (plate)

Fibrous digital sheath

Palmar ligament (plate)

Fibrous digital sheath

Flexor digitorum profundus

Flexor digitorum superficialis (split tendon)

Attachment of palmar aponeurosis to palmar ligament

* Deep transverse metacarpal ligament

D1

D2

Palmar ligament (plate)

D3   P1

Collateral ligament

Twig to joint

D4

Twig to 4th lumbrical

P3   P 2

Radial artery

Deep branch of ulnar nerve

Three perforating branches of deep palmar arch

Hook of hamate

Articular capsule of carpometacarpal joint of thumb

Ligaments [Pisometacarpal
Pisohamate

Tubercle of trapezium

Flexor retinaculum (transverse palmar ligament)

Pisiform

Median nerve

Palmar radiocarpal ligament

Ulnar nerve

Superficial branch of ulnar nerve

Flexor carpi ulnaris

Flexor carpi radialis

Abductor pollicis longus

Pronator quadratus

Brachioradialis

Anterior View

DEEP DISSECTION OF PALM AND DIGITS WITH DEEP BRANCH OF ULNAR NERVE

2.78

- Three unipennate palmar (*P1–P3*) and four bipennate dorsal (*D1–D4*) interosseous muscles are illustrated; the palmar interossei adduct the fingers, and the dorsal interossei abduct the fingers in relation to the axial line, an imaginary line through the long axis of the 3rd digit (see Table 2.13).

- The deep transverse metacarpal ligaments unite the palmar ligaments; the lumbricals pass anterior to the deep transverse metacarpal ligament, and the interossei pass posterior to the ligament.
- The pisohamate and pisometacarpal ligaments form the distal attachment of flexor carpi ulnaris.

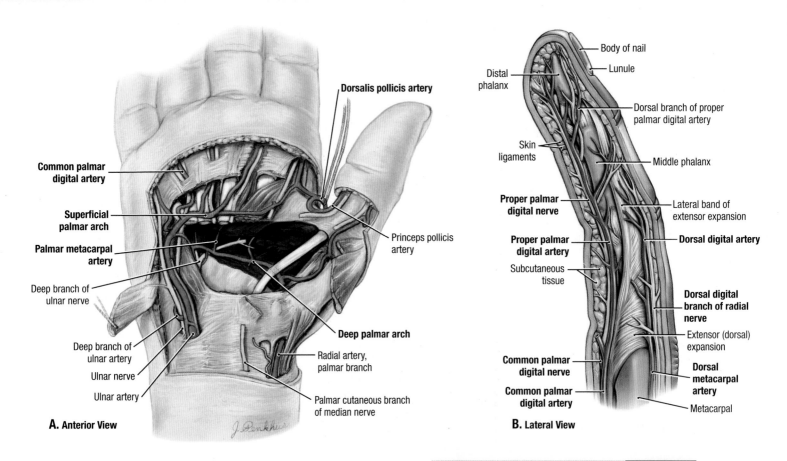

Dorsalis pollicis artery

Body of nail

Lunule

Distal phalanx

Dorsal branch of proper palmar digital artery

Common palmar digital artery

Skin ligaments

Middle phalanx

Superficial palmar arch

Proper palmar digital nerve

Lateral band of extensor expansion

Palmar metacarpal artery

Proper palmar digital artery

Dorsal digital artery

Subcutaneous tissue

Princeps pollicis artery

Dorsal digital branch of radial nerve

Deep branch of ulnar nerve

Extensor (dorsal) expansion

Deep branch of ulnar artery

Common palmar digital nerve

Dorsal metacarpal artery

Ulnar nerve

Deep palmar arch

Common palmar digital artery

Ulnar artery

Radial artery, palmar branch

Metacarpal

Palmar cutaneous branch of median nerve

A. Anterior View

J. Pinkhus

B. Lateral View

2.79 ARTERIAL SUPPLY OF HAND

A. Dissection of palmar arterial arches. **B.** Digital vessels and nerves. **C.** Arteriogram of the hand.

Note that the superficial palmar arch is usually completed by the superficial palmar branch of the radial artery, but in this specimen, the dorsalis pollicis artery completes the arch.

The **superficial and deep palmar (arterial) arches** are not palpable, but their surface markings are visible. The superficial palmar arch occurs at the level of the distal border of the fully extended thumb. The deep palmar arch lies approximately 1 cm proximal to the superficial palmar arch. The location of these arches should be borne in mind in wounds of the palm and when palmar incisions are made.

Intermittent bilateral attacks of **ischemia of the digits**, marked by cyanosis and often accompanied by paresthesia and pain, are characteristically brought on by cold and emotional stimuli. The condition may result from an anatomical abnormality or an underlying disease. When the cause of the condition is idiopathic (unknown) or primary, it is called **Raynaud syndrome** (disease). Since arteries receive innervation from postsynaptic fibers from the sympathetic ganglia, it may be necessary to perform a cervicodorsal presynaptic sympathectomy to dilate the digital arteries.

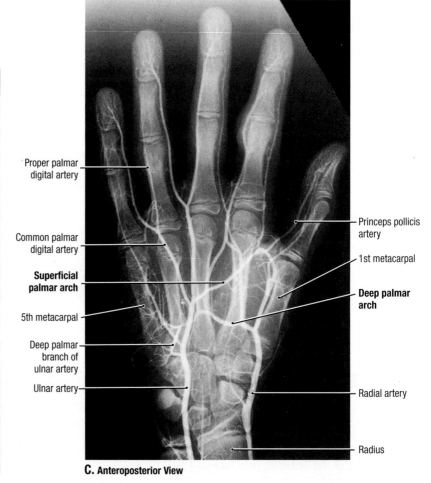

Proper palmar digital artery

Common palmar digital artery

Princeps pollicis artery

Superficial palmar arch

1st metacarpal

5th metacarpal

Deep palmar arch

Deep palmar branch of ulnar artery

Ulnar artery

Radial artery

Radius

C. Anteroposterior View

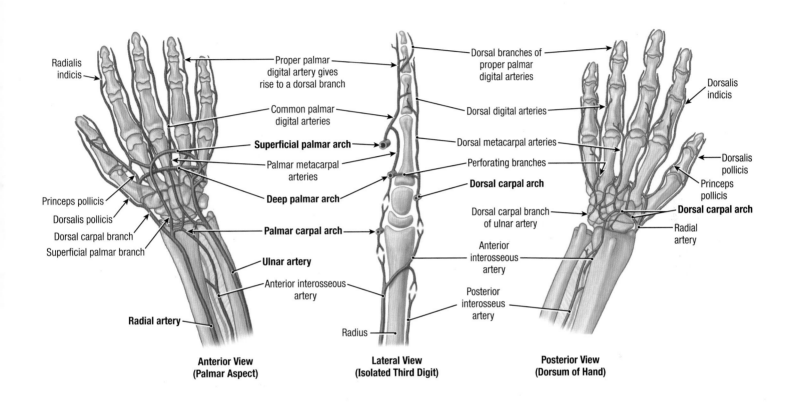

Radialis indicis	Proper palmar digital artery gives rise to a dorsal branch	Dorsal branches of proper palmar digital arteries
	Common palmar digital arteries	Dorsalis indicis
	Superficial palmar arch	Dorsal digital arteries
	Palmar metacarpal arteries	Dorsal metacarpal arteries
	Deep palmar arch	Perforating branches
Princeps pollicis		Dorsalis pollicis
Dorsalis pollicis	**Palmar carpal arch**	**Dorsal carpal arch**
Dorsal carpal branch		Princeps pollicis
Superficial palmar branch	**Ulnar artery**	Dorsal carpal branch of ulnar artery
	Anterior interosseous artery	**Dorsal carpal arch**
Radial artery		Radial artery
	Anterior interosseous artery	
	Posterior interosseus artery	
	Radius	

**Anterior View
(Palmar Aspect)**

**Lateral View
(Isolated Third Digit)**

**Posterior View
(Dorsum of Hand)**

ARTERIAL OF SUPPLY HAND | 2.80

Since hand is placed and held in many different positions, it requires an abundance of highly branched and anastomosing arteries so that oxygenated blood is available in all positions.

TABLE 2.14 | ARTERIES OF HAND

Artery	Origin	Course
Superficial palmar arch	Direct continuation of ulnar artery; arch is completed on lateral side by superficial branch of radial artery or another of its branches	Curves laterally deep to palmar aponeurosis and superficial to long flexor tendons; curve of arch lies across palm at level of distal border of extended thumb
Deep palmar arch	Direct continuation of radial artery; arch is completed on medial side by deep branch of ulnar artery	Curves medially, deep to long flexor tendons and is in contact with bases of metacarpals
Common palmar digital	Superficial palmar arch	Pass directly on lumbricals to webbings of digits
Proper palmar digital	Common palmar digital arteries	Run along sides of digits 2–5
Princeps pollicis	Radial artery as it turns into palm	Descends on palmar aspect of 1st metacarpal and divides at the base of proximal phalanx into two branches that run along sides of thumb
Radialis indicis	Radial artery but may arise from princeps pollicis artery	Passes along lateral side of index finger to its distal end
Dorsal carpal arch	Radial and ulnar arteries	Arches within fascia on dorsum of hand

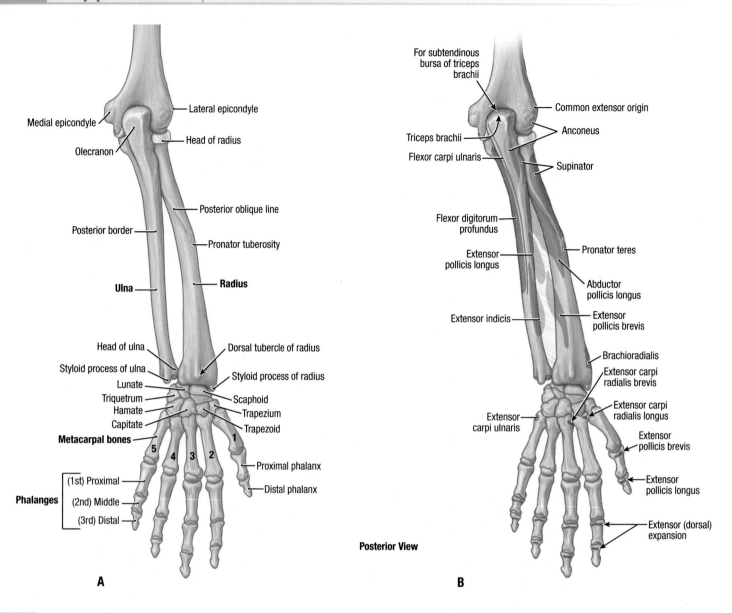

A

Medial epicondyle
Olecranon
Lateral epicondyle
Head of radius
Posterior oblique line
Posterior border
Pronator tuberosity
Ulna
Radius
Head of ulna
Styloid process of ulna
Lunate
Triquetrum
Hamate
Capitate
Dorsal tubercle of radius
Styloid process of radius
Scaphoid
Trapezium
Trapezoid
Metacarpal bones
5 4 3 2 1
Proximal phalanx
Distal phalanx
Phalanges
(1st) Proximal
(2nd) Middle
(3rd) Distal

B

For subtendinous bursa of triceps brachii
Common extensor origin
Anconeus
Triceps brachii
Flexor carpi ulnaris
Supinator
Flexor digitorum profundus
Extensor pollicis longus
Pronator teres
Abductor pollicis longus
Extensor indicis
Extensor pollicis brevis
Brachioradialis
Extensor carpi radialis brevis
Extensor carpi radialis longus
Extensor carpi ulnaris
Extensor pollicis brevis
Extensor pollicis longus
Extensor (dorsal) expansion

Posterior View

2.81 BONES AND MUSCLE ATTACHMENTS ON POSTERIOR FOREARM AND HAND

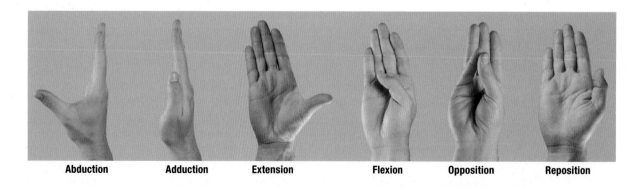

Abduction   **Adduction**   **Extension**   **Flexion**   **Opposition**   **Reposition**

2.82 MOVEMENTS OF THUMB

The thumb is rotated 90 degrees compared to the other digits. Abduction and adduction at the MCP joint occur in a sagittal plane; flexion and extension at the MCP and IP joints occur in frontal planes, opposite to these movements at other joints.

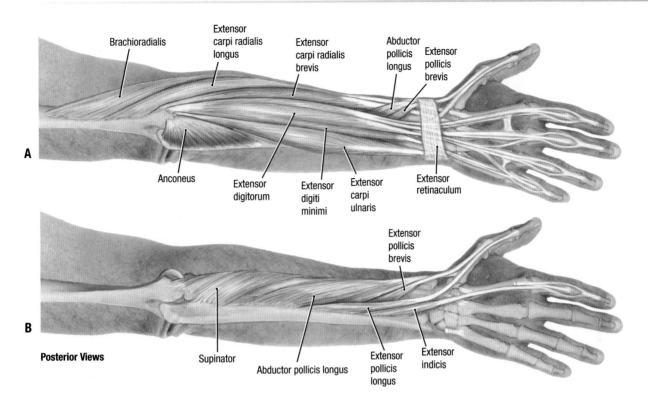

A

B

Posterior Views

MUSCLES OF POSTERIOR FOREARM

2.83

A. Superficial. **B.** Deep.

TABLE 2.15	**MUSCLES OF POSTERIOR SURFACE OF FOREARM**			
Muscle	*Proximal Attachment*	*Distal Attachment*	*Innervation*	*Main Actions*
Brachioradialis	Proximal two thirds of lateral supra-epicondylar ridge of humerus	Lateral surface of distal end of radius	Radial nerve (C5, **C6**, and C7)	Flexes elbow joint
Extensor carpi radialis longus	Lateral supra-epicondylar ridge of humerus	Base of 2nd metacarpal bone	Radial nerve (C6 and C7)	Extend and abduct wrist joint
Extensor carpi radialis brevis		Base of 3rd metacarpal bone	Deep branch of radial nerve (**C7** and C8)	
Extensor digitorum	Lateral epicondyle of humerus	Extensor expansions of medial four digits	Posterior interosseous nerve (C7 and C8), a branch of the radial nerve	Extends medial four metacarpophalangeal joints; extends wrist joint
Extensor digiti minimi		Extensor expansion of 5th digit		Extends MCP and IP joints of 5th digit; extends wrist joint
Extensor carpi ulnaris	Lateral epicondyle of humerus and posterior border of ulna	Base of 5th metacarpal bone		Extends and adducts wrist joint
Anconeus	Lateral epicondyle of humerus	Lateral surface of olecranon and superior part of posterior surface of ulna	Radial nerve (C7, C8, and T1)	Assists triceps brachii in extending elbow joint; stabilizes elbow joint; abducts ulna during pronation
Supinator	Lateral epicondyle of humerus, radial collateral and anular ligaments, supinator fossa, and crest of ulna	Lateral, posterior, and anterior surfaces of proximal third of radius	Deep branch of radial nerve (C5 and **C6**)	Supinates forearm
Abductor pollicis longus	Posterior surface of ulna, radius, and interosseous membrane	Base of 1st metacarpal bone	Posterior interosseous nerve (C7 and **C8**)	Abducts and extends carpometacarpal joint of thumb
Extensor pollicis brevis	Posterior surface of radius and interosseous membrane	Base of proximal phalanx of thumb		Extends MCP joint of thumb; extends wrist joint
Extensor pollicis longus	Posterior surface of middle third of ulna and interosseous membrane	Base of distal phalanx of thumb		Extends MCP and IP joints of thumb; extends wrist joint
Extensor indicis	Posterior surface of ulna and interosseous membrane	Extensor expansion of 2nd digit		Extends MCP and IP joints of 2nd digit; extends wrist joint

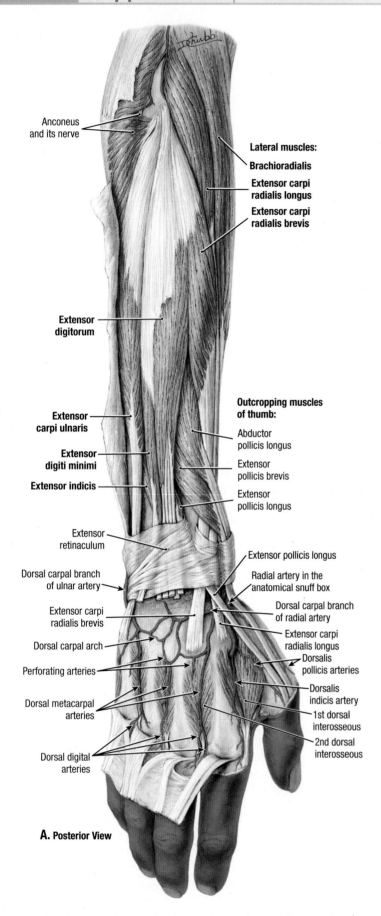

A. Anconeus and its nerve

Lateral muscles:

Brachioradialis

Extensor carpi radialis longus

Extensor carpi radialis brevis

Extensor digitorum

Extensor carpi ulnaris

Extensor digiti minimi

Extensor indicis

Outcropping muscles of thumb:

Abductor pollicis longus

Extensor pollicis brevis

Extensor pollicis longus

Extensor retinaculum

Dorsal carpal branch of ulnar artery

Extensor carpi radialis brevis

Dorsal carpal arch

Perforating arteries

Dorsal metacarpal arteries

Dorsal digital arteries

Extensor pollicis longus

Radial artery in the anatomical snuff box

Dorsal carpal branch of radial artery

Extensor carpi radialis longus

Dorsalis pollicis arteries

Dorsalis indicis artery

1st dorsal interosseous

2nd dorsal interosseous

A. Posterior View

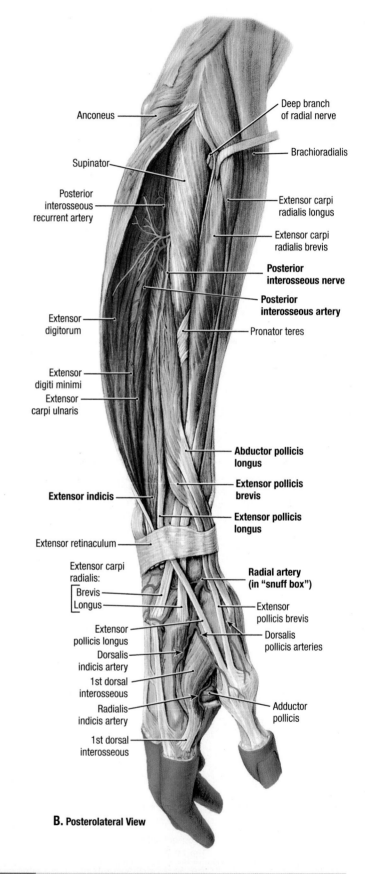

Anconeus

Supinator

Posterior interosseous recurrent artery

Extensor digitorum

Extensor digiti minimi

Extensor carpi ulnaris

Deep branch of radial nerve

Brachioradialis

Extensor carpi radialis longus

Extensor carpi radialis brevis

Posterior interosseous nerve

Posterior interosseous artery

Pronator teres

Abductor pollicis longus

Extensor pollicis brevis

Extensor pollicis longus

Extensor indicis

Extensor retinaculum

Extensor carpi radialis:
[Brevis
[Longus

Extensor pollicis longus

Dorsalis indicis artery

1st dorsal interosseous

Radialis indicis artery

1st dorsal interosseous

Radial artery (in "snuff box")

Extensor pollicis brevis

Dorsalis pollicis arteries

Adductor pollicis

B. Posterolateral View

2.84 EXTENSOR MUSCLES OF FOREARM

A. Superficial dissection. **B.** Deep dissection.

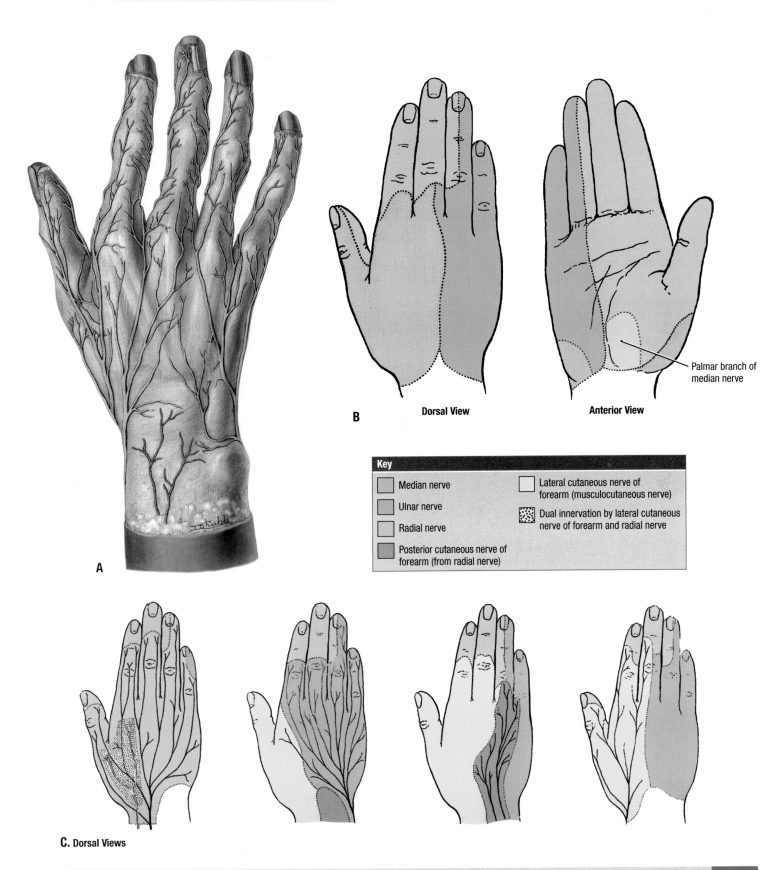

Dorsal View

Anterior View

Palmar branch of
median nerve

B

Key

Median nerve

Ulnar nerve

Radial nerve

Posterior cutaneous nerve of
forearm (from radial nerve)

Lateral cutaneous nerve of
forearm (musculocutaneous nerve)

Dual innervation by lateral cutaneous
nerve of forearm and radial nerve

A

C. Dorsal Views

CUTANEOUS INNERVATION OF HAND

2.85

A. Dissection of nerves of dorsum of hand. **B.** Distribution of the cutaneous nerves to the palm and dorsum of the hand, schematic illustration. **C.** Variations in pattern of cutaneous nerves in dorsum of hand.

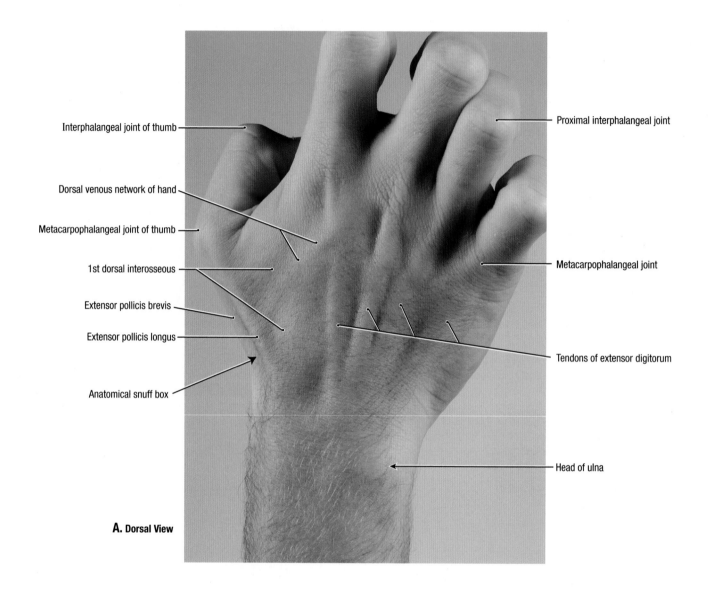

Interphalangeal joint of thumb

Dorsal venous network of hand

Metacarpophalangeal joint of thumb

1st dorsal interosseous

Extensor pollicis brevis

Extensor pollicis longus

Anatomical snuff box

Proximal interphalangeal joint

Metacarpophalangeal joint

Tendons of extensor digitorum

Head of ulna

A. Dorsal View

2.86 DORSUM OF HAND

A. Surface anatomy. The interphalangeal joints are flexed, and the metacarpophalangeal joints are hyperextended to demonstrate the extensor digitorum tendons. **B.** Tendinous (synovial) sheaths distended with blue fluid. **C.** Transverse section of distal forearm. Numbers refer to structures **(B)**. **D.** Sites of bony attachments.

- Six tendinous sheaths occupy the six osseofibrous tunnels deep to the extensor retinaculum. They contain nine tendons: tendons for the thumb in sheaths 1 and 3, tendons for the extensors of the wrist in sheaths 2 and 6, and tendons for the extensors of the wrist and fingers in sheaths 4 and 5.

- The tendon of the extensor pollicis longus hooks around the dorsal tubercle of radius to pass obliquely across the tendons of the extensor carpi radialis longus and brevis to the thumb.

The tendons of the abductor pollicis longus and extensor pollicis brevis are in the same tendinous sheath on the dorsum of the wrist. Excessive friction of these tendons results in fibrous thickening of the sheath and stenosis of the osseofibrous tunnel, **Quervain tenovaginitis stenosans**. This condition causes pain in the wrist that radiates proximally to the forearm and distally to the thumb.

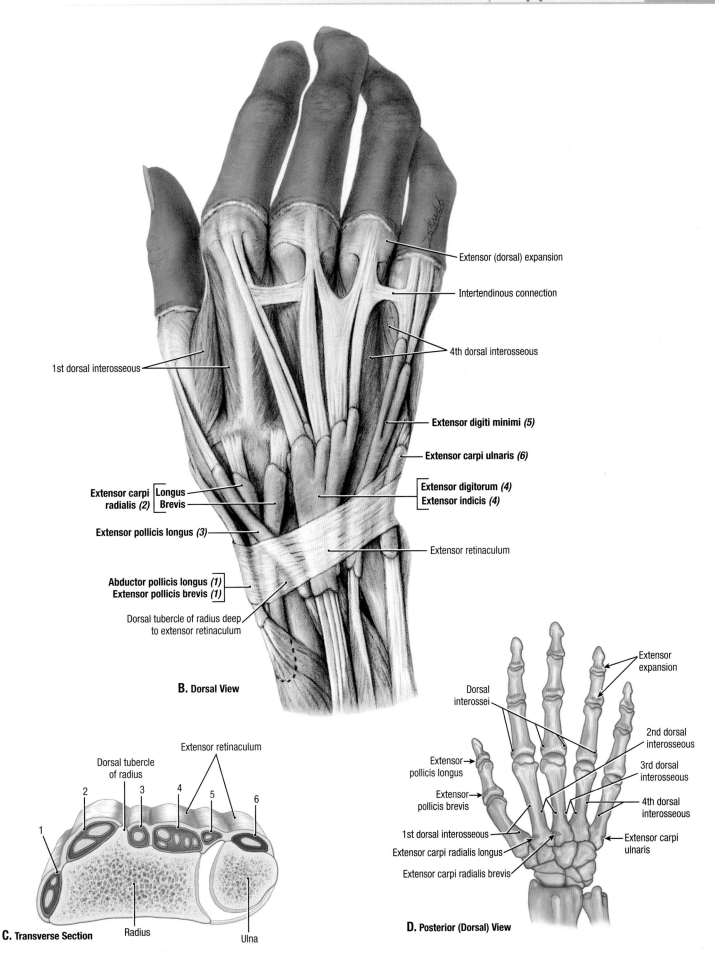

Extensor (dorsal) expansion

Intertendinous connection

4th dorsal interosseous

1st dorsal interosseous

Extensor digiti minimi *(5)*

Extensor carpi ulnaris *(6)*

Extensor carpi radialis *(2)* — Longus / Brevis

Extensor digitorum *(4)*
Extensor indicis *(4)*

Extensor pollicis longus *(3)*

Extensor retinaculum

Abductor pollicis longus *(1)*
Extensor pollicis brevis *(1)*

Dorsal tubercle of radius deep to extensor retinaculum

B. Dorsal View

Extensor retinaculum

Dorsal tubercle of radius

2 3 4 5 6

1

C. Transverse Section   Radius   Ulna

Extensor expansion

Dorsal interossei

2nd dorsal interosseous

Extensor pollicis longus

3rd dorsal interosseous

Extensor pollicis brevis

4th dorsal interosseous

1st dorsal interosseous

Extensor carpi ulnaris

Extensor carpi radialis longus

Extensor carpi radialis brevis

D. Posterior (Dorsal) View

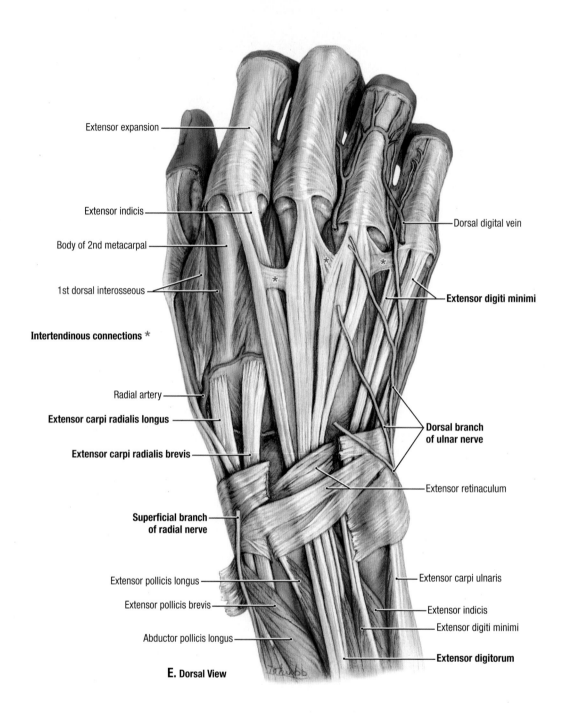

Extensor expansion

Extensor indicis

Body of 2nd metacarpal

1st dorsal interosseous

Intertendinous connections *

Radial artery

Extensor carpi radialis longus

Extensor carpi radialis brevis

Superficial branch of radial nerve

Extensor pollicis longus

Extensor pollicis brevis

Abductor pollicis longus

Dorsal digital vein

Extensor digiti minimi

Dorsal branch of ulnar nerve

Extensor retinaculum

Extensor carpi ulnaris

Extensor indicis

Extensor digiti minimi

Extensor digitorum

E. Dorsal View

2.86   DORSUM OF HAND (*continued*)

E. Tendons on dorsum of hand and extensor retinaculum.
• The deep fascia is thickened to form the extensor retinaculum.
• Proximal to the knuckles, intertendinous connections extend between the tendons of the digital extensors and, thereby, restrict the independent action of the fingers.

Ganglion cyst. Sometimes a nontender cystic swelling appears on the hand, most commonly on the dorsum of the wrist. The thin-walled cyst contains clear mucinous fluid. Clinically, this type of swelling is called a ganglion (a swelling or knot). These synovial cysts are close to and often communicate with the synovial sheaths. The distal attachment of the extensor carpi radialis brevis tendon is a common site for such a cyst.

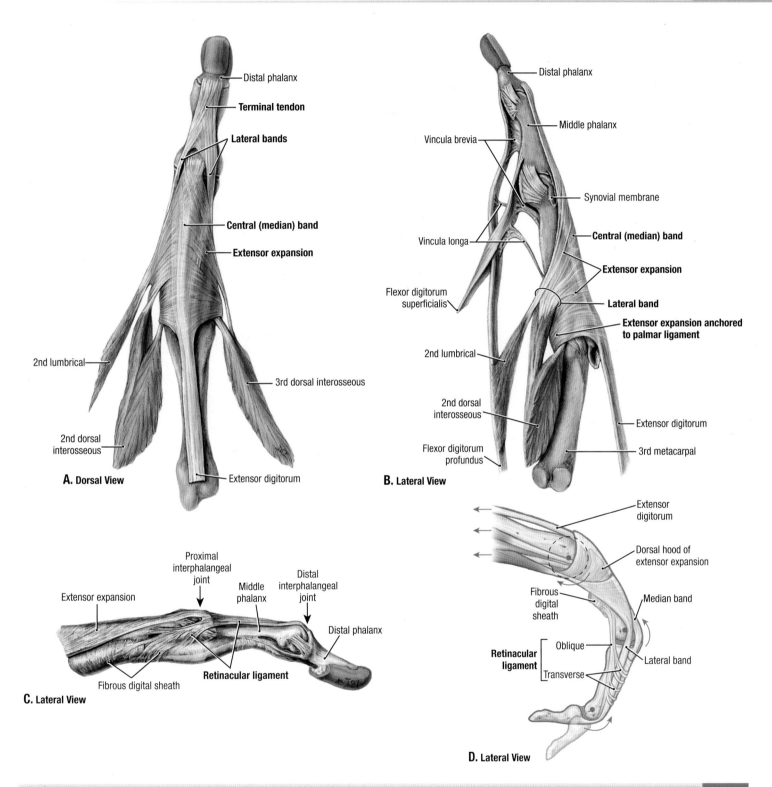

A. Dorsal View

Distal phalanx
Terminal tendon
Lateral bands
Central (median) band
Extensor expansion
2nd lumbrical
3rd dorsal interosseous
2nd dorsal interosseous
Extensor digitorum

B. Lateral View

Distal phalanx
Vincula brevia
Middle phalanx
Synovial membrane
Central (median) band
Vincula longa
Extensor expansion
Flexor digitorum superficialis
Lateral band
Extensor expansion anchored to palmar ligament
2nd lumbrical
2nd dorsal interosseous
Extensor digitorum
Flexor digitorum profundus
3rd metacarpal

C. Lateral View

Proximal interphalangeal joint
Middle phalanx
Distal interphalangeal joint
Extensor expansion
Distal phalanx
Retinacular ligament
Fibrous digital sheath

D. Lateral View

Extensor digitorum
Dorsal hood of extensor expansion
Fibrous digital sheath
Median band
Retinacular ligament Oblique
Lateral band
Transverse

EXTENSOR (DORSAL) EXPANSION OF THIRD DIGIT

2.87

A. Dorsal view. **B.** Lateral view. **C.** Retinacular ligaments of extended digit. **D.** Retinacular ligaments of flexed digit.
- The hood covering the head of the metacarpal is attached to the palmar ligament.
- Contraction of the muscles attaching to the lateral band will produce flexion of the metacarpophalangeal joint and extension of the interphalangeal joints.

- The retinacular ligament is a fibrous band that runs from the proximal phalanx and fibrous digital sheath obliquely across the middle phalanx and two interphalangeal joints to join the extensor (dorsal) expansion and then to the distal phalanx.
- On flexion of the distal interphalangeal joint, the retinacular ligament becomes taut and pulls the proximal joint into flexion; on extension of the proximal joint, the distal joint is pulled by the ligament into nearly complete extension.

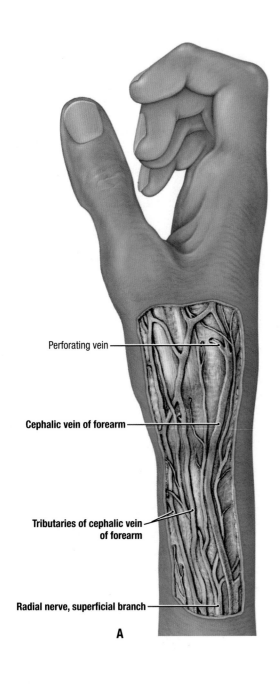

Perforating vein

Cephalic vein of forearm

Tributaries of cephalic vein of forearm

Radial nerve, superficial branch

A

Lateral Views

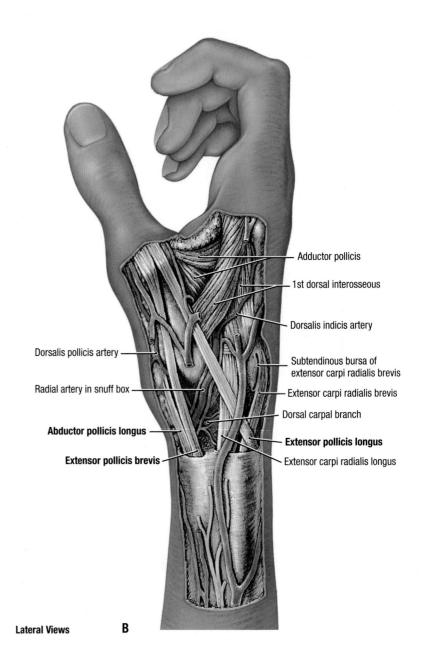

Adductor pollicis

1st dorsal interosseous

Dorsalis indicis artery

Dorsalis pollicis artery

Subtendinous bursa of extensor carpi radialis brevis

Radial artery in snuff box

Extensor carpi radialis brevis

Dorsal carpal branch

Abductor pollicis longus

Extensor pollicis longus

Extensor pollicis brevis

Extensor carpi radialis longus

B

2.88 LATERAL WRIST AND HAND

A. Anatomical snuff box—I.
- The depression at the base of the thumb, the "anatomical snuff box," retains its name from an archaic habit.
- Note the superficial veins, including the cephalic vein of forearm and/or its tributaries, and cutaneous nerves crossing the snuff box.

B. Anatomical snuff box—II.
- Three long tendons of the thumb form the boundaries of the snuff box; the extensor pollicis longus forms the medial boundary and the abductor pollicis longus and extensor pollicis brevis the lateral boundary.
- The radial artery crosses the floor of the snuff box and travels between the two heads of the 1st dorsal interosseous.

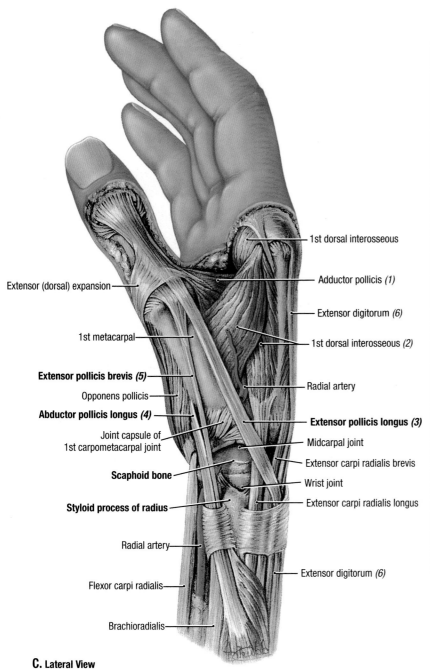

Extensor (dorsal) expansion

1st dorsal interosseous

Adductor pollicis *(1)*

Extensor digitorum *(6)*

1st dorsal interosseous *(2)*

1st metacarpal

Extensor pollicis brevis *(5)*

Opponens pollicis

Radial artery

Abductor pollicis longus *(4)*

Extensor pollicis longus *(3)*

Joint capsule of
1st carpometacarpal joint

Midcarpal joint

Extensor carpi radialis brevis

Scaphoid bone

Wrist joint

Styloid process of radius

Extensor carpi radialis longus

Radial artery

Flexor carpi radialis

Extensor digitorum *(6)*

Brachioradialis

C. Lateral View

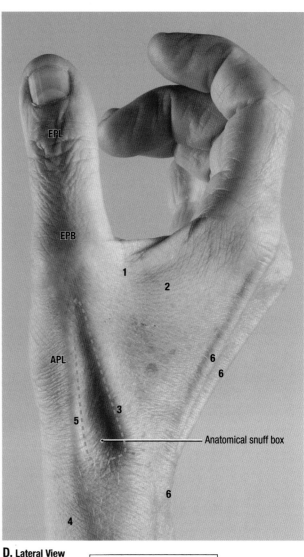

D. Lateral View

Distal Extent of:	
EPL	Extensor pollicis longus
EPB	Extensor pollicis brevis
APL	Abductor pollicis longus

LATERAL WRIST AND HAND (*continued*)   2.88

C. Anatomical snuff box—III. Note the scaphoid bone, the wrist joint proximal to the scaphoid, and the midcarpal joint distal to it.
D. Surface anatomy.

Fracture of the scaphoid often results from a fall on the palm with the hand abducted. The fracture occurs across the narrow part ("waist") of the scaphoid. Pain occurs primarily on the lateral side of the wrist, especially during dorsiflexion and abduction of the hand. Initial radiographs of the wrist may not reveal a fracture, but radiographs taken 10 to 14 days later reveal a fracture because bone resorption has occurred. Owing to the poor blood supply to the proximal part of the scaphoid, union of the fractured parts may take several months. **Avascular necrosis of the proximal fragment of the scaphoid** (pathological death of bone resulting from poor blood supply) may occur and produce degenerative joint disease of the wrist.

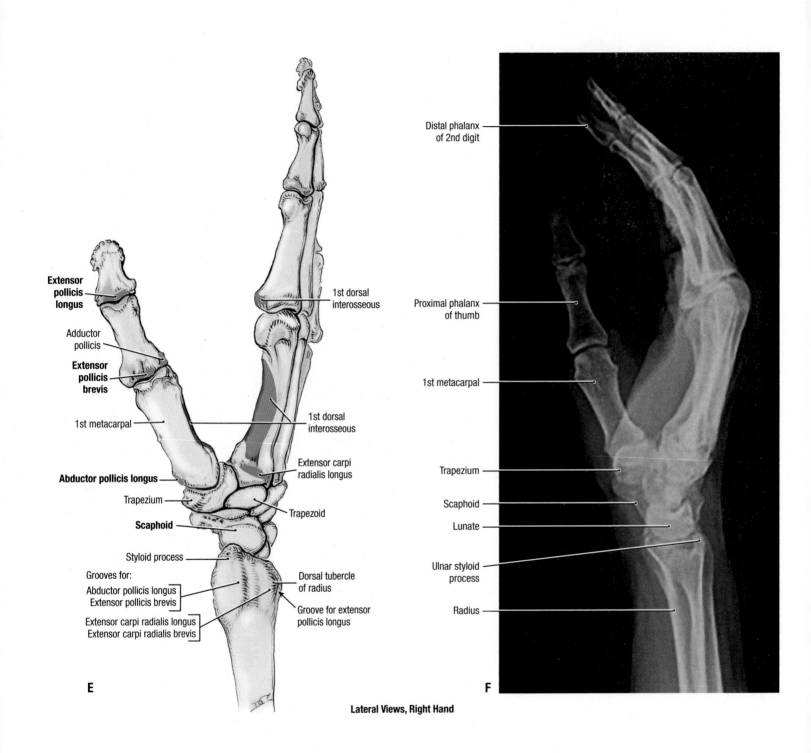

Extensor
pollicis
longus

1st dorsal
interosseous

Adductor
pollicis

Extensor
pollicis
brevis

1st metacarpal

1st dorsal
interosseous

Abductor pollicis longus

Extensor carpi
radialis longus

Trapezium

Trapezoid

Scaphoid

Styloid process

Grooves for:

Dorsal tubercle
of radius

Abductor pollicis longus
Extensor pollicis brevis

Extensor carpi radialis longus
Extensor carpi radialis brevis

Groove for extensor
pollicis longus

E

Distal phalanx
of 2nd digit

Proximal phalanx
of thumb

1st metacarpal

Trapezium

Scaphoid

Lunate

Ulnar styloid
process

Radius

F

Lateral Views, Right Hand

2.88 **LATERAL WRIST AND HAND** (*continued*)

E. Bony hand showing muscle attachments. **F.** Radiograph.
Note that the anatomical snuff box is limited proximally
by the styloid process of the radius and distally by the base of

the 1st metacarpal; parts of the two lateral bones of the carpus
(scaphoid and trapezium) form the floor of the snuff box.

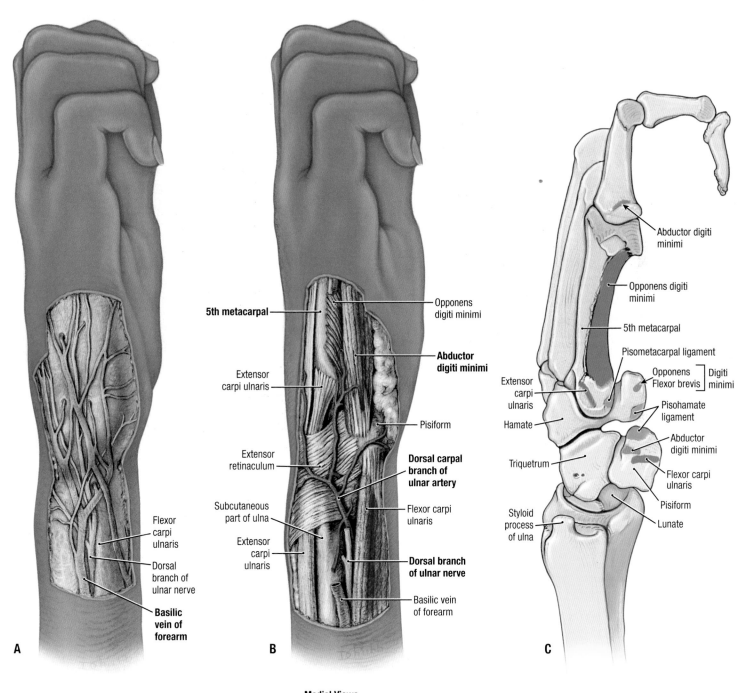

5th metacarpal

Extensor
carpi ulnaris

Extensor
retinaculum

Subcutaneous
part of ulna

Extensor
carpi
ulnaris

Flexor
carpi
ulnaris

Dorsal
branch of
ulnar nerve

**Basilic
vein of
forearm**

Opponens
digiti minimi

**Abductor
digiti minimi**

Pisiform

**Dorsal carpal
branch of
ulnar artery**

Flexor carpi
ulnaris

**Dorsal branch
of ulnar nerve**

Basilic vein
of forearm

Abductor digiti
minimi

Opponens digiti
minimi

5th metacarpal

Pisometacarpal ligament

Opponens ⎤ Digiti
Flexor brevis ⎦ minimi

Extensor
carpi
ulnaris

Hamate

Triquetrum

Styloid
process
of ulna

Pisohamate
ligament

Abductor
digiti minimi

Flexor carpi
ulnaris

Pisiform

Lunate

A B C

Medial Views

MEDIAL WRIST AND HAND

2.89

A. Superficial dissection. **B.** Deep dissection. **C.** Bony hand showing sites of muscular and ligamentous attachments. The extensor carpi ulnaris is inserted directly into the base of the 5th metacarpal, but the flexor carpi ulnaris inserts indirectly to the base of the 5th metacarpal via the pisiform and pisohamate and pisometacarpal ligaments. These ligaments are often considered to be a part of the distal attachment of flexor carpi ulnaris.

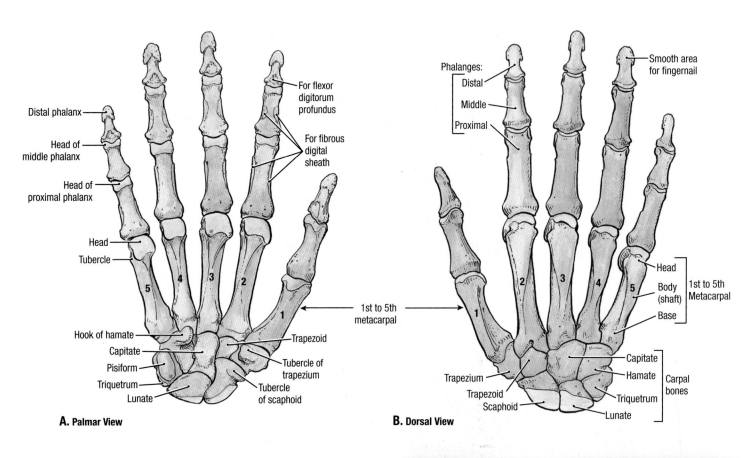

Distal phalanx

Head of middle phalanx

Head of proximal phalanx

Head

Tubercle

5 4 3 2

1

Hook of hamate

Capitate

Pisiform

Triquetrum

Lunate

For flexor digitorum profundus

For fibrous digital sheath

1st to 5th metacarpal

Trapezoid

Tubercle of trapezium

Tubercle of scaphoid

A. Palmar View

Phalanges:

Distal

Middle

Proximal

Smooth area for fingernail

2 3 4 5

1

Head

Body (shaft)

Base

1st to 5th Metacarpal

Trapezium

Trapezoid

Scaphoid

Lunate

Capitate

Hamate

Triquetrum

Carpal bones

B. Dorsal View

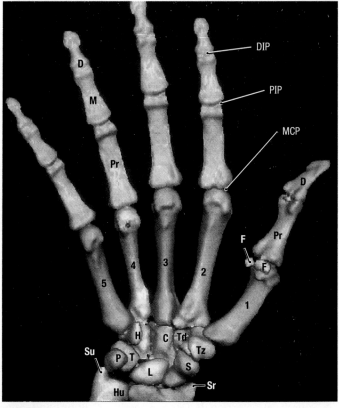

DIP

PIP

MCP

D

M

Pr

D

F

Pr

F

5 4 3 2

1

Su

P T

H C Td

Tz

S

L

Sr

Hu

C. Anterior View

2.90 BONES AND IMAGING OF WRIST AND HAND

A. Palmar view. **B.** Dorsal view. **C.** Three-dimensional computer-generated image of wrist and hand. Letters refer to structures **(D)**.

The eight carpal bones form two rows: in the distal row, the hamate, capitate, trapezoid, and trapezium, the trapezium forming a saddle-shaped joint with the 1st metacarpal; and in the proximal row, the scaphoid, lunate, and pisiform, the pisiform being superimposed on the triquetrum.

Severe **crushing injuries of the hand** may produce multiple metacarpal fractures, resulting in instability of the hand. Similar injuries of the distal phalanges are common (e.g., when a finger is caught in a car door). A **fracture of a distal phalanx** is usually comminuted, and a painful **hematoma** (collection of blood) develops. **Fractures of the proximal and middle phalanges** are usually the result of crushing or hypertension injuries.

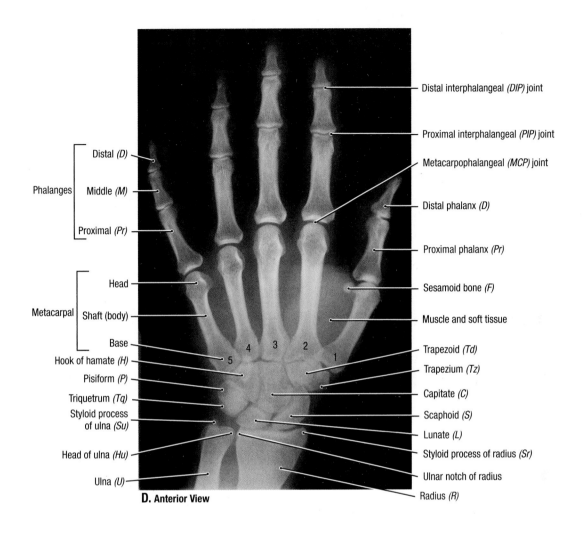

Phalanges
- Distal *(D)*
- Middle *(M)*
- Proximal *(Pr)*

Metacarpal
- Head
- Shaft (body)
- Base

Hook of hamate *(H)*
Pisiform *(P)*
Triquetrum *(Tq)*
Styloid process of ulna *(Su)*
Head of ulna *(Hu)*
Ulna *(U)*

Distal interphalangeal *(DIP)* joint
Proximal interphalangeal *(PIP)* joint
Metacarpophalangeal *(MCP)* joint
Distal phalanx *(D)*
Proximal phalanx *(Pr)*
Sesamoid bone *(F)*
Muscle and soft tissue
Trapezoid *(Td)*
Trapezium *(Tz)*
Capitate *(C)*
Scaphoid *(S)*
Lunate *(L)*
Styloid process of radius *(Sr)*
Ulnar notch of radius
Radius *(R)*

D. Anterior View

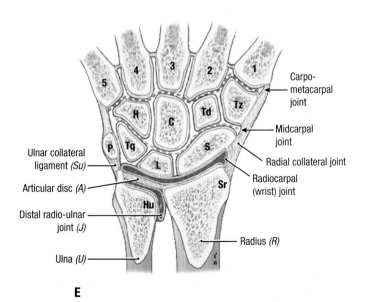

Carpo-metacarpal joint
Midcarpal joint
Radial collateral joint
Radiocarpal (wrist) joint

Ulnar collateral ligament *(Su)*
Articular disc *(A)*
Distal radio-ulnar joint *(J)*
Ulna *(U)*
Radius *(R)*

E

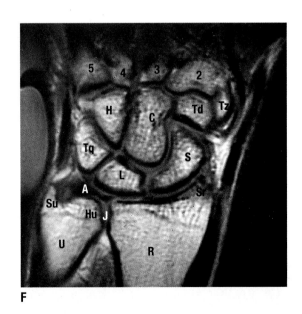

F

BONES AND IMAGING OF WRIST AND HAND (*continued*) | **2.90**

D. Radiograph. **E.** Coronal section. **F.** Coronal MRI. Letters refer to structures (**D**).

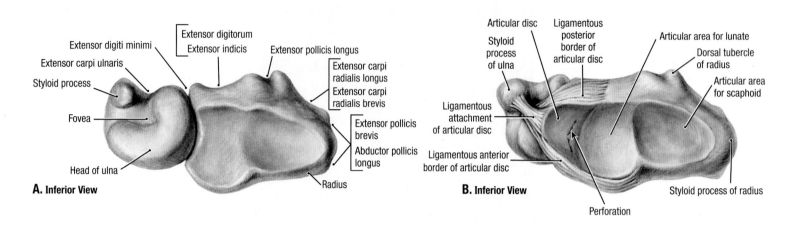

Extensor digiti minimi

Extensor carpi ulnaris

Styloid process

Fovea

Head of ulna

A. Inferior View

Extensor digitorum
Extensor indicis

Extensor pollicis longus

Extensor carpi radialis longus
Extensor carpi radialis brevis

Extensor pollicis brevis

Abductor pollicis longus

Radius

Articular disc

Styloid process of ulna

Ligamentous posterior border of articular disc

Ligamentous attachment of articular disc

Ligamentous anterior border of articular disc

Articular area for lunate

Dorsal tubercle of radius

Articular area for scaphoid

Styloid process of radius

Perforation

B. Inferior View

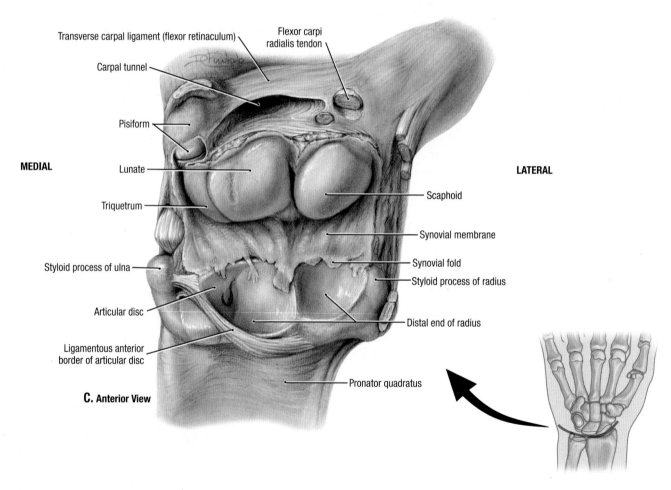

Transverse carpal ligament (flexor retinaculum)

Carpal tunnel

Pisiform

MEDIAL

Lunate

Triquetrum

Styloid process of ulna

Articular disc

Ligamentous anterior border of articular disc

C. Anterior View

Flexor carpi radialis tendon

LATERAL

Scaphoid

Synovial membrane

Synovial fold

Styloid process of radius

Distal end of radius

Pronator quadratus

| **2.91** | RADIOCARPAL (WRIST) JOINT |

A. Distal ends of radius and ulna showing grooves for tendons on the posterior aspects. **B.** Articular disc. The articular disc unites the distal ends of the radius and ulna; it is fibrocartilaginous at the triangular area between the head of the ulna and the lunate bone but ligamentous and pliable elsewhere. The cartilaginous part of the articular disc commonly has a fissure or perforation, as shown here, associated with a roughened surface of the lunate. **C.** Articular surface of the radiocarpal joint, which is opened anteriorly. The lunate articulates with the radius and articular disc; only during adduction of the wrist does the triquetrum come into articulation with the disc.

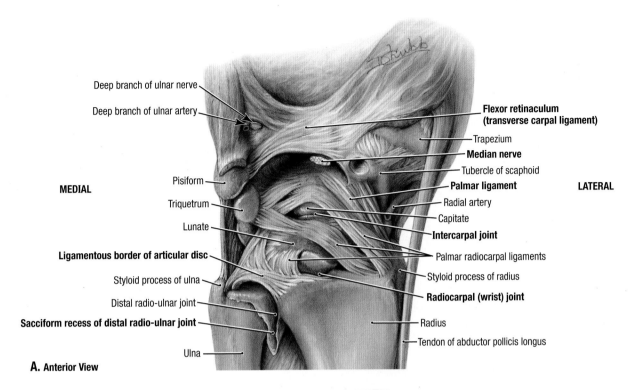

Deep branch of ulnar nerve

Deep branch of ulnar artery

Flexor retinaculum (transverse carpal ligament)

Trapezium

Median nerve

Tubercle of scaphoid

MEDIAL

Pisiform

Palmar ligament

LATERAL

Triquetrum

Radial artery

Capitate

Lunate

Intercarpal joint

Ligamentous border of articular disc

Palmar radiocarpal ligaments

Styloid process of ulna

Styloid process of radius

Distal radio-ulnar joint

Radiocarpal (wrist) joint

Sacciform recess of distal radio-ulnar joint

Radius

Tendon of abductor pollicis longus

Ulna

A. Anterior View

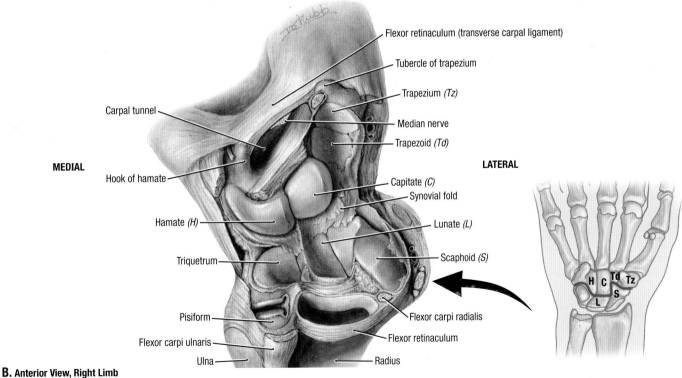

Flexor retinaculum (transverse carpal ligament)

Tubercle of trapezium

Trapezium *(Tz)*

Carpal tunnel

Median nerve

Trapezoid *(Td)*

MEDIAL

LATERAL

Hook of hamate

Capitate *(C)*

Synovial fold

Hamate *(H)*

Lunate *(L)*

Triquetrum

Scaphoid *(S)*

Pisiform

Flexor carpi radialis

Flexor carpi ulnaris

Flexor retinaculum

Ulna

Radius

B. Anterior View, Right Limb

RADIOCARPAL (WRIST) AND MIDCARPAL (TRANSVERSE CARPAL) JOINT — 2.92

A. Ligaments. The hand is forcibly extended. The palmar radio-carpal ligaments pass from the radius to the two rows of carpal bones; they are strong and directed so that the hand moves with the radius during supination. **B.** Articular surfaces of midcarpal (transverse carpal) joint, opened anteriorly.

Note that the flexor retinaculum (transverse carpal ligament) is cut. The proximal part of the ligament, which spans from the pisiform to the scaphoid, is relatively weak; the distal part, which passes from the hook of the hamate to the tubercle of the trapezium, is strong.

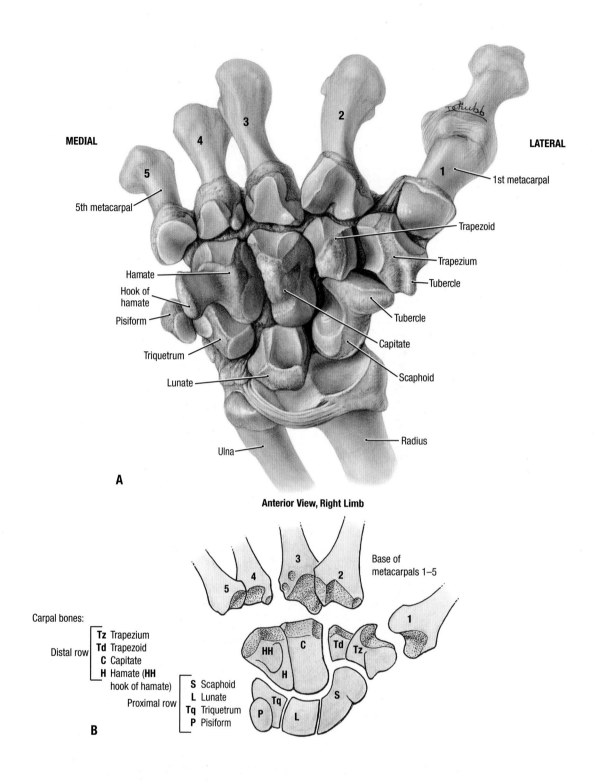

MEDIAL

LATERAL

3

4

2

5

1

1st metacarpal

5th metacarpal

Trapezoid

Trapezium

Tubercle

Hamate

Hook of hamate

Tubercle

Pisiform

Capitate

Triquetrum

Scaphoid

Lunate

Radius

Ulna

A

Anterior View, Right Limb

Carpal bones:

Distal row

Tz Trapezium
Td Trapezoid
C Capitate
H Hamate (**HH** hook of hamate)

Proximal row

S Scaphoid
L Lunate
Tq Triquetrum
P Pisiform

Base of metacarpals 1–5

3

4

5

2

1

HH

C

Td

Tz

H

Tq

S

P

L

B

2.93   CARPAL BONES AND BASES OF METACARPALS

A. Open intercarpal and carpometacarpal (CMC) joints. The dorsal ligaments remain intact and all the joints have been hyperextended. **B.** Articular surfaces of the CMC joints.

Note that the 1st CMC joint is saddle-shaped and especially mobile, allowing opposition of the thumb; the 2nd and 3rd CMC joints have interlocking surfaces and are practically immobile; and the 4th and 5th are hinge-shaped synovial joints with limited movement.

Anterior dislocation of the lunate is a serious injury that usually results from a fall on the extended wrist. The lunate is pushed to the palmar surface of the wrist and may compress the median nerve and lead to carpal tunnel syndrome. Because of poor blood supply, **avascular necrosis of the lunate** may occur.

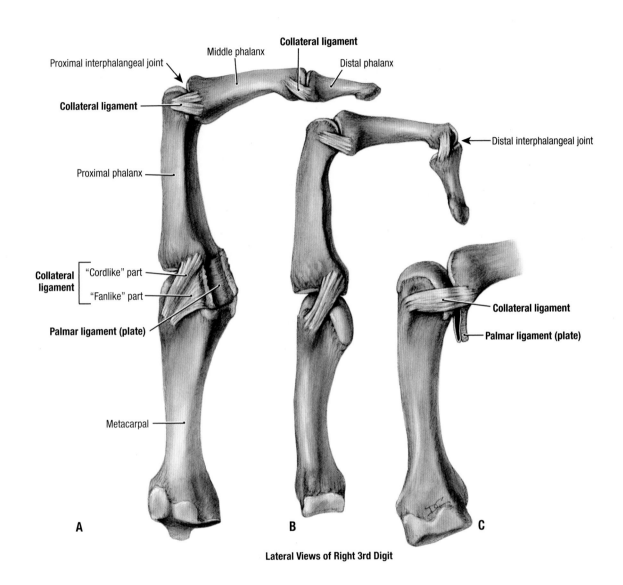

Proximal interphalangeal joint

Middle phalanx

Collateral ligament

Distal phalanx

Collateral ligament

Proximal phalanx

Distal interphalangeal joint

Collateral ligament
  "Cordlike" part
  "Fanlike" part

Palmar ligament (plate)

Collateral ligament

Palmar ligament (plate)

Metacarpal

A   B   C

Lateral Views of Right 3rd Digit

COLLATERAL LIGAMENTS OF METACARPOPHALANGEAL AND INTERPHALANGEAL JOINTS OF THIRD DIGIT | 2.94

A. Extended metacarpophalangeal (MCP) and distal interphalangeal (IP) joints. **B.** Flexed interphalangeal joints. **C.** Flexed MCP joint.

- A fibrocartilaginous plate, the palmar ligament, hangs from the base of the proximal phalanx; is fixed to the head of the metacarpal by the weaker, fanlike part of the collateral ligament **(A)**; and moves like a visor across the metacarpal head **(C)**. The IP joints have similar palmar ligaments.
- The extremely strong, cordlike parts of the collateral ligaments of this joint **(A and B)** are eccentrically attached to

the metacarpal heads; they are slack during extension and taut during flexion **(C)**, so the fingers cannot be spread (abducted) unless the hand is open; the IP joints have similar collateral ligaments.

Skier's thumb refers to the rupture or chronic laxity of the collateral ligament of the 1st metacarpophalangeal joint. The injury results from hyperextension of the joint, which occurs when the thumb is held by the ski pole while the rest lof the hand hits the ground or enters the snow.

TABLE 2.16   **LESIONS OF NERVES OF UPPER LIMB**

Nerve Injury	Injury Description	Impairments	Clinical Aspect
Long thoracic nerve	Stab wound Mastectomy	Abduction of shoulder joint and protraction of the scapula is compromised	Test: Pushing against a wall causes winging of scapula
Axillary nerve	Surgical neck fracture of humerus Anterior dislocation of shoulder joint	Abduction of shoulder joint to horizontal is compromised; sensory loss on lateral side of upper arm	Test: Abduct shoulder joint to horizontal and ask patient to hold position against a downward pull on the distal arm
Radial nerve	Midshaft fracture of humerus Badly fitted crutch Arm draped over a chair	Extension at wrist and joints of digits is lost; supination of forearm is compromised; sensory loss on posterior arm and forearm, and lateral aspect of dorsum of hand	Wrist drop
Median nerve at elbow	Supra-epicondylar fracture of humerus	Flexion of wrist joint is weakened; hand will deviate to ulnar side during flexion of wrist joint; flexion of DIP, PIP, and MCP joints of index and middle digits is lost; abduction, opposition and flexion of thumb joints are lost; sensory loss on palmar and dorsal aspects of index, middle, and lateral half of ring fingers and palmar aspect of thumb	Absence of thumb opposition Lagging 2nd and 3rd digits when making a fist Atrophy of thenar eminence, thumb adducted and extended **A. Inability to oppose thumb** (movement occurs at carpo-metacarpal joint) **B. Simian hand**
Median nerve at wrist	Wrist laceration Carpal tunnel syndrome	Weakened flexion of MCP joints of index and middle fingers; opposition and abduction of CMC and MCP joint of thumb lost; sensory loss same as for median nerve injury at elbow	Test: Make a "O" with thumb and index finger
Ulnar nerve at elbow	Fracture of medial epicondyle of humerus	Hand will deviate to radial side during flexion of wrist joint; flexion of DIP joints of ring and little finger lost; flexion at MCP joint and extension at PIP and DIP joints of little and ring finger are lost; adduction and abduction of MCP joints of digits 2–5 lost; adduction of thumb lost; sensory loss on palmar and dorsal aspects of little and medial half of ring fingers	Claw hand Palmar digital branches Palmar branch **A. Claw hand** **B. Sensory distribution of ulnar nerve**
Ulnar nerve at wrist	Wrist laceration	Flexion at MCP joint and extension at PIP and DIP joints of little and ring fingers lost; adduction and abduction of MCP joints of digits 2–5 lost; adduction of thumb lost; sensory loss same as for ulnar nerve injury at elbow	Test: Hold paper between middle and ring fingers.

CMC, carpometacarpal joint; DIP, distal interphalangeal joint; MCP, metacarpophalangeal joint; PIP, proximal interphalangeal joint.

A. Lateral View

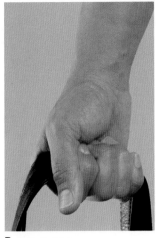

B. Anterior View

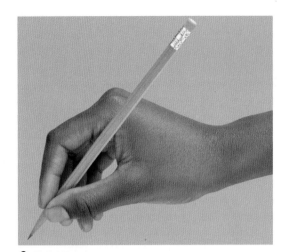

C. Medial View

D. Medial View

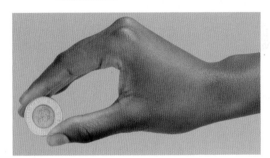

E. Medial View

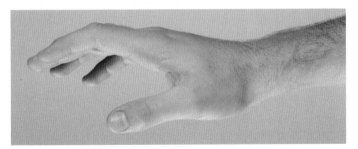

F. Medial View

I. Lateral View

G. Anterior View

H. Anterior View

FUNCTIONAL POSITIONS OF HAND

2.95

A. Cylindrical (power) grasp. When grasping an object, the metacarpophalangeal and interphalangeal joints are flexed, but the radiocarpal joints are extended. Without wrist extension, the grip is weak and insecure. **B.** Hook grasp. This grasp involves primarily the long flexors of the fingers, which are flexed to a varying degree depending on the size of the object. **C.** Tripod (three-jaw chuck) pinch. **D.** and **E.** Fingertip pinch. **F.** Rest position of hand. Casts for fractures are applied most often with the hand in this position. **G.** Loose cylindrical grasp. **H.** Firm cylindrical (power) grasp. **I.** Disc (power) grasp.

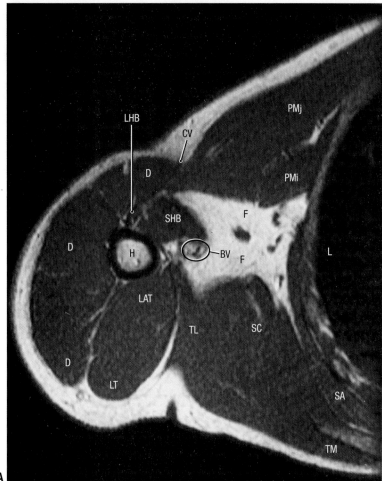

A. ANTERIOR / POSTERIOR

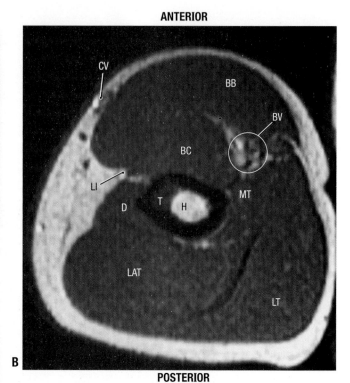

B. ANTERIOR / POSTERIOR

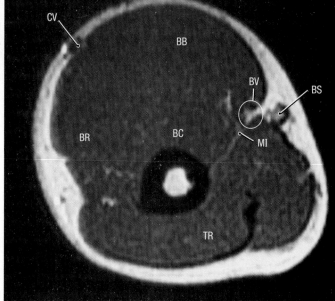

C. ANTERIOR / POSTERIOR

Key for A, B, and C:	
BB	Biceps brachii
BC	Brachialis
BR	Brachioradialis
BS	Basilic Vein
BV	Brachial vessels and nerves
CV	Cephalic vein
D	Deltoid
F	Fat in axilla
H	Humerus
L	Lung
LAT	Lateral head of triceps brachii
LHB	Long head of biceps brachii
LI	Lateral intermuscular septum
LT	Long head of triceps brachii
MI	Medial intermuscular septum
MT	Medial head of triceps brachii
PMi	Pectoralis minor
PMj	Pectoralis major
SA	Serratus anterior
SC	Subscapularis
SHB	Short head of biceps brachii
T	Deltoid tuberosity
TL	Teres major and latissimus dorsi
TM	Teres minor
TR	Triceps brachii

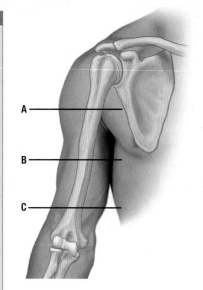

2.96 TRANSVERSE (AXIAL) MRIs OF ARM

A. Transverse MRI through the proximal arm. **B.** Transverse MRI through the middle of the arm. **C.** Transverse MRI through the distal arm.

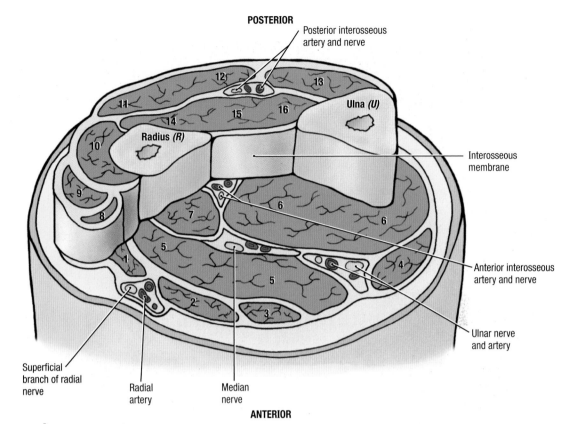

Key

▢	Extensor–supinator compartment
▢	Flexor–pronator compartment

Flexors:

1. Pronator teres
2. Flexor carpi radialis
3. Palmaris longus
4. Flexor carpi ulnaris
5. Flexor digitorum superficialis
6. Flexor digitorum profundus
7. Flexor pollicis longus

Extensors:

8. Brachioradialis
9. Extensor carpi radialis longus
10. Extensor carpi radialis brevis
11. Extensor digitorum
12. Extensor digiti minimi
13. Extensor carpi ulnaris
14. Abductor pollicis longus
15. Extensor pollicis brevis
16. Extensor pollicis longus (extensor indicis)

A. Anterosuperior View

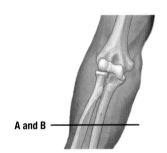

A and B

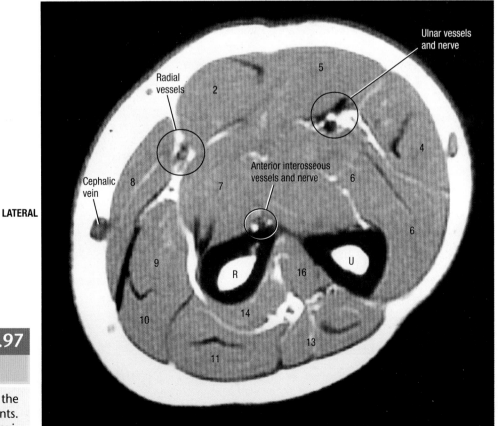

B. Transverse MRI

TRANSVERSE SECTIONS AND TRANSVERSE (AXIAL) MRIs OF FOREARM **2.97**

A. Stepped transverse sections of the anterior and posterior compartments. **B.** Transverse MRI through the proximal forearm.

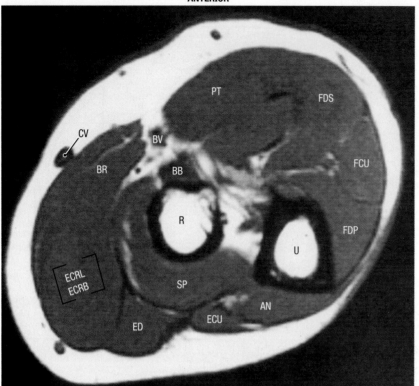

ANTERIOR

LATERAL

MEDIAL

POSTERIOR

C. Transverse MRI

Key for C and D:	
AN	Anconeus
APL	Abductor pollicis longus
BB	Biceps brachii
BR	Brachioradialis
BV	Brachial vessels
CV	Cephalic vein
ECRB	Extensor carpi radialis brevis
ECRL	Extensor carpi radialis longus
ECU	Extensor carpi ulnaris
ED	Extensor digitorum
EPB	Extensor pollicis brevis
EPL	Extensor pollicis longus
FCR	Flexor carpi radialis
FCU	Flexor carpi ulnaris
FDP	Flexor digitorum profundus
FDS	Flexor digitorum superficialis
FPL	Flexor pollicis longus
PQ	Pronator quadratus
PT	Pronator teres
R	Radius
SP	Supinator
U	Ulna
UV	Ulnar vessels and nerve

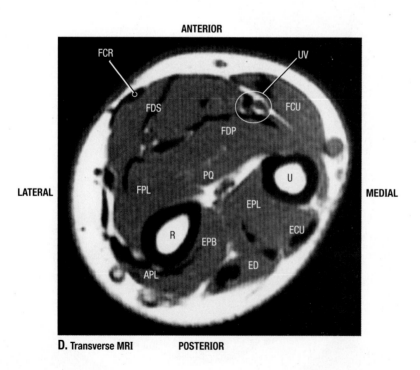

ANTERIOR

LATERAL

MEDIAL

POSTERIOR

D. Transverse MRI

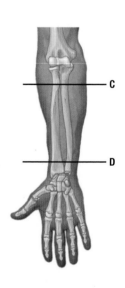

2.97 TRANSVERSE SECTIONS AND TRANSVERSE (AXIAL) MRIs OF FOREARM (*continued*)

C. Transverse MRI through the middle forearm. **D.** Transverse MRI through the distal forearm.

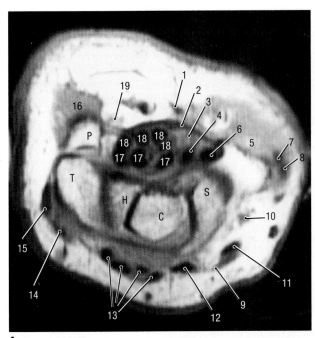

A. Transverse MRI

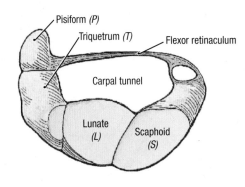

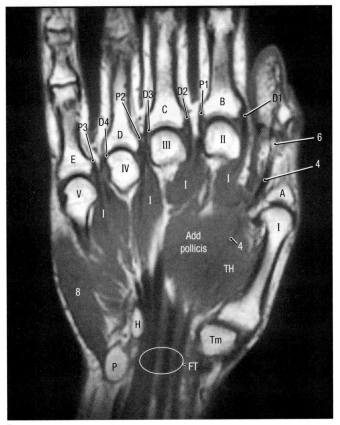

B. Coronal MRI

TRANSVERSE (AXIAL) SECTION AND MRIs THROUGH CARPAL TUNNEL

A. Transverse MRI through the proximal carpal tunnel. Numbers and letters in MRI refer to structures **(D)**. **B.** Coronal MRI of wrist and hand showing the course of the long flexor tendons in the carpal tunnel. Numbers and letters in MRI refer to structures **(D)**.

A–E, proximal phalanges; *FT,* long flexor tendons; *H,* hook of hamate; *I,* interossei; *P,* pisiform; *TH,* thenar muscles; *Tm,* trapezium; *1–5,* heads of metacarpals.

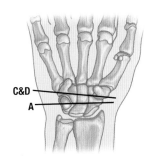

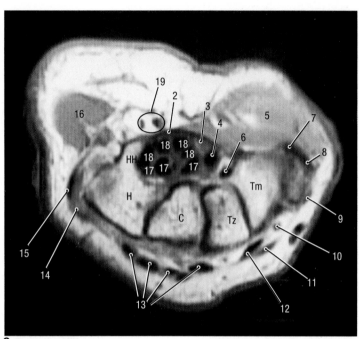

C. Transverse MRI

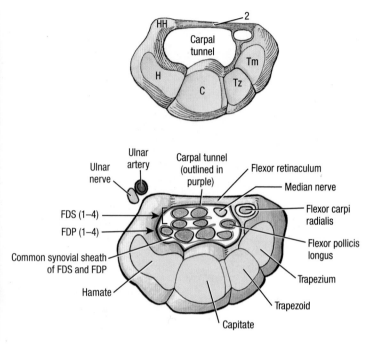

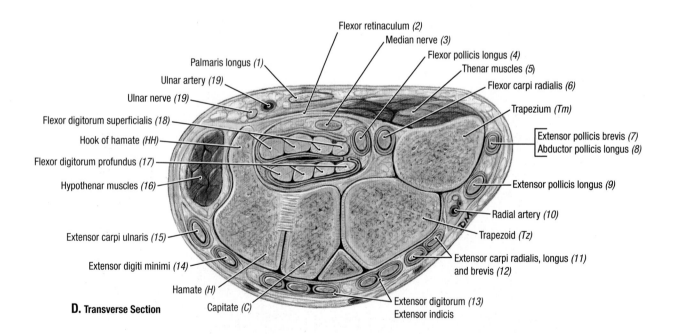

D. Transverse Section

Flexor retinaculum (2)
Median nerve (3)
Palmaris longus (1)
Flexor pollicis longus (4)
Ulnar artery (19)
Thenar muscles (5)
Ulnar nerve (19)
Flexor carpi radialis (6)
Flexor digitorum superficialis (18)
Trapezium (Tm)
Hook of hamate (HH)
Extensor pollicis brevis (7)
Abductor pollicis longus (8)
Flexor digitorum profundus (17)
Extensor pollicis longus (9)
Hypothenar muscles (16)
Radial artery (10)
Extensor carpi ulnaris (15)
Trapezoid (Tz)
Extensor digiti minimi (14)
Extensor carpi radialis, longus (11) and brevis (12)
Hamate (H)
Extensor digitorum (13)
Capitate (C)
Extensor indicis

2.98 | TRANSVERSE (AXIAL) SECTION AND MRIs THROUGH CARPAL TUNNEL (*continued*)

C. Transverse MRI through the distal carpal tunnel. Numbers and letters in MRI refer to structures **(D)**. **D.** Transverse section of carpal tunnel through the distal row of carpal bones.

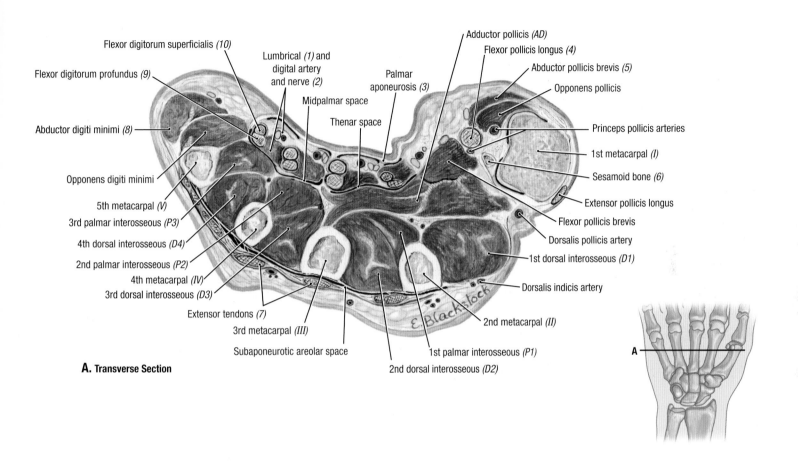

Flexor digitorum superficialis *(10)*
Flexor digitorum profundus *(9)*
Abductor digiti minimi *(8)*
Opponens digiti minimi
5th metacarpal *(V)*
3rd palmar interosseous *(P3)*
4th dorsal interosseous *(D4)*
2nd palmar interosseous *(P2)*
4th metacarpal *(IV)*
3rd dorsal interosseous *(D3)*
Extensor tendons *(7)*
3rd metacarpal *(III)*
Subaponeurotic areolar space

Lumbrical *(1)* and digital artery and nerve *(2)*
Midpalmar space
Thenar space
Palmar aponeurosis *(3)*

Adductor pollicis *(AD)*
Flexor pollicis longus *(4)*
Abductor pollicis brevis *(5)*
Opponens pollicis
Princeps pollicis arteries
1st metacarpal *(I)*
Sesamoid bone *(6)*
Extensor pollicis longus
Flexor pollicis brevis
Dorsalis pollicis artery
1st dorsal interosseous *(D1)*
Dorsalis indicis artery
2nd metacarpal *(II)*
1st palmar interosseous *(P1)*
2nd dorsal interosseous *(D2)*

E. Blackstock

A. Transverse Section

A

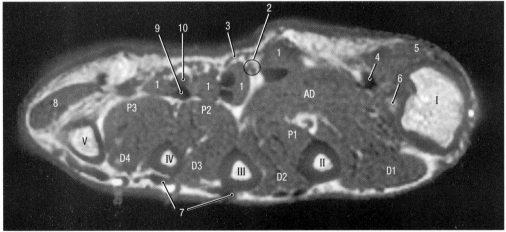

B. Transverse MRI

A. Anatomic section. **B.** MRI

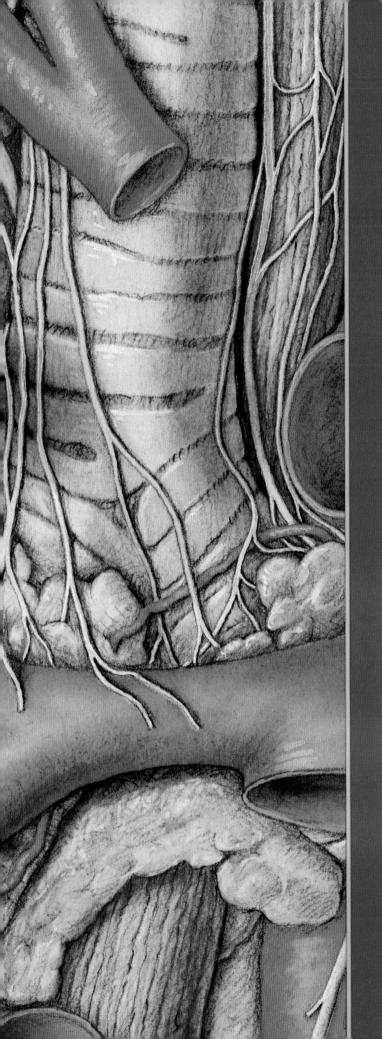

CHAPTER 3

Thorax

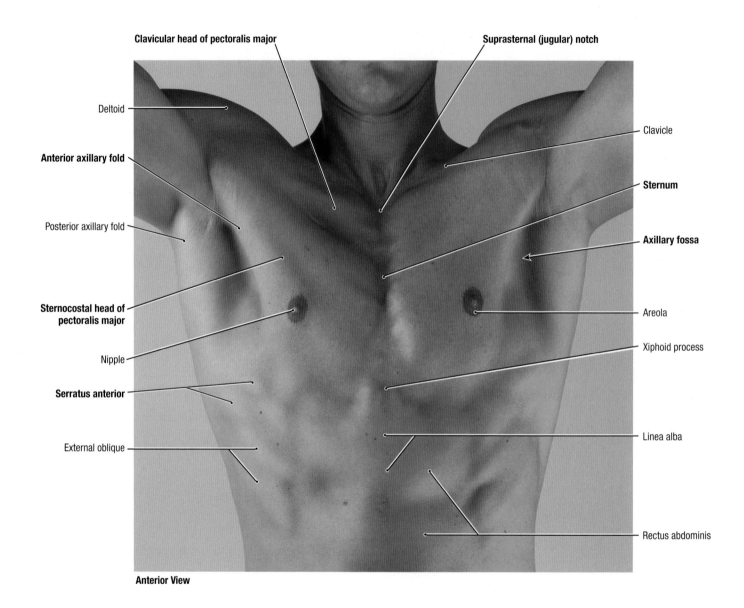

Clavicular head of pectoralis major

Suprasternal (jugular) notch

Deltoid

Clavicle

Anterior axillary fold

Sternum

Posterior axillary fold

Axillary fossa

Sternocostal head of pectoralis major

Areola

Nipple

Xiphoid process

Serratus anterior

External oblique

Linea alba

Rectus abdominis

Anterior View

3.1 SURFACE ANATOMY OF MALE PECTORAL REGION

- The subject is adducting the shoulders against resistance to demonstrate the pectoralis major muscle.
- The sternum (breastbone) lies subcutaneously in the anterior median line and is palpable throughout its length.
- The suprasternal notch can be palpated between the prominent medial ends of the clavicles.
- The pectoralis major muscle has two parts, the sternocostal and clavicular heads.
- The inferior border of the sternocostal head of the pectoralis major muscle forms the anterior axillary fold. The axillary fossa ("armpit") is a surface feature overlying a fat-filled space, the axilla, posterior to the anterior fold.
- The male nipple overlies the 4th intercostal space.

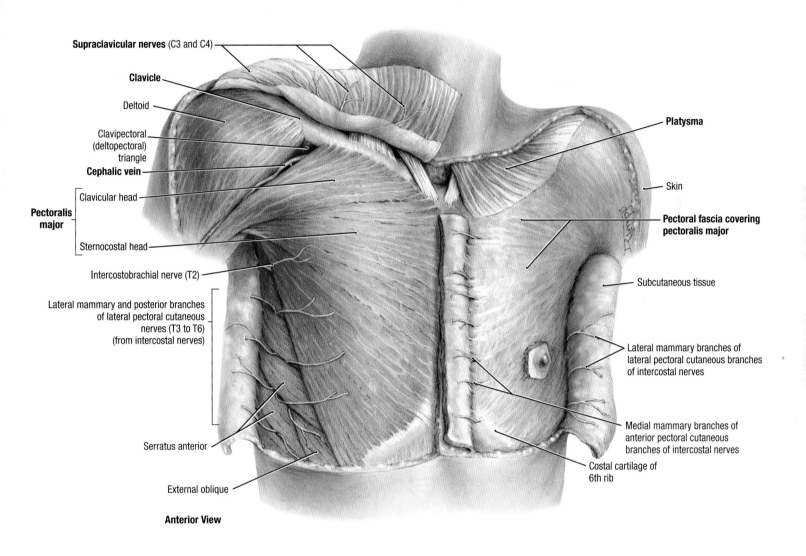

Supraclavicular nerves (C3 and C4)

Clavicle

Deltoid

Clavipectoral (deltopectoral) triangle

Cephalic vein

Pectoralis major — Clavicular head

Sternocostal head

Intercostobrachial nerve (T2)

Lateral mammary and posterior branches of lateral pectoral cutaneous nerves (T3 to T6) (from intercostal nerves)

Serratus anterior

External oblique

Platysma

Skin

Pectoral fascia covering pectoralis major

Subcutaneous tissue

Lateral mammary branches of lateral pectoral cutaneous branches of intercostal nerves

Medial mammary branches of anterior pectoral cutaneous branches of intercostal nerves

Costal cartilage of 6th rib

Anterior View

SUPERFICIAL DISSECTION, MALE PECTORAL REGION   3.2

- The platysma muscle, which descends to the 2nd or 3rd rib, is cut short on both sides of the specimen; together with the supraclavicular nerves, it is reflected superiorly on the right side.
- The pectoral fascia covers the pectoralis major.
- The clavicle lies deep to the subcutaneous tissue and the platysma muscle.
- The cephalic vein passes deeply in the clavipectoral (deltopectoral) triangle to join the axillary vein.

- Supraclavicular (C3 and C4) and upper thoracic nerves (T2 to T6) supply cutaneous innervation to the pectoral region.
- The clavipectoral (deltopectoral) triangle, bounded by the clavicle superiorly, the deltoid muscle laterally, and the clavicular head of the pectoralis major muscle medially, underlies a surface depression called the infraclavicular fossa.

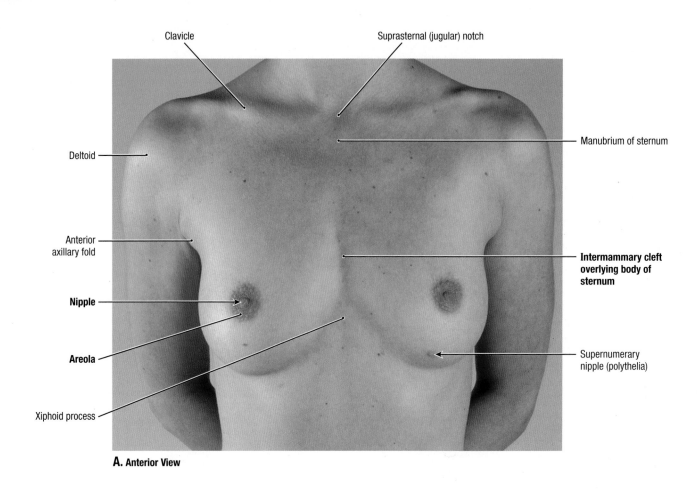

Clavicle

Suprasternal (jugular) notch

Manubrium of sternum

Deltoid

Anterior axillary fold

Intermammary cleft overlying body of sternum

Nipple

Areola

Supernumerary nipple (polythelia)

Xiphoid process

A. Anterior View

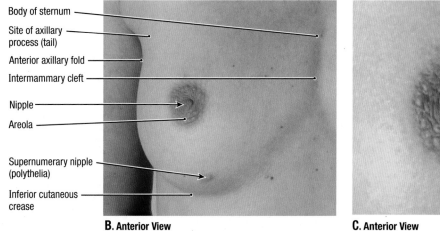

Body of sternum

Site of axillary process (tail)

Anterior axillary fold

Intermammary cleft

Nipple

Areola

Supernumerary nipple (polythelia)

Inferior cutaneous crease

B. Anterior View

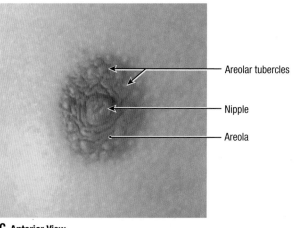

Areolar tubercles

Nipple

Areola

C. Anterior View

3.3   SURFACE ANATOMY OF FEMALE PECTORAL REGION

A. Overview. **B.** Breast. The roughly circular base of the female breast extends transversely from the lateral border of the sternum to the midaxillary line and vertically from the 2nd to 6th ribs. A small part of the breast may extend along the inferolateral edge of the pectoralis major muscle toward the axillary fossa, forming an axillary process or tail (of Spence). **C.** Areola and nipple.

Polymastia (supernumerary breasts) or **polythelia** (accessory nipples) may occur superior or inferior to the normal pair, occasionally developing in the axillary fossa or anterior abdominal wall. Supernumerary breasts usually consist of only a rudimentary nipple and areola, which may be mistaken for a mole (nevus) until they change pigmentation with the normal nipples during pregnancy.

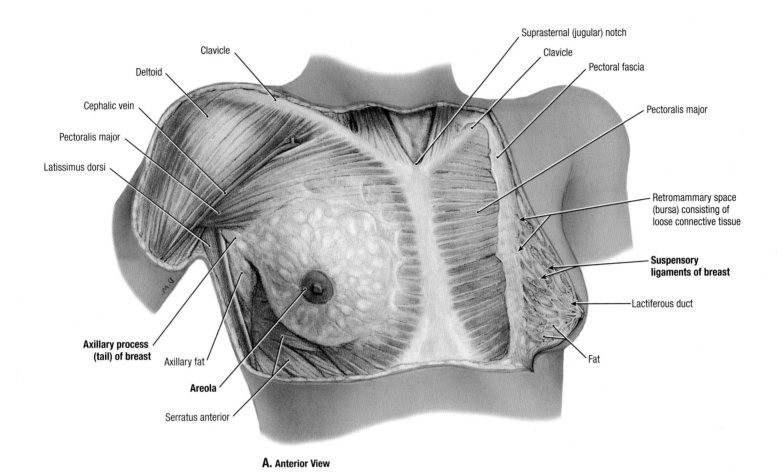

A. Anterior View

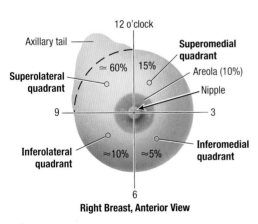

B. Quadrants of Breast: Percentage of Malignant Tumors

SUPERFICIAL DISSECTION, FEMALE PECTORAL REGION

3.4

A. Dissection.
- On the specimen's right side, the skin is removed; on the left side, the breast is sagittally sectioned.
- Two thirds of the breast rests on the pectoral fascia covering the pectoralis major; the other third rests on the fascia covering the serratus anterior muscle.
- The region of loose connective tissue between the pectoral fascia and the deep surface of the breast, the retromammary space (bursa), permits the breast to move on the deep fascia.

Cancer can spread by contiguity (invasion of adjacent tissue). When breast cancer cells invade the retromammary space, attach to or invade the pectoral fascia overlying the pectoralis major, or metastasize to the interpectoral nodes (Fig. 3.7), the breast elevates when the muscle contracts. This movement is a clinical sign of **advanced cancer of the breast**.

B. Breast quadrants. For the anatomical location and description of tumors and cysts, the surface of the breast is divided into four quadrants. For example, "A hard irregular mass was felt in the superior medial quadrant of the breast at the 2 o'clock position, approximately 2.5 cm from the margin of the areola."

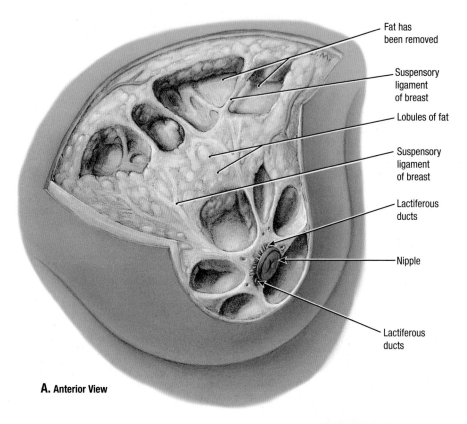

Fat has been removed

Suspensory ligament of breast

Lobules of fat

Suspensory ligament of breast

Lactiferous ducts

Nipple

Lactiferous ducts

A. Anterior View

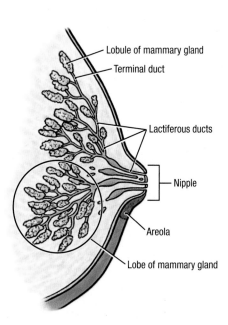

Lobule of mammary gland

Terminal duct

Lactiferous ducts

Nipple

Areola

Lobe of mammary gland

B. Schematic Sagittal Section

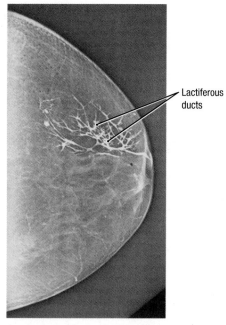

Lactiferous ducts

C. Galactogram

3.5 FEMALE MAMMARY GLAND

A. Dissection. Areas of subcutaneous fat were removed to show the suspensory ligaments of the breast. **B.** Sagittal section. The glandular tissue consists of 15 to 20 lobes, each composed of lobules. Each lobe has a lactiferous duct that widens to form the lactiferous sinus before opening on the nipple. **C.** Galactogram. This is used to image the duct system of the breast. Contrast material is injected into the ducts and mammograms are then taken.

Interference with the lymphatic drainage by cancer may cause **lymphedema** (edema, excess fluid in the subcutaneous tissue), which in turn may result in deviation of the nipple and a leathery, thickened appearance of the breast skin. Prominent (puffy) skin between dimpled pores may develop, which gives the skin an orange-peel appearance (*peau d'orange* sign). Larger dimples may form if pulled by cancerous invasion of the suspensory ligaments of the breast.

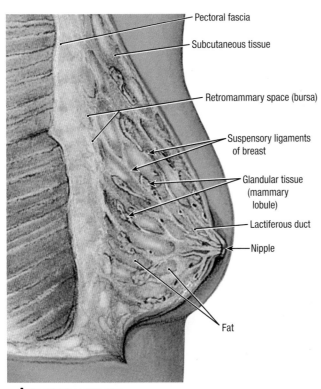

Pectoral fascia

Subcutaneous tissue

Retromammary space (bursa)

Suspensory ligaments of breast

Glandular tissue (mammary lobule)

Lactiferous duct

Nipple

Fat

A. Sagittal Breast Section

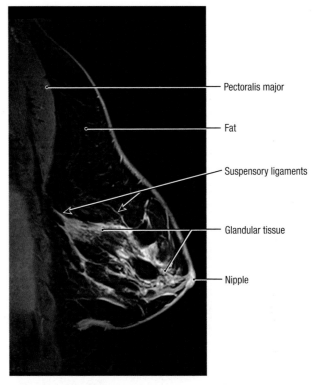

Pectoralis major

Fat

Suspensory ligaments

Glandular tissue

Nipple

B. Sagittal Breast MRI

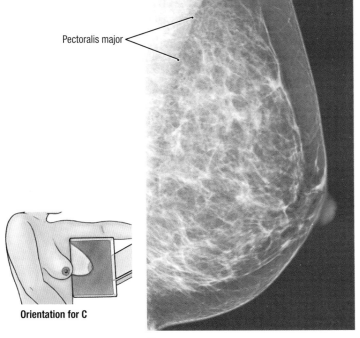

Pectoralis major

Orientation for C

C. MLO Mammogram

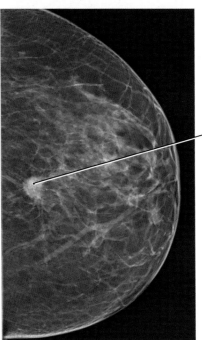

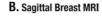

Cancer

Orientation for D

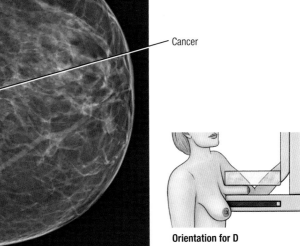

D. CC Mammogram

IMAGING OF BREAST

3.6

A. Illustration of sagittal section of breast. **B.** Sagittal MRI of breast showing many of the features visible in **(A)**. In this MRI, fat appears very dark, whereas glandular tissue is brighter and the linear suspensory ligaments clearly visible. The pectoralis major is also apparent as is the pectoralis minor posterior to it. **C.** and **D.** Scanning mammograms, which use x-rays, are done with a mediolateral oblique (MLO) and a craniocaudal (CC) orientation. These two orientations allow the entire breast to be imaged. A speculated mass (cancer) is identified in **(D)**.

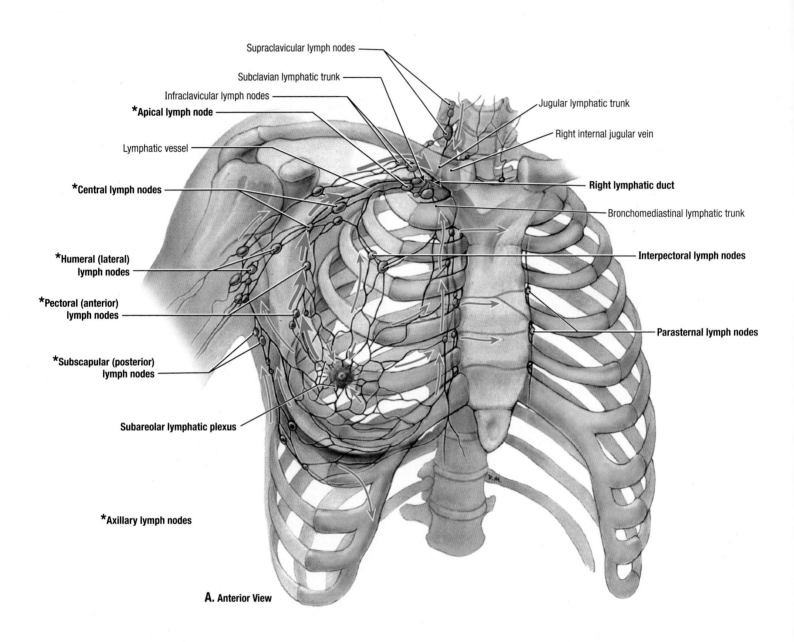

Supraclavicular lymph nodes

Subclavian lymphatic trunk

Infraclavicular lymph nodes

*Apical lymph node

Lymphatic vessel

*Central lymph nodes

*Humeral (lateral) lymph nodes

*Pectoral (anterior) lymph nodes

*Subscapular (posterior) lymph nodes

Subareolar lymphatic plexus

*Axillary lymph nodes

Jugular lymphatic trunk

Right internal jugular vein

Right lymphatic duct

Bronchomediastinal lymphatic trunk

Interpectoral lymph nodes

Parasternal lymph nodes

A. Anterior View

3.7 LYMPHATIC DRAINAGE OF BREAST

A. Overview. Lymph drained from the upper limb and breast passes through nodes arranged irregularly in groups of axillary lymph nodes: (1) pectoral, along the inferior border of the pectoralis minor muscle; (2) subscapular, along the subscapular artery and veins; (3) humeral, along the distal part of the axillary vein; (4) central, at the base of the axilla, embedded in axillary fat; and (5) apical, along the axillary vein between the clavicle and the pectoralis minor muscle. Most of the breast drains via the pectoral, central, and apical axillary nodes to the subclavian lymph trunk, which joins the venous system at the junction of the subclavian and internal jugular veins. The medial part of the breast drains to the parasternal nodes, which are located along the internal thoracic vessels.

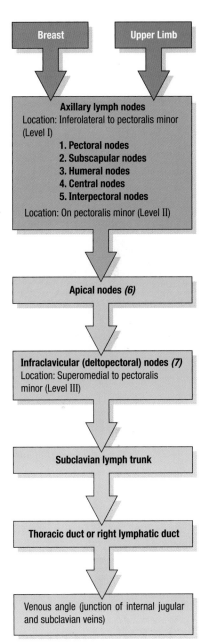

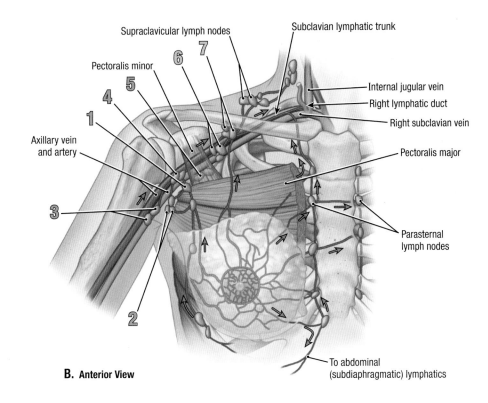

B. Anterior View

Supraclavicular lymph nodes

Subclavian lymphatic trunk

Pectoralis minor

Internal jugular vein

Right lymphatic duct

Right subclavian vein

Axillary vein and artery

Pectoralis major

Parasternal lymph nodes

To abdominal (subdiaphragmatic) lymphatics

C. Flow of Lymph from the Breast and Upper Limb to the Venous Angle.

LYMPHATIC DRAINAGE OF BREAST (*continued*)

3.7

B. Pattern of lymphatic drainage. **Breast cancer** typically spreads by means of lymphatic vessels (lymphogenic metastasis), which carry cancer cells from the breast to the lymph nodes, chiefly those in the axilla. The cells lodge in the nodes, producing nests of tumor cells (metastases). Abundant communications among lymphatic pathways and among axillary, cervical, and parasternal nodes may also cause metastases from the breast to develop in the supraclavicular lymph nodes, the opposite breast, or the abdomen. The prognosis of breast cancer has been correlated with the level of metastasis (Level I, II, or III in **C**) and to the number of involved axillary lymph nodes. **C.** Flow of lymph from the breast and upper limb to the venous angle.

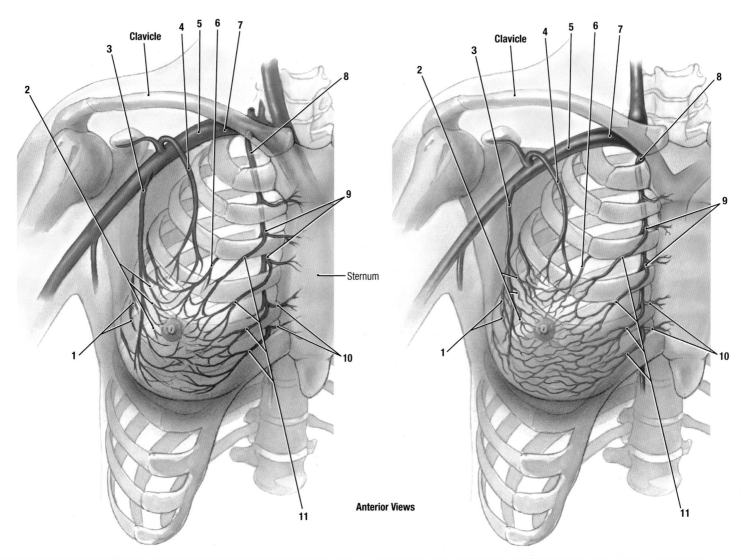

Anterior Views

Arteries
1. Lateral mammary branches of lateral cutaneous branches of posterior intercostal arteries
2. Lateral mammary branches of lateral thoracic artery
3. Lateral thoracic artery
4. Pectoral branch of thoraco-acromial artery
5. Axillary artery
6. Mammary branch of anterior intercostal artery
7. Subclavian artery
8. Internal thoracic artery
9. Perforating branches
10. Sternal branches
11. Medial mammary branches

Veins
1. Lateral mammary branches of lateral cutaneous branches of posterior intercostal veins
2. Lateral mammary branches of lateral thoracic vein
3. Lateral thoracic vein
4. Pectoral branch of thoraco-acromial vein
5. Axillary vein
6. Mammary branch of anterior intercostal vein
7. Subclavian vein
8. Internal thoracic vein
9. Perforating branches
10. Sternal branches
11. Medial mammary veins

3.8   ARTERIAL SUPPLY AND VENOUS DRAINAGE OF BREAST

Arteries enter and veins drain the breast from its superomedial and superolateral aspects; vessels also penetrate the deep surface of the breast. The vessels branch profusely and anastomose with each other.

Breast incisions are placed in the inferior breast quadrants when possible because these quadrants are less vascular than the superior ones.

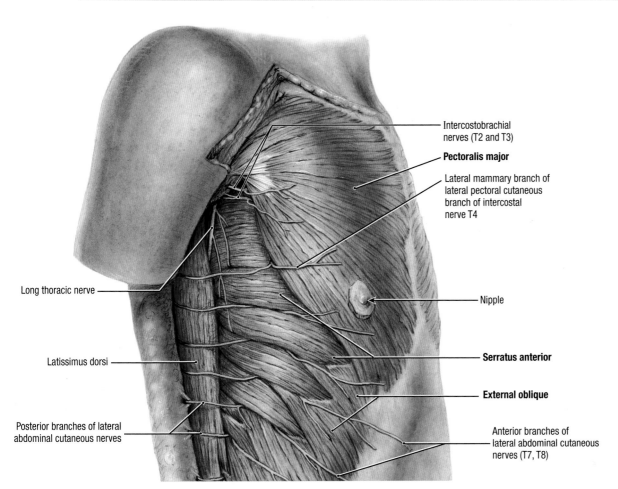

Intercostobrachial
nerves (T2 and T3)

Pectoralis major

Lateral mammary branch of
lateral pectoral cutaneous
branch of intercostal
nerve T4

Nipple

Serratus anterior

External oblique

Anterior branches of
lateral abdominal cutaneous
nerves (T7, T8)

Long thoracic nerve

Latissimus dorsi

Posterior branches of lateral
abdominal cutaneous nerves

A. Anterolateral View (Male)

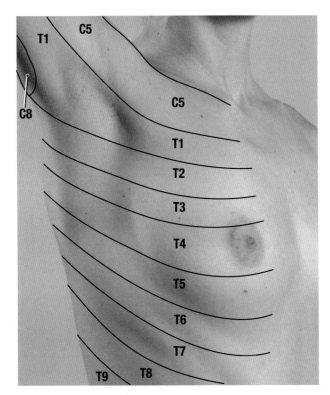

C5

T1

C5

C8

C5

T1

T2

T3

T4

T5

T6

T7

T8

T9

B. Anterolateral View (Female)

MUSCLES AND NERVES OF BED OF BREAST   3.9

A. Muscles comprising bed and cutaneous nerves. **B.** Dermatomes.

Local anesthesia of an intercostal space (intercostal nerve block) is produced by injecting a local anesthetic agent around the intercostal nerves between the paravertebral line and the area of required anesthesia. Because any particular area of skin usually receives innervation from two adjacent nerves, considerable overlapping of contiguous dermatomes occurs. Therefore, complete loss of sensation usually does not occur unless two or more intercostal nerves are anesthetized.

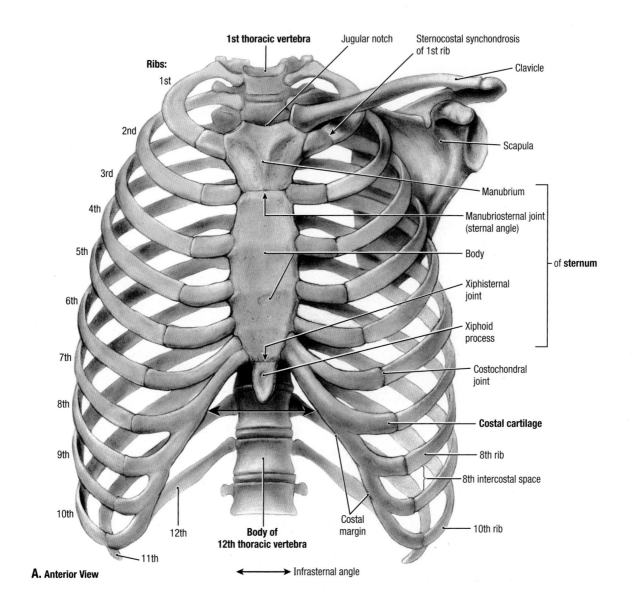

Ribs:
1st

2nd

3rd

4th

5th

6th

7th

8th

9th

10th

12th

11th

1st thoracic vertebra

Jugular notch

Sternocostal synchondrosis of 1st rib

Clavicle

Scapula

Manubrium

Manubriosternal joint (sternal angle)

Body — of **sternum**

Xiphisternal joint

Xiphoid process

Costochondral joint

Costal cartilage

8th rib

8th intercostal space

10th rib

Costal margin

Body of 12th thoracic vertebra

Infrasternal angle

A. Anterior View

3.10   BONY THORAX

- The thoracic cage consists of 12 thoracic vertebrae, 12 pairs of ribs and costal cartilages, and the sternum.
- Anteriorly, the superior seven costal cartilages articulate with the sternum; the 8th, 9th, and 10th cartilages articulate with the cartilage above forming the costal margin; the 11th and 12th are "floating" ribs, that is, their cartilages do not articulate anteriorly.
- The clavicle lies over the 1st rib, making it difficult to palpate. The 2nd rib is easily palpable because its costal cartilage articulates

with the sternum at the sternal angle, located at the junction of the manubrium and body of the sternum.
- The 3rd to 10th ribs can be palpated in sequence inferolaterally from the 2nd rib; the fused costal cartilages of the 7th to 10th ribs form the costal arch (margin), and the tips of the 11th and 12th ribs can be palpated posterolaterally.
- A **rib dislocation** is the displacement of a costal cartilage from the sternum; a **rib separation** refers to dislocation of the costochondral joint.

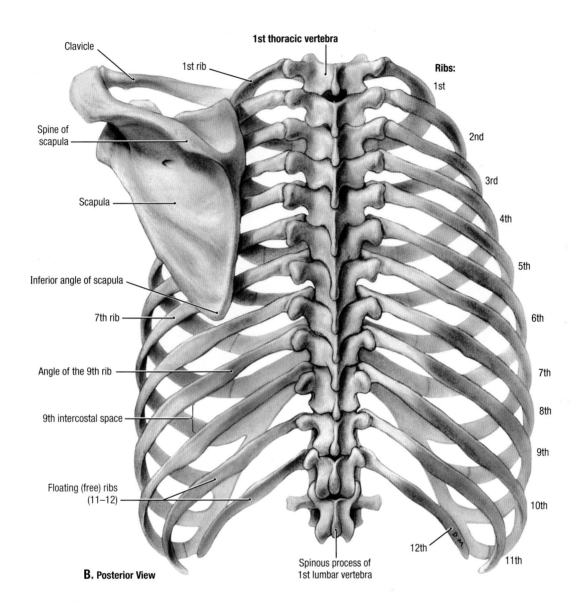

Clavicle

1st rib

1st thoracic vertebra

Spine of scapula

Scapula

Inferior angle of scapula

7th rib

Angle of the 9th rib

9th intercostal space

Floating (free) ribs (11–12)

Ribs:

1st

2nd

3rd

4th

5th

6th

7th

8th

9th

10th

11th

12th

Spinous process of 1st lumbar vertebra

B. Posterior View

BONY THORAX (*continued*)

- The superior thoracic aperture (thoracic inlet) is the doorway between the thoracic cavity and the neck region; it is bounded by the 1st thoracic vertebra, the 1st ribs and their cartilages, and the manubrium of the sternum.
- Each rib articulates posteriorly with the vertebral column.
- Posteriorly, all ribs angle inferiorly; anteriorly, the 3rd to 10th costal cartilages angle superiorly.
- The scapula is suspended from the clavicle and extends across the 2nd to 7th ribs posteriorly.

- When clinicians refer to the superior thoracic aperture as the thoracic "outlet," they are emphasizing the important nerves and arteries that pass through this aperture into the lower neck and upper limb. Hence, various types of **thoracic outlet syndromes** exist, such as the costoclavicular syndrome—pallor and coldness of the skin of the upper limb and diminished radial pulse—resulting from compression of the subclavian artery between the clavicle and the 1st rib.

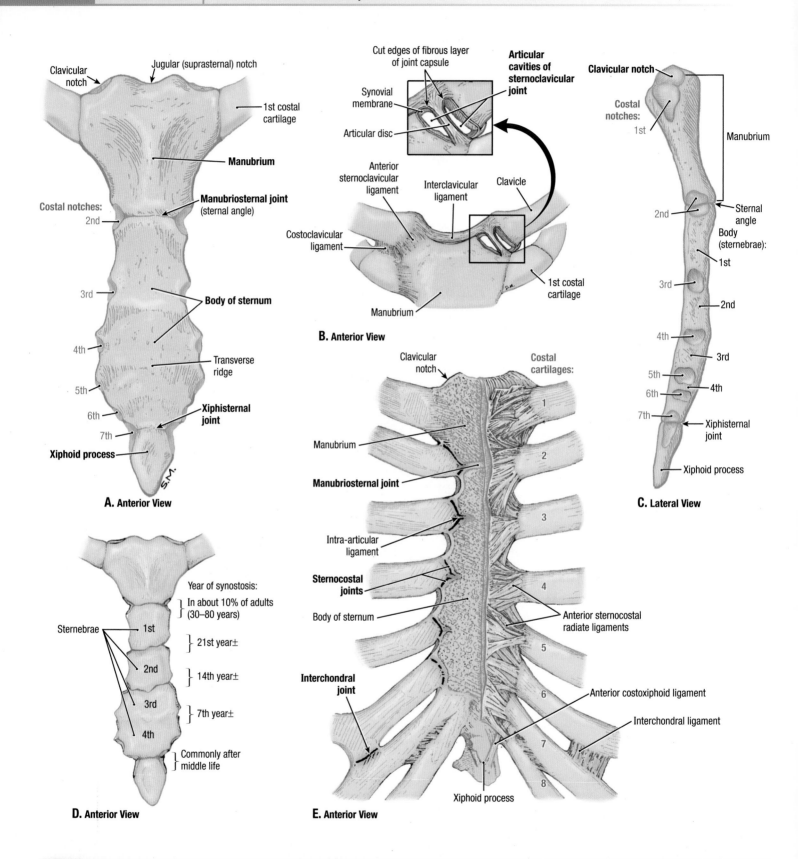

A. Anterior View

Clavicular notch
Jugular (suprasternal) notch
1st costal cartilage
Manubrium
Manubriosternal joint (sternal angle)
Costal notches:
2nd
3rd
4th
5th
6th
7th
Body of sternum
Transverse ridge
Xiphisternal joint
Xiphoid process

B. Anterior View

Cut edges of fibrous layer of joint capsule
Articular cavities of sternoclavicular joint
Synovial membrane
Articular disc
Anterior sternoclavicular ligament
Interclavicular ligament
Clavicle
Costoclavicular ligament
1st costal cartilage
Manubrium

C. Lateral View

Clavicular notch
Costal notches:
1st
Manubrium
2nd
Sternal angle
Body (sternebrae):
1st
3rd
2nd
4th
3rd
5th
4th
6th
7th
Xiphisternal joint
Xiphoid process

D. Anterior View

Sternebrae
1st
2nd
3rd
4th
Year of synostosis:
In about 10% of adults (30–80 years)
21st year±
14th year±
7th year±
Commonly after middle life

E. Anterior View

Clavicular notch
Costal cartilages:
1
2
3
4
5
6
7
8
Manubrium
Manubriosternal joint
Intra-articular ligament
Sternocostal joints
Body of sternum
Interchondral joint
Anterior sternocostal radiate ligaments
Anterior costoxiphoid ligament
Interchondral ligament
Xiphoid process

3.11 **STERNUM AND ASSOCIATED JOINTS**

A. Parts of sternum. **B.** Sternoclavicular joint. **C.** Features of the lateral aspect of the sternum. **D.** Ages of ossification of sternum. **E.** Sternocostal, manubriosternal, and interchondral joints.

On the right side of the specimen, the cortex of the sternum and the external surface of the costal cartilages have been shaved away.

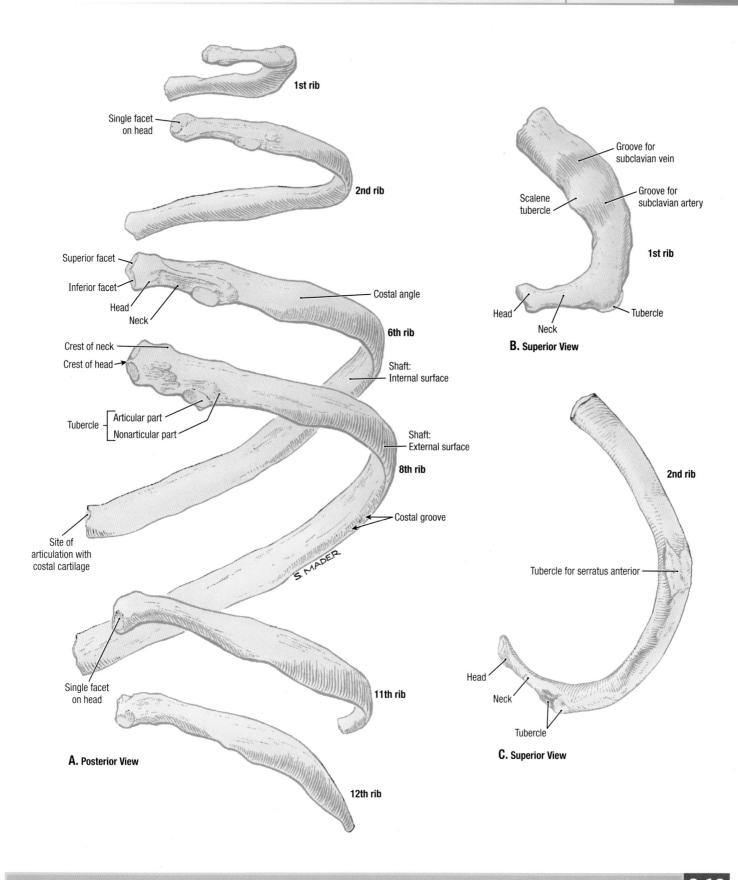

1st rib

Single facet on head

2nd rib

Superior facet

Inferior facet

Head

Neck

Costal angle

6th rib

Shaft: Internal surface

Crest of neck

Crest of head

Shaft: External surface

Tubercle — Articular part / Nonarticular part

8th rib

Site of articulation with costal cartilage

Costal groove

S. MADER

Single facet on head

11th rib

A. Posterior View

12th rib

Groove for subclavian vein

Scalene tubercle

Groove for subclavian artery

1st rib

Head

Neck

Tubercle

B. Superior View

2nd rib

Tubercle for serratus anterior

Head

Neck

Tubercle

C. Superior View

RIBS

3.12

A. "Typical" (6th and 8th) and "atypical" (1st and 2nd and 11th and 12th) ribs. **B.** First rib. **C.** Second rib.

Rib fractures. The weakest part of a rib is immediately anterior to its angle. The middle ribs are most commonly fractured.

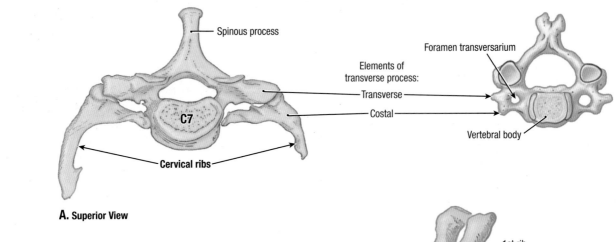

A. Superior View

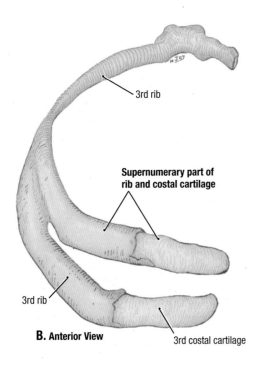

B. Anterior View

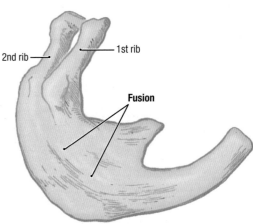

C. Superior View

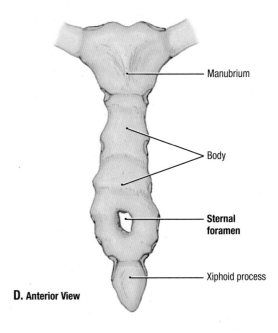

D. Anterior View

3.13 RIB AND STERNUM ANOMALIES

A. Cervical ribs. People usually have 12 ribs on each side, but the number may be increased by the presence of cervical and/or lumbar ribs (supernumerary ribs) or decreased by a failure of the 12th pair to form. **Cervical ribs** (present in up to 1% of people) articulate with the C7 vertebra and are clinically significant because they may compress spinal nerves C8 and T1 or the inferior trunk of the brachial plexus supplying the upper limb. Tingling and numbness may occur along the medial border of the forearm. They may also compress the subclavian artery, resulting in **ischemic muscle pain** (caused by poor blood supply) in the upper limb. **Lumbar ribs** are less common than cervical ribs but have clinical significance in that they may confuse the identity of vertebral levels in diagnostic images. **B.** Bifid rib. The superior component of this 3rd rib is supernumerary and articulated with the lateral aspect of the 1st sternebra. The inferior component articulated at the junction of the 1st and 2nd sternebrae. **C.** Bicipital rib. In this specimen, there has been partial fusion of the first two thoracic ribs. **D.** Sternal foramen.

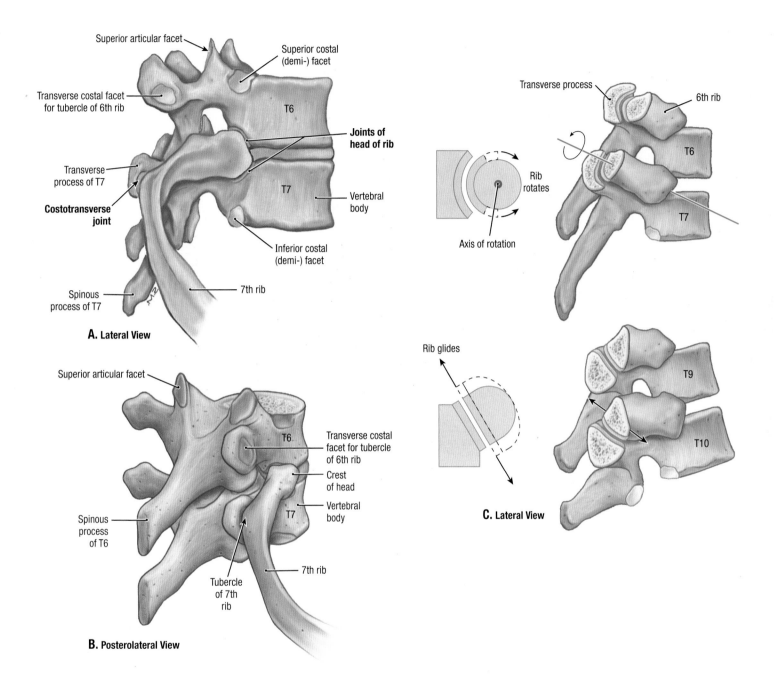

A. Lateral View

- Superior articular facet
- Superior costal (demi-) facet
- Transverse costal facet for tubercle of 6th rib
- T6
- **Joints of head of rib**
- Transverse process of T7
- T7
- **Costotransverse joint**
- Vertebral body
- Inferior costal (demi-) facet
- Spinous process of T7
- 7th rib

B. Posterolateral View

- Superior articular facet
- T6
- Transverse costal facet for tubercle of 6th rib
- Crest of head
- Vertebral body
- T7
- Spinous process of T6
- 7th rib
- Tubercle of 7th rib

- Transverse process
- 6th rib
- Rib rotates
- Axis of rotation
- T6
- T7

- Rib glides
- T9
- T10

C. Lateral View

COSTOVERTEBRAL ARTICULATIONS

3.14

A. and **B.** Articulating structures.

- There are two articular facets on the head of the rib: a larger, inferior costal facet for articulation with the vertebral body of its own number, and a smaller, superior costal facet for articulation with the vertebral body of the vertebra superior to the rib.
- The crest of the head of the rib separates the superior and inferior costal facets.

- The smooth articular part of the tubercle of the rib, the transverse costal facet, articulates with the transverse process of the same numbered vertebra at the costotransverse joint.

C. Movements at the costotransverse joints. At the 1st to 7th costotransverse joints, the ribs rotate, increasing the anteroposterior diameter of the thorax; at the 8th, 9th, and 10th, they glide, increasing the transverse diameter of the upper abdomen.

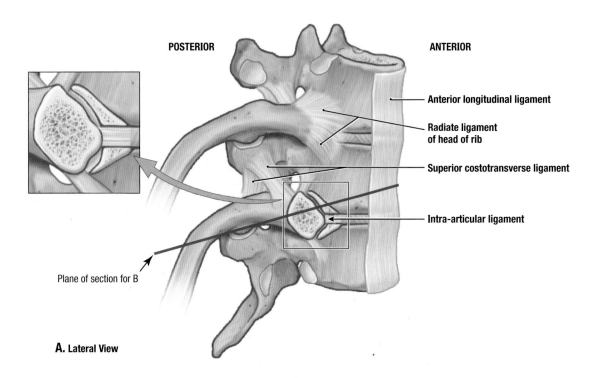

POSTERIOR

ANTERIOR

Anterior longitudinal ligament

Radiate ligament
of head of rib

Superior costotransverse ligament

Intra-articular ligament

Plane of section for B

A. Lateral View

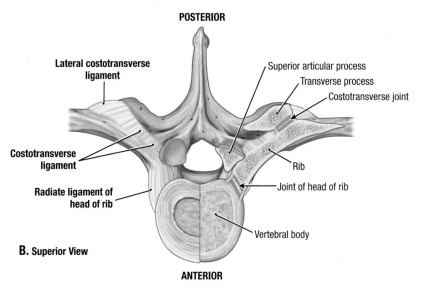

POSTERIOR

Lateral costotransverse
ligament

Superior articular process

Transverse process

Costotransverse joint

Costotransverse
ligament

Radiate ligament of
head of rib

Rib

Joint of head of rib

Vertebral body

B. Superior View

ANTERIOR

3.15 LIGAMENTS OF COSTOVERTEBRAL ARTICULATIONS

A. External and internal ligaments.
- The radiate ligament joins the head of the rib to two vertebral bodies and the interposed intervertebral disc.
- The superior costotransverse ligament joins the crest of the neck of the rib to the transverse process above.
- The intra-articular ligament joins the crest of the head of the rib to the intervertebral disc.

B. Transverse section.
- The vertebral body, transverse processes, superior articulating processes, and posterior elements of the articulating ribs have been transversely sectioned to visualize the joint surfaces and ligaments.
- The costotransverse ligament joins the posterior aspect of the neck of the rib to the adjacent transverse process.
- The lateral costotransverse ligament joins the nonarticulating part of the tubercle of the rib to the tip (apex) of the transverse process.

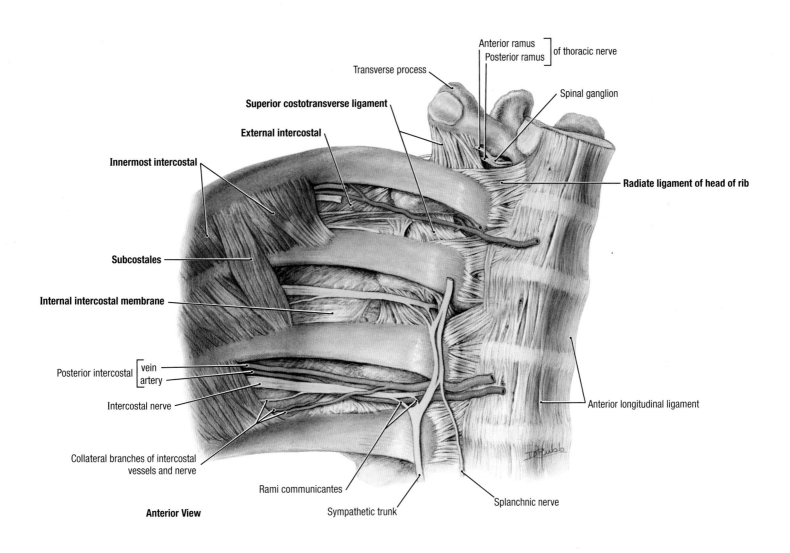

Anterior ramus ⎤ of thoracic nerve
Posterior ramus ⎦

Transverse process

Spinal ganglion

Superior costotransverse ligament

External intercostal

Innermost intercostal

Radiate ligament of head of rib

Subcostales

Internal intercostal membrane

Posterior intercostal ⎡ vein
                      ⎣ artery

Intercostal nerve

Anterior longitudinal ligament

Collateral branches of intercostal vessels and nerve

Rami communicantes

Splanchnic nerve

Sympathetic trunk

Anterior View

| **VERTEBRAL ENDS OF INTERNAL ASPECT OF INTERCOSTAL SPACES** | **3.16** |

- Portions of the innermost intercostal muscle that bridge two intercostal spaces are called subcostales muscles.
- The internal intercostal membrane, in the middle space, is continuous medially with the superior costotransverse ligament.
- Note the order of the structures in the most inferior space: posterior intercostal vein and artery, and intercostal nerve; note also their collateral branches.

- The anterior ramus crosses anterior to the superior costotransverse ligament; the posterior ramus is posterior to it.
- The intercostal nerves attach to the sympathetic trunk by rami communicantes; the splanchnic nerve is a visceral branch of the trunk.

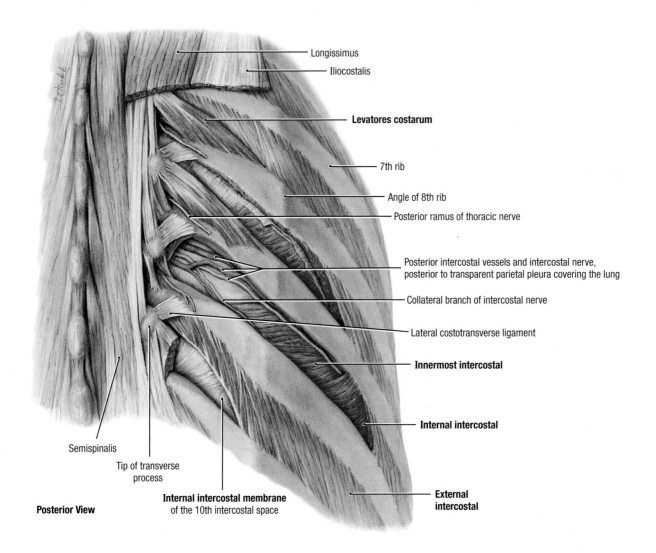

Longissimus
Iliocostalis
Levatores costarum
7th rib
Angle of 8th rib
Posterior ramus of thoracic nerve
Posterior intercostal vessels and intercostal nerve, posterior to transparent parietal pleura covering the lung
Collateral branch of intercostal nerve
Lateral costotransverse ligament
Innermost intercostal
Internal intercostal
Semispinalis
Tip of transverse process
Posterior View
Internal intercostal membrane of the 10th intercostal space
External intercostal

3.17   VERTEBRAL ENDS OF EXTERNAL ASPECT OF INFERIOR INTERCOSTAL SPACES

- The iliocostalis and longissimus muscles have been removed, exposing the levatores costarum muscle. Of the five intercostal spaces shown, the superior two (6th and 7th) are intact. In the 8th and 10th spaces, varying portions of the external intercostal muscle have been removed to reveal the underlying internal intercostal membrane, which is continuous with the internal intercostal muscle. In the 9th space, the levatores costarum muscle has been removed to show the posterior intercostal vessels and intercostal nerve.
- The intercostal vessels and nerve disappear laterally between the internal and innermost intercostal muscles.
- The intercostal nerve is the most inferior of the neurovascular trio (posterior intercostal vein and artery and intercostal nerve) and the least sheltered in the intercostal groove; a collateral branch arises near the angle of the rib.
- **Thoracocentesis.** Sometimes it is necessary to insert a hypodermic needle through an intercostal space into the pleural cavity (see Fig. 3.27) to obtain a sample of pleural fluid or to remove blood or pus. To avoid damage to the intercostal nerve and vessels, the needle is inserted superior to the rib, high enough to avoid the collateral branches.

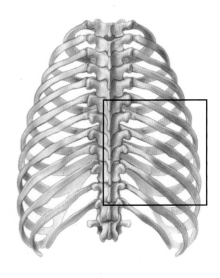

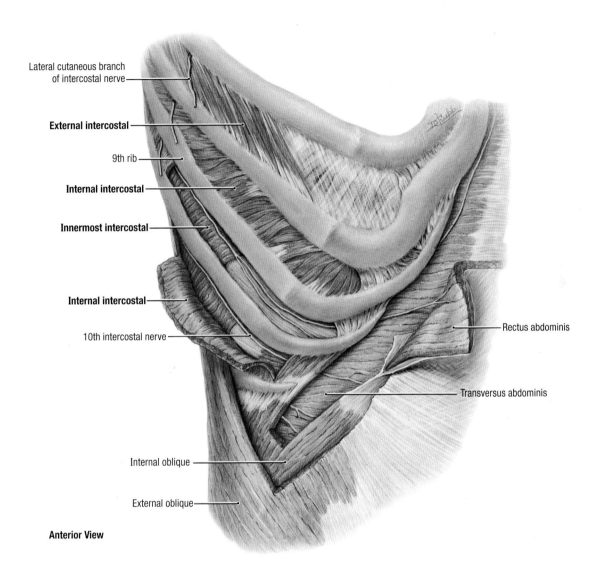

Lateral cutaneous branch of intercostal nerve

External intercostal

9th rib

Internal intercostal

Innermost intercostal

Internal intercostal

10th intercostal nerve

Internal oblique

External oblique

Anterior View

Rectus abdominis

Transversus abdominis

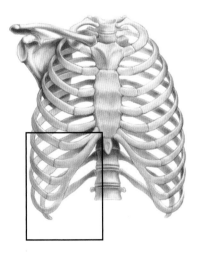

ANTERIOR ENDS OF INFERIOR INTERCOSTAL SPACES

3.18

- The fibers of the external intercostal and external oblique muscles run inferomedially.
- The internal intercostal and internal oblique muscles are in continuity at the ends of the 9th, 10th, and 11th intercostal spaces.
- The intercostal nerves lie deep to the internal intercostal muscle but superficial to the innermost intercostal muscle; anteriorly, these nerves lie superficial to the transversus thoracis or transversus abdominis muscles.
- Intercostal nerves run parallel to the ribs and costal cartilages; on reaching the abdominal wall, nerves T7 and T8 continue superiorly, T9 continues nearly horizontally, and T10 continues inferomedially toward the umbilicus. These nerves provide cutaneous innervation in overlapping segmental bands.

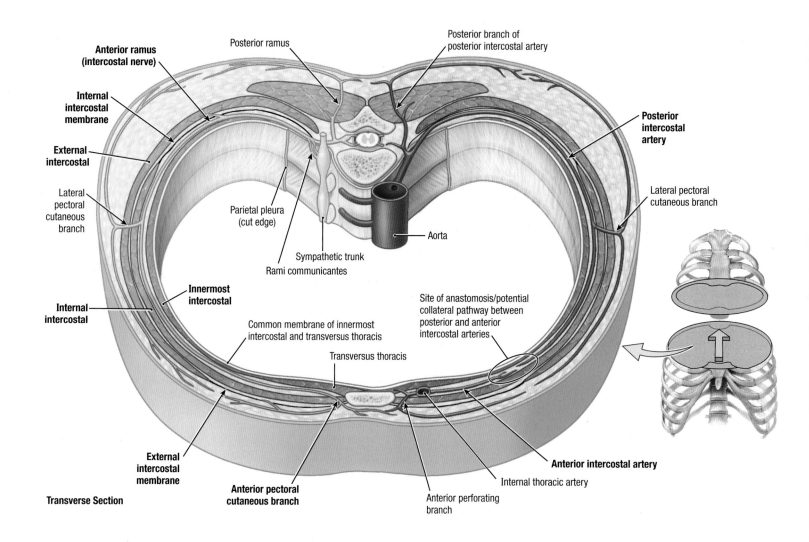

Anterior ramus
(intercostal nerve)

Posterior ramus

Posterior branch of
posterior intercostal artery

Internal
intercostal
membrane

**Posterior
intercostal
artery**

External
intercostal

Lateral
pectoral
cutaneous
branch

Parietal pleura
(cut edge)

Lateral pectoral
cutaneous branch

Aorta

Sympathetic trunk

Rami communicantes

Innermost
intercostal

Internal
intercostal

Site of anastomosis/potential
collateral pathway between
posterior and anterior
intercostal arteries

Common membrane of innermost
intercostal and transversus thoracis

Transversus thoracis

External
intercostal
membrane

Anterior intercostal artery

Internal thoracic artery

Transverse Section

Anterior pectoral
cutaneous branch

Anterior perforating
branch

3.19 CONTENTS OF INTERCOSTAL SPACE, TRANSVERSE SECTION

- The diagram is simplified by showing nerves on the right and arteries on the left.
- The three musculomembranous layers are the external intercostal muscle and membrane, internal intercostal muscle and membrane, and the innermost intercostal muscle, transversus thoracis muscle, and the membrane connecting them.
- The intercostal nerves are the anterior rami of spinal nerves T1 to T11; the anterior ramus of T12 is the subcostal nerve.

- Posterior intercostal arteries are branches of the aorta (the superior two spaces are supplied from the superior intercostal branch of the costocervical trunk); the anterior intercostal arteries are branches of the internal thoracic artery or its branch, the musculophrenic artery.
- The posterior rami innervate the deep back muscles and skin adjacent to the vertebral column.

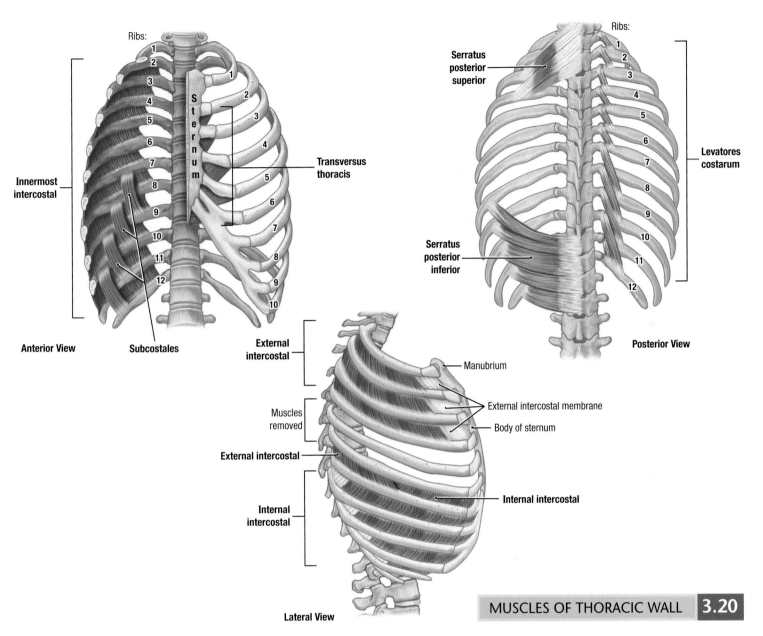

Ribs: 1 2 3 4 5 6 7 8 9 10 11 12

Sternum

Innermost intercostal

Transversus thoracis

Anterior View

Subcostales

Ribs: 1 2 3 4 5 6 7 8 9 10 11 12

Serratus posterior superior

Serratus posterior inferior

Levatores costarum

Posterior View

External intercostal

Muscles removed

External intercostal

Internal intercostal

Manubrium

External intercostal membrane

Body of sternum

Internal intercostal

Lateral View

MUSCLES OF THORACIC WALL | **3.20**

TABLE 3.1 MUSCLES OF THORACIC WALL

Muscles	Superior Attachment	Inferior Attachment	Innervation	Main Action[a]
External intercostal	Inferior border of ribs	Superior border of ribs below	Intercostal nerve	During forced inspiration: elevate ribs[a]
Internal intercostal				During forced respiration: interosseous part depresses ribs; interchondral part elevates ribs[a]
Innermost intercostal				
Transversus thoracis	Posterior surface of lower sternum	Internal surface of costal cartilages 2–6		Weakly depress ribs
Subcostales	Internal surface of lower ribs near their angles	Superior borders of 2nd or 3rd rib below		Probably act in same manner as internal intercostal muscles
Levatores costarum	Transverse processes of C7–T11	Subjacent ribs between tubercle and angle	Posterior rami of C8–T11 nerves	Elevate ribs
Serratus posterior superior	Nuchal ligament, spinous processes of C7–T3 vertebrae	2nd–4th ribs near their angles	2nd–5th intercostal nerves	Elevate ribs[b]
Serratus posterior inferior	Spinous processes of T11–L2 vertebrae	Inferior borders of 8th–12th ribs near their angles	9th–11th intercostal nerves, subcostal (T12) nerve	Depress ribs[b]

[a]The tonus of the intercostal muscles keep intercostal spaces rigid, thereby preventing them from billowing (bulging) out during expiration and from being drawn in during inspiration. The role of individual intercostal muscles and accessory muscles of respiration in moving the ribs is difficult to interpret despite many electromyographic studies.

[b]Action traditionally assigned on the basis of attachments; these muscles appear to be largely proprioceptive in function.

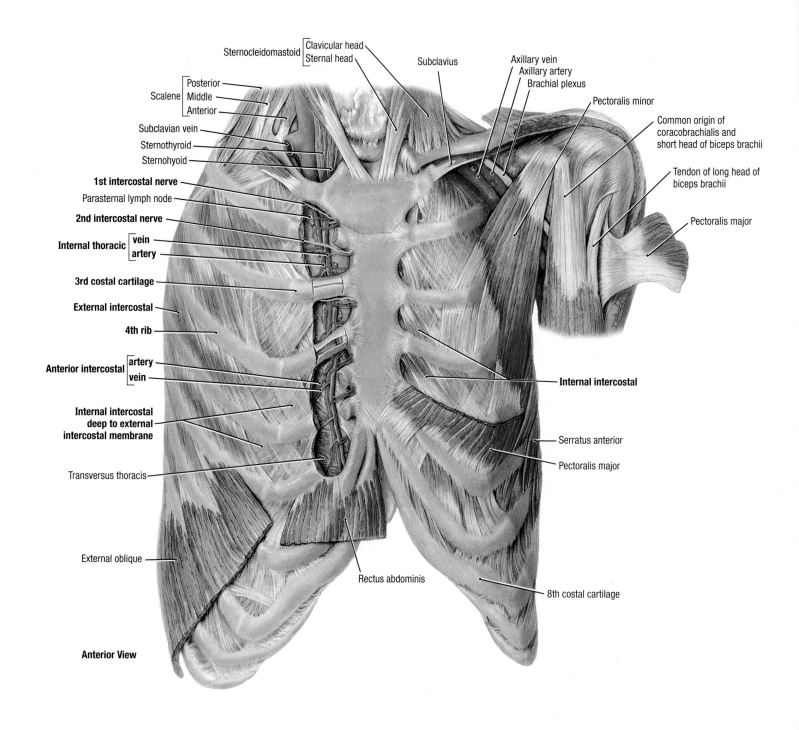

Sternocleidomastoid
Clavicular head
Sternal head
Subclavius
Axillary vein
Axillary artery
Brachial plexus
Pectoralis minor
Scalene — Posterior / Middle / Anterior
Common origin of coracobrachialis and short head of biceps brachii
Subclavian vein
Sternothyroid
Sternohyoid
Tendon of long head of biceps brachii
1st intercostal nerve
Parasternal lymph node
2nd intercostal nerve
Pectoralis major
Internal thoracic — vein / artery
3rd costal cartilage
External intercostal
4th rib
Anterior intercostal — artery / vein
Internal intercostal
Internal intercostal deep to external intercostal membrane
Transversus thoracis
Serratus anterior
Pectoralis major
External oblique
Rectus abdominis
8th costal cartilage

Anterior View

3.21 EXTERNAL ASPECT OF THORACIC WALL

- H-shaped cuts were made through the perichondrium of the 3rd and 4th costal cartilages to shell out segments of cartilage.
- During surgery, **retaining perichondrium** promotes regrowth of removed cartilages.
- The internal thoracic (internal mammary) vessels run inferiorly deep to the costal cartilages and just lateral to the edge of the sternum, providing anterior intercostal branches.

- The parasternal lymph nodes *(green)* receive lymphatic vessels from the anterior parts of intercostal spaces, the costal pleura and diaphragm, and the medial part of the breast.
- The subclavian vessels are "sandwiched" between the 1st rib and clavicle and are "padded" by the subclavius.
- **Surgical access to thorax.** To gain access to the thoracic cavity for surgical procedures, the sternum is divided in the median plane (median sternotomy) and retracted (spread apart). After surgery, the halves of the sternum are held together with wire sutures.

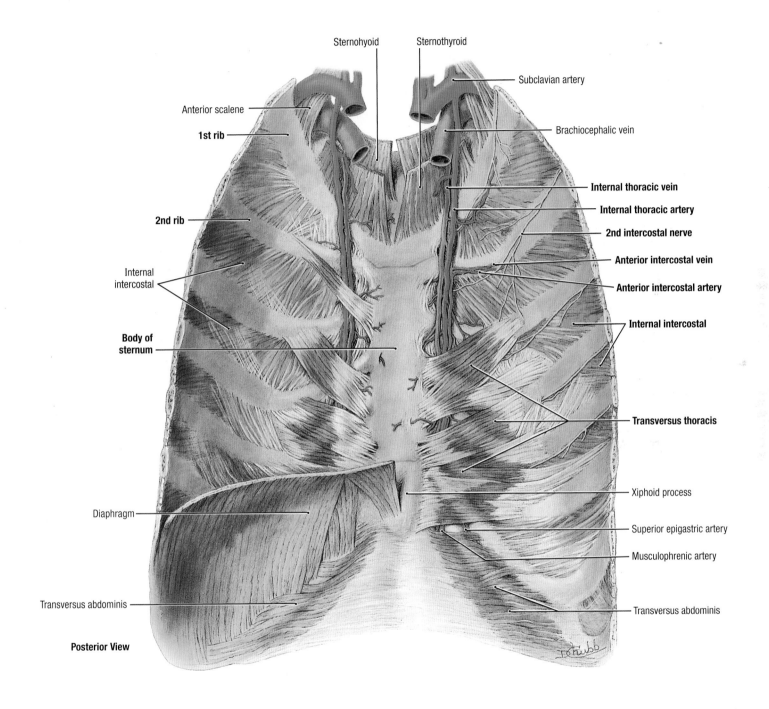

Sternohyoid

Sternothyroid

Subclavian artery

Anterior scalene

Brachiocephalic vein

1st rib

Internal thoracic vein

Internal thoracic artery

2nd rib

2nd intercostal nerve

Anterior intercostal vein

Internal intercostal

Anterior intercostal artery

Internal intercostal

Body of sternum

Transversus thoracis

Xiphoid process

Diaphragm

Superior epigastric artery

Musculophrenic artery

Transversus abdominis

Transversus abdominis

Posterior View

INTERNAL ASPECT OF THE ANTERIOR THORACIC WALL

3.22

- The inferior portions of the internal thoracic vessels are covered posteriorly by the transversus thoracis muscle; the superior portions are in contact with the parietal pleura (removed).
- The transversus thoracis muscle (superior to diaphragm) is continuous with the transversus abdominis muscle (inferior to diaphragm); these form the innermost layer of the three flat muscles of the thoracoabdominal wall.

- The internal thoracic (internal mammary) artery arises from the subclavian artery and is accompanied by two venae comitantes up to the 2nd costal cartilage in this specimen and, superior to this, by the single internal thoracic vein, which drains into the brachiocephalic vein.

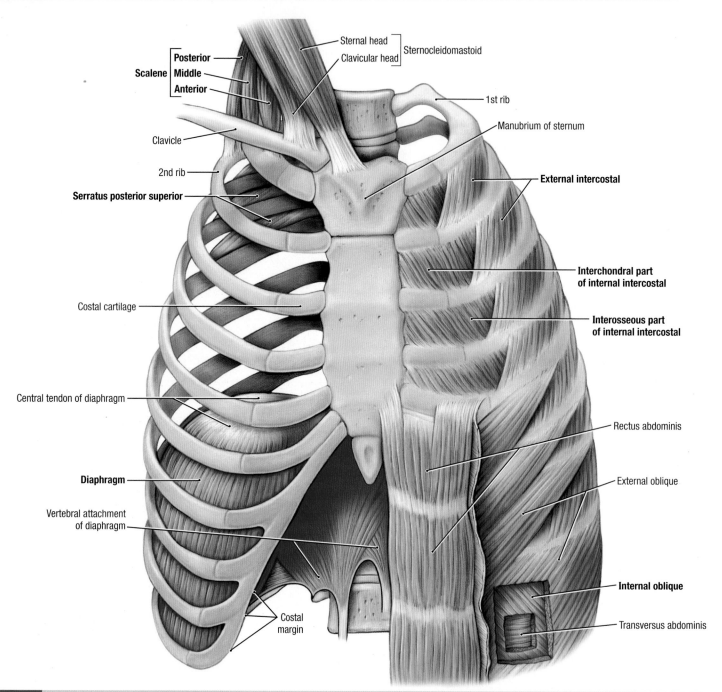

Sternocleidomastoid
- Sternal head
- Clavicular head

Scalene
- **Posterior**
- **Middle**
- **Anterior**

Clavicle

2nd rib

Serratus posterior superior

Costal cartilage

Central tendon of diaphragm

Diaphragm

Vertebral attachment of diaphragm

Costal margin

1st rib

Manubrium of sternum

External intercostal

Interchondral part of internal intercostal

Interosseous part of internal intercostal

Rectus abdominis

External oblique

Internal oblique

Transversus abdominis

3.23 MUSCLES OF RESPIRATION

TABLE 3.2		**MUSCLES OF RESPIRATION**	
		Inspiration	*Expiration*
Normal (Quiet)	Major	Diaphragm (active contraction)	Passive (elastic) recoil of lungs and thoracic cage
	Minor	*Tonic contraction* of external intercostals and interchondral portion of internal intercostals to resist negative pressure	*Tonic contraction* of muscles of anterolateral abdominal walls (rectus abdominis, external and internal obliques, transversus abdominis) to antagonize diaphragm by maintaining intra-abdominal pressure
Active (Forced)		In addition to the above, *active contraction* of sternocleidomastoid, descending (superior) trapezius, pectoralis minor, and scalenes, to elevate and fix upper rib cage	In addition to the above, *active contraction* of muscles of anterolateral abdominal wall (antagonizing diaphragm by increasing intra-abdominal pressure and by pulling inferiorly and fixing inferior costal margin): rectus abdominis, external and internal obliques, and transversus abdominis
		External intercostals, interchondral portion of internal intercostals, subcostales, levatores costarum, and serratus posterior superior[a] to elevate ribs	Internal intercostal (interosseous part) and serratus posterior inferior[a] to depress ribs

[a]Recent studies indicate that the serratus posterior superior and inferior muscles may serve primarily as organs of proprioception rather than motion.

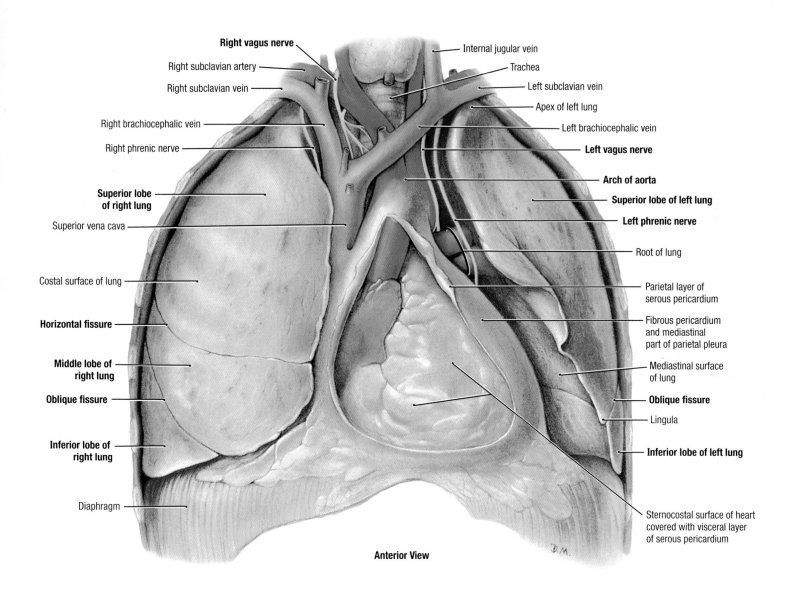

Right vagus nerve

Right subclavian artery

Right subclavian vein

Right brachiocephalic vein

Right phrenic nerve

Superior lobe of right lung

Superior vena cava

Costal surface of lung

Horizontal fissure

Middle lobe of right lung

Oblique fissure

Inferior lobe of right lung

Diaphragm

Internal jugular vein

Trachea

Left subclavian vein

Apex of left lung

Left brachiocephalic vein

Left vagus nerve

Arch of aorta

Superior lobe of left lung

Left phrenic nerve

Root of lung

Parietal layer of serous pericardium

Fibrous pericardium and mediastinal part of parietal pleura

Mediastinal surface of lung

Oblique fissure

Lingula

Inferior lobe of left lung

Sternocostal surface of heart covered with visceral layer of serous pericardium

Anterior View

D.M.

THORACIC CONTENTS *IN SITU*

- The fibrous pericardium, lined by the parietal layer of serous pericardium, is removed anteriorly to expose the heart and great vessels.
- The right lung has three lobes; the superior lobe is separated from the middle lobe by the horizontal fissure, and the middle lobe is separated from the inferior lobe by the oblique fissure.

The left lung has two lobes, superior and inferior, separated by the oblique fissure.
- The anterior border of the left lung is reflected laterally to visualize the phrenic nerve passing anterior to the root of the lung and the vagus nerve lying anterior to the arch of the aorta and then passing posterior to the root of the lung.

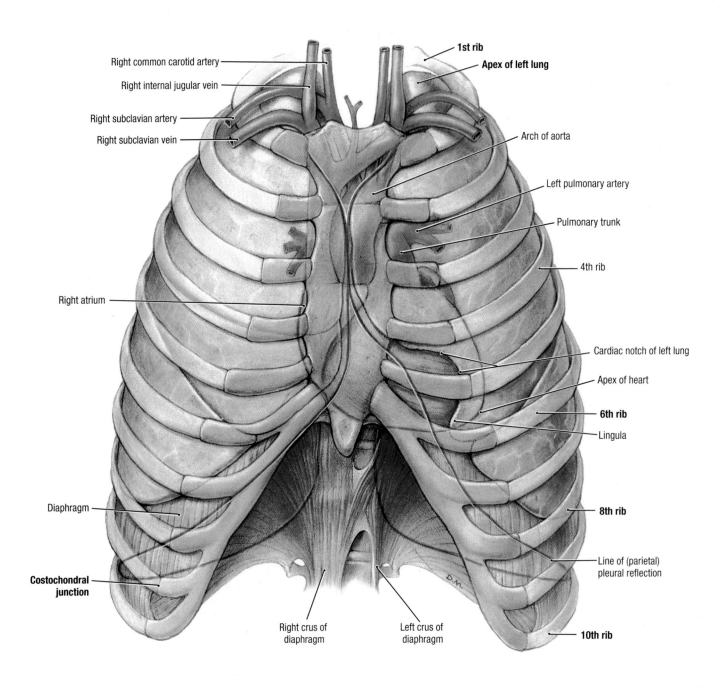

Right common carotid artery

Right internal jugular vein

Right subclavian artery

Right subclavian vein

Right atrium

Diaphragm

Costochondral junction

Right crus of diaphragm

Left crus of diaphragm

1st rib

Apex of left lung

Arch of aorta

Left pulmonary artery

Pulmonary trunk

4th rib

Cardiac notch of left lung

Apex of heart

6th rib

Lingula

8th rib

Line of (parietal) pleural reflection

10th rib

3.25 | TOPOGRAPHY OF THE LUNGS AND MEDIASTINUM

- The mediastinum is located between the pleural cavities and is occupied by the heart and the tissues anterior, posterior, and superior to the heart.
- The apex of the lungs is at the level of the neck of the 1st rib, and the inferior border of the lungs is at the 6th rib in the left midclavicular line and the 8th rib at the lateral aspect of the bony thorax at the midaxillary line.
- The cardiac notch of the left lung and the corresponding deviation of the parietal pleura are away from the median plane toward the left side.

- The inferior reflection of parietal pleura is at the 8th costochondral junction in the midclavicular line, at the 10th rib in the midaxillary line.
- The apex of the heart is in the 5th intercostal space at the left midclavicular line.
- The right atrium forms the right border of the heart and extends just beyond the lateral margin of the sternum.
- The branches of the great vessels pass through the superior thoracic aperture.

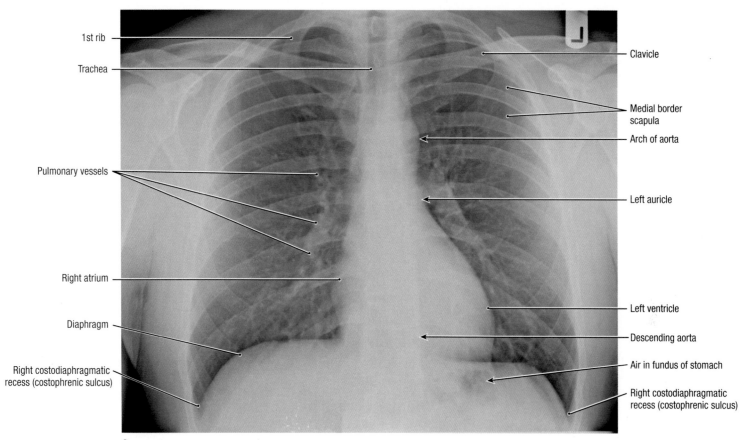

1st rib
Trachea
Pulmonary vessels
Right atrium
Diaphragm
Right costodiaphragmatic recess (costophrenic sulcus)

Clavicle
Medial border scapula
Arch of aorta
Left auricle
Left ventricle
Descending aorta
Air in fundus of stomach
Right costodiaphragmatic recess (costophrenic sulcus)

A. Postero-anterior View

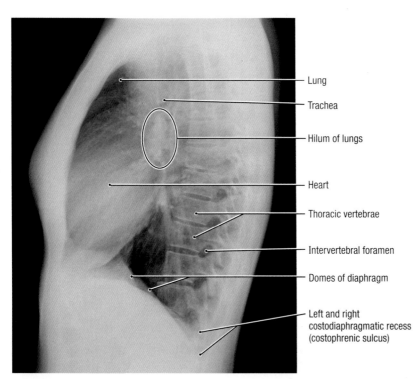

Lung
Trachea
Hilum of lungs
Heart
Thoracic vertebrae
Intervertebral foramen
Domes of diaphragm
Left and right costodiaphragmatic recess (costophrenic sulcus)

B. Lateral View

RADIOGRAPH OF CHEST 3.26

A. Posterior-anterior (PA) radiograph.
- Unless a patient is bedridden, a chest radiograph is done with the x-ray beam traversing the patient from posterior to anterior (PA) because this minimizes distortion. The scapula is protracted and not in the main field of view.
- Right atrium is the primary discernible structure along the right border of the heart.
- Within the dark gray (radiolucent) regions of both sides that show air in the lung, most of the linear denser (whiter) elements are pulmonary veins.
- Along the upper left mediastinal border, the arch of aorta visible, and the aorta can be followed inferiorly.
- Left auricle is often visible along the left border of the heart; inferiorly is the border of the left ventricle.
- In a standing PA radiograph, air is often seen in the fundus of the stomach.

B. Lateral radiograph.
- Note that the left and right are not precisely superimposed on one another.
- Notice how well the heart is shown relative to the aerated lungs, which are radio-opaque because they do not block many photons. A loss of this clear differentiation is known as the silhouette sign and suggests lung disease.
- Any structure in the mediastinum may contribute to **pathological widening of the mediastinal silhouette** (e.g., after trauma that produces hemorrhage into the mediastinum), malignant lymphoma (cancer of lymphatic tissue) that produces massive enlargement of mediastinal lymph nodes, or enlargement (hypertrophy) of the heart occurring with congestive heart failure.

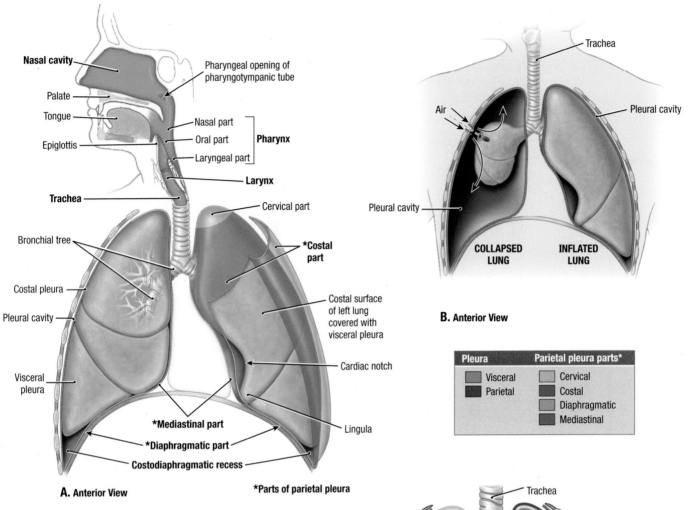

A. Anterior View

*Parts of parietal pleura

B. Anterior View

Pleura	Parietal pleura parts*
Visceral	Cervical
Parietal	Costal
	Diaphragmatic
	Mediastinal

3.27 RESPIRATORY SYSTEM AND PLEURA

A. Overview. **B.** Pleural cavity and pleura. **C.** Diagrammatic section through heart and lungs with pulmonary vessels and tracheobronchial tree.

- The lungs invaginate a continuous membranous pleural sac; the visceral (pulmonary) pleura covers the lungs, and the parietal pleura lines the thoracic cavity; the visceral and parietal pleurae are continuous around the root of the lung.
- The parietal pleura can be divided regionally into the costal, diaphragmatic, mediastinal, and cervical parts; note the costodiaphragmatic recess.
- The pleural cavity is a potential space between the visceral and parietal pleurae that contains a thin layer of fluid. If a sufficient amount of air enters the pleural cavity, the surface tension adhering visceral to parietal pleura (lung to thoracic wall) is broken, and the lung collapses **(atelectasis)** because of its inherent elasticity (elastic recoil). When a lung collapses, the pleural cavity—normally a potential space—becomes a real space **(B)** and may contain air **(pneumothorax)**, blood **(hemothorax)**, etc.

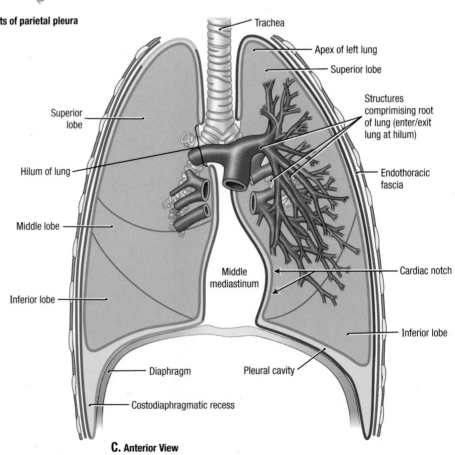

C. Anterior View

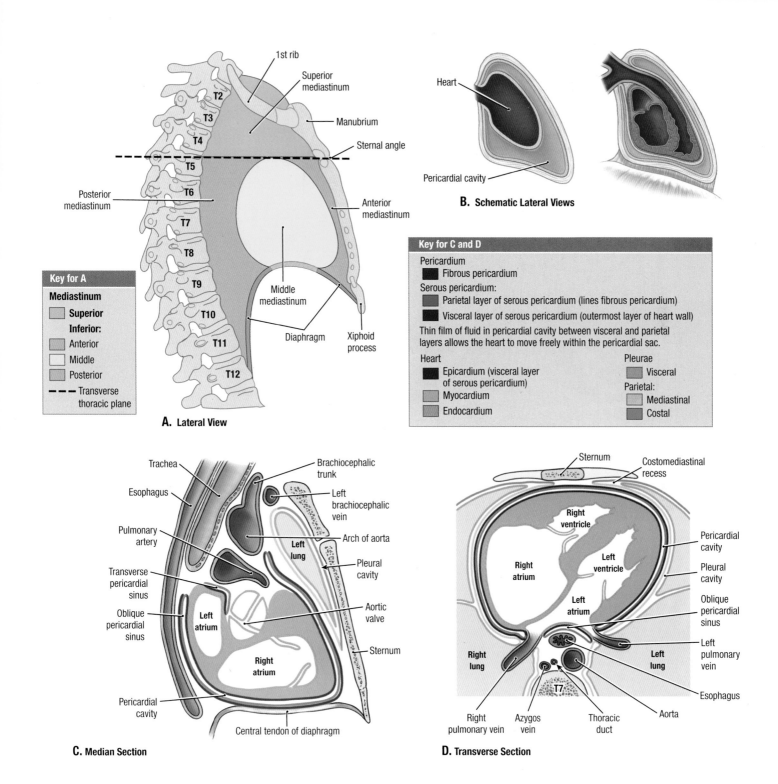

A. Lateral View

B. Schematic Lateral Views

Key for A

Mediastinum

Superior

Inferior:

Anterior

Middle

Posterior

- - - Transverse thoracic plane

Key for C and D

Pericardium

Fibrous pericardium

Serous pericardium:

Parietal layer of serous pericardium (lines fibrous pericardium)

Visceral layer of serous pericardium (outermost layer of heart wall)

Thin film of fluid in pericardial cavity between visceral and parietal layers allows the heart to move freely within the pericardial sac.

Heart

Epicardium (visceral layer of serous pericardium)

Myocardium

Endocardium

Pleurae

Visceral

Parietal:

Mediastinal

Costal

C. Median Section

D. Transverse Section

MEDIASTINUM AND PERICARDIUM

3.28

A. Subdivisions of mediastinum. **B.** Development of pericardial cavity. The embryonic heart invaginates the wall of the serous sac *(left)* and soon practically obliterates the pericardial cavity, leaving only a potential space between the layers of serous pericardium *(right)*. **C.** and **D.** Layers of pericardium and heart in sectional views.

Cardiac tamponade (heart compression) is a potentially lethal condition because heart volume is increasingly compromised by the fluid outside the heart but inside the pericardial cavity. The heart is increasingly compressed and circulation fails. Blood in the pericardial cavity, **hemopericardium**, produces cardiac tamponade.

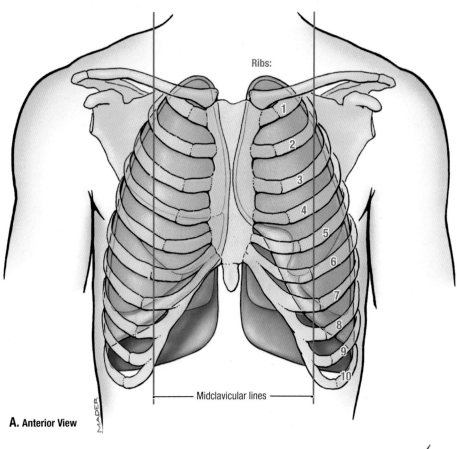

Ribs:

Midclavicular lines

A. Anterior View

3.29 EXTENT OF PARIETAL PLEURA AND LUNGS

Auscultation of lungs. Note the position of the fissures in relation to overlying ribs. To auscultate the upper lobes, place the stethoscope on the anterior thoracic wall superior to the 4th rib on the right and 6th rib on the left; for the middle lobe, place it medial to the right nipple; for the inferior lobes, place it on the posterior thoracic wall below the 3rd rib.

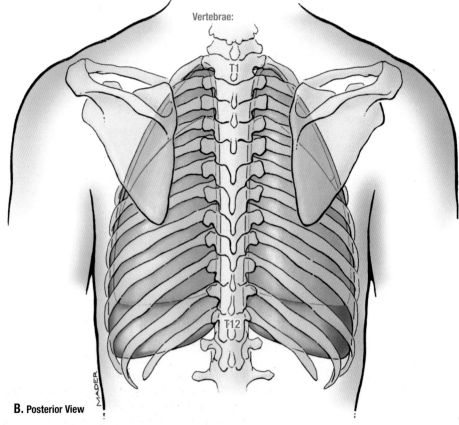

Vertebrae:

T1

T12

B. Posterior View

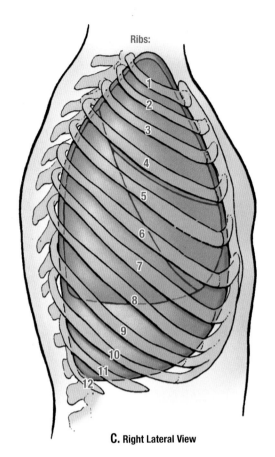

Ribs:

C. Right Lateral View

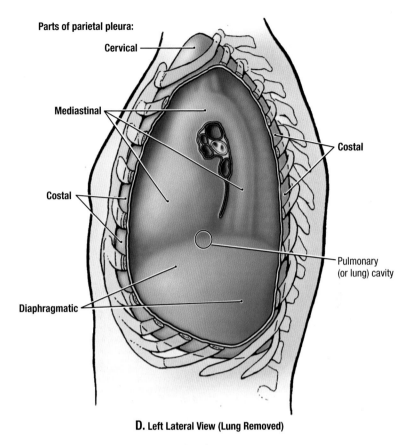

Parts of parietal pleura:

Cervical

Mediastinal

Costal

Costal

Diaphragmatic

Pulmonary (or lung) cavity

D. Left Lateral View (Lung Removed)

EXTENT OF PARIETAL PLEURA AND LUNGS (*continued*)

3.29

TABLE 3.3	SURFACE MARKINGS OF PARIETAL PLEURA (BLUE)	
Level	*Left Pleura*	*Right Pleura*
Apex	About 4 cm superior to middle of clavicle	About 4 cm superior to middle of clavicle
4th costal cartilage	Midline (anteriorly)	Midline (anteriorly)
6th costal cartilage	Lateral margin of sternum	Midline (anteriorly)
8th costal cartilage	Midclavicular line	Midclavicular line
10th rib	Midaxillary line	Midaxillary line
11th rib	Line of inferior angle of scapula	Line of inferior angle of scapula
12th rib	Lateral border of erector spinae to T12 spinous process (slightly lower level than right pleura)	Lateral border of erector spinae to T12 spinous process

SURFACE MARKINGS OF LUNGS COVERED WITH VISCERAL PLEURA (PINK)		
Level	*Left Lung*	*Right Lung*
Apex	About 4 cm superior to middle of clavicle	About 4 cm superior to middle of clavicle
2nd costal cartilage	Midline (anteriorly)	Midline (anteriorly)
4th costal cartilage	Leaves lateral margin of sternum, follows 4th costal cartilage	Lateral margin of sternum
6th costal cartilage	Turns inferiorly to 6th costal cartilage in the midclavicular line (cardiac notch)	Follows 6th costal cartilage to midclavicular line
8th rib	Midaxillary line	Midaxillary line
10th rib	Line of inferior angle of scapula to T10 spinous process	Line of inferior angle of scapula to T10 spinous process

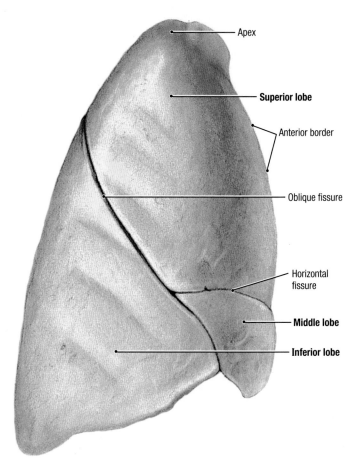

B. Lateral View

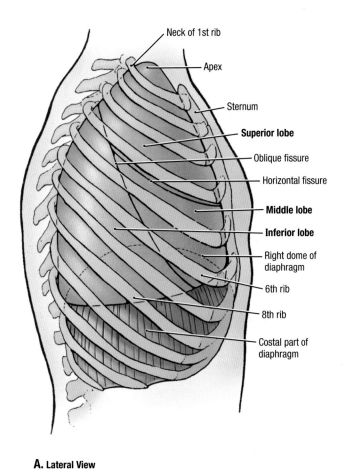

A. Lateral View

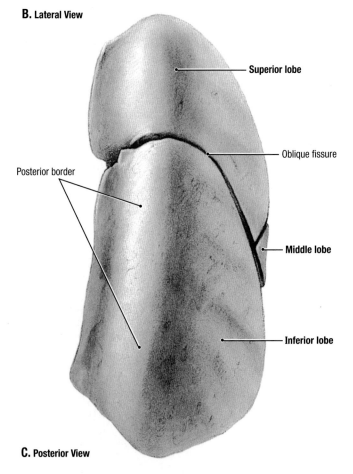

C. Posterior View

3.30   RIGHT LUNG

- The oblique and horizontal fissures divide the right lung into three lobes: superior, middle, and inferior.
- The right lung is larger and heavier than the left but is shorter and wider because the right dome of the diaphragm is higher and the heart bulges more to the left.
- Cadaveric lungs may be shrunken, firm, and discolored, whereas healthy lungs in living people are normally soft, light, and spongy.
- Each lung has an apex and base, three surfaces (costal, mediastinal, and diaphragmatic), and three borders (anterior, inferior, and posterior).

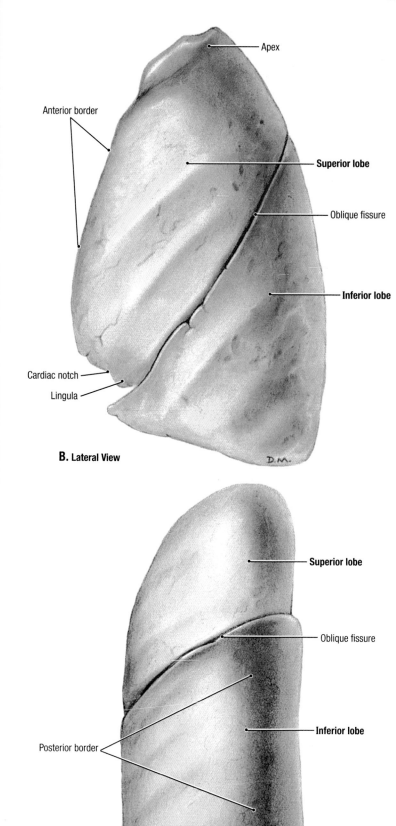

Apex

Anterior border

Superior lobe

Oblique fissure

Inferior lobe

Cardiac notch

Lingula

B. Lateral View

D.M.

Superior lobe

Oblique fissure

Inferior lobe

Posterior border

C. Posterior View

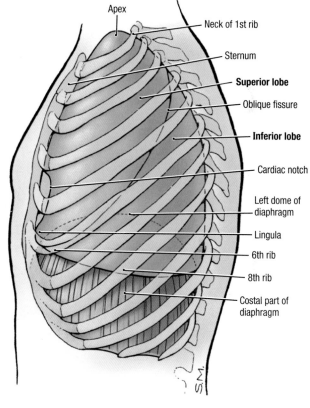

Apex

Neck of 1st rib

Sternum

Superior lobe

Oblique fissure

Inferior lobe

Cardiac notch

Left dome of diaphragm

Lingula

6th rib

8th rib

Costal part of diaphragm

S.M.

A. Lateral View

LEFT LUNG   3.31

- The left lung has two lobes, superior and inferior, separated by the oblique fissure.
- The anterior border has a deep cardiac notch that indents the antero-inferior aspect of the superior lobe.
- The lingula, a tonguelike process of the superior lobe, extends below the cardiac notch and slides in and out of the costome-diastinal recess during inspiration and expiration.
- The lungs of an embalmed cadaver usually retain impressions of structures that lie adjacent to them, such as the ribs and heart.

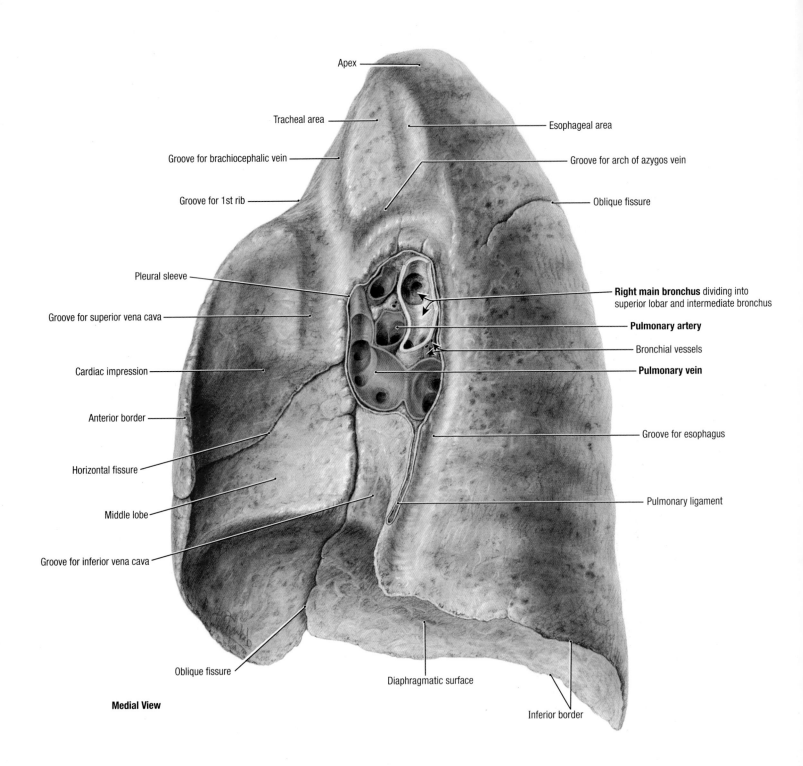

Apex

Tracheal area — Esophageal area

Groove for brachiocephalic vein — Groove for arch of azygos vein

Groove for 1st rib — Oblique fissure

Pleural sleeve — **Right main bronchus** dividing into superior lobar and intermediate bronchus

Groove for superior vena cava — **Pulmonary artery**

Bronchial vessels

Cardiac impression — **Pulmonary vein**

Anterior border — Groove for esophagus

Horizontal fissure

Middle lobe — Pulmonary ligament

Groove for inferior vena cava

Oblique fissure

Diaphragmatic surface

Medial View

Inferior border

3.32 MEDIASTINAL (MEDIAL) SURFACE AND HILUM OF RIGHT LUNG

The embalmed lung shows impressions of the structures with which it comes into contact, clearly demarcated as surface features; the base is contoured by the domes of the diaphragm; the costal surface bears the impressions of the ribs; distended vessels leave their mark, but nerves do not. The oblique fissure is incomplete here.

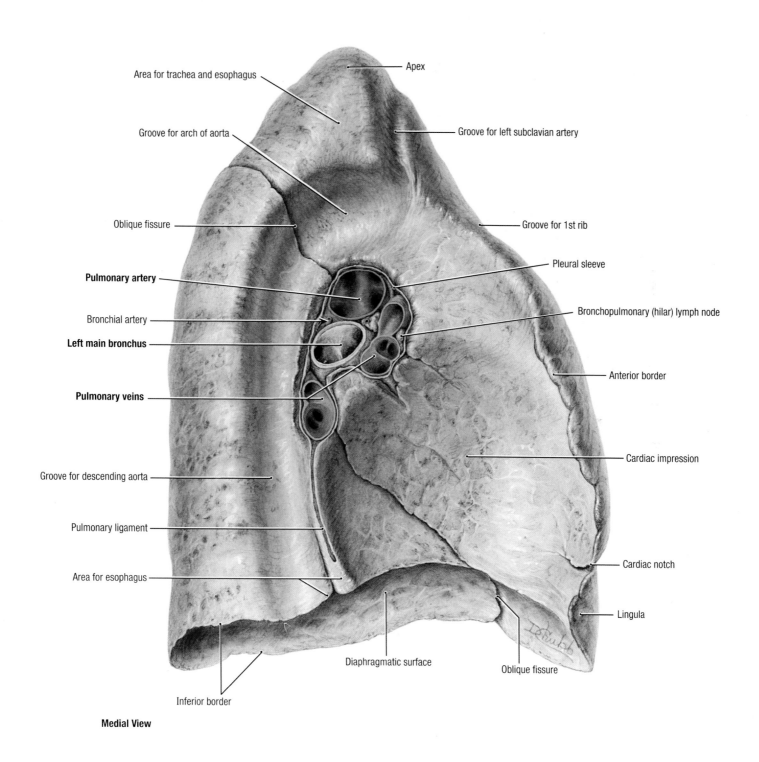

Area for trachea and esophagus

Apex

Groove for arch of aorta

Groove for left subclavian artery

Oblique fissure

Groove for 1st rib

Pleural sleeve

Pulmonary artery

Bronchial artery

Bronchopulmonary (hilar) lymph node

Left main bronchus

Anterior border

Pulmonary veins

Cardiac impression

Groove for descending aorta

Pulmonary ligament

Cardiac notch

Area for esophagus

Lingula

Diaphragmatic surface

Oblique fissure

Inferior border

Medial View

MEDIASTINAL (MEDIAL) SURFACE AND HILUM OF LEFT LUNG

3.33

Note the site of contact with esophagus, between the descending aorta and the inferior end of the pulmonary ligament. In the right and left roots, the artery is superior, the bronchus is posterior, one vein is anterior, and the other is inferior; in the right root, the bronchus to the superior lobe (eparterial bronchus) is the most superior structure.

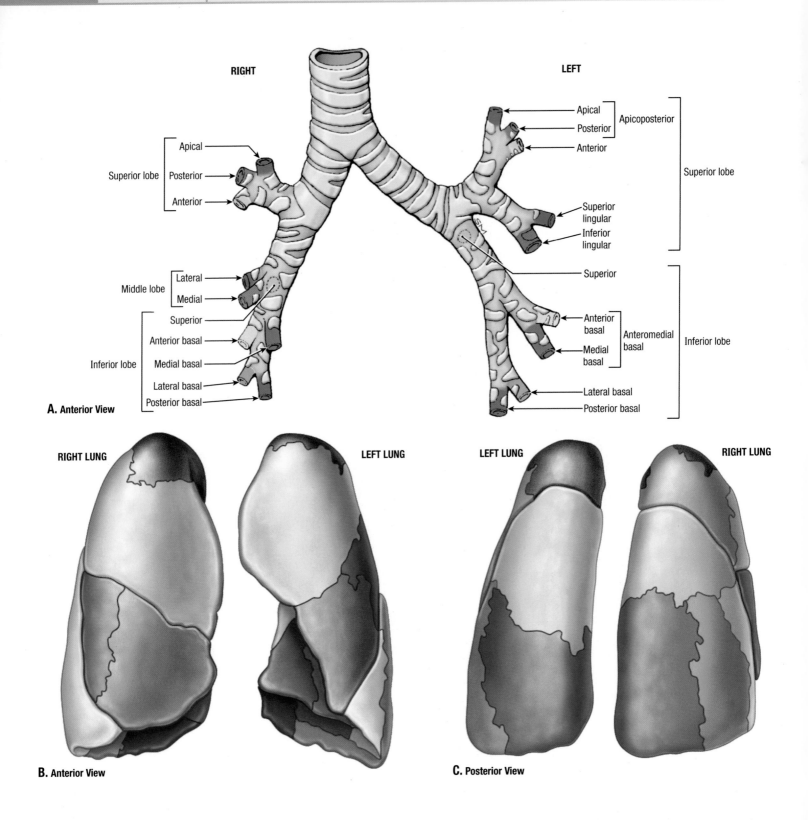

RIGHT

LEFT

Apical ⎤
Posterior ⎦ Apicoposterior

Anterior

Superior lobe

Apical ⎤
Superior lobe — Posterior ⎥
Anterior ⎦

Superior
lingular
Inferior
lingular

Lateral ⎤
Middle lobe — Medial ⎦

Superior

Superior

Anterior
basal
Anteromedial
basal

Inferior lobe

Anterior basal
Inferior lobe — Medial basal
Lateral basal
Posterior basal

Medial
basal

Lateral basal

Posterior basal

A. Anterior View

RIGHT LUNG

LEFT LUNG

LEFT LUNG

RIGHT LUNG

B. Anterior View

C. Posterior View

3.34 SEGMENTAL BRONCHI AND BRONCHOPULMONARY SEGMENTS

A. There are 10 right tertiary or segmental bronchi and 8 left. Note that in the left lung, the apical and posterior bronchi arise from a single stem, as do the anterior basal and medial basal. **B–F.** A bronchopulmonary segment consists of a tertiary bronchus, pulmonary vein and artery, and the portion of lung they serve. These structures are surgically separable to allow segmental resection of the lung. To prepare these specimens, the tertiary bronchi of fresh lungs were isolated within the hilum and injected with latex of various colors. Minor variations in the branching of the bronchi result in variations in the surface patterns.

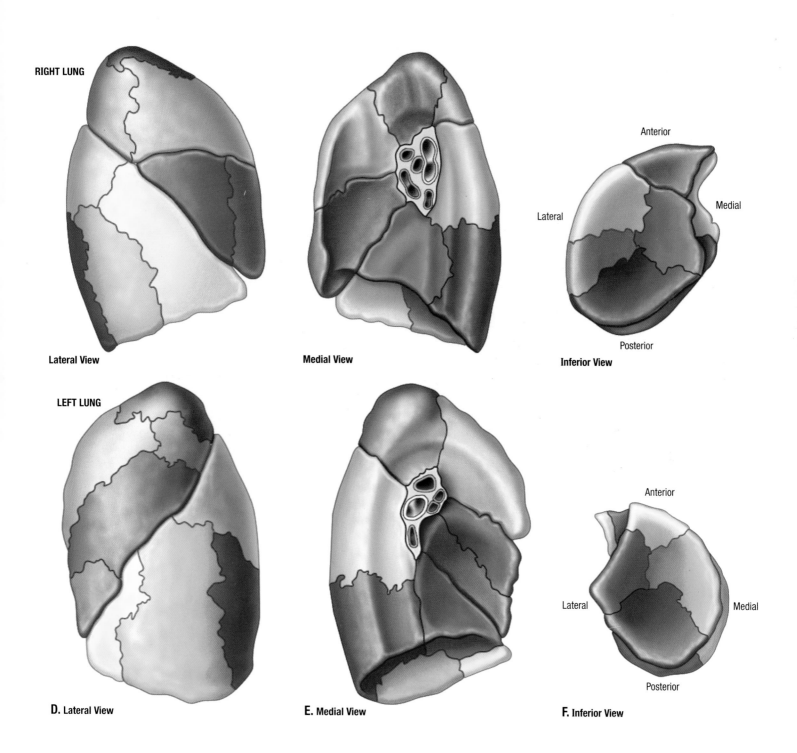

RIGHT LUNG

Lateral View

Medial View

Anterior

Lateral

Medial

Posterior

Inferior View

LEFT LUNG

D. Lateral View

E. Medial View

Anterior

Lateral

Medial

Posterior

F. Inferior View

SEGMENTAL BRONCHI AND BRONCHOPULMONARY SEGMENTS (continued)   3.34

Knowledge of the anatomy of the bronchopulmonary segments is essential for precise interpretations of diagnostic images of the lungs and for surgical resection (removal) of diseased segments. During the treatment of lung cancer, the surgeon may remove a whole lung (**pneumonectomy**), a lobe (**lobectomy**), or one or more bronchopulmonary segments (**segmentectomy**). Knowledge and understanding of the bronchopulmonary segments and their relationship to the bronchial tree are also essential for planning drainage and clearance techniques used in physical therapy for enhancing drainage from specific areas (e.g., in patients with pneumonia or cystic fibrosis).

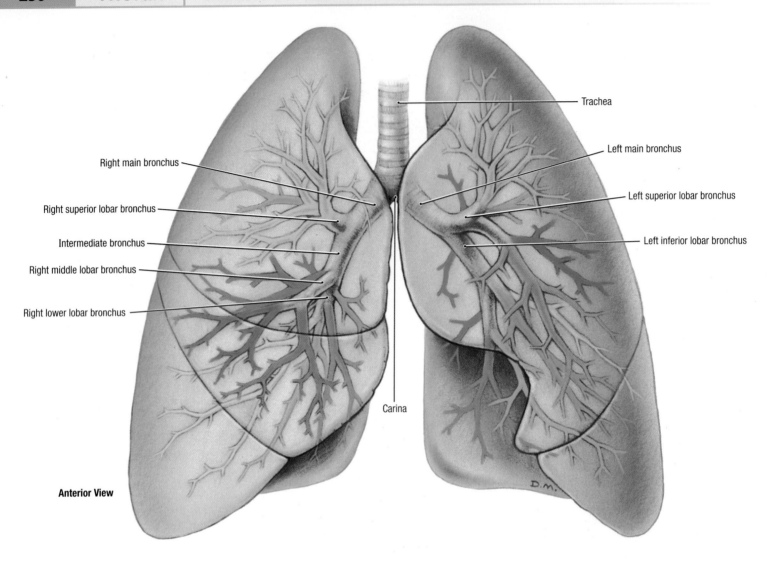

Trachea

Right main bronchus

Right superior lobar bronchus

Intermediate bronchus

Right middle lobar bronchus

Right lower lobar bronchus

Left main bronchus

Left superior lobar bronchus

Left inferior lobar bronchus

Carina

Anterior View

D.M.

3.35   TRACHEA AND BRONCHI *IN SITU*

- The segmental (tertiary) bronchi are color-coded.
- The trachea bifurcates into right and left main (primary) bronchi; the right main bronchus is shorter, wider, and more vertical than the left.
- Therefore, it is more likely that **aspirated foreign bodies** will enter and lodge in the right main bronchus or one of its descending branches.
- The right main bronchus gives off the right superior lobe bronchus (eparterial bronchus) before entering the hilum (hilus) of the lung; after entering the hilum, the continuing intermediate bronchus divides into the right middle and inferior lobar bronchi.
- The left main bronchus divides at the hilum into the left superior and left inferior lobar bronchi; the lobar bronchi further divide into segmental (tertiary) bronchi.

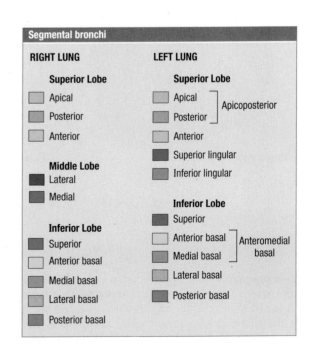

Segmental bronchi

RIGHT LUNG	LEFT LUNG
Superior Lobe	**Superior Lobe**
☐ Apical	☐ Apical ⎤ Apicoposterior
☐ Posterior	☐ Posterior ⎦
☐ Anterior	☐ Anterior
	■ Superior lingular
Middle Lobe	☐ Inferior lingular
■ Lateral	
■ Medial	**Inferior Lobe**
	■ Superior
Inferior Lobe	☐ Anterior basal ⎤ Anteromedial
■ Superior	■ Medial basal ⎦ basal
☐ Anterior basal	■ Lateral basal
■ Medial basal	■ Posterior basal
☐ Lateral basal	
■ Posterior basal	

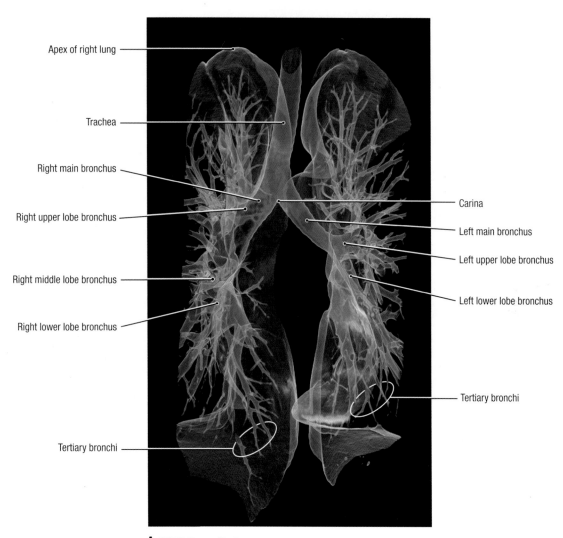

Apex of right lung

Trachea

Right main bronchus

Right upper lobe bronchus

Right middle lobe bronchus

Right lower lobe bronchus

Tertiary bronchi

Carina

Left main bronchus

Left upper lobe bronchus

Left lower lobe bronchus

Tertiary bronchi

A. CT 3D Airway Study

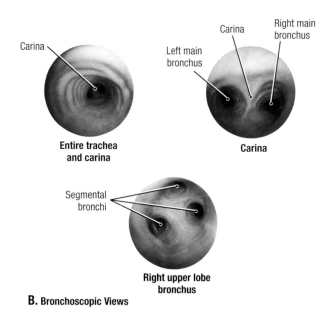

Carina

Entire trachea and carina

Left main bronchus

Carina

Right main bronchus

Carina

Segmental bronchi

Right upper lobe bronchus

B. Bronchoscopic Views

IMAGING OF LUNGS

3.36

A. Normal CT 3D airway study. CT imaging data can be reformatted to demonstrate specific anatomical structures as shown here for the bronchi. **B.** Bronchoscopy.

When examining the bronchi with a **bronchoscope**—an endoscope for inspecting the interior of the tracheobronchial tree for diagnostic purposes—one can observe a ridge, the carina, between the orifices of the main bronchi. If the tracheobronchial lymph nodes in the angle between the main bronchi are enlarged (e.g., because cancer cells have metastasized from a **bronchogenic carcinoma**) the carina is distorted, widened posteriorly, and immobile.

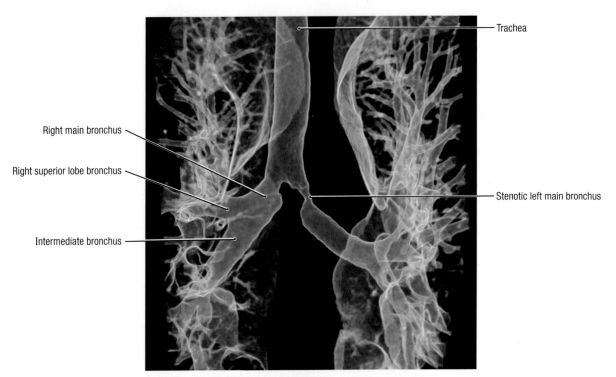

Trachea

Right main bronchus

Right superior lobe bronchus

Stenotic left main bronchus

Intermediate bronchus

C. 3D Airway Study Showing Airway Stenosis

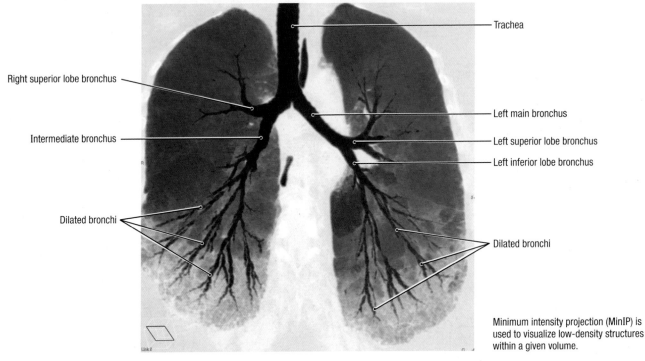

Trachea

Right superior lobe bronchus

Left main bronchus

Intermediate bronchus

Left superior lobe bronchus

Left inferior lobe bronchus

Dilated bronchi

Dilated bronchi

Minimum intensity projection (MinIP) is used to visualize low-density structures within a given volume.

D. CT Minimum Intensity Projection (MinIP) Showing Bronchiectasis

3.36 IMAGING OF LUNGS (*continued*)

C. Stenotic main bronchi. This patient complained of difficulty breathing. A stent was inserted into the bronchus to widen it. **D.** CT MinIP is used to reveal abnormally dilated bronchi, a condition called **bronchiectasis**. The abnormal dilation of these bronchi interferes with mucous removal and is associated with repeated pulmonary infections.

Medial Views

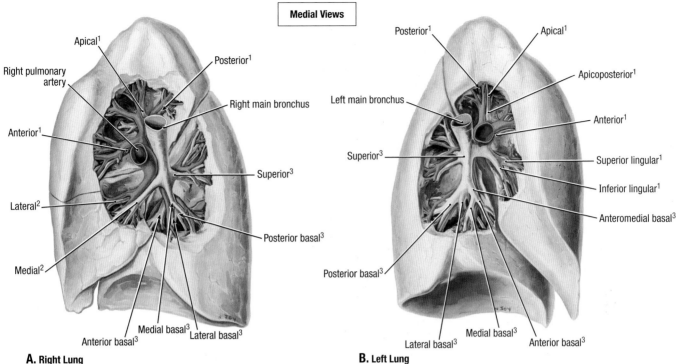

Apical[1]

Posterior[1]

Right pulmonary artery

Right main bronchus

Anterior[1]

Superior[3]

Lateral[2]

Posterior basal[3]

Medial[2]

Medial basal[3]   Lateral basal[3]

Anterior basal[3]

A. Right Lung

Posterior[1]

Apical[1]

Apicoposterior[1]

Left main bronchus

Anterior[1]

Superior[3]

Superior lingular[1]

Inferior lingular[1]

Anteromedial basal[3]

Posterior basal[3]

Lateral basal[3]   Medial basal[3]   Anterior basal[3]

B. Left Lung

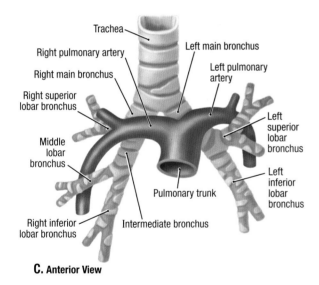

Trachea

Right pulmonary artery

Left main bronchus

Right main bronchus

Left pulmonary artery

Right superior lobar bronchus

Left superior lobar bronchus

Middle lobar bronchus

Right inferior lobar bronchus

Pulmonary trunk

Left inferior lobar bronchus

Intermediate bronchus

C. Anterior View

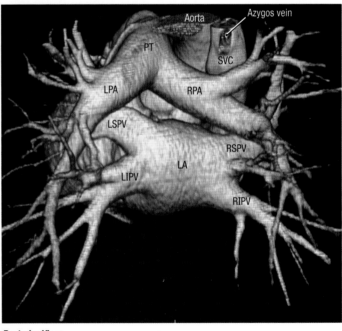

Posterior View

RELATIONSHIP OF BRONCHI AND PULMONARY ARTERIES 3.37

A. Right lung. **B.** Left lung. **C.** Pulmonary arteries and main bronchi. Superscripts indicate segmental bronchi to the [1]superior lobe, [2]middle lobe, and [3]inferior lobe. The pulmonary arteries of fresh lungs were filled with latex; the bronchi were inflated with air. The tissues surrounding the bronchi and vessels were removed.

Obstruction of a pulmonary artery by a blood clot **(pulmonary embolism)** results in partial or complete obstruction of blood flow to the lung.

3D VOLUME RECONSTRUCTION (3DVR) OF PULMONARY ARTERIES AND VEINS AND LEFT ATRIUM 3.38

The pulmonary trunk *(PT)* divides into a longer right pulmonary artery *(RPA)* and shorter left pulmonary artery *(LPA)*; the left superior *(LSPV)* and inferior *(LIPV)* and the right superior *(RSPV)* and inferior *(RIPV)* pulmonary veins drain into the left atrium *(LA). SVC*, superior vena cava.

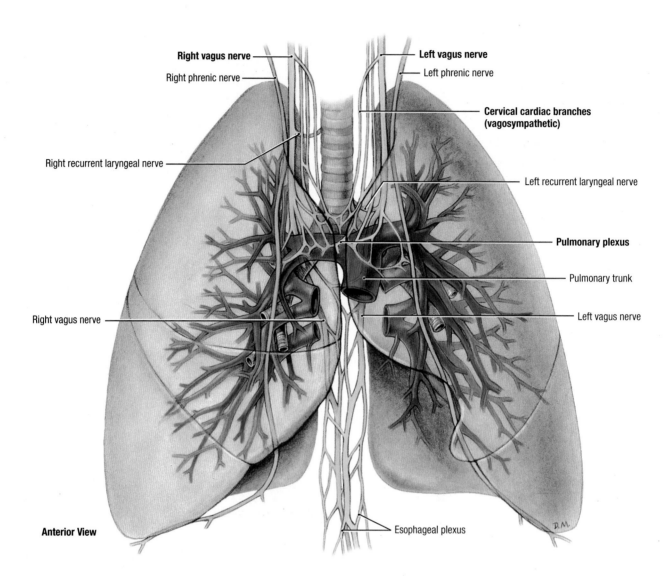

Right vagus nerve

Right phrenic nerve

Right recurrent laryngeal nerve

Right vagus nerve

Anterior View

Left vagus nerve

Left phrenic nerve

Cervical cardiac branches (vagosympathetic)

Left recurrent laryngeal nerve

Pulmonary plexus

Pulmonary trunk

Left vagus nerve

Esophageal plexus

D.M.

3.39   INNERVATION OF LUNGS

- The pulmonary plexuses, located anterior and posterior to the roots of the lungs, receive sympathetic contributions from the right and left sympathetic trunks (2nd to 5th thoracic ganglia, not shown) and parasympathetic contributions from the right and left vagus nerves; cell bodies of postsynaptic parasympathetic neurons are in the pulmonary plexuses and along the branches of the pulmonary tree.
- The right and left vagus nerves continue inferiorly from the posterior pulmonary plexus to contribute fibers to the esophageal plexus.

- The phrenic nerves pass anterior to the root of the lung on their way to the diaphragm.
- Visceral pleura is insensitive to pain. The autonomic nerves reach the visceral pleura in company with the bronchial vessels. The visceral pleura receives no nerves of general sensation.
- Parietal pleura is richly supplied by branches of the somatic intercostal and phrenic nerves. Irritation of the parietal pleura **pleuritus** produces local pain **pleurisy** and referred pain to the areas sharing innervation by the same segments of the spinal cord.

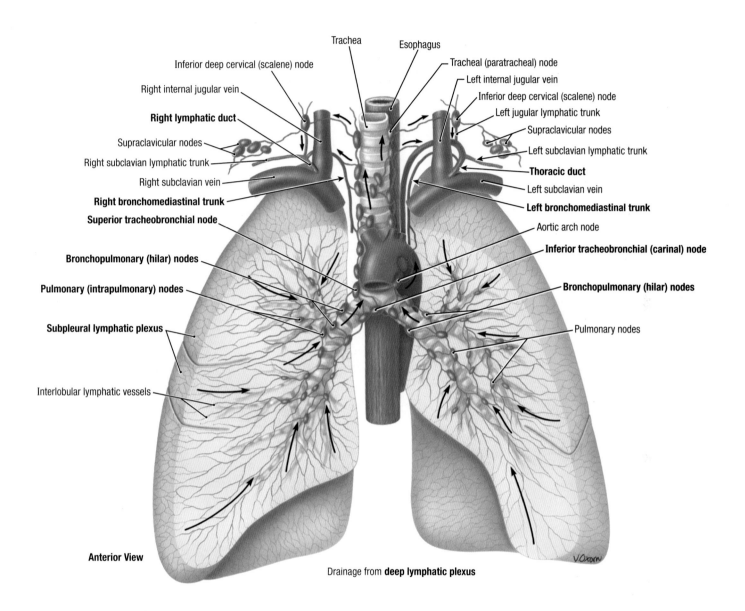

Trachea

Esophagus

Inferior deep cervical (scalene) node

Tracheal (paratracheal) node

Right internal jugular vein

Left internal jugular vein

Inferior deep cervical (scalene) node

Right lymphatic duct

Left jugular lymphatic trunk

Supraclavicular nodes

Supraclavicular nodes

Right subclavian lymphatic trunk

Left subclavian lymphatic trunk

Right subclavian vein

Thoracic duct

Left subclavian vein

Right bronchomediastinal trunk

Left bronchomediastinal trunk

Superior tracheobronchial node

Aortic arch node

Inferior tracheobronchial (carinal) node

Bronchopulmonary (hilar) nodes

Pulmonary (intrapulmonary) nodes

Bronchopulmonary (hilar) nodes

Subpleural lymphatic plexus

Pulmonary nodes

Interlobular lymphatic vessels

Anterior View

V.Oxorn

Drainage from **deep lymphatic plexus**

LYMPHATIC DRAINAGE OF LUNGS

3.40

- Lymphatic vessels originate in the subpleural (superficial) and deep lymphatic plexuses.
- The subpleural lymphatic plexus is superficial, lying deep to the visceral pleura, and drains lymph from the surface of the lung to the bronchopulmonary (hilar) nodes.
- The deep lymphatic plexus is in the lung and follows the bronchi and pulmonary vessels to the pulmonary, and then bronchopulmonary, nodes located at the root of the lung.
- All lymph from the lungs enters the inferior (carinal) and superior tracheobronchial nodes and then continues to the right and left bronchomediastinal trunks to drain into the venous system via the right lymphatic and thoracic ducts; lymph from the left inferior lobe passes largely to the right side.
- Lymph from the parietal pleura drains into lymph nodes of the thoracic wall (Fig. 3.71).

Lung cancer (carcinoma) metastasizes early to the bronchopulmonary lymph nodes and subsequently to the other thoracic lymph nodes. Common sites of **hematogenous metastases** (spreading through the blood) of cancer cells from a bronchogenic carcinoma are the brain, bones, lungs, and suprarenal glands. Often the lymph nodes superior to the clavicle—the supraclavicular lymph nodes—are enlarged when lung (bronchogenic) carcinoma develops owing to metastasis of cancer cells from the tumor. Consequently, the supraclavicular nodes were once referred to as sentinel lymph nodes. More recently, the term sentinel lymph node has been applied to a node or nodes that first receive lymph drainage from a cancer-containing area, regardless of location, following injection of blue dye containing radioactive tracer (technetium-99).

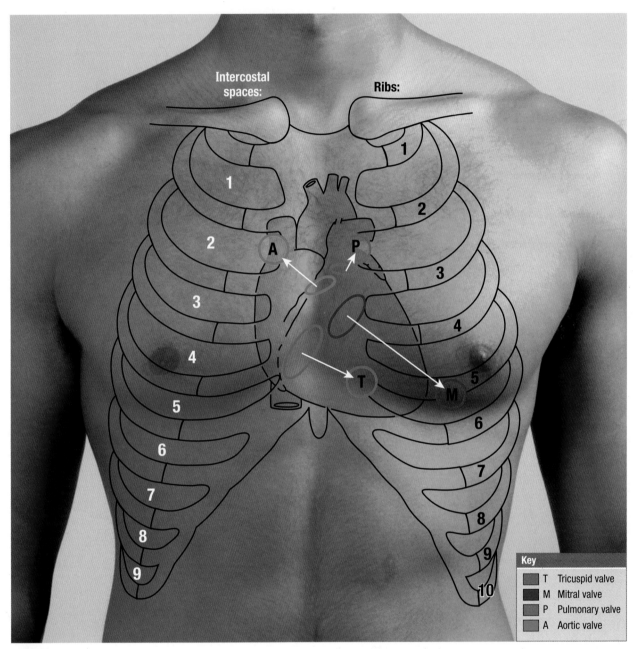

Intercostal spaces:

Ribs:

Key

	T	Tricuspid valve
	M	Mitral valve
	P	Pulmonary valve
	A	Aortic valve

Anterior View

3.41   SURFACE PROJECTIONS OF THE HEART, HEART VALVES, AND THEIR AUSCULTATION AREAS

- The location of each heart valve *in situ* is indicated by a colored oval and the area of auscultation of the valve is indicated as a circle of the same color containing the first letter of the valve name.
- The **auscultation areas** are sites where the sounds of each of the heart's valves can be heard most distinctly through a stethoscope (**cardiac auscultation**).

- The aortic *(A)* and pulmonary *(P)* auscultation areas are in the 2nd intercostal space to the right and left of the sternal border; the tricuspid area *(T)* is near the left sternal border in the 5th or 6th intercostal space; the mitral valve *(M)* is heard best near the apex of the heart in the 5th intercostal space in the midclavicular line.

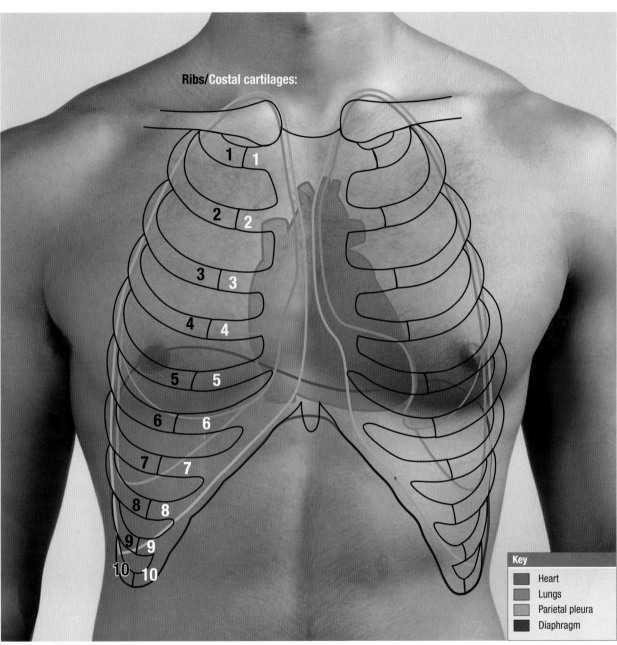

Ribs/Costal cartilages:

Key
- Heart
- Lungs
- Parietal pleura
- Diaphragm

Anterior View

SURFACE MARKINGS OF THE HEART, LUNGS, AND DIAPHRAGM

3.42

- The superior border of the heart is represented by a slightly oblique line joining the 3rd costal cartilages; the convex right side of the heart projects lateral to the sternum and inferiorly, lying at the 6th or 7th costochondral junction; the inferior border of the heart is lying superior to the central tendon of the diaphragm and sloping slightly inferiorly to the apex at the 5th interspace at the midclavicular line.

- The right dome of the diaphragm is higher than the left because of the large size of the liver inferior to the dome; during expiration, the right dome reaches as high as the 5th rib and the left dome ascends to the 5th intercostal space.

- The left pleural cavity is smaller than the right because of the projection of the heart to the left side.

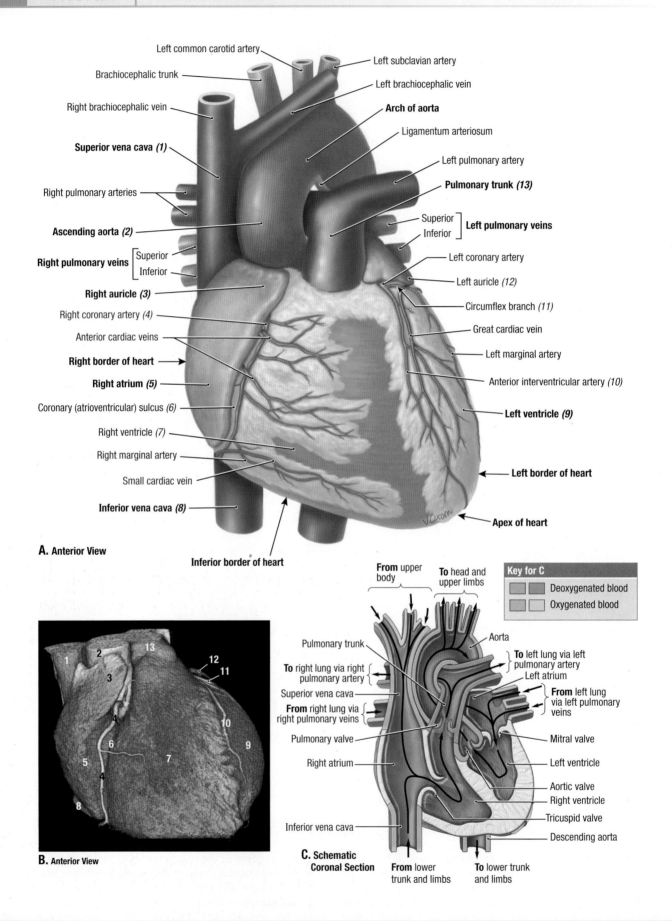

Left common carotid artery

Brachiocephalic trunk

Right brachiocephalic vein

Superior vena cava (1)

Right pulmonary arteries

Ascending aorta (2)

Right pulmonary veins [Superior / Inferior]

Right auricle (3)

Right coronary artery (4)

Anterior cardiac veins

Right border of heart

Right atrium (5)

Coronary (atrioventricular) sulcus (6)

Right ventricle (7)

Right marginal artery

Small cardiac vein

Inferior vena cava (8)

Left subclavian artery

Left brachiocephalic vein

Arch of aorta

Ligamentum arteriosum

Left pulmonary artery

Pulmonary trunk (13)

Superior / Inferior] **Left pulmonary veins**

Left coronary artery

Left auricle (12)

Circumflex branch (11)

Great cardiac vein

Left marginal artery

Anterior interventricular artery (10)

Left ventricle (9)

Left border of heart

Apex of heart

A. Anterior View

Inferior border of heart

B. Anterior View

From upper body

To head and upper limbs

Key for C	
	Deoxygenated blood
	Oxygenated blood

Pulmonary trunk

Aorta

To right lung via right pulmonary artery

To left lung via left pulmonary artery

Left atrium

Superior vena cava

From left lung via left pulmonary veins

From right lung via right pulmonary veins

Pulmonary valve

Mitral valve

Right atrium

Left ventricle

Aortic valve

Right ventricle

Tricuspid valve

Inferior vena cava

Descending aorta

C. Schematic Coronal Section

From lower trunk and limbs

To lower trunk and limbs

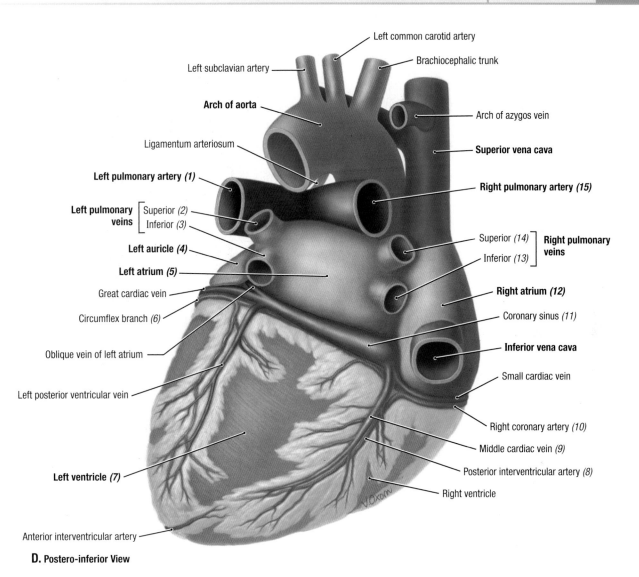

Left common carotid artery

Left subclavian artery

Brachiocephalic trunk

Arch of aorta

Arch of azygos vein

Ligamentum arteriosum

Superior vena cava

Left pulmonary artery *(1)*

Right pulmonary artery *(15)*

Left pulmonary veins Superior *(2)* / Inferior *(3)*

Superior *(14)* / Inferior *(13)* **Right pulmonary veins**

Left auricle *(4)*

Left atrium *(5)*

Right atrium *(12)*

Great cardiac vein

Circumflex branch *(6)*

Coronary sinus *(11)*

Oblique vein of left atrium

Inferior vena cava

Left posterior ventricular vein

Small cardiac vein

Right coronary artery *(10)*

Middle cardiac vein *(9)*

Left ventricle *(7)*

Posterior interventricular artery *(8)*

Right ventricle

Anterior interventricular artery

D. Postero-inferior View

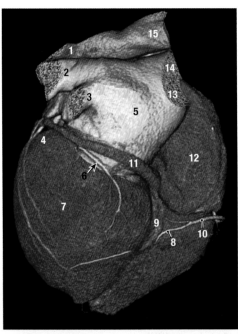

E. Postero-inferior View

HEART AND GREAT VESSELS (*continued*) 3.43

A. Anatomical specimen.
- The right border of the heart, formed by the right atrium, is slightly convex and almost in line with the superior vena cava.
- The inferior border is formed primarily by the right ventricle and part of the left ventricle.
- The left border is formed primarily by the left ventricle and part of the left auricle.

B. 3D volume reconstruction from MRI of heart and coronary vessels (living patient). Numbers refer to structures **(A)**.

C. Circulation of blood through the heart.

D. Anatomical specimen, posterior view.
- Most of the left atrium and left ventricle are visible in this postero-inferior view.
- The right and left pulmonary veins open into the left atrium.
- The arch of the aorta extends superiorly, posteriorly and to the left, in a nearly sagittal plane.

E. 3D volume reconstruction from MRI of heart and coronary vessels. Numbers refer to structures **(D)**.

Right vagus nerve

Right common carotid artery

Trachea

Left common carotid artery

Left vagus nerve

Right internal jugular vein

Left internal jugular vein

Right phrenic nerve

Left phrenic nerve

Right subclavian vein

Left subclavian vein

Brachiocephalic trunk

Left brachiocephalic vein

Right brachiocephalic vein

Manubrium

Right phrenic nerve

Internal thoracic artery

Superior vena cava

Sternal angle at manubriosternal joint (divided)

2nd costal cartilage

Root of lung

Internal thoracic artery

Left phrenic nerve

Right lung

Left lung

Right phrenic nerve

Pericardium

Body of sternum

Right dome of diaphragm

Left dome of diaphragm

Left phrenic nerve

Xiphisternal joint

7th costal cartilage

Xiphoid process

Anterior View

3.44	PERICARDIUM IN RELATION TO STERNUM

- The pericardium lies posterior to the body of the sternum, extending from just superior to the sternal angle to the level of the xiphisternal joint; approximately two thirds lies to the left of the median plane.
- The heart lies between the sternum and the anterior mediastinum anteriorly and the vertebral column and the posterior mediastinum posteriorly.

- In **cardiac compression**, the sternum is depressed 4 to 5 cm, forcing blood out of the heart and into the great vessels.
- Internal thoracic arteries arise from the subclavian arteries and descend posterior to the costal cartilages, running lateral to the sternum and anterior to the pleura.

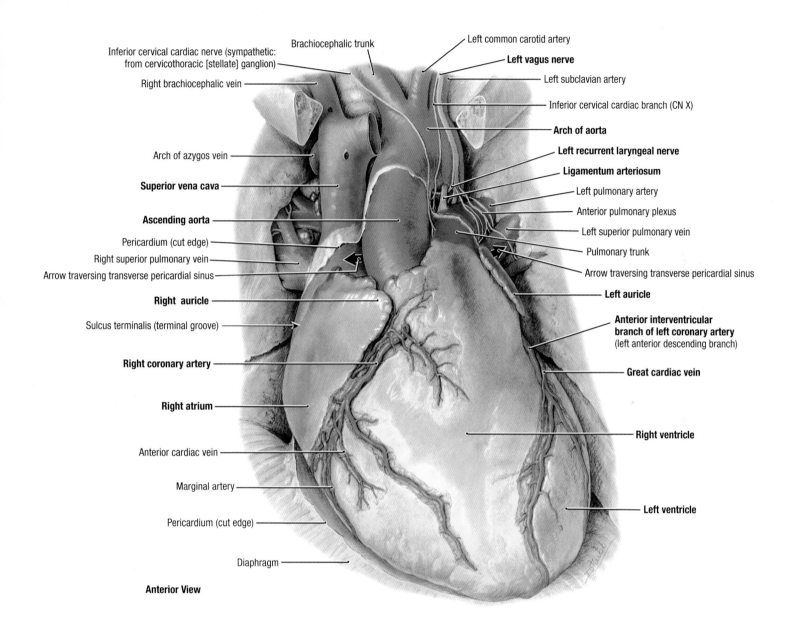

Inferior cervical cardiac nerve (sympathetic: from cervicothoracic [stellate] ganglion)

Right brachiocephalic vein

Brachiocephalic trunk

Left common carotid artery

Left vagus nerve

Left subclavian artery

Inferior cervical cardiac branch (CN X)

Arch of aorta

Arch of azygos vein

Superior vena cava

Ascending aorta

Pericardium (cut edge)

Right superior pulmonary vein

Arrow traversing transverse pericardial sinus

Right auricle

Sulcus terminalis (terminal groove)

Right coronary artery

Right atrium

Anterior cardiac vein

Marginal artery

Pericardium (cut edge)

Diaphragm

Left recurrent laryngeal nerve

Ligamentum arteriosum

Left pulmonary artery

Anterior pulmonary plexus

Left superior pulmonary vein

Pulmonary trunk

Arrow traversing transverse pericardial sinus

Left auricle

Anterior interventricular branch of left coronary artery (left anterior descending branch)

Great cardiac vein

Right ventricle

Left ventricle

Anterior View

STERNOCOSTAL (ANTERIOR) SURFACE OF HEART AND GREAT VESSELS *IN SITU*   **3.45**

- The right ventricle forms most of the sternocostal surface.
- The entire right auricle and much of the right atrium are visible anteriorly, but only a small portion of the left auricle is visible; the auricles, like a closing claw, grasp the origins of the pulmonary trunk and ascending aorta from a posterior approach.
- The ligamentum arteriosum passes from the origin of the left pulmonary artery to the arch of the aorta.
- The right coronary artery courses in the anterior atrioventricular groove, and the anterior interventricular branch of the left coronary artery (anterior descending branch) courses in or parallel to the anterior interventricular groove (see Fig. 3.43B).
- The left vagus nerve passes lateral to the arch of the aorta and then posterior to the root of the lung; the left recurrent laryngeal nerve passes inferior to the arch of the aorta posterior to the ligamentum arteriosum.
- The great cardiac vein ascends beside the anterior interventricular branch of the left coronary artery to drain into the coronary sinus posteriorly.

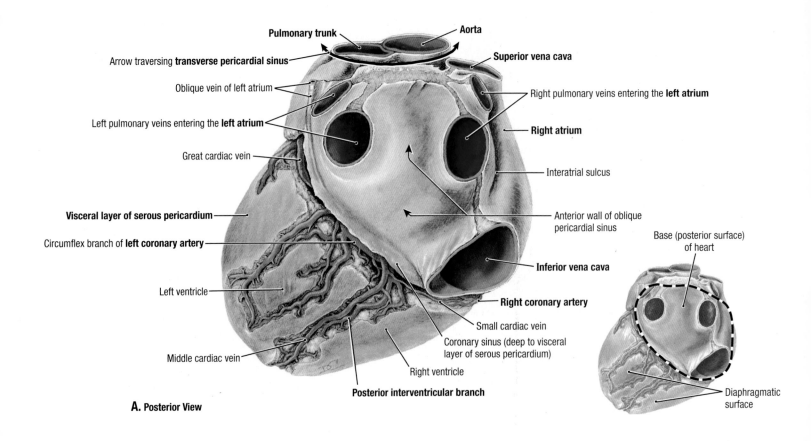

Pulmonary trunk — Aorta
Arrow traversing **transverse pericardial sinus** —
— **Superior vena cava**
Oblique vein of left atrium —
— Right pulmonary veins entering the **left atrium**
Left pulmonary veins entering the **left atrium** —
— **Right atrium**
Great cardiac vein —
— Interatrial sulcus
Visceral layer of serous pericardium —
Circumflex branch of **left coronary artery** —
— Anterior wall of oblique pericardial sinus
Left ventricle —
— **Inferior vena cava**
— **Right coronary artery**
Middle cardiac vein —
— Small cardiac vein
Coronary sinus (deep to visceral layer of serous pericardium)
Right ventricle
Posterior interventricular branch

A. Posterior View

Base (posterior surface) of heart
Diaphragmatic surface

3.46   HEART AND PERICARDIUM

- This heart **(A)** was removed from the interior of the pericardial sac **(B)**.
- The entire base, or posterior surface, and part of the diaphragmatic or inferior surface of the heart are in view *(inset)*.
- The superior vena cava and larger inferior vena cava join the superior and inferior aspects of the right atrium.
- The left atrium forms the greater part of the base (posterior surface) of the heart *(inset)*.
- The left coronary artery in this specimen is dominant, since it supplies the posterior interventricular branch.
- Most branches of cardiac veins cross branches of the coronary arteries superficially.
- The visceral layer of serous pericardium (epicardium) covers the surface of the heart and reflects onto the great vessels; from around the great vessels, the serous pericardium reflects to line the internal aspect of the fibrous pericardium as the parietal layer of serous pericardium. The fibrous pericardium and the parietal layer of serous pericardium form the pericardial sac that encases the heart.
- Note the cut edges of the reflections of serous pericardia around the arterial vessels (the pulmonary trunk and aorta) and venous vessels (the superior and inferior venae cavae and the pulmonary veins).
- **Surgical isolation of cardiac outflow**. The transverse pericardial sinus is especially important to cardiac surgeons. After the pericardial sac has been opened anteriorly, a finger can be passed through the transverse pericardial sinus posterior to the aorta and pulmonary trunk. By passing a surgical clamp or placing a ligature around these vessels, inserting the tubes of a coronary bypass machine, and then tightening the ligature, surgeons can stop or divert the circulation of blood in these large arteries while performing cardiac surgery.

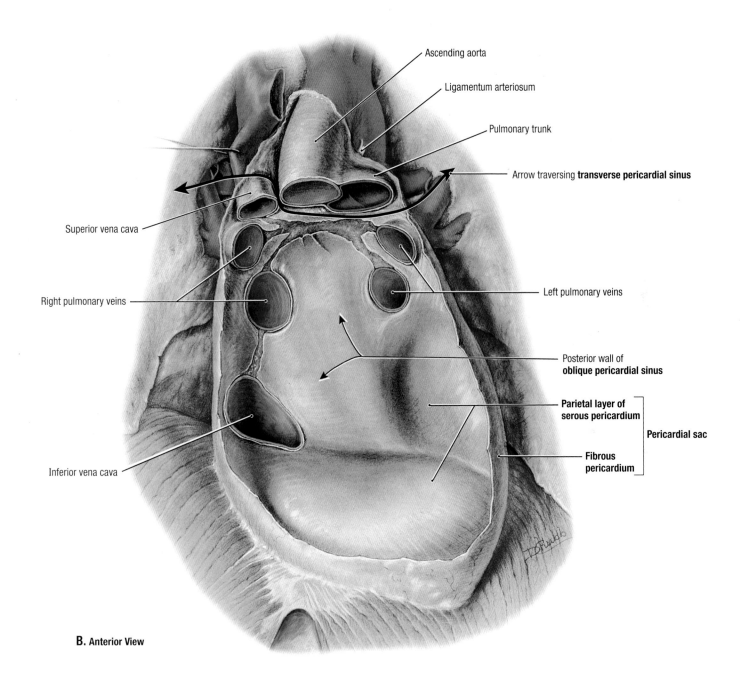

Ascending aorta

Ligamentum arteriosum

Pulmonary trunk

Arrow traversing **transverse pericardial sinus**

Superior vena cava

Right pulmonary veins

Left pulmonary veins

Posterior wall of
oblique pericardial sinus

**Parietal layer of
serous pericardium**

Pericardial sac

**Fibrous
pericardium**

Inferior vena cava

B. Anterior View

HEART AND PERICARDIUM (*continued*)

3.46

- Interior of pericardial sac. Eight vessels were severed to excise the heart: superior and inferior venae cavae, four pulmonary veins, and two pulmonary arteries.
- The oblique sinus is bounded anteriorly by the visceral layer of serous pericardium covering the left atrium **(A)**, posteriorly by the parietal layer of serous pericardium lining the fibrous pericardium, and superiorly and laterally by the reflection of serous pericardium around the four pulmonary veins and the superior and inferior venae cavae **(B)**.
- The transverse sinus is bounded anteriorly by the serous pericardium covering the posterior aspect of the pulmonary trunk and

aorta and posteriorly by the visceral pericardium reflecting from the atria **(A)** inferiorly and the superior vena cava superiorly on the right.
- Blood in the pericardial cavity, **hemopericardium**, produces **cardiac tamponade**. Hemopericardium may result from perforation of a weakened area of the heart muscle owing to a previous **myocardial infarction (MI)** or heart attack, from bleeding into the pericardial cavity after cardiac operations, or from stab wounds. Heart volume is increasingly compromised and circulation fails.

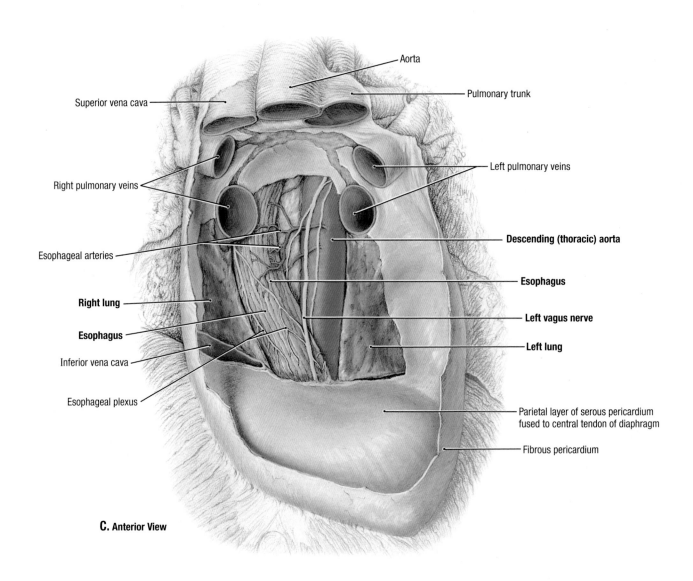

Aorta

Superior vena cava

Pulmonary trunk

Right pulmonary veins

Left pulmonary veins

Esophageal arteries

Descending (thoracic) aorta

Esophagus

Right lung

Left vagus nerve

Esophagus

Left lung

Inferior vena cava

Esophageal plexus

Parietal layer of serous pericardium
fused to central tendon of diaphragm

Fibrous pericardium

C. Anterior View

3.46 HEART AND PERICARDIUM (*continued*)

C. Posterior relationships; dissection. The fibrous and parietal layers of serous pericardium have been removed from posterior and lateral to the oblique sinus. The esophagus in this specimen is deflected to the right; it usually lies in contact with the aorta, forming primary posterior relationships of the heart.

Surgical exposure of venae cavae. After ascending through the diaphragm, the entire thoracic part of the inferior vena cava (IVC) (approximately 2 cm) is enclosed by the pericardium. Consequently, the pericardial sac must be opened to expose the terminal part of the IVC. The same is true for the terminal part of the superior vena cava (SVC), which is partly inside and partly outside the pericardial sac.

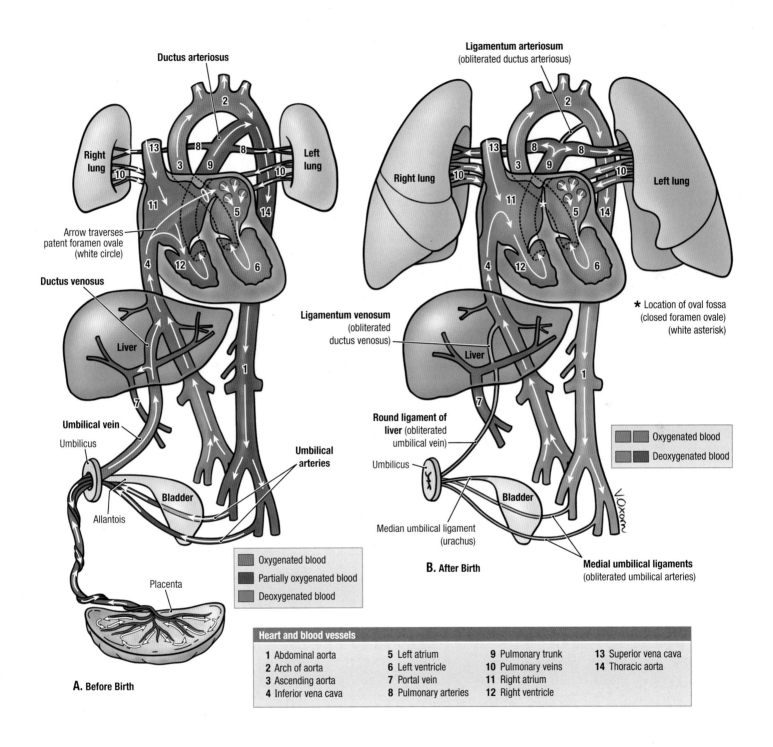

A. Before Birth

Ductus arteriosus

Right lung

Arrow traverses patent foramen ovale (white circle)

Ductus venosus

Liver

Umbilical vein

Umbilicus

Allantois

Placenta

Left lung

Umbilical arteries

Bladder

Oxygenated blood
Partially oxygenated blood
Deoxygenated blood

B. After Birth

Ligamentum arteriosum (obliterated ductus arteriosus)

Right lung

Left lung

★ Location of oval fossa (closed foramen ovale) (white asterisk)

Ligamentum venosum (obliterated ductus venosus)

Liver

Round ligament of liver (obliterated umbilical vein)

Umbilicus

Median umbilical ligament (urachus)

Bladder

Oxygenated blood
Deoxygenated blood

Medial umbilical ligaments (obliterated umbilical arteries)

Heart and blood vessels

1 Abdominal aorta	5 Left atrium	9 Pulmonary trunk	13 Superior vena cava
2 Arch of aorta	6 Left ventricle	10 Pulmonary veins	14 Thoracic aorta
3 Ascending aorta	7 Portal vein	11 Right atrium	
4 Inferior vena cava	8 Pulmonary arteries	12 Right ventricle	

PRE- AND POSTNATAL CIRCULATION

3.47

A. Before birth. **B.** After birth. At birth, two major changes take place: (1) pulmonary respiration starts and (2) after the umbilical cord is ligated, the umbilical arteries (except the most proximal part), umbilical vein, and ductus venosus are occluded and become the medial umbilical ligament, round ligament of liver, and the ligamentum venosum, respectively.

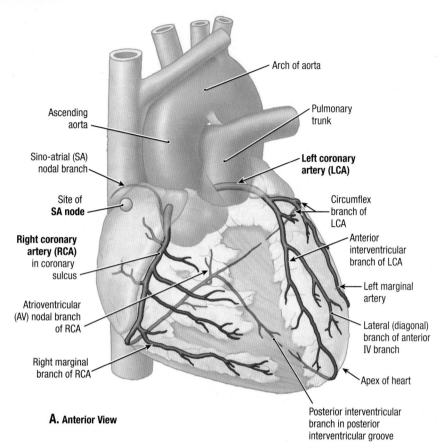

Arch of aorta

Ascending aorta

Pulmonary trunk

Sino-atrial (SA) nodal branch

Left coronary artery (LCA)

Site of **SA node**

Circumflex branch of LCA

Right coronary artery (RCA) in coronary sulcus

Anterior interventricular branch of LCA

Left marginal artery

Atrioventricular (AV) nodal branch of RCA

Lateral (diagonal) branch of anterior IV branch

Right marginal branch of RCA

Apex of heart

Posterior interventricular branch in posterior interventricular groove

A. Anterior View

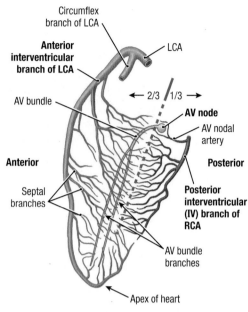

Circumflex branch of LCA

LCA

Anterior interventricular branch of LCA

AV bundle

2/3 1/3

AV node

AV nodal artery

Anterior

Posterior

Septal branches

Posterior interventricular (IV) branch of RCA

AV bundle branches

Apex of heart

C. Arteries of Isolated Interventricular Septum (from Left Side)

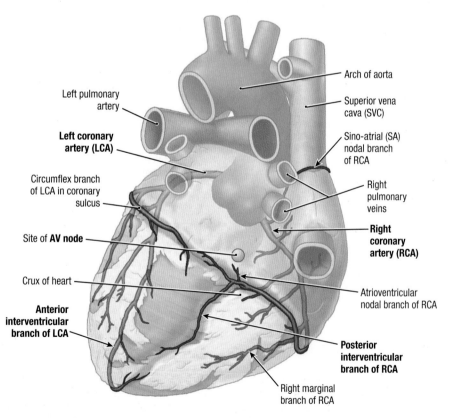

Arch of aorta

Left pulmonary artery

Superior vena cava (SVC)

Left coronary artery (LCA)

Sino-atrial (SA) nodal branch of RCA

Circumflex branch of LCA in coronary sulcus

Right pulmonary veins

Site of **AV node**

Right coronary artery (RCA)

Crux of heart

Atrioventricular nodal branch of RCA

Anterior interventricular branch of LCA

Posterior interventricular branch of RCA

Right marginal branch of RCA

B. Postero-inferior View

3.48   CORONARY ARTERIES

A. Anterior view. **B.** Postero-inferior view. **C.** Arteries of interventricular septum.

- In the most common pattern, the right coronary artery travels in the coronary sulcus to reach the posterior surface of the heart, where it anastomoses with the circumflex branch of the left coronary artery. Early in its course, it gives off the right atrial branch, which supplies the sino-atrial (SA) node via its sino-atrial nodal branch. Major branches are a marginal branch supplying much of the anterior wall of the right ventricle, an atrioventricular (AV) nodal branch given off near the posterior border of the interventricular septum, and a posterior interventricular branch in the interventricular groove that anastomoses with the anterior interventricular branch of the left coronary artery.
- The left coronary artery divides into a circumflex branch that passes posteriorly to anastomose with the right coronary artery on the posterior aspect of the heart and an anterior interventricular branch in the interventricular groove; the origin of the SA nodal branch is variable and may be a branch of the left coronary artery.
- The interventricular septum receives its blood supply from septal branches of the two interventricular (descending) branches: typically the anterior two thirds from the left coronary, and the posterior one third from the right **(C)**.

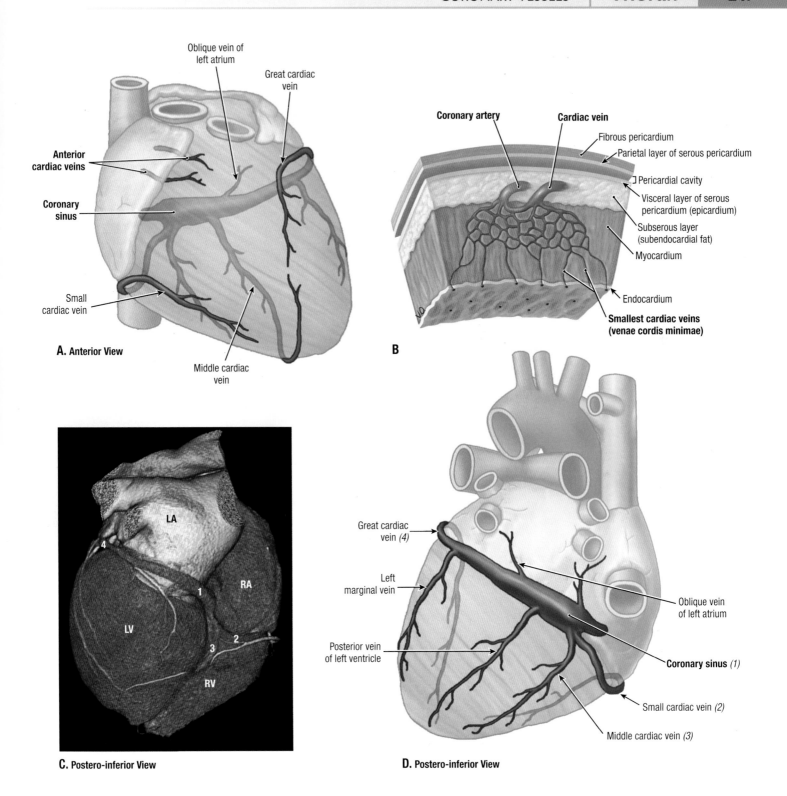

A. Anterior View

Oblique vein of left atrium

Great cardiac vein

Anterior cardiac veins

Coronary sinus

Small cardiac vein

Middle cardiac vein

B

Coronary artery

Cardiac vein

Fibrous pericardium

Parietal layer of serous pericardium

Pericardial cavity

Visceral layer of serous pericardium (epicardium)

Subserous layer (subendocardial fat)

Myocardium

Endocardium

Smallest cardiac veins (venae cordis minimae)

C. Postero-inferior View

LA

RA

LV

RV

D. Postero-inferior View

Great cardiac vein (4)

Left marginal vein

Posterior vein of left ventricle

Oblique vein of left atrium

Coronary sinus (1)

Small cardiac vein (2)

Middle cardiac vein (3)

CARDIAC VEINS

3.49

A. Anterior aspect. **B.** Smallest cardiac veins. **C.** 3D volume reconstruction. Numbers refer to veins in **D**. *LA,* left atrium; *RA,* right atrium; *LV,* left ventricle; *RV,* right ventricle. **D.** Postero-inferior aspect.

The coronary sinus is the major venous drainage vessel of the heart; it is located posteriorly in the atrioventricular (coronary) groove and drains into the right atrium. The great, middle, and small cardiac veins; the oblique vein of the left atrium; and the posterior vein of the left ventricle are the principal vessels draining into the coronary sinus. The anterior cardiac veins drain directly into the right atrium. The smallest cardiac veins (venae cordis minimae) drain the myocardium directly into the atria and ventricles **(B)**. The cardiac veins accompany the coronary arteries and their branches.

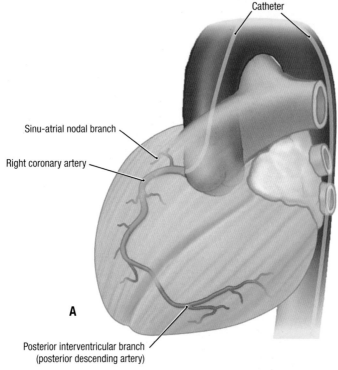

Catheter

Sinu-atrial nodal branch

Right coronary artery

A

Posterior interventricular branch
(posterior descending artery)

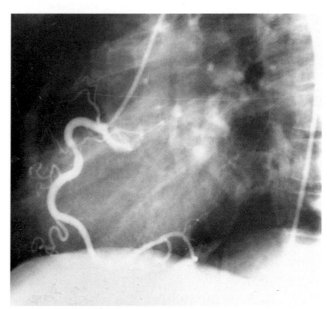

B. Left Anterior Oblique View

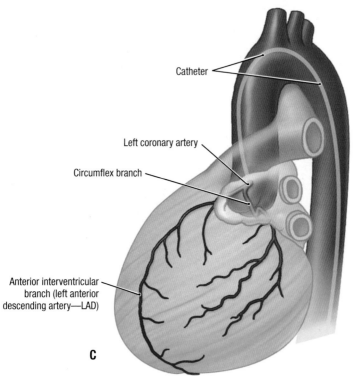

Catheter

Left coronary artery

Circumflex branch

Anterior interventricular
branch (left anterior
descending artery—LAD)

C

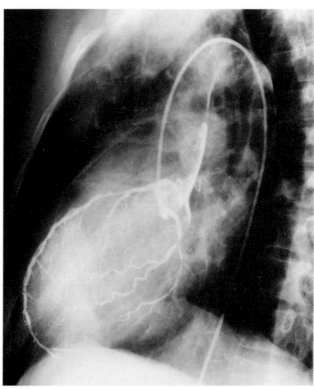

D. Left Anterior Oblique View

3.50 | CORONARY ARTERIOGRAMS WITH ORIENTATION DRAWINGS

Right (**A** and **B**) and left (**C** and **D**) coronary arteriograms.

Coronary artery disease (CAD), one of the leading causes of death, results in a reduced blood supply to the vital myocardial tissue. The three most common sites of coronary artery occlusion and the approximate percentage of occlusions involving each artery are the (1) anterior interventricular (clinically referred to as LAD) branch of the left coronary artery (LCA) (40% to 50%), (2) right coronary artery (RCA) (30% to 40%), and (3) circumflex branch of the LCA (15% to 20%).

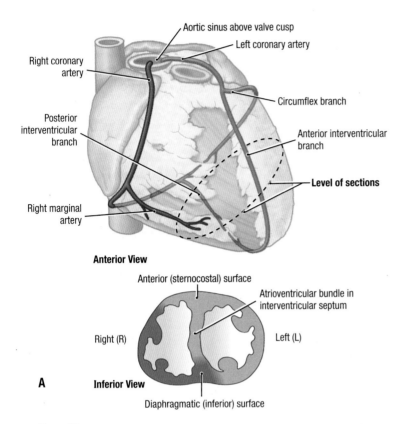

Anterior View

Inferior View

A.

A. and B. Most common pattern (67%). Right coronary artery is dominant, giving rise to the posterior interventricular branch.

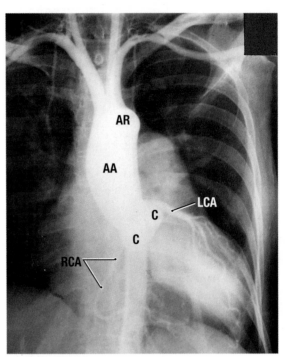

B. Coronary Angiogram, Anteroposterior View

Key for B			
AA	Ascending aorta	LCA	Left coronary artery
AR	Arch of aorta	RCA	Right coronary artery
C	Cusp of aortic valve		

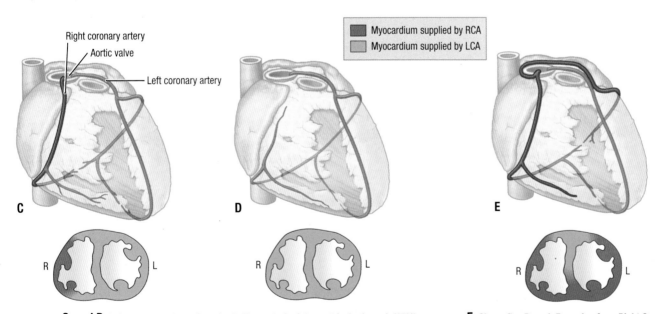

Myocardium supplied by RCA
Myocardium supplied by LCA

C.   **D.**   **E.**

C. and D. Left coronary artery gives rise to the posterior interventricular branch (15%).

E. Circumflex Branch Emerging from Right Coronary Sinus.

VARIATIONS IN DISTRIBUTION OF CORONARY ARTERIES   **3.51**

A. Most common pattern. **B.** Coronary angiogram of most common pattern. **C–E.** Less common patterns.

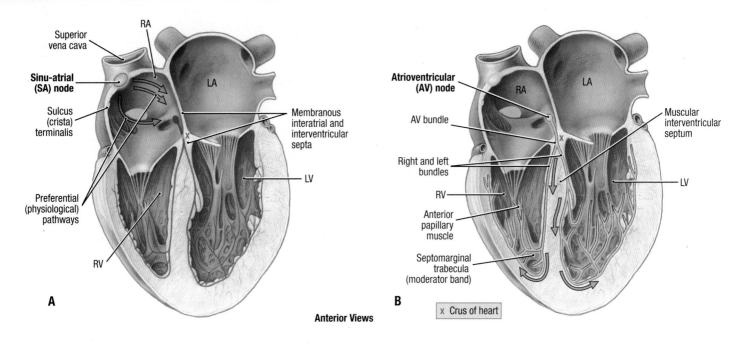

Anterior Views

x Crus of heart

3.52 CONDUCTION SYSTEM OF HEART, CORONAL SECTION

A. Impulses *(arrows)* initiated at the sino-atrial node. **B.** Atrioventricular (AV) node, AV bundle, and bundle branches. **C.** Echocardiogram, apical four-chamber view.

- The sino-atrial (SA) node is in the wall of the right atrium near the superior end of the sulcus terminalis (internally crista terminalis) at the opening of the superior vena cava. The SA node is the "pacemaker" of the heart because it initiates muscle contraction and determines the heart rate. It is supplied by the sino-atrial nodal artery, usually a branch of the right atrial branch of the right coronary artery, but it may arise from the left coronary artery.
- Contraction spreads through the atrial wall (myogenic induction) until it reaches the AV node in the interatrial septum, superomedial to the opening of the coronary sinus. The AV node is supplied by the AV nodal artery, usually arising from the right coronary artery posteriorly at the inferior margin of the interatrial septum.
- The AV bundle, usually supplied by the right coronary artery, passes from the AV node in the membranous part of the interventricular septum, dividing into right and left bundle branches on either side of the muscular part of the interventricular septum.
- The right bundle branch travels inferiorly in the interventricular septum to the anterior wall of the ventricle, with part passing via the septomarginal trabecula to the anterior papillary muscle; excitation spreads throughout the right ventricular wall through a network of subendocardial branches (Purkinje fibers) from the right bundle.
- The left bundle branch lies beneath the endocardium on the left side of the interventricular septum and branches to enter the anterior and posterior papillary muscles and the wall of the left ventricle; further branching into a plexus of subendocardial branches allows the impulses to be conveyed throughout the left ventricular wall. The bundle branches are mostly supplied by the left coronary artery except the posterior limb of the left bundle branch, which is supplied by both coronary arteries.
- **Damage to the cardiac conduction system** (often by compromised blood supply as in coronary artery disease) leads to disturbances of muscle contraction. Damage to the AV node results in "heart block" because the atrial excitation wave does not reach the ventricles, which begin to contract independently at their own slower rate. Damage to one of the bundle branches results in "bundle branch block," in which excitation goes down the unaffected branch to cause systole of that ventricle; the impulse then spreads to the other ventricle, producing later asynchronous contraction.

Key for A and B:

LA Left atrium
LV Left ventricle
RA Right atrium
RV Right ventricle

For this ultrasound image, the transducer is usually placed on the chest wall in the left 5th intercostal space and aimed so that the beam obliquely transects the heart and penetrates all four chambers.

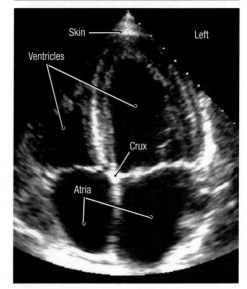

C. Echocardiogram. Apical Four-Chamber View

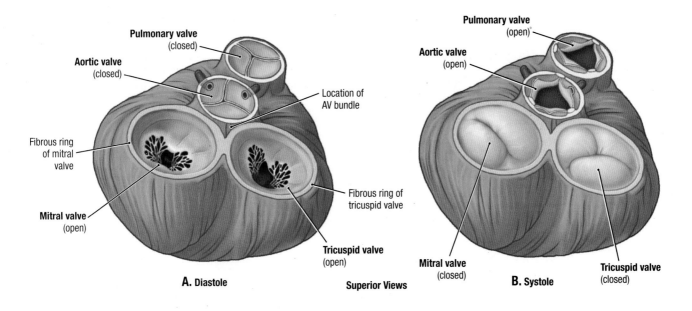

Pulmonary valve
(closed)

Aortic valve
(closed)

Location of
AV bundle

Fibrous ring
of mitral
valve

Mitral valve
(open)

Fibrous ring of
tricuspid valve

Tricuspid valve
(open)

A. Diastole

Superior Views

Pulmonary valve
(open)

Aortic valve
(open)

Mitral valve
(closed)

Tricuspid valve
(closed)

B. Systole

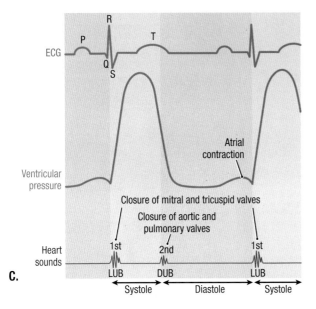

ECG

P

R

T

Q
S

Ventricular
pressure

Atrial
contraction

Closure of mitral and tricuspid valves

Closure of aortic and
pulmonary valves

Heart
sounds

1st   2nd   1st
LUB   DUB   LUB

Systole   Diastole   Systole

C.

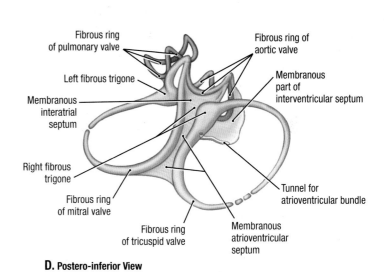

Fibrous ring
of pulmonary valve

Fibrous ring of
aortic valve

Left fibrous trigone

Membranous
part of
interventricular septum

Membranous
interatrial
septum

Right fibrous
trigone

Fibrous ring
of mitral valve

Fibrous ring
of tricuspid valve

Membranous
atrioventricular
septum

Tunnel for
atrioventricular bundle

D. Postero-inferior View

CARDIAC CYCLE AND CARDIAC SKELETON

3.53

A. Ventricular diastole. **B.** Ventricular systole. **C.** Correlation of ventricular pressure, electrocardiogram (ECG), and heart sounds. The cardiac cycle describes the complete movement of the heart or heartbeat and includes the period from the beginning of one heartbeat to the beginning of the next one. The cycle consists of diastole (ventricular relaxation and filling) and systole (ventricular contraction and emptying). The right heart is the pump for the pulmonary circuit; the left heart is the pump for the systemic circuit (see Fig. 3.43C). **D.** Cardiac skeleton. The fibrous framework of dense collagen forms four fibrous rings, which provide attachment for the leaflets and cusps of the valves, and two fibrous trigones that connect the rings, and the membranous parts of the interatrial and interventricular septa. The fibrous skeleton keeps the orifices of the valves patent and separates the myenterically conducted impulses of the atria.

Disorders involving the valves of the heart disturb the pumping efficiency of the heart. **Valvular heart disease** produces either stenosis (narrowing) or insufficiency. **Valvular stenosis** is the failure of a valve to open fully, slowing blood flow from a chamber. **Valvular insufficiency**, or regurgitation, is the failure of the valve to close completely, usually owing to nodule formation on (or scarring and contraction of) the cusps so that the edges do not meet or align. This allows a variable amount of blood (depending on the severity) to flow back into the chamber it was just ejected from. Both stenosis and insufficiency result in an increased workload for the heart. Because valvular diseases are mechanical problems, damaged or defective cardiac valves are often replaced surgically in a procedure called **valvuloplasty**.

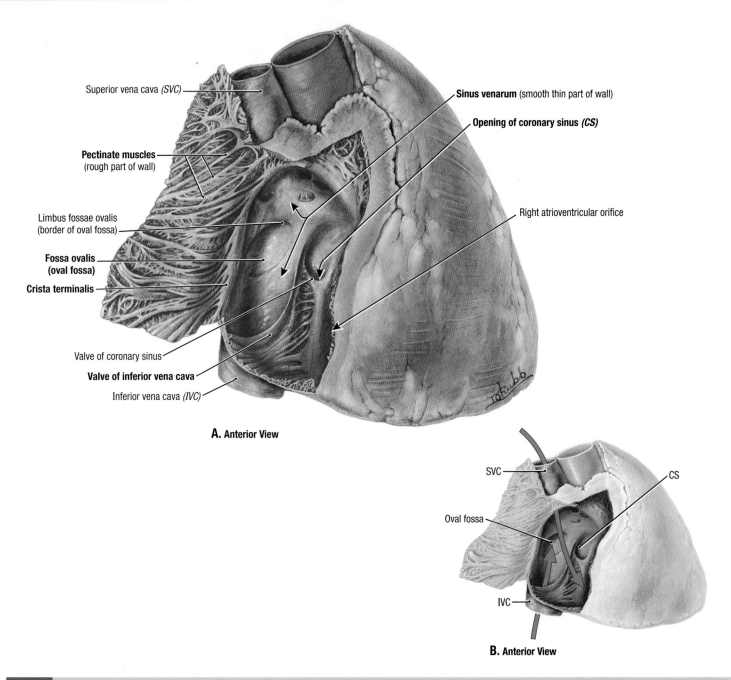

Superior vena cava *(SVC)*

Pectinate muscles
(rough part of wall)

Limbus fossae ovalis
(border of oval fossa)

Fossa ovalis
(oval fossa)

Crista terminalis

Valve of coronary sinus

Valve of inferior vena cava

Inferior vena cava *(IVC)*

Sinus venarum (smooth thin part of wall)

Opening of coronary sinus *(CS)*

Right atrioventricular orifice

A. Anterior View

SVC

Oval fossa

CS

IVC

B. Anterior View

3.54 | RIGHT ATRIUM

A. Interior of right atrium. The anterior wall of the right atrium is reflected. **B.** Blood flow into atrium from the superior and inferior venae cavae.

- The smooth part of the atrial wall is formed by the absorption of the right horn of the sinus venosus, and the rough part is formed from the primitive atrium.
- Crista terminalis, the valve of the inferior vena cava, and the valve of the coronary sinus separate the smooth part from the rough part.
- The pectinate muscle passes anteriorly from the crista terminalis; the crista underlies the sulcus terminalis (not shown), a groove visible externally on the posterolateral surface of the right atrium between the superior and inferior venae cavae.
- The superior and inferior venae cavae and the coronary sinus open onto the smooth part of the right atrium; the anterior cardiac veins and venae cordis minimae (not visible) also open into the atrium.

- The floor of the fossa ovalis is the remnant of the fetal septum primum; the crescent-shaped ridge (limbus fossae ovalis) partially surrounding the fossa is the remnant of the septum secundum.
- Inflow from the superior vena cava is directed toward the tricuspid orifice, whereas blood from the inferior vena cava is directed toward the fossa ovalis **(B)**.
- Congenital anomalies of the interatrial septum, most often incomplete closure of the oval foramen (patent foramen ovale), are **atrial septal defects (ASDs)**. A probe-size patency is present in the superior part of the oval fossa in 15% to 25% of adults (Moore et al., 2012). These small openings, by themselves, cause no hemodynamic abnormalities. Large ASDs allow oxygenated blood from the lungs to be shunted from the left atrium through the ASD into the right atrium, causing enlargement of the right atrium and ventricle and dilation of the pulmonary trunk.

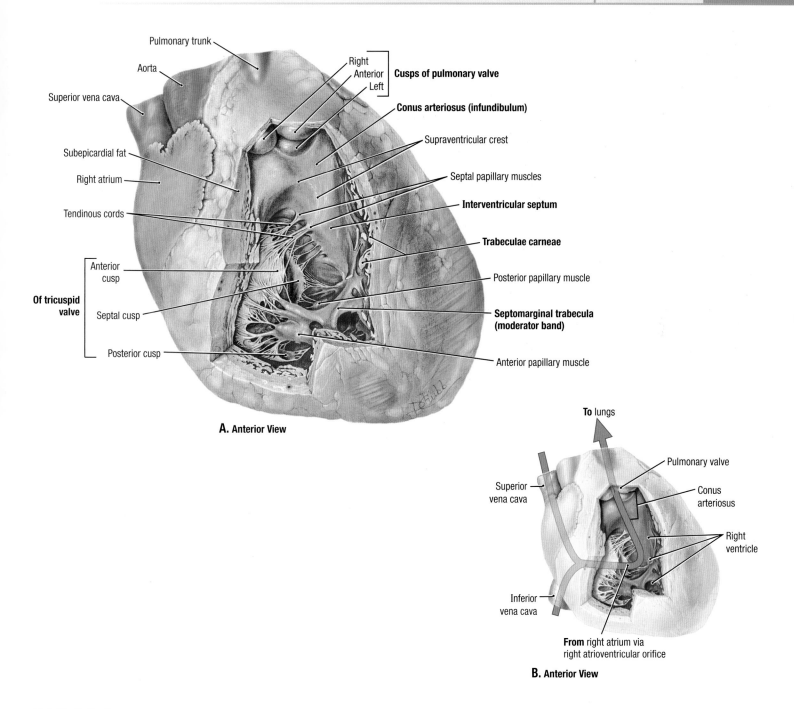

Pulmonary trunk

Aorta

Superior vena cava

Subepicardial fat

Right atrium

Tendinous cords

Of tricuspid valve
- Anterior cusp
- Septal cusp
- Posterior cusp

Right
Anterior
Left
Cusps of pulmonary valve

Conus arteriosus (infundibulum)

Supraventricular crest

Septal papillary muscles

Interventricular septum

Trabeculae carneae

Posterior papillary muscle

Septomarginal trabecula (moderator band)

Anterior papillary muscle

A. Anterior View

To lungs

Superior vena cava

Pulmonary valve

Conus arteriosus

Right ventricle

Inferior vena cava

From right atrium via right atrioventricular orifice

B. Anterior View

RIGHT VENTRICLE

3.55

A. Interior of right ventricle. **B.** Blood flow through right heart.
- The entrance to this chamber, the right atrioventricular or tricuspid orifice, is situated posteriorly; the exit, the orifice of the pulmonary trunk, is superior.
- The outflow portion of the chamber inferior to the pulmonary orifice (conus arteriosus or infundibulum) has a smooth, funnel-shaped wall; the remainder of the ventricle is rough with fleshy trabeculae.
- There are three types of trabeculae: mere ridges, bridges attached only at each end, and fingerlike projections called papillary muscles. The anterior papillary muscle rises from the anterior wall, the posterior (papillary muscle) from the posterior wall, and a series of small septal papillae from the septal wall.

- The septomarginal trabecula, here thick, extends from the septum to the base of the anterior papillary muscle.
- The membranous part of the interventricular septum develops separately from the muscular part and has a complex embryological origin (Moore et al., 2012). Consequently, this part is the common site of **ventricular septal defects (VSDs)**, although defects also occur in the muscular part. VSDs rank first on all lists of cardiac defects. The size of the defect varies from 1 to 25 mm. A VSD causes a left-to-right shunt of blood through the defect. A large shunt increases pulmonary blood flow, which causes severe pulmonary disease (**pulmonary hypertension**, or increased blood pressure) and may cause **cardiac failure**.

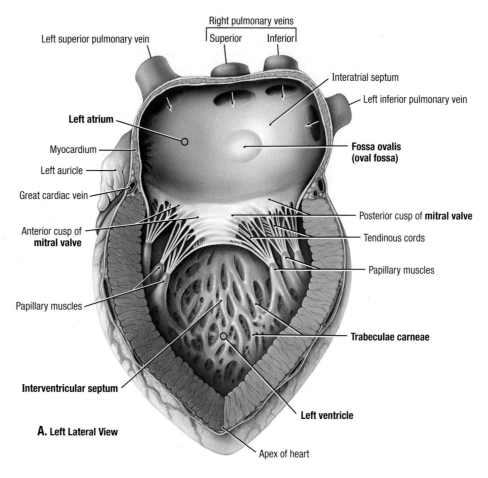

Left superior pulmonary vein

Right pulmonary veins
Superior   Inferior

Interatrial septum

Left inferior pulmonary vein

Left atrium

Myocardium

Fossa ovalis (oval fossa)

Left auricle

Great cardiac vein

Posterior cusp of **mitral valve**

Anterior cusp of **mitral valve**

Tendinous cords

Papillary muscles

Papillary muscles

Trabeculae carneae

Interventricular septum

Left ventricle

A. Left Lateral View

Apex of heart

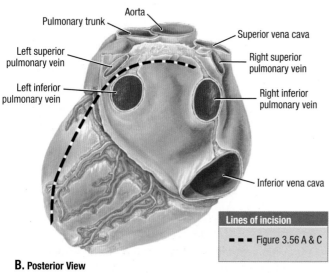

Aorta

Pulmonary trunk

Superior vena cava

Left superior pulmonary vein

Right superior pulmonary vein

Left inferior pulmonary vein

Right inferior pulmonary vein

Inferior vena cava

Lines of incision

- - - Figure 3.56 A & C

B. Posterior View

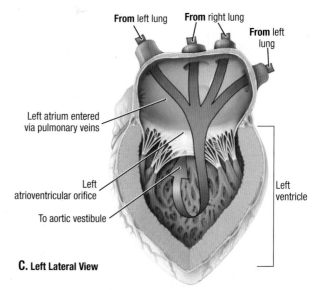

From left lung   **From** right lung   **From** left lung

Left atrium entered via pulmonary veins

Left atrioventricular orifice

To aortic vestibule

Left ventricle

C. Left Lateral View

3.56 LEFT ATRIUM AND LEFT VENTRICLE

A. Interior of left heart. **B.** Line of incision *(black dashed line)* for parts A and C. **C.** Blood flow through the left heart.
- A diagonal cut was made from the base of the heart to the apex, passing between the superior and inferior pulmonary veins and through the posterior cusp of the mitral valve, followed by retraction (spreading) of the left heart wall on each side of the incision.

- The entrances (pulmonary veins) to the left atrium are posterior, and the exit (left atrioventricular or mitral orifice) is anterior.
- The left side of the fossa ovalis is also seen on the left side of the interatrial septum, although the left side is not usually as distinct as the right side is within the right atrium.
- Except for that of the auricle, the atrial wall is smooth.

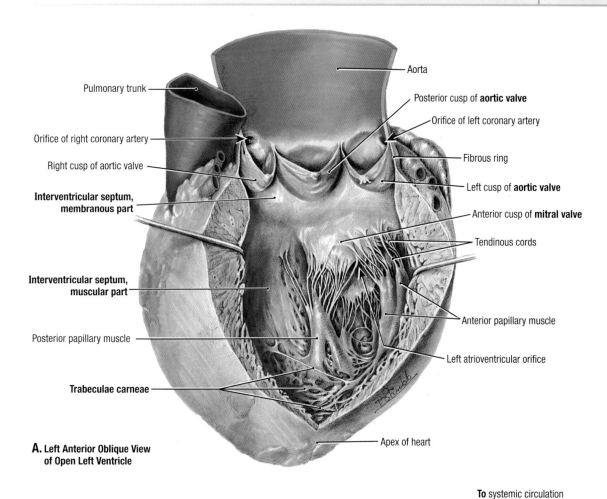

Aorta

Pulmonary trunk

Posterior cusp of **aortic valve**

Orifice of left coronary artery

Orifice of right coronary artery

Fibrous ring

Right cusp of aortic valve

Left cusp of **aortic valve**

Interventricular septum, membranous part

Anterior cusp of **mitral valve**

Tendinous cords

Interventricular septum, muscular part

Posterior papillary muscle

Anterior papillary muscle

Left atrioventricular orifice

Trabeculae carneae

Apex of heart

A. Left Anterior Oblique View of Open Left Ventricle

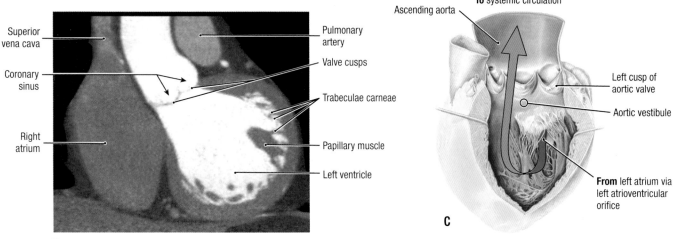

Superior vena cava

Pulmonary artery

Coronary sinus

Valve cusps

Trabeculae carneae

Right atrium

Papillary muscle

Left ventricle

B. Coronal CT

To systemic circulation

Ascending aorta

Left cusp of aortic valve

Aortic vestibule

From left atrium via left atrioventricular orifice

C

LEFT VENTRICLE   3.57

A. Interior of left ventricle. **B.** Coronal CT image from coronary CT arteriography study. The patient was injected with an intravenous (IV) contrast agent and a series of CT images was taken as the contrast material traveled through the heart. For this image, the material has mainly passed through the right side of the heart and is primarily now in the left ventricle and aorta. **C.** Blood flow through the left ventricle.

- A cut was made from the apex along the left margin of the heart, passing posterior to the pulmonary trunk, to open the aortic vestibule and ascending aorta.

- The entrance (left atrioventricular, bicuspid, or mitral orifice) is situated posteriorly, and the exit (aortic orifice) is superior.
- The left ventricular wall is thin and muscular near the apex, thick and muscular superiorly, and thin and fibrous (nonelastic) at the aortic orifice.
- Two large papillary muscles, the anterior from the anterior wall and the posterior from the posterior wall, control the adjacent halves of two cusps of the mitral valve with tendinous cords (chordae tendineae).

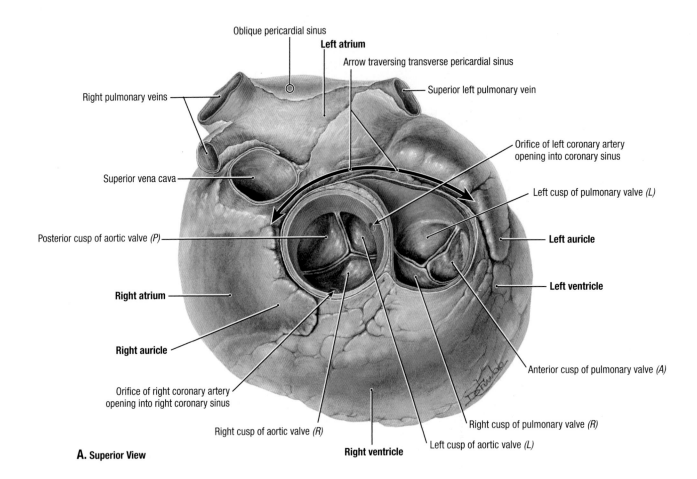

Oblique pericardial sinus

Left atrium

Arrow traversing transverse pericardial sinus

Right pulmonary veins

Superior left pulmonary vein

Superior vena cava

Orifice of left coronary artery opening into coronary sinus

Left cusp of pulmonary valve *(L)*

Posterior cusp of aortic valve *(P)*

Left auricle

Right atrium

Left ventricle

Right auricle

Anterior cusp of pulmonary valve *(A)*

Orifice of right coronary artery opening into right coronary sinus

Right cusp of pulmonary valve *(R)*

Right cusp of aortic valve *(R)*

Left cusp of aortic valve *(L)*

A. Superior View

Right ventricle

3.58 VALVES OF HEART

A. Excised heart.
- The ventricles are positioned anteriorly and to the left, the atria posteriorly and to the right.
- The roots of the aorta and pulmonary artery, which conduct blood from the ventricles, are placed anterior to the atria.
- The aorta and pulmonary artery are enclosed within a common tube of serous pericardium and partly embraced by the auricles of the atria.
- The transverse pericardial sinus curves posterior to the enclosed stems of the aorta and pulmonary trunk and anterior to the superior vena cava and upper limits of the atria.

B. Developmental basis for naming of pulmonary and aortic valve cusps. The truncus arteriosus with four cusps **(I)** splits to form two valves, each with three cusps **(II)**. The heart undergoes partial rotation to the left on its axis, resulting in the arrangement of cusps shown in **(III)**.

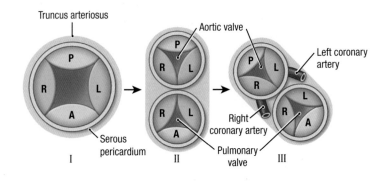

Truncus arteriosus

Aortic valve

Left coronary artery

Serous pericardium

Right coronary artery

Pulmonary valve

I   II   III

B

Semilunar Valves/Cusps	
R Right	**A** Anterior
L Left	**P** Posterior

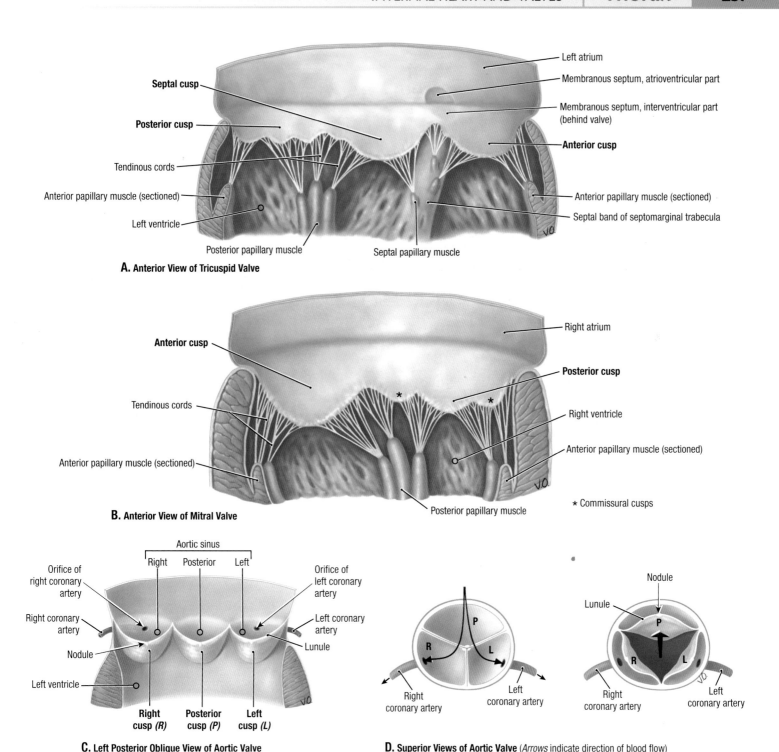

A. Anterior View of Tricuspid Valve

Left atrium
Membranous septum, atrioventricular part
Membranous septum, interventricular part (behind valve)
Anterior cusp
Anterior papillary muscle (sectioned)
Septal band of septomarginal trabecula

Septal cusp
Posterior cusp
Tendinous cords
Anterior papillary muscle (sectioned)
Left ventricle
Posterior papillary muscle
Septal papillary muscle

B. Anterior View of Mitral Valve

Right atrium
Posterior cusp
Right ventricle
Anterior papillary muscle (sectioned)

Anterior cusp
Tendinous cords
Anterior papillary muscle (sectioned)
Posterior papillary muscle

★ Commissural cusps

C. Left Posterior Oblique View of Aortic Valve

Aortic sinus
Right   Posterior   Left
Orifice of right coronary artery
Orifice of left coronary artery
Right coronary artery
Left coronary artery
Nodule
Lunule
Left ventricle
Right cusp (R)   **Posterior cusp (P)**   **Left cusp (L)**

D. Superior Views of Aortic Valve (*Arrows* indicate direction of blood flow)

Nodule
Lunule
Right coronary artery
Left coronary artery

VALVES OF THE HEART

A. and **B.** Atrioventricular valves. **C.** and **D.** Semilunar valves.

Tendinous cords pass from the tips of the papillary muscles to the free margins and ventricular surfaces of the cusps of the tricuspid (**A**) and mitral (**B**) valves. Each papillary muscle or muscle group controls the adjacent sides of two cusps, resisting valve prolapse during systole. In **C** the anulus of the aortic valve has been incised between the right and left cusps and spread open. Each cusp of the semilunar valves bears a nodule in the midpoint of its free edge, flanked by thin connective tissue areas (lunules). When the ventricles relax to fill (diastole), backflow of blood from aortic recoil or pulmonary resistance fills the sinus (space between cusp and dilated part of the aortic or pulmonary wall), causing the nodules and lunules to meet centrally, closing the valve (**D,** *left*). Filling of the coronary arteries occurs during diastole (when ventricular walls are relaxed) as backflow "inflates" the cusps to close the valve.

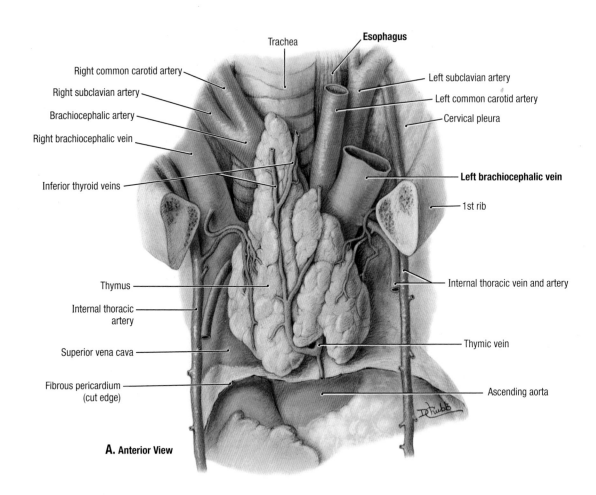

Trachea
Esophagus
Right common carotid artery
Left subclavian artery
Right subclavian artery
Left common carotid artery
Brachiocephalic artery
Cervical pleura
Right brachiocephalic vein
Left brachiocephalic vein
Inferior thyroid veins
1st rib
Thymus
Internal thoracic vein and artery
Internal thoracic artery
Superior vena cava
Thymic vein
Fibrous pericardium (cut edge)
Ascending aorta

A. Anterior View

3.60   SUPERIOR MEDIASTINUM I AND II: SUPERFICIAL DISSECTIONS

A. Dissection I: Thymus *in situ*. The sternum and ribs have been excised and the pleurae removed. It is unusual in an adult to see such a discrete thymus, which is large during puberty but subsequently regresses and is for the most part replaced by fat and fibrous tissue. **B.** Dissection II: Thymus removed. **C.** Relationship of nerves and vessels. The right vagus nerve (CN X) crosses anterior to the right subclavian artery and gives off the right recurrent laryngeal nerve, which passes medially to reach the trachea and esophagus. The left recurrent laryngeal nerve passes inferior and then posterior to the arch of the aorta and ascends between the trachea and esophagus to the larynx.

The distal part of the ascending aorta receives a strong thrust of blood when the left ventricle contracts. Because its wall is not reinforced by fibrous pericardium (the fibrous pericardium blends with the aortic adventitia at the beginning of the arch), an aneurysm may develop. An **aortic aneurysm** is evident on chest film (radiograph of the thorax) or a magnetic resonance angiogram as an enlarged area of the ascending aorta silhouette. Individuals with

an aneurysm usually complain of chest pain that radiates to the back. The aneurysm may exert pressure on the trachea, esophagus, and recurrent laryngeal nerve, causing difficulty in breathing and swallowing.

Mediastinal compression. The recurrent laryngeal nerves supply all the intrinsic muscles of the larynx, except the cricothyroid. Consequently, any investigative procedure or disease process in the superior mediastinum may involve these nerves and affect the voice. Because the left recurrent laryngeal nerve hooks around the arch of the aorta and ascends between the trachea and the esophagus, it may be involved when there is a bronchial or esophageal carcinoma, enlargement of mediastinal lymph nodes, or an aneurysm of the arch of the aorta.

The thymus is a prominent feature during infancy and childhood. In some infants, the thymus may compress the trachea. The thymus plays an important role in the development and maintenance of the immune system. As puberty is reached, the thymus begins to diminish in relative size. By adulthood, it is replaced by adipose tissue.

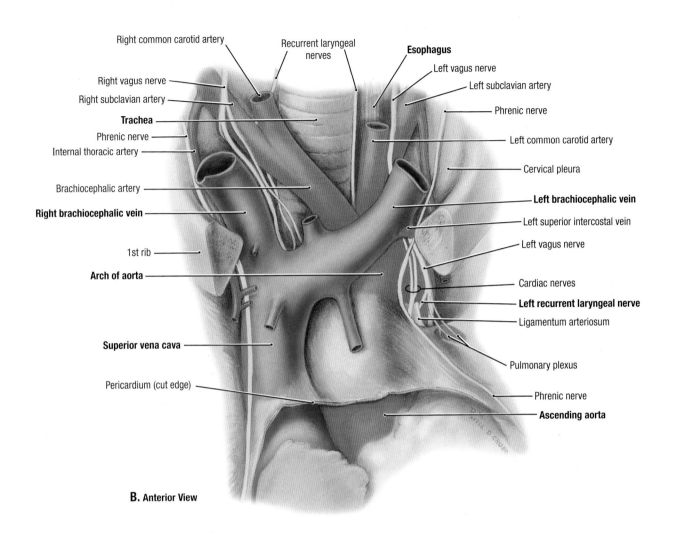

Right common carotid artery

Recurrent laryngeal nerves

Esophagus

Right vagus nerve

Right subclavian artery

Left vagus nerve

Left subclavian artery

Trachea

Phrenic nerve

Phrenic nerve

Internal thoracic artery

Left common carotid artery

Brachiocephalic artery

Cervical pleura

Right brachiocephalic vein

Left brachiocephalic vein

Left superior intercostal vein

1st rib

Left vagus nerve

Arch of aorta

Cardiac nerves

Left recurrent laryngeal nerve

Ligamentum arteriosum

Superior vena cava

Pulmonary plexus

Pericardium (cut edge)

Phrenic nerve

Ascending aorta

B. Anterior View

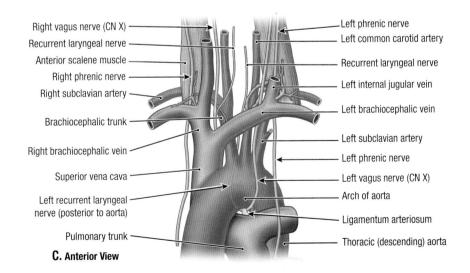

Right vagus nerve (CN X)

Left phrenic nerve

Recurrent laryngeal nerve

Left common carotid artery

Anterior scalene muscle

Recurrent laryngeal nerve

Right phrenic nerve

Left internal jugular vein

Right subclavian artery

Brachiocephalic trunk

Left brachiocephalic vein

Right brachiocephalic vein

Left subclavian artery

Left phrenic nerve

Superior vena cava

Left vagus nerve (CN X)

Left recurrent laryngeal nerve (posterior to aorta)

Arch of aorta

Ligamentum arteriosum

Pulmonary trunk

Thoracic (descending) aorta

C. Anterior View

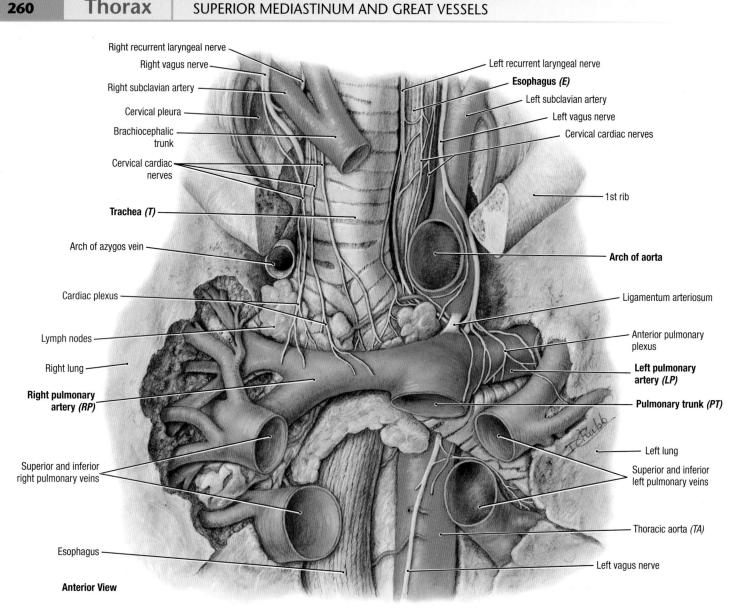

Right recurrent laryngeal nerve
Right vagus nerve
Right subclavian artery
Cervical pleura
Brachiocephalic trunk
Cervical cardiac nerves
Trachea (T)
Arch of azygos vein
Cardiac plexus
Lymph nodes
Right lung
Right pulmonary artery (RP)
Superior and inferior right pulmonary veins
Esophagus

Left recurrent laryngeal nerve
Esophagus (E)
Left subclavian artery
Left vagus nerve
Cervical cardiac nerves
1st rib
Arch of aorta
Ligamentum arteriosum
Anterior pulmonary plexus
Left pulmonary artery (LP)
Pulmonary trunk (PT)
Left lung
Superior and inferior left pulmonary veins
Thoracic aorta (TA)
Left vagus nerve

Anterior View

| **3.61** | SUPERIOR MEDIASTINUM III: CARDIAC PLEXUS AND PULMONARY ARTERIES |

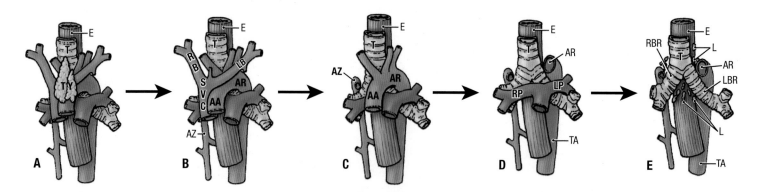

| **3.62** | RELATIONS OF GREAT VESSELS AND TRACHEA |

Observe, from superficial to deep: **(A)** Thymus (TY); **(B)** The right (RB) and left (LB) brachiocephalic veins form the superior vena cava (SVC) and receive the arch of the azygos vein (AZ) posteriorly; **(C)** The ascending aorta (AA) and arch of the aorta (AR) arch over the right pulmonary artery and left main bronchus; **(D)** The right and left pulmonary arteries (RP and LP); and **(E)** The tracheobronchial lymph nodes (L) at the tracheal bifurcation (T). E, esophagus; LBR, left main bronchus; RBR, right main bronchus; TA, thoracic aorta.

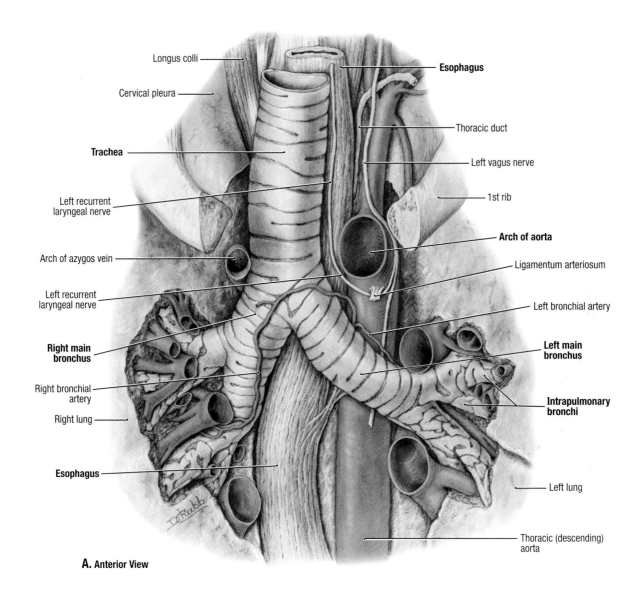

A. Anterior View

- Longus colli
- Cervical pleura
- **Trachea**
- Left recurrent laryngeal nerve
- Arch of azygos vein
- Left recurrent laryngeal nerve
- **Right main bronchus**
- Right bronchial artery
- Right lung
- **Esophagus**
- **Esophagus**
- Thoracic duct
- Left vagus nerve
- 1st rib
- **Arch of aorta**
- Ligamentum arteriosum
- Left bronchial artery
- **Left main bronchus**
- **Intrapulmonary bronchi**
- Left lung
- Thoracic (descending) aorta

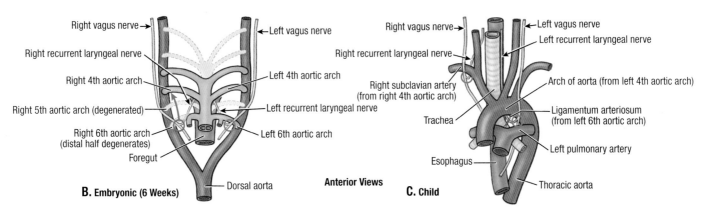

B. Embryonic (6 Weeks)

- Right vagus nerve
- Right recurrent laryngeal nerve
- Right 4th aortic arch
- Right 5th aortic arch (degenerated)
- Right 6th aortic arch (distal half degenerates)
- Foregut
- Left vagus nerve
- Left 4th aortic arch
- Left recurrent laryngeal nerve
- Left 6th aortic arch
- Dorsal aorta

Anterior Views

C. Child

- Right vagus nerve
- Right recurrent laryngeal nerve
- Right subclavian artery (from right 4th aortic arch)
- Trachea
- Esophagus
- Left vagus nerve
- Left recurrent laryngeal nerve
- Arch of aorta (from left 4th aortic arch)
- Ligamentum arteriosum (from left 6th aortic arch)
- Left pulmonary artery
- Thoracic aorta

SUPERIOR MEDIASTINUM IV: TRACHEAL BIFURCATION AND BRONCHI

3.63

A. Dissection. **B.** and **C.** Asymmetrical course of right and left recurrent laryngeal nerves. Arch VI disappears on the right, leaving the right recurrent laryngeal nerve to pass under arch IV, which becomes the right subclavian artery. Arch VI becomes part of the ductus arteriosus on the left side, and arch IV "descends" to become the arch of the aorta; thus, the left recurrent laryngeal nerve is pulled into the thorax.

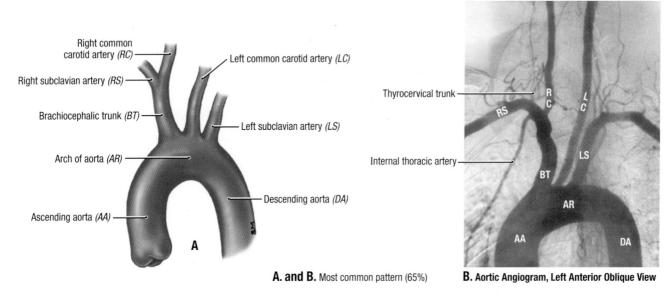

A. and B. Most common pattern (65%)

B. Aortic Angiogram, Left Anterior Oblique View

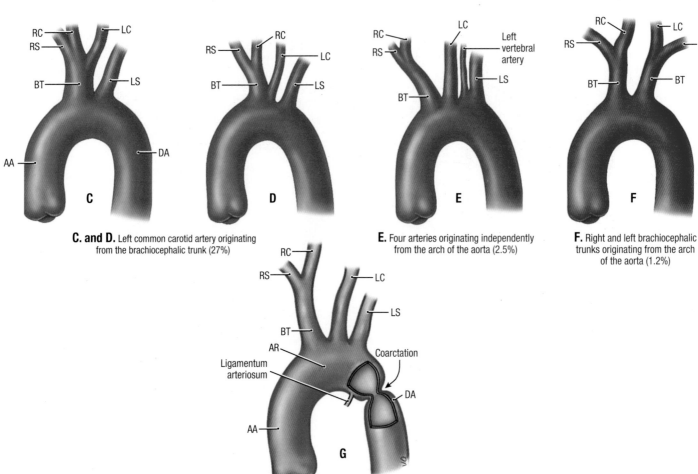

C. and D. Left common carotid artery originating from the brachiocephalic trunk (27%)

E. Four arteries originating independently from the arch of the aorta (2.5%)

F. Right and left brachiocephalic trunks originating from the arch of the aorta (1.2%)

3.64 BRANCHES OF AORTIC ARCH

A. and **B.** Most common pattern (65%). **C–F.** Variations. **G.** In co-arctation of the aorta, the arch or descending aorta has an abnormal narrowing (stenosis) that diminishes the caliber of the aortic lumen, producing an obstruction to blood flow. The most common site is near the ligamentum arteriosum. When the coarctation is inferior to this site (**postductal coarctation**), a good collateral circulation usually develops between the proximal and distal parts of the aorta through the intercostal and internal thoracic arteries.

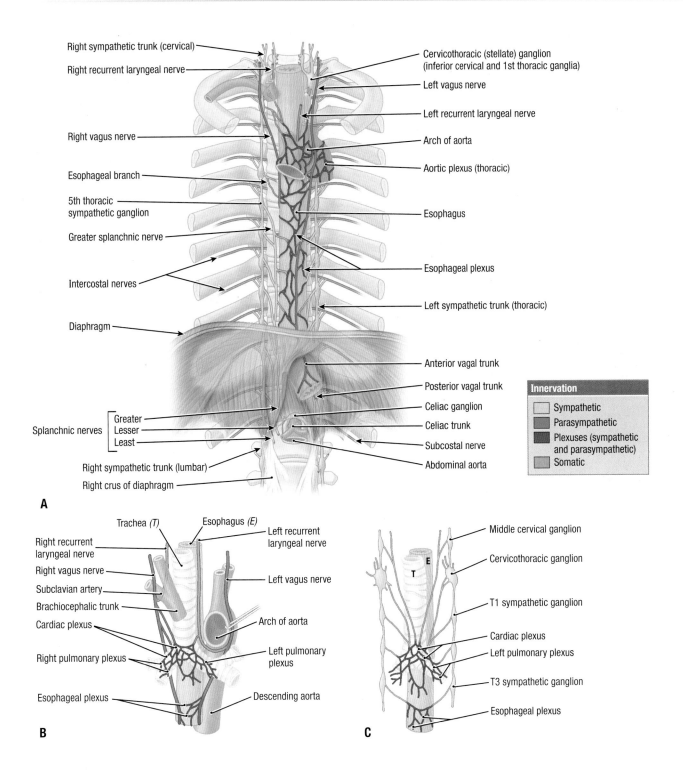

A.

Right sympathetic trunk (cervical)
Right recurrent laryngeal nerve
Right vagus nerve
Esophageal branch
5th thoracic sympathetic ganglion
Greater splanchnic nerve
Intercostal nerves
Diaphragm
Splanchnic nerves — Greater / Lesser / Least
Right sympathetic trunk (lumbar)
Right crus of diaphragm

Cervicothoracic (stellate) ganglion (inferior cervical and 1st thoracic ganglia)
Left vagus nerve
Left recurrent laryngeal nerve
Arch of aorta
Aortic plexus (thoracic)
Esophagus
Esophageal plexus
Left sympathetic trunk (thoracic)
Anterior vagal trunk
Posterior vagal trunk
Celiac ganglion
Celiac trunk
Subcostal nerve
Abdominal aorta

Innervation
☐ Sympathetic
☐ Parasympathetic
☐ Plexuses (sympathetic and parasympathetic)
☐ Somatic

B.
Trachea (T)   **Esophagus (E)**
Right recurrent laryngeal nerve
Right vagus nerve
Subclavian artery
Brachiocephalic trunk
Cardiac plexus
Right pulmonary plexus
Esophageal plexus

Left recurrent laryngeal nerve
Left vagus nerve
Arch of aorta
Left pulmonary plexus
Descending aorta

C.
Middle cervical ganglion
Cervicothoracic ganglion
T1 sympathetic ganglion
Cardiac plexus
Left pulmonary plexus
T3 sympathetic ganglion
Esophageal plexus

CARDIAC AND PULMONARY PLEXUSES

3.65

A. Overview. **B.** Parasympathetic contribution. **C.** Sympathetic contribution.

Heart. Sympathetic stimulation increases the heart's rate and the force of its contractions. Parasympathetic stimulation slows the heart rate, reduces the force of contraction, and constricts the coronary arteries, saving energy between periods of increased demand. While the cardiac plexus is shown in relation to the bifurcation of the trachea, note that it lies directly posterior to the superior margin of the heart (see Fig. 3.28C) and in close proximity to the nodal tissue and origins of the coronary arteries.

Lungs. Sympathetic fibers are inhibitory to the bronchial muscle (bronchodilator), motor to pulmonary vessels (vasoconstrictor), and inhibitory to the alveolar glands of the bronchial tree. Parasympathetic fibers from CN X are bronchoconstrictors, secretory to the glands of the bronchial tree (secretomotor).

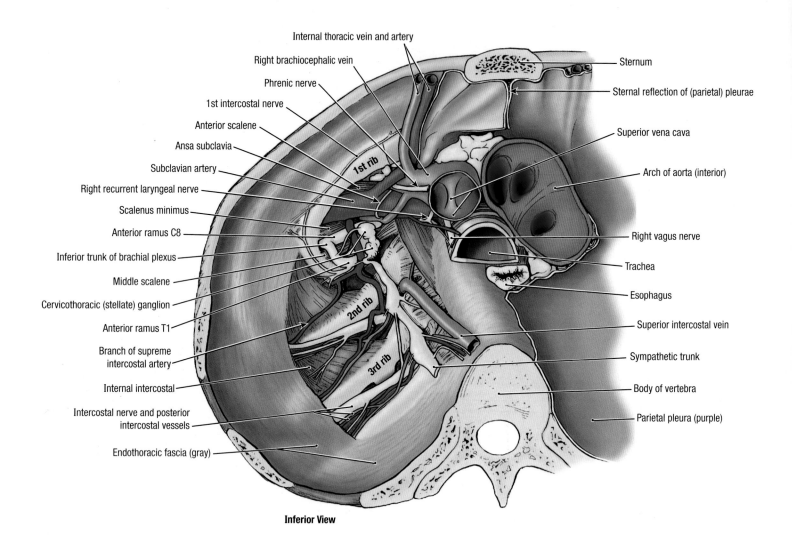

Internal thoracic vein and artery

Right brachiocephalic vein

Phrenic nerve

1st intercostal nerve

Anterior scalene

Ansa subclavia

Subclavian artery

Right recurrent laryngeal nerve

Scalenus minimus

Anterior ramus C8

Inferior trunk of brachial plexus

Middle scalene

Cervicothoracic (stellate) ganglion

Anterior ramus T1

Branch of supreme intercostal artery

Internal intercostal

Intercostal nerve and posterior intercostal vessels

Endothoracic fascia (gray)

1st rib

2nd rib

3rd rib

Sternum

Sternal reflection of (parietal) pleurae

Superior vena cava

Arch of aorta (interior)

Right vagus nerve

Trachea

Esophagus

Superior intercostal vein

Sympathetic trunk

Body of vertebra

Parietal pleura (purple)

Inferior View

3.66 SUPERIOR MEDIASTINUM AND ROOF OF PLEURAL CAVITY

- The cervical, costal, and mediastinal parietal pleura *(purple)* and portions of the endothoracic fascia *(gray)* have been removed from the right side of the specimen to demonstrate structures traversing the superior thoracic aperture.
- The first part of the subclavian artery disappears as it crosses the 1st rib anterior to the anterior scalene muscle.
- The ansa subclavia from the sympathetic trunk and right recurrent laryngeal nerve from the vagus are seen looping inferior to the subclavian artery.
- The anterior rami of C8 and T1 merge to form the inferior trunk of the brachial plexus, which crosses the 1st rib posterior to the anterior scalene muscle.

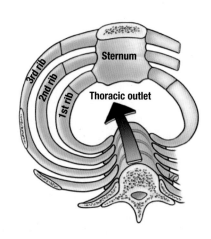

3rd rib

2nd rib

1st rib

Sternum

Thoracic outlet

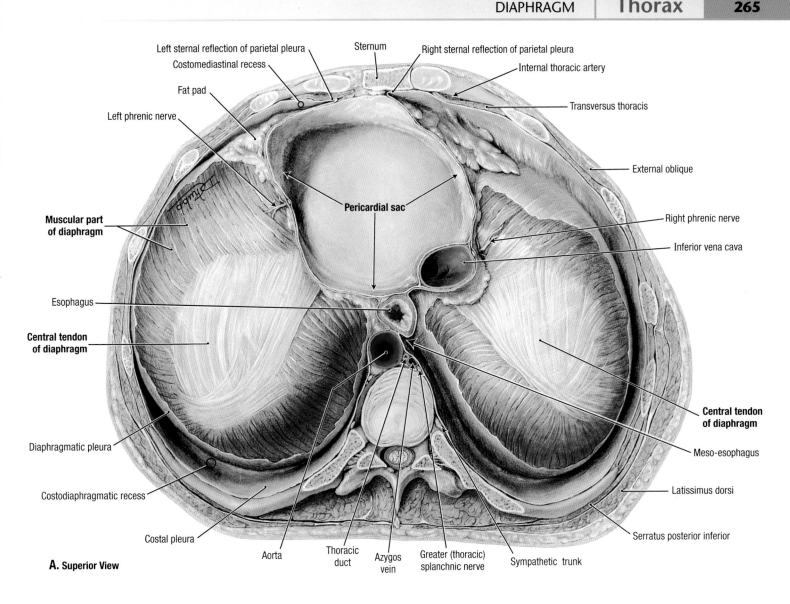

Left sternal reflection of parietal pleura
Costomediastinal recess
Fat pad
Left phrenic nerve
Muscular part of diaphragm
Esophagus
Central tendon of diaphragm
Diaphragmatic pleura
Costodiaphragmatic recess
Costal pleura
Aorta
Thoracic duct
Azygos vein
Greater (thoracic) splanchnic nerve
Sympathetic trunk
Sternum
Right sternal reflection of parietal pleura
Internal thoracic artery
Transversus thoracis
External oblique
Right phrenic nerve
Inferior vena cava
Pericardial sac
Central tendon of diaphragm
Meso-esophagus
Latissimus dorsi
Serratus posterior inferior

A. Superior View

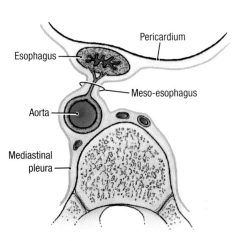

Esophagus
Pericardium
Aorta
Meso-esophagus
Mediastinal pleura

B. Superior View

DIAPHRAGM AND PERICARDIAL SAC   3.67

A. The diaphragmatic pleura is mostly removed. The pericardial sac is situated on the anterior half of the diaphragm; one third is to the right of the median plane, and two thirds to the left. Note also that anterior to the pericardium, the sternal reflection of the left pleural sac approaches but fails to meet that of the right sac in the median plane; and on reaching the vertebral column, the costal pleura becomes the mediastinal pleura.

Irritation of the parietal pleura produces local pain and referred pain to the areas sharing innervation by the same segments of the spinal cord. **Irritation of the costal and peripheral parts of the diaphragmatic pleura** results in local pain and referred pain along the intercostal nerves to the thoracic and abdominal walls. **Irritation of the mediastinal and central diaphragmatic parts of the parietal pleura** results in pain that is referred to the root of the neck and over the shoulder (C3–C5 dermatomes). **B.** Between the inferior part of the esophagus and the aorta, the right and left layers of mediastinal pleura form a dorsal meso-esophagus, especially when the body is in the prone position.

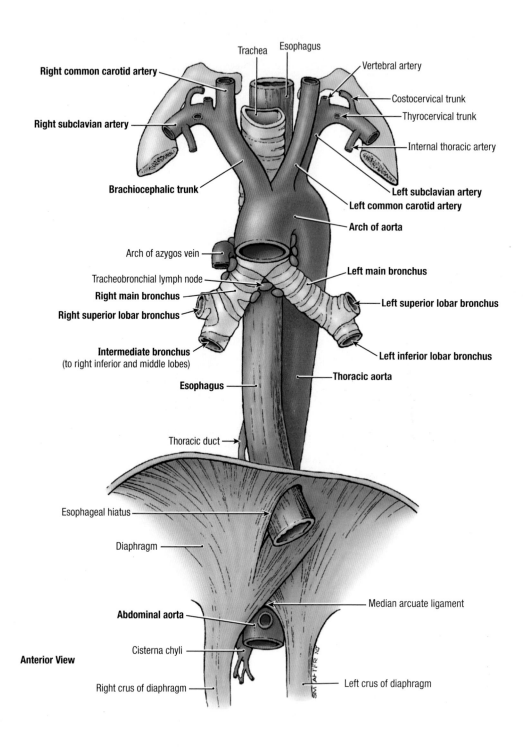

Trachea
Esophagus

Right common carotid artery

Vertebral artery

Costocervical trunk

Thyrocervical trunk

Right subclavian artery

Internal thoracic artery

Brachiocephalic trunk

Left subclavian artery

Left common carotid artery

Arch of aorta

Arch of azygos vein

Tracheobronchial lymph node

Left main bronchus

Right main bronchus

Right superior lobar bronchus

Left superior lobar bronchus

Intermediate bronchus
(to right inferior and middle lobes)

Left inferior lobar bronchus

Esophagus

Thoracic aorta

Thoracic duct

Esophageal hiatus

Diaphragm

Median arcuate ligament

Abdominal aorta

Anterior View

Cisterna chyli

Right crus of diaphragm

Left crus of diaphragm

3.68 ESOPHAGUS, TRACHEA, AND AORTA

- The anterior relations of the thoracic part of the esophagus from superior to inferior are the trachea (from origin at cricoid cartilage to bifurcation), right and left bronchi, inferior tracheobronchial lymph nodes, pericardium (not shown) and, finally, the diaphragm.
- The arch of the aorta passes posterior to the left of these four structures as it arches over the left main bronchus; the arch of the azygos vein passes anterior to their right as it arches over the right main bronchus.

- **Esophageal impressions.** The impressions produced in the esophagus by adjacent structures (aorta, left main bronchus, and esophageal hiatus) are of clinical interest because of the slower passage of substances at these sites. The impressions indicate where swallowed foreign bodies are most likely to lodge and where a stricture may develop after the accidental drinking of a caustic liquid such as lye.

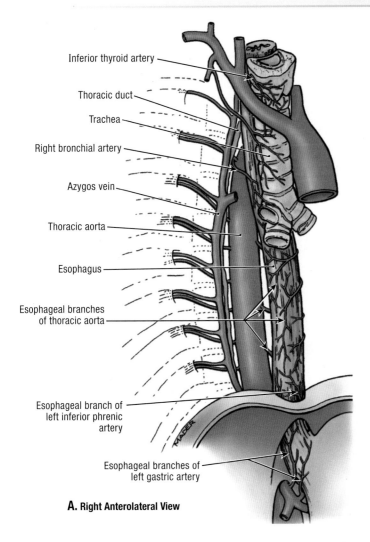

Inferior thyroid artery

Thoracic duct

Trachea

Right bronchial artery

Azygos vein

Thoracic aorta

Esophagus

Esophageal branches
of thoracic aorta

Esophageal branch of
left inferior phrenic
artery

Esophageal branches of
left gastric artery

A. Right Anterolateral View

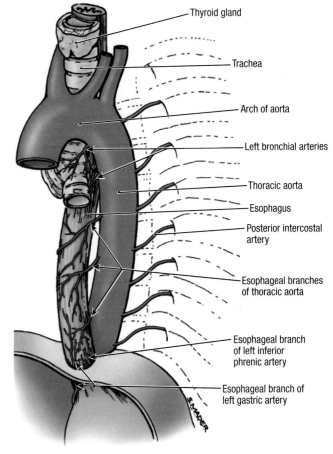

Thyroid gland

Trachea

Arch of aorta

Left bronchial arteries

Thoracic aorta

Esophagus

Posterior intercostal
artery

Esophageal branches
of thoracic aorta

Esophageal branch
of left inferior
phrenic artery

Esophageal branch of
left gastric artery

B. Left Anterolateral View

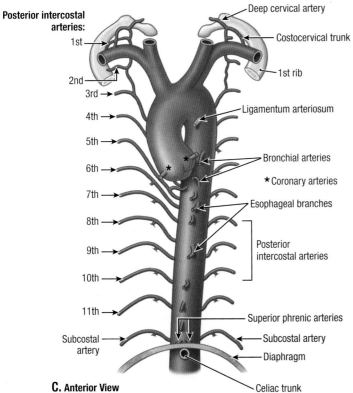

Posterior intercostal
arteries:

1st

2nd

3rd

4th

5th

6th

7th

8th

9th

10th

11th

Subcostal
artery

Deep cervical artery

Costocervical trunk

1st rib

Ligamentum arteriosum

Bronchial arteries

★ Coronary arteries

Esophageal branches

Posterior
intercostal arteries

Superior phrenic arteries

Subcostal artery

Diaphragm

Celiac trunk

C. Anterior View

ARTERIAL SUPPLY TO TRACHEA AND ESOPHAGUS

3.69

A. and **B.** Arteries of trachea and esophagus. The continuous anastomotic chain of arteries on the esophagus is formed (1) by branches of the right and left inferior thyroid and right supreme intercostal arteries superiorly, (2) by the unpaired median aortic (bronchial and esophageal) branches, and (3) by branches of the left gastric and left inferior phrenic arteries inferiorly. The right bronchial artery usually arises from the superior left bronchial or 3rd right posterior intercostal artery (here the 5th) or from the aorta directly. The unpaired median aortic branches also supply the trachea and bronchi. **C.** Branches of the thoracic aorta.

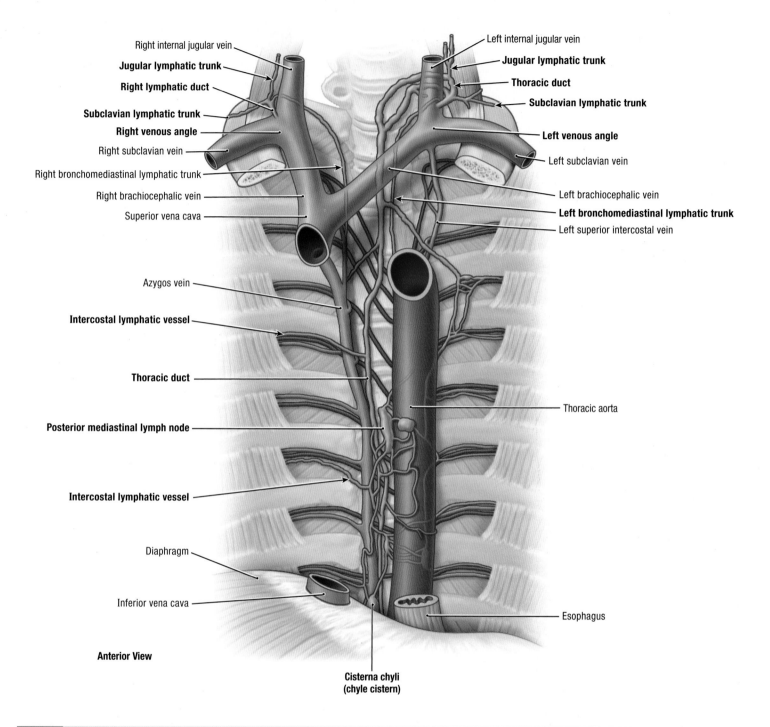

Right internal jugular vein

Jugular lymphatic trunk

Right lymphatic duct

Subclavian lymphatic trunk

Right venous angle

Right subclavian vein

Right bronchomediastinal lymphatic trunk

Right brachiocephalic vein

Superior vena cava

Azygos vein

Intercostal lymphatic vessel

Thoracic duct

Posterior mediastinal lymph node

Intercostal lymphatic vessel

Diaphragm

Inferior vena cava

Left internal jugular vein

Jugular lymphatic trunk

Thoracic duct

Subclavian lymphatic trunk

Left venous angle

Left subclavian vein

Left brachiocephalic vein

Left bronchomediastinal lymphatic trunk

Left superior intercostal vein

Thoracic aorta

Esophagus

Anterior View

**Cisterna chyli
(chyle cistern)**

3.70 THORACIC DUCT

- The descending aorta is located to the left, and the azygos vein slightly to the right of the midline.
- The thoracic duct (1) originates from the cisterna chyli at the T12 vertebral level, (2) ascends on the vertebral column between the azygos vein and the descending aorta, (3) passes to the left at the junction of the posterior and superior mediastina, and continues its ascent to the neck, where (4) it arches laterally to enter the venous system near or at the angle of union of the left internal jugular and subclavian veins (left venous angle).
- The thoracic duct is commonly plexiform (resembling a network) in the posterior mediastinum.

- The termination of the thoracic duct typically receives the left jugular, subclavian, and bronchomediastinal trunks.
- The right lymph duct is short and formed by the union of the right jugular, subclavian, and bronchomediastinal trunks.
- Because the thoracic duct is thin-walled and may be colorless, it may not be easily identified. Consequently, it is vulnerable to inadvertent injury during investigative and/or surgical procedures in the posterior mediastinum. **Laceration of the thoracic duct** results in chyle escaping into the thoracic cavity. Chyle may also enter the pleural cavity, producing chylothorax.

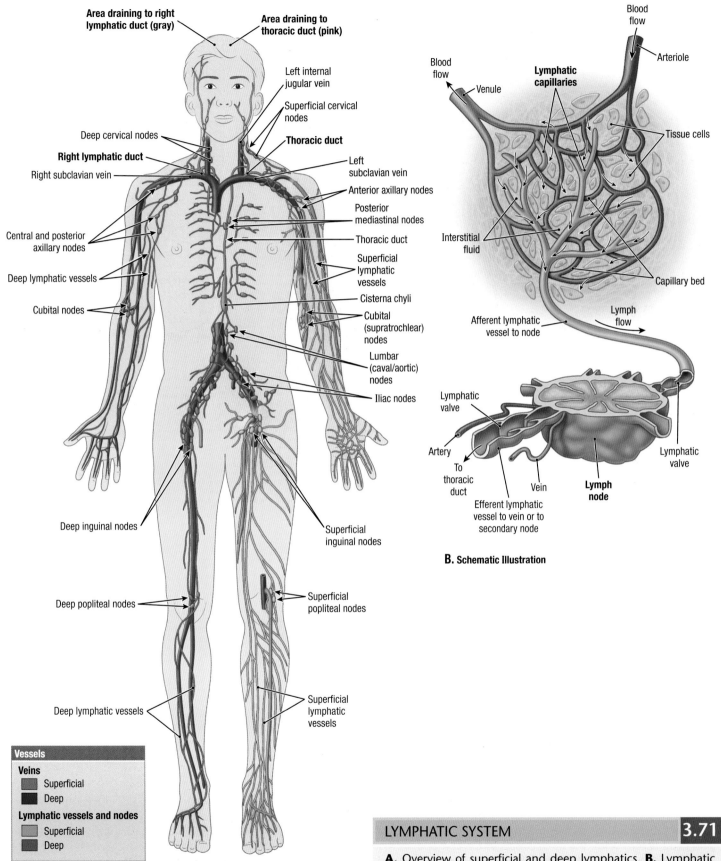

Area draining to right lymphatic duct (gray)

Area draining to thoracic duct (pink)

Left internal jugular vein

Superficial cervical nodes

Deep cervical nodes

Thoracic duct

Right lymphatic duct

Right subclavian vein

Left subclavian vein

Anterior axillary nodes

Posterior mediastinal nodes

Central and posterior axillary nodes

Thoracic duct

Superficial lymphatic vessels

Deep lymphatic vessels

Cisterna chyli

Cubital nodes

Cubital (supratrochlear) nodes

Lumbar (caval/aortic) nodes

Iliac nodes

Deep inguinal nodes

Superficial inguinal nodes

Deep popliteal nodes

Superficial popliteal nodes

Deep lymphatic vessels

Superficial lymphatic vessels

Vessels

Veins
- Superficial
- Deep

Lymphatic vessels and nodes
- Superficial
- Deep

A. Anterior View

Blood flow

Arteriole

Blood flow

Lymphatic capillaries

Venule

Tissue cells

Interstitial fluid

Capillary bed

Afferent lymphatic vessel to node

Lymph flow

Lymphatic valve

Lymphatic valve

Artery

To thoracic duct

Vein

Lymph node

Efferent lymphatic vessel to vein or to secondary node

B. Schematic Illustration

LYMPHATIC SYSTEM | **3.71**

A. Overview of superficial and deep lymphatics. **B.** Lymphatic capillaries, vessels, and nodes. Arrows *(black)* indicate the flow (leaking of interstitial fluid out of blood vessels and absorption) into the lymphatic capillaries.

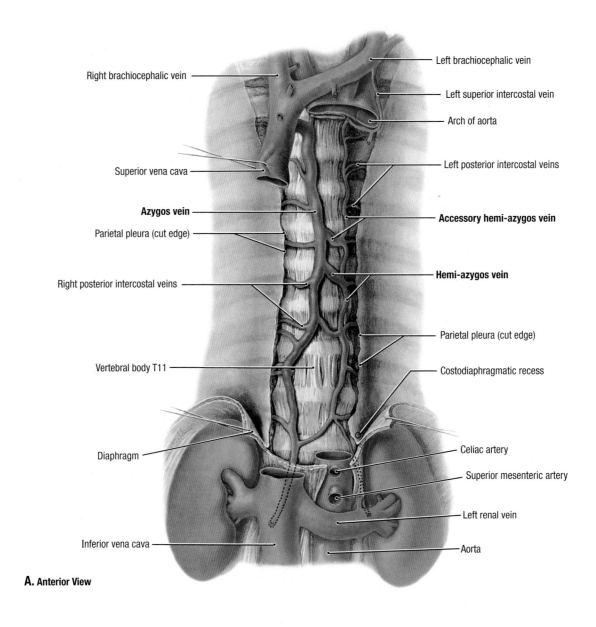

Right brachiocephalic vein

Superior vena cava

Azygos vein

Parietal pleura (cut edge)

Right posterior intercostal veins

Vertebral body T11

Diaphragm

Inferior vena cava

A. Anterior View

Left brachiocephalic vein

Left superior intercostal vein

Arch of aorta

Left posterior intercostal veins

Accessory hemi-azygos vein

Hemi-azygos vein

Parietal pleura (cut edge)

Costodiaphragmatic recess

Celiac artery

Superior mesenteric artery

Left renal vein

Aorta

3.72 AZYGOS SYSTEM OF VEINS

A. Dissection. **B.** Schematic illustration. The ascending lumbar veins connect the common iliac veins to the lumbar veins and join the subcostal veins to become the lateral roots of the azygos and hemi-azygos veins; the medial roots of the azygos and hemi-azygos veins are usually from the inferior vena cava and left renal vein, if present. Typically, the upper four left posterior intercostal veins drain into the left brachiocephalic vein, directly and via the left superior intercostal veins.

The hemi-azygos, accessory hemi-azygos, and left superior inter-costals veins are continuous in **A**, but most commonly, they are discontinuous as in **B**. The hemi-azygos vein crosses the vertebral column at approximately T9, and the accessory hemi-azygos vein crosses at T8, to enter the azygos vein **(B)**. In contrast, there are four cross-connecting channels between the azygos and hemi-azygos systems in **A**. The azygos vein arches superior to the root of the right lung at T4 to drain into the superior vena cava.

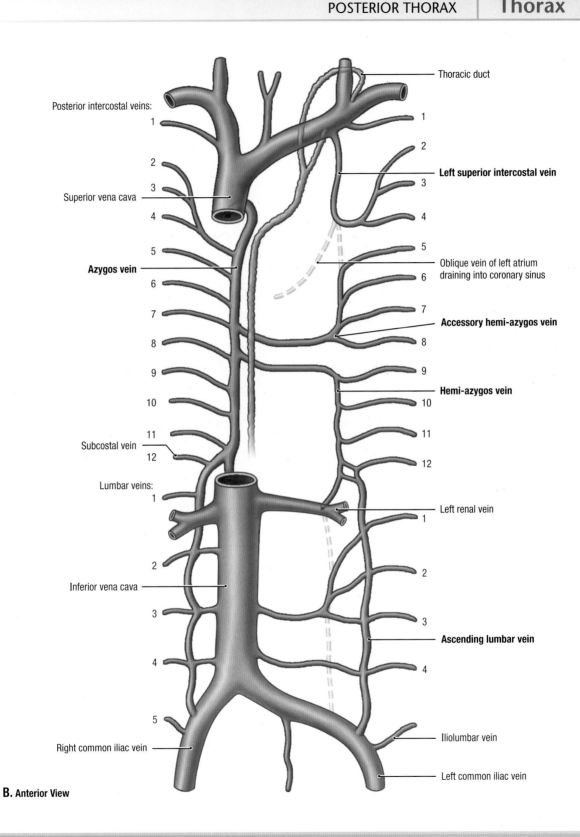

Posterior intercostal veins:
1
2
3
4
5
6
7
8
9
10
11
12

Superior vena cava

Azygos vein

Subcostal vein

Lumbar veins:
1
2

Inferior vena cava

3

4

5

Right common iliac vein

B. Anterior View

Thoracic duct
1
2
Left superior intercostal vein
3
4
5
Oblique vein of left atrium draining into coronary sinus
6
7
Accessory hemi-azygos vein
8
9
Hemi-azygos vein
10
11
12

Left renal vein
1
2
3
Ascending lumbar vein
4

Iliolumbar vein

Left common iliac vein

AZYGOS SYSTEM OF VEINS (*continued*)

The azygos, hemi-azygos, and accessory hemi-azygos veins offer alternate means of venous drainage from the thoracic, abdominal, and back regions when **obstruction of the IVC** occurs. In some people, an accessory azygos vein parallels the main azygos vein on the right side. Other people have no hemi-azygos system of veins. A clinically important variation, although uncommon, is when the azygos system receives all the blood from the IVC, except that from the liver. In these people, the azygos system drains nearly all the blood inferior to the diaphragm, except that from the digestive tract. When **obstruction of the SVC** occurs superior to the entrance of the azygos vein, blood can drain inferiorly into the veins of the abdominal wall and return to the right atrium through the IVC and azygos system of veins.

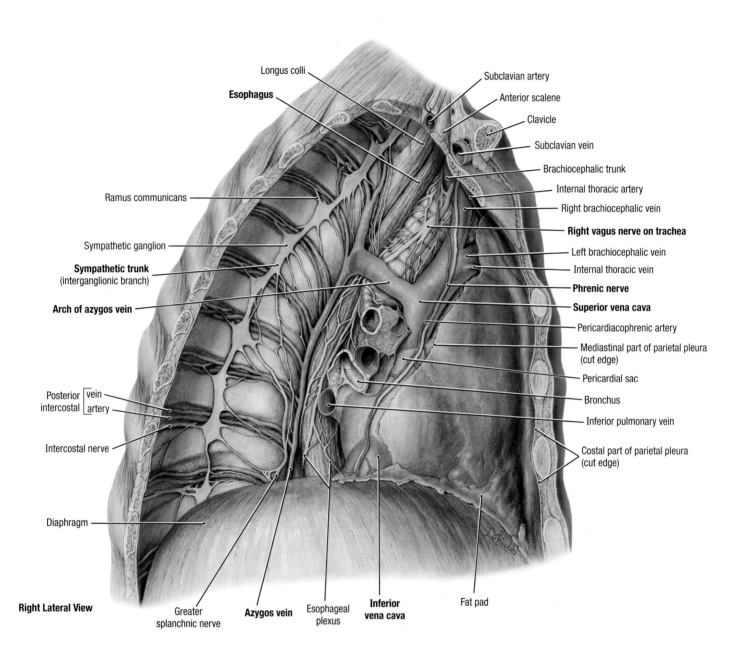

Longus colli

Esophagus

Ramus communicans

Sympathetic ganglion

Sympathetic trunk
(interganglionic branch)

Arch of azygos vein

Posterior ⌈ vein
intercostal ⌊ artery

Intercostal nerve

Diaphragm

Right Lateral View

Greater
splanchnic nerve

Azygos vein

Esophageal
plexus

**Inferior
vena cava**

Fat pad

Subclavian artery

Anterior scalene

Clavicle

Subclavian vein

Brachiocephalic trunk

Internal thoracic artery

Right brachiocephalic vein

Right vagus nerve on trachea

Left brachiocephalic vein

Internal thoracic vein

Phrenic nerve

Superior vena cava

Pericardiacophrenic artery

Mediastinal part of parietal pleura
(cut edge)

Pericardial sac

Bronchus

Inferior pulmonary vein

Costal part of parietal pleura
(cut edge)

3.73　MEDIASTINUM, RIGHT SIDE

- The costal and mediastinal pleurae have mostly been removed, exposing the underlying structures. Compare with the mediastinal surface of the right lung in Figure 3.32.
- The right side of the mediastinum is the "blue side," dominated by the arch of the azygos vein and the superior vena cava.
- Both the trachea and the esophagus are visible from the right side.

- The right vagus nerve descends on the medial surface of the trachea, passes medial to the arch of the azygos vein, posterior to the root of the lung, and then enters the esophageal plexus.
- The right phrenic nerve passes anterior to the root of the lung lateral to both venae cavae.

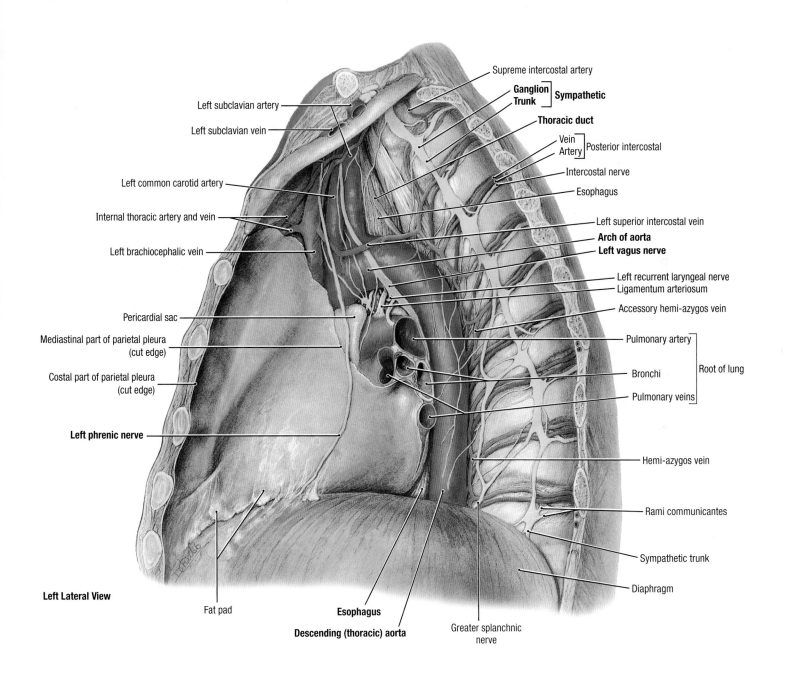

Supreme intercostal artery

Ganglion Trunk **Sympathetic**

Thoracic duct

Vein
Artery Posterior intercostal

Intercostal nerve

Esophagus

Left superior intercostal vein

Arch of aorta

Left vagus nerve

Left recurrent laryngeal nerve

Ligamentum arteriosum

Accessory hemi-azygos vein

Pulmonary artery

Bronchi Root of lung

Pulmonary veins

Hemi-azygos vein

Rami communicantes

Sympathetic trunk

Diaphragm

Left subclavian artery

Left subclavian vein

Left common carotid artery

Internal thoracic artery and vein

Left brachiocephalic vein

Pericardial sac

Mediastinal part of parietal pleura
(cut edge)

Costal part of parietal pleura
(cut edge)

Left phrenic nerve

Left Lateral View

Fat pad

Esophagus

Descending (thoracic) aorta

Greater splanchnic
nerve

MEDIASTINUM, LEFT SIDE

3.74

- Compare with the mediastinal surface of the left lung in Figure 3.33.
- The left side of the mediastinum is the "red side," dominated by the arch and descending portion of the aorta, the left common carotid and subclavian arteries; the latter obscure the trachea from view.
- The thoracic duct can be seen on the left side of the esophagus.
- The left vagus nerve passes posterior to the root of the lung, sending its recurrent laryngeal branch around the ligamentum arteriosum inferior and then medial to the aortic arch.
- The phrenic nerve passes anterior to the root of the lung and penetrates the diaphragm more anteriorly than on the right side.

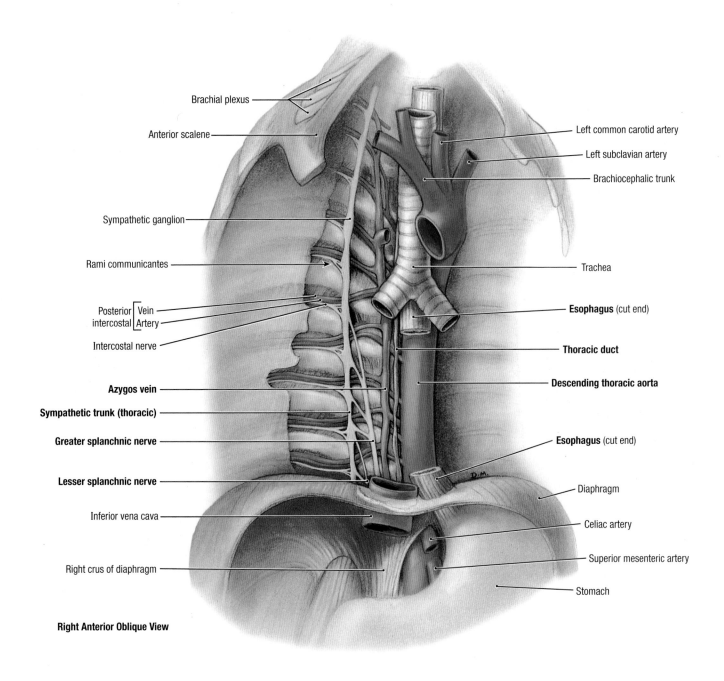

Brachial plexus

Anterior scalene

Sympathetic ganglion

Rami communicantes

Posterior [Vein
intercostal [Artery

Intercostal nerve

Azygos vein

Sympathetic trunk (thoracic)

Greater splanchnic nerve

Lesser splanchnic nerve

Inferior vena cava

Right crus of diaphragm

Left common carotid artery

Left subclavian artery

Brachiocephalic trunk

Trachea

Esophagus (cut end)

Thoracic duct

Descending thoracic aorta

Esophagus (cut end)

Diaphragm

Celiac artery

Superior mesenteric artery

Stomach

Right Anterior Oblique View

3.75   STRUCTURES OF POSTERIOR MEDIASTINUM I

- In this specimen, the parietal pleura is intact on the left side and partially removed on the right side. A portion of the esophagus, between the bifurcation of the trachea and the diaphragm, is also removed.
- The thoracic sympathetic trunk is connected to each intercostal nerve by rami communicantes.

- The greater splanchnic nerve is formed by fibers from the 5th to 10th thoracic sympathetic ganglia, and the lesser splanchnic nerve receives fibers from the 10th and 11th thoracic ganglia. Both nerves contain presynaptic and visceral afferent fibers.
- The azygos vein ascends anterior to the intercostal vessels and to the right of the thoracic duct and aorta and drains into the superior vena cava.

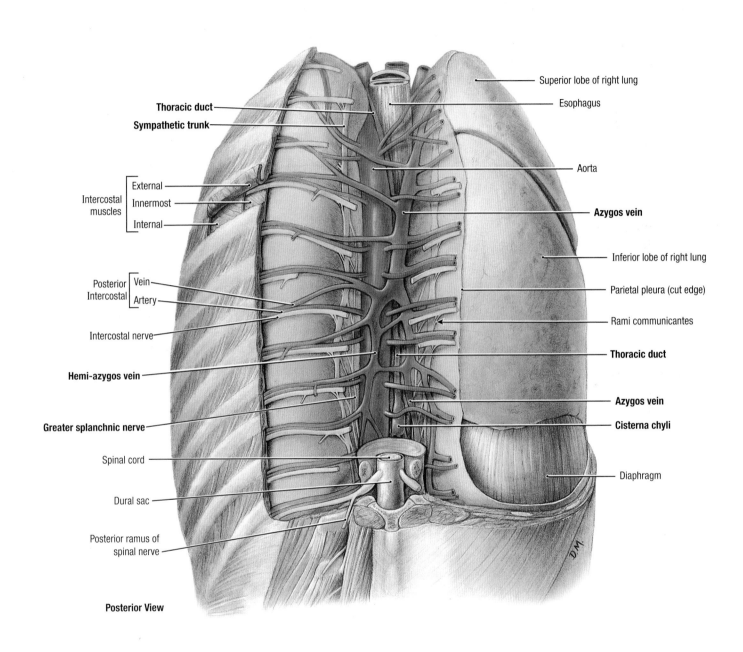

Thoracic duct
Sympathetic trunk

Intercostal muscles
- External
- Innermost
- Internal

Posterior Intercostal
- Vein
- Artery

Intercostal nerve

Hemi-azygos vein

Greater splanchnic nerve

Spinal cord

Dural sac

Posterior ramus of spinal nerve

Superior lobe of right lung
Esophagus

Aorta

Azygos vein

Inferior lobe of right lung

Parietal pleura (cut edge)

Rami communicantes

Thoracic duct

Azygos vein
Cisterna chyli

Diaphragm

Posterior View

STRUCTURES OF POSTERIOR MEDIASTINUM II

3.76

- The thoracic vertebral column and thoracic cage are removed on the right. On the left, the ribs and intercostal musculature are removed posteriorly as far laterally as the angles of the ribs. The parietal pleura is intact on the left side but partially removed on the right to reveal the visceral pleura covering the right lung.

- The azygos vein is on the right side, and the hemi-azygos vein is on the left, crossing the midline (usually at T9 but higher in this specimen) to join the azygos vein. The accessory hemi-azygos vein is absent in this specimen; instead, three most superior posterior intercostal veins drain directly into the azygos vein.

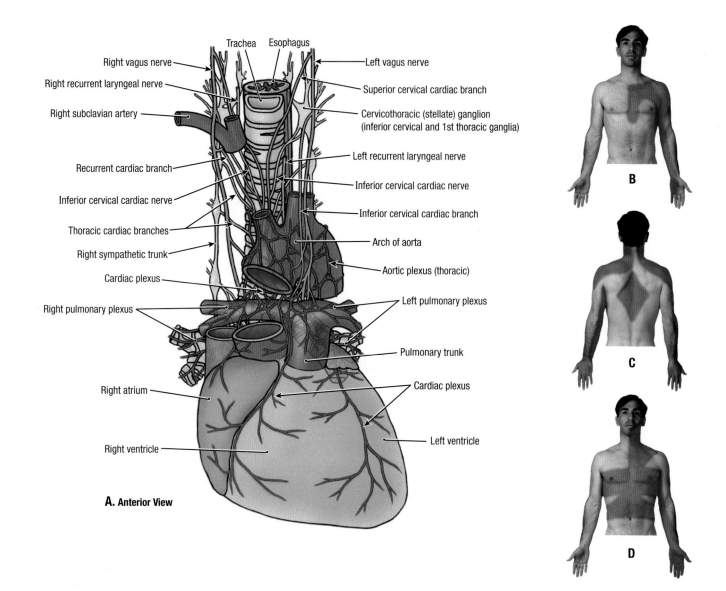

A. Anterior View

3.77   OVERVIEW OF AUTONOMIC AND VISCERAL AFFERENT INNERVATION OF THE THORAX

A. Innervation of heart. **B–D.** Areas of cardiac referred pain *(red)*. **E.** Innervation of posterior and superior mediastina.

The heart is insensitive to touch, cutting, cold, and heat; however, ischemia and the accumulation of metabolic products stimulate pain endings in the myocardium. The afferent pain fibers run centrally in the middle and inferior cervical branches and especially in the thoracic cardiac branches of the sympathetic trunk. The axons of these primary sensory neurons enter spinal cord segments T1 through T4 or T5, especially on the left side.

Cardiac referred pain is a phenomenon whereby noxious stimuli originating in the heart are perceived by a person as pain arising from a superficial part of the body—the skin on the left upper limb, for example. Visceral referred pain is transmitted by visceral afferent fibers accompanying sympathetic fibers and is typically referred to somatic structures or areas such as a limb having afferent fibers with cell bodies in the same spinal ganglion, and central

processes that enter the spinal cord through the same posterior roots (Hardy & Naftel, 2001).

Anginal pain is commonly felt as radiating from the substernal and left pectoral regions to the left shoulder and the medial aspect of the left upper limb **(B)**. This part of the limb is supplied by the medial cutaneous nerve of the arm. Often, the lateral cutaneous branches of the 2nd and 3rd intercostal nerves (the intercostobrachial nerves) join or overlap in their distribution with the medial cutaneous nerve of the arm. Consequently, cardiac pain is referred to the upper limb because the spinal cord segments of these cutaneous nerves (T1–T3) are also common to the visceral afferent terminations for the coronary arteries. Synaptic contacts may also be made with commissural (connector) neurons, which conduct impulses to neurons on the right side of comparable areas of the spinal cord. This occurrence explains why pain of cardiac origin, although usually referred to the left side, may be referred to the right side, both sides, or the back **(C** and **D).**

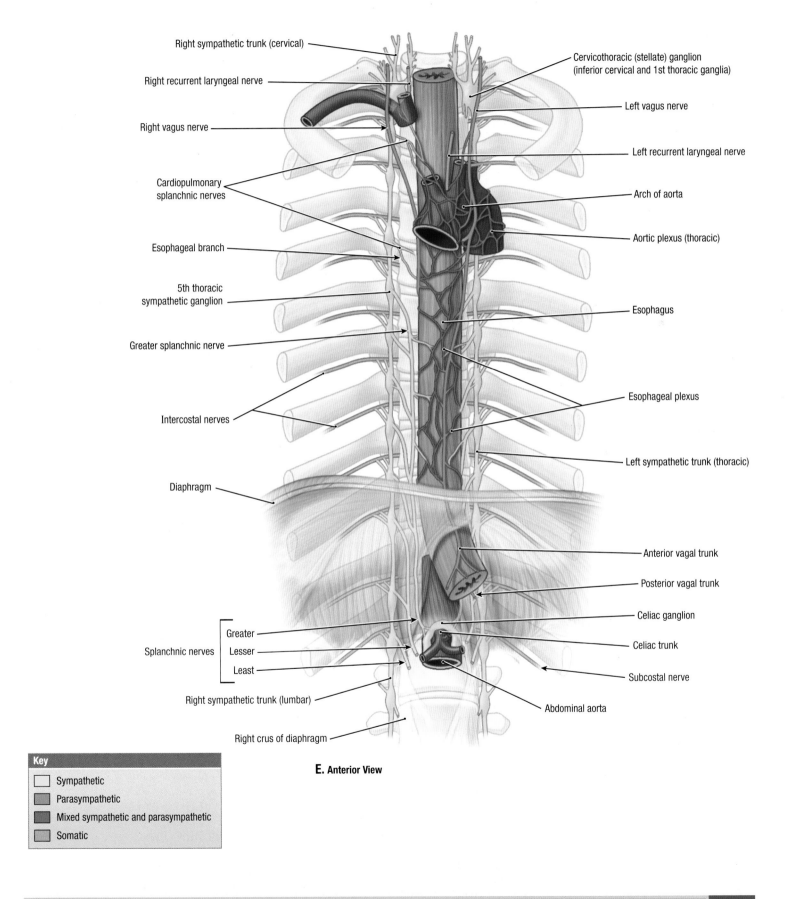

Right sympathetic trunk (cervical)

Right recurrent laryngeal nerve

Right vagus nerve

Cardiopulmonary splanchnic nerves

Esophageal branch

5th thoracic sympathetic ganglion

Greater splanchnic nerve

Intercostal nerves

Diaphragm

Splanchnic nerves
- Greater
- Lesser
- Least

Right sympathetic trunk (lumbar)

Right crus of diaphragm

Cervicothoracic (stellate) ganglion (inferior cervical and 1st thoracic ganglia)

Left vagus nerve

Left recurrent laryngeal nerve

Arch of aorta

Aortic plexus (thoracic)

Esophagus

Esophageal plexus

Left sympathetic trunk (thoracic)

Anterior vagal trunk

Posterior vagal trunk

Celiac ganglion

Celiac trunk

Subcostal nerve

Abdominal aorta

E. Anterior View

Key
- ☐ Sympathetic
- ☐ Parasympathetic
- ☐ Mixed sympathetic and parasympathetic
- ☐ Somatic

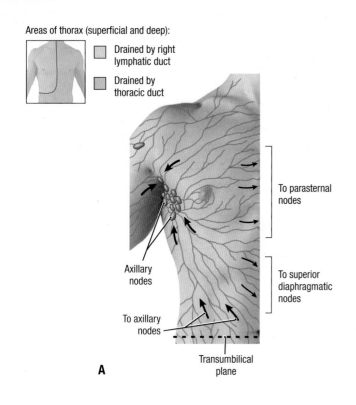

Areas of thorax (superficial and deep):

☐ Drained by right lymphatic duct

☐ Drained by thoracic duct

To parasternal nodes

To superior diaphragmatic nodes

Axillary nodes

To axillary nodes

Transumbilical plane

A

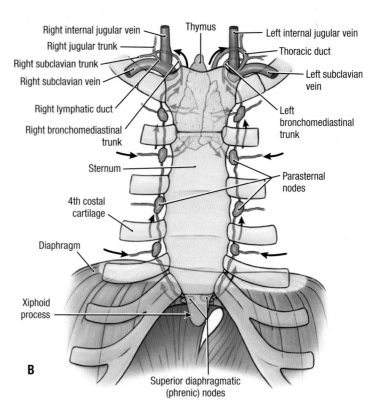

Right internal jugular vein — Thymus — Left internal jugular vein

Right jugular trunk — Thoracic duct

Right subclavian trunk — Left subclavian vein

Right subclavian vein

Right lymphatic duct — Left bronchomediastinal trunk

Right bronchomediastinal trunk

Sternum — Parasternal nodes

4th costal cartilage

Diaphragm

Xiphoid process

Superior diaphragmatic (phrenic) nodes

B

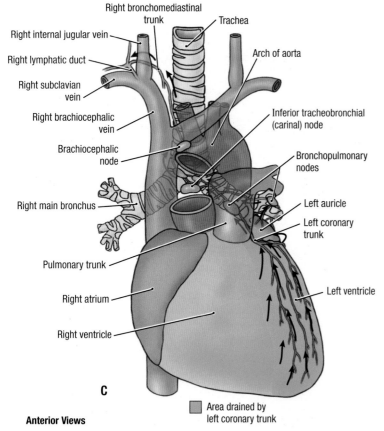

Right bronchomediastinal trunk — Trachea

Right internal jugular vein — Arch of aorta

Right lymphatic duct

Right subclavian vein — Inferior tracheobronchial (carinal) node

Right brachiocephalic vein

Brachiocephalic node — Bronchopulmonary nodes

Right main bronchus — Left auricle

— Left coronary trunk

Pulmonary trunk

Right atrium — Left ventricle

Right ventricle

C

Anterior Views

☐ Area drained by left coronary trunk

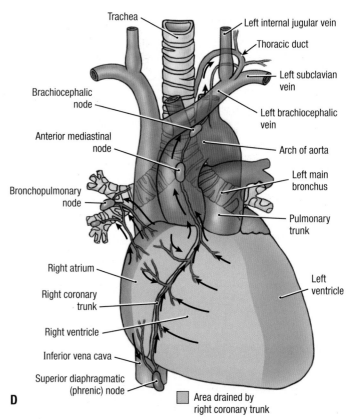

Trachea — Left internal jugular vein

— Thoracic duct

Brachiocephalic node — Left subclavian vein

Anterior mediastinal node — Left brachiocephalic vein

— Arch of aorta

Bronchopulmonary node — Left main bronchus

— Pulmonary trunk

Right atrium

Right coronary trunk — Left ventricle

Right ventricle

Inferior vena cava

Superior diaphragmatic (phrenic) node

D

☐ Area drained by right coronary trunk

3.78   OVERVIEW OF LYMPHATIC DRAINAGE OF THORAX

A. Superficial lymphatic drainage. **B.** Deep lymphatic drainage of parasternal nodes. **C.** Lymphatic drainage of left side of heart. **D.** Lymphatic drainage of right side of heart.

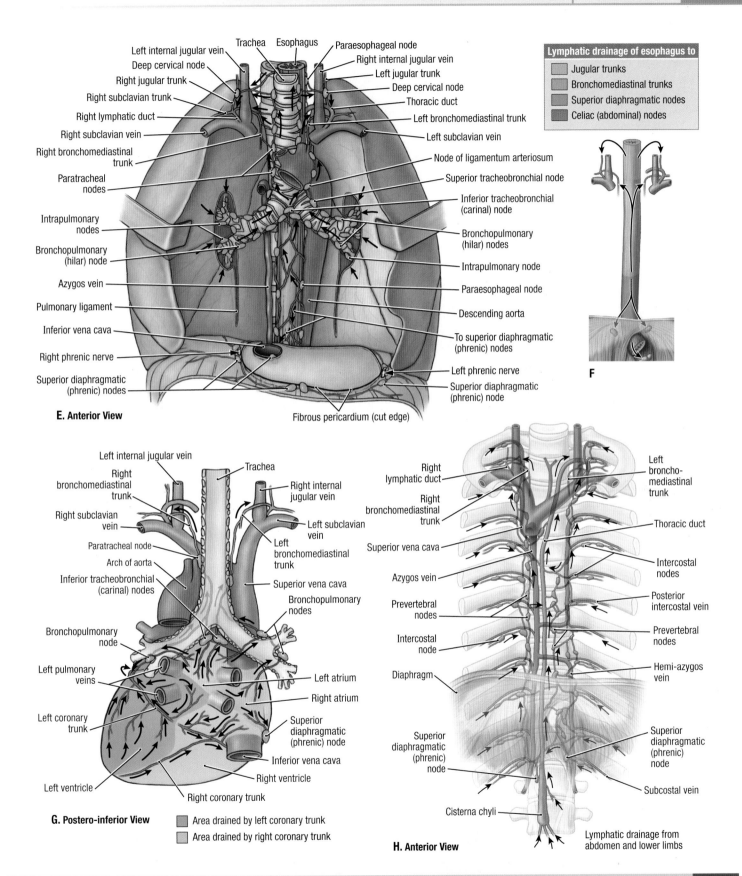

Lymphatic drainage of esophagus to

- Jugular trunks
- Bronchomediastinal trunks
- Superior diaphragmatic nodes
- Celiac (abdominal) nodes

E. Anterior View

Left internal jugular vein
Deep cervical node
Right jugular trunk
Right subclavian trunk
Right lymphatic duct
Right subclavian vein
Right bronchomediastinal trunk
Paratracheal nodes
Intrapulmonary nodes
Bronchopulmonary (hilar) node
Azygos vein
Pulmonary ligament
Inferior vena cava
Right phrenic nerve
Superior diaphragmatic (phrenic) nodes

Trachea Esophagus Paraesophageal node
Right internal jugular vein
Left jugular trunk
Deep cervical node
Thoracic duct
Left bronchomediastinal trunk
Left subclavian vein
Node of ligamentum arteriosum
Superior tracheobronchial node
Inferior tracheobronchial (carinal) node
Bronchopulmonary (hilar) nodes
Intrapulmonary node
Paraesophageal node
Descending aorta
To superior diaphragmatic (phrenic) nodes
Left phrenic nerve
Superior diaphragmatic (phrenic) node
Fibrous pericardium (cut edge)

F

G. Postero-inferior View

Left internal jugular vein
Right bronchomediastinal trunk
Right subclavian vein
Paratracheal node
Arch of aorta
Inferior tracheobronchial (carinal) nodes
Bronchopulmonary node
Left pulmonary veins
Left coronary trunk
Left ventricle
Right coronary trunk

Trachea
Right internal jugular vein
Left subclavian vein
Left bronchomediastinal trunk
Superior vena cava
Bronchopulmonary nodes
Left atrium
Right atrium
Superior diaphragmatic (phrenic) node
Inferior vena cava
Right ventricle

Area drained by left coronary trunk
Area drained by right coronary trunk

H. Anterior View

Right lymphatic duct
Right bronchomediastinal trunk
Superior vena cava
Azygos vein
Prevertebral nodes
Intercostal node
Diaphragm
Superior diaphragmatic (phrenic) node
Cisterna chyli

Left broncho-mediastinal trunk
Thoracic duct
Intercostal nodes
Posterior intercostal vein
Prevertebral nodes
Hemi-azygos vein
Superior diaphragmatic (phrenic) node
Subcostal vein
Lymphatic drainage from abdomen and lower limbs

OVERVIEW OF LYMPHATIC DRAINAGE OF THORAX (*continued*)   **3.78**

E. Lymphatic drainage of lungs, esophagus, and superior surface of diaphragm. **F.** Lymphatic drainage of esophagus. **G.** Lymphatic drainage of posterior and inferior surfaces of heart. **H.** Lymphatic drainage of posterior mediastinum.

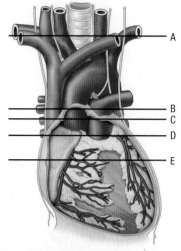

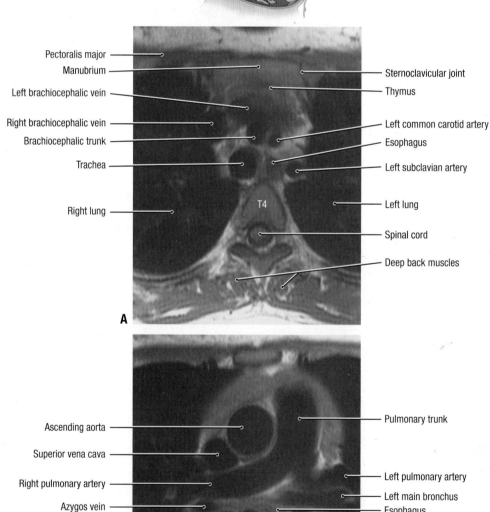

Pectoralis major

Manubrium

Left brachiocephalic vein

Right brachiocephalic vein

Brachiocephalic trunk

Trachea

Right lung

Sternoclavicular joint

Thymus

Left common carotid artery

Esophagus

Left subclavian artery

Left lung

Spinal cord

Deep back muscles

T4

A

Ascending aorta

Superior vena cava

Right pulmonary artery

Azygos vein

Right lung

Pulmonary trunk

Left pulmonary artery

Left main bronchus

Esophagus

Descending aorta

Spinal cord

Deep back muscles

T7

B

3.79 TRANSVERSE (AXIAL) MRIs OF THORAX (A–E)

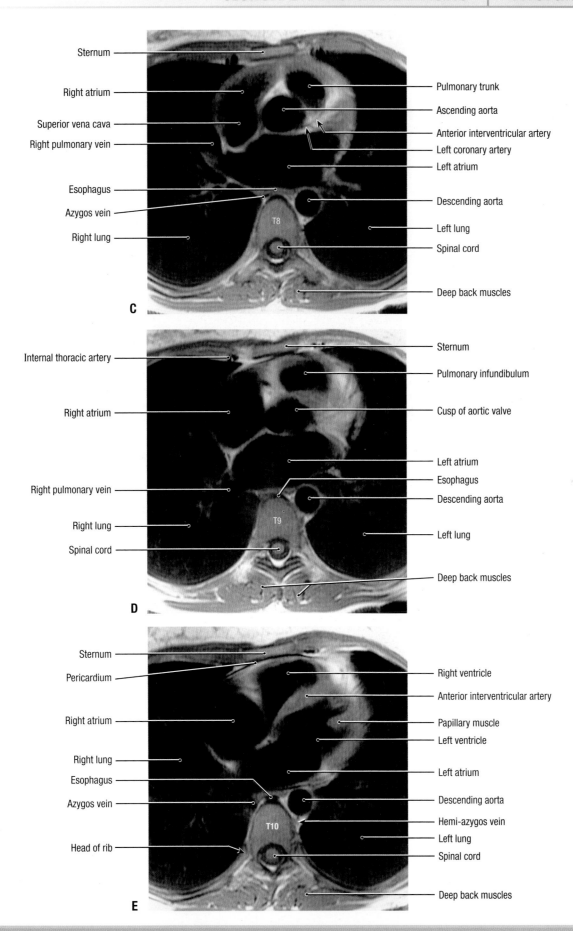

Sternum
Right atrium
Superior vena cava
Right pulmonary vein
Esophagus
Azygos vein
Right lung

Pulmonary trunk
Ascending aorta
Anterior interventricular artery
Left coronary artery
Left atrium
Descending aorta
Left lung
Spinal cord
Deep back muscles

T8

C

Internal thoracic artery
Right atrium
Right pulmonary vein
Right lung
Spinal cord

Sternum
Pulmonary infundibulum
Cusp of aortic valve
Left atrium
Esophagus
Descending aorta
Left lung
Deep back muscles

T9

D

Sternum
Pericardium
Right atrium
Right lung
Esophagus
Azygos vein
Head of rib

Right ventricle
Anterior interventricular artery
Papillary muscle
Left ventricle
Left atrium
Descending aorta
Hemi-azygos vein
Left lung
Spinal cord
Deep back muscles

T10

E

TRANSVERSE (AXIAL) MRIs OF THORAX (*continued*)

3.79

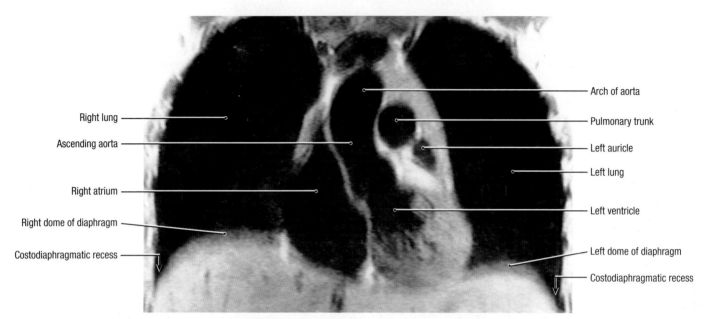

Arch of aorta

Right lung

Ascending aorta

Pulmonary trunk

Left auricle

Left lung

Right atrium

Left ventricle

Right dome of diaphragm

Costodiaphragmatic recess

Left dome of diaphragm

Costodiaphragmatic recess

A. Coronal MRI through Ascending and Arch of Aorta

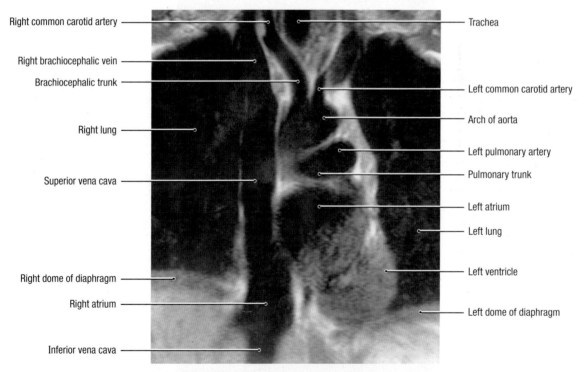

Right common carotid artery

Trachea

Right brachiocephalic vein

Brachiocephalic trunk

Left common carotid artery

Arch of aorta

Right lung

Left pulmonary artery

Superior vena cava

Pulmonary trunk

Left atrium

Left lung

Right dome of diaphragm

Left ventricle

Right atrium

Left dome of diaphragm

Inferior vena cava

B. Coronal MRI through Superior and Inferior Vena Cava

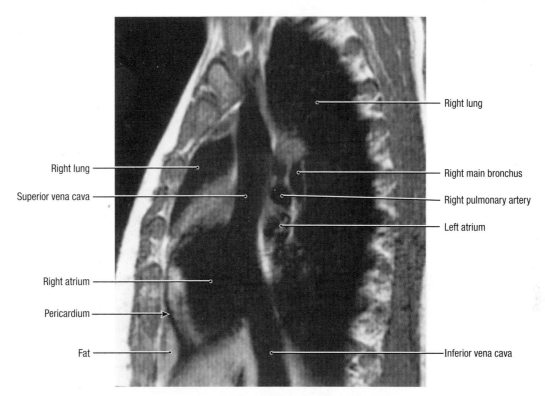

Right lung

Right lung

Superior vena cava

Right main bronchus

Right pulmonary artery

Left atrium

Right atrium

Pericardium

Fat

Inferior vena cava

A. Sagittal MRI through Superior and Inferior Vena Cava

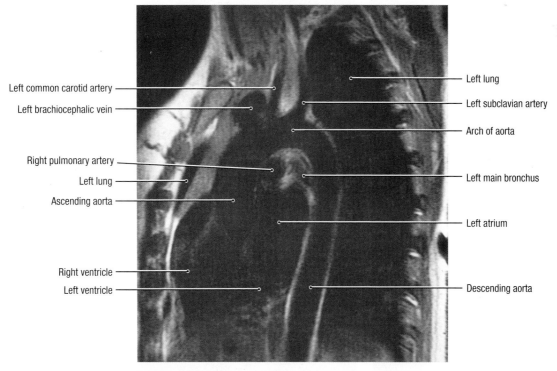

Left common carotid artery

Left brachiocephalic vein

Right pulmonary artery

Left lung

Ascending aorta

Right ventricle

Left ventricle

Left lung

Left subclavian artery

Arch of aorta

Left main bronchus

Left atrium

Descending aorta

B. Sagittal MRI through Arch of Aorta

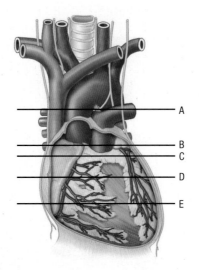

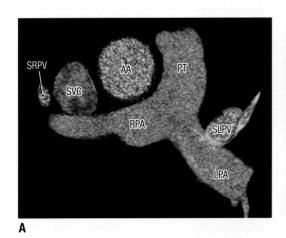

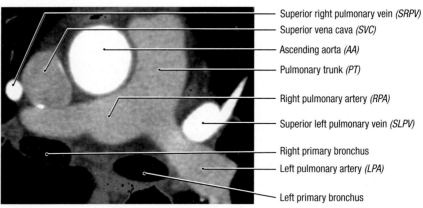

- Superior right pulmonary vein (SRPV)
- Superior vena cava (SVC)
- Ascending aorta (AA)
- Pulmonary trunk (PT)
- Right pulmonary artery (RPA)
- Superior left pulmonary vein (SLPV)
- Right primary bronchus
- Left pulmonary artery (LPA)
- Left primary bronchus

A

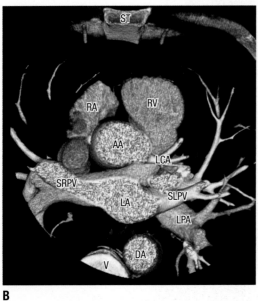

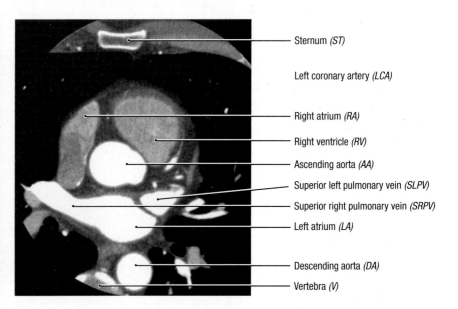

- Sternum (ST)
- Left coronary artery (LCA)
- Right atrium (RA)
- Right ventricle (RV)
- Ascending aorta (AA)
- Superior left pulmonary vein (SLPV)
- Superior right pulmonary vein (SRPV)
- Left atrium (LA)
- Descending aorta (DA)
- Vertebra (V)

B

3.82   TRANSVERSE OR HORIZONTAL (AXIAL) 3D VOLUME RECONSTRUCTIONS (*LEFT SIDE OF PAGE*) AND CT ANGIOGRAMS OF THORAX (*A–E*)

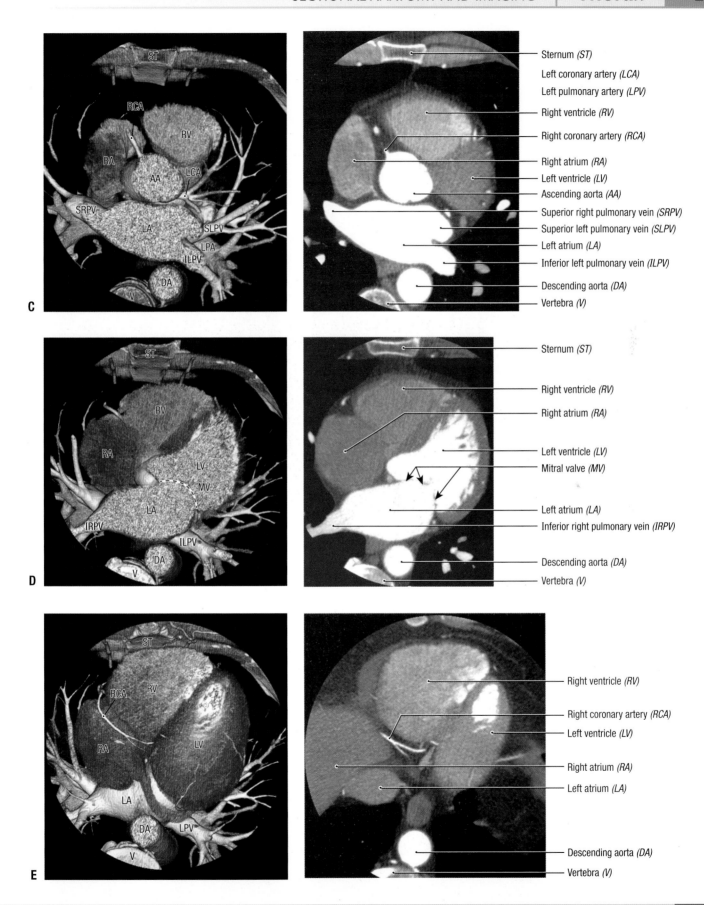

Sternum (ST)
Left coronary artery (LCA)
Left pulmonary artery (LPV)
Right ventricle (RV)
Right coronary artery (RCA)
Right atrium (RA)
Left ventricle (LV)
Ascending aorta (AA)
Superior right pulmonary vein (SRPV)
Superior left pulmonary vein (SLPV)
Left atrium (LA)
Inferior left pulmonary vein (ILPV)
Descending aorta (DA)
Vertebra (V)

Sternum (ST)
Right ventricle (RV)
Right atrium (RA)
Left ventricle (LV)
Mitral valve (MV)
Left atrium (LA)
Inferior right pulmonary vein (IRPV)
Descending aorta (DA)
Vertebra (V)

Right ventricle (RV)
Right coronary artery (RCA)
Left ventricle (LV)
Right atrium (RA)
Left atrium (LA)
Descending aorta (DA)
Vertebra (V)

TRANSVERSE OR HORIZONTAL (AXIAL) 3D VOLUME RECONSTRUCTIONS (*LEFT SIDE OF PAGE*) AND CT ANGIOGRAMS OF THORAX (A–E) (*continued*)

3.82

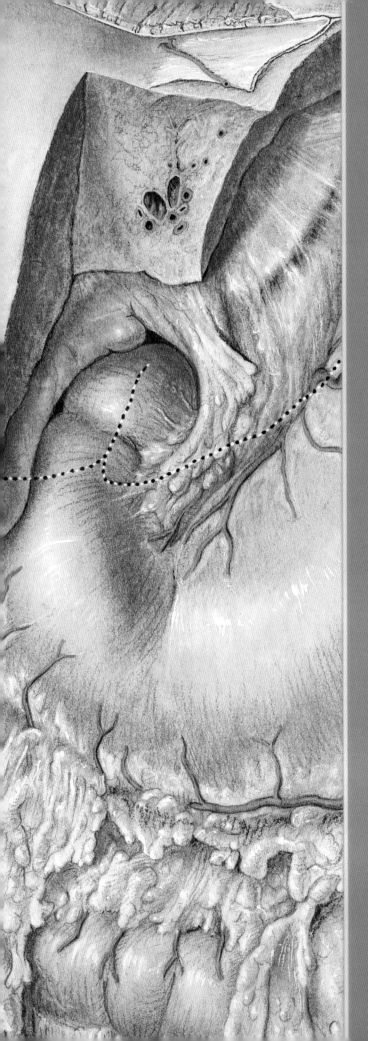

CHAPTER 4

Abdomen

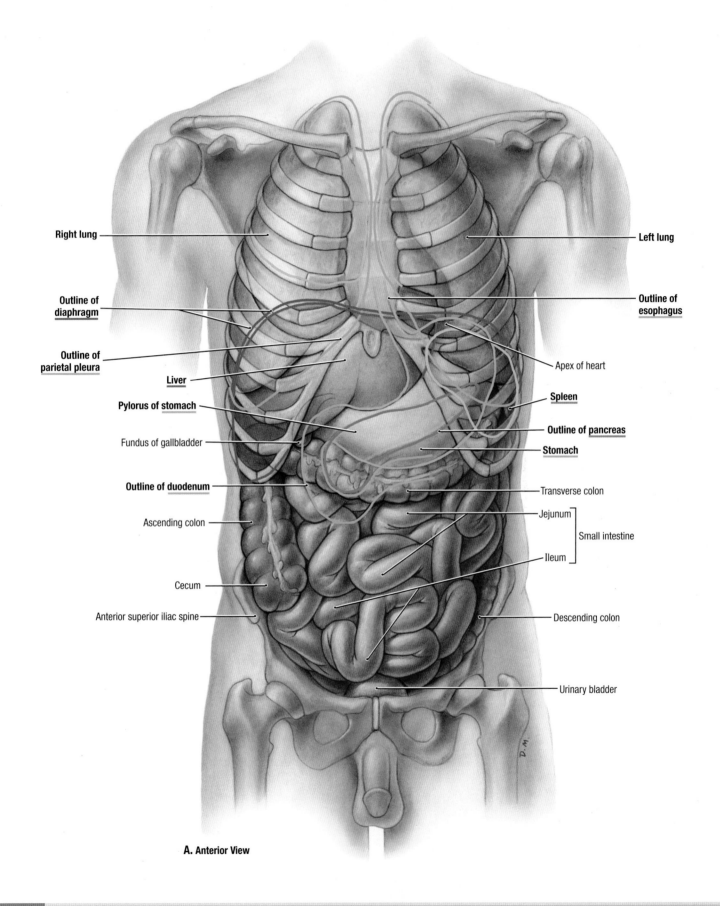

Right lung

Left lung

Outline of diaphragm

Outline of esophagus

Outline of parietal pleura

Apex of heart

Liver

Spleen

Pylorus of stomach

Outline of pancreas

Fundus of gallbladder

Stomach

Outline of duodenum

Transverse colon

Ascending colon

Jejunum

Small intestine

Ileum

Cecum

Anterior superior iliac spine

Descending colon

Urinary bladder

A. Anterior View

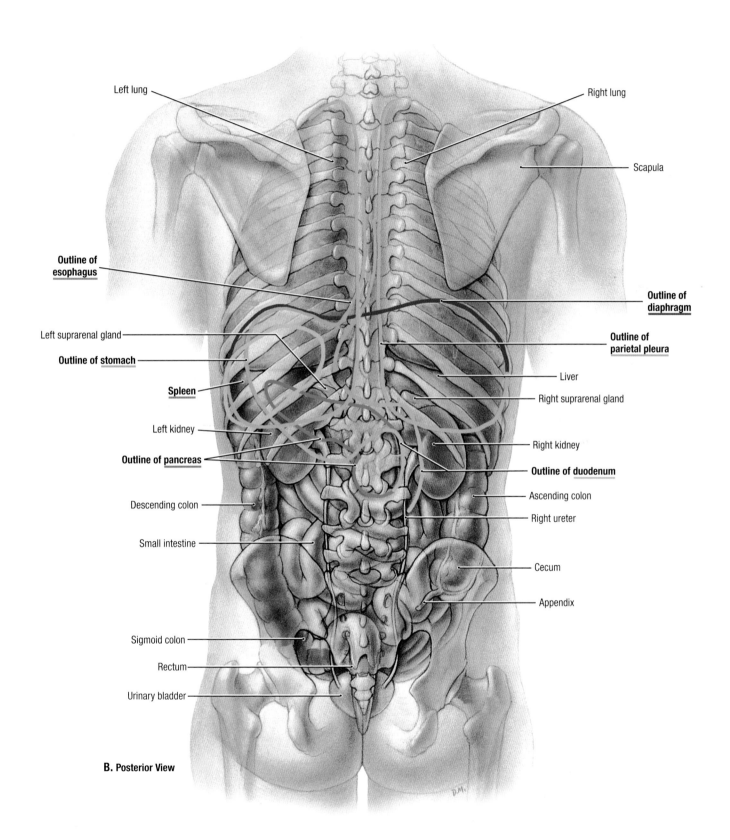

Left lung

Right lung

Scapula

Outline of esophagus

Outline of diaphragm

Left suprarenal gland

Outline of parietal pleura

Outline of stomach

Liver

Spleen

Right suprarenal gland

Left kidney

Right kidney

Outline of pancreas

Outline of duodenum

Descending colon

Ascending colon

Right ureter

Small intestine

Cecum

Appendix

Sigmoid colon

Rectum

Urinary bladder

B. Posterior View

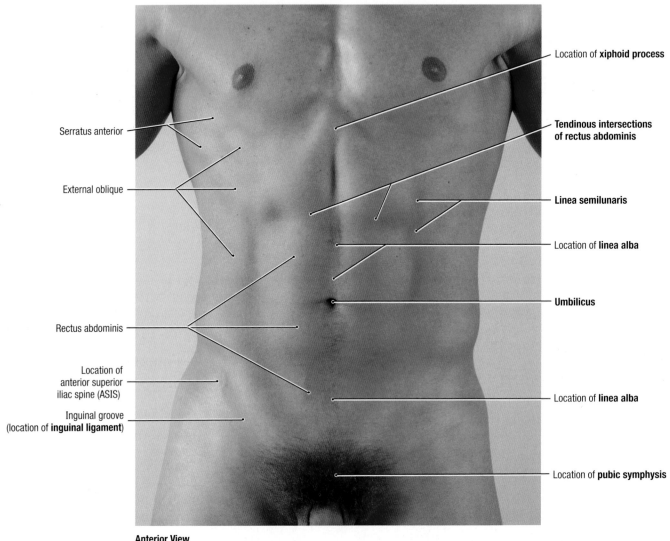

Serratus anterior

External oblique

Rectus abdominis

Location of anterior superior iliac spine (ASIS)

Inguinal groove (location of **inguinal ligament**)

Location of **xiphoid process**

Tendinous intersections of rectus abdominis

Linea semilunaris

Location of **linea alba**

Umbilicus

Location of **linea alba**

Location of **pubic symphysis**

Anterior View

4.2   SURFACE ANATOMY

Surface features.

- The umbilicus is where the umbilical cord entered the fetus and indicates the anterior level of the T10 dermatome. Typically, the umbilicus lies at the level of the intervertebral disc between the L3 and L4 vertebrae.
- The linea alba is a fibrous band formed by the fusion of the right and left abdominal aponeuroses between the xiphoid process and the pubic symphysis demarcated superficially by a midline vertical skin groove.

- A curved skin groove, the linea semilunaris, demarcates the lateral border of the right and left rectus abdominis muscles and rectus sheath.
- In lean individuals with good muscle development, three to four transverse skin grooves overlie the tendinous intersections of the rectus abdominis muscle.
- The site of the inguinal ligament is indicated by a skin crease, the inguinal groove, just inferior and parallel to the ligament, marking the division between the anterolateral abdominal wall and the thigh.

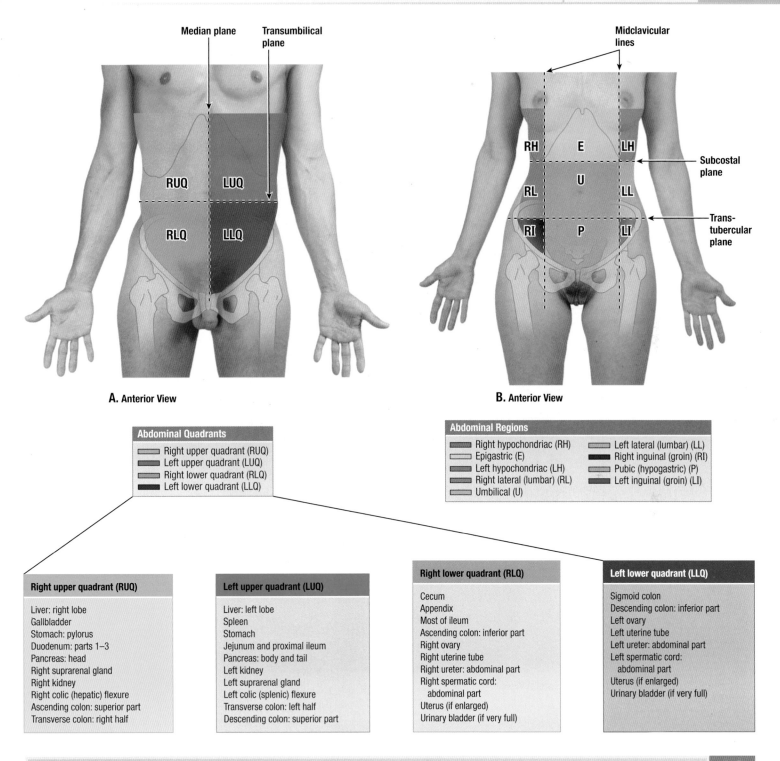

Median plane   Transumbilical plane

Midclavicular lines

RH   **E**   **LH**

RUQ   **LUQ**

RL   **U**   **LL**

RLQ   **LLQ**

RI   **P**   **LI**

Subcostal plane

Transtubercular plane

A. Anterior View

B. Anterior View

Abdominal Quadrants

- Right upper quadrant (RUQ)
- Left upper quadrant (LUQ)
- Right lower quadrant (RLQ)
- Left lower quadrant (LLQ)

Abdominal Regions

- Right hypochondriac (RH)
- Epigastric (E)
- Left hypochondriac (LH)
- Right lateral (lumbar) (RL)
- Umbilical (U)
- Left lateral (lumbar) (LL)
- Right inguinal (groin) (RI)
- Pubic (hypogastric) (P)
- Left inguinal (groin) (LI)

Right upper quadrant (RUQ)

Liver: right lobe
Gallbladder
Stomach: pylorus
Duodenum: parts 1–3
Pancreas: head
Right suprarenal gland
Right kidney
Right colic (hepatic) flexure
Ascending colon: superior part
Transverse colon: right half

Left upper quadrant (LUQ)

Liver: left lobe
Spleen
Stomach
Jejunum and proximal ileum
Pancreas: body and tail
Left kidney
Left suprarenal gland
Left colic (splenic) flexure
Transverse colon: left half
Descending colon: superior part

Right lower quadrant (RLQ)

Cecum
Appendix
Most of ileum
Ascending colon: inferior part
Right ovary
Right uterine tube
Right ureter: abdominal part
Right spermatic cord:
   abdominal part
Uterus (if enlarged)
Urinary bladder (if very full)

Left lower quadrant (LLQ)

Sigmoid colon
Descending colon: inferior part
Left ovary
Left uterine tube
Left ureter: abdominal part
Left spermatic cord:
   abdominal part
Uterus (if enlarged)
Urinary bladder (if very full)

ABDOMINAL REGIONS AND QUADRANTS

4.3

A. Quadrants. **B.** Regions. It is important to know what organs are located in each abdominal region or quadrant so that one knows where to auscultate, percuss, and palpate them and to record the locations of findings during a physical exam.

The six common causes of **abdominal protrusion** begin with the letter F: food, fluid, fat, feces, flatus, and fetus. Eversion of the umbilicus may be a sign of increased intra-abdominal pressure, usually resulting from ascites (abdominal accumulation of serous fluid in the peritoneal cavity), or a large mass (e.g., a tumor, fetus, or enlarged organ such as the liver [hepatomegaly]).

Warm hands are important when palpating the abdominal wall because cold hands make the anterolateral abdominal muscles tense, producing involuntary muscle spasms known as guarding. Intense guarding, boardlike reflexive muscular rigidity that cannot be willfully suppressed, occurs during palpation when an organ (such as the appendix) is inflamed and in itself constitutes a clinically significant sign of **acute abdomen**. The involuntary muscular spasms attempt to protect the viscera from pressure, which is painful when an abdominal infection is present. The common nerve supply of the skin and muscles of the wall explains why these spasms occur.

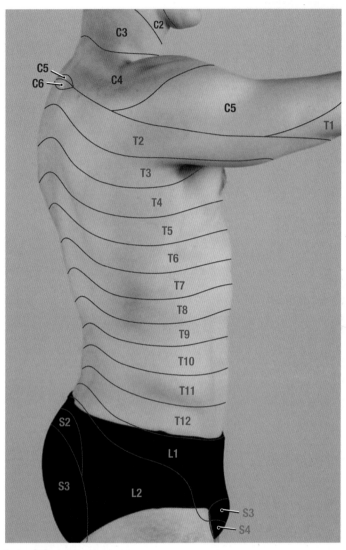

Lateral View

4.4 DERMATOMES

The thoraco-abdominal (T7–T11) nerves run between the external and internal oblique muscles to supply sensory innervation to the overlying skin. The T10 nerve supplies the region of the umbilicus. The subcostal nerve (T12) runs along the inferior border of the 12th rib to supply the skin over the anterior superior iliac spine and hip. The iliohypogastric nerve (L1) innervates the skin over the iliac crest and lower pubic region and the ilio-inguinal nerve (L1) innervates the skin of the medial aspect of the thigh, the scrotum or labium majus, and mons pubis.

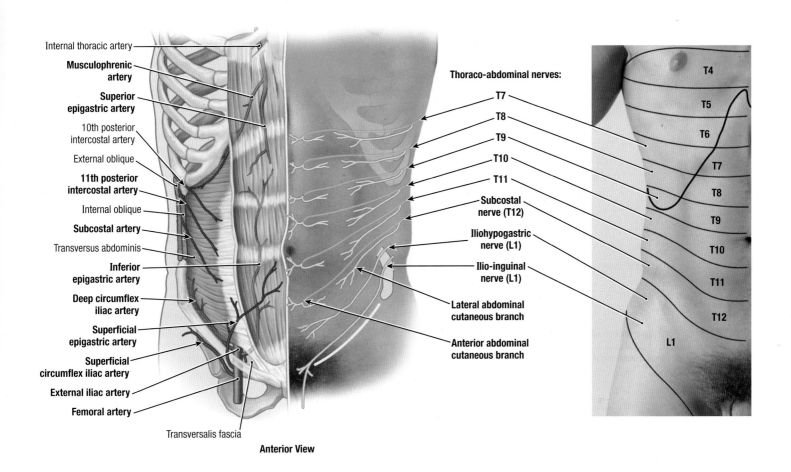

Internal thoracic artery
Musculophrenic artery
Superior epigastric artery
10th posterior intercostal artery
External oblique
11th posterior intercostal artery
Internal oblique
Subcostal artery
Transversus abdominis
Inferior epigastric artery
Deep circumflex iliac artery
Superficial epigastric artery
Superficial circumflex iliac artery
External iliac artery
Femoral artery
Transversalis fascia

Thoraco-abdominal nerves:
T7
T8
T9
T10
T11
Subcostal nerve (T12)
Iliohypogastric nerve (L1)
Ilio-inguinal nerve (L1)
Lateral abdominal cutaneous branch
Anterior abdominal cutaneous branch

T4
T5
T6
T7
T8
T9
T10
T11
T12
L1

Anterior View

ARTERIES AND NERVES OF ANTEROLATERAL ABDOMINAL WALL

4.5

The skin and muscles of the anterolateral abdominal wall are supplied mainly by the:

- Thoraco-abdominal nerves: distal, abdominal parts of the anterior rami of the inferior six thoracic spinal nerves (T7–T11), which have muscular branches and anterior and lateral abdominal cutaneous branches. The anterior abdominal cutaneous branches pierce the rectus sheath a short distance from the median plane, after the rectus abdominis muscle has been supplied. Spinal nerves T7–T9 supply the skin superior to the umbilicus; T10 innervates the skin around the umbilicus.
- Spinal nerve T11, plus the cutaneous branches of the subcostal (T12), iliohypogastric, and ilio-inguinal (L1) nerves: supply the skin inferior to the umbilicus.
- Subcostal nerve: large anterior ramus of spinal nerve T12.

The blood vessels of the anterolateral abdominal wall are the:

- Superior epigastric vessels and branches of the musculophrenic vessels, the terminal branches of the internal thoracic vessels.

- Inferior epigastric and deep circumflex iliac vessels from the external iliac vessels.
- Superficial circumflex iliac and superficial epigastric vessels from the femoral artery and great saphenous vein.
- Posterior intercostal vessels in the 11th intercostal space and anterior branches of subcostal vessels.

Incisional nerve injury. The inferior thoracic spinal nerves (T7–T12) and the iliohypogastric and ilio-inguinal nerves (L1) approach the abdominal musculature separately to provide the multisegmental innervation of the abdominal muscles. Thus, they are distributed across the anterolateral abdominal wall, where they run oblique but mostly horizontal courses. They are susceptible to injury in surgical incisions or from trauma at any level of the abdominal wall. Injury to them may result in weakening of the muscles. In the inguinal region, such a weakness may predispose an individual to development of an inguinal hernia.

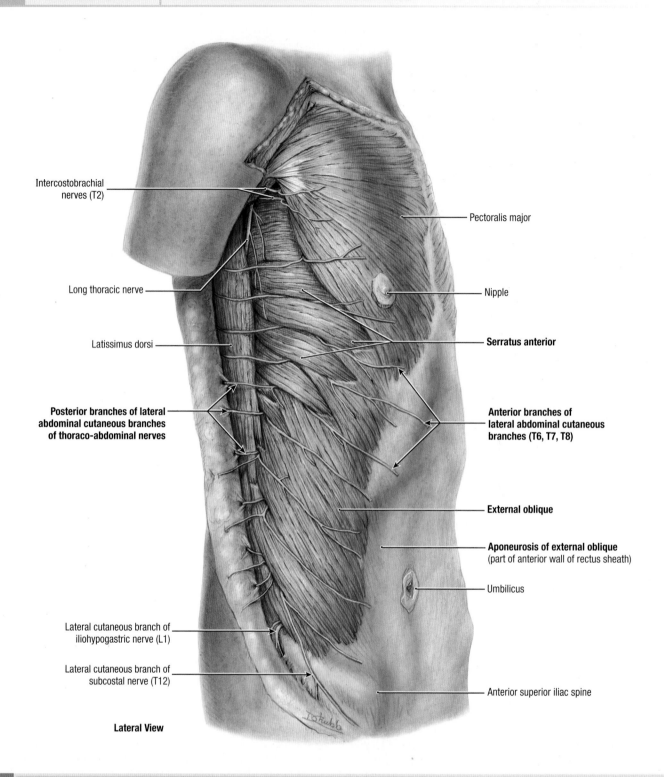

Intercostobrachial nerves (T2)

Long thoracic nerve

Latissimus dorsi

Posterior branches of lateral abdominal cutaneous branches of thoraco-abdominal nerves

Lateral cutaneous branch of iliohypogastric nerve (L1)

Lateral cutaneous branch of subcostal nerve (T12)

Pectoralis major

Nipple

Serratus anterior

Anterior branches of lateral abdominal cutaneous branches (T6, T7, T8)

External oblique

Aponeurosis of external oblique (part of anterior wall of rectus sheath)

Umbilicus

Anterior superior iliac spine

Lateral View

4.6 ANTEROLATERAL ABDOMINAL WALL, SUPERFICIAL DISSECTION

The muscular portion of the external oblique muscle interdigitates with slips of the serratus anterior muscle, and the aponeurotic portion contributes to the anterior wall of the rectus sheath. The anterior and posterior branches of the lateral abdominal cutaneous branches of the thoraco-abdominal nerves course superficially in the subcutaneous tissue.

- **Umbilical hernias** are usually small protrusions of extraperitoneal fat and/or peritoneum and omentum and sometimes bowel. They result from increased intra-abdominal pressure in the presence of weakness or incomplete closure of the anterior abdominal wall after ligation of the umbilical cord at birth, or may be acquired later, most commonly in women and obese people.

- The lines along which the fibers of the abdominal aponeurosis interlace (see Fig. 4.10A, B, and D) are also potential sites of herniation. These gaps may be congenital, the result of the stresses of obesity and aging, or the consequence of surgical or traumatic wounds.

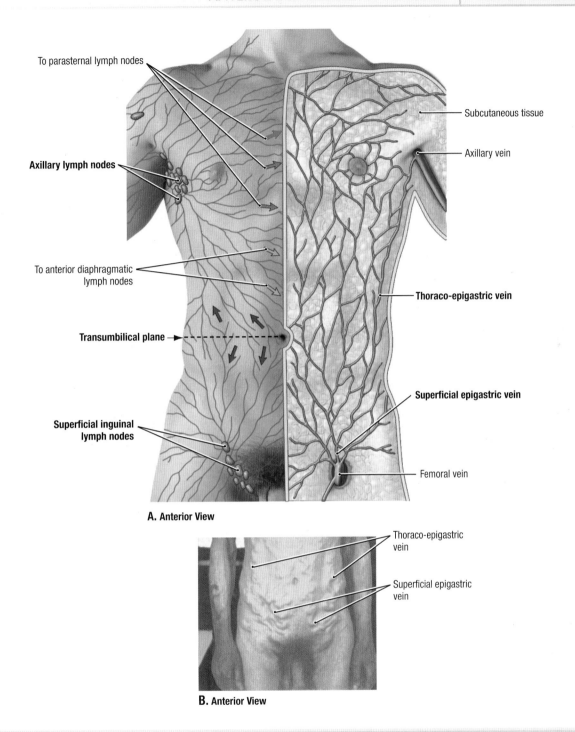

To parasternal lymph nodes

Axillary lymph nodes

To anterior diaphragmatic lymph nodes

Transumbilical plane →

Superficial inguinal lymph nodes

Subcutaneous tissue

Axillary vein

Thoraco-epigastric vein

Superficial epigastric vein

Femoral vein

A. Anterior View

Thoraco-epigastric vein

Superficial epigastric vein

B. Anterior View

LYMPHATIC DRAINAGE AND SUBCUTANEOUS (SUPERFICIAL) VENOUS DRAINAGE OF ANTEROLATERAL ABDOMINAL WALL

4.7

A. Overview.

- The skin and subcutaneous tissue of the abdominal wall are served by an intricate subcutaneous venous plexus, draining superiorly to the internal thoracic vein medially and the lateral thoracic vein laterally, and inferiorly to the superficial and inferior epigastric veins, tributaries of the femoral and external iliac veins, respectively.

- Superficial lymphatic vessels accompany the subcutaneous veins; those superior to the transumbilical plane drain mainly to the axillary lymph nodes; however, a few drain to the parasternal lymph nodes. Superficial lymphatic vessels inferior to the transumbilical plane drain to the superficial inguinal lymph nodes.

B. Enlargement of subcutaneous veins.

- **Liposuction** is a surgical method for removing unwanted subcutaneous fat using a percutaneously placed suction tube and high vacuum pressure. The tubes are inserted subdermally through small skin incisions.

- When flow in the superior or inferior vena cava is obstructed, anastomoses between the tributaries of these systemic veins, such as the thoraco-epigastric vein, may provide **collateral pathways** by which the obstruction may be bypassed, allowing blood to return to the heart. The veins become enlarged and tortuous **(B)**.

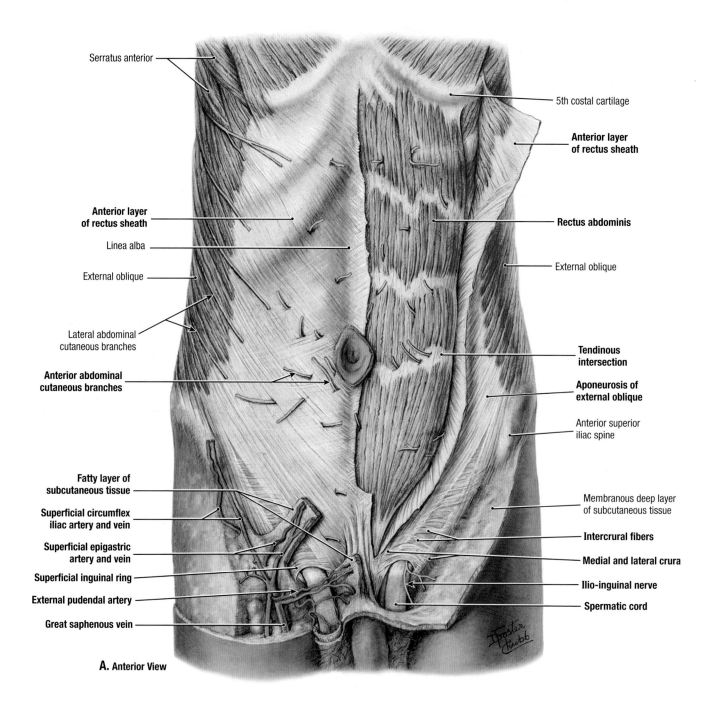

Serratus anterior

5th costal cartilage

**Anterior layer
of rectus sheath**

**Anterior layer
of rectus sheath**

Rectus abdominis

Linea alba

External oblique

External oblique

Lateral abdominal
cutaneous branches

**Tendinous
intersection**

**Anterior abdominal
cutaneous branches**

**Aponeurosis of
external oblique**

Anterior superior
iliac spine

**Fatty layer of
subcutaneous tissue**

Membranous deep layer
of subcutaneous tissue

**Superficial circumflex
iliac artery and vein**

Intercrural fibers

**Superficial epigastric
artery and vein**

Medial and lateral crura

Superficial inguinal ring

Ilio-inguinal nerve

External pudendal artery

Spermatic cord

Great saphenous vein

A. Anterior View

4.8   ANTERIOR ABDOMINAL WALL

A. Superficial dissection demonstrating the relationship of the cutaneous nerves and superficial vessels to the musculoaponeurotic structures. The anterior wall of the left rectus sheath is reflected, revealing the rectus abdominis muscle, segmented by tendinous intersections.

- After the T7 to T12 spinal nerves supply the muscles, their anterior abdominal cutaneous branches emerge from the rectus abdominis muscle and pierce the anterior wall of its sheath.
- The three superficial inguinal branches of the femoral artery (superficial circumflex iliac artery, superficial epigastric artery,

and external pudendal artery) and the great saphenous vein lie in the fatty layer of subcutaneous tissue.

- The fibers of the external oblique aponeurosis separate into medial and lateral crura, which, with the intercrural fibers that unite them, form the superficial inguinal ring. The spermatic cord of the male (shown here), or round ligament of the female, exits the inguinal canal through the superficial inguinal ring along with the ilio-inguinal nerve.

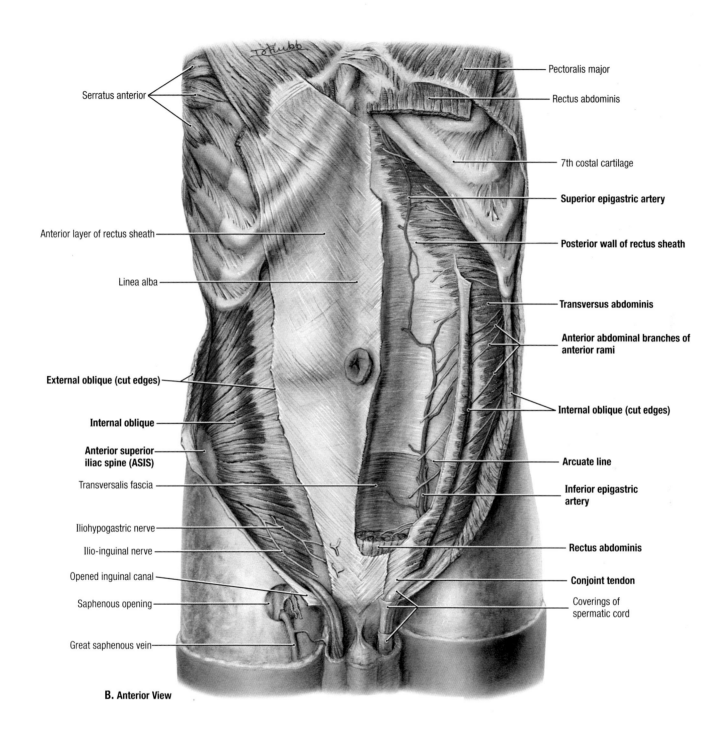

Serratus anterior

Anterior layer of rectus sheath

Linea alba

External oblique (cut edges)

Internal oblique

Anterior superior iliac spine (ASIS)

Transversalis fascia

Iliohypogastric nerve

Ilio-inguinal nerve

Opened inguinal canal

Saphenous opening

Great saphenous vein

Pectoralis major

Rectus abdominis

7th costal cartilage

Superior epigastric artery

Posterior wall of rectus sheath

Transversus abdominis

Anterior abdominal branches of anterior rami

Internal oblique (cut edges)

Arcuate line

Inferior epigastric artery

Rectus abdominis

Conjoint tendon

Coverings of spermatic cord

B. Anterior View

ANTERIOR ABDOMINAL WALL (*continued*) 4.8

B. Deep dissection. On the right side of the specimen, most of the external oblique muscle is excised. On the left, the internal oblique muscle is divided and the rectus abdominis muscle is excised, revealing the posterior wall of the rectus sheath.

- The fibers of the internal oblique muscle run horizontally at the level of the anterior superior iliac spine (ASIS), obliquely upward superior to the ASIS, and obliquely downward inferior to the ASIS.
- The arcuate line is at the level of the ASIS; inferior to the line, transversalis fascia lies immediately posterior to the rectus abdominis muscle.

- Initially, the anterior abdominal branches of the anterior rami course between the internal oblique and transversus abdominis muscles.
- The anastomosis between the superior and inferior epigastric arteries indirectly unites the subclavian artery of the upper limb to the external iliac arteries of the lower limb. The anastomosis can become functionally patent in response to **slowly developing occlusion of the aorta.**

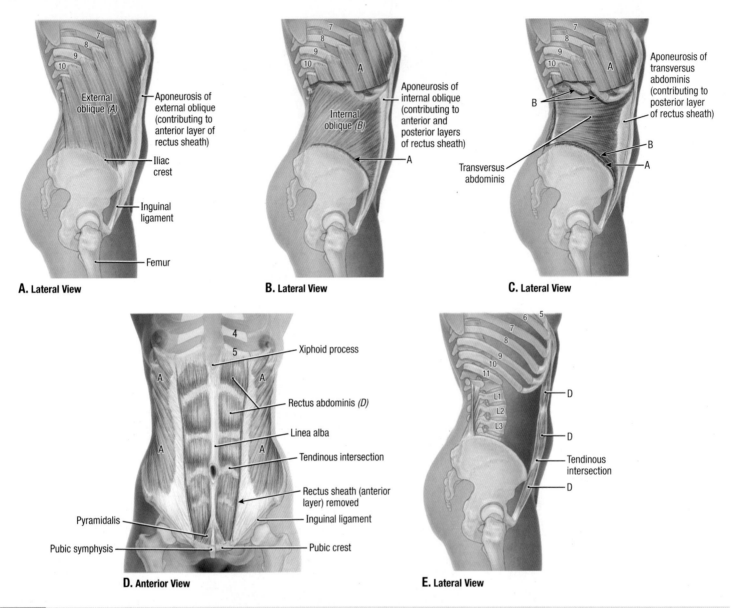

A. Lateral View

B. Lateral View

C. Lateral View

D. Anterior View

E. Lateral View

4.9   MUSCLES OF ANTEROLATERAL ABDOMINAL WALL

A. External oblique. **B.** Internal oblique. **C.** Transversus abdominis. **D.** and **E.** Rectus abdominis and pyramidalis.

TABLE 4.1	PRINCIPAL MUSCLES OF ANTEROLATERAL ABDOMINAL WALL			
Muscles[a]	**Origin**	**Insertion**	**Innervation**	**Action(s)**
External oblique **(A)**	External surfaces of 5th–12th ribs	Linea alba, pubic tubercle, and anterior half of iliac crest	Thoraco-abdominal nerves (anterior rami of T7–T11) and subcostal nerve	Compresses and supports abdominal viscera; flexes and rotates trunk
Internal oblique **(B)**	Thoracolumbar fascia, anterior two thirds of iliac crest, and connective tissue deep to inguinal ligament	Inferior borders of 10th–12th ribs, linea alba, and pubis via conjoint tendon	Thoraco-abdominal nerves (anterior rami of T7–T11), subcostal nerve, and first lumbar nerve	
Transversus abdominis **(C)**	Internal surfaces of 7th–12th costal cartilages, thoracolumbar fascia, iliac crest, and connective tissue deep to inguinal ligament (iliopsoas fascia)	Linea alba with aponeurosis of internal oblique, pubic crest, and pectin pubis via conjoint tendon		Compresses and supports abdominal viscera (with external oblique ipsilaterally, internal oblique contralaterally)
Rectus abdominis **(D)**	Pubic symphysis and pubic crest	Xiphoid process and 5th–7th costal cartilages	Thoraco-abdominal nerves (T7–T11) and subcostal nerve	Flexes trunk and compresses abdominal viscera[b]; stabilizes and controls tilt of pelvis

[a]Approximately 80% of people have a *pyramidalis* muscle, which is located in the rectus sheath anterior to the most inferior part of the rectus abdominis. It extends from the pubic crest of the hip bone to the linea alba. This small muscle tenses the linea alba.

[b]In so doing, these muscles act as antagonists of the diaphragm to produce expiration.

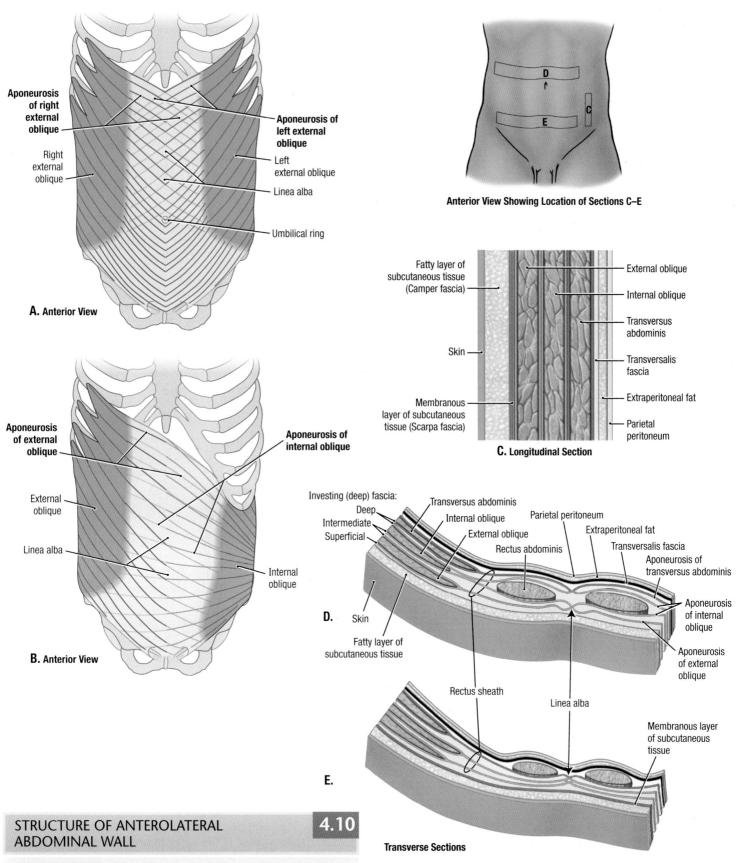

A. Anterior View

Aponeurosis of right external oblique

Aponeurosis of left external oblique

Right external oblique

Left external oblique

Linea alba

Umbilical ring

B. Anterior View

Aponeurosis of external oblique

Aponeurosis of internal oblique

External oblique

Linea alba

Internal oblique

Anterior View Showing Location of Sections C–E

C. Longitudinal Section

Fatty layer of subcutaneous tissue (Camper fascia)

Skin

Membranous layer of subcutaneous tissue (Scarpa fascia)

External oblique

Internal oblique

Transversus abdominis

Transversalis fascia

Extraperitoneal fat

Parietal peritoneum

D.

Investing (deep) fascia:
Deep
Intermediate
Superficial

Transversus abdominis

Internal oblique

External oblique

Rectus abdominis

Parietal peritoneum

Extraperitoneal fat

Transversalis fascia

Aponeurosis of transversus abdominis

Aponeurosis of internal oblique

Aponeurosis of external oblique

Skin

Fatty layer of subcutaneous tissue

Rectus sheath

Linea alba

E.

Membranous layer of subcutaneous tissue

Transverse Sections

STRUCTURE OF ANTEROLATERAL ABDOMINAL WALL

4.10

A. Interdigitation of the aponeuroses of the right and left external oblique muscles. **B.** Interdigitation of the aponeuroses of the contralateral external and internal oblique muscles. **C–E.** Layers of the abdominal wall and the rectus sheath.

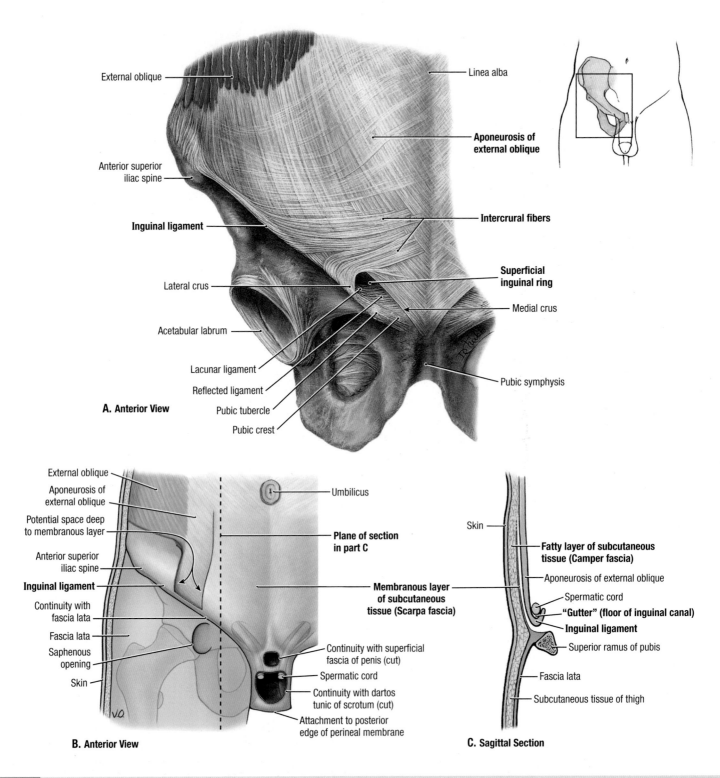

External oblique

Linea alba

Aponeurosis of
external oblique

Anterior superior
iliac spine

Inguinal ligament

Intercrural fibers

Superficial
inguinal ring

Lateral crus

Medial crus

Acetabular labrum

Lacunar ligament

Reflected ligament

Pubic symphysis

Pubic tubercle

A. Anterior View

Pubic crest

External oblique

Aponeurosis of
external oblique

Umbilicus

Potential space deep
to membranous layer

Skin

Fatty layer of subcutaneous
tissue (Camper fascia)

**Plane of section
in part C**

Anterior superior
iliac spine

Aponeurosis of external oblique

Inguinal ligament

Spermatic cord

Continuity with
fascia lata

**Membranous layer
of subcutaneous
tissue (Scarpa fascia)**

"Gutter" (floor of inguinal canal)

Inguinal ligament

Fascia lata

Superior ramus of pubis

Saphenous
opening

Skin

Continuity with superficial
fascia of penis (cut)

Fascia lata

Spermatic cord

Subcutaneous tissue of thigh

Continuity with dartos
tunic of scrotum (cut)

Attachment to posterior
edge of perineal membrane

B. Anterior View

C. Sagittal Section

4.11   INGUINAL REGION OF MALE I

A. Formations of the aponeurosis of the external oblique muscle. **B.** and **C.** Membranous (deep) layer of subcutaneous tissue. Inferior to the umbilicus, the subcutaneous tissue is composed of two layers: a superficial fatty layer and a deep membranous layer. Laterally, the membranous layer fuses with the fascia lata of the thigh about a finger's breadth inferior to the inguinal ligament. Medially, it fuses with the linea alba and pubic symphysis in the midline, and inferiorly, it continues as the membranous layer of the subcutaneous tissue of the perineum and penis and the dartos fascia of the scrotum. The inferior margin of the external oblique aponeurosis is thickened and turned internally forming the inguinal ligament. The superior surface of the in-turning inguinal ligament forms a shallow trough or "gutter" that is the floor of the inguinal canal.

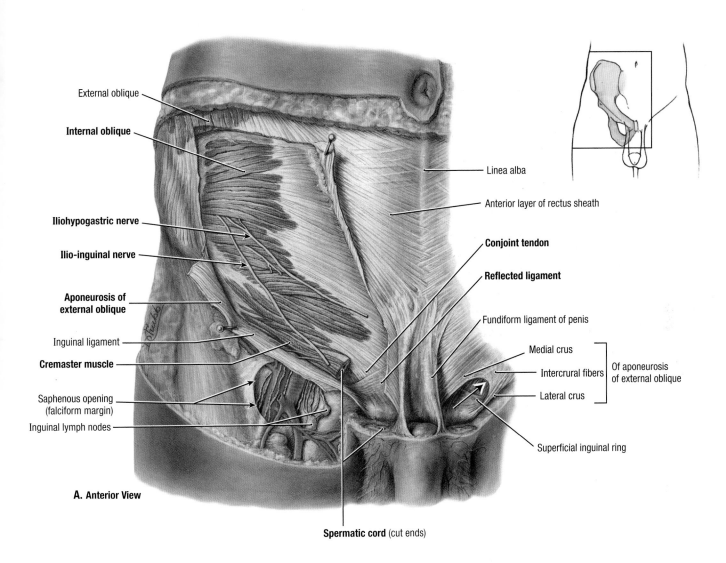

External oblique

Internal oblique

Linea alba

Anterior layer of rectus sheath

Iliohypogastric nerve

Ilio-inguinal nerve

Conjoint tendon

Reflected ligament

**Aponeurosis of
external oblique**

Fundiform ligament of penis

Inguinal ligament

Medial crus

Cremaster muscle

Intercrural fibers

Of aponeurosis
of external oblique

Lateral crus

Saphenous opening
(falciform margin)

Inguinal lymph nodes

Superficial inguinal ring

A. Anterior View

Spermatic cord (cut ends)

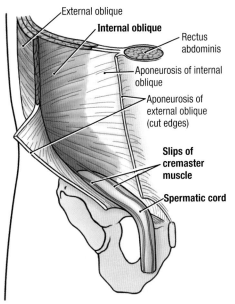

External oblique

Internal oblique

Rectus
abdominis

Aponeurosis of internal
oblique

Aponeurosis of
external oblique
(cut edges)

**Slips of
cremaster
muscle**

Spermatic cord

B. Anterior View

INGUINAL REGION OF MALE II

4.12

A. Internal oblique and cremaster muscle. Part of the aponeurosis of the external oblique muscle is cut away, and the spermatic cord is cut short. **B.** Schematic illustration.

- The cremaster fascia covers the spermatic cord. Cremaster muscle is dispersed within the cremasteric fascia.
- The reflected ligament is formed by aponeurotic fibers of the external oblique muscle and lies anterior to the conjoint tendon. The conjoint tendon is formed by the fusion of the inferior most parts of the aponeurosis of the internal oblique and transversus abdominis muscles.
- The cutaneous branches of the iliohypogastric and ilio-inguinal nerves (L1) course between the internal and external oblique muscles and must be avoided when an **appendectomy (gridiron) incision** is made in this region.

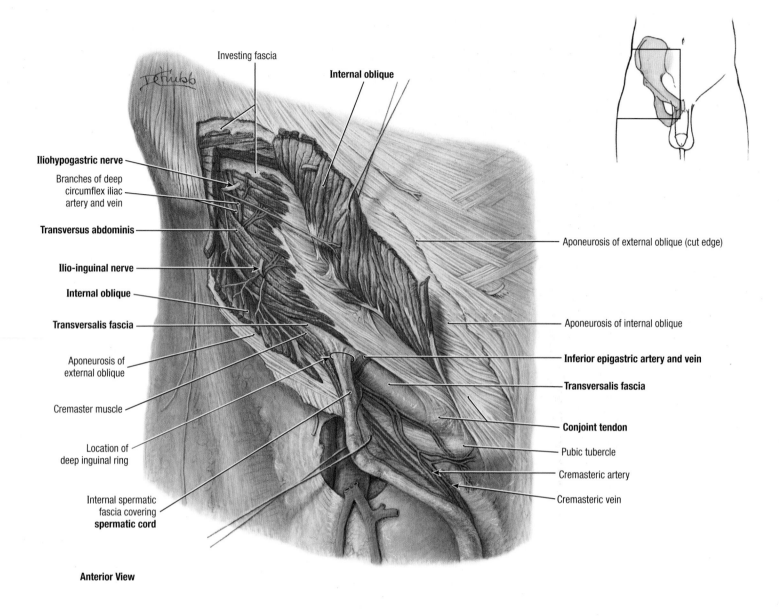

Investing fascia

Internal oblique

Iliohypogastric nerve

Branches of deep circumflex iliac artery and vein

Transversus abdominis

Ilio-inguinal nerve

Internal oblique

Transversalis fascia

Aponeurosis of external oblique

Cremaster muscle

Location of deep inguinal ring

Internal spermatic fascia covering **spermatic cord**

Aponeurosis of external oblique (cut edge)

Aponeurosis of internal oblique

Inferior epigastric artery and vein

Transversalis fascia

Conjoint tendon

Pubic tubercle

Cremasteric artery

Cremasteric vein

Anterior View

4.13   INGUINAL REGION OF MALE III

The internal oblique muscle is reflected, and the spermatic cord is retracted.
- The internal oblique muscle portion of the conjoint tendon is attached to the pubic crest, and the transversus abdominis portion to the pectineal line.

- The iliohypogastric and ilio-inguinal nerves (L1) supply the internal oblique and transversus abdominis muscles.
- The transversalis fascia is evaginated to form the tubular internal spermatic fascia. The mouth of the tube, called the deep inguinal ring, is situated lateral to the inferior epigastric vessels.

TABLE 4.2	**BOUNDARIES OF INGUINAL CANAL**		
Boundary	*Deep Ring/Lateral Third*	*Middle Third*	*Lateral Third/Superficial Ring*
Posterior wall	Transversalis fascia	Transversalis fascia	Inguinal falx (conjoint tendon) plus reflected inguinal ligament
Anterior wall	Internal oblique plus lateral crus of aponeurosis of external oblique	Aponeurosis of external oblique (lateral crus and intercrural fibers)	Aponeurosis of external oblique (intercrural fibers), with fascia of external oblique continuing onto cord as external spermatic fascia
Roof	Transversalis fascia	Musculo-aponeurotic arches of internal oblique and transversus abdominis	Medial crus of aponeurosis of external oblique
Floor	Iliopubic tract	Inguinal ligament	Lacunar ligament

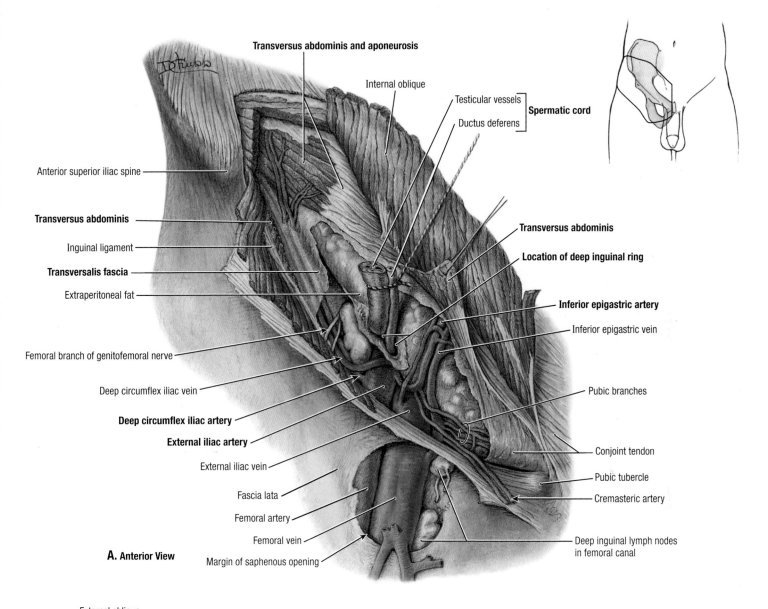

Transversus abdominis and aponeurosis

Internal oblique

Testicular vessels ⎤
Ductus deferens ⎦ **Spermatic cord**

Anterior superior iliac spine

Transversus abdominis

Inguinal ligament

Transversalis fascia

Extraperitoneal fat

Transversus abdominis

Location of deep inguinal ring

Inferior epigastric artery

Inferior epigastric vein

Femoral branch of genitofemoral nerve

Deep circumflex iliac vein

Pubic branches

Deep circumflex iliac artery

External iliac artery

External iliac vein

Fascia lata

Femoral artery

Femoral vein

Margin of saphenous opening

Conjoint tendon

Pubic tubercle

Cremasteric artery

Deep inguinal lymph nodes in femoral canal

A. Anterior View

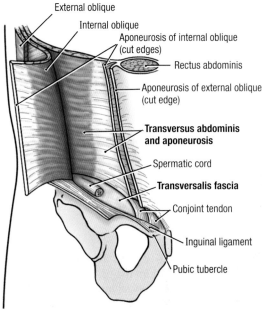

External oblique

Internal oblique

Aponeurosis of internal oblique (cut edges)

Rectus abdominis

Aponeurosis of external oblique (cut edge)

Transversus abdominis and aponeurosis

Spermatic cord

Transversalis fascia

Conjoint tendon

Inguinal ligament

Pubic tubercle

B. Anterior View

INGUINAL REGION OF MALE IV

4.14

A. The inguinal part of the transversus abdominis muscle and transversalis fascia is partially cut away, the spermatic cord is excised, and the ductus deferens is retracted. **B.** Schematic illustration.

- The deep inguinal ring is located superior to the inguinal ligament at the midpoint between the anterior superior iliac spine and pubic tubercle.
- The external iliac artery has two branches, the deep circumflex iliac and inferior epigastric arteries. Note also the cremasteric artery and pubic branch arising from the latter.

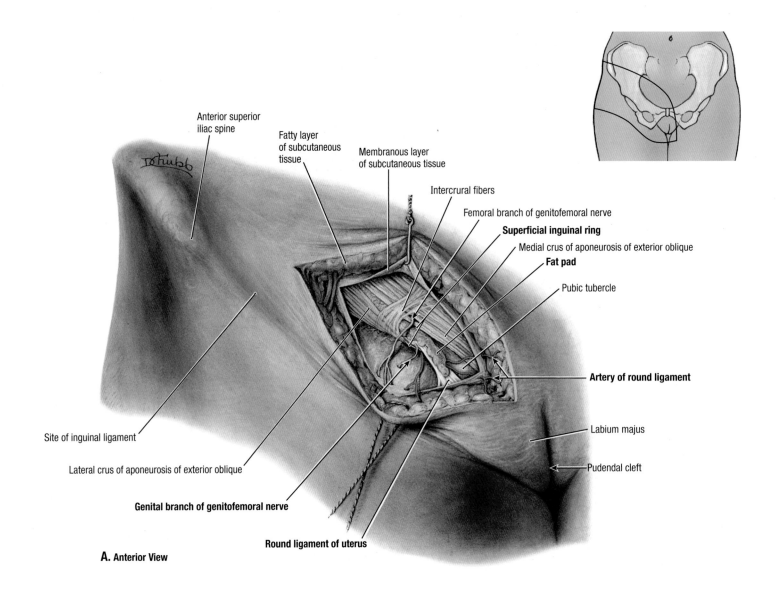

Anterior superior
iliac spine

Fatty layer
of subcutaneous
tissue

Membranous layer
of subcutaneous tissue

Intercrural fibers

Femoral branch of genitofemoral nerve

Superficial inguinal ring

Medial crus of aponeurosis of exterior oblique

Fat pad

Pubic tubercle

Artery of round ligament

Labium majus

Pudendal cleft

Site of inguinal ligament

Lateral crus of aponeurosis of exterior oblique

Genital branch of genitofemoral nerve

Round ligament of uterus

A. Anterior View

4.15 INGUINAL CANAL OF FEMALE

Progressive dissections of the female inguinal canal.
- The superficial inguinal ring is small **(A)**. Passing through the superficial inguinal ring are the round ligament of the uterus, a closely applied fat pad, the genital branch of the genitofemoral nerve, and the artery of the round ligament of the uterus **(B)**.

- The round ligament breaks up into strands as it leaves the inguinal canal and approaches the labium majus. The ilio-inguinal nerve may also pass through the superficial inguinal ring **(C)**.
- The external iliac artery and vein are exposed deep to the inguinal canal by excising the transversalis fascia **(D)**.

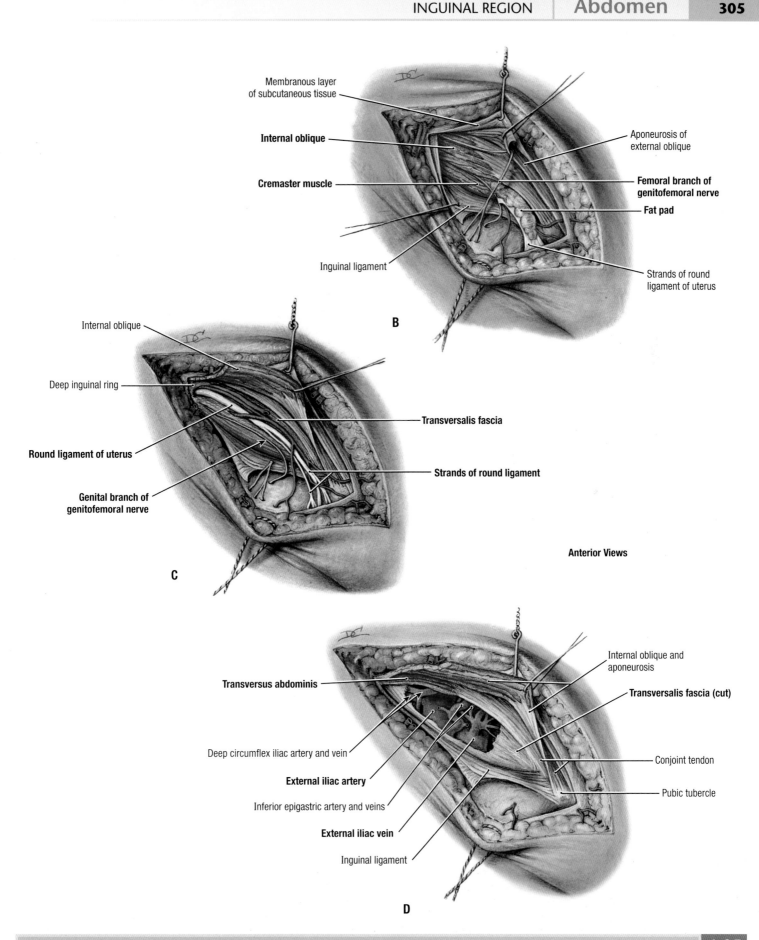

Membranous layer
of subcutaneous tissue

Internal oblique

Cremaster muscle

Inguinal ligament

Aponeurosis of
external oblique

**Femoral branch of
genitofemoral nerve**

Fat pad

Strands of round
ligament of uterus

B

Internal oblique

Deep inguinal ring

Round ligament of uterus

**Genital branch of
genitofemoral nerve**

Transversalis fascia

Strands of round ligament

Anterior Views

C

Transversus abdominis

Deep circumflex iliac artery and vein

External iliac artery

Inferior epigastric artery and veins

External iliac vein

Inguinal ligament

Internal oblique and
aponeurosis

Transversalis fascia (cut)

Conjoint tendon

Pubic tubercle

D

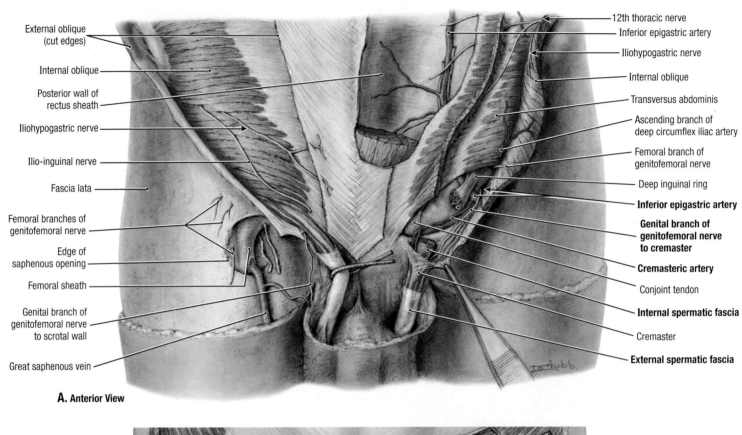

External oblique (cut edges)

Internal oblique

Posterior wall of rectus sheath

Iliohypogastric nerve

Ilio-inguinal nerve

Fascia lata

Femoral branches of genitofemoral nerve

Edge of saphenous opening

Femoral sheath

Genital branch of genitofemoral nerve to scrotal wall

Great saphenous vein

12th thoracic nerve

Inferior epigastric artery

Iliohypogastric nerve

Internal oblique

Transversus abdominis

Ascending branch of deep circumflex iliac artery

Femoral branch of genitofemoral nerve

Deep inguinal ring

Inferior epigastric artery

Genital branch of genitofemoral nerve to cremaster

Cremasteric artery

Conjoint tendon

Internal spermatic fascia

Cremaster

External spermatic fascia

A. Anterior View

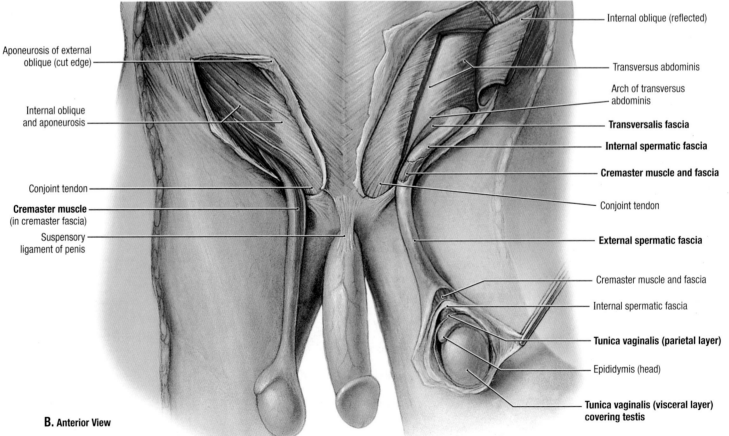

Aponeurosis of external oblique (cut edge)

Internal oblique and aponeurosis

Conjoint tendon

Cremaster muscle (in cremaster fascia)

Suspensory ligament of penis

Internal oblique (reflected)

Transversus abdominis

Arch of transversus abdominis

Transversalis fascia

Internal spermatic fascia

Cremaster muscle and fascia

Conjoint tendon

External spermatic fascia

Cremaster muscle and fascia

Internal spermatic fascia

Tunica vaginalis (parietal layer)

Epididymis (head)

Tunica vaginalis (visceral layer) covering testis

B. Anterior View

4.16 INGUINAL CANAL, SPERMATIC CORD, AND TESTIS

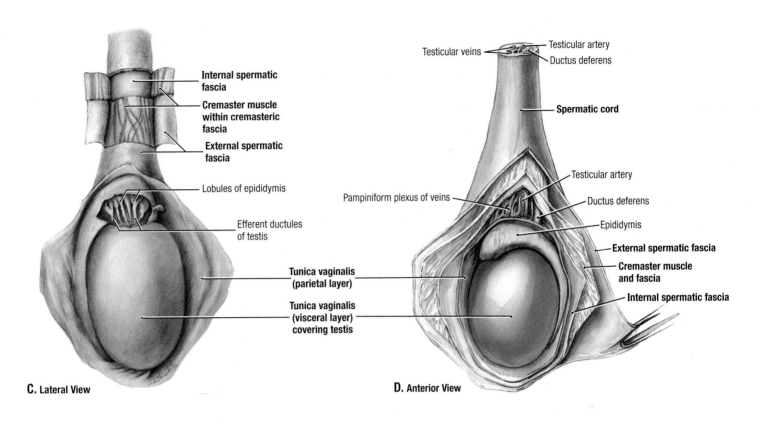

C. Lateral View

Internal spermatic fascia

Cremaster muscle within cremasteric fascia

External spermatic fascia

Lobules of epididymis

Efferent ductules of testis

Tunica vaginalis (parietal layer)

Tunica vaginalis (visceral layer) covering testis

D. Anterior View

Testicular veins

Testicular artery

Ductus deferens

Spermatic cord

Pampiniform plexus of veins

Testicular artery

Ductus deferens

Epididymis

External spermatic fascia

Cremaster muscle and fascia

Internal spermatic fascia

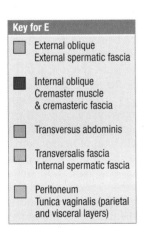

Key for E

External oblique
External spermatic fascia

Internal oblique
Cremaster muscle
& cremasteric fascia

Transversus abdominis

Transversalis fascia
Internal spermatic fascia

Peritoneum
Tunica vaginalis (parietal and visceral layers)

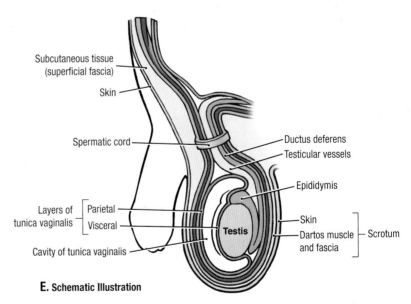

Subcutaneous tissue (superficial fascia)

Skin

Spermatic cord

Ductus deferens

Testicular vessels

Epididymis

Layers of tunica vaginalis — Parietal / Visceral

Skin

Testis

Dartos muscle and fascia

Scrotum

Cavity of tunica vaginalis

E. Schematic Illustration

INGUINAL CANAL, SPERMATIC CORD, AND TESTIS (*continued*) 4.16

A. Dissection of inguinal canal. **B.** Dissection of inguinal region and coverings of the spermatic cord and testis. **C–E.** Coverings of spermatic cord and testis. The cavity of the tunica vaginalis is normally a potential space.

Male

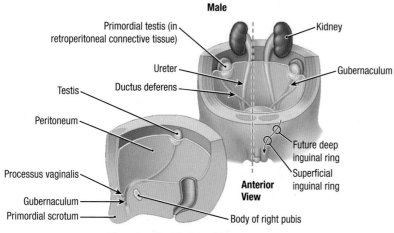

Primordial testis (in retroperitoneal connective tissue)
Kidney
Ureter
Ductus deferens
Gubernaculum
Testis
Peritoneum
Future deep inguinal ring
Superficial inguinal ring
Processus vaginalis
Gubernaculum
Primordial scrotum
Body of right pubis
Anterior View

Diagrammatic oblique sagittal section to right of midline

A. Seventh Week

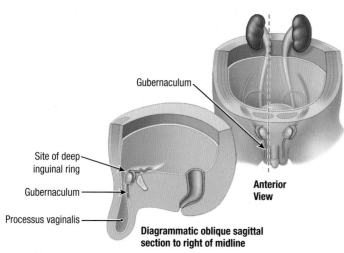

Gubernaculum
Site of deep inguinal ring
Gubernaculum
Processus vaginalis
Anterior View

Diagrammatic oblique sagittal section to right of midline

B. Seventh Month

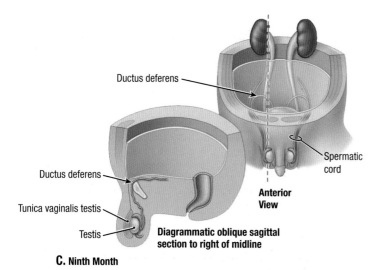

Ductus deferens
Ductus deferens
Tunica vaginalis testis
Testis
Spermatic cord
Anterior View

Diagrammatic oblique sagittal section to right of midline

C. Ninth Month

Female

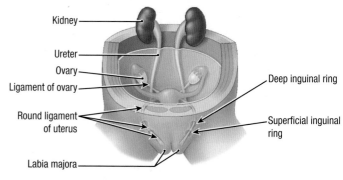

Primordial ovaries
Upper gubernaculum (inguinal fold–becomes ligament of ovary)
Paramesonephric duct
Developing kidney
Lower gubernaculum (becomes round ligament of uterus)
Mesonephric duct
Peritoneum

D. 2 Months

Kidney
Ureter
Ovary
Ligament of ovary
Deep inguinal ring
Round ligament of uterus
Superficial inguinal ring
Labia majora

E. 15 Weeks

4.17   RELOCATION OF GONADS

The inguinal canals in females are narrower than those in males, and the canals in infants of both sexes are shorter and much less oblique than in adults. For a complete description of the embryology of the inguinal region, see Moore et al. (2012).

The fetal testes relocate from the dorsal abdominal wall in the superior lumbar region to the deep inguinal rings during the 9th to 12th fetal weeks. This movement probably results from the growth of the vertebral column and pelvis. The male gubernaculum, attached to the caudal pole of the testis and accompanied by an outpouching of peritoneum, the processus vaginalis, projects into the scrotum. The testis passes posterior to the processus vaginalis. The inferior remnant of the processus vaginalis forms the tunica vaginalis covering the testis. The ductus deferens, testicular vessels, nerves, and lymphatics accompany the testis. The final descent of the testis usually occurs before or shortly after birth.

The fetal ovaries also relocate from the dorsal abdominal wall in the superior lumbar region during the 12th week but pass into the lesser pelvis. The female gubernaculum attaches to the caudal pole of the ovary and projects into the labia majora, attaching en route to the uterus; the part passing from the uterus to the ovary forms the ovarian ligament, and the remainder of it becomes the round ligament of the uterus. Because of the attachment of the ovarian ligaments to the uterus, the ovaries do not relocate to the inguinal region; however, the round ligament passes through the inguinal canal and attaches to the subcutaneous tissue of the labium majus.

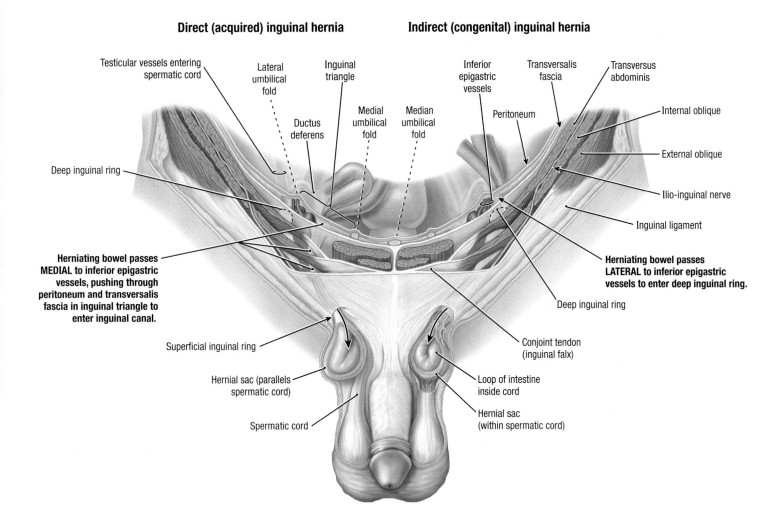

Direct (acquired) inguinal hernia   **Indirect (congenital) inguinal hernia**

Testicular vessels entering spermatic cord

Lateral umbilical fold

Inguinal triangle

Ductus deferens

Medial umbilical fold

Median umbilical fold

Inferior epigastric vessels

Transversalis fascia

Transversus abdominis

Peritoneum

Internal oblique

External oblique

Ilio-inguinal nerve

Inguinal ligament

Deep inguinal ring

Herniating bowel passes MEDIAL to inferior epigastric vessels, pushing through peritoneum and transversalis fascia in inguinal triangle to enter inguinal canal.

Herniating bowel passes LATERAL to inferior epigastric vessels to enter deep inguinal ring.

Deep inguinal ring

Superficial inguinal ring

Conjoint tendon (inguinal falx)

Hernial sac (parallels spermatic cord)

Loop of intestine inside cord

Spermatic cord

Hernial sac (within spermatic cord)

COURSE OF DIRECT AND INDIRECT INGUINAL HERNIAS

4.18

An **inguinal hernia** is a protrusion of parietal peritoneum and viscera, such as the small intestine, through the abdominal wall in the inguinal region. There are two major categories of inguinal hernia: indirect and direct. More than two thirds are indirect hernias, most commonly occurring in males.

TABLE 4.3   CHARACTERISTICS OF INGUINAL HERNIAS

Characteristics	Direct (Acquired)	Indirect (Congenital)
Predisposing factors	Weakness of anterior abdominal wall in inguinal triangle (e.g., owing to distended superficial ring, narrow conjoint tendon, or attenuation of aponeurosis in males >40 years of age)	Patency of processus vaginalis (complete or at least of superior part) in younger persons, the great majority of whom are males
Frequency	Less common (one third to one fourth of inguinal hernias)	More common (two thirds to three fourths of inguinal hernias)
Coverings at exit from abdominal cavity	Peritoneum plus transversalis fascia (lies outside inner one or two fascial coverings, parallel to cord)	Peritoneum of persistent processus vaginalis plus all three fascial coverings of cord/round ligament
Course	Usually traverses only medial third of inguinal canal, external and parallel to vestige of processus vaginalis	Traverses inguinal canal (entire canal if it is sufficient size) within processus vaginalis
Exit from anterior abdominal wall	Via superficial ring, lateral to cord; rarely enters scrotum	Via superficial ring inside cord, commonly passing into scrotum/labium majus

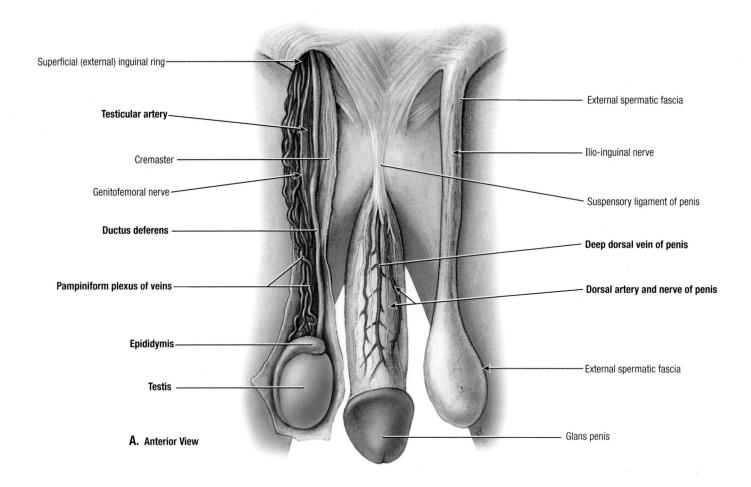

Superficial (external) inguinal ring

Testicular artery

Cremaster

Genitofemoral nerve

Ductus deferens

Pampiniform plexus of veins

Epididymis

Testis

A. Anterior View

External spermatic fascia

Ilio-inguinal nerve

Suspensory ligament of penis

Deep dorsal vein of penis

Dorsal artery and nerve of penis

External spermatic fascia

Glans penis

4.19 SPERMATIC CORD, TESTIS, AND EPIDIDYMIS

A. Dissection of spermatic cord. The subcutaneous tissue (dartos fascia) covering the penis has been removed and the deep fascia rendered transparent to demonstrate the median deep dorsal vein and the bilateral dorsal arteries and nerves of the penis. On the specimen's right, the coverings of the spermatic cord and testis are reflected, and the contents of the cord are separated. The testicular artery has been separated from the pampiniform plexus of veins that surrounds it as it courses parallel to the ductus deferens. Lymphatic vessels and autonomic nerve fibers (not shown) are also present. **B.** The tunica vaginalis has been incised longitudinally to expose its cavity, surrounding the testis anteriorly and laterally, and extending between the testis and epididymis at the sinus of the epididymis. The epididymis is located posterolateral to the testis, that is, toward the right side of the right testis and the left side of the left testis. The appendices of the testis and epididymis may be observed in some specimens. These structures are small remnants of the embryonic genital (paramesonephric) duct.

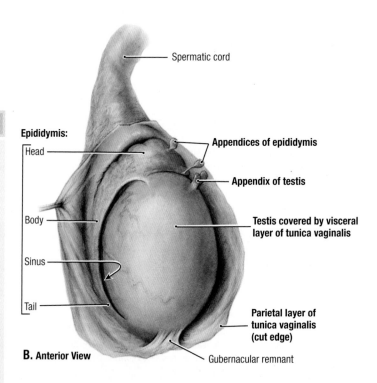

Spermatic cord

Epididymis:

Head

Body

Sinus

Tail

Appendices of epididymis

Appendix of testis

Testis covered by visceral layer of tunica vaginalis

Parietal layer of tunica vaginalis (cut edge)

Gubernacular remnant

B. Anterior View

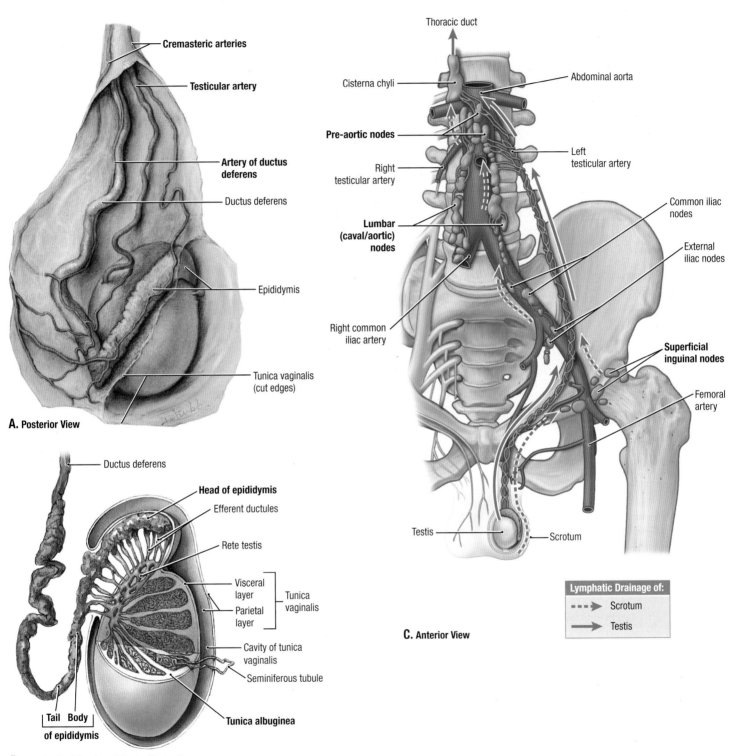

A. Posterior View

Cremasteric arteries

Testicular artery

Artery of ductus deferens

Ductus deferens

Epididymis

Tunica vaginalis (cut edges)

B. Longitudinal Section of Tunica Vaginalis; Testis Sectioned in Sagittal and Transverse Planes

Ductus deferens

Head of epididymis

Efferent ductules

Rete testis

Visceral layer

Parietal layer

Tunica vaginalis

Cavity of tunica vaginalis

Seminiferous tubule

Tunica albuginea

Tail Body of epididymis

C. Anterior View

Thoracic duct

Cisterna chyli

Abdominal aorta

Pre-aortic nodes

Right testicular artery

Left testicular artery

Lumbar (caval/aortic) nodes

Common iliac nodes

External iliac nodes

Right common iliac artery

Superficial inguinal nodes

Femoral artery

Testis

Scrotum

Lymphatic Drainage of:
- - - → Scrotum
——→ Testis

BLOOD SUPPLY AND LYMPHATIC DRAINAGE OF TESTIS **4.20**

A. Blood supply. **B.** Internal structure. **C.** Lymphatic drainage.

Because the testes relocate from the posterior abdominal wall into the scrotum during fetal development, their lymphatic drainage differs from that of the scrotum, which is an outpouching of the abdominal skin. Consequently, **cancer of the testis** metastasizes initially to the lumbar lymph nodes, and **cancer of the scrotum** metastasizes initially to the superficial inguinal lymph nodes.

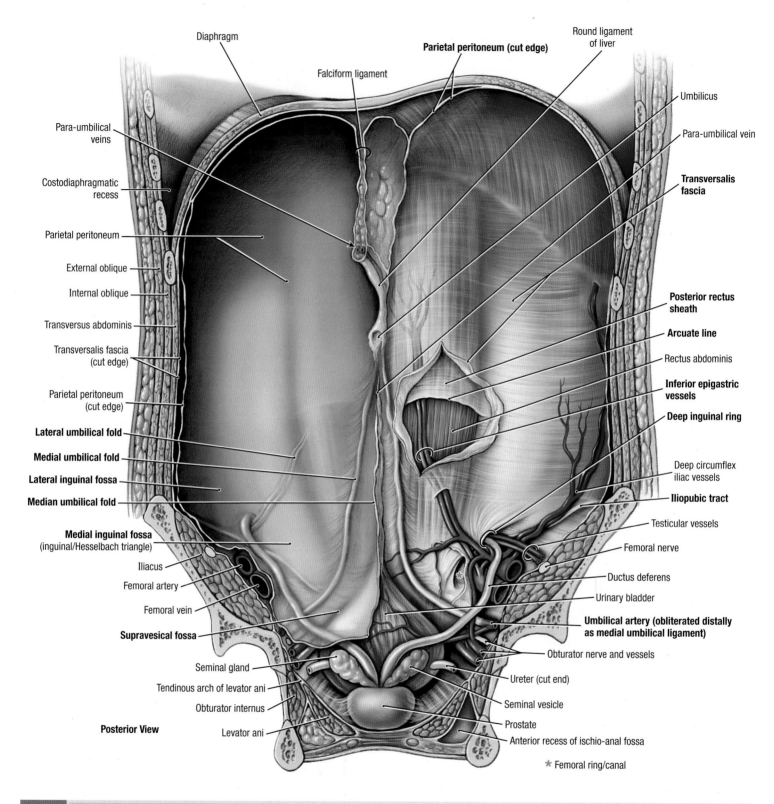

Diaphragm

Falciform ligament

Round ligament of liver

Parietal peritoneum (cut edge)

Umbilicus

Para-umbilical veins

Para-umbilical vein

Costodiaphragmatic recess

Transversalis fascia

Parietal peritoneum

External oblique

Internal oblique

Transversus abdominis

Transversalis fascia (cut edge)

Posterior rectus sheath

Arcuate line

Rectus abdominis

Parietal peritoneum (cut edge)

Inferior epigastric vessels

Lateral umbilical fold

Deep inguinal ring

Medial umbilical fold

Deep circumflex iliac vessels

Lateral inguinal fossa

Median umbilical fold

Iliopubic tract

Testicular vessels

Medial inguinal fossa
(inguinal/Hesselbach triangle)

Femoral nerve

Iliacus

Ductus deferens

Femoral artery

Femoral vein

Urinary bladder

Umbilical artery (obliterated distally as medial umbilical ligament)

Supravesical fossa

Obturator nerve and vessels

Seminal gland

Ureter (cut end)

Tendinous arch of levator ani

Seminal vesicle

Obturator internus

Prostate

Posterior View

Levator ani

Anterior recess of ischio-anal fossa

✳ Femoral ring/canal

4.21   POSTERIOR ASPECT OF THE ANTEROLATERAL ABDOMINAL WALL

Umbilical folds (median, medial, and lateral) are reflections of the parietal peritoneum that are raised from the body wall by underlying structures. The median umbilical fold extends from the urinary bladder to the umbilicus and covers the median umbilical ligament (the remnant of the urachus). The two medial umbilical folds cover the medial umbilical ligaments (occluded remnants of the fetal umbilical arteries). Two lateral umbilical folds cover the inferior epigastric vessels. The supravesical fossae are between the median and medial umbilical folds, the medial inguinal fossae (inguinal triangles) are between the medial and lateral umbilical folds, and the lateral inguinal fossae and deep inguinal rings are lateral to the lateral umbilical folds.

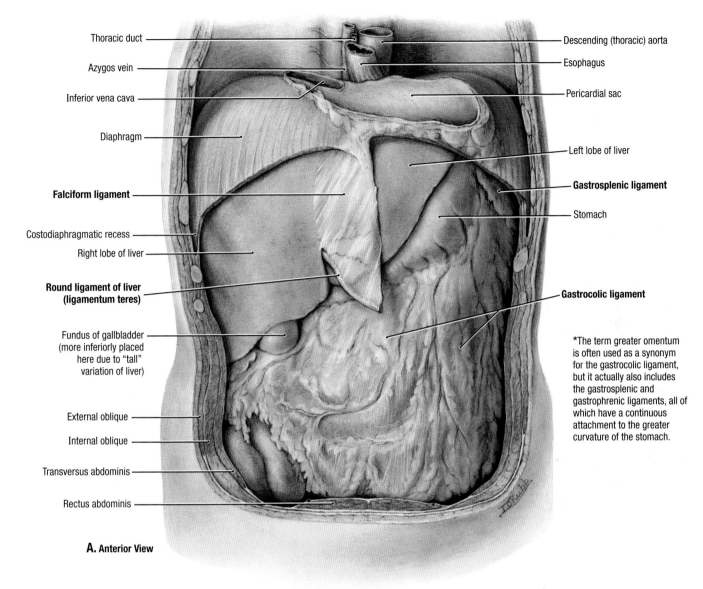

Thoracic duct

Azygos vein

Inferior vena cava

Diaphragm

Falciform ligament

Costodiaphragmatic recess

Right lobe of liver

**Round ligament of liver
(ligamentum teres)**

Fundus of gallbladder
(more inferiorly placed
here due to "tall"
variation of liver)

External oblique

Internal oblique

Transversus abdominis

Rectus abdominis

Descending (thoracic) aorta

Esophagus

Pericardial sac

Left lobe of liver

Gastrosplenic ligament

Stomach

Gastrocolic ligament

*The term greater omentum
is often used as a synonym
for the gastrocolic ligament,
but it actually also includes
the gastrosplenic and
gastrophrenic ligaments, all of
which have a continuous
attachment to the greater
curvature of the stomach.

A. Anterior View

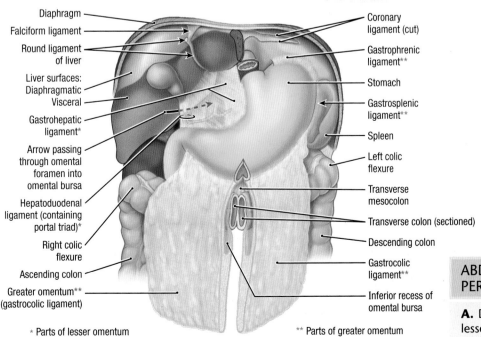

Diaphragm

Falciform ligament

Round ligament
of liver

Liver surfaces:
Diaphragmatic
Visceral

Gastrohepatic
ligament*

Arrow passing
through omental
foramen into
omental bursa

Hepatoduodenal
ligament (containing
portal triad)*

Right colic
flexure

Ascending colon

Greater omentum**
(gastrocolic ligament)

Coronary
ligament (cut)

Gastrophrenic
ligament**

Stomach

Gastrosplenic
ligament**

Spleen

Left colic
flexure

Transverse
mesocolon

Transverse colon (sectioned)

Descending colon

Gastrocolic
ligament**

Inferior recess of
omental bursa

* Parts of lesser omentum

** Parts of greater omentum

B. Anterior View

ABDOMINAL CONTENTS AND
PERITONEUM

4.22

A. Dissection. **B.** Components of greater and
lesser omentum.

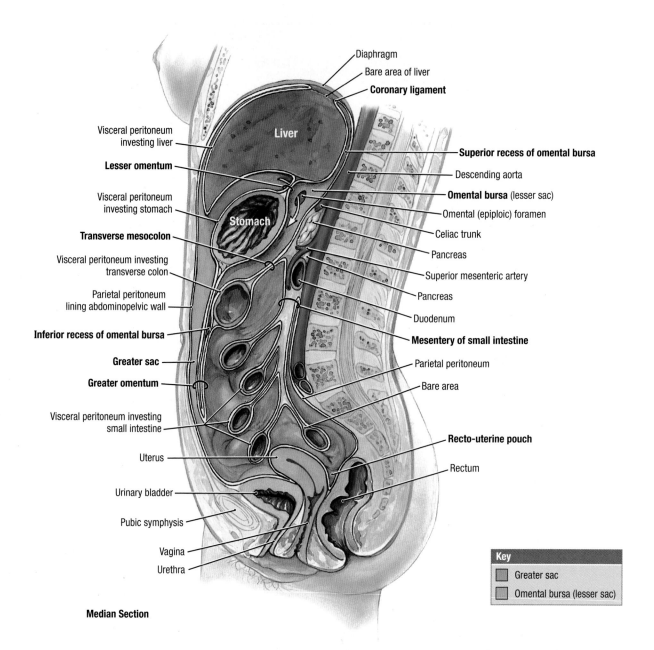

Diaphragm
Bare area of liver
Coronary ligament

Visceral peritoneum investing liver

Liver

Superior recess of omental bursa

Descending aorta

Lesser omentum

Omental bursa (lesser sac)

Omental (epiploic) foramen

Visceral peritoneum investing stomach

Stomach

Celiac trunk

Transverse mesocolon

Pancreas

Superior mesenteric artery

Visceral peritoneum investing transverse colon

Pancreas

Parietal peritoneum lining abdominopelvic wall

Duodenum

Mesentery of small intestine

Inferior recess of omental bursa

Greater sac

Parietal peritoneum

Greater omentum

Bare area

Visceral peritoneum investing small intestine

Uterus

Recto-uterine pouch

Urinary bladder

Rectum

Pubic symphysis

Vagina

Urethra

Median Section

Key	
▮	Greater sac
▮	Omental bursa (lesser sac)

4.23 PERITONEAL FORMATIONS AND BARE AREAS

Various terms are used to describe the parts of the peritoneum that connect organs with other organs or to the abdominal wall and to describe the compartments and recesses that are formed as a consequence. The *arrow* passes through the omental (epiploic) foramen.

TABLE 4.4	TERMS USED TO DESCRIBE PARTS OF PERITONEUM
Term	*Definition*
Peritoneal ligament	Double layer of peritoneum that connects an organ with another organ or to the abdominal wall.
Mesentery	Double layer of peritoneum that occurs as a result of the invagination of the peritoneum by one or more organs and constitutes a continuity of the visceral and parietal peritoneum.
Omentum	Double-layered extension of peritoneum passing from the proximal duodenum and/or stomach and to adjacent organs. The greater omentum extends from the greater curvature of the stomach and the proximal duodenum; the lesser omentum from the lesser curvature.
Bare area	Every organ must have an area, the bare area, that is not covered with visceral peritoneum, to allow the entrance and exit of neurovascular structures. Bare areas are formed in relation to the attachments of mesenteries, omenta, and ligaments. Named bare areas (e.g., bare area of liver) are especially extensive.

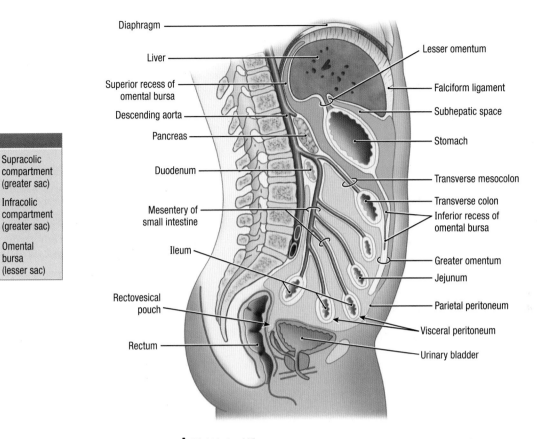

A. Right Lateral View

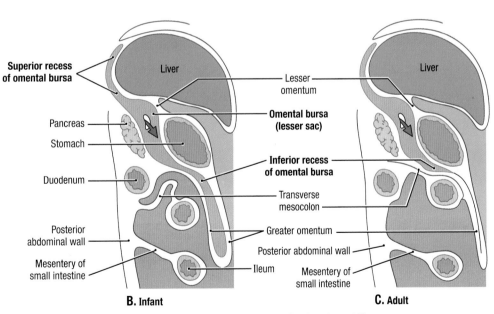

B. Infant   **C.** Adult

Schematic Sagittal Sections, Lateral View

SUBDIVISIONS OF PERITONEAL CAVITY

4.24

A. Sagittal section. **B.** In an infant, the omental bursa (lesser sac) is an isolated part of the peritoneal cavity, lying posterior to the stomach and extending superiorly between the liver and diaphragm (superior recess of the omental bursa) and inferiorly between the layers of the greater omentum (inferior recess of the omental bursa). **C.** In an adult, after fusion of the layers of the greater omentum, the inferior recess of the omental bursa now extends inferiorly only as far as the transverse colon. The *arrows (red)* pass from the greater sac through the omental (epiploic) foramen into the omental bursa.

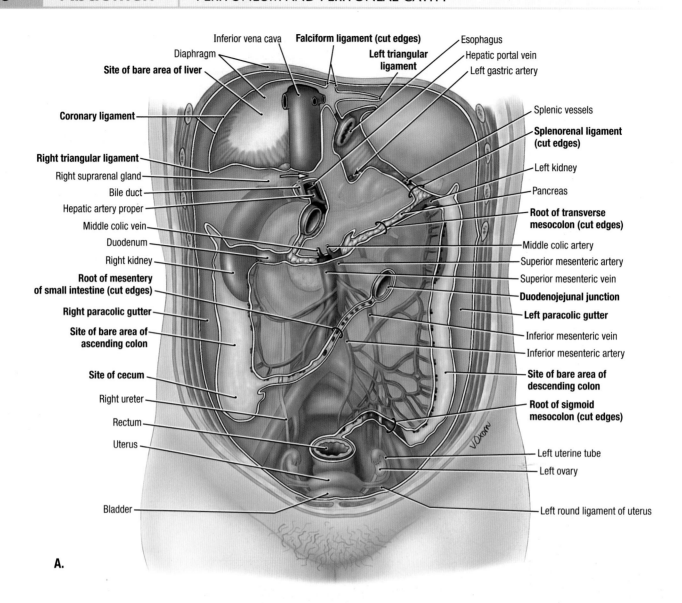

Inferior vena cava

Diaphragm

Falciform ligament (cut edges)

Left triangular ligament

Esophagus

Hepatic portal vein

Left gastric artery

Site of bare area of liver

Coronary ligament

Splenic vessels

Splenorenal ligament (cut edges)

Right triangular ligament

Right suprarenal gland

Bile duct

Hepatic artery proper

Middle colic vein

Duodenum

Right kidney

Left kidney

Pancreas

Root of transverse mesocolon (cut edges)

Middle colic artery

Superior mesenteric artery

Superior mesenteric vein

Root of mesentery of small intestine (cut edges)

Right paracolic gutter

Site of bare area of ascending colon

Duodenojejunal junction

Left paracolic gutter

Inferior mesenteric vein

Inferior mesenteric artery

Site of cecum

Right ureter

Rectum

Uterus

Site of bare area of descending colon

Root of sigmoid mesocolon (cut edges)

Left uterine tube

Left ovary

Bladder

Left round ligament of uterus

A.

Anterior Views

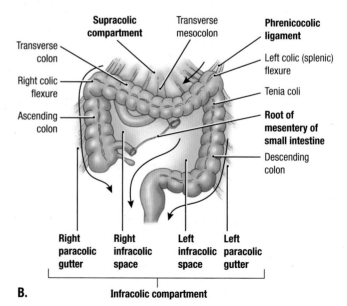

Supracolic compartment

Transverse mesocolon

Phrenicocolic ligament

Transverse colon

Right colic flexure

Ascending colon

Left colic (splenic) flexure

Tenia coli

Root of mesentery of small intestine

Descending colon

Right paracolic gutter

Right infracolic space

Left infracolic space

Left paracolic gutter

B.

Infracolic compartment

4.25 POSTERIOR WALL OF PERITONEAL CAVITY

A. Roots of the peritoneal reflections. The peritoneal reflections from the posterior abdominal wall (mesenteries and reflections surrounding bare areas of liver and secondarily retroperitoneal organs) have been cut at their roots, and the intraperitoneal and secondarily retroperitoneal viscera have been removed. The *arrow (white)* passes through the omental (epiploic) foramen. **B.** Supracolic and infracolic compartments of the greater sac.

The infracolic spaces and paracolic gutters are of clinical importance because they determine the paths *(black arrows)* for the **flow of ascitic fluid with changes in position**, and the spread of intraperitoneal infections.

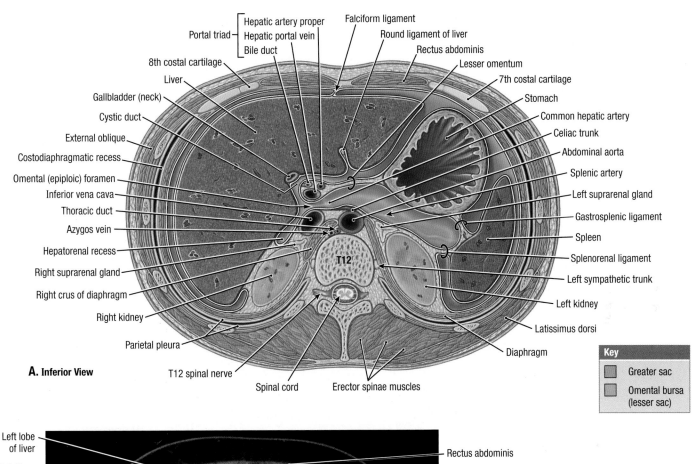

A. Inferior View

Key
- Greater sac
- Omental bursa (lesser sac)

Labels (A. Inferior View):
Portal triad — Hepatic artery proper, Hepatic portal vein, Bile duct
Falciform ligament
Round ligament of liver
Rectus abdominis
Lesser omentum
7th costal cartilage
Stomach
Common hepatic artery
Celiac trunk
Abdominal aorta
Splenic artery
Left suprarenal gland
Gastrosplenic ligament
Spleen
Splenorenal ligament
Left sympathetic trunk
Left kidney
Latissimus dorsi
Diaphragm
8th costal cartilage
Liver
Gallbladder (neck)
Cystic duct
External oblique
Costodiaphragmatic recess
Omental (epiploic) foramen
Inferior vena cava
Thoracic duct
Azygos vein
Hepatorenal recess
Right suprarenal gland
Right crus of diaphragm
Right kidney
Parietal pleura
T12 spinal nerve
Spinal cord
Erector spinae muscles

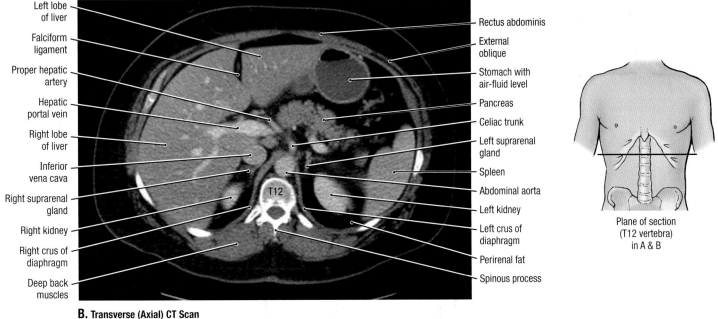

B. Transverse (Axial) CT Scan

Labels (B. Transverse CT Scan):
Left lobe of liver
Falciform ligament
Proper hepatic artery
Hepatic portal vein
Right lobe of liver
Inferior vena cava
Right suprarenal gland
Right kidney
Right crus of diaphragm
Deep back muscles
Rectus abdominis
External oblique
Stomach with air-fluid level
Pancreas
Celiac trunk
Left suprarenal gland
Spleen
Abdominal aorta
Left kidney
Left crus of diaphragm
Perirenal fat
Spinous process

Plane of section (T12 vertebra) in A & B

TRANSVERSE SECTION AND AXIAL CT IMAGE THROUGH GREATER SAC AND OMENTAL BURSA 4.26

- When bacterial contamination occurs or when the gut is traumatically penetrated or ruptured as the result of infection and inflammation, gas, fecal matter, and bacteria enter the peritoneal cavity. The result is infection and inflammation of the peritoneum, called **peritonitis**.
- Under certain pathological conditions such as peritonitis, the peritoneal cavity may be distended with abnormal fluid, **ascites**.

Widespread metastases (spread) of cancer cells to the abdominal viscera cause exudation (escape) of fluid that is often blood stained. Thus, the peritoneal cavity may be distended with several liters of abnormal fluid. Surgical puncture of the peritoneal cavity for the aspiration of drainage of fluid is called **paracentesis**.

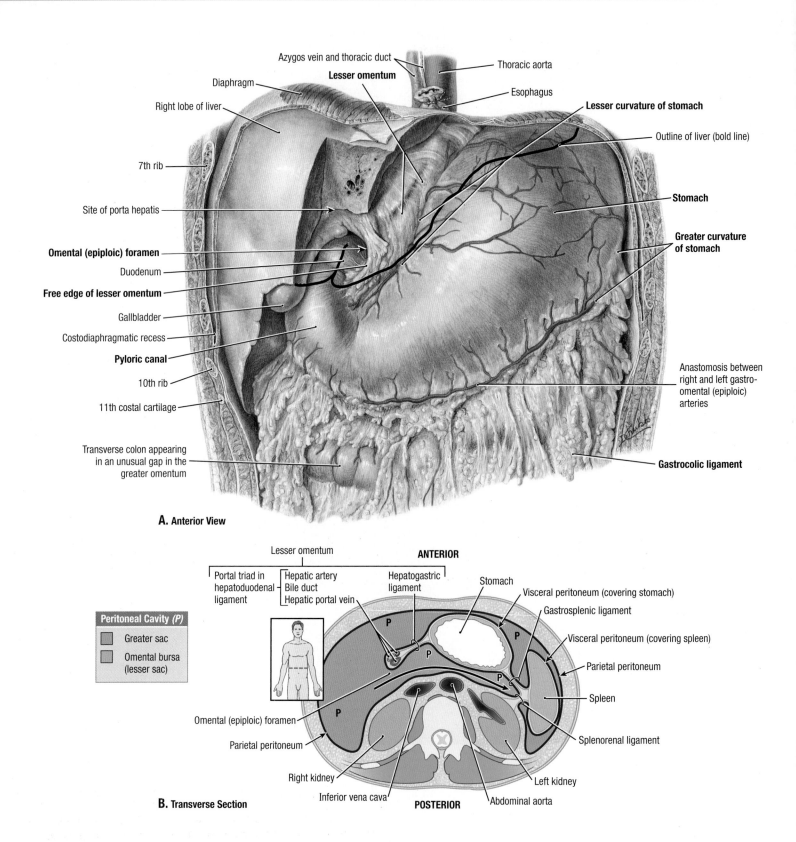

Azygos vein and thoracic duct
Lesser omentum
Thoracic aorta
Diaphragm
Esophagus
Right lobe of liver
Lesser curvature of stomach
Outline of liver (bold line)
7th rib
Site of porta hepatis
Stomach
Omental (epiploic) foramen
Greater curvature of stomach
Duodenum
Free edge of lesser omentum
Gallbladder
Costodiaphragmatic recess
Pyloric canal
Anastomosis between right and left gastro-omental (epiploic) arteries
10th rib
11th costal cartilage
Transverse colon appearing in an unusual gap in the greater omentum
Gastrocolic ligament

A. Anterior View

Lesser omentum
ANTERIOR
Portal triad in hepatoduodenal ligament
- Hepatic artery
- Bile duct
- Hepatic portal vein
Hepatogastric ligament
Stomach
Visceral peritoneum (covering stomach)
Gastrosplenic ligament

Peritoneal Cavity (P)
Greater sac
Omental bursa (lesser sac)

Visceral peritoneum (covering spleen)
Parietal peritoneum
Spleen
Omental (epiploic) foramen
Parietal peritoneum
Splenorenal ligament
Right kidney
Left kidney
Inferior vena cava
Abdominal aorta
POSTERIOR

B. Transverse Section

4.27   STOMACH AND OMENTA

A. Lesser and greater omenta. The stomach is inflated with air, and the left part of the liver is cut away. The gallbladder, followed superiorly, leads to the free margin of the lesser omentum and serves as a guide to the omental (epiploic) foramen, which lies posterior to that free margin. **B.** Omental bursa (lesser sac), schematic transverse section. Arrow is traversing omental foramen and bursa.

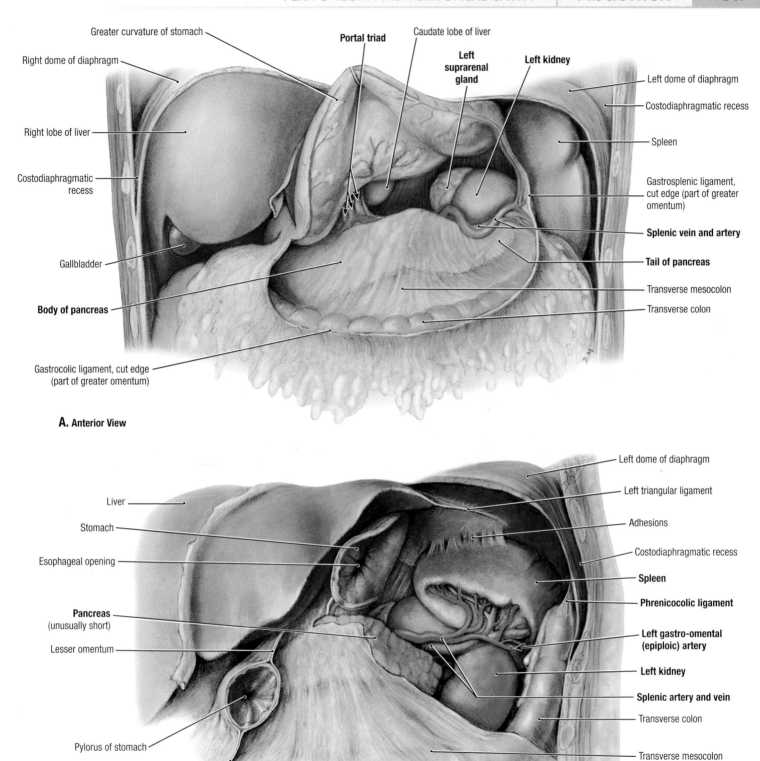

A. Anterior View

Greater curvature of stomach

Right dome of diaphragm

Right lobe of liver

Costodiaphragmatic recess

Gallbladder

Body of pancreas

Gastrocolic ligament, cut edge (part of greater omentum)

Portal triad

Caudate lobe of liver

Left suprarenal gland

Left kidney

Left dome of diaphragm

Costodiaphragmatic recess

Spleen

Gastrosplenic ligament, cut edge (part of greater omentum)

Splenic vein and artery

Tail of pancreas

Transverse mesocolon

Transverse colon

B. Anterior View

Liver

Stomach

Esophageal opening

Pancreas (unusually short)

Lesser omentum

Pylorus of stomach

Gastrocolic ligament (cut edge)

Left dome of diaphragm

Left triangular ligament

Adhesions

Costodiaphragmatic recess

Spleen

Phrenicocolic ligament

Left gastro-omental (epiploic) artery

Left kidney

Splenic artery and vein

Transverse colon

Transverse mesocolon

POSTERIOR RELATIONSHIPS OF OMENTAL BURSA (LESSER SAC)

4.28

A. Opened omental bursa. The greater omentum has been cut along the greater curvature of the stomach; the stomach is reflected superiorly. Peritoneum of the posterior wall of the bursa is partially removed. **B.** Stomach bed. The stomach is excised. Peritoneum covering the stomach bed and inferior part of the kidney and pancreas is largely removed. **Adhesions** binding intraperitoneal organs, such as the spleen to the diaphragm are pathological, but not unusual.

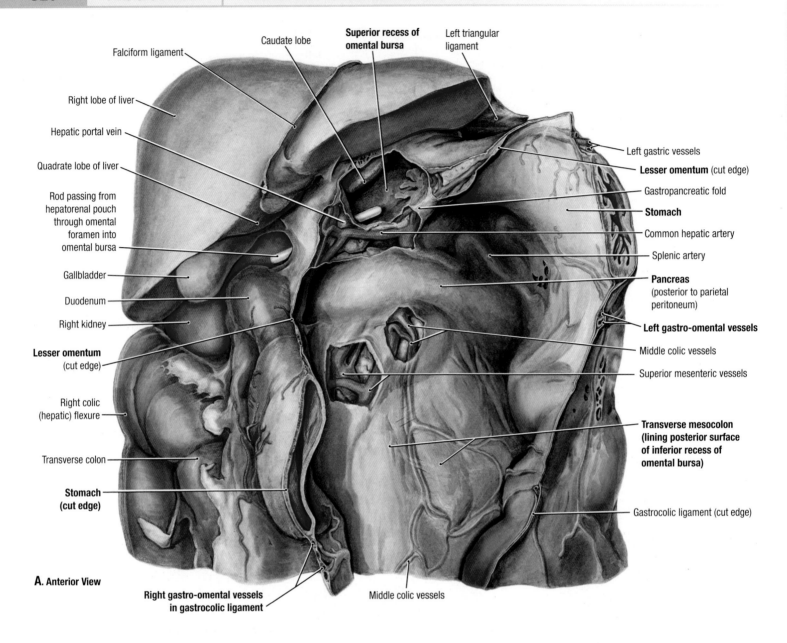

A. Anterior View

Falciform ligament

Right lobe of liver

Hepatic portal vein

Quadrate lobe of liver

Rod passing from hepatorenal pouch through omental foramen into omental bursa

Gallbladder

Duodenum

Right kidney

Lesser omentum (cut edge)

Right colic (hepatic) flexure

Transverse colon

Stomach (cut edge)

Caudate lobe

Superior recess of omental bursa

Left triangular ligament

Left gastric vessels

Lesser omentum (cut edge)

Gastropancreatic fold

Stomach

Common hepatic artery

Splenic artery

Pancreas (posterior to parietal peritoneum)

Left gastro-omental vessels

Middle colic vessels

Superior mesenteric vessels

Transverse mesocolon (lining posterior surface of inferior recess of omental bursa)

Gastrocolic ligament (cut edge)

Right gastro-omental vessels in gastrocolic ligament

Middle colic vessels

4.29   OMENTAL BURSA (LESSER SAC), OPENED

A. Dissection. **B.** Line of incision **(A)**. The anterior wall of the omental bursa, consisting of the stomach, lesser omentum, anterior layer of the greater omentum, and vessels along the curvatures of the stomach, has been sectioned sagittally. The two halves have been retracted to the left and right: the body of the stomach on the left side, and the pyloric part of the stomach and first part of the duodenum on the right. The right kidney forms the posterior wall of the hepatorenal pouch (part of greater sac), and the pancreas lies horizontally on the posterior wall of the main compartment of the omental bursa (lesser sac). The gastrocolic ligament forms the anterior wall and the lower part of the posterior wall of the inferior recess of the omental bursa. The transverse mesocolon forms the upper part of the posterior wall of the inferior recess of the omental bursa.

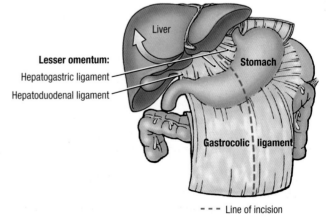

Liver

Lesser omentum:
Hepatogastric ligament

Hepatoduodenal ligament

Stomach

Gastrocolic ligament

- - - Line of incision

B. Anterior View

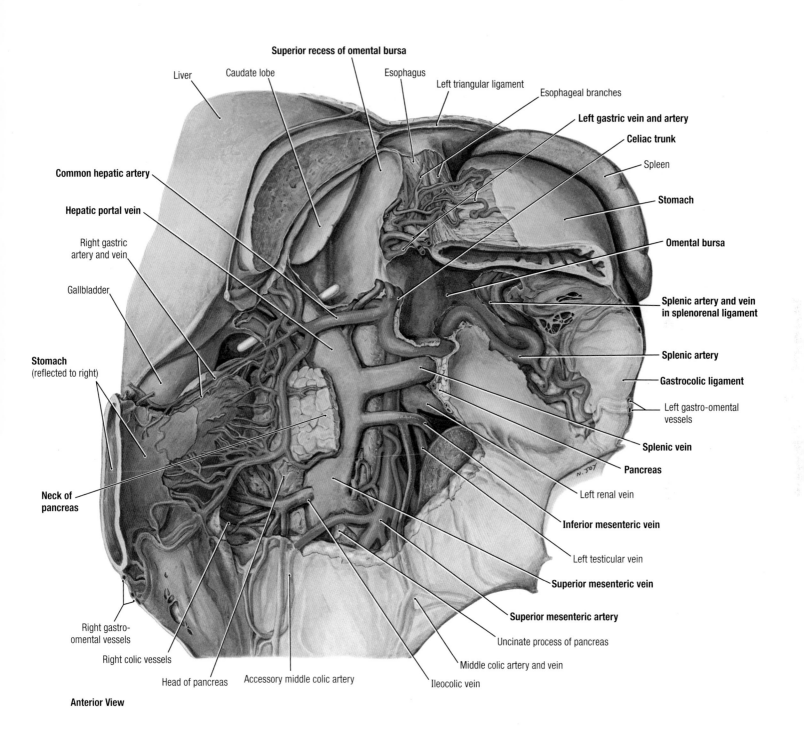

Superior recess of omental bursa

Liver · Caudate lobe · Esophagus · Left triangular ligament · Esophageal branches

Left gastric vein and artery

Celiac trunk

Spleen

Common hepatic artery

Stomach

Hepatic portal vein

Omental bursa

Right gastric artery and vein

Gallblader

Splenic artery and vein in splenorenal ligament

Splenic artery

Stomach (reflected to right)

Gastrocolic ligament

Left gastro-omental vessels

Splenic vein

Neck of pancreas

Pancreas

Left renal vein

Inferior mesenteric vein

Left testicular vein

Superior mesenteric vein

Right gastro-omental vessels

Superior mesenteric artery

Uncinate process of pancreas

Right colic vessels

Middle colic artery and vein

Head of pancreas · Accessory middle colic artery · Ileocolic vein

Anterior View

POSTERIOR WALL OF OMENTAL BURSA

4.30

The parietal peritoneum of the posterior wall of the omental bursa has been mostly removed, and a section of the pancreas has been excised. The rod passes through the omental foramen.

- The celiac trunk gives rise to the left gastric artery, the splenic artery that runs tortuously to the left, and the common hepatic artery that runs to the right, passing anterior to the hepatic portal vein.
- The hepatic portal vein is formed posterior to the neck of the pancreas by the union of the superior mesenteric and splenic

veins, with the inferior mesenteric vein joining at or near the angle of union.
- The left testicular vein usually drains into the left renal vein. Both are systemic veins.
- **Inflammation of the parietal peritoneum** can occur due to an enlarged organ or by the escape of fluid from an organ. The area becomes inflamed and causes pain over the affected region.
- **Rebound tenderness** is a pain that is elicited after pressure over the inflamed area is released.

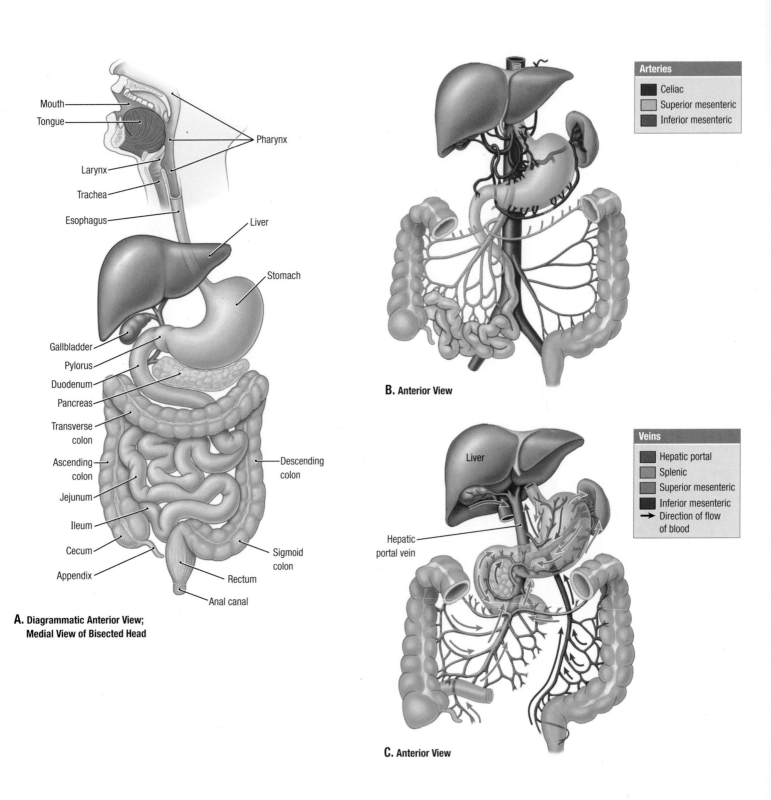

A. Diagrammatic Anterior View;
Medial View of Bisected Head

Mouth
Tongue
Pharynx
Larynx
Trachea
Esophagus
Liver
Stomach
Gallbladder
Pylorus
Duodenum
Pancreas
Transverse colon
Ascending colon
Descending colon
Jejunum
Ileum
Cecum
Sigmoid colon
Appendix
Rectum
Anal canal

B. Anterior View

Arteries
- Celiac
- Superior mesenteric
- Inferior mesenteric

C. Anterior View

Liver
Hepatic portal vein

Veins
- Hepatic portal
- Splenic
- Superior mesenteric
- Inferior mesenteric
- → Direction of flow of blood

4.31 ALIMENTARY SYSTEM

A. Overview. The alimentary system extends from the lips to the anus. Associated organs include the liver, gallbladder, and pancreas. **B.** Overview of arterial supply. **C.** Overview of portal venous drainage.

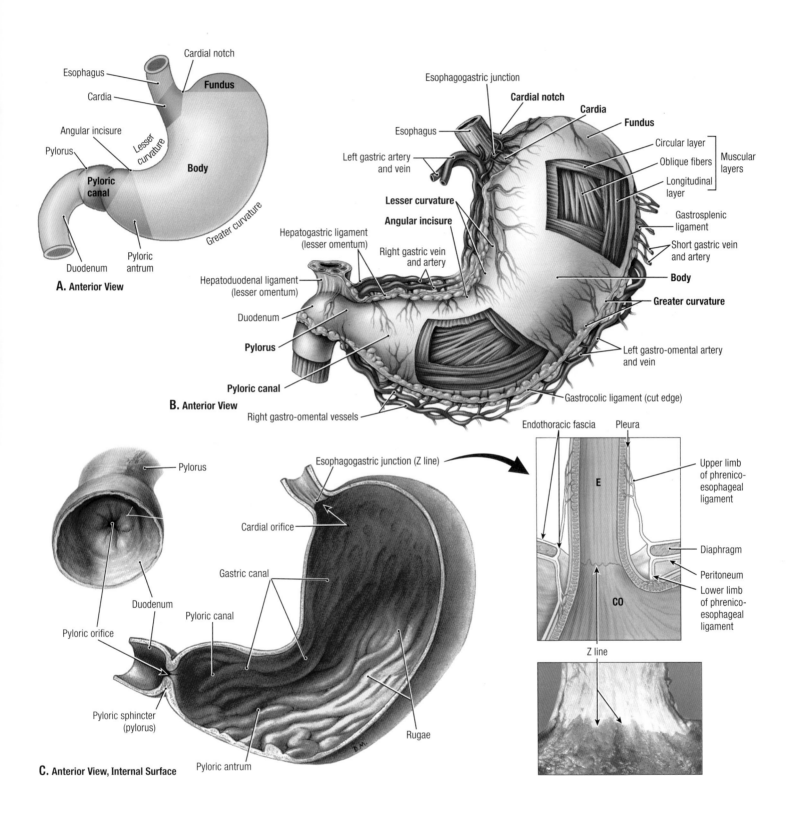

A. Anterior View

B. Anterior View

C. Anterior View, Internal Surface

STOMACH

4.32

A. Parts. **B.** External surface. **C.** Internal surface (mucous membrane), anterior wall removed. *Insets*: Left side of page—pylorus, viewed from the duodenum. Right side of page—details of the esophagogastric junction. The Z line is where the stratified squamous epithelium of the esophagus (white portion in photograph) changes to the simple columnar epithelium of the stomach (dark portion).

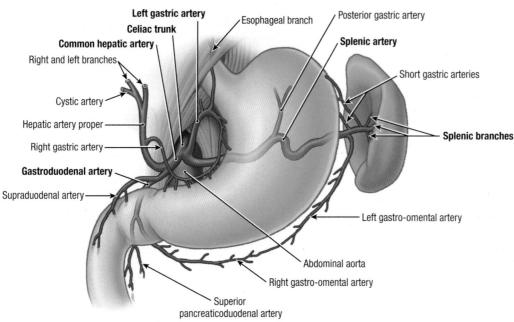

Left gastric artery

Celiac trunk

Common hepatic artery

Right and left branches

Cystic artery

Hepatic artery proper

Right gastric artery

Gastroduodenal artery

Supraduodenal artery

Esophageal branch

Posterior gastric artery

Splenic artery

Short gastric arteries

Splenic branches

Left gastro-omental artery

Abdominal aorta

Right gastro-omental artery

Superior pancreaticoduodenal artery

A. Anterior View

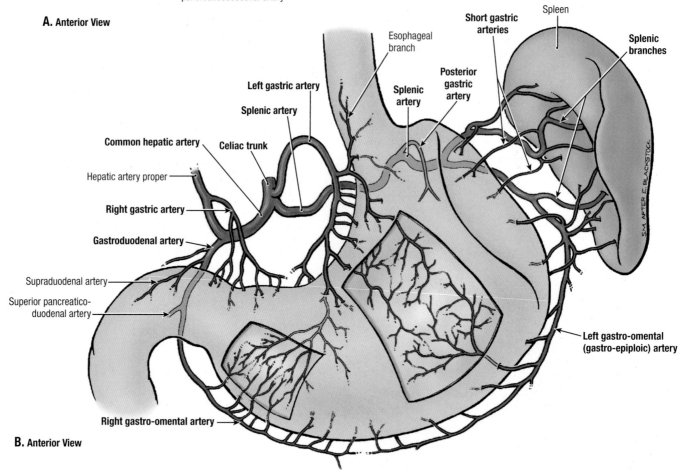

Esophageal branch

Short gastric arteries

Spleen

Splenic branches

Left gastric artery

Splenic artery

Posterior gastric artery

Common hepatic artery

Celiac trunk

Splenic artery

Hepatic artery proper

Right gastric artery

Gastroduodenal artery

Supraduodenal artery

Superior pancreatico-duodenal artery

Left gastro-omental (gastro-epiploic) artery

Right gastro-omental artery

B. Anterior View

4.33 CELIAC ARTERY

A. Branches of celiac trunk. The celiac trunk is a branch of the abdominal aorta, arising immediately inferior to the aortic hiatus of the diaphragm (T12 vertebral level). The vessel is usually 1 to 2 cm long and divides into the left gastric, common hepatic, and splenic arteries. The celiac trunk supplies the liver, gallbladder, inferior esophagus, stomach, pancreas, spleen, and duodenum. **B.** Arteries of stomach and spleen. The serous and muscular coats are removed from two areas of the stomach, revealing anastomotic networks in the submucous coat.

Five main sites where
esophagus is constricted:

1. Junction of pharynx
   and esophagus
   (in neck)

2. Aortic arch

3. Left main bronchus
   (at tracheal bifurcation)

4. Left atrium

5. Esophageal hiatus

A. Lateral View

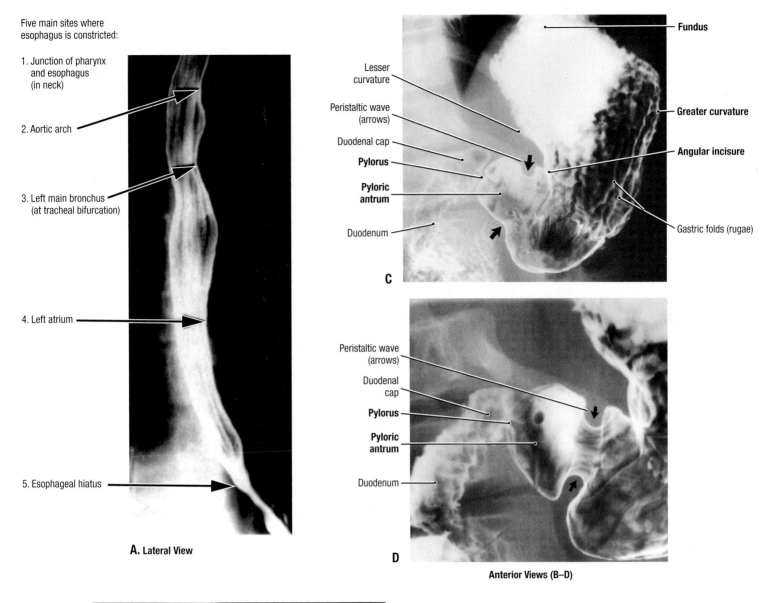

Fundus

Lesser
curvature

Greater curvature

Peristaltic wave
(arrows)

Duodenal cap

Pylorus

Angular incisure

**Pyloric
antrum**

Duodenum

Gastric folds (rugae)

C

Peristaltic wave
(arrows)

Duodenal
cap

Pylorus

**Pyloric
antrum**

Duodenum

D

Anterior Views (B–D)

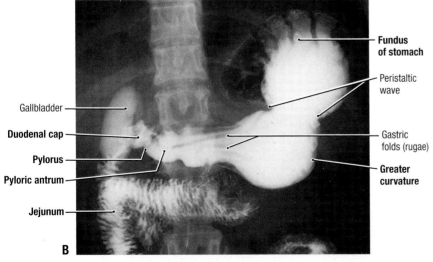

**Fundus
of stomach**

Peristaltic
wave

Gallbladder

Duodenal cap

Gastric
folds (rugae)

Pylorus

Pyloric antrum

**Greater
curvature**

Jejunum

B

RADIOGRAPHS OF ESOPHAGUS, STOMACH, DUODENUM (BARIUM SWALLOW)   4.34

A. Five sites of normal esophageal constriction. **B.** Stomach, small intestine, and gallbladder. Note additional contrast medium in gallbladder. **C.** Stomach and duodenum. **D.** Pyloric antrum and duodenal cap.

Blockage of esophagus. The impressions produced in the esophagus by adjacent structures are of clinical interest because of the slower passage of substances at these sites. The impressions indicate where swallowed foreign objects are most likely to lodge and where a stricture may develop, for example, after the accidental drinking of a caustic liquid, such as lye.

A hiatal (hiatus) hernia is a protrusion of a part of the stomach into the mediastinum through the esophageal hiatus of the diaphragm. The hernias occur most often in people after middle age, possibly because of weakening of the muscular part of the diaphragm and widening of the esophageal hiatus.

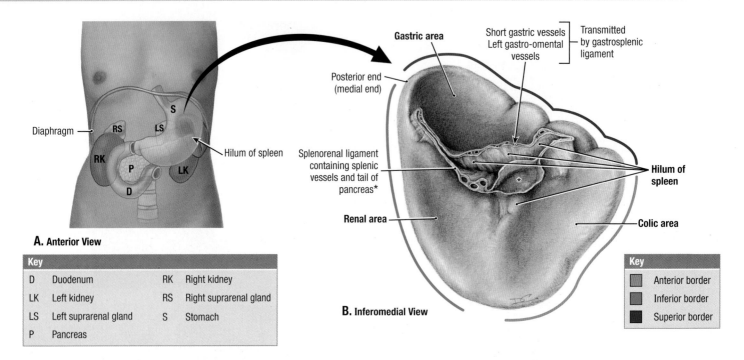

A. Anterior View

Key			
D	Duodenum	RK	Right kidney
LK	Left kidney	RS	Right suprarenal gland
LS	Left suprarenal gland	S	Stomach
P	Pancreas		

B. Inferomedial View

Key	
	Anterior border
	Inferior border
	Superior border

4.35   SPLEEN

A. The surface anatomy of the spleen. The spleen lies superficially in the left upper abdominal quadrant between the 9th and 11th ribs.
B. Note the impressions (colic, renal, and gastric areas) made by structures in contact with the spleen's visceral surface. Its superior border is notched.

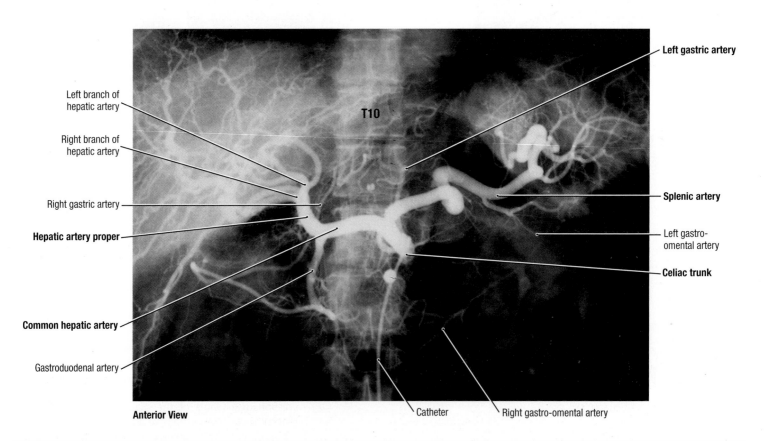

Anterior View

4.36   CELIAC ARTERIOGRAM

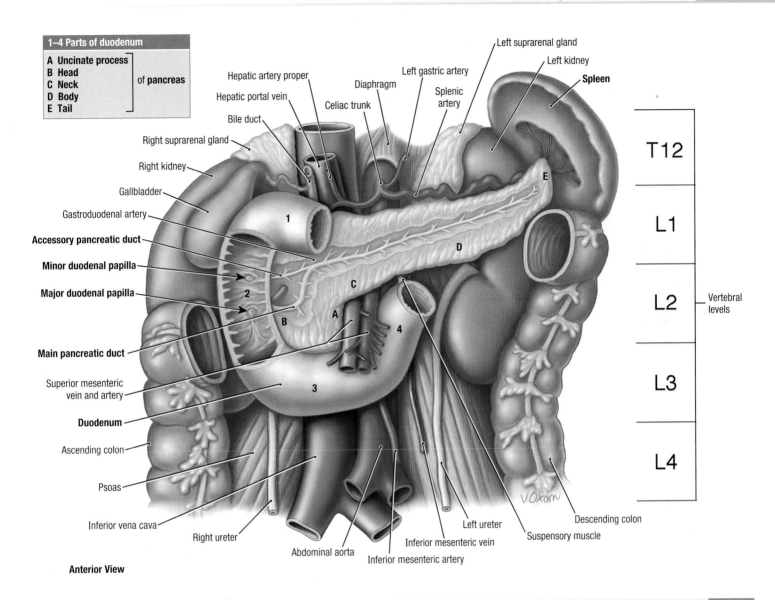

1–4 Parts of duodenum

A Uncinate process	
B Head	
C Neck	of **pancreas**
D Body	
E Tail	

Hepatic artery proper
Hepatic portal vein
Bile duct
Right suprarenal gland
Right kidney
Gallbladder
Gastroduodenal artery
Accessory pancreatic duct
Minor duodenal papilla
Major duodenal papilla
Main pancreatic duct
Superior mesenteric vein and artery
Duodenum
Ascending colon
Psoas
Inferior vena cava
Right ureter
Abdominal aorta
Inferior mesenteric artery
Inferior mesenteric vein
Left ureter
Suspensory muscle
Descending colon
Celiac trunk
Diaphragm
Left gastric artery
Splenic artery
Left suprarenal gland
Left kidney
Spleen

V. Oxorn

Vertebral levels: T12, L1, L2, L3, L4

Anterior View

PARTS AND RELATIONSHIPS OF PANCREAS AND DUODENUM   4.37

Pancreas and duodenum *in situ*.

TABLE 4.5	**PARTS AND RELATIONSHIPS OF DUODENUM**					
Part of Duodenum	*Anterior*	*Posterior*	*Medial*	*Superior*	*Inferior*	*Vertebral Level*
Superior (1st) part	Peritoneum Gallbladder Quadrate lobe of liver	Bile duct Gastroduodenal artery Hepatic portal vein IVC		Neck of gallbladder	Neck of pancreas	Anterolateral to L1 vertebra
Descending (2nd) part	Transverse colon Transverse mesocolon Coils of small intestine	Hilum of right kidney Renal vessels Ureter Psoas major	Head of pancreas Pancreatic duct Bile duct			Right of L2–L3 vertebrae
Inferior (horizontal or 3rd) part	Superior mesenteric artery Superior mesenteric vein Coils of small intestine	Right psoas major IVC Aorta Right ureter		Head and uncinate process of pancreas Superior mesenteric artery and vein		Anterior to L3 vertebra
Ascending (4th) part	Beginning of root of mesentery Coils of jejunum	Left psoas major Left margin of aorta	Superior mesenteric artery and vein	Body of pancreas		Left of L3 vertebra

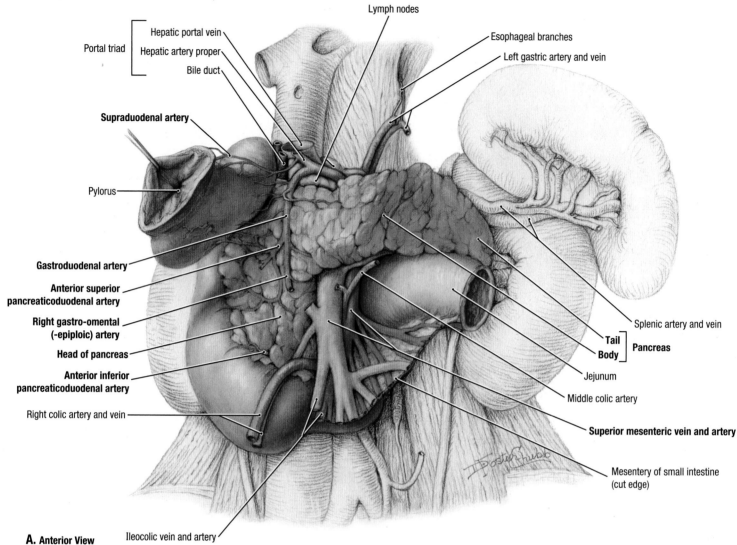

Lymph nodes

Esophageal branches

Left gastric artery and vein

Portal triad
Hepatic portal vein
Hepatic artery proper
Bile duct

Supraduodenal artery

Pylorus

Gastroduodenal artery

Anterior superior pancreaticoduodenal artery

Right gastro-omental (-epiploic) artery

Head of pancreas

Anterior inferior pancreaticoduodenal artery

Right colic artery and vein

Splenic artery and vein

Tail
Body } **Pancreas**

Jejunum

Middle colic artery

Superior mesenteric vein and artery

Mesentery of small intestine (cut edge)

Ileocolic vein and artery

A. Anterior View

4.38 VASCULAR RELATIONSHIPS OF PANCREAS AND DUODENUM

A. Anterior relationships. The gastroduodenal artery descends anterior to the neck of the pancreas.
B. Posterior relationships. The splenic artery and vein course on the posterior aspect of the pancreatic tail, which usually extends to the spleen. The pancreas "loops" around the right side of the superior mesenteric vessels so that its neck is anterior, its head is to the right, and its uncinate process is posterior to the vessels. The splenic and superior mesenteric veins unite posterior to the neck to form the hepatic portal vein. The bile duct descends in a fissure (opened up) in the posterior part of the head of the pancreas.

Most inflammatory erosions of the duodenal wall, **duodenal (peptic) ulcers**, are in the posterior wall of the superior (1st) part of the duodenum within 3 cm of the pylorus.

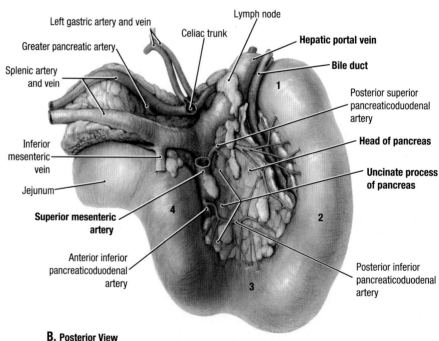

Left gastric artery and vein

Greater pancreatic artery

Celiac trunk

Lymph node

Hepatic portal vein

Splenic artery and vein

Bile duct

1

Posterior superior pancreaticoduodenal artery

Inferior mesenteric vein

Head of pancreas

Uncinate process of pancreas

Jejunum

2

Superior mesenteric artery

4

3

Anterior inferior pancreaticoduodenal artery

Posterior inferior pancreaticoduodenal artery

B. Posterior View

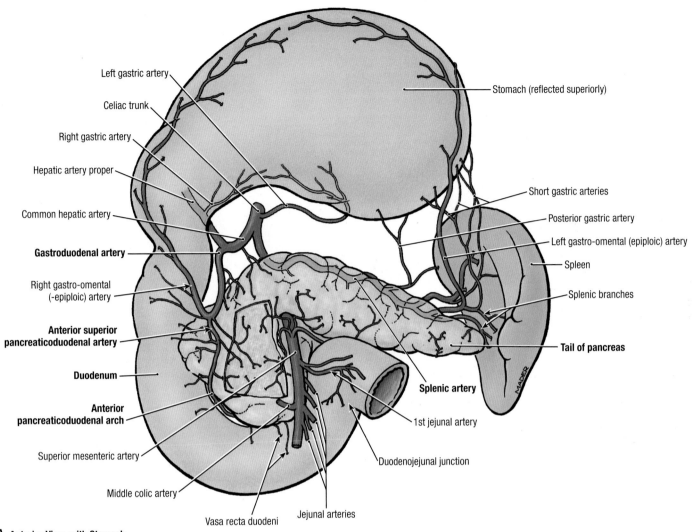

Left gastric artery

Celiac trunk

Right gastric artery

Hepatic artery proper

Common hepatic artery

Gastroduodenal artery

Right gastro-omental
(-epiploic) artery

**Anterior superior
pancreaticoduodenal artery**

Duodenum

**Anterior
pancreaticoduodenal arch**

Superior mesenteric artery

Middle colic artery

Vasa recta duodeni

Jejunal arteries

Stomach (reflected superiorly)

Short gastric arteries

Posterior gastric artery

Left gastro-omental (epiploic) artery

Spleen

Splenic branches

Tail of pancreas

Splenic artery

1st jejunal artery

Duodenojejunal junction

**A. Anterior View, with Stomach
Reflected Superiorly**

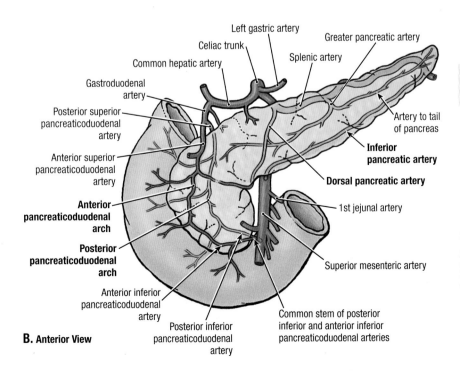

Left gastric artery

Celiac trunk

Common hepatic artery

Gastroduodenal
artery

Posterior superior
pancreaticoduodenal
artery

Anterior superior
pancreaticoduodenal
artery

**Anterior
pancreaticoduodenal
arch**

**Posterior
pancreaticoduodenal
arch**

Anterior inferior
pancreaticoduodenal
artery

Posterior inferior
pancreaticoduodenal
artery

Greater pancreatic artery

Splenic artery

Artery to tail
of pancreas

**Inferior
pancreatic artery**

Dorsal pancreatic artery

1st jejunal artery

Superior mesenteric artery

Common stem of posterior
inferior and anterior inferior
pancreaticoduodenal arteries

B. Anterior View

BLOOD SUPPLY TO THE PANCREAS, DUODENUM, AND SPLEEN `4.39`

A. Celiac trunk and superior mesenteric artery.

B. Pancreatic and pancreaticoduodenal arteries.

- The anterior superior pancreaticoduodenal artery from the gastroduodenal artery and the anterior inferior pancreaticoduodenal artery of the superior mesenteric artery form the anterior pancreaticoduodenal arch anterior to the head of the pancreas. The posterior superior and posterior inferior branches of the same two arteries form the posterior pancreaticoduodenal arch posterior to the pancreas. The anterior and posterior inferior arteries often arise from a common stem.

- Arteries supplying the pancreas are derived from the common hepatic artery, gastroduodenal artery, pancreaticoduodenal arches, splenic artery, and superior mesenteric artery.

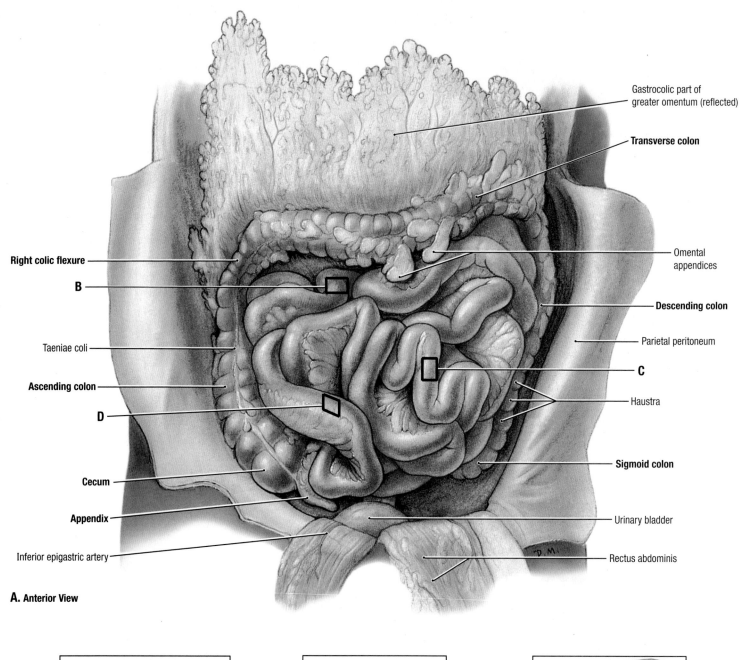

Gastrocolic part of
greater omentum (reflected)

Transverse colon

Right colic flexure

B

Taeniae coli

Ascending colon

D

Cecum

Appendix

Inferior epigastric artery

Omental
appendices

Descending colon

Parietal peritoneum

C

Haustra

Sigmoid colon

Urinary bladder

Rectus abdominis

A. Anterior View

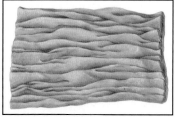

B. Proximal Jejunum

C. Proximal Ileum

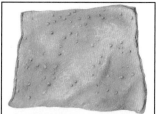

D. Distal Ileum

4.40   INTESTINES *IN SITU*, INTERIOR OF SMALL INTESTINE

A. Intestines *in situ,* greater omentum reflected. The ileum is reflected to expose the appendix. The appendix usually lies posterior to the cecum (retrocecal) or, as in this case, projects over the
pelvic brim. The features of the large intestines are the taeniae
coli, haustra, and omental appendices. **B.** Proximal jejunum. The
circular folds are tall, closely packed, and commonly branched.
C. Proximal ileum. The circular folds are low and becoming sparse.
The caliber of the gut is reduced, and the wall is thinner. **D.** Distal
ileum. Circular folds are absent, and solitary lymph nodules stud
the wall.

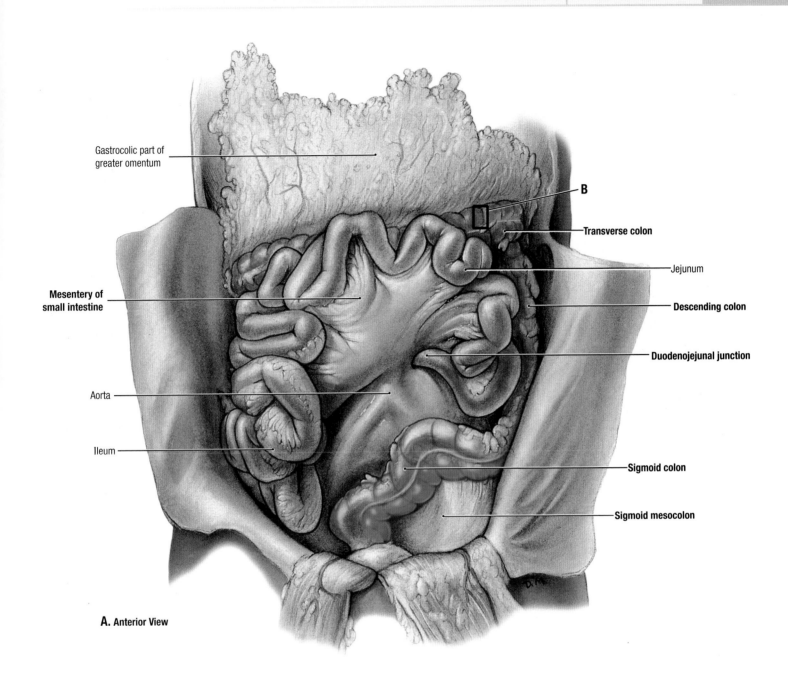

Gastrocolic part of
greater omentum

B

Transverse colon

Jejunum

Mesentery of
small intestine

Descending colon

Duodenojejunal junction

Aorta

Ileum

Sigmoid colon

Sigmoid mesocolon

A. Anterior View

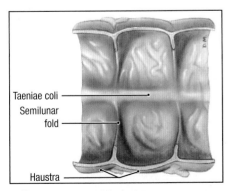

Taeniae coli
Semilunar
fold

Haustra

B. Transverse Colon

SIGMOID MESOCOLON AND MESENTERY OF SMALL INTESTINE, INTERIOR OF TRANSVERSE COLON

4.41

A. Sigmoid mesocolon and mesentery of the small intestine.
- The duodenojejunal junction is situated to the left of the median plane.
- The mesentery of the small intestine fans out extensively from its short root to accommodate the length of jejunum and ileum (~6 m).
- The descending colon is the narrowest part of the large intestine and is retroperitoneal. The sigmoid colon has a mesentery, the sigmoid mesocolon; the sigmoid colon is continuous with the rectum at the point at which the sigmoid mesocolon ends.

B. Transverse colon. The semilunar folds and taeniae coli form prominent features on the smooth-surfaced wall.

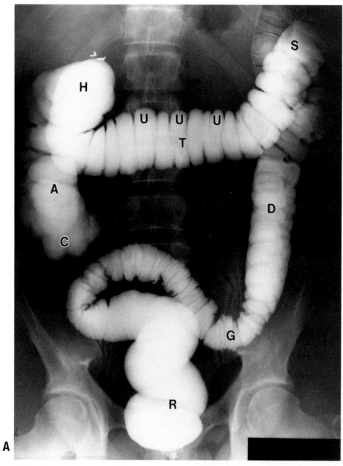

A

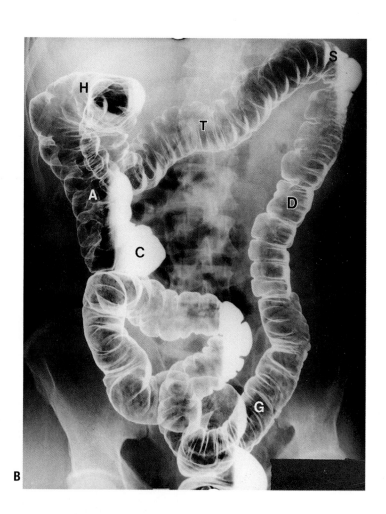

B

Postero-anterior Radiographs

Key					
A	Ascending colon	G	Sigmoid colon	S	Splenic flexure
C	Cecum	H	Hepatic flexure	T	Transverse colon
D	Descending colon	R	Rectum	U	Haustra

4.42   BARIUM ENEMA AND COLONOSCOPY OF COLON

A. Single-contrast study. A barium enema has filled the colon. **B.** Double-contrast study. Barium can be seen coating the walls of the colon, which is distended with air, providing a vivid view of the mucosal relief and haustra. **C.** The interior of the colon can be observed with an elongated endoscope, usually a fiber-optic flexible colonoscope. The endoscope is a tube that inserts into the colon through the anus and rectum. **D.** Diverticulo-sis of the colon can be photographed through a colonoscope. **E.** Diverticulosis is a disorder in which multiple false diverticula (external evaginations or outpocketings of the mucosa of the colon) develop along the intestine. It primarily affects middle-aged and elderly people. Diverticulosis is commonly (60%) found in the sigmoid colon. Diverticula are subject to infection and rupture, leading to **diverticulitis**, and they can distort and erode the nutrient arteries, leading to hemorrhage.

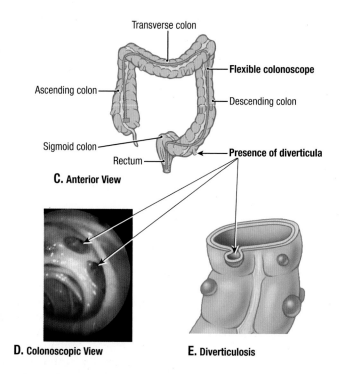

Transverse colon

Ascending colon

Sigmoid colon

Rectum

Flexible colonoscope

Descending colon

Presence of diverticula

C. Anterior View

D. Colonoscopic View

E. Diverticulosis

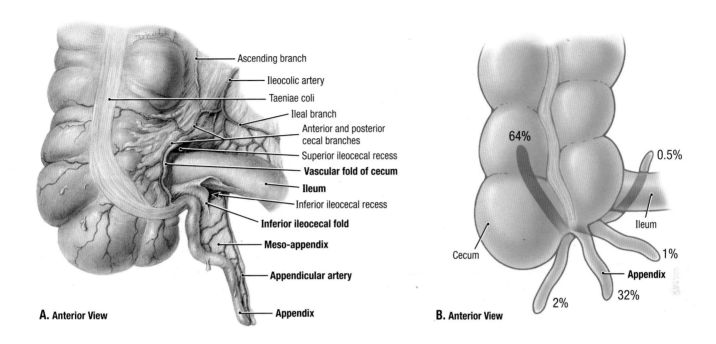

A. Anterior View

- Ascending branch
- Ileocolic artery
- Taeniae coli
- Ileal branch
- Anterior and posterior cecal branches
- Superior ileocecal recess
- **Vascular fold of cecum**
- **Ileum**
- Inferior ileocecal recess
- **Inferior ileocecal fold**
- **Meso-appendix**
- **Appendicular artery**
- **Appendix**

B. Anterior View

64%
0.5%
Ileum
Cecum
1%
Appendix
2%
32%

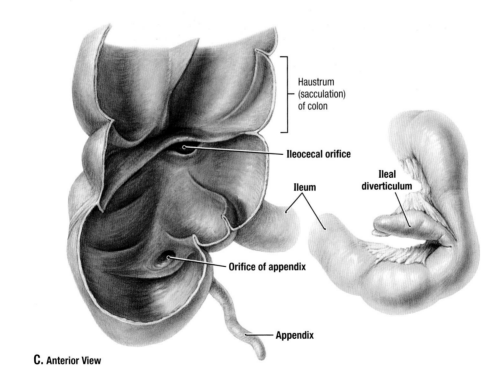

C. Anterior View

- Haustrum (sacculation) of colon
- **Ileocecal orifice**
- **Ileal diverticulum**
- **Ileum**
- **Orifice of appendix**
- **Appendix**

ILEOCECAL REGION AND APPENDIX

4.43

A. Blood supply. The appendicular artery is located in the free edge of the meso-appendix. The inferior ileocecal fold is bloodless, whereas the superior ileocecal fold is called the vascular fold of the cecum. **B.** The approximate incidence of various positions of the appendix. **C.** Interior of a dried cecum and ileal diverticulum (of Meckel). This cecum was filled with air until dry, opened, and varnished. Ileal diverticulum is a congenital anomaly that occurs in 1% to 2% of persons. It is a pouchlike remnant (3 to 6 cm long) of the proximal part of the yolk stalk, typically within 50 cm of the ileocecal junction. It sometimes becomes inflamed and produces pain that may mimic that produced by appendicitis.

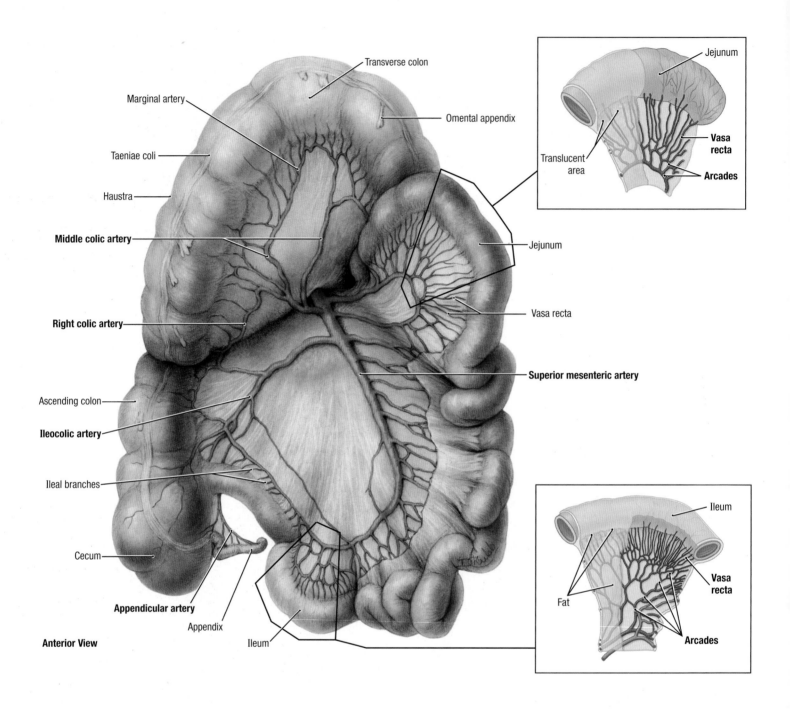

Transverse colon

Marginal artery

Taeniae coli

Haustra

Middle colic artery

Right colic artery

Ascending colon

Ileocolic artery

Ileal branches

Cecum

Appendicular artery

Appendix

Anterior View

Ileum

Omental appendix

Jejunum

Vasa recta

Superior mesenteric artery

Jejunum

Vasa recta

Arcades

Translucent area

Ileum

Vasa recta

Arcades

Fat

4.44   SUPERIOR MESENTERIC ARTERY AND ARTERIAL ARCADES

The peritoneum is partially stripped off.
- The superior mesenteric artery ends by anastomosing with one of its own branches, the ileal branch of the ileocolic artery.
- On the inset drawings of jejunum and ileum, compare the diameter, thickness of wall, number of arterial arcades, long or short vasa recta, presence of translucent (fat-free) areas at the mesenteric border, and fat encroaching on the wall of the gut between the jejunum and ileum.

- **Acute inflammation of the appendix** is a common cause of an acute abdomen (severe abdominal pain arising suddenly). The pain of appendicitis usually commences as a vague pain in the peri-umbilical region because afferent pain fibers enter the spinal cord at the T10 level. Later, severe pain in the right lower quadrant results from irritation of the parietal peritoneum lining the posterior abdominal wall.

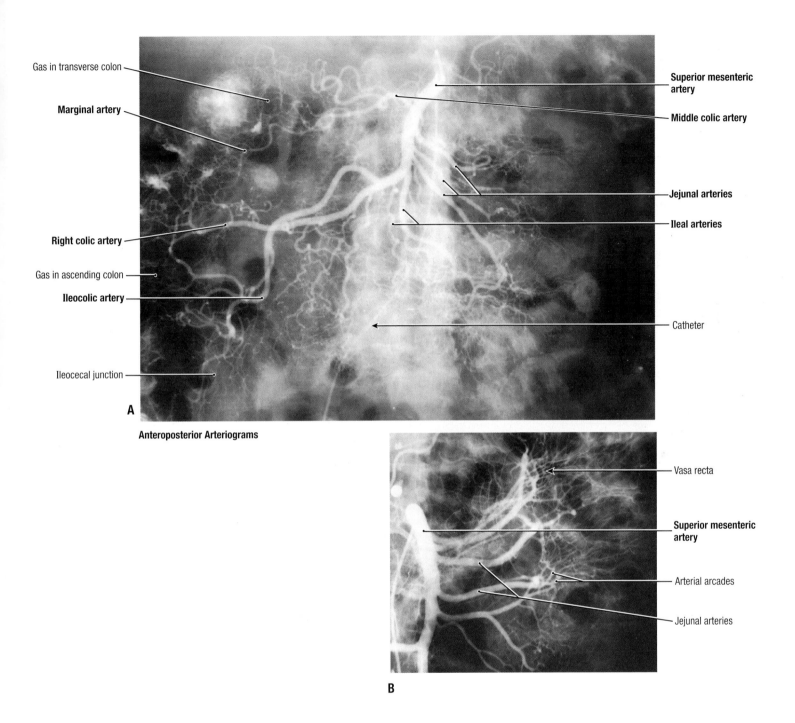

Gas in transverse colon

Marginal artery

Right colic artery

Gas in ascending colon

Ileocolic artery

Ileocecal junction

A

Anteroposterior Arteriograms

Superior mesenteric artery

Middle colic artery

Jejunal arteries

Ileal arteries

Catheter

Vasa recta

Superior mesenteric artery

Arterial arcades

Jejunal arteries

B

SUPERIOR MESENTERIC ARTERIOGRAMS

A. Branches of superior mesenteric artery. Consult Figure 4.44 to identify the branches. **B.** Enlargement to show the jejunal arteries, arterial arcades, and vasa recta.

- The branches of the superior mesenteric artery include, from its left side, 12 or more jejunal and ileal arteries that anastomose to form arcades from which vasa recta pass to the small intestine and, from its right side, the middle colic, ileocolic, and commonly (but not here) an independent right colic artery that anastomose to form a marginal artery that parallels the mesenteric border at the colon and from which vasa recta pass to the large intestine.

- **Occlusion of the vasa recta** by emboli results in ischemia of the part of the intestine concerned. If the ischemia is severe, necrosis of the involved segment results and **ileus** (obstruction of the intestine) of the paralytic type occurs. Ileus is accompanied by a severe colicky pain, along with abdominal distension, vomiting, and often fever and dehydration. If the condition is diagnosed early (e.g., using a superior mesenteric arteriogram), the obstructed part of the vessel may be cleared surgically.

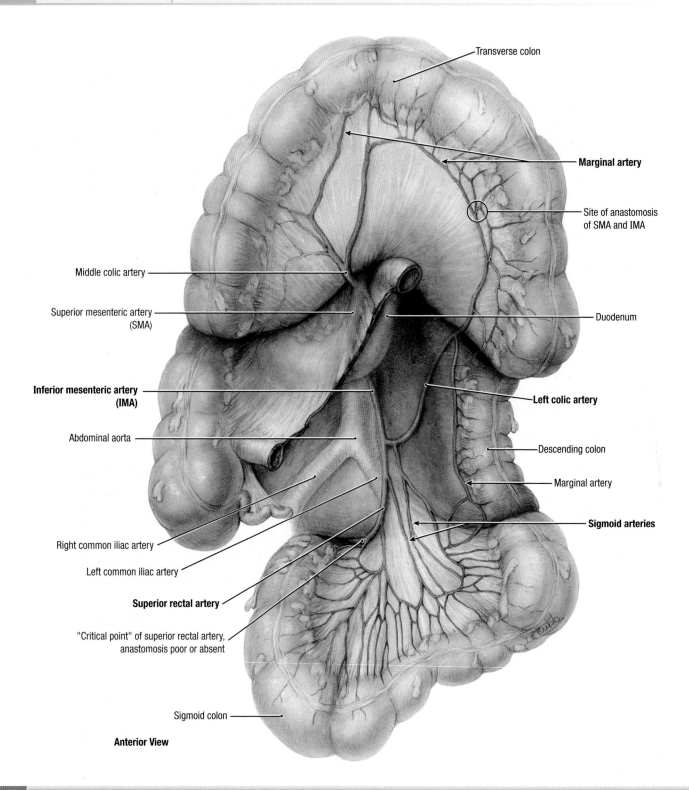

Transverse colon

Marginal artery

Site of anastomosis of SMA and IMA

Middle colic artery

Superior mesenteric artery (SMA)

Duodenum

Inferior mesenteric artery (IMA)

Left colic artery

Abdominal aorta

Descending colon

Marginal artery

Sigmoid arteries

Right common iliac artery

Left common iliac artery

Superior rectal artery

"Critical point" of superior rectal artery, anastomosis poor or absent

Sigmoid colon

Anterior View

4.46 | INFERIOR MESENTERIC ARTERY

The mesentery of the small intestine has been cut at its root.
- The inferior mesenteric artery arises posterior to the ascending part of the duodenum, about 4 cm superior to the bifurcation of the aorta; on crossing the left common iliac artery, it becomes the superior rectal artery.
- The branches of the inferior mesenteric artery include the left colic artery and several sigmoid arteries; the inferior two sigmoid arteries branch from the superior rectal artery.

- The point at which the last sigmoidal artery branches from the superior rectal artery is known as the "critical point" of the superior rectal artery; distal to this point, there are poor or no anastomotic connections between the superior rectal artery and the marginal artery.

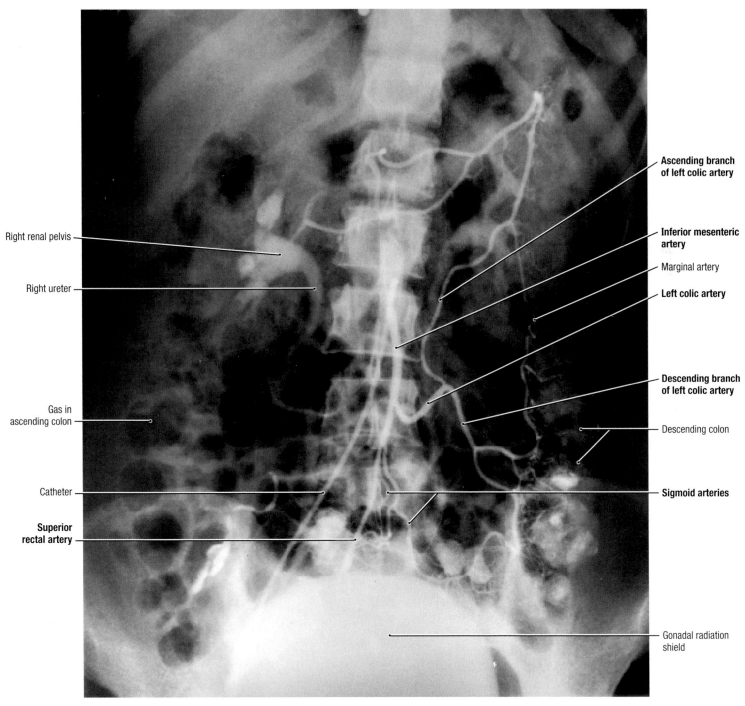

Ascending branch
of left colic artery

Inferior mesenteric
artery

Marginal artery

Left colic artery

Descending branch
of left colic artery

Descending colon

Sigmoid arteries

Gonadal radiation
shield

Right renal pelvis

Right ureter

Gas in
ascending colon

Catheter

Superior
rectal artery

Postero-anterior Arteriogram

INFERIOR MESENTERIC ARTERIOGRAM

4.47

- The left colic artery courses to the left toward the descending colon and splits into ascending and descending branches.
- The sigmoid arteries, two to four in number, supply the sigmoid colon.
- The superior rectal artery, which is the continuation of the inferior mesenteric artery, supplies the rectum; the superior rectal anastomoses are formed by branches of the middle and inferior rectal arteries (from the internal iliac artery).

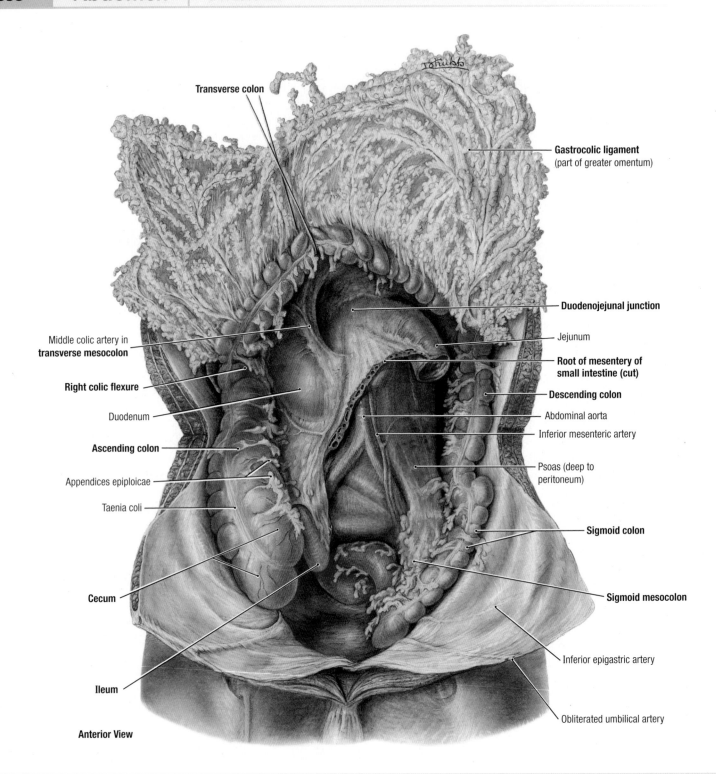

Transverse colon

Gastrocolic ligament
(part of greater omentum)

Duodenojejunal junction

Jejunum

**Root of mesentery of
small intestine (cut)**

Descending colon

Abdominal aorta

Inferior mesenteric artery

Psoas (deep to
peritoneum)

Sigmoid colon

Sigmoid mesocolon

Inferior epigastric artery

Obliterated umbilical artery

Middle colic artery in
transverse mesocolon

Right colic flexure

Duodenum

Ascending colon

Appendices epiploicae

Taenia coli

Cecum

Ileum

Anterior View

4.48 PERITONEUM OF POSTERIOR ABDOMINAL CAVITY

The gastrocolic ligament is retracted superiorly, along with the transverse colon and transverse mesocolon. The appendix had been surgically removed. This dissection is continued in Figure 4.49.
- The root of the mesentery of the small intestine, approximately 15 to 20 cm in length, extends between the duodenojejunal junction and ileocecal junction.
- The large intestine forms 3½ sides of a square, "framing" the jejunum and ileum. On the right are the cecum and ascending colon, superior is the transverse colon, on the left

is the descending and sigmoid colon, and inferiorly is the sigmoid colon.
- **Chronic inflammation of the colon (ulcerative colitis, Crohn disease)** is characterized by severe inflammation and ulceration of the colon and rectum. In some patients, a colectomy is performed, during which the terminal ileum and colon as well as the rectum and anal canal are removed. An ileostomy is then constructed to establish an artificial cutaneous opening between the ileum and the skin of the anterolateral abdominal wall.

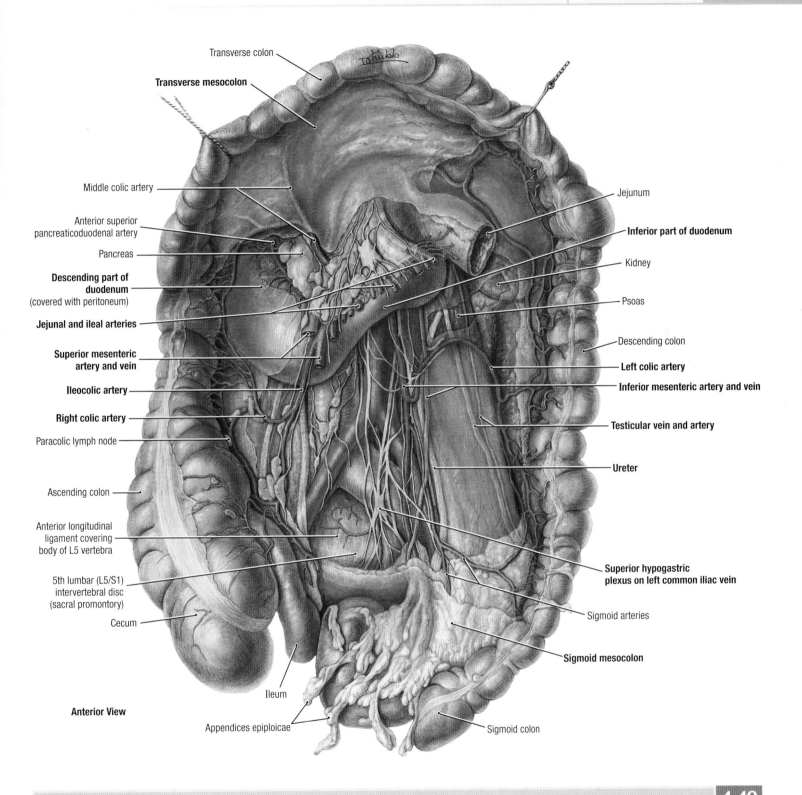

Transverse colon

Transverse mesocolon

Middle colic artery

Anterior superior pancreaticoduodenal artery

Pancreas

Descending part of duodenum (covered with peritoneum)

Jejunal and ileal arteries

Superior mesenteric artery and vein

Ileocolic artery

Right colic artery

Paracolic lymph node

Ascending colon

Anterior longitudinal ligament covering body of L5 vertebra

5th lumbar (L5/S1) intervertebral disc (sacral promontory)

Cecum

Anterior View

Appendices epiploicae

Ileum

Jejunum

Inferior part of duodenum

Kidney

Psoas

Descending colon

Left colic artery

Inferior mesenteric artery and vein

Testicular vein and artery

Ureter

Superior hypogastric plexus on left common iliac vein

Sigmoid arteries

Sigmoid mesocolon

Sigmoid colon

POSTERIOR ABDOMINAL CAVITY WITH PERITONEUM REMOVED

4.49

The jejunal and ileal branches (cut) pass from the left side of the superior mesenteric artery. The right colic artery here is a branch of the ileocolic artery. This is the same specimen as in Figure 4.48.
- The duodenum is larger in diameter before crossing the superior mesenteric vessels and narrow afterward.
- On the right side, there are lymph nodes on the colon, paracolic nodes beside the colon, and nodes along the ileocolic artery, which drain into nodes anterior to the pancreas.

- The intestines and intestinal vessels lie on a resectable plane (remnant of the embryological dorsal mesentery) anterior to that of the testicular vessels; these, in turn, lie anterior to the plane of the kidney, its vessels, and the ureter.
- The superior hypogastric plexus lie inferior to the bifurcation of the aorta and anterior to the left common iliac vein, the body of the 5th lumbar vertebra, and the 5th intervertebral disc.

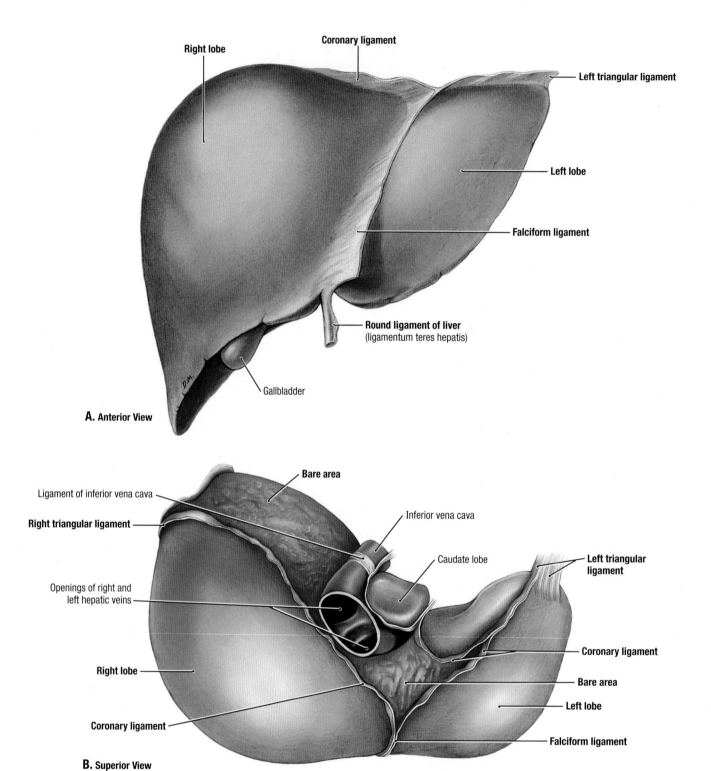

A. Anterior View

Right lobe

Coronary ligament

Left triangular ligament

Left lobe

Falciform ligament

Round ligament of liver
(ligamentum teres hepatis)

Gallbladder

B. Superior View

Bare area

Ligament of inferior vena cava

Right triangular ligament

Inferior vena cava

Caudate lobe

Left triangular
ligament

Openings of right and
left hepatic veins

Right lobe

Coronary ligament

Bare area

Left lobe

Coronary ligament

Falciform ligament

4.50 DIAPHRAGMATIC (ANTERIOR AND SUPERIOR) SURFACE OF LIVER

A. The falciform ligament has been severed close to its attachment to the diaphragm and anterior abdominal wall and demarcates the right and left lobes of the liver. The round ligament of the liver (ligamentum teres) lies within the free edge of the falciform ligament.

B. The two layers of peritoneum that form the falciform ligament separate over the superior aspect (surrounding the bare area) of the liver to form the superior layer of the coronary ligament and the right and left triangular ligaments.

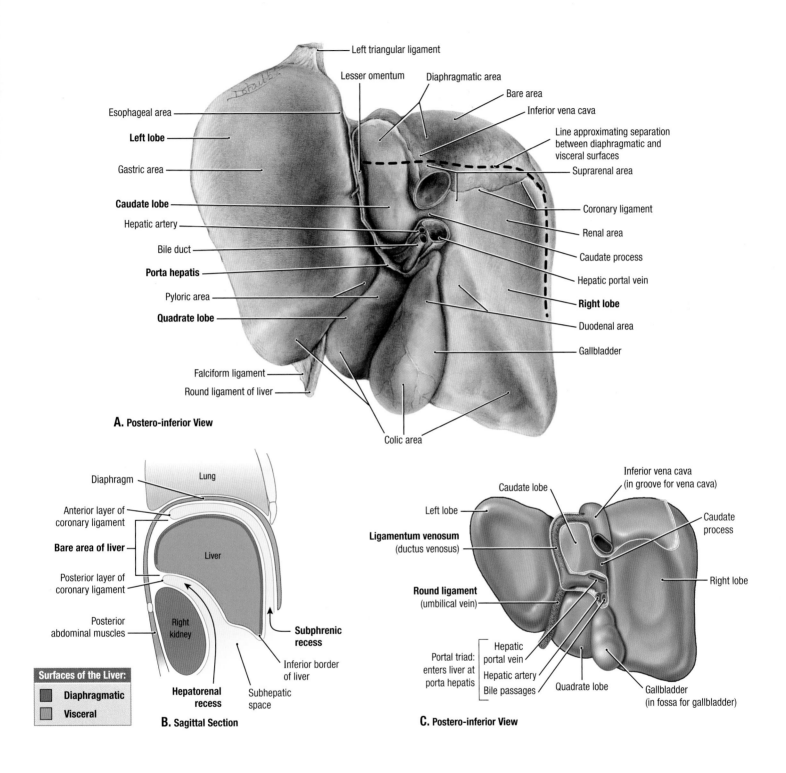

A. Postero-inferior View

Left triangular ligament
Lesser omentum
Diaphragmatic area
Bare area
Inferior vena cava
Esophageal area
Left lobe
Gastric area
Caudate lobe
Hepatic artery
Bile duct
Porta hepatis
Pyloric area
Quadrate lobe
Falciform ligament
Round ligament of liver
Line approximating separation between diaphragmatic and visceral surfaces
Suprarenal area
Coronary ligament
Renal area
Caudate process
Hepatic portal vein
Right lobe
Duodenal area
Gallbladder
Colic area

B. Sagittal Section

Diaphragm
Lung
Anterior layer of coronary ligament
Bare area of liver
Liver
Posterior layer of coronary ligament
Posterior abdominal muscles
Right kidney
Subphrenic recess
Inferior border of liver
Hepatorenal recess
Subhepatic space

Surfaces of the Liver:
Diaphragmatic
Visceral

C. Postero-inferior View

Caudate lobe
Inferior vena cava (in groove for vena cava)
Left lobe
Caudate process
Ligamentum venosum (ductus venosus)
Right lobe
Round ligament (umbilical vein)
Portal triad: enters liver at porta hepatis
Hepatic portal vein
Hepatic artery
Bile passages
Quadrate lobe
Gallbladder (in fossa for gallbladder)

VISCERAL (POSTERO-INFERIOR) SURFACE OF LIVER

4.51

A. Isolated specimen demonstrating lobes, and impressions of adjacent viscera. **B.** Hepatic surfaces and peritoneal recesses. **C.** Round ligament of liver and ligamentum venosum. The round ligament of liver includes the obliterated remains of the umbilical vein that carried well-oxygenated blood from the placenta to the fetus. The ligamentum venosum is the fibrous remnant of the fetal ductus venosus that shunted blood from the umbilical vein to the inferior vena cava by passing the liver. Hepatic tissue may be obtained for diagnostic purposes by **liver biopsy**. The needle puncture is commonly made through the right 10th intercostal space in the midaxillary line. Before the physician takes the biopsy, the person is asked to hold his or her breath in full expiration to minimize the costodiaphragmatic recess and to lessen the possibility of damaging the lung and contaminating the pleural cavity.

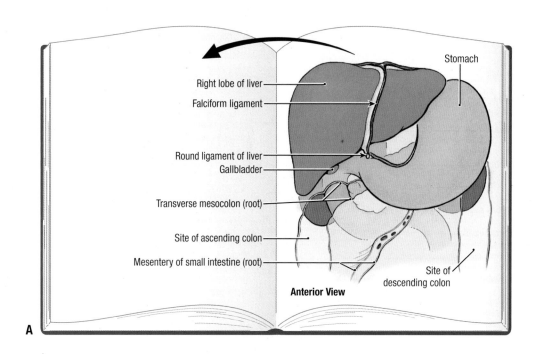

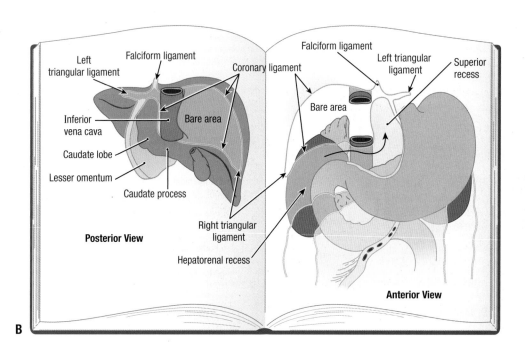

LIVER AND ITS POSTERIOR RELATIONS, SCHEMATIC ILLUSTRATION

A. Liver *in situ*. The jejunum, ileum, and the ascending, transverse, and descending colons have been removed. **B.** The liver is drawn schematically on a page in a book, so that as the page is turned (*arrow* in **A**), the liver is reflected to the right to reveal its posterior surface, and on the facing page, the posterior relations that compose the bed of the liver are viewed. The *arrow* **(B)** traverses the omental (epiploic) foramen to enter the omental bursa and its superior recess (*arrowhead*).

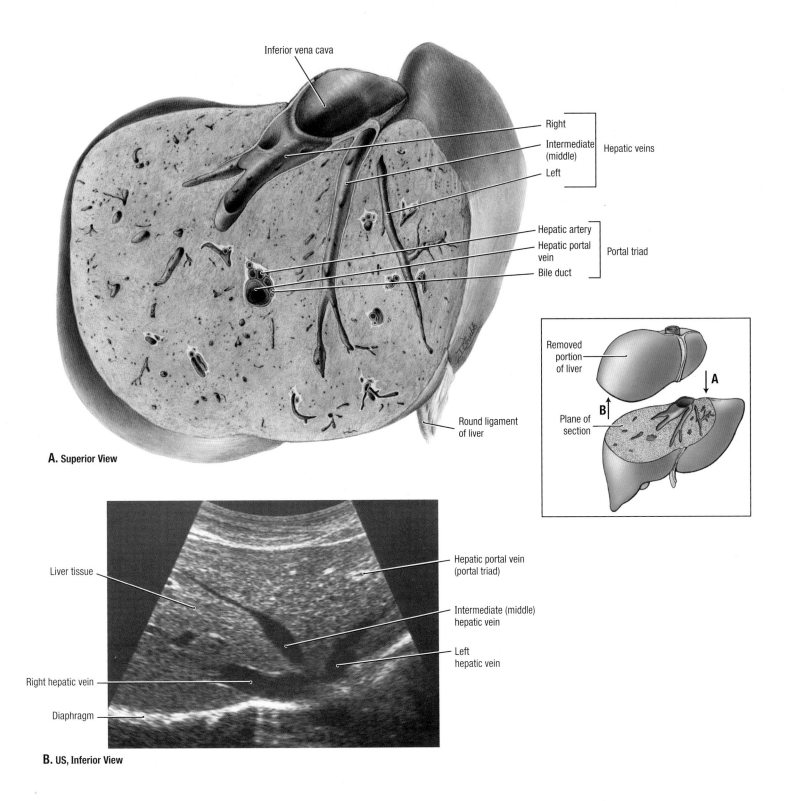

Inferior vena cava

Right

Intermediate (middle)

Left

Hepatic veins

Hepatic artery

Hepatic portal vein

Bile duct

Portal triad

Round ligament of liver

A. Superior View

Removed portion of liver

Plane of section

Liver tissue

Hepatic portal vein (portal triad)

Intermediate (middle) hepatic vein

Left hepatic vein

Right hepatic vein

Diaphragm

B. US, Inferior View

HEPATIC VEINS

4.53

A. Approximately horizontal section of liver with the posterior aspect at the top of page. Note the multiple perivascular fibrous capsules sectioned throughout the cut surface, each containing a portal triad (the hepatic portal vein, hepatic artery, bile ductules) plus lymph vessels. Interdigitating with these are branches of the three main hepatic veins (right, intermediate, and left), which, unaccompanied and lacking capsules, converge on the inferior vena cava. **B.** Ultrasound scan. The transducer was placed under the costal margin and directed posteriorly, producing an inverted image **(A)**.

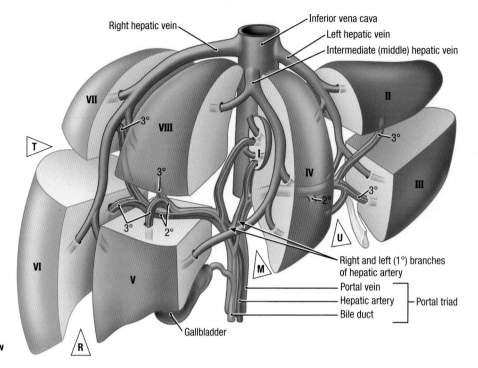

Right hepatic vein

Inferior vena cava

Left hepatic vein

Intermediate (middle) hepatic vein

VII

VIII

3°

T

I

II

IV

III

3°

3°

2°

3°

2°

U

VI

V

M

R

Key

M = Main portal fissure
R = Right portal fissure
T = Transverse hepatic plane
U = Umbilical fissure
2° = Secondary branches of portal triad structures
3° = Tertiary branches of portal triad structures

Right and left (1°) branches of hepatic artery

Portal vein ⎤
Hepatic artery ⎬ Portal triad
Bile duct ⎦

Gallbladder

A. Anterior View

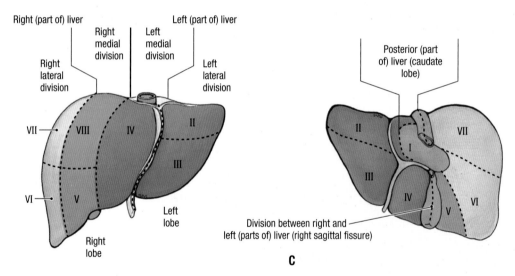

Right (part of) liver

Right medial division

Left medial division

Left (part of) liver

Right lateral division

Left lateral division

VII

VIII

IV

II

III

VI

V

Left lobe

Right lobe

B

Posterior (part of) liver (caudate lobe)

II

I

VII

III

IV

V

VI

Division between right and left (parts of) liver (right sagittal fissure)

C

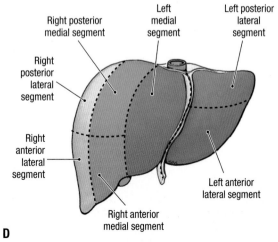

Right posterior medial segment

Left medial segment

Left posterior lateral segment

Right posterior lateral segment

Right anterior lateral segment

Left anterior lateral segment

Right anterior medial segment

D

Left posterior lateral segment

Posterior (caudate) segment

Right posterior lateral segment

Right anterior lateral segment

Left anterior lateral segment

Left medial segment

Right anterior medial segment

E

Anterior Views (B, D)   **Postero-inferior Views (C, E)**

HEPATIC SEGMENTATION (continued)

Each segment is supplied by a secondary or tertiary branch of the hepatic artery, bile duct, and portal vein. The hepatic veins interdigitate between the portal triads and are intersegmental in that they drain adjacent segments. Since the right and left hepatic arteries and ducts and branches of the right and left portal veins do not communicate, it is possible to perform **hepatic lobectomies** (removal of the right or left part of the liver) and **segmentectomies**. Each segment can be identified numerically or by name (Table 4.6).

TABLE 4.6	SCHEMA OF TERMINOLOGY FOR SUBDIVISIONS OF LIVER						
Anatomical Term	*Right Lobe*		*Left Lobe*			*Caudate Lobe*	
Functional/surgical term*a*	Right (part of) liver [Right portal lobe*b*]		Left (part of) liver [Left portal lobe*c*]			Posterior (part of) liver	
	Right lateral division	Right medial division	Left medial division		Left lateral division	[Right caudate lobe*b*]	[Left caudate lobe*c*]
	Posterior lateral segment **Segment VII** [Posterior superior area]	Posterior medial segment **Segment VIII** [Anterior superior area]	[Medial superior area]	Lateral segment **Segment II** [Lateral superior area]		Posterior segment **Segment I**	
			Left medial segment **Segment IV**				
	Right anterior lateral segment **Segment VI** [Posterior inferior area]	Anterior medial segment **Segment V** [Anterior inferior area]	[Medial inferior area = quadrate lobe]	Left anterior lateral segment **Segment III** [Lateral inferior area]			

*a*The labels in the table and figure above reflect the *Terminologia Anatomica: International Anatomical Terminology*. Previous terminology is in brackets.
*b,c*Under the schema of the previous terminology, the caudate lobe was divided into right and left halves, and *b*the right half of the caudate lobe was considered a subdivision of the right portal lobe; *c*the left half of the caudate lobe was considered a subdivision of the left portal lobe.

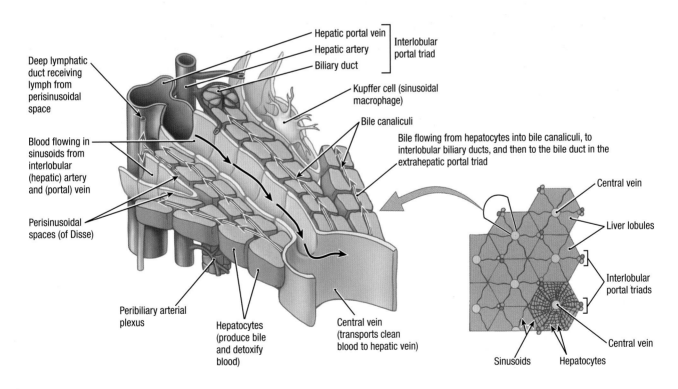

Deep lymphatic duct receiving lymph from perisinusoidal space

Blood flowing in sinusoids from interlobular (hepatic) artery and (portal) vein

Perisinusoidal spaces (of Disse)

Peribiliary arterial plexus

Hepatocytes (produce bile and detoxify blood)

Central vein (transports clean blood to hepatic vein)

Hepatic portal vein
Hepatic artery
Biliary duct
} Interlobular portal triad

Kupffer cell (sinusoidal macrophage)

Bile canaliculi

Bile flowing from hepatocytes into bile canaliculi, to interlobular biliary ducts, and then to the bile duct in the extrahepatic portal triad

Central vein

Liver lobules

Interlobular portal triads

Central vein

Sinusoids   Hepatocytes

FLOW OF BLOOD AND BILE IN THE LIVER

This small part of a liver lobule shows the components of the interlobular portal triad and the positioning of the sinusoids and bile canaliculi (*right*). The cut surface of the liver shows the hexagonal pattern of the lobules.

- With the exception of lipids, every substance absorbed by the alimentary tract is received first by the liver via the hepatic portal vein. In addition to its many metabolic activities, the liver stores glycogen and secretes bile.
- There is progressive destruction of hepatocytes in **cirrhosis of the liver** and replacement of them by fibrous tissue. This tissue surrounds the intrahepatic blood vessels and biliary ducts, making the liver firm and impeding circulation of blood through it.

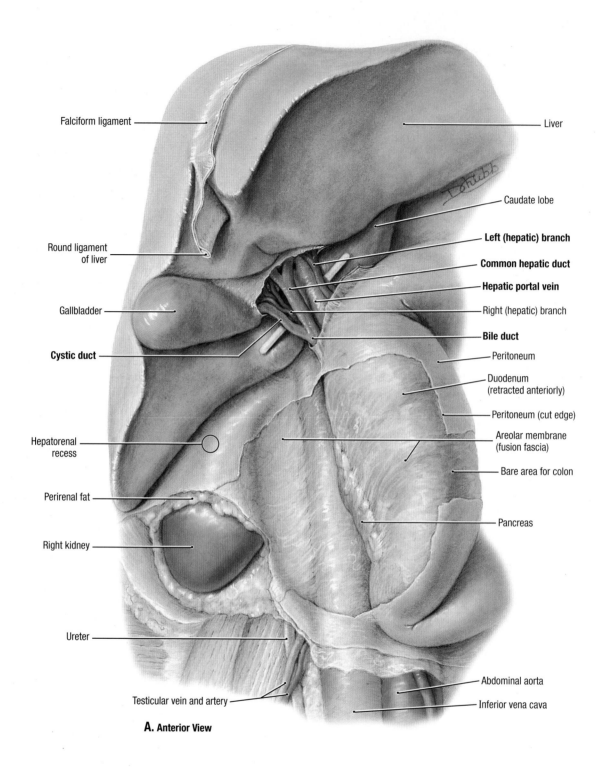

Falciform ligament

Liver

Caudate lobe

Round ligament
of liver

Left (hepatic) branch

Common hepatic duct

Hepatic portal vein

Gallbladder

Right (hepatic) branch

Cystic duct

Bile duct

Peritoneum

Duodenum
(retracted anteriorly)

Peritoneum (cut edge)

Hepatorenal
recess

Areolar membrane
(fusion fascia)

Bare area for colon

Perirenal fat

Pancreas

Right kidney

Ureter

Abdominal aorta

Testicular vein and artery

Inferior vena cava

A. Anterior View

4.56 EXPOSURE OF THE PORTAL TRIAD IN HEPATODUODENAL LIGAMENT

A. The hepatoduodenal ligament (hepatic pedicle) includes the portal triad consisting of the hepatic portal vein (posteriorly), the hepatic artery proper (ascending from the left), and the bile passages (descending to the right). Here, the hepatic artery proper is replaced by a left hepatic branch, arising directly from the common hepatic artery, and a right hepatic branch, arising from the superior mesenteric artery (a common variation). A rod traverses the omental (epiploic) foramen. The lesser omentum and transverse colon are removed, and the peritoneum is cut along the right border of the duodenum; this part of the duodenum is retracted anteriorly. The space opened up reveals two smooth areolar membranes (fusion fascia) normally applied to each other that are vestiges of the embryonic peritoneum originally covering these surfaces.

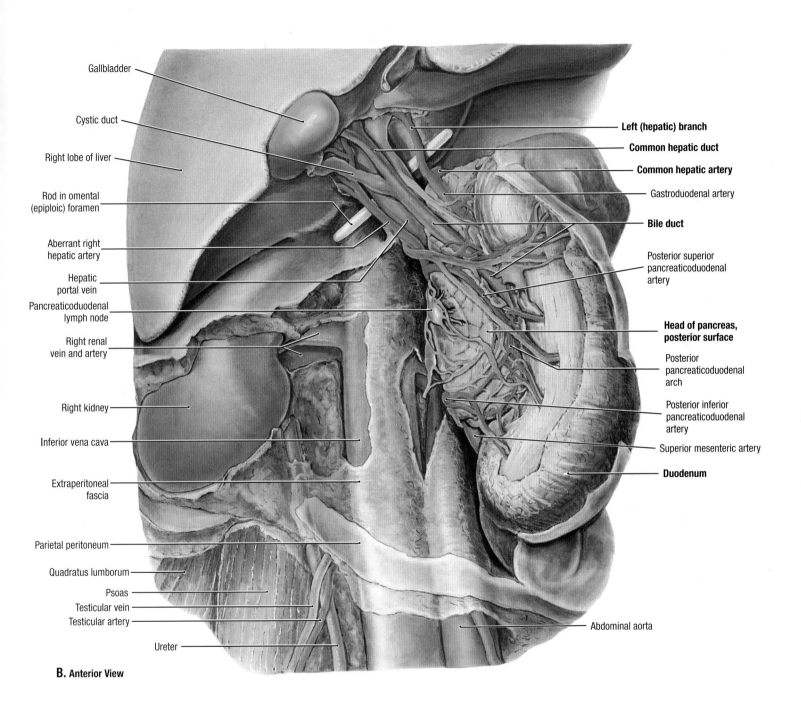

Gallbladder

Cystic duct

Right lobe of liver

Rod in omental (epiploic) foramen

Aberrant right hepatic artery

Hepatic portal vein

Pancreaticoduodenal lymph node

Right renal vein and artery

Right kidney

Inferior vena cava

Extraperitoneal fascia

Parietal peritoneum

Quadratus lumborum

Psoas

Testicular vein

Testicular artery

Ureter

Left (hepatic) branch

Common hepatic duct

Common hepatic artery

Gastroduodenal artery

Bile duct

Posterior superior pancreaticoduodenal artery

Head of pancreas, posterior surface

Posterior pancreaticoduodenal arch

Posterior inferior pancreaticoduodenal artery

Superior mesenteric artery

Duodenum

Abdominal aorta

B. Anterior View

EXPOSURE OF THE PORTAL TRIAD IN HEPATODUODENAL LIGAMENT (*continued*) | 4.56

B. Continuing the dissection, the secondarily retroperitoneal viscera (duodenum and head of the pancreas) are retracted anteriorly and to the left. The areolar membrane (fusion fascia) covering the posterior aspect of the pancreas and duodenum is largely removed, and that covering the anterior aspect of the great vessels is partly removed.

A common method for **reducing portal hypertension** is to divert blood from the portal venous system to the systemic venous system by creating a communication between the portal vein and the inferior vena cava (IVC). This **portacaval anastomosis** or **portosystemic shunt** may be created where these vessels lie close to each other posterior to the liver.

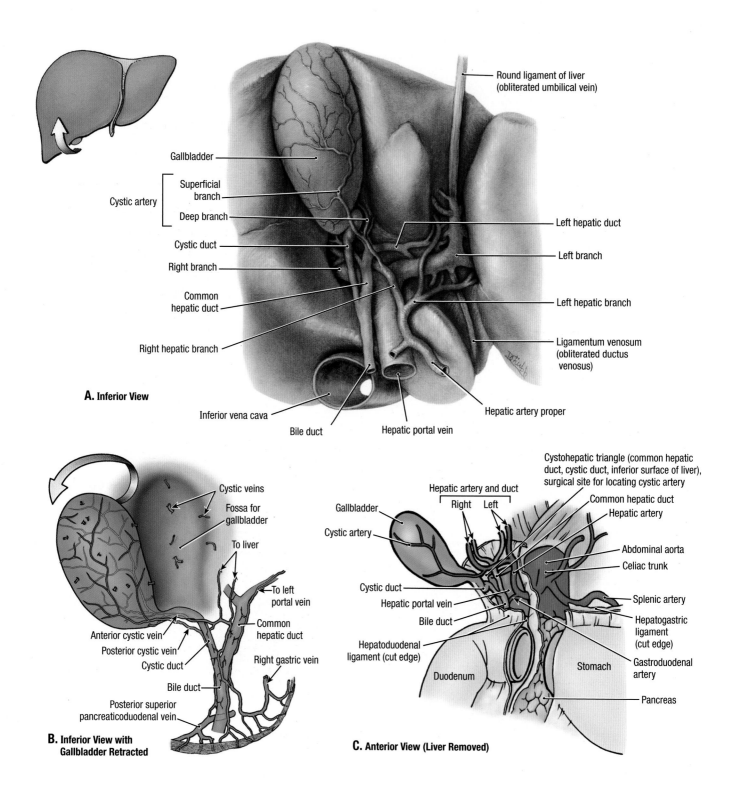

A. Inferior View

Round ligament of liver (obliterated umbilical vein)

Gallbladder

Cystic artery
- Superficial branch
- Deep branch

Cystic duct

Right branch

Common hepatic duct

Right hepatic branch

Left hepatic duct

Left branch

Left hepatic branch

Ligamentum venosum (obliterated ductus venosus)

Inferior vena cava

Bile duct

Hepatic portal vein

Hepatic artery proper

B. Inferior View with Gallbladder Retracted

Cystic veins

Fossa for gallbladder

To liver

To left portal vein

Common hepatic duct

Right gastric vein

Anterior cystic vein

Posterior cystic vein

Cystic duct

Bile duct

Posterior superior pancreaticoduodenal vein

C. Anterior View (Liver Removed)

Cystohepatic triangle (common hepatic duct, cystic duct, inferior surface of liver), surgical site for locating cystic artery

Hepatic artery and duct
- Right
- Left

Common hepatic duct

Hepatic artery

Gallbladder

Cystic artery

Cystic duct

Hepatic portal vein

Bile duct

Hepatoduodenal ligament (cut edge)

Duodenum

Abdominal aorta

Celiac trunk

Splenic artery

Hepatogastric ligament (cut edge)

Gastroduodenal artery

Stomach

Pancreas

4.57 GALLBLADDER AND STRUCTURES OF PORTA HEPATIS

A. Gallbladder, cystic artery, and extrahepatic bile ducts. The inferior border of the liver is elevated to demonstrate its visceral surface (as in orientation figure). **B.** Venous drainage of the gallbladder and extrahepatic ducts. Most veins are tributaries of the hepatic portal vein, but some drain directly to the liver. **C.** Portal triad within the hepatoduodenal ligament (free edge of lesser omentum).

Gallstones are concretions in the gallbladder or extrahepatic biliary ducts. The cystohepatic (hepatobiliary) triangle (Calot), between the common hepatic duct, cystic duct, and liver, is an important endoscopic landmark for locating the cystic artery during **cholecystectomy**.

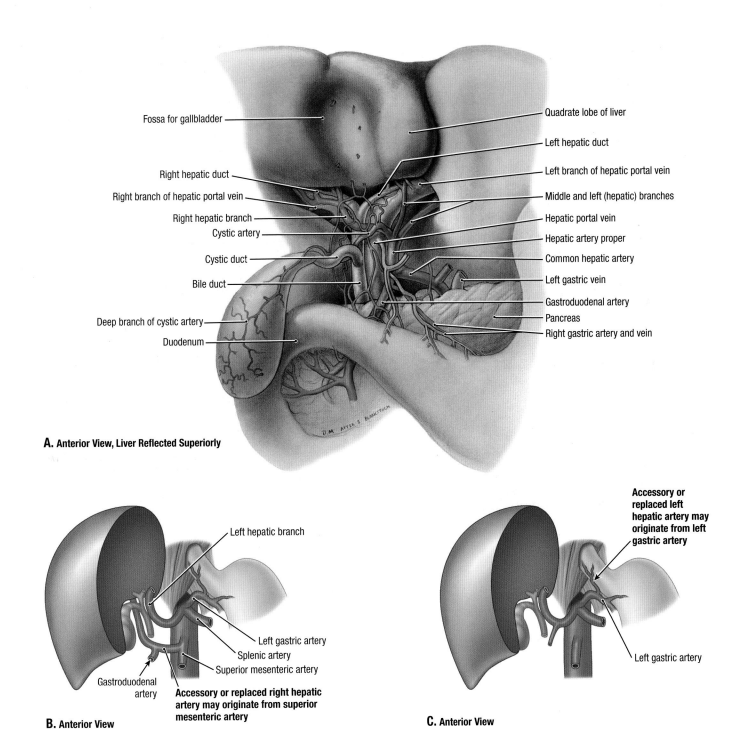

Fossa for gallbladder

Right hepatic duct

Right branch of hepatic portal vein

Right hepatic branch

Cystic artery

Cystic duct

Bile duct

Deep branch of cystic artery

Duodenum

Quadrate lobe of liver

Left hepatic duct

Left branch of hepatic portal vein

Middle and left (hepatic) branches

Hepatic portal vein

Hepatic artery proper

Common hepatic artery

Left gastric vein

Gastroduodenal artery

Pancreas

Right gastric artery and vein

A. Anterior View, Liver Reflected Superiorly

Left hepatic branch

Left gastric artery

Splenic artery

Superior mesenteric artery

Gastroduodenal artery

Accessory or replaced right hepatic artery may originate from superior mesenteric artery

B. Anterior View

Accessory or replaced left hepatic artery may originate from left gastric artery

Left gastric artery

C. Anterior View

VESSELS IN PORTA HEPATIS

A. Hepatic and cystic vessels. The liver is reflected superiorly. The gallbladder, freed from its bed or fossa, has remained nearly in its anatomical position, pulled slightly to the right. The deep branch of the cystic artery on the deep, or attached, surface of the gallbladder anastomoses with branches of the superficial branch of the cystic artery and sends twigs into the bed of the gallbladder.

Veins (not all shown) accompany most arteries. **B.** Aberrant (accessory or replaced) right hepatic artery. **C.** Aberrant left hepatic artery.

Awareness of the variations in arteries and bile duct formation is important for surgeons when they ligate the cystic duct during **cholecystectomy** (removal of the gallbladder).

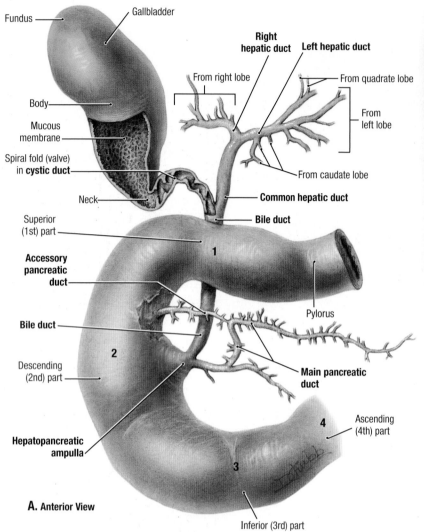

Fundus
Gallbladder
Body
Mucous membrane
Spiral fold (valve) in **cystic duct**
Neck
Superior (1st) part
Accessory pancreatic duct
Bile duct
Descending (2nd) part
Hepatopancreatic ampulla

Right hepatic duct
Left hepatic duct
From right lobe
From quadrate lobe
From left lobe
From caudate lobe
Common hepatic duct
Bile duct
1
Pylorus
2
Main pancreatic duct
4
Ascending (4th) part
3
Inferior (3rd) part

A. Anterior View

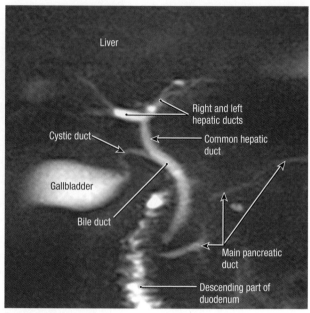

Liver
Right and left hepatic ducts
Cystic duct
Common hepatic duct
Gallbladder
Bile duct
Main pancreatic duct
Descending part of duodenum

C. Magnetic Resonance Cholangiopancreatography (MRCP)

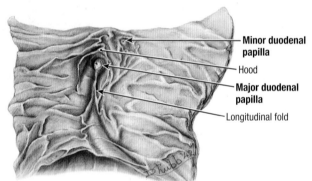

Minor duodenal papilla
Hood
Major duodenal papilla
Longitudinal fold

D. Internal View

Accessory pancreatic duct
Bile duct
Bile duct
1
Minor duodenal papilla
2 wall removed
Major duodenal papilla
B
A
4
Main pancreatic duct
Superior mesenteric vein and artery
E
D
C
3

B. Anterior View

Key

1–4 Parts of duodenum
Parts of pancreas:
A Uncinate process (extends posterior to superior mesenteric vein)
B Head   **D** Body
C Neck   **E** Tail

4.59 BILE AND PANCREATIC DUCTS

A. and **B.** Extrahepatic bile passages and pancreatic ducts. **C.** Magnetic resonance cholangiopancreatography (MRCP) demonstrating the bile and pancreatic ducts. The right and left hepatic ducts collect bile from the liver; the common hepatic duct unites with the cystic duct superior to the duodenum to form the bile duct, which descends posterior to the superior (1st) part of the duodenum. **D.** Interior of the descending (2nd) part of the duodenum. The bile duct joins the main pancreatic duct, forming the hepatopancreatic ampulla, which opens on the major duodenal papilla. This opening is the narrowest part of the biliary passages and is the common site for **impaction of a gallstone**. Gallstones may produce biliary colic (pain in the epigastric region). The accessory pancreatic duct opens on the minor duodenal papilla.

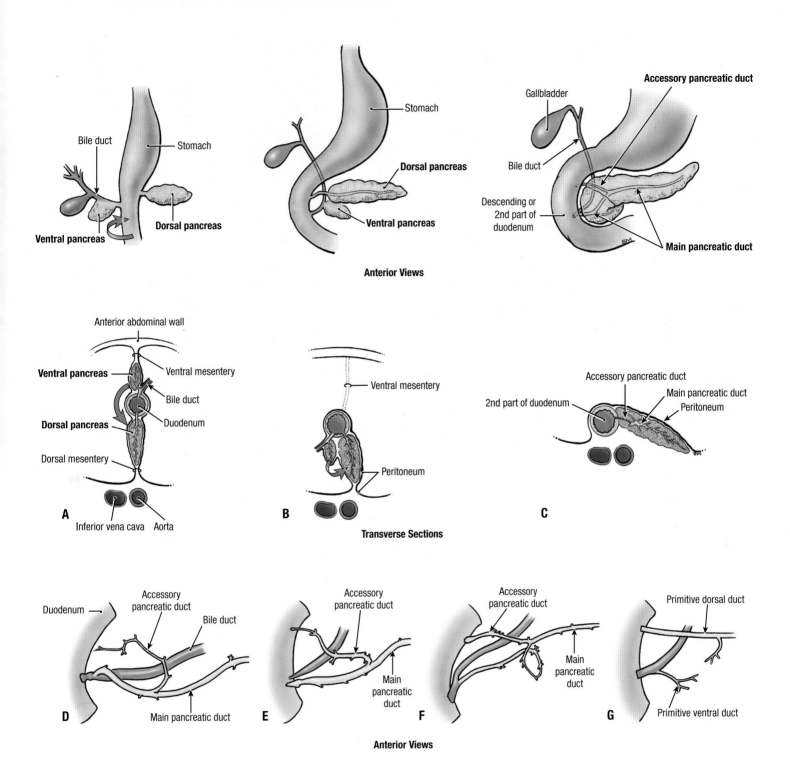

Anterior Views

Transverse Sections

Anterior Views

DEVELOPMENT AND VARIABILITY OF THE PANCREATIC DUCTS

4.60

A–C. Anterior views *(upper row)* and transverse sections *(middle row)* of the stages in the development of the pancreas. **A.** The small, primitive ventral bud arises in common with the bile duct, and a larger, primitive dorsal bud arises independently from the duodenum. **B.** The 2nd, or descending, part of the duodenum rotates on its long axis, which brings the ventral bud and bile duct posterior to the dorsal bud. **C.** A connecting segment unites the dorsal duct to the ventral duct, whereupon the duodenal end of the dorsal duct atrophies, and the direction of flow within it is reversed. **D–G.** Common variations of the pancreatic duct. **D.** An accessory duct that has lost its connection with the duodenum. **E.** An accessory duct that is large enough to relieve an obstructed main duct. **F.** An accessory duct that could probably substitute for the main duct. **G.** A persisting primitive dorsal duct unconnected to the primitive ventral duct.

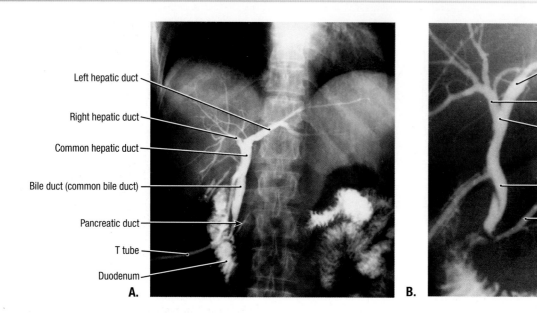

Left hepatic duct

Right hepatic duct

Common hepatic duct

Bile duct (common bile duct)

Pancreatic duct

T tube

Duodenum

A.

Left hepatic duct

Right hepatic duct

Common hepatic duct

Bile duct

Pancreatic duct (partially filled)

B.

4.61 RADIOGRAPHS OF BILIARY PASSAGES

After a cholecystectomy (removal of the gallbladder), contrast medium was injected with a T tube inserted into the bile passages. The biliary passages are visualized in the superior abdomen (**A**) and are more localized in **B**.

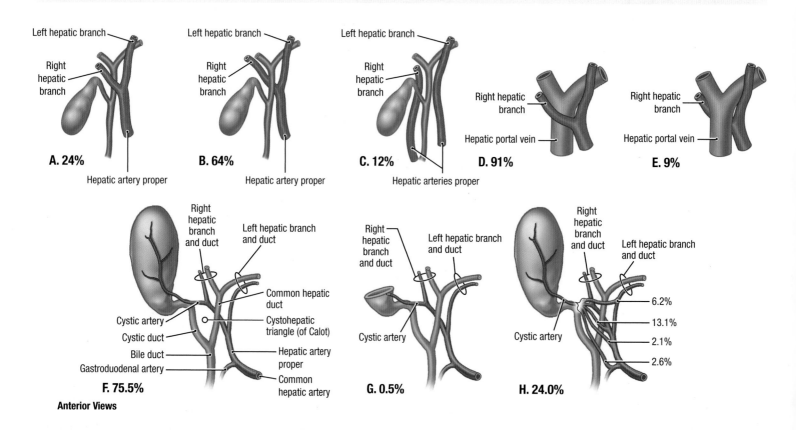

Left hepatic branch

Right hepatic branch

Hepatic artery proper

A. 24%

Left hepatic branch

Right hepatic branch

Hepatic artery proper

B. 64%

Left hepatic branch

Right hepatic branch

Hepatic arteries proper

C. 12%

Right hepatic branch

Hepatic portal vein

D. 91%

Right hepatic branch

Hepatic portal vein

E. 9%

Right hepatic branch and duct

Left hepatic branch and duct

Common hepatic duct

Cystohepatic triangle (of Calot)

Hepatic artery proper

Common hepatic artery

Cystic artery

Cystic duct

Bile duct

Gastroduodenal artery

F. 75.5%

Right hepatic branch and duct

Left hepatic branch and duct

Cystic artery

G. 0.5%

Right hepatic branch and duct

Left hepatic branch and duct

Cystic artery

6.2%

13.1%

2.1%

2.6%

H. 24.0%

Anterior Views

4.62 VARIATIONS IN HEPATIC AND CYSTIC ARTERIES

In a study of 165 cadavers in Dr. Grant's laboratory, five patterns were observed. **A.** Right hepatic artery crossing anterior to bile passages, 24%. **B.** Right hepatic artery crossing posterior to bile passages, 64%. **C.** Aberrant artery arising from the superior mesenteric artery, 12%. The artery crossed anterior (**D**) to the portal vein in 91% and posterior (**E**) in 9%. The cystic artery usually arises from the right hepatic artery in the angle between the common hepatic duct and cystic duct (see cystohepatic triangle, Fig. 4.57A), without crossing the common hepatic duct (**F** and **G**). However, when it arises on the left of the bile passages, it almost always crosses anterior to the passages (**H**).

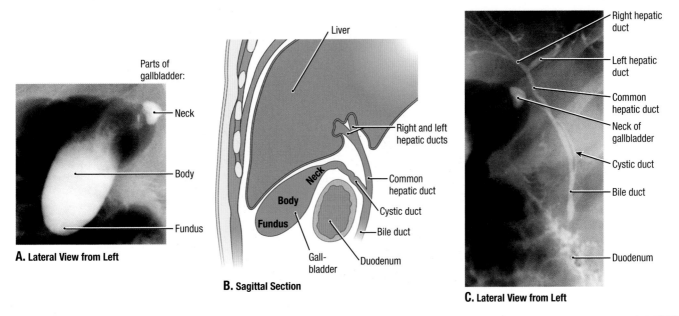

A. Lateral View from Left

Parts of gallbladder:
— Neck
— Body
— Fundus

B. Sagittal Section

Liver

Right and left hepatic ducts

Neck

Body

Fundus

Common hepatic duct

Cystic duct

Bile duct

Gall-bladder

Duodenum

C. Lateral View from Left

Right hepatic duct

Left hepatic duct

Common hepatic duct

Neck of gallbladder

Cystic duct

Bile duct

Duodenum

GALLBLADDER AND EXTRAHEPATIC BILIARY DUCTS

4.63

A. Gallbladder demonstrated by endoscopic retrograde cholangiography (ERCP). **B.** Relationships to superior part of duodenum. **C.** ERCP of bile passages.

Endoscopic retrograde cholangiography (ERCP) is done by first passing a fiberoptic endoscope through the mouth, esophagus, and stomach. Then the duodenum is entered, and a cannula is inserted into the major duodenal papilla and advanced under fluoroscopic guidance into the duct of choice (bile duct or pancreatic duct) for injection of radiographic contrast medium.

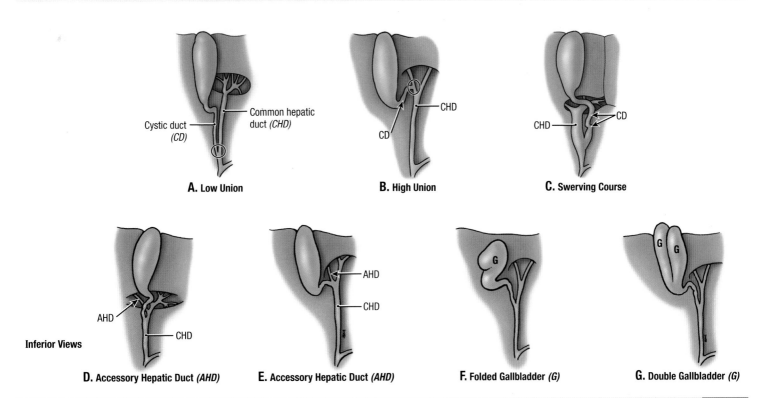

Cystic duct (CD) — Common hepatic duct (CHD)

A. Low Union

CHD — CD

B. High Union

CHD — CD

C. Swerving Course

Inferior Views

AHD — CHD

D. Accessory Hepatic Duct (AHD)

AHD — CHD

E. Accessory Hepatic Duct (AHD)

G

F. Folded Gallbladder (G)

G G

G. Double Gallbladder (G)

VARIATIONS OF CYSTIC AND HEPATIC DUCTS AND GALLBLADDER

4.64

The cystic duct usually lies on the right side of the common hepatic duct, joining it just above the superior (first) part of the duodenum, but this varies **(A–C)**. Of 95 gallbladders and bile passages studied in Dr. Grant's laboratory, 7 had accessory ducts. Of these, four joined the common hepatic duct near the cystic duct **(D)**, two joined the cystic duct **(E)**, and one was an anastomosing duct connecting the cystic with the common hepatic duct. **F.** Folded gallbladder. **G.** Double gallbladder.

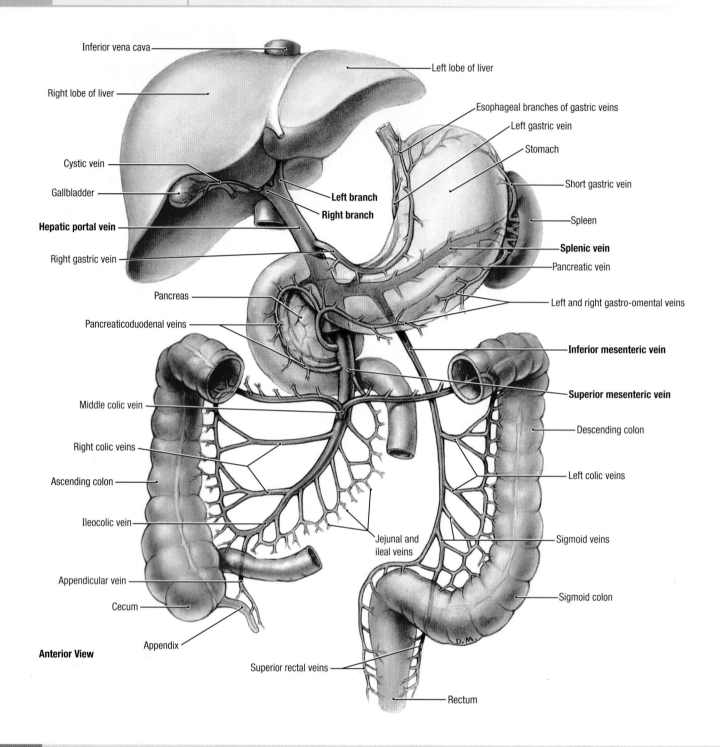

Inferior vena cava

Right lobe of liver

Cystic vein

Gallbladder

Hepatic portal vein

Right gastric vein

Pancreas

Pancreaticoduodenal veins

Middle colic vein

Right colic veins

Ascending colon

Ileocolic vein

Appendicular vein

Cecum

Appendix

Anterior View

Left lobe of liver

Esophageal branches of gastric veins

Left gastric vein

Stomach

Short gastric vein

Left branch

Right branch

Spleen

Splenic vein

Pancreatic vein

Left and right gastro-omental veins

Inferior mesenteric vein

Superior mesenteric vein

Descending colon

Left colic veins

Jejunal and ileal veins

Sigmoid veins

Sigmoid colon

Superior rectal veins

Rectum

4.65   PORTAL VENOUS SYSTEM

- The hepatic portal vein drains venous blood from the gastrointestinal tract, spleen, pancreas, and gallbladder to the sinusoids of the liver; from here, the blood is conveyed to the systemic venous system by the hepatic veins that drain directly to the inferior vena cava.
- The hepatic portal vein forms posterior to the neck of the pancreas by the union of the superior mesenteric and splenic veins, with the inferior mesenteric vein joining at or near the angle of union.
- The splenic vein drains blood from the inferior mesenteric, left gastro-omental (epiploic), short gastric, and pancreatic veins.

- The right gastro-omental, pancreaticoduodenal, jejunal, ileal, right, and middle colic veins drain into the superior mesenteric vein.
- The inferior mesenteric vein commences in the rectal plexus as the superior rectal vein and, after crossing the common iliac vessels, becomes the inferior mesenteric vein; branches include the sigmoid and left colic veins.
- The hepatic portal vein divides into right and left branches at the porta hepatis. The left branch carries mainly, but not exclusively, blood from the inferior mesenteric, gastric, and splenic veins, and the right branch carries blood mainly from the superior mesenteric vein.

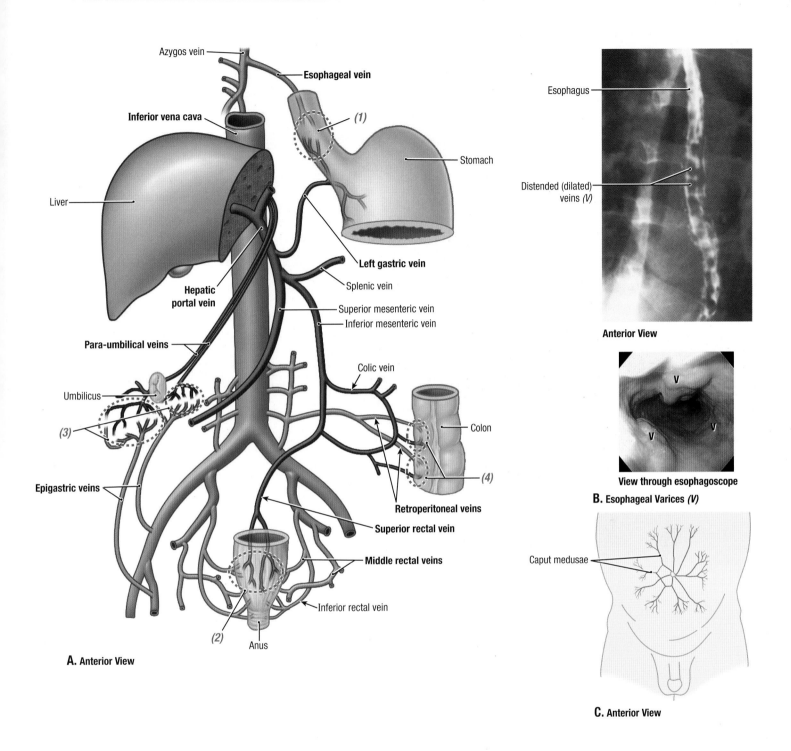

A. Anterior View

Anterior View

View through esophagoscope

B. Esophageal Varices *(V)*

C. Anterior View

PORTACAVAL SYSTEM

4.66

A. Portacaval system. In this diagram, portal tributaries *(dark blue)*, and systemic tributaries and communicating veins *(light blue)*. In **portal hypertension** (as in hepatic cirrhosis), the portal blood cannot pass freely through the liver, and the portocaval anastomoses become engorged, dilated, or even varicose; as a consequence, these veins may rupture. The sites of the portocaval anastomosis shown are between *(1)* esophageal veins draining into the azygos vein (systemic) and left gastric vein (portal), which when dilated are esophageal varices; *(2)* the inferior and middle rectal veins, draining into the inferior vena cava (systemic) and the superior rectal vein continuing as the inferior mesenteric vein (portal) (hemorrhoids result if the vessels are dilated); *(3)* paraumbilical veins (portal) and small epigastric veins of the anterior abdominal wall (systemic), which when varicose form "caput medusae" (so named because of the resemblance of the radiating veins to the serpents on the head of Medusa, a character in Greek mythology); and *(4)* twigs of colic veins (portal) anastomosing with systemic retroperitoneal veins. **B.** Esophageal varices. **C.** Caput medusae.

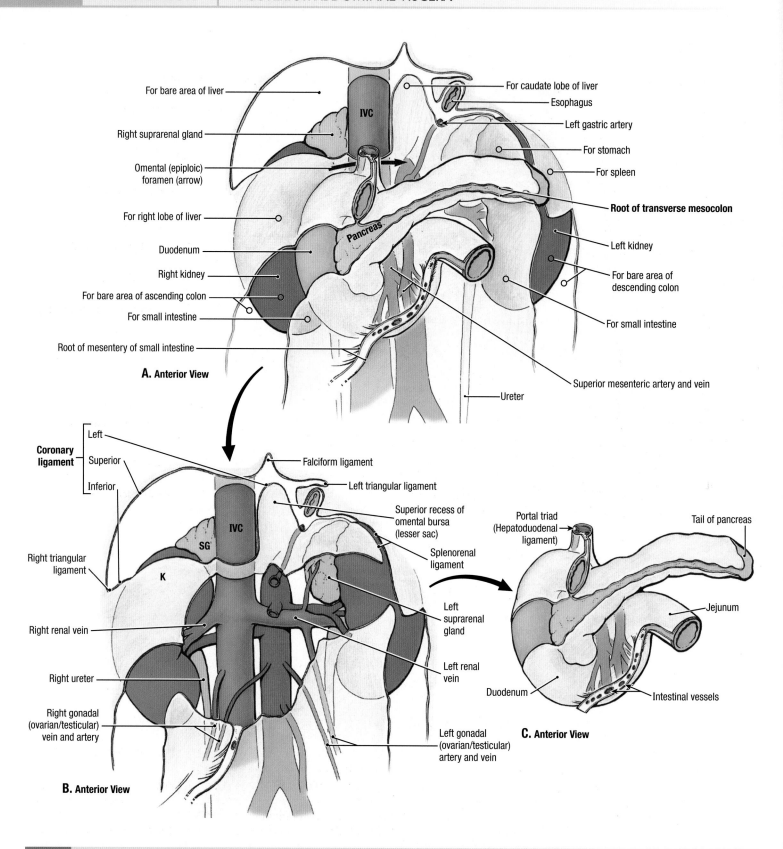

For bare area of liver

Right suprarenal gland

Omental (epiploic) foramen (arrow)

For right lobe of liver

Duodenum

Right kidney

For bare area of ascending colon

For small intestine

Root of mesentery of small intestine

IVC

Pancreas

For caudate lobe of liver

Esophagus

Left gastric artery

For stomach

For spleen

Root of transverse mesocolon

Left kidney

For bare area of descending colon

For small intestine

Superior mesenteric artery and vein

Ureter

A. Anterior View

Coronary ligament

Left

Superior

Inferior

Right triangular ligament

Right renal vein

Right ureter

Right gonadal (ovarian/testicular) vein and artery

IVC

SG

K

Falciform ligament

Left triangular ligament

Superior recess of omental bursa (lesser sac)

Splenorenal ligament

Left suprarenal gland

Left renal vein

Left gonadal (ovarian/testicular) artery and vein

B. Anterior View

Portal triad (Hepatoduodenal ligament)

Tail of pancreas

Jejunum

Intestinal vessels

Duodenum

C. Anterior View

4.67   POSTERIOR ABDOMINAL VISCERA AND THEIR ANTERIOR RELATIONS

A. Duodenum and pancreas *in situ.* Note the line of attachment of the root of the transverse mesocolon is to the body and tail of the pancreas. The viscera contacting specific regions are indicated by the term "for." The omental (epiploic) foramen is traversed by an arrow. **B.** After removal of duodenum and pancreas. The three parts of the coronary ligament are attached to the diaphragm, except where the inferior vena cava (IVC), suprarenal gland (SG), and kidney (K) intervene. **C.** Pancreas and duodenum removed from **A.**

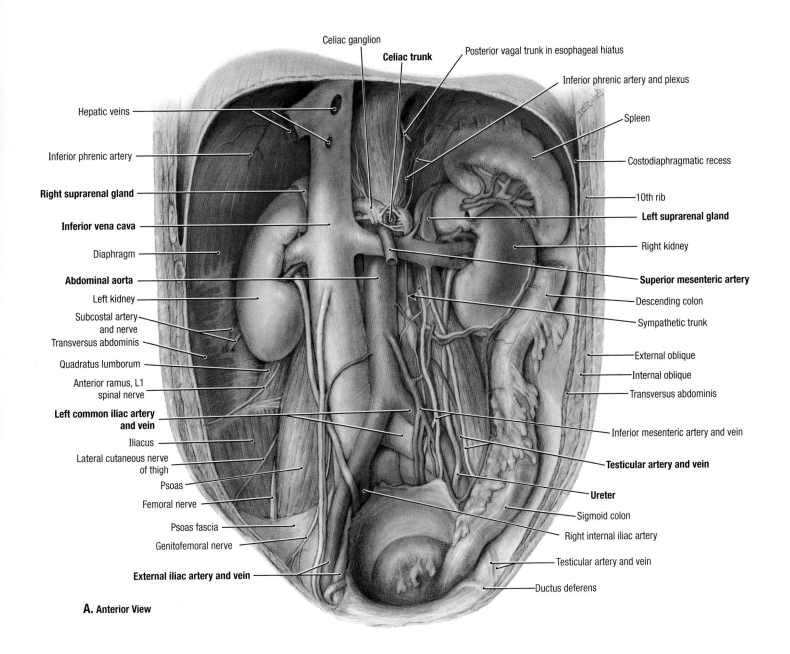

Celiac ganglion

Celiac trunk

Posterior vagal trunk in esophageal hiatus

Inferior phrenic artery and plexus

Hepatic veins

Inferior phrenic artery

Right suprarenal gland

Inferior vena cava

Diaphragm

Abdominal aorta

Left kidney

Subcostal artery and nerve

Transversus abdominis

Quadratus lumborum

Anterior ramus, L1 spinal nerve

Left common iliac artery and vein

Iliacus

Lateral cutaneous nerve of thigh

Psoas

Femoral nerve

Psoas fascia

Genitofemoral nerve

External iliac artery and vein

Spleen

Costodiaphragmatic recess

10th rib

Left suprarenal gland

Right kidney

Superior mesenteric artery

Descending colon

Sympathetic trunk

External oblique

Internal oblique

Transversus abdominis

Inferior mesenteric artery and vein

Testicular artery and vein

Ureter

Sigmoid colon

Right internal iliac artery

Testicular artery and vein

Ductus deferens

A. Anterior View

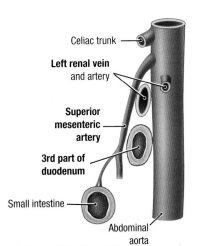

Celiac trunk

Left renal vein and artery

Superior mesenteric artery

3rd part of duodenum

Small intestine

Abdominal aorta

B. Lateral View (from Left)

VISCERA AND VESSELS OF POSTERIOR ABDOMINAL WALL 4.68

A. Great vessels, kidneys, and suprarenal glands. **B.** Relationships of left renal vein and inferior (third) part of duodenum to aorta and superior mesenteric artery.

- The abdominal aorta is shorter and smaller in caliber than the inferior vena cava.
- The inferior mesenteric artery arises about 4 cm superior to the aortic bifurcation and crosses the left common iliac vessels to become the superior rectal artery.
- The left renal vein drains the left testis, left suprarenal gland, and left kidney; the renal arteries are posterior to the renal veins.
- The ureter crosses the external iliac artery just beyond the common iliac bifurcation.
- The testicular vessels cross anterior to the ureter and join the ductus deferens at the deep inguinal ring.
- The left renal vein and duodenum (and uncinate process of pancreas—not shown) pass between the aorta posteriorly and the superior mesenteric artery anteriorly; they may be compressed like nuts in a nutcracker (**B**).

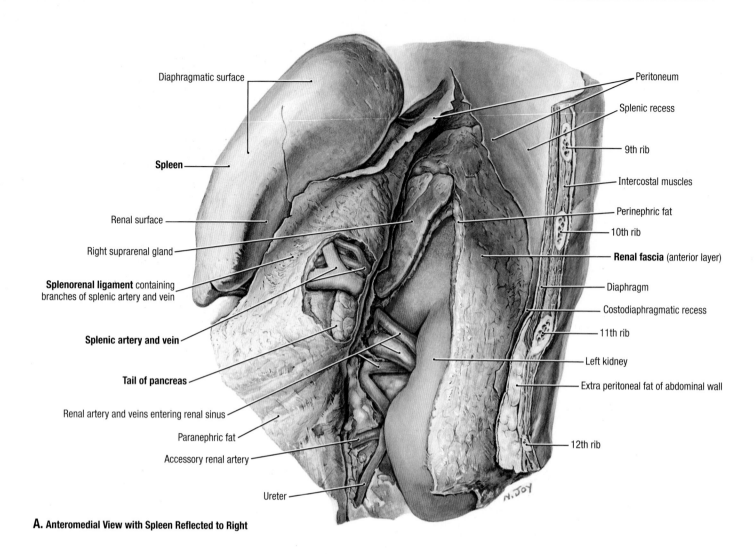

Diaphragmatic surface

Spleen

Renal surface

Right suprarenal gland

Splenorenal ligament containing branches of splenic artery and vein

Splenic artery and vein

Tail of pancreas

Renal artery and veins entering renal sinus

Paranephric fat

Accessory renal artery

Ureter

Peritoneum

Splenic recess

9th rib

Intercostal muscles

Perinephric fat

10th rib

Renal fascia (anterior layer)

Diaphragm

Costodiaphragmatic recess

11th rib

Left kidney

Extra peritoneal fat of abdominal wall

12th rib

N.JOY

A. Anteromedial View with Spleen Reflected to Right

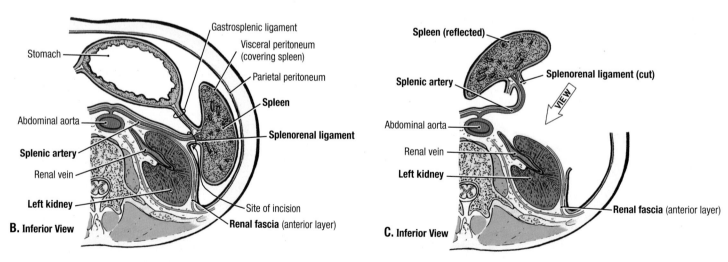

Gastrosplenic ligament

Stomach

Visceral peritoneum (covering spleen)

Parietal peritoneum

Spleen

Splenorenal ligament

Abdominal aorta

Splenic artery

Renal vein

Left kidney

Site of incision

Renal fascia (anterior layer)

B. Inferior View

Spleen (reflected)

Splenic artery

Splenorenal ligament (cut)

VIEW

Abdominal aorta

Renal vein

Left kidney

Renal fascia (anterior layer)

C. Inferior View

4.69	EXPOSURE OF THE LEFT KIDNEY AND SUPRARENAL GLAND

A. Dissection. **B.** Schematic section with spleen and splenorenal ligament intact. **C.** Procedure used in A to expose the kidney. The spleen and splenorenal ligament are reflected anteriorly, with the splenic vessels and tail of the pancreas. Part of the renal fascia of the kidney is removed. Note the proximity of the splenic vein and left renal vein, enabling a **splenorenal shunt** to be established surgically to relieve portal hypertension.

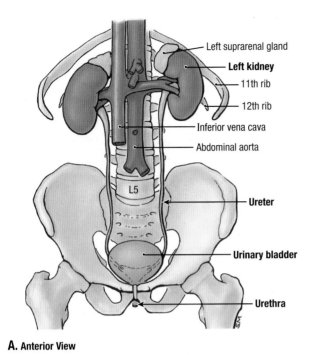

Left suprarenal gland
Left kidney
11th rib
12th rib
Inferior vena cava
Abdominal aorta

L5

Ureter

Urinary bladder

Urethra

A. Anterior View

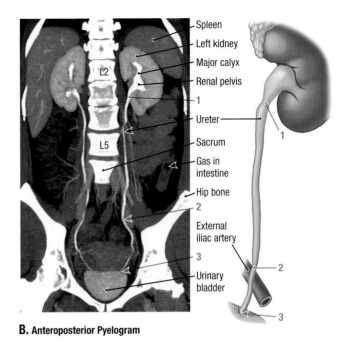

Spleen
Left kidney
Major calyx
Renal pelvis
1
Ureter
Sacrum
Gas in intestine
Hip bone
2
External iliac artery
3
Urinary bladder

L2

L5

1

2

3

B. Anteroposterior Pyelogram

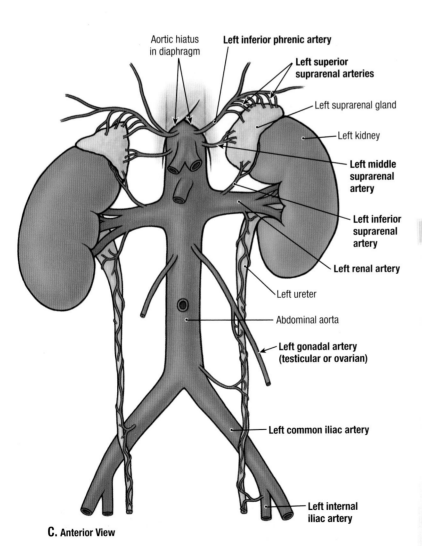

Aortic hiatus in diaphragm
Left inferior phrenic artery

Left superior suprarenal arteries

Left suprarenal gland

Left kidney

Left middle suprarenal artery

Left inferior suprarenal artery

Left renal artery

Left ureter

Abdominal aorta

Left gonadal artery (testicular or ovarian)

Left common iliac artery

Left internal iliac artery

C. Anterior View

KIDNEYS AND SUPRARENAL GLANDS 　4.70

A. Overview of urinary system. **B.** Retrograde pyelogram. Contrast medium was injected into the ureters from a flexible endoscope (urethroscope) in the bladder. Note the papillae bulging into the minor calices, which empty into a major calyx that opens, in turn, into the renal pelvis drained by the ureter. Sites at which relative constrictions in the ureters normally appear: (1) ureteropelvic junction; (2) crossing external iliac vessels or pelvic brim; and (3) as ureter traverses bladder wall. These constricted areas are potential sites of obstruction by ureteric (kidney) stones. **C.** Arterial supply of the suprarenal glands, kidneys, and ureters.

Renal transplantation is now an established operation for the treatment of selected cases of chronic renal failure. The kidney can be removed from the donor without damaging the suprarenal gland because of the weak septum of renal fascia that separates the kidney from this gland. The site for transplanting a kidney is in the iliac fossa of the greater pelvis. The renal artery and vein are joined to the external iliac artery and vein, respectively, and the ureter is sutured into the urinary bladder.

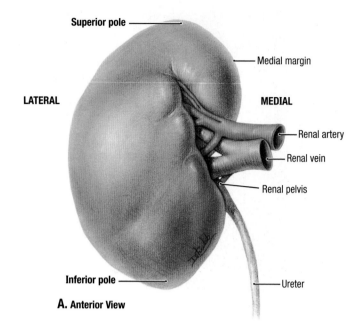

Superior pole

Medial margin

LATERAL

MEDIAL

Renal artery

Renal vein

Renal pelvis

Inferior pole

Ureter

A. Anterior View

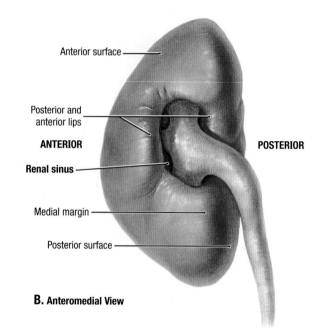

Anterior surface

Posterior and anterior lips

ANTERIOR

POSTERIOR

Renal sinus

Medial margin

Posterior surface

B. Anteromedial View

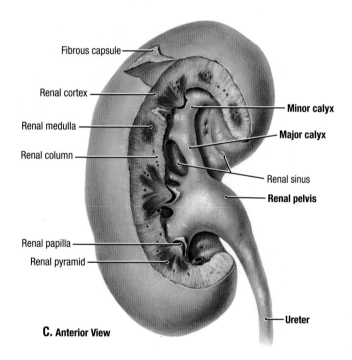

Fibrous capsule

Renal cortex

Renal medulla

Renal column

Minor calyx

Major calyx

Renal sinus

Renal pelvis

Renal papilla

Renal pyramid

Ureter

C. Anterior View

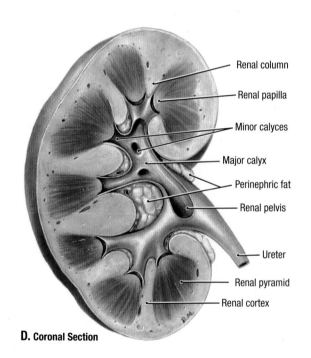

Renal column

Renal papilla

Minor calyces

Major calyx

Perinephric fat

Renal pelvis

Ureter

Renal pyramid

Renal cortex

D. Coronal Section

4.71   STRUCTURE OF KIDNEY

A. External features. The superior pole of the kidney is closer to the median plane than the inferior pole. Approximately 25% of kidneys may have a 2nd, 3rd, and even 4th accessory renal artery branching from the aorta. These multiple vessels enter through the renal sinus or at the superior or inferior pole (polar arteries). **B.** Renal sinus. The renal sinus is a vertical "pocket" opening on the medial side of the kidney. Tucked into the pocket are the renal pelvis and renal vessels in a matrix of perinephric fat. **C.** Renal calices. The anterior wall of the renal sinus has been cut away to expose the renal pelvis and the calices. **D.** Internal features. **Cysts in the kidney,** multiple or solitary, are common and usually benign findings during ultrasound examinations and dissection of cadavers. **Adult polycystic disease** of the kidneys, however, is an important cause of renal failure.

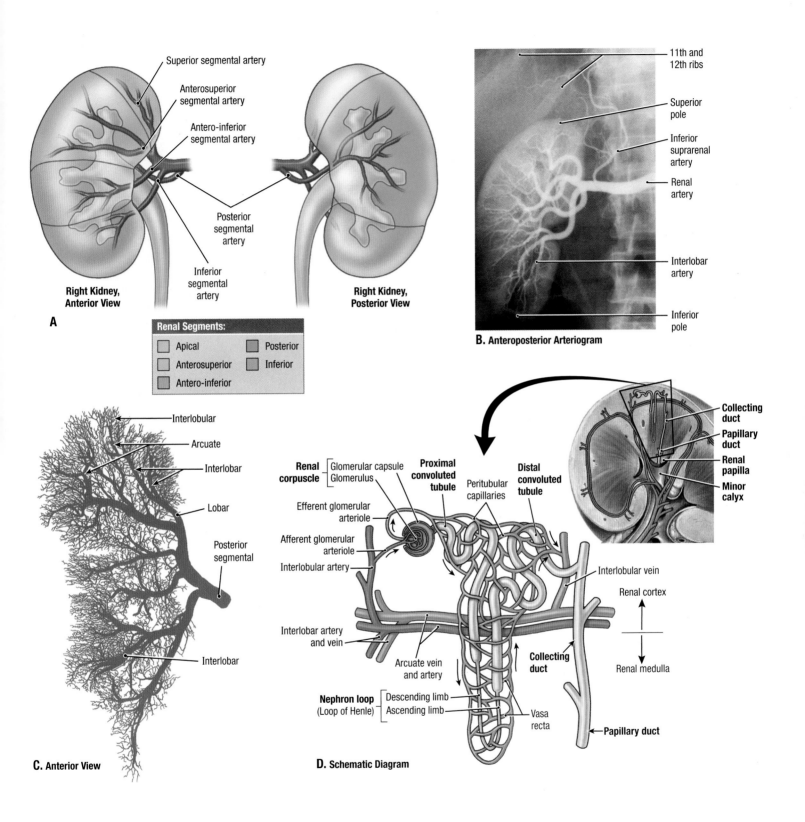

A. Segmental arteries. Superior segmental artery, Anterosuperior segmental artery, Antero-inferior segmental artery, Posterior segmental artery, Inferior segmental artery. Right Kidney, Anterior View. Right Kidney, Posterior View.

Renal Segments:
- Apical
- Anterosuperior
- Antero-inferior
- Posterior
- Inferior

B. Anteroposterior Arteriogram. 11th and 12th ribs, Superior pole, Inferior suprarenal artery, Renal artery, Interlobar artery, Inferior pole.

C. Anterior View. Interlobular, Arcuate, Interlobar, Lobar, Posterior segmental, Interlobar artery and vein, Interlobar.

D. Schematic Diagram. Renal corpuscle [Glomerular capsule, Glomerulus], Proximal convoluted tubule, Peritubular capillaries, Distal convoluted tubule, Efferent glomerular arteriole, Afferent glomerular arteriole, Interlobular artery, Interlobar artery and vein, Interlobular vein, Renal cortex, Renal medulla, Arcuate vein and artery, Collecting duct, Nephron loop (Loop of Henle) [Descending limb, Ascending limb], Vasa recta, Papillary duct, Collecting duct, Papillary duct, Renal papilla, Minor calyx.

SEGMENTS OF THE KIDNEYS

4.72

A. Segmental arteries. Segmental arteries do not anastomose significantly with other segmental arteries; they are end arteries. The area supplied by each segmented artery is an independent, surgically respectable unit or **renal segment**. **B.** Renal arteriogram. **C.** Corrosion cast of posterior segmental artery of kidney. **D.** The nephron is the functional unit of the kidney consisting of a renal corpuscle, proximal tubule, nephron loop, and distal tubule. Papillary ducts open onto renal papillae, emptying into minor calices.

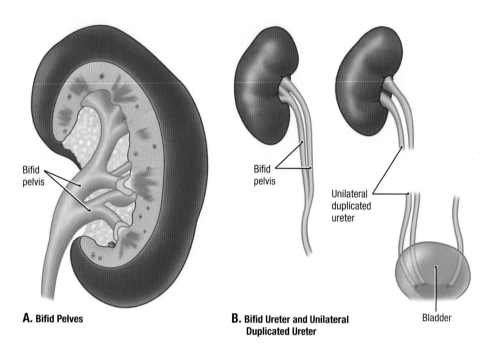

A. Bifid Pelves

Bifid pelvis

B. Bifid Ureter and Unilateral Duplicated Ureter

Bifid pelvis

Unilateral duplicated ureter

Bladder

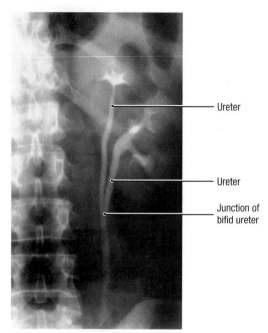

Anteroposterior Pyelogram

Ureter

Ureter

Junction of bifid ureter

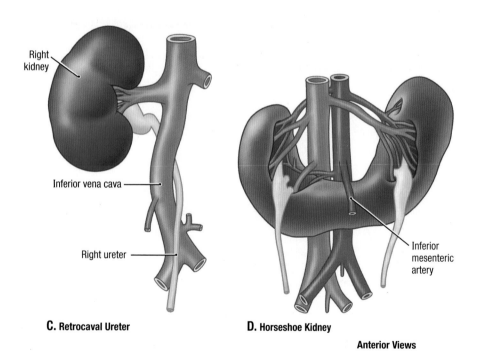

C. Retrocaval Ureter

Right kidney

Inferior vena cava

Right ureter

D. Horseshoe Kidney

Inferior mesenteric artery

Anterior Views

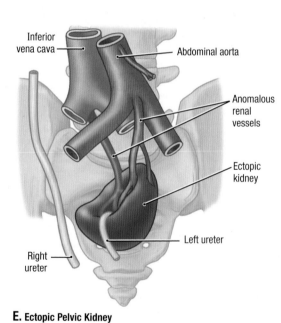

E. Ectopic Pelvic Kidney

Inferior vena cava

Abdominal aorta

Anomalous renal vessels

Ectopic kidney

Left ureter

Right ureter

4.73 ANOMALIES OF KIDNEY AND URETER

A. Bifid pelves. The pelves are almost replaced by two long major calices, which extend outside the sinus. **B.** Duplicated, or bifid, ureters. These can be unilateral or bilateral and complete or incomplete. **C.** Retrocaval ureter. The ureter courses posterior and then anterior to the inferior vena cava. **D.** Horseshoe kidney. The right and left kidneys are fused in the midline. **E.** Ectopic pelvic kidney. Pelvic kidneys have no fatty capsule and can be unilateral or bilateral. During childbirth, they may cause obstruction and suffer injury.

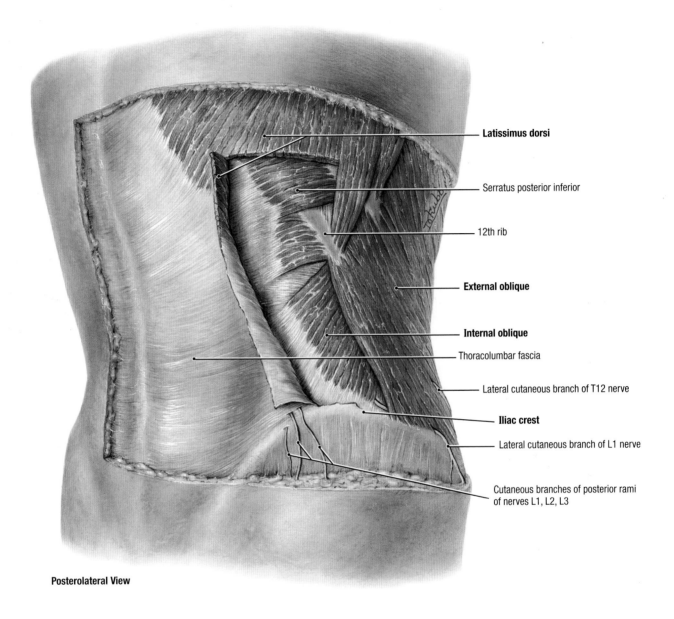

Latissimus dorsi

Serratus posterior inferior

12th rib

External oblique

Internal oblique

Thoracolumbar fascia

Lateral cutaneous branch of T12 nerve

Iliac crest

Lateral cutaneous branch of L1 nerve

Cutaneous branches of posterior rami of nerves L1, L2, L3

Posterolateral View

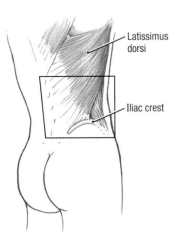

Latissimus dorsi

Iliac crest

POSTEROLATERAL ABDOMINAL WALL: EXPOSURE OF KIDNEY I | 4.74

The latissimus dorsi is partially reflected.

- The external oblique muscle has an oblique, free posterior border that extends from the tip of the 12th rib to the midpoint of the iliac crest.
- The internal oblique muscle extends posteriorly beyond the border of the external oblique muscle.

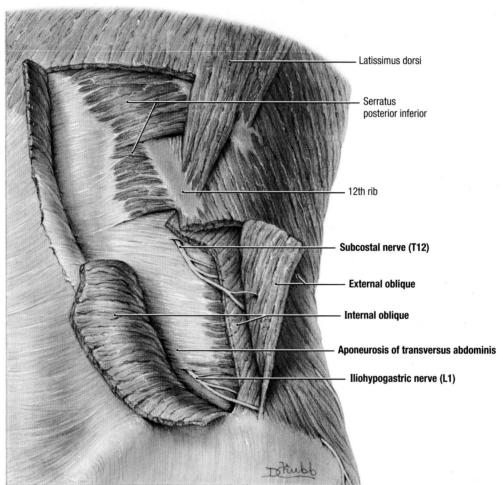

Latissimus dorsi

Serratus posterior inferior

12th rib

Subcostal nerve (T12)

External oblique

Internal oblique

Aponeurosis of transversus abdominis

Iliohypogastric nerve (L1)

Posterolateral View

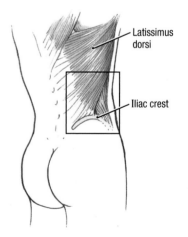

Latissimus dorsi

Iliac crest

4.75 POSTEROLATERAL ABDOMINAL WALL: EXPOSURE OF KIDNEY II

The external oblique muscle is incised and reflected laterally, and the internal oblique muscle is incised and reflected medially; the transversus abdominis muscle and its posterior aponeurosis are exposed where pierced by the subcostal (T12) and iliohypogastric (L1) nerves. These nerves give off motor twigs and lateral cutaneous branches and continue anteriorly between the internal oblique and transversus abdominis muscles.

4.76 POSTEROLATERAL ABDOMINAL WALL: EXPOSURE OF KIDNEY III AND RENAL FASCIA (*next page*)

A. The posterior aponeurosis of the transversus abdominis muscle is divided between the subcostal and iliohypogastric nerves and lateral to the oblique lateral border of the quadratus lumborum muscle; the retroperitoneal fat surrounding the kidney is exposed.

B. Renal fascia and retroperitoneal fat, schematic transverse section. The renal fascia is within this fat; fat internal to the renal fascia is termed perinephric fat (perirenal fat capsule), and the fat immediately external is paranephric fat (pararenal fat body).

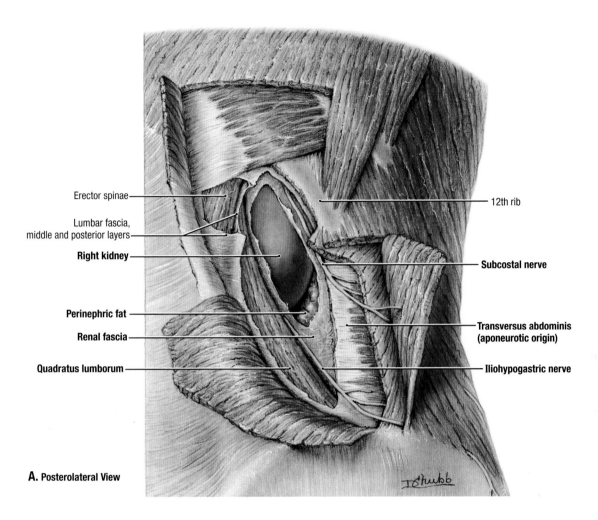

Erector spinae

Lumbar fascia, middle and posterior layers

Right kidney

Perinephric fat

Renal fascia

Quadratus lumborum

12th rib

Subcostal nerve

Transversus abdominis (aponeurotic origin)

Iliohypogastric nerve

A. Posterolateral View

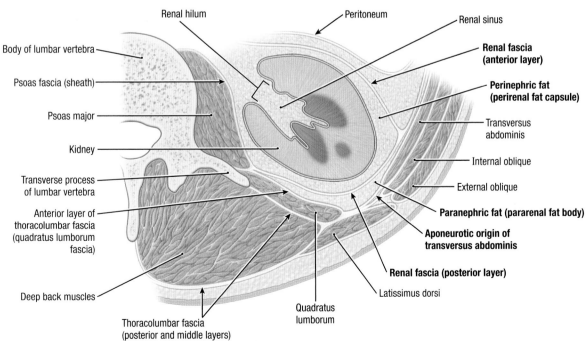

Renal hilum

Peritoneum

Renal sinus

Body of lumbar vertebra

Psoas fascia (sheath)

Psoas major

Kidney

Transverse process of lumbar vertebra

Anterior layer of thoracolumbar fascia (quadratus lumborum fascia)

Deep back muscles

Renal fascia (anterior layer)

Perinephric fat (perirenal fat capsule)

Transversus abdominis

Internal oblique

External oblique

Paranephric fat (pararenal fat body)

Aponeurotic origin of transversus abdominis

Renal fascia (posterior layer)

Latissimus dorsi

Thoracolumbar fascia (posterior and middle layers)

Quadratus lumborum

B. Transverse Section, Inferior View

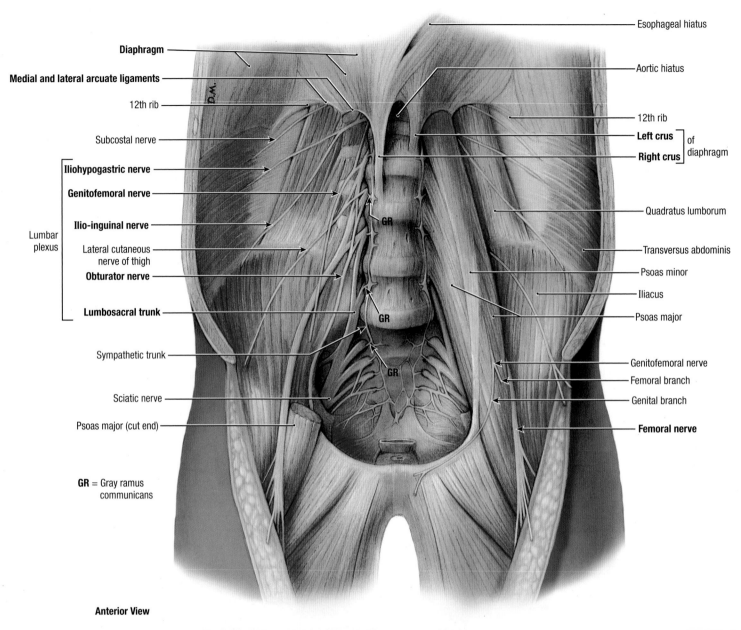

Esophageal hiatus

Diaphragm

Medial and lateral arcuate ligaments

Aortic hiatus

12th rib

12th rib

Subcostal nerve

Left crus | of
Right crus | diaphragm

Iliohypogastric nerve

Genitofemoral nerve

Quadratus lumborum

Ilio-inguinal nerve

Transversus abdominis

Lateral cutaneous nerve of thigh

Psoas minor

Obturator nerve

Iliacus

Lumbar plexus

Psoas major

Lumbosacral trunk

GR

GR

Sympathetic trunk

Genitofemoral nerve

Femoral branch

GR

Sciatic nerve

Genital branch

Psoas major (cut end)

Femoral nerve

GR = Gray ramus communicans

Anterior View

4.77 LUMBAR PLEXUS AND VERTEBRAL ATTACHMENT OF DIAPHRAGM

TABLE 4.7	PRINCIPAL MUSCLES OF POSTERIOR ABDOMINAL WALL			
Muscle	Superior Attachments	Inferior Attachments	Innervation	Actions
Psoas major[a,b]	Transverse processes of lumbar vertebrae; sides of bodies of T12–L5 vertebrae and intervening intervertebral discs	By a strong tendon to lesser trochanter of femur	Anterior rami of lumbar nerves (**L1**[c], **L2**[c], L3)	Acting inferiorly with iliacus, it flexes thigh at hip; acting superiorly, it flexes vertebral column laterally; it is used to balance the trunk; during sitting, it acts inferiorly with iliacus to flex trunk
Iliacus[a]	Superior two thirds of iliac fossa, ala of sacrum, and anterior sacro-iliac ligaments	Lesser trochanter of femur and shaft inferior to it, and to psoas major tendon	Femoral nerve (**L2**[c], L3, L4)	Flexes thigh and stabilizes hip joint; acts with psoas major
Quadratus lumborum	Medial half of inferior border of 12th rib and tips of lumbar transverse processes	Iliolumbar ligament and internal lip of iliac crest	Anterior rami of T12 and L1–L4 nerves	Extends and laterally flexes vertebral column; fixes 12th rib during inspiration

[a]Psoas major and iliacus muscles are often described together as the iliopsoas muscle when flexion of the hip joint is discussed.
[b]Psoas minor attaches proximally to the sides of bodies of T12–L1 vertebrae and intervertebral disc and distally to the pectineal line and iliopectineal eminence via the iliopectineal arch; it does not cross the hip joint.
    It is used to balance the trunk, in conjunction with psoas major. Innervation is from the anterior rami of lumbar nerves (L1, L2).
[c]Primary segment(s) of innervation are boldface type.

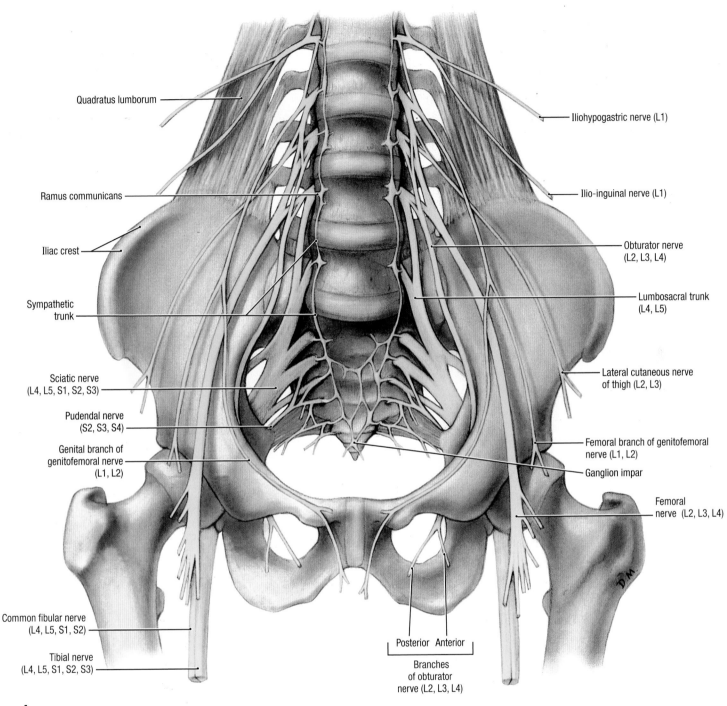

Quadratus lumborum

Ramus communicans

Iliac crest

Sympathetic trunk

Sciatic nerve (L4, L5, S1, S2, S3)

Pudendal nerve (S2, S3, S4)

Genital branch of genitofemoral nerve (L1, L2)

Common fibular nerve (L4, L5, S1, S2)

Tibial nerve (L4, L5, S1, S2, S3)

Iliohypogastric nerve (L1)

Ilio-inguinal nerve (L1)

Obturator nerve (L2, L3, L4)

Lumbosacral trunk (L4, L5)

Lateral cutaneous nerve of thigh (L2, L3)

Femoral branch of genitofemoral nerve (L1, L2)

Ganglion impar

Femoral nerve (L2, L3, L4)

Posterior   Anterior

Branches of obturator nerve (L2, L3, L4)

A. Anterior View

NERVES OF LUMBAR PLEXUS

4.78

The lumbar plexus of nerves is composed of the anterior rami of L1–L4 nerves:

- Ilio-inguinal and iliohypogastric nerves (L1) enter the abdomen posterior to the medial arcuate ligaments; they run between the transversus abdominis and internal oblique to supply the skin of the suprapubic and inguinal regions.
- Lateral cutaneous nerve of thigh (L2, L3) enters the thigh posterior to the inguinal ligament, just medial to the anterior superior iliac spine; it supplies the skin on the anterolateral surface of the thigh.

- Femoral nerve (L2–L4) emerges from the lateral border of the psoas; innervates the iliacus muscle and the extensor muscles of the knee.
- Genitofemoral nerve (L1, L2) pierces the anterior surface of the psoas major muscle; divides into femoral and genital branches.
- Obturator nerve (L2–L4) emerges from the medial border of the psoas to supply the adductor muscles of the thigh.
- Lumbosacral trunk (L4, L5) passes over the ala of the sacrum to join the sacral plexus.

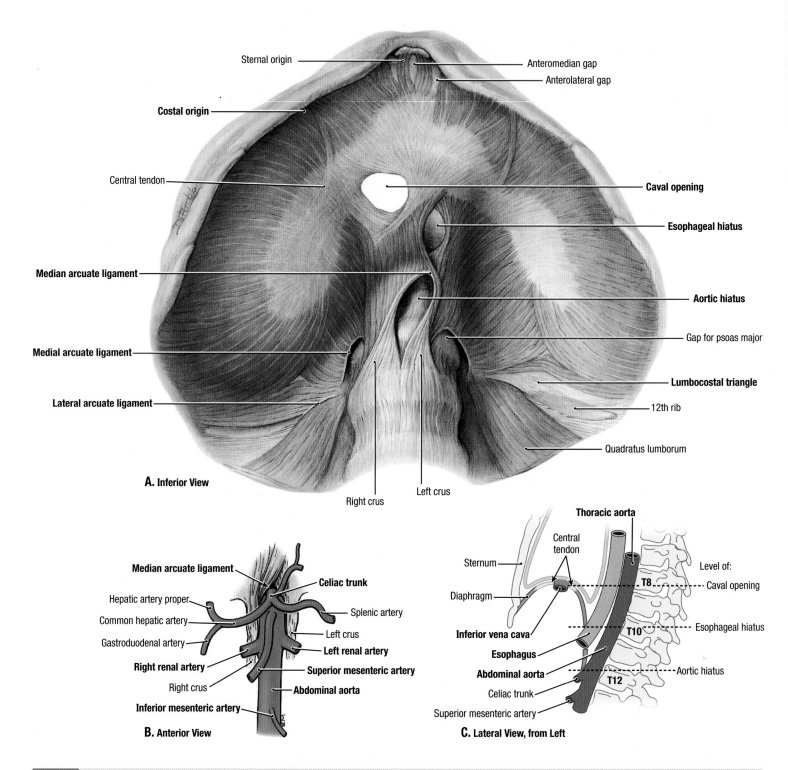

A. Inferior View

- Sternal origin
- Anteromedian gap
- Anterolateral gap
- **Costal origin**
- Central tendon
- **Caval opening**
- **Esophageal hiatus**
- **Median arcuate ligament**
- **Aortic hiatus**
- Gap for psoas major
- **Medial arcuate ligament**
- **Lumbocostal triangle**
- 12th rib
- **Lateral arcuate ligament**
- Quadratus lumborum
- Right crus
- Left crus

B. Anterior View

- **Median arcuate ligament**
- **Celiac trunk**
- Hepatic artery proper
- Common hepatic artery
- Splenic artery
- Gastroduodenal artery
- Left crus
- **Right renal artery**
- **Left renal artery**
- Right crus
- **Superior mesenteric artery**
- **Abdominal aorta**
- **Inferior mesenteric artery**

C. Lateral View, from Left

- **Thoracic aorta**
- Central tendon
- Sternum
- Level of:
- T8
- Caval opening
- Diaphragm
- **Inferior vena cava**
- T10
- Esophageal hiatus
- **Esophagus**
- **Abdominal aorta**
- Aortic hiatus
- Celiac trunk
- T12
- Superior mesenteric artery

4.79   DIAPHRAGM

A. Dissection. The clover-shaped central tendon is the aponeurotic insertion of the muscle. **Diaphragmatic hernia.** The diaphragm in this specimen fails to arise from the left lateral arcuate ligament, leaving a potential opening, the lumbocostal triangle, through which abdominal contents may be herniated into the thoracic cavity following a sudden increase in intra-thoracic or intra-abdominal pressure. A **hiatal hernia** is a protrusion of part of the stomach into the thorax through the esophageal hiatus.

B. Median arcuate ligament and branches of the aorta. **C.** Openings of the diaphragm. There are three major openings: (1) the caval opening for the inferior vena cava, most anterior, at the T8 vertebral level to the right of the midline; (2) the esophageal hiatus, intermediate, at T10 level and to the left; and (3) the aortic hiatus, which allows the aorta to pass posterior to the vertebral attachment of the diaphragm in the midline at T12.

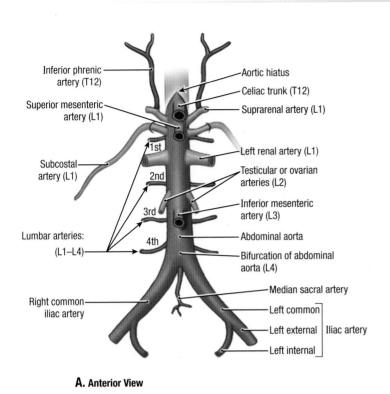

A. Anterior View

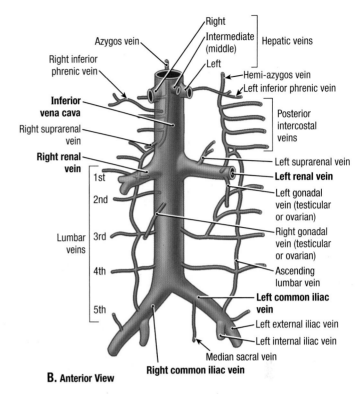

B. Anterior View

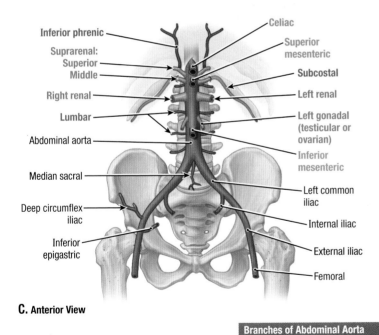

C. Anterior View

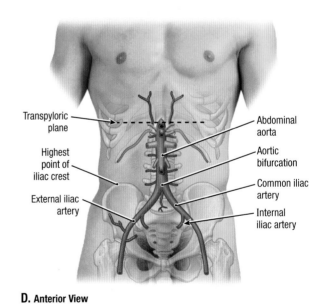

D. Anterior View

Branches of Abdominal Aorta		
▢ Anterior midline	▢ Lateral	▢ Posterolateral

ABDOMINAL AORTA AND INFERIOR VENA CAVA AND THEIR BRANCHES

4.80

A. Branches (and their vertebral levels) of abdominal aorta. **B.** Tributaries of the inferior vena cava (IVC). **C.** Arteries of posterior abdominal wall, branches of aorta. **D.** Surface anatomy.

Rupture of an **aortic aneurysm** (localized enlargement of the abdominal aorta) causes severe pain in the abdomen or back. If unrecognized, a ruptured aneurysm has a mortality of nearly 90%

because of heavy blood loss. Surgeons can repair an aneurysm by opening it, inserting a prosthetic graft (such as one made of Dacron), and sewing the wall of the aneurysmal aorta over the graft to protect it. Aneurysms may also be treated by endovascular catheterization procedures.

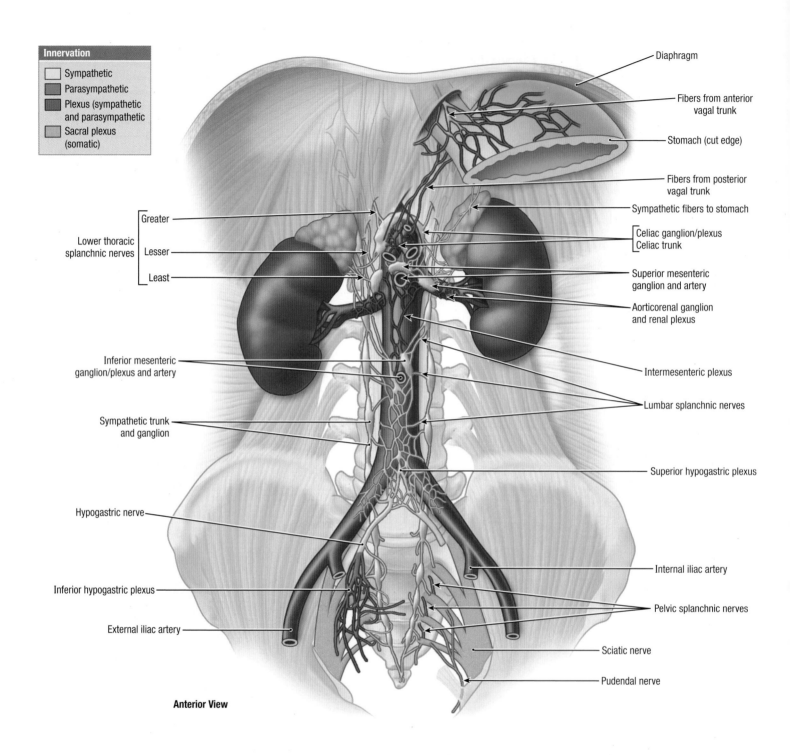

Innervation
- Sympathetic
- Parasympathetic
- Plexus (sympathetic and parasympathetic
- Sacral plexus (somatic)

Diaphragm

Fibers from anterior vagal trunk

Stomach (cut edge)

Fibers from posterior vagal trunk

Sympathetic fibers to stomach

Celiac ganglion/plexus
Celiac trunk

Superior mesenteric ganglion and artery

Aorticorenal ganglion and renal plexus

Intermesenteric plexus

Lumbar splanchnic nerves

Superior hypogastric plexus

Internal iliac artery

Pelvic splanchnic nerves

Sciatic nerve

Pudendal nerve

Lower thoracic splanchnic nerves
- Greater
- Lesser
- Least

Inferior mesenteric ganglion/plexus and artery

Sympathetic trunk and ganglion

Hypogastric nerve

Inferior hypogastric plexus

External iliac artery

Anterior View

4.81 ABDOMINOPELVIC NERVE PLEXUSES AND GANGLIA

The sympathetic part of the autonomic nervous system in the abdomen consists of:
- *Abdominopelvic splanchnic nerves* from the thoracic and abdominal sympathetic trunks.
- Prevertebral sympathetic ganglia.

- *Abdominal aortic plexus* and its extensions, the peri-arterial plexuses.

  The plexuses are mixed, shared with the parasympathetic nervous system and visceral afferent fibers.

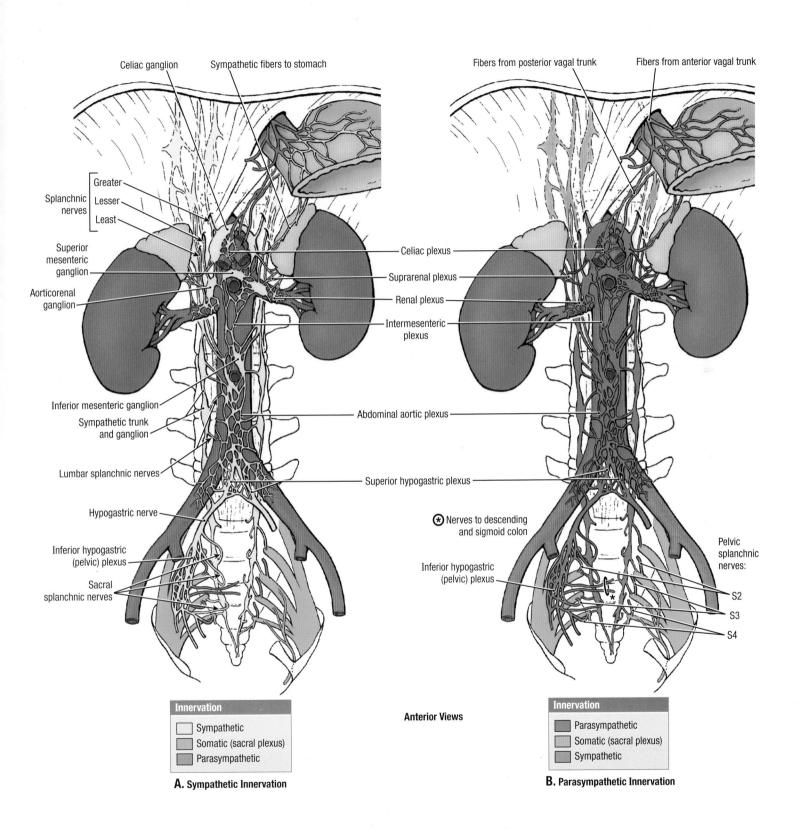

Celiac ganglion
Sympathetic fibers to stomach
Fibers from posterior vagal trunk
Fibers from anterior vagal trunk

Greater
Splanchnic nerves — Lesser
Least

Superior mesenteric ganglion

Aorticorenal ganglion

Celiac plexus
Suprarenal plexus
Renal plexus
Intermesenteric plexus

Inferior mesenteric ganglion

Sympathetic trunk and ganglion

Abdominal aortic plexus

Lumbar splanchnic nerves

Superior hypogastric plexus

Hypogastric nerve

✱ Nerves to descending and sigmoid colon

Inferior hypogastric (pelvic) plexus

Inferior hypogastric (pelvic) plexus

Sacral splanchnic nerves

Pelvic splanchnic nerves:

S2
S3
S4

Anterior Views

Innervation
☐ Sympathetic
☐ Somatic (sacral plexus)
☐ Parasympathetic

Innervation
☐ Parasympathetic
☐ Somatic (sacral plexus)
☐ Sympathetic

A. Sympathetic Innervation

B. Parasympathetic Innervation

OVERVIEW OF AUTONOMIC NERVOUS SYSTEM

4.82

A. Sympathetic. **B.** Parasympathetic.

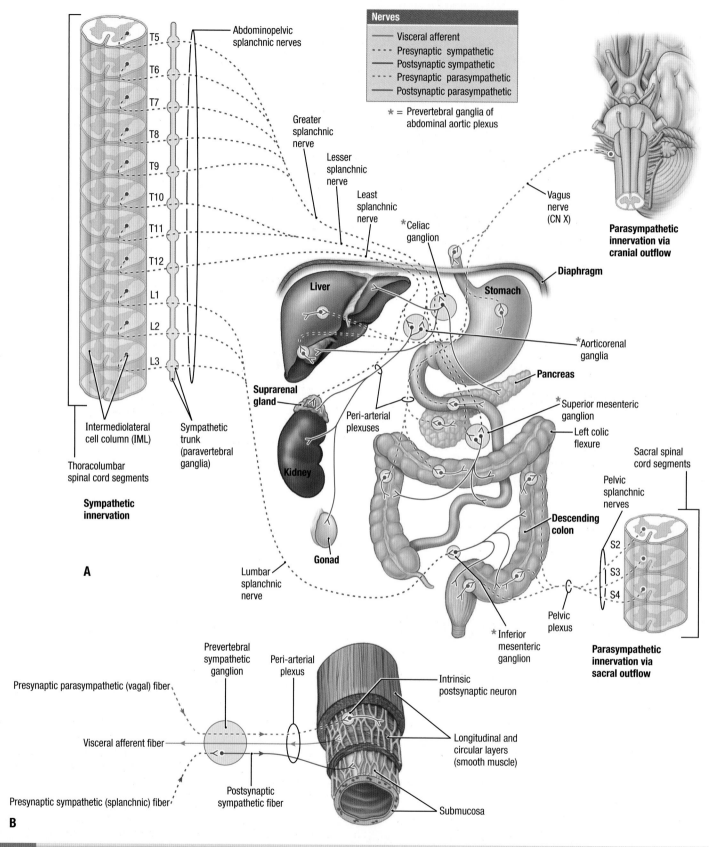

Nerves
— Visceral afferent
---- Presynaptic sympathetic
— Postsynaptic sympathetic
---- Presynaptic parasympathetic
— Postsynaptic parasympathetic

* = Prevertebral ganglia of abdominal aortic plexus

Abdominopelvic splanchnic nerves

Greater splanchnic nerve

Lesser splanchnic nerve

Least splanchnic nerve

*Celiac ganglion

Vagus nerve (CN X)

Parasympathetic innervation via cranial outflow

T5, T6, T7, T8, T9, T10, T11, T12, L1, L2, L3

Intermediolateral cell column (IML)

Sympathetic trunk (paravertebral ganglia)

Thoracolumbar spinal cord segments

Sympathetic innervation

A

Liver

Stomach

Diaphragm

*Aorticorenal ganglia

Pancreas

*Superior mesenteric ganglion

Left colic flexure

Suprarenal gland

Peri-arterial plexuses

Kidney

Gonad

Lumbar splanchnic nerve

Descending colon

Sacral spinal cord segments

Pelvic splanchnic nerves

S2, S3, S4

Pelvic plexus

*Inferior mesenteric ganglion

Parasympathetic innervation via sacral outflow

Prevertebral sympathetic ganglion

Peri-arterial plexus

Presynaptic parasympathetic (vagal) fiber

Intrinsic postsynaptic neuron

Visceral afferent fiber

Longitudinal and circular layers (smooth muscle)

Presynaptic sympathetic (splanchnic) fiber

Postsynaptic sympathetic fiber

Submucosa

B

4.83 ORIGIN AND DISTRIBUTION OF PRESYNAPTIC AND POSTSYNAPTIC SYMPATHETIC AND PARASYMPATHETIC FIBERS, AND GANGLIA INVOLVED IN SUPPLYING ABDOMINAL VISCERA

A. Overview. **B.** Fibers supplying the intrinsic plexuses of abdominal viscera.

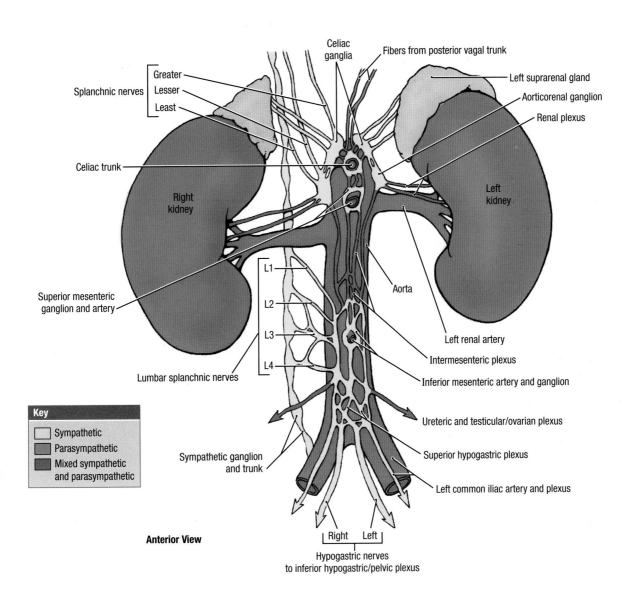

Celiac ganglia

Fibers from posterior vagal trunk

Splanchnic nerves
- Greater
- Lesser
- Least

Left suprarenal gland

Aorticorenal ganglion

Renal plexus

Celiac trunk

Right kidney

Left kidney

Superior mesenteric ganglion and artery

L1

L2

L3

L4

Aorta

Left renal artery

Intermesenteric plexus

Inferior mesenteric artery and ganglion

Lumbar splanchnic nerves

Ureteric and testicular/ovarian plexus

Superior hypogastric plexus

Left common iliac artery and plexus

Key
- ☐ Sympathetic
- ▨ Parasympathetic
- ▨ Mixed sympathetic and parasympathetic

Sympathetic ganglion and trunk

Anterior View

Right Left

Hypogastric nerves to inferior hypogastric/pelvic plexus

ABDOMINAL NERVE PLEXUSES AND GANGLIA

4.84

TABLE 4.8	AUTONOMIC INNERVATION OF ABDOMINAL VISCERA (SPLANCHNIC NERVES)				
Splanchnic Nerves	*Autonomic Fiber Type[a]*	*System*	*Origin*		*Destination*
A. Cardiopulmonary (Cervical and upper thoracic)	Postsynaptic	Sympathetic	Cervical and upper thoracic sympathetic trunk		Thoracic cavity (viscera superior to the level of diaphragm)

B. Abdominopelvic
  1. Lower thoracic
    a. Greater
    b. Lesser
    c. Least
  2. Lumbar
  3. Sacral
	Presynaptic		Lower thoracic and abdominopelvic sympathetic trunk:
			1. Thoracic sympathetic trunk:
			  a. T5–T9 or T10 level
			  b. T10–T11 level
			  c. T12 level
			2. Abdominal sympathetic trunk
			3. Pelvic (sacral) sympathetic trunk
					Abdominopelvic cavity (prevertebral ganglia serving viscera and suprarenal glands inferior to the level of diaphragm)
					1. Abdominal prevertebral ganglia:
					  a. Celiac ganglia
					  b. Aorticorenal ganglia
					  c. & 2. Other abdominal prevertebral ganglia (superior and inferior mesenteric and of intermesenteric/hypogastric plexuses)
					3. Pelvic prevertebral ganglia

C. Pelvic	Presynaptic	Parasympathetic	Anterior rami of S2–S4 spinal nerves		Intrinsic ganglia of descending and sigmoid colon, rectum, and pelvic viscera

[a]Splanchnic nerves also convey visceral afferent fibers, which are not part of the autonomic nervous system.

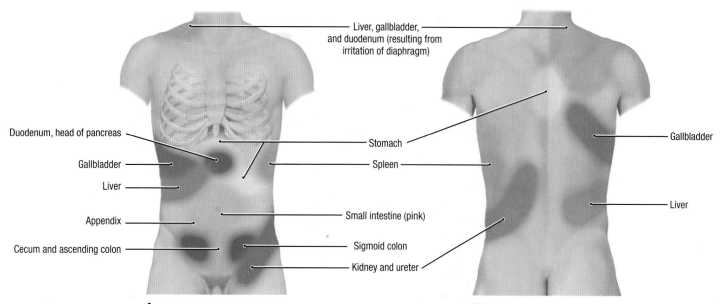

Liver, gallbladder, and duodenum (resulting from irritation of diaphragm)

Duodenum, head of pancreas

Gallbladder

Liver

Appendix

Cecum and ascending colon

Stomach

Spleen

Small intestine (pink)

Sigmoid colon

Kidney and ureter

Gallbladder

Liver

A. Anterior View

B. Posterior View

4.85   SURFACE PROJECTIONS OF VISCERAL PAIN

A. and **B.** Sites of visceral referred pain. **C.** Approximate spinal cord segments and spinal sensory ganglia involved in sympathetic and visceral afferent (pain) innervation of abdominal viscera.

Pain is an unpleasant sensation associated with actual or potential tissue damage, mediated by specific nerve fibers to the brain, where its conscious appreciation may be modified. Organic pain arising from an organ such as the stomach varies from dull to severe; however, the pain is poorly localized. It radiates to the dermatome level served by the corresponding sensory ganglion, which receives the visceral afferent fibers from the organ concerned. **Visceral referred pain** from a gastric ulcer, for example, is referred to the epigastric region because the stomach is supplied by pain afferents that reach the T7 and T8 spinal (sensory) ganglia and spinal cord segments through the greater splanchnic nerve. The brain interprets the pain as though the irritation occurred in the skin of the epigastric region, which is also supplied by the same sensory ganglia and spinal cord segments.

Pain arising from the parietal peritoneum is of the somatic type and is usually severe. The site of its origin may be localized. The anatomical basis for this localization of pain is that the parietal peritoneum is supplied by somatic sensory fibers through thoracic nerves, whereas a viscus such as the appendix is supplied by visceral afferent fibers in the lesser splanchnic nerve. Inflamed parietal peritoneum is extremely sensitive to stretching. When digital pressure is applied to the anterolateral abdominal wall over the site of inflammation, the parietal peritoneum is stretched. When the fingers are suddenly removed, extreme localized pain is usually felt, known as **rebound tenderness**.

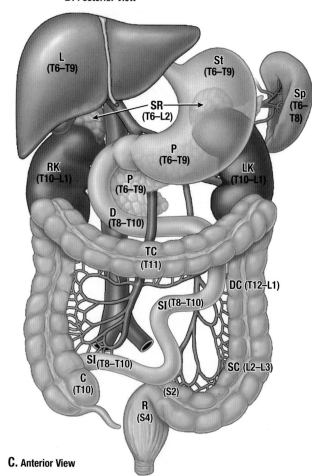

L (T6–T9)

St (T6–T9)

Sp (T6–T8)

SR (T6–L2)

P (T6–T9)

RK (T10–L1)

P (T6–T9)

LK (T10–L1)

D (T8–T10)

TC (T11)

DC (T12–L1)

SI (T8–T10)

SI (T8–T10)

SC (L2–L3)

C (T10)

(S2)

R (S4)

C. Anterior View

Key					
C	Cecum	P	Pancreas	Sp	Spleen
D	Duodenum	R	Rectum	SR	Suprarenal glands
DC	Descending colon	RK	Right kidney	St	Stomach
L	Liver	SC	Sigmoid colon	TC	Transverse colon
LK	Left kidney	SI	Small intestine		

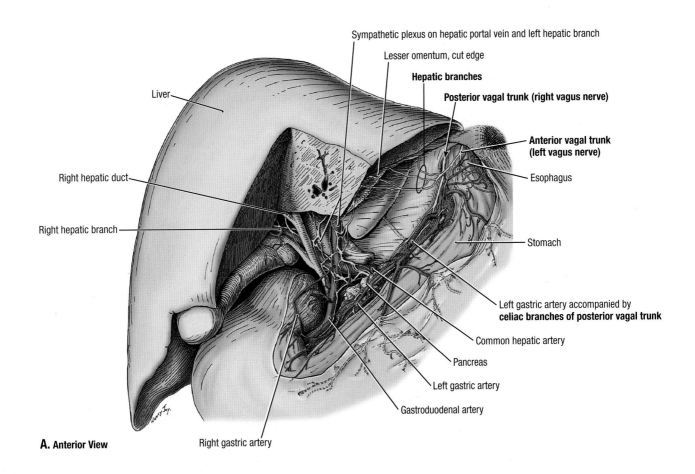

Sympathetic plexus on hepatic portal vein and left hepatic branch

Lesser omentum, cut edge

Hepatic branches

Posterior vagal trunk (right vagus nerve)

Anterior vagal trunk (left vagus nerve)

Esophagus

Liver

Right hepatic duct

Right hepatic branch

Stomach

Left gastric artery accompanied by **celiac branches of posterior vagal trunk**

Common hepatic artery

Pancreas

Left gastric artery

Gastroduodenal artery

Right gastric artery

A. Anterior View

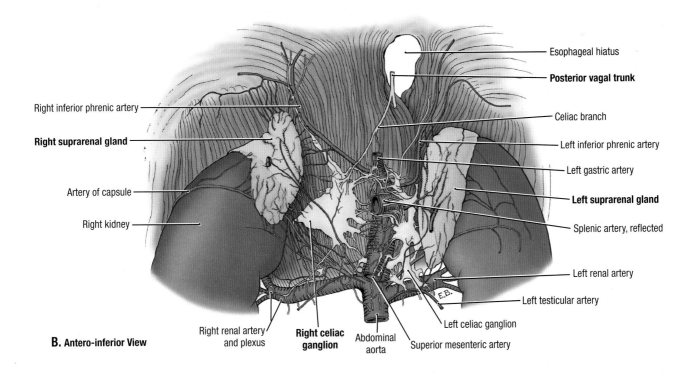

Esophageal hiatus

Posterior vagal trunk

Right inferior phrenic artery

Celiac branch

Right suprarenal gland

Left inferior phrenic artery

Left gastric artery

Artery of capsule

Left suprarenal gland

Right kidney

Splenic artery, reflected

Left renal artery

Left testicular artery

Right renal artery and plexus

Right celiac ganglion

Abdominal aorta

Superior mesenteric artery

Left celiac ganglion

B. Antero-inferior View

VAGUS NERVES IN ABDOMEN

4.86

A. Anterior and posterior vagal trunks. **B.** Celiac plexus and ganglia and suprarenal glands.

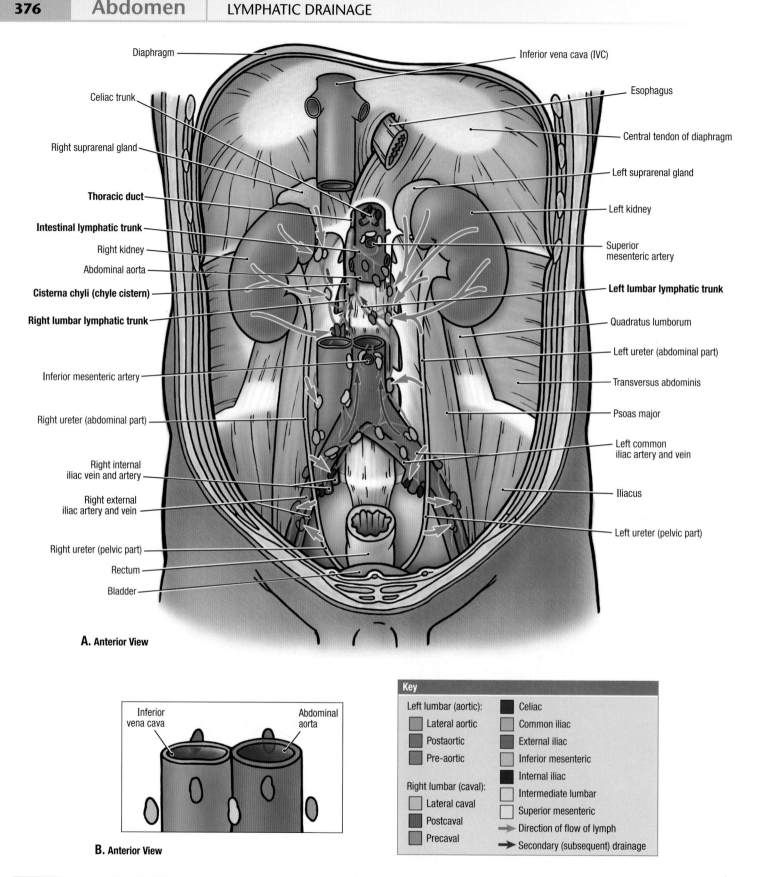

Diaphragm

Inferior vena cava (IVC)

Celiac trunk

Esophagus

Right suprarenal gland

Central tendon of diaphragm

Thoracic duct

Left suprarenal gland

Intestinal lymphatic trunk

Left kidney

Right kidney

Superior mesenteric artery

Abdominal aorta

Cisterna chyli (chyle cistern)

Left lumbar lymphatic trunk

Right lumbar lymphatic trunk

Quadratus lumborum

Left ureter (abdominal part)

Inferior mesenteric artery

Transversus abdominis

Right ureter (abdominal part)

Psoas major

Left common iliac artery and vein

Right internal iliac vein and artery

Right external iliac artery and vein

Iliacus

Right ureter (pelvic part)

Left ureter (pelvic part)

Rectum

Bladder

A. Anterior View

Key

Left lumbar (aortic):
- Lateral aortic
- Postaortic
- Pre-aortic

Right lumbar (caval):
- Lateral caval
- Postcaval
- Precaval

- Celiac
- Common iliac
- External iliac
- Inferior mesenteric
- Internal iliac
- Intermediate lumbar
- Superior mesenteric
- Direction of flow of lymph
- Secondary (subsequent) drainage

Inferior vena cava

Abdominal aorta

B. Anterior View

4.87 LYMPHATIC DRAINAGE OF SUPRARENAL GLANDS, KIDNEYS, AND URETERS

Lymphatic vessels from the suprarenal glands, kidneys, and upper ureters drain to the lumbar nodes. Lymphatic vessels from the middle part of the ureter usually drain into the **common iliac lymph nodes**, whereas vessels from its inferior part drain into the common, external, or internal **iliac lymph nodes**.

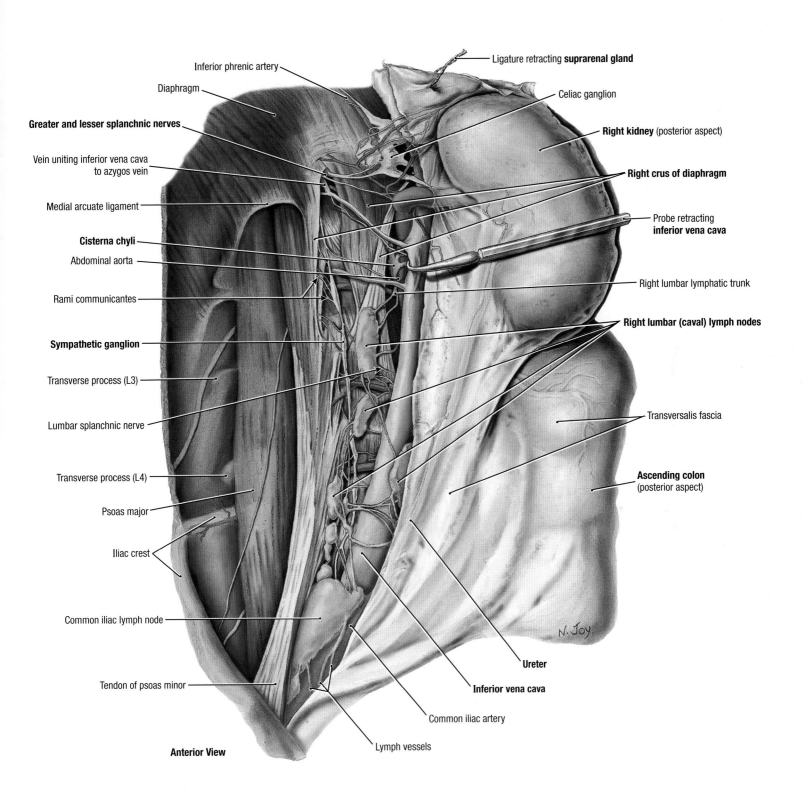

Ligature retracting **suprarenal gland**

Inferior phrenic artery

Diaphragm

Celiac ganglion

Greater and lesser splanchnic nerves

Right kidney (posterior aspect)

Vein uniting inferior vena cava to azygos vein

Right crus of diaphragm

Medial arcuate ligament

Probe retracting **inferior vena cava**

Cisterna chyli

Abdominal aorta

Right lumbar lymphatic trunk

Rami communicantes

Right lumbar (caval) lymph nodes

Sympathetic ganglion

Transverse process (L3)

Lumbar splanchnic nerve

Transversalis fascia

Transverse process (L4)

Ascending colon (posterior aspect)

Psoas major

Iliac crest

Common iliac lymph node

Ureter

Tendon of psoas minor

Inferior vena cava

Common iliac artery

Anterior View

Lymph vessels

N. Joy

LUMBAR LYMPH NODES, SYMPATHETIC TRUNK, NERVES, AND GANGLIA

4.88

The right suprarenal gland, kidney, ureter, and colon are reflected to the left along with the transversalis fascia covering their posterior aspects. The inferior vena cava is pulled medially, and the third and fourth lumbar veins are removed. In this specimen, the greater and lesser splanchnic nerves, the sympathetic trunk, and a communicating vein pass through an unusually wide cleft in the right crus. The splanchnic nerves convey preganglionic fibers arising from the cell bodies in the (thoracolumbar) sympathetic trunk. The greater splanchnic nerve is from thoracic ganglia 5 to 9, and the lesser from thoracic ganglia 10 and 11.

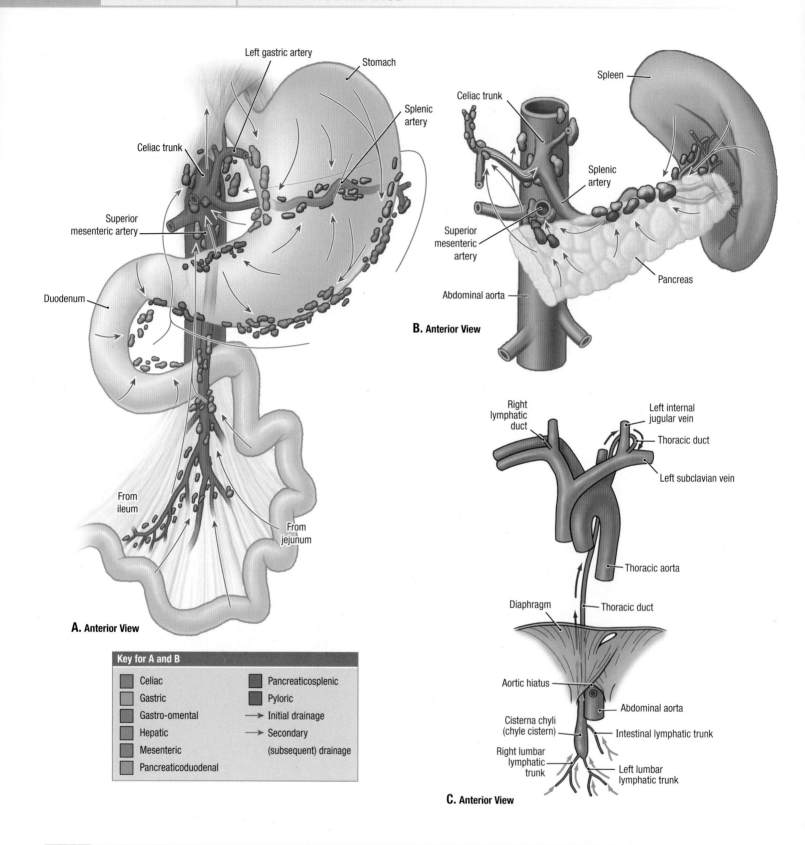

Left gastric artery

Stomach

Splenic artery

Celiac trunk

Superior mesenteric artery

Duodenum

From ileum

From jejunum

A. Anterior View

Celiac trunk

Spleen

Splenic artery

Superior mesenteric artery

Pancreas

Abdominal aorta

B. Anterior View

Right lymphatic duct

Left internal jugular vein

Thoracic duct

Left subclavian vein

Thoracic aorta

Diaphragm

Thoracic duct

Aortic hiatus

Abdominal aorta

Cisterna chyli (chyle cistern)

Intestinal lymphatic trunk

Right lumbar lymphatic trunk

Left lumbar lymphatic trunk

C. Anterior View

Key for A and B	
▮ Celiac	▮ Pancreaticosplenic
▮ Gastric	▮ Pyloric
▮ Gastro-omental	→ Initial drainage
▮ Hepatic	→ Secondary
▮ Mesenteric	(subsequent) drainage
▮ Pancreaticoduodenal	

4.89 LYMPHATIC DRAINAGE

A. Stomach and small intestine. **B.** Spleen and pancreas. **C.** Drainage from lumbar and intestinal lymphatic trunks. The *arrows* indicate the direction of lymph flow; each group of lymph nodes is color-coded. Lymph from the abdominal nodes drains into the cisterna chyli, origin of the inferior end of the thoracic duct. The thoracic duct receives all lymph that forms inferior to the diaphragm and left upper quadrant (thorax and left upper limb) and empties into the junction of the left subclavian and left internal jugular veins.

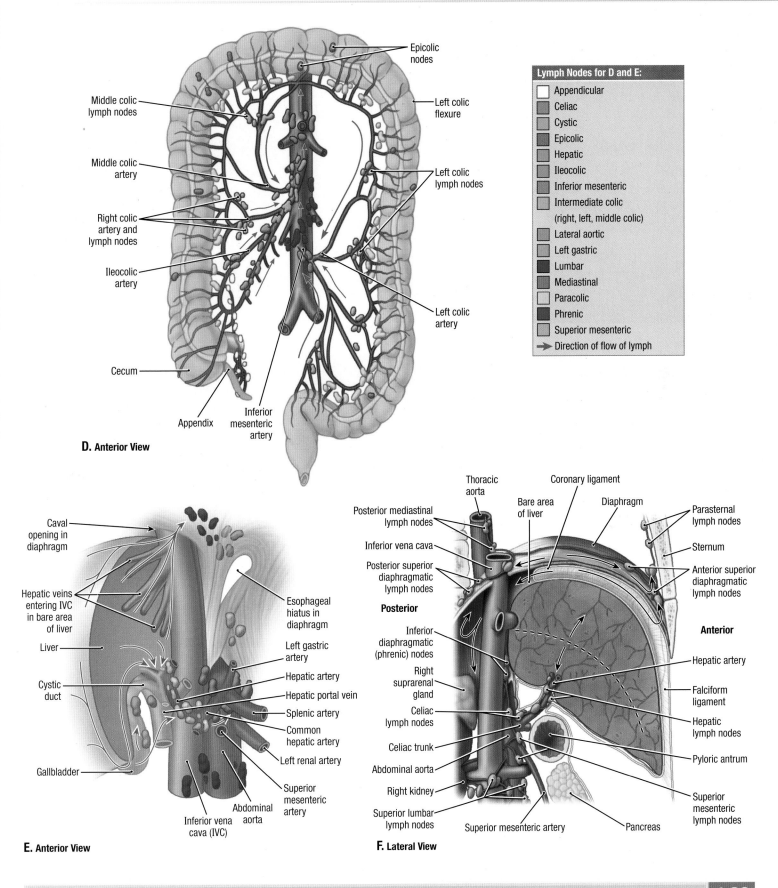

Epicolic nodes

Middle colic lymph nodes

Left colic flexure

Middle colic artery

Left colic lymph nodes

Right colic artery and lymph nodes

Ileocolic artery

Left colic artery

Cecum

Appendix

Inferior mesenteric artery

D. Anterior View

Lymph Nodes for D and E:

- Appendicular
- Celiac
- Cystic
- Epicolic
- Hepatic
- Ileocolic
- Inferior mesenteric
- Intermediate colic (right, left, middle colic)
- Lateral aortic
- Left gastric
- Lumbar
- Mediastinal
- Paracolic
- Phrenic
- Superior mesenteric
- → Direction of flow of lymph

Caval opening in diaphragm

Hepatic veins entering IVC in bare area of liver

Liver

Cystic duct

Gallbladder

Esophageal hiatus in diaphragm

Left gastric artery

Hepatic artery

Hepatic portal vein

Splenic artery

Common hepatic artery

Left renal artery

Superior mesenteric artery

Inferior vena cava (IVC)

Abdominal aorta

E. Anterior View

Thoracic aorta

Coronary ligament

Diaphragm

Bare area of liver

Parasternal lymph nodes

Posterior mediastinal lymph nodes

Inferior vena cava

Sternum

Posterior superior diaphragmatic lymph nodes

Anterior superior diaphragmatic lymph nodes

Posterior

Anterior

Inferior diaphragmatic (phrenic) nodes

Hepatic artery

Right suprarenal gland

Falciform ligament

Celiac lymph nodes

Hepatic lymph nodes

Celiac trunk

Abdominal aorta

Pyloric antrum

Right kidney

Superior mesenteric lymph nodes

Superior lumbar lymph nodes

Superior mesenteric artery

Pancreas

F. Lateral View

LYMPHATIC DRAINAGE (*continued*)

4.89

D. Large intestine. **E.** Liver and gallbladder. **F.** Liver.

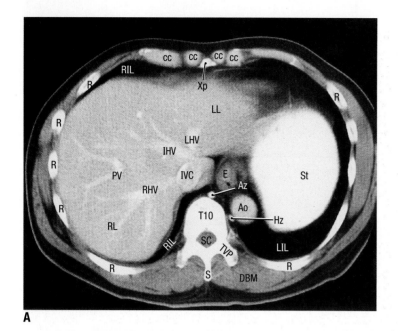

A

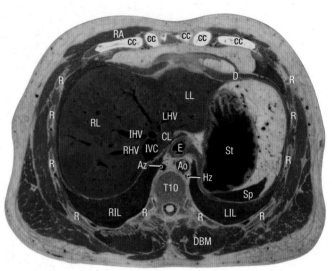

B

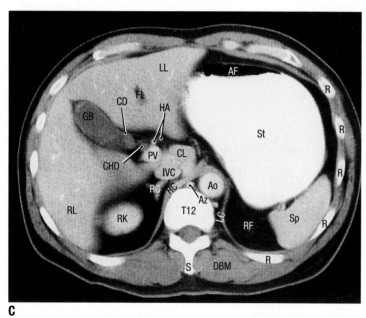

C

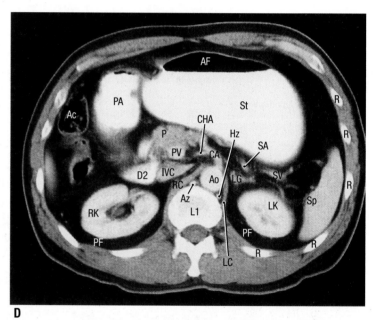

D

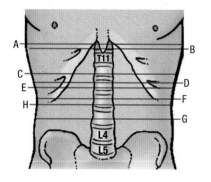

Key					
Ac	Ascending colon	Dc	Descending colon	LG	Left suprarenal gland
AF	Air-fluid level of stomach	D2	Descending part of duodenum	LHV	Left hepatic vein
Ao	Aorta	D3	Inferior part of duodenum	LIL	Left inferior lobe of lung
Az	Azygos vein	E	Esophagus	LK	Left kidney
CA	Celiac artery	FL	Falciform ligament	LL	Left lobe of liver
cc	Costal cartilage	GB	Gallbladder	LRV	Left renal vein
CD	Cystic duct	HA	Hepatic artery	LU	Left ureter
CHA	Common hepatic artery	Hz	Hemi-azygos vein	P	Pancreas
CHD	Common hepatic duct	IHV	Intermediate hepatic vein	PA	Pyloric antrum of stomach
CL	Caudate lobe of liver	IMV	Inferior mesenteric vein	PB	Body of pancreas
D	Diaphragm	IVC	Inferior vena cava	PC	Portal confluence
DBM	Deep back muscles	LC	Left crus of diaphragm		

4.90 TRANSVERSE (AXIAL) MRIs OF ABDOMEN

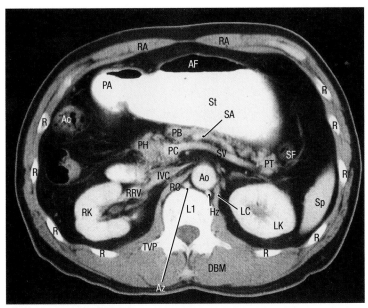

E

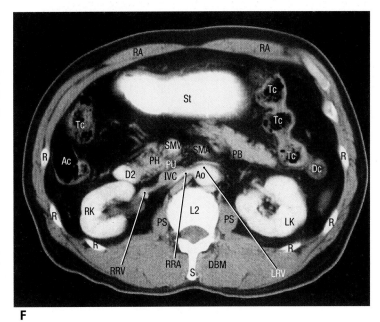

F

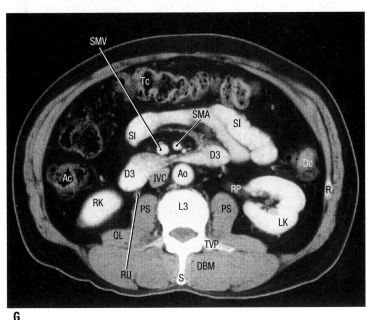

G

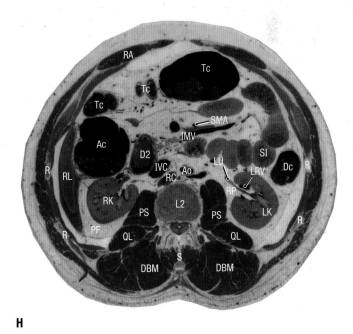

H

Key (continued)

PF	Perinephric fat	RC	Right crus of diaphragm	RRV	Right renal vein	Sp	Spleen
PH	Head of pancreas	RF	Retroperitoneal fat	RU	Right ureter	St	Stomach
PS	Psoas muscle	RG	Right suprarenal gland	S	Spinous process	SV	Splenic vein
PT	Tail of pancreas	RHV	Right hepatic vein	SA	Splenic artery	Tc	Transverse colon
PU	Uncinate process of pancreas	RIL	Right inferior lobe of lung	SC	Spinal cord	TVP	Transverse process
PV	Hepatic portal vein	RK	Right kidney	SF	Splenic flexure	Xp	Xiphoid process
QL	Quadratus lumborum	RL	Right lobe of liver	SI	Small intestine		
R	Rib	RP	Renal pelvis	SMA	Superior mesenteric artery		
RA	Rectus abdominis	RRA	Right renal artery	SMV	Superior mesenteric vein		

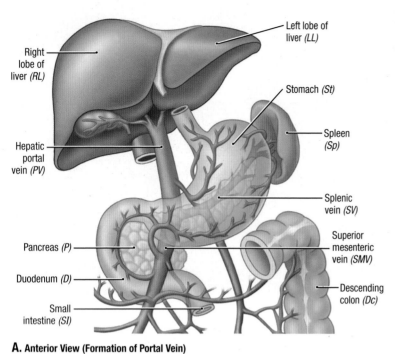

A. Anterior View (Formation of Portal Vein)

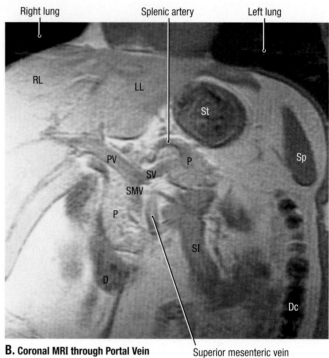

B. Coronal MRI through Portal Vein

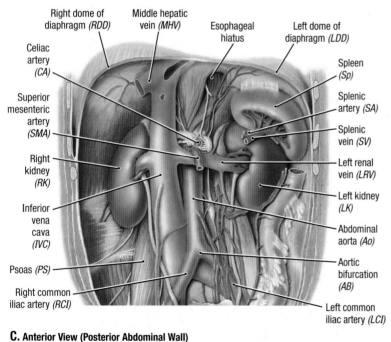

C. Anterior View (Posterior Abdominal Wall)

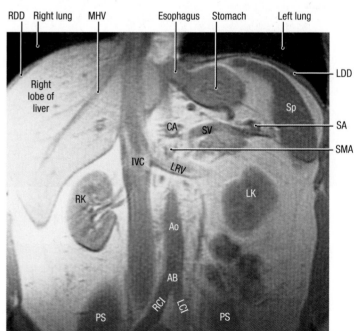

D. Coronal MRI through Inferior Vena Cava

4.91   CORONAL MRIs OF ABDOMEN

A. Illustration of formation of the hepatic portal vein. **B.** Coronal MRI through hepatic portal vein. **C.** Illustration of posterior abdominal wall. **D.** Coronal MRI through inferior vena cava and right and left kidneys.

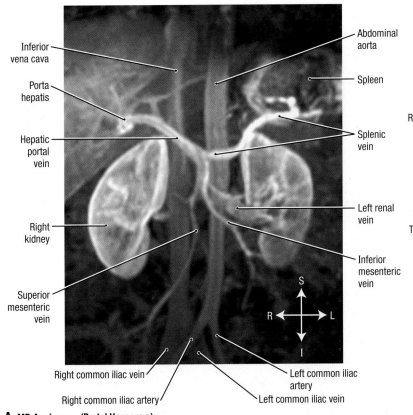

A. MR Angiogram (Portal Venogram)

Labels: Inferior vena cava, Porta hepatis, Hepatic portal vein, Right kidney, Superior mesenteric vein, Abdominal aorta, Spleen, Splenic vein, Left renal vein, Inferior mesenteric vein, Right common iliac vein, Right common iliac artery, Left common iliac artery, Left common iliac vein

Orientation: S, I, R, L

C. Sagittal MRI through Aorta and Celiac and Superior Mesenteric Arteries

Labels: LIL, LL, GE, RC, Ao, T12, SV, P, CA, St, SMA, L1, LRV, Tc, Do, L2, L3, L4

Key for C:			
Ao	Aorta	P	Pancreas
CA	Celiac artery	RC	Right crus
Do	Duodenum	SMA	Superior mesenteric artery
GE	Gastro-esophageal junction	St	Stomach
LIL	Inferior lobe of left lung	SV	Splenic vein
LL	Left lobe of liver	Tc	Transverse colon
LRV	Left renal vein		

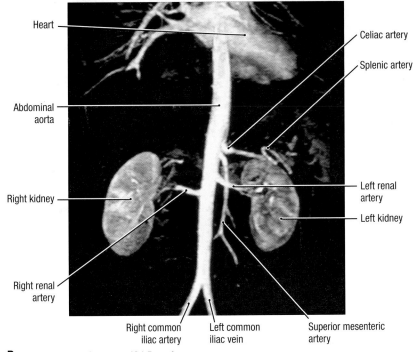

B. MR Angiogram of Aorta and Its Branches

Labels: Heart, Abdominal aorta, Right kidney, Right renal artery, Right common iliac artery, Celiac artery, Splenic artery, Left renal artery, Left kidney, Left common iliac vein, Superior mesenteric artery

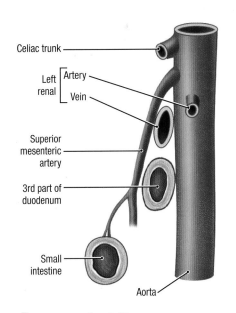

D. Lateral View (from Left)

Labels: Celiac trunk, Left renal Artery / Vein, Superior mesenteric artery, 3rd part of duodenum, Small intestine, Aorta

MR ANGIOGRAMS AND SAGITTAL MRI OF ABDOMEN

4.92

A. Magnetic resonance angiogram (portal venogram) demonstrating the tributaries and formation of the hepatic portal vein. **B.** MR angiogram of aorta and branches. **C.** Sagittal MRI through aorta showing the relationships of the celiac and superior mesenteric arteries to surrounding structures. **D.** Schematic illustration of relationships of superior mesenteric artery.

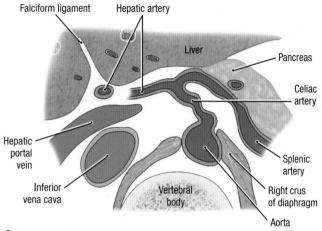

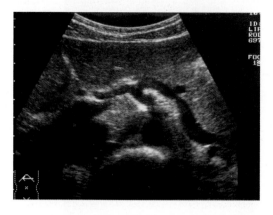

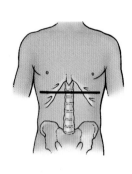

A. Transverse US Scan through Celiac Axis (Area of Branching)

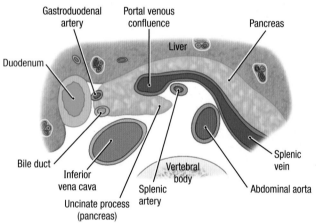

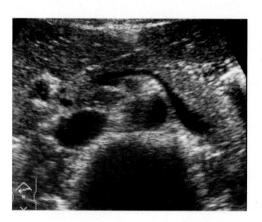

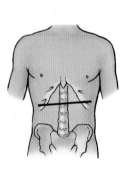

B. Transverse US Scan through Splenic View

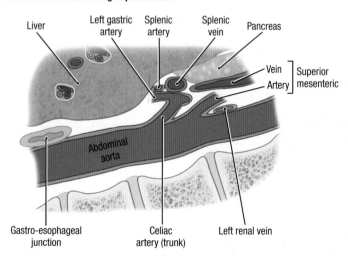

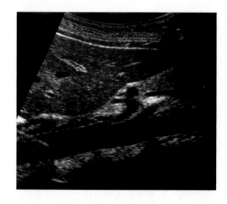

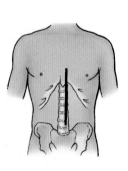

C. Midsagittal US Scan through Abdominal Aorta

4.93 ULTRASOUND SCANS OF ABDOMEN

A. Transverse ultrasound scan through celiac artery (axis). **B.** Transverse ultrasound scan through pancreas. **C.** and **D.** Sagittal ultrasound scans through the aorta, celiac trunk, and superior mesenteric artery (**D** with Doppler). **E.** Transverse ultrasound scan at hilum of left kidney with the left renal artery and vein (with Doppler). **F.** Sagittal ultrasound scan of the right kidney.

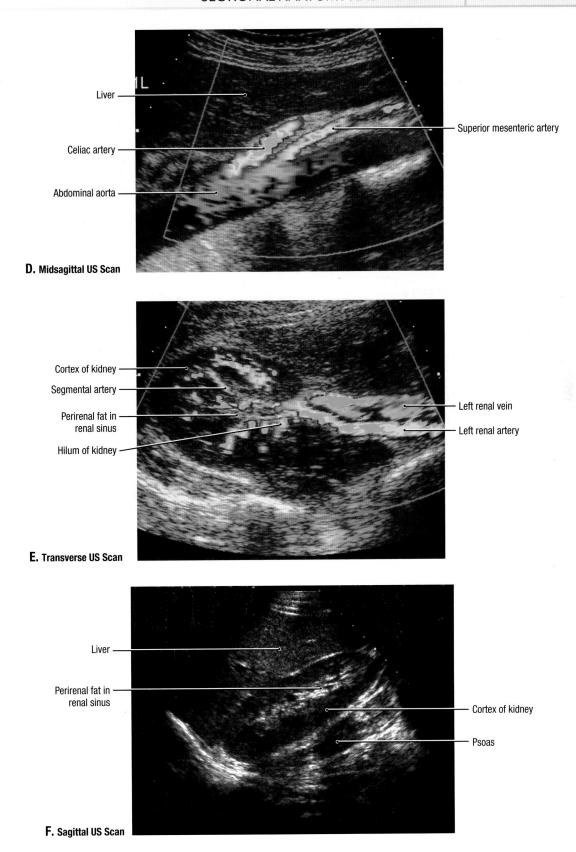

D. Midsagittal US Scan

Liver

Celiac artery

Abdominal aorta

Superior mesenteric artery

E. Transverse US Scan

Cortex of kidney

Segmental artery

Perirenal fat in renal sinus

Hilum of kidney

Left renal vein

Left renal artery

F. Sagittal US Scan

Liver

Perirenal fat in renal sinus

Cortex of kidney

Psoas

ULTRASOUND SCANS OF ABDOMEN (continued)

4.93

A major advantage of ultrasonography is its ability to produce real-time images, demonstrating motion of structures and flow within blood vessels. In Doppler ultrasonography (**D** and **E**), the shifts in frequency between emitted ultrasonic waves and their echoes are used to measure the velocities of moving objects. This technique is based on the principle of the Doppler effect. Blood flow through vessels is displayed in color, superimposed on the two-dimensional cross-sectional image (slow flow: *blue*, fast flow: *orange*).

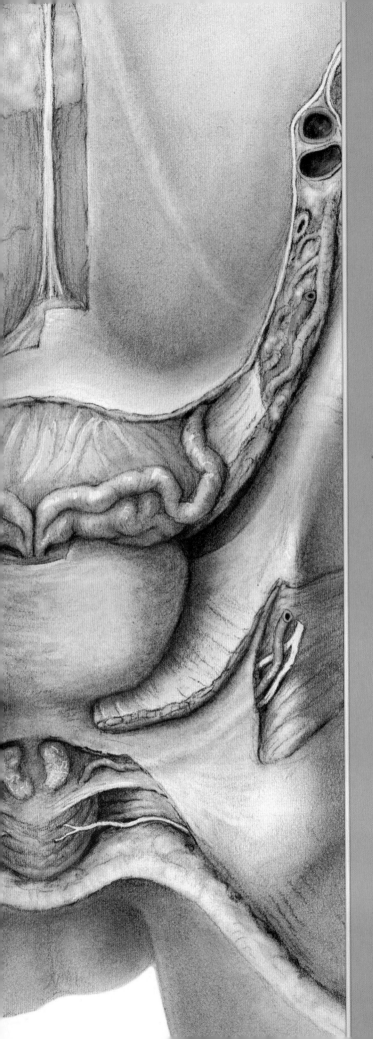

CHAPTER 5

Pelvis and Perineum

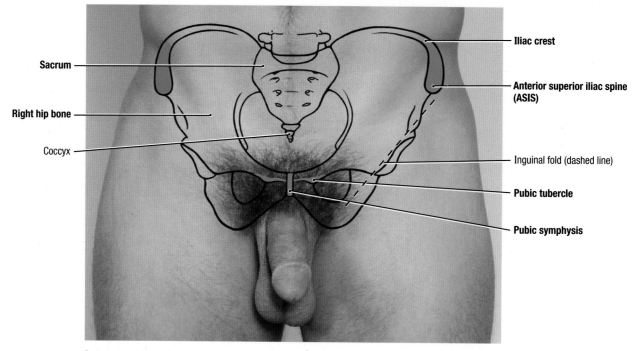

Iliac crest

Sacrum

Anterior superior iliac spine (ASIS)

Right hip bone

Coccyx

Inguinal fold (dashed line)

Pubic tubercle

Pubic symphysis

A. Anterior View

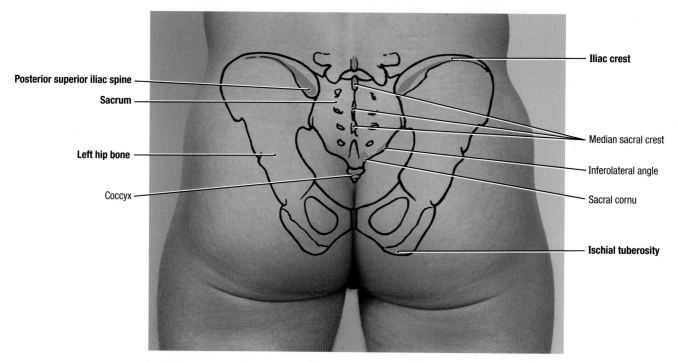

Iliac crest

Posterior superior iliac spine

Sacrum

Median sacral crest

Left hip bone

Inferolateral angle

Coccyx

Sacral cornu

Ischial tuberosity

B. Posterior View

5.1   SURFACE ANATOMY OF MALE PELVIC GIRDLE

The pelvic girdle (bony pelvis) is a basin-shaped ring of three bones (right and left hip bones and sacrum) that connects the vertebral column to the femora. **Palpable features** *(green)* **should be symmetrical across the midline. A.** The anterior third of the iliac crests are subcutaneous and usually easily palpable. The remainder of the crests may also be palpable, depending on the thickness of the overlying subcutaneous tissue (fat). The inguinal ligament spans between the palpable anterior superior iliac spine (ASIS) and pubic tubercle, located superior to the lateral and medial ends of the inguinal fold. **B.** The posterior superior iliac spine (PSIS) is usually palpable and often lies deep to a visible dimple, indicating the S2 vertebral level. The ischial tuberosities may be palpated when the hip joint is flexed.

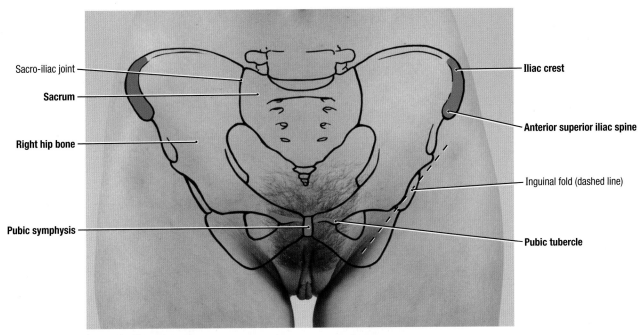

Sacro-iliac joint

Sacrum

Right hip bone

Pubic symphysis

Iliac crest

Anterior superior iliac spine

Inguinal fold (dashed line)

Pubic tubercle

A. Anterior View

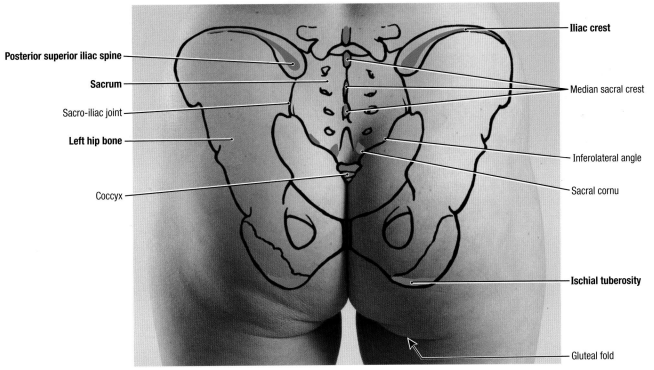

Posterior superior iliac spine

Sacrum

Sacro-iliac joint

Left hip bone

Coccyx

Iliac crest

Median sacral crest

Inferolateral angle

Sacral cornu

Ischial tuberosity

Gluteal fold

B. Posterior View

SURFACE ANATOMY OF FEMALE PELVIC GIRDLE

5.2

The female pelvic girdle is relatively wider and shallower than that of the male, related to its additional roles of bearing the weight of the gravid uterus in late **pregnancy** and allowing passage of the fetus through the pelvic outlet during childbirth **(parturition)**. **A.** Palpable features *(green)*: The hip bones are joined anteriorly at the pubic symphysis. The presence of a thick overlying pubic fat pad forming the mons pubis may interfere with palpation of the pubic tubercles and symphysis. **B.** Posteriorly, the hip bones articulate with the sacrum at the sacro-iliac joints.

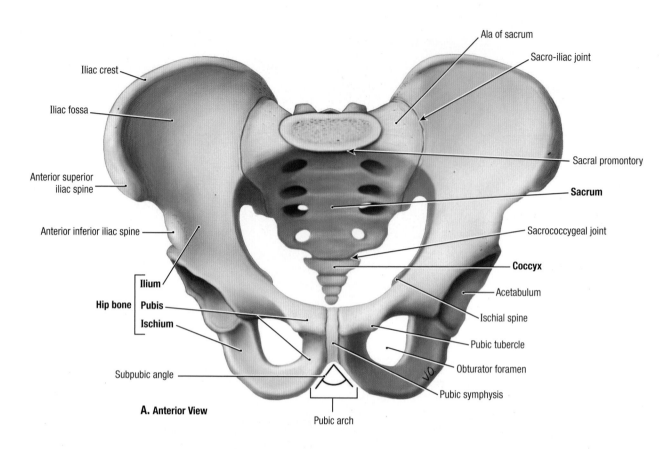

A. Anterior View

Iliac crest

Iliac fossa

Anterior superior iliac spine

Anterior inferior iliac spine

Ilium

Hip bone **Pubis**

Ischium

Subpubic angle

Pubic arch

Ala of sacrum

Sacro-iliac joint

Sacral promontory

Sacrum

Sacrococcygeal joint

Coccyx

Acetabulum

Ischial spine

Pubic tubercle

Obturator foramen

Pubic symphysis

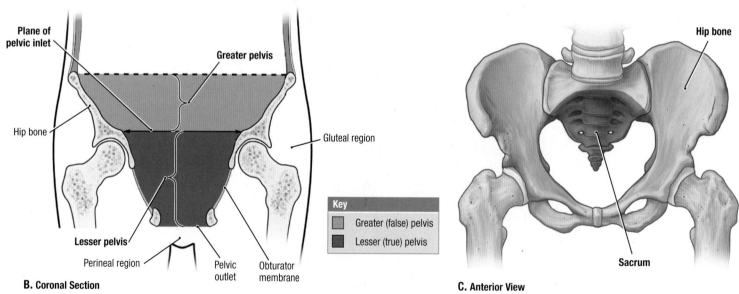

B. Coronal Section

Plane of pelvic inlet

Greater pelvis

Hip bone

Lesser pelvis

Perineal region

Pelvic outlet

Obturator membrane

Gluteal region

Key
Greater (false) pelvis
Lesser (true) pelvis

C. Anterior View

Hip bone

Sacrum

5.3 BONES AND DIVISIONS OF PELVIS

A. Bones of pelvis. The three bones composing the pelvis are the pubis, ischium, and ilium. **B.** and **C.** Lesser and greater pelvis, schematic illustrations. The plane of the pelvic inlet (*double-headed arrow* in **B**) separates the greater pelvis (part of the abdominal cavity) from the lesser pelvis (pelvic cavity).

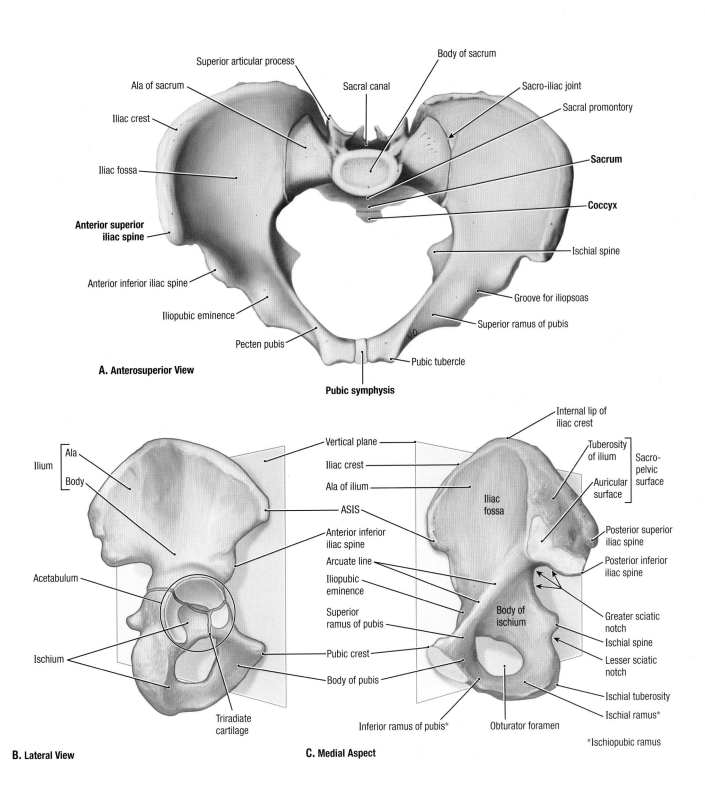

A. Anterosuperior View

- Superior articular process
- Body of sacrum
- Ala of sacrum
- Sacral canal
- Sacro-iliac joint
- Iliac crest
- Sacral promontory
- Iliac fossa
- **Sacrum**
- **Anterior superior iliac spine**
- **Coccyx**
- Anterior inferior iliac spine
- Ischial spine
- Iliopubic eminence
- Groove for iliopsoas
- Pecten pubis
- Superior ramus of pubis
- Pubic tubercle
- **Pubic symphysis**

B. Lateral View

- Ilium
  - Ala
  - Body
- Acetabulum
- Ischium
- Triradiate cartilage

C. Medial Aspect

- Vertical plane
- Internal lip of iliac crest
- Iliac crest
- Tuberosity of ilium
- Ala of ilium
- Sacro-pelvic surface
- ASIS
- Iliac fossa
- Auricular surface
- Anterior inferior iliac spine
- Arcuate line
- Posterior superior iliac spine
- Iliopubic eminence
- Posterior inferior iliac spine
- Superior ramus of pubis
- Body of ischium
- Pubic crest
- Greater sciatic notch
- Body of pubis
- Ischial spine
- Lesser sciatic notch
- Inferior ramus of pubis*
- Obturator foramen
- Ischial tuberosity
- Ischial ramus*
- *Ischiopubic ramus

PELVIS, ANATOMICAL POSITION

5.4

A. Pelvic girdle. **B.** Placement of hip bone in anatomical position. In the anatomical position, (1) the anterior superior iliac spine *(ASIS)* and the anterior aspect of the pubis lie in the same vertical plane and (2) the sacrum is located superiorly, the coccyx posteriorly, and the pubic symphysis antero-inferiorly. **C.** Features of hip bone.

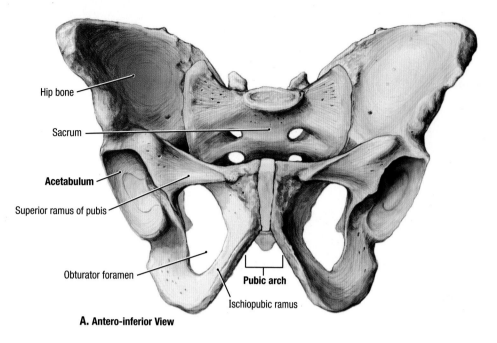

Hip bone

Sacrum

Acetabulum

Superior ramus of pubis

Obturator foramen

Pubic arch

Ischiopubic ramus

A. Antero-inferior View

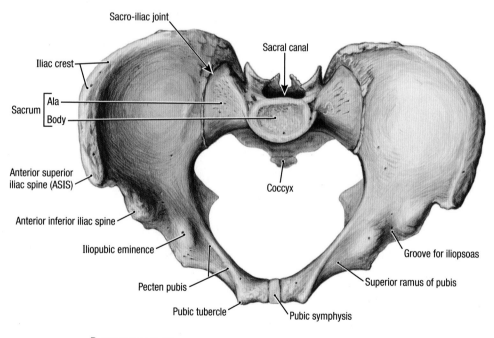

Sacro-iliac joint

Sacral canal

Iliac crest

Sacrum — Ala
          Body

Anterior superior
iliac spine (ASIS)

Anterior inferior iliac spine

Iliopubic eminence

Pecten pubis

Pubic tubercle

Coccyx

Groove for iliopsoas

Superior ramus of pubis

Pubic symphysis

B. Anterosuperior View

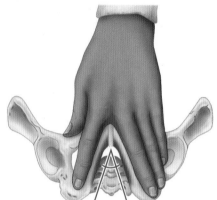

C. Subpubic Angle
"V" shaped

5.5   MALE PELVIC GIRDLE

TABLE 5.1	DIFFERENCES BETWEEN MALE AND FEMALE PELVES	
Bony Pelvis	*Male*	*Female*
General structure	Thicker and heavier	Thinner and lighter
Greater pelvis (pelvis major)	Deeper	Shallower
Lesser pelvis (pelvis minor)	Narrower and deeper, tapering	Wider and shallower, cylindrical
Pelvic inlet (superior pelvic aperture)	Heart shaped, narrower	More oval or rounded, wider
Sacrum/coccyx	More curved	Less curved

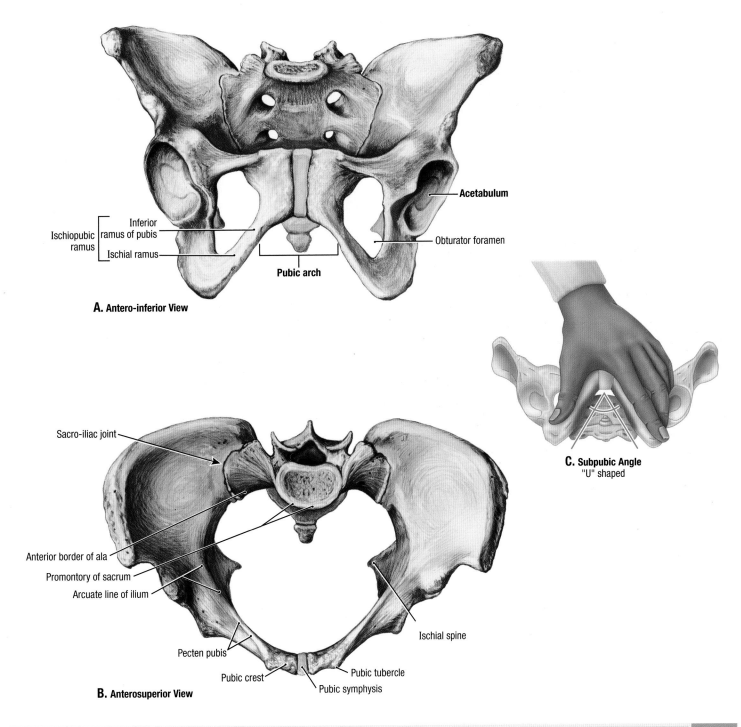

A. Antero-inferior View

Inferior
ramus of pubis
Ischiopubic
ramus
Ischial ramus

Acetabulum

Obturator foramen

Pubic arch

C. Subpubic Angle
"U" shaped

Sacro-iliac joint

Anterior border of ala

Promontory of sacrum

Arcuate line of ilium

Pecten pubis

Ischial spine

Pubic crest

Pubic tubercle

Pubic symphysis

B. Anterosuperior View

FEMALE PELVIC GIRDLE

5.6

TABLE 5.1	DIFFERENCES BETWEEN MALE AND FEMALE PELVES (*continued*)		
Bony Pelvis	*Male*		*Female*
Pelvic outlet (inferior pelvic aperture)	Comparatively small		Comparatively large
Pubic arch and subpubic angle	Narrower		Wider
Obturator foramen	Round		Oval
Acetabulum	Large		Small

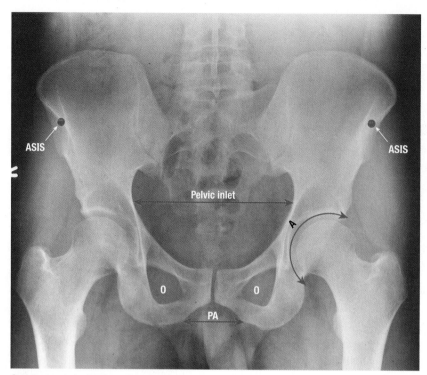

A. Anteroposterior Radiograph, Male Pelvis

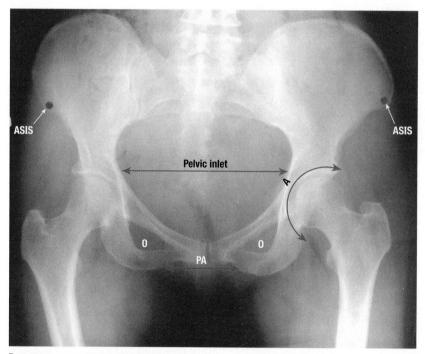

B. Anteroposterior Radiograph, Female Pelvis

5.7   RADIOGRAPHS OF PELVIS

A. Male. **B.** Female. Some of the main differences of male and female pelves are listed in Table 5.1. The radiographs highlight some of these differences. *A*, acetabulum; *ASIS*, anterior superior iliac spine; *O*, obturator foramen; *PA*, pubic arch.

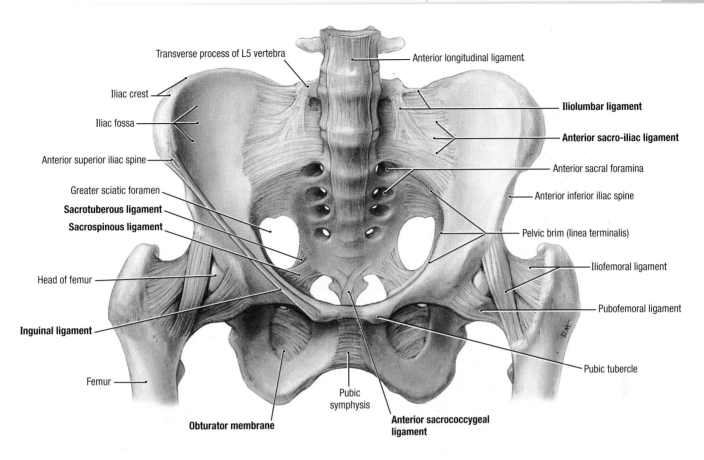

Transverse process of L5 vertebra

Iliac crest

Iliac fossa

Anterior superior iliac spine

Greater sciatic foramen

Sacrotuberous ligament

Sacrospinous ligament

Head of femur

Inguinal ligament

Femur

Anterior longitudinal ligament

Iliolumbar ligament

Anterior sacro-iliac ligament

Anterior sacral foramina

Anterior inferior iliac spine

Pelvic brim (linea terminalis)

Iliofemoral ligament

Pubofemoral ligament

Pubic tubercle

Obturator membrane

Pubic symphysis

Anterior sacrococcygeal ligament

A. Anterior View

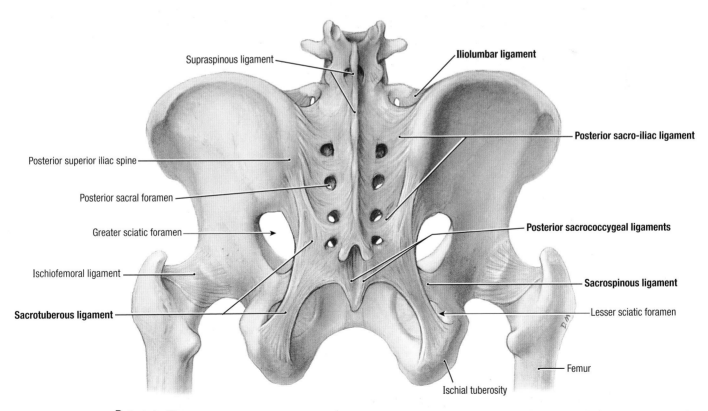

Supraspinous ligament

Iliolumbar ligament

Posterior superior iliac spine

Posterior sacral foramen

Greater sciatic foramen

Ischiofemoral ligament

Sacrotuberous ligament

Posterior sacro-iliac ligament

Posterior sacrococcygeal ligaments

Sacrospinous ligament

Lesser sciatic foramen

Femur

Ischial tuberosity

B. Posterior View

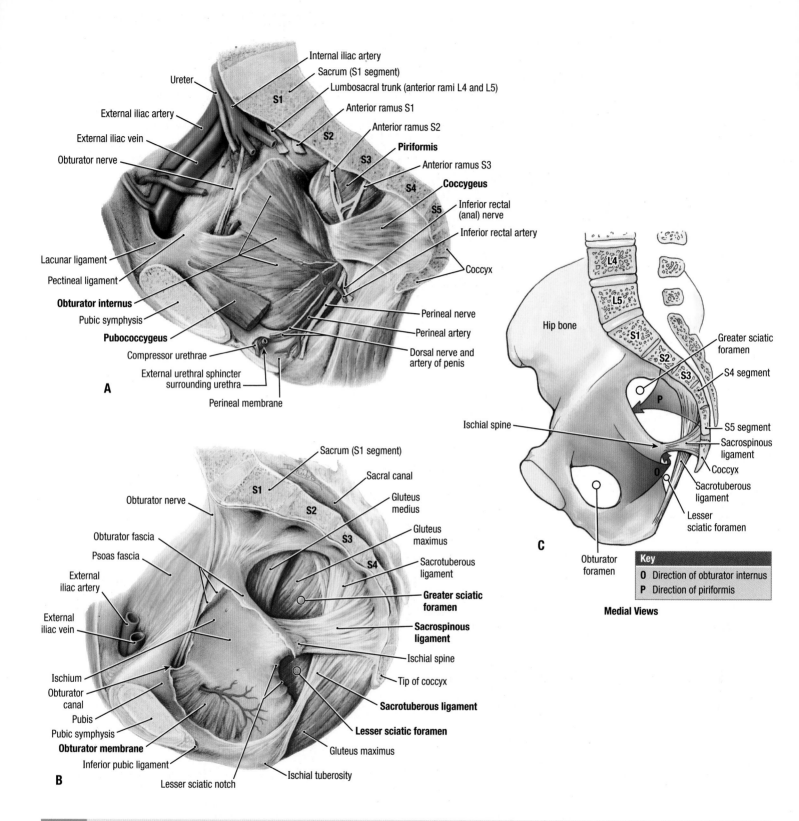

Internal iliac artery
Sacrum (S1 segment)
Ureter
Lumbosacral trunk (anterior rami L4 and L5)
External iliac artery
Anterior ramus S1
External iliac vein
Anterior ramus S2
Obturator nerve
Piriformis
Anterior ramus S3
Coccygeus
Inferior rectal (anal) nerve
Inferior rectal artery
Lacunar ligament
Coccyx
Pectineal ligament
Obturator internus
Perineal nerve
Pubic symphysis
Perineal artery
Pubococcygeus
Dorsal nerve and artery of penis
Compressor urethrae
External urethral sphincter surrounding urethra
Perineal membrane

A

Sacrum (S1 segment)
Sacral canal
Obturator nerve
Gluteus medius
Obturator fascia
Gluteus maximus
Psoas fascia
External iliac artery
Sacrotuberous ligament
External iliac vein
Greater sciatic foramen
Sacrospinous ligament
Ischium
Ischial spine
Obturator canal
Tip of coccyx
Pubis
Pubic symphysis
Sacrotuberous ligament
Obturator membrane
Lesser sciatic foramen
Inferior pubic ligament
Gluteus maximus
Lesser sciatic notch
Ischial tuberosity

B

Hip bone
Greater sciatic foramen
S4 segment
Ischial spine
S5 segment
Sacrospinous ligament
Coccyx
Sacrotuberous ligament
Lesser sciatic foramen
Obturator foramen

C

Key

O Direction of obturator internus
P Direction of piriformis

Medial Views

5.9 OBTURATOR INTERNUS AND PIRIFORMIS

- On the lateral pelvic wall, the obturator foramen is closed by the obturator membrane except for the obturator canal; the obturator internus muscle attaches to the obturator membrane and surrounding bone and exits the lesser pelvis through the lesser sciatic foramen; obturator fascia lies on the medial surface of the muscle.

- Piriformis lies on the posterolateral pelvic wall and leaves the lesser pelvis through the greater sciatic foramen.

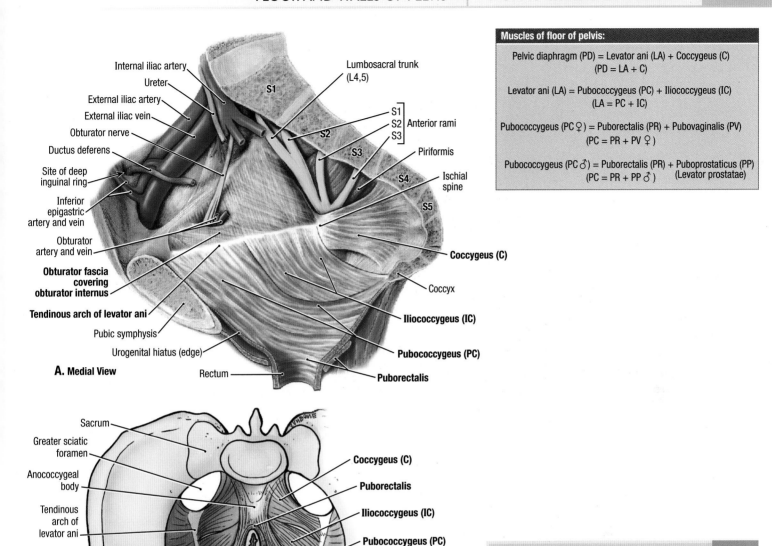

Muscles of floor of pelvis:

Pelvic diaphragm (PD) = Levator ani (LA) + Coccygeus (C)
(PD = LA + C)

Levator ani (LA) = Pubococcygeus (PC) + Iliococcygeus (IC)
(LA = PC + IC)

Pubococcygeus (PC ♀) = Puborectalis (PR) + Pubovaginalis (PV)
(PC = PR + PV ♀)

Pubococcygeus (PC ♂) = Puborectalis (PR) + Puboprostaticus (PP)
(PC = PR + PP ♂)   (Levator prostatae)

A. Medial View

B. Anterosuperior View

MUSCLES OF PELVIC DIAPHRAGM 5.10

A. The pelvic floor is formed by the funnel- or bowl-shaped pelvic diaphragm. The funnel shape can be seen in a medial view of a median section. **B.** The bowl shape from a superior view.

TABLE 5.2	**MUSCLES OF PELVIC WALLS AND FLOOR**				
Boundary	*Muscle*	*Proximal Attachment*	*Distal Attachment*	*Innervation*	*Main Action*
Lateral wall	Obturator internus	Pelvic surfaces of ilium and ischium, obturator membrane	Greater trochanter of femur	Nerve to obturator internus (L5, S1, S2)	Rotates hip joint laterally; assists in holding head of femur in acetabulum
Posterolateral wall	Piriformis	Pelvic surface of S2–S4 segments, superior margin of greater sciatic notch, sacrotuberous ligament		Anterior rami of S1 and S2	Rotates hip joint laterally; abducts hip joint; assists in holding head of femur in acetabulum
Floor	Levator ani (pubococcygeus, puborectalis, and iliococcygeus)	Body of pubis, tendinous arch of obturator fascia, ischial spine	Perineal body, coccyx, anococcygeal ligament, walls of prostate or vagina, rectum, and anal canal	Nerve to levator ani (branches of S4), inferior anal (rectal) nerve, and coccygeal plexus	Forms most of pelvic diaphragm that helps support pelvic viscera and resists increases in intra-abdominal pressure
	Coccygeus (ischiococcygeus)	Ischial spine	Inferior end of sacrum and coccyx	Branches of S4 and S5 spinal nerves	Forms small part of pelvic diaphragm that supports pelvic viscera; flexes sacrococcygeal joints

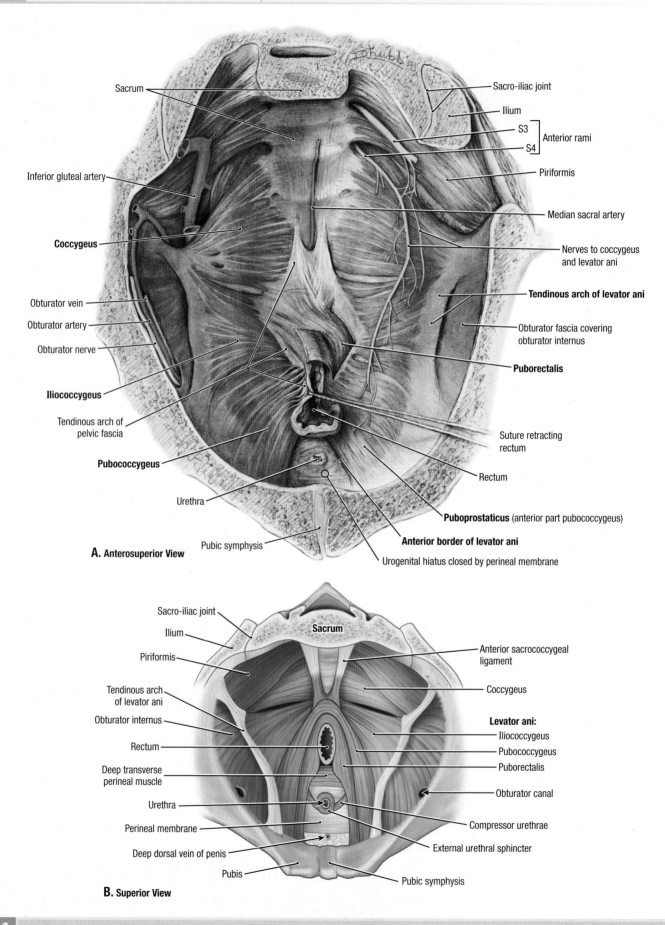

Sacrum

Sacro-iliac joint

Ilium

S3
S4 } Anterior rami

Piriformis

Inferior gluteal artery

Median sacral artery

Coccygeus

Nerves to coccygeus and levator ani

Tendinous arch of levator ani

Obturator vein

Obturator artery

Obturator fascia covering obturator internus

Obturator nerve

Puborectalis

Iliococcygeus

Tendinous arch of pelvic fascia

Suture retracting rectum

Pubococcygeus

Rectum

Urethra

Puboprostaticus (anterior part pubococcygeus)

A. Anterosuperior View

Pubic symphysis

Anterior border of levator ani

Urogenital hiatus closed by perineal membrane

Sacro-iliac joint

Ilium

Piriformis

Sacrum

Anterior sacrococcygeal ligament

Coccygeus

Tendinous arch of levator ani

Obturator internus

Levator ani:
Iliococcygeus
Pubococcygeus
Puborectalis

Rectum

Deep transverse perineal muscle

Obturator canal

Urethra

Compressor urethrae

Perineal membrane

External urethral sphincter

Deep dorsal vein of penis

Pubis

B. Superior View

Pubic symphysis

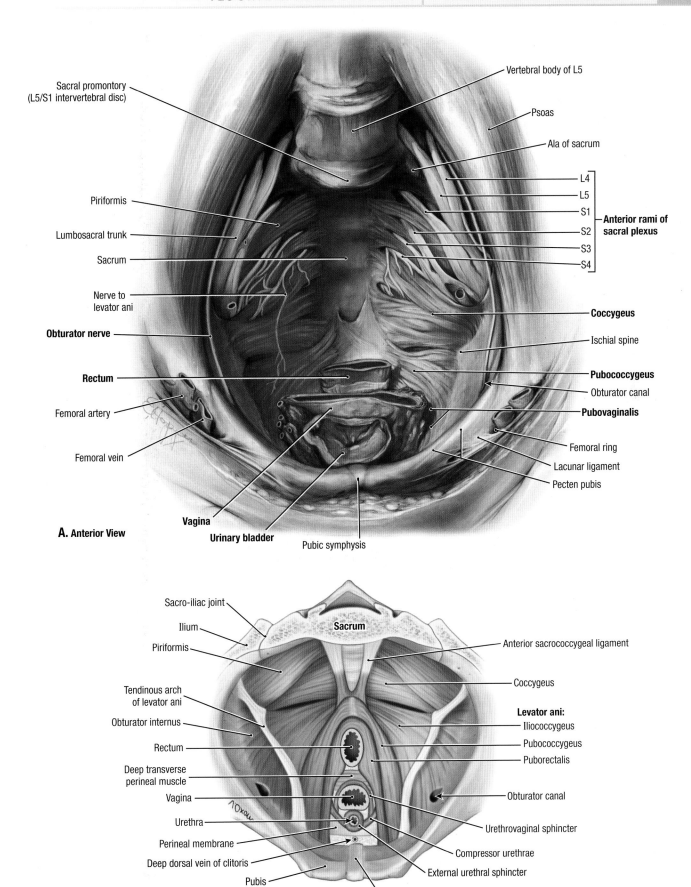

A. Anterior View

Sacral promontory
(L5/S1 intervertebral disc)

Piriformis

Lumbosacral trunk

Sacrum

Nerve to
levator ani

Obturator nerve

Rectum

Femoral artery

Femoral vein

Vagina

Urinary bladder

Pubic symphysis

Vertebral body of L5

Psoas

Ala of sacrum

L4
L5
S1
S2   **Anterior rami of
S3   sacral plexus**
S4

Coccygeus

Ischial spine

Pubococcygeus

Obturator canal

Pubovaginalis

Femoral ring

Lacunar ligament

Pecten pubis

B. Superior View

Sacro-iliac joint

Ilium

Piriformis

Tendinous arch
of levator ani

Obturator internus

Rectum

Deep transverse
perineal muscle

Vagina

Urethra

Perineal membrane

Deep dorsal vein of clitoris

Pubis

Pubic symphysis

Sacrum

Anterior sacrococcygeal ligament

Coccygeus

Levator ani:
Iliococcygeus

Pubococcygeus

Puborectalis

Obturator canal

Urethrovaginal sphincter

Compressor urethrae

External urethral sphincter

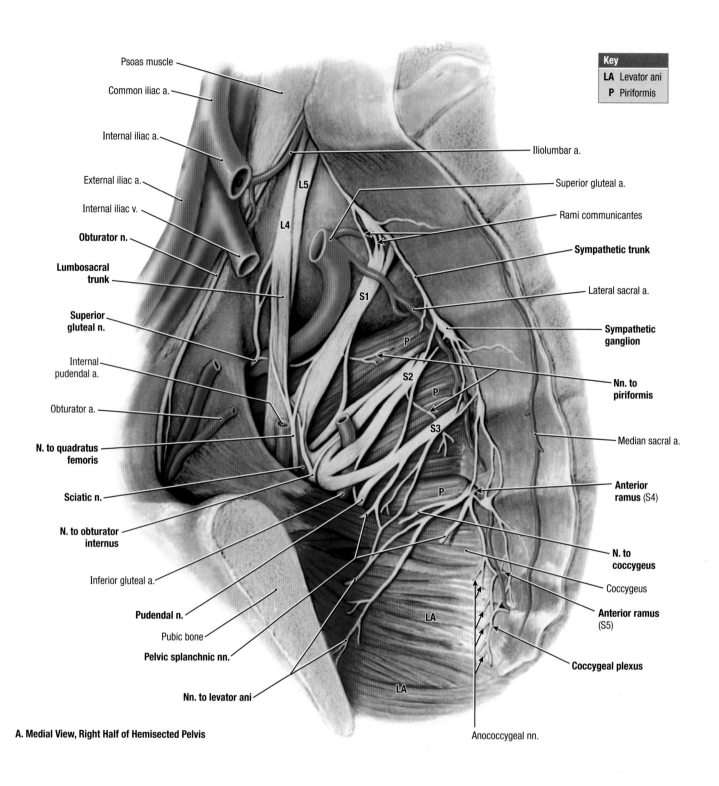

Psoas muscle

Common iliac a.

Internal iliac a.

External iliac a.

Internal iliac v.

Obturator n.

Lumbosacral trunk

Superior gluteal n.

Internal pudendal a.

Obturator a.

N. to quadratus femoris

Sciatic n.

N. to obturator internus

Inferior gluteal a.

Pudendal n.

Pubic bone

Pelvic splanchnic nn.

Nn. to levator ani

L5

L4

S1

P

S2

P

S3

P

LA

LA

Key
LA Levator ani
P Piriformis

Iliolumbar a.

Superior gluteal a.

Rami communicantes

Sympathetic trunk

Lateral sacral a.

Sympathetic ganglion

Nn. to piriformis

Median sacral a.

Anterior ramus (S4)

N. to coccygeus

Coccygeus

Anterior ramus (S5)

Coccygeal plexus

Anococcygeal nn.

A. Medial View, Right Half of Hemisected Pelvis

5.13 SACRAL AND COCCYGEAL NERVE PLEXUSES

A. Dissection.

• The sympathetic trunk or its ganglia send rami communicantes to each sacral and coccygeal nerve.

• The anterior ramus from L4 joins that of L5 to form the lumbosacral trunk.

• The sciatic nerve arises from anterior rami of L4, L5, S1, S2, and S3; the pudendal nerve from S2, S3, and S4; and the coccygeal plexus from S4, S5, and coccygeal segments.

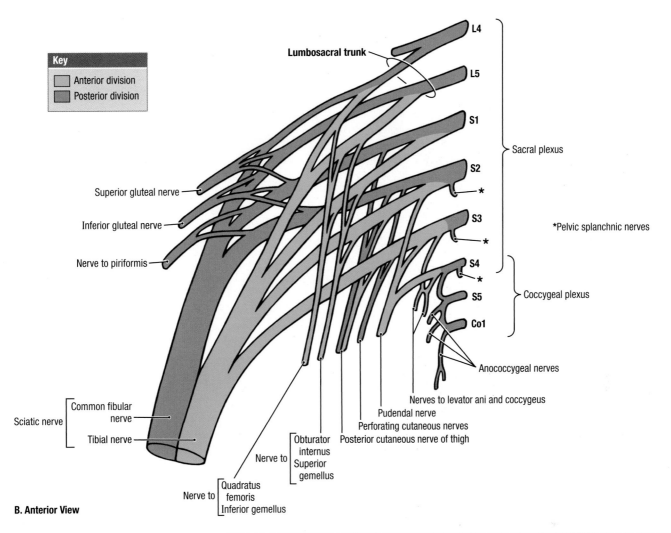

Key
- Anterior division
- Posterior division

Lumbosacral trunk

L4
L5
S1
S2
*
S3
*
S4
*
S5
Co1

Sacral plexus

*Pelvic splanchnic nerves

Coccygeal plexus

Superior gluteal nerve
Inferior gluteal nerve
Nerve to piriformis

Anococcygeal nerves
Nerves to levator ani and coccygeus
Pudendal nerve
Perforating cutaneous nerves
Posterior cutaneous nerve of thigh

Common fibular nerve
Sciatic nerve
Tibial nerve

Obturator internus
Nerve to
Superior gemellus

Nerve to
Quadratus femoris
Inferior gemellus

B. Anterior View

SACRAL AND COCCYGEAL NERVE PLEXUSES (*continued*)

5.13

B. Branches of anterior and posterior divisions of sacral and coccygeal plexuses.

TABLE 5.3	**NERVES OF SACRAL AND COCCYGEAL PLEXUSES**	
Nerve	*Origin*	*Distribution*
Sciatic: 1. Common fibular 2. Tibial	L4, L5, S1, S2 L4, L5, S1, S2, S3	Articular branches to hip joint and muscular branches to flexors of knee joint in thigh and all muscles in leg and foot
3. Superior gluteal	L4, L5, S1	Gluteus medius and gluteus minimus muscles
4. Nerve to quadratus femoris and inferior gemellus	L4, L5, S1	Quadratus femoris and inferior gemellus muscles
5. Inferior gluteal	L5, S1, S2	Gluteus maximus muscle
6. Nerve to obturator internus and superior gemellus	L5, S1, S2	Obturator internus and superior gemellus muscles
7. Nerve to piriformis	S1, S2	Piriformis muscle
8. Posterior cutaneous nerve of thigh	S1, S2, S3	Cutaneous branches to buttock and uppermost medial and posterior surfaces of thigh
9. Perforating cutaneous	S2, S3	Cutaneous branches to medial part of buttock
10. Pudendal	S2, S3, S4	Structures in perineum, sensory to genitalia, muscular branches to perineal muscles, external urethral sphincter, and external anal sphincter
11. Pelvic splanchnic	S2, S3, S4	Pelvic viscera via inferior hypogastric and pelvic plexuses
12. Nerves to levator ani and coccygeus	S3, S4	Levator ani and coccygeus muscles
13. Anococcygeal nerve	S4, S5, Co1	Penetrate coccygeal attachments of sacrospinous/sacrotuberous ligaments to supply overlying skin

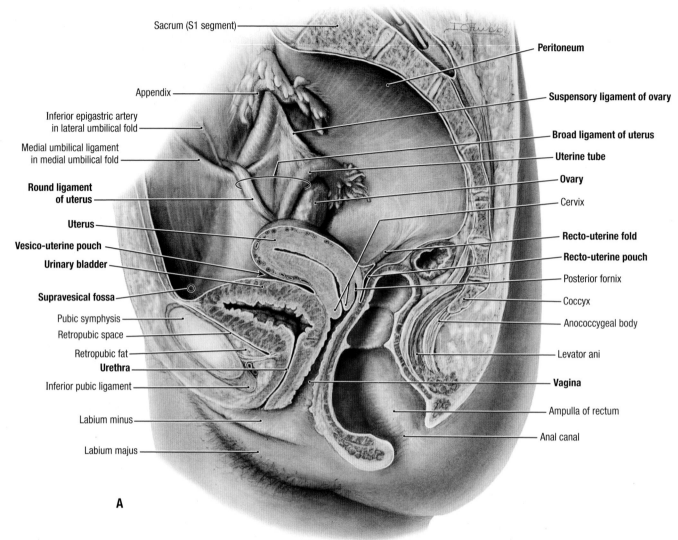

Sacrum (S1 segment)

Appendix

Inferior epigastric artery
in lateral umbilical fold

Medial umbilical ligament
in medial umbilical fold

**Round ligament
of uterus**

Uterus

Vesico-uterine pouch

Urinary bladder

Supravesical fossa

Pubic symphysis

Retropubic space

Retropubic fat

Urethra

Inferior pubic ligament

Labium minus

Labium majus

Peritoneum

Suspensory ligament of ovary

Broad ligament of uterus

Uterine tube

Ovary

Cervix

Recto-uterine fold

Recto-uterine pouch

Posterior fornix

Coccyx

Anococcygeal body

Levator ani

Vagina

Ampulla of rectum

Anal canal

A

(B) Peritoneal reflections in females

Peritoneum passes:
1. From the anterior abdominal wall
2. Superior to the pubic bone, forming supravesical fossa
3. On the superior surface of the urinary bladder
4. From the bladder to mid-uterus, forming the vesico-uterine pouch
5. On the fundus and body of the uterus, and posterior fornix of the vagina
6. Between the rectum and uterus, forming the recto-uterine pouch
7. On the anterior and lateral sides of the rectum
8. Posteriorly to become the sigmoid mesocolon

**Medial Views of Right Half of
Hemisected Female Pelvis**

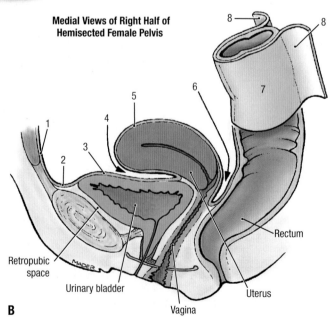

Retropubic
space

Urinary bladder

Vagina

Rectum

Uterus

5.14 PERITONEUM COVERING FEMALE PELVIC ORGANS

A. Organs *in situ* with peritoneal reflections. **B.** Schematic illustration of peritoneal reflections. The level of the supravesical fossa changes with filling and emptying of bladder.

B

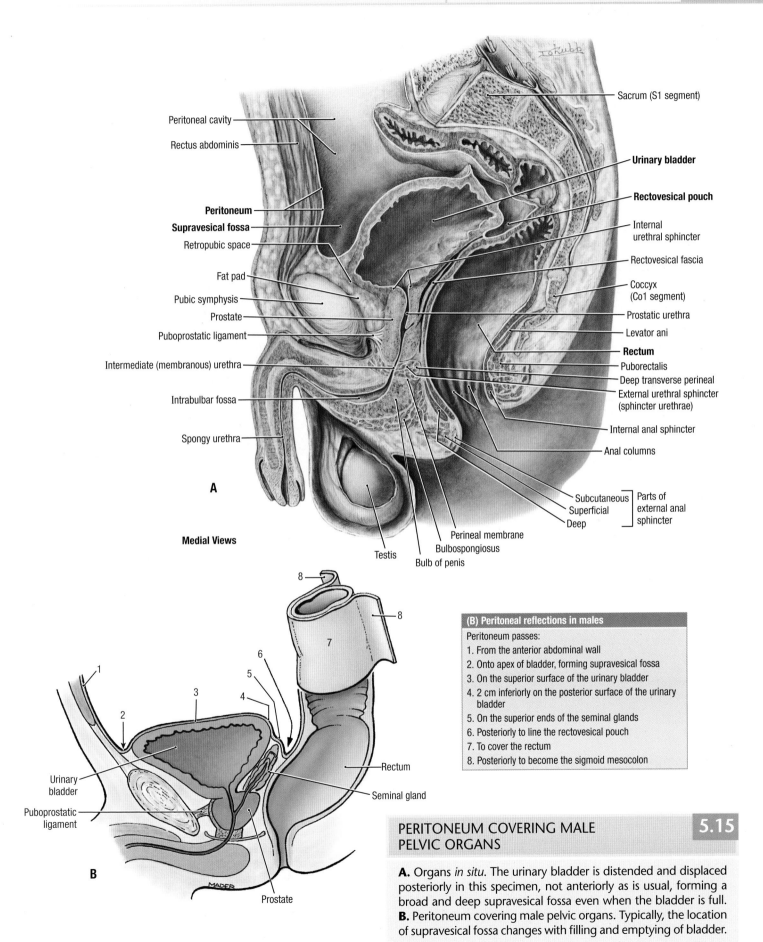

Sacrum (S1 segment)

Peritoneal cavity

Rectus abdominis

Urinary bladder

Rectovesical pouch

Peritoneum

Supravesical fossa

Internal
urethral sphincter

Retropubic space

Rectovesical fascia

Fat pad

Coccyx
(Co1 segment)

Pubic symphysis

Prostate

Prostatic urethra

Puboprostatic ligament

Levator ani

Intermediate (membranous) urethra

Rectum

Puborectalis

Deep transverse perineal

External urethral sphincter
(sphincter urethrae)

Intrabulbar fossa

Internal anal sphincter

Spongy urethra

Anal columns

A

Subcutaneous | Parts of
Superficial | external anal
Deep | sphincter

Medial Views

Perineal membrane

Bulbospongiosus

Testis

Bulb of penis

8

8

7

6

5

3

4

2

1

(B) Peritoneal reflections in males

Peritoneum passes:
1. From the anterior abdominal wall
2. Onto apex of bladder, forming supravesical fossa
3. On the superior surface of the urinary bladder
4. 2 cm inferiorly on the posterior surface of the urinary bladder
5. On the superior ends of the seminal glands
6. Posteriorly to line the rectovesical pouch
7. To cover the rectum
8. Posteriorly to become the sigmoid mesocolon

Urinary
bladder

Puboprostatic
ligament

Rectum

Seminal gland

B

Prostate

MADER

PERITONEUM COVERING MALE PELVIC ORGANS

5.15

A. Organs *in situ.* The urinary bladder is distended and displaced posteriorly in this specimen, not anteriorly as is usual, forming a broad and deep supravesical fossa even when the bladder is full. **B.** Peritoneum covering male pelvic organs. Typically, the location of supravesical fossa changes with filling and emptying of bladder.

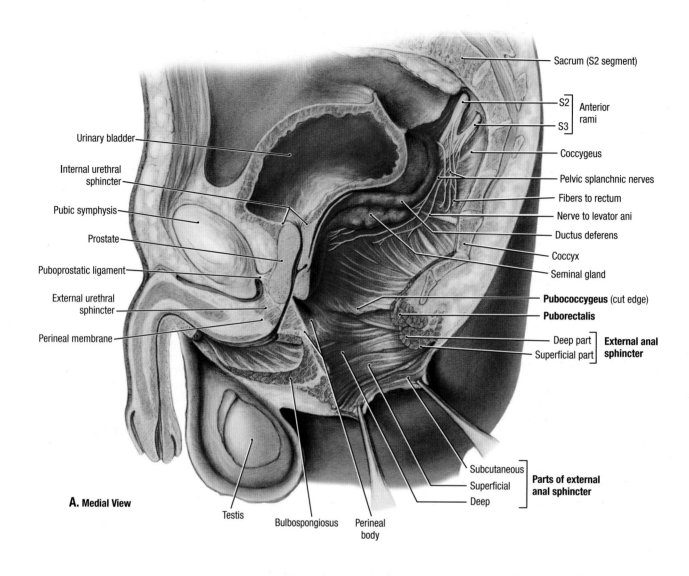

Urinary bladder

Internal urethral sphincter

Pubic symphysis

Prostate

Puboprostatic ligament

External urethral sphincter

Perineal membrane

Sacrum (S2 segment)

S2
S3 } Anterior rami

Coccygeus

Pelvic splanchnic nerves

Fibers to rectum

Nerve to levator ani

Ductus deferens

Coccyx

Seminal gland

Pubococcygeus (cut edge)

Puborectalis

Deep part
Superficial part } **External anal sphincter**

Subcutaneous
Superficial
Deep } **Parts of external anal sphincter**

A. Medial View

Testis   Bulbospongiosus   Perineal body

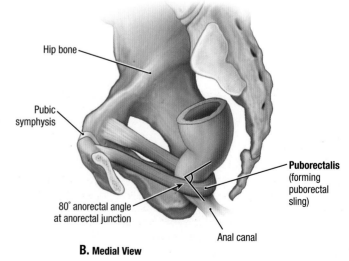

Hip bone

Pubic symphysis

80° anorectal angle at anorectal junction

Puborectalis (forming puborectal sling)

Anal canal

B. Medial View

5.16   ANAL SPHINCTERS AND ANAL CANAL

A. Levator ani, in right half of hemisected pelvis.

- The subcutaneous fibers of the external anal sphincter and overlying skin are reflected with forceps. The pubococcygeus muscle is cut to reveal the anal canal, to which it is, in part, attached.

B. Puborectalis.

- The innermost part of the levator ani/pubococcygeus muscle, the puborectalis, forms a U-shaped muscular "sling" around the anorectal junction, which maintains the anorectal (perineal) flexure.

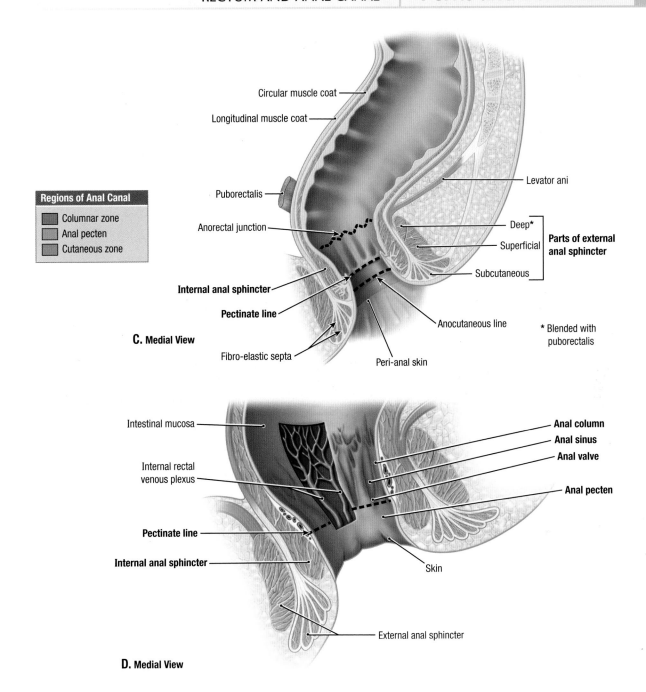

Regions of Anal Canal
- Columnar zone
- Anal pecten
- Cutaneous zone

Circular muscle coat

Longitudinal muscle coat

Puborectalis

Anorectal junction

Internal anal sphincter

Pectinate line

C. Medial View

Fibro-elastic septa

Levator ani

Deep*
Superficial — **Parts of external anal sphincter**
Subcutaneous

Anocutaneous line

* Blended with puborectalis

Peri-anal skin

Intestinal mucosa

Internal rectal venous plexus

Pectinate line

Internal anal sphincter

Anal column
Anal sinus
Anal valve

Anal pecten

Skin

D. Medial View

External anal sphincter

ANAL SPHINCTERS AND ANAL CANAL (*continued*) 5.16

C. External and internal anal sphincters.
- The internal anal sphincter is a thickening of the inner, circular muscular coat of the anal canal.
- The external anal sphincter has three often indistinct continuous zones: deep, superficial, and subcutaneous; the deep part intermingles with the puborectalis muscle posteriorly.
- The longitudinal muscle layer of the rectum separates the internal and external anal sphincters and terminates in the subcutaneous tissue and skin around the anus.

D. Features of the anal canal.
- The anal columns are 5 to 10 vertical folds of mucosa separated by anal sinuses and valves; they contain portions of the rectal venous plexus.

- The pecten is a smooth area of hairless stratified epithelium that lies between the anal valves superiorly and the inferior border of the internal anal sphincter inferiorly.
- The pectinate line is an irregular line at the base of the anal valves where the intestinal mucosa is continuous with the pecten; this indicates the junction of the superior part of the anal canal (derived from embryonic hindgut) and the inferior part of the anal canal (derived from the anal pit [proctodeum]). Innervation is visceral proximal to the line and somatic distally; lymphatic drainage is to the pararectal nodes proximally and to the superficial inguinal nodes distally.

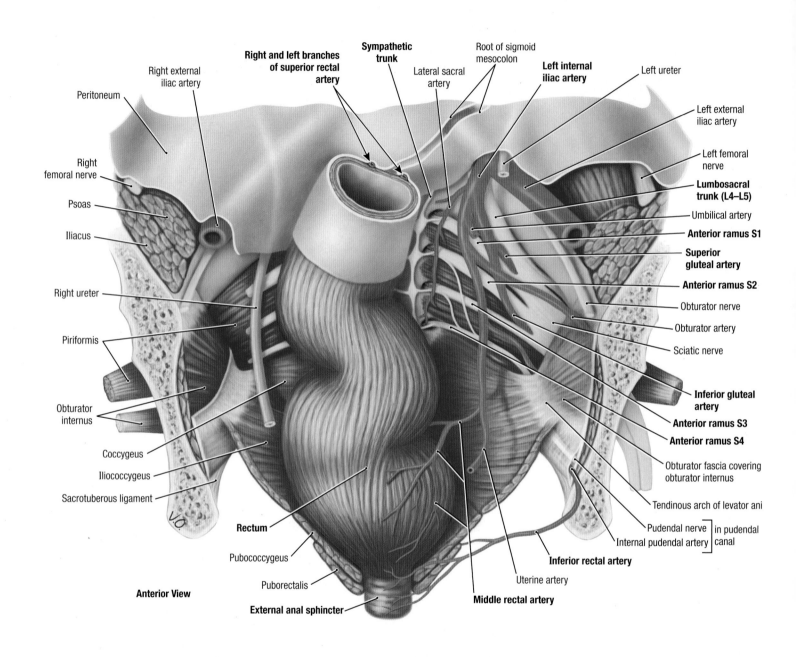

Right external
iliac artery

Peritoneum

Right
femoral nerve

Psoas

Iliacus

Right ureter

Piriformis

Obturator
internus

Coccygeus

Iliococcygeus

Sacrotuberous ligament

Anterior View

Rectum

Pubococcygeus

Puborectalis

External anal sphincter

**Right and left branches
of superior rectal
artery**

**Sympathetic
trunk**

Lateral sacral
artery

Root of sigmoid
mesocolon

**Left internal
iliac artery**

Left ureter

Left external
iliac artery

Left femoral
nerve

**Lumbosacral
trunk (L4–L5)**

Umbilical artery

Anterior ramus S1

**Superior
gluteal artery**

Anterior ramus S2

Obturator nerve

Obturator artery

Sciatic nerve

**Inferior gluteal
artery**

Anterior ramus S3

Anterior ramus S4

Obturator fascia covering
obturator internus

Tendinous arch of levator ani

Pudendal nerve ⎤ in pudendal
Internal pudendal artery ⎦ canal

Inferior rectal artery

Uterine artery

Middle rectal artery

5.17 RECTUM, ANAL CANAL, AND NEUROVASCULAR STRUCTURES OF POSTERIOR PELVIS

The pelvis is coronally bisected anterior to the rectum and anal canal. The superior gluteal artery often passes posteriorly between the anterior rami of L5 and S1, and the inferior gluteal artery between S2 and S3.

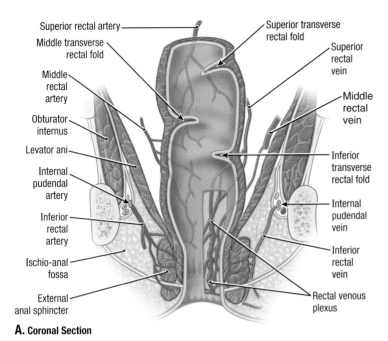

Superior rectal artery
Middle transverse rectal fold
Middle rectal artery
Obturator internus
Levator ani
Internal pudendal artery
Inferior rectal artery
Ischio-anal fossa
External anal sphincter

Superior transverse rectal fold
Superior rectal vein
Middle rectal vein
Inferior transverse rectal fold
Internal pudendal vein
Inferior rectal vein
Rectal venous plexus

A. Coronal Section

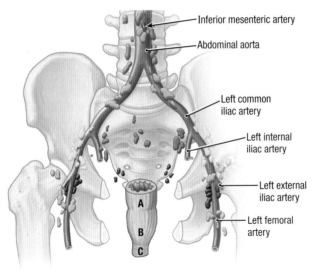

Inferior mesenteric artery
Abdominal aorta
Left common iliac artery
Left internal iliac artery
Left external iliac artery
Left femoral artery

A
B
C

B. Anterior View

Key for B

A	Superior half of rectum
B	Inferior half of rectum
C	Anal canal
	Lumbar
	Inferior mesenteric
	Common iliac
	Internal iliac
	External iliac
	Superficial inguinal
	Deep inguinal
	Sacral
→	Direction of flow of lymph

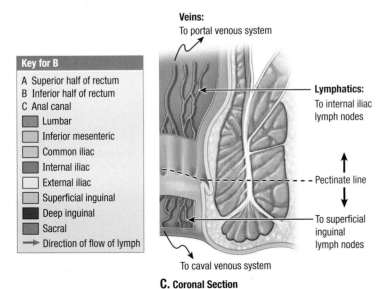

Veins:
To portal venous system
Lymphatics:
To internal iliac lymph nodes
Pectinate line
To superficial inguinal lymph nodes
To caval venous system

C. Coronal Section

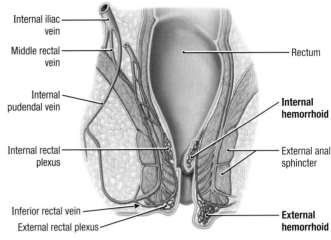

Internal iliac vein
Middle rectal vein
Internal pudendal vein
Internal rectal plexus
Inferior rectal vein
External rectal plexus

Rectum
Internal hemorrhoid
External anal sphincter
External hemorrhoid

D. Anterior Views of Coronal Section

VASCULATURE AND LYMPHATIC DRAINAGE OF RECTUM

5.18

A. Arterial and venous drainage. **B.** Lymphatic drainage. **C.** Venous and lymphatic drainage superior and inferior to the pectinate line. **D.** Hemorrhoids. **Internal hemorrhoids** (piles) are prolapses of rectal mucosa containing the normally dilated veins of the internal rectal venous plexus. Internal hemorrhoids are thought to result from a breakdown of the muscularis mucosae, a smooth muscle layer deep to the mucosa. Internal hemorrhoids that prolapse through the anal canal are often compressed by the contracted sphincters, impeding blood flow. As a result, they tend to strangulate and ulcerate. Because of the presence of abundant arteriovenous anastomoses, bleeding from internal hemorrhoids is characteristically bright red. The current practice is to treat only prolapsed, ulcerated internal hemorrhoids.

External hemorrhoids are thromboses (blood clots) in the veins of the external rectal venous plexus and are covered by skin. Predisposing factors for hemorrhoids include pregnancy, chronic constipation, and any disorder that impedes venous return including increased intra-abdominal pressure. The superior rectal vein drains into the inferior mesenteric vein, whereas the middle and inferior rectal veins drain through the systemic system into the inferior vena cava. Any abnormal increase in pressure in the valveless portal system or veins of the trunk may cause enlargement of the superior rectal veins, resulting in an increase in blood flow or stasis in the internal rectal venous plexus. In **portal hypertension** that occurs in relation to **hepatic cirrhosis**, the portacaval anastomosis (e.g., esophageal) may become varicose and rupture. Note that the veins of the rectal plexuses normally appear varicose (dilated and tortuous), even in newborns, and that internal hemorrhoids occur most commonly in the absence of portal hypertension.

Regarding pain from and the treatment of hemorrhoids, note that the anal canal superior to the pectinate line is visceral; thus, it is innervated by visceral afferent pain fibers, so that an incision or needle insertion into this region is painless. Internal hemorrhoids are not painful and can be treated without anesthesia. Inferior to the pectinate line, the anal canal is somatic, supplied by the inferior anal (rectal) nerves containing somatic sensory fibers. Therefore, it is sensitive to painful stimuli (e.g., to the prick of a hypodermic needle). External hemorrhoids can be painful but often resolve in a few days.

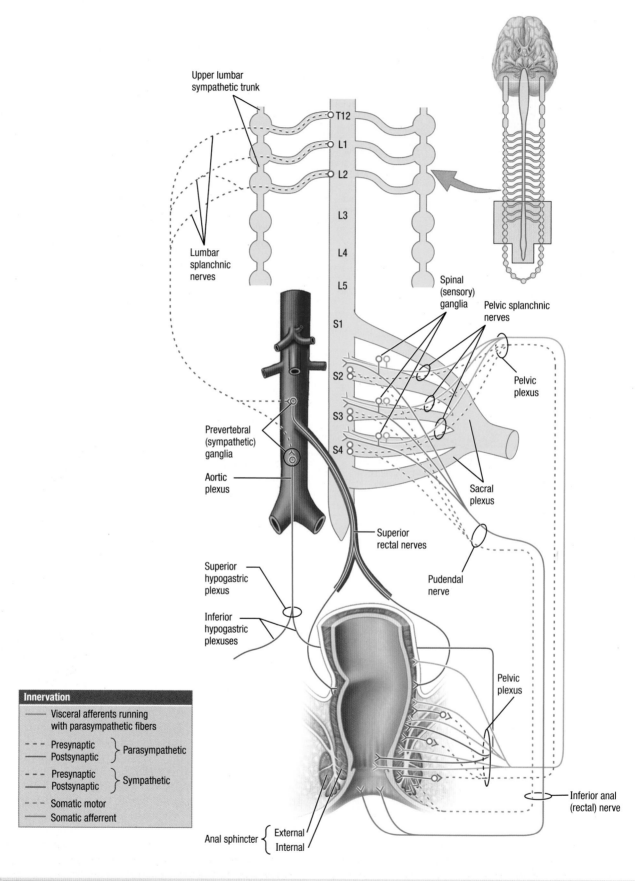

Upper lumbar
sympathetic trunk

Lumbar
splanchnic
nerves

Spinal
(sensory)
ganglia

Pelvic splanchnic
nerves

Pelvic
plexus

Prevertebral
(sympathetic)
ganglia

Aortic
plexus

Sacral
plexus

Superior
rectal nerves

Superior
hypogastric
plexus

Inferior
hypogastric
plexuses

Pudendal
nerve

Pelvic
plexus

Inferior anal
(rectal) nerve

Innervation

— Visceral afferents running
with parasympathetic fibers

- - - Presynaptic } Parasympathetic
— Postsynaptic

- - - Presynaptic } Sympathetic
— Postsynaptic

- - - Somatic motor
— Somatic afferent

Anal sphincter { External
Internal

5.19 INNERVATION OF RECTUM AND ANAL CANAL

The lumbar and pelvic spinal nerves and hypogastric plexuses have been retracted laterally for clarity.

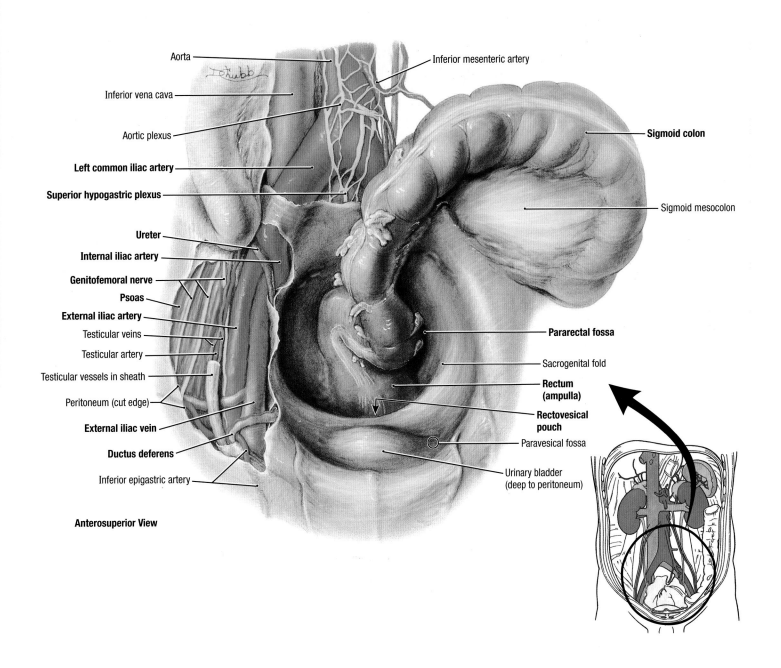

Aorta

Inferior mesenteric artery

Inferior vena cava

Aortic plexus

Left common iliac artery

Superior hypogastric plexus

Ureter

Internal iliac artery

Genitofemoral nerve

Psoas

External iliac artery

Testicular veins

Testicular artery

Testicular vessels in sheath

Peritoneum (cut edge)

External iliac vein

Ductus deferens

Inferior epigastric artery

Sigmoid colon

Sigmoid mesocolon

Pararectal fossa

Sacrogenital fold

Rectum (ampulla)

Rectovesical pouch

Paravesical fossa

Urinary bladder (deep to peritoneum)

Anterosuperior View

RECTUM *IN SITU*   5.20

- The sigmoid colon begins at the left pelvic brim and becomes the rectum anterior to the third sacral segment in the midline.
- The superior hypogastric plexus lies inferior to the bifurcation of the aorta and anterior to the left common iliac vein.
- The ureter adheres to the external aspect of the peritoneum, crosses the external iliac vessels, and descends anterior to the

internal iliac artery. The ductus deferens and its artery also adhere to the peritoneum, cross the external iliac vessels, and then hook around the inferior epigastric artery to join the other components of the spermatic cord.
- The genitofemoral nerve lies on the psoas.

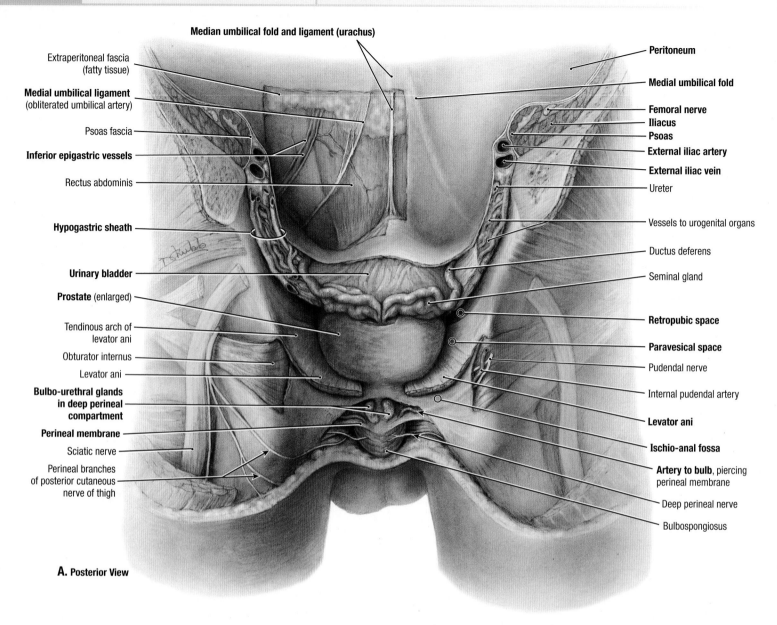

Median umbilical fold and ligament (urachus)

Extraperitoneal fascia (fatty tissue)

Medial umbilical ligament (obliterated umbilical artery)

Psoas fascia

Inferior epigastric vessels

Rectus abdominis

Hypogastric sheath

Urinary bladder

Prostate (enlarged)

Tendinous arch of levator ani

Obturator internus

Levator ani

Bulbo-urethral glands in deep perineal compartment

Perineal membrane

Sciatic nerve

Perineal branches of posterior cutaneous nerve of thigh

Peritoneum

Medial umbilical fold

Femoral nerve
Iliacus
Psoas
External iliac artery
External iliac vein
Ureter

Vessels to urogenital organs

Ductus deferens

Seminal gland

Retropubic space

Paravesical space

Pudendal nerve

Internal pudendal artery

Levator ani

Ischio-anal fossa

Artery to bulb, piercing perineal membrane

Deep perineal nerve

Bulbospongiosus

A. Posterior View

5.21 **POSTERIOR APPROACH TO ANTERIOR PELVIC AND PERINEAL STRUCTURES AND SPACES**

A. Dissection. The rectovesical septum and all pelvic and perineal structures posterior to it have been removed. **B.** Schematic coronal section through the anterior pelvis (plane of urinary bladder and prostate) demonstrating pelvic fascia.

- The inferior epigastric artery and accompanying veins enter the rectus sheath, covered posteriorly with peritoneum to form the lateral umbilical fold. The medial umbilical fold is formed by peritoneum overlying the medial umbilical ligament (obliterated umbilical artery), and the median umbilical fold is formed by the median umbilical ligament (urachus).
- The pelvic genito-urinary organs are subperitoneal. Near the bladder, the ureter accompanies a "leash" of internal iliac vessels and derivatives within the hypogastric sheath, a fibro-areolar structure.

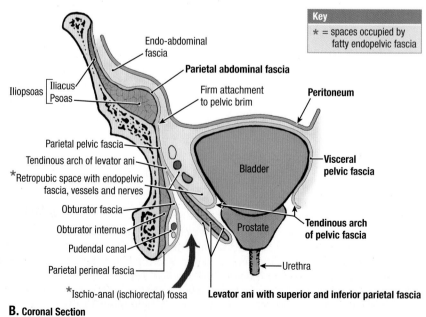

Key
* = spaces occupied by fatty endopelvic fascia

Endo-abdominal fascia

Parietal abdominal fascia

Firm attachment to pelvic brim

Peritoneum

Iliopsoas {Iliacus / Psoas}

Parietal pelvic fascia

Tendinous arch of levator ani

*Retropubic space with endopelvic fascia, vessels and nerves

Obturator fascia

Obturator internus

Pudendal canal

Parietal perineal fascia

*Ischio-anal (ischiorectal) fossa

Bladder

Prostate

Visceral pelvic fascia

Tendinous arch of pelvic fascia

Urethra

Levator ani with superior and inferior parietal fascia

B. Coronal Section

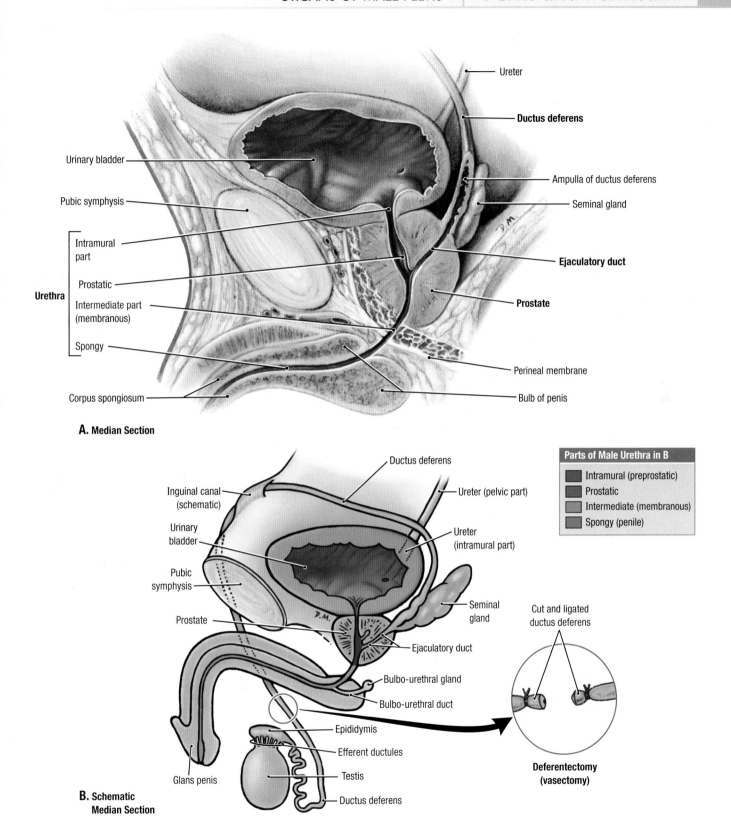

Ureter

Ductus deferens

Urinary bladder

Ampulla of ductus deferens

Pubic symphysis

Seminal gland

Intramural part

Prostatic

Ejaculatory duct

Urethra

Intermediate part (membranous)

Prostate

Spongy

Corpus spongiosum

Perineal membrane

Bulb of penis

A. Median Section

Ductus deferens

Inguinal canal (schematic)

Ureter (pelvic part)

Urinary bladder

Ureter (intramural part)

Pubic symphysis

Prostate

Seminal gland

Ejaculatory duct

Bulbo-urethral gland

Bulbo-urethral duct

Epididymis

Efferent ductules

Glans penis

Testis

Ductus deferens

B. Schematic Median Section

Cut and ligated ductus deferens

Deferentectomy (vasectomy)

Parts of Male Urethra in B
Intramural (preprostatic)
Prostatic
Intermediate (membranous)
Spongy (penile)

URINARY BLADDER, PROSTATE, SEMINAL GLANDS, AND DUCTUS DEFERENS

5.22

A. Dissection. The ejaculatory duct (~2 cm in length) is formed by the union of the ductus deferens and duct of the seminal gland; it passes anteriorly and inferiorly through the substance of the prostate to enter the prostatic urethra. **B.** Overview of urogenital system, schematic illustration. The common method of sterilizing males is a **deferentectomy**, popularly called **vasectomy**. During this procedure, part of the ductus deferens is ligated and/or excised through an incision in the superior part of the scrotum. Hence, the subsequent ejaculated fluid contains no sperms.

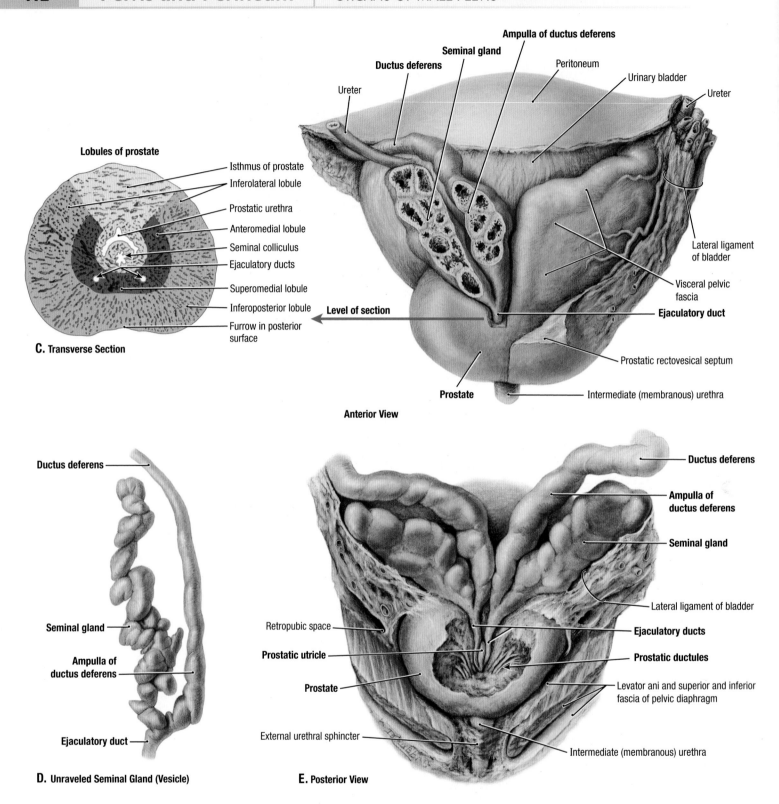

Lobules of prostate

Isthmus of prostate
Inferolateral lobule
Prostatic urethra
Anteromedial lobule
Seminal colliculus
Ejaculatory ducts
Superomedial lobule
Inferoposterior lobule
Furrow in posterior surface

C. Transverse Section

Ureter
Ductus deferens
Seminal gland
Ampulla of ductus deferens
Peritoneum
Urinary bladder
Ureter

Lateral ligament of bladder
Visceral pelvic fascia
Ejaculatory duct
Level of section
Prostatic rectovesical septum
Prostate
Intermediate (membranous) urethra

Anterior View

Ductus deferens

Seminal gland

Ampulla of ductus deferens

Ejaculatory duct

D. Unraveled Seminal Gland (Vesicle)

Ductus deferens

Ampulla of ductus deferens

Seminal gland

Lateral ligament of bladder
Ejaculatory ducts
Prostatic ductules
Levator ani and superior and inferior fascia of pelvic diaphragm

Retropubic space
Prostatic utricle
Prostate
External urethral sphincter
Intermediate (membranous) urethra

E. Posterior View

5.22 URINARY BLADDER, PROSTATE, SEMINAL GLANDS, AND DUCTUS DEFERENS (*continued*)

C. Bladder, ductus deferens, seminal glands (vesicles), and lobules of prostate. The left seminal gland and ampulla of the ductus deferens are dissected and opened; part of the prostate is cut away to expose the ejaculatory duct. **D.** Seminal gland unraveled. The gland is a tortuous tube with numerous dilatations. The ampulla of the ductus deferens has similar dilatations. **E.** Prostate, dissected posteriorly. The ejaculatory duct enters the prostatic urethra on the seminal colliculus. The prostatic utricle lies between the ends of the two ejaculatory ducts. The prostatic ductules mostly open onto the prostatic sinus.

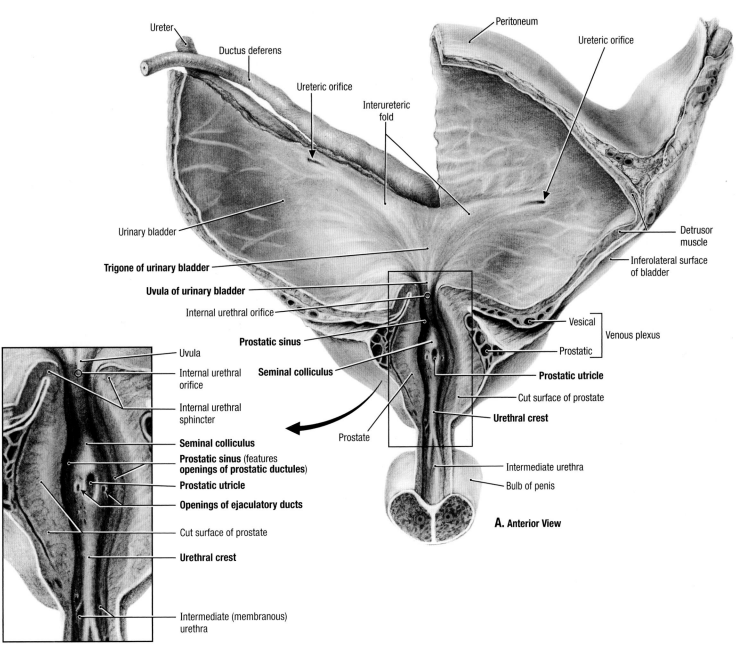

A. Anterior View

Labels on part A:
Ureter, Ductus deferens, Ureteric orifice, Interureteric fold, Peritoneum, Ureteric orifice, Urinary bladder, **Trigone of urinary bladder**, **Uvula of urinary bladder**, Internal urethral orifice, **Prostatic sinus**, **Seminal colliculus**, Prostate, Detrusor muscle, Inferolateral surface of bladder, Vesical, Prostatic, Venous plexus, **Prostatic utricle**, Cut surface of prostate, **Urethral crest**, Intermediate urethra, Bulb of penis

Labels on part B:
Uvula, Internal urethral orifice, Internal urethral sphincter, **Seminal colliculus**, **Prostatic sinus** (features openings of prostatic ductules), **Prostatic utricle**, **Openings of ejaculatory ducts**, Cut surface of prostate, **Urethral crest**, Intermediate (membranous) urethra

B. Anterior View

INTERIOR OF MALE URINARY BLADDER AND PROSTATIC URETHRA

A. Dissection. The anterior walls of the bladder, prostate, and urethra were cut away. **B.** Features of the prostatic urethra.

- The mucous membrane is smooth over the trigone of the urinary bladder (triangular region demarcated by ureteric and internal urethral orifices) but folded elsewhere, especially when the bladder is empty.
- The opening of the vestigial prostatic utricle is in the seminal colliculus on the urethral crest; there is an orifice of an ejaculatory duct on each side of the prostatic utricle. The prostatic fascia encloses the prostatic venous plexus.

The prostate is of considerable medical interest because enlargement or **benign hypertrophy of the prostate (BHP)** is common after middle age, affecting virtually every male who lives long enough. An enlarged prostate projects into the urinary bladder and impedes urination by distorting the prostatic urethra. The middle lobule usually enlarges the most and obstructs the internal urethral orifice. The more the person strains, the more the valvelike prostatic mass occludes the urethra.

BHP is a common cause of urethral obstruction, leading to **nocturia** (needing to void during the night), **dysuria** (difficulty and/or pain during urination), and **urgency** (sudden desire to void). BHP also increases the risk of bladder infections (**cystitis**) as well as kidney damage.

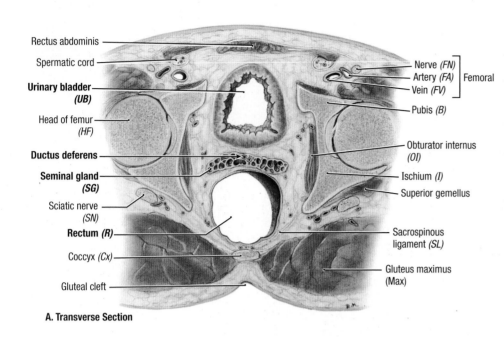

Rectus abdominis
Spermatic cord
Urinary bladder (UB)
Head of femur (HF)
Ductus deferens
Seminal gland (SG)
Sciatic nerve (SN)
Rectum (R)
Coccyx (Cx)
Gluteal cleft

Nerve (FN)
Artery (FA) — Femoral
Vein (FV)
Pubis (B)
Obturator internus (OI)
Ischium (I)
Superior gemellus
Sacrospinous ligament (SL)
Gluteus maximus (Max)

A. Transverse Section

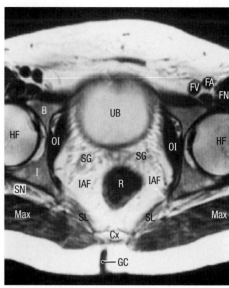

Transverse MRI

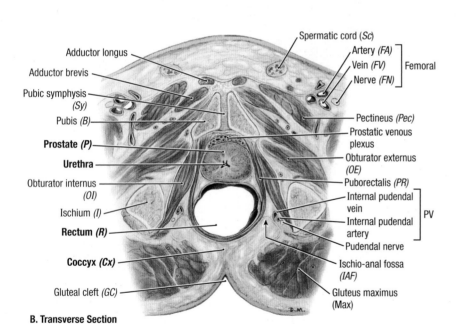

Adductor longus
Adductor brevis
Pubic symphysis (Sy)
Pubis (B)
Prostate (P)
Urethra
Obturator internus (OI)
Ischium (I)
Rectum (R)
Coccyx (Cx)
Gluteal cleft (GC)

Spermatic cord (Sc)
Artery (FA)
Vein (FV) — Femoral
Nerve (FN)
Pectineus (Pec)
Prostatic venous plexus
Obturator externus (OE)
Puborectalis (PR)
Internal pudendal vein
Internal pudendal artery — PV
Pudendal nerve
Ischio-anal fossa (IAF)
Gluteus maximus (Max)

B. Transverse Section

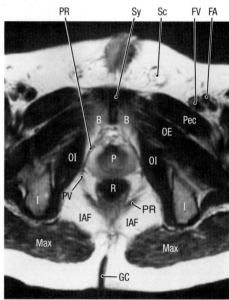

Transverse MRI

5.24 MALE PELVIS, TRANSVERSE SECTIONS AND MRI

A. Transverse section and MRI through urinary bladder, seminal gland, and rectum. **B.** Transverse section and MRI through prostate and rectum. **C.** Digital rectal examination.

The prostate is examined for enlargement and tumors (focal masses or asymmetry) by **digital rectal examination**. A full bladder offers resistance, holding the gland in place and making it more readily palpable. The malignant prostate feels hard and often irregular.

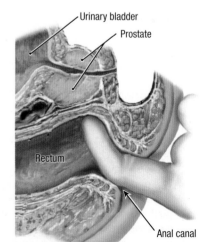

Urinary bladder
Prostate
Rectum
Anal canal

C. Sagittal Section

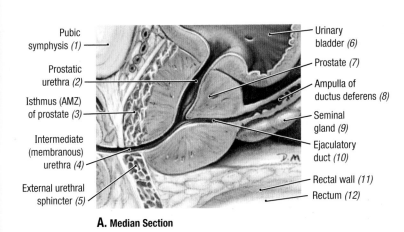

Pubic symphysis (1)

Prostatic urethra (2)

Isthmus (AMZ) of prostate (3)

Intermediate (membranous) urethra (4)

External urethral sphincter (5)

Urinary bladder (6)

Prostate (7)

Ampulla of ductus deferens (8)

Seminal gland (9)

Ejaculatory duct (10)

Rectal wall (11)

Rectum (12)

D.M

A. Median Section

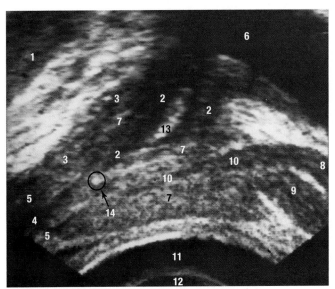

Longitudinal (Median) US

Key for US Scan	
12	Site of transducer in rectum
13	Concretions surrounding distended and collapsed urethra
14	Calcification in seminal colliculus

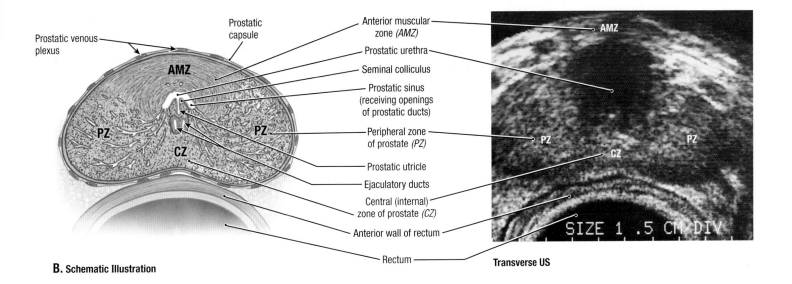

Prostatic venous plexus

Prostatic capsule

AMZ

PZ

PZ

CZ

Anterior muscular zone (AMZ)

Prostatic urethra

Seminal colliculus

Prostatic sinus (receiving openings of prostatic ducts)

Peripheral zone of prostate (PZ)

Prostatic utricle

Ejaculatory ducts

Central (internal) zone of prostate (CZ)

Anterior wall of rectum

Rectum

B. Schematic Illustration

AMZ

PZ

CZ

PZ

SIZE 1 .5 CM/DIV

Transverse US

TRANSRECTAL ULTRASOUND SCANS OF MALE PELVIS

5.25

A. Longitudinal scan. **B.** Transverse scan. The probe was inserted into the rectum to scan the anteriorly located prostate. The ducts of the glands in the peripheral zone open into the prostatic sinuses, whereas the ducts of the glands in the central (internal) zone open into the prostatic sinuses and onto the seminal colliculus.

Because of the close relationship of the prostate to the prostatic urethra, obstructions of the urethra may be relieved endoscopically. The instrument is inserted transurethrally through the external urethral orifice and spongy urethra into the prostatic urethra. All or part of the prostate, or just the hypertrophied part, is removed by

transurethral resection of the prostate (TURP). In more serious cases, the entire prostate is removed along with the seminal glands, ejaculatory ducts, and terminal parts of the deferent ducts **(radical prostatectomy)**.

TURP and improved operative techniques (laparoscopic or robotic surgery) attempt to preserve the nerves and blood vessels associated with the capsule of the prostate and adjacent to the seminal vesicles as they pass to and from the penis, increasing the possibility for patients to retain sexual function after surgery as well as restoring normal urinary control.

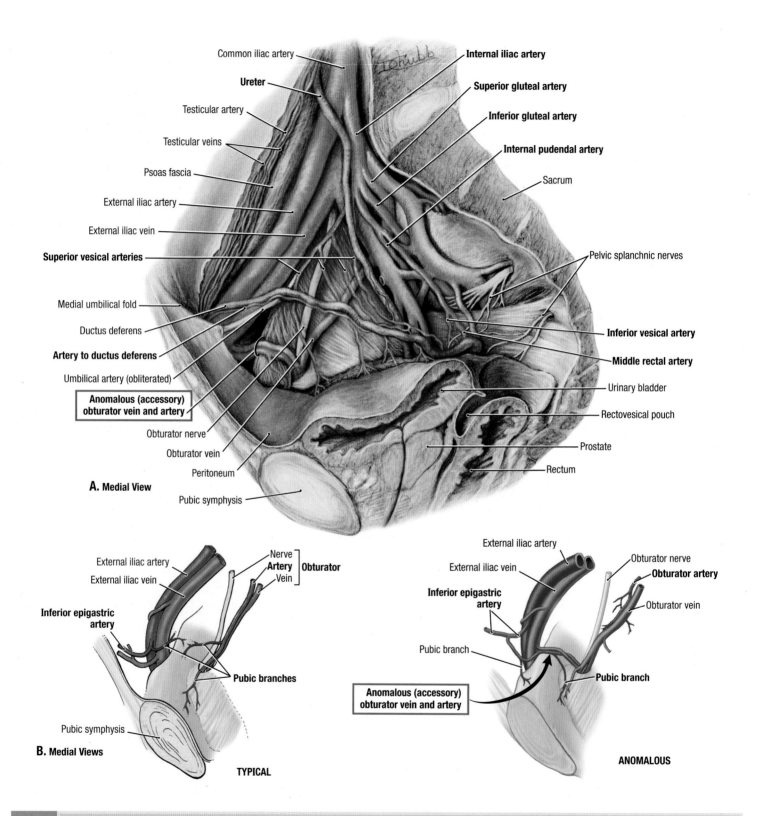

Common iliac artery

Internal iliac artery

Ureter

Testicular artery

Superior gluteal artery

Testicular veins

Inferior gluteal artery

Psoas fascia

Internal pudendal artery

External iliac artery

Sacrum

External iliac vein

Superior vesical arteries

Pelvic splanchnic nerves

Medial umbilical fold

Ductus deferens

Artery to ductus deferens

Inferior vesical artery

Umbilical artery (obliterated)

Middle rectal artery

Anomalous (accessory) obturator vein and artery

Urinary bladder

Obturator nerve

Rectovesical pouch

Obturator vein

Prostate

Peritoneum

Rectum

A. Medial View

Pubic symphysis

External iliac artery

Nerve

External iliac artery

External iliac vein

Artery Obturator

Obturator nerve

Vein

Obturator artery

Inferior epigastric artery

External iliac vein

Obturator vein

Inferior epigastric artery

Pubic branch

Pubic branches

Pubic branch

Anomalous (accessory) obturator vein and artery

Pubic symphysis

B. Medial Views

TYPICAL

ANOMALOUS

5.26 PELVIC VESSELS *IN SITU*; LATERAL PELVIC WALL

A. Dissection of lateral pelvic wall. The ureter crosses the external iliac artery at its origin (common iliac bifurcation), and the ductus deferens crosses the external iliac artery at its termination (deep inguinal ring). In this specimen, an anomalous (accessory) obturator artery branches from the inferior epigastric artery. **B.** Typical and anomalous obturator arteries. Surgeons performing hernia repairs must keep this common variation in mind.

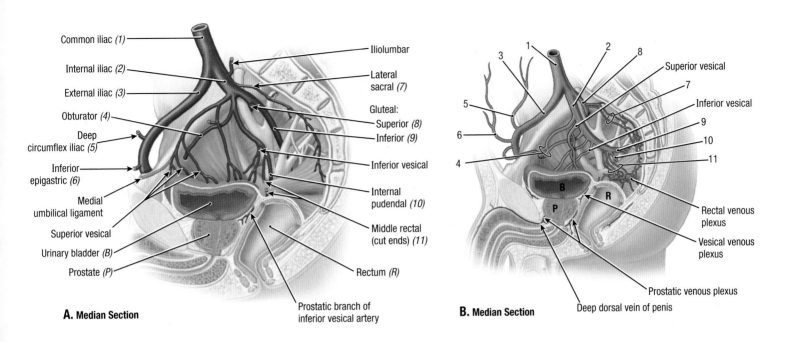

A. Median Section

Common iliac *(1)*
Internal iliac *(2)*
External iliac *(3)*
Obturator *(4)*
Deep circumflex iliac *(5)*
Inferior epigastric *(6)*
Medial umbilical ligament
Superior vesical
Urinary bladder *(B)*
Prostate *(P)*

Iliolumbar
Lateral sacral *(7)*
Gluteal:
Superior *(8)*
Inferior *(9)*
Inferior vesical
Internal pudendal *(10)*
Middle rectal (cut ends) *(11)*
Rectum *(R)*
Prostatic branch of inferior vesical artery

B. Median Section

Superior vesical
Inferior vesical
Rectal venous plexus
Vesical venous plexus
Prostatic venous plexus
Deep dorsal vein of penis

ARTERIES AND VEINS OF MALE PELVIS

5.27

A. Arteries. **B.** Veins.

The neurovascular structures of the pelvis lie extraperitoneally. When dissecting from the pelvic cavity toward the pelvic walls, the pelvic arteries are encountered first, followed by the associated pelvic veins, and then the somatic nerves of the pelvis.

TABLE 5.4	ARTERIES OF MALE PELVIS		
Artery	**Origin**	**Course**	**Distribution**
Internal iliac	Common iliac artery	Passes medially over pelvic brim and descends into pelvic cavity; often forms anterior and posterior divisions	Main blood supply to pelvic organs, gluteal muscles, and perineum
Anterior division of internal iliac artery	Internal iliac artery	Passes laterally along lateral wall of pelvis, dividing into visceral, obturator, and internal pudendal arteries	Pelvic viscera, perineum, and muscles of superior medial thigh
Umbilical	Anterior division of internal iliac artery	Short pelvic course; gives off superior vesical arteries, then obliterates, becoming medial umbilical ligament	Urinary bladder and, in some males, ductus deferens
Superior vesical	Patent part of umbilical artery	Usually multiple; pass to superior aspect of urinary bladder	Superior aspect of urinary bladder and distal ureter
Artery to ductus deferens	Superior or inferior vesical artery	Runs subperitoneally to ductus deferens	Ductus deferens
Obturator	Anterior division of internal iliac artery	Runs antero-inferiorly on lateral pelvic wall	Pelvic muscles, nutrient artery to head of femur and medial compartment of thigh
Inferior vesical		Passes subperitoneally giving rise to prostatic artery and occasionally the artery to the ductus deferens	Inferior aspect of urinary bladder, pelvic ureter, seminal glands, and prostate
Middle rectal		Descends in pelvis to rectum	Seminal glands, prostate, and inferior part of rectum
Internal pudendal		Exits pelvis through greater sciatic foramen and enters perineum via lesser sciatic foramen	Main artery to perineum, including muscles and skin of anal and urogenital triangles; erectile bodies
Posterior division of internal iliac artery	Internal iliac artery	Passes posteriorly and gives rise to parietal branches	Pelvic wall and gluteal region
Iliolumbar	Posterior division of internal iliac artery	Ascends anterior to sacro-iliac joint and posterior to common iliac vessels and psoas major	Iliacus, psoas major, quadratus lumborum muscles, and cauda equina in vertebral canal
Lateral sacral (superior and inferior)		Run on anteromedial aspect of piriformis to send branches into pelvic sacral foramina	Piriformis muscle, structures in sacral canal and erector spinae muscles
Testicular (gonadal)	Abdominal aorta	Descends retroperitoneally; traverses inguinal canal and enters scrotum	Abdominal ureter, testis and epididymis

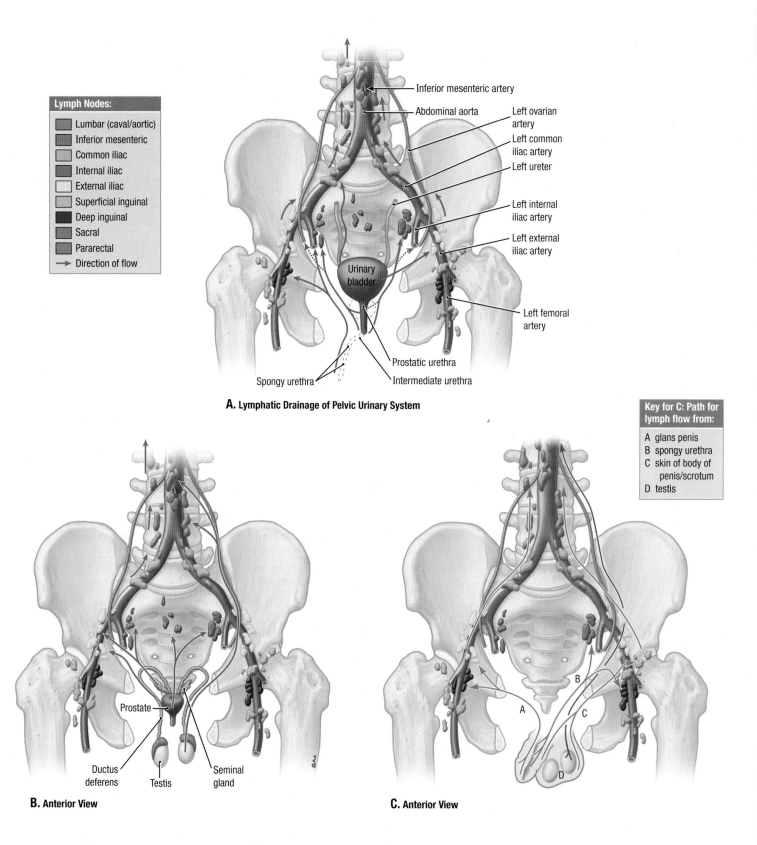

Lymph Nodes:
- Lumbar (caval/aortic)
- Inferior mesenteric
- Common iliac
- Internal iliac
- External iliac
- Superficial inguinal
- Deep inguinal
- Sacral
- Pararectal
- → Direction of flow

Inferior mesenteric artery
Abdominal aorta
Left ovarian artery
Left common iliac artery
Left ureter
Left internal iliac artery
Left external iliac artery
Urinary bladder
Left femoral artery
Prostatic urethra
Spongy urethra
Intermediate urethra

A. Lymphatic Drainage of Pelvic Urinary System

Key for C: Path for lymph flow from:
A glans penis
B spongy urethra
C skin of body of penis/scrotum
D testis

Prostate
Ductus deferens
Testis
Seminal gland

B. Anterior View

C. Anterior View

5.28 LYMPHATIC DRAINAGE OF MALE PELVIS AND PERINEUM

A. Pelvic urinary system. **B.** Internal genital organs. **C.** Penis, spongy urethra, scrotum and testis.

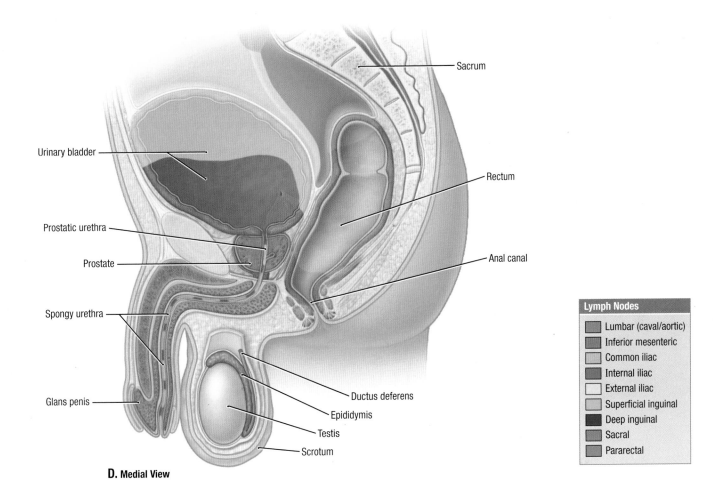

Sacrum

Urinary bladder

Rectum

Prostatic urethra

Prostate

Anal canal

Spongy urethra

Ductus deferens

Epididymis

Glans penis

Testis

Scrotum

D. Medial View

Lymph Nodes

- Lumbar (caval/aortic)
- Inferior mesenteric
- Common iliac
- Internal iliac
- External iliac
- Superficial inguinal
- Deep inguinal
- Sacral
- Pararectal

LYMPHATIC DRAINAGE OF MALE PELVIS AND PERINEUM (*continued*)

5.28

D. Zones of pelvis and perineum initially draining into specific groups of lymph nodes.

TABLE 5.5	LYMPHATIC DRAINAGE OF MALE PELVIS AND PERINEUM
Lymph Node Group	**Structures Typically Draining to Lymph Node Group**
Lumbar	Gonads and associated structures (including testicular vessels), urethra, testis, epididymis, common iliac nodes
Inferior mesenteric nodes	Superiormost rectum, sigmoid colon, descending colon, pararectal nodes
Common iliac nodes	External and internal iliac lymph nodes
Internal iliac nodes	Inferior pelvic structures, deep perineal structures, sacral nodes, prostatic urethra, prostate, base of bladder, inferior part of pelvic ureter, inferior part of seminal glands, cavernous bodies, anal canal (above pectinate line), inferior rectum
External iliac nodes	Anterosuperior pelvic structures, deep inguinal nodes, superior aspect of bladder, superior part of pelvic ureter, upper part of seminal gland, pelvic part of ductus deferens, intermediate and spongy urethra
Superficial inguinal nodes	Lower limb, superficial drainage of inferolateral quadrant of trunk, including anterior abdominal wall inferior to umbilicus, gluteal region, superficial perineal structures, skin of perineum including skin and prepuce of penis, scrotum, peri-anal skin, anal canal inferior to pectinate line
Deep inguinal nodes	Glans of penis, distal spongy urethra, superficial inguinal nodes
Sacral nodes	Postero-inferior pelvic structures, inferior rectum
Pararectal nodes	Superior rectum

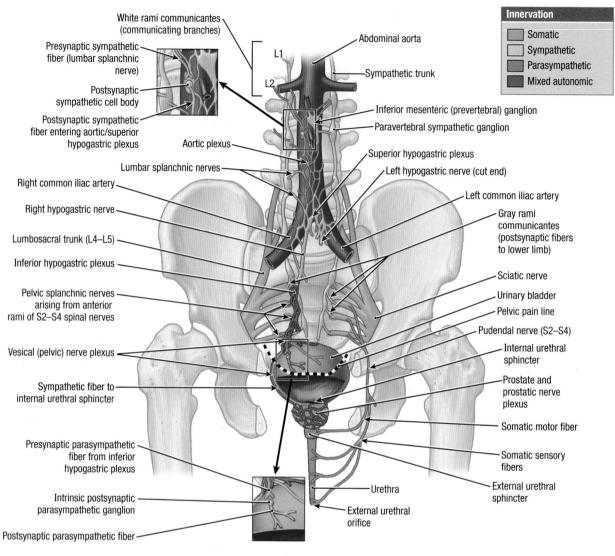

Innervation
- Somatic
- Sympathetic
- Parasympathetic
- Mixed autonomic

White rami communicantes (communicating branches)

Presynaptic sympathetic fiber (lumbar splanchnic nerve)

Postsynaptic sympathetic cell body

Postsynaptic sympathetic fiber entering aortic/superior hypogastric plexus

Aortic plexus

Lumbar splanchnic nerves

Right common iliac artery

Right hypogastric nerve

Lumbosacral trunk (L4–L5)

Inferior hypogastric plexus

Pelvic splanchnic nerves arising from anterior rami of S2–S4 spinal nerves

Vesical (pelvic) nerve plexus

Sympathetic fiber to internal urethral sphincter

Presynaptic parasympathetic fiber from inferior hypogastric plexus

Intrinsic postsynaptic parasympathetic ganglion

Postsynaptic parasympathetic fiber

L1
L2

Abdominal aorta

Sympathetic trunk

Inferior mesenteric (prevertebral) ganglion

Paravertebral sympathetic ganglion

Superior hypogastric plexus

Left hypogastric nerve (cut end)

Left common iliac artery

Gray rami communicantes (postsynaptic fibers to lower limb)

Sciatic nerve

Urinary bladder

Pelvic pain line

Pudendal nerve (S2–S4)

Internal urethral sphincter

Prostate and prostatic nerve plexus

Somatic motor fiber

Somatic sensory fibers

External urethral sphincter

Urethra

External urethral orifice

A. Anterior View

5.29 INNERVATION OF MALE PELVIS AND PERINEUM

A. Overview.

TABLE 5.6	EFFECT OF SYMPATHETIC AND PARASYMPATHETIC STIMULATION ON URINARY TRACT, GENITAL SYSTEM, AND RECTUM	
Organ, Tract, or System	*Effect of Sympathetic Stimulation*	*Effect of Parasympathetic Stimulation*
Urinary tract	Vasoconstriction of renal vessels slows urine formation; internal sphincter of male bladder contracted to prevent retrograde ejaculation and maintain urinary continence	Inhibits contraction of internal sphincter of bladder in males; contracts detrusor muscle of the bladder wall causing urination
Genital system	Causes ejaculation and vasoconstriction resulting in remission of erection	Produces engorgement (erection) of erectile tissues of the external genitals
Rectum	Maintains tonus of internal anal sphincter; inhibits peristalsis of rectum	Rectal contraction (peristalsis) for defecation; inhibition of contraction of internal anal sphincter

The parasympathetic system is restricted in its distribution to the head, neck, and body cavities (except for erectile tissues of genitalia); otherwise, parasympathetic fibers are never found in the body wall and limbs. Sympathetic fibers, by comparison, are distributed to all vascularized portions of the body.

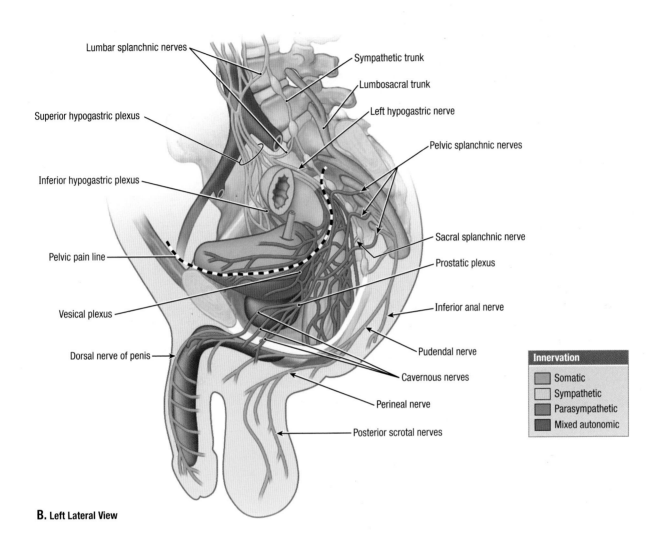

Lumbar splanchnic nerves

Superior hypogastric plexus

Inferior hypogastric plexus

Pelvic pain line

Vesical plexus

Dorsal nerve of penis

Sympathetic trunk

Lumbosacral trunk

Left hypogastric nerve

Pelvic splanchnic nerves

Sacral splanchnic nerve

Prostatic plexus

Inferior anal nerve

Pudendal nerve

Cavernous nerves

Perineal nerve

Posterior scrotal nerves

Innervation
- Somatic
- Sympathetic
- Parasympathetic
- Mixed autonomic

B. Left Lateral View

INNERVATION OF MALE PELVIS AND PERINEUM (*continued*)

5.29

B. Innervation of prostate and external genitalia.
- The primary function of the sacral sympathetic trunks is to provide postsynaptic fibers to the sacral plexus for sympathetic innervation of the lower limb.
- The peri-arterial plexuses of the ovarian, superior rectal, and internal iliac arteries are minor routes by which sympathetic fibers enter the pelvis. Their primary function is vasomotion of the arteries they accompany.
- The hypogastric plexuses (superior and inferior) are networks of sympathetic and visceral afferent nerve fibers.
- The superior hypogastric plexus carries fibers conveyed to and from the aortic (intermesenteric) plexus by the L3 and L4 splanchnic nerves. The superior hypogastric plexus divides into right and left hypogastric nerves that merge with the parasympathetic pelvic splanchnic nerves to form the inferior hypogastric plexuses.

- The fibers of the inferior hypogastric plexuses continue to the pelvic viscera on which they form pelvic plexuses (e.g., prostatic nerve plexus).
- The pelvic splanchnic nerves convey presynaptic parasympathetic fibers from the S2–S4 spinal cord segments, which make up the sacral outflow of the parasympathetic system.
- Visceral afferents conveying unconscious reflex sensation follow the course of the parasympathetic fibers retrogradely to the spinal sensory ganglia of S2–S4, as do those transmitting pain sensations from the viscera inferior to the pelvic pain line (structures that do not contact the peritoneum plus the distal sigmoid colon and rectum). Visceral afferent fibers conducting pain from structures superior to the pelvic pain line (structures in contact with the peritoneum, except for the distal sigmoid colon and rectum) follow the sympathetic fibers retrogradely to inferior thoracic and superior lumbar spinal ganglia.

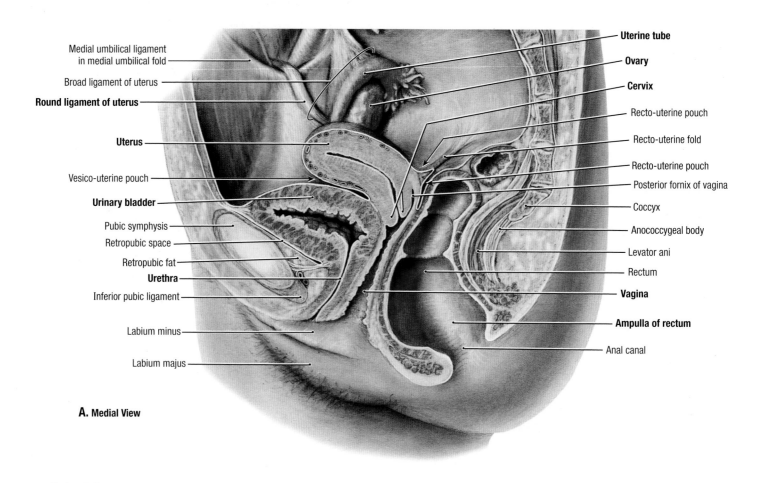

Medial umbilical ligament in medial umbilical fold

Broad ligament of uterus

Round ligament of uterus

Uterus

Vesico-uterine pouch

Urinary bladder

Pubic symphysis

Retropubic space

Retropubic fat

Urethra

Inferior pubic ligament

Labium minus

Labium majus

Uterine tube

Ovary

Cervix

Recto-uterine pouch

Recto-uterine fold

Recto-uterine pouch

Posterior fornix of vagina

Coccyx

Anococcygeal body

Levator ani

Rectum

Vagina

Ampulla of rectum

Anal canal

A. Medial View

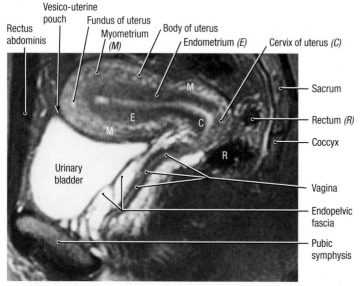

B. Midsagittal US

C. Longitudinal (Median) Transabdominal US

| 5.30 | FEMALE PELVIC ORGANS *IN SITU* |

A. Median section. The adult uterus is typically *anteverted* (tipped anterosuperiorly relative to the axis of the vagina) and *anteflexed* (flexed or bent anteriorly relative to the cervix, creating the *angle of flexion*) so that its mass lies over the bladder. The cervix, opening on the anterior wall of the vagina, has a short, round, anterior lip and a long, thin, posterior lip. **B.** Midsagittal MRI of uterus. **C.** Median (transabdominal) ultrasound image. The urinary bladder is distended to displace the loops of bowel from the pelvis.

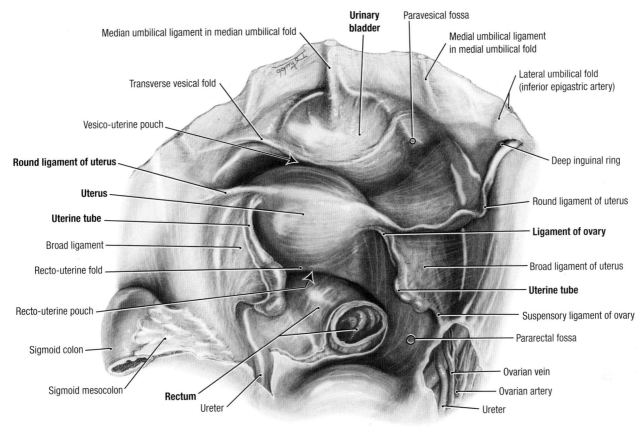

Median umbilical ligament in median umbilical fold

Transverse vesical fold

Vesico-uterine pouch

Round ligament of uterus

Uterus

Uterine tube

Broad ligament

Recto-uterine fold

Recto-uterine pouch

Sigmoid colon

Sigmoid mesocolon

Rectum

Ureter

Urinary bladder

Paravesical fossa

Medial umbilical ligament in medial umbilical fold

Lateral umbilical fold (inferior epigastric artery)

Deep inguinal ring

Round ligament of uterus

Ligament of ovary

Broad ligament of uterus

Uterine tube

Suspensory ligament of ovary

Pararectal fossa

Ovarian vein

Ovarian artery

Ureter

D. Superior View

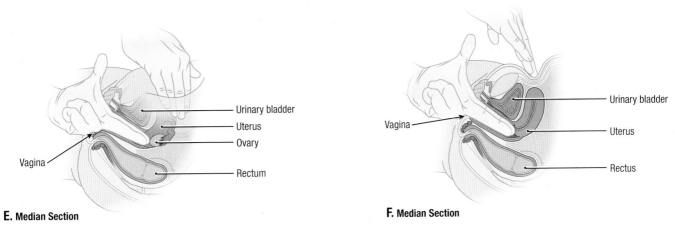

Vagina

Urinary bladder

Uterus

Ovary

Rectum

E. Median Section

Vagina

Urinary bladder

Uterus

Rectus

F. Median Section

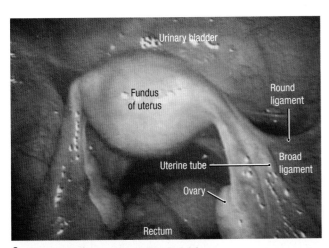

Urinary bladder

Fundus of uterus

Round ligament

Uterine tube

Broad ligament

Ovary

Rectum

G. Laparoscopic View of Normal Female Pelvis

FEMALE PELVIC ORGANS *IN SITU* (continued)   5.30

D. True pelvis with peritoneum intact, viewed from above. The uterus is usually asymmetrically placed. The round ligament of the female takes the same subperitoneal course as the ductus deferens of the male. **E. Bimanual palpation of uterine adnexa** (accessory structures, e.g., ovaries) **F. Bimanual palpation of uterus. G. Laparoscopy** involves inserting a laparoscope into the peritoneal cavity through a small incision below the umbilicus. Insufflation of inert gas creates a pneumoperitoneum to provide space to visualize the pelvic organs. Additional openings (ports) can be made to introduce other instruments for manipulation or to enable therapeutic procedures (e.g., ligation of the uterine tubes).

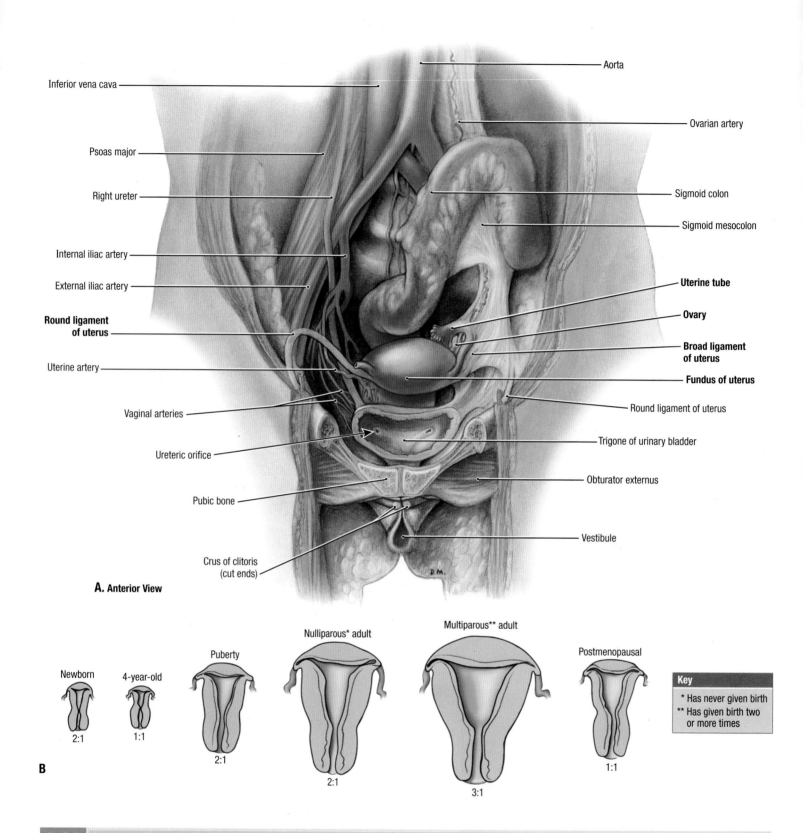

Aorta

Inferior vena cava

Ovarian artery

Psoas major

Right ureter

Sigmoid colon

Sigmoid mesocolon

Internal iliac artery

External iliac artery

Uterine tube

Ovary

Round ligament of uterus

Broad ligament of uterus

Uterine artery

Fundus of uterus

Round ligament of uterus

Vaginal arteries

Trigone of urinary bladder

Ureteric orifice

Obturator externus

Pubic bone

Vestibule

Crus of clitoris (cut ends)

A. Anterior View

Newborn
2:1

4-year-old
1:1

Puberty
2:1

Nulliparous* adult
2:1

Multiparous** adult
3:1

Postmenopausal
1:1

Key

* Has never given birth
** Has given birth two or more times

B

5.31 FEMALE GENITAL ORGANS

A. Dissection. Part of the pubic bones, the anterior aspect of the bladder, and—on the specimen's right side—the uterine tube, ovary, broad ligament, and peritoneum covering the lateral wall of the pelvis have been removed. **B.** Lifetime changes in uterine size and proportion (body to cervical ratio, e.g., 2:1). All these stages represent normal anatomy for the particular age and reproductive status of the woman.

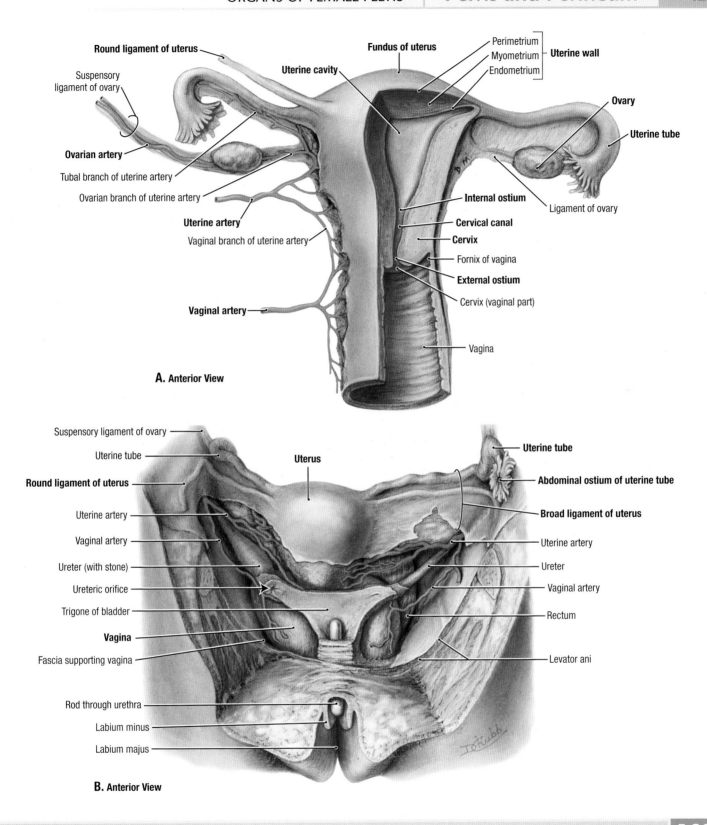

A. Anterior View

Round ligament of uterus

Suspensory ligament of ovary

Ovarian artery

Tubal branch of uterine artery

Ovarian branch of uterine artery

Uterine artery

Vaginal branch of uterine artery

Vaginal artery

Fundus of uterus

Uterine cavity

Perimetrium
Myometrium — Uterine wall
Endometrium

Ovary

Uterine tube

Internal ostium

Cervical canal

Cervix

Fornix of vagina

External ostium

Cervix (vaginal part)

Vagina

Ligament of ovary

B. Anterior View

Suspensory ligament of ovary

Uterine tube

Round ligament of uterus

Uterine artery

Vaginal artery

Ureter (with stone)

Ureteric orifice

Trigone of bladder

Vagina

Fascia supporting vagina

Rod through urethra

Labium minus

Labium majus

Uterus

Uterine tube

Abdominal ostium of uterine tube

Broad ligament of uterus

Uterine artery

Ureter

Vaginal artery

Rectum

Levator ani

UTERUS AND ITS ADNEXA

5.32

A. Blood supply. On the specimen's left side, part of the uterine wall with the round ligament and the vaginal wall have been cut away to expose the cervix, uterine cavity, and thick muscular wall of the uterus, the myometrium. On the specimen's right side, the ovarian artery (from the aorta) and uterine artery (from the internal iliac) supply the ovary, uterine tube, and uterus and anastomose in the broad ligament along the lateral aspect of the uterus. The uterine artery sends a uterine branch to supply the uterine body and fundus and a vaginal branch to supply the cervix and vagina. **B.** Uterus and broad ligament. The pubic bones and bladder, trigone excepted, are removed, as a continued dissection from Figure 5.31A.

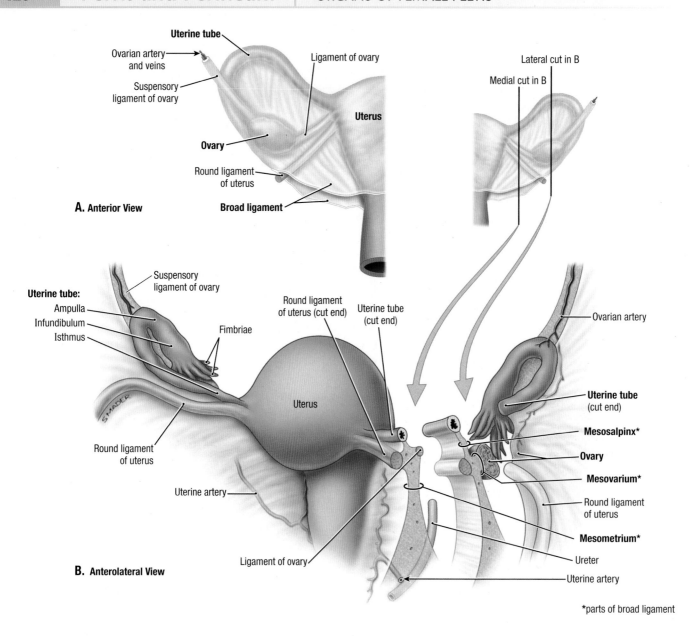

A. Anterior View

B. Anterolateral View

*parts of broad ligament

5.33 UTERUS AND BROAD LIGAMENT

A. and **B.** Two paramedian sections show "mesenteries" with the prefix meso-. "Salpinx" is the Greek word for trumpet or tube, and "metro" for uterus. The mesentery of the uterus and uterine tube is called the broad ligament. The major part of the broad ligament, the *mesometrium*, is attached to the uterus. The ovary is attached to the broad ligament by a mesentery of its own, called the *mesovarium*, to the uterus by the ligament of the ovary, and near the pelvic brim, by the suspensory ligament of the ovary containing the ovarian vessels. The part of the broad ligament superior to the level of the mesovarium is called the *mesosalpinx*. **C. Hysterectomy** (excision of the uterus) is performed through the lower anterior abdominal wall or through the vagina. Because the uterine artery crosses superior to the ureter near the lateral fornix of the vagina, the ureter is in danger of being inadvertently clamped or severed when the uterine artery is tied off during a hysterectomy.

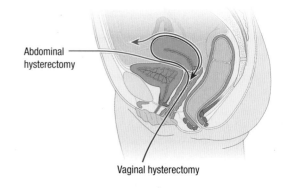

Abdominal hysterectomy

Vaginal hysterectomy

C. Medial View

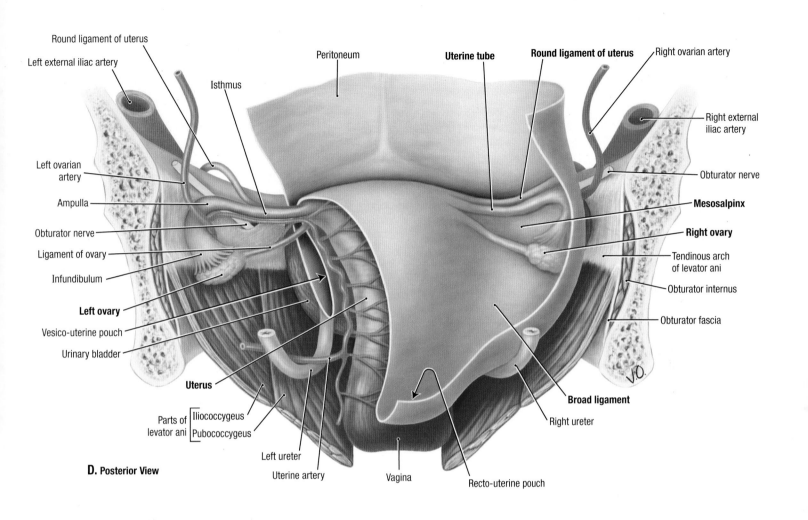

Round ligament of uterus
Left external iliac artery
Isthmus
Peritoneum
Uterine tube
Round ligament of uterus
Right ovarian artery
Right external iliac artery
Left ovarian artery
Ampulla
Obturator nerve
Ligament of ovary
Infundibulum
Left ovary
Vesico-uterine pouch
Urinary bladder
Uterus
Parts of levator ani { Iliococcygeus / Pubococcygeus }
Left ureter
Uterine artery
Vagina
Recto-uterine pouch
Right ureter
Broad ligament
Obturator internus
Obturator fascia
Tendinous arch of levator ani
Right ovary
Mesosalpinx
Obturator nerve

D. Posterior View

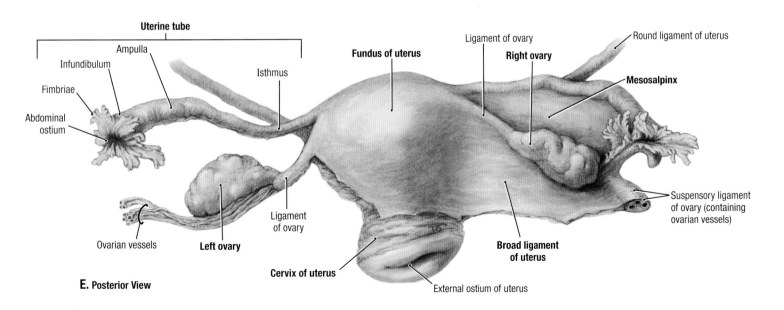

Uterine tube
Ampulla
Infundibulum
Fimbriae
Abdominal ostium
Isthmus
Fundus of uterus
Ligament of ovary
Right ovary
Round ligament of uterus
Mesosalpinx
Ligament of ovary
Left ovary
Ovarian vessels
Cervix of uterus
External ostium of uterus
Broad ligament of uterus
Suspensory ligament of ovary (containing ovarian vessels)

E. Posterior View

UTERUS AND BROAD LIGAMENT (*continued*)

D. Uterus *in situ*. **E.** Uterus and adnexa, removed from cadaver.

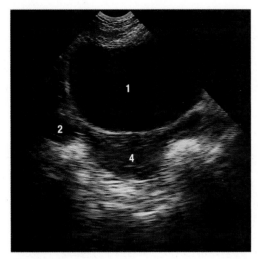

A. Transverse (Axial) US

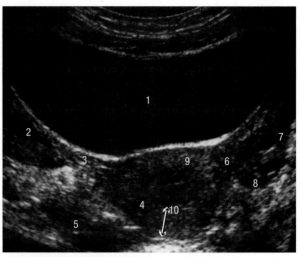

B. Transverse (Axial) US

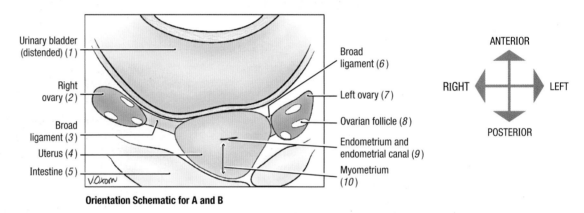

Urinary bladder (distended) (1)

Right ovary (2)

Broad ligament (3)

Uterus (4)

Intestine (5)

Broad ligament (6)

Left ovary (7)

Ovarian follicle (8)

Endometrium and endometrial canal (9)

Myometrium (10)

ANTERIOR

RIGHT · LEFT

POSTERIOR

Orientation Schematic for A and B

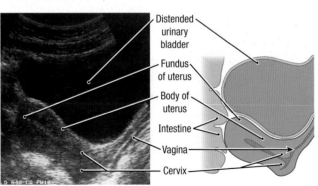

Distended urinary bladder

Fundus of uterus

Body of uterus

Intestine

Vagina

Cervix

C. Longitudinal (Median) US

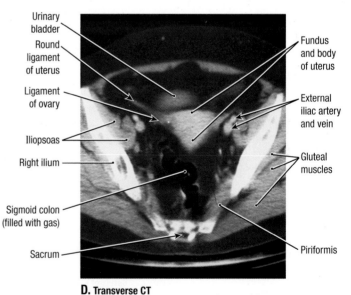

Urinary bladder

Round ligament of uterus

Ligament of ovary

Iliopsoas

Right ilium

Sigmoid colon (filled with gas)

Sacrum

Fundus and body of uterus

External iliac artery and vein

Gluteal muscles

Piriformis

D. Transverse CT

5.34 IMAGING OF UTERUS AND UTERINE ADNEXA

A. and **B.** Transverse ultrasound images. **C.** Longitudinal ultrasound image. Temporary retroversion and retroflexion result when a fully distended urinary bladder temporarily retroverts the uterus and decreases the angle of flexion. **D.** Transverse (axial) CT.

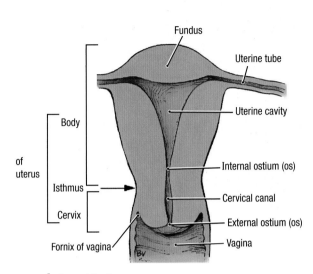

Fundus

Uterine tube

Uterine cavity

Body

of uterus

Internal ostium (os)

Isthmus

Cervical canal

Cervix

External ostium (os)

Fornix of vagina

Vagina

A. Coronal Section

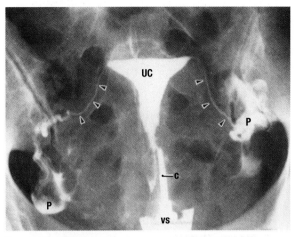

B. Hysterosalpingogram of Normal Uterus, Anteroposterior View

Key for B					
▲▲	Uterine tubes	P	Peritoneal cavity	VS	Vaginal speculum
C	Catheter in cervical canal	UC	Uterine cavity		

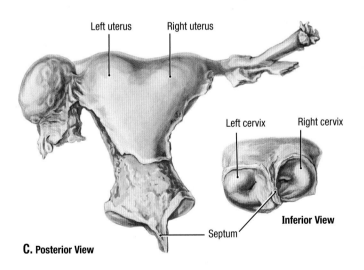

Left uterus

Right uterus

Left cervix

Right cervix

Inferior View

Septum

C. Posterior View

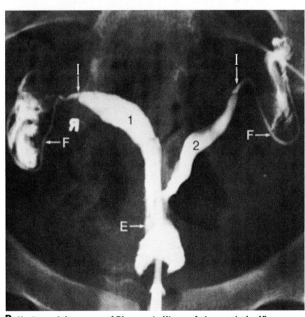

D. Hysterosalpingogram of Bicornuate Uterus, Anteroposterior View

Key for D			
1 and 2	Uterine cavities	F	Uterine tubes
E	Cervical canal	I	Isthmus of uterine tubes

RADIOGRAPH OF UTERUS AND UTERINE TUBES (HYSTEROSALPINGOGRAM)

5.35

A. Parts of uterus and superior vagina. **B.** During **hysterosalpingography**, radiopaque material is injected into the uterus through external os of the uterus. If normal, contrast medium travels through the triangular uterine cavity (*UC*) and uterine tubes (*arrowheads*) and passes into the pararectal fossae (*P*) of the peritoneal cavity. The female genital tract is in direct communication with the peritoneal cavity and is, therefore, a potential pathway for the spread of an infection from the vagina and uterus. **C.** Illustration of duplicated uterus. **D.** Hysterosalpingogram of a bicornuate ("two-horned") uterus.

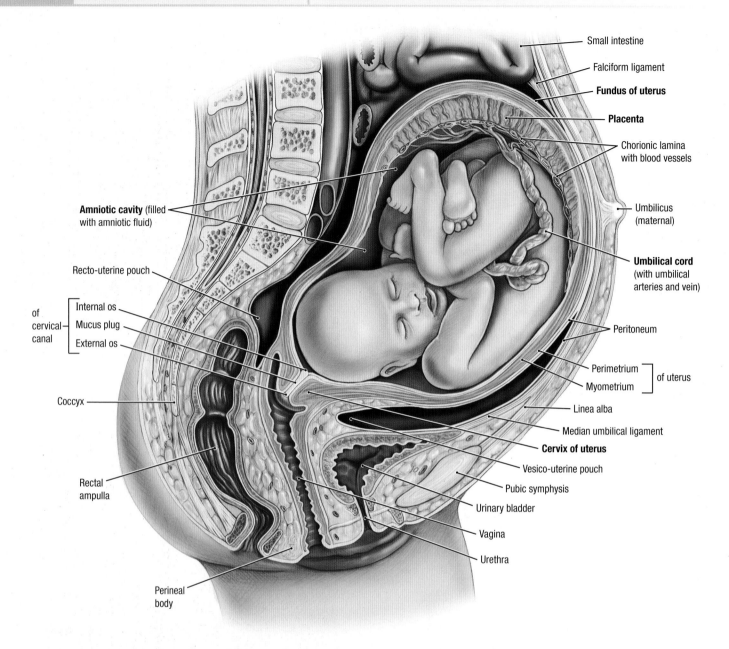

Small intestine

Falciform ligament

Fundus of uterus

Placenta

Chorionic lamina
with blood vessels

Amniotic cavity (filled
with amniotic fluid)

Umbilicus
(maternal)

Recto-uterine pouch

Umbilical cord
(with umbilical
arteries and vein)

of
cervical
canal { Internal os
Mucus plug
External os

Peritoneum

Perimetrium }
Myometrium } of uterus

Coccyx

Linea alba

Median umbilical ligament

Cervix of uterus

Vesico-uterine pouch

Rectal
ampulla

Pubic symphysis

Urinary bladder

Vagina

Urethra

Perineal
body

A. Median Section

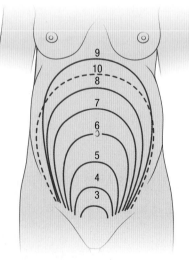

9
10
8
7
6
5
4
3

B. Anterior View

5.36   PREGNANT UTERUS

A. Median section; fetus is intact. **B. Monthly changes in size of uterus during pregnancy.** Over the 9 months of pregnancy, the *gravid* uterus expands greatly to accommodate the fetus, becoming larger and increasingly thin walled. At the end of pregnancy, the fetus "drops," as the head becomes engaged in the lesser pelvis. The uterus becomes nearly membranous, with the fundus dropping below its highest level (achieved in the 9th month), at which time it extends superiorly to the costal margin, occupying most of the abdominopelvic cavity.

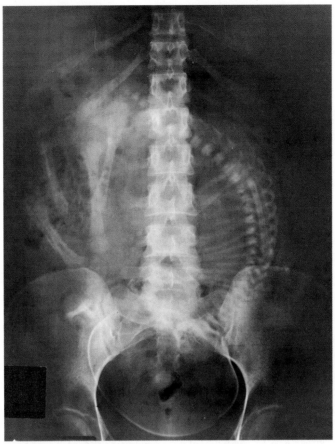

C. Anteroposterior View

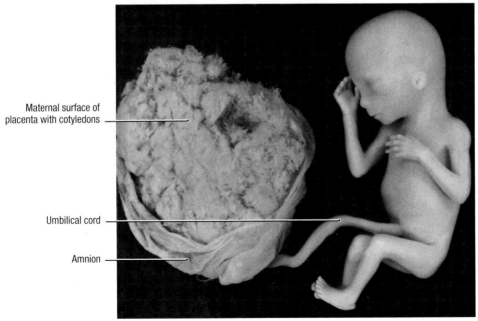

Maternal surface of placenta with cotyledons

Umbilical cord

Amnion

D. Maternal Surface of Placenta

PREGNANT UTERUS (*continued*)

5.36

C. Radiograph of fetus. **D.** Photograph of an 18-week-old fetus connected to the placenta by the umbilical cord.

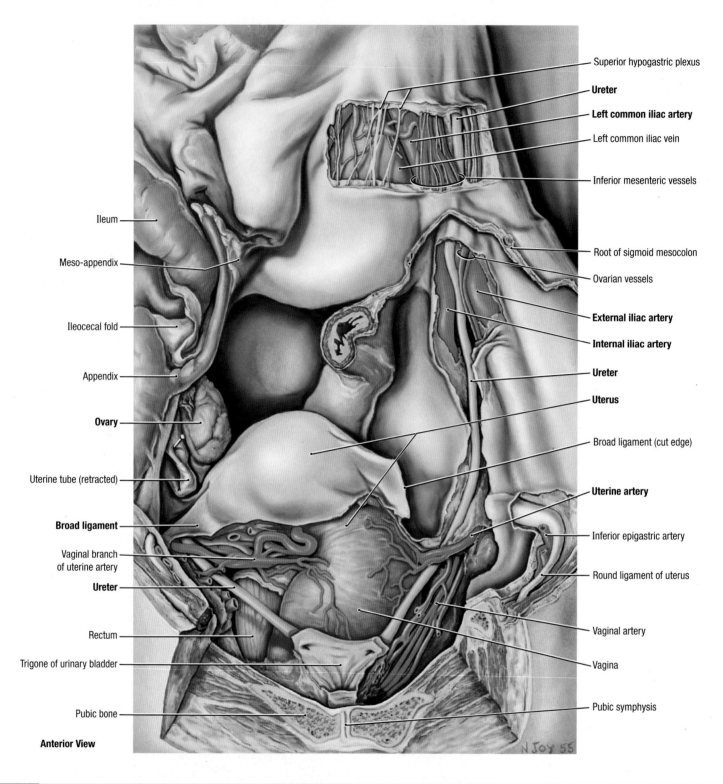

Superior hypogastric plexus

Ureter

Left common iliac artery

Left common iliac vein

Inferior mesenteric vessels

Root of sigmoid mesocolon

Ovarian vessels

External iliac artery

Internal iliac artery

Ureter

Uterus

Broad ligament (cut edge)

Uterine artery

Inferior epigastric artery

Round ligament of uterus

Vaginal artery

Vagina

Pubic symphysis

Ileum

Meso-appendix

Ileocecal fold

Appendix

Ovary

Uterine tube (retracted)

Broad ligament

Vaginal branch of uterine artery

Ureter

Rectum

Trigone of urinary bladder

Pubic bone

Anterior View

N JOY 55

5.37 URETER AND RELATIONSHIP TO UTERINE ARTERY

- Most of the pubic symphysis and most of the bladder (except the trigone) have been removed.
- The left ureter is crossed by the ovarian vessels and nerves; the apex of the inverted V-shaped root of the sigmoid mesocolon is situated anterior to the left ureter.

- The left ureter crosses the external iliac artery at the bifurcation of the common iliac artery and then descends anterior to the internal iliac artery; its course is subperitoneal from where it enters the pelvis to where it passes deep to the broad ligament and is crossed by the uterine artery. **Injury of the ureter** may occur in this region when the uterine artery is ligated and cut during hysterectomy.

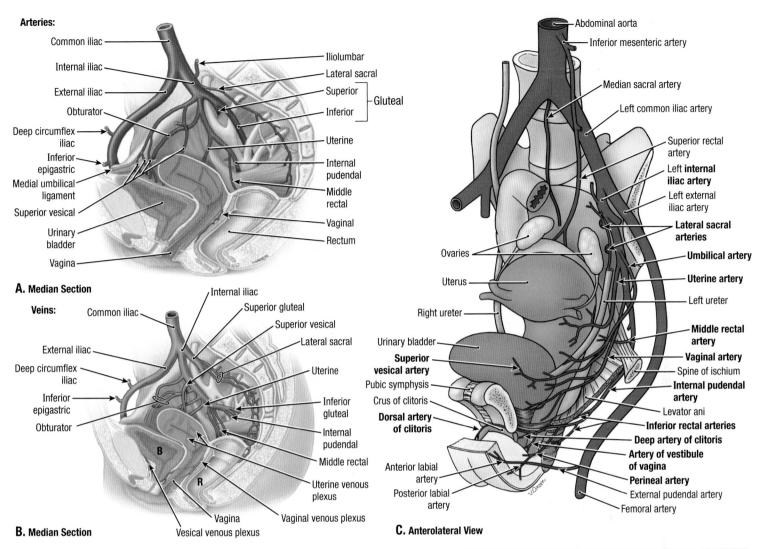

Arteries:

A. Median Section

- Common iliac
- Internal iliac
- External iliac
- Obturator
- Deep circumflex iliac
- Inferior epigastric
- Medial umbilical ligament
- Superior vesical
- Urinary bladder
- Vagina
- Iliolumbar
- Lateral sacral
- Superior — Gluteal
- Inferior — Gluteal
- Uterine
- Internal pudendal
- Middle rectal
- Vaginal
- Rectum

Veins:

B. Median Section

- Common iliac
- External iliac
- Deep circumflex iliac
- Inferior epigastric
- Obturator
- Internal iliac
- Superior gluteal
- Superior vesical
- Lateral sacral
- Uterine
- Inferior gluteal
- Internal pudendal
- Middle rectal
- Uterine venous plexus
- Vaginal venous plexus
- Vaginal
- Vesical venous plexus

C. Anterolateral View

- Abdominal aorta
- Inferior mesenteric artery
- Median sacral artery
- Left common iliac artery
- Superior rectal artery
- **Left internal iliac artery**
- Left external iliac artery
- **Lateral sacral arteries**
- **Umbilical artery**
- **Uterine artery**
- Left ureter
- **Middle rectal artery**
- **Vaginal artery**
- Spine of ischium
- **Internal pudendal artery**
- Levator ani
- **Inferior rectal arteries**
- **Deep artery of clitoris**
- **Artery of vestibule of vagina**
- **Perineal artery**
- External pudendal artery
- Femoral artery
- Ovaries
- Uterus
- Right ureter
- Urinary bladder
- **Superior vesical artery**
- Pubic symphysis
- Crus of clitoris
- **Dorsal artery of clitoris**
- Anterior labial artery
- Posterior labial artery

ARTERIES AND VEINS OF FEMALE PELVIS

5.38

TABLE 5.7	**ARTERIES OF FEMALE PELVIS (DERIVATIVES OF INTERNAL ILIAC ARTERY [IIA])**		
Artery	*Origin*	*Course*	*Distribution*
Anterior division of IIA	Internal iliac artery	Passes anteriorly along lateral wall of pelvis, dividing into visceral and obturator arteries	Pelvic viscera and muscles of superior medial thigh and perineum
Umbilical	Anterior div. IIA	Short pelvic course, gives off superior vesical arteries	Superior aspect of urinary bladder
Superior vesical artery	Patent umbilical a.	Usually multiple, pass to superior aspect of urinary bladder	Superior aspect of urinary bladder
Obturator	Anterior division of internal iliac artery	Runs antero-inferiorly on lateral pelvic wall	Pelvic muscles, ilium, femoral head, medial thigh
Uterine		Runs anteromedially between broad and cardinal ligs.; crosses ureter superiorly to lateral aspect of uterine cervix	Uterus, ligaments of uterus, medial parts of uterine tube and ovary, and superior vagina
Vaginal		Divides into vaginal and inferior vesical branches	Vaginal branch: lower vagina, vestibular bulb, and adjacent rectum; inferior vesical branch: fundus of urinary bladder
Middle rectal		Descends in pelvis to inferior part of rectum	Inferior part of rectum
Internal pudendal		Exits pelvis via greater sciatic foramen and enters perineum (ischio-anal fossa) via lesser sciatic foramen	Main artery to perineum including muscles of anal canal and perineum, skin and urogenital triangle and erectile bodies
Posterior division of IIA	Internal iliac artery	Passes posteriorly and gives rise to parietal branches	Pelvic wall and gluteal region
Iliolumbar	Posterior division of internal iliac artery	Ascends anterior to sacro-iliac joint and posterior to common iliac vessels and psoas major muscle	Iliacus, psoas major, quadratus lumborum muscles, and cauda equina in vertebral canal
Lateral sacral		Runs on anteromedial aspect of piriformis muscle	Piriformis and erector spinae muscles, structures in sacral canal
Ovarian	Abdominal aorta	Crosses pelvic brim and descends in suspensory ligament to ovary	Abdominal and/or pelvic ureter, ovary, and ampullary end of uterine tube

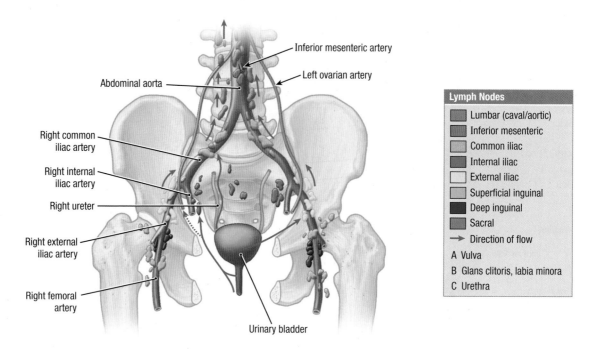

Inferior mesenteric artery

Left ovarian artery

Abdominal aorta

Right common
iliac artery

Right internal
iliac artery

Right ureter

Right external
iliac artery

Right femoral
artery

Urinary bladder

Lymph Nodes

- Lumbar (caval/aortic)
- Inferior mesenteric
- Common iliac
- Internal iliac
- External iliac
- Superficial inguinal
- Deep inguinal
- Sacral
- → Direction of flow
- A Vulva
- B Glans clitoris, labia minora
- C Urethra

A. Pelvic Urinary System

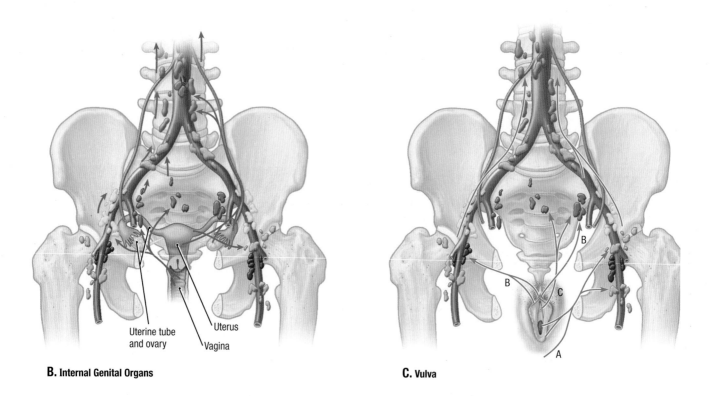

Uterine tube
and ovary

Uterus

Vagina

B. Internal Genital Organs

B

B

C

A

C. Vulva

Anterior Views

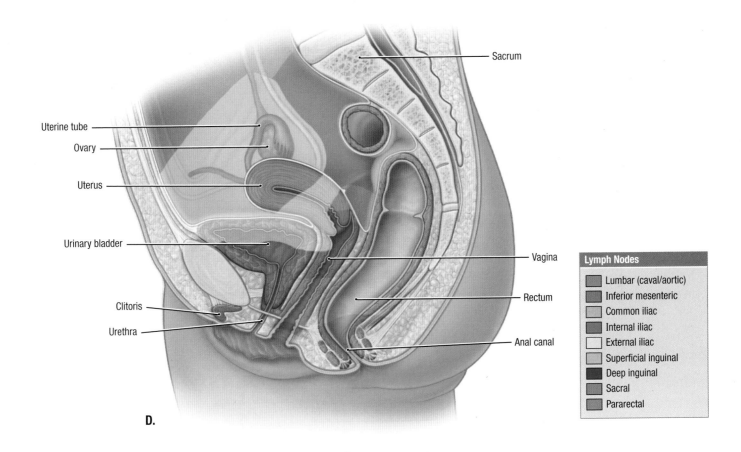

Sacrum

Uterine tube

Ovary

Uterus

Urinary bladder

Clitoris

Urethra

Vagina

Rectum

Anal canal

Lymph Nodes

	Lumbar (caval/aortic)
	Inferior mesenteric
	Common iliac
	Internal iliac
	External iliac
	Superficial inguinal
	Deep inguinal
	Sacral
	Pararectal

D.

LYMPHATIC DRAINAGE OF FEMALE PELVIS AND PERINEUM (*continued*)   5.39

D. Zones of pelvis and perineum initially draining to specific groups of regional nodes.

TABLE 5.8	**LYMPHATIC DRAINAGE OF STRUCTURES OF FEMALE PELVIS AND PERINEUM**
Lymph Node Group	*Structures Typically Draining to Lymph Node Group*
Lumbar	Gonads and associated structures (along ovarian vessels), ovary, uterine tube (except isthmus and intra-uterine parts), fundus of uterus, common iliac nodes
Inferior mesenteric	Superiormost rectum, sigmoid colon, descending colon, pararectal nodes
Common iliac	External and internal iliac lymph nodes
Internal iliac	Inferior pelvic structures, deep perineal structures, sacral nodes, base of bladder, inferior pelvic ureter, anal canal (above pectinate line), inferior rectum, middle and upper vagina, cervix, body of uterus, sacral nodes
External iliac	Anterosuperior pelvic structures, deep inguinal nodes, superior bladder, superior pelvic ureter, upper vagina, cervix, lower body of uterus
Superficial inguinal	Lower limb, superficial drainage of inferolateral quadrant of trunk, including anterior abdominal wall inferior to umbilicus, gluteal region, superolateral uterus (near attachment of round ligament), skin of perineum including vulva, ostium of vagina (inferior to hymen), prepuce of clitoris, peri-anal skin, anal canal inferior to pectinate line
Deep inguinal	Glans of clitoris, superficial inguinal nodes
Sacral	Postero-inferior pelvic structures, inferior rectum, inferior vagina
Pararectal	Superior rectum

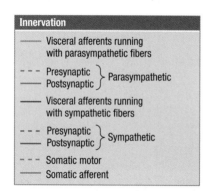

Innervation

——	Visceral afferents running with parasympathetic fibers
- - -	Presynaptic ⎫ Parasympathetic
——	Postsynaptic ⎭
——	Visceral afferents running with sympathetic fibers
- - -	Presynaptic ⎫ Sympathetic
——	Postsynaptic ⎭
- - -	Somatic motor
——	Somatic afferent

5.40   INNERVATION OF FEMALE PELVIC VISCERA

- Pelvic splanchnic nerves (S2–S4) supply parasympathetic motor fibers to the uterus and vagina (and vasodilator fibers to the erectile tissue of the clitoris and bulb of the vestibule; not shown).
- Presynaptic sympathetic fibers pass through the lumbar splanchnic nerves to synapse in prevertebral ganglia; the postsynaptic fibers travel through the superior and inferior hypogastric plexuses to reach the pelvic viscera.
- Visceral afferent fibers conducting pain from intraperitoneal viscera travel with the sympathetic fibers to the T12–L2 spinal ganglia. Visceral afferent fibers conducting pain from subperitoneal viscera travel with parasympathetic fibers to the S2–S4 spinal ganglia.
- Somatic sensation from the opening of the vagina also passes to the S2–S4 spinal ganglia via the pudendal nerve.
- Muscular contractions of the uterus are hormonally induced.

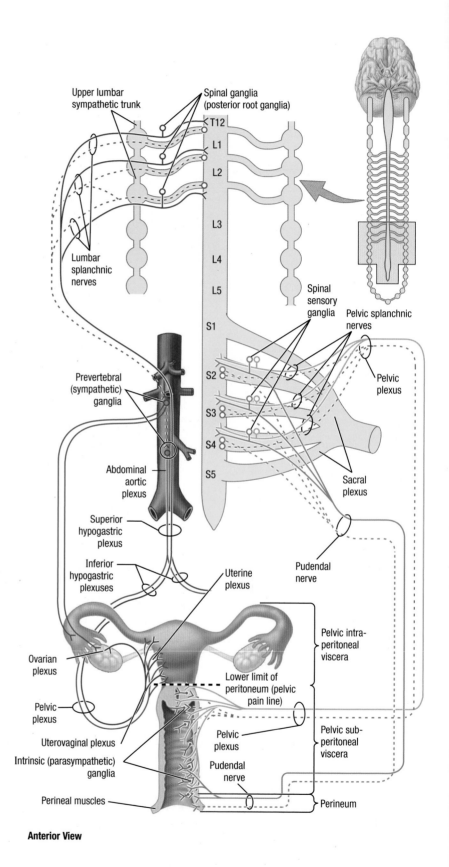

Anterior View

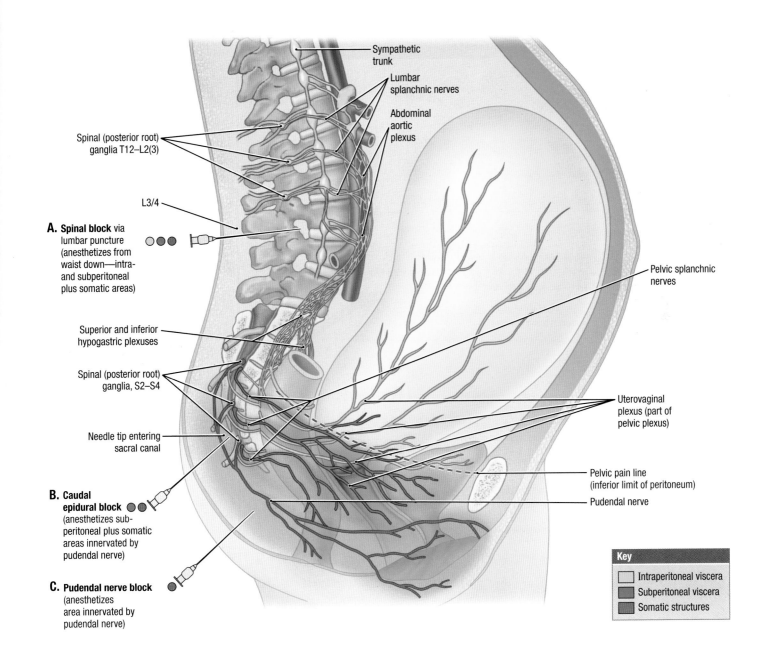

Sympathetic trunk

Lumbar splanchnic nerves

Abdominal aortic plexus

Spinal (posterior root) ganglia T12–L2(3)

L3/4

A. Spinal block via lumbar puncture (anesthetizes from waist down—intra- and subperitoneal plus somatic areas)

Superior and inferior hypogastric plexuses

Spinal (posterior root) ganglia, S2–S4

Needle tip entering sacral canal

B. Caudal epidural block (anesthetizes sub-peritoneal plus somatic areas innervated by pudendal nerve)

C. Pudendal nerve block (anesthetizes area innervated by pudendal nerve)

Pelvic splanchnic nerves

Uterovaginal plexus (part of pelvic plexus)

Pelvic pain line (inferior limit of peritoneum)

Pudendal nerve

Key	
☐	Intraperitoneal viscera
▨	Subperitoneal viscera
▦	Somatic structures

INNERVATION OF PELVIC VISCERA—OBSTETRICAL NERVE BLOCKS

5.41

- A **spinal block**, in which the anesthetic agent is introduced with a needle into the spinal subarachnoid space at the L3–L4 vertebral level, produces complete anesthesia inferior to approximately the waist level. The perineum, pelvic floor, and birth canal are anesthetized, and motor and sensory functions of the entire lower limbs, as well as sensation of uterine contractions, are temporarily eliminated.
- With the **caudal epidural block**, the anesthetic agent is administered using an in-dwelling catheter in the sacral canal. The entire

birth canal, pelvic floor, and most of the perineum are anesthetized, but the lower limbs are not usually affected. The mother is aware of her uterine contractions.

- A **pudendal nerve block** is a peripheral nerve block that provides local anesthesia over the S2–S4 dermatomes (most of the perineum) and the inferior quarter of the vagina. It does not block pain from the superior birth canal (uterine cervix and superior vagina), so the mother is able to feel uterine contractions.

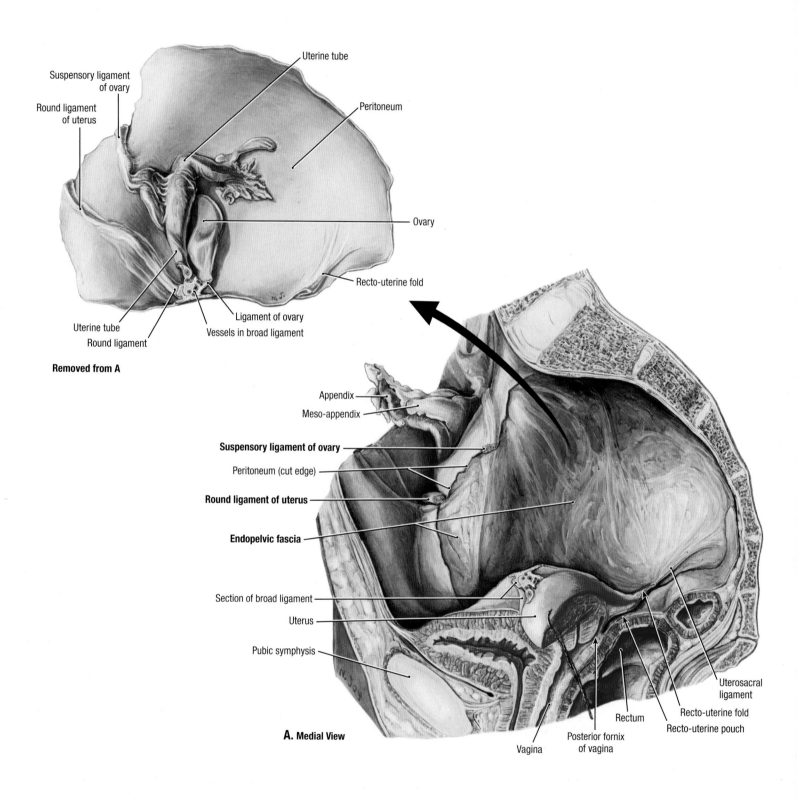

Suspensory ligament of ovary

Round ligament of uterus

Uterine tube

Peritoneum

Ovary

Recto-uterine fold

Uterine tube

Round ligament

Ligament of ovary

Vessels in broad ligament

Removed from A

Appendix

Meso-appendix

Suspensory ligament of ovary

Peritoneum (cut edge)

Round ligament of uterus

Endopelvic fascia

Section of broad ligament

Uterus

Pubic symphysis

Uterosacral ligament

Recto-uterine fold

Recto-uterine pouch

Rectum

Posterior fornix of vagina

Vagina

A. Medial View

5.42 SERIAL DISSECTION OF AUTONOMIC NERVES OF FEMALE PELVIS

A. Broad ligament and peritoneum of the lateral wall of the pelvic cavity have been removed to expose the endopelvic fascia.

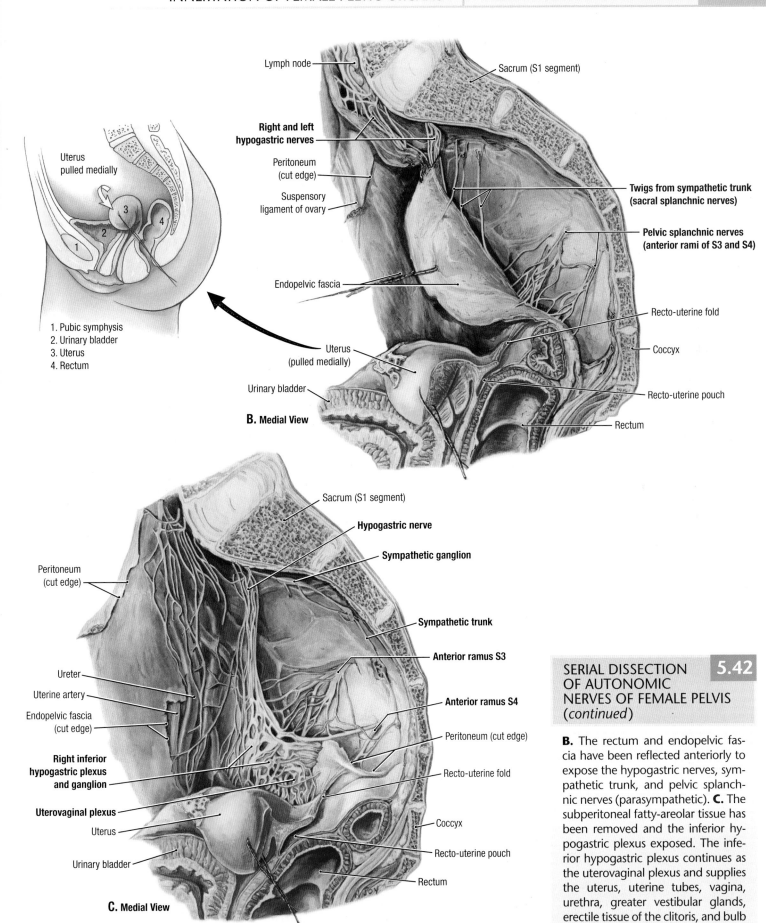

Uterus pulled medially

1. Pubic symphysis
2. Urinary bladder
3. Uterus
4. Rectum

Lymph node

Sacrum (S1 segment)

Right and left hypogastric nerves

Peritoneum (cut edge)

Suspensory ligament of ovary

Twigs from sympathetic trunk (sacral splanchnic nerves)

Pelvic splanchnic nerves (anterior rami of S3 and S4)

Endopelvic fascia

Recto-uterine fold

Coccyx

Uterus (pulled medially)

Urinary bladder

Recto-uterine pouch

B. Medial View

Rectum

Sacrum (S1 segment)

Hypogastric nerve

Sympathetic ganglion

Peritoneum (cut edge)

Sympathetic trunk

Anterior ramus S3

Ureter

Uterine artery

Endopelvic fascia (cut edge)

Anterior ramus S4

Peritoneum (cut edge)

Right inferior hypogastric plexus and ganglion

Recto-uterine fold

Uterovaginal plexus

Uterus

Coccyx

Urinary bladder

Recto-uterine pouch

Rectum

C. Medial View

SERIAL DISSECTION OF AUTONOMIC NERVES OF FEMALE PELVIS (*continued*)

5.42

B. The rectum and endopelvic fascia have been reflected anteriorly to expose the hypogastric nerves, sympathetic trunk, and pelvic splanchnic nerves (parasympathetic). **C.** The subperitoneal fatty-areolar tissue has been removed and the inferior hypogastric plexus exposed. The inferior hypogastric plexus continues as the uterovaginal plexus and supplies the uterus, uterine tubes, vagina, urethra, greater vestibular glands, erectile tissue of the clitoris, and bulb of the vestibule.

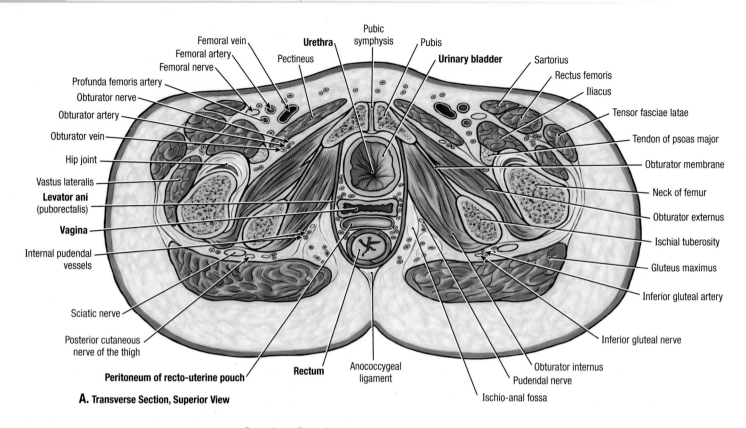

A. Transverse Section, Superior View

Femoral vein
Femoral artery
Femoral nerve
Profunda femoris artery
Obturator nerve
Obturator artery
Obturator vein
Hip joint
Vastus lateralis
Levator ani (puborectalis)
Vagina
Internal pudendal vessels
Sciatic nerve
Posterior cutaneous nerve of the thigh
Peritoneum of recto-uterine pouch
Pectineus
Pubic symphysis
Urethra
Pubis
Urinary bladder
Sartorius
Rectus femoris
Iliacus
Tensor fasciae latae
Tendon of psoas major
Obturator membrane
Neck of femur
Obturator externus
Ischial tuberosity
Gluteus maximus
Inferior gluteal artery
Inferior gluteal nerve
Obturator internus
Pudendal nerve
Ischio-anal fossa
Rectum
Anococcygeal ligament

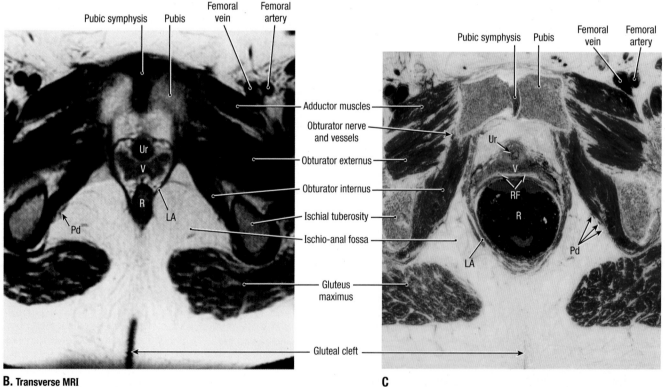

B. Transverse MRI

Pubic symphysis
Pubis
Femoral vein
Femoral artery
Ur
V
R
LA
Pd

Adductor muscles
Obturator nerve and vessels
Obturator externus
Obturator internus
Ischial tuberosity
Ischio-anal fossa
Gluteus maximus
Gluteal cleft

C

Pubic symphysis
Pubis
Femoral vein
Femoral artery
Ur
V
RF
R
Pd
LA

Key for B and C	
LA	Levator ani
Pd	Pudendal nerve and vessels
R	Rectum
RF	Recto-uterine fold
Ur	Urethra
V	Vagina

5.43 TRANSVERSE SECTIONS AND MRIs THROUGH FEMALE PELVIS

A. Transverse section through the ischial tuberosities, bladder, vagina, rectum, and recto-uterine pouch. **B.** Transverse (axial) MRI. **C.** Sectioned specimen.

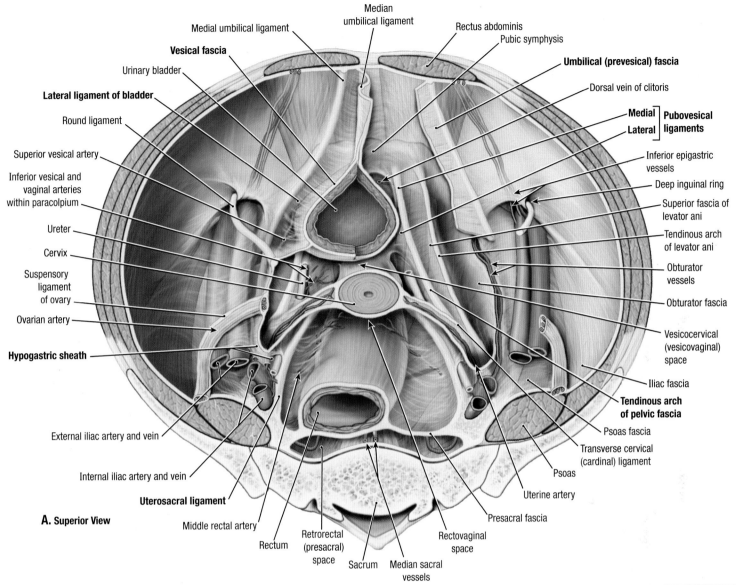

Median umbilical ligament

Medial umbilical ligament

Vesical fascia

Urinary bladder

Lateral ligament of bladder

Round ligament

Superior vesical artery

Inferior vesical and vaginal arteries within paracolpium

Ureter

Cervix

Suspensory ligament of ovary

Ovarian artery

Hypogastric sheath

External iliac artery and vein

Internal iliac artery and vein

Uterosacral ligament

A. Superior View

Middle rectal artery

Rectum

Retrorectal (presacral) space

Sacrum

Median sacral vessels

Rectus abdominis

Pubic symphysis

Umbilical (prevesical) fascia

Dorsal vein of clitoris

Medial ⎤ **Pubovesical**
Lateral ⎦ **ligaments**

Inferior epigastric vessels

Deep inguinal ring

Superior fascia of levator ani

Tendinous arch of levator ani

Obturator vessels

Obturator fascia

Vesicocervical (vesicovaginal) space

Iliac fascia

Tendinous arch of pelvic fascia

Psoas fascia

Transverse cervical (cardinal) ligament

Psoas

Uterine artery

Presacral fascia

Rectovaginal space

PELVIC FASCIA AND SUPPORTING MECHANISM OF CERVIX AND UPPER VAGINA — **5.44**

A. Greater and lesser pelvis demonstrating pelvic viscera and endopelvic fascia. **B.** Schematic illustration of fascial ligaments and areolar spaces at the level of tendinous arch of pelvic fascia.

- Note the parietal pelvic fascia covering the obturator internus and levator ani muscles and the visceral pelvic fasciae are continuous where the organs penetrate the pelvic floor, forming a tendinous arch of pelvic fascia bilaterally.
- The endopelvic fascia lies between, and is continuous with, both visceral and parietal layers of pelvic fascia. The loose, areolar portions of the endopelvic fascia have been removed; the fibrous, condensed portions remain. Note the condensation of this fascia into the hypogastric sheath, containing the vessels to the pelvic viscera, the ureters, and (in the male) the ductus deferens.
- Observe the ligamentous extensions of the hypogastric sheath: the lateral ligament of the urinary bladder, the transverse cervical ligament at the base of the broad ligament, and a less prominent lamina posteriorly containing the middle rectal vessels.

Key

▨ Tendinous arch of pelvic fascia

ANTERIOR

Pubic symphysis

Retropubic space (opened)

Pubovesical ligament

Vesical fascia

Tendinous arch of levator ani

Transverse cervical ligament

Recto-uterine pouch

Uterosacral (recto-uterine ligament)

Rectal fascia

Presacral space (opened)

Urinary bladder

Cervix

Rectum

Sacrum

B. Superior View

POSTERIOR

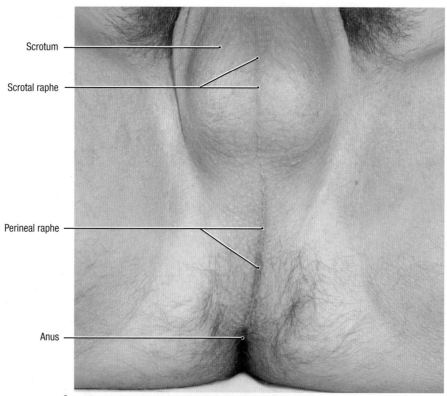

Scrotum

Scrotal raphe

Perineal raphe

Anus

A. Inferior View, penis/scrotum retracted anteriorly

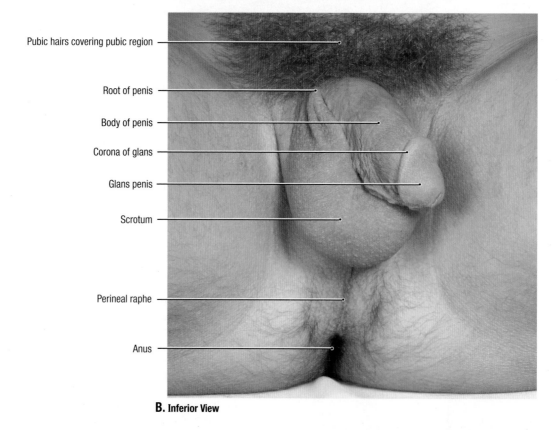

Pubic hairs covering pubic region

Root of penis

Body of penis

Corona of glans

Glans penis

Scrotum

Perineal raphe

Anus

B. Inferior View

5.45 SURFACE ANATOMY OF MALE PERINEUM

A. Center of male perineal region. **B.** Penis, scrotum, and anal region.

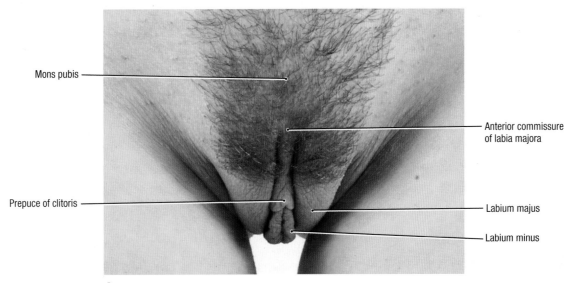

Mons pubis

Anterior commissure of labia majora

Prepuce of clitoris

Labium majus

Labium minus

A. Anterior View

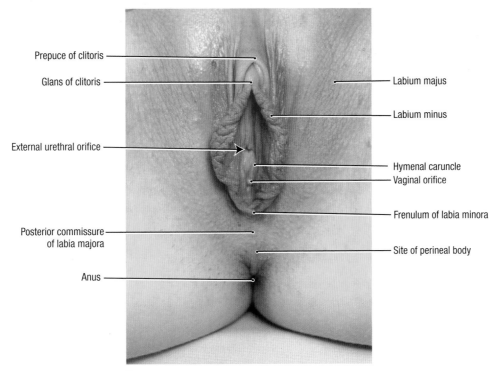

Prepuce of clitoris

Glans of clitoris

Labium majus

Labium minus

External urethral orifice

Hymenal caruncle

Vaginal orifice

Frenulum of labia minora

Posterior commissure of labia majora

Site of perineal body

Anus

B. Antero-inferior View (Lithotomy Position)

SURFACE ANATOMY OF THE FEMALE PERINEUM

5.46

A. External genitalia (pudendum; vulva), standing position. **B.** Vestibule of vagina and the external urethral and vaginal orifices opening into it (recumbent position).

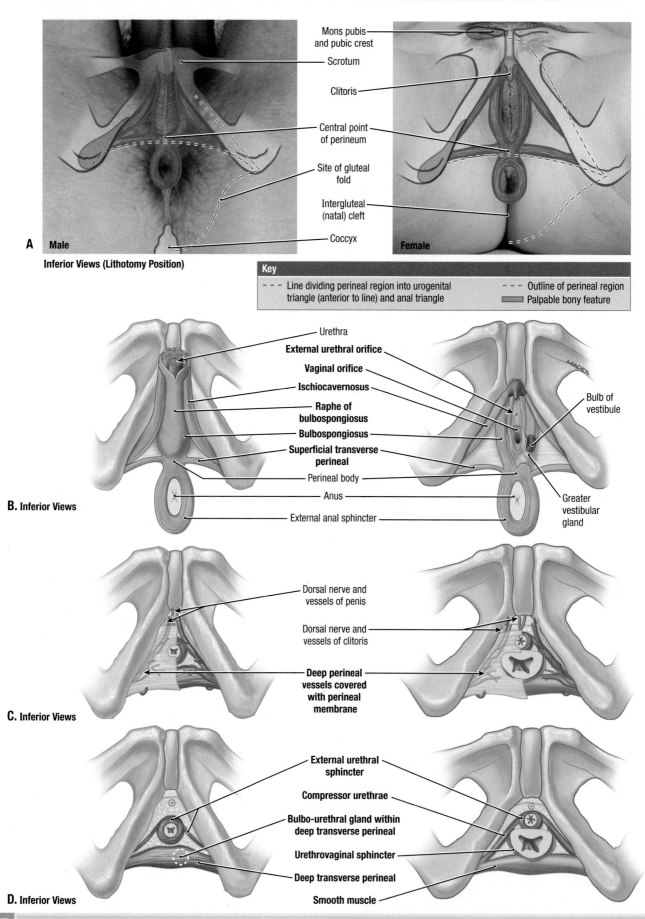

A Male

Inferior Views (Lithotomy Position)

Mons pubis and pubic crest

Scrotum

Clitoris

Central point of perineum

Site of gluteal fold

Intergluteal (natal) cleft

Coccyx

Female

Key

- - - Line dividing perineal region into urogenital triangle (anterior to line) and anal triangle

- - - Outline of perineal region

Palpable bony feature

B. Inferior Views

Urethra

External urethral orifice

Vaginal orifice

Ischiocavernosus

Raphe of bulbospongiosus

Bulbospongiosus

Superficial transverse perineal

Perineal body

Anus

External anal sphincter

Bulb of vestibule

Greater vestibular gland

C. Inferior Views

Dorsal nerve and vessels of penis

Dorsal nerve and vessels of clitoris

Deep perineal vessels covered with perineal membrane

D. Inferior Views

External urethral sphincter

Compressor urethrae

Bulbo-urethral gland within deep transverse perineal

Urethrovaginal sphincter

Deep transverse perineal

Smooth muscle

5.47   LAYERS OF PERINEUM

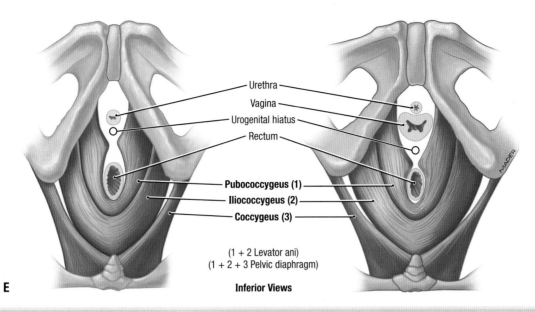

Urethra
Vagina
Urogenital hiatus
Rectum
Pubococcygeus (1)
Iliococcygeus (2)
Coccygeus (3)

(1 + 2 Levator ani)
(1 + 2 + 3 Pelvic diaphragm)

E

Inferior Views

LAYERS OF PERINEUM (*continued*)

5.47

A–E. The layers are shown from superficial to deep.

TABLE 5.9	**MUSCLES OF PERINEUM**			
Muscle	*Origin*	*Course and Insertion*	*Innervation*	*Main Action*
External anal sphincter	Skin and fascia surrounding anus; coccyx via anococcygeal ligament	Passes around lateral aspects of anal canal; insertion into perineal body	Inferior anal (rectal) nerve, a branch of pudendal nerve (S2–S4)	Constricts anal canal during peristalsis, resisting defecation; supports and fixes perineal body and pelvic floor
Bulbospongiosus	*Male:* median raphe on ventral surface of bulb of penis; perineal body	*Male:* surrounds lateral aspects of bulb of penis and most proximal part of body of penis, inserting into perineal membrane, dorsal aspect of corpora spongiosum and cavernosa, and fascia of bulb of penis	Muscular (deep) branch of perineal nerve, a branch of the pudendal nerve (S2–S4)	*Male:* supports and fixes perineal body/pelvic floor; compresses bulb of penis to expel last drops of urine/semen; assists erection by compressing outflow via deep perineal vein and by pushing blood from bulb into body of penis
	Female: perineal body	*Female:* passes on each side of lower vagina, enclosing bulb and greater vestibular gland; inserts onto pubic arch and fascia of corpora cavernosa of clitoris		*Female:* supports and fixes perineal body/pelvic floor; "sphincter" of vagina; assists in erection of clotiris (and perhaps bulb of vestibule); compresses greater vestibular gland
Ischiocavernosus	Internal surface of ischiopubic ramus and ischial tuberosity	Embraces crus of penis or clitoris, inserting onto the inferior and medial aspects of the crus and to the perineal membrane medial to the crus		Maintains erection of penis or clitoris by-compressing outflow veins and pushing blood from the root of penis or clitoris into the body of penis or clitoris
Superficial transverse perineal	Internal surface of ischiopubic ramus and ischial tuberosity	Passes along inferior aspect of posterior border of perineal membrane to perineal body		Supports and fixes perineal body (pelvic floor) to support abdominopelvic viscera and resist increased intra-abdominal pressure
Deep transverse perineal (male only)		Passes along superior aspect of posterior border of perineal membrane to perineal body, and external anal sphincter	Muscular (deep) branch of perineal nerve	
Smooth muscle (female only)		Passes to lateral wall of urethra and vagina	Autonomic nerves	Quantity of smooth muscle increases with age; function uncertain
External urethral sphincter	Ischiopubic rami	Surrounds urethra superior to perineal membrane; in males, also ascends anterior aspect of prostate	Dorsal nerve of penis or clitoris, terminal branch of pudendal nerve (S2–S4)	Compresses urethra to maintain urinary continence
Compressor urethrae (females only)	Internal surface of ischiopubic ramus	Continuous with external urethral sphincter		Compresses urethra; with pelvic diaphragm assists in elongation of urethra
Urethrovaginal sphincter (females only)	Anterior side of urethra	Continuous with compressor urethrae; extends posteriorly on lateral wall of urethra and vagina to interdigitate with fibers from opposite side of perineal body		Compresses urethra and vagina

Oelrich TM. The urethral sphincter muscle in the male. *Am J Anat* 1980;158:229–246.
Oelrich TM. The striated urogenital sphincter muscle in the female. *Anat Rec* 1983;205:223–232.
Mirilas P, Skandalakis JE. Urogenital diaphragm: an erroneous concept casting its shadow over the sphincter urethrae and deep perineal space. *J Am Coll Surg* 2004;198:279–290.
DeLancey JO. Correlative study of paraurethral anatomy. *Obstet Gynecol* 1986;68:91–97.

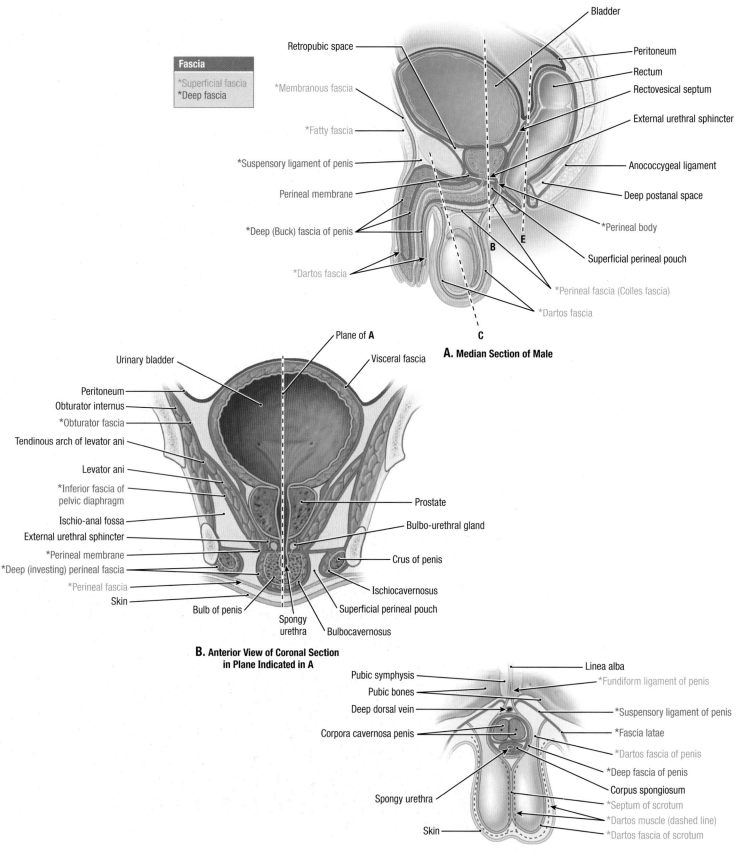

Fascia
*Superficial fascia
*Deep fascia

Retropubic space
*Membranous fascia
*Fatty fascia
*Suspensory ligament of penis
Perineal membrane
*Deep (Buck) fascia of penis
*Dartos fascia

Bladder
Peritoneum
Rectum
Rectovesical septum
External urethral sphincter
Anococcygeal ligament
Deep postanal space
*Perineal body
Superficial perineal pouch
*Perineal fascia (Colles fascia)
*Dartos fascia

Plane of **A**
Visceral fascia

A. Median Section of Male

Urinary bladder
Peritoneum
Obturator internus
*Obturator fascia
Tendinous arch of levator ani
Levator ani
*Inferior fascia of pelvic diaphragm
Ischio-anal fossa
External urethral sphincter
*Perineal membrane
*Deep (investing) perineal fascia
*Perineal fascia
Skin
Bulb of penis
Spongy urethra
Bulbocavernosus

Prostate
Bulbo-urethral gland
Crus of penis
Ischiocavernosus
Superficial perineal pouch

B. Anterior View of Coronal Section in Plane Indicated in A

Pubic symphysis
Pubic bones
Deep dorsal vein
Corpora cavernosa penis
Spongy urethra
Skin

Linea alba
*Fundiform ligament of penis
*Suspensory ligament of penis
*Fascia latae
*Dartos fascia of penis
*Deep fascia of penis
Corpus spongiosum
*Septum of scrotum
*Dartos muscle (dashed line)
*Dartos fascia of scrotum

C. Anterior View of Coronal Section in Plane Indicated in A

5.48 FASCIAE OF PERINEUM

A–C. Male perineum.

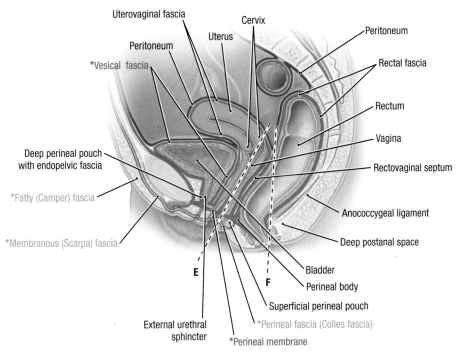

Uterovaginal fascia
Cervix
Uterus
Peritoneum
Peritoneum
Rectal fascia
*Vesical fascia
Rectum
Vagina
Deep perineal pouch with endopelvic fascia
Rectovaginal septum
*Fatty (Camper) fascia
Anococcygeal ligament
*Membranous (Scarpa) fascia
Deep postanal space
Bladder
E
F
Perineal body
Superficial perineal pouch
External urethral sphincter
*Perineal fascia (Colles fascia)
*Perineal membrane

D. Median Section of Female

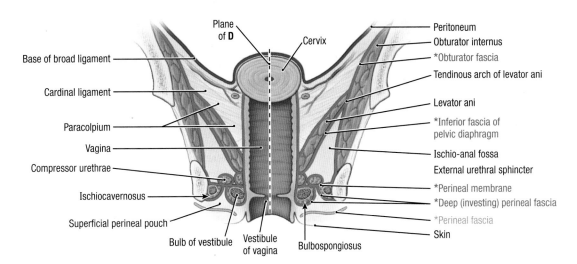

Plane of **D**
Cervix
Peritoneum
Obturator internus
Base of broad ligament
*Obturator fascia
Tendinous arch of levator ani
Cardinal ligament
Levator ani
Paracolpium
*Inferior fascia of pelvic diaphragm
Vagina
Ischio-anal fossa
Compressor urethrae
External urethral sphincter
Ischiocavernosus
*Perineal membrane
*Deep (investing) perineal fascia
Superficial perineal pouch
*Perineal fascia
Bulb of vestibule
Vestibule of vagina
Bulbospongiosus
Skin

E. Anterior View of Coronal Section in Plane Indicated in D

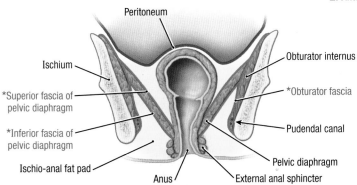

Peritoneum
Ischium
Obturator internus
*Superior fascia of pelvic diaphragm
*Obturator fascia
*Inferior fascia of pelvic diaphragm
Pudendal canal
Ischio-anal fat pad
Pelvic diaphragm
Anus
External anal sphincter

F. Anterior View of Coronal Section in Plane Indicated in D

FASCIAE OF PERINEUM (continued)

5.48

D–F. Female perineum.

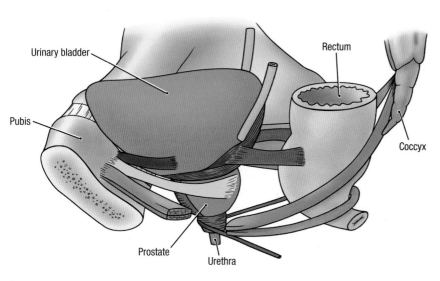

A. Left Lateral View, Male

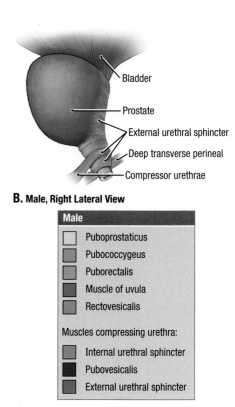

B. Male, Right Lateral View

Male	
☐	Puboprostaticus
▨	Pubococcygeus
☐	Puborectalis
▨	Muscle of uvula
▨	Rectovesicalis
Muscles compressing urethra:	
▨	Internal urethral sphincter
■	Pubovesicalis
▨	External urethral sphincter

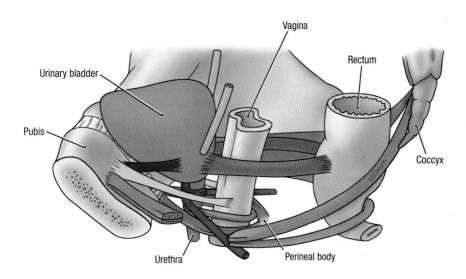

C. Left Lateral View, Female

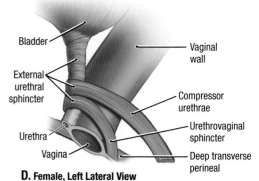

D. Female, Left Lateral View

Female	
■	Pubovesicalis
▨	Pubococcygeus
☐	Puborectalis
▨	Rectovesicalis
Muscles compressing urethra:	
▨	Compressor urethrae
▨	External urethral sphincter
Muscles compressing vagina:	
☐	Pubovaginalis
▨	Urethrovaginal sphincter (part of external urethral sphincter)
■	Bulbospongiosus

5.49   SUPPORTING AND COMPRESSOR/SPHINCTERIC MUSCLES OF PELVIS

A. Male. **B.** Male urethral sphincters. **C.** Female. **D.** Female urethral sphincters.

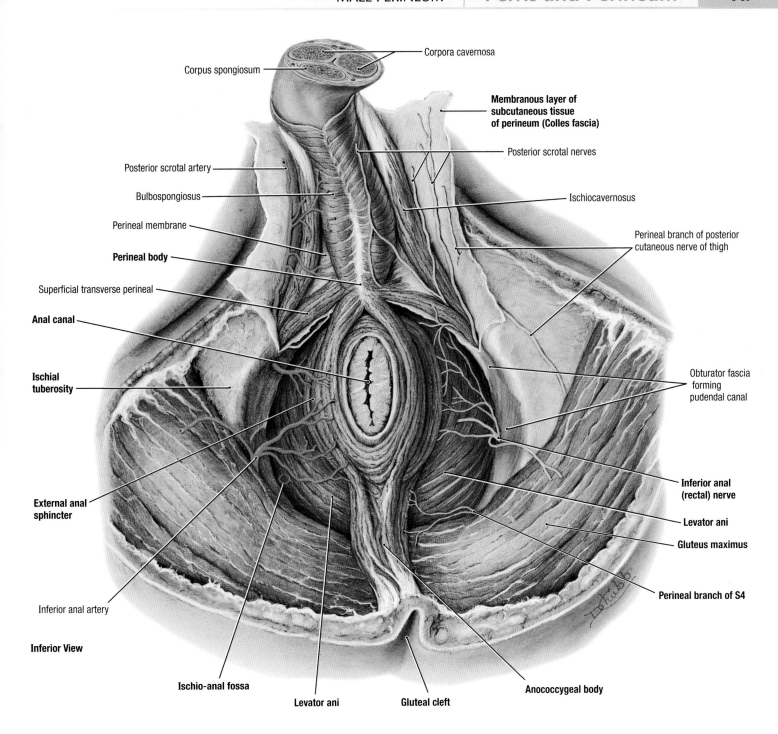

Corpora cavernosa

Corpus spongiosum

Membranous layer of subcutaneous tissue of perineum (Colles fascia)

Posterior scrotal nerves

Posterior scrotal artery

Bulbospongiosus

Ischiocavernosus

Perineal membrane

Perineal branch of posterior cutaneous nerve of thigh

Perineal body

Superficial transverse perineal

Anal canal

Ischial tuberosity

Obturator fascia forming pudendal canal

External anal sphincter

Inferior anal (rectal) nerve

Levator ani

Gluteus maximus

Inferior anal artery

Perineal branch of S4

Inferior View

Ischio-anal fossa

Levator ani

Gluteal cleft

Anococcygeal body

DISSECTION OF MALE PERINEUM I

5.50

Superficial dissection.
- The membranous layer of subcutaneous tissue of the perineum was incised and reflected, opening the subcutaneous perineal compartment (pouch) in which the cutaneous nerves course.
- The perineal membrane is exposed between the three paired muscles of the superficial compartment; although not evident here, the muscles are individually ensheathed with investing fascia.
- The anal canal is surrounded by the external anal sphincter. The superficial fibers of the sphincter anchor the anal canal anteriorly to the perineal body and posteriorly, via the anococcygeal body (ligament), to the coccyx and skin of the gluteal cleft.

- Ischio-anal (ischiorectal) fossae, from which fat bodies have been removed, lie on each side of the external anal sphincter. The fossae are also bound medially and superiorly by the levator ani, laterally by the ischial tuberosities and obturator internus fascia, and posteriorly by the gluteus maximus overlying the sacrotuberous ligaments. An anterior recess of each ischio-anal fossa extends superior to the perineal membrane.
- In the lateral wall of the fossa, the inferior anal (rectal) nerve emerges from the pudendal canal and, with the perineal branch of S4, supplies the voluntary external anal sphincter and perianal skin; most cutaneous twigs have been removed.

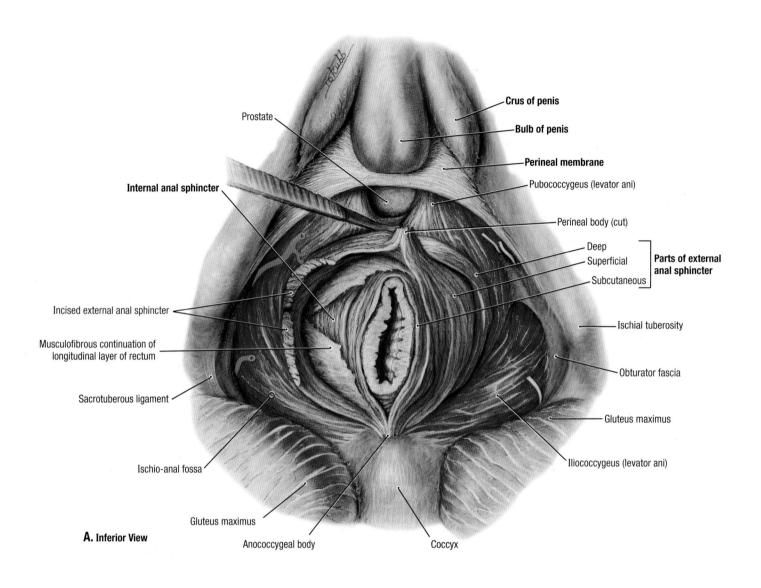

Prostate

Crus of penis

Bulb of penis

Perineal membrane

Pubococcygeus (levator ani)

Internal anal sphincter

Perineal body (cut)

Deep
Superficial } **Parts of external anal sphincter**
Subcutaneous

Incised external anal sphincter

Ischial tuberosity

Musculofibrous continuation of longitudinal layer of rectum

Obturator fascia

Sacrotuberous ligament

Gluteus maximus

Ischio-anal fossa

Iliococcygeus (levator ani)

Gluteus maximus

Anococcygeal body

Coccyx

A. Inferior View

| 5.51 | DISSECTION OF THE MALE PERINEUM II |

A. The superficial perineal muscles have been removed, revealing the roots of the erectile bodies (crura and bulb) of the penis, attached to the ischiopubic rami and perineal membrane. On the left side, the superficial and deep parts of the external anal sphincter were incised and reflected; the underlying musculofibrous continuation of the outer longitudinal layer of the muscular layer of the rectum is cut to reveal thickening of the inner circular layer that comprises the internal anal sphincter.
B. Rupture of the spongy urethra in the bulb of the penis results in extravasation (abnormal passage) of urine into the subcutaneous perineal compartment. The attachments of the membranous layer of subcutaneous tissue determine the direction and restrictions of flow of the extravasated urine. Urine and blood may pass deep to the continuations of the membranous layer in the scrotum, penis, and inferior abdominal wall. The urine cannot pass laterally and inferiorly into the thighs because the membranous layer fuses with the fascia lata (deep fascia of the thigh) nor posteriorly into the anal triangle due to continuity with the perineal membrane and perineal body.

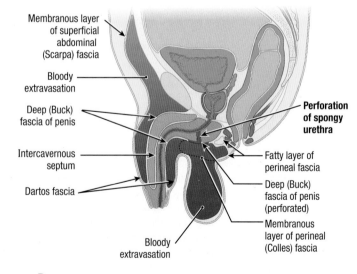

Membranous layer of superficial abdominal (Scarpa) fascia

Bloody extravasation

Deep (Buck) fascia of penis

Intercavernous septum

Dartos fascia

Perforation of spongy urethra

Fatty layer of perineal fascia

Deep (Buck) fascia of penis (perforated)

Membranous layer of perineal (Colles) fascia

Bloody extravasation

B. Medial View (from Left)

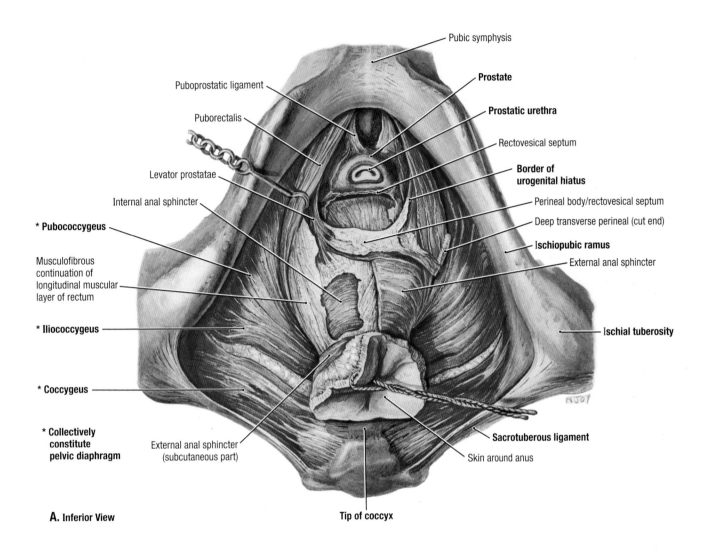

Pubic symphysis

Puboprostatic ligament

Puborectalis

Prostate

Prostatic urethra

Levator prostatae

Rectovesical septum

Internal anal sphincter

Border of urogenital hiatus

* Pubococcygeus

Perineal body/rectovesical septum

Deep transverse perineal (cut end)

Musculofibrous continuation of longitudinal muscular layer of rectum

Ischiopubic ramus

External anal sphincter

* Iliococcygeus

Ischial tuberosity

* Coccygeus

* Collectively constitute pelvic diaphragm

External anal sphincter (subcutaneous part)

Sacrotuberous ligament

Skin around anus

A. Inferior View

Tip of coccyx

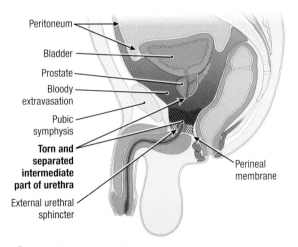

Peritoneum

Bladder

Prostate

Bloody extravasation

Pubic symphysis

Torn and separated intermediate part of urethra

External urethral sphincter

Perineal membrane

B. Medial View (from Left)

DISSECTION OF THE MALE PERINEUM III 5.52

A. The perineal membrane and structures superficial to it have been removed. The prostatic urethra, base of the prostate, and rectum are visible through the urogenital hiatus of the pelvic diaphragm. The osseofibrous boundaries are demonstrated. **B. Rupture of the intermediate part of the urethra** results in extravasation of urine and blood into the deep perineal compartment. The fluid may pass superiorly through the urogenital hiatus and distribute extraperitoneally around the prostate and bladder.

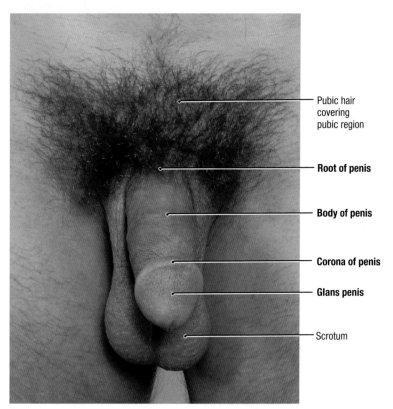

Pubic hair covering pubic region

Root of penis

Body of penis

Corona of penis

Glans penis

Scrotum

A. Anterior View

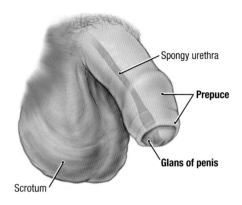

Spongy urethra

Prepuce

Glans of penis

Scrotum

B. Right Anterolateral View

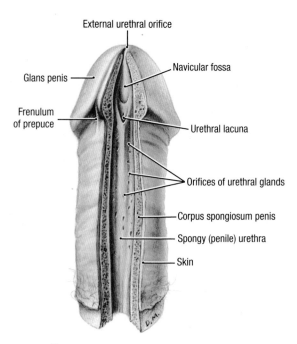

External urethral orifice

Glans penis

Navicular fossa

Frenulum of prepuce

Urethral lacuna

Orifices of urethral glands

Corpus spongiosum penis

Spongy (penile) urethra

Skin

D. Urethral Aspect of Distal Penis

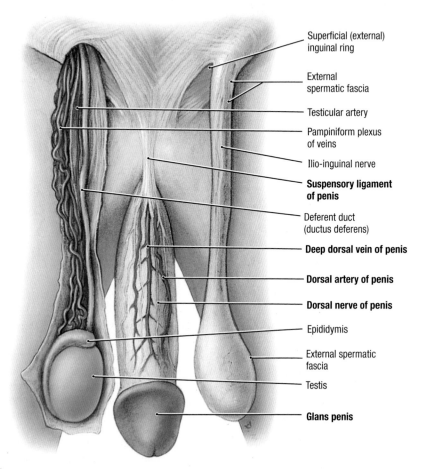

Superficial (external) inguinal ring

External spermatic fascia

Testicular artery

Pampiniform plexus of veins

Ilio-inguinal nerve

Suspensory ligament of penis

Deferent duct (ductus deferens)

Deep dorsal vein of penis

Dorsal artery of penis

Dorsal nerve of penis

Epididymis

External spermatic fascia

Testis

Glans penis

C. Anterior View

5.53 GLANS, PREPUCE, AND NEUROVASCULAR BUNDLE OF PENIS

A. Surface anatomy, penis circumcised. **B.** Uncircumcised penis. **C.** Vessels and nerves of penis and contents of spermatic cord. The superficial and deep fasciae covering the penis are removed to expose the midline deep dorsal vein and the bilateral dorsal arteries and nerves of the penis. **D.** Spongy urethra, interior. A longitudinal incision was made on the urethral surface of the penis and carried through the floor of the urethra, allowing a view of the dorsal surface of the interior of the urethra.

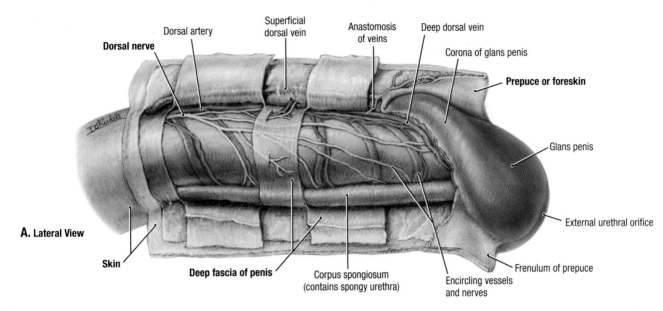

A. Lateral View

Dorsal nerve — Dorsal artery — Superficial dorsal vein — Anastomosis of veins — Deep dorsal vein — Corona of glans penis — **Prepuce or foreskin** — Glans penis — External urethral orifice — Frenulum of prepuce — Encircling vessels and nerves — Corpus spongiosum (contains spongy urethra) — **Deep fascia of penis** — **Skin**

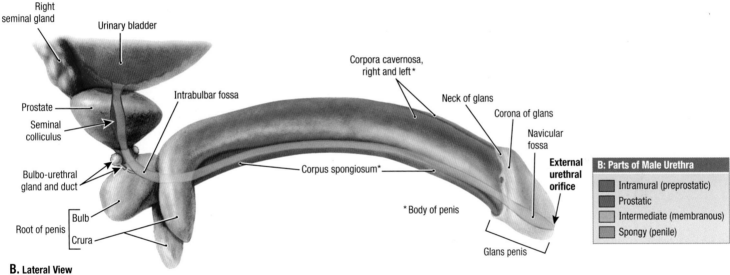

B. Lateral View

Right seminal gland — Urinary bladder — Prostate — Seminal colliculus — Bulbo-urethral gland and duct — Root of penis — Bulb — Crura — Intrabulbar fossa — Corpora cavernosa, right and left * — Neck of glans — Corona of glans — Navicular fossa — **External urethral orifice** — Corpus spongiosum* — * Body of penis — Glans penis

B: Parts of Male Urethra

- Intramural (preprostatic)
- Prostatic
- Intermediate (membranous)
- Spongy (penile)

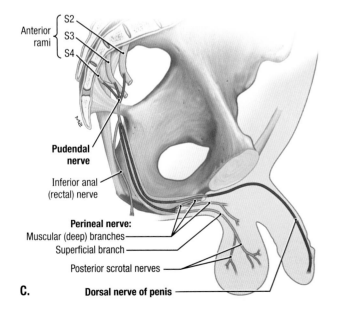

C.

Anterior rami { S2, S3, S4 } — **Pudendal nerve** — Inferior anal (rectal) nerve — **Perineal nerve:** — Muscular (deep) branches — Superficial branch — Posterior scrotal nerves — **Dorsal nerve of penis**

C: Pudendal Nerve and Branches, by Region

- Deep perineal pouch
- Dorsum of penis
- Superficial perineum
- Pelvis
- Gluteal region
- Pudendal canal

URETHRA, LAYERS, AND NERVES OF PENIS 5.54

A. Dissection. The skin, subcutaneous tissue, and deep fascia of the penis and prepuce are reflected separately. **B.** Parts of male urethra. **C.** Distribution of pudendal nerve, right hemipelvis. Five regions transversed by the nerve are demonstrated.

An uncircumcised prepuce covers all or most of the glans penis. The prepuce is usually sufficiently elastic to allow retraction over the glans. In some males, it is tight and cannot be retracted easily (phimosis), if at all. Secretions (smegma) may accumulate in the preputial sac, located between the glans penis and prepuce, causing irritation. **Circumcision** exposes most, or all, of the glans.

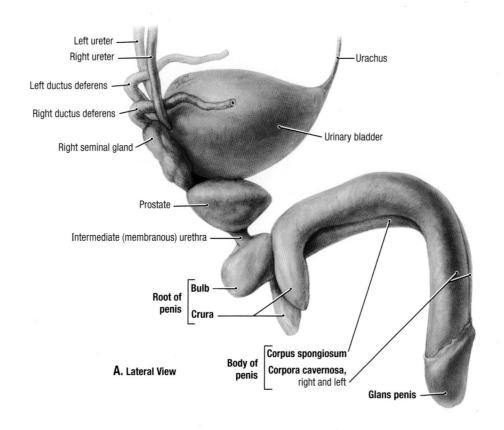

A. Lateral View

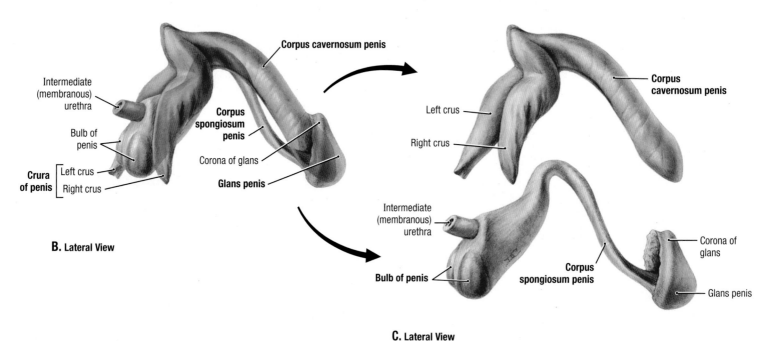

B. Lateral View

C. Lateral View

5.55 MALE UROGENITAL SYSTEM, ERECTILE BODIES

A. Pelvic components of genital and urinary tracts and erectile bodies of perineum. **B.** Dissection of male erectile bodies (corpora cavernosa and corpus spongiosum). **C.** Corpus spongiosum and corpora cavernosa, separated. The erectile bodies are flexed where the penis is suspended by the suspensory ligament of the penis from the pubic symphysis. The corpus spongiosum extends posteriorly as the bulb of the penis and terminates anteriorly as the glans.

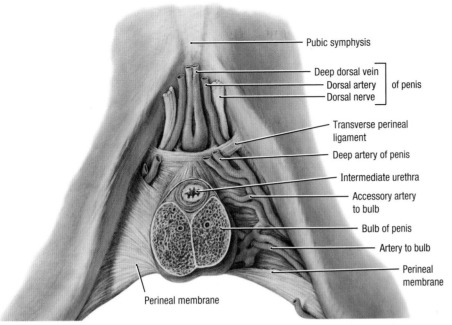

A. Anterior/Inferior View

Labels: Pubic symphysis; Deep dorsal vein / Dorsal artery / Dorsal nerve (of penis); Transverse perineal ligament; Deep artery of penis; Intermediate urethra; Accessory artery to bulb; Bulb of penis; Artery to bulb; Perineal membrane; Perineal membrane

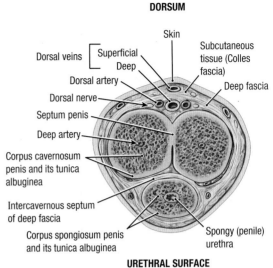

DORSUM

C. Transverse Section

Labels: Skin; Subcutaneous tissue (Colles fascia); Deep fascia; Dorsal veins (Superficial / Deep); Dorsal artery; Dorsal nerve; Septum penis; Deep artery; Corpus cavernosum penis and its tunica albuginea; Intercavernous septum of deep fascia; Corpus spongiosum penis and its tunica albuginea; Spongy (penile) urethra; URETHRAL SURFACE

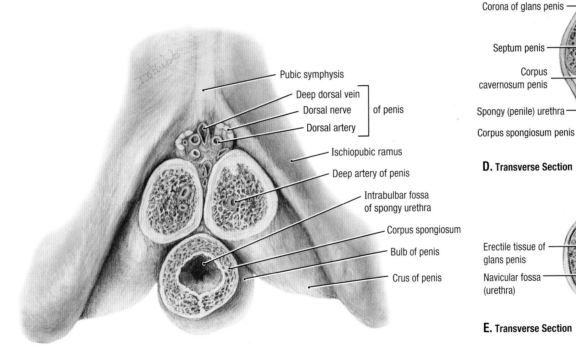

B. Anterior View

Labels: Pubic symphysis; Deep dorsal vein / Dorsal nerve / Dorsal artery (of penis); Ischiopubic ramus; Deep artery of penis; Intrabulbar fossa of spongy urethra; Corpus spongiosum; Bulb of penis; Crus of penis

D. Transverse Section

Labels: Corona of glans penis; Septum penis; Corpus cavernosum penis; Spongy (penile) urethra; Corpus spongiosum penis

E. Transverse Section

Labels: Erectile tissue of glans penis; Navicular fossa (urethra)

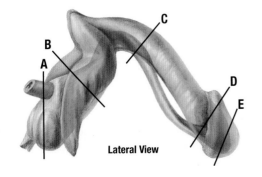

Lateral View

CROSS SECTIONS OF PENIS 5.56

A. Transverse section through bulb of penis with crura removed. The bulb is cut posterior to the entry of the intermediate urethra. On the left side, the perineal membrane is partially removed, opening the deep perineal compartment. **B.** The crura and bulb of penis have been sectioned obliquely. The spongy urethra is dilated within the bulb of the penis. **C.** Transverse section through body of penis. **D.** Transverse section through the proximal part of the glans penis. **E.** Transverse section through the distal part of the glans penis.

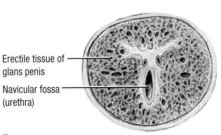

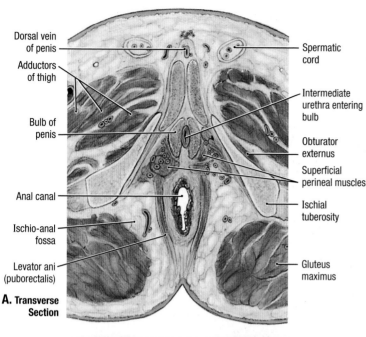

Dorsal vein of penis
Adductors of thigh
Bulb of penis
Anal canal
Ischio-anal fossa
Levator ani (puborectalis)

Spermatic cord
Intermediate urethra entering bulb
Obturator externus
Superficial perineal muscles
Ischial tuberosity
Gluteus maximus

A. Transverse Section

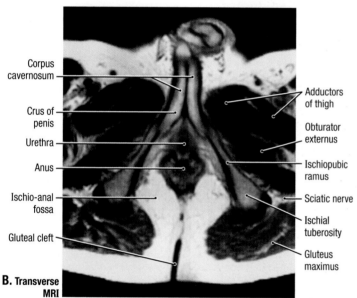

Corpus cavernosum
Crus of penis
Urethra
Anus
Ischio-anal fossa
Gluteal cleft

Adductors of thigh
Obturator externus
Ischiopubic ramus
Sciatic nerve
Ischial tuberosity
Gluteus maximus

B. Transverse MRI

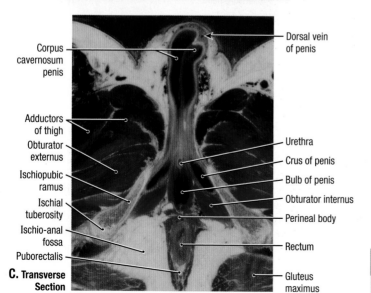

Corpus cavernosum penis
Adductors of thigh
Obturator externus
Ischiopubic ramus
Ischial tuberosity
Ischio-anal fossa
Puborectalis

Dorsal vein of penis
Urethra
Crus of penis
Bulb of penis
Obturator internus
Perineal body
Rectum
Gluteus maximus

C. Transverse Section

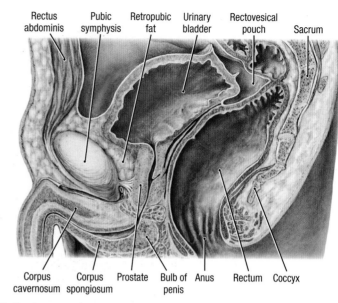

Rectus abdominis
Pubic symphysis
Retropubic fat
Urinary bladder
Rectovesical pouch
Sacrum

Corpus cavernosum
Corpus spongiosum
Prostate
Bulb of penis
Anus
Rectum
Coccyx

D. Median Section, Male

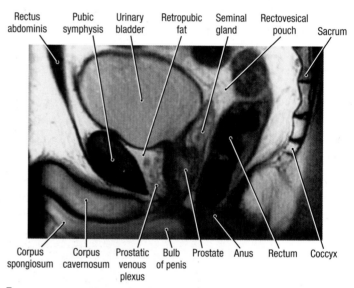

Rectus abdominis
Pubic symphysis
Urinary bladder
Retropubic fat
Seminal gland
Rectovesical pouch
Sacrum

Corpus spongiosum
Corpus cavernosum
Prostatic venous plexus
Bulb of penis
Prostate
Anus
Rectum
Coccyx

E. Median MRI, Prostate

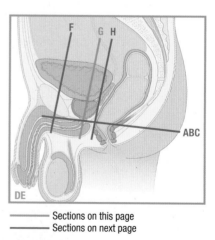

F G H
ABC
DE

Sections on this page
Sections on next page

5.57 IMAGING OF MALE PELVIS AND PERINEUM

Coronal MRIs:

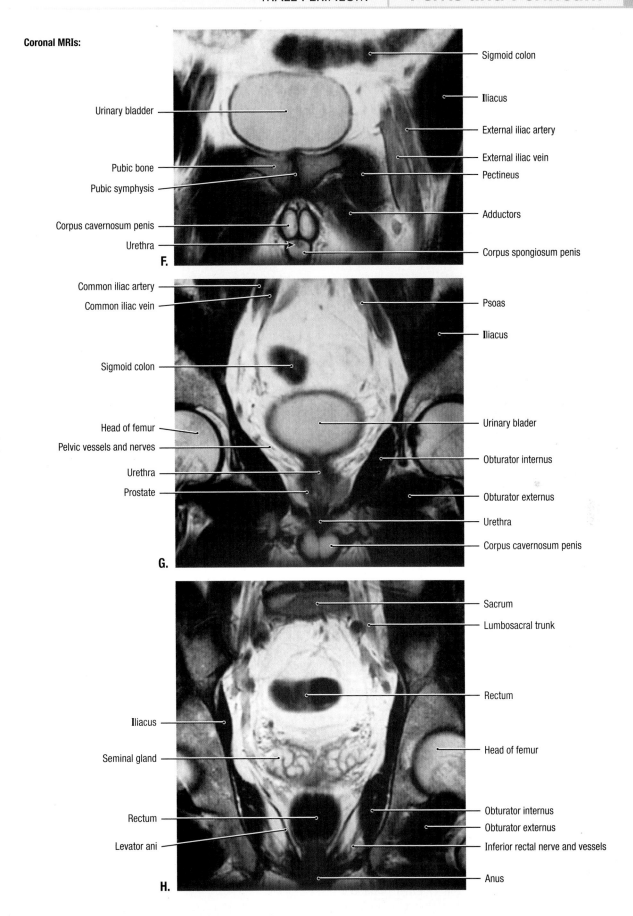

F.

- Sigmoid colon
- Urinary bladder
- Iliacus
- External iliac artery
- External iliac vein
- Pubic bone
- Pubic symphysis
- Pectineus
- Corpus cavernosum penis
- Adductors
- Urethra
- Corpus spongiosum penis

G.

- Common iliac artery
- Common iliac vein
- Psoas
- Iliacus
- Sigmoid colon
- Head of femur
- Pelvic vessels and nerves
- Urinary blader
- Urethra
- Obturator internus
- Prostate
- Obturator externus
- Urethra
- Corpus cavernosum penis

H.

- Sacrum
- Lumbosacral trunk
- Rectum
- Iliacus
- Seminal gland
- Head of femur
- Rectum
- Obturator internus
- Levator ani
- Obturator externus
- Inferior rectal nerve and vessels
- Anus

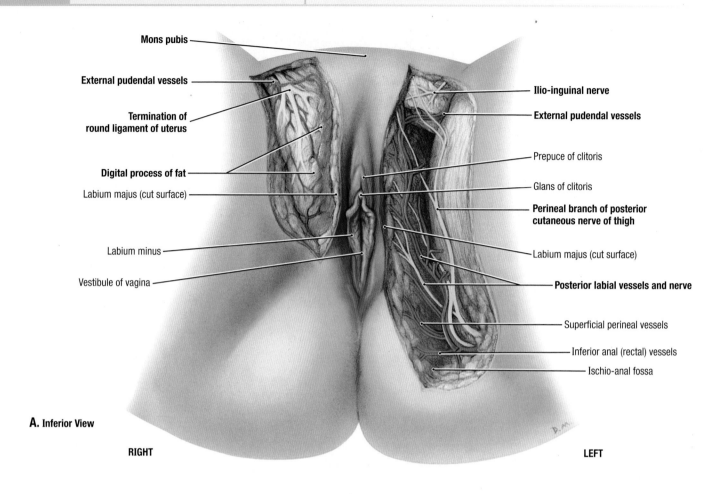

Mons pubis

External pudendal vessels

Termination of
round ligament of uterus

Digital process of fat

Labium majus (cut surface)

Labium minus

Vestibule of vagina

Ilio-inguinal nerve

External pudendal vessels

Prepuce of clitoris

Glans of clitoris

Perineal branch of posterior
cutaneous nerve of thigh

Labium majus (cut surface)

Posterior labial vessels and nerve

Superficial perineal vessels

Inferior anal (rectal) vessels

Ischio-anal fossa

A. Inferior View

RIGHT   LEFT

5.58   FEMALE PERINEUM I

A. Superficial dissection.

On the right side of the specimen:

- A long digital process of fat lies deep to the fatty subcutaneous tissue and descends into the labium majus.
- The round ligament of the uterus ends as a branching band of fascia that spreads out superficial to the fatty digital process.

On the left side of the specimen:

- Most of the fatty digital process is removed.
- The mons pubis is the rounded fatty prominence anterior to the pubic symphysis and bodies of the pubic bones.
- The posterior labial vessels and nerves (S2, S3) are joined by the perineal branch of the posterior cutaneous nerve of thigh (S1, S2, S3) and run anterior to the mons pubis. At the mons pubis, the vessels anastomose with the external pudendal vessels, and the nerves overlap in supply with the ilio-inguinal nerve (L1).

B. Cutaneous zones of innervation.

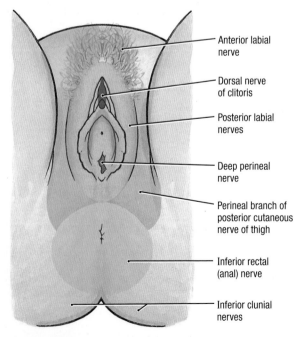

Anterior labial
nerve

Dorsal nerve
of clitoris

Posterior labial
nerves

Deep perineal
nerve

Perineal branch of
posterior cutaneous
nerve of thigh

Inferior rectal
(anal) nerve

Inferior clunial
nerves

B. Inferior View

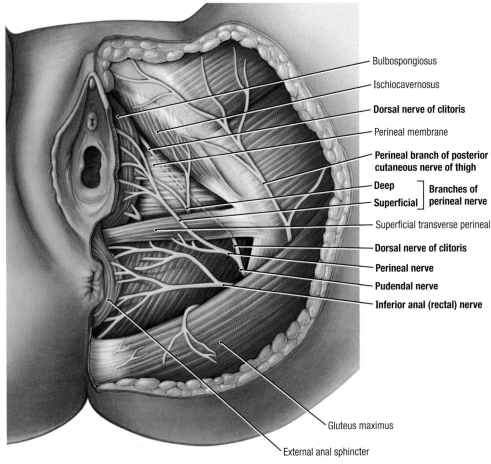

Bulbospongiosus

Ischiocavernosus

Dorsal nerve of clitoris

Perineal membrane

Perineal branch of posterior cutaneous nerve of thigh

Deep ⎤ Branches of
Superficial ⎦ perineal nerve

Superficial transverse perineal

Dorsal nerve of clitoris

Perineal nerve

Pudendal nerve

Inferior anal (rectal) nerve

Gluteus maximus

External anal sphincter

A. Inferior View

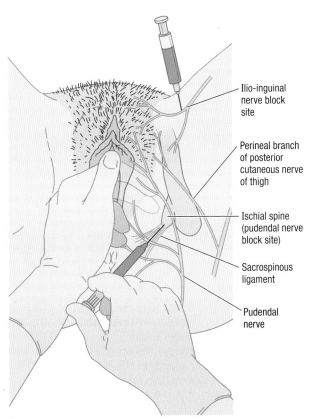

Ilio-inguinal nerve block site

Perineal branch of posterior cutaneous nerve of thigh

Ischial spine (pudendal nerve block site)

Sacrospinous ligament

Pudendal nerve

B. Inferior View (Lithotomy Position)

INNERVATION OF THE FEMALE PERINEUM   5.59

A. Dissection of perineal nerves. The anterior aspect of the perineum is supplied by anterior labial nerves, derived from the ilio-inguinal nerve and genital branch of the genitofemoral nerve. The pudendal nerve is the main nerve of the perineum. Posterior labial nerves, derived from the superficial perineal nerve, supply most of the vulva. The deep perineal nerve supplies the orifice of the vagina and superficial perineal muscles; and the dorsal nerve of the clitoris supplies deep perineal muscles and sensations to the clitoris. The inferior anal (rectal) nerve, also from the pudendal nerve, innervates the external anal sphincter and the peri-anal skin. The lateral perineum is supplied by the perineal branch of the posterior cutaneous nerve of the thigh. **B.** To relieve the pain experienced during childbirth, **pudendal nerve block anesthesia** may be performed by injecting a local anesthetic agent into the tissue surrounding the pudendal nerve, near the ischial spine. A pudendal nerve block does not abolish sensations from the anterior and lateral parts of the perineum. Therefore, **an anesthetic block of the ilio-inguinal and/or perineal branch of the posterior cutaneous nerve of the thigh** may also need to be performed.

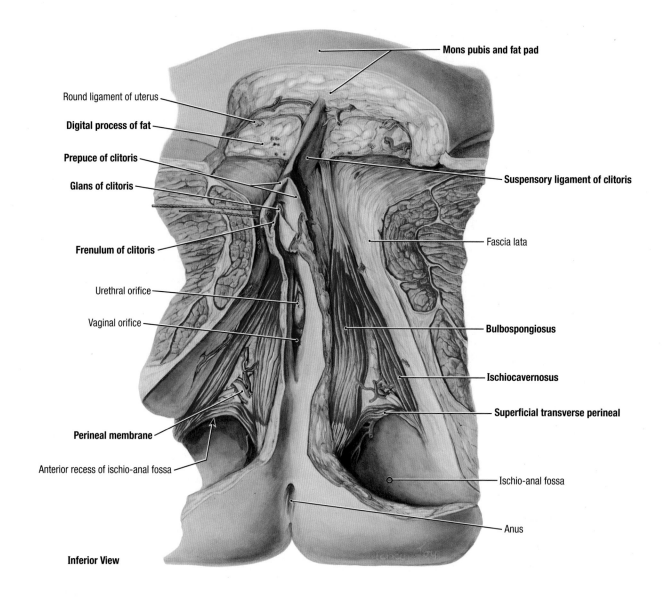

Mons pubis and fat pad

Round ligament of uterus

Digital process of fat

Prepuce of clitoris

Glans of clitoris

Frenulum of clitoris

Urethral orifice

Vaginal orifice

Perineal membrane

Anterior recess of ischio-anal fossa

Suspensory ligament of clitoris

Fascia lata

Bulbospongiosus

Ischiocavernosus

Superficial transverse perineal

Ischio-anal fossa

Anus

Inferior View

5.60 FEMALE PERINEUM II

- Note the thickness of the subcutaneous fatty tissue of the mons pubis and the encapsulated digital process of fat deep to this. The suspensory ligament of the clitoris descends from the linea alba.
- Anteriorly, each labium minus forms two laminae or folds: The lateral laminae of the labia pass on each side of the glans clitoris and unite, forming a hood that partially or completely covers the glans, the prepuce (foreskin) of the clitoris. The medial laminae of the labia merge posterior to the glans, forming the frenulum of the clitoris.

- There are three muscles on each side: bulbospongiosus, ischiocavernosus, and superficial transverse perineal; the perineal membrane is visible between them.
- The bulbospongiosus muscle overlies the bulb of the vestibule and the great vestibular gland. In the male, the muscles of the two sides are united by a median raphe; in the female, the orifice of the vagina separates the right from the left.

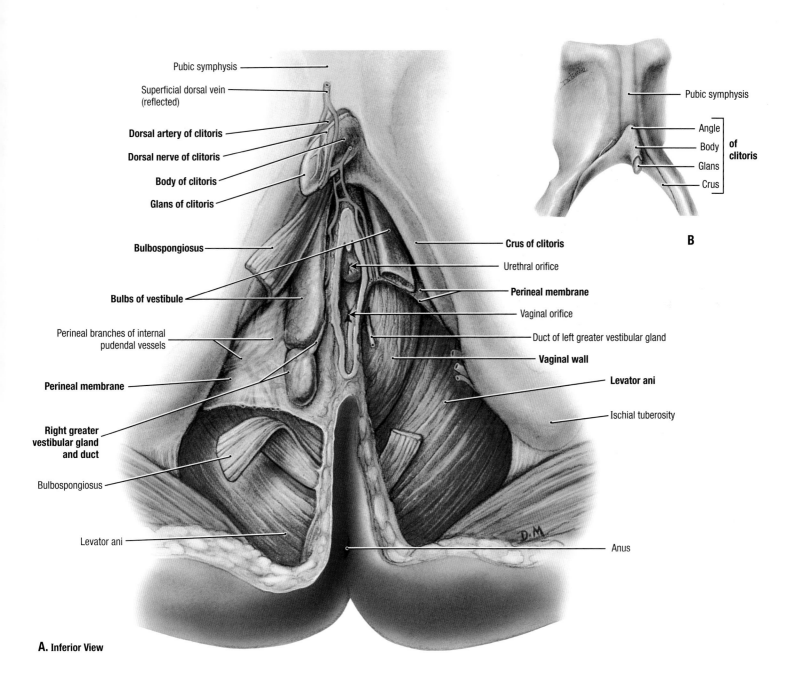

Pubic symphysis

Superficial dorsal vein (reflected)

Dorsal artery of clitoris

Dorsal nerve of clitoris

Body of clitoris

Glans of clitoris

Bulbospongiosus

Bulbs of vestibule

Perineal branches of internal pudendal vessels

Perineal membrane

Right greater vestibular gland and duct

Bulbospongiosus

Levator ani

Pubic symphysis

Angle
Body
Glans
Crus

of clitoris

B

Crus of clitoris

Urethral orifice

Perineal membrane

Vaginal orifice

Duct of left greater vestibular gland

Vaginal wall

Levator ani

Ischial tuberosity

Anus

A. Inferior View

FEMALE PERINEUM III

5.61

A. Deeper dissection. **B.** Clitoris.

In **A**:

- The bulbospongiosus muscle is reflected on the right side and mostly removed on the left side; the posterior portion of the bulb of the vestibule and the greater vestibular gland have been removed on the left side.
- The glans and body of the clitoris is displaced to the right so that the distribution of the dorsal vessels and nerve of the clitoris can be seen.
- Homologues of the bulb of the penis, the bulbs of the vestibule exist as two masses of elongated erectile tissue that lie along the

sides of the vaginal orifice; veins connect the bulbs of the vestibule to the glans of the clitoris.
- On the specimen's right side, the greater vestibular gland is situated at the posterior end of the bulb; both structures are covered by bulbospongiosus muscle.
- On the specimen's left side, the bulb, gland, and perineal membrane are cut away, thereby revealing the external aspect of the vaginal wall.

In **B**:

- The body of the clitoris, composed of two crura (corpora cavernosa), is capped by the glans.

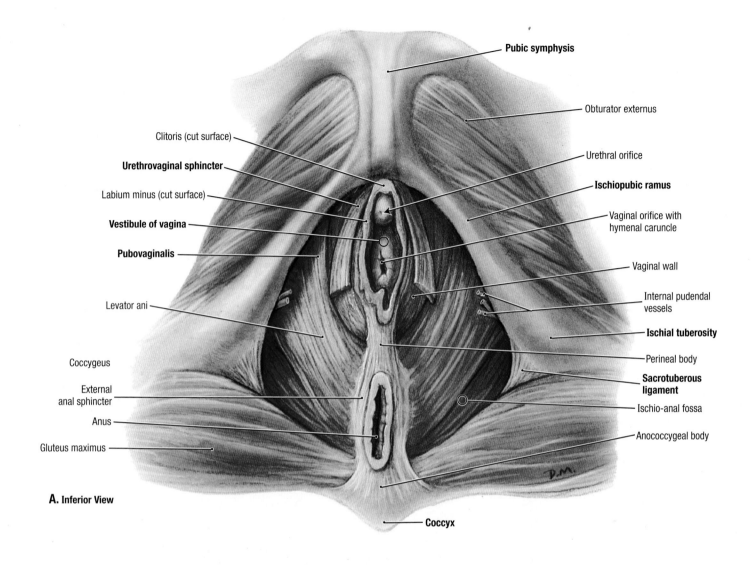

Pubic symphysis

Obturator externus

Clitoris (cut surface)

Urethrovaginal sphincter

Labium minus (cut surface)

Urethral orifice

Ischiopubic ramus

Vestibule of vagina

Vaginal orifice with hymenal caruncle

Pubovaginalis

Vaginal wall

Levator ani

Internal pudendal vessels

Ischial tuberosity

Coccygeus

Perineal body

External anal sphincter

Sacrotuberous ligament

Anus

Ischio-anal fossa

Gluteus maximus

Anococcygeal body

A. Inferior View

Coccyx

5.62 FEMALE PERINEUM IV

A. Deep perineal compartment. The perineal membrane and smooth muscle corresponding in position to the deep transverse perineal muscle in the male have been removed.

- The most anterior and medial part of the levator ani muscle, the pubovaginalis, passes posterior to the vaginal orifice.
- The urethrovaginal sphincter, part of the external urethral sphincter of the female, rests on the urethra and straddles the vagina.
- The labia minora (cut short here) bound the vestibule of the vagina.

  A. and **B.** The osseoligamentous boundaries of the diamond-shaped perineum are the pubic symphysis, ischiopubic rami, ischial tuberosities, sacrotuberous ligaments, and coccyx. For descriptive purposes, a transverse line connecting the ischial tuberosities subdivides the diamond into urogenital and anal triangles.

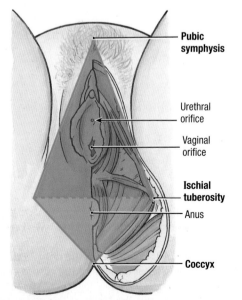

Pubic symphysis

Urethral orifice

Vaginal orifice

Ischial tuberosity

Anus

Coccyx

B. Inferior View

Key	
	Urogenital triangle
	Anal triangle

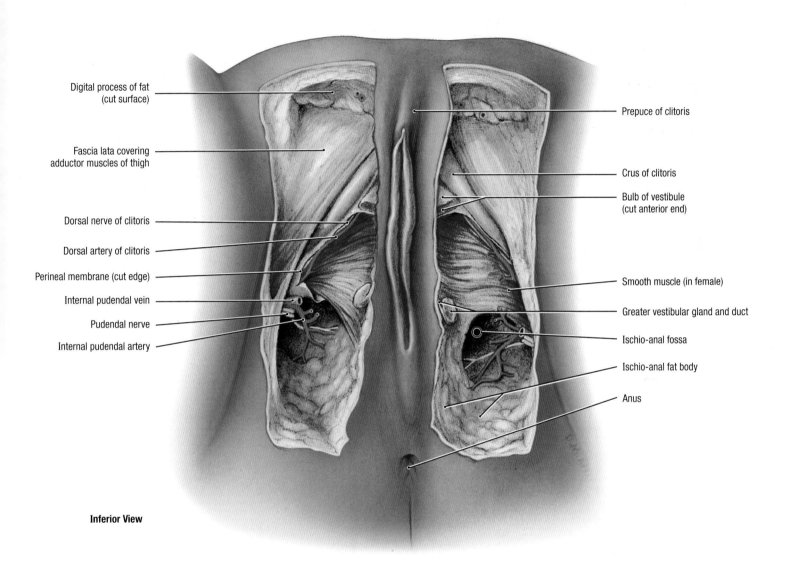

Digital process of fat (cut surface)

Fascia lata covering adductor muscles of thigh

Dorsal nerve of clitoris

Dorsal artery of clitoris

Perineal membrane (cut edge)

Internal pudendal vein

Pudendal nerve

Internal pudendal artery

Prepuce of clitoris

Crus of clitoris

Bulb of vestibule (cut anterior end)

Smooth muscle (in female)

Greater vestibular gland and duct

Ischio-anal fossa

Ischio-anal fat body

Anus

Inferior View

FEMALE PERINEUM V

5.63

This is a different dissection than the previous series, with the vulva undissected centrally but the perineum dissected deeply on each side. Although most of the perineal membrane and bulbs of the vestibule have been removed, the greater vestibular glands (structures of the superficial perineal compartment) have been left in place. The development and extent of the smooth muscle layer corresponding in position to the voluntary deep transverse perineal muscles of the male are highly variable, being relatively extensive in this case, blending centrally with voluntary fibers of the external urethral sphincter and the perineal body.

The greater vestibular glands are usually not palpable but are so when infected. Occlusion of the vestibular gland duct can predispose the individual to **infection of the vestibular gland**. The gland is the site or origin of most **vulvar adenocarcinomas** (cancers). **Bartholinitis**, inflammation of the greater vestibular (Bartholin) glands, may result from a number of pathogenic organisms. Infected glands may enlarge to a diameter of 4 to 5 cm and impinge on the wall of the rectum. Occlusion of the vestibular gland duct without infection can result in the accumulation of mucin (**Bartholin cyst**).

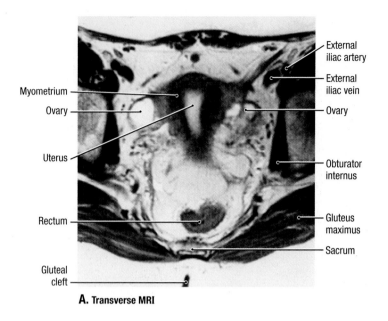

External iliac artery

External iliac vein

Ovary

Obturator internus

Gluteus maximus

Sacrum

Myometrium

Ovary

Uterus

Rectum

Gluteal cleft

A. Transverse MRI

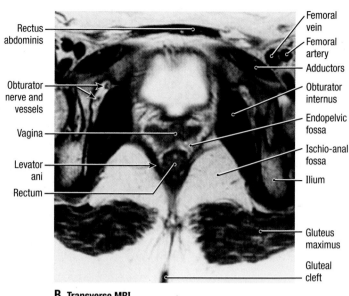

Rectus abdominis

Obturator nerve and vessels

Vagina

Levator ani

Rectum

Femoral vein

Femoral artery

Adductors

Obturator internus

Endopelvic fossa

Ischio-anal fossa

Ilium

Gluteus maximus

Gluteal cleft

B. Transverse MRI

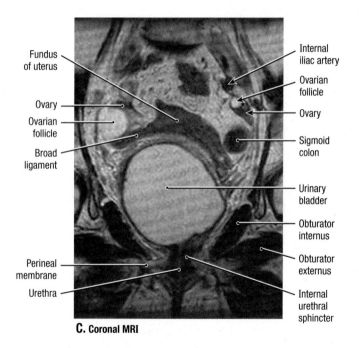

Fundus of uterus

Ovary

Ovarian follicle

Broad ligament

Perineal membrane

Urethra

Internal iliac artery

Ovarian follicle

Ovary

Sigmoid colon

Urinary bladder

Obturator internus

Obturator externus

Internal urethral sphincter

C. Coronal MRI

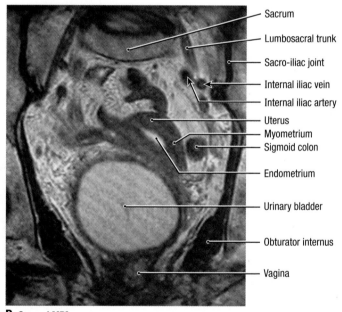

Sacrum

Lumbosacral trunk

Sacro-iliac joint

Internal iliac vein

Internal iliac artery

Uterus

Myometrium

Sigmoid colon

Endometrium

Urinary bladder

Obturator internus

Vagina

D. Coronal MRI

5.64 IMAGING OF FEMALE PELVIS AND PERINEUM

A. and **B.** Transverse (axial) MRIs of female pelvis. **C.** and **D.** Coronal MRIs. **E–H.** Transverse anatomical sections and corresponding MRIs of female perineum.

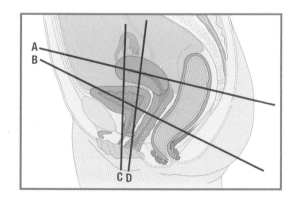

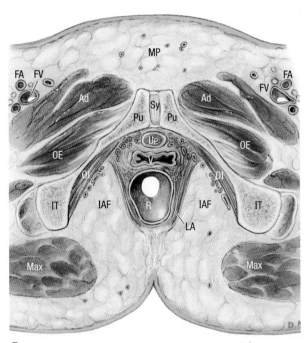

E. Anatomical Transverse Section

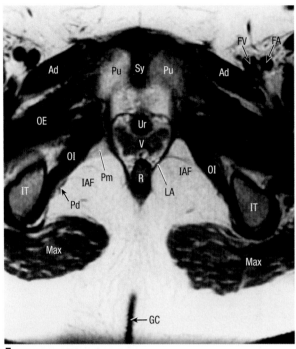

F. Transverse MRI

Key					
AC	Anal canal	LA	Levator ani	PR	Puborectalis
Ad	Adductor muscles	LM	Labium majus	Pu	Pubic bone
CC	Crus of clitoris	Max	Gluteus maximus	QF	Quadratus femoris
FA	Femoral artery	MP	Mons pubis	R	Rectum
FV	Femoral vein	OE	Obturator externus	Sy	Pubic symphysis
GC	Gluteal cleft	OI	Obturator internus	Ur	Urethra
IAF	Ischio-anal fossa	Pd	Pudendal canal	V	Vagina
IPR	Ischiopubic ramus	Pec	Pectineus	Ve	Vestibule of the vagina
IT	Ischial tuberosity	Pm	Perineal membrane		

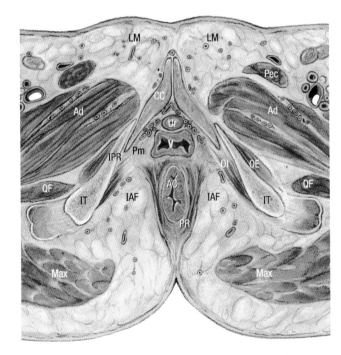

G. Anatomical Transverse Section

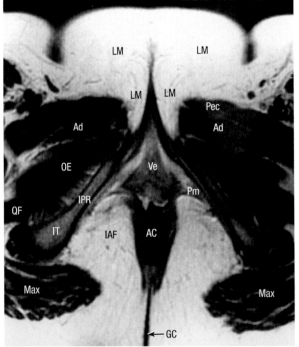

H. Transverse MRI

IMAGING OF FEMALE PELVIS AND PERINEUM (*continued*)

5.64

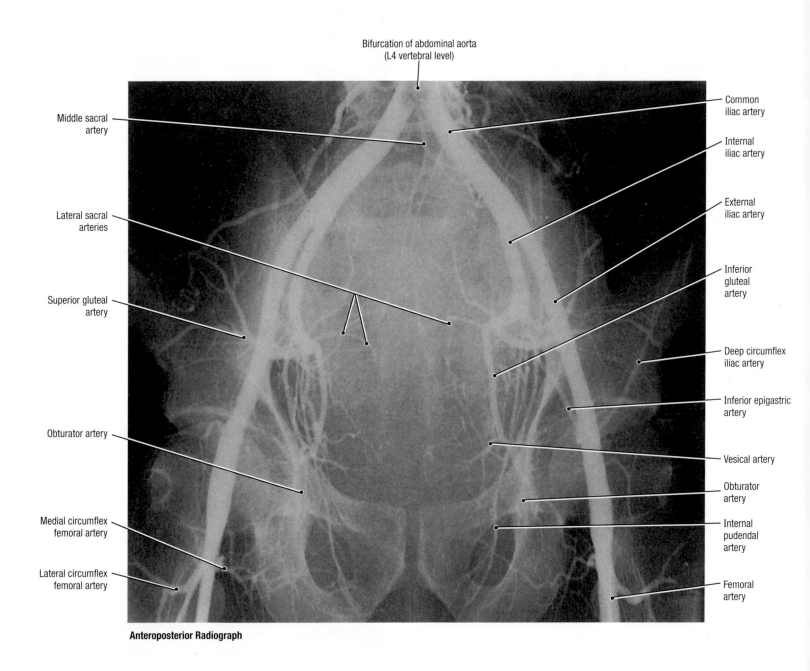

Bifurcation of abdominal aorta
(L4 vertebral level)

Middle sacral artery

Lateral sacral arteries

Superior gluteal artery

Obturator artery

Medial circumflex femoral artery

Lateral circumflex femoral artery

Common iliac artery

Internal iliac artery

External iliac artery

Inferior gluteal artery

Deep circumflex iliac artery

Inferior epigastric artery

Vesical artery

Obturator artery

Internal pudendal artery

Femoral artery

Anteroposterior Radiograph

5.65 PELVIC ANGIOGRAPHY

Radiopaque dye released into the aorta of this male patient entered the branches of the external and internal iliac arteries at the time this radiograph was produced.

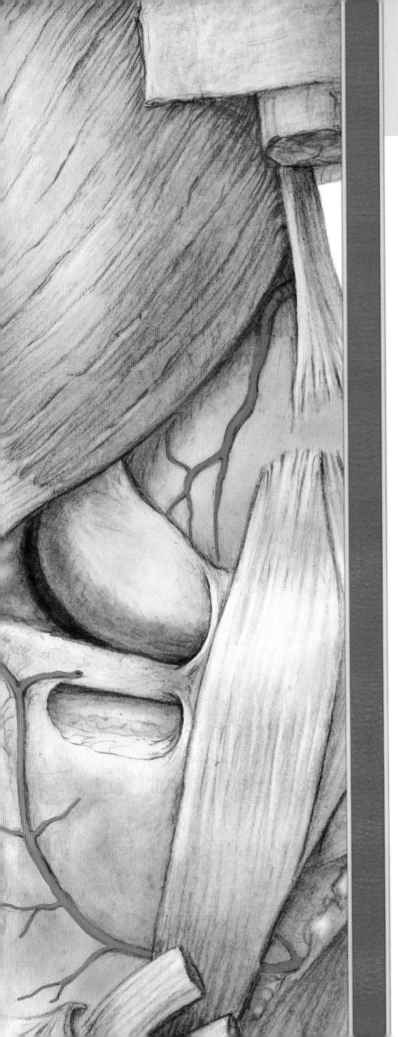

CHAPTER 6

Lower Limb

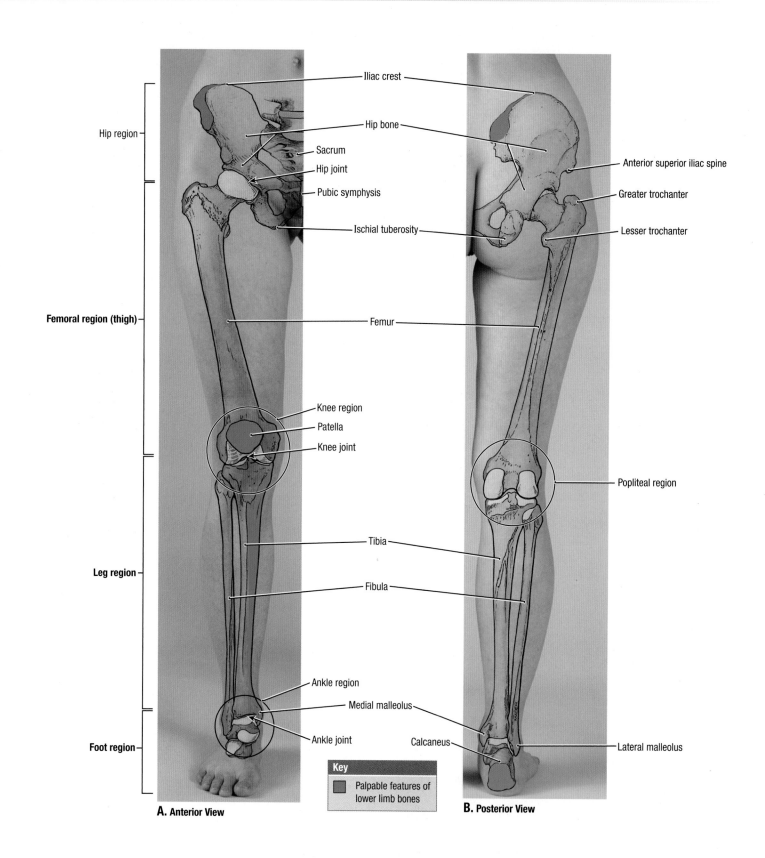

Hip region

Femoral region (thigh)

Leg region

Foot region

Iliac crest

Hip bone

Sacrum

Hip joint

Pubic symphysis

Ischial tuberosity

Femur

Knee region

Patella

Knee joint

Tibia

Fibula

Ankle region

Medial malleolus

Ankle joint

Calcaneus

Anterior superior iliac spine

Greater trochanter

Lesser trochanter

Popliteal region

Lateral malleolus

Key

Palpable features of lower limb bones

A. Anterior View

B. Posterior View

6.1　　REGIONS, BONES, AND MAJOR JOINTS OF LOWER LIMB

The hip bones meet anteriorly at the pubic symphysis and articulate with the sacrum posteriorly. The femur articulates with the hip bone proximally and the tibia distally. The tibia and fibula are the bones of the leg that join the foot at the ankle.

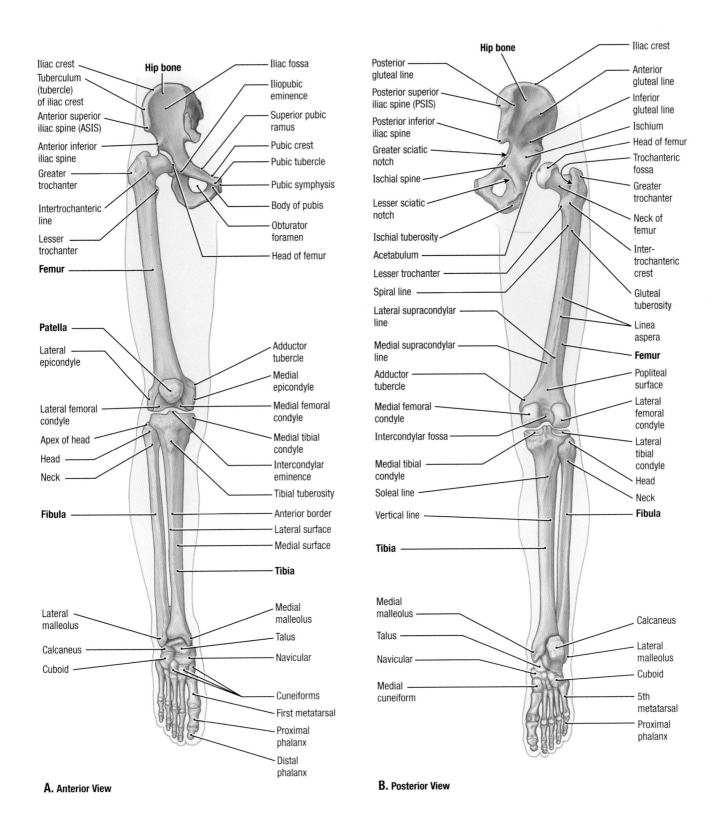

Iliac crest
Tuberculum (tubercle) of iliac crest
Anterior superior iliac spine (ASIS)
Anterior inferior iliac spine
Greater trochanter
Intertrochanteric line
Lesser trochanter
Femur

Patella
Lateral epicondyle
Lateral femoral condyle
Apex of head
Head
Neck
Fibula

Lateral malleolus
Calcaneus
Cuboid

Hip bone

Iliac fossa
Iliopubic eminence
Superior pubic ramus
Pubic crest
Pubic tubercle
Pubic symphysis
Body of pubis
Obturator foramen
Head of femur

Adductor tubercle
Medial epicondyle
Medial femoral condyle
Medial tibial condyle
Intercondylar eminence
Tibial tuberosity
Anterior border
Lateral surface
Medial surface
Tibia

Medial malleolus
Talus
Navicular

Cuneiforms
First metatarsal
Proximal phalanx
Distal phalanx

A. Anterior View

Hip bone

Posterior gluteal line
Posterior superior iliac spine (PSIS)
Posterior inferior iliac spine
Greater sciatic notch
Ischial spine
Lesser sciatic notch
Ischial tuberosity
Acetabulum
Lesser trochanter
Spiral line
Lateral supracondylar line
Medial supracondylar line
Adductor tubercle
Medial femoral condyle
Intercondylar fossa
Medial tibial condyle
Soleal line
Vertical line
Tibia

Medial malleolus
Talus
Navicular
Medial cuneiform

Iliac crest
Anterior gluteal line
Inferior gluteal line
Ischium
Head of femur
Trochanteric fossa
Greater trochanter
Neck of femur
Inter-trochanteric crest
Gluteal tuberosity
Linea aspera
Femur
Popliteal surface
Lateral femoral condyle
Lateral tibial condyle
Head
Neck
Fibula

Calcaneus
Lateral malleolus
Cuboid
5th metatarsal
Proximal phalanx

B. Posterior View

FEATURES OF BONES OF LOWER LIMB

6.2

The foot is in full plantar flexion. The hip joint is disarticulated **(B)** to demonstrate the acetabulum of the hip bone and the entire head of the femur.

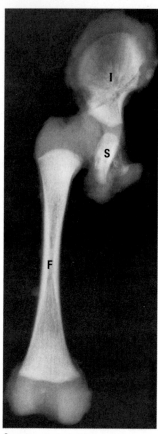

A. Anteroposterior View

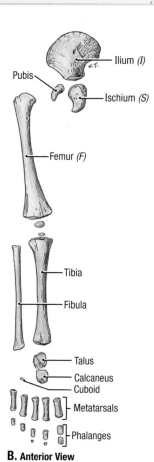

Ilium (I)
Pubis
Ischium (S)
Femur (F)
Tibia
Fibula
Talus
Calcaneus
Cuboid
Metatarsals
Phalanges

B. Anterior View

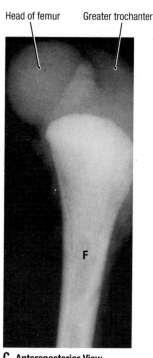

Head of femur   Greater trochanter

C. Anteroposterior View

6.3 POSTNATAL LOWER LIMB DEVELOPMENT

A. and **C.** Anteroposterior radiographs of normal postmortem specimens of newborns show the bony *(white)* and cartilaginous *(gray)* components of the femur and hip bone. **B.** Ossified portions of bones of lower limb at birth. The hip bone can be divided into three primary parts: ilium, ischium, and pubis. The diaphyses (bodies) of the long bones are well ossified. Some epiphyses (growth plates) and tarsal bones have begun to ossify. **D.** Foot of child age 4.

Dislocated epiphysis of femoral head. In older children and adolescents (10 to 17 years of age), the epiphysis of the femoral head may slip away from the femoral neck because of weakness of the epiphyseal plate. This injury may be caused by acute trauma or repetitive microtraumas that place increased shearing stress on the epiphysis, especially with abduction and lateral rotation.

Fractures involving epiphyseal plates. The primary ossification center for the superior end of the tibia appears shortly after birth and joins the shaft of the tibia during adolescence (usually 16 to 18 years of age). Tibial fractures in children are more serious if they involve the epiphyseal plates because continued normal growth of bone may be jeopardized. Disruption of the epiphyseal plate at the tibial tuberosity may cause inflammation of the tuberosity and chronic recurring pain during adolescence (Osgood-Schlatter disease), especially in young athletes.

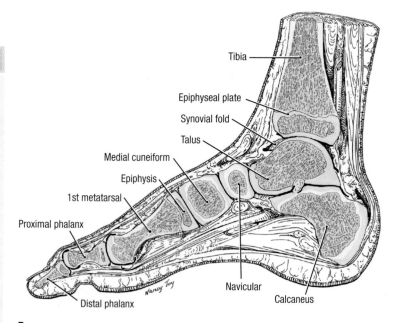

Tibia
Epiphyseal plate
Synovial fold
Talus
Medial cuneiform
Epiphysis
1st metatarsal
Proximal phalanx
Distal phalanx
Navicular
Calcaneus

D. Sagittal Section

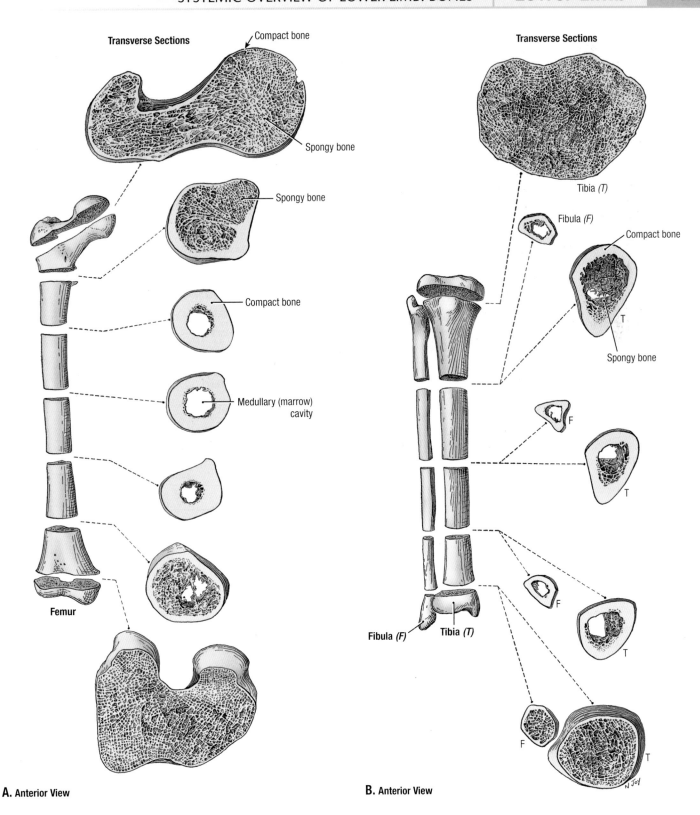

Transverse Sections

Compact bone

Spongy bone

Spongy bone

Compact bone

Medullary (marrow) cavity

Femur

A. Anterior View

Transverse Sections

Tibia (T)

Fibula (F)

Compact bone

T

Spongy bone

F

T

F

T

Fibula (F)   Tibia (T)

F   T

B. Anterior View

TRANSVERSE SECTIONS THROUGH FEMUR, TIBIA AND FIBULA

6.4

A. Femur. **B.** Tibia and fibula. Note the differences in thickness of the compact and spongy bone and in the width of the medullary (marrow) cavity. Compact and spongy bones are distinguished by the relative amount of solid matter and by the number and size of the spaces they contain. All bones have a superficial thin layer of compact bone around a central mass of spongy bone, except where the latter is replaced by the medullary (marrow) cavity. Within the medullary cavity of adult bones and between the spicules (trabeculae) of spongy bone, yellow (fatty) or red (blood cell and platelet-forming) bone marrow or both are found. This is significant for MRIs where the compact bone is seen as a thin black line surrounding the whiter spongy bone with its abundant fatty marrow.

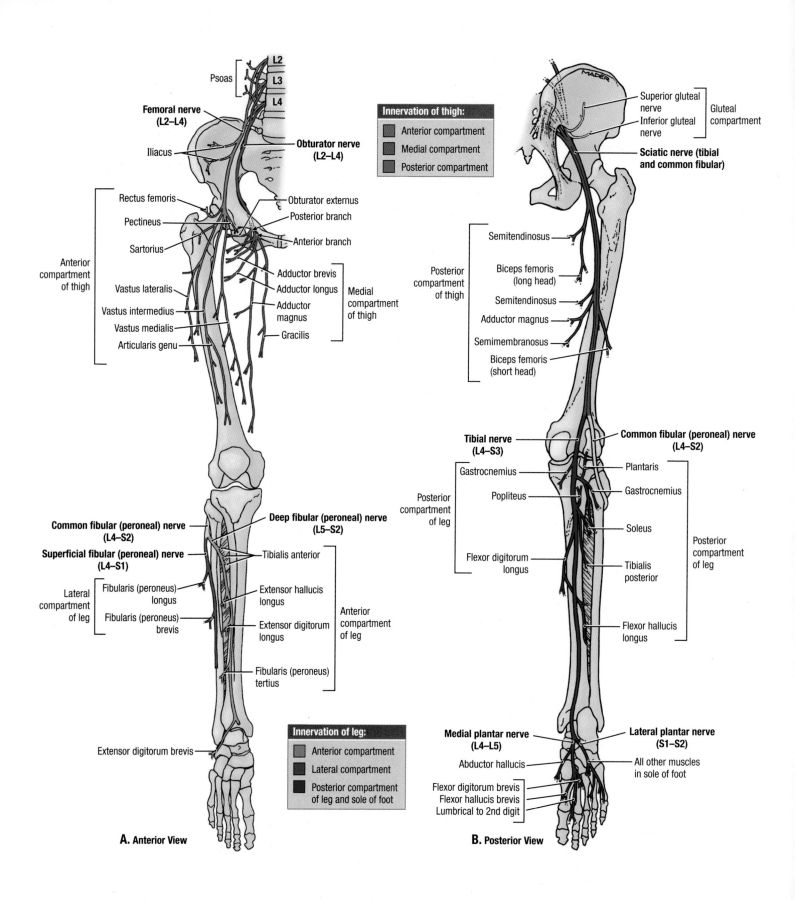

Innervation of thigh:
- Anterior compartment
- Medial compartment
- Posterior compartment

Innervation of leg:
- Anterior compartment
- Lateral compartment
- Posterior compartment of leg and sole of foot

A. Anterior View

L2
L3
L4
Psoas
Femoral nerve (L2–L4)
Iliacus
Obturator nerve (L2–L4)
Rectus femoris
Pectineus
Sartorius
Obturator externus
Posterior branch
Anterior branch
Adductor brevis
Adductor longus
Adductor magnus
Gracilis
Anterior compartment of thigh
Vastus lateralis
Vastus intermedius
Vastus medialis
Articularis genu
Medial compartment of thigh
Common fibular (peroneal) nerve (L4–S2)
Superficial fibular (peroneal) nerve (L4–S1)
Deep fibular (peroneal) nerve (L5–S2)
Tibialis anterior
Lateral compartment of leg
Fibularis (peroneus) longus
Fibularis (peroneus) brevis
Extensor hallucis longus
Extensor digitorum longus
Anterior compartment of leg
Fibularis (peroneus) tertius
Extensor digitorum brevis

B. Posterior View

Superior gluteal nerve
Inferior gluteal nerve
Gluteal compartment
Sciatic nerve (tibial and common fibular)
Posterior compartment of thigh
Semitendinosus
Biceps femoris (long head)
Semitendinosus
Adductor magnus
Semimembranosus
Biceps femoris (short head)
Tibial nerve (L4–S3)
Common fibular (peroneal) nerve (L4–S2)
Gastrocnemius
Plantaris
Popliteus
Gastrocnemius
Posterior compartment of leg
Soleus
Flexor digitorum longus
Tibialis posterior
Posterior compartment of leg
Flexor hallucis longus
Medial plantar nerve (L4–L5)
Lateral plantar nerve (S1–S2)
Abductor hallucis
All other muscles in sole of foot
Flexor digitorum brevis
Flexor hallucis brevis
Lumbrical to 2nd digit

6.5 | OVERVIEW OF MOTOR INNERVATION OF LOWER LIMB

TABLE 6.1	MOTOR NERVES OF LOWER LIMB		
Nerve	**Origin**	**Course**	**Distribution**
Femoral	Lumbar plexus (L2–L4)	Passes deep to midpoint of inguinal ligament, lateral to femoral vessels, dividing into muscular and cutaneous branches in femoral triangle	Anterior thigh muscles
Obturator		Traverses lesser pelvis to enter thigh via obturator foramen and then divides; its anterior branch descends between adductor longus and adductor brevis; its posterior branch descends between adductor brevis and adductor magnus	*Anterior branch:* adductor longus, adductor brevis, gracilis, and pectineus *Posterior branch:* obturator externus and adductor magnus
Sciatic	Sacral plexus (L4–S3)	Enters gluteal region through greater sciatic foramen, usually passing inferior to piriformis, descends in posterior compartment of thigh, bifurcating at apex of popliteal fossa into tibial and common fibular (peroneal) nerves	Muscles of posterior thigh, leg and sole and dorsum of foot
Tibial	Sciatic nerve	Terminal branch of sciatic nerve arising at apex of popliteal fossa; descends through popliteal fossa with popliteal vessels, continuing in deep posterior compartment of leg with posterior tibial vessels; bifurcates into medial and lateral plantar nerves	Hamstring muscles of posterior compartment of thigh, muscles of posterior compartment of leg, and sole of foot
Common fibular (peroneal)		Terminal branch of sciatic nerve arising at apex of popliteal fossa; follows medial border of biceps femoris and its tendon to wind around neck of fibula deep to fibularis longus, where it bifurcates into superficial and deep fibular nerves	Short head of biceps femoris, muscles of anterior and lateral compartments of leg, and dorsum of foot
Superficial fibular (peroneal)	Common fibular nerve	Arises deep to fibularis longus on neck of fibula and descends in lateral compartment of the leg; pierces crural fascia in distal third of leg to become cutaneous	Muscles of lateral compartment of leg
Deep fibular (peroneal)		Arises deep to fibularis longus on neck of fibula; passes through extensor digitorum longus into anterior compartment, descending on interosseous membrane; crosses ankle joint and enters dorsum of foot	Muscles of anterior compartment of leg and dorsum of foot

TABLE 6.2	NERVE LESIONS		
Injured Nerve	**Injury Description**	**Impairments**	**Clinical Aspects**
Femoral nerve	Trauma at femoral triangle Pelvic fracture	Flexion of thigh is weakened Extension of leg is lost Sensory loss on anterior thigh and medial leg	Loss of knee jerk reflex Anesthesia on anterior thigh
Obturator nerve	Anterior hip dislocation Radical retropubic prostatectomy	Adduction of thigh is lost Variable sensory loss on medial thigh	Rare injury due to protected position
Superior gluteal nerve	Surgery Posterior hip dislocation Poliomyelitis	Gluteus medius and minimus function is lost Ability to pull contralateral pelvis up to level and abduction of thigh are lost	**Superior gluteal nerve palsy** Gluteus medius limp or "waddling gait" Positive Trendelenburg sign
Inferior gluteal nerve	Surgery Posterior hip dislocation	Gluteus maximus function is lost Ability to rise from a seated position, climb stairs or incline, or jump is lost	**Inferior gluteal nerve palsy** Patient will lean the body trunk backward at heel strike
Common fibular nerve	Blow to lateral aspect of leg Fracture of neck of fibula	Eversion of foot is lost Dorsiflexion of foot is lost Extension of toes is lost Sensory loss on anterolateral leg and dorsum of foot	**Common fibular nerve palsy** Patient will present with foot plantar flexed ("footdrop") and inverted Patient cannot stand on heels "Foot slap"
Tibial nerve at popliteal fossa	Trauma at popliteal fossa	Inversion of foot is weakened Plantar flexion of foot is lost Sensory loss on sole of foot	Patient will present with foot dorsiflexed and everted Patient cannot stand on toes

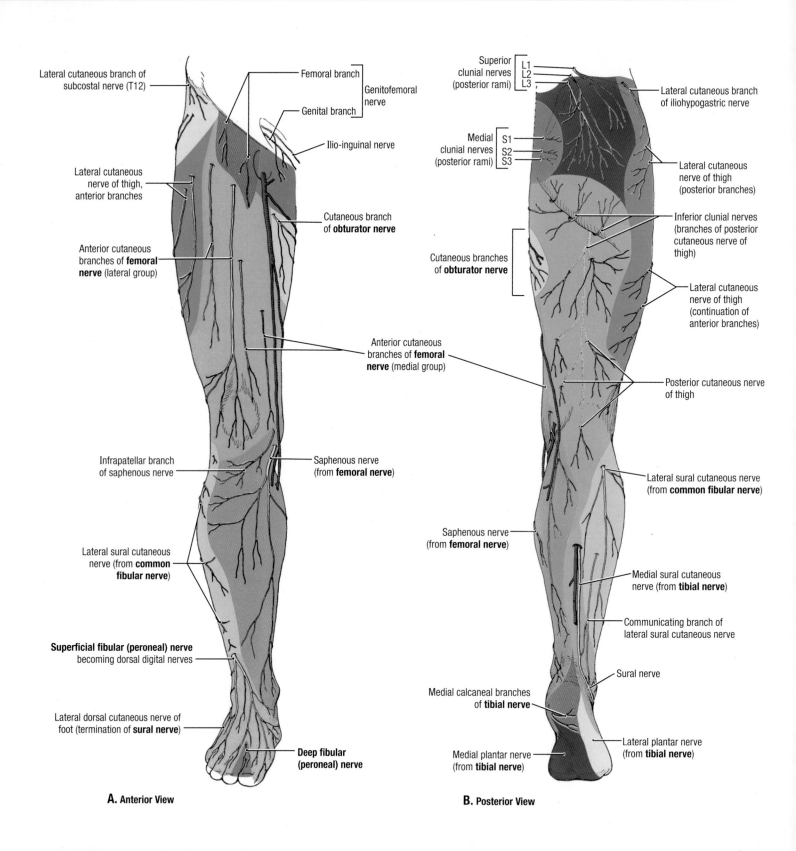

Lateral cutaneous branch of subcostal nerve (T12)

Lateral cutaneous nerve of thigh, anterior branches

Anterior cutaneous branches of **femoral nerve** (lateral group)

Infrapatellar branch of saphenous nerve

Lateral sural cutaneous nerve (from **common fibular** nerve)

Superficial fibular (peroneal) nerve becoming dorsal digital nerves

Lateral dorsal cutaneous nerve of foot (termination of **sural nerve**)

Femoral branch

Genital branch

Genitofemoral nerve

Ilio-inguinal nerve

Cutaneous branch of **obturator nerve**

Anterior cutaneous branches of **femoral nerve** (medial group)

Saphenous nerve (from **femoral nerve**)

Deep fibular (peroneal) nerve

A. Anterior View

Superior clunial nerves (posterior rami)   L1 L2 L3

Medial clunial nerves (posterior rami)   S1 S2 S3

Cutaneous branches of **obturator nerve**

Saphenous nerve (from **femoral nerve**)

Medial calcaneal branches of **tibial nerve**

Medial plantar nerve (from **tibial nerve**)

Lateral cutaneous branch of iliohypogastric nerve

Lateral cutaneous nerve of thigh (posterior branches)

Inferior clunial nerves (branches of posterior cutaneous nerve of thigh)

Lateral cutaneous nerve of thigh (continuation of anterior branches)

Posterior cutaneous nerve of thigh

Lateral sural cutaneous nerve (from **common fibular nerve**)

Medial sural cutaneous nerve (from **tibial nerve**)

Communicating branch of lateral sural cutaneous nerve

Sural nerve

Lateral plantar nerve (from **tibial nerve**)

B. Posterior View

CUTANEOUS NERVES OF LOWER LIMB

6.6

Cutaneous nerves in the subcutaneous tissue supply the skin of the lower limb. In the posterior view, the medial sural cutaneous nerve (*sural* is Latin for calf) is joined between the popliteal fossa and posterior aspect of the ankle by a communicating branch of the lateral sural cutaneous nerve to form the sural nerve. The level of the junction is variable and is low in this specimen.

TABLE 6.3	CUTANEOUS NERVES OF LOWER LIMB		
Nerve	*Origin (Contributing Spinal Nerves)*	*Course*	*Distribution to Skin of Lower Limb*
Subcostal (lateral cutaneous branch)	T12 anterior ramus	Descends over iliac crest	Hip region inferior to anterior part of iliac crest and anterior to greater trochanter
Iliohypogastric	Lumbar plexus (L1; occasionally T12)	Parallels iliac crest	Lateral cutaneous branch supplies superolateral quadrant of buttock
Ilio-inguinal	Lumbar plexus (L1; occasionally T12)	Passes through inguinal canal	Inguinal fold; femoral branch supplies skin over medial femoral triangle
Genitofemoral	Lumbar plexus (L1–L2)	Descends anterior surface of psoas major	Femoral branch supplies skin over lateral part of femoral triangle; genital branch supplies anterior scrotum or labia majora
Lateral cutaneous nerve of thigh	Lumbar plexus (L2–L3)	Passes deep to inguinal ligament, ~1 cm medial to anterior superior iliac spine	Skin on anterior and lateral aspects of thigh
Anterior cutaneous branches	Lumbar plexus via femoral nerve (L2–L4)	Arise in femoral triangle; pierce fascia lata along the path of sartorius muscle	Skin of anterior and medial aspects of thigh
Cutaneous branch of obturator nerve	Lumbar plexus via obturator nerve (L2–L4)	Following its descent between adductors longus and brevis, obturator nerve pierces fascia lata to reach the skin of thigh	Skin of middle part of medial thigh
Posterior cutaneous nerve of thigh	Sacral plexus (S1–S3)	Enters gluteal region via greater sciatic foramen deep to gluteus maximus; then descends deep to fascia lata; terminal branches pierce fascia lata	Skin of posterior thigh and popliteal fossa
Saphenous nerve	Lumbar plexus via femoral nerve (L3–L4)	Traverses adductor canal but does not pass through adductor hiatus	Skin on medial side of leg and foot
Superficial fibular nerve	Common fibular nerve (L4–S1)	After supplying fibular muscles, perforates deep fascia of leg	Skin of anterolateral leg and dorsum of foot
Deep fibular nerve	Common fibular nerve (L5)	After supplying muscles on dorsum of foot, pierces deep fascia superior to heads of 1st and 2nd metatarsals	Skin of web between great and 2nd toes
Sural nerve	Tibial and common fibular nerves (S1–S2)	Medial sural cutaneous branch of tibial nerve and lateral sural cutaneous branch of common fibular nerve merge at varying levels on posterior leg	Skin of posterolateral leg and lateral margin of foot
Medial plantar nerve	Tibial nerve (L4–L5)	Passes between first and second layers of plantar muscles	Skin of medial side of sole, and plantar aspect, sides, and nail beds of medial 3½ toes
Lateral plantar nerve	Tibial nerve (S1–S2)	Passes between first and second layers of plantar muscles	Skin of lateral sole, and plantar aspect, sides, and nail beds of lateral 1½ toes
Calcaneal nerves	Tibial and sural nerves (S1–S2)	Branches over calcaneal tuberosity	Skin of heel
Superior clunial nerves	L1–L3 posterior rami	Course laterally/inferiorly in subcutaneous tissue	Skin overlying superior and central parts of buttock
Medial clunial nerves	S1–S3 posterior rami	From dorsal sacral foramina; enter overlying subcutaneous tissue	Skin of medial buttock and intergluteal cleft
Inferior clunial nerves	Posterior cutaneous nerve of thigh (S2–S3)	Arise deep to gluteus maximus; emerge from beneath inferior border of muscle	Skin of inferior buttock (overlying gluteal fold)

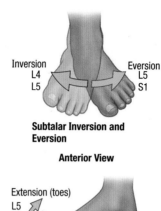
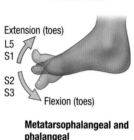
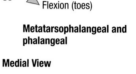
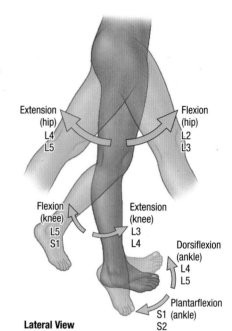

A

Myotatic (Deep Tendon) Reflex	Spinal Cord Segments
Quadriceps (knee jerk)	L3/L4
Calcaneal (Achilles; ankle jerk)	S1/S2

B

6.7 | MYOTOMES AND DEEP TENDON REFLEXES

A. Myotomes. Somatic motor (general somatic efferent) fibers transmit impulses to skeletal (voluntary) muscles. The unilateral muscle mass receiving innervation from the somatic motor fibers conveyed by a single spinal nerve is a myotome. Each skeletal muscle is usually innervated by the somatic motor fibers of several spinal nerves; therefore, the muscle myotome will consist of several segments. The muscle myotomes have been grouped by joint movement to facilitate clinical testing. **B.** Myotatic (deep tendon) reflexes. A myotatic (stretch) reflex is an involuntary contraction of a muscle in response to being stretched. Deep tendon reflexes (e.g., "knee jerk") are monosynaptic stretch reflexes that are elicited by briskly tapping the tendon with a reflex hammer. Each tendon reflex is mediated by specific spinal nerves. Stretch reflexes control muscle tone (e.g., in antigravity, muscles that keep the body upright against gravity).

TABLE 6.4 | NERVE ROOT (ANTERIOR RAMUS) LESIONS

Compressed Nerve Root	Dermatome Affected	Muscles Affected	Movement Weakness/Deficit	Nerve and Reflex Involved
L4	L4: medial surface of leg; big toe	Quadriceps	Extension of knee	Femoral nerve ↓ Knee jerk
L5	L5: lateral surface of leg; dorsum of foot	Tibialis anterior Extensor hallucis longus Extensor digitorum longus	Dorsiflexion of ankle (patient cannot stand on heels) Extension of toes	Common fibular nerve No reflex loss
S1	S1: posterior surface of lower limb; little toe	Gastrocnemius Soleus	Plantar flexion of ankle (patient cannot stand on toes) Flexion of toes	Tibial nerve ↓ Ankle jerk

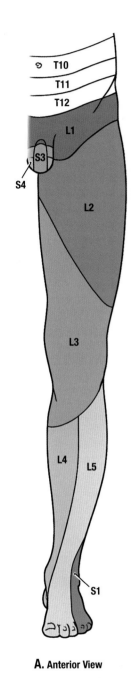

A. Anterior View

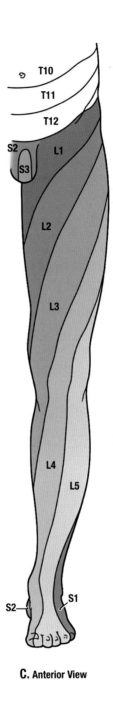

B. Posterior View

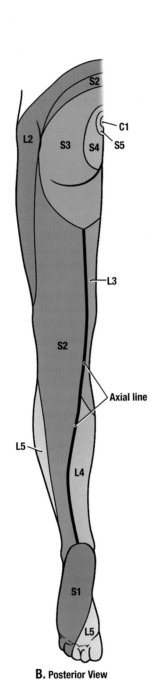

C. Anterior View

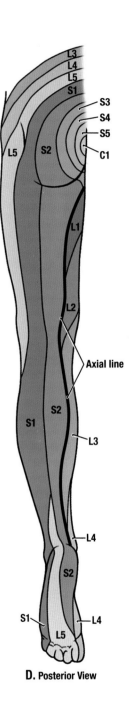

D. Posterior View

DERMATOMES OF LOWER LIMB

6.8

The dermatomal, or segmental, pattern of distribution of sensory nerve fibers persists despite the merging of spinal nerves in plexus formation during development. Two different dermatome maps are commonly used. **A.** and **B.** The dermatome pattern of the lower limb according to Foerster (1933) is preferred by many because of its correlation with clinical findings. **C.** and **D.** The dermatome pattern of the lower limb according to Keegan and Garrett (1948) is preferred by others for its aesthetic uniformity and obvious correlation with development. Although depicted as distinct zones, adjacent dermatomes overlap considerably, except along the axial line.

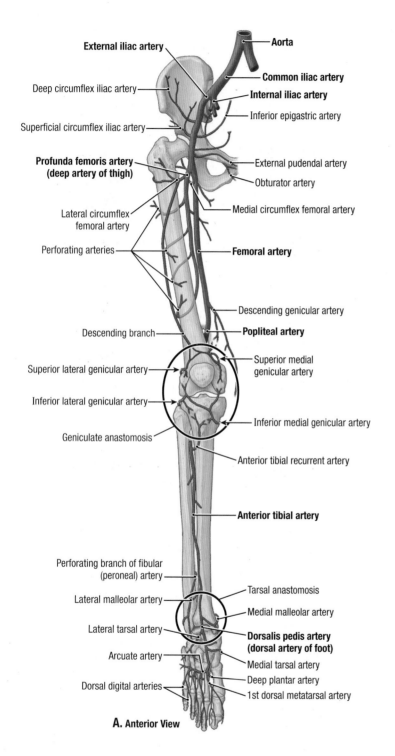

A. Anterior View

Aorta

External iliac artery

Deep circumflex iliac artery

Common iliac artery

Internal iliac artery

Inferior epigastric artery

Superficial circumflex iliac artery

Profunda femoris artery
(deep artery of thigh)

External pudendal artery

Obturator artery

Lateral circumflex
femoral artery

Medial circumflex femoral artery

Perforating arteries

Femoral artery

Descending genicular artery

Descending branch

Popliteal artery

Superior medial
genicular artery

Superior lateral genicular artery

Inferior lateral genicular artery

Inferior medial genicular artery

Geniculate anastomosis

Anterior tibial recurrent artery

Anterior tibial artery

Perforating branch of fibular
(peroneal) artery

Tarsal anastomosis

Lateral malleolar artery

Medial malleolar artery

Lateral tarsal artery

Dorsalis pedis artery
(dorsal artery of foot)

Arcuate artery

Medial tarsal artery

Deep plantar artery

Dorsal digital arteries

1st dorsal metatarsal artery

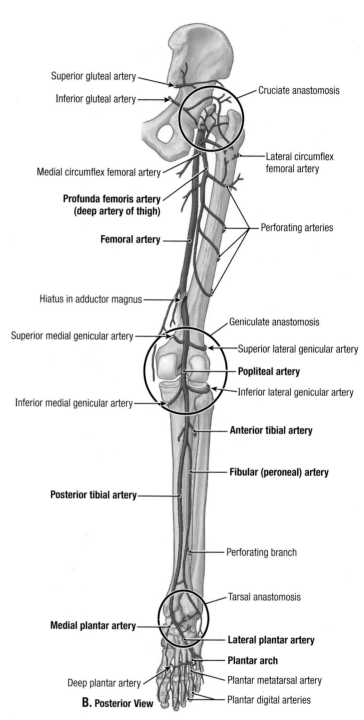

B. Posterior View

Superior gluteal artery

Inferior gluteal artery

Cruciate anastomosis

Lateral circumflex
femoral artery

Medial circumflex femoral artery

Profunda femoris artery
(deep artery of thigh)

Perforating arteries

Femoral artery

Hiatus in adductor magnus

Geniculate anastomosis

Superior medial genicular artery

Superior lateral genicular artery

Popliteal artery

Inferior lateral genicular artery

Inferior medial genicular artery

Anterior tibial artery

Fibular (peroneal) artery

Posterior tibial artery

Perforating branch

Tarsal anastomosis

Medial plantar artery

Lateral plantar artery

Plantar arch

Deep plantar artery

Plantar metatarsal artery

Plantar digital arteries

6.9 OVERVIEW OF ARTERIES OF LOWER LIMB

The arteries often anastomose or communicate to form networks to ensure blood supply distal to the joint throughout the range of movement (cruciate, genicule and tarsal anastomoses).

If a main channel is slowly occluded, the smaller alternate channels can usually increase in size, providing a **collateral circulation** that ensures the blood supply to structures distal to the blockage.

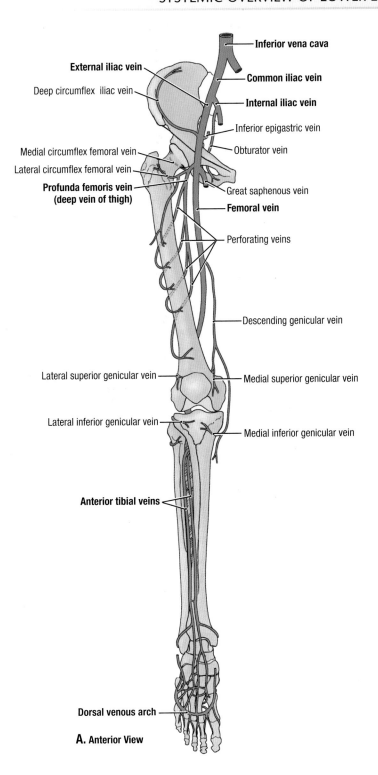

- Inferior vena cava
- External iliac vein
- Deep circumflex iliac vein
- Common iliac vein
- Internal iliac vein
- Inferior epigastric vein
- Obturator vein
- Medial circumflex femoral vein
- Lateral circumflex femoral vein
- **Profunda femoris vein (deep vein of thigh)**
- Great saphenous vein
- **Femoral vein**
- Perforating veins
- Descending genicular vein
- Lateral superior genicular vein
- Medial superior genicular vein
- Lateral inferior genicular vein
- Medial inferior genicular vein
- **Anterior tibial veins**
- **Dorsal venous arch**

A. Anterior View

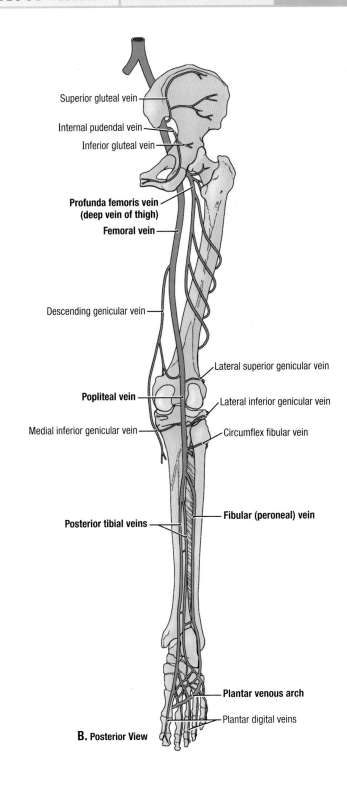

- Superior gluteal vein
- Internal pudendal vein
- Inferior gluteal vein
- **Profunda femoris vein (deep vein of thigh)**
- **Femoral vein**
- Descending genicular vein
- Lateral superior genicular vein
- **Popliteal vein**
- Lateral inferior genicular vein
- Medial inferior genicular vein
- Circumflex fibular vein
- **Posterior tibial veins**
- **Fibular (peroneal) vein**
- **Plantar venous arch**
- Plantar digital veins

B. Posterior View

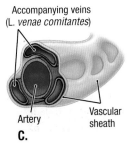

- Accompanying veins (L. *venae comitantes*)
- Artery
- Vascular sheath

C.

DEEP VEINS OF LOWER LIMB | 6.10

A. and **B.** Deep veins lie internal to the deep fascia. Although only the anterior and posterior tibial veins are depicted as paired structures in this schematic illustration, typically in the limbs deep veins occur as multiple, generally parallel, continually interanastomosing accompanying veins (L. *venae comitantes*) surrounding and sharing the name of the artery they accompany. **C.** Accompanying veins.

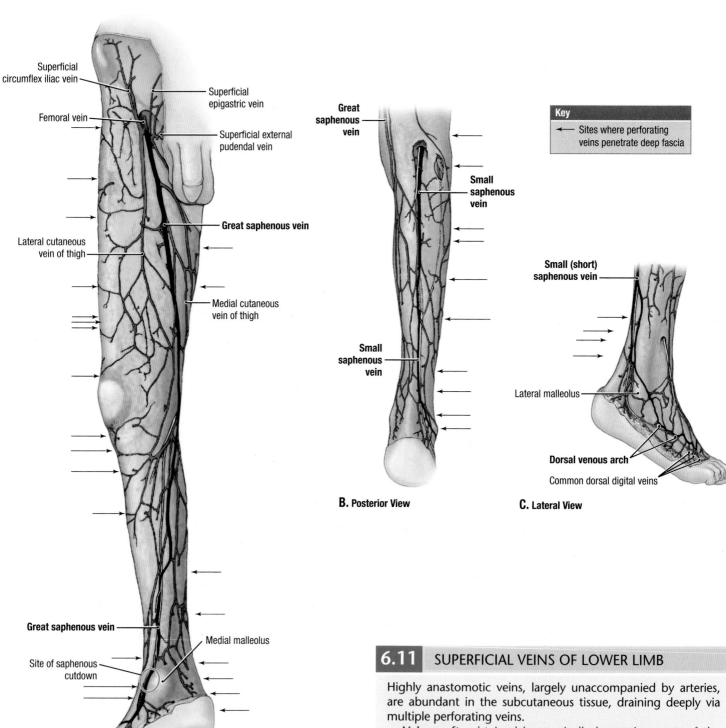

Superficial circumflex iliac vein

Superficial epigastric vein

Femoral vein

Superficial external pudendal vein

Great saphenous vein

Lateral cutaneous vein of thigh

Medial cutaneous vein of thigh

Great saphenous vein

Site of saphenous cutdown

Medial malleolus

A. Anteromedial View

Great saphenous vein

Small saphenous vein

Small saphenous vein

B. Posterior View

Key
← Sites where perforating veins penetrate deep fascia

Small (short) saphenous vein

Lateral malleolus

Dorsal venous arch

Common dorsal digital veins

C. Lateral View

6.11 SUPERFICIAL VEINS OF LOWER LIMB

Highly anastomotic veins, largely unaccompanied by arteries, are abundant in the subcutaneous tissue, draining deeply via multiple perforating veins.

Vein grafts obtained by surgically harvesting parts of the great saphenous vein are used to bypass obstructions in blood vessels (e.g., a coronary artery). When used as a bypass, the vein is reversed so that the valves do not obstruct blood flow. Because there are so many anastomosing leg veins, removal of the great saphenous vein rarely affects circulation seriously, provided the deep veins are intact.

Saphenous cut down. The great saphenous vein can be located by making a skin incision anterior to the medial malleolus. This procedure is used to insert a cannula for prolonged administration of blood, electrolytes, drugs, etc.

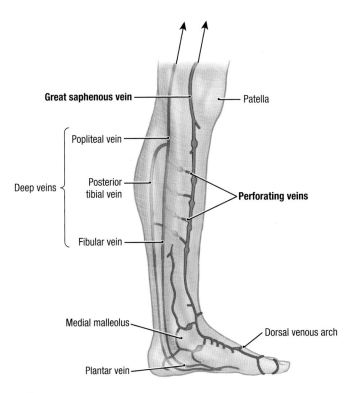

A. Medial View

Great saphenous vein — Patella

Deep veins {
Popliteal vein
Posterior tibial vein — Perforating veins
Fibular vein
}

Medial malleolus — Dorsal venous arch

Plantar vein

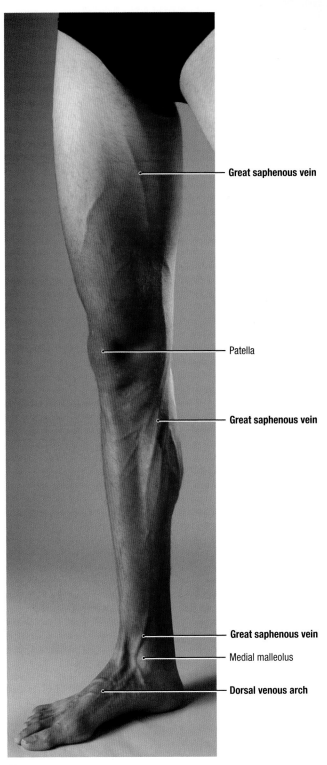

C. Anteromedial View, Normal Veins

Great saphenous vein

Great saphenous vein

Patella

Great saphenous vein

Great saphenous vein
Medial malleolus
Dorsal venous arch

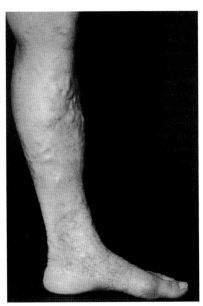

B. Medial View, Varicose Veins

DRAINAGE AND SURFACE ANATOMY OF SUPERFICIAL VEINS OF LOWER LIMB

6.12

A. Schematic diagram of drainage of superficial veins. Blood is shunted from the superficial veins (e.g., great saphenous vein) to the deep veins (e.g., fibular and posterior tibial veins) via perforating veins that penetrate the deep fascia. Muscular compression of deep veins assists return of blood to the heart against gravity. **B.** Varicose veins form when either the deep fascia or the valves of the perforating veins are incompetent. This allows the muscular compression that normally propels blood toward the heart to push blood from the deep to the superficial veins. Consequently, superficial veins become enlarged and tortuous. **C.** Normal veins distended following exercise.

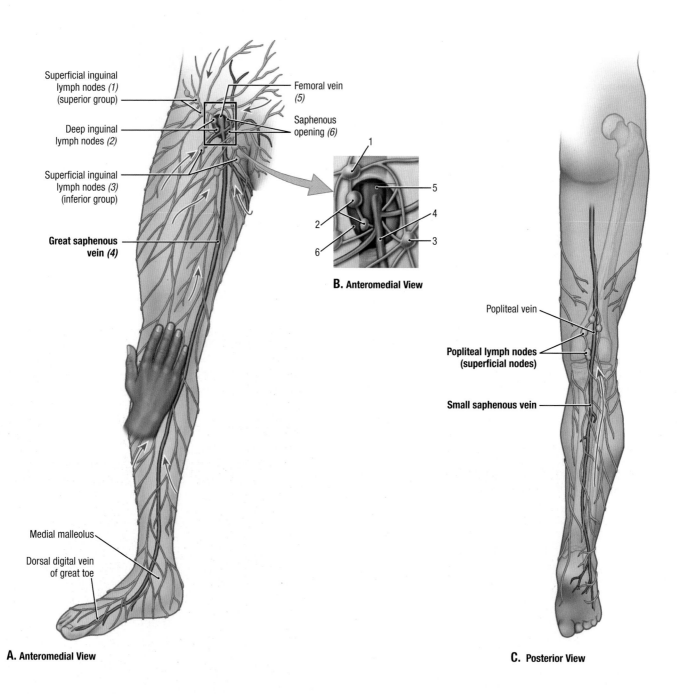

Superficial inguinal lymph nodes *(1)* (superior group)

Femoral vein *(5)*

Deep inguinal lymph nodes *(2)*

Saphenous opening *(6)*

Superficial inguinal lymph nodes *(3)* (inferior group)

Great saphenous vein *(4)*

Medial malleolus

Dorsal digital vein of great toe

A. Anteromedial View

B. Anteromedial View

Popliteal vein

Popliteal lymph nodes (superficial nodes)

Small saphenous vein

C. Posterior View

6.13   SUPERFICIAL LYMPHATIC DRAINAGE OF LOWER LIMB

The superficial lymphatic vessels accompany the saphenous veins and their tributaries in the superficial fascia. The lymphatic vessels along the great saphenous vein drain into the superficial inguinal lymph nodes; those along the small saphenous vein drain into the popliteal lymph nodes. Lymph from the superficial inguinal nodes drains to the deep inguinal and external iliac nodes. Lymph from the popliteal nodes ascends through deep lymphatic vessels accompanying the deep blood vessels to the deep inguinal nodes.

Note that the great saphenous vein lies anterior to the medial malleolus and a hand's breadth posterior to the medial border of the patella. **Lymph nodes enlarge** when diseased. Abrasions and minor sepsis, caused by pathogenic micro-organisms or their toxins, may produce slight enlargement of the superficial inguinal nodes (lymphadenopathy) in otherwise healthy people. Malignancies (e.g., of the external genitalia and uterus) and perineal abscesses also result in enlargement of these nodes.

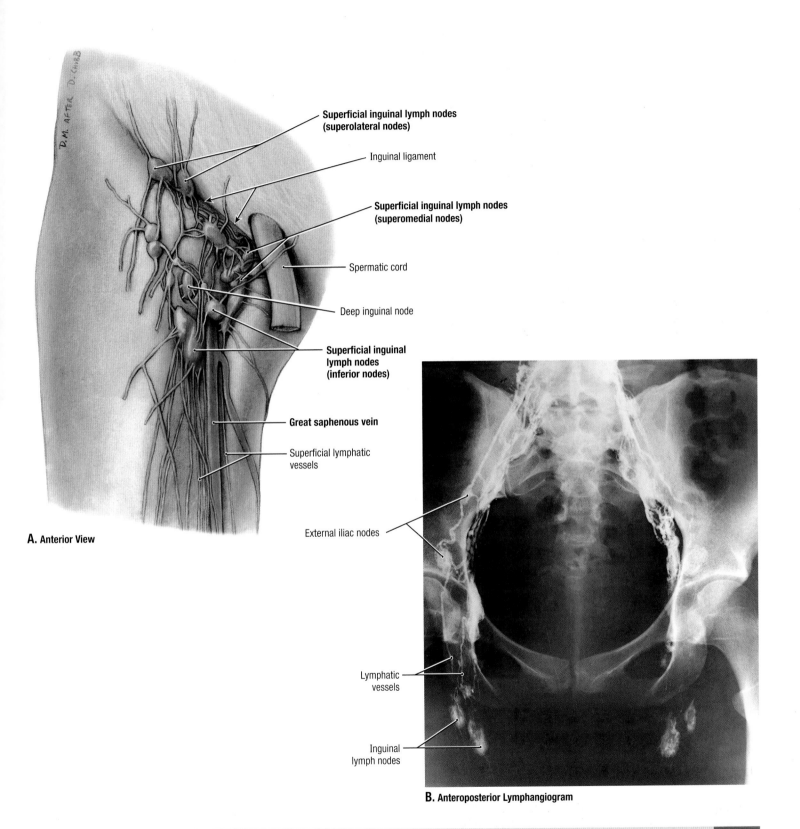

Superficial inguinal lymph nodes (superolateral nodes)

Inguinal ligament

Superficial inguinal lymph nodes (superomedial nodes)

Spermatic cord

Deep inguinal node

Superficial inguinal lymph nodes (inferior nodes)

Great saphenous vein

Superficial lymphatic vessels

A. Anterior View

External iliac nodes

Lymphatic vessels

Inguinal lymph nodes

B. Anteroposterior Lymphangiogram

INGUINAL LYMPH NODES

6.14

A. Dissection. **B.** Lymphangiogram.

- **Observe the arrangement of the nodes:** a proximal chain parallel to the inguinal ligament (superolateral and superomedial superficial inguinal lymph nodes) and a distal chain on the sides of the great saphenous vein (inferior superficial inguinal lymph nodes). Efferent vessels leave these nodes and pass deep to the inguinal ligament to enter the deep inguinal and external iliac nodes.
- Note the anastomosis between the lymph vessels.

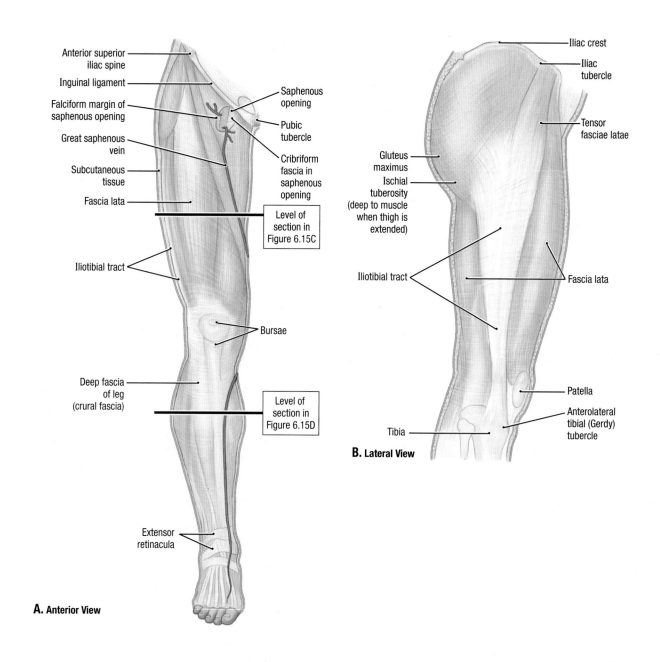

A. Anterior View

- Anterior superior iliac spine
- Inguinal ligament
- Falciform margin of saphenous opening
- Great saphenous vein
- Subcutaneous tissue
- Fascia lata
- Saphenous opening
- Pubic tubercle
- Cribriform fascia in saphenous opening
- Level of section in Figure 6.15C
- Iliotibial tract
- Bursae
- Deep fascia of leg (crural fascia)
- Level of section in Figure 6.15D
- Extensor retinacula

B. Lateral View

- Iliac crest
- Iliac tubercle
- Tensor fasciae latae
- Gluteus maximus
- Ischial tuberosity (deep to muscle when thigh is extended)
- Iliotibial tract
- Fascia lata
- Patella
- Anterolateral tibial (Gerdy) tubercle
- Tibia

6.15 FASCIA AND MUSCULOFASCIAL COMPARTMENTS OF LOWER LIMB

A. Anterior skin and subcutaneous tissue have been removed to reveal the deep fascia of the thigh (fascia lata) and leg (crural fascia). **B.** Lateral skin and subcutaneous tissue have been removed to reveal the fascia lata. The fascia lata is thick laterally and forms the iliotibial tract. The iliotibial tract serves as a common aponeurosis for the gluteus maximus and tensor fasciae latae muscles. One of the most common causes of lateral knee pain in endurance athletes

(e.g., runners, cyclers, hikers) is **iliotibial tract (band) syndrome (ITBS)**. Friction of the IT tract against the lateral epicondyle of the femur with flexion and extension of the knee (e.g., during running) may result in the inflammation of the IT tract over the lateral aspect of the knee or its attachment to the dorsolateral tubercle (Gerdy tubercle). ITBS may also occur in the hip region, especially in older individuals.

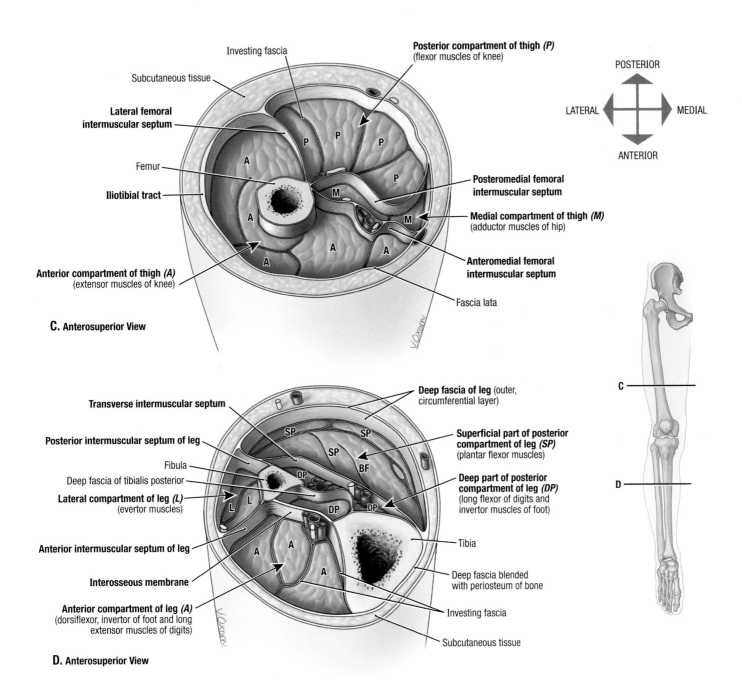

C. Anterosuperior View

Investing fascia

Subcutaneous tissue

Lateral femoral intermuscular septum

Femur

Iliotibial tract

Anterior compartment of thigh (A)
(extensor muscles of knee)

Posterior compartment of thigh (P)
(flexor muscles of knee)

POSTERIOR

LATERAL — MEDIAL

ANTERIOR

Posteromedial femoral intermuscular septum

Medial compartment of thigh (M)
(adductor muscles of hip)

Anteromedial femoral intermuscular septum

Fascia lata

D. Anterosuperior View

Transverse intermuscular septum

Posterior intermuscular septum of leg

Fibula

Deep fascia of tibialis posterior

Lateral compartment of leg (L)
(evertor muscles)

Anterior intermuscular septum of leg

Interosseous membrane

Anterior compartment of leg (A)
(dorsiflexor, invertor of foot and long
extensor muscles of digits)

Deep fascia of leg (outer,
circumferential layer)

**Superficial part of posterior
compartment of leg (SP)**
(plantar flexor muscles)

**Deep part of posterior
compartment of leg (DP)**
(long flexor of digits and
invertor muscles of foot)

Tibia

Deep fascia blended
with periosteum of bone

Investing fascia

Subcutaneous tissue

FASCIA AND MUSCULOFASCIAL COMPARTMENTS OF LOWER LIMB (*continued*) | **6.15**

C. and **D.** The fascial compartments of the thigh (**C**) and leg (**D**) are demonstrated in transverse section. The fascial compartments contain muscles that generally perform common functions and share common innervation and contain the spread of infection. While both thigh and leg have anterior and posterior compartments, the thigh also includes a medial compartment and the leg a lateral compartment. Trauma to muscles and/or vessels in the compartments may produce hemorrhage, edema, and inflammation of the muscles. Because the septa, deep fascia, and bony attachments firmly bound the compartments, increased volume resulting from these processes raises intracompartmental pressure. In **compartment syndromes**, structures within or distal to the compressed area become ischemic and may become permanently injured (e.g., compression of capillary beds results in denervation and consequent paralysis of muscles). A **fasciotomy** (incision of bounding fascia or septum) may be performed to relieve the pressure in the compartment and restore circulation.

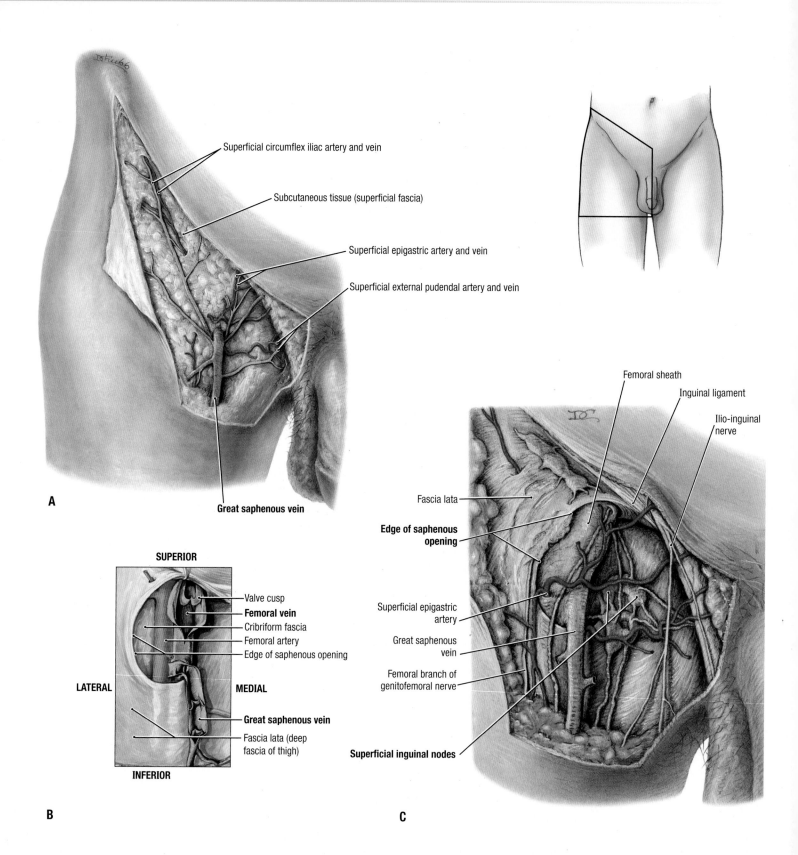

A. Superficial inguinal vessels.

Superficial circumflex iliac artery and vein

Subcutaneous tissue (superficial fascia)

Superficial epigastric artery and vein

Superficial external pudendal artery and vein

Great saphenous vein

A

SUPERIOR

Valve cusp
Femoral vein
Cribriform fascia
Femoral artery
Edge of saphenous opening

LATERAL

MEDIAL

Great saphenous vein

Fascia lata (deep fascia of thigh)

INFERIOR

B

Femoral sheath

Inguinal ligament

Ilio-inguinal nerve

Fascia lata

Edge of saphenous opening

Superficial epigastric artery

Great saphenous vein

Femoral branch of genitofemoral nerve

Superficial inguinal nodes

C

6.16 SUPERFICIAL INGUINAL VESSELS AND SAPHENOUS OPENING

A. Superficial inguinal vessels. The arteries are branches of the femoral artery, and the veins are tributaries of the great saphenous vein. **B.** Valves of the proximal part of femoral and great saphenous veins. **C.** Saphenous opening.

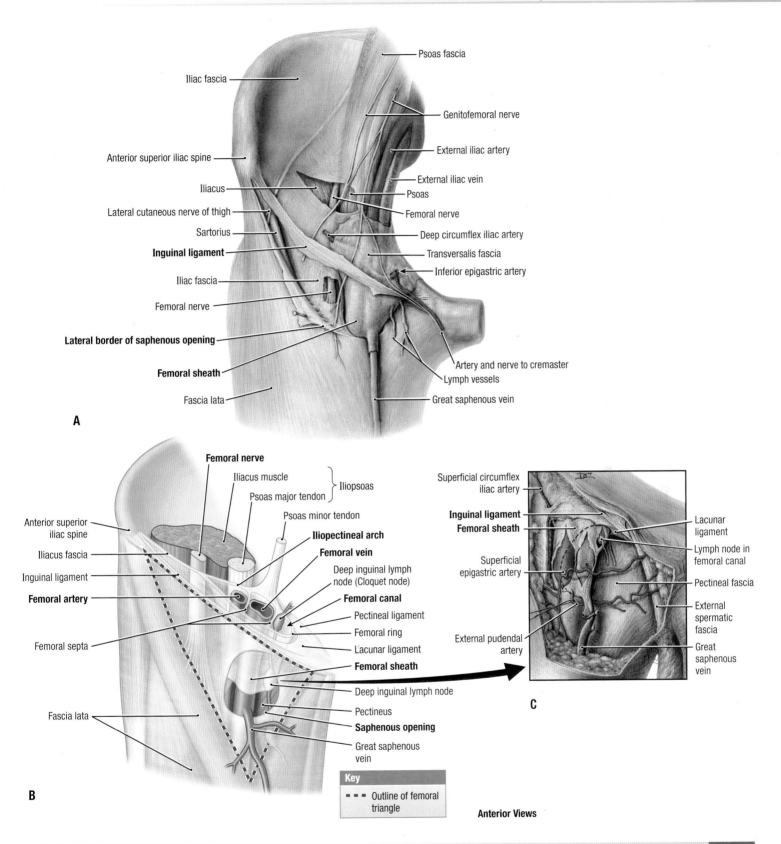

A. Dissection.

Psoas fascia
Iliac fascia
Genitofemoral nerve
Anterior superior iliac spine
External iliac artery
External iliac vein
Iliacus
Psoas
Lateral cutaneous nerve of thigh
Femoral nerve
Sartorius
Deep circumflex iliac artery
Inguinal ligament
Transversalis fascia
Iliac fascia
Inferior epigastric artery
Femoral nerve
Lateral border of saphenous opening
Femoral sheath
Artery and nerve to cremaster
Lymph vessels
Fascia lata
Great saphenous vein

A

Femoral nerve
Iliacus muscle
Psoas major tendon } Iliopsoas
Psoas minor tendon
Iliopectineal arch
Femoral vein
Anterior superior iliac spine
Deep inguinal lymph node (Cloquet node)
Iliacus fascia
Femoral canal
Inguinal ligament
Pectineal ligament
Femoral artery
Femoral ring
Lacunar ligament
Femoral septa
Femoral sheath
Deep inguinal lymph node
Fascia lata
Pectineus
Saphenous opening
Great saphenous vein

Superficial circumflex iliac artery
Inguinal ligament
Femoral sheath
Lacunar ligament
Lymph node in femoral canal
Superficial epigastric artery
Pectineal fascia
External spermatic fascia
External pudendal artery
Great saphenous vein

C

Key
- - - Outline of femoral triangle

Anterior Views

B

FEMORAL SHEATH AND INGUINAL LIGAMENT

6.17

A. Dissection. **B.** Schematic illustration. The femoral sheath contains the femoral artery, vein, and lymph vessels, but the femoral nerve, lying posterior to the iliacus fascia, is outside the femoral sheath. **C.** Femoral sheath and femoral ring. The three compartments of the femoral sheath are for the femoral artery, vein, and femoral canal. The femoral canal has a small proximal opening at its abdominal end, the femoral ring, closed by extraperitoneal fatty tissue.

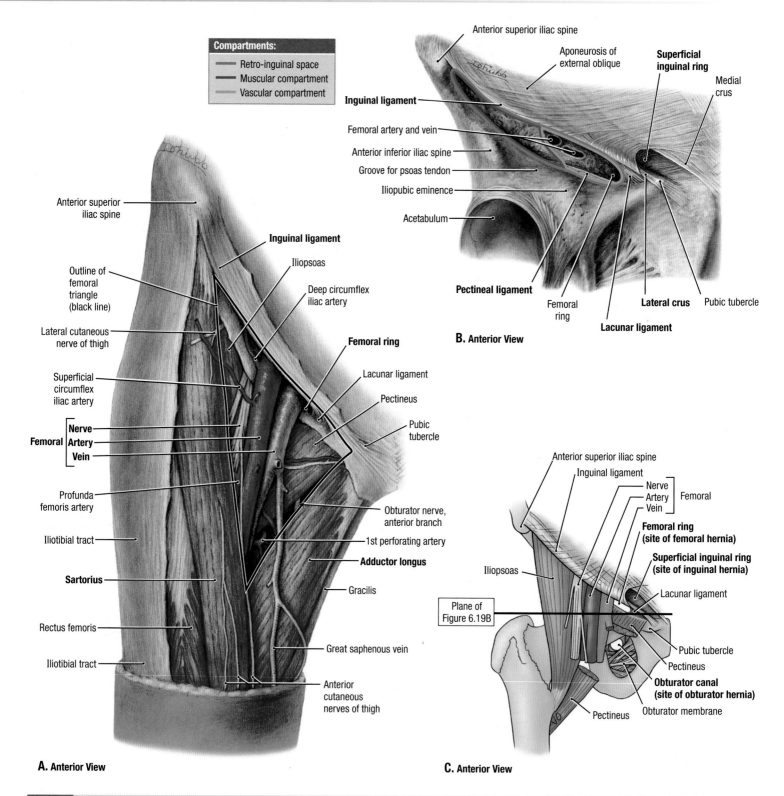

Compartments:
- Retro-inguinal space
- Muscular compartment
- Vascular compartment

B. Anterior View

Labels (figure B): Anterior superior iliac spine · Aponeurosis of external oblique · **Superficial inguinal ring** · Medial crus · Inguinal ligament · Femoral artery and vein · Anterior inferior iliac spine · Groove for psoas tendon · Iliopubic eminence · Acetabulum · **Pectineal ligament** · Femoral ring · **Lateral crus** · **Lacunar ligament** · Pubic tubercle

A. Anterior View

Labels (figure A): Anterior superior iliac spine · **Inguinal ligament** · Iliopsoas · Outline of femoral triangle (black line) · Deep circumflex iliac artery · Lateral cutaneous nerve of thigh · **Femoral ring** · Superficial circumflex iliac artery · Lacunar ligament · Pectineus · **Femoral** [Nerve, Artery, Vein] · Pubic tubercle · Profunda femoris artery · Iliotibial tract · Obturator nerve, anterior branch · 1st perforating artery · **Adductor longus** · **Sartorius** · Gracilis · Rectus femoris · Iliotibial tract · Great saphenous vein · Anterior cutaneous nerves of thigh

C. Anterior View

Labels (figure C): Anterior superior iliac spine · Inguinal ligament · Nerve, Artery, Vein [Femoral] · Iliopsoas · **Femoral ring (site of femoral hernia)** · **Superficial inguinal ring (site of inguinal hernia)** · Lacunar ligament · Plane of Figure 6.19B · Pubic tubercle · Pectineus · **Obturator canal (site of obturator hernia)** · Obturator membrane · Pectineus

6.18 STRUCTURES PASSING TO/FROM FEMORAL TRIANGLE VIA RETRO-INGUINAL PASSAGE

A. Dissection. The boundaries of the femoral triangle are the inguinal ligament superiorly (base of triangle), the medial border of the sartorius (lateral side), and the lateral border of the adductor longus (medial side). The point at which the lateral and medial sides converge inferiorly forms the apex. The femoral triangle is bisected by the femoral vessels. **B.** Retro-inguinal passage between the inguinal ligament anteriorly and the bony pelvis posteriorly.

C. The iliopsoas muscle, the femoral nerve, artery, and vein, and the lymphatic vessels draining the inguinal nodes pass deep to the inguinal ligament to enter the anterior thigh or return to the trunk. Three potential sites for **hernia formation** are indicated. **Pulsations of the femoral artery** can be felt distal to the inguinal ligament, midway between the anterior superior iliac spine and the pubic tubercle.

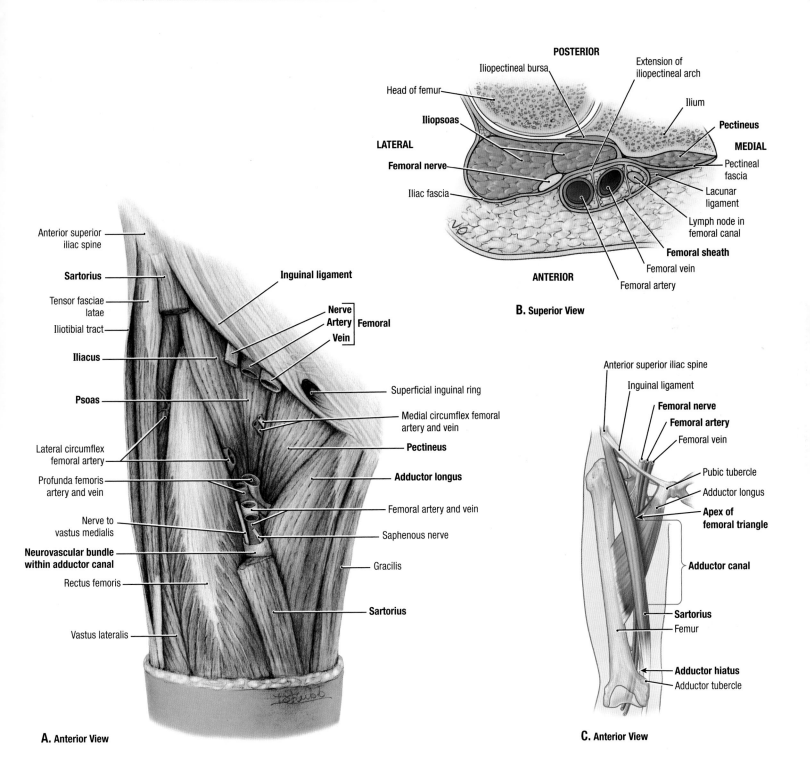

POSTERIOR

Iliopectineal bursa

Head of femur

Iliopsoas

LATERAL

Femoral nerve

Iliac fascia

Extension of
iliopectineal arch

Ilium

Pectineus

MEDIAL

Pectineal
fascia

Lacunar
ligament

Lymph node in
femoral canal

Femoral sheath

Femoral vein

Femoral artery

ANTERIOR

B. Superior View

Anterior superior
iliac spine

Sartorius

Tensor fasciae
latae

Iliotibial tract

Iliacus

Psoas

Lateral circumflex
femoral artery

Profunda femoris
artery and vein

Nerve to
vastus medialis

**Neurovascular bundle
within adductor canal**

Rectus femoris

Vastus lateralis

Inguinal ligament

Nerve
Artery } **Femoral**
Vein

Superficial inguinal ring

Medial circumflex femoral
artery and vein

Pectineus

Adductor longus

Femoral artery and vein

Saphenous nerve

Gracilis

Sartorius

A. Anterior View

Anterior superior iliac spine

Inguinal ligament

Femoral nerve

Femoral artery

Femoral vein

Pubic tubercle

Adductor longus

**Apex of
femoral triangle**

Adductor canal

Sartorius

Femur

Adductor hiatus

Adductor tubercle

C. Anterior View

FLOOR OF FEMORAL CANAL AND RETRO-INGUINAL PASSAGE

6.19

A. Dissection. Portions of the sartorius muscle, femoral vessels, and femoral nerve have been removed revealing the floor of the femoral triangle, formed by the iliopsoas laterally and the pectineus medially. At the apex of the triangle, the femoral vessels, saphenous nerve, and the nerve to the vastus medialis pass deep to the sartorius into the adductor (subsartorial) canal. **B.** Transverse section of the femoral triangle at the level of head of femur. The iliopsoas and femoral nerve traverse the retro-inguinal passage and femoral triangle in a fascial sheath separate from the femoral vessels, which are contained within the femoral sheath (see Fig. 6.18C for level of section). **C.** Schematic illustration of course of femoral vessels. The adductor canal extends from the apex of the femoral triangle to the adductor hiatus by which the vessels enter and leave the popliteal fossa.

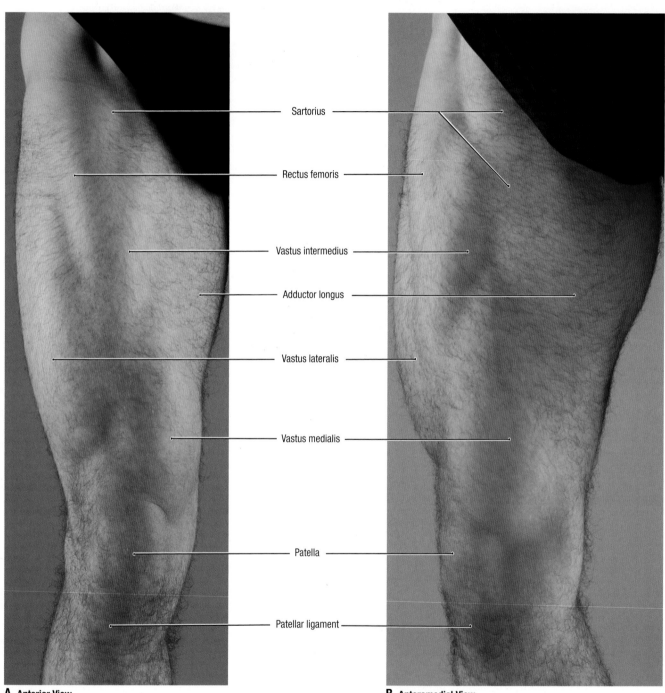

Sartorius

Rectus femoris

Vastus intermedius

Adductor longus

Vastus lateralis

Vastus medialis

Patella

Patellar ligament

A. Anterior View

B. Anteromedial View

6.20   SURFACE ANATOMY OF ANTERIOR AND MEDIAL ASPECTS OF THIGH

Patellar tendinitis (jumper's knee) is caused by continuous over-loading of the knee extensor mechanism, resulting in microtears of the tendon. The most vulnerable site is where the patellar ligament (tendon) attaches to the patella. This overuse injury can result in degeneration and tearing of the tendon.

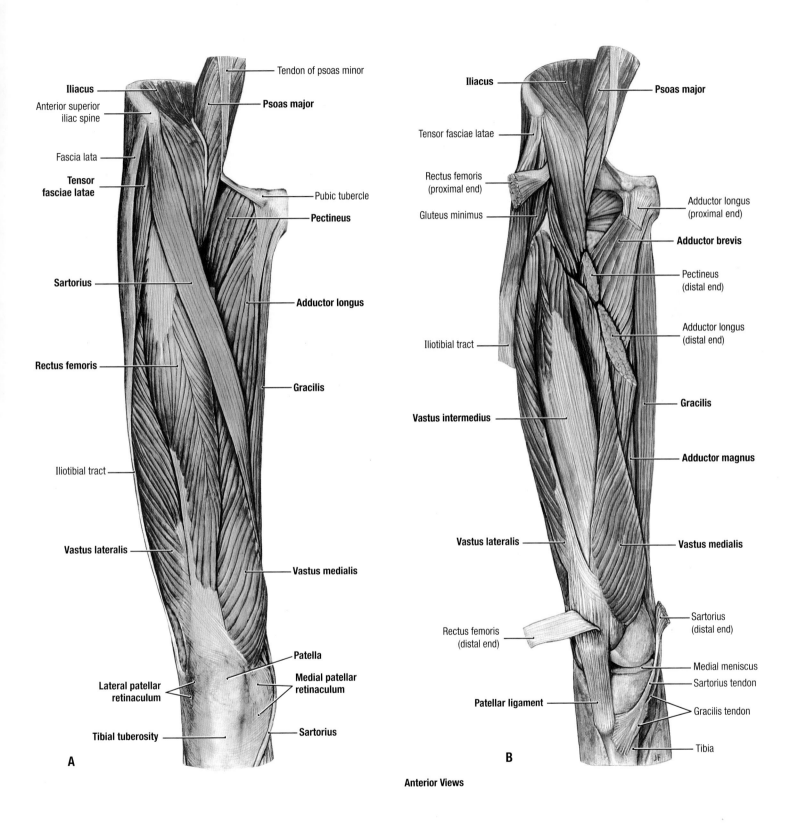

A. Superficial dissection. **B.** Deep dissection. The central portions of the muscle bellies of the sartorius, rectus femoris, pectineus, and adductor longus muscles have been removed. **Weakness of**

the vastus medialis or vastus lateralis, resulting from arthritis or trauma to the knee joint, for example, can result in abnormal patellar movement and loss of joint stability.

Anterior Views

ANTERIOR AND MEDIAL THIGH MUSCLES, SUPERFICIAL AND DEEP DISSECTIONS

6.21

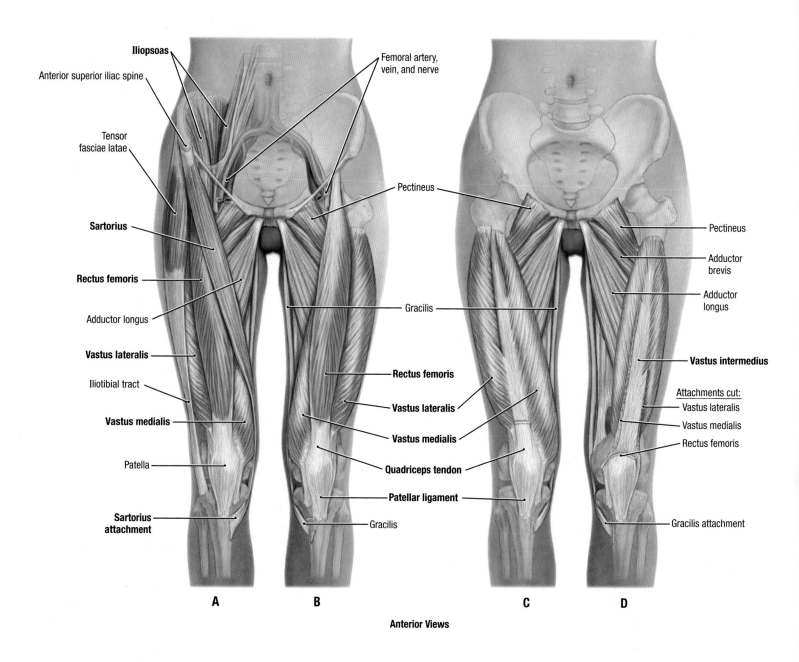

Anterior Views

6.22 ANTERIOR AND MEDICAL THIGH MUSCLES, SCHEMATIC ILLUSTRATIONS

A–D. Sequential views from superficial to deep.

A "hip pointer," which is a **contusion of the iliac crest,** usually occurs at its anterior part (e.g., where the sartorius attaches to the anterior superior iliac spine). This is one of the most common injuries to the hip region, usually occurring in association with collision sports. Contusions cause bleeding from ruptured capillaries and infiltration of blood into the muscles, tendons, and other soft tissues. The term hip pointer may also refer to avulsion of bony muscle attachments, for example, of the sartorius or rectus femoris from the anterior superior or inferior iliac spines or of the iliopsoas from the lesser trochanter of the femur. However, these injuries should be called **avulsion fractures.**

A person with a **paralyzed quadriceps** cannot extend the leg against resistance and usually presses on the distal end of the thigh during walking to prevent inadvertent flexion of the knee joint.

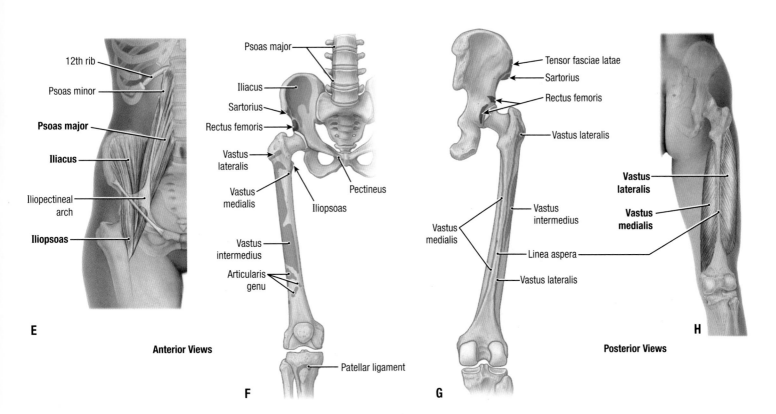

E. Iliopsoas.

Anterior Views

Anterior Views

Posterior Views

E

F

G

H

ANTERIOR AND MEDICAL THIGH MUSCLES, SCHEMATIC ILLUSTRATIONS (*continued*)   6.22

E. Iliopsoas. **F.** and **G.** Attachments of anterior muscles of thigh. **H.** Posterior attachment of vastus medialis and lateralis.

TABLE 6.5	**MUSCLES OF ANTERIOR THIGH**			
Muscle	*Proximal Attachment[a]*	*Distal Attachment[a]*	*Innervation[b]*	*Main Actions*
Iliopsoas				
Psoas major	Lateral aspects of T12–L5 vertebrae and intervertebral discs; transverse processes of all lumbar vertebrae	Lesser trochanter of femur	Anterior rami of lumbar nerves (**L1**, **L2**, and L3)	Flexes and stabilizes[c] hip joint
Iliacus	Iliac crest, iliac fossa, ala of sacrum and anterior sacro-iliac ligaments	Tendon of psoas major, lesser trochanter, and femur distal to it	Femoral nerve (L2 and L3)	
Tensor fasciae latae	Anterior superior iliac spine and anterior part of iliac crest	Iliotibial tract that attaches to lateral condyle of tibia	Superior gluteal (L4 and L5)	Abducts, medially rotates, and flexes hip joint; helps to keep knee extended; steadies trunk on thigh
Sartorius	Anterior superior iliac spine and superior part of notch inferior to it	Superior part of medial surface of tibia	Femoral nerve (L2 and L3)	Flexes, abducts, and laterally rotates hip joint; flexes knee joint[d]
Quadriceps femoris				
Rectus femoris	Anterior inferior iliac spine and ilium superior to acetabulum	Base of patella and by patellar ligament to tibial tuberosity; medial and lateral vasti also attach to tibia and patella via aponeuroses (medial and lateral patellar retinacula)	Femoral nerve (L2, **L3**, and **L4**)	Extends knee joint; rectus femoris also steadies hip joint and helps iliopsoas to flex hip joint
Vastus lateralis	Greater trochanter and lateral lip of linea aspera of femur			
Vastus medialis	Intertrochanteric line and medial lip of linea aspera of femur			
Vastus intermedius	Anterior and lateral surfaces of body of femur			

[a]See also Figure 6.22 for muscle attachments.
[b]Numbers indicate spinal cord segmental innervation of nerves (e.g., L1, L2, and L3 indicate that nerves supplying psoas major are derived from first three lumbar segments of the spinal cord; boldface type [e.g., **L1**, **L2**] indicates main segmental innervation). Damage to one or more of these spinal cord segments or to motor nerve roots arising from these segments results in paralysis of the muscles concerned.
[c]Psoas major is also a postural muscle that helps control deviation of trunk and is active during standing.
[d]Four actions of sartorius (L. *sartor*, tailor) produce the once-common cross-legged sitting position used by tailors—hence the name.

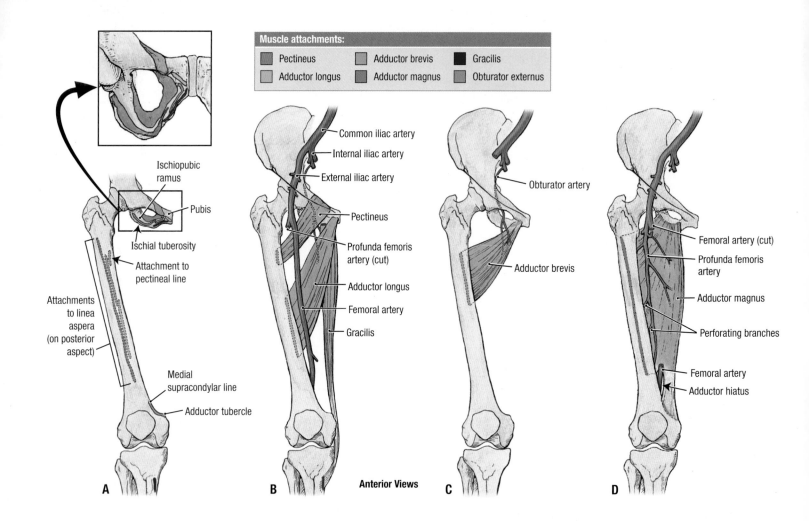

Muscle attachments:

- Pectineus
- Adductor brevis
- Gracilis
- Adductor longus
- Adductor magnus
- Obturator externus

A. Overview of attachments. B. Pectineus, adductor longus, and gracilis. C. Adductor brevis. D. Adductor magnus.

6.23 ATTACHMENTS OF MUSCLES OF MEDIAL ASPECT OF THIGH

TABLE 6.6	MUSCLES OF MEDIAL THIGH			
Muscle	*Proximal Attachment*	*Distal Attachment*[a]	*Innervation*[b]	*Main Actions*
Pectineus	Superior pubic ramus	Pectineal line of femur, just inferior to lesser trochanter	Femoral nerve (**L2** and L3) may receive a branch from obturator nerve	Adducts and flexes hip joint; assists with medial rotation of hip joint
Adductor longus	Body of pubis inferior to pubic crest	Middle third of linea aspera of femur	Obturator nerve, (L2, **L3**, and L4)	Adducts hip joint
Adductor brevis	Body of pubis and inferior pubic ramus	Pectineal line and proximal part of linea aspera of femur	Obturator nerve (L2, **L3**, and L4)	Adducts hip joint and, to some extent, flexes it
Adductor magnus	Inferior pubic ramus, ramus of ischium (adductor part), and ischial tuberosity	Gluteal tuberosity, linea aspera, medial supracondylar line (adductor part), and adductor tubercle of femur (hamstring part)	*Adductor part:* obturator nerve (L2, **L3**, and **L4**) *Hamstring part:* tibial part of sciatic nerve (**L4**)	Adducts hip joint; its adductor part also flexes hip joint, and its hamstring part extends it
Gracilis	Body of pubis and inferior pubic ramus	Superior part of medial surface of tibia	Obturator nerve (**L2** and L3)	Adducts hip joint, flexes knee joint, and helps rotate it medially
Obturator externus	Margins of obturator foramen and obturator membrane	Trochanteric fossa of femur	Obturator nerve (L3 and **L4**)	Laterally rotates hip joint; steadies head of femur in acetabulum

Collectively, the first five muscles listed are the adductors of the thigh, but their actions are more complex (e.g., they act as flexors of the hip joint during flexion of the knee joint and are active during walking).
[a]See Figure 6.22 for muscle attachments.
[b]See Table 6.1 for explanation of segmental innervation. Numbers indicate spinal cord segmental innervation of nerves (e.g., L2, L3, and L4 indicate that the obturator nerve supplying adductor longus is derived from lumbar segments of the spinal cord; boldface type **[L3]** indicates main segmental innervation). Damage to one or more of these spinal cord segments or to motor nerve roots arising from these segments results in paralysis of the muscles concerned.

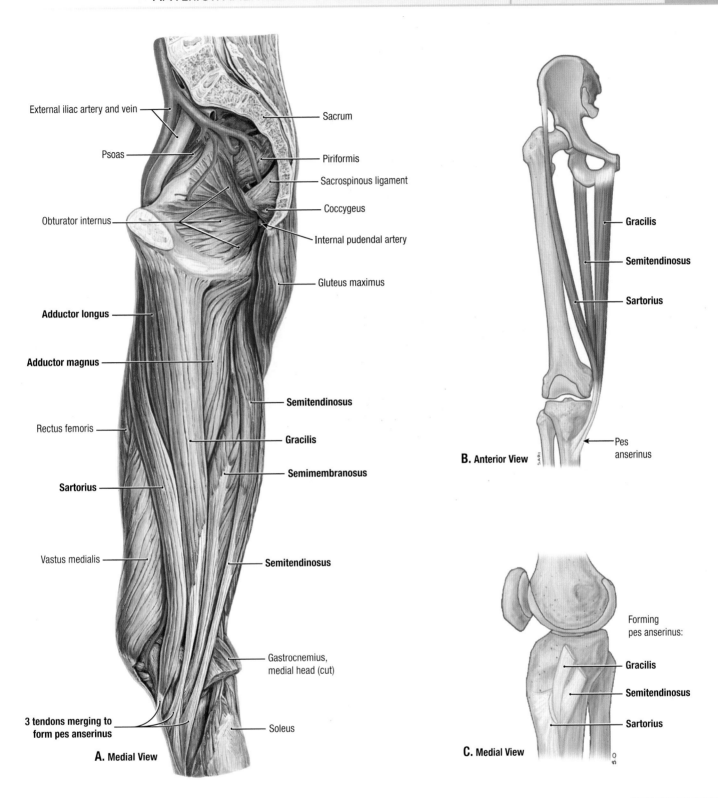

External iliac artery and vein

Psoas

Obturator internus

Adductor longus

Adductor magnus

Rectus femoris

Sartorius

Vastus medialis

3 tendons merging to form pes anserinus

A. Medial View

Sacrum

Piriformis

Sacrospinous ligament

Coccygeus

Internal pudendal artery

Gluteus maximus

Semitendinosus

Gracilis

Semimembranosus

Semitendinosus

Gastrocnemius, medial head (cut)

Soleus

Gracilis

Semitendinosus

Sartorius

Pes anserinus

B. Anterior View

Forming pes anserinus:

Gracilis

Semitendinosus

Sartorius

C. Medial View

MUSCLES OF MEDIAL ASPECT OF THIGH

6.24

A. Dissection. **B.** Muscular tripod. The sartorius, gracilis, and semitendinosus muscles form an inverted tripod arising from three different components of the hip bone. These muscles course within three different compartments, perform three different functions, and are innervated by three different nerves yet share a common distal attachment. **C.** Distal attachment of sartorius, gracilis, and semitendinosus muscles. All three tendons become thin and aponeurotic and are collectively referred to as the pes anserinus.

The gracilis is a relatively weak member of the adductor group and hence can be removed without noticeable loss of its actions on the leg. Surgeons often **transplant the gracilis**, or part of it, with its nerve and blood vessels to replace a damaged muscle, in the hand, for example.

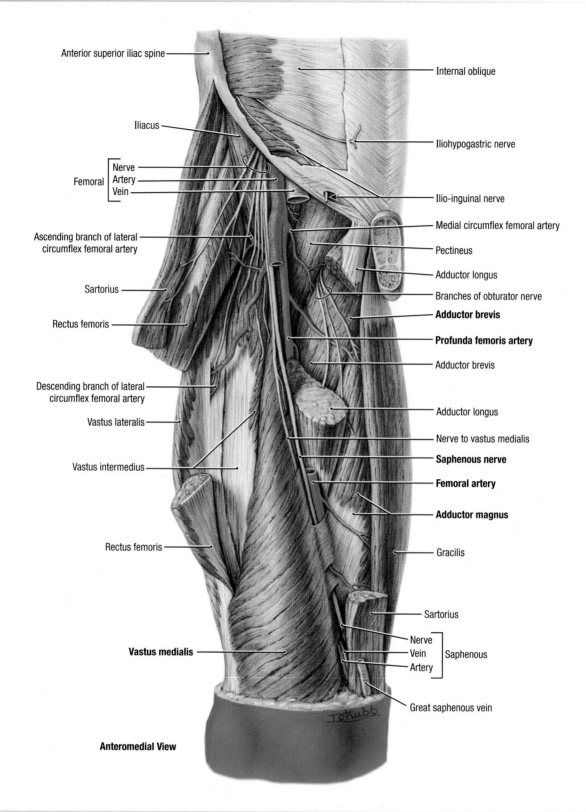

Anterior superior iliac spine

Internal oblique

Iliacus

Iliohypogastric nerve

Femoral — Nerve / Artery / Vein

Ilio-inguinal nerve

Medial circumflex femoral artery

Ascending branch of lateral circumflex femoral artery

Pectineus

Adductor longus

Sartorius

Branches of obturator nerve

Rectus femoris

Adductor brevis

Profunda femoris artery

Adductor brevis

Descending branch of lateral circumflex femoral artery

Adductor longus

Vastus lateralis

Nerve to vastus medialis

Saphenous nerve

Vastus intermedius

Femoral artery

Adductor magnus

Rectus femoris

Gracilis

Sartorius

Nerve / Vein / Artery } Saphenous

Vastus medialis

Great saphenous vein

Anteromedial View

6.25 ANTEROMEDIAL ASPECT OF THIGH

- The limb is rotated laterally.
- The femoral nerve breaks up into multiple nerves on entering the thigh.
- The femoral artery lies between two motor territories: that of the obturator nerve, which is medial, and that of the femoral nerve, which is lateral.
- The nerve to the vastus medialis muscle and the saphenous nerve accompany the femoral artery into the adductor canal.
- The profunda femoris artery (deep artery of thigh) is the largest branch of the femoral artery and the chief artery to the thigh.

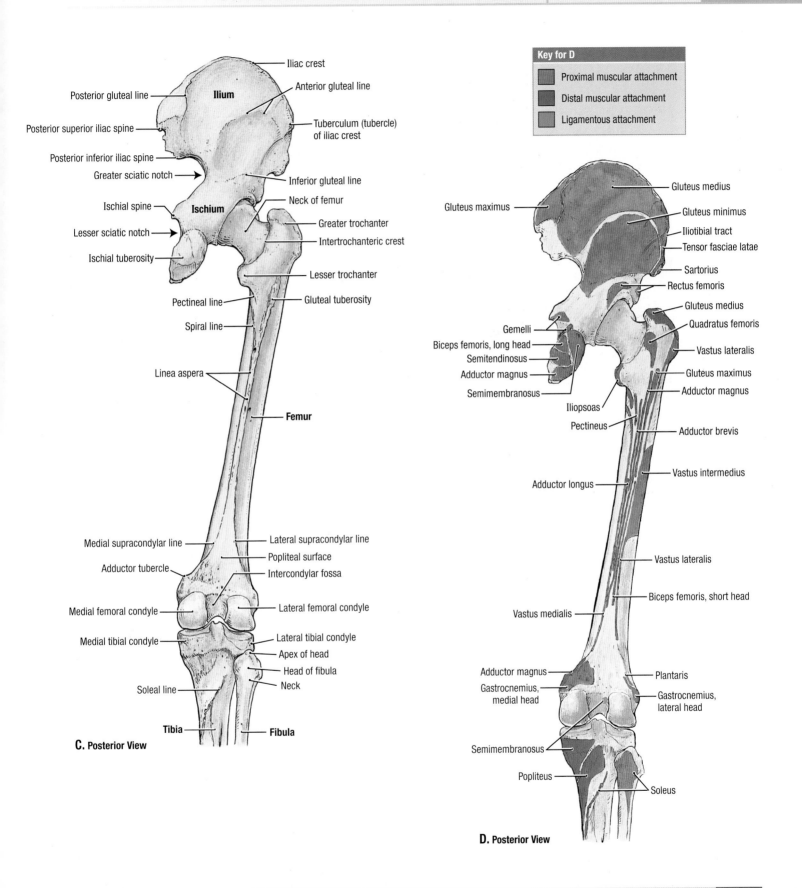

Key for D

Proximal muscular attachment

Distal muscular attachment

Ligamentous attachment

Iliac crest

Anterior gluteal line

Posterior gluteal line

Ilium

Posterior superior iliac spine

Tuberculum (tubercle) of iliac crest

Posterior inferior iliac spine

Greater sciatic notch

Inferior gluteal line

Ischial spine

Neck of femur

Ischium

Lesser sciatic notch

Greater trochanter

Ischial tuberosity

Intertrochanteric crest

Lesser trochanter

Pectineal line

Gluteal tuberosity

Spiral line

Linea aspera

Femur

Medial supracondylar line

Lateral supracondylar line

Adductor tubercle

Popliteal surface

Medial femoral condyle

Intercondylar fossa

Medial tibial condyle

Lateral femoral condyle

Lateral tibial condyle

Apex of head

Head of fibula

Soleal line

Neck

Tibia

Fibula

C. Posterior View

Gluteus maximus

Gluteus medius

Gluteus minimus

Iliotibial tract

Tensor fasciae latae

Sartorius

Rectus femoris

Gluteus medius

Gemelli

Quadratus femoris

Biceps femoris, long head

Vastus lateralis

Semitendinosus

Adductor magnus

Gluteus maximus

Semimembranosus

Adductor magnus

Iliopsoas

Pectineus

Adductor brevis

Vastus intermedius

Adductor longus

Vastus lateralis

Biceps femoris, short head

Vastus medialis

Adductor magnus

Plantaris

Gastrocnemius, medial head

Gastrocnemius, lateral head

Semimembranosus

Popliteus

Soleus

D. Posterior View

BONES OF THE THIGH AND PROXIMAL LEG (continued)

6.27

C. Bony features. **D.** Muscle attachment sites.

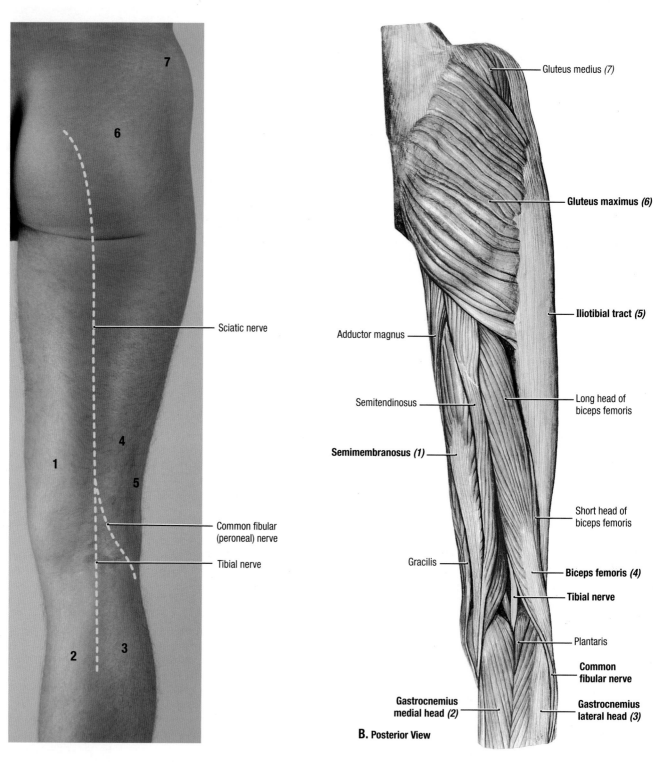

A. Posterior View

- Sciatic nerve
- Common fibular (peroneal) nerve
- Tibial nerve

Gluteus medius *(7)*

Gluteus maximus *(6)*

Iliotibial tract *(5)*

Adductor magnus

Semitendinosus

Long head of biceps femoris

Semimembranosus *(1)*

Short head of biceps femoris

Gracilis

Biceps femoris *(4)*

Tibial nerve

Plantaris

Common fibular nerve

Gastrocnemius medial head *(2)*

Gastrocnemius lateral head *(3)*

B. Posterior View

| 6.28 | MUSCLES OF THE GLUTEAL REGION AND POSTERIOR THIGH I |

A. Surface anatomy. Numbers refer to structures labeled in **(B)**.
B. Superficial dissection. Muscles of gluteal region and posterior thigh (hamstring muscles consist of semimembranosus, semitendinosus, and biceps femoris).

Hamstring strains (pulled and/or torn hamstrings) are common in running, jumping, and quick-start sports. The muscular exertion required to excel in these sports may tear part of the proximal attachments of the hamstrings from the ischial tuberosity.

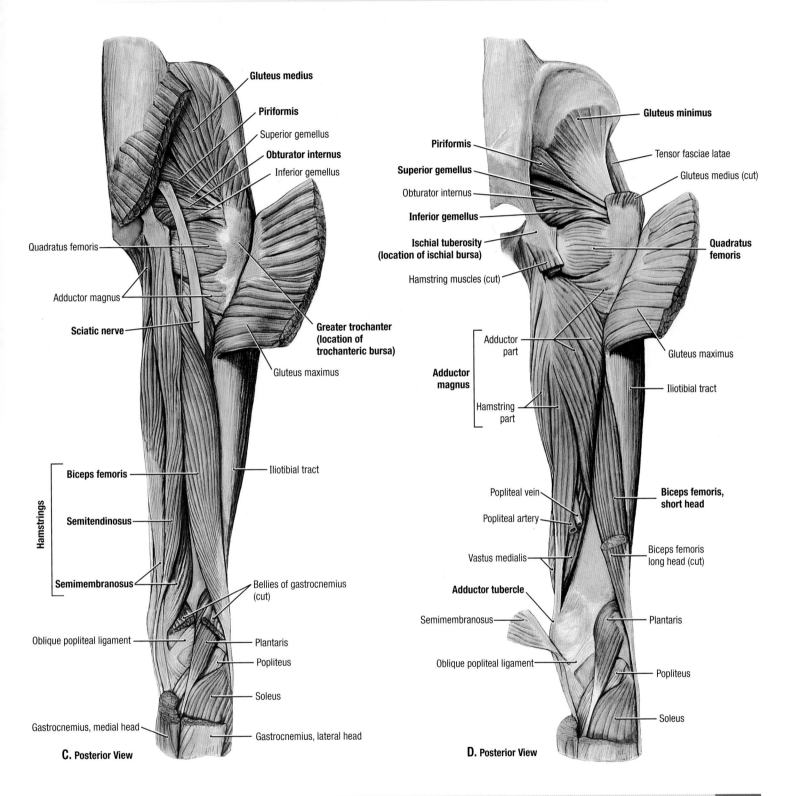

C. Posterior View

D. Posterior View

MUSCLES OF THE GLUTEAL REGION AND POSTERIOR THIGH (*continued*) II AND III

6.28

C. Muscles of gluteal region and posterior thigh with gluteus maximus reflected. **D.** Adductor magnus muscle. The adductor magnus has two parts: one belongs to the adductor group, innervated by the obturator nerve and the other to the hamstring group, innervated by the tibial portion of the sciatic nerve. The trochanteric bursa separates the superior fibers of the gluteus maximus from the greater trochanter of the femur, and the ischial bursa

separates the inferior part of the gluteus maximus from the ischial tuberosity.

Diffuse deep pain in the lateral thigh region (e.g., during stair climbing) may be caused by **trochanteric bursitis**. It is characterized by point tenderness over the greater trochanter, with pain radiating along the iliotibial tract. **Ischial bursitis** results from excessive friction between the ischial bursae and ischial tuberosities (e.g., as from cycling).

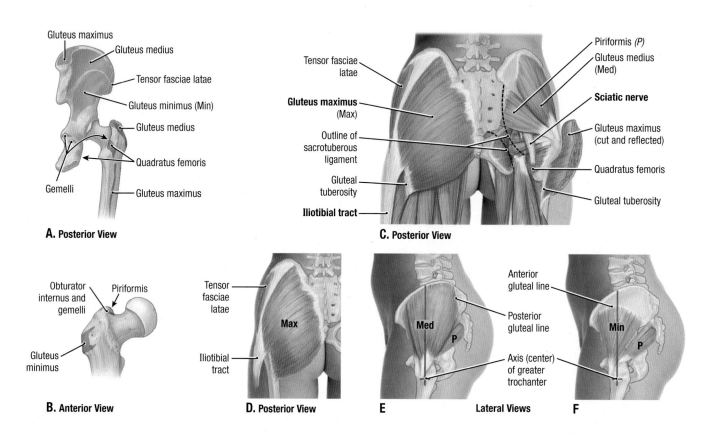

A. Posterior View

B. Anterior View

C. Posterior View

D. Posterior View

E **Lateral Views** **F**

6.29 MUSCLES OF GLUTEAL REGION

A. and **B.** Attachments. **C.** Relationship of gluteal muscles. **D.** Gluteus maximus and tensor fasciae latae. **E.** Gluteus medius. **F.** Gluteus minimus.

TABLE 6.7	**MUSCLES OF GLUTEAL REGION**			
Muscle	*Proximal Attachment[a] (Red)*	*Distal Attachment[a] (Blue)*	*Innervation[b]*	*Main Actions*
Gluteus maximus	Ilium posterior to posterior gluteal line, dorsal surface of sacrum and coccyx, sacro-tuberous ligament	Iliotibial tract that inserts into lateral condyle of tibia; lower, deep fibers to gluteal tuberosity	Inferior gluteal nerve (L5, **S1**, **S2**)	Extends hip joint and assists in lateral rotation; steadies thigh and assists in raising trunk from flexed position
Gluteus medius	External surface of ilium between anterior and posterior gluteal lines; gluteal fascia	Lateral surface of greater trochanter of femur	Superior gluteal nerve (**L5**, S1)	Abducts and medially rotates hip joint[c]; keeps pelvis level when opposite leg is off ground and advances pelvis during swing phase of gait; TFL also contributes to stability of extended knee
Gluteus minimus	External surface of ilium between anterior and inferior gluteal lines	Anterior surface of greater trochanter of femur		
Tensor fasciae latae (TFL)	Anterior superior iliac spine and iliac crest	Iliotibial tract that attaches to lateral condyle (Gerdy tubercle) of tibia		
Piriformis	Anterior surface of sacrum and sacrotuberous ligament	Superior border of greater trochanter of femur	Anterior rami of S1 and S2	Laterally rotate extended hip joint and abduct flexed hip joint; steady femoral head in acetabulum
Obturator internus	Pelvic surface of obturator membrane and surrounding bones	Medial surface (trochanteric fossa) of greater trochanter of femur by common tendons	Nerve to obturator internus (L5, S1)	
Superior gemellus	Ischial spine			
Inferior gemellus	Ischial tuberosity			
Quadratus femoris	Lateral border of ischial tuberosity	Quadrate tubercle on intertrochanteric crest of femur	Nerve to quadratus femoris (L5, S1)	Laterally rotates hip joint,[d] steadies femoral head in acetabulum

[a]See Figure 6.22 for muscle attachments.

[b]Numbers indicate spinal cord segmental innervation of nerves (e.g., L5, S1, and S2 indicate that the inferior gluteal nerve supplying gluteus maximus is derived from three segments of the spinal cord; boldface type [**S1, S2**] indicates main segmental innervation). Damage to one or more of these spinal cord segments or to motor nerve roots arising from these segments results in paralysis of the muscles concerned.

[c]Gluteus medius and minimus: anterior fibers medially rotate hip joint and posterior fibers laterally rotate hip joint.

[d]There are six lateral rotators of the hip joint: piriformis, obturator internus, gemelli (superior and inferior), quadratus femoris, and obturator externus. These muscles also stabilize the hip joint.

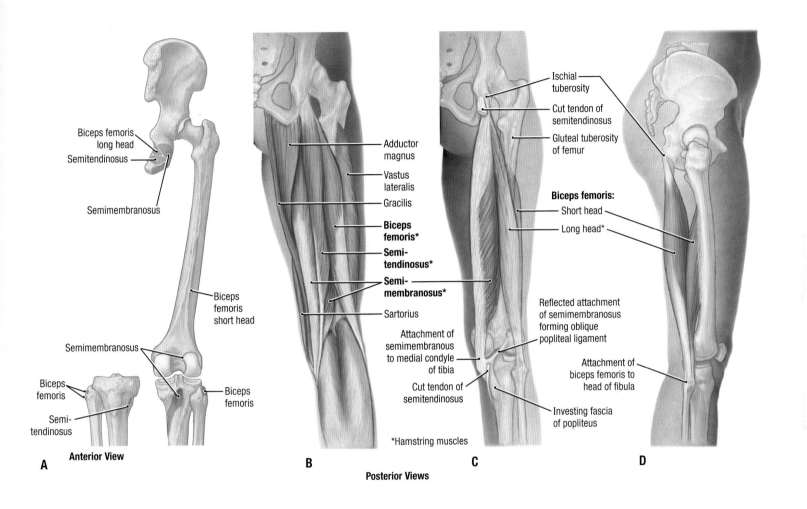

MUSCLES OF POSTERIOR THIGH

6.30

A. Attachments. **B.** Superficial layer. **C.** Intermediate layer. **D.** Deep layer.

TABLE 6.8	**MUSCLES OF POSTERIOR THIGH (HAMSTRING)**			
Muscle[a]	Proximal Attachment[a] (Red)	Distal Attachment[a] (Blue)	Innervation[b]	Main Actions
Semitendinosus	Ischial tuberosity	Medial surface of superior part of tibia	Tibial division of sciatic nerve (L5, S1, and S2)	Extend hip joint; flex knee joint and rotate it medially; when hip and knee joints are flexed, can extend trunk
Semimembranosus		Posterior part of medial condyle of tibia; reflected attachment forms oblique popliteal ligament to lateral femoral condyle		
Biceps femoris	*Long head:* ischial tuberosity *Short head:* linea aspera and lateral supracondylar line of femur	Lateral side of head of fibula; tendon is split at this site by fibular collateral ligament of knee	*Long head:* tibial division of sciatic nerve (L5, S1, and S2) *Short head:* common fibular (peroneal) division of sciatic nerve (L5, S1, and S2)	Flexes knee joint and rotates it laterally; extends hip joint (e.g., when initiating a walking gait)

[a]See Figure 6.22 for muscle attachments.
[b]See Table 6.1 for explanation of segmental innervation.

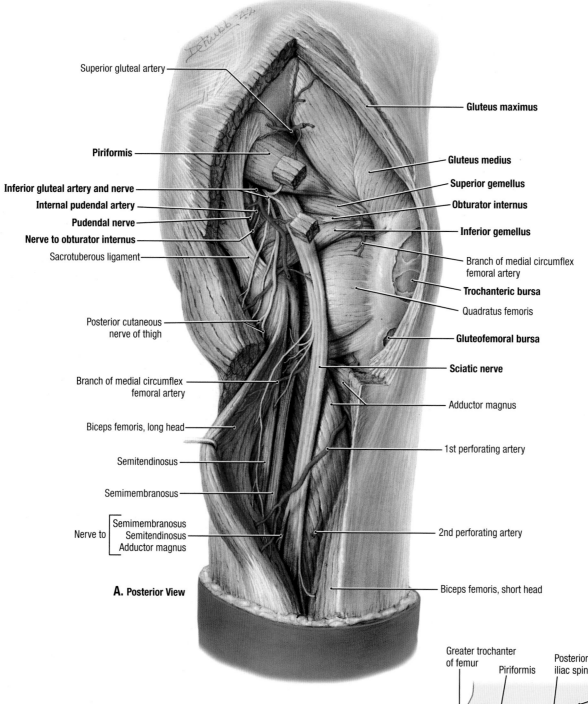

Superior gluteal artery

Piriformis

Inferior gluteal artery and nerve

Internal pudendal artery

Pudendal nerve

Nerve to obturator internus

Sacrotuberous ligament

Posterior cutaneous
nerve of thigh

Branch of medial circumflex
femoral artery

Biceps femoris, long head

Semitendinosus

Semimembranosus

Nerve to { Semimembranosus
Semitendinosus
Adductor magnus

Gluteus maximus

Gluteus medius

Superior gemellus

Obturator internus

Inferior gemellus

Branch of medial circumflex
femoral artery

Trochanteric bursa

Quadratus femoris

Gluteofemoral bursa

Sciatic nerve

Adductor magnus

1st perforating artery

2nd perforating artery

Biceps femoris, short head

A. Posterior View

6.31 MUSCLES OF GLUTEAL REGION AND POSTERIOR THIGH IV

A. Dissection. The gluteus maximus muscle is split superiorly and inferiorly, and the middle part is excised; two cubes remain to identify its nerve. The gluteus maximus is the only muscle to cover the greater trochanter; it is aponeurotic and has underlying bursae where it glides on the trochanter (trochanteric bursa) and the aponeurosis of the vastus lateralis muscle (gluteofemoral bursa). **B.** Intragluteal injection. Injections can be made safely only into the superolateral part of the buttock to avoid injury to the sciatic and gluteal nerves. This site has a rich vascular network from the superior gluteal vessels that lie between the gluteus medius and minimus muscles.

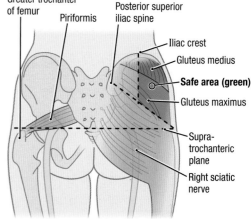

Greater trochanter
of femur

Piriformis

Posterior superior
iliac spine

Iliac crest

Gluteus medius

Safe area (green)

Gluteus maximus

Supra-
trochanteric
plane

Right sciatic
nerve

B. Posterior View, Intragluteal Injection

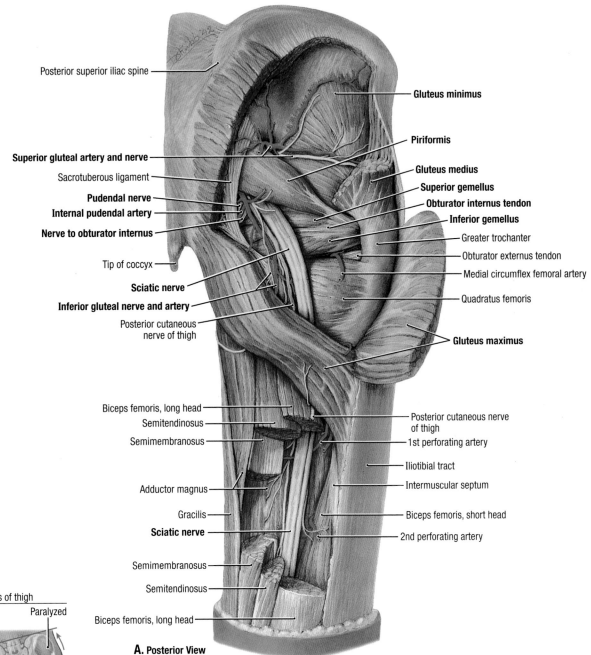

Posterior superior iliac spine

Gluteus minimus

Piriformis

Superior gluteal artery and nerve

Gluteus medius

Sacrotuberous ligament

Superior gemellus

Pudendal nerve

Obturator internus tendon

Internal pudendal artery

Inferior gemellus

Nerve to obturator internus

Greater trochanter

Obturator externus tendon

Tip of coccyx

Medial circumflex femoral artery

Sciatic nerve

Quadratus femoris

Inferior gluteal nerve and artery

Posterior cutaneous
nerve of thigh

Gluteus maximus

Biceps femoris, long head

Posterior cutaneous nerve
of thigh

Semitendinosus

1st perforating artery

Semimembranosus

Iliotibial tract

Intermuscular septum

Adductor magnus

Gracilis

Biceps femoris, short head

Sciatic nerve

2nd perforating artery

Semimembranosus

Semitendinosus

Biceps femoris, long head

A. Posterior View

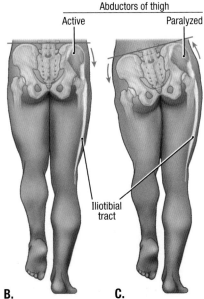

Abductors of thigh

Active

Paralyzed

Iliotibial
tract

B.

C.

Posterior Views

MUSCLES OF GLUTEAL REGION AND POSTERIOR THIGH V | 6.32

A. The proximal three quarters of the gluteus maximus muscle is reflected, and parts of the gluteus medius and the three hamstring muscles are excised. The superior gluteal vessels and nerves emerge superior to the piriformis muscle; all other vessels and nerves emerge inferior to it. **B.** When the weight is borne by one limb, the muscles on the supported side fix the pelvis so that it does not sag to the unsupported side, keeping the pelvis level. **C.** When the right **abductors are paralyzed**, owing to a lesion of the right superior gluteal nerve, fixation by these muscles is lost and the pelvis tilts to the unsupported left side (positive Trendelenburg sign).

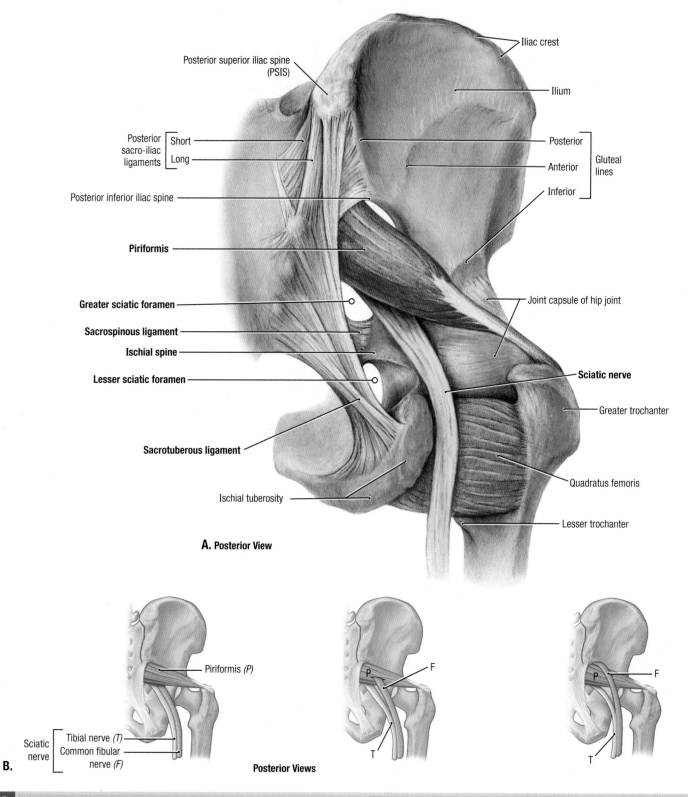

Iliac crest

Posterior superior iliac spine (PSIS)

Ilium

Posterior sacro-iliac ligaments — Short
Long

Posterior
Anterior — Gluteal lines
Inferior

Posterior inferior iliac spine

Piriformis

Greater sciatic foramen

Joint capsule of hip joint

Sacrospinous ligament

Ischial spine

Sciatic nerve

Lesser sciatic foramen

Greater trochanter

Sacrotuberous ligament

Quadratus femoris

Ischial tuberosity

Lesser trochanter

A. Posterior View

Piriformis *(P)*

F

F

P

P

Sciatic nerve — Tibial nerve *(T)*
Common fibular nerve *(F)*

T

T

B.

Posterior Views

| **6.33** | LATERAL ROTATORS OF HIP, SCIATIC NERVE, AND LIGAMENTS OF GLUTEAL REGION |

A. Piriformis and quadratus femoris. **B.** Relationship of sciatic nerve to piriformis muscle. Of 640 limbs studied in Dr. Grant's laboratory, in 87%, the tibial and fibular (peroneal) divisions passed inferior to the piriformis *(left)*; in 12.2%, the fibular (peroneal) division passed through the piriformis *(center)*; and in 0.5%, the fibular (peroneal) division passed superior to the piriformis *(right)*.

Sciatic nerve block. Sensation conveyed by the sciatic nerve can be blocked by injecting an anesthetic agent a few centimeters inferior to the midpoint of the line joining the PSIS and the superior border of the greater trochanter. Paresthesia radiates to the foot because of anesthesia of the plantar nerves, which are terminal branches of the tibial nerve derived from the sciatic nerve.

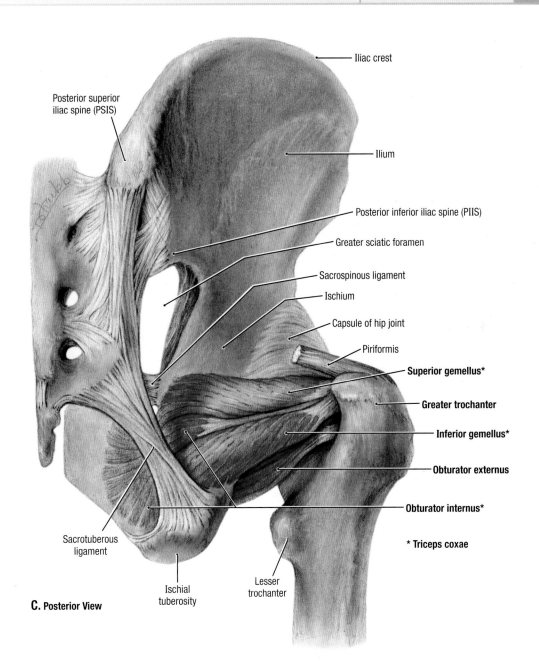

Iliac crest

Posterior superior iliac spine (PSIS)

Ilium

Posterior inferior iliac spine (PIIS)

Greater sciatic foramen

Sacrospinous ligament

Ischium

Capsule of hip joint

Piriformis

Superior gemellus*

Greater trochanter

Inferior gemellus*

Obturator externus

Obturator internus*

* Triceps coxae

Sacrotuberous ligament

Ischial tuberosity

Lesser trochanter

C. Posterior View

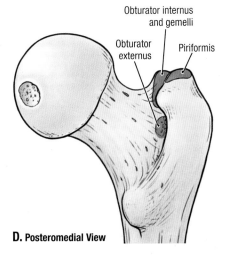

Obturator internus and gemelli

Obturator externus

Piriformis

D. Posteromedial View

LATERAL ROTATORS OF HIP, SCIATIC NERVE, AND LIGAMENTS OF GLUTEAL REGION (continued)

6.33

- **C.** Obturator internus, obturator externus, and superior and inferior gemelli.
- **D.** Muscle attachments of the posterior aspect of the proximal femur.
- The obturator internus is located partly in the pelvis, where it covers most of the lateral wall of the lesser pelvis. It leaves the pelvis through the lesser sciatic foramen, makes a right-angle turn, becomes tendinous, and receives the distal attachments of the gemelli before attaching to the medial surface of the greater trochanter (trochanteric fossa).
- The obturator externus extends from the external surface of the obturator membrane and surrounding bone of the pelvis to the posterior aspect of the greater trochanter, passing directly under the acetabulum and neck of the femur.
- **Common fibular nerve compression at piriformis.** In the approximately 12% of people in whom the common fibular division of the sciatic nerve passes through the piriformis, this muscle may compress the nerve.

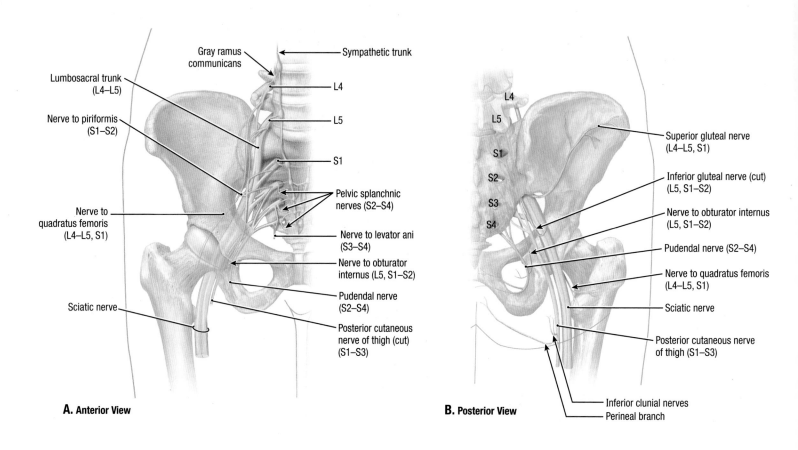

A. Anterior View

B. Posterior View

6.34 NERVES OF GLUTEAL REGION

The muscles of the gluteal region are innervated by the sacral plexus.

TABLE 6.9	NERVES OF GLUTEAL REGION		
Nerve	*Origin*	*Course*	*Distribution in Gluteal Region*
Clunial (superior, middle, and inferior)	*Superior:* posterior rami of L1–L3 nerves *Middle:* posterior rami of S1–S3 nerves *Inferior:* posterior cutaneous nerve of thigh	*Superior nerves* cross iliac crest; *middle nerves* exit through posterior sacral foramina and enter gluteal region; *inferior nerves* curve around inferior border of gluteus maximus	Gluteal region as far laterally as greater trochanter
Sciatic	Sacral plexus (L4–S3)	Exits pelvis via greater sciatic foramen inferior to piriformis to enter gluteal region	No muscles in gluteal region
Posterior cutaneous nerve of thigh	Sacral plexus (S1–S3)	Exits pelvis via greater sciatic foramen inferior to piriformis, emerges from inferior border of gluteus maximus coursing deep to fascia lata	Skin of buttock via inferior cluneal branches, skin over posterior thigh and popliteal fossa; skin of lateral perineum and upper medial thigh via perineal branch
Superior gluteal	Anterior rami of L4–S1 nerves	Exits pelvis via greater sciatic foramen superior to piriformis; courses between gluteus medius and minimus	Gluteus medius, gluteus minimus, and tensor fasciae latae
Inferior gluteal	Anterior rami of L5–S2 nerves	Exits pelvis via greater sciatic foramen inferior to piriformis, dividing into multiple branches	Gluteus maximus
Nerve to quadratus femoris	Anterior rami of L4–S1 nerves	Exits pelvis via greater sciatic foramen deep to sciatic nerve	Posterior hip joint, inferior gemellus, and quadratus femoris
Pudendal	Anterior rami of S2–S4 nerves	Exits pelvis via greater sciatic foramen inferior to piriformis; descends posterior to sacrospinous ligament; enters perineum (pudendal canal) through lesser sciatic foramen	No structures in gluteal region (supplies most of perineum)
Nerve to obturator internus	Anterior rami of L5–S2 nerves	Exits pelvis via greater sciatic foramen inferior to piriformis; descends posterior to ischial spine; enters lesser sciatic foramen and passes to obturator internus	Superior gemellus and obturator internus

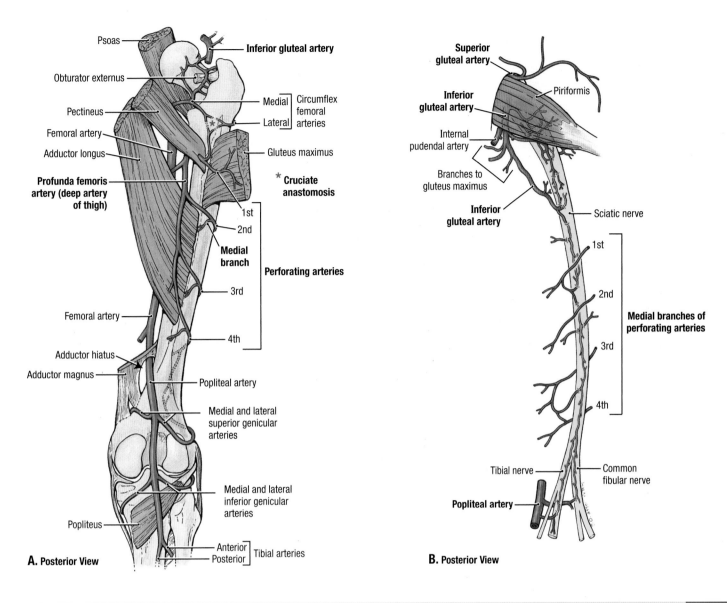

A. Posterior View

Psoas
Obturator externus
Pectineus
Femoral artery
Adductor longus
Profunda femoris artery (deep artery of thigh)
Femoral artery
Adductor hiatus
Adductor magnus
Popliteus

Inferior gluteal artery
Medial / Lateral — Circumflex femoral arteries
Gluteus maximus
*** Cruciate anastomosis**
1st
2nd
Medial branch
Perforating arteries
3rd
4th
Popliteal artery
Medial and lateral superior genicular arteries
Medial and lateral inferior genicular arteries
Anterior / Posterior — Tibial arteries

B. Posterior View

Superior gluteal artery
Inferior gluteal artery
Internal pudendal artery
Branches to gluteus maximus
Inferior gluteal artery
Piriformis
Sciatic nerve
1st
2nd
Medial branches of perforating arteries
3rd
4th
Tibial nerve
Common fibular nerve
Popliteal artery

ARTERIES OF GLUTEAL REGION AND POSTERIOR THIGH

6.35

TABLE 6.10	ARTERIES OF GLUTEAL REGION AND POSTERIOR THIGH		
Artery	*Origin*	*Course*	*Distribution*
Superior gluteal		Enters gluteal region through greater sciatic foramen superior to piriformis; divides into superficial and deep branches; anastomoses with inferior gluteal and medial circumflex femoral arteries	*Superficial branch:* superior gluteus maximus *Deep branch:* runs between gluteus medius and minimus, supplying both and tensor fasciae latae
Inferior gluteal	Internal iliac	Enters gluteal region through greater sciatic foramen inferior to piriformis; descends on medial side of sciatic nerve; anastomoses with superior gluteal artery and participates in cruciate anastomosis of thigh	Inferior gluteus maximus, obturator internus, quadratus femoris, and superior parts of hamstring muscles
Internal pudendal		Enters gluteal region through greater sciatic foramen; descends posterior to ischial spine; exits gluteal region via lesser sciatic foramen to perineum	No structures in gluteal region (supplies external genitalia and muscles in perineal region)
Perforating arteries		Perforate aponeurotic portion of adductor magnus attachment and medial intermuscular septum to enter and supply muscular branches to posterior compartment; then pierce lateral intermuscular septum to enter posterolateral aspect of anterior compartment	Hamstring muscles in posterior compartment; posterior portion of vastus lateralis in anterior compartment; femur (via femoral nutrient arteries); reinforce arterial supply of sciatic nerve
Lateral circumflex femoral	Profunda femoris (may arise from femoral)	Passes laterally deep to sartorius and rectus femoris; enter gluteal region	Anterior part of gluteal region
Medial circumflex femoral		Passes medially and posteriorly between pectineus and iliopsoas; enters gluteal region	Supplies most blood to head and neck of femur; hip region

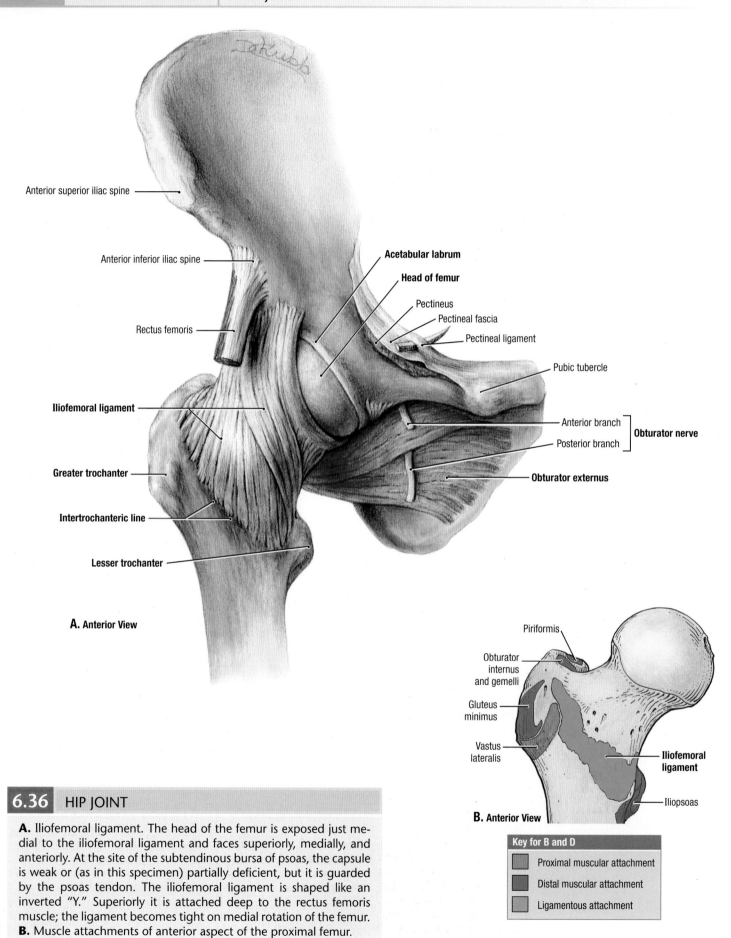

Anterior superior iliac spine

Anterior inferior iliac spine

Rectus femoris

Iliofemoral ligament

Greater trochanter

Intertrochanteric line

Lesser trochanter

Acetabular labrum

Head of femur

Pectineus

Pectineal fascia

Pectineal ligament

Pubic tubercle

Anterior branch

Posterior branch

Obturator nerve

Obturator externus

A. Anterior View

Piriformis

Obturator internus and gemelli

Gluteus minimus

Vastus lateralis

Iliofemoral ligament

Iliopsoas

B. Anterior View

| 6.36 | HIP JOINT |

A. Iliofemoral ligament. The head of the femur is exposed just medial to the iliofemoral ligament and faces superiorly, medially, and anteriorly. At the site of the subtendinous bursa of psoas, the capsule is weak or (as in this specimen) partially deficient, but it is guarded by the psoas tendon. The iliofemoral ligament is shaped like an inverted "Y." Superiorly it is attached deep to the rectus femoris muscle; the ligament becomes tight on medial rotation of the femur. **B.** Muscle attachments of anterior aspect of the proximal femur.

Key for B and D	
	Proximal muscular attachment
	Distal muscular attachment
	Ligamentous attachment

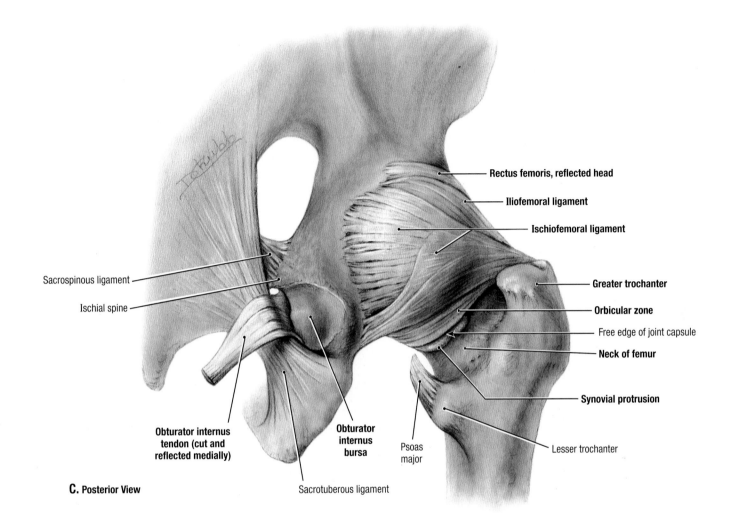

Rectus femoris, reflected head

Iliofemoral ligament

Ischiofemoral ligament

Greater trochanter

Orbicular zone

Free edge of joint capsule

Neck of femur

Synovial protrusion

Lesser trochanter

Psoas major

Sacrotuberous ligament

Obturator internus bursa

Obturator internus tendon (cut and reflected medially)

Ischial spine

Sacrospinous ligament

C. Posterior View

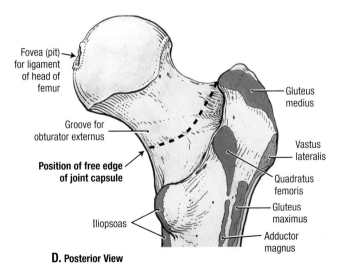

Fovea (pit) for ligament of head of femur

Groove for obturator externus

Position of free edge of joint capsule

Iliopsoas

Gluteus medius

Vastus lateralis

Quadratus femoris

Gluteus maximus

Adductor magnus

D. Posterior View

HIP JOINT (*continued*) **6.36**

C. Ischiofemoral ligament. The fibers of the capsule spiral to become taut during extension and medial rotation of the femur. The synovial membrane protrudes inferior to the fibrous capsule and forms a bursa for the tendon of the obturator externus muscle. Note the large subtendinous bursa of the obturator internus at the lesser sciatic notch, where the tendon turns 90 degrees to attach to the greater trochanter. **D.** Muscle attachments onto the posterior aspect of proximal femur.

Osteoarthritis of the hip joint, characterized by pain, edema, limitation of motion, and erosion of articular cartilage, is a common cause of disability. During **hip replacement**, a metal prosthesis anchored to the person's femur by bone cement replaces the femoral head and neck. A plastic socket cemented to the hip bone replaces the acetabulum.

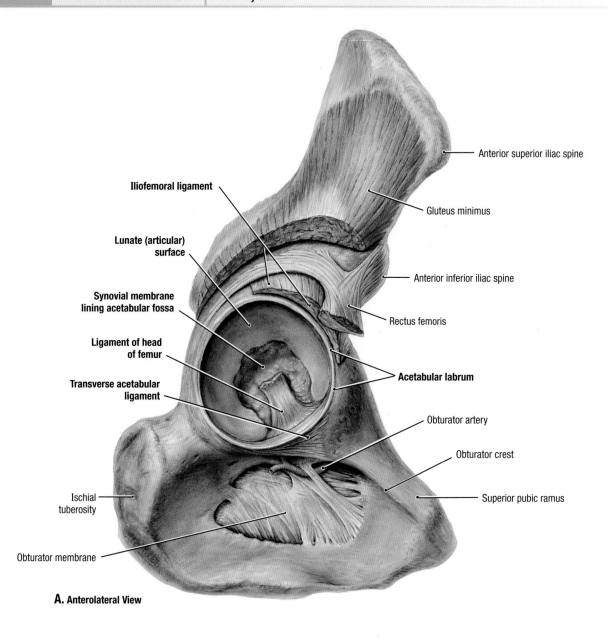

Iliofemoral ligament

Lunate (articular) surface

Synovial membrane lining acetabular fossa

Ligament of head of femur

Transverse acetabular ligament

Ischial tuberosity

Obturator membrane

Anterior superior iliac spine

Gluteus minimus

Anterior inferior iliac spine

Rectus femoris

Acetabular labrum

Obturator artery

Obturator crest

Superior pubic ramus

A. Anterolateral View

6.37   ACETABULAR REGION

A. Dissection of acetabulum. **B.** Muscle attachments of acetabular region.

In **A**:

- The transverse acetabular ligament bridges the acetabular notch.
- The acetabular labrum is attached to the acetabular rim and transverse acetabular ligament and forms a complete ring around the head of the femur.
- The ligament of the head of the femur lies between the head of the femur and the acetabulum. These fibers are attached superiorly to the pit (fovea) on the head of the femur and inferiorly to the transverse acetabular ligament and the margins of the acetabular notch. The artery of the ligament of the head of the femur passes through the acetabular notch and into the ligament of the head of the femur.

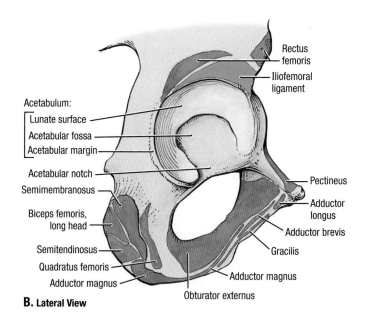

Rectus femoris

Iliofemoral ligament

Acetabulum:
  Lunate surface
  Acetabular fossa
  Acetabular margin

Acetabular notch

Semimembranosus

Biceps femoris, long head

Semitendinosus

Quadratus femoris

Adductor magnus

Pectineus

Adductor longus

Adductor brevis

Gracilis

Adductor magnus

Obturator externus

B. Lateral View

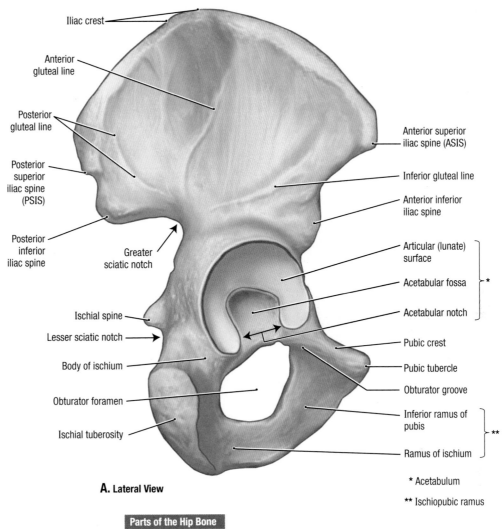

Iliac crest
Anterior gluteal line
Posterior gluteal line
Posterior superior iliac spine (PSIS)
Posterior inferior iliac spine
Greater sciatic notch
Ischial spine
Lesser sciatic notch
Body of ischium
Obturator foramen
Ischial tuberosity

Anterior superior iliac spine (ASIS)
Inferior gluteal line
Anterior inferior iliac spine
Articular (lunate) surface
Acetabular fossa   *
Acetabular notch
Pubic crest
Pubic tubercle
Obturator groove
Inferior ramus of pubis   **
Ramus of ischium

* Acetabulum
** Ischiopubic ramus

A. Lateral View

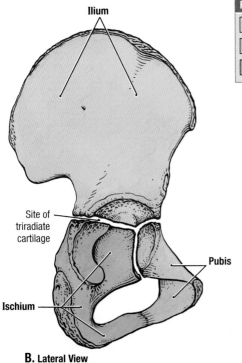

Ilium
Site of triradiate cartilage
Pubis
Ischium

B. Lateral View

Parts of the Hip Bone
- Ilium
- Pubis
- Ischium

HIP BONE   6.38

A. Features of the lateral aspect. In the anatomical position, the anterior superior iliac spine and pubic tubercle are in the same coronal plane, and the ischial spine and superior end of the pubic symphysis are in the same horizontal plane; the internal aspect of the body of the pubis faces superiorly, and the acetabulum faces inferolaterally. **B.** Hip bone in youth. The three parts of the hip bone (ilium, ischium, and pubis) meet in the acetabulum at the triradiate synchondrosis. One or more primary centers of ossification appear in the triradiate cartilage at approximately the 12th year. Secondary centers of ossification appear along the length of the iliac crest, at the anterior inferior iliac spine, the ischial tuberosity, and the pubic symphysis at about puberty; fusion is usually complete by age 23.

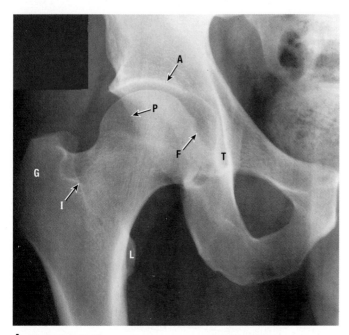

A. Anteroposterior View

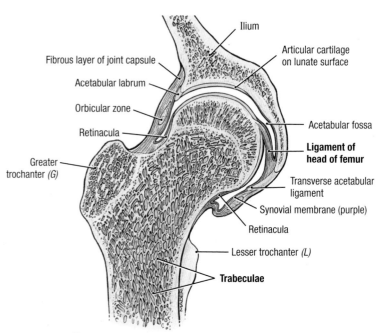

B. Coronal Section

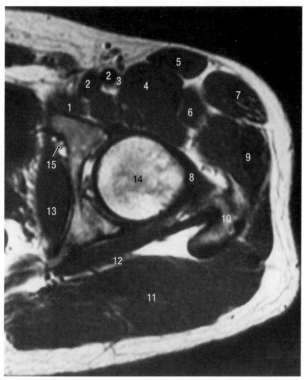

C. Transverse MRI

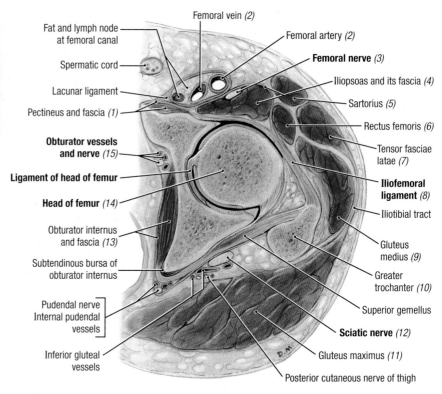

D. Transverse Section, Inferior View

6.39 RADIOGRAPH AND TRANSVERSE MRI OF HIP JOINT

A. Radiograph. On the femur, note the greater (G) and lesser (L) trochanters, the intertrochanteric crest (I), and the pit or fovea (F) for the ligament of the head. On the pelvis, note the roof (A) and posterior rim (P) of the acetabulum and the "teardrop" appearance (T) caused by the superimposition of structures at the inferior margin of the acetabulum. **B.** Coronal section. **C.** MRI. Numbers refer to structures labeled in **(D). D.** Transverse section.

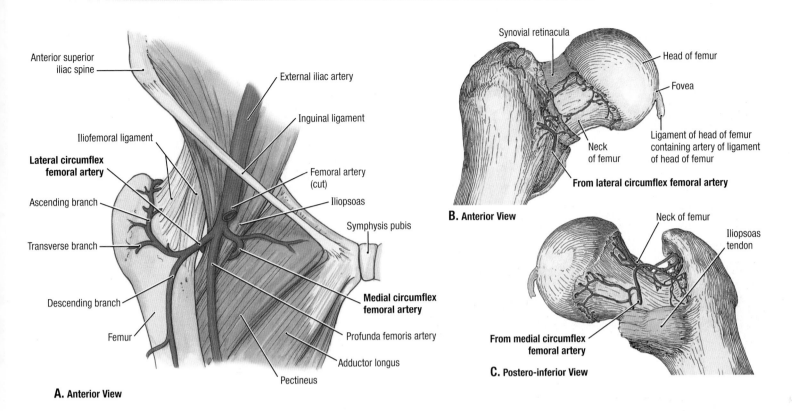

Anterior superior iliac spine

External iliac artery

Iliofemoral ligament

Inguinal ligament

Lateral circumflex femoral artery

Femoral artery (cut)

Ascending branch

Iliopsoas

Symphysis pubis

Transverse branch

Medial circumflex femoral artery

Descending branch

Profunda femoris artery

Femur

Adductor longus

Pectineus

A. Anterior View

Synovial retinacula

Head of femur

Fovea

Ligament of head of femur containing artery of ligament of head of femur

Neck of femur

From lateral circumflex femoral artery

B. Anterior View

Neck of femur

Iliopsoas tendon

From medial circumflex femoral artery

C. Postero-inferior View

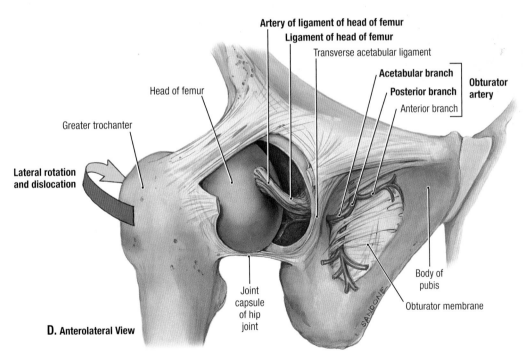

Artery of ligament of head of femur

Ligament of head of femur

Transverse acetabular ligament

Head of femur

Acetabular branch

Obturator artery

Posterior branch

Anterior branch

Greater trochanter

Lateral rotation and dislocation

Body of pubis

Joint capsule of hip joint

Obturator membrane

D. Anterolateral View

BLOOD SUPPLY TO HEAD OF FEMUR

6.40

A. Medial and lateral circumflex femoral arteries in femoral triangle. **B.** Branches of lateral circumflex femoral artery. **C.** Branches of medial circumflex femoral artery. **D.** Obturator artery. The artery of the ligament of the head of the femur is a branch of the acetabular artery and can be seen traveling in the ligament to the head of the femur.

Fractures of the femoral neck often disrupt the blood supply to the head of the femur. The medial circumflex femoral artery supplies most of the blood to the head and neck of the femur and is often torn when the femoral neck is fractured. In some cases, the blood supplied by the artery of the ligament of the head may be the only blood received by the proximal fragment of the femoral head, which may be inadequate. If the blood vessels are ruptured, the fragment of bone may receive no blood and undergo aseptic avascular necrosis.

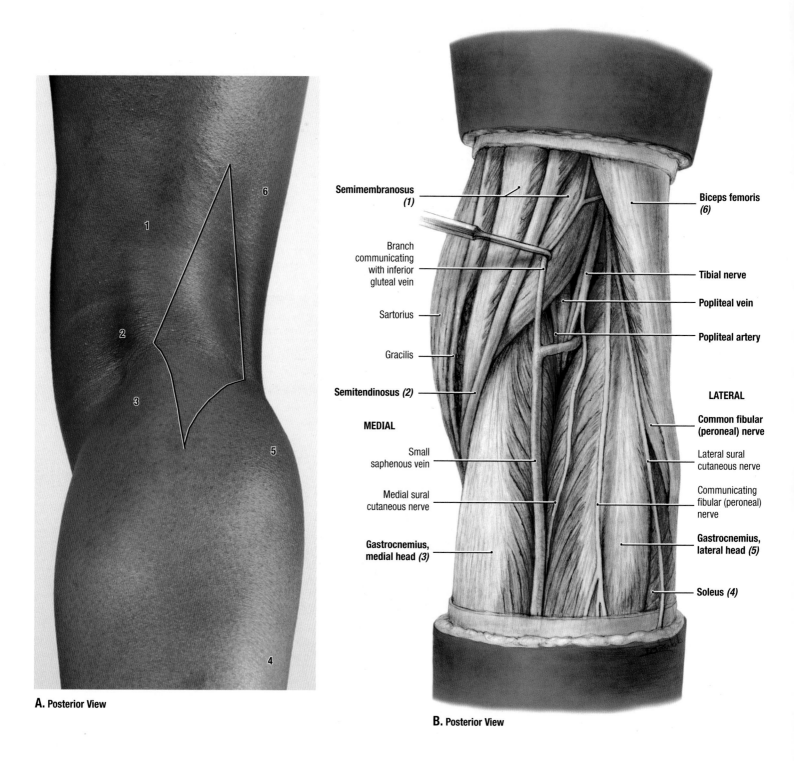

A. Posterior View

Semimembranosus (1)

Branch communicating with inferior gluteal vein

Sartorius

Gracilis

Semitendinosus (2)

MEDIAL

Small saphenous vein

Medial sural cutaneous nerve

Gastrocnemius, medial head (3)

Biceps femoris (6)

Tibial nerve

Popliteal vein

Popliteal artery

LATERAL

Common fibular (peroneal) nerve

Lateral sural cutaneous nerve

Communicating fibular (peroneal) nerve

Gastrocnemius, lateral head (5)

Soleus (4)

B. Posterior View

6.41 POPLITEAL FOSSA

A. Surface anatomy. Numbers refer to structures labeled in **(B)**.
B. Superficial dissection.

Because the popliteal artery is deep in the popliteal fossa, it may be difficult to feel the **popliteal pulse**. Palpation of this pulse is commonly performed by placing the person in the prone position with the knee flexed to relax the popliteal fascia and hamstrings. The pulsations are best felt in the inferior part of the fossa. Weakening or loss of the popliteal pulse is a sign of femoral artery obstruction.

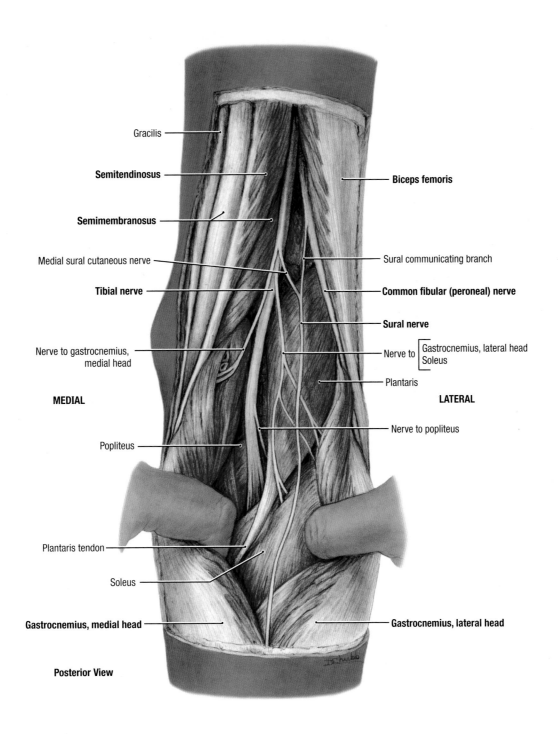

Gracilis

Semitendinosus

Semimembranosus

Medial sural cutaneous nerve

Tibial nerve

Nerve to gastrocnemius, medial head

MEDIAL

Popliteus

Plantaris tendon

Soleus

Gastrocnemius, medial head

Posterior View

Biceps femoris

Sural communicating branch

Common fibular (peroneal) nerve

Sural nerve

Nerve to ⌈ Gastrocnemius, lateral head
         ⌊ Soleus

Plantaris

LATERAL

Nerve to popliteus

Gastrocnemius, lateral head

NERVES OF POPLITEAL FOSSA

6.42

The two heads of the gastrocnemius muscle are separated. A cutaneous branch of the tibial nerve joins a communicating branch of the common fibular (peroneal) nerve to form the sural nerve. In this specimen, the junction is high; usually it is 5 to 8 cm proximal to the ankle.

All motor branches in this region emerge from the tibial nerve, one branch from its medial side and the others from its lateral side; hence, it is safer to dissect on the medial side.

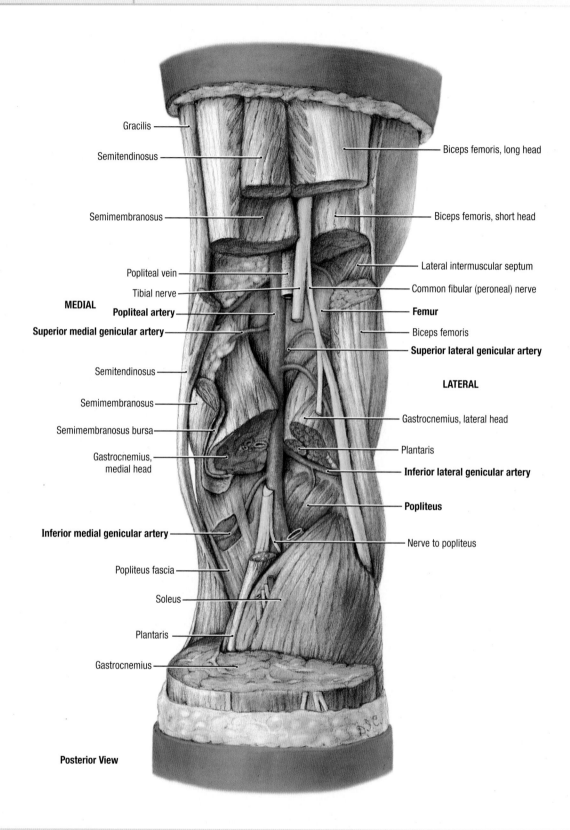

Gracilis

Semitendinosus

Semimembranosus

Popliteal vein

Tibial nerve

MEDIAL

Popliteal artery

Superior medial genicular artery

Semitendinosus

Semimembranosus

Semimembranosus bursa

Gastrocnemius, medial head

Inferior medial genicular artery

Popliteus fascia

Soleus

Plantaris

Gastrocnemius

Biceps femoris, long head

Biceps femoris, short head

Lateral intermuscular septum

Common fibular (peroneal) nerve

Femur

Biceps femoris

Superior lateral genicular artery

LATERAL

Gastrocnemius, lateral head

Plantaris

Inferior lateral genicular artery

Popliteus

Nerve to popliteus

Posterior View

6.43 DEEP DISSECTION OF POPLITEAL FOSSA

The common fibular (peroneal) nerve follows the posterior border of the biceps femoris muscle, formed centrally by the fibrous capsule of the knee joint. The popliteal artery lies on the floor of the popliteal fossa. The floor is formed by the femur, capsule of the knee joint, and popliteus muscle and fascia. The popliteal artery gives off genicular branches that also lie on the floor of the fossa. A **popliteal aneurysm** (abnormal dilation of all or part of the popliteal artery) usually causes edema (swelling) and pain in the popliteal fossa. If the femoral artery has to be ligated, blood can bypass the occlusion through the genicular anastomosis and reach the popliteal artery distal to the ligation.

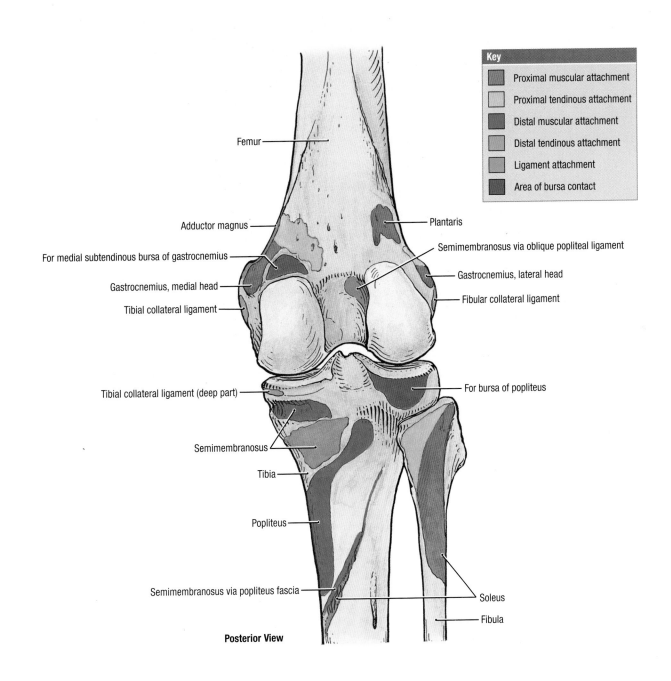

Key
- Proximal muscular attachment
- Proximal tendinous attachment
- Distal muscular attachment
- Distal tendinous attachment
- Ligament attachment
- Area of bursa contact

Femur

Adductor magnus

For medial subtendinous bursa of gastrocnemius

Gastrocnemius, medial head

Tibial collateral ligament

Plantaris

Semimembranosus via oblique popliteal ligament

Gastrocnemius, lateral head

Fibular collateral ligament

Tibial collateral ligament (deep part)

For bursa of popliteus

Semimembranosus

Tibia

Popliteus

Semimembranosus via popliteus fascia

Soleus

Fibula

Posterior View

ATTACHMENT OF MUSCLES OF POPLITEAL REGION

6.44

Lighter tones are secondary attachments.

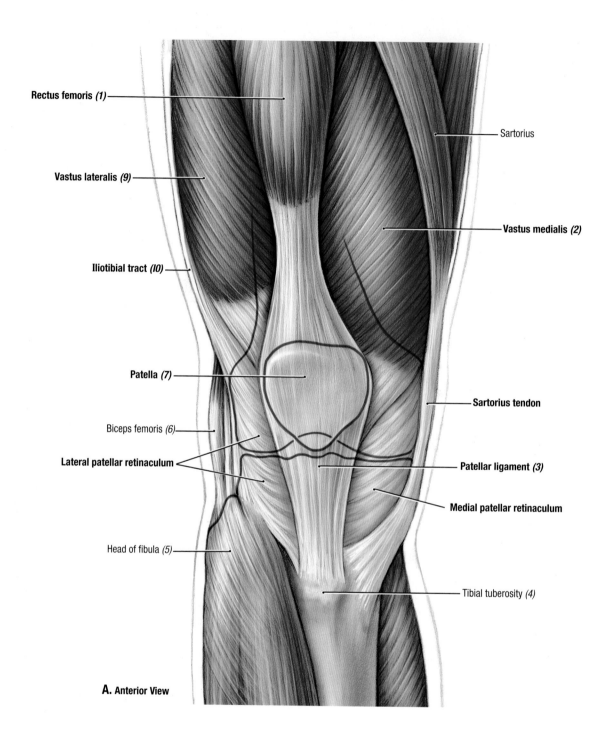

Rectus femoris *(1)*

Vastus lateralis *(9)*

Iliotibial tract *(10)*

Patella *(7)*

Biceps femoris *(6)*

Lateral patellar retinaculum

Head of fibula *(5)*

Sartorius

Vastus medialis *(2)*

Sartorius tendon

Patellar ligament *(3)*

Medial patellar retinaculum

Tibial tuberosity *(4)*

A. Anterior View

6.45 ANTERIOR ASPECT OF KNEE

A. Distal thigh and knee regions. Note that the tendons of the four parts of the quadriceps unite to form the quadriceps tendon, a broad band that attaches to the patella. The patellar ligament, a continuation of the quadriceps tendon, attaches the patella to the tibial tuberosity. The lateral and medial patellar retinacula, formed largely by continuation of the iliotibial tract and investing fascia of the vasti muscles, maintains alignment of the patella and patellar ligament. The retinacula also form the anterolateral and anteromedial portions of the fibrous layer of the joint capsule of the knee.

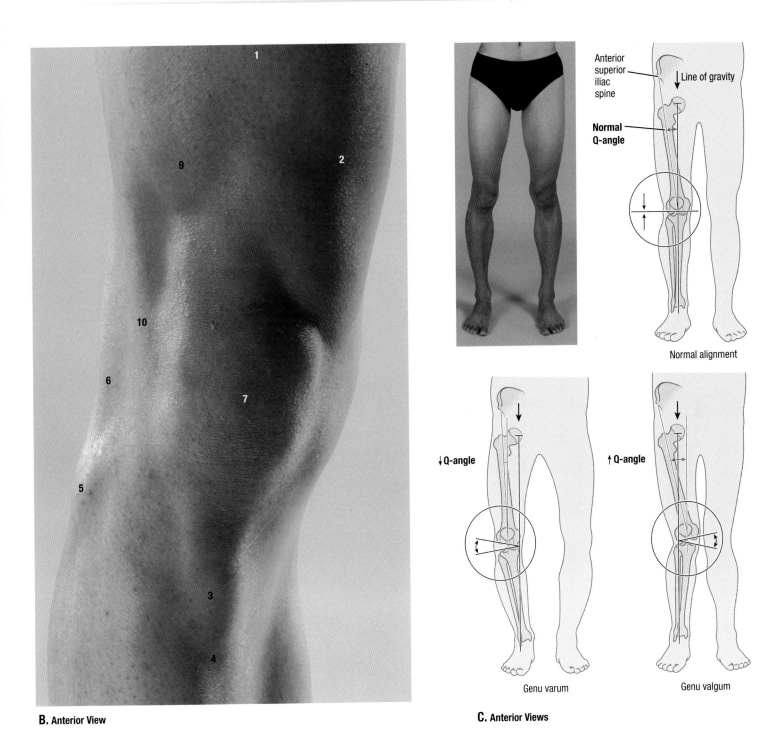

B. Anterior View

C. Anterior Views

Normal alignment

Genu varum

Genu valgum

ANTERIOR ASPECT OF KNEE (*continued*)

B. Surface anatomy. Numbers refer to structures labeled in (**A**). The femur is placed diagonally within the thigh, whereas the tibia is almost vertical within the leg, creating an angle at the knee between the long axes of the bones. The angle between the two bones, referred to clinically as the **Q-angle**, is assessed by drawing a line from the anterior superior iliac spine to the middle of the patella and extrapolating a second (vertical) line passing through the middle of the patella and tibial tuberosity. The Q-angle is typically greater in adult females, owing to their wider

pelves. **C.** Genu valgum and genu varum. A medial angulation of the leg in relation to the thigh, in which the femur is abnormally vertical and the Q-angle is small, is a deformity called **genu varum** (bowleg) that causes unequal weight bearing resulting in arthrosis (destruction of knee cartilages), and an overstressed fibular collateral ligament. A lateral angulation of the leg (large Q-angle, >17 degrees) in relation to the thigh is called **genu valgum** (knock-knee). This results in excess stress and degeneration of the lateral structures of the knee joint.

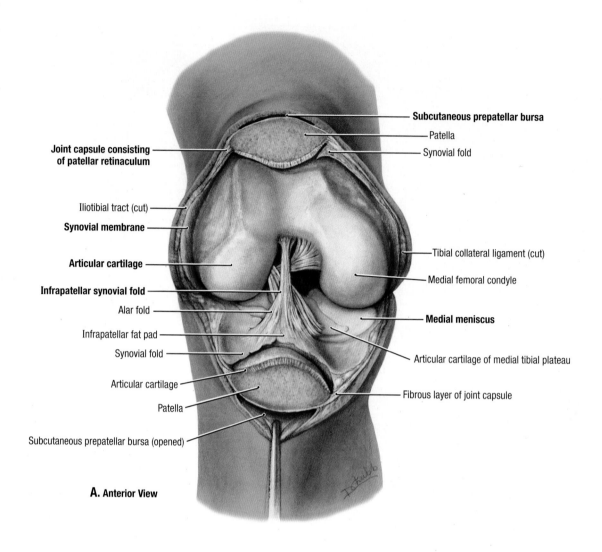

Subcutaneous prepatellar bursa
Patella
Synovial fold

Joint capsule consisting of patellar retinaculum

Iliotibial tract (cut)
Synovial membrane

Articular cartilage

Infrapatellar synovial fold
Alar fold
Infrapatellar fat pad
Synovial fold
Articular cartilage
Patella
Subcutaneous prepatellar bursa (opened)

Tibial collateral ligament (cut)
Medial femoral condyle

Medial meniscus

Articular cartilage of medial tibial plateau
Fibrous layer of joint capsule

A. Anterior View

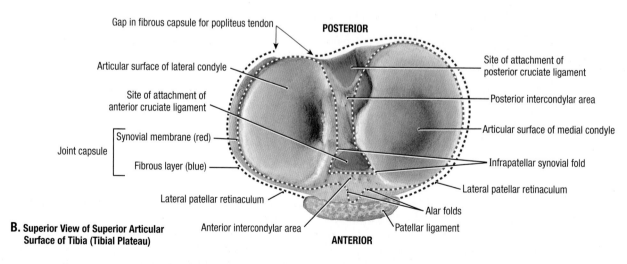

Gap in fibrous capsule for popliteus tendon
POSTERIOR

Articular surface of lateral condyle
Site of attachment of anterior cruciate ligament

Joint capsule — Synovial membrane (red)
Fibrous layer (blue)

Lateral patellar retinaculum
Anterior intercondylar area
ANTERIOR

Site of attachment of posterior cruciate ligament
Posterior intercondylar area
Articular surface of medial condyle
Infrapatellar synovial fold
Lateral patellar retinaculum
Alar folds
Patellar ligament

B. Superior View of Superior Articular Surface of Tibia (Tibial Plateau)

6.46 FIBROUS LAYER AND SYNOVIAL MEMBRANE OF JOINT CAPSULE

A. Dissection. **B.** Attachment of the layers of the joint capsule to the tibia. The fibrous layer (blue dotted line) and synovial membrane (red dotted line) are adjacent on each side, but they part company centrally to accommodate intercondylar and infrapatellar structures that are intracapsular (inside the fibrous layer) but extra-articular (excluded from the articular cavity by synovial membrane).

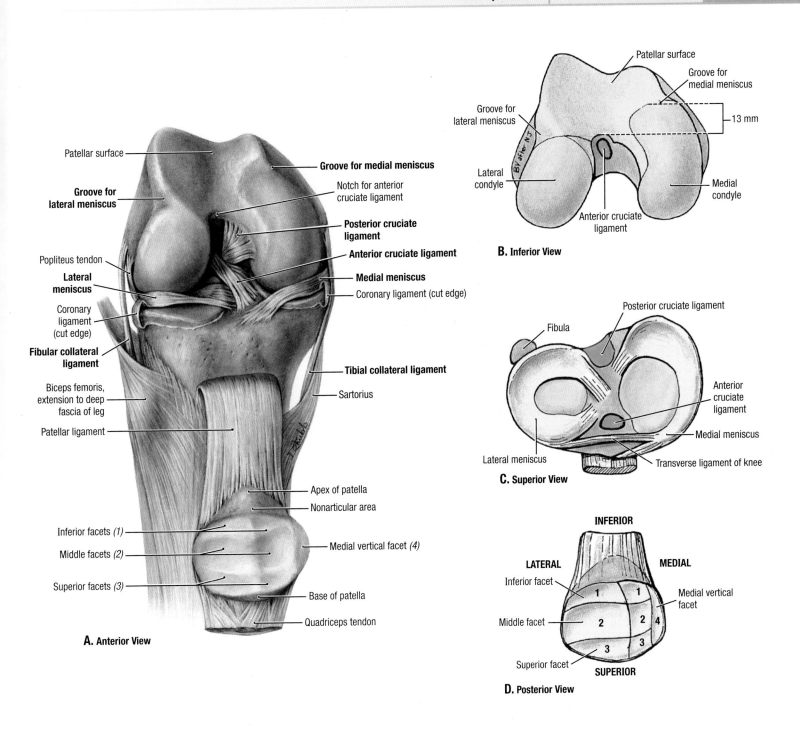

Patellar surface

Groove for medial meniscus

Groove for lateral meniscus

Lateral condyle

Anterior cruciate ligament

13 mm

Medial condyle

B. Inferior View

Patellar surface

Groove for medial meniscus

Notch for anterior cruciate ligament

Groove for lateral meniscus

Posterior cruciate ligament

Popliteus tendon

Anterior cruciate ligament

Lateral meniscus

Medial meniscus

Coronary ligament (cut edge)

Coronary ligament (cut edge)

Fibular collateral ligament

Tibial collateral ligament

Biceps femoris, extension to deep fascia of leg

Sartorius

Patellar ligament

Apex of patella

Nonarticular area

Inferior facets *(1)*

Middle facets *(2)*

Medial vertical facet *(4)*

Superior facets *(3)*

Base of patella

Quadriceps tendon

A. Anterior View

Posterior cruciate ligament

Fibula

Anterior cruciate ligament

Medial meniscus

Lateral meniscus

Transverse ligament of knee

C. Superior View

INFERIOR

LATERAL

MEDIAL

Inferior facet

Medial vertical facet

Middle facet

Superior facet

SUPERIOR

D. Posterior View

ARTICULAR SURFACES AND LIGAMENTS OF KNEE JOINT

6.47

A. Flexed knee joint with patella reflected. There are indentations on the sides of the femoral condyles at the junction of the patellar and tibial articular areas. The lateral tibial articular area is shorter than the medial one. The notch at the anterolateral part of the intercondylar notch is for the anterior cruciate ligament on full extension. **B.** Distal femur. **C.** Tibial plateaus. **D.** Articular surfaces of patella. The three paired facets (superior, middle, and inferior) on the posterior surface of the patella articulate with the patellar surface of the femur successively during (*1*) extension, (*2*) slight flexion, (*3*) flexion, and the most medial vertical facet on the patella (*4*) articulates during full flexion with the

crescentic facet on the medial margin of the intercondylar notch of the femur.

When **patellar dislocation** occurs, it nearly always dislocates laterally. The tendency toward lateral dislocation is normally counterbalanced by the medial, more horizontal pull of the powerful vastus medialis. In addition, the more anterior projection of the lateral femoral condyle and deeper slope for the large lateral patellar facet provides a mechanical deterrent to lateral dislocation. An imbalance of the lateral pull and the mechanisms resisting it result in abnormal tracking of the patella within the patellar groove and chronic patellar pain, even if actual dislocation does not occur.

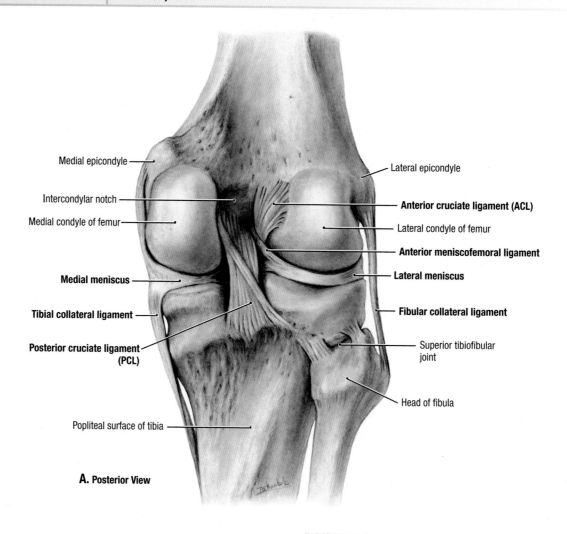

Medial epicondyle

Intercondylar notch

Medial condyle of femur

Medial meniscus

Tibial collateral ligament

Posterior cruciate ligament (PCL)

Popliteal surface of tibia

Lateral epicondyle

Anterior cruciate ligament (ACL)

Lateral condyle of femur

Anterior meniscofemoral ligament

Lateral meniscus

Fibular collateral ligament

Superior tibiofibular joint

Head of fibula

A. Posterior View

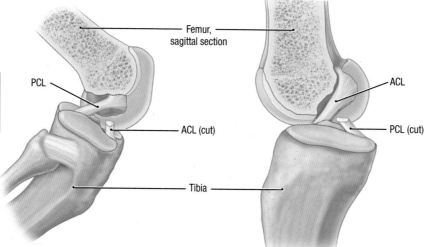

Femur, sagittal section

PCL

ACL (cut)

Tibia

ACL

PCL (cut)

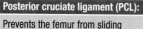

Posterior cruciate ligament (PCL):

Prevents the femur from sliding anteriorly on the tibia, particularly when the knee is flexed

Anterior cruciate ligament (ACL):

Prevents the femur from sliding posteriorly on the tibia, preventing hyperextension of the knee, and limits medial rotation of the femur when the foot is planted (leg is fixed)

B. Lateral View

C. Medial View

6.48 LIGAMENTS OF KNEE JOINT

A. Posterior aspect of joint. **B.** Anterior cruciate ligament (ACL). **C.** Posterior cruciate ligament (PCL). In each illustration, half the femur is sagittally sectioned and removed with the proximal part of the corresponding cruciate ligament. **Injury to the knee joint** is frequently caused by a blow to the lateral side of the extended knee or excessive lateral twisting of the flexed knee, which disrupts the tibial collateral ligament and concomitantly tears and/or detaches the medial meniscus from the joint capsule. This injury is common in athletes who twist their flexed knees while running (e.g., in football and soccer). The ACL, which serves as a pivot for rotary movements of the knee, is taut during flexion and may also tear subsequent to the rupture of the tibial collateral ligament.

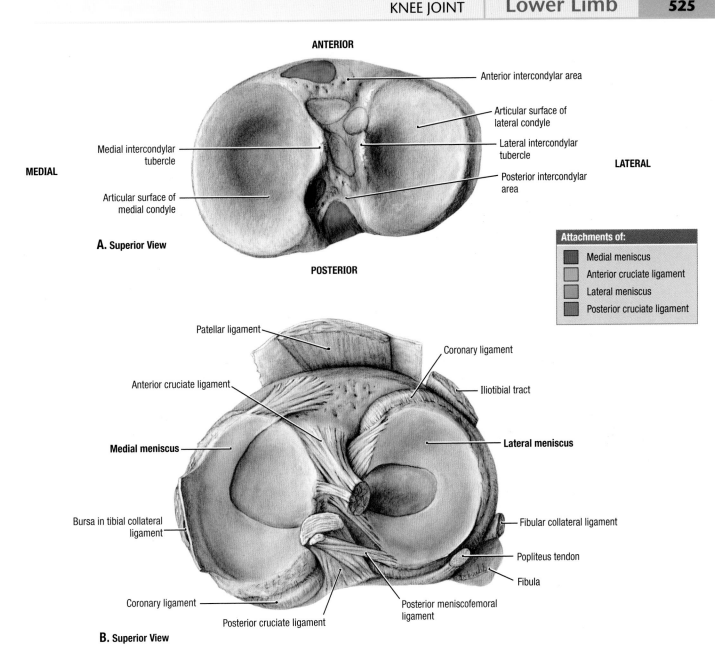

ANTERIOR

Anterior intercondylar area

Articular surface of
lateral condyle

Medial intercondylar
tubercle

MEDIAL

Lateral intercondylar
tubercle

LATERAL

Posterior intercondylar
area

Articular surface of
medial condyle

A. Superior View

POSTERIOR

Attachments of:
■ Medial meniscus
■ Anterior cruciate ligament
■ Lateral meniscus
■ Posterior cruciate ligament

Patellar ligament

Coronary ligament

Anterior cruciate ligament

Iliotibial tract

Medial meniscus

Lateral meniscus

Bursa in tibial collateral
ligament

Fibular collateral ligament

Popliteus tendon

Fibula

Coronary ligament

Posterior meniscofemoral
ligament

Posterior cruciate ligament

B. Superior View

CRUCIATE LIGAMENTS AND MENISCI   **6.49**

A. Attachments sites on tibia. **B.** Menisci *in situ.*
- The lateral tibial condyle is flatter, shorter from anterior to posterior, and more circular. The medial condyle is concave, longer from anterior to posterior, and more oval.
- The menisci conform to the shapes of the surfaces on which they rest. Because the horns of the lateral meniscus are attached close together and its coronary ligament is slack, this meniscus can slide anteriorly and posteriorly on the (flat) condyle; because the horns of the medial meniscus are attached further apart, its movements on the (concave) condyle are restricted.

C. Arthroscopy of knee joint.

  Arthroscopy is an endoscopic examination that allows visualization of the interior of the knee joint cavity with minimal disruption of tissue. The arthroscope and one (or more) additional cannula(e) are inserted through tiny incisions, known as portals. The second cannula is for passage of specialized tools. This technique allows removal of torn menisci, loose bodies in the joint such as bone chips, and debridement (the excision of devitalized articular cartilaginous material). Ligament repair or replacement may also be performed using an arthroscope.

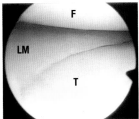

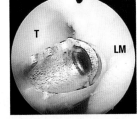

Normal lateral meniscus

Trimming torn lateral meniscus

C. Femoral condyle *(F)*, Tibial plateau *(T)*, Lateral meniscus *(LM)*

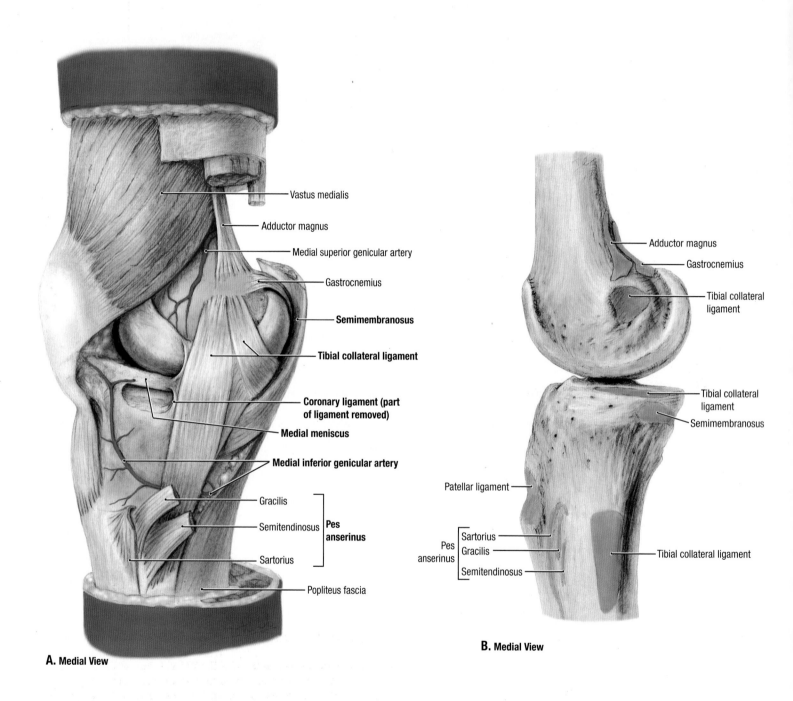

Vastus medialis

Adductor magnus

Medial superior genicular artery

Gastrocnemius

Semimembranosus

Tibial collateral ligament

Coronary ligament (part of ligament removed)

Medial meniscus

Medial inferior genicular artery

Gracilis ⎤
Semitendinosus ⎬ **Pes anserinus**
Sartorius ⎦

Popliteus fascia

A. Medial View

Adductor magnus

Gastrocnemius

Tibial collateral ligament

Tibial collateral ligament

Semimembranosus

Patellar ligament

Sartorius ⎤
Pes anserinus Gracilis ⎬
Semitendinosus ⎦

Tibial collateral ligament

B. Medial View

6.50 MEDIAL ASPECT OF KNEE

A. Dissection. The bandlike part of the tibial collateral ligament attaches to the medial epicondyle of the femur, bridges superficial to the insertion of the semimembranosus muscle, and crosses the medial inferior genicular artery. Distally, the ligament is crossed by the three tendons forming the pes anserinus (sartorius, gracilis, and semitendinosus). **B.** Muscle and ligament attachment sites.

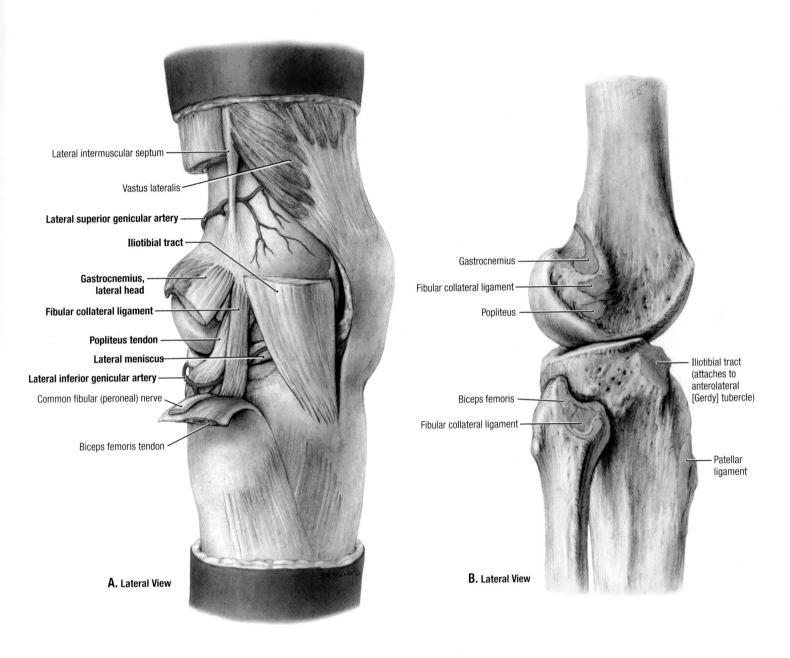

Lateral intermuscular septum

Vastus lateralis

Lateral superior genicular artery

Iliotibial tract

Gastrocnemius, lateral head

Fibular collateral ligament

Popliteus tendon

Lateral meniscus

Lateral inferior genicular artery

Common fibular (peroneal) nerve

Biceps femoris tendon

A. Lateral View

Gastrocnemius

Fibular collateral ligament

Popliteus

Biceps femoris

Fibular collateral ligament

Iliotibial tract (attaches to anterolateral [Gerdy] tubercle)

Patellar ligament

B. Lateral View

LATERAL ASPECT OF KNEE

6.51

A. Dissection. Three structures arise from the lateral epicondyle and are uncovered by reflecting the biceps femoris tendon: the gastrocnemius muscle is posterosuperior; the popliteus muscle is antero-inferior; and the fibular collateral ligament is in between, crossing superficial to the popliteus muscle. The lateral inferior genicular artery courses along the lateral meniscus. **B.** Muscle and ligament attachments.

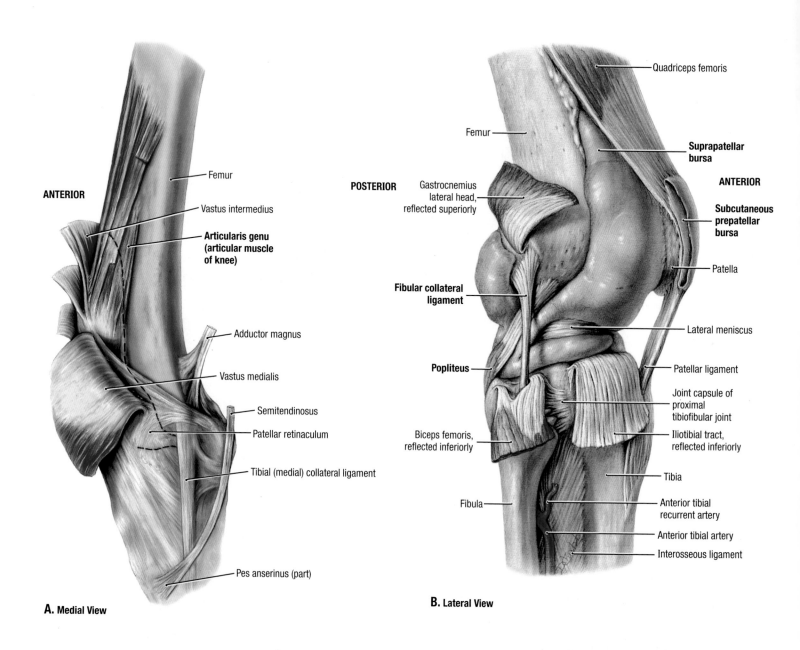

ANTERIOR

— Femur

— Vastus intermedius

**Articularis genu
(articular muscle
of knee)**

— Adductor magnus

— Vastus medialis

— Semitendinosus

— Patellar retinaculum

— Tibial (medial) collateral ligament

— Pes anserinus (part)

A. Medial View

POSTERIOR

Femur —

Gastrocnemius
lateral head,
reflected superiorly

**Fibular collateral
ligament**

Popliteus —

Biceps femoris,
reflected inferiorly

Fibula —

Quadriceps femoris

**Suprapatellar
bursa**

ANTERIOR

**Subcutaneous
prepatellar
bursa**

Patella

Lateral meniscus

Patellar ligament

Joint capsule of
proximal
tibiofibular joint

Iliotibial tract,
reflected inferiorly

Tibia

Anterior tibial
recurrent artery

Anterior tibial artery

Interosseous ligament

B. Lateral View

6.52 ARTICULARIS GENU AND BURSAE OF KNEE REGION

A. Articularis genu (articular muscle of the knee). This muscle lies deep to the vastus intermedius muscle and consists of fibers arising from the anterior surface of the femur proximally and attaching into the synovial membrane distally. The articularis genu pulls the synovial membrane of the suprapatellar bursa *(dotted line)* superiorly during extension of the knee so that it will not be caught between the patella and femur within the knee joint. **B.** Lateral aspect of knee. Latex was injected into the articular cavity and fixed with acetic acid. The distended synovial membrane was exposed and cleaned. The gastrocnemius muscle was reflected proximally, and the biceps femoris muscle and the iliotibial tract were reflected distally. The extent of the synovial capsule: superiorly, it rises superior to the patella, where it rests on a layer of fat that allows it to glide

freely with movements of the joint—this superior part is called the suprapatellar bursa; posteriorly, it rises as high as the origin of the gastrocnemius muscle; laterally, it curves inferior to the lateral femoral epicondyle, where the popliteus tendon and fibular collateral ligament are attached; and inferiorly, it bulges inferior to the lateral meniscus, overlapping the tibia (the coronary ligament is removed to show this). **Prepatellar bursitis** (housemaid's knee) is usually a friction bursitis caused by friction between the skin and the patella. The suprapatellar bursa communicates with the articular cavity of the knee joint; consequently, abrasions or penetrating wounds superior to the patella may result in **suprapatellar bursitis** caused by bacteria entering the bursa from the torn skin. The infection may spread to the knee joint. **C.** Posterior aspect of knee.

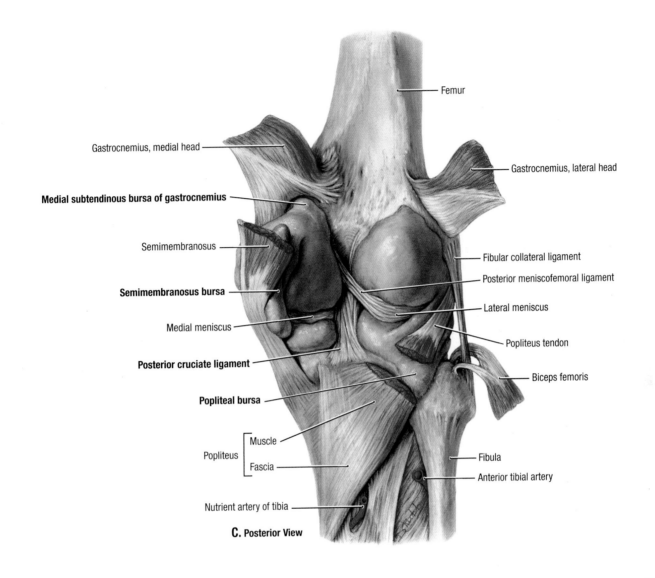

Femur

Gastrocnemius, medial head

Gastrocnemius, lateral head

Medial subtendinous bursa of gastrocnemius

Semimembranosus

Fibular collateral ligament

Posterior meniscofemoral ligament

Semimembranosus bursa

Lateral meniscus

Medial meniscus

Popliteus tendon

Posterior cruciate ligament

Biceps femoris

Popliteal bursa

Popliteus { Muscle / Fascia }

Fibula

Anterior tibial artery

Nutrient artery of tibia

C. Posterior View

BURSAE OF KNEE REGION (*continued*)

6.52

TABLE 6.11	**BURSAE AROUND KNEE**	
Bursa	*Location*	*Structural Features or Functions*
Suprapatellar	Located between femur and tendon of quadriceps femoris	Held in position by articular muscle of knee; superior extension of synovial cavity of knee joint
Popliteus	Located between tendon of popliteus and lateral condyle of tibia	Opens into synovial cavity of knee joint, inferior to lateral meniscus
Anserine	Separates tendons of sartorius, gracilis, and semitendinosus from tibia and tibial collateral ligament	Area where tendons of these muscles attach to tibia (pes anserinus) resembles the foot of a goose (L. *pes*, foot; L. *anser*, goose)
Medial subtendinous bursa of gastrocnemius	Lies deep to proximal attachment of tendon of medial head of gastrocnemius	Extension of synovial cavity of knee joint
Semimembranosus	Located between medial head of gastrocnemius and semimembranosus tendon	Related to the distal attachment of semimembranosus
Subcutaneous prepatellar	Lies between skin and anterior surface of patella	Allows free movement of skin over patella during movements of leg
Subcutaneous infrapatellar	Located between skin and tibial tuberosity	Helps knee to withstand pressure when kneeling*
Deep infrapatellar	Lies between patellar ligament and anterior surface of tibia	Separated from knee joint by infrapatellar fat pad*

*See Figure 6.56.

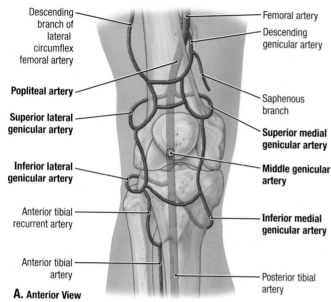

Descending branch of lateral circumflex femoral artery

Femoral artery

Descending genicular artery

Popliteal artery

Saphenous branch

Superior lateral genicular artery

Superior medial genicular artery

Inferior lateral genicular artery

Middle genicular artery

Anterior tibial recurrent artery

Inferior medial genicular artery

Anterior tibial artery

Posterior tibial artery

A. Anterior View

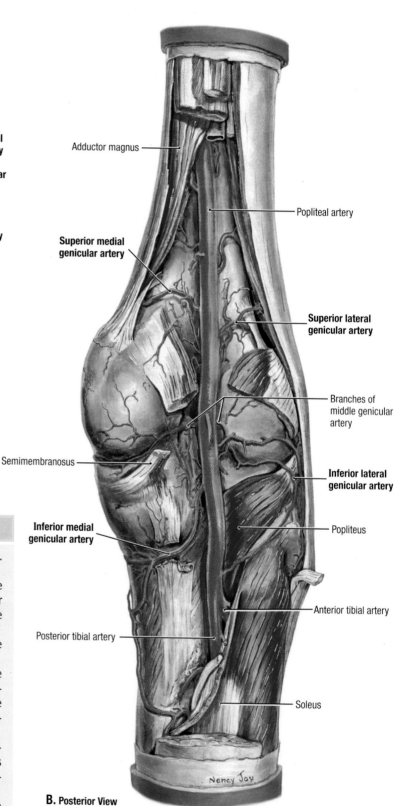

Adductor magnus

Popliteal artery

Superior medial genicular artery

Superior lateral genicular artery

Branches of middle genicular artery

Semimembranosus

Inferior lateral genicular artery

Inferior medial genicular artery

Popliteus

Posterior tibial artery

Anterior tibial artery

Soleus

Nancy Joy

B. Posterior View

6.53 ANASTOMOSES AROUND KNEE

A. Genicular anastomosis on the anterior aspect of the knee.
B. Popliteal artery in popliteal fossa.

- The popliteal artery runs from the adductor hiatus (in the adductor magnus muscle) proximally to the inferior border of the popliteus muscle distally, where it bifurcates into the anterior and posterior tibial arteries.
- The three anterior relations of the popliteal artery include the femur, joint capsule of the knee, and the popliteus muscle.
- The genicular arteries participate in the formation of the peri-articular genicular anastomosis, a network of vessels surrounding the knee that provides collateral circulation capable of maintaining blood supply to the leg during full knee flexion, which may kink the popliteal artery.
- Five genicular branches of the popliteal artery supply the capsule and ligaments of the knee joint. The genicular arteries are the superior lateral, superior medial, middle, inferior lateral, and inferior medial genicular arteries.
- Other contributors are the descending genicular artery, a branch of the femoral artery, superomedially; descending branch of the lateral circumflex femoral artery, superolaterally; and anterior tibial recurrent artery, a branch of the anterior tibial artery, inferolaterally.

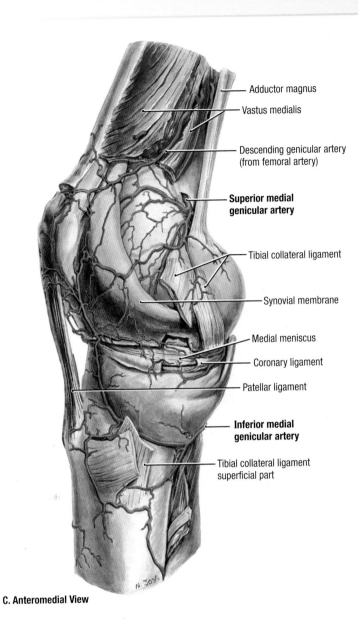

Adductor magnus

Vastus medialis

Descending genicular artery
(from femoral artery)

**Superior medial
genicular artery**

Tibial collateral ligament

Synovial membrane

Medial meniscus

Coronary ligament

Patellar ligament

**Inferior medial
genicular artery**

Tibial collateral ligament
superficial part

C. Anteromedial View

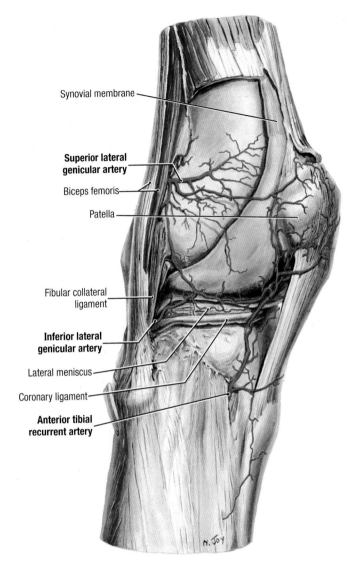

Synovial membrane

**Superior lateral
genicular artery**

Biceps femoris

Patella

Fibular collateral
ligament

**Inferior lateral
genicular artery**

Lateral meniscus

Coronary ligament

**Anterior tibial
recurrent artery**

E. Anterolateral View

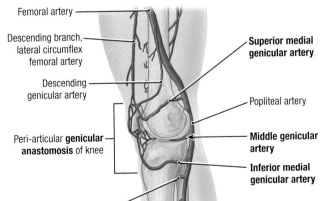

Femoral artery

Descending branch,
lateral circumflex
femoral artery

Descending
genicular artery

Peri-articular **genicular
anastomosis** of knee

Anterior tibial
artery

**Superior medial
genicular artery**

Popliteal artery

**Middle genicular
artery**

**Inferior medial
genicular artery**

Posterior
tibial artery

D. Medial View

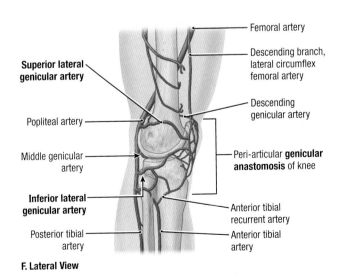

Femoral artery

Descending branch,
lateral circumflex
femoral artery

Descending
genicular artery

**Superior lateral
genicular artery**

Popliteal artery

Middle genicular
artery

**Inferior lateral
genicular artery**

Posterior tibial
artery

Peri-articular **genicular
anastomosis** of knee

Anterior tibial
recurrent artery

Anterior tibial
artery

F. Lateral View

6.53

ANASTOMOSES AROUND KNEE (*continued*)

C. and **D.** Medial aspect of the knee showing superior and inferior medial genicular arteries. **E.** and **F.** Lateral aspect of the knee showing superior and inferior lateral genicular arteries.

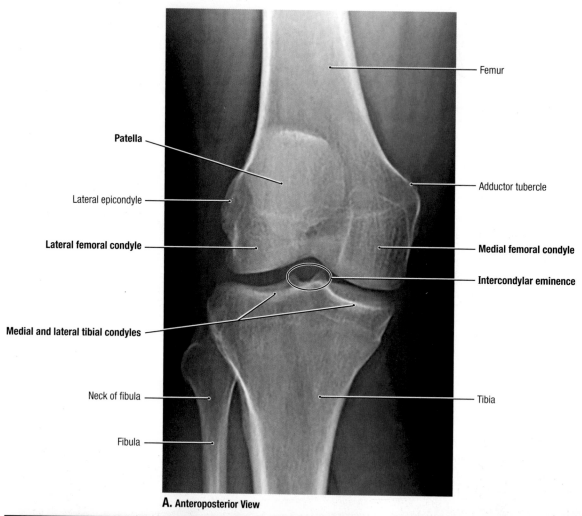

Femur

Patella

Lateral epicondyle

Lateral femoral condyle

Medial and lateral tibial condyles

Neck of fibula

Fibula

Adductor tubercle

Medial femoral condyle

Intercondylar eminence

Tibia

A. Anteroposterior View

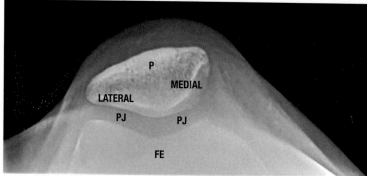

B. Skyline (Merchant) View (Knee in Flexion)

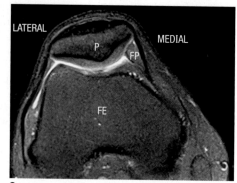

C. Transverse MRI

6.54 IMAGING OF THE KNEE AND PATELLOFEMORAL ARTICULATION

A. Anteroposterior radiograph of knee. **B.** Radiograph of patella (knee joint flexed). **C.** Transverse MRI showing the patellofemoral joint. *FE*, femur; *FP*, fat pad; *P*, patella; *PJ*, patellofemoral joint.

Pain deep to the patella often results from excessive running; hence, this type of pain is often called "runner's knee." The pain results from repetitive microtrauma caused by abnormal tracking of the patella relative to the patellar surface of the femur, a condition known as the **patellofemoral syndrome**. This syndrome may also

result from a direct blow to the patella and from osteoarthritis of the patellofemoral compartment (degenerative wear and tear of articular cartilages). In some cases, strengthening of the vastus medialis corrects patellofemoral dysfunction. This muscle tends to prevent lateral dislocation of the patella resulting from the Q-angle because the vastus medialis attaches to and pulls on the medial border of the patella. Hence, weakness of the vastus medialis predisposes the individual to patellofemoral dysfunction and patellar dislocation.

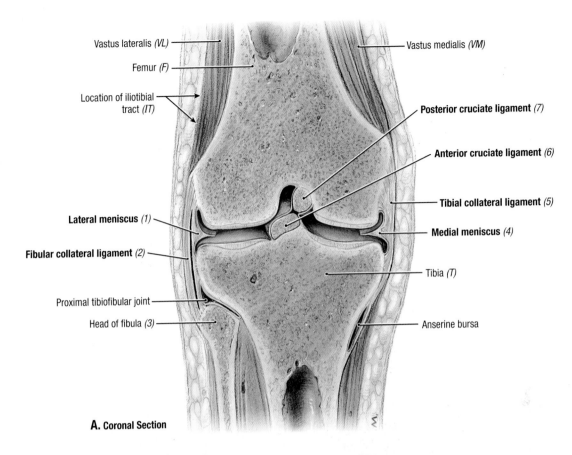

Vastus lateralis *(VL)*
Femur *(F)*
Location of iliotibial tract *(IT)*

Lateral meniscus *(1)*

Fibular collateral ligament *(2)*

Proximal tibiofibular joint
Head of fibula *(3)*

Vastus medialis *(VM)*

Posterior cruciate ligament *(7)*

Anterior cruciate ligament *(6)*

Tibial collateral ligament *(5)*

Medial meniscus *(4)*

Tibia *(T)*

Anserine bursa

A. Coronal Section

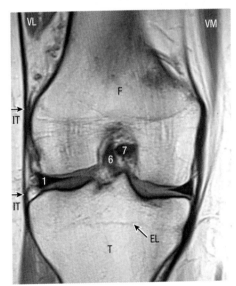

B. Coronal MRI

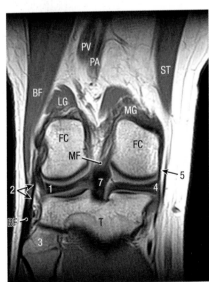

C. Coronal MRI

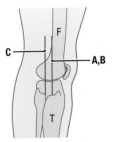

Lateral View

CORONAL SECTION AND MRI OF KNEE | 6.55

A. Section through intercondylar notch of femur, tibia, and fibula. **B.** MRI through intercondylar notch of femur and tibia. **C.** MRI through femoral condyles tibia and fibula. Numbers in MRIs refer to structures **(A)**. *BF,* biceps femoris; *EL,* epiphyseal line; *F,* fat in popliteal fossa; *FC,* femoral condyle; *IT,* iliotibial tract; *LG,* lateral head of gastrocnemius; *MF,* meniscofemoral ligament; *MG,* medial head of gastrocnemius; *PA,* popliteal artery; *PV,* popliteal vein; *ST,* semitendinosus; *T,* tibia; *VL,* vastus lateralis; *VM,* vastus medialis.

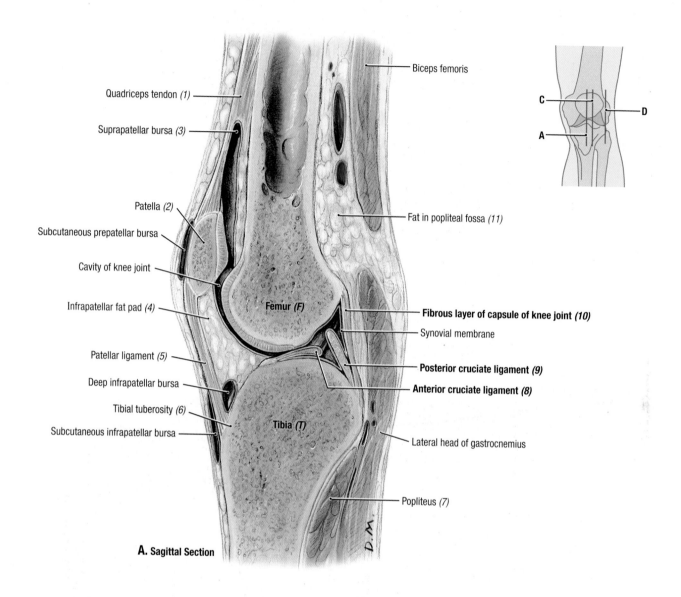

Quadriceps tendon *(1)*

Suprapatellar bursa *(3)*

Patella *(2)*

Subcutaneous prepatellar bursa

Cavity of knee joint

Infrapatellar fat pad *(4)*

Patellar ligament *(5)*

Deep infrapatellar bursa

Tibial tuberosity *(6)*

Subcutaneous infrapatellar bursa

Femur *(F)*

Tibia *(T)*

Biceps femoris

Fat in popliteal fossa *(11)*

Fibrous layer of capsule of knee joint *(10)*

Synovial membrane

Posterior cruciate ligament *(9)*

Anterior cruciate ligament *(8)*

Lateral head of gastrocnemius

Popliteus *(7)*

A. Sagittal Section

6.56 SAGITTAL SECTION AND IMAGING OF KNEE

A. Illustration of section through lateral aspect of intercondylar notch of femur.

Fractures of the distal end of the femur, or lacerations of the anterior thigh, may involve the suprapatellar bursa and result in infection of the knee joint. When the knee joint is infected and inflamed, the amount of synovial fluid may increase. **Joint effusions**, the escape of fluid from blood or lymphatic vessels, result in increased amounts of fluid in the joint cavity. Because the suprapatellar bursa is a superior continuation of the synovial cavity of the knee joint, fullness of the thigh in the region of the bursa may indicate increased synovial fluid. This bursa can be aspirated to remove the fluid for examination. Direct **aspiration of the knee joint** is usually performed with the patient sitting on a table with the knee flexed. The joint is approached laterally, using three bony points as landmarks for needle insertion: the anterolateral tibial (Gerdy) tubercle, the lateral epicondyle of the femur, and the apex of the patella. In addition, this triangular area also is used for drug injection for treating pathology of the knee joint.

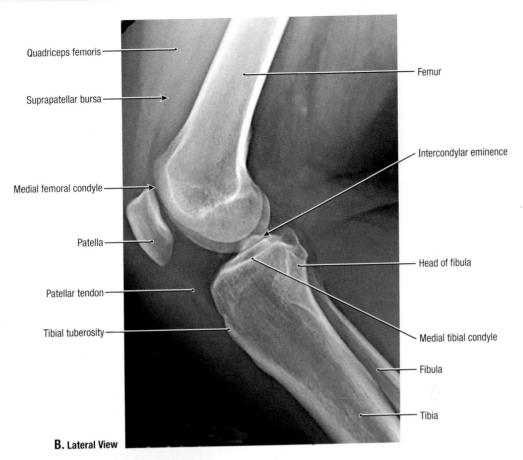

Quadriceps femoris

Suprapatellar bursa

Medial femoral condyle

Patella

Patellar tendon

Tibial tuberosity

Femur

Intercondylar eminence

Head of fibula

Medial tibial condyle

Fibula

Tibia

B. Lateral View

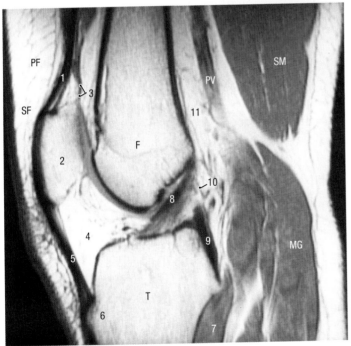

C. Sagittal MRI

D. Sagittal MRI

6.56

SAGITTAL SECTION AND IMAGING OF KNEE (*continued*)

B. Lateral radiograph of flexed knee. The fabella is an inconsistent sesamoid bone in the lateral head of gastrocnemius muscle. **C.** MRI through medial aspect of intercondylar notch of femur showing cruciate ligaments. **D.** MRI through medial femoral and tibial condyles. Numbers in MRIs refer to structures labeled in **(A)**. *AM*, anterior horn of medial meniscus; *F*, fibula; *MG*, medial head of gastrocnemius; *PF*, prefemoral fat; *PM*, posterior horn of medial meniscus; *PV*, popliteal vessels; *SF*, suprapatellar fat; *SM*, semimembranosus; *ST*, semitendinosus; *T*, tibia; *VM*, vastus medialis.

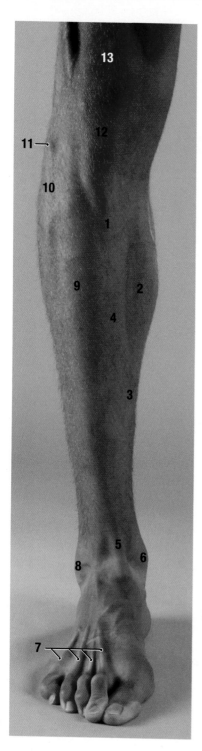

A. Anterior View

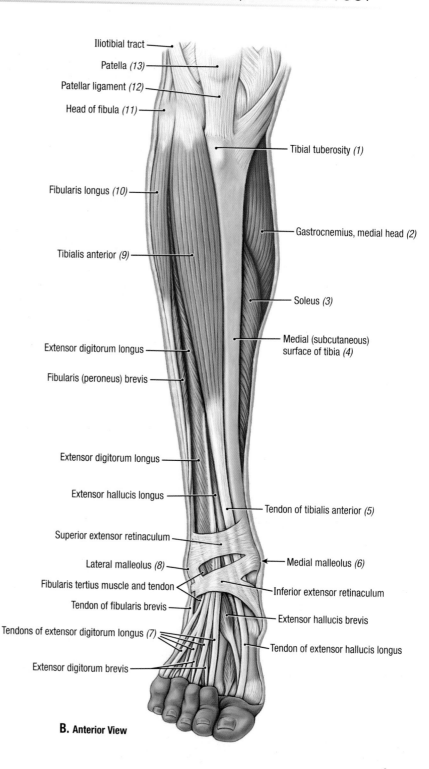

Iliotibial tract

Patella *(13)*

Patellar ligament *(12)*

Head of fibula *(11)*

Tibial tuberosity *(1)*

Fibularis longus *(10)*

Gastrocnemius, medial head *(2)*

Tibialis anterior *(9)*

Soleus *(3)*

Medial (subcutaneous) surface of tibia *(4)*

Extensor digitorum longus

Fibularis (peroneus) brevis

Extensor digitorum longus

Extensor hallucis longus

Tendon of tibialis anterior *(5)*

Superior extensor retinaculum

Lateral malleolus *(8)*

Medial malleolus *(6)*

Fibularis tertius muscle and tendon

Inferior extensor retinaculum

Tendon of fibularis brevis

Extensor hallucis brevis

Tendons of extensor digitorum longus *(7)*

Tendon of extensor hallucis longus

Extensor digitorum brevis

B. Anterior View

6.57 ANTERIOR LEG: SUPERFICIAL MUSCLES

A. Surface anatomy. Numbers refer to structures labeled in **(B)**. **B.** Dissection. The muscles of the anterior compartment are ankle dorsiflexors/toe extensors. They are active in walking as they concentrically contract to raise the forefoot to clear the ground during the swing phase of the gait cycle and eccentrically contract to lower the forefoot to the ground after the heel strike of the stance phase.

Shin splints, edema, and pain in the area of the distal third of the tibia result from repetitive microtrauma of the anterior compartment muscles, especially the tibialis anterior. This produces a mild form of **anterior compartment syndrome**. The pain commonly occurs during traumatic injury or athletic overexertion of the muscles. Edema and muscle-tendon inflammation causes swelling that reduces blood flow to the muscles. Swollen ischemic muscles are painful and tender to pressure.

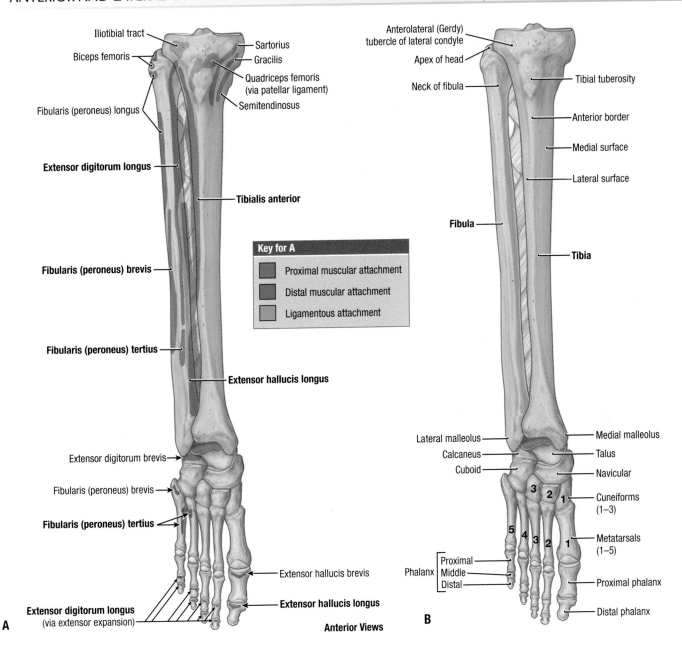

A. Attachments. B. Features of bones.

Key for A
- Proximal muscular attachment
- Distal muscular attachment
- Ligamentous attachment

Anterior Views

ANTERIOR LEG AND DORSUM OF FOOT: FEATURES OF BONES AND MUSCLE ATTACHMENTS

6.58

TABLE 6.12 — MUSCLES OF ANTERIOR COMPARTMENT OF LEG

Muscle	Proximal Attachment	Distal Attachment	Innervation[a]	Main Actions
Tibialis anterior	Lateral condyle and superior half of lateral surface of tibia	Medial and inferior surfaces of medial cuneiform and base of 1st metatarsal	Deep fibular (peroneal) nerve (L4–L5)	Dorsiflexes ankle joint and inverts foot
Extensor hallucis longus	Middle part of anterior surface of fibula and interosseous membrane	Dorsal aspect of base of distal phalanx of great toe (hallux)	Deep fibular (peroneal) nerve (L5–S1)	Extends great toe and dorsiflexes ankle joint
Extensor digitorum longus	Lateral condyle of tibia and superior three fourths of anterior surface of interosseous membrane	Middle and distal phalanges of lateral four digits		Extends lateral four digits and dorsiflexes ankle joint
Fibularis (peroneus) tertius	Inferior third of anterior surface of fibula and interosseous membrane	Dorsum of base of 5th metatarsal		Dorsiflexes ankle joint and aids in eversion of foot

[a]See Table 6.1 for explanation of segmental innervation.

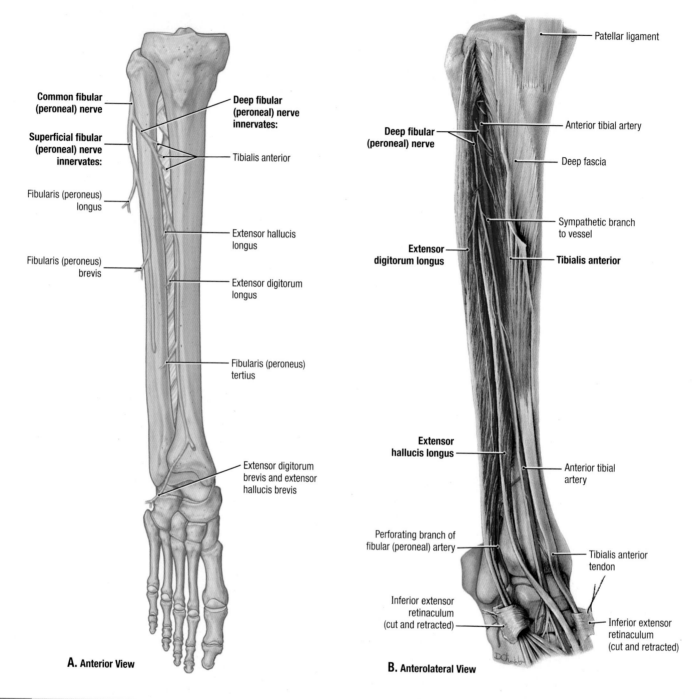

A. Anterior View

- Common fibular (peroneal) nerve
- Superficial fibular (peroneal) nerve innervates:
- Fibularis (peroneus) longus
- Fibularis (peroneus) brevis
- Deep fibular (peroneal) nerve innervates:
- Tibialis anterior
- Extensor hallucis longus
- Extensor digitorum longus
- Fibularis (peroneus) tertius
- Extensor digitorum brevis and extensor hallucis brevis

B. Anterolateral View

- Patellar ligament
- Deep fibular (peroneal) nerve
- Anterior tibial artery
- Deep fascia
- Sympathetic branch to vessel
- Extensor digitorum longus
- Tibialis anterior
- Extensor hallucis longus
- Anterior tibial artery
- Perforating branch of fibular (peroneal) artery
- Tibialis anterior tendon
- Inferior extensor retinaculum (cut and retracted)
- Inferior extensor retinaculum (cut and retracted)

6.59 ANTERIOR LEG: MUSCLES, NERVES, AND VESSELS

TABLE 6.13	COMMON, SUPERFICIAL, AND DEEP FIBULAR (PERONEAL) NERVES		
Nerve	*Origin*	*Course*	*Distribution/Structure(s) Supplied*
Common fibular	Sciatic nerve	Forms as sciatic nerve bifurcates at the apex of popliteal fossa and follows medial border of biceps femoris; winds around neck of fibula, dividing into superficial and deep fibular nerves	Skin on lateral part of posterior aspect of leg via the lateral sural cutaneous nerve; lateral aspect of knee joint via its articular branch
Superficial fibular	Common fibular nerve	Arises deep to fibularis longus and descends in lateral compartment of leg; pierces crural fascia at distal third of leg to become cutaneous	Fibularis longus and brevis and skin on distal third of anterolateral surface of leg and dorsum of foot
Deep fibular	Common fibular nerve	Arises deep to fibularis longus; passes through extensor digitorum longus, descends on interosseous membrane, and continues on dorsum of foot	Anterior muscles of leg, dorsum of foot, and skin of first interdigital cleft; dorsal aspect of joints crossed via articular branches

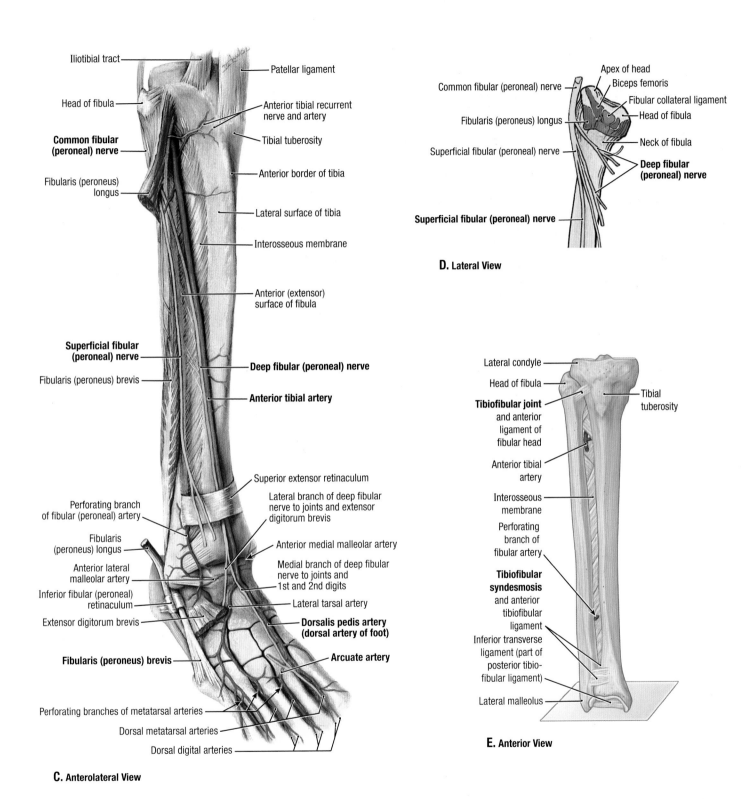

C. Anterolateral View

D. Lateral View

E. Anterior View

ANTERIOR LEG: MUSCLES, NERVES, AND VESSELS (*continued*) | **6.59**

A. Overview of motor innervation. **B.** Deep dissection of the anterior compartment of leg. The muscles are separated to display anterior tibial artery and deep fibular nerve. **C.** Neurovascular structures of lateral compartment and dorsum of foot. **D.** Relations of common fibular nerve and branches to the proximal fibula. **E.** Interosseous membrane.

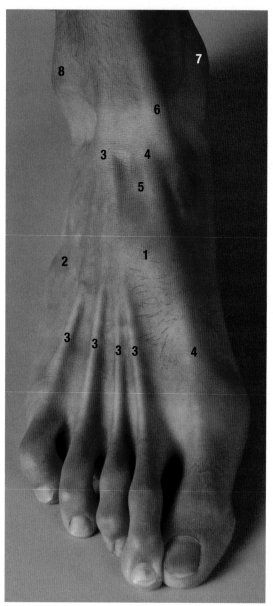

A. Superior View

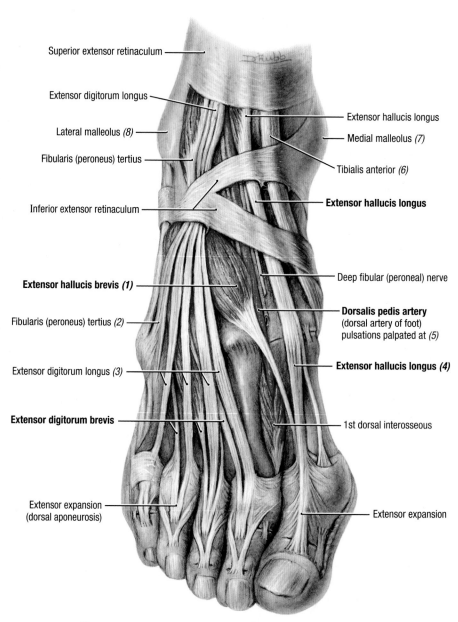

Superior extensor retinaculum

Extensor digitorum longus

Extensor hallucis longus

Lateral malleolus (8)

Medial malleolus (7)

Fibularis (peroneus) tertius

Tibialis anterior (6)

Inferior extensor retinaculum

Extensor hallucis longus

Extensor hallucis brevis (1)

Deep fibular (peroneal) nerve

Fibularis (peroneus) tertius (2)

Dorsalis pedis artery
(dorsal artery of foot)
pulsations palpated at (5)

Extensor digitorum longus (3)

Extensor hallucis longus (4)

Extensor digitorum brevis

1st dorsal interosseous

Extensor expansion
(dorsal aponeurosis)

Extensor expansion

B. Superior View

6.60 | DORSUM OF FOOT

A. Surface anatomy. Numbers refer to structures labeled in **(B)**.
B. Dissection. The dorsal vein of foot and deep fibular nerve are cut.

At the ankle, the dorsalis pedis artery (dorsal artery of foot) and deep fibular nerve lie midway between the malleoli. On the dorsum of the foot, the dorsal artery of foot is crossed by the extensor hallucis brevis muscle and disappears between the two heads of the first dorsal interosseous muscle.

Clinically, knowing the location of the belly of the extensor digitorum brevis is important for distinguishing this muscle from abnormal edema. Contusion and tearing of the muscle fibers and associated blood vessels result in a **hematoma in extensor digitorum brevis**, producing edema anteromedial to the lateral malleolus. Most people who have not seen this inflamed muscle assume they have a severely sprained ankle.

The **dorsalis pedis pulse** may be palpated with the feet slightly dorsiflexed. The pulse is usually easy to palpate because the dorsal arteries of the foot are subcutaneous and pass along a line from the extensor retinaculum to a point just lateral to the extensor hallucis longus tendon. A diminished or absent dorsalis pedis pulse usually suggests vascular insufficiency resulting from arterial disease.

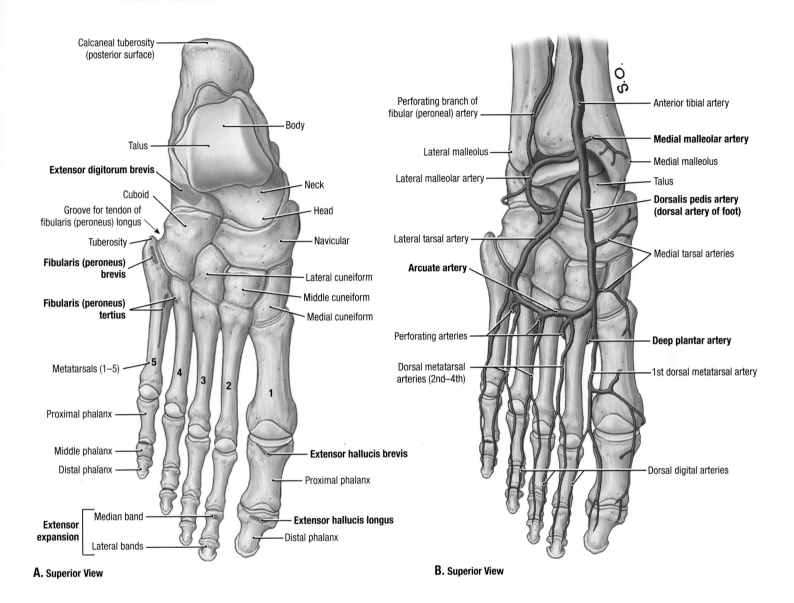

Calcaneal tuberosity (posterior surface)
Body
Talus
Extensor digitorum brevis
Cuboid
Groove for tendon of fibularis (peroneus) longus
Tuberosity
Fibularis (peroneus) brevis
Fibularis (peroneus) tertius
Neck
Head
Navicular
Lateral cuneiform
Middle cuneiform
Medial cuneiform
Metatarsals (1–5)
5 4 3 2 1
Proximal phalanx
Middle phalanx
Distal phalanx
Extensor hallucis brevis
Proximal phalanx
Extensor expansion
Median band
Lateral bands
Extensor hallucis longus
Distal phalanx

A. Superior View

Perforating branch of fibular (peroneal) artery
Lateral malleolus
Lateral malleolar artery
Lateral tarsal artery
Arcuate artery
Perforating arteries
Dorsal metatarsal arteries (2nd–4th)
Anterior tibial artery
Medial malleolar artery
Medial malleolus
Talus
Dorsalis pedis artery (dorsal artery of foot)
Medial tarsal arteries
Deep plantar artery
1st dorsal metatarsal artery
Dorsal digital arteries

B. Superior View

MUSCLE ATTACHMENTS AND ARTERIES OF DORSUM OF FOOT
6.61

A. Attachments. **B.** Arterial supply.

TABLE 6.14	ARTERIAL SUPPLY TO DORSUM OF FOOT		
Artery	*Origin*	*Course*	*Distribution*
Dorsalis pedis (dorsal artery of foot)	Continuation of anterior tibial artery distal to talocrural joint	Descends anteromedially to 1st interosseous space and divides into deep plantar and arcuate arteries	Dorsal surface of hind foot
Lateral tarsal artery	From dorsalis pedis artery (dorsal artery of foot)	Runs an arched course laterally beneath extensor digitorum brevis to anastomose with branches of arcuate artery	
Arcuate artery		Runs laterally from 1st interosseous space across bases of lateral four metatarsals, deep to extensor tendons	
Deep plantar artery		Passes to sole of foot and joins plantar arch	Sole of foot
Metatarsal arteries: 1st	From deep plantar artery	Run between metatarsals to clefts of toes where each vessel divides into two dorsal digital arteries	Dorsal surface of forefoot
2nd to 4th	From arcuate artery	Perforating arteries connect to plantar arch and plantar metatarsal arteries.	
Dorsal digital arteries	From metatarsal arteries	Pass to sides of adjoining digits	Proximal dorsal digits

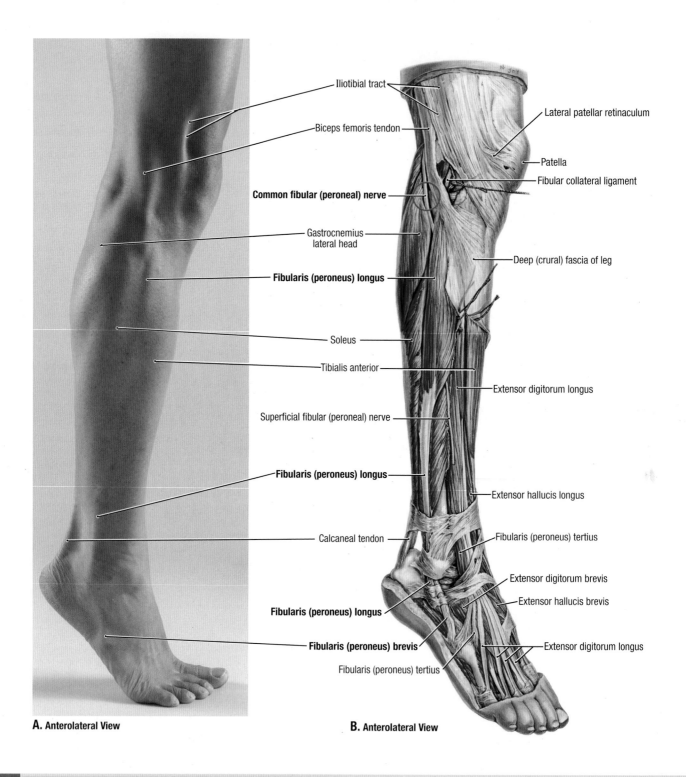

Iliotibial tract

Lateral patellar retinaculum

Biceps femoris tendon

Patella

Common fibular (peroneal) nerve

Fibular collateral ligament

Gastrocnemius lateral head

Deep (crural) fascia of leg

Fibularis (peroneus) longus

Soleus

Tibialis anterior

Extensor digitorum longus

Superficial fibular (peroneal) nerve

Extensor hallucis longus

Fibularis (peroneus) longus

Fibularis (peroneus) tertius

Calcaneal tendon

Extensor digitorum brevis

Extensor hallucis brevis

Fibularis (peroneus) longus

Extensor digitorum longus

Fibularis (peroneus) brevis

Fibularis (peroneus) tertius

A. Anterolateral View

B. Anterolateral View

6.62   LATERAL LEG AND FOOT: MUSCLES

A. Surface anatomy. **B.** Dissection.

- The two fibular (peroneal) muscles both attach to two thirds of the fibula, the fibularis (peroneus) longus muscle to the proximal two thirds, and the fibularis (peroneus) brevis muscle to the distal two thirds. Where they overlap, the fibularis brevis muscle lies anteriorly.
- The fibularis (peroneus) longus muscle enters the foot by hooking around the cuboid and traveling medially to the base of the 1st metatarsal and medial cuneiform.

- **Common fibular (peroneal) nerve lesion.** The nerve lies in contact with the neck of the fibula deep to the fibularis longus muscle, where it is vulnerable to injury (red circle). This injury may have serious implications because the nerve supplies the extensor and everter muscle groups, with loss of function resulting in **footdrop** (inability to dorsiflex the ankle) and difficulty in everting the foot.

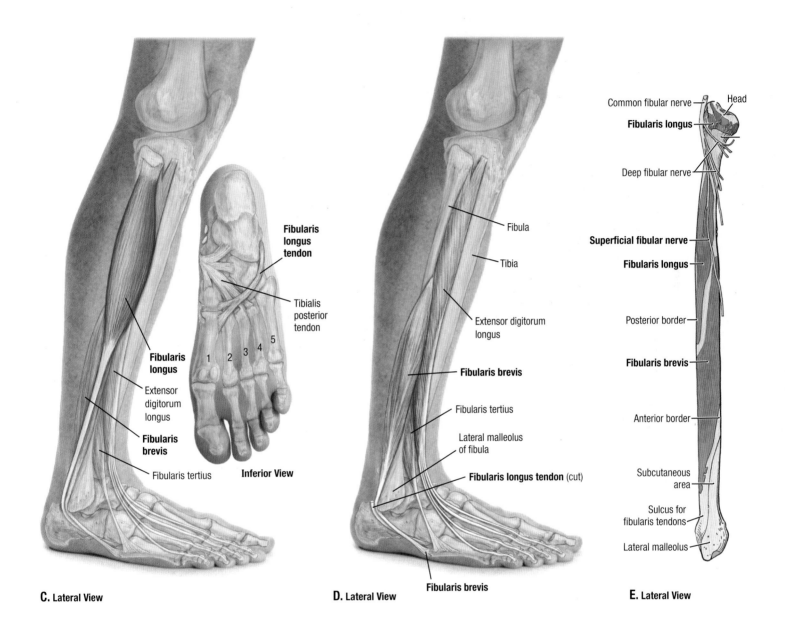

Fibularis longus tendon

Tibialis posterior tendon

5 4 3 2 1

Fibularis longus

Extensor digitorum longus

Fibularis brevis

Fibularis tertius

Inferior View

C. Lateral View

Fibula

Tibia

Extensor digitorum longus

Fibularis brevis

Fibularis tertius

Lateral malleolus of fibula

Fibularis longus tendon (cut)

Fibularis brevis

D. Lateral View

Common fibular nerve — Head

Fibularis longus

Deep fibular nerve

Superficial fibular nerve

Fibularis longus

Posterior border

Fibularis brevis

Anterior border

Subcutaneous area

Sulcus for fibularis tendons

Lateral malleolus

E. Lateral View

LATERAL LEG AND FOOT: MUSCLES (*continued*)

6.62

C. Fibularis (peroneus) longus. **D.** Fibularis (peroneus) brevis. **E.** Attachments sites on fibula.

TABLE 6.15	MUSCLES OF LATERAL COMPARTMENT OF LEG			
Muscle	*Proximal Attachment*	*Distal Attachment*	*Innervation[a]*	*Main Actions*
Fibularis (peroneus) longus	Head and superior two thirds of lateral surface of fibula	Base of 1st metatarsal and medial cuneiform	Superficial fibular (peroneal) nerve (L5, S1, and S2)	Evert foot and weakly plantar flex ankle joint reflexively resist inadvertent inversion of foot
Fibularis (peroneus) brevis	Inferior two thirds of lateral surface of fibula	Dorsal surface of tuberosity on lateral side of base of 5th metatarsal		

[a]See Table 6.1 for explanation of segmental innervation.

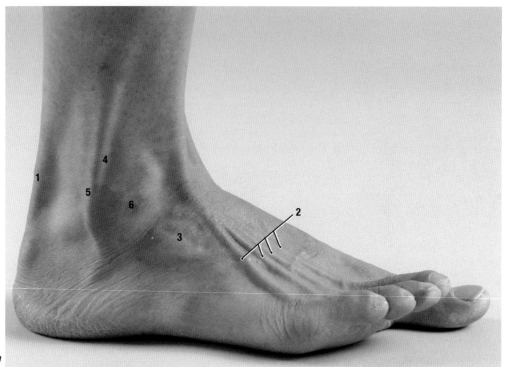

A. Lateral View

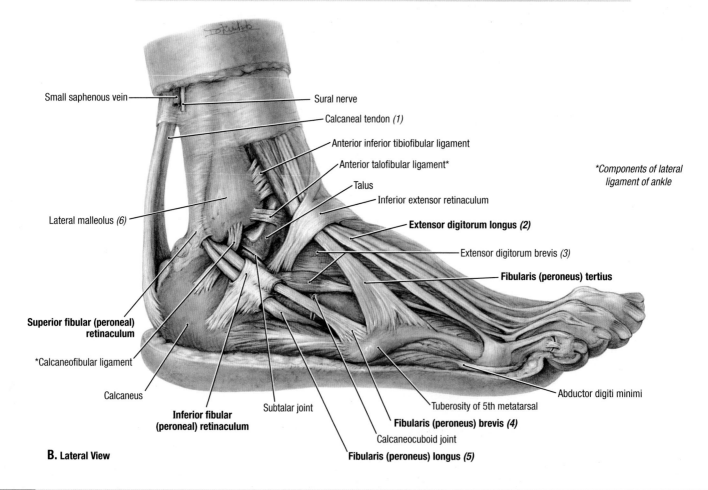

Small saphenous vein

Lateral malleolus *(6)*

Superior fibular (peroneal) retinaculum

*Calcaneofibular ligament

Calcaneus

Inferior fibular (peroneal) retinaculum

Subtalar joint

Sural nerve

Calcaneal tendon *(1)*

Anterior inferior tibiofibular ligament

Anterior talofibular ligament*

Talus

Inferior extensor retinaculum

Extensor digitorum longus *(2)*

Extensor digitorum brevis *(3)*

Fibularis (peroneus) tertius

Abductor digiti minimi

Tuberosity of 5th metatarsal

Fibularis (peroneus) brevis *(4)*

Calcaneocuboid joint

Fibularis (peroneus) longus *(5)*

Components of lateral ligament of ankle

B. Lateral View

6.63   SYNOVIAL SHEATHS AND TENDONS AT ANKLE

A. Surface anatomy. Numbers refer to structures labeled in **(B)**. **B.** Tendons at the lateral aspect of the ankle.

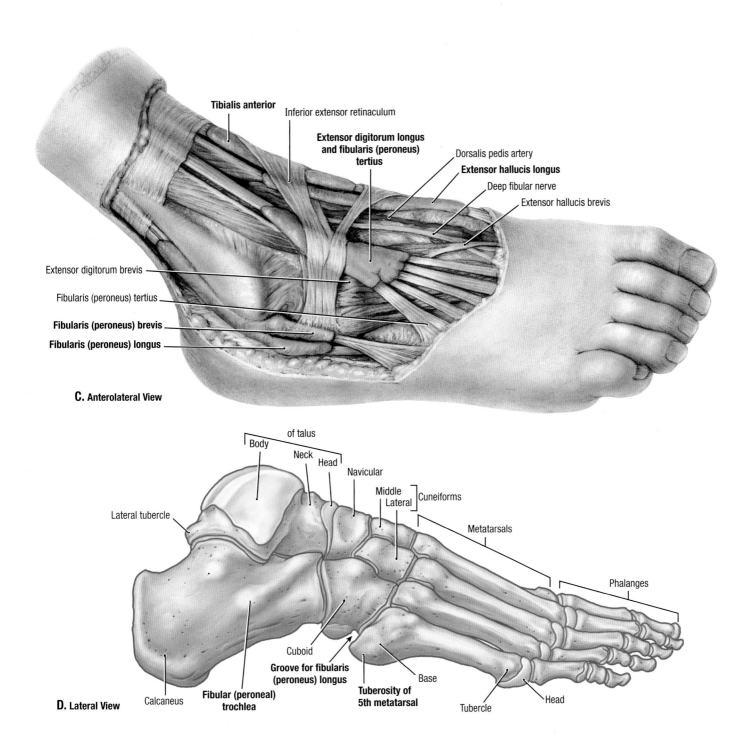

Tibialis anterior
Inferior extensor retinaculum
Extensor digitorum longus and fibularis (peroneus) tertius
Dorsalis pedis artery
Extensor hallucis longus
Deep fibular nerve
Extensor hallucis brevis

Extensor digitorum brevis
Fibularis (peroneus) tertius
Fibularis (peroneus) brevis
Fibularis (peroneus) longus

C. Anterolateral View

of talus
Body
Neck
Head
Navicular
Middle
Lateral
Cuneiforms
Metatarsals

Lateral tubercle

Phalanges

Cuboid
Groove for fibularis (peroneus) longus
Fibular (peroneal) trochlea
Base
Tuberosity of 5th metatarsal
Head
Tubercle
Calcaneus

D. Lateral View

SYNOVIAL SHEATHS AND TENDONS AT ANKLE (*continued*)   **6.63**

C. Synovial sheaths of tendons on the anterolateral aspect of the ankle. The tendons of the fibularis (peroneus) longus and fibularis (peroneus) brevis muscles are enclosed in a common synovial sheath posterior to the lateral malleolus. This sheath splits into two, one for each tendon, posterior to the fibular (peroneal) trochlea. **D.** Lateral aspect of bones of foot.

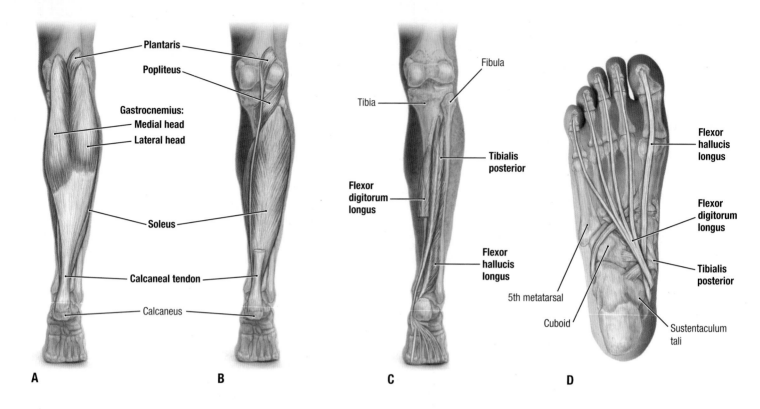

A   **B**   **C**   **D**

6.64   POSTERIOR LEG: MUSCLES

A. and **B.** Muscles of superficial compartment. **C.** and **D.** Muscles of deep compartment.

TABLE 6.16	MUSCLES OF POSTERIOR COMPARTMENT OF LEG			
Muscle	*Proximal Attachment*	*Distal Attachment*	*Innervation[a]*	*Main Actions*
Superficial Muscles				
Gastrocnemius	*Lateral head:* lateral aspect of lateral condyle of femur	Posterior surface of calcaneus via calcaneal tendon (tendocalcaneus)	Tibial nerve (S1 and S2)	Plantar flexes ankle joint when knee joint is extended; raises heel during walking, and flexes knee joint
	Medial head: popliteal surface of femur, superior to medial condyle			
Soleus	Posterior aspect of head of fibula, superior fourth of posterior surface of fibula, soleal line and medial border of tibia			Plantar flexes ankle joint (independent of knee position) and steadies leg on foot
Plantaris	Inferior end of lateral supracondylar line of femur and oblique popliteal ligament			Weakly assists gastrocnemius in plantar flexing ankle joint and flexing knee joint
Deep Muscles				
Popliteus	Lateral surface of lateral condyle of femur and lateral meniscus	Posterior surface of tibia, superior to soleal line	Tibial nerve (**L4**, L5, and S1)	Unlocks fully extended knee joint (laterally rotates femur 5 degrees on planted tibia); weakly flexes knee joint
Flexor hallucis longus	Inferior two thirds of posterior surface of fibula and inferior part of interosseous membrane	Base of distal phalanx of great toe (hallux)		Flexes great toe at all joints and plantar flexes ankle joint; supports medial longitudinal arch of foot
Flexor digitorum longus	Medial part of posterior surface of tibia inferior to soleal line, and by a broad tendon to fibula	Bases of distal phalanges of lateral four digits	Tibial nerve (**S2** and S3)	Flexes lateral four digits and plantar flexes ankle joint; supports longitudinal arches of foot
Tibialis posterior	Interosseous membrane, posterior surface of tibia inferior to soleal line and posterior surface of fibula	Tuberosity of navicular, cuneiform, and cuboid and bases of metatarsals 2–4	Tibial nerve (L4 and L5)	Plantar flexes ankle joint and inverts foot

[a]Numbers indicate spinal cord segmental innervation of nerves (e.g., S2, and S3 indicate that the part of the tibial nerve supplying flexor digitorum longus is derived from two segments of the spinal cord; boldface type [**S2**] indicates main segmental innervation). Damage to one or more of these spinal cord segments or to motor nerve roots arising from these segments results in paralysis of the muscles concerned.

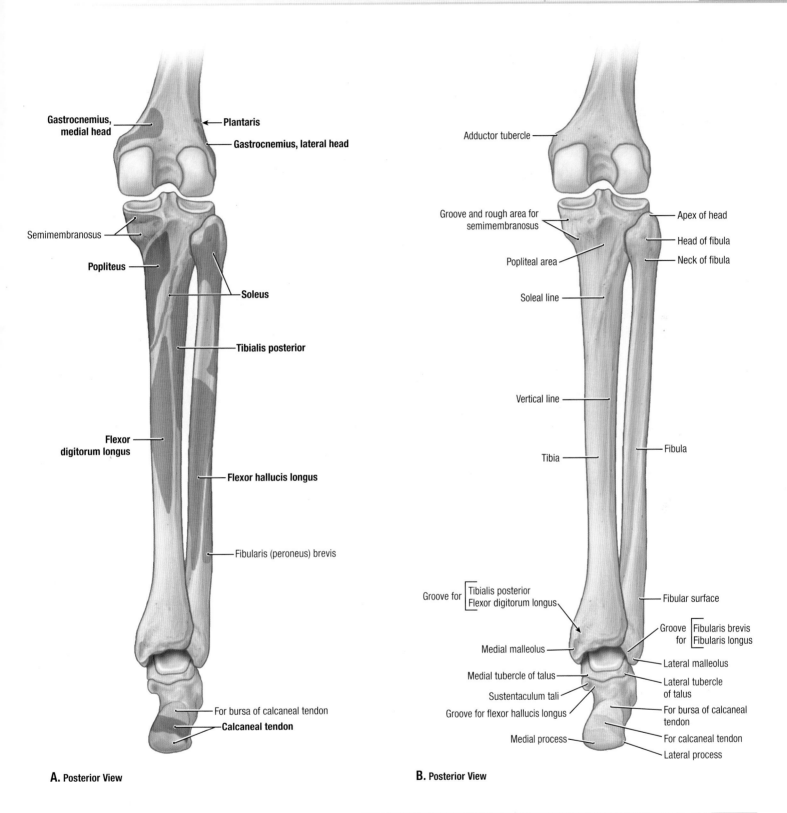

Gastrocnemius, medial head
Plantaris
Gastrocnemius, lateral head
Semimembranosus
Popliteus
Soleus
Tibialis posterior
Flexor digitorum longus
Flexor hallucis longus
Fibularis (peroneus) brevis
For bursa of calcaneal tendon
Calcaneal tendon

A. Posterior View

Adductor tubercle
Groove and rough area for semimembranosus
Popliteal area
Soleal line
Vertical line
Tibia
Apex of head
Head of fibula
Neck of fibula
Fibula
Groove for | Tibialis posterior | Flexor digitorum longus
Medial malleolus
Medial tubercle of talus
Sustentaculum tali
Groove for flexor hallucis longus
Medial process
Fibular surface
Groove for | Fibularis brevis | Fibularis longus
Lateral malleolus
Lateral tubercle of talus
For bursa of calcaneal tendon
For calcaneal tendon
Lateral process

B. Posterior View

POSTERIOR LEG: BONES

6.65

A. Muscle attachments. **B.** Features of bones.

Tibial fractures. The tibial shaft is narrowest at the junction of its middle and inferior thirds, which is the most frequent site of fracture. Unfortunately, this area of the bone also has the poorest blood supply.

Fibular fractures. These commonly occur 2 to 6 cm proximal to the distal end of the lateral malleolus and are often associated with fracture/dislocations of the ankle joint, which are combined with tibial fractures. When a person slips and the foot is forced into an excessively inverted position, the ankle ligaments tear, forcibly tilting the talus against the lateral malleolus and shearing it off.

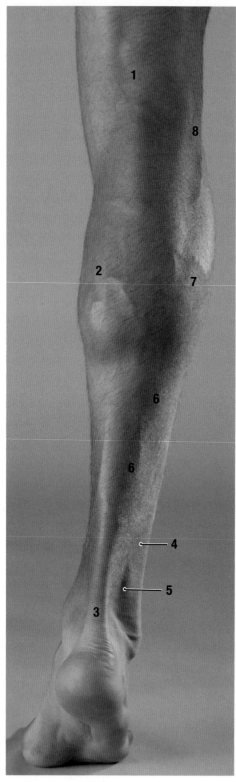

A. Posterior View

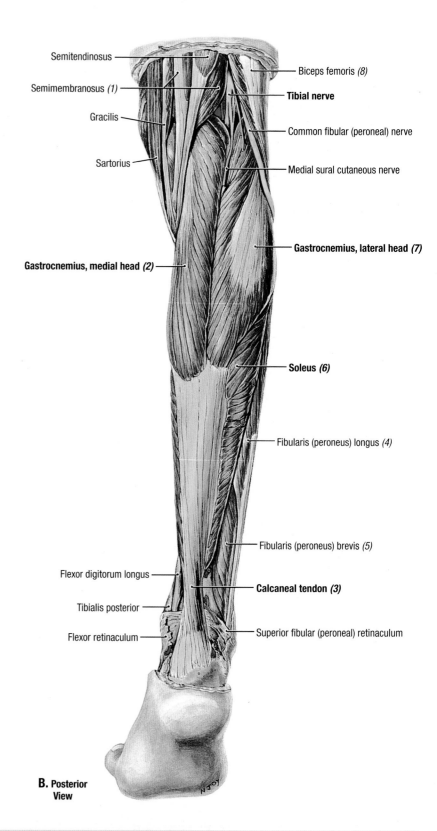

Semitendinosus

Semimembranosus (1)

Gracilis

Sartorius

Gastrocnemius, medial head (2)

Flexor digitorum longus

Tibialis posterior

Flexor retinaculum

Biceps femoris (8)

Tibial nerve

Common fibular (peroneal) nerve

Medial sural cutaneous nerve

Gastrocnemius, lateral head (7)

Soleus (6)

Fibularis (peroneus) longus (4)

Fibularis (peroneus) brevis (5)

Calcaneal tendon (3)

Superior fibular (peroneal) retinaculum

B. Posterior View

6.66 POSTERIOR LEG: SUPERFICIAL MUSCLES OF POSTERIOR COMPARTMENT

A. Surface anatomy. Numbers refer to structures labeled in **(B)**.
B. Dissection.
  Gastrocnemius strain (tennis leg) is a painful calf injury resulting from partial tearing of the medial belly of the muscle at or near its musculotendinous junction. It is caused by overstretching the muscle during simultaneous full extension of the knee joint and dorsiflexion of the ankle joint.

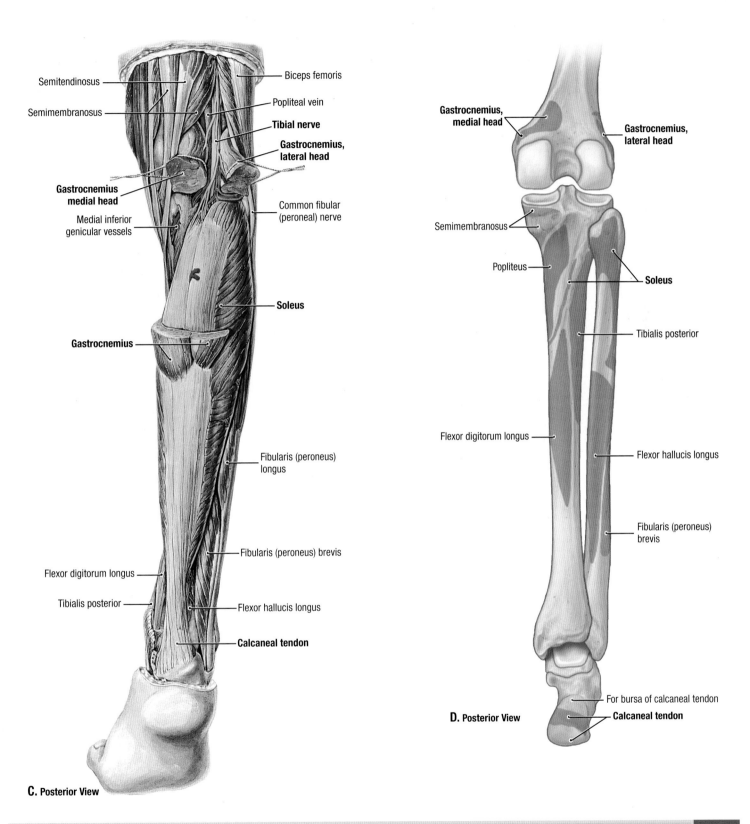

C. Posterior View

D. Posterior View

Labels (C, left figure):
- Semitendinosus
- Semimembranosus
- **Gastrocnemius medial head**
- Medial inferior genicular vessels
- **Gastrocnemius**
- Flexor digitorum longus
- Tibialis posterior
- Biceps femoris
- Popliteal vein
- **Tibial nerve**
- **Gastrocnemius, lateral head**
- Common fibular (peroneal) nerve
- **Soleus**
- Fibularis (peroneus) longus
- Fibularis (peroneus) brevis
- Flexor hallucis longus
- **Calcaneal tendon**

Labels (D, right figure):
- **Gastrocnemius, medial head**
- Semimembranosus
- Popliteus
- Flexor digitorum longus
- **Gastrocnemius, lateral head**
- **Soleus**
- Tibialis posterior
- Flexor hallucis longus
- Fibularis (peroneus) brevis
- For bursa of calcaneal tendon
- **Calcaneal tendon**

POSTERIOR LEG: SUPERFICIAL MUSCLES OF POSTERIOR COMPARTMENT (*continued*)   **6.66**

C. Dissection revealing soleus. **D.** Bones of leg showing muscle attachments.

Inflammation of the calcaneal tendon due to microscopic tears of collagen fibers in the tendon, particularly just superior to its attachment to the calcaneus, results in **calcaneal tendinitis**, which causes pain during walking. **Calcaneal tendon rupture** is probably the most severe acute muscular problem of the leg. Following complete rupture of the tendon, passive dorsiflexion is excessive, and the person cannot plantar flex against resistance.

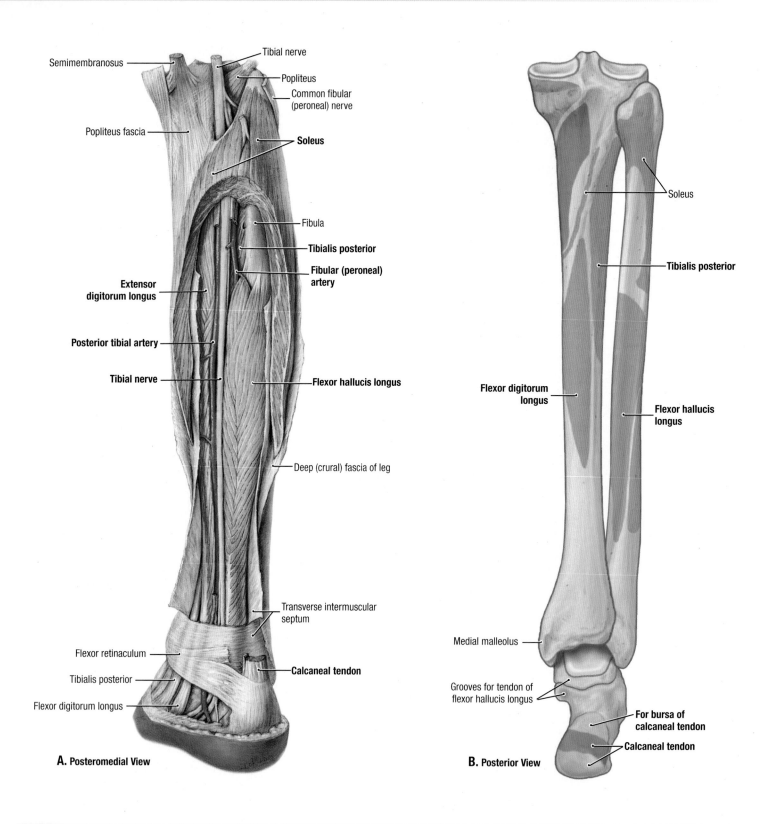

A. Posteromedial View

Semimembranosus
Tibial nerve
Popliteus
Common fibular (peroneal) nerve
Popliteus fascia
Soleus
Fibula
Tibialis posterior
Extensor digitorum longus
Fibular (peroneal) artery
Posterior tibial artery
Tibial nerve
Flexor hallucis longus
Deep (crural) fascia of leg
Transverse intermuscular septum
Flexor retinaculum
Tibialis posterior
Calcaneal tendon
Flexor digitorum longus

B. Posterior View

Soleus
Tibialis posterior
Flexor digitorum longus
Flexor hallucis longus
Medial malleolus
Grooves for tendon of flexor hallucis longus
For bursa of calcaneal tendon
Calcaneal tendon

6.67 POSTERIOR LEG: DEEP MUSCLES OF POSTERIOR COMPARTMENT

A. Superficial dissection. The calcaneal (Achilles) tendon is cut, the gastrocnemius muscle is removed, and only a horseshoe-shaped proximal part of the soleus muscle remains in place. **B.** Bones of leg showing muscle attachments. **Tarsal tunnel syndrome**, the entrapment and compression of the tibial nerve, occurs when there is edema and tightness in the ankle involving the synovial sheaths of the tendons of muscles in the posterior compartment of the leg. The area involved is from the medial malleolus to the calcaneus. The heel pain results from compression of the tibial nerve by the flexor retinaculum.

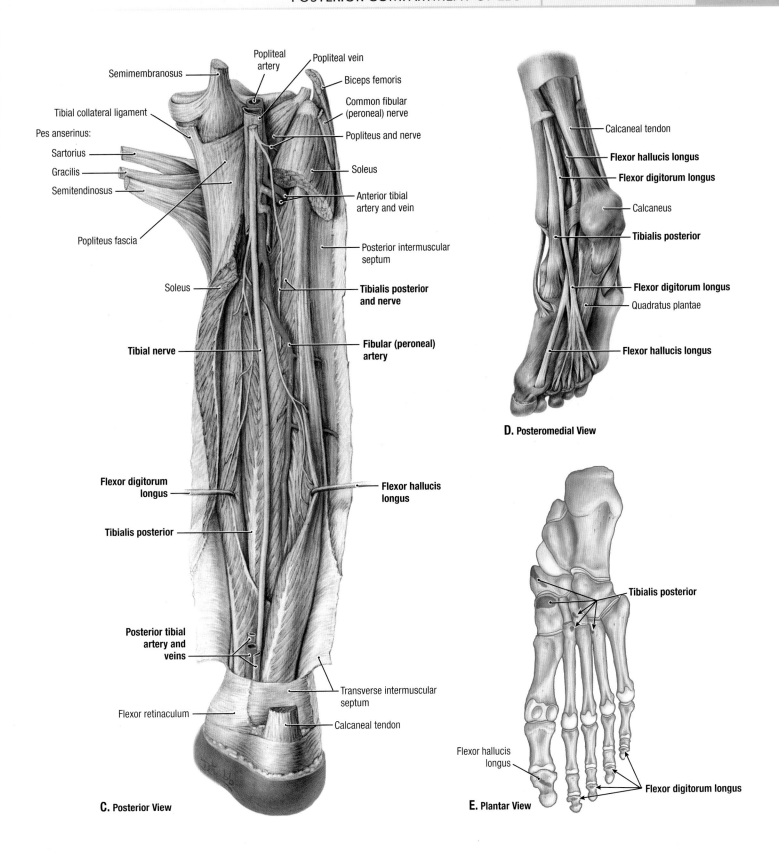

Semimembranosus

Tibial collateral ligament

Pes anserinus:

Sartorius

Gracilis

Semitendinosus

Popliteus fascia

Soleus

Tibial nerve

Flexor digitorum longus

Tibialis posterior

Posterior tibial artery and veins

Flexor retinaculum

C. Posterior View

Popliteal artery

Popliteal vein

Biceps femoris

Common fibular (peroneal) nerve

Popliteus and nerve

Soleus

Anterior tibial artery and vein

Posterior intermuscular septum

Tibialis posterior and nerve

Fibular (peroneal) artery

Flexor hallucis longus

Transverse intermuscular septum

Calcaneal tendon

Calcaneal tendon

Flexor hallucis longus

Flexor digitorum longus

Calcaneus

Tibialis posterior

Flexor digitorum longus

Quadratus plantae

Flexor hallucis longus

D. Posteromedial View

Tibialis posterior

Flexor hallucis longus

Flexor digitorum longus

E. Plantar View

POSTERIOR LEG: DEEP MUSCLES OF POSTERIOR COMPARTMENT (*continued*) | **6.67**

C. Deeper dissection. The flexor hallucis longus and flexor digitorum longus are pulled apart, and the posterior tibial artery is partly excised. The tibialis posterior lies deep to the two long digital flexors. **D.** Crossing of muscles (tendons) of the deep compartment superoposterior to the medial malleolus and in the sole of the foot. **E.** Bones of foot showing muscle attachments.

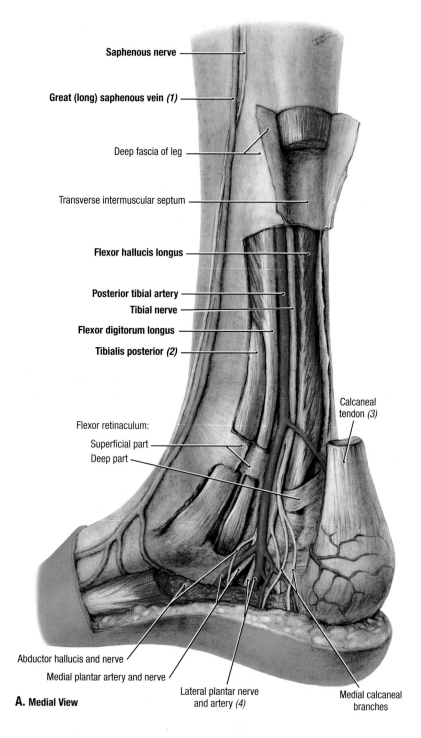

Saphenous nerve

Great (long) saphenous vein *(1)*

Deep fascia of leg

Transverse intermuscular septum

Flexor hallucis longus

Posterior tibial artery

Tibial nerve

Flexor digitorum longus

Tibialis posterior *(2)*

Flexor retinaculum:

Superficial part

Deep part

Calcaneal tendon *(3)*

Abductor hallucis and nerve

Medial plantar artery and nerve

Lateral plantar nerve and artery *(4)*

Medial calcaneal branches

A. Medial View

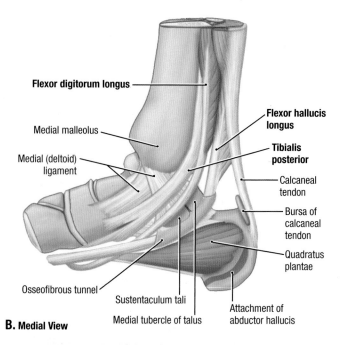

Flexor digitorum longus

Medial malleolus

Medial (deltoid) ligament

Flexor hallucis longus

Tibialis posterior

Calcaneal tendon

Bursa of calcaneal tendon

Quadratus plantae

Osseofibrous tunnel

Sustentaculum tali

Medial tubercle of talus

Attachment of abductor hallucis

B. Medial View

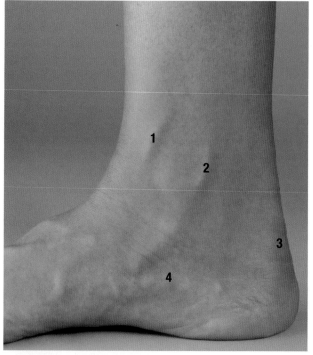

C. Medial View

6.68 MEDIAL ANKLE REGION

A. Dissection. The calcaneal tendon and posterior part of the abductor hallucis were excised. **B.** Schematic illustration of the tendons passing posterior to medial malleolus. **C.** Surface anatomy. Numbers refer to structures labeled in **(A)**.

• The posterior tibial artery and the tibial nerve lie between the flexor digitorum longus land flexor hallucis longus muscles and divide into medial and lateral plantar branches.

• The tibialis posterior and flexor digitorum longus tendons occupy separate osseofibrous tunnels posterior to the medial malleolus.

• The **posterior tibial pulse** can usually be palpated between the posterior surface of the medial malleolus and the medial border of the calcaneal tendon.

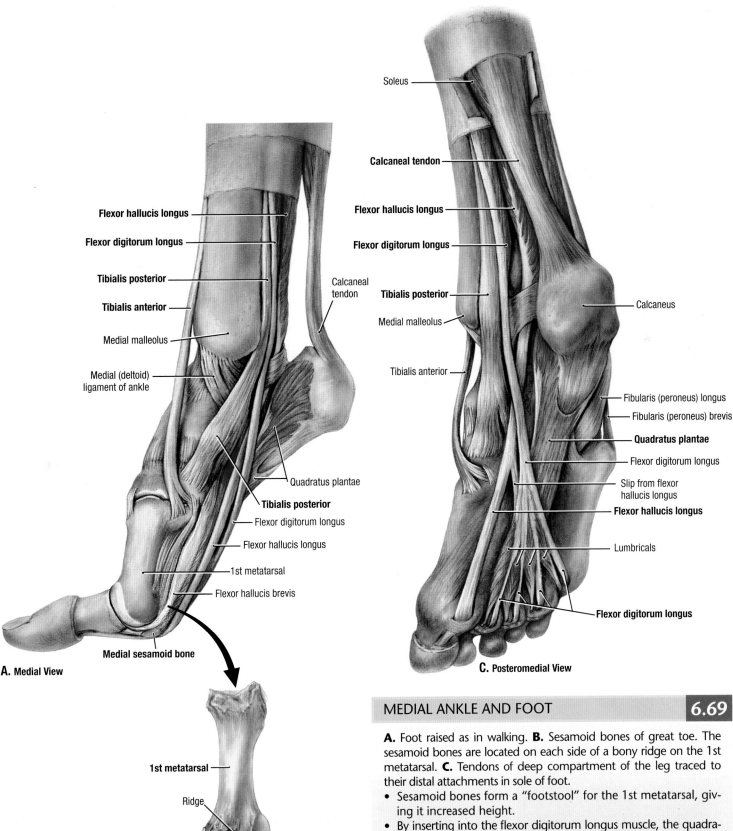

A. Medial View

Flexor hallucis longus
Flexor digitorum longus
Tibialis posterior
Tibialis anterior
Medial malleolus
Medial (deltoid) ligament of ankle
Calcaneal tendon
Quadratus plantae
Tibialis posterior
Flexor digitorum longus
Flexor hallucis longus
1st metatarsal
Flexor hallucis brevis
Medial sesamoid bone

B. Plantar Surface

1st metatarsal
Ridge
Medial sesamoid
Lateral sesamoid
Sheath of flexor hallucis longus tendon

C. Posteromedial View

Soleus
Calcaneal tendon
Flexor hallucis longus
Flexor digitorum longus
Tibialis posterior
Medial malleolus
Tibialis anterior
Calcaneus
Fibularis (peroneus) longus
Fibularis (peroneus) brevis
Quadratus plantae
Flexor digitorum longus
Slip from flexor hallucis longus
Flexor hallucis longus
Lumbricals
Flexor digitorum longus

MEDIAL ANKLE AND FOOT   6.69

A. Foot raised as in walking. **B.** Sesamoid bones of great toe. The sesamoid bones are located on each side of a bony ridge on the 1st metatarsal. **C.** Tendons of deep compartment of the leg traced to their distal attachments in sole of foot.

- Sesamoid bones form a "footstool" for the 1st metatarsal, giving it increased height.
- By inserting into the flexor digitorum longus muscle, the quadratus plantae muscle modifies the oblique pull of flexor tendons.
- The flexor hallucis longus muscle uses three pulleys: grooves on the posterior aspect of the distal end of the tibia, on the posterior aspect of the talus, and inferior to the sustentaculum tali.
- The flexor digitorum longus muscle crosses superficial to the tibialis posterior, superoposterior to the medial malleolus.

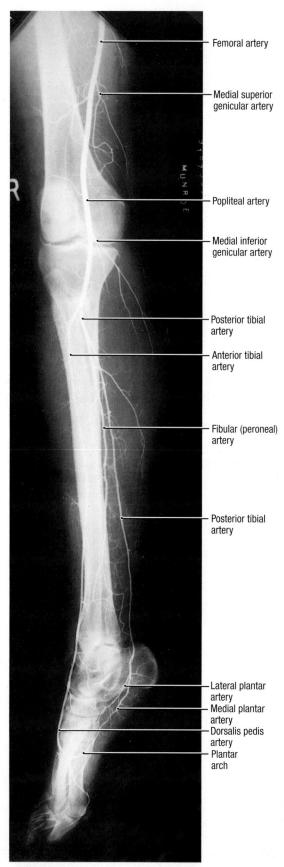

A. Medial View

Femoral artery

Medial superior genicular artery

Popliteal artery

Medial inferior genicular artery

Posterior tibial artery

Anterior tibial artery

Fibular (peroneal) artery

Posterior tibial artery

Lateral plantar artery

Medial plantar artery

Dorsalis pedis artery

Plantar arch

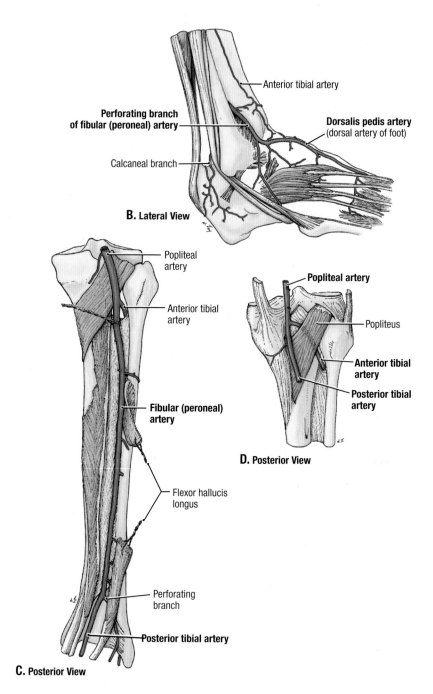

Anterior tibial artery

Perforating branch of fibular (peroneal) artery

Calcaneal branch

Dorsalis pedis artery (dorsal artery of foot)

B. Lateral View

Popliteal artery

Anterior tibial artery

Fibular (peroneal) artery

Flexor hallucis longus

Perforating branch

Posterior tibial artery

C. Posterior View

Popliteal artery

Popliteus

Anterior tibial artery

Posterior tibial artery

D. Posterior View

6.70 POPLITEAL ARTERIOGRAM AND ARTERIAL ANOMALIES

A. Popliteal arteriogram. The femoral artery becomes the popliteal artery at the adductor hiatus. The anterior tibial artery continues as the dorsalis pedis (dorsal artery of the foot). The posterior tibial artery terminates as the medial and lateral plantar arteries; its major branch is the fibular artery. **B.** Anomalous dorsalis pedis artery. The perforating branch of the fibular artery rarely continues as the dorsalis pedis artery, but when it does, the anterior tibial artery ends proximal to the ankle or is a slender vessel. **C.** Absence of posterior tibial artery. Compensatory enlargement of the fibular artery was found to occur in approximately 5% of limbs. **D.** High division of popliteal artery, along with the anterior tibial artery descending anterior to the popliteus muscle. This anomaly was found to occur in approximately 2% of limbs.

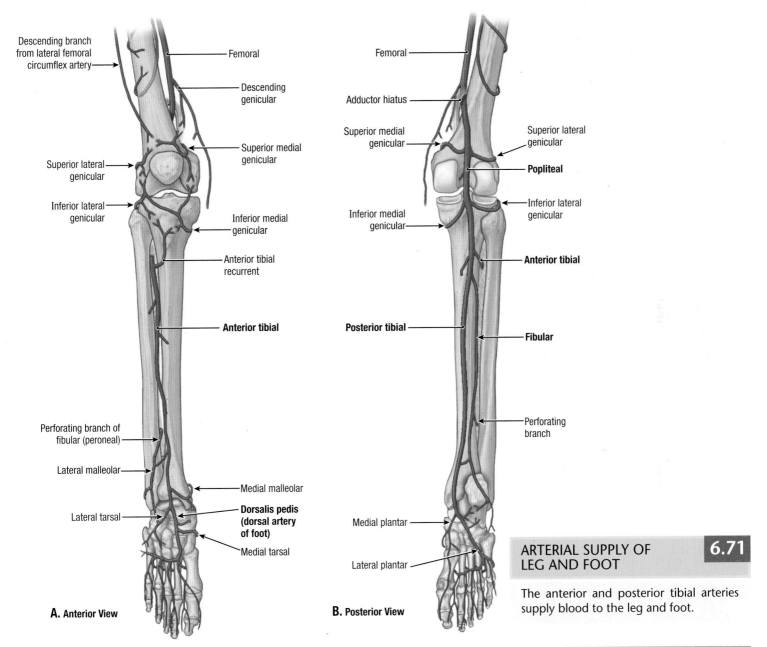

Descending branch from lateral femoral circumflex artery

Femoral

Descending genicular

Superior medial genicular

Superior lateral genicular

Inferior lateral genicular

Inferior medial genicular

Anterior tibial recurrent

Anterior tibial

Perforating branch of fibular (peroneal)

Lateral malleolar

Lateral tarsal

Medial malleolar

Dorsalis pedis (dorsal artery of foot)

Medial tarsal

A. Anterior View

Femoral

Adductor hiatus

Superior medial genicular

Superior lateral genicular

Popliteal

Inferior medial genicular

Inferior lateral genicular

Anterior tibial

Posterior tibial

Fibular

Perforating branch

Medial plantar

Lateral plantar

B. Posterior View

ARTERIAL SUPPLY OF LEG AND FOOT | 6.71

The anterior and posterior tibial arteries supply blood to the leg and foot.

TABLE 6.17	ARTERIAL SUPPLY OF LEG AND FOOT		
Artery	Origin	Course	Distribution in Leg
Popliteal	Continuation of femoral artery at adductor hiatus	Passes through popliteal fossa to leg; divides into anterior and posterior tibial arteries at lower border of popliteus	All aspects of knee via genicular arteries
Anterior tibial	From popliteal	Passes between tibia and fibula into anterior compartment through gap superior to interosseous membrane; descends between tibialis anterior and extensor digitorum longus muscles	Anterior compartment of leg
Dorsalis pedis (dorsal artery of foot)	Continuation of anterior tibial artery distal to talocrural joint	Descends to first interosseous space; pierces first dorsal interosseous muscle as deep plantar artery; joins deep plantar arch	Muscles on dorsum of foot
Posterior tibial	From popliteal	Passes through posterior compartment; divides into medial and lateral plantar arteries posterior to medial malleolus	Posterior and lateral compartments of leg, nutrient artery passes to tibia
Fibular (peroneal)		Descends in posterior compartment adjacent to posterior intermuscular septum	Posterior compartment: perforating branches supply lateral compartment
Medial plantar	From posterior tibial	In foot between abductor hallucis and flexor digitorum brevis muscles	Supplies mainly muscles of great toe and skin on medial side of sole of foot
Lateral plantar		Runs anterolaterally deep to abductor hallucis and flexor digitorum brevis, and then arches medially to form deep plantar arch	Supplies lateral aspect of sole of foot

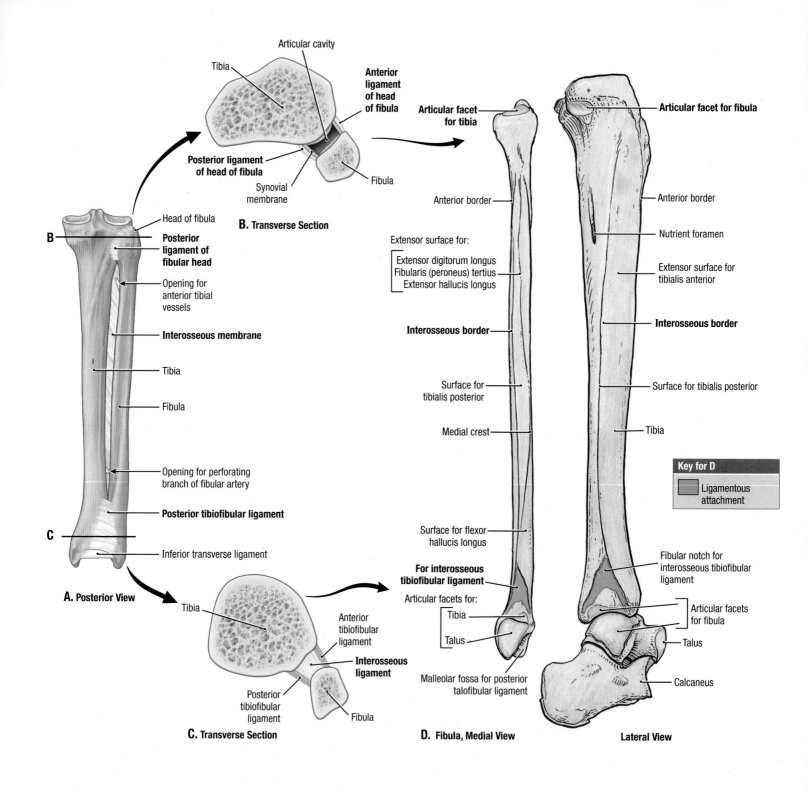

Articular cavity

Tibia

Anterior ligament of head of fibula

Articular facet for tibia

Articular facet for fibula

Posterior ligament of head of fibula

Synovial membrane

Fibula

B. Transverse Section

Head of fibula

B

Posterior ligament of fibular head

Opening for anterior tibial vessels

Interosseous membrane

Tibia

Fibula

Opening for perforating branch of fibular artery

Posterior tibiofibular ligament

C

Inferior transverse ligament

A. Posterior View

Anterior border

Extensor surface for:
Extensor digitorum longus
Fibularis (peroneus) tertius
Extensor hallucis longus

Interosseous border

Surface for tibialis posterior

Medial crest

Surface for flexor hallucis longus

For interosseous tibiofibular ligament

Articular facets for:
Tibia
Talus

Malleolar fossa for posterior talofibular ligament

D. Fibula, Medial View

Anterior border

Nutrient foramen

Extensor surface for tibialis anterior

Interosseous border

Surface for tibialis posterior

Tibia

Key for D
Ligamentous attachment

Fibular notch for interosseous tibiofibular ligament

Articular facets for fibula

Talus

Calcaneus

Lateral View

Tibia

Anterior tibiofibular ligament

Interosseous ligament

Posterior tibiofibular ligament

Fibula

C. Transverse Section

6.72 TIBIOFIBULAR JOINT AND TIBIOFIBULAR SYNDESMOSIS

A. Overview. **B.** Tibiofibular joint. **C.** Tibiofibular syndesmosis. **D.** Tibia and fibula, disarticulated.

- The superior tibiofibular joint (proximal tibiofibular joint) is a plane type of synovial joint between the flat facet on the fibular head and a similar facet located posterolaterally on the lateral tibial condyle. The tense joint capsule surrounds the joint and attaches to the margins of the articular surfaces of the fibula and tibia.

- The tibiofibular syndesmosis is a fibrous joint. This articulation is essential for stability of the ankle joint because it keeps the lateral malleolus firmly against the lateral surface of the talus. The strong interosseous tibiofibular ligament is continuous superiorly with the interosseous membrane and forms the principal connection between the distal ends of the tibia and fibula.

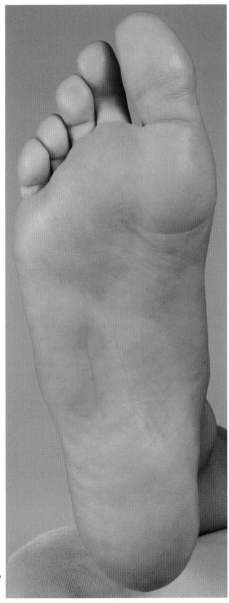

A. Plantar View

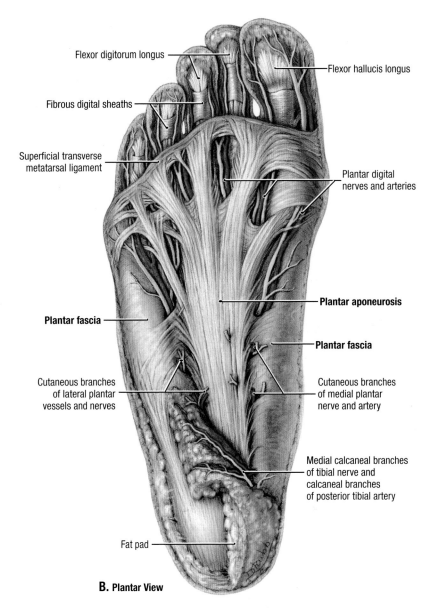

Flexor digitorum longus

Fibrous digital sheaths

Superficial transverse metatarsal ligament

Flexor hallucis longus

Plantar digital nerves and arteries

Plantar aponeurosis

Plantar fascia

Plantar fascia

Cutaneous branches of lateral plantar vessels and nerves

Cutaneous branches of medial plantar nerve and artery

Medial calcaneal branches of tibial nerve and calcaneal branches of posterior tibial artery

Fat pad

B. Plantar View

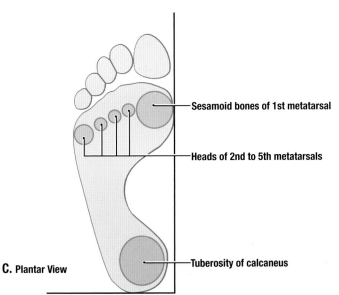

Sesamoid bones of 1st metatarsal

Heads of 2nd to 5th metatarsals

Tuberosity of calcaneus

C. Plantar View

SOLE OF FOOT, SUPERFICIAL

6.73

A. Surface anatomy. **B.** Dissection. Plantar aponeurosis and fascia, with neurovascular structures. **C.** Weight-bearing areas. The weight of the body is transmitted to the talus from the tibia and fibula. It is then transmitted to the tuberosity of the calcaneus, the heads of the 2nd to 5th metatarsals, and the sesamoid bones of the first digit.

Plantar fasciitis, strain and inflammation of the plantar aponeurosis may result from running and high-impact aerobics, especially when inappropriate footwear is worn. It causes pain on the plantar surface of the heel and on the medial aspect of the foot. Point tenderness is located at the proximal attachment of the plantar aponeurosis to the medial tubercle of the calcaneus and on the medial surface of this bone. The pain increases with passive extension of the great toe and may be further exacerbated by dorsiflexion of the ankle and/or weight bearing.

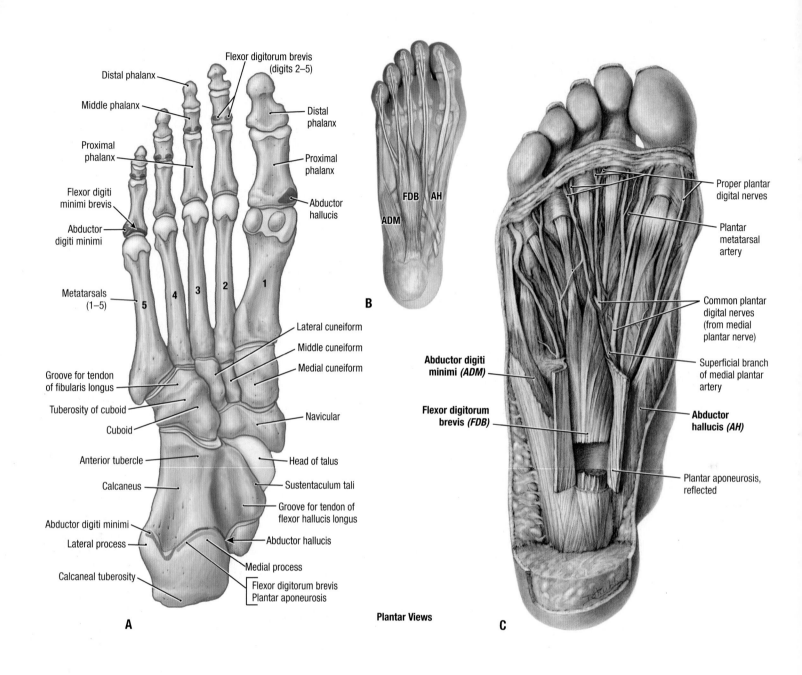

Flexor digitorum brevis (digits 2–5)

Distal phalanx

Middle phalanx

Proximal phalanx

Flexor digiti minimi brevis

Abductor digiti minimi

Metatarsals (1–5)

Groove for tendon of fibularis longus

Tuberosity of cuboid

Cuboid

Anterior tubercle

Calcaneus

Abductor digiti minimi

Lateral process

Calcaneal tuberosity

Distal phalanx

Proximal phalanx

Abductor hallucis

Lateral cuneiform

Middle cuneiform

Medial cuneiform

Navicular

Head of talus

Sustentaculum tali

Groove for tendon of flexor hallucis longus

Abductor hallucis

Medial process

Flexor digitorum brevis Plantar aponeurosis

A

B

FDB   AH

ADM

Proper plantar digital nerves

Plantar metatarsal artery

Common plantar digital nerves (from medial plantar nerve)

Superficial branch of medial plantar artery

Abductor hallucis (AH)

Plantar aponeurosis, reflected

Abductor digiti minimi (ADM)

Flexor digitorum brevis (FDB)

Plantar Views

C

6.74 FIRST LAYER OF MUSCLES OF SOLE OF FOOT

A. Bones. **B.** Overview. **C.** Dissection. Muscles and neurovascular structures.

TABLE 6.18	MUSCLES IN SOLE OF FOOT—FIRST LAYER			
Muscle	Proximal Attachment	Distal Attachment	Innervation	Actions[a]
Abductor hallucis	Medial process of tuberosity of calcaneus, flexor retinaculum, and plantar aponeurosis	Medial side of base of proximal phalanx of first digit	Medial plantar nerve (S2–S3)	Abducts and flexes first digit
Flexor digitorum brevis	Medial process of tuberosity of calcaneus, plantar aponeurosis, and intermuscular septa	Both sides of middle phalanges of lateral four digits		Flexes lateral four digits
Abductor digiti minimi	Medial and lateral processes of tuberosity of calcaneus, plantar aponeurosis, and intermuscular septa	Lateral side of base of proximal phalanx of fifth digit	Lateral plantar nerve (S2–S3)	Abducts and flexes fifth digit

[a]Although individual actions are described, the primary function of the intrinsic muscles of the foot is to act collectively to resist forces that stress (attempt to flatten) the arches of the foot.

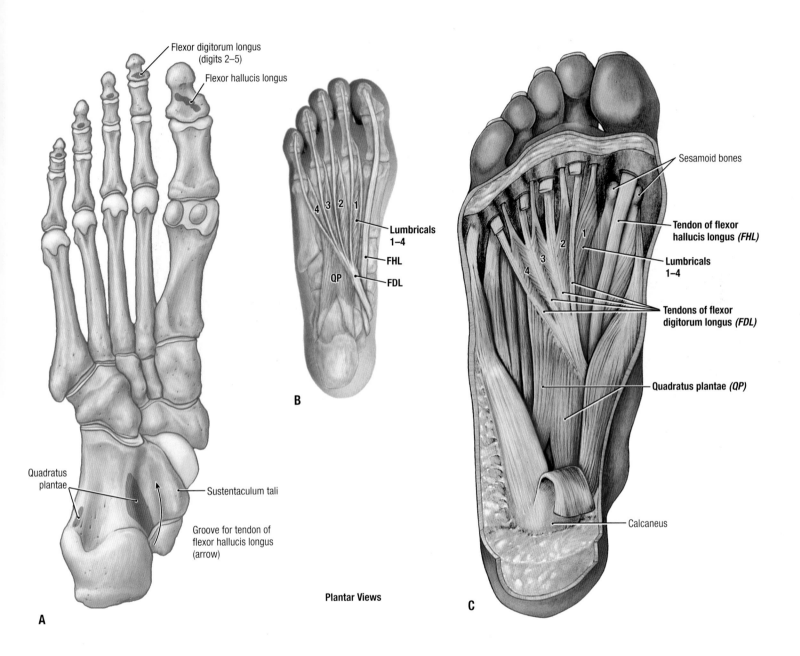

Flexor digitorum longus (digits 2–5)

Flexor hallucis longus

Quadratus plantae

Sustentaculum tali

Groove for tendon of flexor hallucis longus (arrow)

A

Lumbricals 1–4

FHL

QP

FDL

Plantar Views

B

Sesamoid bones

Tendon of flexor hallucis longus *(FHL)*

Lumbricals 1–4

Tendons of flexor digitorum longus *(FDL)*

Quadratus plantae *(QP)*

Calcaneus

C

SECOND LAYER OF MUSCLES OF SOLE OF FOOT

6.75

A. Bony attachments. **B.** Overview. **C.** Dissection.

TABLE 6.19	MUSCLES IN SOLE OF FOOT—SECOND LAYER			
Muscle	*Proximal Attachment*	*Distal Attachment*	*Innervation*	*Actions[a]*
Quadratus plantae	Medial surface and lateral margin of plantar surface of calcaneus	Posterolateral margin of tendon of flexor digitorum longus	Lateral plantar nerve (S2–S3)	Assists flexor digitorum longus in flexing lateral four digits
Lumbricals	Tendons of flexor digitorum longus	Medial aspect of extensor expansion over lateral four digits	*Medial one:* medial plantar nerve (S2–S3) *Lateral three:* lateral plantar nerve (S2–S3)	Flex proximal phalanges and extend middle and distal phalanges of lateral four digits

[a]Although individual actions are described, the primary function of the intrinsic muscles of the foot is to act collectively to resist forces that stress (attempt to flatten) the arches of the foot.

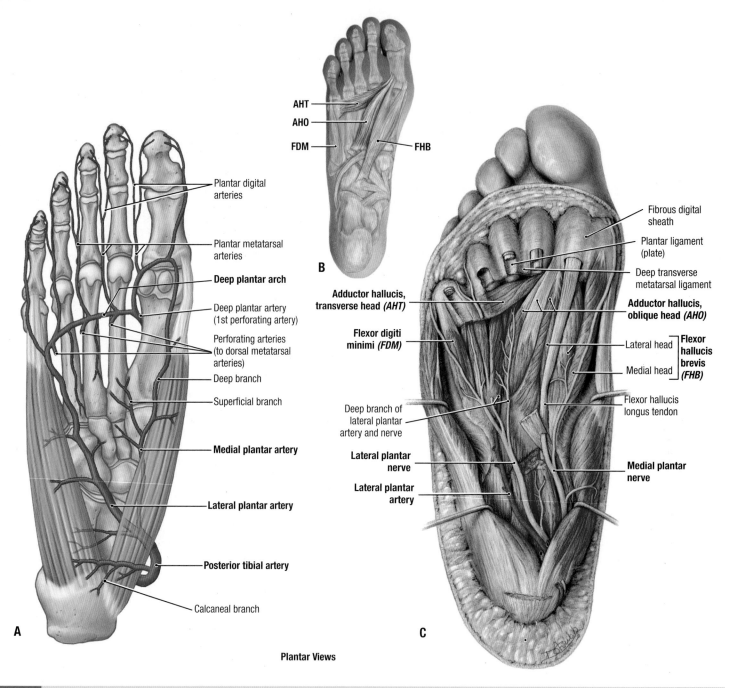

Plantar digital arteries

Plantar metatarsal arteries

Deep plantar arch

Deep plantar artery (1st perforating artery)

Perforating arteries (to dorsal metatarsal arteries)

Deep branch

Superficial branch

Medial plantar artery

Lateral plantar artery

Posterior tibial artery

Calcaneal branch

AHT
AHO
FDM — — FHB

B

Adductor hallucis, transverse head **(AHT)**

Flexor digiti minimi **(FDM)**

Deep branch of lateral plantar artery and nerve

Lateral plantar nerve

Lateral plantar artery

Fibrous digital sheath

Plantar ligament (plate)

Deep transverse metatarsal ligament

Adductor hallucis, oblique head (AHO)

Lateral head **Flexor hallucis brevis (FHB)**
Medial head

Flexor hallucis longus tendon

Medial plantar nerve

A   **C**

Plantar Views

6.76 THIRD LAYER OF MUSCLES AND ARTERIAL SUPPLY OF SOLE OF FOOT

A. Arterial supply. **B.** Overview. **C.** Dissection. Muscles and neurovascular structures.

TABLE 6.20	MUSCLES IN SOLE OF FOOT—THIRD LAYER			
Muscle	*Proximal Attachment*	*Distal Attachment*	*Innervation*	*Actions[a]*
Flexor hallucis brevis	Plantar surfaces of cuboid and lateral cuneiforms	Both sides of base of proximal phalanx of first digit	Medial plantar nerve (S2–S3)	Flexes proximal phalanx of first digit
Adductor hallucis	*Oblique head:* bases of metatarsals 2–4 *Transverse head:* plantar ligaments of metatarsophalangeal joints	Tendons of both heads attach to lateral side of base of proximal phalanx of first digit	Deep branch of lateral plantar nerve (S2–S3)	Adducts first digit; assists in maintaining transverse arch of foot
Flexor digiti minimi	Base of 5th metatarsal	Base of proximal phalanx of fifth digit	Superficial branch of lateral plantar nerve (S2–S3)	Flexes proximal phalanx of fifth digit, thereby assisting with its flexion

[a]Although individual actions are described, the primary function of the intrinsic muscles of the foot is to act collectively to resist forces that stress (attempt to flatten) the arches of the foot.

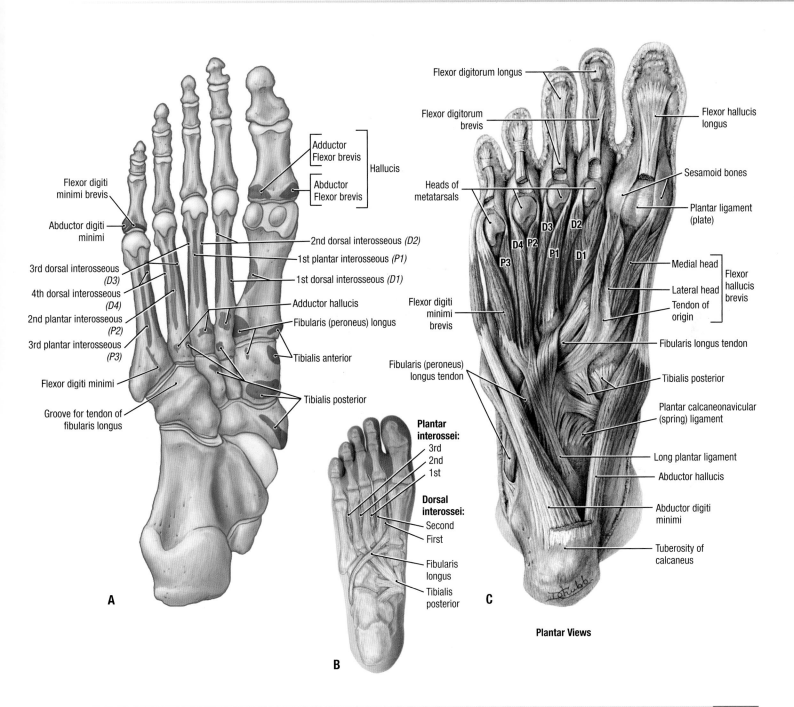

A. Bony attachments of muscles of third and fourth layers. **B.** Overview. **C.** Dissection. Muscles and ligaments.

FOURTH LAYER OF MUSCLES OF SOLE OF FOOT

6.77

TABLE 6.21	MUSCLES IN SOLE OF FOOT—FOURTH LAYER			
Muscle	*Proximal Attachment*	*Distal Attachment*	*Innervation*	*Actions[a]*
Plantar interossei (three muscles; P1–P3)	Plantar aspect of medial sides of shafts of metatarsals 3–5	Medial sides of bases of proximal phalanges of third to fifth digits	Lateral plantar nerve (S2–S3)	Adduct digits 3–5 and flex metatarsophalangeal joints
Dorsal interossei (four muscles; D1–D4)	Adjacent sides of shafts of metatarsals 1–5	First: medial side of proximal phalanx of second digit Second to fourth: lateral sides of second to fourth digits		Abduct digits 2–4 and flex metatarsophalangeal joints

[a]Although individual actions are described, the primary function of the intrinsic muscles of the foot is to act collectively to resist forces that stress (attempt to flatten) the arches of the foot.

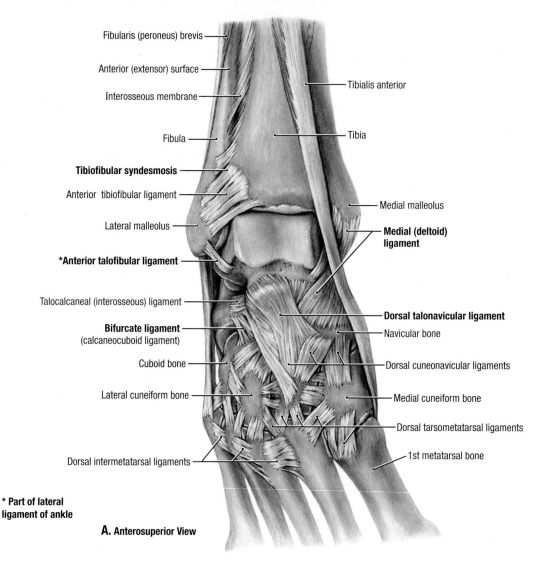

Fibularis (peroneus) brevis

Anterior (extensor) surface

Interosseous membrane

Fibula

Tibiofibular syndesmosis

Anterior tibiofibular ligament

Lateral malleolus

*Anterior talofibular ligament

Talocalcaneal (interosseous) ligament

Bifurcate ligament
(calcaneocuboid ligament)

Cuboid bone

Lateral cuneiform bone

Dorsal intermetatarsal ligaments

Tibialis anterior

Tibia

Medial malleolus

**Medial (deltoid)
ligament**

Dorsal talonavicular ligament

Navicular bone

Dorsal cuneonavicular ligaments

Medial cuneiform bone

Dorsal tarsometatarsal ligaments

1st metatarsal bone

*** Part of lateral
ligament of ankle**

A. Anterosuperior View

| 6.78 | ANKLE JOINT AND LIGAMENTS OF DORSUM OF FOOT |

A. Dissection. The ankle joint is plantar flexed, and its anterior capsular fibers are removed. Note that the bifurcate ligament, a Y-shaped ligament consisting of calcaneocuboid and calcaneonavicular ligaments, and the dorsal talonavicular ligament are the primary dorsal ligaments of the transverse tarsal joint. **B.** Ankle joint with joint cavity distended with injected latex. Note the relations of the tendons to the sustentaculum tali: the flexor hallucis longus inferior to it, flexor digitorum longus along its medial aspect, and tibialis posterior superior to it and in contact with the medial (deltoid) ligament.

A **Pott fracture-dislocation of the ankle** occurs when the foot is forcibly everted. This action pulls on the extremely strong medial (deltoid) ligament, often avulsing the medial malleolus and compressing the lateral malleolus against the talus, shearing off the malleolus or, more often, fracturing the fibula superior to the tibiofibular syndesmosis.

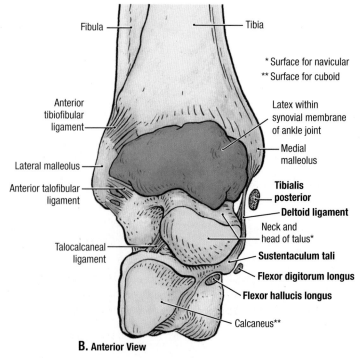

Fibula

Tibia

* Surface for navicular
** Surface for cuboid

Anterior
tibiofibular
ligament

Lateral malleolus

Anterior talofibular
ligament

Talocalcaneal
ligament

Latex within
synovial membrane
of ankle joint

Medial
malleolus

**Tibialis
posterior**

Deltoid ligament

Neck and
head of talus*

Sustentaculum tali

Flexor digitorum longus

Flexor hallucis longus

Calcaneus**

B. Anterior View

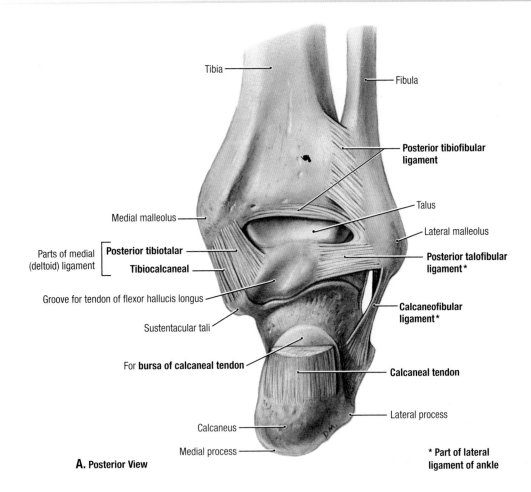

Tibia

Fibula

Posterior tibiofibular ligament

Talus

Medial malleolus

Lateral malleolus

Parts of medial (deltoid) ligament — [**Posterior tibiotalar**

Tibiocalcaneal

Posterior talofibular ligament *

Groove for tendon of flexor hallucis longus

Calcaneofibular ligament *

Sustentacular tali

For **bursa of calcaneal tendon**

Calcaneal tendon

Lateral process

Calcaneus

Medial process

A. Posterior View

* Part of lateral ligament of ankle

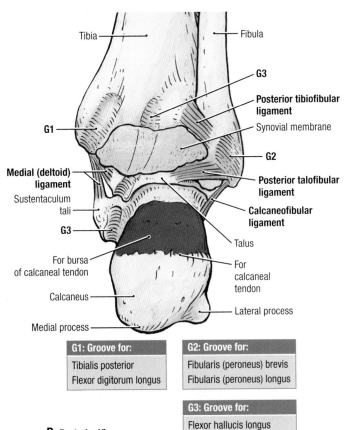

Tibia

Fibula

G3

Posterior tibiofibular ligament

Synovial membrane

G1

G2

Medial (deltoid) ligament

Posterior talofibular ligament

Sustentaculum tali

Calcaneofibular ligament

G3

Talus

For bursa of calcaneal tendon

For calcaneal tendon

Calcaneus

Lateral process

Medial process

G1: Groove for:	**G2: Groove for:**
Tibialis posterior	Fibularis (peroneus) brevis
Flexor digitorum longus	Fibularis (peroneus) longus

G3: Groove for:
Flexor hallucis longus

B. Posterior View

POSTERIOR ASPECT OF ANKLE JOINT 6.79

A. Dissection. **B.** Ankle joint with joint cavity distended with latex. Observe the grooves for the flexor hallucis longus muscle, which crosses the middle of the ankle joint posteriorly, the two tendons posterior to the medial malleolus, and the two tendons posterior to the lateral malleolus.

- The posterior aspect of the ankle joint is strengthened by the transversely oriented posterior tibiofibular and posterior talofibular ligaments.
- The calcaneofibular ligament stabilizes the joint laterally, and the posterior tibiotalar and tibiocalcanean parts of the medial (deltoid) ligament stabilize it medially.
- The groove for the flexor hallucis tendon is between the medial and lateral tubercles of the talus and continues inferior to the sustentaculum tali.

Calcaneal bursitis results from inflammation of the bursa of the calcaneal tendon located between the calcaneal tendon and the superior part of the posterior surface of the calcaneus. Calcaneal bursitis causes pain posterior to the heel and occurs commonly during long-distance running, basketball, and tennis. It is caused by excessive friction on the bursa as the calcaneal tendon continuously slides over it.

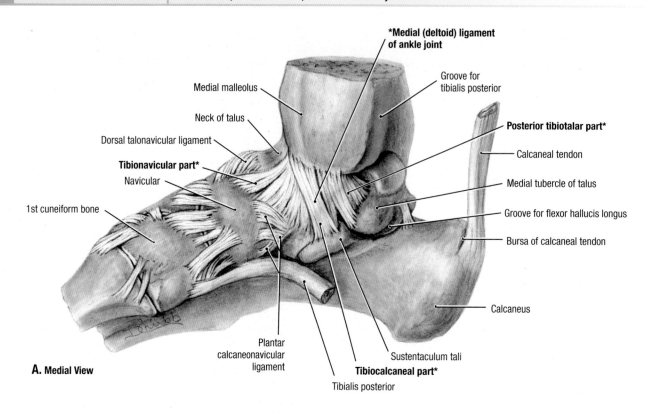

*Medial (deltoid) ligament of ankle joint

Medial malleolus

Neck of talus

Dorsal talonavicular ligament

Tibionavicular part*

Navicular

1st cuneiform bone

Groove for tibialis posterior

Posterior tibiotalar part*

Calcaneal tendon

Medial tubercle of talus

Groove for flexor hallucis longus

Bursa of calcaneal tendon

Calcaneus

Plantar calcaneonavicular ligament

Sustentaculum tali

Tibiocalcaneal part*

Tibialis posterior

A. Medial View

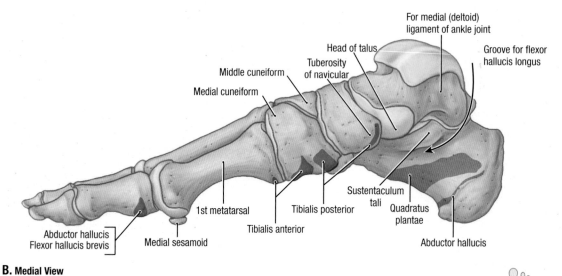

For medial (deltoid) ligament of ankle joint

Head of talus

Tuberosity of navicular

Middle cuneiform

Medial cuneiform

Groove for flexor hallucis longus

1st metatarsal

Tibialis posterior

Tibialis anterior

Sustentaculum tali

Quadratus plantae

Abductor hallucis

Abductor hallucis
Flexor hallucis brevis

Medial sesamoid

B. Medial View

6.80 MEDIAL LIGAMENTS OF ANKLE REGION

A. Dissection. **B.** Bones. The joint capsule of the ankle joint is re-inforced medially by the large, strong medial (deltoid) ligament that attaches proximally to the medial malleolus and fans out from it to attach distally to the talus, calcaneus, and navicular via four adjacent and continuous parts: the tibionavicular part, the tibiocalcaneal part, and the anterior and posterior tibiotalar parts. The medial ligament stabilizes the ankle joint during eversion of the foot and prevents subluxation (partial dislocation) of the ankle joint. **C.** Normal medial longitudinal arch.

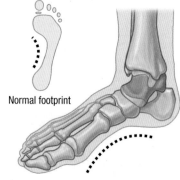

Normal footprint

C. Normal Medial Longitudinal Arch

Key for A, B and C

A	Calcaneal (Achilles) tendon
Ca	Calcaneus
Cb	Cuboid
Cu	Cuneiforms
F	Fat
L	Lateral malleolus
M	Medial malleolus
MT	Metatarsal
N	Navicular
S	Sustentaculum tali
Su	Superimposed tibia and fibula
T	Talus
TF	Tibiofibular syndesmosis
TH	Head of talus
TN	Neck of talus
TS	Tarsal sinus

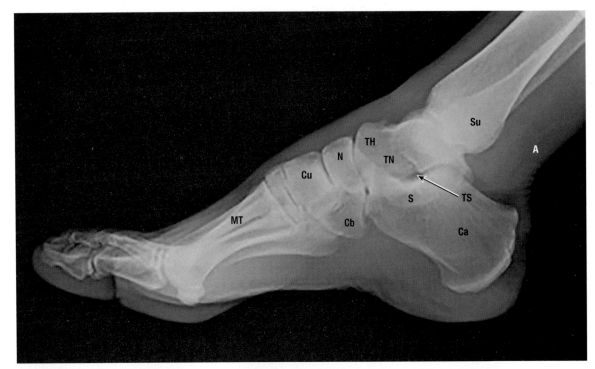

A. Medial Radiograph

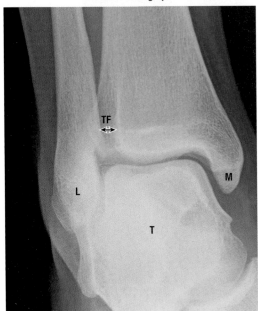

B. Anteroposterior Radiograph

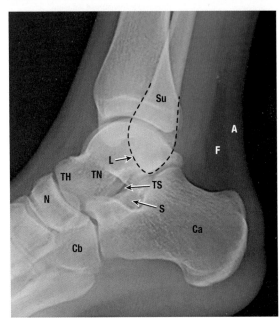

C. Lateral Radiograph

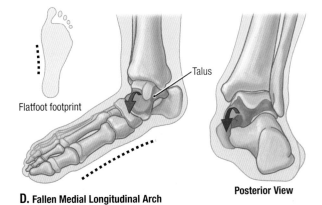

D. Fallen Medial Longitudinal Arch

Posterior View

RADIOGRAPHS OF ANKLE AND FOOT | 6.81

A–C. Imaging of ankle region and tarsal bones. **D. Pes planus** (flatfeet). Acquired flatfeet ("fallen arches") are likely to be secondary to dysfunction of the tibialis posterior due to trauma, degeneration with age, or denervation. In the absence of normal passive or dynamic support, the plantar calcaneonavicular ligament fails to support the head of the talus. Consequently, the head of the talus displaces inferomedially. As a result, flattening of the medial longitudinal arch occurs along with lateral deviation of the forefoot. Flatfeet are common in older people, particularly if they undertake much unaccustomed standing or gain weight rapidly, adding stress on the muscles and increasing strain on the ligaments supporting the arches.

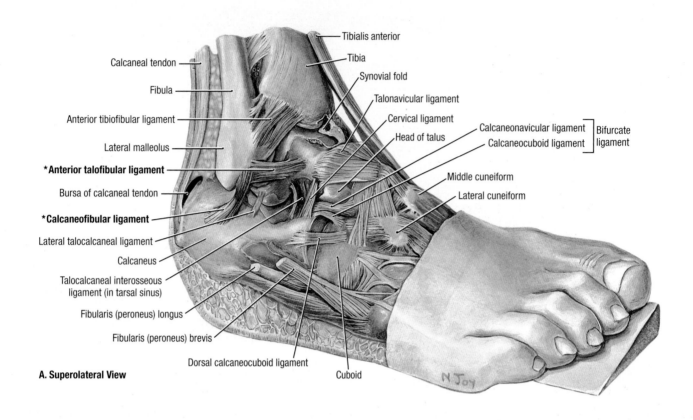

Tibialis anterior
Tibia
Synovial fold
Talonavicular ligament
Cervical ligament
Head of talus
Calcaneonavicular ligament ⎤ Bifurcate
Calcaneocuboid ligament ⎦ ligament
Middle cuneiform
Lateral cuneiform

Calcaneal tendon
Fibula
Anterior tibiofibular ligament
Lateral malleolus
*Anterior talofibular ligament
Bursa of calcaneal tendon
*Calcaneofibular ligament
Lateral talocalcaneal ligament
Calcaneus
Talocalcaneal interosseous ligament (in tarsal sinus)
Fibularis (peroneus) longus
Fibularis (peroneus) brevis
Dorsal calcaneocuboid ligament
Cuboid

A. Superolateral View

*Parts of lateral ligament of ankle

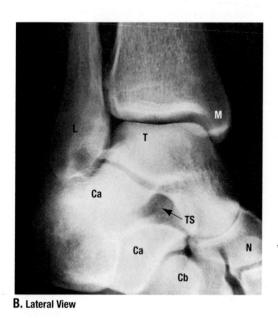

B. Lateral View

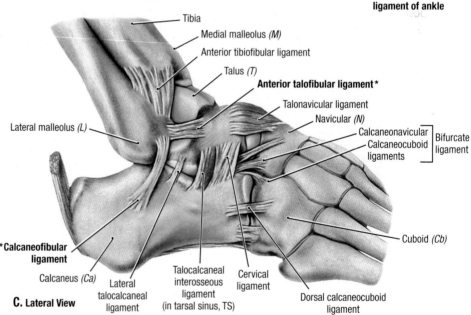

Tibia
Medial malleolus (M)
Anterior tibiofibular ligament
Talus (T)
Anterior talofibular ligament *
Talonavicular ligament
Navicular (N)
Calcaneonavicular ⎤ Bifurcate
Calcaneocuboid ⎦ ligament
ligaments

Lateral malleolus (L)

Cuboid (Cb)

*Calcaneofibular ligament
Calcaneus (Ca)
Lateral talocalcaneal ligament
Talocalcaneal interosseous ligament (in tarsal sinus, TS)
Cervical ligament
Dorsal calcaneocuboid ligament

C. Lateral View

6.82 | LATERAL LIGAMENTS OF ANKLE REGION

A. Dissection with foot inverted by underlying wedge. **B.** Lateral radiograph. Abbreviations refer to structures labeled in **(C). C.** Dissection.

The lateral ligament of the ankle consists of three separate ligaments: (1) anterior talofibular ligament, (2) calcaneofibular ligament, and (3) posterior talofibular ligament (see Fig. 6.79A).

Ankle sprains (partial or fully torn ligaments) are common injuries. Ankle sprains nearly always result from forceful inversion of the weight-bearing plantar flexed foot. The anterior talofibular ligament is most commonly injured, resulting in instability of the ankle. The calcaneofibular is also often torn.

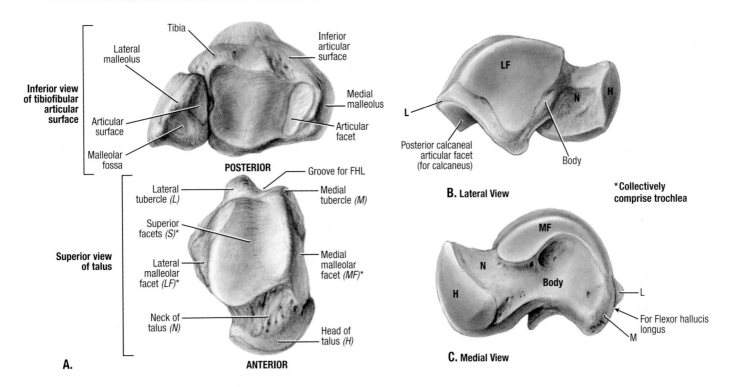

A.

Inferior view of tibiofibular articular surface

Tibia

Lateral malleolus

Inferior articular surface

Medial malleolus

Articular surface

Articular facet

Malleolar fossa

POSTERIOR

Superior view of talus

Groove for FHL

Lateral tubercle (L)

Medial tubercle (M)

Superior facets (S)*

Lateral malleolar facet (LF)*

Medial malleolar facet (MF)*

Neck of talus (N)

Head of talus (H)

ANTERIOR

LF

L

Posterior calcaneal articular facet (for calcaneus)

N

H

Body

B. Lateral View

*Collectively comprise trochlea

MF

N

Body

H

L

For Flexor hallucis longus

M

C. Medial View

ARTICULAR SURFACES OF ANKLE JOINT

6.83

A. Superior view of talus separated from distal ends of tibia and fibula. The superior articular surface of the talus is broader anteriorly than posteriorly. The fully dorsiflexed position is stable compared with the fully plantar flexed position. In plantar flexion, when the tibia and fibula articulate with the narrower posterior part of the superior articular surface of the talus, some side-to-side movement of the joint is allowed, accounting for the instability of the joint in this position. **B.** Lateral view of talus. The triangular lateral facet is for articulation with the lateral malleolus. **C.** Medial view of talus. The comma-shaped medial facet is for articulation with the medial malleolus.

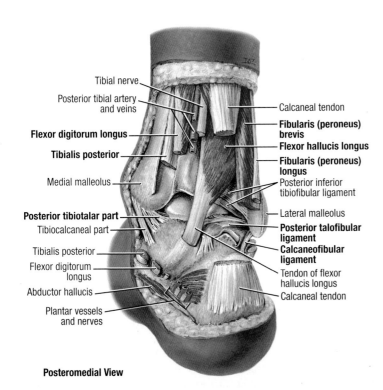

Tibial nerve

Posterior tibial artery and veins

Flexor digitorum longus

Tibialis posterior

Medial malleolus

Posterior tibiotalar part

Tibiocalcaneal part

Tibialis posterior

Flexor digitorum longus

Abductor hallucis

Plantar vessels and nerves

Calcaneal tendon

Fibularis (peroneus) brevis

Flexor hallucis longus

Fibularis (peroneus) longus

Posterior inferior tibiofibular ligament

Lateral malleolus

Posterior talofibular ligament

Calcaneofibular ligament

Tendon of flexor hallucis longus

Calcaneal tendon

Posteromedial View

RELATIONSHIP OF ANKLE LIGAMENTS TO MUSCULAR AND NEUROVASCULAR STRUCTURES

6.84

- The flexor hallucis longus muscle is midway between the medial and lateral malleoli; the tendons of the flexor digitorum and tibialis posterior are medial to it, and the tendons of the fibularis longus and brevis are lateral to it.
- The strongest parts of the ligaments of the ankle are those that prevent anterior displacement of the leg bones, namely, the posterior part of the medial ligament (posterior tibiotalar), the posterior talofibular, and calcaneofibular and tibiocalcaneal parts.

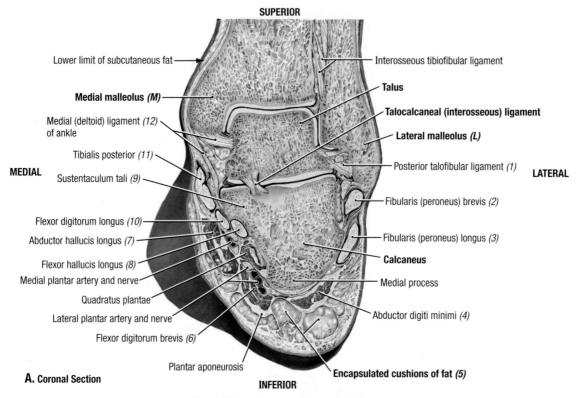

SUPERIOR

Lower limit of subcutaneous fat

Interosseous tibiofibular ligament

Medial malleolus (M)

Talus

Medial (deltoid) ligament (12) of ankle

Talocalcaneal (interosseous) ligament

Tibialis posterior (11)

Lateral malleolus (L)

MEDIAL

Sustentaculum tali (9)

Posterior talofibular ligament (1)

LATERAL

Flexor digitorum longus (10)

Fibularis (peroneus) brevis (2)

Abductor hallucis longus (7)

Fibularis (peroneus) longus (3)

Flexor hallucis longus (8)

Calcaneus

Medial plantar artery and nerve

Medial process

Quadratus plantae

Lateral plantar artery and nerve

Abductor digiti minimi (4)

Flexor digitorum brevis (6)

A. Coronal Section

Plantar aponeurosis

INFERIOR

Encapsulated cushions of fat (5)

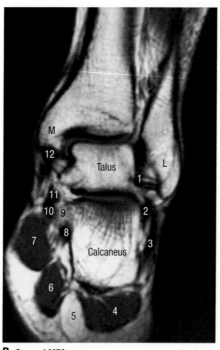

B. Coronal MRI

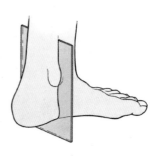

6.85 CORONAL SECTION AND MRI THROUGH ANKLE

A. Coronal section. **B.** Coronal MRI. Numbers in **(B)** refer to labeled structures in **(A)**.

- The tibia rests on the talus, and the talus rests on the calcaneus; between the calcaneus and the skin are several encapsulated cushions of fat.

- The lateral malleolus descends farther inferiorly than the medial malleolus.

- The talocalcaneal (interosseous) ligament between the talus and calcaneus separates the subtalar, or posterior talocalcaneal joint from the talocalcaneonavicular joint.

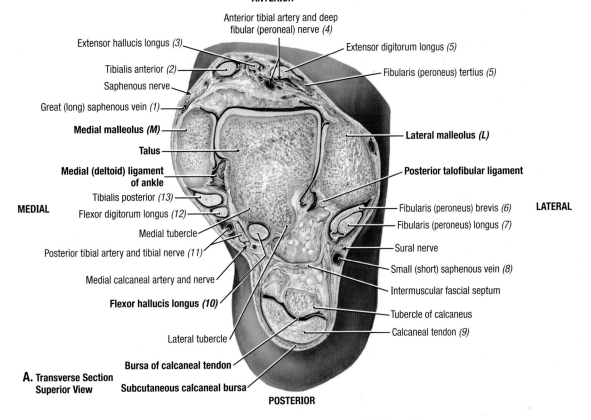

ANTERIOR

Anterior tibial artery and deep fibular (peroneal) nerve *(4)*

Extensor hallucis longus *(3)*

Tibialis anterior *(2)*

Saphenous nerve

Great (long) saphenous vein *(1)*

Medial malleolus *(M)*

Talus

Medial (deltoid) ligament of ankle

Tibialis posterior *(13)*

Flexor digitorum longus *(12)*

Medial tubercle

Posterior tibial artery and tibial nerve *(11)*

Medial calcaneal artery and nerve

Flexor hallucis longus *(10)*

Lateral tubercle

Bursa of calcaneal tendon

A. Transverse Section Superior View

Subcutaneous calcaneal bursa

POSTERIOR

Extensor digitorum longus *(5)*

Fibularis (peroneus) tertius *(5)*

Lateral malleolus *(L)*

Posterior talofibular ligament

Fibularis (peroneus) brevis *(6)*

Fibularis (peroneus) longus *(7)*

Sural nerve

Small (short) saphenous vein *(8)*

Intermuscular fascial septum

Tubercle of calcaneus

Calcaneal tendon *(9)*

MEDIAL

LATERAL

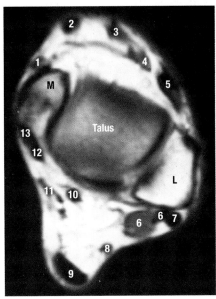

B. Transverse MRI

TRANSVERSE SECTION AND MRI THROUGH ANKLE

6.86

A. Transverse section. **B.** Transverse MRI. Numbers in **(B)** refer to labeled structures in **(A)**.

- The body of the talus is wedge-shaped and positioned between the malleoli, which are bound to it by the medial (deltoid) and posterior talofibular ligaments.

- The flexor hallucis longus muscle lies within its osseofibrous sheath between the medial and lateral tubercles of the talus.

- There is a small, inconstant subcutaneous bursa superficial to the calcaneal tendon and a large, constant bursa of calcaneal tendon deep to it.

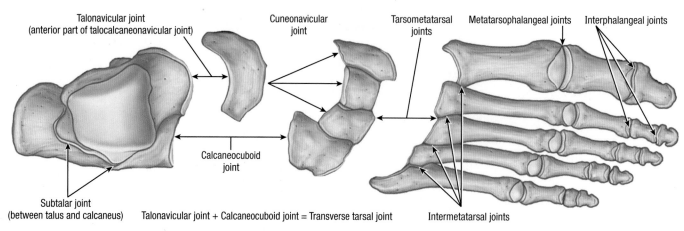

A. Superior View

Talonavicular joint (anterior part of talocalcaneonavicular joint)

Cuneonavicular joint

Tarsometatarsal joints

Metatarsophalangeal joints

Interphalangeal joints

Calcaneocuboid joint

Subtalar joint (between talus and calcaneus)

Talonavicular joint + Calcaneocuboid joint = Transverse tarsal joint

Intermetatarsal joints

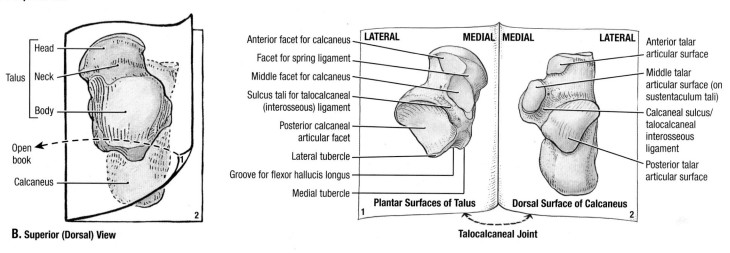

B. Superior (Dorsal) View

Talus — Head, Neck, Body

Open book

Calcaneus

Anterior facet for calcaneus
Facet for spring ligament
Middle facet for calcaneus
Sulcus tali for talocalcaneal (interosseous) ligament
Posterior calcaneal articular facet
Lateral tubercle
Groove for flexor hallucis longus
Medial tubercle

LATERAL MEDIAL MEDIAL LATERAL

Plantar Surfaces of Talus

Dorsal Surface of Calcaneus

Anterior talar articular surface
Middle talar articular surface (on sustentaculum tali)
Calcaneal sulcus/talocalcaneal interosseous ligament
Posterior talar articular surface

Talocalcaneal Joint

6.87 JOINTS OF FOOT

A. Overview. **B.** Talocalcaneal joint.

TABLE 6.22	**JOINTS OF FOOT**				
Joint	*Type*	*Articular Surface*	*Joint Capsule*	*Ligaments*	*Movements*
Subtalar	Synovial (plane) joint	Inferior surface of body of talus articulates with superior surface of calcaneus	Attached to margins of articular surfaces	Medial, lateral, and posterior talocalcaneal ligaments support capsule; talocalcaneal (interosseous) ligament binds bones together	Inversion and eversion of foot
Talocalcaneonavicular	Synovial joint; talonavicular part is a pivot joint	Head of talus articulates with calcaneus and navicular bones	Incompletely encloses joint	Plantar calcaneonavicular ("spring") ligament supports head of talus	Gliding and rotary movements
Calcaneocuboid	Synovial (plane) joint	Anterior end of calcaneus articulates with posterior surface of cuboid	Encloses joint	Dorsal calcaneocuboid ligament, plantar calcaneocuboid ligament, and long plantar ligament support joint capsule	Inversion and eversion of foot
Cuneonavicular	Synovial (plane) joint	Anterior navicular articulates with posterior surface of cuneiforms	Common joint capsule	Dorsal and plantar ligaments	Limited gliding movement
Tarsometatarsal	Synovial (plane) joint	Anterior tarsal bones articulate with bases of metatarsal bones	Encloses joint	Dorsal, plantar, and interosseous ligaments	Gliding or sliding
Intermetatarsal	Synovial (plane) joint	Bases of metatarsal bones articulate with each other	Encloses each joint	Dorsal, plantar, and interosseous ligaments bind bones together	Little individual movement
Metatarsophalangeal	Synovial (condyloid) joint	Heads of metatarsal bones articulate with bases of proximal phalanges	Encloses each joint	Collateral ligaments support capsule on each side; plantar ligament supports plantar part of capsule	Flexion, extension, and some abduction, adduction and circumduction
Interphalangeal	Synovial (hinge) joint	Head of proximal or middle phalanx articulates with base of phalanx distal to it	Encloses each joint	Collateral and plantar ligaments support joints	Flexion and extension

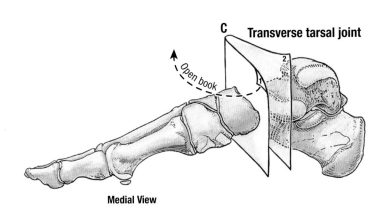

C Transverse tarsal joint

Open book

Medial View

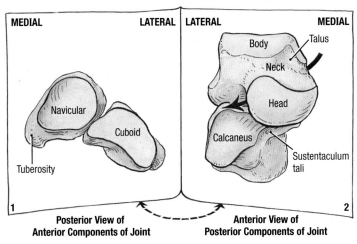

C. Transverse Tarsal Joint

MEDIAL — LATERAL | LATERAL — MEDIAL

Body — Talus
Neck
Navicular
Cuboid
Head
Calcaneus
Tuberosity
Sustentaculum tali

1 Posterior View of Anterior Components of Joint — Anterior View of Posterior Components of Joint **2**

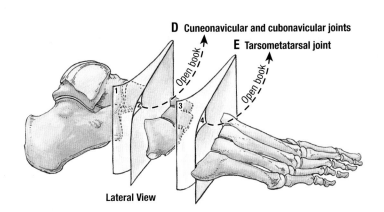

D Cuneonavicular and cubonavicular joints
E Tarsometatarsal joint

Open book
Open book

Lateral View

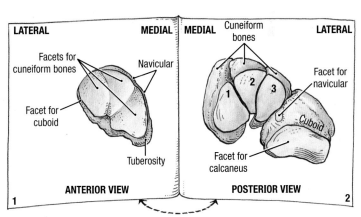

D. Cuneonavicular and Cubonavicular Joints

LATERAL — MEDIAL | MEDIAL — LATERAL

Cuneiform bones
Facets for cuneiform bones
Navicular
Facet for cuboid
Facet for navicular
Cuboid
Tuberosity
Facet for calcaneus

1 ANTERIOR VIEW — POSTERIOR VIEW **2**

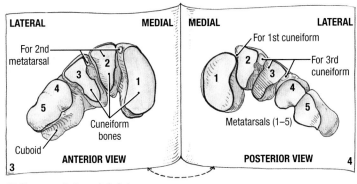

E. Tarsometatarsal Joints

LATERAL — MEDIAL | MEDIAL — LATERAL

For 2nd metatarsal
For 1st cuneiform
For 3rd cuneiform
Cuneiform bones
Metatarsals (1–5)
Cuboid

3 ANTERIOR VIEW — POSTERIOR VIEW **4**

JOINTS OF FOOT (continued)   6.87

C. Transverse tarsal joint. The *black arrow* traverses the tarsal sinus, in which the talocalcaneal (interosseous) ligament is located.
D. Cuneonavicular and cubonavicular joints. **E.** Tarsometatarsal joints.
- The joints of inversion and eversion are the subtalar (posterior talocalcaneal) joint, talocalcaneonavicular joint, and transverse tarsal (combined calcaneocuboid and talonavicular) joint.

- The talus participates in the ankle joint, of the posterior and anterior talocalcaneal joints, and of the talonavicular joint.
Metatarsal fractures (dancer's fracture) usually occur when the dancer loses balance, putting full body weight on the metatarsal.
Fatigue fractures of the metatarsals, usually transverse, may result from prolonged walking with repeated stress on the metatarsals.

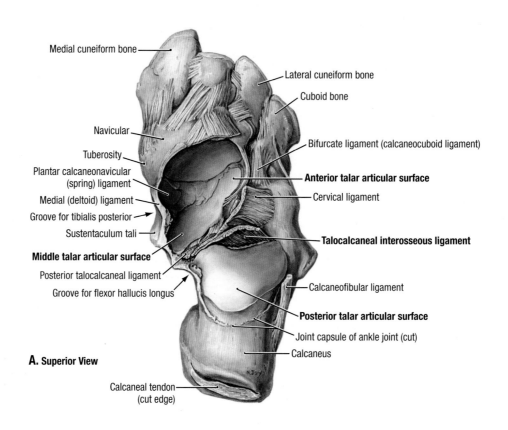

Medial cuneiform bone

Lateral cuneiform bone

Cuboid bone

Navicular

Bifurcate ligament (calcaneocuboid ligament)

Tuberosity

Plantar calcaneonavicular (spring) ligament

Anterior talar articular surface

Medial (deltoid) ligament

Cervical ligament

Groove for tibialis posterior

Sustentaculum tali

Talocalcaneal interosseous ligament

Middle talar articular surface

Posterior talocalcaneal ligament

Groove for flexor hallucis longus

Calcaneofibular ligament

Posterior talar articular surface

Joint capsule of ankle joint (cut)

Calcaneus

A. Superior View

Calcaneal tendon (cut edge)

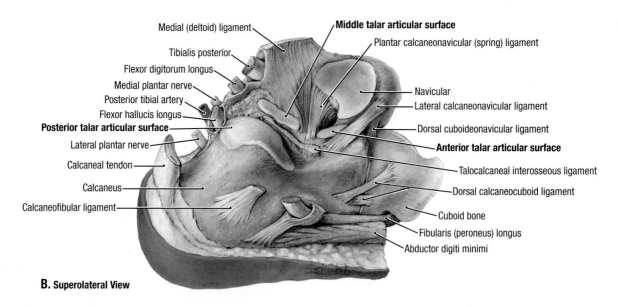

Medial (deltoid) ligament

Middle talar articular surface

Tibialis posterior

Plantar calcaneonavicular (spring) ligament

Flexor digitorum longus

Medial plantar nerve

Navicular

Posterior tibial artery

Lateral calcaneonavicular ligament

Flexor hallucis longus

Dorsal cuboideonavicular ligament

Posterior talar articular surface

Anterior talar articular surface

Lateral plantar nerve

Calcaneal tendon

Talocalcaneal interosseous ligament

Calcaneus

Dorsal calcaneocuboid ligament

Calcaneofibular ligament

Cuboid bone

Fibularis (peroneus) longus

Abductor digiti minimi

B. Superolateral View

6.88 JOINTS OF INVERSION AND EVERSION

The joints of inversion and eversion are the subtalar (posterior talocalcaneal), talocalcaneonavicular, and transverse tarsal (combined calcaneocuboid and talonavicular) joints. **A.** Posterior and middle parts of foot with talus removed. **B.** Posterior part of foot with talus removed. The convex posterior talar facet is separated from the concave middle, and anterior facets by the talocalcaneal (interosseous) ligament within the tarsal sinus. The posterior and anterior talocalcaneal joints are separated from each other by the sulcus tali and calcaneal sulcus, which, when the talus and calcaneus are in articulation, become the tarsal sinus.

Calcaneal fractures. A hard fall onto the heel (e.g., from a ladder) may fracture the calcaneus into several pieces, resulting in a comminuted fracture. A calcaneal fracture is usually disabling because it disrupts the subtalar (talocalcaneal) joint.

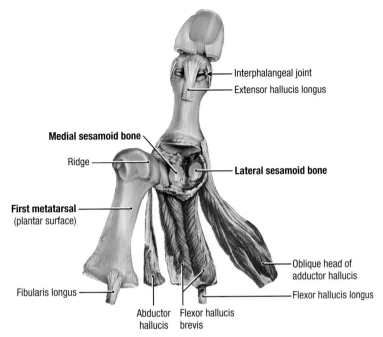

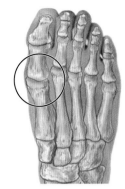

1st metatarsophalangeal
joint (circled)

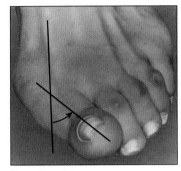

B. Hallux Valgus

A. Superior View of Phalanges and Nail, Right Great Toe; Medial View of First Metatarsal; Superior View of Sesamoid Bones

METATARSOPHALANGEAL JOINT OF GREAT TOE | 6.89

A. First metatarsal and sesamoid bones of the right great toe. The 1st metatarsal has been reflected medially. **B. Hallux valgus** is a foot deformity caused by pressure from footwear and degenerative joint disease. It is characterized by lateral deviation of the base of the 1st metatarsal and base of the proximal phalanx of the great toe (L. *hallux*). In some people, the deviation is so great that the 1st toe overlaps the 2nd toe. These individuals are unable to move their 1st digit away from their 2nd digit because the sesamoid bones under the head of the 1st metatarsal are displaced and lie in the space between the heads of the 1st and 2nd metatarsals. In addition, a subcutaneous bursa may form owing to pressure and friction against the shoe. When tender and inflamed, the bursa is called a **bunion**.

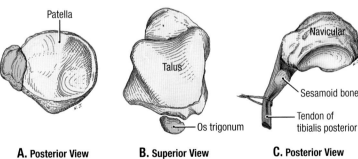

A. Posterior View   **B. Superior View**   **C. Posterior View**

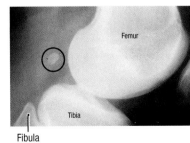

D. Lateral View   **E. Lateral View (sesamoid bone circled)**

BONY ANOMALIES | 6.90

A. Bipartite patella. Occasionally, the superolateral angle of the patella ossifies independently and remains discrete. **B. Os trigonum.** The lateral (posterior) tubercle of the talus has a separate center of ossification that appears from the ages of 7 to 13 years; when this fails to fuse with the body of the talus, as in the left bone of this pair, it is called an os trigonum. It was found in Dr. Grant's lab in 7.7% of 558 adult feet; 22 were paired, and 21 were unpaired. **C.** Sesamoid bone in the tendon of tibialis posterior. A sesamoid bone was found in 23% of 348 adults. **D.** Sesamoid bone in the tendon of fibularis (peroneus) longus. A sesamoid bone was found in 26% of 92 specimens. In this specimen, it is bipartite, and the fibularis (peroneus) longus muscle has an additional attachment to the 5th metatarsal bone. **E. Fabella.** A sesamoid bone in the lateral head of the gastrocnemius muscle was present in 21.6% of 116 limbs.

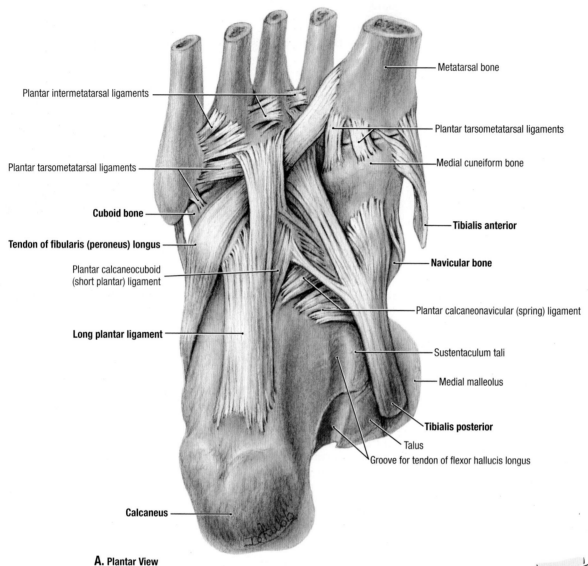

Metatarsal bone

Plantar intermetatarsal ligaments

Plantar tarsometatarsal ligaments

Medial cuneiform bone

Plantar tarsometatarsal ligaments

Cuboid bone

Tibialis anterior

Tendon of fibularis (peroneus) longus

Navicular bone

Plantar calcaneocuboid (short plantar) ligament

Plantar calcaneonavicular (spring) ligament

Long plantar ligament

Sustentaculum tali

Medial malleolus

Tibialis posterior

Talus

Groove for tendon of flexor hallucis longus

Calcaneus

A. Plantar View

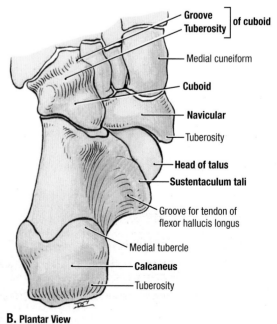

Groove
Tuberosity } of cuboid

Medial cuneiform

Cuboid

Navicular

Tuberosity

Head of talus

Sustentaculum tali

Groove for tendon of flexor hallucis longus

Medial tubercle

Calcaneus

Tuberosity

B. Plantar View

6.91 LIGAMENTS OF SOLE OF FOOT

A. Dissection of superficial ligaments. **B.** Bones lying deep to ligaments. The head of the talus is exposed between the sustentaculum tali of the calcaneus and the navicular.

In **A**:

- Note the insertions of three long tendons: fibularis (peroneus) longus, tibialis anterior, and tibialis posterior.
- The tendon of the fibularis (peroneus) longus muscle crosses the sole of the foot in the groove anterior to the tuberosity of the cuboid, is bridged by some fibers of the long plantar ligament, and inserts into the base of the 1st metatarsal.
- Observe the slips of the tibialis posterior tendon extending to the bones anterior to the transverse tarsal joint.

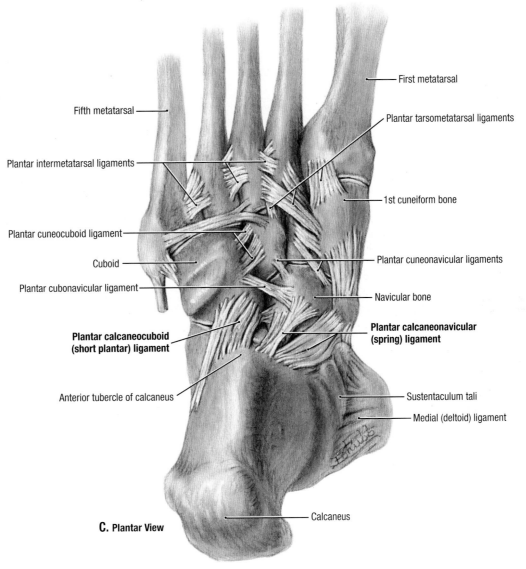

First metatarsal

Fifth metatarsal

Plantar tarsometatarsal ligaments

Plantar intermetatarsal ligaments

1st cuneiform bone

Plantar cuneocuboid ligament

Cuboid

Plantar cuneonavicular ligaments

Plantar cubonavicular ligament

Navicular bone

Plantar calcaneocuboid (short plantar) ligament

Plantar calcaneonavicular (spring) ligament

Anterior tubercle of calcaneus

Sustentaculum tali

Medial (deltoid) ligament

Calcaneus

C. Plantar View

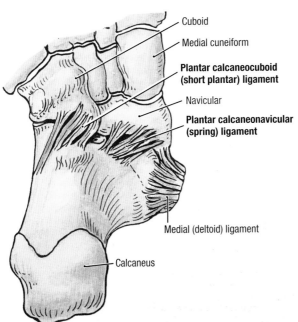

Cuboid

Medial cuneiform

Plantar calcaneocuboid (short plantar) ligament

Navicular

Plantar calcaneonavicular (spring) ligament

Medial (deltoid) ligament

Calcaneus

D. Plantar View

LIGAMENTS OF SOLE OF FOOT (*continued*)   6.91

C. Dissection of the deep ligaments. **D.** Support for head of talus. The head of the talus is supported by the plantar calcaneonavicular ligament (spring ligament) and the tendon of the tibialis posterior.

- The plantar calcaneocuboid (short plantar) and plantar calcaneonavicular (spring) ligaments are the primary plantar ligaments of the transverse tarsal joint.
- The ligaments of the anterior foot diverge laterally and posteriorly from each side of the long axis of the 3rd metatarsal and 3rd cuneiform; hence, a posterior thrust received by the 1st metatarsal, as when rising on the big toe while in walking, is transmitted directly to the navicular and talus by the first cuneiform and indirectly by the 2nd metatarsal, 2nd cuneiform, 3rd metatarsal, and 3rd cuneiform.
- A posterior thrust received by the 4th and 5th metatarsals is transmitted directly to the cuboid and calcaneus.

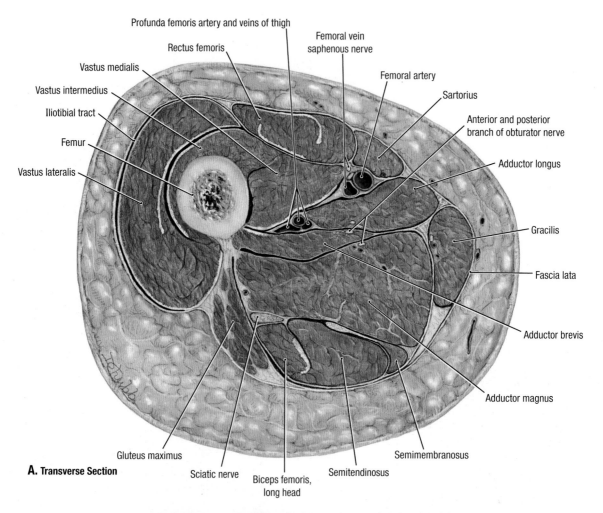

Profunda femoris artery and veins of thigh

Rectus femoris

Vastus medialis

Vastus intermedius

Iliotibial tract

Femur

Vastus lateralis

Femoral vein
saphenous nerve

Femoral artery

Sartorius

Anterior and posterior
branch of obturator nerve

Adductor longus

Gracilis

Fascia lata

Adductor brevis

Adductor magnus

Gluteus maximus

Sciatic nerve

Biceps femoris,
long head

Semitendinosus

Semimembranosus

A. Transverse Section

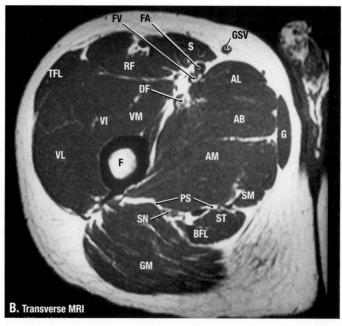

B. Transverse MRI

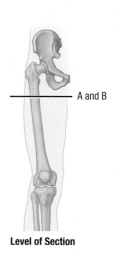

A and B

Level of Section

6.92 TRANSVERSE SECTIONS AND MRIs OF THIGH

A. Anatomical section of proximal thigh. **B.** Transverse MRI of proximal thigh.

Compartments of thigh

- ☐ Anterior
- ☐ Medial
- ☐ Posterior

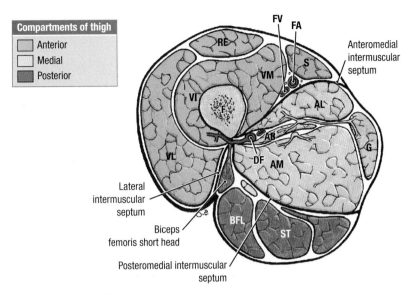

C

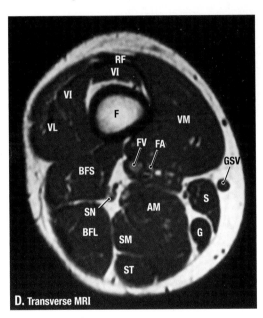

D. Transverse MRI

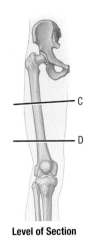

Level of Section

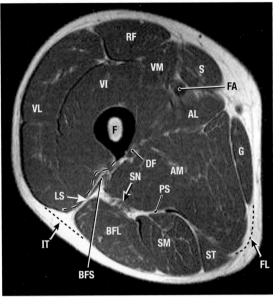

Transverse MRI

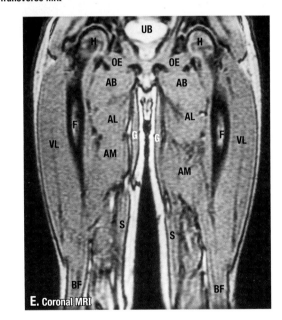

E. Coronal MRI

Key

AB	Adductor brevis	H	Head of femur
AL	Adductor longus	IT	Iliotibial tract
AM	Adductor magnus	LS	Lateral intermuscular septum
AS	Anteromedial intermuscular septum	OE	Obturator externus
BF	Biceps femoris	PS	Posterior intermuscular septum
BFL	Long head of biceps femoris	RF	Rectus femoris
BFS	Short head of biceps femoris	S	Sartorius
DF	Profunda femoris artery	SM	Semimembranosus
F	Femur	SN	Sciatic nerve
FA	Femoral artery	ST	Semitendinosus
FL	Fascia lata	TFL	Tensor fasciae latae
FV	Femoral vein	UB	Urinary bladder
G	Gracilis	VI	Vastus intermedius
GM	Gluteus maximus	VL	Vastus lateralis
GSV	Great saphenous vein	VM	Vastus medialis

TRANSVERSE SECTIONS AND MRIs OF THIGH (continued)

6.92

C. Diagrammatic anatomical section and transverse (axial) MRI of midthigh. **D.** Transverse (axial) MRI of distal thigh. **E.** Coronal MRI.

The thigh has three compartments, each with its own nerve supply and primary function: anterior group extends the knee and is supplied by the femoral nerve; medial group adducts the hip and is supplied by the obturator nerve; and posterior group flexes the knee and is supplied by the sciatic nerve.

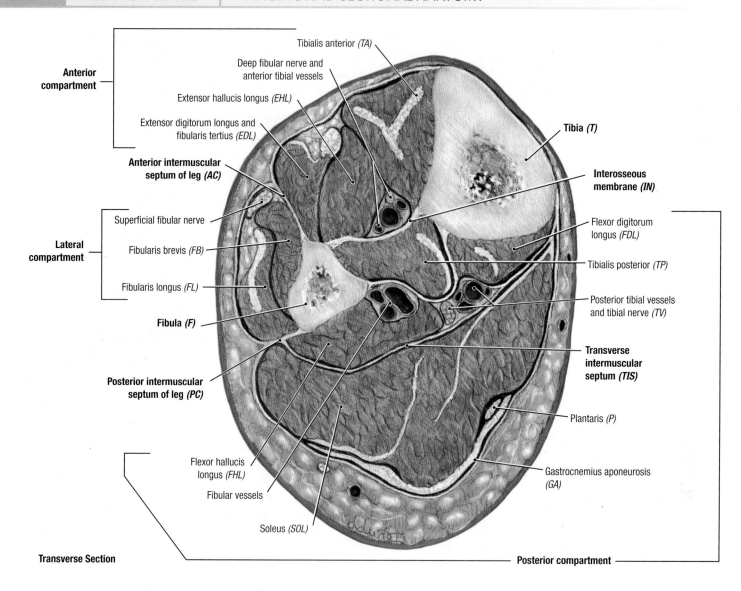

Anterior compartment

- Tibialis anterior (TA)
- Deep fibular nerve and anterior tibial vessels
- Extensor hallucis longus (EHL)
- Extensor digitorum longus and fibularis tertius (EDL)
- **Anterior intermuscular septum of leg (AC)**

Lateral compartment

- Superficial fibular nerve
- Fibularis brevis (FB)
- Fibularis longus (FL)
- **Fibula (F)**

Posterior intermuscular septum of leg (PC)

- Flexor hallucis longus (FHL)
- Fibular vessels
- Soleus (SOL)

- **Tibia (T)**
- **Interosseous membrane (IN)**
- Flexor digitorum longus (FDL)
- Tibialis posterior (TP)
- Posterior tibial vessels and tibial nerve (TV)
- **Transverse intermuscular septum (TIS)**
- Plantaris (P)
- Gastrocnemius aponeurosis (GA)

Transverse Section

Posterior compartment

6.93   TRANSVERSE SECTION OF LEG

Boundaries of anterior, lateral, and posterior compartments of leg. Anterior compartment: tibia, interosseous membrane, fibula, anterior intermuscular septum, and crural fascia. Lateral compartment: fibula, anterior and posterior intermuscular septa, and the crural fascia. Posterior compartment: tibia, interosseous membrane, fibula, posterior intermuscular septum, and crural fascia. The posterior compartment is subdivided by the transverse intermuscular septum into superficial and deep subcompartments.

Compartmental infections in the leg. Because the septa and deep fascia forming the boundaries of the leg compartments are strong, the increased volume consequent to infection with suppuration (formation of pus) increases intracompartmental pressure. Inflammation within the anterior and posterior compartments spreads chiefly in a distal direction; however, a purulent infection in the lateral compartment can ascend proximally into the popliteal fossa, presumably along the course of the fibular nerve. **Fasciotomy** may be necessary to relieve compartmental pressure and debride (remove by scraping) pockets of infection.

Level of Section

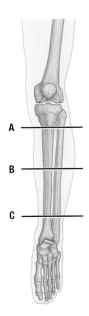

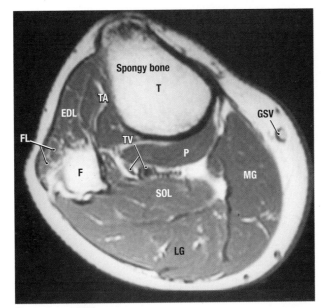

A. Transverse MRI

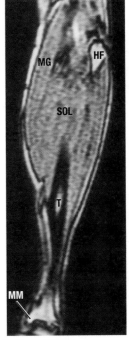

D. Coronal MRI

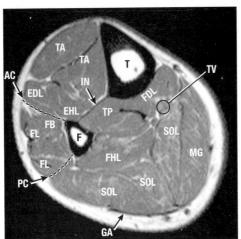

B. Transverse Section and MRI

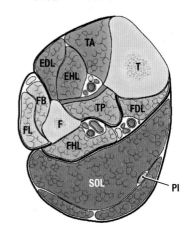

Key

AC	Anterior intermuscular septum
AV	Anterior tibial vessels and deep fibular nerve
EDL	Extensor digitorum longus
EHL	Extensor hallucis longus
F	Fibula
FB	Fibularis brevis
FDL	Flexor digitorum longus
FHL	Flexor hallucis longus
FL	Fibularis longus
GA	Gastrocnemius aponeurosis
GSV	Great saphenous vein
HF	Head of fibula
IN	Interosseous membrane
LG	Lateral head of gastrocnemius
MG	Medial head of gastrocnemius
MM	Medial malleolus
P	Popliteus
PI	Plantaris
PC	Posterior intermuscular septum
SOL	Soleus
SSV	Small saphenous vein
T	Tibia
TA	Tibialis anterior
TC	Calcaneal tendon
TP	Tibialis posterior
TV	Tibial nerve and posterior tibial vessels

Key for B

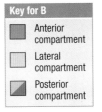

- Anterior compartment
- Lateral compartment
- Posterior compartment

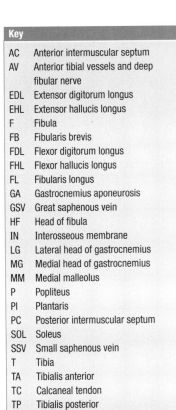

C. Transverse MRI

MRIs OF LEG | 6.94

A–C. Transverse (axial) MRIs. **D.** Coronal MRI.

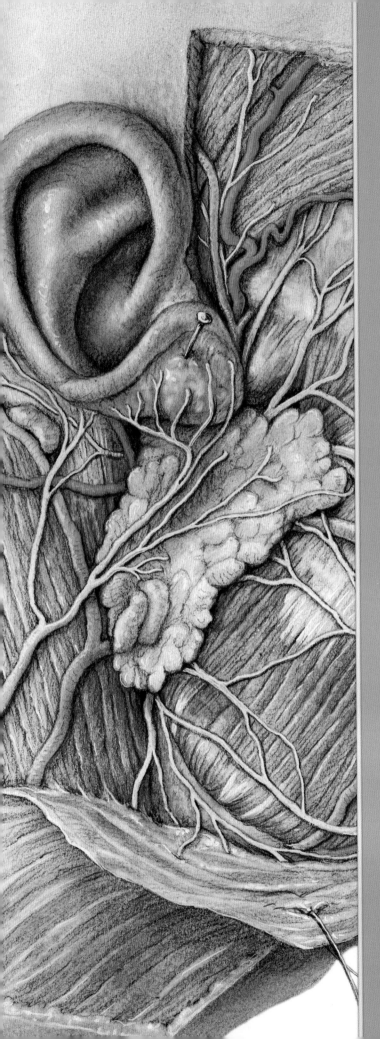

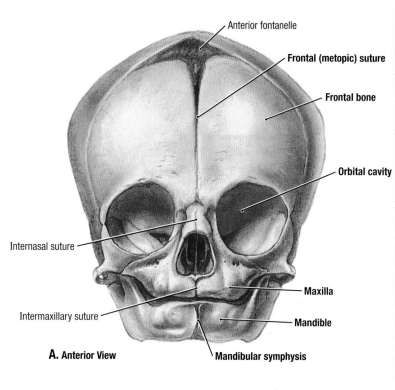

Anterior fontanelle

Frontal (metopic) suture

Frontal bone

Orbital cavity

Internasal suture

Maxilla

Intermaxillary suture

Mandible

A. Anterior View

Mandibular symphysis

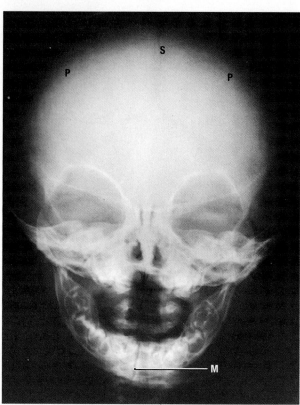

B. Anteroposterior View

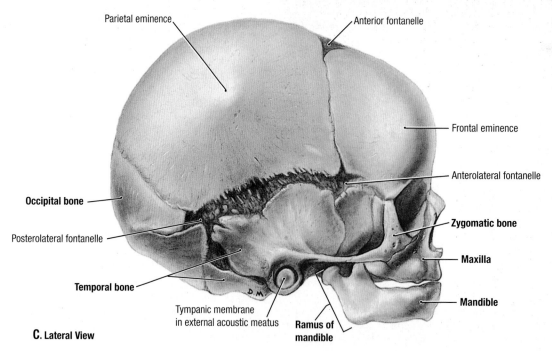

Parietal eminence

Anterior fontanelle

Frontal eminence

Anterolateral fontanelle

Occipital bone

Zygomatic bone

Posterolateral fontanelle

Maxilla

Temporal bone

Mandible

Tympanic membrane
in external acoustic meatus

**Ramus of
mandible**

C. Lateral View

7.1   CRANIUM AT BIRTH AND IN EARLY CHILDHOOD

A. Cranium at birth, anterior aspect. **B.** Radiograph of 6½-month-old child. **C.** Cranium at birth, lateral aspect.
Compared with the adult skull (Figs. 7.2 to 7.4):
• The maxilla and mandible are proportionately small.
• The mandibular symphysis, which closes during the second year, and the frontal suture, which closes during the sixth year are still open (unfused).

• The orbital cavities are proportionately large, but the face is small; the facial skeleton forming only one eighth of the whole cranium, while in the adult, it forms one third.

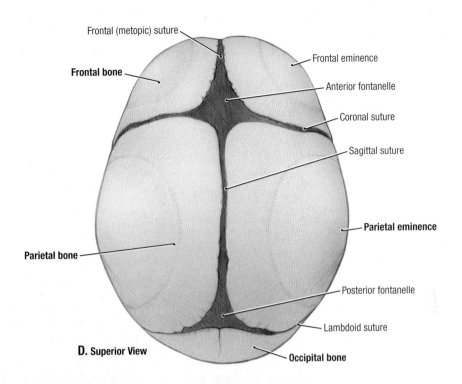

Frontal (metopic) suture

Frontal bone

Parietal bone

Frontal eminence

Anterior fontanelle

Coronal suture

Sagittal suture

Parietal eminence

Posterior fontanelle

Lambdoid suture

Occipital bone

D. Superior View

Key for B, E, and F

A	Angle of mandible
B	Body of mandible
C	Coronal suture
F	Frontal bone
L	Lambdoid suture
M	Mandibular symphysis
O	Occipital bone
P	Parietal eminence
S	Sagittal suture
SP	Sphenoid
T	Temporal bone
X	Maxilla
Y	Mastoid process
Z	Zygomatic bone

Arrowheads = Membranous outline
of parietal bone

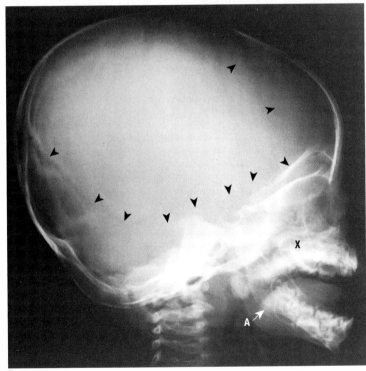

E. Lateral View

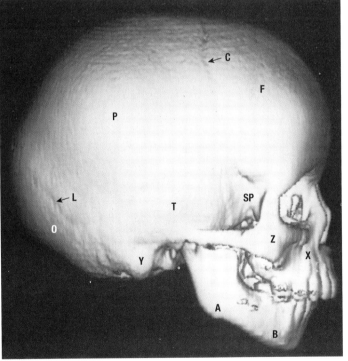

F. Lateral View

CRANIUM AT BIRTH AND IN EARLY CHILDHOOD *(continued)*

7.1

D. Cranium at birth, superior aspect. **E.** Radiograph of 6½-month-old child. **F.** Three-dimensional computer-generated image of 3-year-old child's cranium.

- The parietal eminence is a shallow, rounded cone. Ossification, which starts at the eminences, has not yet reached the ultimate four angles of the parietal bone; accordingly, these regions are membranous, and the membrane is blended with the pericranium externally and the dura mater internally to form the fontanelles. The fontanelles are usually closed by the second year. There is no mastoid process until the second year.

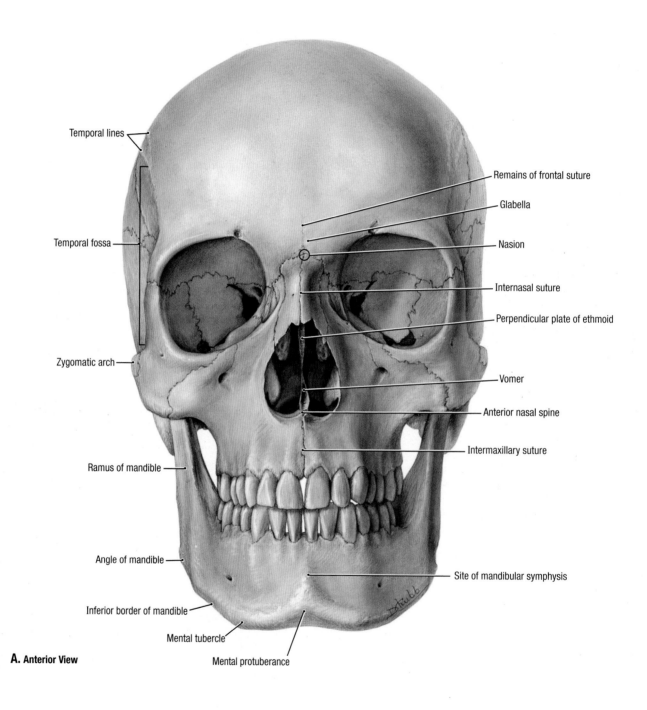

Temporal lines

Remains of frontal suture

Glabella

Temporal fossa

Nasion

Internasal suture

Perpendicular plate of ethmoid

Zygomatic arch

Vomer

Anterior nasal spine

Intermaxillary suture

Ramus of mandible

Angle of mandible

Site of mandibular symphysis

Inferior border of mandible

Mental tubercle

A. Anterior View

Mental protuberance

7.2 CRANIUM, FACIAL (FRONT) ASPECT

A. Formations of the bony cranium. **B.** Bones of cranium and their features. The individual bones forming the cranium are color-coded. For the orbital cavity, see also Figure 7.36A.

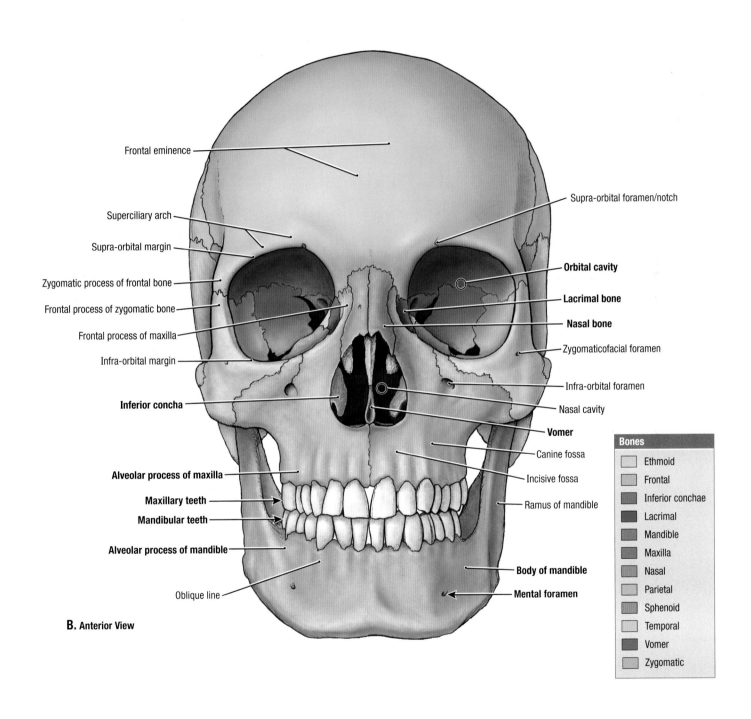

Frontal eminence

Superciliary arch

Supra-orbital margin

Zygomatic process of frontal bone

Frontal process of zygomatic bone

Frontal process of maxilla

Infra-orbital margin

Inferior concha

Alveolar process of maxilla

Maxillary teeth

Mandibular teeth

Alveolar process of mandible

Oblique line

B. Anterior View

Supra-orbital foramen/notch

Orbital cavity

Lacrimal bone

Nasal bone

Zygomaticofacial foramen

Infra-orbital foramen

Nasal cavity

Vomer

Canine fossa

Incisive fossa

Ramus of mandible

Body of mandible

Mental foramen

Bones	
	Ethmoid
	Frontal
	Inferior conchae
	Lacrimal
	Mandible
	Maxilla
	Nasal
	Parietal
	Sphenoid
	Temporal
	Vomer
	Zygomatic

CRANIUM, FACIAL (FRONT) ASPECT (*continued*) 7.2

Extraction of teeth causes the alveolar bone to resorb in the affected region(s). Following complete loss or extraction of teeth, the sockets begin to fill in with bone, and the alveolar processes begin to resorb. The mental foramen may eventually lie near the superior border of the body of the mandible. In some cases, mental foramina resorption may extend to the mental nerves, exposing them to injury.

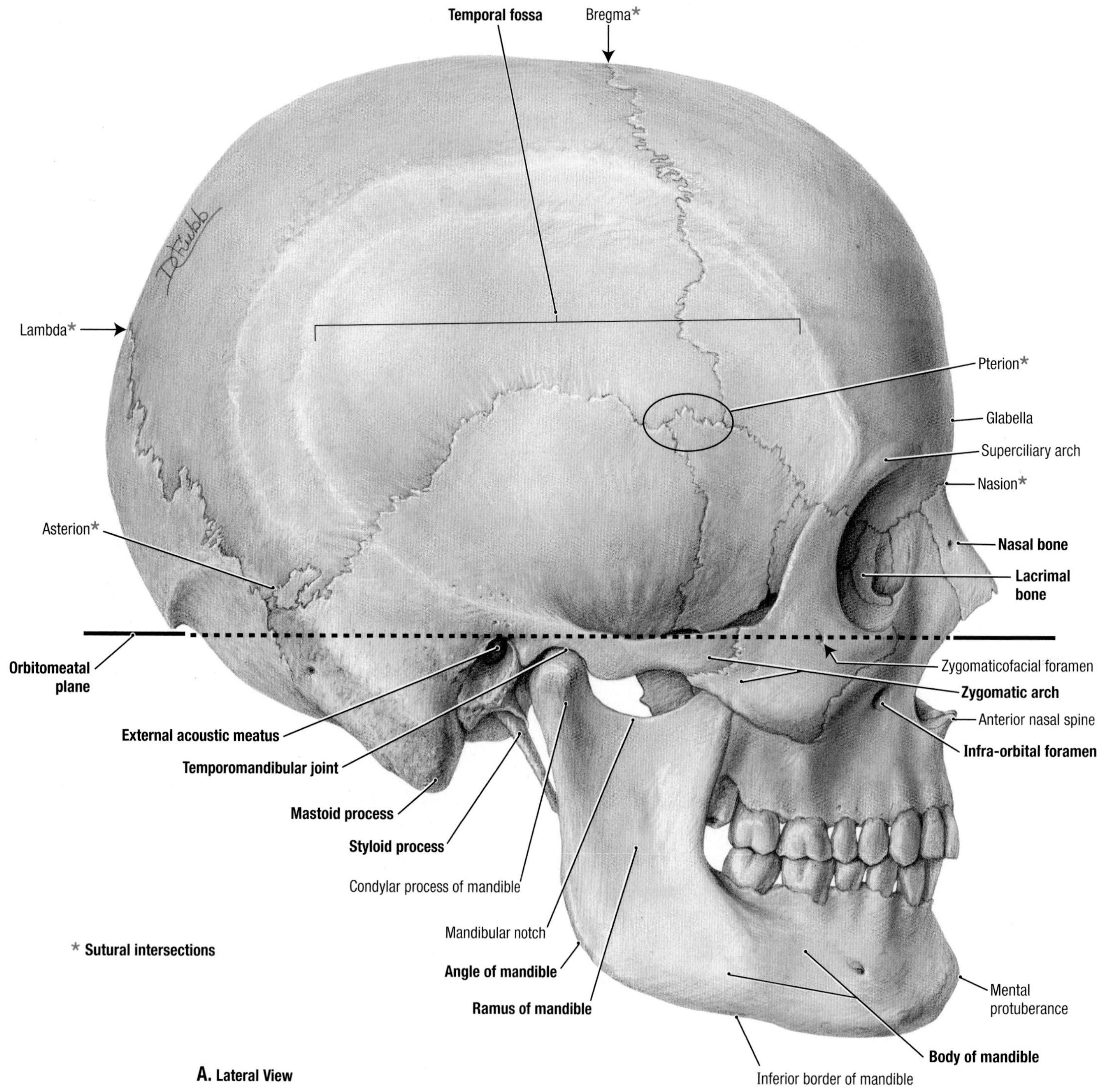

A. Lateral View

* **Sutural intersections**

7.3 CRANIUM, LATERAL ASPECT

A. Bony cranium. **B.** Cranium with bones color-coded. The cranium is in the anatomical position when the orbitomeatal plane is horizontal. **C.** Buttresses of cranium. The buttresses are thicker portions of cranial bones that transfer forces around the weaker regions of the orbits and nasal cavity.

The convexity of the neurocranium (braincase) distributes and thereby minimizes the effects of a blow to it. However, hard blows to the head in thin areas of the cranium (e.g., in the temporal fossa) are likely to produce **depressed fractures**, in which a fragment of bone is depressed inward, compressing and/or injuring the brain. In **comminuted fractures**, the bone is broken into several pieces. **Linear fractures**, the most frequent type, usually occur at the point of impact, but fracture lines often radiate away from it in two or more directions.

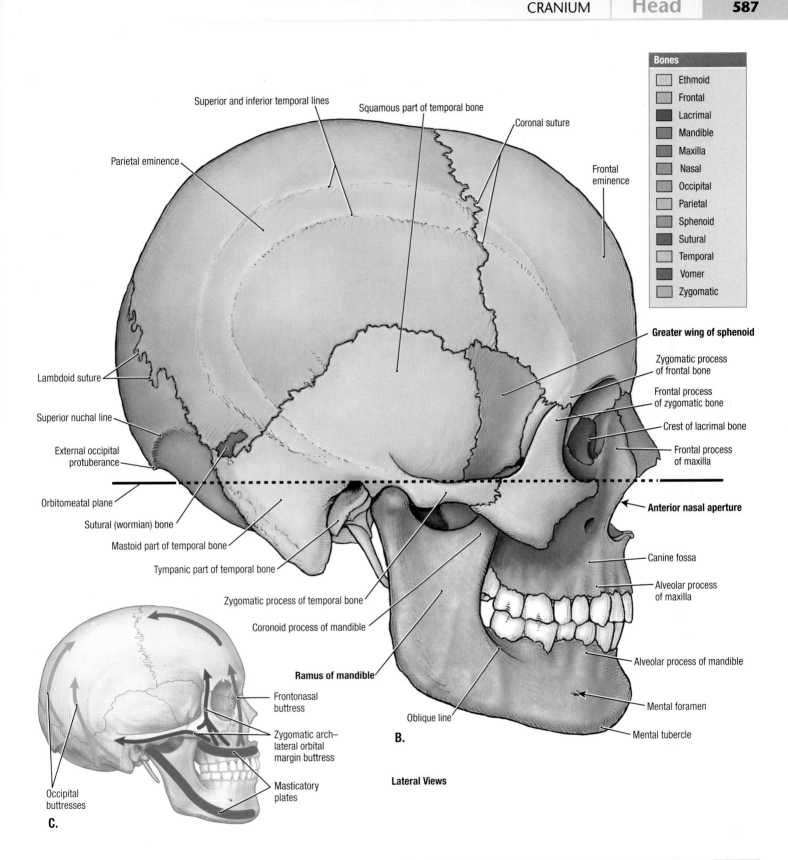

Bones
- Ethmoid
- Frontal
- Lacrimal
- Mandible
- Maxilla
- Nasal
- Occipital
- Parietal
- Sphenoid
- Sutural
- Temporal
- Vomer
- Zygomatic

Superior and inferior temporal lines

Squamous part of temporal bone

Coronal suture

Parietal eminence

Frontal eminence

Greater wing of sphenoid

Zygomatic process of frontal bone

Frontal process of zygomatic bone

Crest of lacrimal bone

Frontal process of maxilla

Lambdoid suture

Superior nuchal line

External occipital protuberance

Orbitomeatal plane

Anterior nasal aperture

Sutural (wormian) bone

Mastoid part of temporal bone

Tympanic part of temporal bone

Canine fossa

Alveolar process of maxilla

Zygomatic process of temporal bone

Coronoid process of mandible

Ramus of mandible

Oblique line

B.

Alveolar process of mandible

Mental foramen

Mental tubercle

Occipital buttresses

Frontonasal buttress

Zygomatic arch– lateral orbital margin buttress

Masticatory plates

C.

Lateral Views

CRANIUM, LATERAL ASPECT (continued)

7.3

If the area of the neurocranium is thick at the site of impact, the bone usually bends inward without fracturing; however, a fracture may occur some distance from the site of direct trauma where the calvaria is thinner. In a **contrecoup (counterblow) fracture**, the fracture occurs on the opposite side of the cranium rather than at the point of impact. One or more sutural (accessory) bones may be located along the lambdoid suture or near the mastoid process.

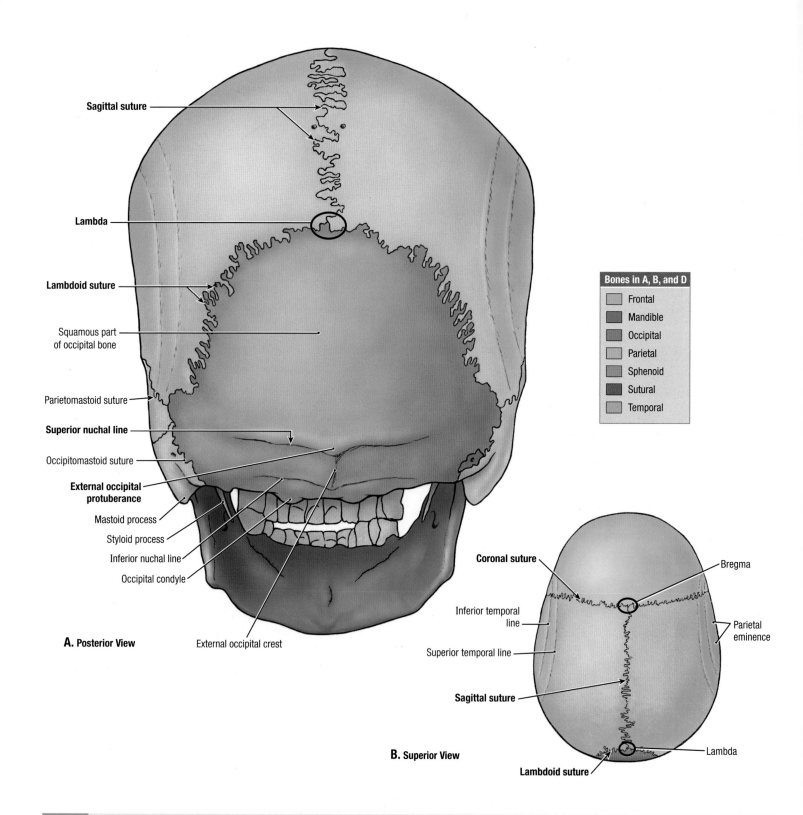

Bones in A, B, and D
- Frontal
- Mandible
- Occipital
- Parietal
- Sphenoid
- Sutural
- Temporal

A. Posterior View

- Sagittal suture
- Lambda
- **Lambdoid suture**
- Squamous part of occipital bone
- Parietomastoid suture
- **Superior nuchal line**
- Occipitomastoid suture
- **External occipital protuberance**
- Mastoid process
- Styloid process
- Inferior nuchal line
- Occipital condyle
- External occipital crest

B. Superior View

- **Coronal suture**
- Inferior temporal line
- Superior temporal line
- **Sagittal suture**
- **Lambdoid suture**
- Bregma
- Parietal eminence
- Lambda

7.4 CRANIUM, OCCIPITAL ASPECT, CALVARIA, AND ANTERIOR PART OF POSTERIOR CRANIAL FOSSA

A. The lambda, near the center of this convex surface, is located at the junction of the sagittal and lambdoid sutures. **B.** The roof of the neurocranium, or calvaria (skullcap), is formed primarily by the paired parietal bones, the frontal bone, and the occipital bone.

Premature closure of the coronal suture results in a high, towerlike cranium, called **oxycephaly** or **turricephaly**. Premature closure of sutures usually does not affect brain development. When premature closure occurs on one side only, the cranium is asymmetrical, a condition known as **plagiocephaly**.

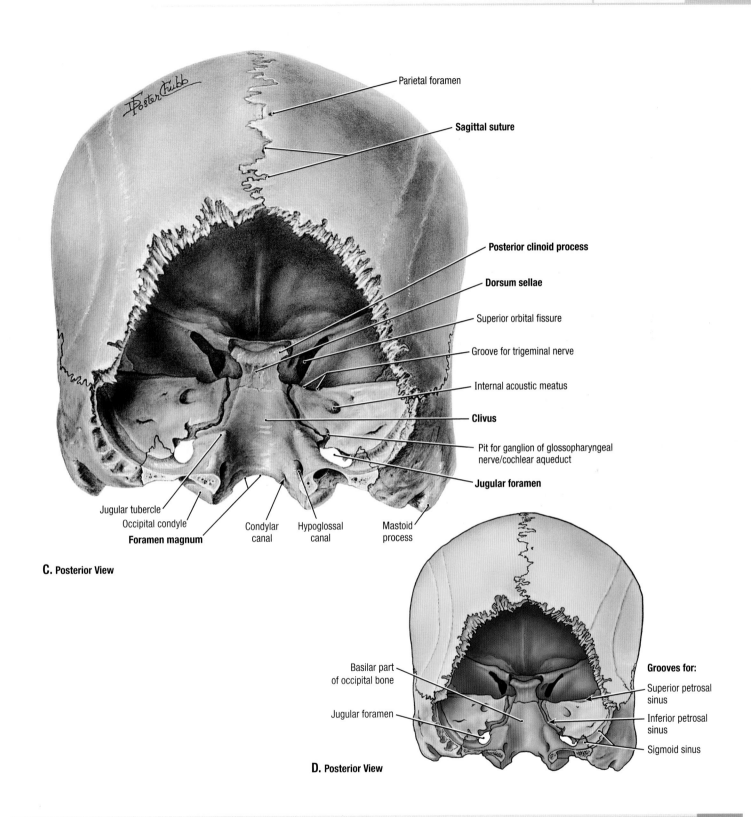

C. Posterior View

Labels (top to bottom, clockwise):
- Parietal foramen
- **Sagittal suture**
- **Posterior clinoid process**
- **Dorsum sellae**
- Superior orbital fissure
- Groove for trigeminal nerve
- Internal acoustic meatus
- **Clivus**
- Pit for ganglion of glossopharyngeal nerve/cochlear aqueduct
- **Jugular foramen**
- Mastoid process
- Jugular tubercle
- Occipital condyle
- **Foramen magnum**
- Condylar canal
- Hypoglossal canal

D. Posterior View

- Basilar part of occipital bone
- Jugular foramen
- Grooves for:
  - Superior petrosal sinus
  - Inferior petrosal sinus
  - Sigmoid sinus

CRANIUM, OCCIPITAL ASPECT, CALVARIA, AND ANTERIOR PART OF POSTERIOR CRANIAL FOSSA (*continued*)

7.4

C. and **D.** Cranium after removal of squamous part of occipital bone.
- The dorsum sellae projects from the body of the sphenoid; the posterior clinoid processes form its superolateral corners.
- The clivus is the slope descending from the dorsum sellae to the foramen magnum.

- The grooves for the sigmoid sinus and inferior petrosal sinus lead inferiorly to the jugular foramen.

   Premature closure of the sagittal suture, in which the anterior fontanelle is small or absent, results in a long, narrow, wedge-shaped cranium, a condition called **scaphocephaly**.

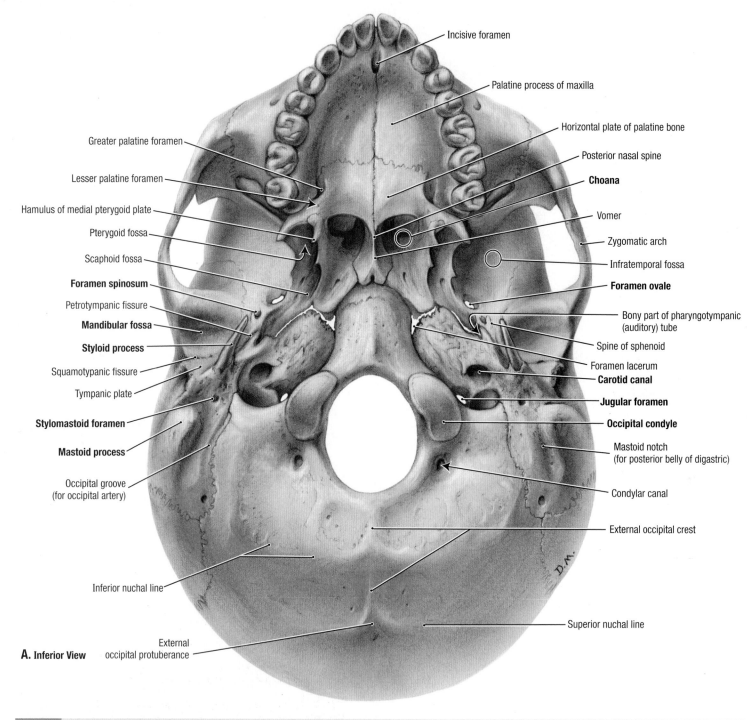

Incisive foramen

Palatine process of maxilla

Horizontal plate of palatine bone

Posterior nasal spine

Choana

Vomer

Zygomatic arch

Infratemporal fossa

Foramen ovale

Bony part of pharyngotympanic (auditory) tube

Spine of sphenoid

Foramen lacerum

Carotid canal

Jugular foramen

Occipital condyle

Mastoid notch (for posterior belly of digastric)

Condylar canal

External occipital crest

Superior nuchal line

Greater palatine foramen

Lesser palatine foramen

Hamulus of medial pterygoid plate

Pterygoid fossa

Scaphoid fossa

Foramen spinosum

Petrotympanic fissure

Mandibular fossa

Styloid process

Squamotympanic fissure

Tympanic plate

Stylomastoid foramen

Mastoid process

Occipital groove (for occipital artery)

Inferior nuchal line

A. Inferior View

External occipital protuberance

7.5   CRANIUM, INFERIOR ASPECT

A. Bony cranium. **B.** Diagram of cranium with bones color-coded.

TABLE 7.1	**FORAMINA AND OTHER APERTURES OF NEUROCRANIUM AND CONTENTS (SEE FIGS. 7.2 TO 7.6)**
Foramen cecum: Nasal emissary vein (1% of population)	Optic canals: Optic nerve (CN II) and ophthalmic arteries
Cribriform plate: Olfactory nerves (CN I)	Superior orbital fissure: Ophthalmic veins; ophthalmic nerve (CN V_1); CN III, IV, and VI; and sympathetic fibers
Anterior and posterior ethmoidal foramina: Vessels and nerves with same names	Foramen rotundum: Maxillary nerve (CN V_2)
Foramen ovale: Mandibular nerve (CN V_3) and accessory meningeal artery	Jugular foramen: CN IX, X, and XI; superior bulb of internal jugular vein; inferior petrosal and sigmoid sinuses; meningeal branches of ascending pharyngeal and occipital arteries

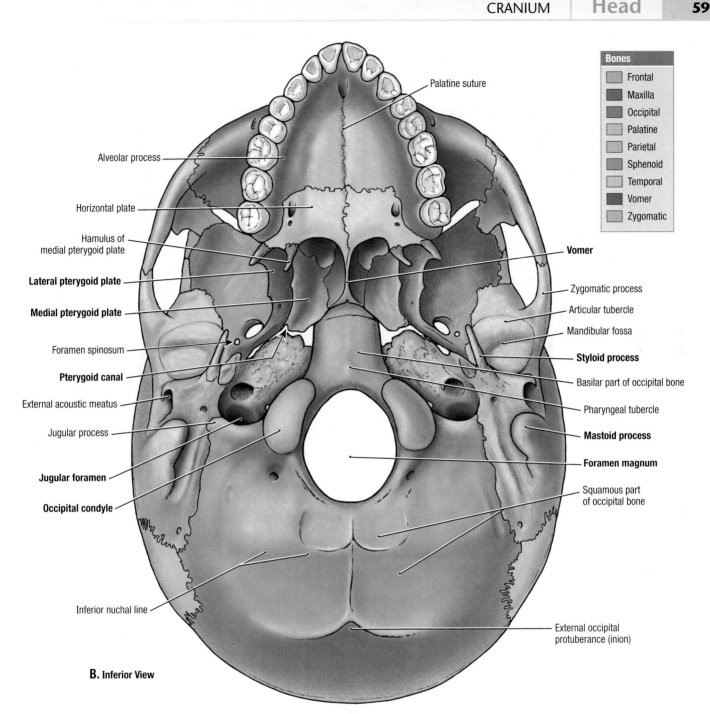

Palatine suture

Alveolar process

Horizontal plate

Hamulus of
medial pterygoid plate

Lateral pterygoid plate

Medial pterygoid plate

Foramen spinosum

Pterygoid canal

External acoustic meatus

Jugular process

Jugular foramen

Occipital condyle

Inferior nuchal line

Vomer

Zygomatic process

Articular tubercle

Mandibular fossa

Styloid process

Basilar part of occipital bone

Pharyngeal tubercle

Mastoid process

Foramen magnum

Squamous part
of occipital bone

External occipital
protuberance (inion)

B. Inferior View

Bones

- Frontal
- Maxilla
- Occipital
- Palatine
- Parietal
- Sphenoid
- Temporal
- Vomer
- Zygomatic

CRANIUM, INFERIOR ASPECT (continued)

7.5

TABLE 7.1	FORAMINA AND OTHER APERTURES OF NEUROCRANIUM AND CONTENTS (SEE FIGS. 7.2 TO 7.6) (continued)
Foramen spinosum: Middle meningeal artery/vein and meningeal branch of CN V3	Hypoglossal canal: Hypoglossal nerve (CN XII)
Foramen lacerum[a]: Deep petrosal nerve, some meningeal arterial branches and small veins	Foramen magnum: Spinal cord; spinal accessory nerve (CN XI); vertebral arteries; internal vertebral venous plexus
Groove of greater petrosal nerve: Greater petrosal nerve and petrosal branch of middle meningeal artery	Condylar canal: Condyloid emissary vein (passes from sigmoid sinus to vertebral veins in neck)
Carotid canal: Internal carotid artery and accompanying sympathetic and venous plexuses	Stylomastoid foramen: Facial nerve (CN VII)
Internal acoustic meatus: Facial nerve/intermediate nerve (CN VII); vestibulocochlear nerve (CN VIII); labyrinthine artery	Mastoid foramina: Mastoid emissary vein from sigmoid sinus and meningeal branch of occipital artery

[a]The internal carotid artery and its accompanying sympathetic and venous plexuses actually pass horizontally across (rather than vertically through) the area of the foramen lacerum, an artifact of dry crania, which is closed by cartilage in life.

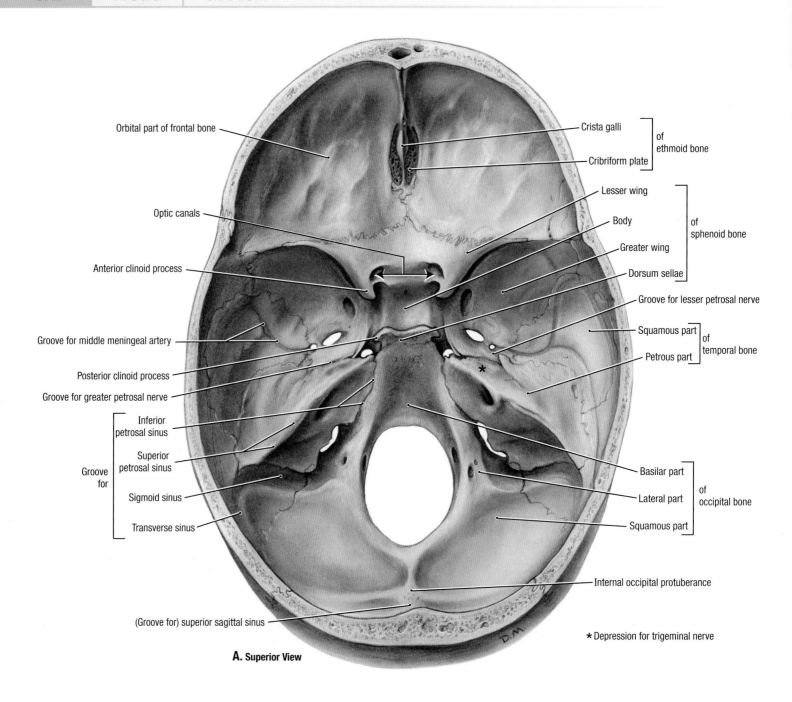

Orbital part of frontal bone

Optic canals

Anterior clinoid process

Groove for middle meningeal artery

Posterior clinoid process

Groove for greater petrosal nerve

Groove for — Inferior petrosal sinus

Superior petrosal sinus

Sigmoid sinus

Transverse sinus

(Groove for) superior sagittal sinus

Crista galli ⎱ of
Cribriform plate ⎰ ethmoid bone

Lesser wing
Body ⎱ of
Greater wing ⎰ sphenoid bone
Dorsum sellae

Groove for lesser petrosal nerve

Squamous part ⎱ of
Petrous part ⎰ temporal bone

Basilar part
Lateral part ⎱ of
Squamous part ⎰ occipital bone

Internal occipital protuberance

★ Depression for trigeminal nerve

A. Superior View

7.6   INTERIOR OF THE CRANIAL BASE

A. Bony cranial base. **B.** Anterior, middle, and posterior cranial fossae. **C.** Diagrammatic cranial base with bones color-coded.

- **Fractures in the floor of the anterior cranial fossa** may involve the cribriform plate of the ethmoid, resulting in leakage of CSF through the nose (CSF rhinorrhea). **CSF rhinorrhea** may be a primary indication of a cranial base fracture which increases the risk of meningitis, because an infection could spread to the meninges from the ear or nose.

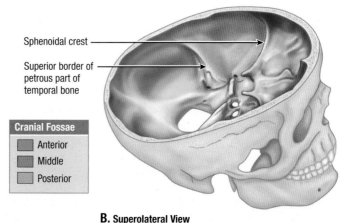

Sphenoidal crest

Superior border of petrous part of temporal bone

Cranial Fossae

	Anterior
	Middle
	Posterior

B. Superolateral View

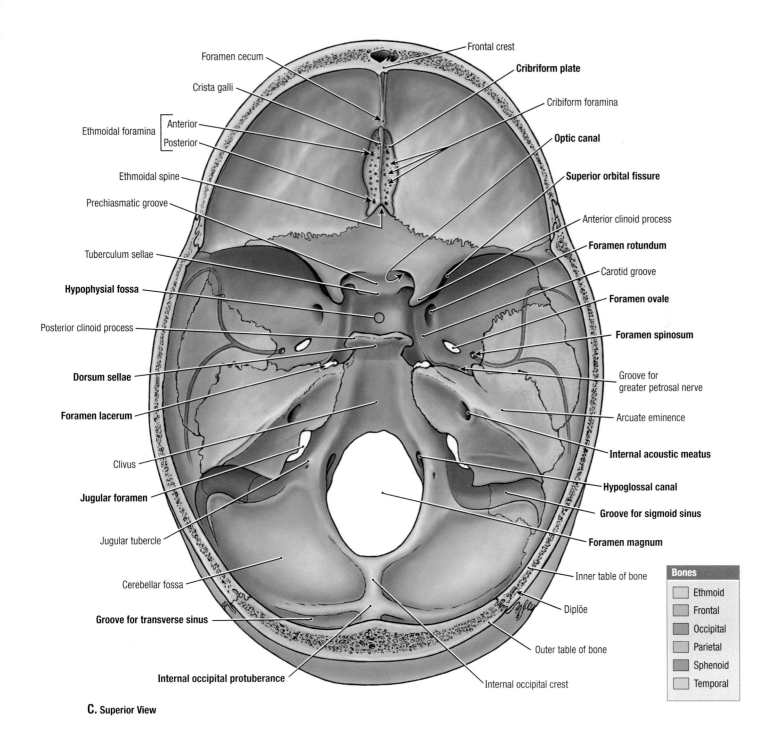

Foramen cecum
Crista galli
Ethmoidal foramina { Anterior / Posterior }
Ethmoidal spine
Prechiasmatic groove
Tuberculum sellae
Hypophysial fossa
Posterior clinoid process
Dorsum sellae
Foramen lacerum
Clivus
Jugular foramen
Jugular tubercle
Cerebellar fossa
Groove for transverse sinus
Internal occipital protuberance

Frontal crest
Cribriform plate
Cribiform foramina
Optic canal
Superior orbital fissure
Anterior clinoid process
Foramen rotundum
Carotid groove
Foramen ovale
Foramen spinosum
Groove for greater petrosal nerve
Arcuate eminence
Internal acoustic meatus
Hypoglossal canal
Groove for sigmoid sinus
Foramen magnum
Inner table of bone
Diplöe
Outer table of bone
Internal occipital crest

Bones	
	Ethmoid
	Frontal
	Occipital
	Parietal
	Sphenoid
	Temporal

C. Superior View

INTERIOR OF THE CRANIAL BASE (continued)   7.6

In **B**, note the following midline features:
- In the anterior cranial fossa, the frontal crest and crista galli for anterior attachment of the falx cerebri have between them the foramen cecum, which, during development, transmits a vein connecting the superior sagittal sinus with the veins of the frontal sinus and root of the nose.
- In the middle cranial fossa, the tuberculum sellae, hypophysial fossa, dorsum sellae, and posterior clinoid processes constitute the sella turcica (L. Turkish saddle).

- In the posterior cranial fossa, note the clivus, foramen magnum, internal occipital crest for attachment of the falx cerebelli, and the internal occipital protuberance, from which the grooves for the transverse sinuses course laterally.

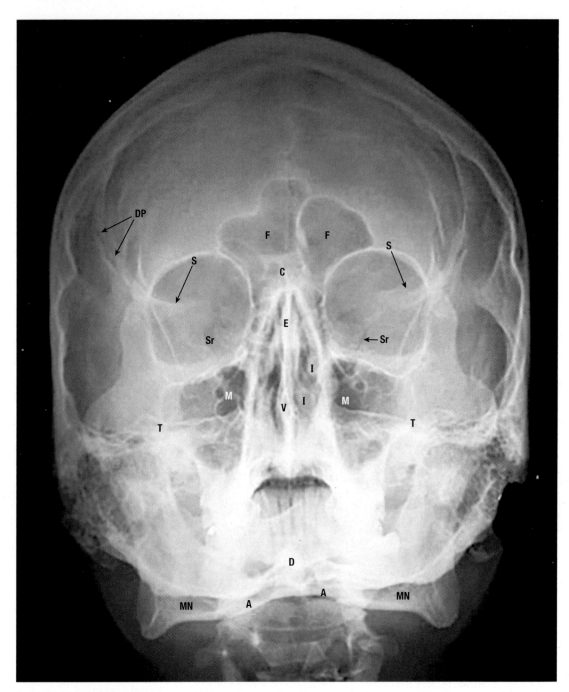

A. Postero-anterior Radiograph

7.7 RADIOGRAPHS OF THE CRANIUM

A. Postero-anterior (*Caldwell*) radiograph. This view places the orbits centrally in the head and is used to examine the orbits and paranasal sinuses.

Observe in **A**:

- The labeled features include the superior orbital fissure (*Sr*), lesser wing of the sphenoid (*S*), superior surface of the petrous part of the temporal bone (*T*), crista galli (*C*), frontal sinus (*F*), mandible (*MN*), maxillary sinus (*M*), and diploic veins (*DP*).
- The nasal septum is formed by the perpendicular plate of the ethmoid (*E*) and the vomer (*V*); note the inferior and middle conchae (*I*) of the lateral wall of the nose.
- Superimposed on the facial skeleton are the dens (*D*) and lateral masses of the atlas (*A*).

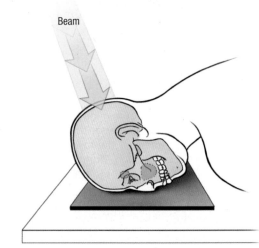

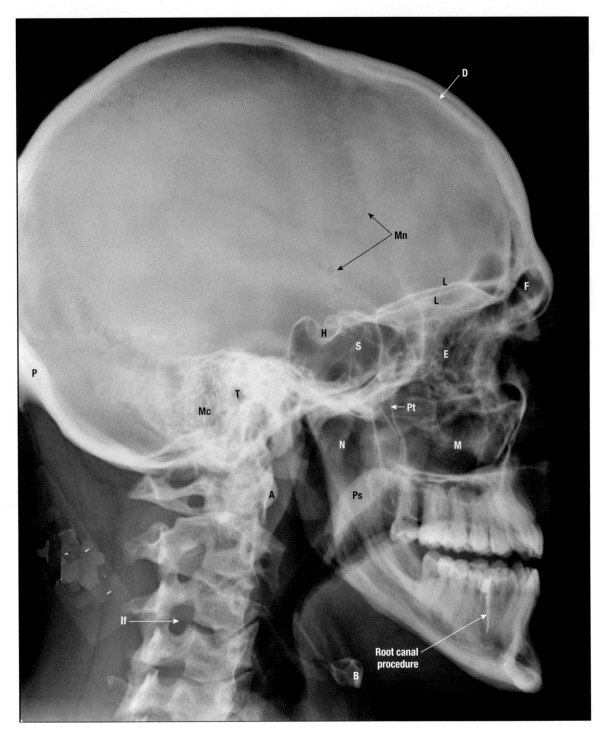

B. Lateral Radiograph

RADIOGRAPHS OF THE CRANIUM (*continued*) 7.7

B. Lateral radiograph of the cranium. Most of the relatively thin bone of the facial skeleton (viscerocranium) is radiolucent (appears *black*).

- The labeled features include the anterior tubercle of atlas (*A*), ethmoidal cells (*E*), frontal (*F*), sphenoidal (*S*) and maxillary (*M*) sinuses, the hypophysial fossa (*H*) for the pituitary gland, the petrous part of the temporal bone (*T*), mastoid cells (*Mc*),

grooves for the branches of the middle meningeal vessels (*Mn*), internal occipital protuberance (*P*), diploe (*D*), pterygopalatine fossa (*Pt*), soft palate (*Ps*), intervertebral foramen (*If*), hyoid (*B*), and the nasopharynx (*N*).

- The right and left orbital plates of the frontal bone are not superimposed; thus, the floor of the anterior cranial fossa appears as two lines (*L*).

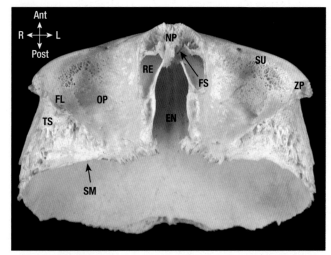

A. Inferior View

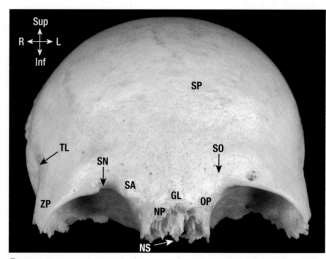

B. Anterior View

Key for A and B: Frontal Bone							
EN	Ethmoidal notch	NP	Nasal part	SA	Superciliary arch	SU	Supra-orbital margin
FL	Fossa for lacrimal gland	NS	Nasal spine	SM	Sphenoidal margin	TL	Temporal line
FS	Opening of frontal sinus	OP	Orbital part	SN	Supra-orbital notch	TS	Temporal surface
GL	Glabella	RE	Root of ethmoid cells	SO	Supra-orbital foramen	ZP	Zygomatic process
				SP	Squamous part		

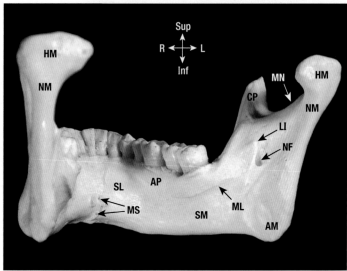

C. Posteromedial View

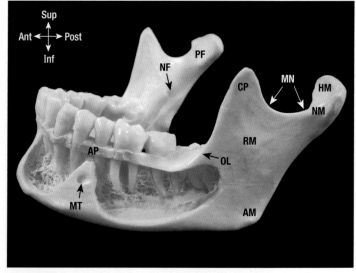

D. Lateral View

Key for C and D: Mandible							
AM	Angle of mandible (gonial angle)	MN	Mandibular notch	PF	Pterygoid fovea
AP	Alveolar part	MS	Mental (genial) spines	RM	Ramus of mandible
CP	Coronoid process	MT	Mental foramen	SL	Sublingual fossa
HM	Head of mandible	NF	Mandibular foramen	SM	Submandibular fossa
LI	Lingula	NM	Neck of mandible		
ML	Mylohyoid groove	OL	Oblique line		

7.8   MANDIBLE, MAXILLA, FRONTAL, ETHMOID, AND LACRIMAL BONES

A. and **B.** Frontal bone. **C.** and **D.** Mandible.

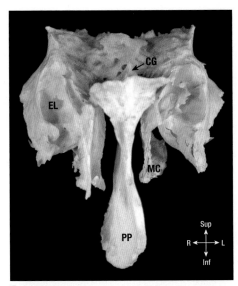

E. Anterior View

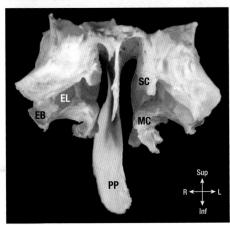

F. Posterior View

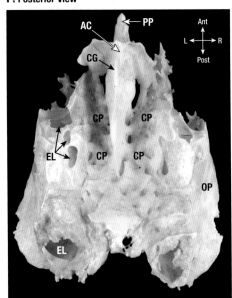

G. Superior View

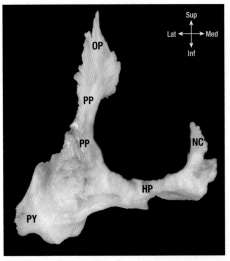

H. Anterior View

Key for H: Palatine Bone

HP	Horizontal plate	PP	Perpendicular plate
NC	Nasal crest	PY	Pyramidal process
OP	Orbital process		

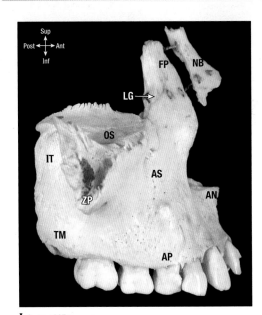

I. Lateral View

Key for I: Maxilla and Nasal Bone

AN	Anterior nasal spine	LG	Lacrimal groove
AP	Alveolar part	NB	Nasal bone
AS	Anterior surface	OS	Orbital surface
FP	Frontal process	TM	Tuberosity
IT	Infratemporal surface	ZP	Zygomatic process

Key for E–G: Ethmoid Bone

AC	Ala of crista galli	EB	Ethmoidal bulla	OP	Orbital plate
CG	Crista galli	EL	Ethmoidal labyrinth (cells)	PP	Perpendicular plate
CP	Cribriform plate	MC	Middle nasal concha	SC	Superior nasal concha

MANDIBLE, MAXILLA, FRONTAL, ETHMOID, AND LACRIMAL BONES (*continued*) 7.8

E–G. Ethmoid bone. **H.** Lacrimal bone. **I.** Maxilla.

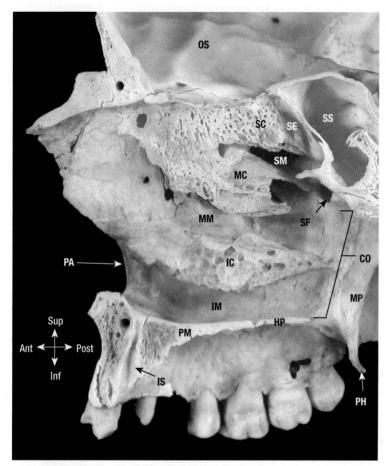

A. Lateral Wall of Nose, Medial View

Key for A: Lateral Wall of Nose	
CO	Choana (posterior nasal aperture)
HP	Horizontal plate of palatine bone
IC	Inferior nasal concha
IM	Inferior nasal meatus
IS	Incisive canal
MC	Middle nasal concha
MM	Middle nasal meatus
MP	Medial pterygoid plate
OS	Orbital surface of frontal bone
PA	Piriform aperture
PH	Pterygoid hamulus
PM	Palatine process of maxilla
SC	Superior nasal concha
SE	Spheno-ethmoidal recess
SF	Sphenopalatine foramen
SM	Superior nasal meatus
SS	Sphenoidal sinus

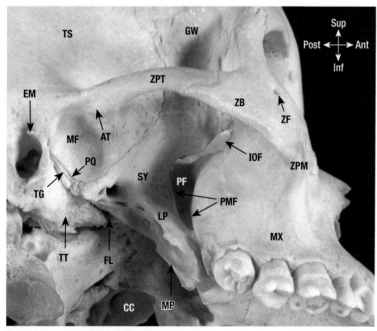

B. Infratemporal Region, Inferolateral View

Key for B: Infratemporal Region	
AT	Articular tubercle
CC	Carotid canal
EM	External acoustic meatus
FL	Foramen lacerum
GW	Greater wing of sphenoid
IOF	Inferior orbital fissure
LP	Lateral pterygoid plate
MF	Mandibular fossa
MP	Medial pterygoid plate
MX	Maxilla
PF	Pterygopalatine fossa
PMF	Pterygomaxillary fissure
PQ	Petrosquamous fissure
SY	Stylomastoid foramen
TG	Tegmen tympani
TS	Temporal bone (squamous part)
TT	Temporal bone (tympanic part)
ZB	Zygomatic bone
ZF	Zygomaticofacial foramen
ZPM	Zygomatic process of maxilla
ZPT	Zygomatic process of temporal bone

7.9   LATERAL WALL OF NOSE AND INFRATEMPORAL REGION

A. Lateral wall of nose. **B.** Infratemporal region.

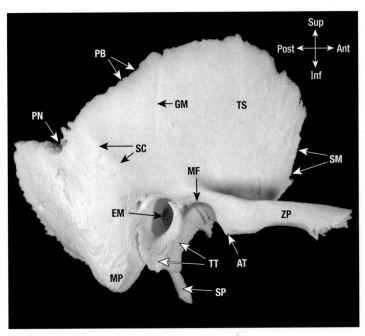

A. Lateral View

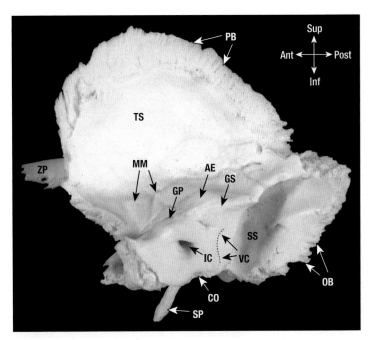

B. Medial View

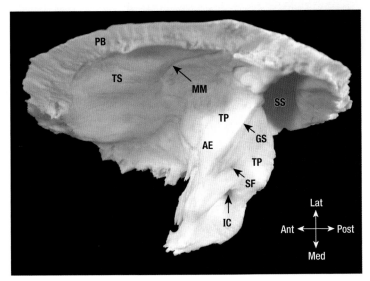

C. Superior View

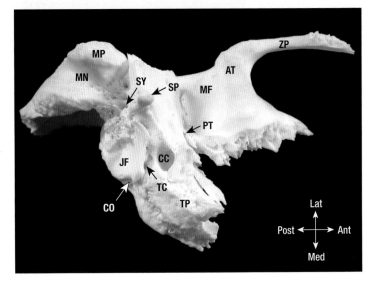

D. Inferior View

Key for A–D: Temporal Bone		
AE	Arcuate eminence	MF	Mandibular fossa	SM	Sphenoid margin
AT	Articular tubercle	MM	Groove for middle meningeal artery	SP	Styloid process
CC	Carotid canal	MN	Mastoid notch	SS	Groove for sigmoid sinus
CO	Cochlear canaliculus	MP	Mastoid process	SY	Stylomastoid foramen
EM	External acoustic meatus	OB	Occipital border	TC	Tympanic canaliculus
GM	Groove for middle temporal artery	PB	Parietal border	TP	Temporal bone (petrous part)
GP	Hiatus for greater petrosal nerve	PN	Parietal notch	TS	Temporal bone (squamous part)
GS	Groove for superior petrosal sinus	PT	Petrotympanic fissure	TT	Temporal bone (tympanic part)
IC	Internal acoustic meatus	SC	Supramastoid crest	VC	Vestibular canaliculus
JF	Jugular fossa	SF	Subarcuate fossa	ZP	Zygomatic process

TEMPORAL BONE

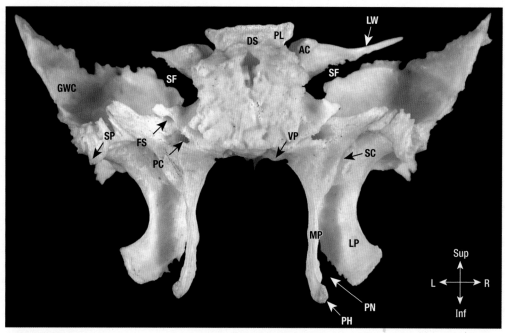

A. Posterior View

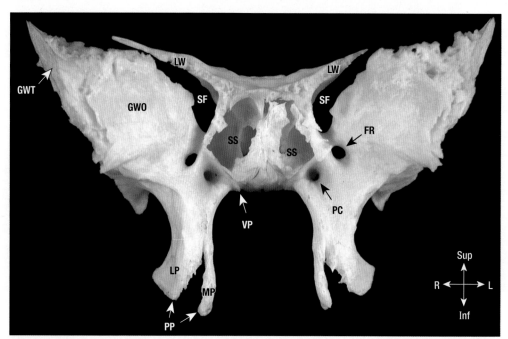

B. Anterior View

Key for A–D: Sphenoid Bone					
AC	Anterior clinoid process	FO	Foramen ovale	GWO	Greater wing (orbital surface)
CG	Carotid sulcus	FR	Foramen rotundum	GWT	Greater wing (temporal surface)
CS	Prechiasmatic sulcus	FS	Foramen spinosum	H	Hypophysial fossa
DS	Dorsum sellae	GWC	Greater wing (cerebral surface)	LP	Lateral pterygoid plate
ES	Ethmoidal spine	GWI	Greater wing (infratemporal surface)	LW	Lesser wing

7.11   SPHENOID BONE

A. Posterior aspect. **B.** Anterior aspect. The sphenoid is an irregular unpaired bone that is wedged between the frontal, temporal, and occipital bones.

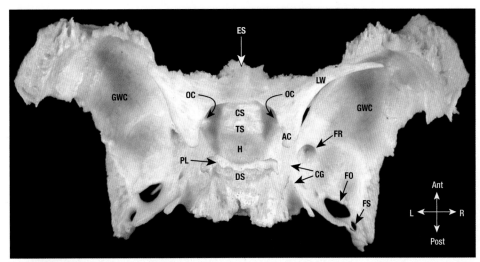

C. Superior View

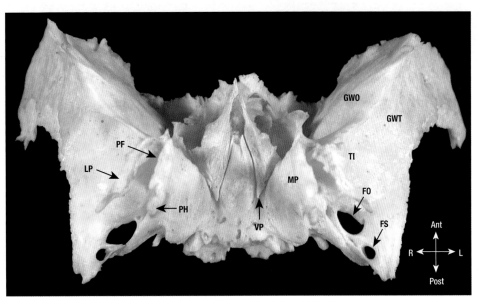

D. Inferior View

Key for A-D: Sphenoid Bone (continued)					
MP	Medial pterygoid plate	PL	Posterior clinoid process	SP	Spine of sphenoid bone
OC	Optic canal	PN	Pterygoid notch	SS	Sphenoidal sinus (in body of sphenoid)
PC	Pterygoid canal	PP	Pterygoid process	TI	Greater wing of sphenoid (infratemporal surface)
PF	Pterygoid fossa	SC	Scaphoid fossa	TS	Tuberculum sellae
PH	Pterygoid hamulus	SF	Superior orbital fissure	VP	Vaginal process

SPHENOID BONE (*continued*) 7.11

C. Superior aspect. **D.** Inferior aspect. It consists of a body and three pairs of processes: greater wings, lesser wings, and pterygoid processes.

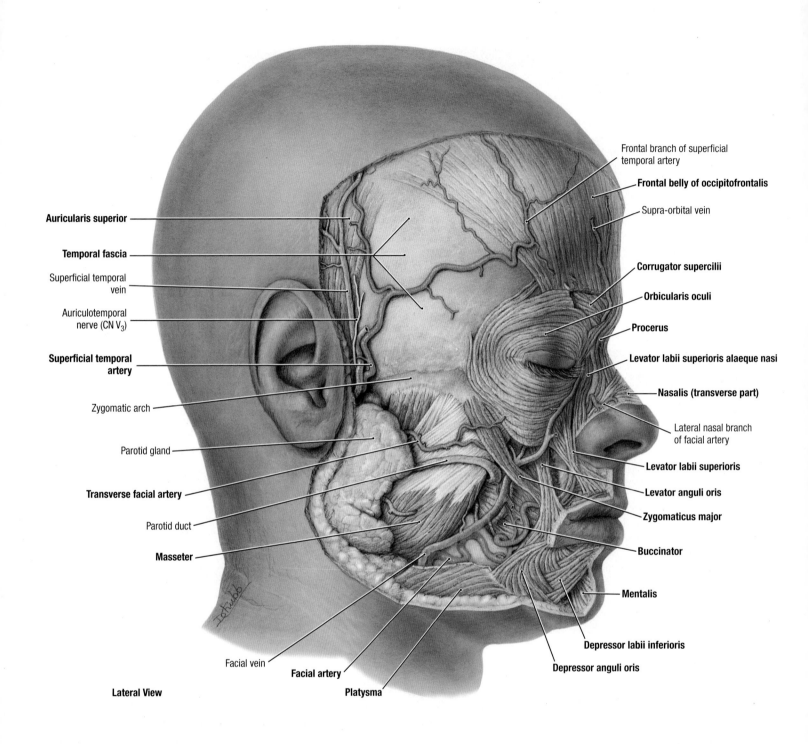

Frontal branch of superficial temporal artery

Frontal belly of occipitofrontalis

Supra-orbital vein

Corrugator supercilii

Orbicularis oculi

Procerus

Levator labii superioris alaeque nasi

Nasalis (transverse part)

Lateral nasal branch of facial artery

Levator labii superioris

Levator anguli oris

Zygomaticus major

Buccinator

Mentalis

Depressor labii inferioris

Depressor anguli oris

Auricularis superior

Temporal fascia

Superficial temporal vein

Auriculotemporal nerve (CN V₃)

Superficial temporal artery

Zygomatic arch

Parotid gland

Transverse facial artery

Parotid duct

Masseter

Facial vein

Facial artery

Platysma

Lateral View

7.12 MUSCLES OF FACIAL EXPRESSION AND ARTERIES OF THE FACE

- The muscles of facial expression are the superficial sphincters and dilators of the openings of the head; all are supplied by the facial nerve (CN VII). The masseter and temporalis (the latter covered here by temporal fascia) are muscles of mastication that are innervated by the trigeminal nerve (CN V).

- **Superficial temporal and facial artery pulses.** Anesthesiologists, usually stationed at the head of the operating table, take these pulses. The superficial temporal pulse is palpated anterior to the auricle as the artery crosses the zygomatic arch. The facial pulse is palpated where the facial artery crosses the inferior border of the mandible immediately anterior to the masseter.

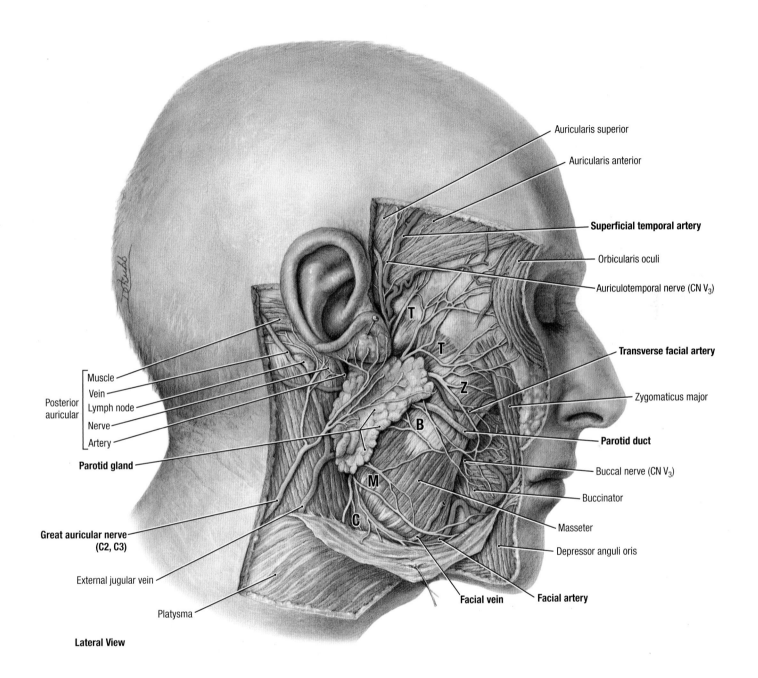

Auricularis superior
Auricularis anterior
Superficial temporal artery
Orbicularis oculi
Auriculotemporal nerve (CN V₃)
Transverse facial artery
Zygomaticus major
Parotid duct
Buccal nerve (CN V₃)
Buccinator
Masseter
Depressor anguli oris
Facial artery
Facial vein

T
T
Z
B
M
C

Muscle
Vein
Posterior auricular { Lymph node
Nerve
Artery
Parotid gland

Great auricular nerve (C2, C3)
External jugular vein
Platysma

Lateral View

RELATIONSHIPS OF BRANCHES OF FACIAL NERVE AND VESSELS TO THE PAROTID GLAND AND DUCT 7.13

- The parotid duct extends across the masseter muscle just inferior to the zygomatic arch; the duct turns medially to pierce the buccinator and opens into the oral vestibule.

- The facial nerve (CN VII) innervates the muscles of facial expression. After emerging from the stylomastoid foramen, the main stem of the facial nerve has posterior auricular, digastric, and stylohyoid branches; the parotid plexus gives rise to temporal (*T*), zygomatic (*Z*), buccal (*B*), marginal mandibular (*M*), cervical (*C*), and posterior auricular branches. These branches form a plexus within the parotid gland, the branches of which radiate over the face, anastomosing with each other and the branches of the trigeminal nerve.

- During **parotidectomy** (surgical excision of the parotid gland), identification, dissection, and preservation of the branches of the facial nerve are critical.

- The parotid gland may become infected by infectious agents that pass through the bloodstream, as occurs in mumps, an acute communicable viral disease. Infection of the gland causes inflammation, **parotiditis**, and swelling of the gland. Severe pain occurs because the parotid sheath, innervated by the great auricular nerve, is distended by swelling.

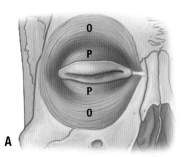

A

B N

C

— Nose *(N)*

Occipitofrontalis

Corrugator supercilii

Procerus + transverse part of nasalis

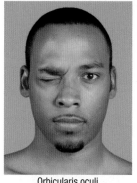

Orbicularis oculi

Lev. labii sup. alaeque nasi + alar part of nasalis

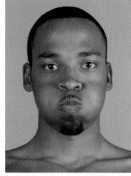

Buccinator + orbicularis oris

Zygomaticus major + minor

Risorius

Risorius + depressor labii inferioris

Levator labii sup. + depressor labii

Dilators of mouth: Risorius plus levator labii superioris + depressor labii inferioris

D

Orbicularis oris

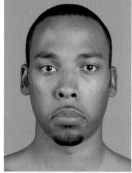

Depressor anguli oris

Mentalis

Platysma

Anterior Views

| 7.14 | MUSCLES OF FACIAL EXPRESSION |

A. Orbicularis oculi: palpebral (*P*) and orbital (*O*) parts. Eyelids close from lateral to medial washing lacrimal fluid across the cornea. **B.** Gentle closure of eyelid—palpebral part. **C.** Tight closure of eyelid—orbital part. **D.** Actions of selected muscles of facial expression.

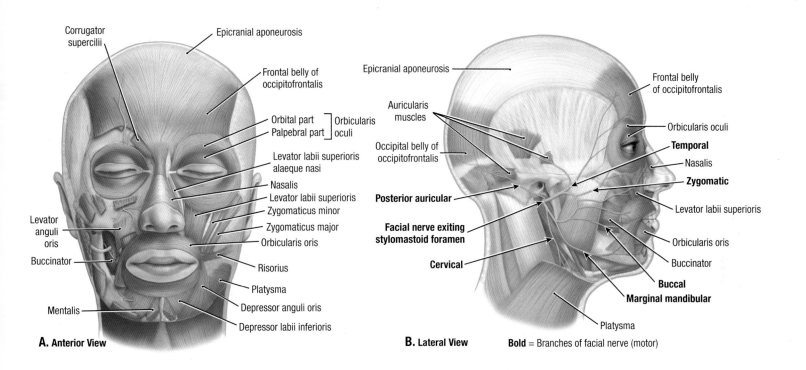

A. Anterior View

B. Lateral View   **Bold** = Branches of facial nerve (motor)

BRANCHES OF FACIAL NERVE AND MUSCLES OF FACIAL EXPRESSION   7.15

A. Muscles. **B.** Branches of facial nerve.

TABLE 7.2	MAIN MUSCLES OF FACIAL EXPRESSION		
Muscle[a]	Origin	Insertion	Action
Occipitofrontalis, frontal belly	Epicranial aponeurosis	Skin of and subcutaneous tissue of eyebrows and forehead	Elevates eyebrows and wrinkles skin of forehead; protracts scalp (indicating surprise or curiosity)
Occipitofrontalis, occipital belly	Lateral two thirds of superior nuchal line	Epicranial aponeurosis	Retracts scalp; increasing effectiveness of frontal belly
Orbicularis oculi	Medial orbital margin, medial palpebral ligament; lacrimal bone	Skin around margin of orbit; superior and inferior tarsal plates	Closes eyelids; palpebral part does so gently; orbital part tightly (winking)
Orbicularis oris	Medial maxilla and mandible; deep surface of perioral skin; angle of mouth (modiolus)	Mucous membrane of lips	Tonus closes oral fissure; phasic contraction compresses and protrudes lips (kissing) or resists distension (when blowing)
Levator labii superioris	Infra-orbital margin (maxilla)	Skin of upper lip	Part of dilators of mouth; retract (elevate) and/or evert upper lip; deepen nasolabial sulcus (showing sadness)
Zygomaticus minor	Anterior aspect, zygomatic bone		
Buccinator	Mandible, alveolar processes of maxilla and mandible; pterygomandibular raphe	Angle of mouth (modiolus); orbicularis oris	Presses cheek against molar teeth; works with tongue to keep food between occlusal surfaces and out of oral vestibule; resists distension (when blowing)
Zygomaticus major	Lateral aspect of zygomatic bone	Angle of mouth (modiolus)	Part of dilators of mouth; elevate labial commissure—bilaterally to smile (happiness); unilaterally to sneer (disdain)
Risorius	Parotid fascia and buccal skin (highly variable)		Part of dilators of mouth; widens oral fissure
Platysma	Subcutaneous tissue of infraclavicular and supraclavicular regions	Base of mandible; skin of cheek and lower lip; angle of mouth (modiolus); orbicularis oris	Depresses mandible (against resistance); tenses skin of inferior face and neck (conveying tension and stress)

[a]All of these muscles are supplied by the facial nerve (CN VII).

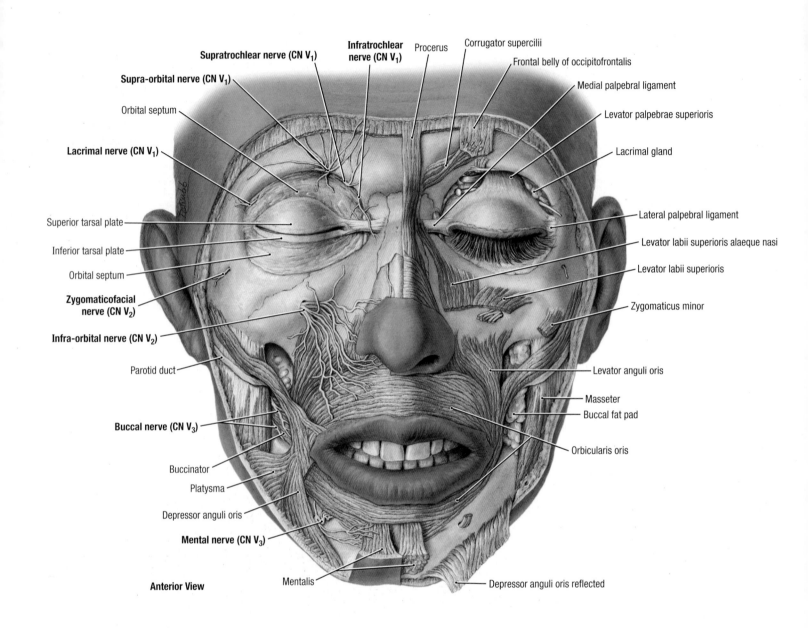

Supratrochlear nerve (CN V₁)

Supra-orbital nerve (CN V₁)

Orbital septum

Lacrimal nerve (CN V₁)

Superior tarsal plate

Inferior tarsal plate

Orbital septum

Zygomaticofacial nerve (CN V₂)

Infra-orbital nerve (CN V₂)

Parotid duct

Buccal nerve (CN V₃)

Buccinator

Platysma

Depressor anguli oris

Mental nerve (CN V₃)

Anterior View

Mentalis

Infratrochlear nerve (CN V₁)

Procerus

Corrugator supercilii

Frontal belly of occipitofrontalis

Medial palpebral ligament

Levator palpebrae superioris

Lacrimal gland

Lateral palpebral ligament

Levator labii superioris alaeque nasi

Levator labii superioris

Zygomaticus minor

Levator anguli oris

Masseter

Buccal fat pad

Orbicularis oris

Depressor anguli oris reflected

7.16 CUTANEOUS BRANCHES OF TRIGEMINAL NERVE, MUSCLES OF FACIAL EXPRESSION, AND EYELID

Injury to the facial nerve (CN VII) or its branches produces paralysis of some or all of the facial muscles on the affected side (Bell facial palsy). The affected area sags, and facial expression is distorted. The loss of tonus causes the inferior lid to evert (fall away from the surface of the eyeball). As a result, the lacrimal fluid is not spread over the cornea, preventing adequate lubrication, hydration, and flushing of the cornea. This makes the cornea vulnerable to ulceration. If the injury weakens or paralyzes the buccinator and orbicularis oris, food will accumulate in the oral vestibule during chewing, usually requiring continual removal with a finger. When the sphincters or dilators of the mouth are affected, displacement of the mouth (drooping of the corner) is produced by gravity and

contraction of unopposed contralateral facial muscles, resulting in food and saliva dribbling out of the side of the mouth. Weakened lip muscles affect speech. Affected people cannot whistle or blow a wind instrument effectively. They frequently dab their eyes and mouth with a handkerchief to wipe the fluid (tears and saliva) that runs from the drooping lid and mouth.

Because the face does not have a distinct layer of deep fascia and the subcutaneous tissue is loose between the attachments of facial muscles, **facial lacerations** tend to gap (part widely). Consequently, the skin must be sutured carefully to prevent scarring. The looseness of the subcutaneous tissue also enables fluid and blood to accumulate in the loose connective tissue causing **bruising of the face.**

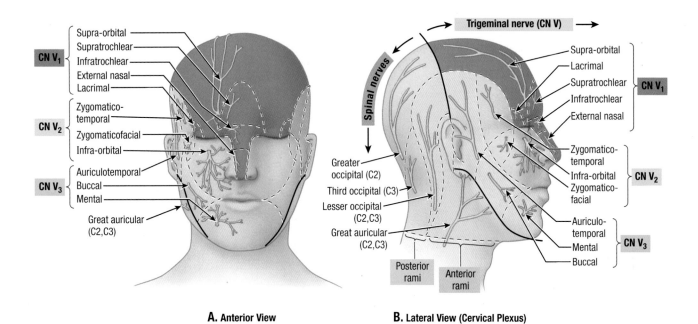

A. Anterior View

B. Lateral View (Cervical Plexus)

NERVES OF FACE AND SCALP

TABLE 7.3	**NERVES OF FACE AND SCALP**		
Nerve	*Origin*	*Course*	*Distribution*
Frontal	Ophthalmic nerve (CN V_1)	Crosses orbit on superior aspect of levator palpebrae superioris; divides into supra-orbital and supratrochlear branches	Skin of forehead, scalp, superior eyelid, and nose; conjunctiva of superior lid and mucosa of frontal sinus
Supra-orbital	Continuation of frontal nerve (CN V_1)	Emerges through supra-orbital notch, or foramen, and breaks up into small branches	Mucous membrane of frontal sinus and conjunctiva (lining) of superior eyelid; skin of forehead as far as vertex
Supratrochlear	Frontal nerve (CN V_1)	Continues anteromedially along roof of orbit, passing lateral to trochlea	Skin in middle of forehead to hairline
Infratrochlear	Nasociliary nerve (CN V_1)	Follows medial wall of orbit passing inferior to trochlea to superior eyelid	Skin and conjunctiva (lining) of superior eyelid
Lacrimal	Ophthalmic nerve (CN V_1)	Passes through palpebral fascia of superior eyelid near lateral angle (canthus) of eye	Lacrimal gland and small area of skin and conjunctiva of lateral part of superior eyelid
External nasal	Anterior ethmoidal nerve (CN V_1)	Runs in nasal cavity and emerges on face between nasal bone and lateral nasal cartilage	Skin on dorsum of nose, including tip of nose
Zygomatic	Maxillary nerve (CN V_2)	Arises in floor of orbit, divides into zygomaticofacial and zygomaticotemporal nerves, which traverse foramina of same name	Skin over zygomatic arch and anterior temporal region
Infra-orbital	Terminal branch of maxillary nerve (CN V_2)	Runs in floor of orbit and emerges at infra-orbital foramen	Skin of cheek, inferior lid, lateral side of nose and inferior septum and superior lip, upper premolar incisors and canine teeth; mucosa of maxillary sinus and superior lip
Auriculotemporal	Mandibular nerve (CN V_3)	From posterior division of CN V_3, it passes between neck of mandible and external acoustic meatus to accompany superficial temporal artery	Skin anterior to ear and posterior temporal region, tragus and part of helix of auricle, and roof of external acoustic meatus and upper tympanic membrane
Buccal	Mandibular nerve (CN V_3)	From the anterior division of CN V_3 in infratemporal fossa, it passes anteriorly to reach cheek	Skin and mucosa of cheek, buccal gingiva adjacent to 2nd and 3rd molar teeth
Mental	Terminal branch of inferior alveolar nerve (CN V_3)	Emerges from mandibular canal at mental foramen	Skin of chin and inferior lip and mucosa of lower lip

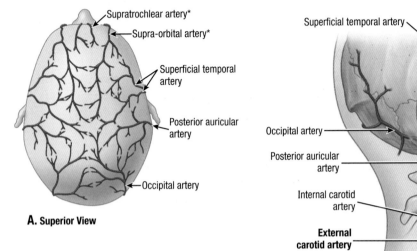

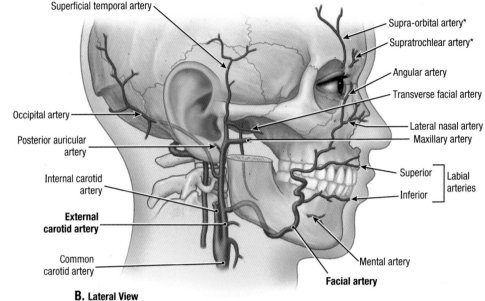

A. Superior View

B. Lateral View

*Source = internal carotid artery (ophthalmic artery); all other labeled
arteries are from external carotid

7.18 ARTERIES OF FACE AND SCALP

Most superficial arteries of the face are branches or derivatives of
the external carotid artery. The facial artery, a branch of the external carotid artery, provides the major arterial supply to the face. The
facial artery winds its way to the inferior border of the mandible,
just anterior to the masseter, and then courses over the face to the
medial angle (canthus) of the eye, where the superior and inferior
eyelids meet.

TABLE 7.4	ARTERIES OF SUPERFICIAL FACE AND SCALP		
Artery	*Origin*	*Course*	*Distribution*
Facial	External carotid artery	Ascends deep to submandibular gland, winds around inferior border of mandible and enters face	Muscles of facial expression and face
Inferior labial	Facial artery near angle of mouth	Runs medially in lower lip	Lower lip and chin
Superior labial		Runs medially in upper lip	Upper lip and ala (side) and septum of nose
Lateral nasal	Facial artery as it ascends alongside nose	Passes to ala of nose	Skin on ala and dorsum of nose
Angular	Terminal branch of facial artery	Passes to medial angle (canthus) of eye	Superior part of cheek and lower eyelid
Occipital	External carotid artery	Passes medial to posterior belly of digastric and mastoid process; accompanies occipital nerve in occipital region	Scalp of back of head, as far as vertex
Posterior auricular		Passes posteriorly, deep to parotid, along styloid process between mastoid and ear	Scalp posterior to auricle and auricle
Superficial temporal	Smaller terminal branch of external carotid artery	Ascends anterior to ear to temporal region and ends in scalp	Facial muscles and skin of frontal and temporal regions
Transverse facial	Superficial temporal artery within parotid gland	Crosses face superficial to masseter and inferior to zygomatic arch	Parotid gland and duct, muscles and skin of face
Mental	Terminal branch of inferior alveolar artery	Emerges from mental foramen and passes to chin	Facial muscles and skin of chin
***Supra-orbital**	Terminal branch of ophthalmic artery, a branch of internal carotid	Passes superiorly from supra-orbital foramen	Muscles and skin of forehead and scalp
***Supratrochlear**		Passes superiorly from supratrochlear notch	Muscles and skin of scalp

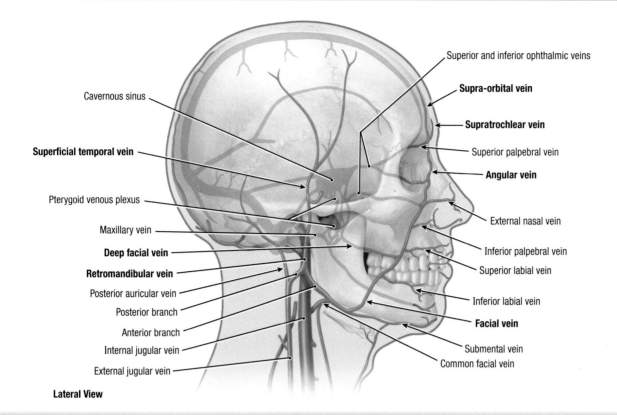

Superior and inferior ophthalmic veins

Supra-orbital vein

Supratrochlear vein

Superior palpebral vein

Angular vein

External nasal vein

Inferior palpebral vein

Superior labial vein

Inferior labial vein

Facial vein

Submental vein

Common facial vein

Cavernous sinus

Superficial temporal vein

Pterygoid venous plexus

Maxillary vein

Deep facial vein

Retromandibular vein

Posterior auricular vein

Posterior branch

Anterior branch

Internal jugular vein

External jugular vein

Lateral View

VEINS OF FACE | 7.19

TABLE 7.5	**VEINS OF FACE**			
Vein	*Origin*	*Course*	*Termination*	*Area Drained*
Supratrochlear	Begins from a venous plexus on the forehead and scalp, through which it communicates with the frontal branch of the superficial temporal vein, its contralateral partner, and the supra-orbital vein	Descends near the midline of the forehead to the root of the nose where it joins the supra-orbital vein	Angular vein at the root of the nose	Anterior part of scalp and forehead
Supra-orbital	Begins in the forehead by anastomosing with a frontal tributary of the superficial temporal vein	Passes medially superior to the orbit and joins the supratrochlear vein; a branch passes through the supra-orbital notch and joins with the superior ophthalmic vein		
Angular	Begins at root of nose by union of supratrochlear and supra-orbital veins	Descends obliquely along the root and side of the nose to the inferior margin of the orbit	Becomes the facial vein at the inferior margin of the orbit	In addition to above, drains upper and lower lids and conjunctiva; may receive drainage from cavernous sinus
Facial	Continuation of angular vein past inferior margin of orbit	Descends along lateral border of the nose, receiving external nasal and inferior palpebral veins, then obliquely across face to mandible; receives anterior division of retromandibular vein, after which it is sometimes called the common facial vein	Internal jugular vein at or inferior to the level of the hyoid bone	Anterior scalp and forehead, eyelids, external nose, and anterior cheek, lips, chin, and submandibular gland
Deep facial	Pterygoid venous plexus	Runs anteriorly on maxilla above buccinator and deep to masseter, emerging medial to anterior border of masseter onto face	Enters posterior aspect of facial vein	Infratemporal fossa (most areas supplied by maxillary artery)
Superficial temporal	Begins from a widespread plexus of veins on the side of the scalp and along the zygomatic arch	Its frontal and parietal tributaries unite anterior to the auricle; it crosses the temporal root of the zygomatic arch to pass from the temporal region and enters the substance of the parotid gland	Joins the maxillary vein posterior to the neck of the mandible to form the retromandibular vein	Side of the scalp, superficial aspect of the temporal muscle, and external ear
Retromandibular	Formed anterior to the ear by the union of the superficial temporal and maxillary veins	Runs posterior and deep to the ramus of the mandible through the substance of the parotid gland; communicates at its inferior end with the facial vein	*Anterior branch* unites with facial vein to form common facial vein; *posterior branch* unites with the posterior auricular vein to form the external jugular vein	Parotid gland and masseter muscle

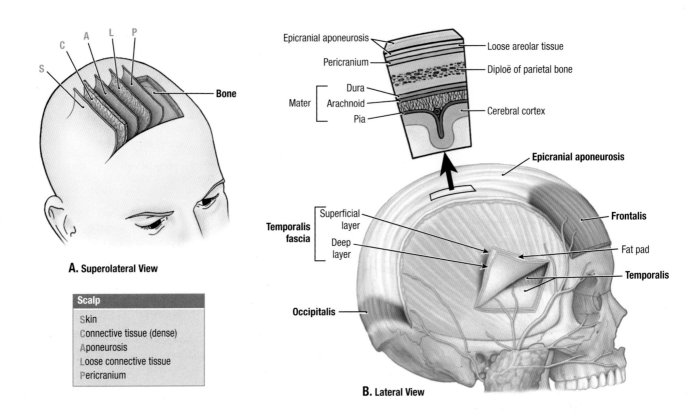

A. Superolateral View

Scalp
Skin
Connective tissue (dense)
Aponeurosis
Loose connective tissue
Pericranium

B. Lateral View

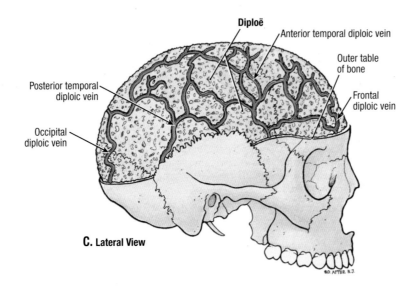

C. Lateral View

7.20 SCALP

A. Layers of scalp. **B.** Epicranial aponeurosis. **C.** Diploic veins. The outer layer of the compact bone of the cranium has been filed away, exposing the channels for the diploic veins in the cancellous bone that composes the diploë.

Scalp injuries and infections. The loose areolar tissue layer is the danger area of the scalp because pus or blood spreads easily in it. Infection in this layer can pass into the cranial cavity through emissary veins, which pass through parietal foramina in the calvaria and reach intracranial structures such as the meninges. An infection cannot pass into the neck because the occipital belly of the occipitofrontalis attaches to the occipital bone and mastoid parts of the temporal bones. Neither can a scalp infection spread laterally beyond the zygomatic arches because the epicranial aponeurosis is continuous with the temporalis fascia that attaches to these arches. An infection or fluid (e.g., pus or blood) can enter the eyelids and the root of the nose because the frontal belly of the occipitofrontalis inserts into the skin and dense subcutaneous tissue and does not attach to the bone. **Ecchymoses**, or purple patches, develop as a result of extravasation of blood into the subcutaneous tissue, skin of the eyelids and surrounding regions.

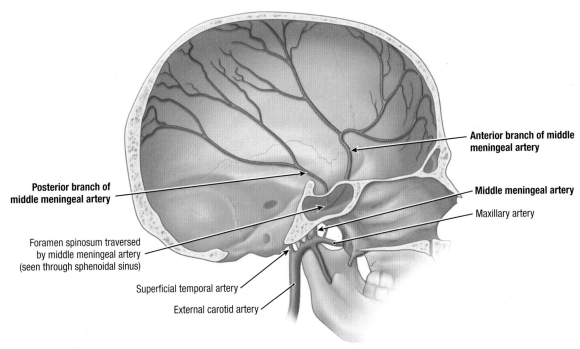

Anterior branch of middle meningeal artery

Posterior branch of middle meningeal artery

Middle meningeal artery

Maxillary artery

Foramen spinosum traversed by middle meningeal artery (seen through sphenoidal sinus)

Superficial temporal artery

External carotid artery

A. Medial View, Left Half of Bisected Cranium

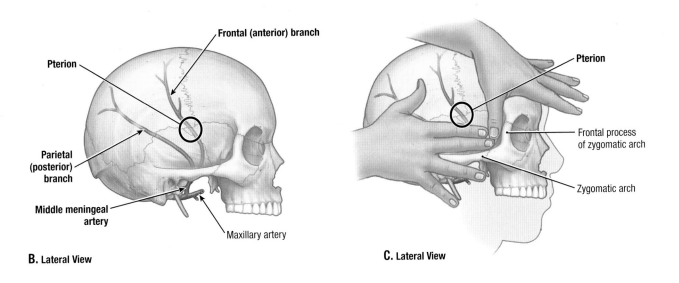

Frontal (anterior) branch

Pterion

Pterion

Parietal (posterior) branch

Middle meningeal artery

Maxillary artery

Frontal process of zygomatic arch

Zygomatic arch

B. Lateral View

C. Lateral View

MIDDLE MENINGEAL ARTERY AND PTERION

7.21

A. Course of the middle meningeal artery in the cranium. **B.** Surface projections of middle meningeal artery. **C.** Locating the pterion. The pterion is located two fingers breadth superior to the zygomatic arch and one thumb breadth posterior to the frontal process of the zygomatic bone (approximately 4 cm superior to the midpoint of the zygomatic arch); the anterior branch of the middle meningeal artery crosses the pterion.

A hard blow to the side of the head may fracture the thin bones forming the pterion, rupturing the anterior branch of the middle meningeal artery crossing the pterion. The resulting **extradural (epidural) hematoma** exerts pressure on the underlying cerebral cortex. Untreated middle meningeal artery hemorrhage may cause death in a few hours.

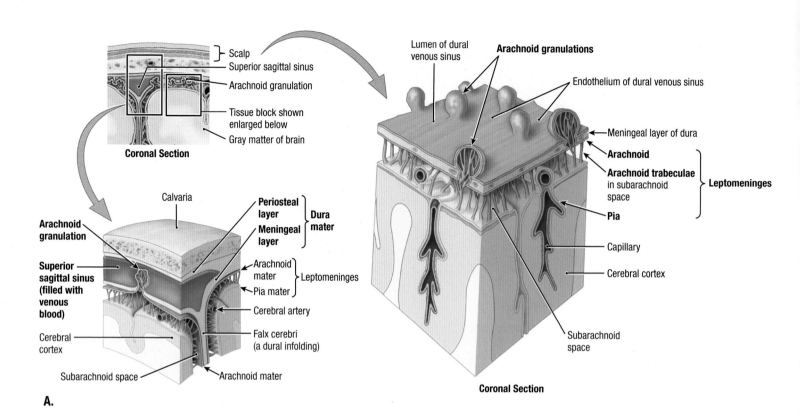

Coronal Section

A.

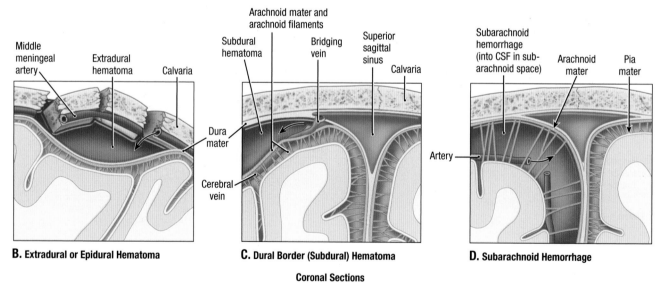

B. Extradural or Epidural Hematoma

C. Dural Border (Subdural) Hematoma

D. Subarachnoid Hemorrhage

Coronal Sections

7.22 MENINGES

A. Cranium and meninges. The three meningeal spaces include the extradural (epidural) space between the cranial bones and dura, which is a potential space normally (it becomes a real space pathologically if blood accumulates in it); the similarly potential subdural space between the dura and arachnoid; and the subarachnoid space, the normal realized space between the arachnoid and pia, which contains cerebrospinal fluid (CSF). **B.** Extradural (epidural) **hematomas** result from bleeding from a torn middle meningeal artery. **C.** Dural border (subdural) **hematomas** commonly result from tearing of a cerebral vein as it enters the superior sagittal sinus. **D.** Subarachnoid hemorrhage results from bleeding within the subarachnoid space (e.g., from rupture of an aneurysm).

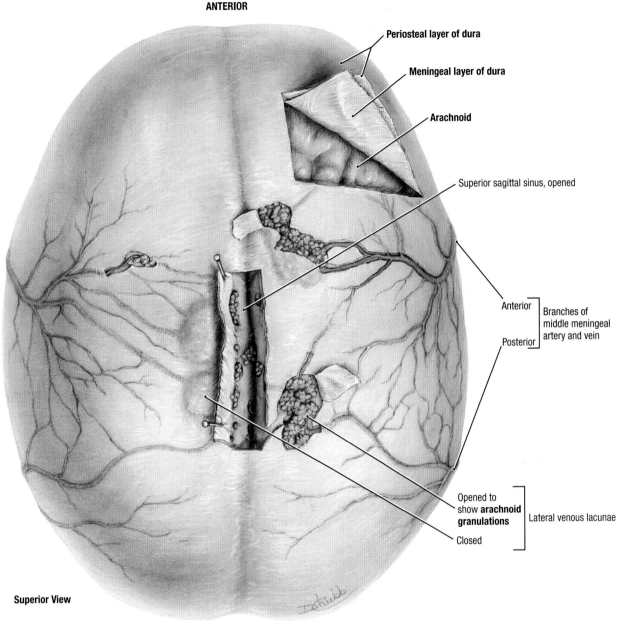

ANTERIOR

Periosteal layer of dura

Meningeal layer of dura

Arachnoid

Superior sagittal sinus, opened

Anterior · Branches of middle meningeal artery and vein

Posterior

Opened to show **arachnoid granulations** · Lateral venous lacunae

Closed

Superior View

POSTERIOR

DURA MATER AND ARACHNOID GRANULATIONS

7.23

- The calvaria is removed. In the median plane, the thick roof of the superior sagittal sinus is partly pinned aside, and laterally, the thin roofs of two lateral lacunae are reflected.
- The middle meningeal artery courses with the middle meningeal veins, which enlarge superiorly and drain into a lateral lacunae. Other channels drain the lateral lacunae into the superior sagittal sinus.
- Arachnoid granulations in the lacunae are responsible for absorption of CSF from the subarachnoid space into the venous system.

- The dura is sensitive to pain, especially where it is related to the dural venous sinuses and meningeal arteries. Although the causes of **headache** are numerous, distention of the scalp or meningeal vessels (or both) is believed to be one cause of headache. Many headaches appear to be dural in origin, such as the headache occurring after a lumbar spinal puncture for removal of CSF. These headaches are thought to result from stimulation of sensory nerve endings in the dura.

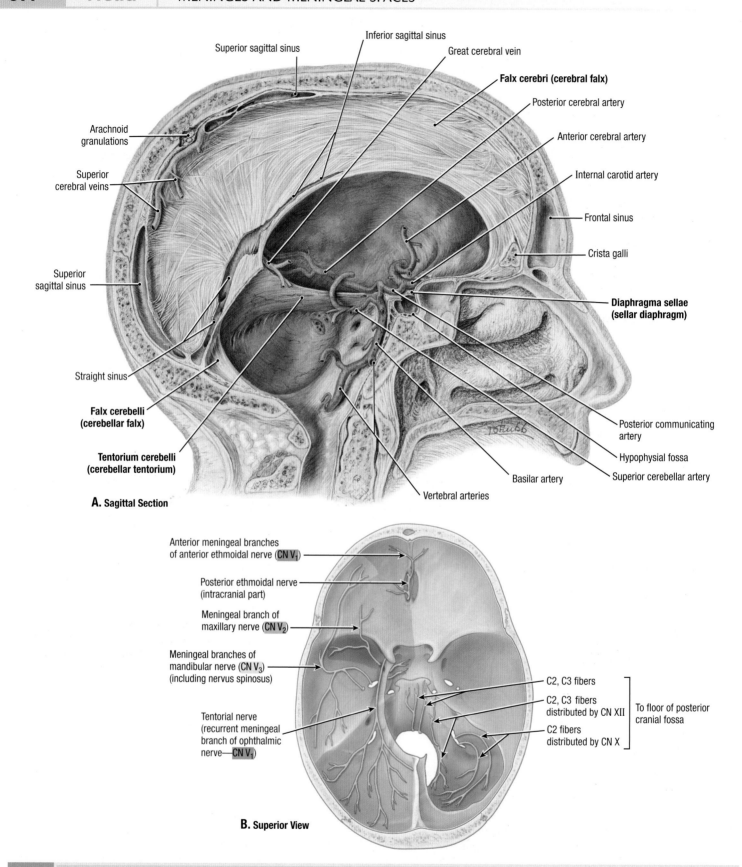

Inferior sagittal sinus
Superior sagittal sinus
Great cerebral vein
Falx cerebri (cerebral falx)
Posterior cerebral artery
Anterior cerebral artery
Internal carotid artery
Frontal sinus
Crista galli
Diaphragma sellae (sellar diaphragm)
Posterior communicating artery
Hypophysial fossa
Superior cerebellar artery
Basilar artery
Vertebral arteries
Arachnoid granulations
Superior cerebral veins
Superior sagittal sinus
Straight sinus
Falx cerebelli (cerebellar falx)
Tentorium cerebelli (cerebellar tentorium)

A. Sagittal Section

Anterior meningeal branches of anterior ethmoidal nerve (CN V₁)
Posterior ethmoidal nerve (intracranial part)
Meningeal branch of maxillary nerve (CN V₂)
Meningeal branches of mandibular nerve (CN V₃) (including nervus spinosus)
Tentorial nerve (recurrent meningeal branch of ophthalmic nerve—CN V₁)
C2, C3 fibers
C2, C3 fibers distributed by CN XII
To floor of posterior cranial fossa
C2 fibers distributed by CN X

B. Superior View

7.24   DURA MATER

A. Reflections of the dura mater. **B.** Innervation of the dura of the cranial base. The dura of the cranial base is innervated by branches of the trigeminal nerve and sensory fibers of cervical spinal nerves (C2, C3) passing directly from those nerves or via meningeal branches of the vagus (CN X) and hypoglossal (CN XII) nerves.

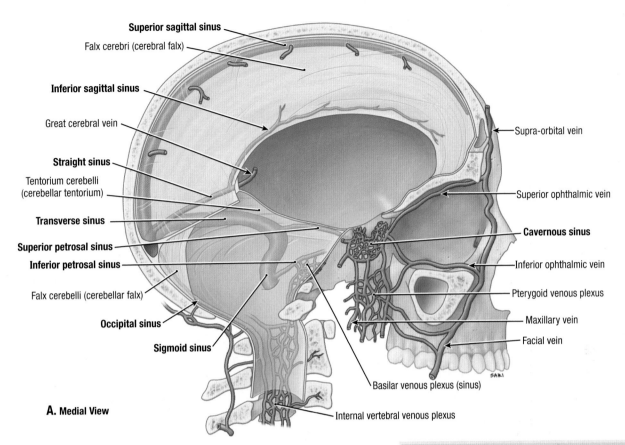

Superior sagittal sinus

Falx cerebri (cerebral falx)

Inferior sagittal sinus

Great cerebral vein

Straight sinus

Tentorium cerebelli (cerebellar tentorium)

Transverse sinus

Superior petrosal sinus

Inferior petrosal sinus

Falx cerebelli (cerebellar falx)

Occipital sinus

Sigmoid sinus

Supra-orbital vein

Superior ophthalmic vein

Cavernous sinus

Inferior ophthalmic vein

Pterygoid venous plexus

Maxillary vein

Facial vein

Basilar venous plexus (sinus)

Internal vertebral venous plexus

A. Medial View

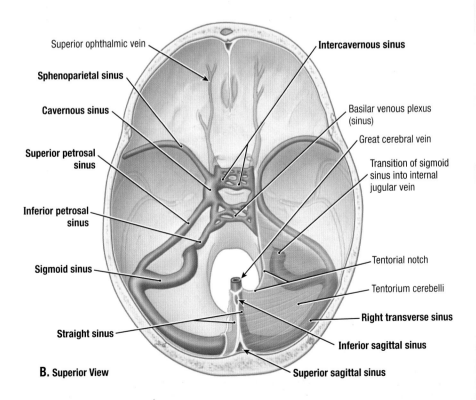

Superior ophthalmic vein

Sphenoparietal sinus

Cavernous sinus

Superior petrosal sinus

Inferior petrosal sinus

Sigmoid sinus

Straight sinus

B. Superior View

Intercavernous sinus

Basilar venous plexus (sinus)

Great cerebral vein

Transition of sigmoid sinus into internal jugular vein

Tentorial notch

Tentorium cerebelli

Right transverse sinus

Inferior sagittal sinus

Superior sagittal sinus

VENOUS SINUSES OF DURA MATER | 7.25

A. Schematic of left half of cranial cavity and right facial skeleton. **B.** Venous sinuses of the cranial base.

- The superior sagittal sinus is at the superior border of the falx cerebri, and the inferior sagittal sinus is in its free border. The great cerebral vein joins the inferior sagittal sinus to form the straight sinus.
- The superior sagittal sinus usually becomes the right transverse sinus, which drains into the right sigmoid sinus, and next into the right internal jugular vein; the straight sinus similarly drains through the left transverse sinus, left sigmoid sinus, and left internal jugular vein.
- The cavernous sinus communicates with the veins of the face through the ophthalmic veins and pterygoid plexus of veins and with the sigmoid sinus through the superior and inferior petrosal sinuses.
- **Metastasis of tumor cells to dural sinuses.** The basilar and occipital sinuses communicate through the foramen magnum with the internal vertebral venous plexuses. Because these venous channels are valveless, increased intra-abdominopelvic or intrathoracic pressure, as occurs during heavy coughing and straining, may force venous blood from these regions into the internal vertebral venous system and from it into the dural venous sinuses. As a result, pus in abscesses and tumor cells in these regions may spread to the vertebrae and brain.

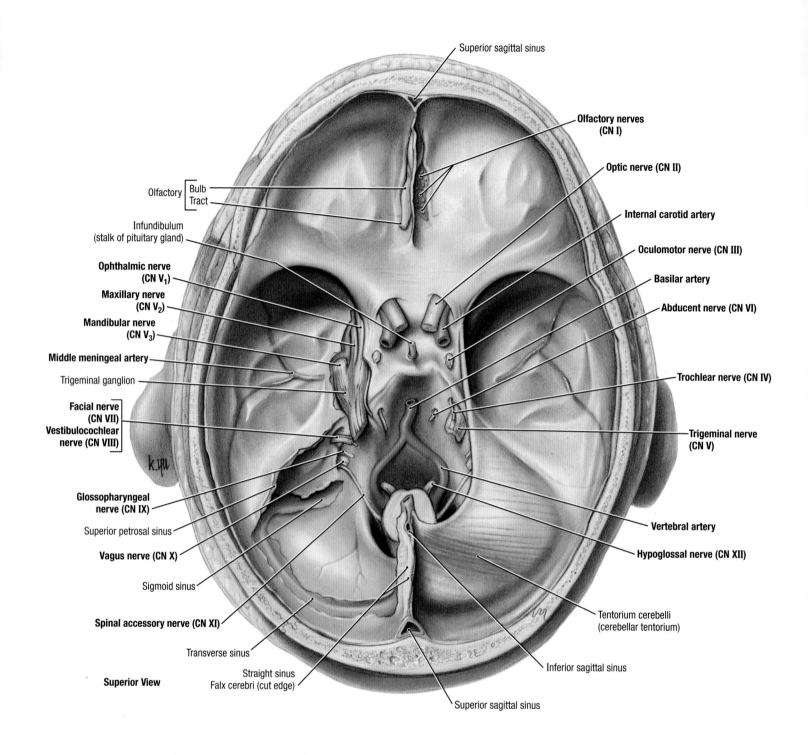

Superior sagittal sinus

Olfactory nerves (CN I)

Optic nerve (CN II)

Internal carotid artery

Oculomotor nerve (CN III)

Basilar artery

Abducent nerve (CN VI)

Trochlear nerve (CN IV)

Trigeminal nerve (CN V)

Vertebral artery

Hypoglossal nerve (CN XII)

Tentorium cerebelli (cerebellar tentorium)

Inferior sagittal sinus

Superior sagittal sinus

Olfactory Bulb / Tract

Infundibulum (stalk of pituitary gland)

Ophthalmic nerve (CN V₁)

Maxillary nerve (CN V₂)

Mandibular nerve (CN V₃)

Middle meningeal artery

Trigeminal ganglion

Facial nerve (CN VII)

Vestibulocochlear nerve (CN VIII)

Glossopharyngeal nerve (CN IX)

Superior petrosal sinus

Vagus nerve (CN X)

Sigmoid sinus

Spinal accessory nerve (CN XI)

Transverse sinus

Straight sinus

Falx cerebri (cut edge)

Superior View

7.26 NERVES AND VESSELS OF THE INTERIOR OF THE BASE OF CRANIUM

- On the left of the specimen, the dura mater forming the roof of the trigeminal cave is cut away to expose the trigeminal ganglion and its three branches. The tentorium cerebelli is removed to reveal the transverse and superior petrosal sinuses.
- The frontal lobes of the cerebrum are located in the anterior cranial fossa, the temporal lobes in the middle cranial fossa, and

the brainstem and cerebellum in the posterior cranial fossa; the occipital lobes rest on the tentorium cerebelli.
- The sites where the 12 cranial nerves and the internal carotid, vertebral, basilar, and middle meningeal arteries penetrate the dura mater are shown.

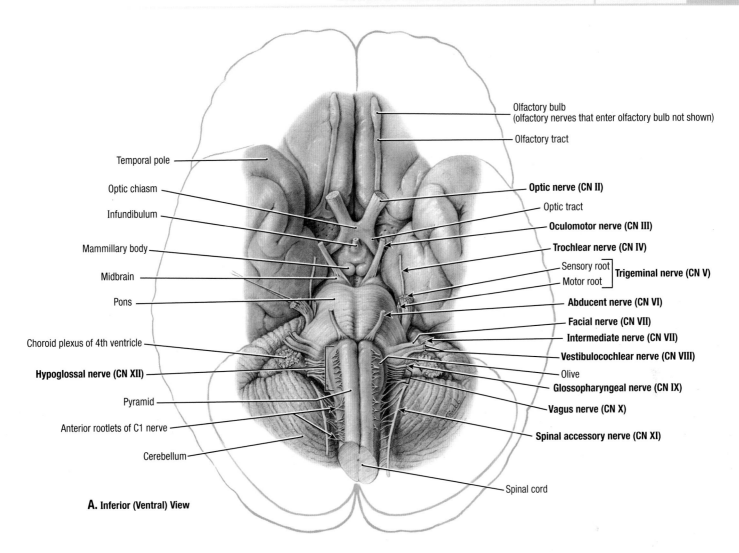

Olfactory bulb (olfactory nerves that enter olfactory bulb not shown)

Olfactory tract

Temporal pole

Optic chiasm

Infundibulum

Mammillary body

Midbrain

Pons

Choroid plexus of 4th ventricle

Hypoglossal nerve (CN XII)

Pyramid

Anterior rootlets of C1 nerve

Cerebellum

Optic nerve (CN II)

Optic tract

Oculomotor nerve (CN III)

Trochlear nerve (CN IV)

Sensory root
Motor root **Trigeminal nerve (CN V)**

Abducent nerve (CN VI)

Facial nerve (CN VII)

Intermediate nerve (CN VII)

Vestibulocochlear nerve (CN VIII)

Olive

Glossopharyngeal nerve (CN IX)

Vagus nerve (CN X)

Spinal accessory nerve (CN XI)

Spinal cord

A. Inferior (Ventral) View

BASE OF BRAIN AND SUPERFICIAL ORIGINS OF CRANIAL NERVES **7.27**

A. Cranial nerves in relation to the base of the brain. **B.** Cranial fossae. Foramina of skull and their associated cranial nerve(s) are listed below.

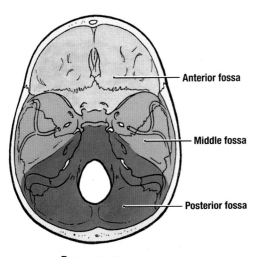

Anterior fossa

Middle fossa

Posterior fossa

B. Superior View

TABLE 7.6	**OPENINGS BY WHICH CRANIAL NERVES EXIT CRANIAL CAVITY**
Foramina/Apertures	*Cranial Nerve*
Anterior cranial fossa	
Cribriform foramina in cribriform plate	Axons of olfactory cells in olfactory epithelium form olfactory nerves (CN I)
Middle cranial fossa	
Optic canal	Optic nerve (CN II)
Superior orbital fissure	Ophthalmic nerve (CN V₁) and branches, oculomotor nerve (CN III), trochlear nerve (CN IV), and abducent nerve (CN VI)
Foramen rotundum	Maxillary nerve (CN V₂)
Foramen ovale	Mandibular nerve (CN V₃)
Posterior cranial fossa	
Foramen magnum	Spinal accessory nerve (CN XI)
Jugular foramen	Glossopharyngeal nerve (CN IX), vagus nerve (CN X), and spinal accessory nerve (CN XI)
Hypoglossal canal	Hypoglossal nerve (CN XII)

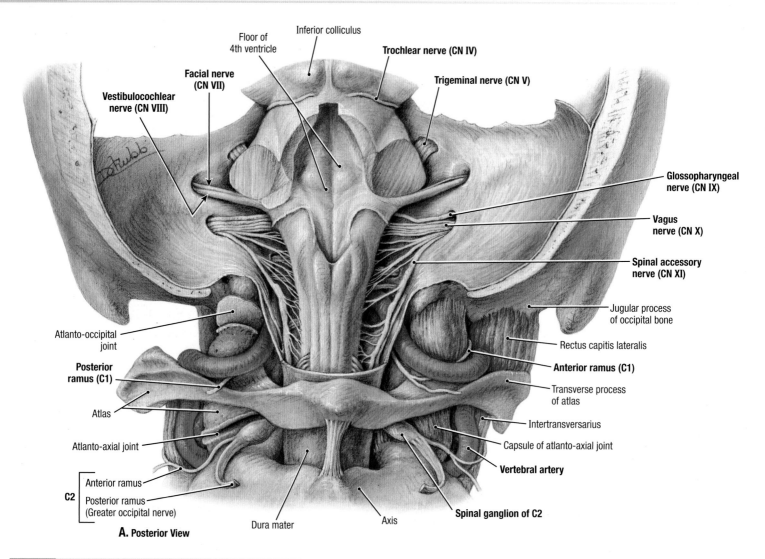

Floor of 4th ventricle
Inferior colliculus
Trochlear nerve (CN IV)
Facial nerve (CN VII)
Trigeminal nerve (CN V)
Vestibulocochlear nerve (CN VIII)
Glossopharyngeal nerve (CN IX)
Vagus nerve (CN X)
Spinal accessory nerve (CN XI)
Jugular process of occipital bone
Atlanto-occipital joint
Rectus capitis lateralis
Posterior ramus (C1)
Anterior ramus (C1)
Atlas
Transverse process of atlas
Atlanto-axial joint
Intertransversarius
Capsule of atlanto-axial joint
Vertebral artery
Anterior ramus
C2
Posterior ramus (Greater occipital nerve)
Spinal ganglion of C2
Dura mater
Axis
A. Posterior View

7.28 POSTERIOR EXPOSURES OF CRANIAL NERVES

A. and **B.** Squamous part of occipital bone has been removed posterior to foramen magnum to reveal posterior cranial fossa. **A.** Brainstem *in situ*. **B.** Brainstem removed *(right side)*. The trochlear nerves (CN IV) arise from the dorsal aspect of the midbrain, just inferior to the inferior colliculi.

- The sensory and motor roots of the trigeminal nerves (CN V) pass anterolaterally to enter the mouth of the trigeminal cave.
- The facial (CN VII) and vestibulocochlear (CN VIII) nerves course laterally to enter the internal acoustic meatus.
- The glossopharyngeal nerve (CN IX) pierces the dura mater separately but passes with the vagus (CN X) and spinal accessory (CN XI) nerves through the jugular foramen.
- An **acoustic neuroma** (neurofibroma) is a slow-growing benign tumor of the neurolemma (Schwann) cells. The tumor begins in the vestibulocochlear nerve (CN VIII) while it is in the internal acoustic meatus. The early symptom of an acoustic neuroma is usually loss of hearing. Dysequilibrium and tinnitus also may occur.

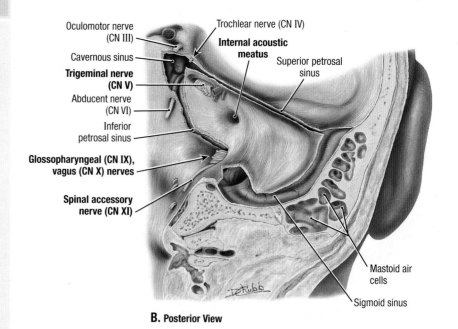

Oculomotor nerve (CN III)
Trochlear nerve (CN IV)
Internal acoustic meatus
Cavernous sinus
Superior petrosal sinus
Trigeminal nerve (CN V)
Abducent nerve (CN VI)
Inferior petrosal sinus
Glossopharyngeal (CN IX), vagus (CN X) nerves
Spinal accessory nerve (CN XI)
Mastoid air cells
Sigmoid sinus

B. Posterior View

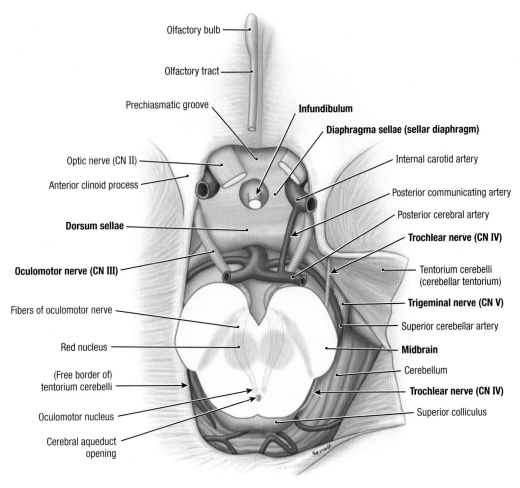

Olfactory bulb

Olfactory tract

Prechiasmatic groove

Infundibulum

Diaphragma sellae (sellar diaphragm)

Optic nerve (CN II)

Anterior clinoid process

Internal carotid artery

Posterior communicating artery

Posterior cerebral artery

Dorsum sellae

Trochlear nerve (CN IV)

Oculomotor nerve (CN III)

Tentorium cerebelli (cerebellar tentorium)

Trigeminal nerve (CN V)

Fibers of oculomotor nerve

Superior cerebellar artery

Red nucleus

Midbrain

(Free border of) tentorium cerebelli

Cerebellum

Trochlear nerve (CN IV)

Oculomotor nucleus

Superior colliculus

Cerebral aqueduct opening

Superior View

TENTORIAL NOTCH

7.29

- The brain has been removed by cutting through the midbrain, revealing the tentorial notch through which the brainstem extends from the posterior into the middle cranial fossa.
- On the right side of the specimen, the tentorium cerebelli is divided and reflected. The trochlear nerve (CN IV) passes around the midbrain under the free edge of the tentorium cerebelli; the roots of the trigeminal nerve (CN V) enter the mouth of the trigeminal cave.
- There is a circular opening in the diaphragma sellae for the infundibulum, the stalk of the pituitary gland.
- The oculomotor nerve (CN III) passes between the posterior cerebral and superior cerebellar arteries and then laterally around the posterior clinoid process.

- The tentorial notch is the opening in the tentorium cerebelli for the brainstem, which is slightly larger than is necessary to accommodate the midbrain. Hence, space-occupying lesions, such as tumors in the supratentorial compartment, produce increased intracranial pressure that may cause part of the adjacent temporal lobe of the brain to herniate through the tentorial notch. During **tentorial herniation**, the temporal lobe may be lacerated by the tough tentorium cerebelli, and the oculomotor nerve (CN III) may be stretched, compressed, or both. Oculomotor lesions may produce paralysis of the extrinsic eye muscles supplied by CN III.

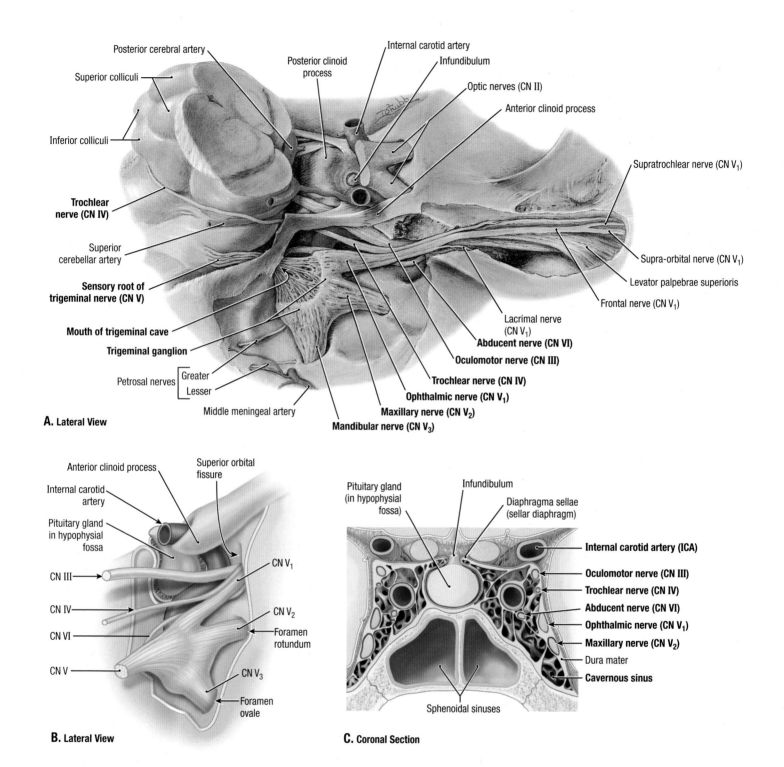

A. Lateral View

- Posterior cerebral artery
- Superior colliculi
- Inferior colliculi
- **Trochlear nerve (CN IV)**
- Superior cerebellar artery
- **Sensory root of trigeminal nerve (CN V)**
- **Mouth of trigeminal cave**
- **Trigeminal ganglion**
- Petrosal nerves [Greater / Lesser]
- Middle meningeal artery
- Posterior clinoid process
- Internal carotid artery
- Infundibulum
- Optic nerves (CN II)
- Anterior clinoid process
- Supratrochlear nerve (CN V₁)
- Supra-orbital nerve (CN V₁)
- Levator palpebrae superioris
- Frontal nerve (CN V₁)
- Lacrimal nerve (CN V₁)
- **Abducent nerve (CN VI)**
- **Oculomotor nerve (CN III)**
- **Trochlear nerve (CN IV)**
- **Ophthalmic nerve (CN V₁)**
- **Maxillary nerve (CN V₂)**
- **Mandibular nerve (CN V₃)**

B. Lateral View

- Anterior clinoid process
- Superior orbital fissure
- Internal carotid artery
- Pituitary gland in hypophysial fossa
- CN III
- CN IV
- CN VI
- CN V
- CN V₁
- CN V₂
- Foramen rotundum
- CN V₃
- Foramen ovale

C. Coronal Section

- Pituitary gland (in hypophysial fossa)
- Infundibulum
- Diaphragma sellae (sellar diaphragm)
- **Internal carotid artery (ICA)**
- **Oculomotor nerve (CN III)**
- **Trochlear nerve (CN IV)**
- **Abducent nerve (CN VI)**
- **Ophthalmic nerve (CN V₁)**
- **Maxillary nerve (CN V₂)**
- Dura mater
- **Cavernous sinus**
- Sphenoidal sinuses

7.30 NERVES AND VESSELS OF MIDDLE CRANIAL FOSSA I

A. Superficial dissection. The tentorium cerebelli is cut away. The dura mater is largely removed from the middle cranial fossa. The roof of the orbit is partly removed. **B.** Relationship of oculomotor, trochlear, trigeminal, and abducent nerves to the internal carotid artery. **C.** Coronal section through the cavernous sinus.

In **fractures of the cranial base**, the internal carotid artery may be torn, producing an arteriovenous fistula within the cavernous sinus. Arterial blood rushes into the sinus, enlarging it and forcing retrograde blood flow into its venous tributaries, especially the ophthalmic veins. As a result, the eyeball protrudes (**exophthalmos**) and the conjunctiva becomes engorged (**chemosis**). Because CN III, CN IV, CN VI, CN V₁, and CN V₂ lie in or close to the lateral wall of the cavernous sinus, these nerves may also be affected.

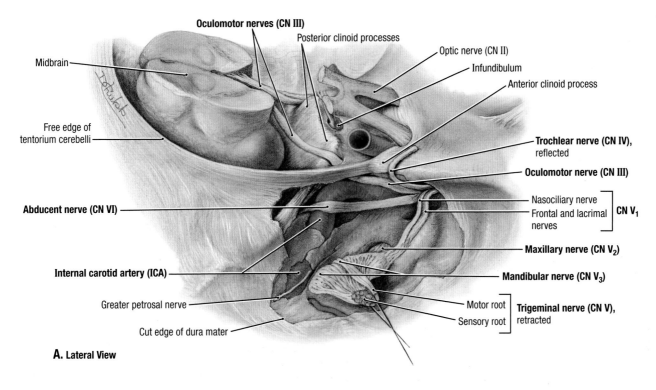

Oculomotor nerves (CN III)
Posterior clinoid processes
Optic nerve (CN II)
Infundibulum
Anterior clinoid process
Midbrain
Trochlear nerve (CN IV), reflected
Free edge of tentorium cerebelli
Oculomotor nerve (CN III)
Nasociliary nerve
Frontal and lacrimal nerves] **CN V₁**
Abducent nerve (CN VI)
Maxillary nerve (CN V₂)
Mandibular nerve (CN V₃)
Internal carotid artery (ICA)
Motor root] **Trigeminal nerve (CN V),**
Greater petrosal nerve
Sensory root] retracted
Cut edge of dura mater

A. Lateral View

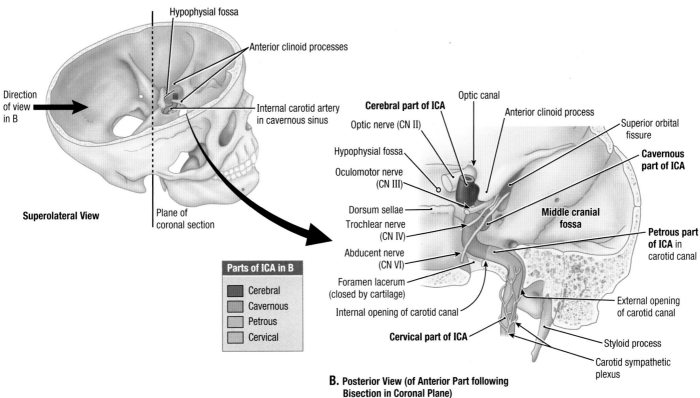

Hypophysial fossa
Anterior clinoid processes

Direction of view in B

Internal carotid artery in cavernous sinus

Optic canal
Cerebral part of ICA
Anterior clinoid process
Optic nerve (CN II)
Superior orbital fissure
Hypophysial fossa
Cavernous part of ICA
Oculomotor nerve (CN III)
Superolateral View
Plane of coronal section
Dorsum sellae
Middle cranial fossa
Trochlear nerve (CN IV)
Petrous part of ICA in carotid canal
Abducent nerve (CN VI)

Parts of ICA in B
Cerebral
Cavernous
Petrous
Cervical

Foramen lacerum (closed by cartilage)
External opening of carotid canal
Internal opening of carotid canal
Cervical part of ICA
Styloid process
Carotid sympathetic plexus

B. Posterior View (of Anterior Part following Bisection in Coronal Plane)

NERVES AND VESSELS OF MIDDLE CRANIAL FOSSA II

A. Deep dissection. The roots of the trigeminal nerve are divided, withdrawn from the mouth of the trigeminal cave, and turned anteriorly. The trochlear nerve is reflected anteriorly. **B.** Course of the internal carotid artery.

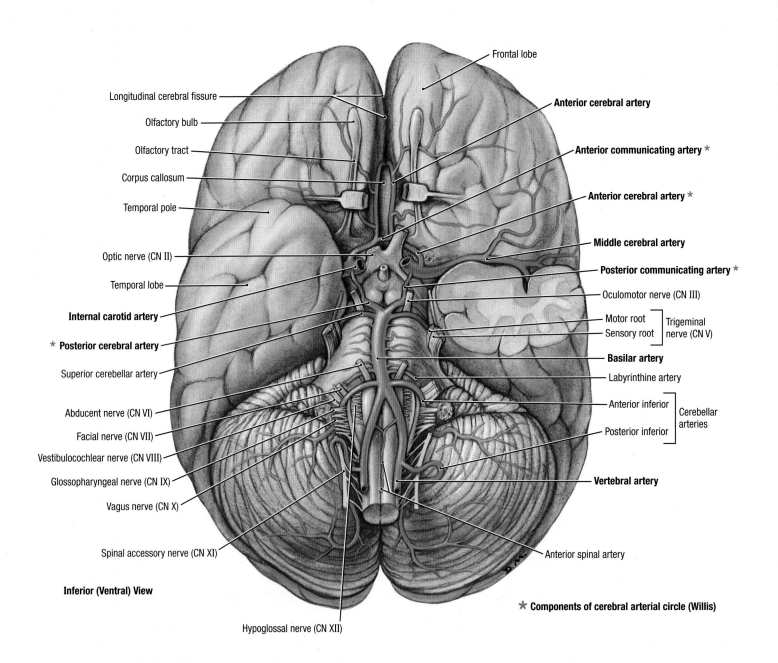

Frontal lobe

Longitudinal cerebral fissure

Olfactory bulb

Olfactory tract

Corpus callosum

Temporal pole

Optic nerve (CN II)

Temporal lobe

Internal carotid artery

* Posterior cerebral artery

Superior cerebellar artery

Abducent nerve (CN VI)

Facial nerve (CN VII)

Vestibulocochlear nerve (CN VIII)

Glossopharyngeal nerve (CN IX)

Vagus nerve (CN X)

Spinal accessory nerve (CN XI)

Inferior (Ventral) View

Anterior cerebral artery

Anterior communicating artery *

Anterior cerebral artery *

Middle cerebral artery

Posterior communicating artery *

Oculomotor nerve (CN III)

Motor root ⎤ Trigeminal
Sensory root ⎦ nerve (CN V)

Basilar artery

Labyrinthine artery

Anterior inferior ⎤ Cerebellar
Posterior inferior ⎦ arteries

Vertebral artery

Anterior spinal artery

* **Components of cerebral arterial circle (Willis)**

Hypoglossal nerve (CN XII)

7.32 BASE OF BRAIN AND CEREBRAL ARTERIAL CIRCLE

The anterior part of the left temporal lobe is removed to enable visualization of the middle cerebral artery in the lateral fissure. The frontal lobes are separated to expose the anterior cerebral arteries and corpus callosum.

An **ischemic stroke** denotes the sudden development of neurological deficits that are consequences of impaired cerebral blood flow. The most common causes of strokes are spontaneous cerebrovascular accidents such as cerebral embolism, thrombosis, or hemorrhage, and subarachnoid hemorrhage (Rowland, 2000). The cerebral arterial circle is an important means of collateral circulation in the event of gradual obstruction of one of the major arteries forming the circle. Sudden occlusion, even if only partial,

results in neurological deficits. In elderly persons, the anastomoses are often inadequate when a large artery (e.g., internal carotid) is occluded, even if the occlusion is gradual. In such cases, function is impaired at least to some degree.

Hemorrhagic stroke follows the rupture of an artery or a saccular aneurysm, a saclike dilation on a weak part of the arterial wall. The most common type of saccular aneurysm is a berry aneurysm, occurring in the vessels of or near the cerebral arterial circle. In time, especially in people with hypertension (high blood pressure), the weak part of the arterial wall expands and may rupture, allowing blood to enter the subarachnoid space.

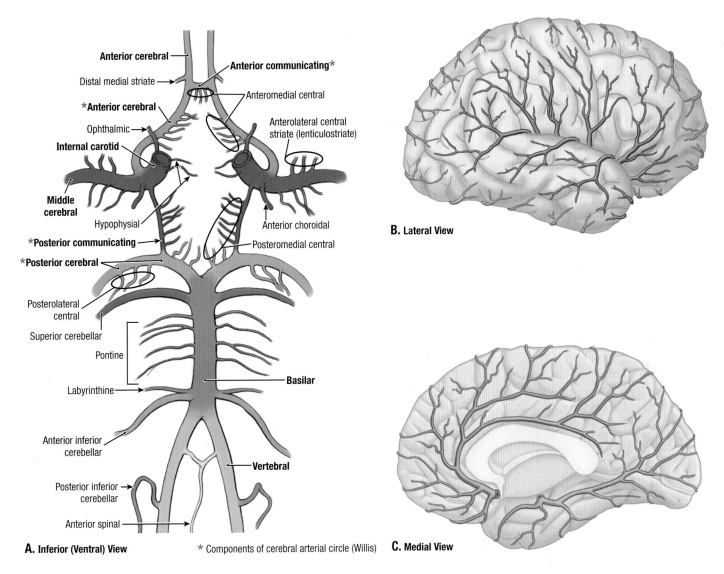

A. Schematic overview. **B.** and **C.** Distribution of anterior, middle, and posterior cerebral arteries.

ARTERIES OF BRAIN

7.33

TABLE 7.7	**ARTERIAL SUPPLY TO BRAIN**	
Artery	*Origin*	*Distribution*
Vertebral	Subclavian artery	Cranial meninges and cerebellum
Posterior inferior cerebellar	Vertebral artery	Postero-inferior aspect of cerebellum
Basilar	Formed by junction of vertebral arteries	Brainstem, cerebellum, and cerebrum
Pontine		Numerous branches to brainstem
Anterior inferior cerebellar	Basilar artery	Inferior aspect of cerebellum
Superior cerebellar		Superior aspect of cerebellum
Internal carotid	Common carotid artery at superior border of thyroid cartilage	Gives branches in cavernous sinus and provides supply to brain
Anterior cerebral	Internal carotid artery	Cerebral hemispheres, except for occipital lobes
Middle cerebral	Continuation of the internal carotid artery distal to anterior cerebral artery	Most of lateral surface of cerebral hemispheres
Posterior cerebral	Terminal branch of basilar artery	Inferior aspect of cerebral hemisphere and occipital lobe
Anterior communicating	Anterior cerebral artery	Cerebral arterial circle
Posterior communicating	Internal carotid artery	

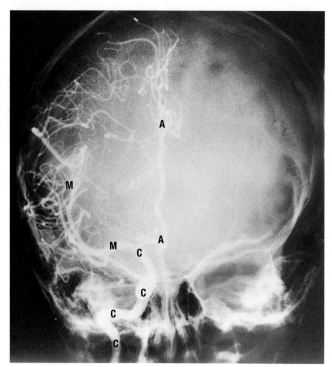

A. Postero-anterior Angiogram

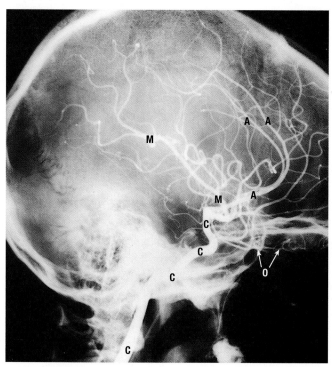

B. Lateral Angiogram

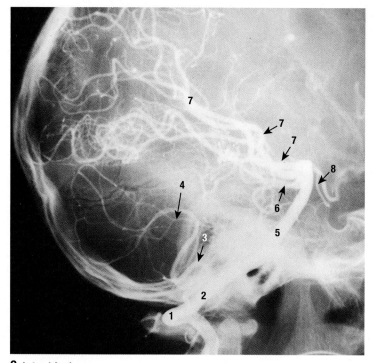

C. Lateral Angiogram

Key for A, B, and C	
A	Anterior cerebral artery
C	Internal carotid artery
M	Middle cerebral artery
O	Ophthalmic artery
1	Vertebral artery on posterior arch of atlas
2	Vertebral artery entering skull through foramen magnum
3	Posterior inferior cerebellar artery
4	Anterior inferior cerebellar artery
5	Basilar artery
6	Superior cerebellar artery
7	Posterior cerebral artery
8	Posterior communicating artery

7.34   **ARTERIOGRAMS**

A. and **B.** Carotid arteriogram. The four Cs indicate the parts of the internal carotid artery: cervical, before entering the cranium; petrous, within the temporal bone; cavernous, within the sinus; and cerebral, within the cranial subarachnoid space. **C.** Vertebral arteriogram. **Transient ischemic attacks (TIAs)** refer to neurological symptoms resulting from ischemia (deficient blood supply) of the brain.

The symptoms of a TIA may be ambiguous: staggering, dizziness, light-headedness, fainting, and paresthesias (e.g., tingling in a limb). Most TIAs last a few minutes, but some persist longer. Individuals with TIAs are at increased risk for myocardial infarction and *ischemic stroke* (Brust, 2000)

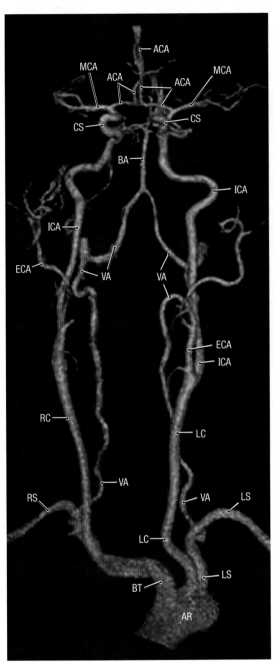

A. Anterior View

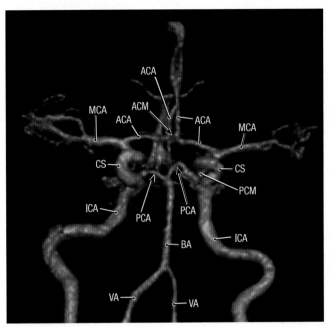

B. Anterior View

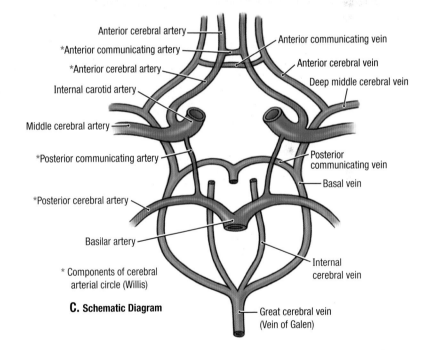

Anterior cerebral artery
*Anterior communicating artery
*Anterior cerebral artery
Internal carotid artery
Middle cerebral artery
*Posterior communicating artery
*Posterior cerebral artery
Basilar artery

Anterior communicating vein
Anterior cerebral vein
Deep middle cerebral vein
Posterior communicating vein
Basal vein
Internal cerebral vein

* Components of cerebral arterial circle (Willis)

Great cerebral vein (Vein of Galen)

C. Schematic Diagram

Key for A and B							
ACA	Anterior cerebral artery	BT	Brachiocephalic trunk	LC	Left common carotid artery	PCM	Posterior communicating artery
ACM	Anterior communicating artery	CS	Carotid siphon	LS	Left subclavian artery	RC	Right common carotid artery
AR	Arch of aorta	ECA	External carotid artery	MCA	Middle cerebral artery	RS	Right subclavian artery
BA	Basilar artery	ICA	Internal carotid artery	PCA	Posterior cerebral artery	VA	Vertebral artery

BLOOD SUPPLY OF HEAD AND NECK

7.35

A. CT angiogram of arteries of head and neck. **B.** CT angiogram of cerebral arterial circle (circle of Willis). **C.** Schematic diagram of cerebral arterial circle and veins of cerebral base.

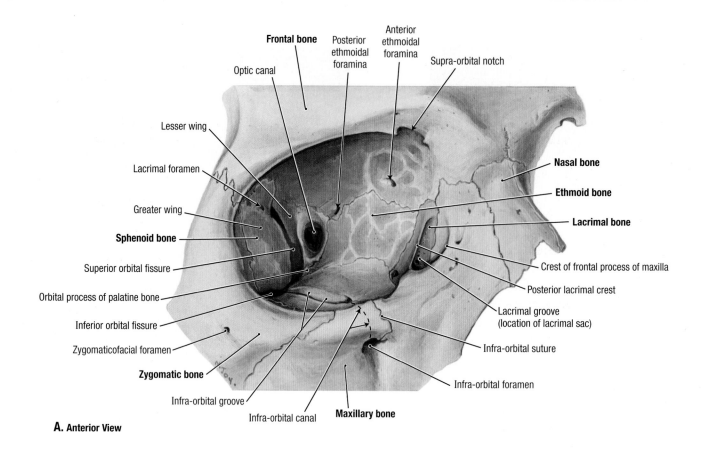

Frontal bone
Posterior ethmoidal foramina
Anterior ethmoidal foramina
Supra-orbital notch
Optic canal
Lesser wing
Lacrimal foramen
Greater wing
Sphenoid bone
Superior orbital fissure
Orbital process of palatine bone
Inferior orbital fissure
Zygomaticofacial foramen
Zygomatic bone
Infra-orbital groove
Infra-orbital canal
Maxillary bone
Nasal bone
Ethmoid bone
Lacrimal bone
Crest of frontal process of maxilla
Posterior lacrimal crest
Lacrimal groove (location of lacrimal sac)
Infra-orbital suture
Infra-orbital foramen

A. Anterior View

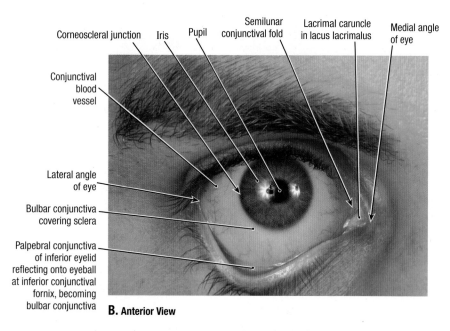

Corneoscleral junction
Iris
Pupil
Semilunar conjunctival fold
Lacrimal caruncle in lacus lacrimalis
Medial angle of eye
Conjunctival blood vessel
Lateral angle of eye
Bulbar conjunctiva covering sclera
Palpebral conjunctiva of inferior eyelid reflecting onto eyeball at inferior conjunctival fornix, becoming bulbar conjunctiva

B. Anterior View

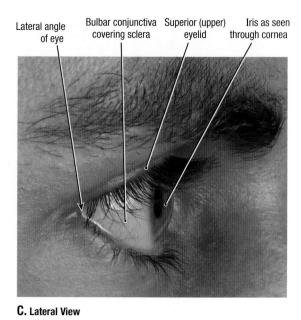

Lateral angle of eye
Bulbar conjunctiva covering sclera
Superior (upper) eyelid
Iris as seen through cornea

C. Lateral View

7.36 ORBITAL CAVITY AND SURFACE ANATOMY OF THE EYE

A. Bones and features of the orbital cavity. **B.** and **C.** Surface anatomy of the eye. The inferior eyelid is everted to demonstrate the palpebral conjunctiva **(B)**. When powerful blows impact directly on the bony rim of the orbit, the resulting **orbital fractures** usually occur at the sutures between the bones forming the orbital margin. Fractures of the inferior wall may involve the maxillary sinus; fractures of the medial wall are less common and may involve the ethmoidal and sphenoidal sinuses. Although the superior wall is stronger, it is thin enough to be translucent and may be readily penetrated. Thus, a sharp object may pass through it into the frontal lobe of the brain. Orbital fractures often result in intra-orbital bleeding, which exerts pressure on the eyeball, causing **exophthalmos** (protrusion of the eyeball).

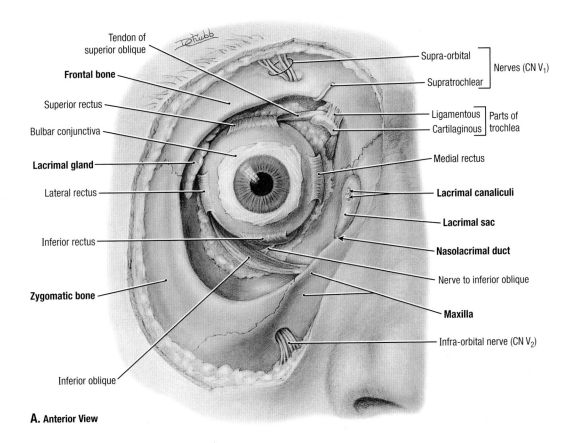

Tendon of superior oblique

Frontal bone

Superior rectus

Bulbar conjunctiva

Lacrimal gland

Lateral rectus

Inferior rectus

Zygomatic bone

Inferior oblique

Supra-orbital
Supratrochlear
} Nerves (CN V₁)

Ligamentous
Cartilaginous
} Parts of trochlea

Medial rectus

Lacrimal canaliculi

Lacrimal sac

Nasolacrimal duct

Nerve to inferior oblique

Maxilla

Infra-orbital nerve (CN V₂)

A. Anterior View

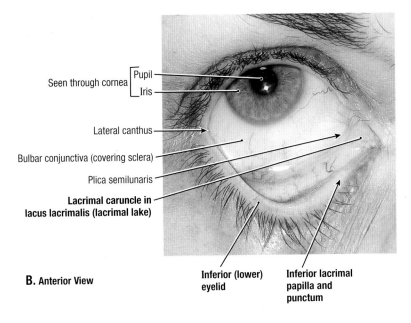

Seen through cornea {
Pupil
Iris

Lateral canthus

Bulbar conjunctiva (covering sclera)

Plica semilunaris

Lacrimal caruncle in lacus lacrimalis (lacrimal lake)

B. Anterior View

Inferior (lower) eyelid

Inferior lacrimal papilla and punctum

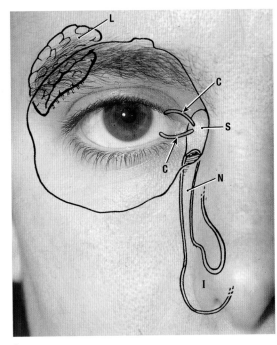

L

C

S

C

N

I

C. Anterior View

EYE AND LACRIMAL APPARATUS

7.37

A. Anterior dissection of orbital cavity. The eyelids, orbital septum, levator palpebrae superioris, and some fat are removed. **B.** Surface features, with the inferior eyelid everted. **C.** Surface projection of lacrimal apparatus. Tears, secreted by the lacrimal gland (*L*) in the superolateral angle of the bony orbit, pass across the eyeball and enter the lacus lacrimalis (lacrimal lake) at the medial angle of the eye; from here they drain through the lacrimal puncta and lacrimal canaliculi (*C*) to the lacrimal sac (*S*). The lacrimal sac drains into the nasolacrimal duct (*N*), which empties into the inferior meatus (*I*) of the nose.

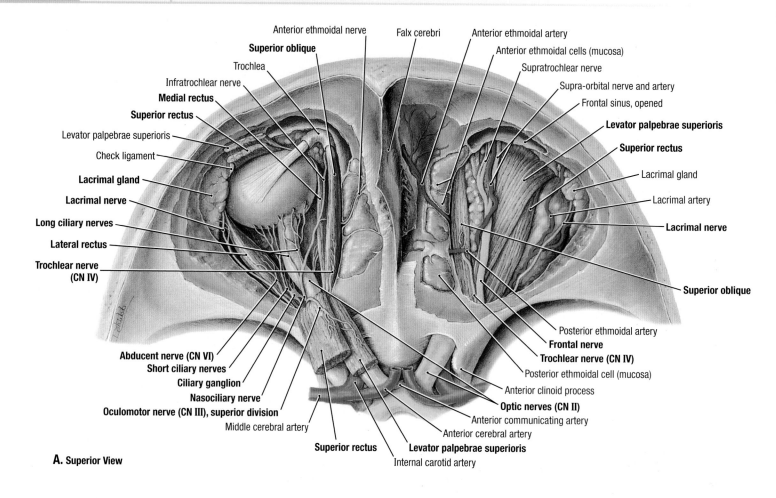

Anterior ethmoidal nerve
Falx cerebri
Anterior ethmoidal artery
Superior oblique
Anterior ethmoidal cells (mucosa)
Trochlea
Supratrochlear nerve
Infratrochlear nerve
Supra-orbital nerve and artery
Medial rectus
Frontal sinus, opened
Superior rectus
Levator palpebrae superioris
Levator palpebrae superioris
Superior rectus
Check ligament
Lacrimal gland
Lacrimal gland
Lacrimal artery
Lacrimal nerve
Lacrimal nerve
Long ciliary nerves
Lateral rectus
Superior oblique
Trochlear nerve (CN IV)
Posterior ethmoidal artery
Frontal nerve
Trochlear nerve (CN IV)
Abducent nerve (CN VI)
Posterior ethmoidal cell (mucosa)
Short ciliary nerves
Anterior clinoid process
Ciliary ganglion
Optic nerves (CN II)
Nasociliary nerve
Anterior communicating artery
Oculomotor nerve (CN III), superior division
Anterior cerebral artery
Middle cerebral artery
Superior rectus
Levator palpebrae superioris
Internal carotid artery

A. Superior View

7.38 ORBITAL CAVITY, SUPERIOR APPROACH

A. Superficial dissection. On the right side of **A**, the orbital plate of the frontal bone is removed. On the left side of **A**, the levator palpebrae and superior rectus muscles are reflected.

- The trochlear nerve (CN IV) lies on the medial side of the superior oblique muscle, and the abducent nerve (CN VI) on the medial side of the lateral rectus muscle.
- The lacrimal nerve runs superior to the lateral rectus muscle supplying sensory fibers to the conjunctiva and skin of the superior eyelid; it receives a communicating branch of the zygomaticotemporal nerve carrying secretory motor fibers from the pterygopalatine ganglion prior to entering or within the lacrimal gland.
- The parasympathetic ciliary ganglion, placed between the lateral rectus muscle and the optic nerve (CN II), gives rise to many short ciliary nerves; the nasociliary nerve gives rise to two long ciliary nerves that anastomose with each other and the short ciliary nerves.

B. Distribution of nerve fibers to ciliary ganglion and eyeball.

Horner syndrome results from interruption of a cervical sympathetic trunk and is manifest by the absence of sympathetically stimulated functions on the ipsilateral side of the head. The syndrome includes the following signs: constriction of the pupil (**miosis**), drooping of the superior eyelid (**ptosis**), redness and increased temperature of the skin (**vasodilatation**), and absence of sweating (**anhydrosis**).

- The ciliary ganglion receives sensory fibers from the nasociliary branches of CN VI, postsynaptic sympathetic fibers from the continuation of the internal carotid plexus extending along the ophthalmic artery, and presynaptic parasympathetic fibers from the inferior branch of the oculomotor nerve; only the latter synapse in the ganglion.

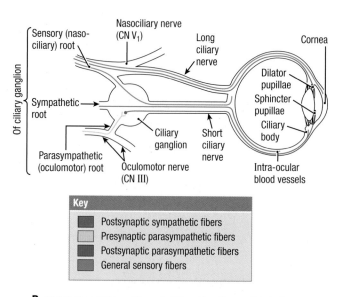

Nasociliary nerve (CN V₁)

Sensory (naso-ciliary) root

Long ciliary nerve

Cornea

Of ciliary ganglion

Sympathetic root

Dilator pupillae
Sphincter pupillae
Ciliary body

Parasympathetic (oculomotor) root

Ciliary ganglion

Short ciliary nerve

Oculomotor nerve (CN III)

Intra-ocular blood vessels

Key

■ Postsynaptic sympathetic fibers
▨ Presynaptic parasympathetic fibers
■ Postsynaptic parasympathetic fibers
■ General sensory fibers

B. Distribution of Nerve Fibers to Ciliary Ganglion and Eyeball

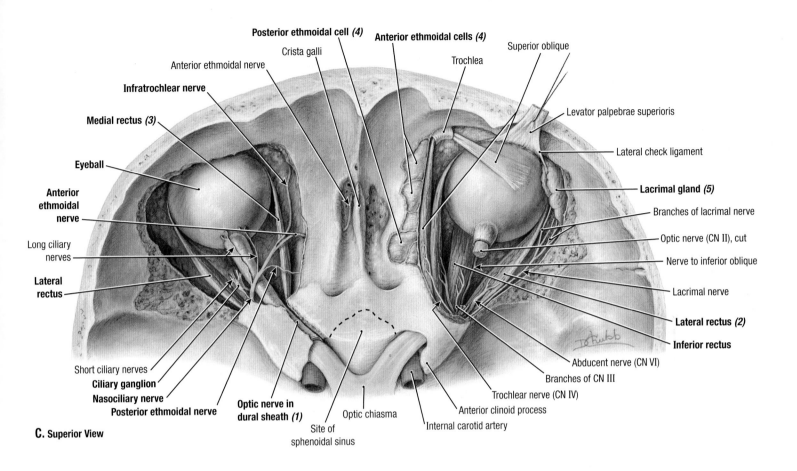

Posterior ethmoidal cell *(4)*
Anterior ethmoidal cells *(4)*
Crista galli
Trochlea
Superior oblique
Anterior ethmoidal nerve
Infratrochlear nerve
Levator palpebrae superioris
Medial rectus *(3)*
Lateral check ligament
Eyeball
Anterior ethmoidal nerve
Lacrimal gland *(5)*
Branches of lacrimal nerve
Long ciliary nerves
Optic nerve (CN II), cut
Nerve to inferior oblique
Lateral rectus
Lacrimal nerve
Lateral rectus *(2)*
Inferior rectus
Short ciliary nerves
Abducent nerve (CN VI)
Ciliary ganglion
Branches of CN III
Nasociliary nerve
Trochlear nerve (CN IV)
Posterior ethmoidal nerve
Optic nerve in dural sheath *(1)*
Anterior clinoid process
Optic chiasma
Internal carotid artery
Site of sphenoidal sinus

C. Superior View

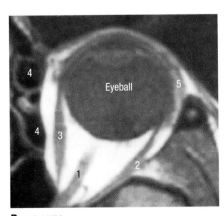

D. Axial MRI

ORBITAL CAVITY, SUPERIOR APPROACH *(continued)* | 7.38

C. Deep dissection before *(left side of specimen)* and after *(right side of specimen)* section of the optic nerve (CN II). **D.** Transverse (axial) MRI of orbital cavity. The numbers refer to structures labeled in **C**.

Observe on the right side of **C**:

- The eyeball occupies the anterior half of the orbital cavity.

Observe on the left of **C**:

- The parasympathetic ciliary ganglion lies posteriorly between the lateral rectus muscle and the sheath of the optic nerve.
- The nasociliary nerve (CN V$_1$) sends a branch to the ciliary ganglion and crosses the optic nerve (CN II), where it gives off two long ciliary nerves (sensory to the eyeball and cornea) and the posterior ethmoidal nerve (to the sphenoidal sinus and posterior ethmoidal cells). The nasociliary nerve then divides into the anterior ethmoidal and infratrochlear nerves.
- Complete **oculomotor nerve palsy** affects four of the six ocular muscles, the levator palpebrae superioris, and the sphincter pupillae. The superior eyelid droops (**ptosis**) and cannot be raised voluntarily because of the unopposed activity of the orbicularis oculi (supplied by the facial nerve). The pupil is also fully dilated and nonreactive because of the unopposed dilator pupillae. The pupil is fully abducted and depressed ("down and out") because of the unopposed activity of the lateral rectus and superior oblique, respectively.
- A **lesion of the abducent nerve** results in loss of lateral gaze to the ipsilateral side because of paralysis of the lateral rectus muscle. On forward gaze, the eye is diverted medially because of the lack of normal resting tone in the lateral rectus, resulting in diplopia (double vision).

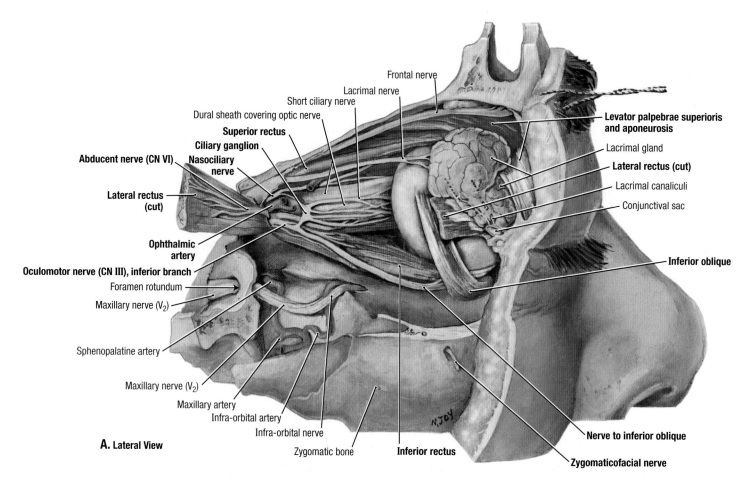

Frontal nerve

Lacrimal nerve

Short ciliary nerve

Dural sheath covering optic nerve

Superior rectus

Ciliary ganglion

Nasociliary nerve

Abducent nerve (CN VI)

Lateral rectus (cut)

Ophthalmic artery

Oculomotor nerve (CN III), inferior branch

Foramen rotundum

Maxillary nerve (V₂)

Sphenopalatine artery

Maxillary nerve (V₂)

Maxillary artery

Infra-orbital artery

Infra-orbital nerve

Zygomatic bone

Inferior rectus

Levator palpebrae superioris and aponeurosis

Lacrimal gland

Lateral rectus (cut)

Lacrimal canaliculi

Conjunctival sac

Inferior oblique

Nerve to inferior oblique

Zygomaticofacial nerve

A. Lateral View

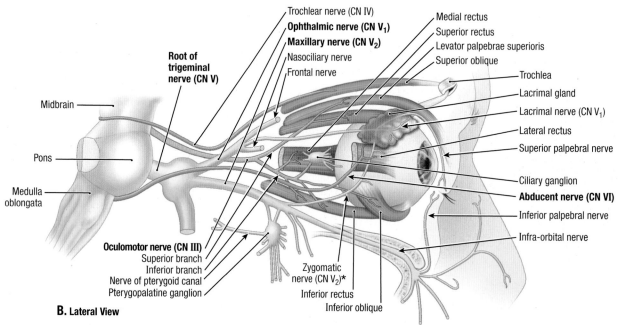

Trochlear nerve (CN IV)

Ophthalmic nerve (CN V₁)

Maxillary nerve (CN V₂)

Nasociliary nerve

Frontal nerve

Root of trigeminal nerve (CN V)

Midbrain

Pons

Medulla oblongata

Medial rectus

Superior rectus

Levator palpebrae superioris

Superior oblique

Trochlea

Lacrimal gland

Lacrimal nerve (CN V₁)

Lateral rectus

Superior palpebral nerve

Ciliary ganglion

Abducent nerve (CN VI)

Inferior palpebral nerve

Infra-orbital nerve

Oculomotor nerve (CN III)

Superior branch

Inferior branch

Nerve of pterygoid canal

Pterygopalatine ganglion

Zygomatic nerve (CN V₂)*

Inferior rectus

Inferior oblique

B. Lateral View

7.39 LATERAL ASPECT OF THE ORBIT AND STRUCTURE OF THE EYELID

A. Dissection. **B.** Nerves. **C.** Sagittal and cross section through optic nerve. The subarachnoid space around the optic nerve is continuous with the subarachnoid space around the brain. **D.** Sagittal MRI. The numbers refer to structures labeled in **C.** circled, optic foramen; *M*, maxillary sinus; *S*, superior ophthalmic vein. **E.** Structure of eyelid.

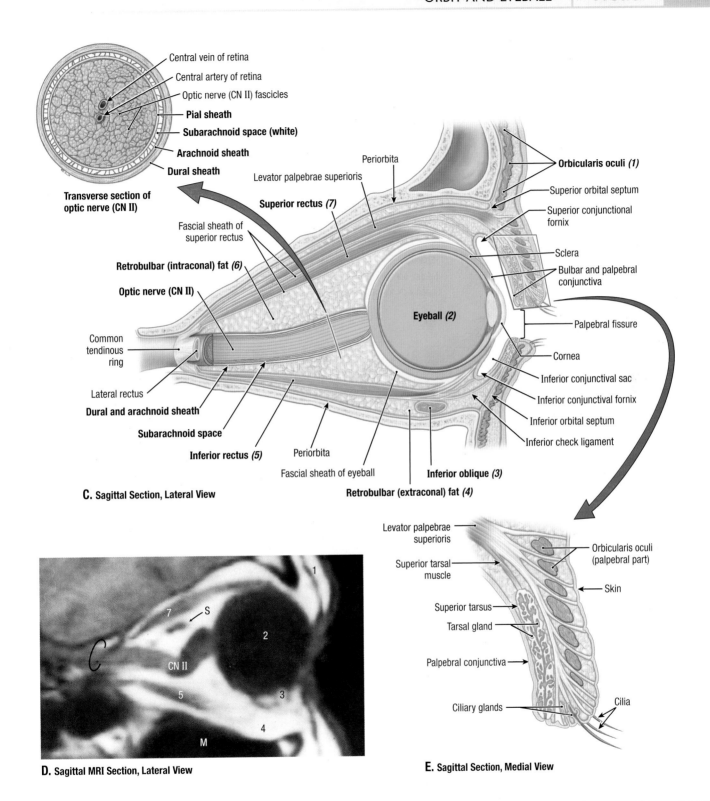

Central vein of retina
Central artery of retina
Optic nerve (CN II) fascicles
Pial sheath
Subarachnoid space (white)
Arachnoid sheath
Dural sheath

Transverse section of optic nerve (CN II)

Periorbita
Levator palpebrae superioris
Superior rectus (7)

Fascial sheath of superior rectus

Retrobulbar (intraconal) fat (6)

Optic nerve (CN II)

Common tendinous ring

Lateral rectus

Dural and arachnoid sheath

Subarachnoid space

Inferior rectus (5)   Periorbita
                Fascial sheath of eyeball
C. Sagittal Section, Lateral View

Orbicularis oculi (1)
Superior orbital septum
Superior conjunctional fornix
Sclera
Bulbar and palpebral conjunctiva
Palpebral fissure

Eyeball (2)

Cornea
Inferior conjunctival sac
Inferior conjunctival fornix
Inferior orbital septum
Inferior check ligament

Inferior oblique (3)
Retrobulbar (extraconal) fat (4)

D. Sagittal MRI Section, Lateral View

Levator palpebrae superioris
Superior tarsal muscle
Superior tarsus
Tarsal gland
Palpebral conjunctiva
Ciliary glands

Orbicularis oculi (palpebral part)
Skin
Cilia

E. Sagittal Section, Medial View

LATERAL ASPECT OF THE ORBIT AND STRUCTURE OF THE EYELID (*continued*)

7.39

- Foreign objects, such as sand or metal filings, produce **corneal abrasions** that cause sudden, stabbing eye pain and tears. Opening and closing the eyelids is also painful. **Corneal lacerations** are caused by sharp objects such as fingernails or the corner of a page of a book.
- Any of the glands in the eyelid may become inflamed and swollen from infection or obstruction of their ducts. If the ducts of

the ciliary glands are obstructed, a painful red suppurative (pus-producing) swelling, a sty (**hordeolum**), develops on the eyelid. **Obstruction of a tarsal gland** produces inflammation, a **tarsal chalazion**, that protrudes toward the eyeball and rubs against it as the eyelids blink.

7.40 EXTRA-OCULAR MUSCLES AND THEIR MOVEMENTS

A. The line of pull of the muscles relative to the eyeball and the axes around which movements occur. The orientation of the orbit is important in understanding the actions of the extra-ocular muscles. The common tendinous ring (origin of the recti), the origin of the inferior oblique, and the trochlea of the superior oblique all lie medial to the eyeball and to the anteroposterior (A-P) and vertical axes. **(a)** The medial and lateral recti are the primary adductors and abductors of the eyeball. However, when movements begin from the primary position (gaze directed anteriorly along the A-P axis): (1) The line of pull of the superior and inferior rectus muscles passes medial and anterior to the vertical axis, resulting in secondary actions of adduction; and (2) the line of pull of the superior and inferior oblique muscles passes medial and posterior to the vertical axis, resulting in secondary actions of abduction. **(b)** Pulling in opposite directions relative to the transverse axis, the superior rectus and inferior oblique muscles are synergistic elevators, and the inferior rectus and superior oblique are synergistic depressors. **(c)** Medial pull produced by the muscles attaching to the superior eyeball (superior rectus and oblique) produces secondary actions of medial rotation (intorsion), and that produced by muscles attaching to the inferior eyeball (inferior rectus and oblique) produces lateral rotation (extorsion).
B. Movements produced by isolated contraction of the four rectus and two oblique muscles, starting from the primary position. *Large arrows* indicate prime movers for the six cardinal movements. Movements in directions between large arrows (e.g., vertical elevation or depression) require synergistic actions of adjacent muscles. Contralaterally paired muscles that work synergistically to direct parallel binocular gaze are called yoke muscles. For example, the right LR and left MR act as yoke muscles in directing gaze to the right.

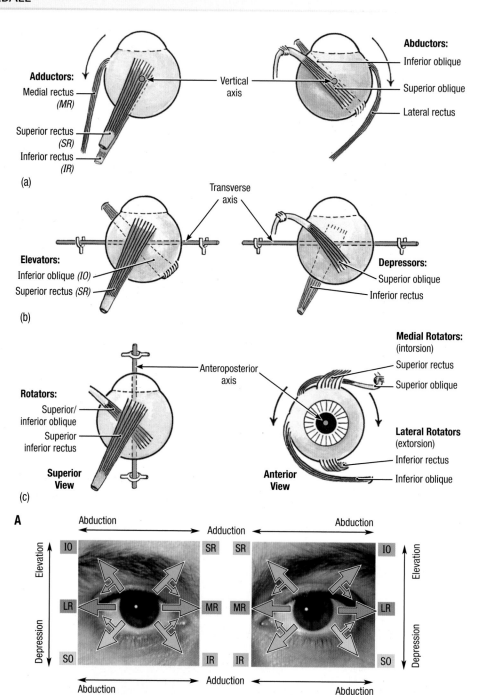

B. Anterior View of Right and Left Eyes

TABLE 7.8	ACTIONS OF MUSCLES OF ORBIT STARTING FROM PRIMARY POSITION[a]		
	Main Action		
Muscle	**Horizontal Axis (A)**	**Vertical Axis (B)**	**Anteroposterior Axis (C)**
Superior rectus (SR)	Elevates	Adducts	Rotates medially (intorsion)
Inferior rectus (IR)	Depresses	Adducts	Rotates laterally (extorsion)
Superior oblique (SO)	Depresses	Abducts	Rotates medially (intorsion)
Inferior oblique (IO)	Elevates	Abducts	Rotates laterally (extorsion)
Medial rectus (MR)	N/A	Adducts	N/A
Lateral rectus (LR)	N/A	Abducts	N/A

[a]Primary position, gaze directed anteriorly.

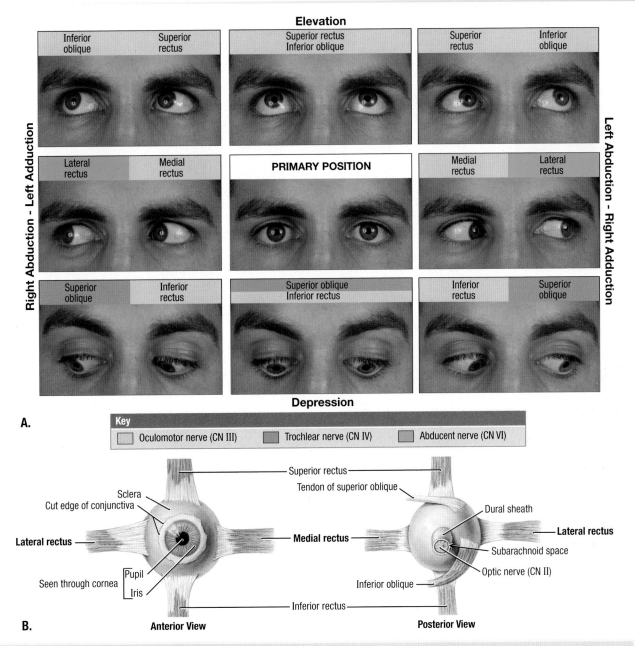

EXTRA-OCULAR MUSCLES AND THEIR MOVEMENTS

7.41

A. Binocular movements of eyeball from primary position, and muscles and nerves producing them. **B.** Muscles of eyeball.

TABLE 7.9	MUSCLES OF ORBIT			
Muscle	*Origin*	*Insertion*	*Innervation*	*Main Action(s)[a]*
Levator palpebrae superioris	Lesser wing of sphenoid bone, superior and anterior to optic canal	Superior tarsus and skin of superior eyelid	Oculomotor nerve; deep layer (superior tarsal muscle) supplied by sympathetic fibers	Elevates superior eyelid
Superior oblique (SO)	Body of sphenoid bone	Tendon passes through trochlea to insert into sclera, deep to SR	Trochlear nerve (CN IV)	Abducts, depresses, and rotates eyeball medially (intorsion)
Inferior oblique (IO)	Anterior part of floor of orbit	Sclera deep to lateral rectus muscle	Oculomotor nerve (CN III)	Abducts, elevates, and rotates eyeball laterally (extorsion)
Superior rectus (SR)	Common tendinous ring	Sclera just posterior to corneoscleral junction	Oculomotor nerve (CN III)	Elevates, adducts, and rotates eyeball medially (intorsion)
Inferior rectus (IR)				Depresses, adducts, and rotates eyeball laterally (extorsion)
Medial rectus (MR)				Adducts eyeball
Lateral rectus (LR)			Abducent nerve (CN VI)	Abducts eyeball

[a]It is essential to appreciate that all muscles are continuously involved in eyeball movements; thus, the individual actions are not usually tested clinically.

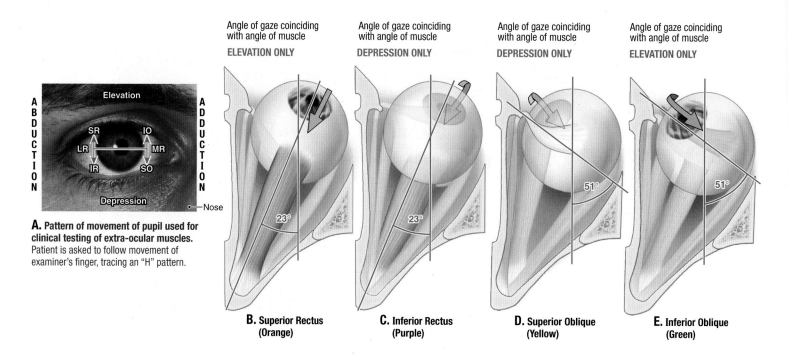

A. Pattern of movement of pupil used for clinical testing of extra-ocular muscles. Patient is asked to follow movement of examiner's finger, tracing an "H" pattern.

B. Superior Rectus (Orange)

C. Inferior Rectus (Purple)

D. Superior Oblique (Yellow)

E. Inferior Oblique (Green)

7.42 CLINICAL TESTING OF EXTRA-OCULAR MUSCLES AND MOTOR NERVES (CN III, IV, AND VI)

Most movements from the primary position involve synergists. When **testing the extra-ocular muscles** (usually to determine the integrity of the involved motor nerve), it is desirable to isolate muscle activity. If the pupil is first adducted (MR—CN III) so that the direction of gaze coincides with the line of pull of the oblique muscles, only the SO (CN IV) can depress and only the IO (CN III) can elevate the pupil. If the pupil is first abducted (LR—CN VI) so that the direction of gaze coincides with the line of pull of the superior and inferior recti, only these muscles can elevate and depress the pupil (superior and inferior divisions of CN III).

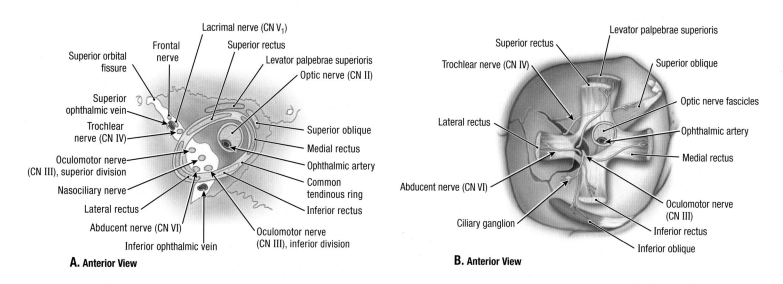

A. Anterior View

B. Anterior View

7.43 NERVES OF ORBIT

A. Overview. **B.** Relationships at apex of orbit.

Orbital tumors. Because of the closeness of the optic nerve to the sphenoidal and posterior ethmoidal sinuses, a malignant tumor in these sinuses may erode the thin bony walls of the orbit and compress the optic nerve and orbital contents.

Tumors in the orbit produce **exophthalmos** (protrusion of eyeball). Tumors in the middle cranial fossa enter the orbital cavity through the superior orbital fissure. Tumors in the temporal or infratemporal fossae enter the orbit through the inferior orbital fissure.

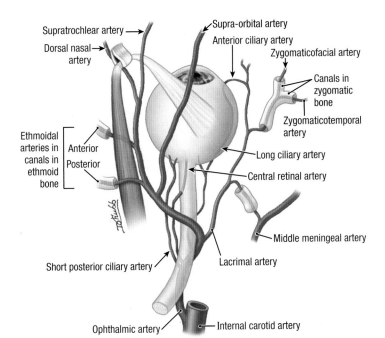

A. Superior View

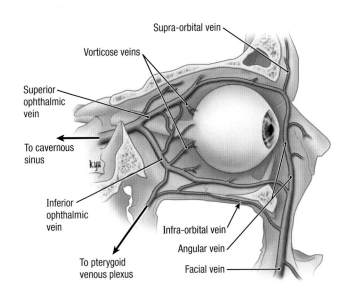

B. Lateral View

ARTERIES AND VEINS OF ORBIT

7.44

A. Arteries.

Blockage of central retinal artery. The terminal branches of the central retinal artery are end arteries. Obstruction of the artery by an embolus results in instant and total blindness. Blockage of the artery is usually unilateral and occurs in older people. **B. Veins.** The superior and inferior ophthalmic veins receive the vorticose veins from the eyeball and drain into the cavernous sinus posteriorly and the pterygoid plexus inferiorly. They communicate with the facial and supra-orbital veins anteriorly.

- The facial veins make clinically important connections with the cavernous sinus through the superior ophthalmic veins. **Cavernous sinus thrombosis** usually results from infections in

the orbit, nasal sinuses, and superior part of the face (the danger triangle). In persons with thrombophlebitis of the facial vein, pieces of an infected thrombus may extend into the cavernous sinus, producing **thrombophlebitis of the cavernous sinus**. The infection usually involves only one sinus initially but may spread to the opposite side through the intercavernous sinuses.

- **Blockade of central retinal vein.** The central retinal vein enters the cavernous sinus. Thrombophlebitis of this sinus may result in passage of a thrombus to the central retinal vein and produce a blockage in one of the small retinal veins. Occlusion of a branch of the central vein of the retina usually results in slow, painless loss of vision.

TABLE 7.10	ARTERIES OF ORBIT	
Artery	*Origin*	*Course and Distribution*
Ophthalmic	Internal carotid artery	Traverses optic foramen to reach orbital cavity
Central retinal		Runs in dural sheath of optic nerve, entering nerve near eyeball; appears at center of optic disc; supplies optic retina (except cones and rods)
Supra-orbital		Passes superiorly and posteriorly from supra-orbital foramen to supply forehead and scalp
Supratrochlear		Passes from supra-orbital margin to forehead and scalp
Lacrimal		Passes along superior border of lateral rectus muscle to supply lacrimal gland, conjunctiva, and eyelids
Dorsal nasal	Ophthalmic artery	Courses along dorsal aspect of nose and supplies its surface
Short posterior ciliary		Pierces sclera at periphery of optic nerve to supply choroid, which, in turn, supplies cones and rods of optic retina
Long posterior ciliary		Pierces sclera to supply ciliary body and iris
Posterior ethmoidal		Passes through posterior ethmoidal foramen to posterior ethmoidal cells
Anterior ethmoidal		Passes through anterior ethmoidal foramen to anterior cranial fossa; supplies anterior and middle ethmoidal cells, frontal sinus, nasal cavity, and skin on dorsum of nose
Anterior ciliary	Muscular rami of the ophthalmic and infra-orbital arteries	Pierces sclera at attachments of rectus muscles and forms network in iris and ciliary body
Infra-orbital	Third part of maxillary artery	Passes along infra-orbital groove and exits through infra-orbital foramen to face

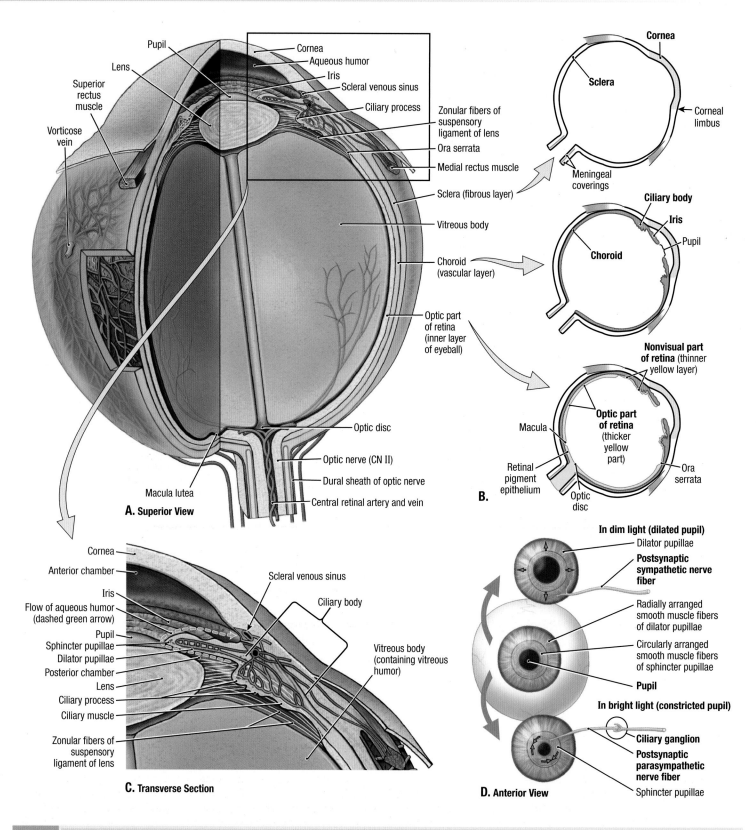

A. Superior View

Pupil
Lens
Superior rectus muscle
Vorticose vein
Macula lutea

Cornea
Aqueous humor
Iris
Scleral venous sinus
Ciliary process
Zonular fibers of suspensory ligament of lens
Ora serrata
Medial rectus muscle
Sclera (fibrous layer)
Vitreous body
Choroid (vascular layer)
Optic part of retina (inner layer of eyeball)
Optic disc
Optic nerve (CN II)
Dural sheath of optic nerve
Central retinal artery and vein

B.

Cornea
Sclera
Corneal limbus
Meningeal coverings

Ciliary body
Iris
Pupil
Choroid

Nonvisual part of retina (thinner yellow layer)
Optic part of retina (thicker yellow part)
Macula
Retinal pigment epithelium
Optic disc
Ora serrata

C. Transverse Section

Cornea
Anterior chamber
Iris
Flow of aqueous humor (dashed green arrow)
Pupil
Sphincter pupillae
Dilator pupillae
Posterior chamber
Lens
Ciliary process
Ciliary muscle
Zonular fibers of suspensory ligament of lens

Scleral venous sinus
Ciliary body
Vitreous body (containing vitreous humor)

D. Anterior View

In dim light (dilated pupil)
Dilator pupillae
Postsynaptic sympathetic nerve fiber
Radially arranged smooth muscle fibers of dilator pupillae
Circularly arranged smooth muscle fibers of sphincter pupillae
Pupil
In bright light (constricted pupil)
Ciliary ganglion
Postsynaptic parasympathetic nerve fiber
Sphincter pupillae

7.45   ILLUSTRATION OF A DISSECTED EYEBALL

A. Parts of the eyeball. **B.** Layers (coats) of eyeball. **C.** Anterior segment. **D.** Structure and function of iris. The aqueous humor is produced by the ciliary processes and provides nutrients for the avascular cornea and lens; the aqueous humor drains into the scleral venous sinus (also called the sinus venosus sclerae or canal of Schlemm). **Glaucoma.** If drainage of the aqueous humor is reduced significantly, pressure builds up in the chambers of the eye (glaucoma). Blindness can result from compression of the inner layer of the retina and retinal arteries if aqueous humor production is not reduced to maintain normal intra-ocular pressure.

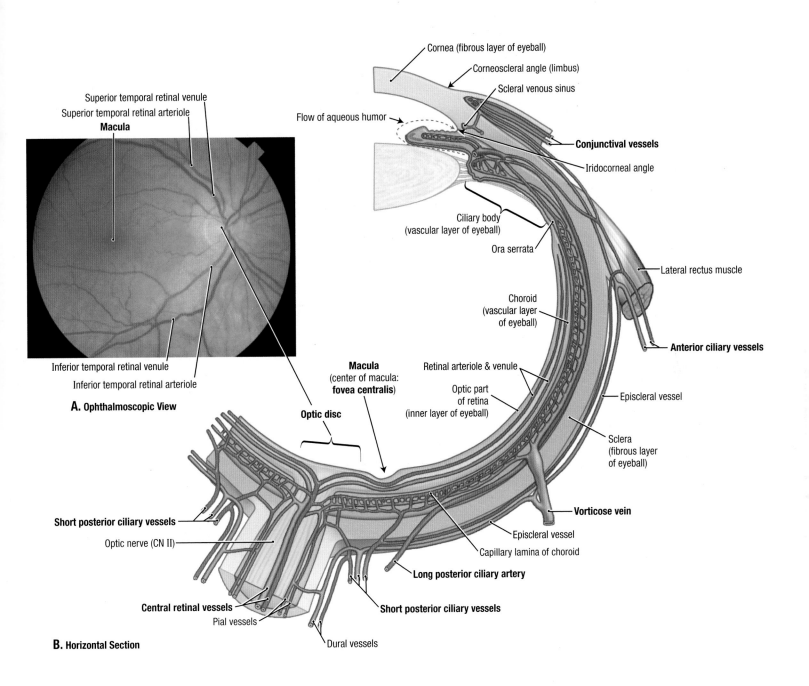

A. Ophthalmoscopic View

- Superior temporal retinal venule
- Superior temporal retinal arteriole
- **Macula**
- Inferior temporal retinal venule
- Inferior temporal retinal arteriole

- Cornea (fibrous layer of eyeball)
- Corneoscleral angle (limbus)
- Scleral venous sinus
- Flow of aqueous humor
- **Conjunctival vessels**
- Iridocorneal angle
- Ciliary body (vascular layer of eyeball)
- Ora serrata
- Lateral rectus muscle
- Choroid (vascular layer of eyeball)
- Retinal arteriole & venule
- **Anterior ciliary vessels**
- Episcleral vessel
- **Macula** (center of macula: **fovea centralis**)
- **Optic disc**
- Optic part of retina (inner layer of eyeball)
- Sclera (fibrous layer of eyeball)
- **Short posterior ciliary vessels**
- Optic nerve (CN II)
- Episcleral vessel
- **Vorticose vein**
- Capillary lamina of choroid
- **Long posterior ciliary artery**
- **Central retinal vessels**
- Pial vessels
- **Short posterior ciliary vessels**
- Dural vessels

B. Horizontal Section

OCULAR FUNDUS AND BLOOD SUPPLY TO THE EYEBALL

A. Right ocular fundus, ophthalmoscopic view. Retinal venules (wider) and retinal arterioles (narrower) radiate from the center of the oval optic disc, formed in relation to the entry of the optic nerve into the eyeball. The round, dark area lateral to the disc is the macula; branches of vessels extend to this area but do not reach its center, the fovea centralis, a depressed spot that is the area of most acute vision. It is avascular but, like the rest of the outermost (cones and rods) layer of the retina, is nourished by the adjacent choriocapillaris. Increased intracranial pressure is transmitted through the CSF in the subarachnoid space surrounding the optic nerve, causing the optic disc to protrude. The protrusion, called **papilledema**, is apparent during ophthalmoscopy. **B.** Blood supply to eyeball.

The eyeball has three layers: (1) the external, fibrous layer is the sclera and cornea; (2) the middle, vascular layer is the choroid, ciliary body, and iris; and (3) the internal, neural layer or retina consists of a pigment cell layer and a neural layer. The central artery of the retina, a branch of the ophthalmic artery, is an end artery. Of the eight posterior ciliary arteries, six are short posterior ciliary arteries and supply the choroid, which in turn nourishes the outer, non-vascular layer of the retina. Two long posterior ciliary arteries, one on each side of the eyeball, run between the sclera and choroid to anastomose with the anterior ciliary arteries, which are derived from muscular branches. The choroid is drained by posterior ciliary veins, and four to five vorticose veins that drain into the ophthalmic veins.

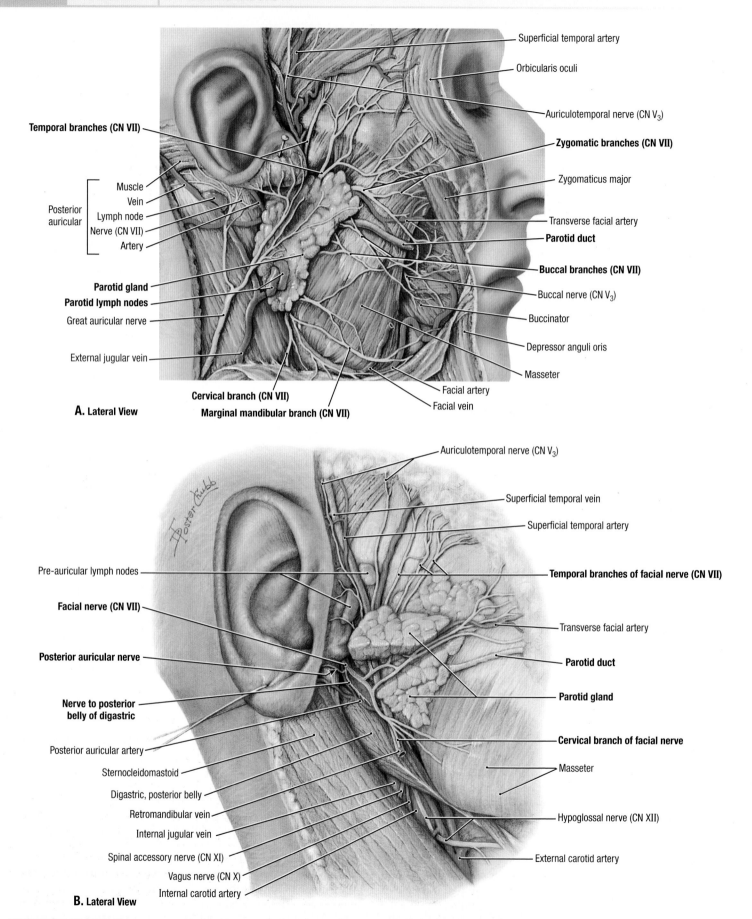

Superficial temporal artery

Orbicularis oculi

Auriculotemporal nerve (CN V₃)

Temporal branches (CN VII)

Zygomatic branches (CN VII)

Zygomaticus major

Posterior auricular
Muscle
Vein
Lymph node
Nerve (CN VII)
Artery

Transverse facial artery

Parotid duct

Parotid gland

Parotid lymph nodes

Great auricular nerve

Buccal branches (CN VII)

Buccal nerve (CN V₃)

Buccinator

Depressor anguli oris

External jugular vein

Masseter

Facial artery

Facial vein

Cervical branch (CN VII)

Marginal mandibular branch (CN VII)

A. Lateral View

Auriculotemporal nerve (CN V₃)

Superficial temporal vein

Superficial temporal artery

Pre-auricular lymph nodes

Temporal branches of facial nerve (CN VII)

Facial nerve (CN VII)

Transverse facial artery

Posterior auricular nerve

Parotid duct

Nerve to posterior belly of digastric

Parotid gland

Posterior auricular artery

Cervical branch of facial nerve

Sternocleidomastoid

Masseter

Digastric, posterior belly

Retromandibular vein

Internal jugular vein

Spinal accessory nerve (CN XI)

Hypoglossal nerve (CN XII)

Vagus nerve (CN X)

External carotid artery

Internal carotid artery

B. Lateral View

7.47 PAROTID REGION

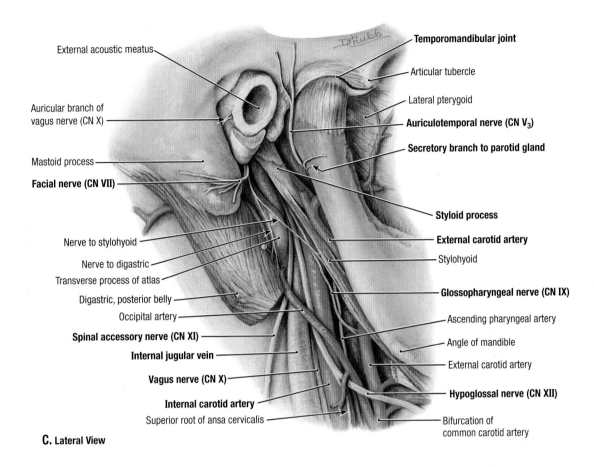

External acoustic meatus

Auricular branch of
vagus nerve (CN X)

Mastoid process

Facial nerve (CN VII)

Nerve to stylohyoid

Nerve to digastric

Transverse process of atlas

Digastric, posterior belly

Occipital artery

Spinal accessory nerve (CN XI)

Internal jugular vein

Vagus nerve (CN X)

Internal carotid artery

Superior root of ansa cervicalis

Temporomandibular joint

Articular tubercle

Lateral pterygoid

Auriculotemporal nerve (CN V₃)

Secretory branch to parotid gland

Styloid process

External carotid artery

Stylohyoid

Glossopharyngeal nerve (CN IX)

Ascending pharyngeal artery

Angle of mandible

External carotid artery

Hypoglossal nerve (CN XII)

Bifurcation of
common carotid artery

C. Lateral View

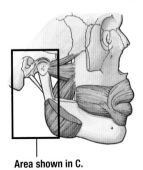

Area shown in C.

PAROTID REGION (*continued*)

7.47

A. Superficial dissection. **B.** Deep dissection with part of the gland removed. During **parotidectomy** (surgical excision of the parotid gland), identification, dissection, and preservation of the facial nerve are critical. The parotid gland has superficial and deep parts. In parotidectomy the superficial part is removed, then the plexus may be retracted to remove the deep part. **C.** Deep dissection following removal of the parotid gland and auricle. The facial nerve, posterior belly of the digastric muscle, and its nerve are retracted; the external carotid artery, stylohyoid muscle, and the nerve to the stylohyoid remain *in situ*. The internal jugular vein, internal carotid artery, and glossopharyngeal (CN IX), vagus (CN X), spinal accessory (CN XI), and hypoglossal (CN XII) nerves cross anterior to the transverse process of the atlas and deep to the styloid process.

Hypoglossal nerve palsy. Trauma, such as a fractured mandible, may injure the hypoglossal nerve (CN XII), resulting in paralysis and eventual atrophy of one side of the tongue. The tongue deviates to the paralyzed side during protrusion.

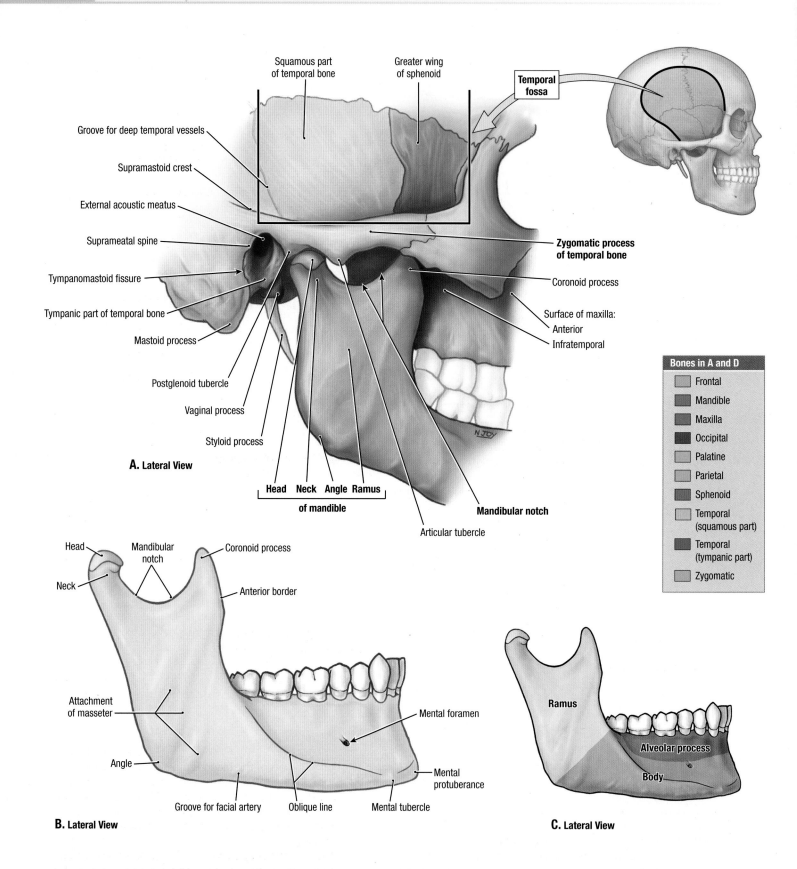

Squamous part of temporal bone

Greater wing of sphenoid

Temporal fossa

Groove for deep temporal vessels

Supramastoid crest

External acoustic meatus

Suprameatal spine

Tympanomastoid fissure

Tympanic part of temporal bone

Mastoid process

Postglenoid tubercle

Vaginal process

Styloid process

A. Lateral View

Head   Neck   Angle   Ramus
of mandible

Zygomatic process of temporal bone

Coronoid process

Surface of maxilla:
Anterior
Infratemporal

Mandibular notch

Articular tubercle

Bones in A and D

	Frontal
	Mandible
	Maxilla
	Occipital
	Palatine
	Parietal
	Sphenoid
	Temporal (squamous part)
	Temporal (tympanic part)
	Zygomatic

Head

Neck

Mandibular notch

Coronoid process

Anterior border

Attachment of masseter

Angle

Groove for facial artery

Oblique line

Mental foramen

Mental protuberance

Mental tubercle

B. Lateral View

Ramus

Alveolar process

Body

C. Lateral View

7.48   TEMPORAL AND INFRATEMPORAL FOSSAE AND MANDIBLE

A. Bones and bony features. Note that superficially the zygomatic process of the temporal bone is the boundary between the temporal fossa superiorly and the infratemporal fossa inferiorly. **B.** External surface of the mandible. **C.** Parts of mandible.

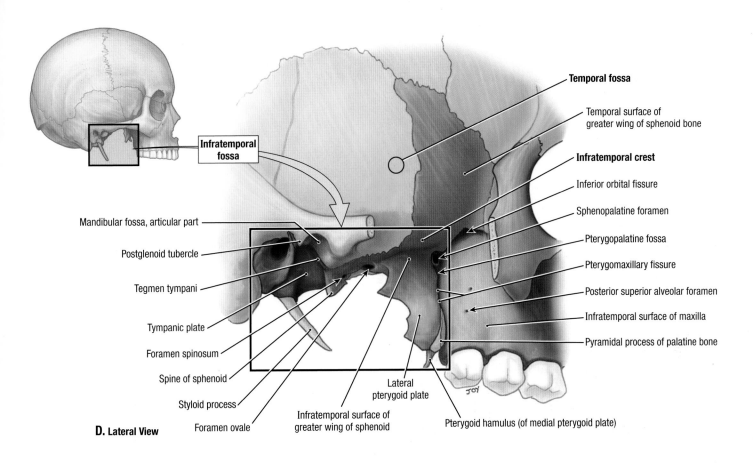

Infratemporal fossa

Temporal fossa

Temporal surface of greater wing of sphenoid bone

Infratemporal crest

Inferior orbital fissure

Sphenopalatine foramen

Pterygopalatine fossa

Pterygomaxillary fissure

Posterior superior alveolar foramen

Infratemporal surface of maxilla

Pyramidal process of palatine bone

Mandibular fossa, articular part

Postglenoid tubercle

Tegmen tympani

Tympanic plate

Foramen spinosum

Spine of sphenoid

Styloid process

Foramen ovale

Infratemporal surface of greater wing of sphenoid

Lateral pterygoid plate

Pterygoid hamulus (of medial pterygoid plate)

D. Lateral View

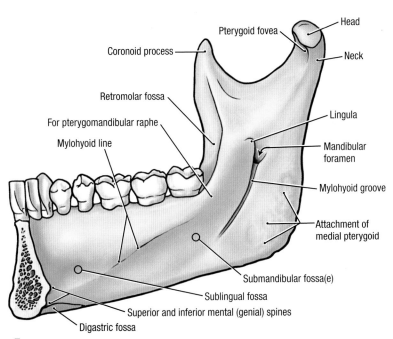

Pterygoid fovea

Head

Coronoid process

Neck

Retromolar fossa

Lingula

For pterygomandibular raphe

Mandibular foramen

Mylohyoid line

Mylohyoid groove

Attachment of medial pterygoid

Submandibular fossa(e)

Sublingual fossa

Superior and inferior mental (genial) spines

Digastric fossa

E. Medial View

7.48 TEMPORAL AND INFRATEMPORAL FOSSAE AND MANDIBLE (*continued*)

D. Bones and bony features of the infratemporal fossa. The mandible and part of the zygomatic arch have been removed. Deeply, the infratemporal crest separates the temporal and infratemporal fossae. **E.** Internal surface of the mandible.

- The temporal region is the region of the head that includes the lateral area of the scalp and the deeper soft tissues overlying the temporal fossa of the cranium, superior to the zygomatic arch. The temporal fossa, occupied primarily by the upper portion of the temporalis muscle, is bounded by the inferior temporal lines (see Fig. 7.3B).
- The infratemporal fossa is an irregularly shaped space deep and inferior to the zygomatic arch, deep to the ramus of the mandible and posterior to the maxilla. It communicates with the temporal fossa through the interval between the zygomatic arch and the cranial bones.

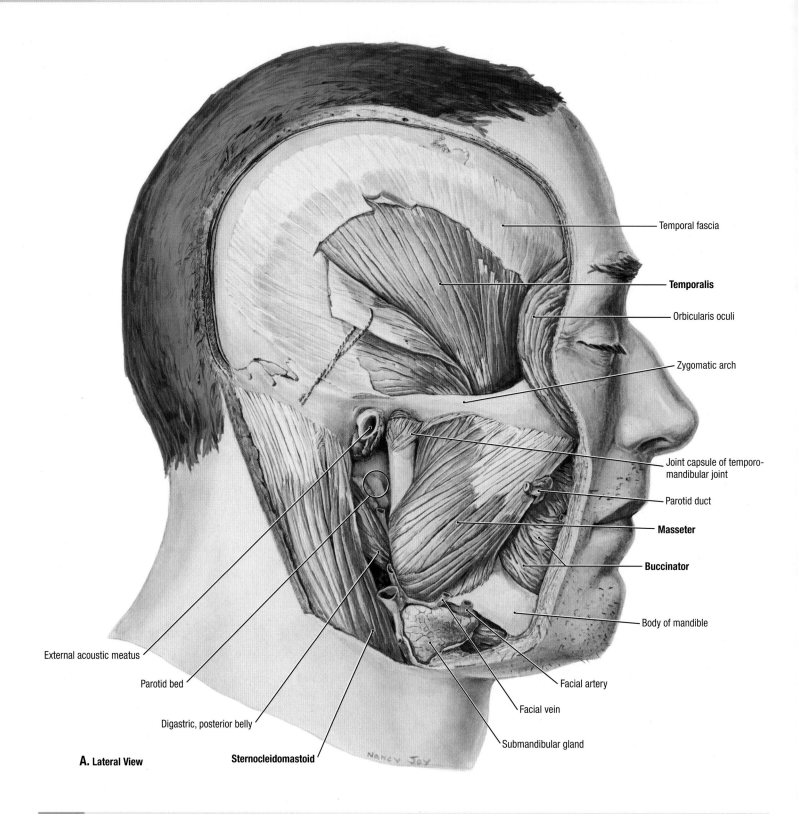

Temporal fascia

Temporalis

Orbicularis oculi

Zygomatic arch

Joint capsule of temporo-mandibular joint

Parotid duct

Masseter

Buccinator

Body of mandible

Facial artery

Facial vein

Submandibular gland

External acoustic meatus

Parotid bed

Digastric, posterior belly

A. Lateral View

Sternocleidomastoid

NANCY JOY

7.49 TEMPORALIS AND MASSETER

A. Superficial dissection.

- The temporalis and masseter muscles are supplied by the mandibular nerve (CN V$_3$), and both elevate the mandible. The buccinator muscle, supplied by the facial nerve (CN VII), functions during chewing to keep food between the teeth but does not act on the mandible.

- The sternocleidomastoid muscle, supplied by the spinal accessory nerve (CN XI), is the chief flexor of the head and neck; it forms the lateral part of the posterior boundary of the parotid region/parotid bed.

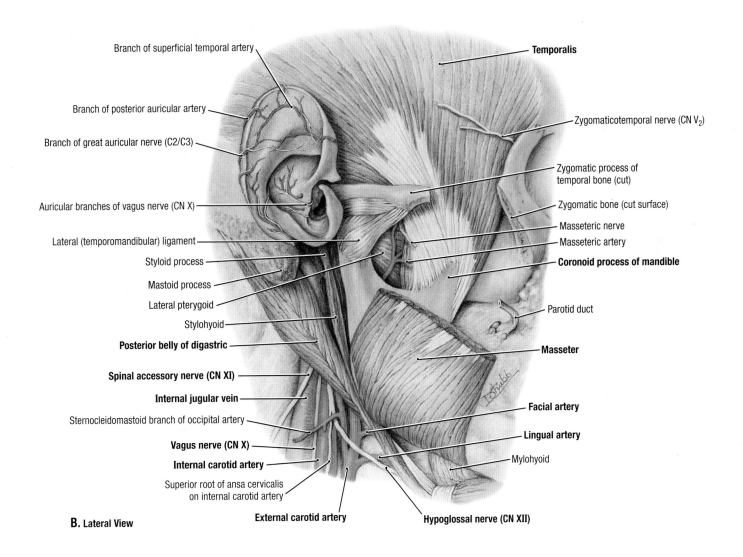

Branch of superficial temporal artery

Branch of posterior auricular artery

Branch of great auricular nerve (C2/C3)

Auricular branches of vagus nerve (CN X)

Lateral (temporomandibular) ligament

Styloid process

Mastoid process

Lateral pterygoid

Stylohyoid

Posterior belly of digastric

Spinal accessory nerve (CN XI)

Internal jugular vein

Sternocleidomastoid branch of occipital artery

Vagus nerve (CN X)

Internal carotid artery

Superior root of ansa cervicalis on internal carotid artery

External carotid artery

B. Lateral View

Temporalis

Zygomaticotemporal nerve (CN V₂)

Zygomatic process of temporal bone (cut)

Zygomatic bone (cut surface)

Masseteric nerve

Masseteric artery

Coronoid process of mandible

Parotid duct

Masseter

Facial artery

Lingual artery

Mylohyoid

Hypoglossal nerve (CN XII)

TEMPORALIS AND MASSETER (*continued*)

B. Deep dissection.
- Parts of the zygomatic arch and masseter muscle have been re-moved to expose the attachment of the temporalis muscle to the coronoid process of the mandible.
- The carotid sheath surrounding the internal jugular vein, internal carotid artery, and the vagus nerve (CN X) has been removed.

The external carotid artery and its lingual, facial, and occipital branches, and the spinal accessory (CN XI) and hypoglossal (CN XII) nerves pass medial to the posterior belly of the digastric muscle.

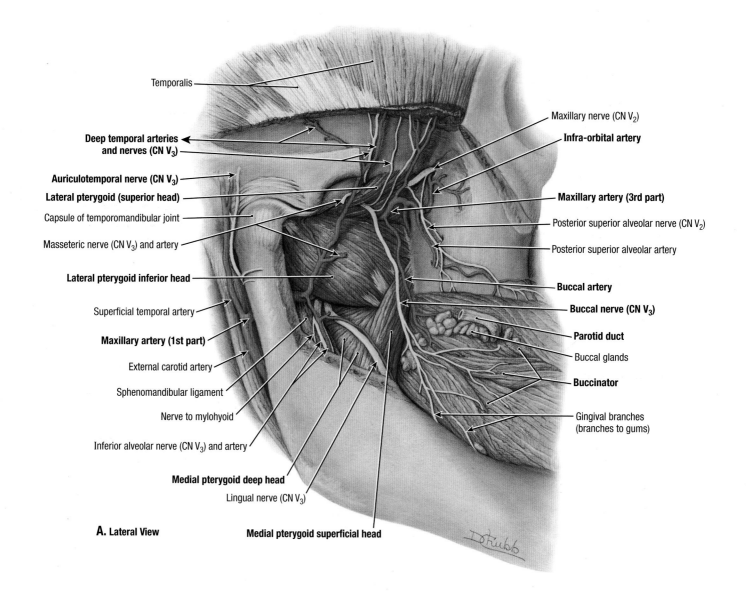

Temporalis

Deep temporal arteries
and nerves (CN V₃)

Auriculotemporal nerve (CN V₃)

Lateral pterygoid (superior head)

Capsule of temporomandibular joint

Masseteric nerve (CN V₃) and artery

Lateral pterygoid inferior head

Superficial temporal artery

Maxillary artery (1st part)

External carotid artery

Sphenomandibular ligament

Nerve to mylohyoid

Inferior alveolar nerve (CN V₃) and artery

Medial pterygoid deep head

Lingual nerve (CN V₃)

Maxillary nerve (CN V₂)

Infra-orbital artery

Maxillary artery (3rd part)

Posterior superior alveolar nerve (CN V₂)

Posterior superior alveolar artery

Buccal artery

Buccal nerve (CN V₃)

Parotid duct

Buccal glands

Buccinator

Gingival branches
(branches to gums)

A. Lateral View

Medial pterygoid superficial head

7.50 INFRATEMPORAL REGION

A. Superficial dissection.
- The maxillary artery, the larger of two terminal branches of the external carotid, is divided into three parts relative to the lateral pterygoid muscle.
- The buccinator is pierced by the parotid duct, the ducts of the buccal glands, and sensory branches of the buccal nerve.
- The lateral pterygoid muscle arises by two heads, one head from the roof, and the other head from the medial wall of the infratemporal fossa; both heads insert in relation to the temporomandibular joint—the superior head attaching primarily to the articular disc of the joint and the inferior head primarily to the anterior aspect of the neck of the mandible (pterygoid fovea).

- Because of the close relationship of the facial and auriculotemporal nerves to the temporomandibular joint (TMJ), care must be taken during **surgical procedures on the temporomandibular joint** to preserve both the branches of the facial nerve overlying it and the articular branches of the auriculotemporal nerve that enter the posterior part of the joint. Injury to articular branches of the auriculotemporal nerve supplying the TMJ—associated with traumatic dislocation and rupture of the joint capsule and lateral ligament—leads to laxity and instability of the TMJ.

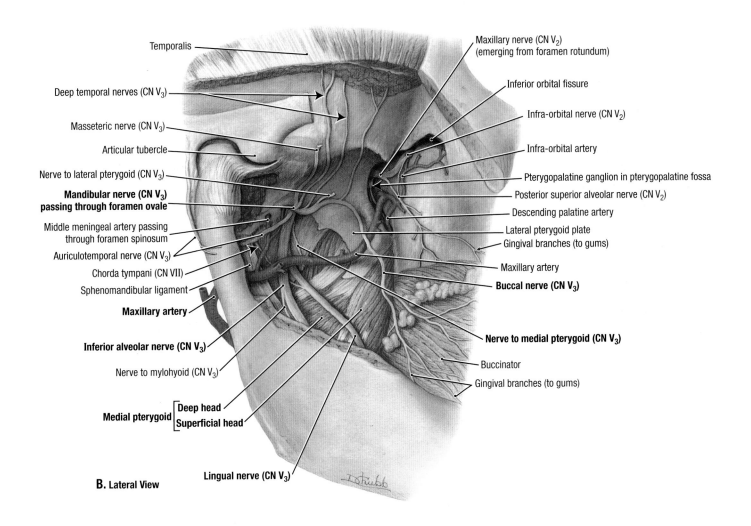

Temporalis

Deep temporal nerves (CN V₃)

Masseteric nerve (CN V₃)

Articular tubercle

Nerve to lateral pterygoid (CN V₃)

Mandibular nerve (CN V₃)
passing through foramen ovale

Middle meningeal artery passing
through foramen spinosum

Auriculotemporal nerve (CN V₃)

Chorda tympani (CN VII)

Sphenomandibular ligament

Maxillary artery

Inferior alveolar nerve (CN V₃)

Nerve to mylohyoid (CN V₃)

Medial pterygoid [Deep head / Superficial head]

B. Lateral View

Maxillary nerve (CN V₂)
(emerging from foramen rotundum)

Inferior orbital fissure

Infra-orbital nerve (CN V₂)

Infra-orbital artery

Pterygopalatine ganglion in pterygopalatine fossa

Posterior superior alveolar nerve (CN V₂)

Descending palatine artery

Lateral pterygoid plate

Gingival branches (to gums)

Maxillary artery

Buccal nerve (CN V₃)

Nerve to medial pterygoid (CN V₃)

Buccinator

Gingival branches (to gums)

Lingual nerve (CN V₃)

INFRATEMPORAL REGION (continued)

B. Deeper dissection.

- The lateral pterygoid muscle and most of the branches of the maxillary artery have been removed to expose the mandibular nerve (CN V₃) entering the infratemporal fossa through the foramen ovale and the middle meningeal artery passing through the foramen spinosum.
- The deep head of the medial pterygoid muscle arises from the medial surface of the lateral pterygoid plate and the pyramidal process of the palatine bone. It has a small, superficial head that arises from the tuberosity of the maxilla.
- The inferior alveolar and lingual nerves descend on the medial pterygoid muscle. The inferior alveolar nerve gives off the nerve to mylohyoid and nerve to anterior belly of the digastric muscle, and the lingual nerve receives the chorda tympani, which carries secretory parasympathetic fibers and fibers of taste.
- Motor nerves arising from CN V₃ supply the four muscles of mastication: the masseter, temporalis, and lateral and medial pterygoids. The buccal nerve from the mandibular nerve is sensory; the buccal branch of the facial nerve is the motor supply to the buccinator muscle.
- To perform a **mandibular nerve block**, an anesthetic agent is injected near the mandibular nerve where it enters the infratemporal fossa. This block usually anesthetizes the auriculotemporal, inferior alveolar, lingual, and buccal branches of the mandibular nerve.

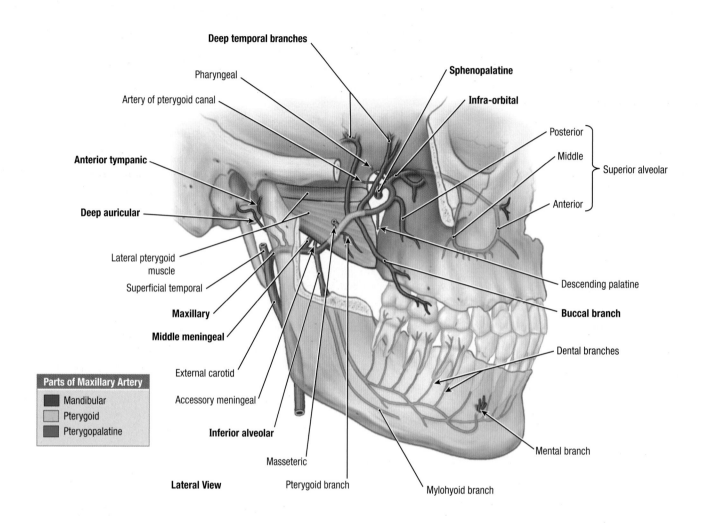

Deep temporal branches

Pharyngeal

Artery of pterygoid canal

Sphenopalatine

Infra-orbital

Posterior

Middle

Superior alveolar

Anterior tympanic

Anterior

Deep auricular

Lateral pterygoid muscle

Descending palatine

Superficial temporal

Maxillary

Buccal branch

Middle meningeal

Dental branches

External carotid

Parts of Maxillary Artery
Mandibular
Pterygoid
Pterygopalatine

Accessory meningeal

Inferior alveolar

Mental branch

Masseteric

Lateral View

Pterygoid branch

Mylohyoid branch

7.51   BRANCHES OF MAXILLARY ARTERY

- The maxillary artery arises at the neck of the mandible and is divided into three parts (mandibular, pterygoid, and pterygopalatine) by the lateral pterygoid muscle; it can pass medial or lateral to the lateral pterygoid.
- The branches of the *first (mandibular) part* pass through foramina or canals: the deep auricular to the external acoustic meatus, the anterior tympanic to the tympanic cavity, the middle and accessory meningeal to the cranial cavity, and the inferior alveolar to the mandible and teeth.

- The branches of the *second (pterygoid) part*, directly related to the lateral pterygoid muscle, supply muscles via the masseteric, deep temporal, pterygoid, and buccal branches.
- The branches of the *third (pterygopalatine) part* (posterior superior alveolar, infra-orbital, descending palatine, and sphenopalatine arteries) arise immediately proximal to and within the pterygopalatine fossa.

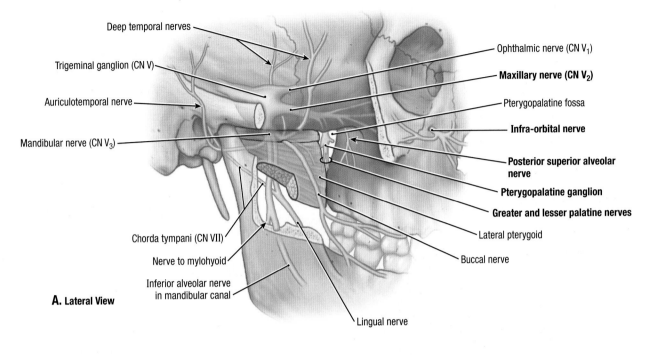

Deep temporal nerves
Trigeminal ganglion (CN V)
Auriculotemporal nerve
Mandibular nerve (CN V₃)

Ophthalmic nerve (CN V₁)
Maxillary nerve (CN V₂)
Pterygopalatine fossa
Infra-orbital nerve
Posterior superior alveolar nerve
Pterygopalatine ganglion
Greater and lesser palatine nerves
Lateral pterygoid
Buccal nerve

Chorda tympani (CN VII)
Nerve to mylohyoid
Inferior alveolar nerve in mandibular canal
Lingual nerve

A. Lateral View

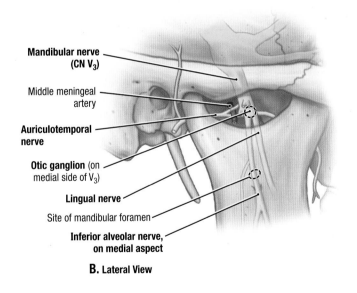

Mandibular nerve (CN V₃)
Middle meningeal artery
Auriculotemporal nerve
Otic ganglion (on medial side of V₃)
Lingual nerve
Site of mandibular foramen
Inferior alveolar nerve, on medial aspect

B. Lateral View

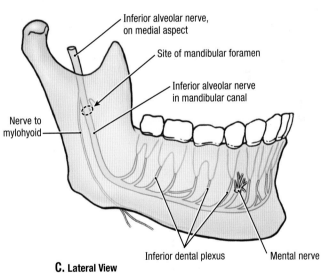

Inferior alveolar nerve, on medial aspect
Site of mandibular foramen
Inferior alveolar nerve in mandibular canal
Nerve to mylohyoid
Inferior dental plexus
Mental nerve

C. Lateral View

BRANCHES OF MAXILLARY AND MANDIBULAR NERVES

7.52

A. Infratemporal region and pterygopalatine fossa. Branches of the maxillary (CN V₂) and mandibular (CN V₃) nerves accompany branches from the three parts of the maxillary artery. **B.** Nerves of infratemporal fossa and otic ganglion. **C.** Mandible and inferior alveolar nerve.

An **alveolar nerve block**—commonly used by dentists when repairing mandibular teeth—anesthetizes the inferior alveolar nerve, a branch of CN V₃. The anesthetic agent is injected around the mandibular foramen, the opening into the mandibular canal on the medial aspect of the ramus of the mandible. This canal gives passage to the inferior alveolar nerve, artery, and vein. When this nerve block is successful, all mandibular teeth are anesthetized to the median plane. The skin and mucous membrane of the lower lip, the labial alveolar mucosa and gingiva, and the skin of the chin are also anesthetized because they are supplied by the mental branch of this nerve.

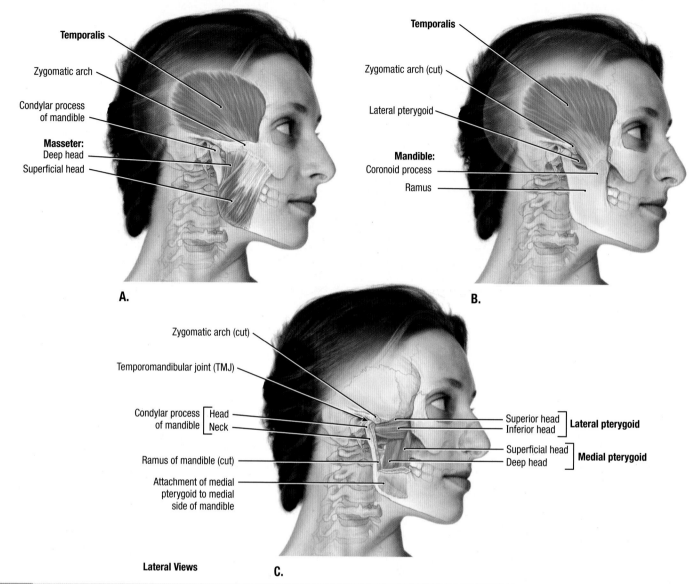

A. Temporalis and masseter. **B.** Temporalis. Zygomatic arch has been removed. **C.** Medial and lateral pterygoid.

Lateral Views

7.53 MUSCLES OF MASTICATION

TABLE 7.11	**MUSCLES OF MASTICATION (ACTING ON TEMPOROMANDIBULAR JOINT)**			
Muscle	*Origin*	*Insertion*	*Innervation*	*Main Action*
Temporalis	Floor of temporal fossa and deep surface of temporal fascia	Tip and medial surface of coronoid process and anterior border of ramus of mandible	Deep temporal branches of mandibular nerve (CN V$_3$)	Elevates mandible, closing jaws; posterior fibers retrude mandible after protrusion
Masseter	Inferior border and medial surface of zygomatic arch	Lateral surface of ramus of mandible and coronoid process	Mandibular nerve (CN V$_3$) through masseteric nerve that enters deep surface of the muscle	Elevates and protrudes mandible, thus closing jaws; deep fibers retrude it
Lateral pterygoid	*Superior head:* infratemporal surface and infratemporal crest of greater wing of sphenoid bone *Inferior head:* lateral surface of lateral pterygoid plate	Neck of mandible, articular disc, and capsule of temporomandibular joint	Mandibular nerve (CN V$_3$) through lateral pterygoid nerve which enters its deep surface	*Acting bilaterally*, protrude mandible and depress chin; *acting unilaterally* alternately, they produce side-to-side movements of mandible
Medial pterygoid	*Deep head:* medial surface of lateral pterygoid plate and pyramidal process of palatine bone *Superficial head:* tuberosity of maxilla	Medial surface of ramus of mandible, inferior to mandibular foramen	Mandibular nerve (CN V$_3$) through medial pterygoid nerve	Helps elevate mandible, closing jaws; *acting bilaterally* protrude mandible; *acting unilaterally*, protrudes side of jaw; acting alternately, they produce a grinding motion

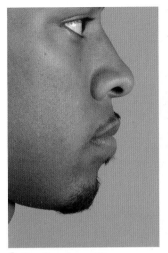

A. Elevation of mandible

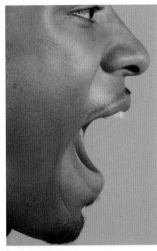

B. Depression of mandible

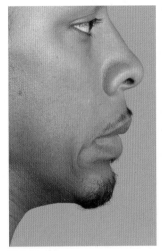

C. Retrusion

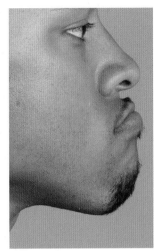

D. Protrusion

Lateral Views

E. Protrusion

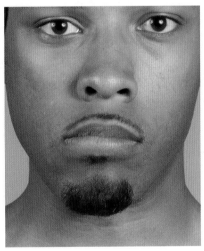

F. Lateral movement to right side

G. Lateral movement to left side

Anterior Views

MOVEMENTS OF TEMPOROMANDIBULAR JOINT

7.54

Temporomandibular joint movements are produced chiefly by the muscles of mastication. These four muscles (temporalis, masseter, and medial and lateral pterygoid muscles) develop from the mesoderm of the first pharyngeal arch; consequently, they are innervated by the nerve of that arch, the motor root of the mandibular nerve (CN V$_3$).

TABLE 7.12	**MOVEMENTS OF TEMPOROMANDIBULAR JOINT**
Movements	*Muscles*
Elevation (close mouth) (A)	Temporalis, masseter, and medial pterygoid
Depression (open mouth) (B)	Lateral pterygoid; suprahyoid and infrahyoid muscles; gravity
Retrusion (retrude chin) (C)	Temporalis (posterior oblique and near horizontal fibers) and masseter
Protrusion (protrude chin) (D and E)	Lateral pterygoid, masseter, and medial pterygoid
Lateral movements (grinding and chewing) (F and G)	Temporalis of same side, pterygoids of opposite side, and masseter

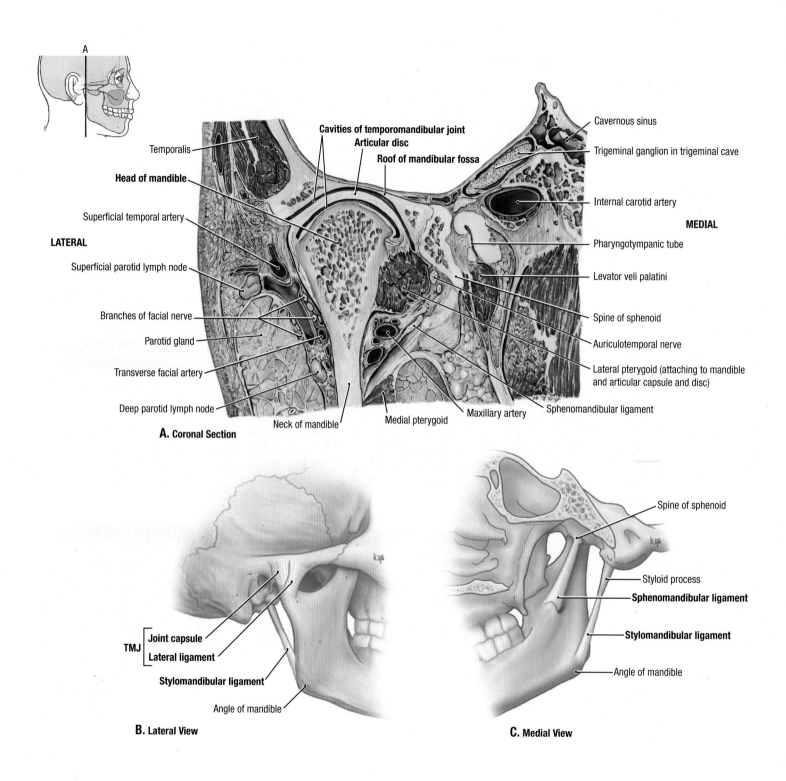

A

A. Coronal Section

Temporalis

Head of mandible

Superficial temporal artery

LATERAL

Superficial parotid lymph node

Branches of facial nerve

Parotid gland

Transverse facial artery

Deep parotid lymph node

Neck of mandible

Medial pterygoid

Maxillary artery

Cavities of temporomandibular joint
Articular disc
Roof of mandibular fossa

Cavernous sinus

Trigeminal ganglion in trigeminal cave

Internal carotid artery

MEDIAL

Pharyngotympanic tube

Levator veli palatini

Spine of sphenoid

Auriculotemporal nerve

Lateral pterygoid (attaching to mandible and articular capsule and disc)

Sphenomandibular ligament

B. Lateral View

TMJ { **Joint capsule** / **Lateral ligament** }

Stylomandibular ligament

Angle of mandible

C. Medial View

Spine of sphenoid

Styloid process

Sphenomandibular ligament

Stylomandibular ligament

Angle of mandible

7.55   TEMPOROMANDIBULAR JOINT

A. Coronal section. **B.** Temporomandibular joint *(TMJ)* and stylomandibular ligament. The joint capsule of the temporomandibular joint attaches to the margins of the mandibular fossa and articular tubercle of the temporal bone and around the neck of the mandible; the lateral (temporomandibular) ligament strengthens the lateral aspect of the joint. **C.** Stylomandibular and sphenomandibular ligaments. The strong sphenomandibular ligament descends from near the spine of the sphenoid to the lingula of the mandible and is the "swinging hinge" by which the mandible is suspended; the weaker stylomandibular ligament is a thickened part of the parotid sheath that joins the styloid process to the angle of the mandible.

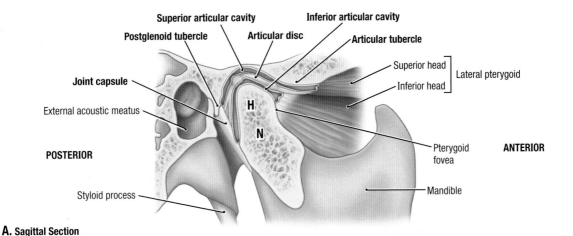

A. Sagittal Section

Labels on figure A:
Superior articular cavity
Postglenoid tubercle
Articular disc
Inferior articular cavity
Articular tubercle
Joint capsule
External acoustic meatus
Superior head
Inferior head
Lateral pterygoid
H
N
POSTERIOR
ANTERIOR
Pterygoid fovea
Styloid process
Mandible

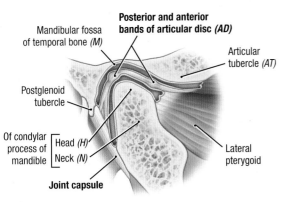

B. Closed Mouth, Sagittal Section

Labels on figure B:
Mandibular fossa of temporal bone (M)
Posterior and anterior bands of articular disc (AD)
Articular tubercle (AT)
Postglenoid tubercle
Of condylar process of mandible — Head (H) / Neck (N)
Joint capsule
Lateral pterygoid

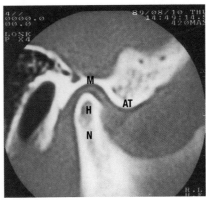

Sagittal CT

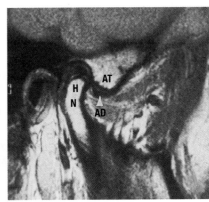

Sagittal MRI

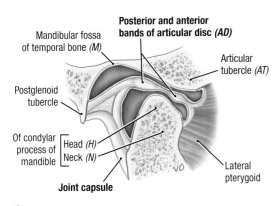

C. Open Mouth, Sagittal Section

Labels on figure C:
Mandibular fossa of temporal bone (M)
Posterior and anterior bands of articular disc (AD)
Articular tubercle (AT)
Postglenoid tubercle
Of condylar process of mandible — Head (H) / Neck (N)
Joint capsule
Lateral pterygoid

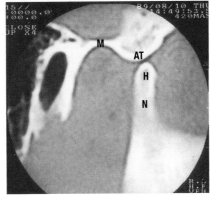

Sagittal CT

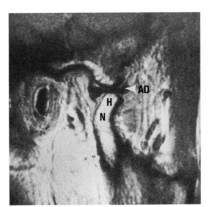

Sagittal MRI

SECTIONAL ANATOMY OF TEMPOROMANDIBULAR JOINT

7.56

A. Temporomandibular joint and related structures, sagittal section. **B.** Sagittal orientation figure, CT, and MRI—mouth closed. **C.** Sagittal orientation figure, CT, and MRI images—mouth opened widely. The articular disc divides the articular cavity into superior and inferior compartments, each lined by a separate synovial membrane.

Dislocation of mandible. During yawning or taking large bites, excessive contraction of the lateral pterygoids can cause the head of the mandible to dislocate (pass anterior to the articular tubercle). In this position, the mouth remains wide open, and the person cannot close it without manual distraction.

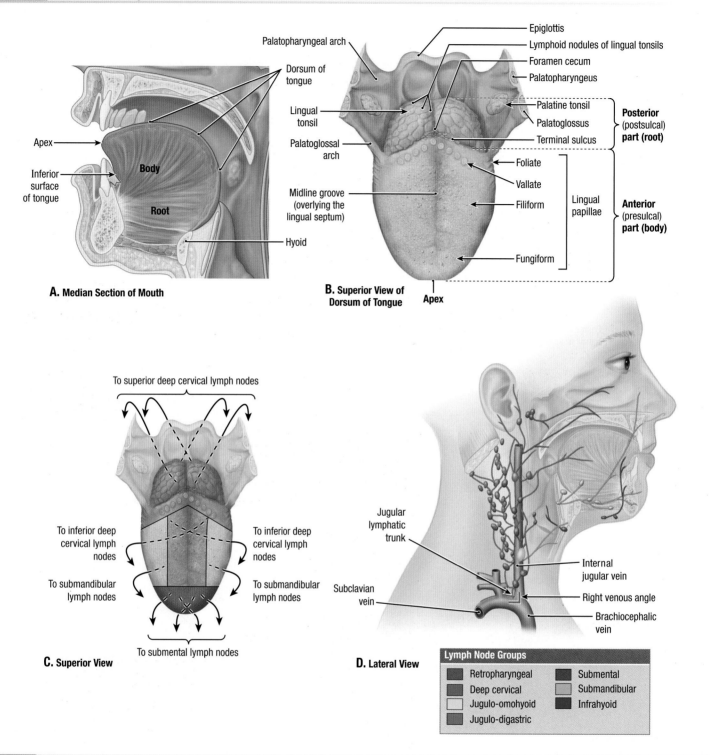

A. Median Section of Mouth

Apex
Inferior surface of tongue
Body
Root
Hyoid

B. Superior View of Dorsum of Tongue

Palatopharyngeal arch
Dorsum of tongue
Lingual tonsil
Palatoglossal arch
Midline groove (overlying the lingual septum)
Hyoid

Epiglottis
Lymphoid nodules of lingual tonsils
Foramen cecum
Palatopharyngeus
Palatine tonsil
Palatoglossus
Terminal sulcus
Foliate
Vallate
Filiform
Fungiform
Apex

Posterior (postsulcal) part (root)
Anterior (presulcal) part (body)
Lingual papillae

C. Superior View

To superior deep cervical lymph nodes
To inferior deep cervical lymph nodes
To inferior deep cervical lymph nodes
To submandibular lymph nodes
To submandibular lymph nodes
To submental lymph nodes

D. Lateral View

Jugular lymphatic trunk
Subclavian vein
Internal jugular vein
Right venous angle
Brachiocephalic vein

Lymph Node Groups
Retropharyngeal
Deep cervical
Jugulo-omohyoid
Jugulo-digastric
Submental
Submandibular
Infrahyoid

7.57   PARTS AND LYMPHATIC DRAINAGE OF TONGUE

A. Parts of tongue. **B.** Features of dorsum of the tongue. The foramen cecum is the upper end of the primitive thyroglossal duct; the arms of the V-shaped terminal sulcus diverge from the foramen, demarcating the posterior third of the tongue from the anterior two thirds. **C.** Lymphatic drainage of dorsum of tongue. **D.** Lymphatic drainage of tongue, mouth, nasal cavity, and nose.

Carcinoma of tongue. Malignant tumors in the posterior part of the tongue metastasize to the superior deep cervical lymph nodes on both sides. In contrast, tumors in the apex and anterolateral parts usually do not metastasize to the inferior deep

cervical nodes until late in the disease. Because the deep nodes are closely related to the internal jugular vein (IJV), metastases from the carcinoma may spread to the submental and submandibular regions and along the IJV into the neck.

Gag reflex. One may touch the anterior part of the tongue without feeling discomfort; however, when the posterior part is touched, one usually gags. CN IX and CN X are responsible for the muscular contraction of each side of the pharynx. Glossopharyngeal branches (CN IX) provide the afferent limb of the gag reflex.

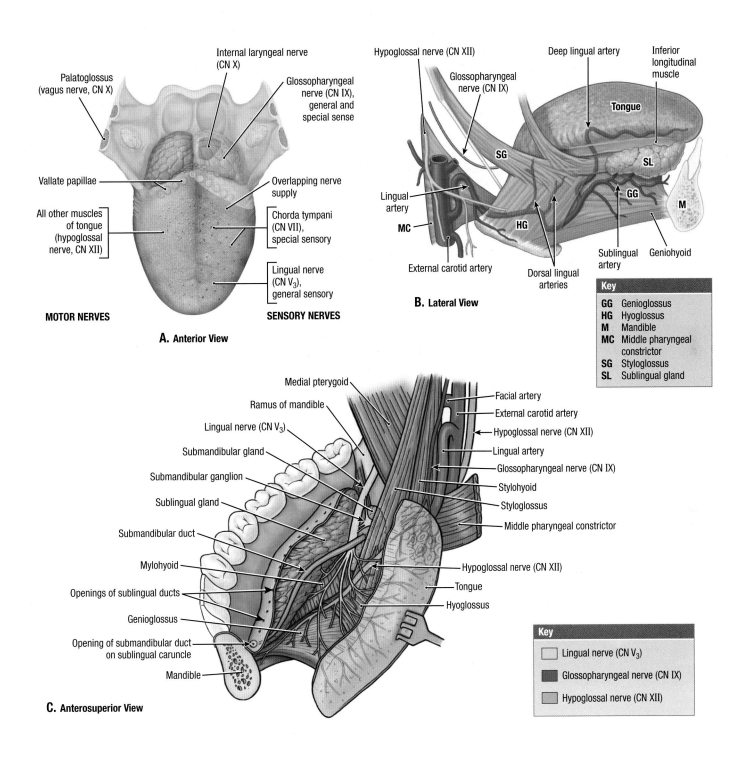

A. Anterior View

Palatoglossus (vagus nerve, CN X)

Internal laryngeal nerve (CN X)

Glossopharyngeal nerve (CN IX), general and special sense

Vallate papillae

All other muscles of tongue (hypoglossal nerve, CN XII)

Overlapping nerve supply

Chorda tympani (CN VII), special sensory

Lingual nerve (CN V₃), general sensory

MOTOR NERVES

SENSORY NERVES

B. Lateral View

Hypoglossal nerve (CN XII)

Glossopharyngeal nerve (CN IX)

Deep lingual artery

Inferior longitudinal muscle

Tongue

SG

SL

Lingual artery

GG

M

MC

HG

External carotid artery

Dorsal lingual arteries

Sublingual artery

Geniohyoid

Key

GG	Genioglossus
HG	Hyoglossus
M	Mandible
MC	Middle pharyngeal constrictor
SG	Styloglossus
SL	Sublingual gland

C. Anterosuperior View

Medial pterygoid

Ramus of mandible

Lingual nerve (CN V₃)

Submandibular gland

Submandibular ganglion

Sublingual gland

Submandibular duct

Mylohyoid

Openings of sublingual ducts

Genioglossus

Opening of submandibular duct on sublingual caruncle

Mandible

Facial artery

External carotid artery

Hypoglossal nerve (CN XII)

Lingual artery

Glossopharyngeal nerve (CN IX)

Stylohyoid

Styloglossus

Middle pharyngeal constrictor

Hypoglossal nerve (CN XII)

Tongue

Hyoglossus

Key

	Lingual nerve (CN V₃)
	Glossopharyngeal nerve (CN IX)
	Hypoglossal nerve (CN XII)

ARTERIES AND NERVES OF THE TONGUE

7.58

A. General sensory, special sensory (taste), and motor innervation of tongue. **B.** Course and distribution of the lingual artery. **C.** Dissection of right side of floor of mouth.

Sialography. The parotid and submandibular salivary glands may be examined radiographically after the injection of a contrast medium into their ducts. This special type of radiograph (sialogram) demonstrates the salivary ducts and some secretory units. Because of the small size and number of sublingual ducts of the sublingual glands, one cannot usually inject contrast medium into them.

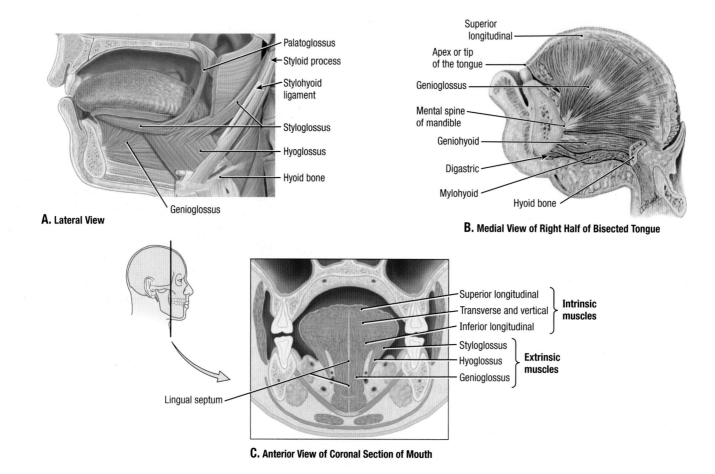

A. Lateral View

- Palatoglossus
- Styloid process
- Stylohyoid ligament
- Styloglossus
- Hyoglossus
- Hyoid bone
- Genioglossus

B. Medial View of Right Half of Bisected Tongue

- Superior longitudinal
- Apex or tip of the tongue
- Genioglossus
- Mental spine of mandible
- Geniohyoid
- Digastric
- Mylohyoid
- Hyoid bone

C. Anterior View of Coronal Section of Mouth

- Superior longitudinal
- Transverse and vertical — **Intrinsic muscles**
- Inferior longitudinal
- Styloglossus
- Hyoglossus — **Extrinsic muscles**
- Genioglossus
- Lingual septum

7.59 MUSCLES OF TONGUE

A. Extrinsic muscles. **B.** Median section. **C.** Coronal section. The extrinsic muscles of the tongue originate outside the tongue and attach to it, whereas the intrinsic muscles have their attachments entirely within the tongue and are not attached to bone.

TABLE 7.13	**MUSCLES OF TONGUE**			
Extrinsic Muscles				
Muscle	*Origin*	*Insertion*	*Innervation*	*Main Action*
Genioglossus	Superior part of mental spine of mandible	Dorsum of tongue and body of hyoid bone		Depresses tongue; its posterior part pulls tongue anteriorly for protrusion[a]
Hyoglossus	Body and greater horn of hyoid bone	Side and inferior aspect of tongue	Hypoglossal nerve (CN XII)	Depresses and retracts tongue
Styloglossus	Styloid process of temporal bone and stylohyoid ligament	Side and inferior aspect of tongue		Retracts tongue and draws it up to create a trough for swallowing
Palatoglossus	Palatine aponeurosis of soft palate	Side of tongue	CN X and pharyngeal plexus	Elevates posterior part of tongue plexus
Intrinsic Muscles				
Muscle	*Origin*	*Insertion*	*Innervation*	*Main Action*
Superior longitudinal	Submucous fibrous layer and lingual septum	Margins and mucous membrane of tongue		Curls tip and sides of tongue superiorly and shortens tongue
Inferior longitudinal	Root of tongue and body of hyoid bone	Apex of tongue	Hypoglossal nerve (CN XII)	Curls tip of tongue inferiorly and shortens tongue
Transverse	Lingual septum	Fibrous tissue at margins of tongue		Narrows and elongates the tongue[a]
Vertical	Superior surface of borders of tongue	Inferior surface of borders of tongue		Flattens and broadens the tongue[a]

[a]Acts simultaneously to protrude tongue.

A. Transverse Section

Palatopharyngeal arch
Longus capitis
Sternocleidomastoid
Carotid sheath
Digastric, posterior belly
Stylohyoid
Stylopharyngeus
Styloglossus
Medial pterygoid
Superior pharyngeal constrictor
Masseter
Inferior alveolar nerve (CN V₃)
Palatoglossus in palatoglossal arch
Site of section B
Tongue
Buccinator and oral muscles

Cavity of pharynx
Longus colli
Axis
Superior constrictor
Retropharyngeal space
Prevertebral fascia
Spinal accessory nerve (CN XI)
Internal jugular vein
Vagus nerve (CN X)
Parotid gland
Hypoglossal nerve (CN XII)
Retromandibular vein
Glossopharyngeal nerve (CN IX)
External carotid artery
Internal carotid artery
Sympathetic ganglion
Lateral pharyngeal space
Palatine tonsil
Ramus of mandible
Lingual nerve (CN V₃)
Facial vein
Buccal glands
Facial artery and branches

B. Anterior View of Coronal Section

Molar tooth
Alveolar mucosa
Superior buccal gingiva (proper)
Oral vestibule
Buccal mucosa*
Buccinator
Bolus of food
Inferior buccal gingiva (proper)
Mandible

C Crown
N Neck } of tooth
R Root

Oral cavity proper
Palatine mucosa*
Superior lingual gingiva (proper)
Tongue
Inferior lingual gingiva (proper)
Oral mucosa* of floor of mouth
* Mucous membrane of mouth

Plane of section

SECTIONS THROUGH MOUTH

7.60

A. The viscerocranium has been sectioned at the C1 vertebral level, the plane of section passing through the oral fissure anteriorly. The retropharyngeal space (opened up in this specimen) allows the pharynx to contract and relax during swallowing; the retropharyngeal space is closed laterally at the carotid sheath and limited posteriorly by the prevertebral fascia. The beds of the parotid glands are also demonstrated. **B.** Schematic coronal section demonstrating how the tongue and buccinator (or, anteriorly, the orbicularis oris) work together to retain food between the teeth when chewing. The buccinator and superior part of the orbicularis oris are innervated by the buccal branch of the facial nerve (CN VII).

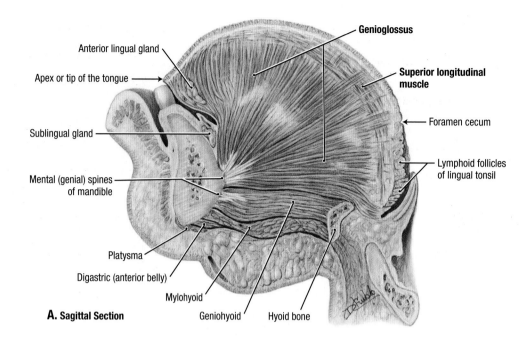

Genioglossus

Superior longitudinal muscle

Anterior lingual gland

Apex or tip of the tongue

Foramen cecum

Sublingual gland

Lymphoid follicles of lingual tonsil

Mental (genial) spines of mandible

Platysma

Digastric (anterior belly)

Mylohyoid

Geniohyoid   Hyoid bone

A. Sagittal Section

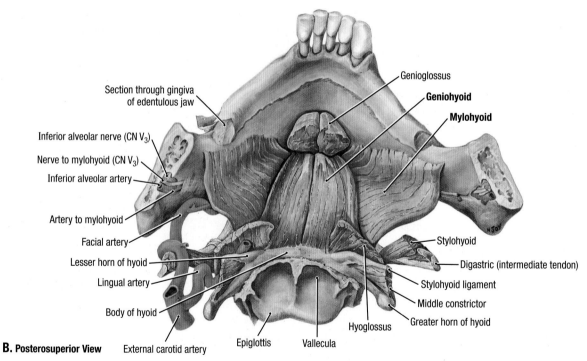

Genioglossus

Geniohyoid

Mylohyoid

Section through gingiva of edentulous jaw

Inferior alveolar nerve (CN V$_3$)

Nerve to mylohyoid (CN V$_3$)

Inferior alveolar artery

Artery to mylohyoid

Facial artery

Lesser horn of hyoid

Lingual artery

Body of hyoid

Stylohyoid

Digastric (intermediate tendon)

Stylohyoid ligament

Middle constrictor

Greater horn of hyoid

Hyoglossus

Epiglottis   Vallecula

B. Posterosuperior View   External carotid artery

7.61 TONGUE AND FLOOR OF MOUTH

A. Median section though the tongue and lower jaw. The tongue is composed mainly of muscle; extrinsic muscles alter the position of the tongue, and intrinsic muscles alter its shape. The genioglossus is the extrinsic muscle apparent in this plane, and the superior longitudinal muscle is the intrinsic muscle. **B.** Muscles of the floor of the mouth viewed posterosuperiorly. The mylohyoid muscle extends between the two mylohyoid lines of the mandible. It has a thick, free posterior border and becomes thinner anteriorly.

Genioglossus paralysis. When the genioglossus is paralyzed, the tongue mass has a tendency to shift posteriorly, obstructing the airway and presenting the risk of suffocation. Total relaxation of the genioglossus muscles occurs during general anesthesia; therefore, the tongue of an anesthetized patient must be prevented from relapsing by inserting an airway.

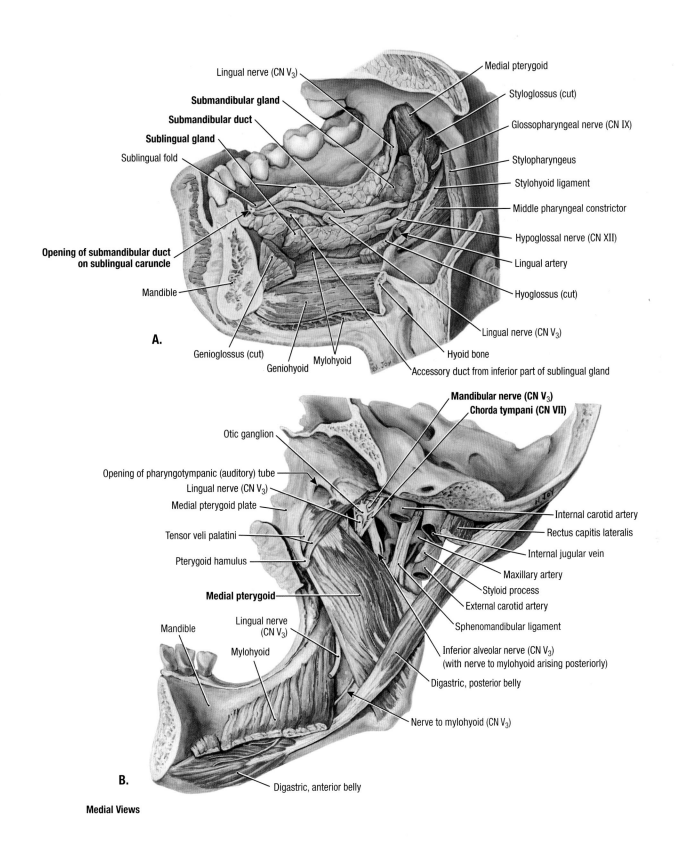

A.

Lingual nerve (CN V₃)

Submandibular gland

Submandibular duct

Sublingual gland

Sublingual fold

**Opening of submandibular duct
on sublingual caruncle**

Mandible

Genioglossus (cut)

Geniohyoid

Mylohyoid

Medial pterygoid

Styloglossus (cut)

Glossopharyngeal nerve (CN IX)

Stylopharyngeus

Stylohyoid ligament

Middle pharyngeal constrictor

Hypoglossal nerve (CN XII)

Lingual artery

Hyoglossus (cut)

Lingual nerve (CN V₃)

Hyoid bone

Accessory duct from inferior part of sublingual gland

B.

Otic ganglion

Opening of pharyngotympanic (auditory) tube

Lingual nerve (CN V₃)

Medial pterygoid plate

Tensor veli palatini

Pterygoid hamulus

Medial pterygoid

Mandible

Lingual nerve
(CN V₃)

Mylohyoid

Digastric, anterior belly

Mandibular nerve (CN V₃)

Chorda tympani (CN VII)

Internal carotid artery

Rectus capitis lateralis

Internal jugular vein

Maxillary artery

Styloid process

External carotid artery

Sphenomandibular ligament

Inferior alveolar nerve (CN V₃)
(with nerve to mylohyoid arising posteriorly)

Digastric, posterior belly

Nerve to mylohyoid (CN V₃)

Medial Views

MUSCLES, GLANDS, AND VESSELS OF FLOOR OF MOUTH AND MEDIAL ASPECT OF MANDIBLE   7.62

A. Sublingual and submandibular glands. The tongue has been excised. **B.** Structures related to the medial surface of the mandible. The otic ganglion lies medial to the mandibular nerve (CN V₃) and between the foramen ovale superiorly and the medial pterygoid muscle inferiorly.

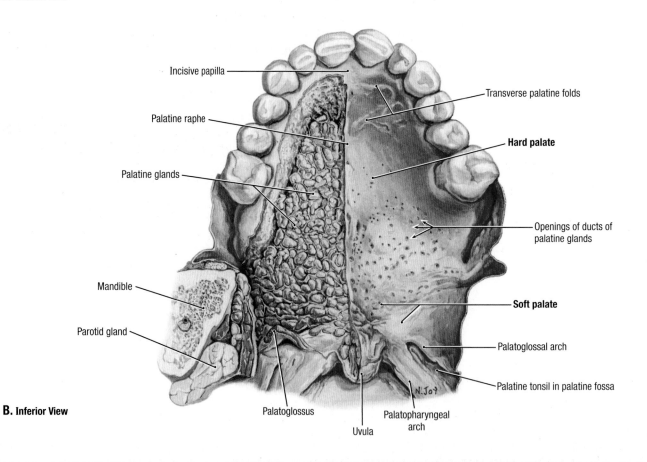

Incisive fossa

Incisive bone (premaxilla)

Maxilla, palatine process

Intermaxillary suture

Groove for greater palatine vessels

Median palatine suture

Greater Lesser **Palatine foramina**

Horizontal plate

Palatine bone

Pyramidal process (tubercle)

Lateral pterygoid plate

Scaphoid fossa

Medial pterygoid plate

Medial pterygoid plate Hamulus Tubercle

Vomer

Posterior nasal spine

A. Inferior View

Incisive papilla

Palatine raphe

Palatine glands

Transverse palatine folds

Hard palate

Openings of ducts of palatine glands

Mandible

Parotid gland

Soft palate

Palatoglossal arch

Palatine tonsil in palatine fossa

B. Inferior View

Palatoglossus

Uvula

Palatopharyngeal arch

7.63	PALATE

A. Bones of the head palate. The palatine aponeurosis, which forms the fibrous "skeleton" of the soft palate, stretches between the hamuli of the medial pterygoid plates. **B.** Mucous membrane and glands of palate.

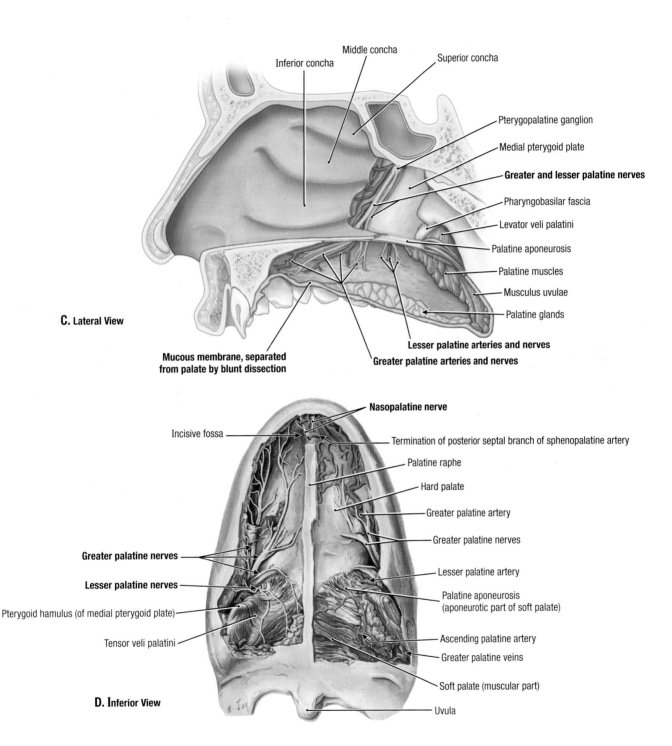

Middle concha
Inferior concha
Superior concha

Pterygopalatine ganglion

Medial pterygoid plate

Greater and lesser palatine nerves

Pharyngobasilar fascia

Levator veli palatini

Palatine aponeurosis

Palatine muscles

Musculus uvulae

Palatine glands

C. Lateral View

**Mucous membrane, separated
from palate by blunt dissection**

Lesser palatine arteries and nerves

Greater palatine arteries and nerves

Nasopalatine nerve

Incisive fossa

Termination of posterior septal branch of sphenopalatine artery

Palatine raphe

Hard palate

Greater palatine artery

Greater palatine nerves

Greater palatine nerves

Lesser palatine artery

Lesser palatine nerves

Palatine aponeurosis
(aponeurotic part of soft palate)

Pterygoid hamulus (of medial pterygoid plate)

Tensor veli palatini

Ascending palatine artery

Greater palatine veins

Soft palate (muscular part)

D. Inferior View

Uvula

PALATE (continued)

7.63

C. Nerves and vessels of palatine canal. The lateral wall of the nasal cavity is shown. The posterior ends of the middle and inferior conchae are excised along with the mucoperiosteum; the thin, perpendicular plate of the palatine bone is removed to expose the palatine nerves and arteries. **D.** Dissection of an edentulous palate. The greater palatine nerve supplies the gingivae and hard palate, the nasopalatine nerves the incisive region, and the lesser palatine nerves the soft palate. **Anesthesia of palatine nerves.** The nasopalatine nerves can be anesthetized by injecting anesthetic into the mouth of the incisive fossa in the hard palate. The anesthetized tissues are the palatal mucosa, the lingual gingivae, the six anterior maxillary teeth, and associated alveolar bone. The greater palatine nerve can be anesthetized by injecting anesthetic into the greater palatine foramen. The nerve emerges between the second and third maxillary molar teeth. This nerve block anesthetizes the palatal mucosa and lingual gingivae posterior to the maxillary canine teeth, and the underlying bone of the palate.

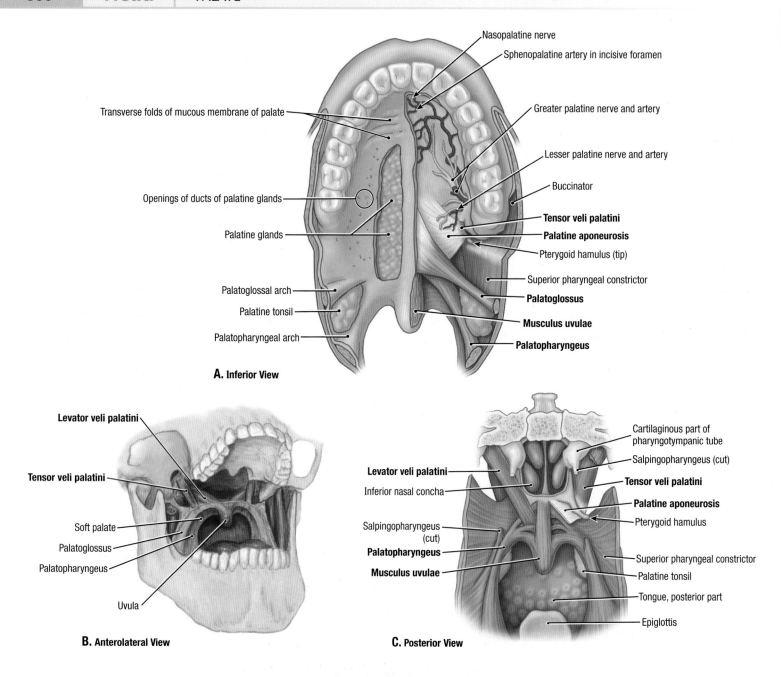

A. Inferior View

B. Anterolateral View

C. Posterior View

7.64   MUSCLES OF SOFT PALATE

TABLE 7.14	**MUSCLES OF SOFT PALATE**				
Muscle	*Superior Attachment*	*Inferior Attachment*	*Innervation*	*Main Action(s)*
Levator veli palatini	Cartilage of pharyngotympanic tube and petrous part of temporal bone	Palatine aponeurosis	Pharyngeal branch of vagus nerve through pharyngeal plexus	Elevates soft palate during swallowing and yawning
Tensor veli palatini	Scaphoid fossa of medial pterygoid plate, spine of sphenoid bone, and cartilage of pharyngotympanic tube		Medial pterygoid nerve (CN V₃) through otic ganglion	Tenses soft palate and opens mouth of pharyngotympanic tube during swallowing and yawning
Palatoglossus	Palatine aponeurosis	Side of tongue	Pharyngeal branch of vagus nerve (CN X) via pharyngeal plexus	Elevates posterior part of tongue and draws soft palate onto tongue
Palatopharyngeus	Hard palate and palatine aponeurosis	Lateral wall of pharynx		Tenses soft palate and pulls walls of pharynx superiorly, anteriorly, and medially during swallowing
Musculus uvulae	Posterior nasal spine and palatine aponeurosis	Mucosa of uvula		Shortens uvula and pulls it superiorly

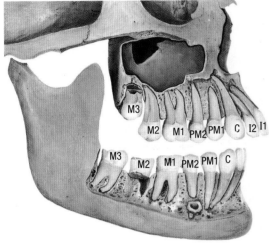

A. Lateral View

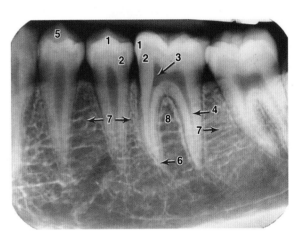

B. Pantomographic Radiograph

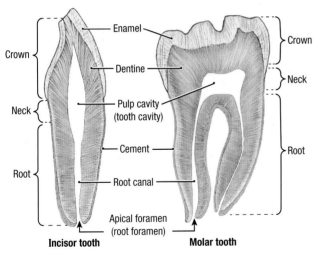

C. Longitudinal Section

Incisor tooth Molar tooth

D. Lateral Radiograph

Key		
1 Enamel	**2** Dentine	**3** Pulp cavity
4 Root canal	**5** Buccal cusp	**6** Root apex
7 Interalveolar septa (alveolar bone)		
8 Interradicular septum (alveolar bone)		

PERMANENT TEETH I

7.65

A. Teeth *in situ* with roots exposed. Incisors (*I1*, *I2*), canine (*C1*), premolars (*PM1*, *PM2*), and molars (*M1*, *M2*, *M3*). The roots of the 2nd lower molar have been removed. **B.** Pantomographic radiograph of mandible and maxilla. The left lower third molar is not present. **C.** Longitudinal sections of an incisor and a molar tooth. **D.** Lateral radiograph.

Decay of the hard tissues of a tooth results in the formation of **dental caries** (cavities). Invasion of the pulp of the tooth by a carious lesion (cavity) results in infection and irritation of the tissues in the pulp cavity. This condition causes an inflammatory process (pulpitis). Because the pulp cavity is a rigid space, the swollen pulpal tissues cause pain (toothache).

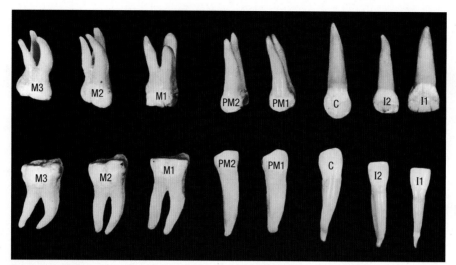

A. Vestibular View

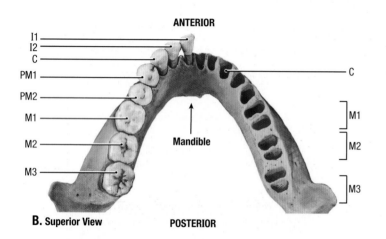

B. Superior View

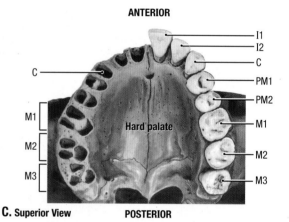

C. Superior View

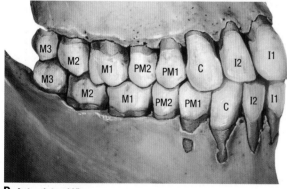

D. Anterolateral View

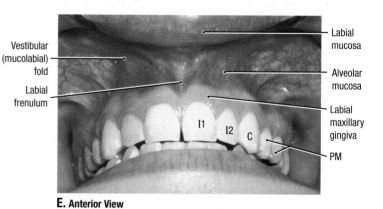

E. Anterior View

7.66 PERMANENT TEETH II

A. Removed teeth, displaying roots. There are 32 permanent teeth; 8 are on each side of each dental arch on the top (maxillary teeth) and bottom (mandibular teeth): 2 incisors (*I1*, *I2*), 1 canine (*C*), 2 premolars (*PM1*, *PM2*), and 3 molars (*M1* to *M3*). **B.** Permanent mandibular teeth and their sockets. **C.** Permanent maxillary teeth and their sockets. **D.** Teeth in occlusion. **E.** Vestibule and gingivae of the maxilla.

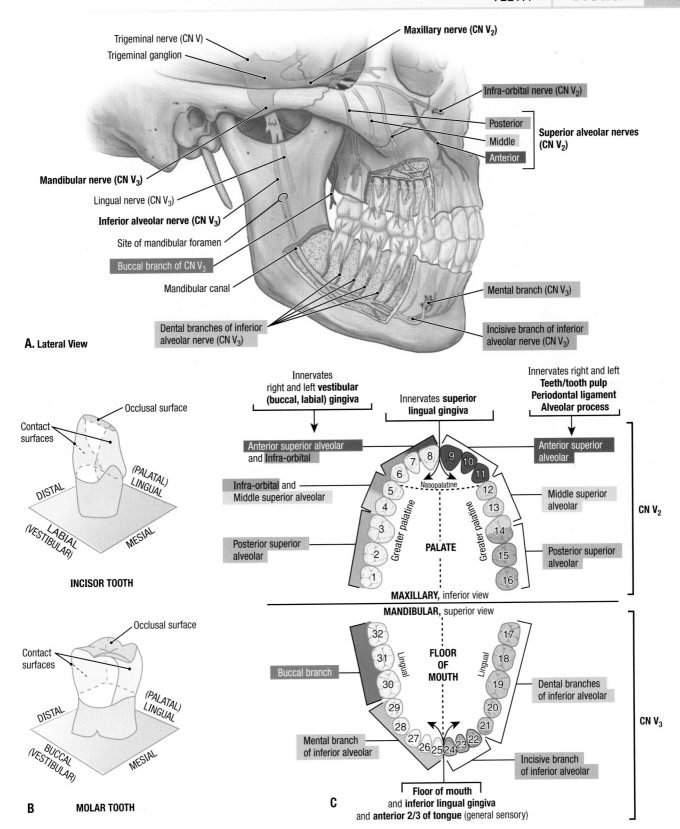

A. Lateral View

Trigeminal nerve (CN V)
Trigeminal ganglion
Maxillary nerve (CN V₂)
Infra-orbital nerve (CN V₂)
Posterior
Middle
Anterior
Superior alveolar nerves (CN V₂)
Mandibular nerve (CN V₃)
Lingual nerve (CN V₃)
Inferior alveolar nerve (CN V₃)
Site of mandibular foramen
Buccal branch of CN V₃
Mandibular canal
Mental branch (CN V₃)
Incisive branch of inferior alveolar nerve (CN V₃)
Dental branches of inferior alveolar nerve (CN V₃)

Occlusal surface
Contact surfaces
(PALATAL) LINGUAL
DISTAL
LABIAL (VESTIBULAR)
MESIAL
INCISOR TOOTH

Occlusal surface
Contact surfaces
(PALATAL) LINGUAL
DISTAL
BUCCAL (VESTIBULAR)
MESIAL
B   MOLAR TOOTH

Innervates right and left **vestibular (buccal, labial) gingiva**
Innervates **superior lingual gingiva**
Innervates right and left **Teeth/tooth pulp Periodontal ligament Alveolar process**

Anterior superior alveolar and Infra-orbital
Anterior superior alveolar
Infra-orbital and Middle superior alveolar
Nasopalatine
Middle superior alveolar
Posterior superior alveolar
Greater palatine
PALATE
Greater palatine
Posterior superior alveolar

MAXILLARY, inferior view
CN V₂

MANDIBULAR, superior view
Buccal branch
FLOOR OF MOUTH
Lingual
Lingual
Dental branches of inferior alveolar
Mental branch of inferior alveolar
Incisive branch of inferior alveolar

C
Floor of mouth and **inferior lingual gingiva** and **anterior 2/3 of tongue** (general sensory)
CN V₃

INNERVATION OF TEETH   7.67

A. Superior and inferior alveolar nerves. **B.** Surfaces of an incisor and molar tooth. **C.** Innervation of the mouth and teeth.

Improper oral hygiene results in food deposits in tooth and gingival crevices, which may cause inflammation of the gingivae, **gingivitis**. If untreated, the disease spreads to other supporting structures (including the alveolar bone), producing **periodontitis**. Periodontitis results in inflammation of the gingivae and may result in absorption of alveolar bone and gingival recession. Gingival recession exposes the sensitive cement of the teeth.

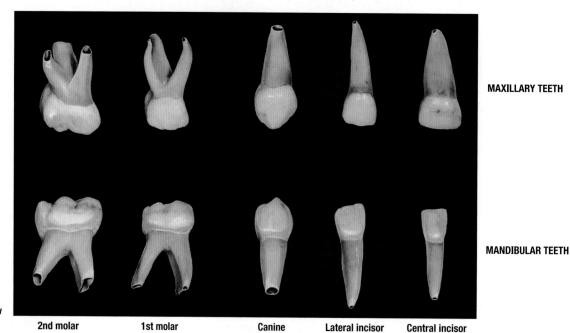

MAXILLARY TEETH

MANDIBULAR TEETH

A. Vestibular View

2nd molar   1st molar   Canine   Lateral incisor   Central incisor

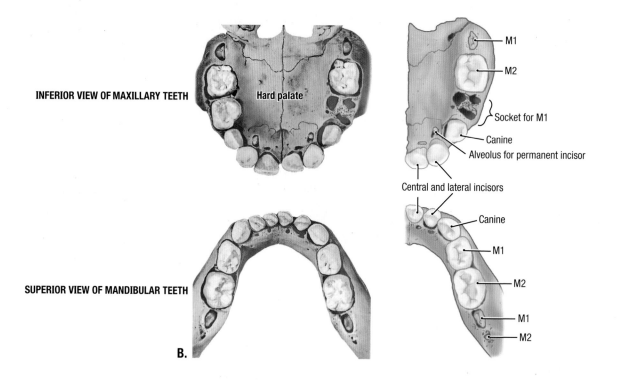

INFERIOR VIEW OF MAXILLARY TEETH

Hard palate

M1

M2

Socket for M1

Canine

Alveolus for permanent incisor

Central and lateral incisors

Canine

SUPERIOR VIEW OF MANDIBULAR TEETH

M1

M2

M1

M2

B.

7.68   PRIMARY TEETH

A. Removed teeth. There are 20 primary (deciduous) teeth, 5 in each half of the mandible and 5 in each maxilla. They are named central incisor, lateral incisor, canine, 1st molar (*M1*), and 2nd molar (*M2*). Primary teeth differ from permanent teeth in that the primary teeth are smaller and whiter; the molars also have more bulbous crowns and more divergent roots. **B.** Teeth *in situ*, younger than 2 years of age. Permanent teeth are colored orange; the crowns of the un-erupted 1st and 2nd permanent molars are partly visible.

TABLE 7.15	PRIMARY AND SECONDARY DENTITION				
Deciduous Teeth	*Central Incisor*	*Lateral Incisor*	*Canine*	*First Molar*	*Second Molar*
Eruption (months)[a]	6–8	8–10	16–20	12–16	20–24
Shedding (years)	6–7	7–8	10–12	9–11	10–12

[a]In some normal infants, the first teeth (medial incisors) may not erupt until 12 to 13 months of age.

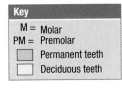

Key

M = Molar
PM = Premolar
◻ Permanent teeth
◻ Deciduous teeth

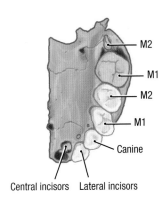

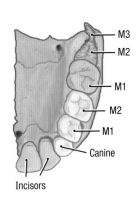

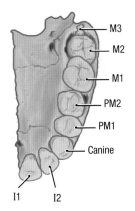

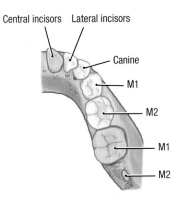

Age: 6–7 years

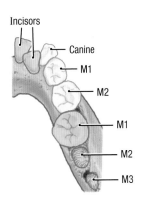

Age: 8 years

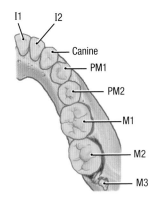

Age: 12 years

TABLE 7.15	PRIMARY AND SECONDARY DENTITION *(continued)*						
Permanent Teeth	*Central Incisor*	*Lateral Incisor*	*Canine*	*First Premolar*	*Second Premolar*	*First Molar*	*Second Molar*	*Third Molar*
Eruption (years)	7–8	8–9	10–12	10–11	11–12	6–7	12	13–25

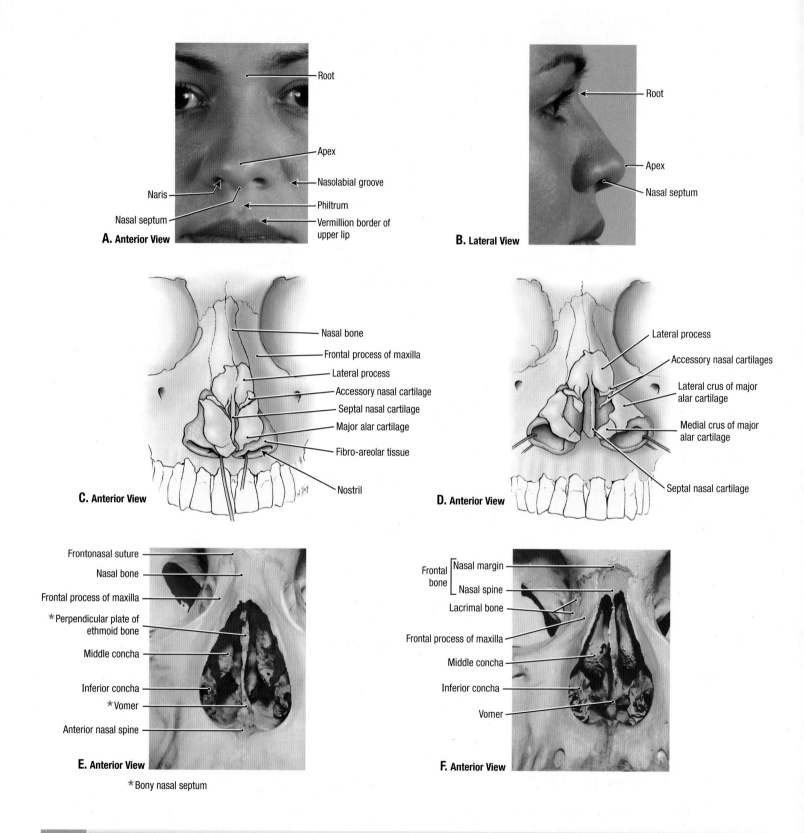

A. Anterior View

- Root
- Apex
- Nasolabial groove
- Naris
- Philtrum
- Nasal septum
- Vermillion border of upper lip

B. Lateral View

- Root
- Apex
- Nasal septum

C. Anterior View

- Nasal bone
- Frontal process of maxilla
- Lateral process
- Accessory nasal cartilage
- Septal nasal cartilage
- Major alar cartilage
- Fibro-areolar tissue
- Nostril

D. Anterior View

- Lateral process
- Accessory nasal cartilages
- Lateral crus of major alar cartilage
- Medial crus of major alar cartilage
- Septal nasal cartilage

E. Anterior View

- Frontonasal suture
- Nasal bone
- Frontal process of maxilla
- *Perpendicular plate of ethmoid bone
- Middle concha
- Inferior concha
- *Vomer
- Anterior nasal spine

*Bony nasal septum

F. Anterior View

- Frontal bone [Nasal margin / Nasal spine]
- Lacrimal bone
- Frontal process of maxilla
- Middle concha
- Inferior concha
- Vomer

7.69 SURFACE ANATOMY, CARTILAGES, AND BONES OF NOSE

A. Surface features of anterior aspect of nose. **B.** Surface features of lateral aspect of nose. **C.** Nasal cartilages, with the septum pulled inferiorly. **D.** Nasal cartilages, separated and retracted laterally. **E.** Lower conchae and bony septum seen through the piriform aperture. The margin of the piriform aperture is sharp and formed by the maxillae and nasal bones. **F.** Nasal bones removed. The areas of the frontal processes of the maxillae *(yellow)* and of the frontal bone *(blue)* that articulate with the nasal bones can be seen.

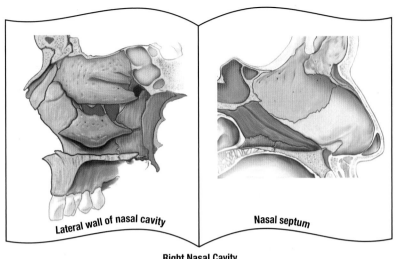

Lateral wall of nasal cavity

Nasal septum

Right Nasal Cavity

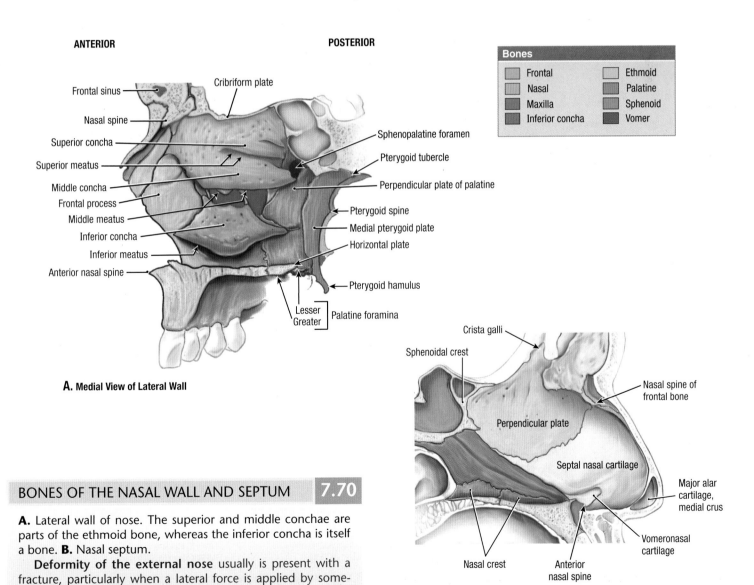

ANTERIOR

POSTERIOR

Frontal sinus

Cribriform plate

Nasal spine

Superior concha

Superior meatus

Middle concha

Frontal process

Middle meatus

Inferior concha

Inferior meatus

Anterior nasal spine

Sphenopalatine foramen

Pterygoid tubercle

Perpendicular plate of palatine

Pterygoid spine

Medial pterygoid plate

Horizontal plate

Pterygoid hamulus

Lesser
Greater } Palatine foramina

A. Medial View of Lateral Wall

Bones	
Frontal	Ethmoid
Nasal	Palatine
Maxilla	Sphenoid
Inferior concha	Vomer

Crista galli

Sphenoidal crest

Nasal spine of frontal bone

Perpendicular plate

Septal nasal cartilage

Major alar cartilage, medial crus

Vomeronasal cartilage

Nasal crest

Anterior nasal spine

B. Lateral View of Nasal Septum

BONES OF THE NASAL WALL AND SEPTUM   7.70

A. Lateral wall of nose. The superior and middle conchae are parts of the ethmoid bone, whereas the inferior concha is itself a bone. **B.** Nasal septum.

Deformity of the external nose usually is present with a fracture, particularly when a lateral force is applied by someone's elbow, for example. When the injury results from a direct blow (e.g., from a hockey stick), the cribriform plate of the ethmoid bone may fracture, resulting in CSF rhinorrhea.

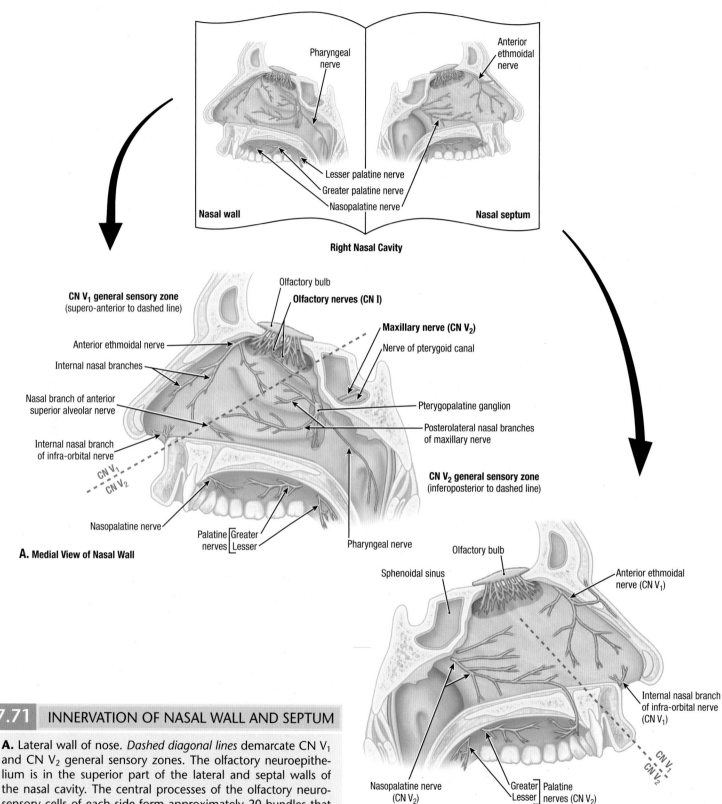

A. Medial View of Nasal Wall

B. Lateral View of Nasal Septum

7.71 INNERVATION OF NASAL WALL AND SEPTUM

A. Lateral wall of nose. *Dashed diagonal lines* demarcate CN V$_1$ and CN V$_2$ general sensory zones. The olfactory neuroepithelium is in the superior part of the lateral and septal walls of the nasal cavity. The central processes of the olfactory neurosensory cells of each side form approximately 20 bundles that together form an olfactory nerve (CN I). **B.** Nasal septum. The nasopalatine nerve from the pterygopalatine ganglion supplies the postero-inferior septum, and the anterior ethmoidal nerve (branch of V$_1$) supplies the anterosuperior septum.

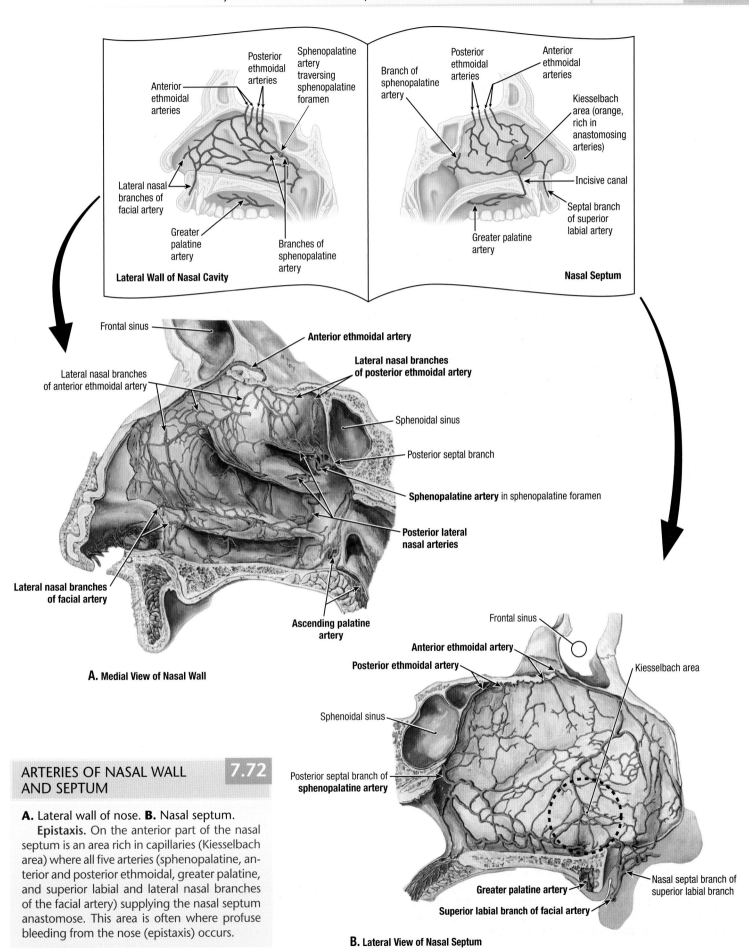

Lateral Wall of Nasal Cavity

Anterior ethmoidal arteries

Posterior ethmoidal arteries

Sphenopalatine artery traversing sphenopalatine foramen

Lateral nasal branches of facial artery

Greater palatine artery

Branches of sphenopalatine artery

Nasal Septum

Branch of sphenopalatine artery

Posterior ethmoidal arteries

Anterior ethmoidal arteries

Kiesselbach area (orange, rich in anastomosing arteries)

Incisive canal

Septal branch of superior labial artery

Greater palatine artery

Frontal sinus

Anterior ethmoidal artery

Lateral nasal branches of anterior ethmoidal artery

Lateral nasal branches of posterior ethmoidal artery

Sphenoidal sinus

Posterior septal branch

Sphenopalatine artery in sphenopalatine foramen

Posterior lateral nasal arteries

Lateral nasal branches of facial artery

Ascending palatine artery

A. Medial View of Nasal Wall

Frontal sinus

Anterior ethmoidal artery

Posterior ethmoidal artery

Kiesselbach area

Sphenoidal sinus

Posterior septal branch of **sphenopalatine artery**

Nasal septal branch of superior labial branch

Greater palatine artery

Superior labial branch of facial artery

B. Lateral View of Nasal Septum

ARTERIES OF NASAL WALL AND SEPTUM   7.72

A. Lateral wall of nose. **B.** Nasal septum.

Epistaxis. On the anterior part of the nasal septum is an area rich in capillaries (Kiesselbach area) where all five arteries (sphenopalatine, anterior and posterior ethmoidal, greater palatine, and superior labial and lateral nasal branches of the facial artery) supplying the nasal septum anastomose. This area is often where profuse bleeding from the nose (epistaxis) occurs.

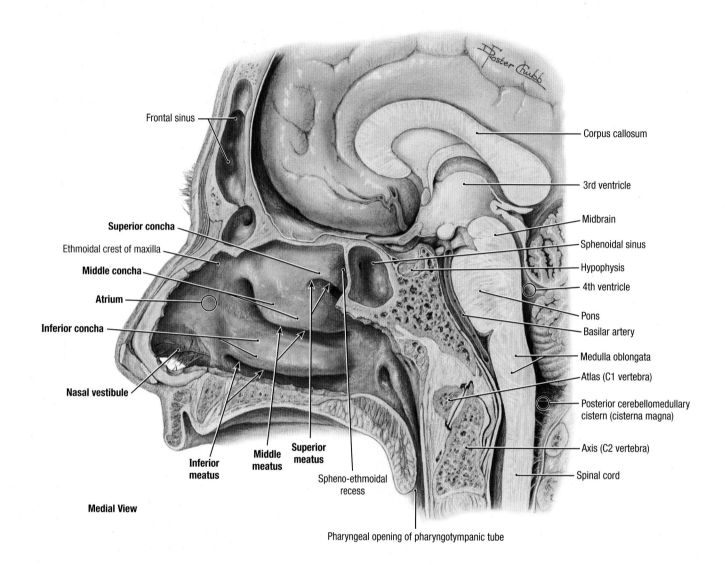

Medial View

Labels (clockwise from top left):
- Frontal sinus
- Corpus callosum
- 3rd ventricle
- Superior concha
- Ethmoidal crest of maxilla
- Midbrain
- Sphenoidal sinus
- Middle concha
- Hypophysis
- Atrium
- 4th ventricle
- Inferior concha
- Pons
- Basilar artery
- Nasal vestibule
- Medulla oblongata
- Atlas (C1 vertebra)
- Posterior cerebellomedullary cistern (cisterna magna)
- Axis (C2 vertebra)
- Spinal cord
- Inferior meatus
- Middle meatus
- Superior meatus
- Spheno-ethmoidal recess
- Pharyngeal opening of pharyngotympanic tube

7.73 RIGHT HALF OF HEMISECTED HEAD DEMONSTRATING UPPER RESPIRATORY TRACT

- The vestibule is superior to the nostril and anterior to the inferior meatus; hairs grow from its skin-lined surface. The atrium is superior to the vestibule and anterior to the middle meatus.
- The inferior and middle conchae curve inferiorly and medially from the lateral wall, dividing it into three nearly equal parts and covering the inferior and middle meatuses, respectively. The middle concha ends inferior to the sphenoidal sinus, and the inferior concha ends inferior to the middle concha, just anterior to the orifice of the auditory tube. The superior concha is small and anterior to the sphenoidal sinus.
- The roof comprises an anterior sloping part corresponding to the bridge of the nose; an intermediate horizontal part; a perpendicular part anterior to the sphenoidal sinus; and a curved part, inferior to the sinus, that is continuous with the roof of the nasopharynx.

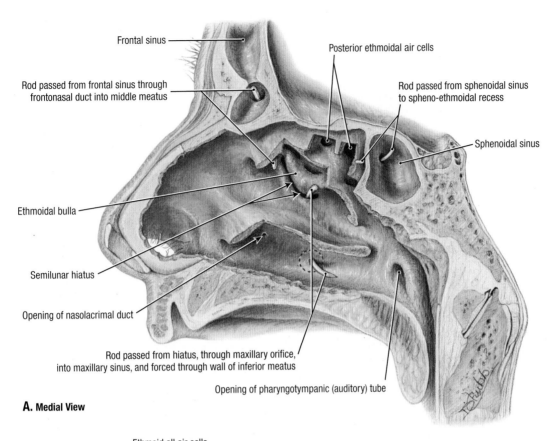

Frontal sinus

Posterior ethmoidal air cells

Rod passed from frontal sinus through frontonasal duct into middle meatus

Rod passed from sphenoidal sinus to spheno-ethmoidal recess

Sphenoidal sinus

Ethmoidal bulla

Semilunar hiatus

Opening of nasolacrimal duct

Rod passed from hiatus, through maxillary orifice, into maxillary sinus, and forced through wall of inferior meatus

Opening of pharyngotympanic (auditory) tube

A. Medial View

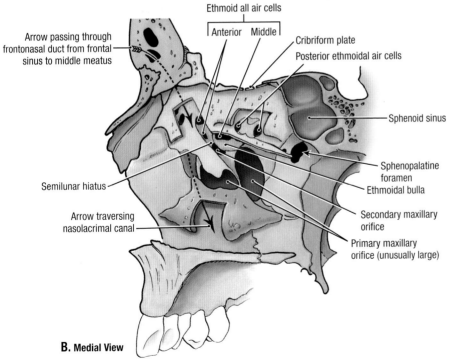

Ethmoid all air cells

Anterior   Middle

Cribriform plate

Posterior ethmoidal air cells

Arrow passing through frontonasal duct from frontal sinus to middle meatus

Sphenoid sinus

Sphenopalatine foramen

Ethmoidal bulla

Semilunar hiatus

Secondary maxillary orifice

Arrow traversing nasolacrimal canal

Primary maxillary orifice (unusually large)

B. Medial View

Bones in B		Soft Tissue
Ethmoid	Maxilla	Lateral wall of maxillary sinus
Frontal	Nasal	
Inferior concha	Palatine	
Lacrimal	Sphenoid	

COMMUNICATIONS THROUGH NASAL WALL

7.74

A. Dissection. Parts of the superior, middle, and inferior conchae are cut away to reveal the openings of the air sinuses. **B.** Diagrams of the bones and openings of the lateral wall of nasal cavity following dissection. Note one arrow passing from the frontal sinus through the frontonasal duct into the middle meatus and another arrow coming from the antero-medial orbit via the nasolacrimal canal.

Rhinitis. The nasal mucosa becomes swollen and inflamed (rhinitis) during upper respiratory infections and allergic reactions (e.g., hay fever). Swelling of this mucous membrane occurs readily because of its vascularity and abundant mucosal glands. Infections of the nasal cavities may spread to the anterior cranial fossa through the cribriform plate, nasopharynx and retropharyngeal soft tissues, middle ear through the pharyngotympanic (auditory) tube, paranasal sinuses, lacrimal apparatus, and conjunctiva.

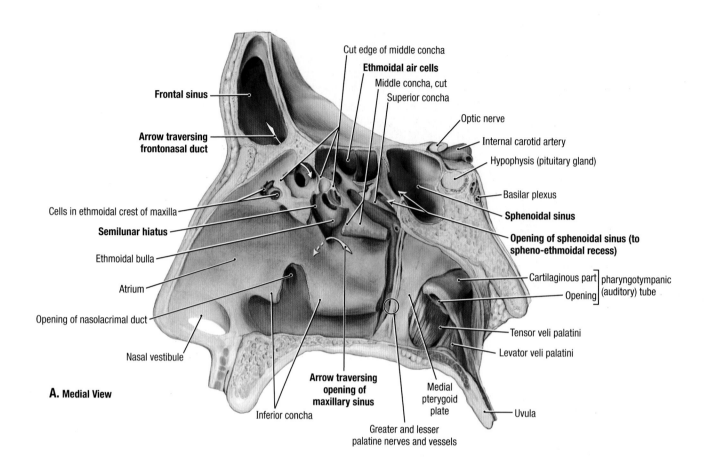

Cut edge of middle concha

Ethmoidal air cells

Middle concha, cut

Superior concha

Optic nerve

Internal carotid artery

Hypophysis (pituitary gland)

Basilar plexus

Sphenoidal sinus

Opening of sphenoidal sinus (to spheno-ethmoidal recess)

Cartilaginous part ⎤ pharyngotympanic

Opening ⎦ (auditory) tube

Tensor veli palatini

Levator veli palatini

Uvula

Medial pterygoid plate

Greater and lesser palatine nerves and vessels

Arrow traversing opening of maxillary sinus

Inferior concha

A. Medial View

Nasal vestibule

Opening of nasolacrimal duct

Atrium

Ethmoidal bulla

Semilunar hiatus

Cells in ethmoidal crest of maxilla

Frontal sinus

Arrow traversing frontonasal duct

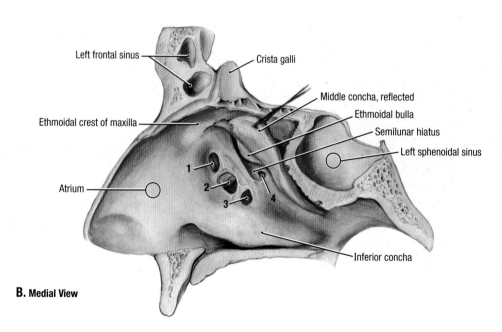

Left frontal sinus

Crista galli

Middle concha, reflected

Ethmoidal bulla

Semilunar hiatus

Left sphenoidal sinus

Ethmoidal crest of maxilla

Atrium

1

2

3

4

Inferior concha

B. Medial View

7.75 PARANASAL SINUSES, OPENINGS, AND PALATINE MUSCLES IN NASAL WALL

A. Dissection. Parts of the middle and inferior conchae and lateral wall of the nasal cavity are cut away to expose the nerves and vessels in the palatine canal and the extrinsic palatine muscles.

B. Accessory maxillary orifices. In addition to the primary, or normal, ostium (not shown), there are four secondary, or acquired, ostia (numbered *1* to *4*).

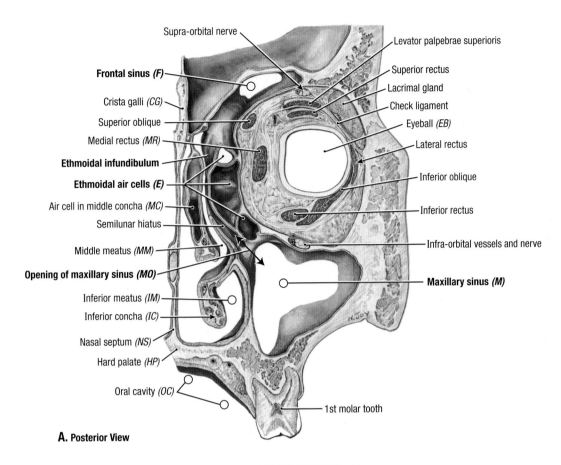

Supra-orbital nerve
Levator palpebrae superioris
Frontal sinus *(F)*
Superior rectus
Crista galli *(CG)*
Lacrimal gland
Superior oblique
Check ligament
Medial rectus *(MR)*
Eyeball *(EB)*
Ethmoidal infundibulum
Lateral rectus
Ethmoidal air cells *(E)*
Air cell in middle concha *(MC)*
Inferior oblique
Semilunar hiatus
Inferior rectus
Middle meatus *(MM)*
Opening of maxillary sinus *(MO)*
Infra-orbital vessels and nerve
Inferior meatus *(IM)*
Maxillary sinus *(M)*
Inferior concha *(IC)*
Nasal septum *(NS)*
Hard palate *(HP)*
Oral cavity *(OC)*
1st molar tooth

A. **Posterior View**

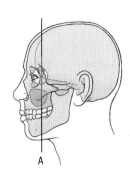

A

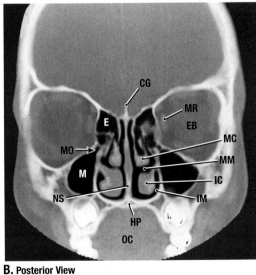

B. **Posterior View**

C. **Anteroposterior View**

PARANASAL SINUSES AND NASAL CAVITY

A. Coronal section of right side of the head. **B.** CT image. **C.** Radiograph of cranium. Letters in **B** and **C** refer to structures labeled in **A**.

If nasal drainage is blocked, **infections of the ethmoidal cells** of the ethmoidal sinuses may break through the fragile medial wall of the orbit. Severe infections from this source may cause blindness but could also affect the dural sheath of the optic nerve, causing **optic neuritis**.

During removal of a maxillary molar tooth, a **fracture of a tooth root** may occur. If proper retrieval methods are not used, a piece of the root may be driven superiorly into the maxillary sinus.

Radiographs/CT images of the frontal sinuses may be used for **forensic identification of unknown individuals**. The frontal sinuses are unique to each person, much like fingerprints.

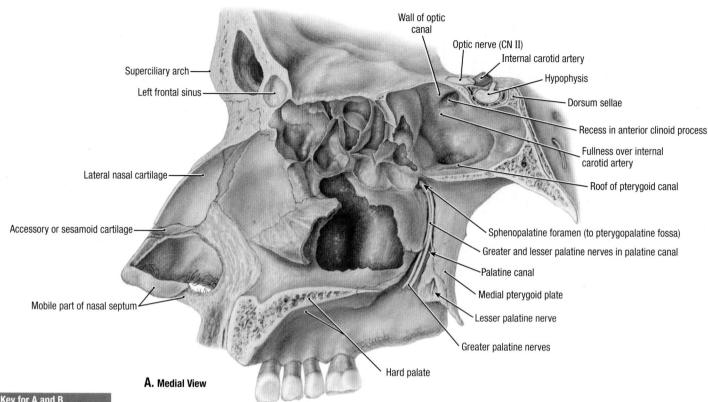

A. Medial View

Key for A and B

Sinuses:

- Ethmoidal air cells *(E)*
- Frontal sinus *(F)*
- Maxillary sinus *(M)*
- Sphenoidal sinus *(S)*

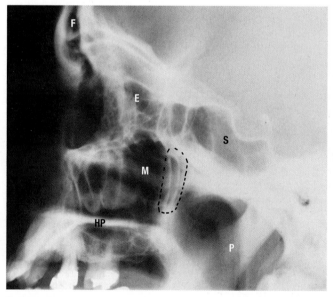

B. Lateral View

7.77 PARANASAL SINUSES

A. Opened sinuses. Sinuses are color-coded. **B.** Radiograph of cranium. *dotted lines,* pterygopalatine fossa; *HP,* hard palate; *P,* pharynx. The maxillary sinuses are the most commonly infected, as their ostia are small and located high on their superomedial walls, a poor location for natural drainage of the sinus. When the mucous membrane of the sinus is congested, the maxillary openings (ostia) often are obstructed. The **maxillary sinusitis** is treated with antibiotics; the sinus can also be cannulated and drained. For chronic maxillary sinusitis, **sinuplasty** or **maxillary antrostomy** are used to improve the drainage of the maxillary sinus by enlarging the opening of the ostia of one or more sinuses.

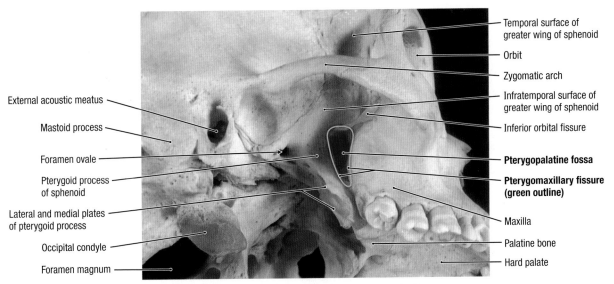

External acoustic meatus
Mastoid process
Foramen ovale
Pterygoid process of sphenoid
Lateral and medial plates of pterygoid process
Occipital condyle
Foramen magnum

Temporal surface of greater wing of sphenoid
Orbit
Zygomatic arch
Infratemporal surface of greater wing of sphenoid
Inferior orbital fissure
Pterygopalatine fossa
Pterygomaxillary fissure (green outline)
Maxilla
Palatine bone
Hard palate

A. Inferolateral and Slightly Posterior View, Looking into Infratemporal and Pterygopalatine Fossae

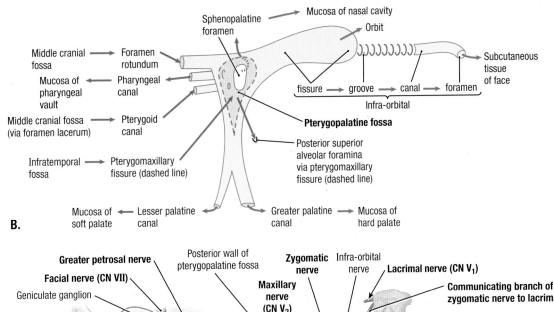

Sphenopalatine foramen
Mucosa of nasal cavity
Orbit

Middle cranial fossa → Foramen rotundum
Mucosa of pharyngeal vault ← Pharyngeal canal
Middle cranial fossa (via foramen lacerum) → Pterygoid canal
Infratemporal fossa → Pterygomaxillary fissure (dashed line)

fissure → groove → canal → foramen
Infra-orbital

Subcutaneous tissue of face

Pterygopalatine fossa

Posterior superior alveolar foramina via pterygomaxillary fissure (dashed line)

Mucosa of soft palate ← Lesser palatine canal
Greater palatine canal → Mucosa of hard palate

B.

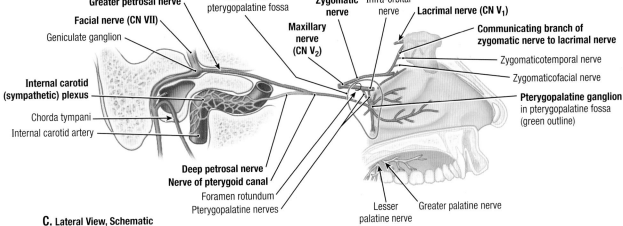

Greater petrosal nerve
Posterior wall of pterygopalatine fossa
Zygomatic nerve
Infra-orbital nerve
Lacrimal nerve (CN V₁)

Facial nerve (CN VII)
Geniculate ganglion
Maxillary nerve (CN V₂)
Communicating branch of zygomatic nerve to lacrimal nerve
Zygomaticotemporal nerve
Zygomaticofacial nerve

Internal carotid (sympathetic) plexus
Chorda tympani
Internal carotid artery
Pterygopalatine ganglion in pterygopalatine fossa (green outline)

Deep petrosal nerve
Nerve of pterygoid canal
Foramen rotundum
Pterygopalatine nerves
Lesser palatine nerve
Greater palatine nerve

C. Lateral View, Schematic

PTERYGOPALATINE FOSSA

7.78

A. Bony relationships. The pterygopalatine fossa is a small pyramidal space inferior to the apex of the orbit. It lies between the pterygoid process of the sphenoid and the posterior aspect of the maxilla anteriorly. **B.** Schematic illustration. (From Paff GH. *Anatomy of the Head and Neck*. Philadelphia, PA: W.B. Saunders Company; 1973.) **C.** Pterygopalatine ganglion and related nerves.

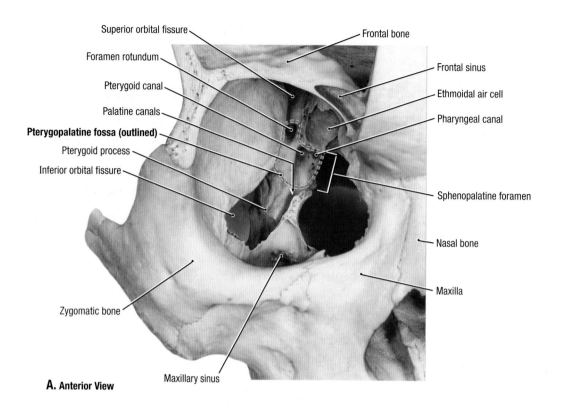

Superior orbital fissure
Foramen rotundum
Pterygoid canal
Palatine canals
Pterygopalatine fossa (outlined)
Pterygoid process
Inferior orbital fissure
Zygomatic bone
Maxillary sinus

Frontal bone
Frontal sinus
Ethmoidal air cell
Pharyngeal canal
Sphenopalatine foramen
Nasal bone
Maxilla

A. Anterior View

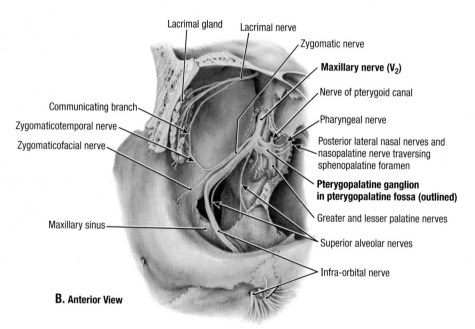

Lacrimal gland
Lacrimal nerve
Zygomatic nerve
Maxillary nerve (V₂)
Nerve of pterygoid canal
Communicating branch
Zygomaticotemporal nerve
Zygomaticofacial nerve
Pharyngeal nerve
Posterior lateral nasal nerves and nasopalatine nerve traversing sphenopalatine foramen
Pterygopalatine ganglion in pterygopalatine fossa (outlined)
Greater and lesser palatine nerves
Maxillary sinus
Superior alveolar nerves
Infra-orbital nerve

B. Anterior View

7.79 NERVES OF THE PTERYGOPALATINE FOSSA

A. Bones and foramina, orbital approach. **B.** Vessels and nerves, orbital approach. In **A** and **B**, the pterygopalatine fossa has been exposed through the maxillary sinus after removal of the floor of the orbit.

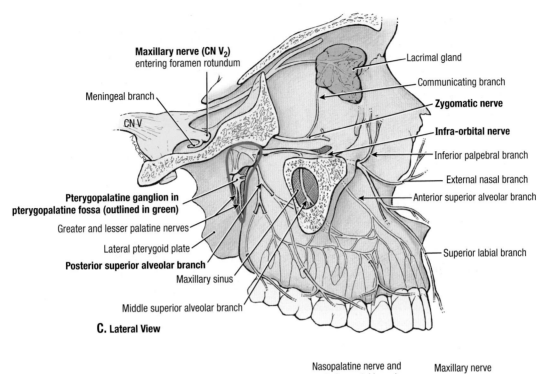

Maxillary nerve (CN V₂) entering foramen rotundum

Meningeal branch

CN V

Pterygopalatine ganglion in pterygopalatine fossa (outlined in green)

Greater and lesser palatine nerves

Lateral pterygoid plate

Posterior superior alveolar branch

Maxillary sinus

Middle superior alveolar branch

Lacrimal gland

Communicating branch

Zygomatic nerve

Infra-orbital nerve

Inferior palpebral branch

External nasal branch

Anterior superior alveolar branch

Superior labial branch

C. Lateral View

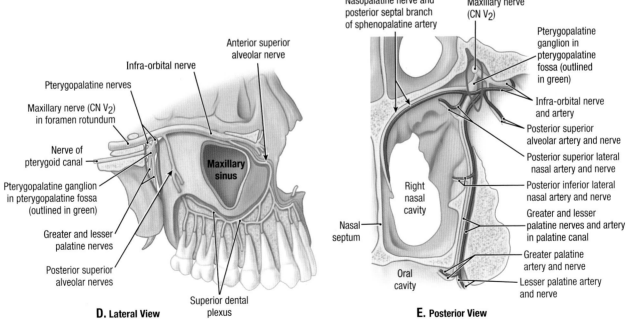

Pterygopalatine nerves

Maxillary nerve (CN V₂) in foramen rotundum

Nerve of pterygoid canal

Pterygopalatine ganglion in pterygopalatine fossa (outlined in green)

Greater and lesser palatine nerves

Posterior superior alveolar nerves

Infra-orbital nerve

Anterior superior alveolar nerve

Maxillary sinus

Superior dental plexus

D. Lateral View

Nasopalatine nerve and posterior septal branch of sphenopalatine artery

Maxillary nerve (CN V₂)

Pterygopalatine ganglion in pterygopalatine fossa (outlined in green)

Infra-orbital nerve and artery

Posterior superior alveolar artery and nerve

Posterior superior lateral nasal artery and nerve

Right nasal cavity

Nasal septum

Oral cavity

Posterior inferior lateral nasal artery and nerve

Greater and lesser palatine nerves and artery in palatine canal

Greater palatine artery and nerve

Lesser palatine artery and nerve

E. Posterior View

NERVES OF THE PTERYGOPALATINE FOSSA (*continued*)

7.79

C. Maxillary nerve (CN V₂) and branches. **D.** The fossa is viewed laterally. Part of the wall of the maxillary sinus has been removed. **E.** Nasopalatine and greater and lesser palatine nerves.

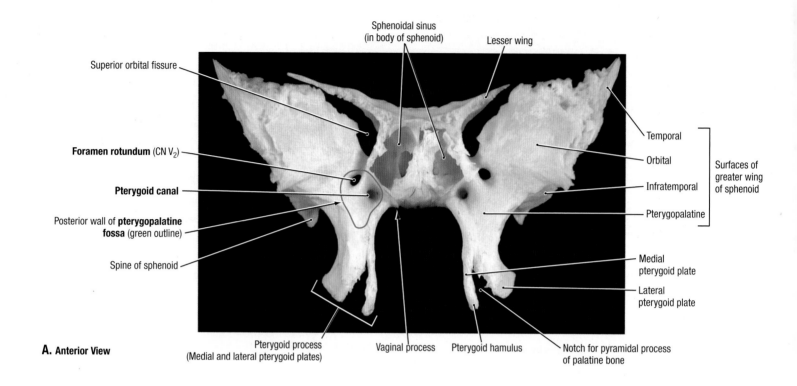

Superior orbital fissure

Sphenoidal sinus (in body of sphenoid)

Lesser wing

Foramen rotundum (CN V₂)

Pterygoid canal

Posterior wall of pterygopalatine fossa (green outline)

Spine of sphenoid

Temporal

Orbital

Infratemporal

Pterygopalatine

Surfaces of greater wing of sphenoid

Medial pterygoid plate

Lateral pterygoid plate

Pterygoid process (Medial and lateral pterygoid plates)

Vaginal process

Pterygoid hamulus

Notch for pyramidal process of palatine bone

A. Anterior View

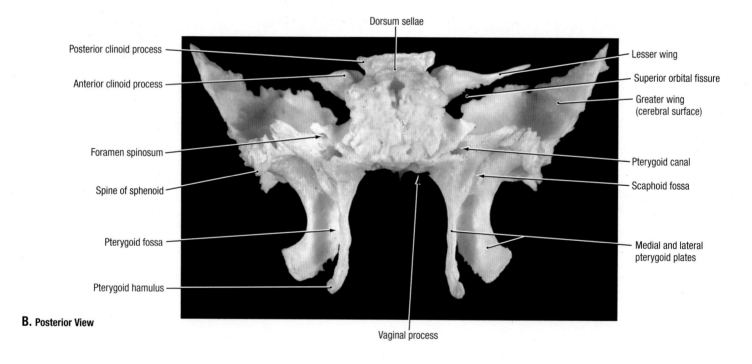

Dorsum sellae

Posterior clinoid process

Anterior clinoid process

Foramen spinosum

Spine of sphenoid

Pterygoid fossa

Pterygoid hamulus

Lesser wing

Superior orbital fissure

Greater wing (cerebral surface)

Pterygoid canal

Scaphoid fossa

Medial and lateral pterygoid plates

Vaginal process

B. Posterior View

7.80 SPHENOID BONE: FEATURES AND RELATIONSHIP TO PTERYGOPALATINE FOSSA

A. The pterygopalatine fossa communicates posterosuperiorly with the middle cranial fossa through the foramen rotundum and pterygoid canal. **B.** Bony features.

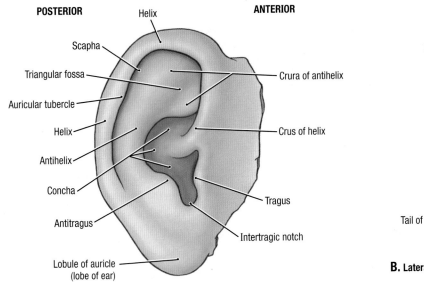

POSTERIOR — Helix — **ANTERIOR**

Scapha

Triangular fossa

Auricular tubercle

Helix

Antihelix

Concha

Antitragus

Lobule of auricle
(lobe of ear)

Crura of antihelix

Crus of helix

Tragus

Intertragic notch

A. Lateral View

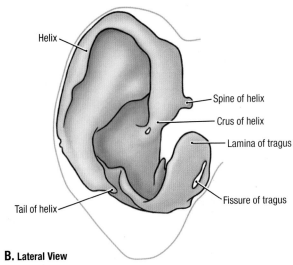

Helix

Spine of helix

Crus of helix

Lamina of tragus

Tail of helix

Fissure of tragus

B. Lateral View

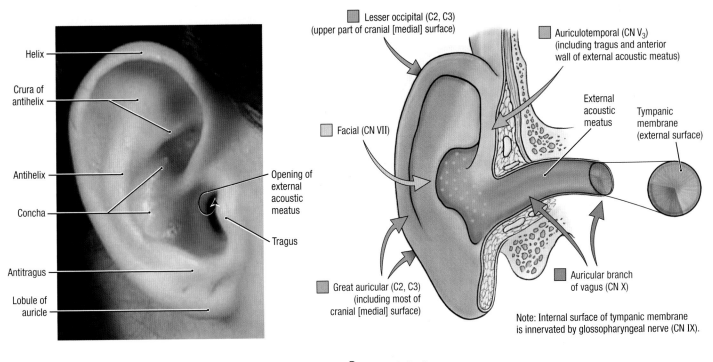

Helix

Crura of
antihelix

Antihelix

Concha

Antitragus

Lobule of
auricle

Opening of
external
acoustic
meatus

Tragus

C. Lateral View

Lesser occipital (C2, C3)
(upper part of cranial [medial] surface)

Auriculotemporal (CN V₃)
(including tragus and anterior
wall of external acoustic meatus)

Facial (CN VII)

External
acoustic
meatus

Tympanic
membrane
(external surface)

Great auricular (C2, C3)
(including most of
cranial [medial] surface)

Auricular branch
of vagus (CN X)

Note: Internal surface of tympanic membrane
is innervated by glossopharyngeal nerve (CN IX).

D. Schematic Section

AURICLE

7.81

A. Features of auricle. **B.** Cartilage of auricle. **C.** Surface anatomy of auricle. **D.** Sensory innervation.

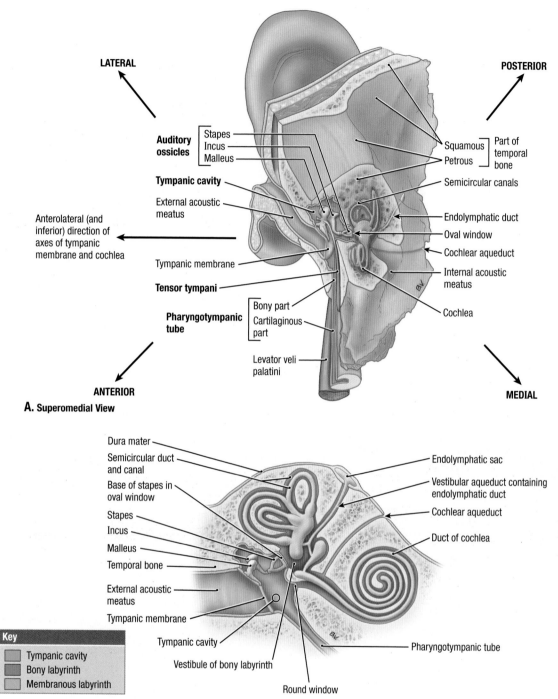

LATERAL

POSTERIOR

Auditory ossicles — Stapes / Incus / Malleus

Tympanic cavity

External acoustic meatus

Anterolateral (and inferior) direction of axes of tympanic membrane and cochlea

Tympanic membrane

Tensor tympani

Pharyngotympanic tube — Bony part / Cartilaginous part

Levator veli palatini

Squamous / Petrous — Part of temporal bone

Semicircular canals

Endolymphatic duct

Oval window

Cochlear aqueduct

Internal acoustic meatus

Cochlea

ANTERIOR

MEDIAL

A. Superomedial View

Dura mater

Semicircular duct and canal

Base of stapes in oval window

Stapes

Incus

Malleus

Temporal bone

External acoustic meatus

Tympanic membrane

Tympanic cavity

Vestibule of bony labyrinth

Round window

Endolymphatic sac

Vestibular aqueduct containing endolymphatic duct

Cochlear aqueduct

Duct of cochlea

Pharyngotympanic tube

Key
- Tympanic cavity
- Bony labyrinth
- Membranous labyrinth

B. Oblique Section of Petrous Temporal Bone

7.82   EXTERNAL, MIDDLE, AND INTERNAL EAR I: OVERVIEWS

A. Right temporal bone and auricle, sectioned in planes of (1) externa acoustic meatus and (2) pharyngotympanic tube.
B. Schematic section of petrous temporal bone.
- The external ear comprises the auricle and external acoustic (auditory) meatus.
- The middle ear (tympanum) lies between the tympanic membrane and internal ear. Three ossicles extend from the lateral to the medial walls of the tympanum. Of these, the malleus is attached to the tympanic membrane. The stapes is attached by the annular ligament to the oval window, and the incus connects to the malleus and stapes. The pharyngotympanic tube, extending from the nasopharynx, opens into the anterior wall of the tympanic cavity.
- The membranous labyrinth comprises a closed system of membranous tubes and bulbs filled with fluid, endolymph and bathed in surrounding fluid, called perilymph; both membranous labyrinth and perilymph are contained within the bony labyrinth.

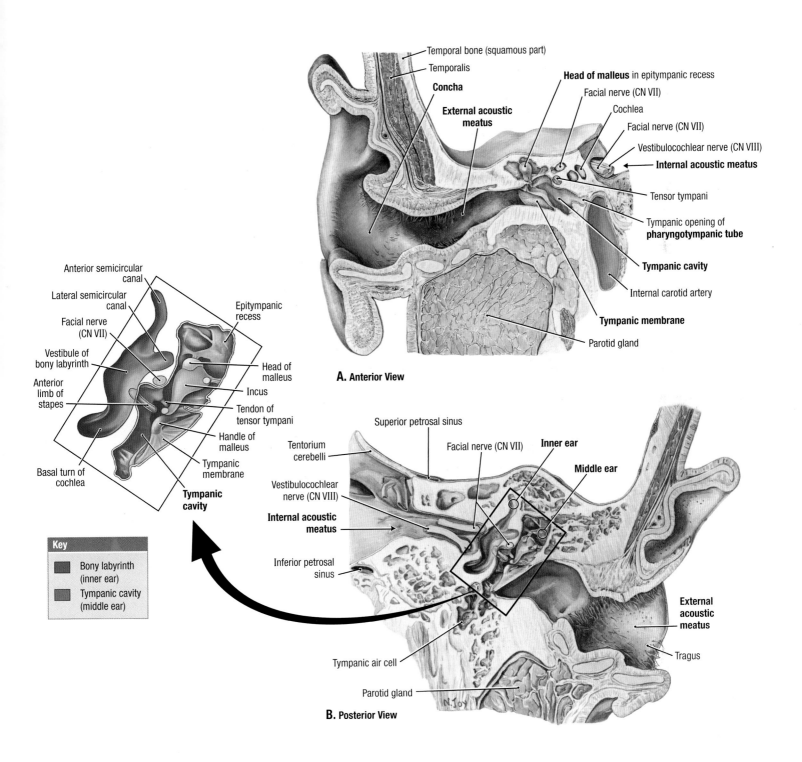

Temporal bone (squamous part)

Temporalis

Concha

External acoustic meatus

Head of malleus in epitympanic recess

Facial nerve (CN VII)

Cochlea

Facial nerve (CN VII)

Vestibulocochlear nerve (CN VIII)

Internal acoustic meatus

Tensor tympani

Tympanic opening of **pharyngotympanic tube**

Tympanic cavity

Internal carotid artery

Tympanic membrane

Parotid gland

A. Anterior View

Anterior semicircular canal

Lateral semicircular canal

Facial nerve (CN VII)

Vestibule of bony labyrinth

Anterior limb of stapes

Epitympanic recess

Head of malleus

Incus

Tendon of tensor tympani

Handle of malleus

Tympanic membrane

Basal turn of cochlea

Tympanic cavity

Superior petrosal sinus

Facial nerve (CN VII)

Inner ear

Middle ear

Tentorium cerebelli

Vestibulocochlear nerve (CN VIII)

Internal acoustic meatus

Inferior petrosal sinus

External acoustic meatus

Tympanic air cell

Tragus

Parotid gland

N. Joy

B. Posterior View

Key

Bony labyrinth (inner ear)

Tympanic cavity (middle ear)

EXTERNAL, MIDDLE, AND INTERNAL EAR II: CORONALLY SECTIONED

7.83

A. Anterior portion. **B.** Posterior portion. The inset *(outlined by the box)* is an enlargement of the structures of the middle and internal ear as they appear in **B**.

- The external acoustic meatus is about 3 cm long; half is cartilaginous and half is bony. It is narrowest at the isthmus, near the junction of the cartilaginous and bony parts.
- The external acoustic meatus is innervated by the auriculotemporal branch of the mandibular nerve (CN V₃) and the auricular

branches of the vagus nerve (CN X); the middle ear is innervated by the glossopharyngeal nerve (CN IX).

- The cartilaginous part of the external acoustic meatus is lined with thick skin; the bony part is lined with thin skin that adheres to the periosteum and forms the outermost layer of the tympanic membrane.

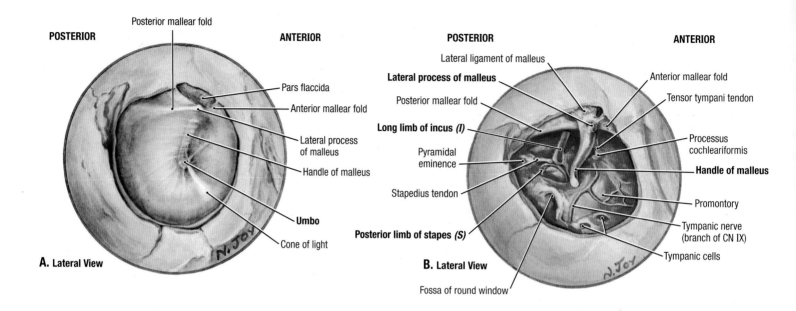

POSTERIOR Posterior mallear fold **ANTERIOR**

Pars flaccida
Anterior mallear fold
Lateral process of malleus
Handle of malleus
Umbo
Cone of light

A. Lateral View

POSTERIOR **ANTERIOR**

Lateral ligament of malleus
Lateral process of malleus
Anterior mallear fold
Posterior mallear fold
Tensor tympani tendon
Long limb of incus (I)
Processus cochleariformis
Pyramidal eminence
Handle of malleus
Stapedius tendon
Promontory
Posterior limb of stapes (S)
Tympanic nerve (branch of CN IX)
Tympanic cells
B. Lateral View
Fossa of round window

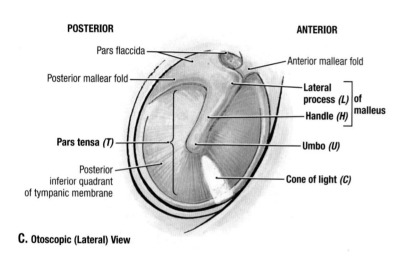

POSTERIOR **ANTERIOR**

Pars flaccida
Anterior mallear fold
Posterior mallear fold
Lateral process (L) of malleus
Handle (H)
Pars tensa (T)
Umbo (U)
Posterior inferior quadrant of tympanic membrane
Cone of light (C)

C. Otoscopic (Lateral) View

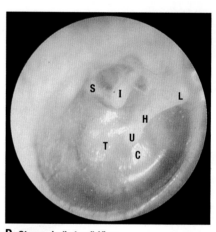

D. Otoscopic (Lateral) View

| 7.84 | TYMPANIC MEMBRANE |

A. External (lateral) surface of tympanic membrane. **B.** Tympanic membrane removed, demonstrating structures that lie medially. **C.** Diagram of otoscopic view of tympanic membrane. **D.** Otoscopic view of tympanic membrane. Letter labels are identified in **B** and **C**.

- The oval tympanic membrane is a shallow cone deepest at the central apex, the umbo, where the membrane is attached to the tip of the handle of the malleus. The handle of the malleus is attached to the membrane along its entire length as it extends anterosuperiorly toward the periphery of the membrane.
- Superior to the lateral process of the malleus, the membrane is thin (pars flaccida); the flaccid part lacks the radial and circular fibers present in the remainder of the membrane (pars tensa).

The junction between the two parts is marked by anterior and posterior mallear folds.
- The lateral surface of the tympanic membrane is innervated by the auricular branch of the auriculotemporal nerve (CN V₃) and the auricular branch of the vagus nerve (CN X); the medial surface is innervated by tympanic branches of CN IX.

Examination of the external acoustic meatus and tympanic membrane begins by straightening the meatus. In adults, the helix is grasped and pulled posterosuperiorly (up, out, and back). These movements reduce the curvature of the external acoustic meatus, facilitating insertion of the otoscope. The external acoustic meatus is relatively short in infants; therefore, extra care must be taken to prevent damage to the tympanic membrane.

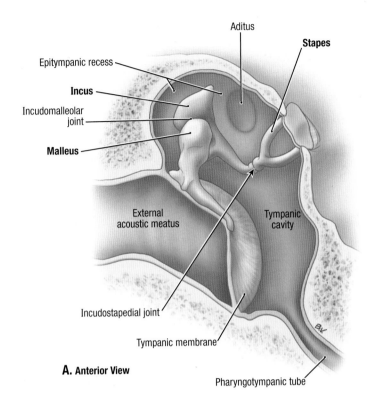

Aditus

Stapes

Epitympanic recess

Incus

Incudomalleolar joint

Malleus

External acoustic meatus

Tympanic cavity

Incudostapedial joint

Tympanic membrane

A. Anterior View

Pharyngotympanic tube

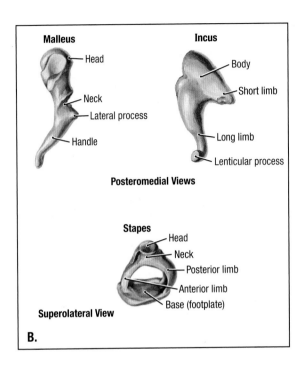

Malleus — Head, Neck, Lateral process, Handle

Incus — Body, Short limb, Long limb, Lenticular process

Posteromedial Views

Stapes — Head, Neck, Posterior limb, Anterior limb, Base (footplate)

Superolateral View

B.

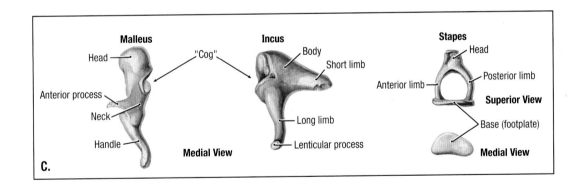

Malleus — Head, "Cog", Anterior process, Neck, Handle

Medial View

Incus — Body, Short limb, Long limb, Lenticular process

Stapes — Head, Anterior limb, Posterior limb

Superior View

Base (footplate)

Medial View

C.

OSSICLES OF THE MIDDLE EAR

A. Ossicles *in situ*, as revealed by a coronal section of the temporal bone. **B.** and **C.** Isolated ossicles.

- The head of the malleus and body and short process of the incus lie in the epitympanic recess, and the handle of the malleus is embedded in the tympanic membrane.
- The saddle-shaped articular surface of the head of the malleus and the reciprocally shaped articular surface of the body of the incus form the incudomalleolar synovial joint.
- A convex articular facet at the end of the long process of the incus articulates with the head of the stapes to compose the incudostapedial synovial joint.

- An earache and bulging red tympanic membrane may indicate pus or fluid in the middle ear, a sign of **otitis media**. Infection of the middle ear often is secondary to upper respiratory infections. Inflammation and swelling of the mucous membrane lining the tympanic cavity may cause partial or complete blockage of the pharyngotympanic tube. The tympanic membrane becomes red and bulges and the person may complain of "ear popping." If untreated, otitis media may produce impaired hearing as the result of scarring of the auditory ossicles, limiting the ability of these bones to move in response to sound.

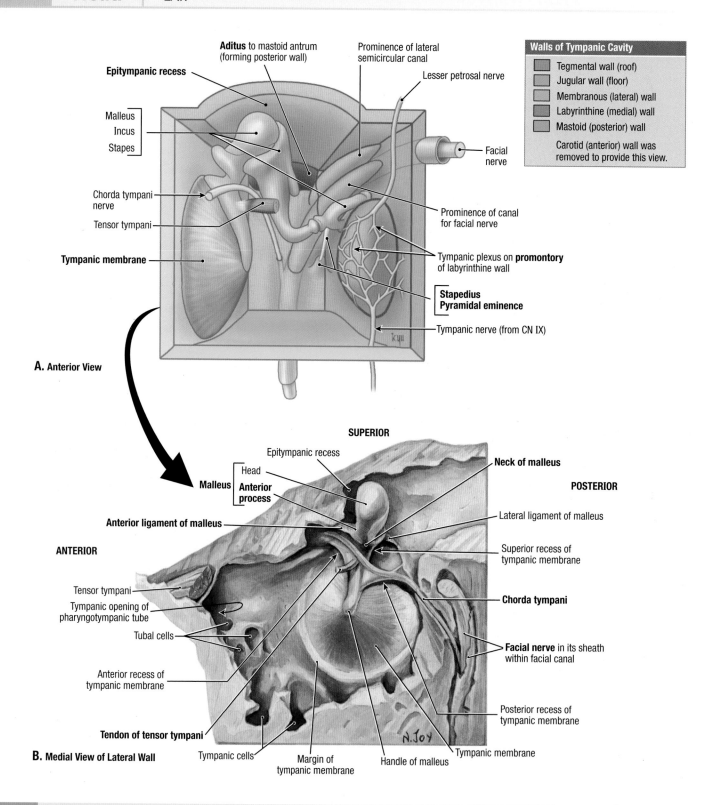

A. Anterior View

Walls of Tympanic Cavity

- Tegmental wall (roof)
- Jugular wall (floor)
- Membranous (lateral) wall
- Labyrinthine (medial) wall
- Mastoid (posterior) wall

Carotid (anterior) wall was removed to provide this view.

A. Anterior View labels: Aditus to mastoid antrum (forming posterior wall); Epitympanic recess; Prominence of lateral semicircular canal; Lesser petrosal nerve; Malleus; Incus; Stapes; Facial nerve; Chorda tympani nerve; Tensor tympani; Tympanic membrane; Prominence of canal for facial nerve; Tympanic plexus on **promontory** of labyrinthine wall; **Stapedius** **Pyramidal eminence**; Tympanic nerve (from CN IX)

B. Medial View of Lateral Wall labels: SUPERIOR; Epitympanic recess; **Neck of malleus**; POSTERIOR; Malleus — Head, **Anterior process**; **Lateral ligament of malleus**; **Anterior ligament of malleus**; Superior recess of tympanic membrane; ANTERIOR; **Chorda tympani**; Tensor tympani; Tympanic opening of pharyngotympanic tube; Tubal cells; Anterior recess of tympanic membrane; **Facial nerve** in its sheath within facial canal; **Tendon of tensor tympani**; Tympanic cells; Margin of tympanic membrane; Handle of malleus; Tympanic membrane; Posterior recess of tympanic membrane

7.86   STRUCTURES OF THE TYMPANIC CAVITY

A. Schematic illustration of the tympanic cavity with the anterior wall removed. **B.** Lateral wall of the tympanic cavity. The facial nerve lies within the facial canal surrounded by a tough periosteal tube; the chorda tympani leaves the facial nerve and lies within two crescentic folds of mucous membrane, crossing the neck of the malleus superior to the tendon of tensor tympani.

Perforation of the tympanic membrane (ruptured eardrum) may result from otitis media. Perforation may also result from foreign bodies in the external acoustic meatus, trauma, or excessive pressure. Because the superior half of the tympanic membrane is much more vascular than the inferior half, incisions are made postero-inferiorly through the membrane. This incision also avoids injury to the chorda tympani nerve and auditory ossicles.

ANTERIOR

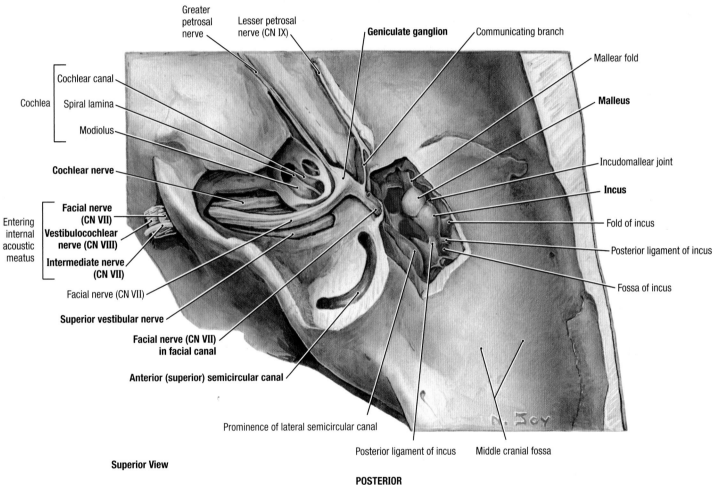

Greater petrosal nerve

Lesser petrosal nerve (CN IX)

Geniculate ganglion

Communicating branch

Mallear fold

Cochlear canal

Cochlea

Spiral lamina

Modiolus

Cochlear nerve

Malleus

Incudomallear joint

Incus

Fold of incus

Entering internal acoustic meatus

Facial nerve (CN VII)

Vestibulocochlear nerve (CN VIII)

Intermediate nerve (CN VII)

Facial nerve (CN VII)

Posterior ligament of incus

Fossa of incus

Superior vestibular nerve

Facial nerve (CN VII) in facial canal

Anterior (superior) semicircular canal

Prominence of lateral semicircular canal

Superior View

Posterior ligament of incus

Middle cranial fossa

POSTERIOR

MIDDLE AND INNER EAR *IN SITU*

7.87

The tegmen tympani has been removed to expose the middle ear. In addition, the arcuate eminence has been removed to reveal the anterior semicircular canal, and the course of the facial and vestibulocochlear nerves through the internal acoustic meatus and internal ear.

At the geniculate ganglion, the facial nerve executes a sharp bend, called the genu, and then curves postero-inferiorly within the bony facial canal; the thin lateral wall of the facial canal separates the facial nerve from the tympanic cavity of the middle ear.

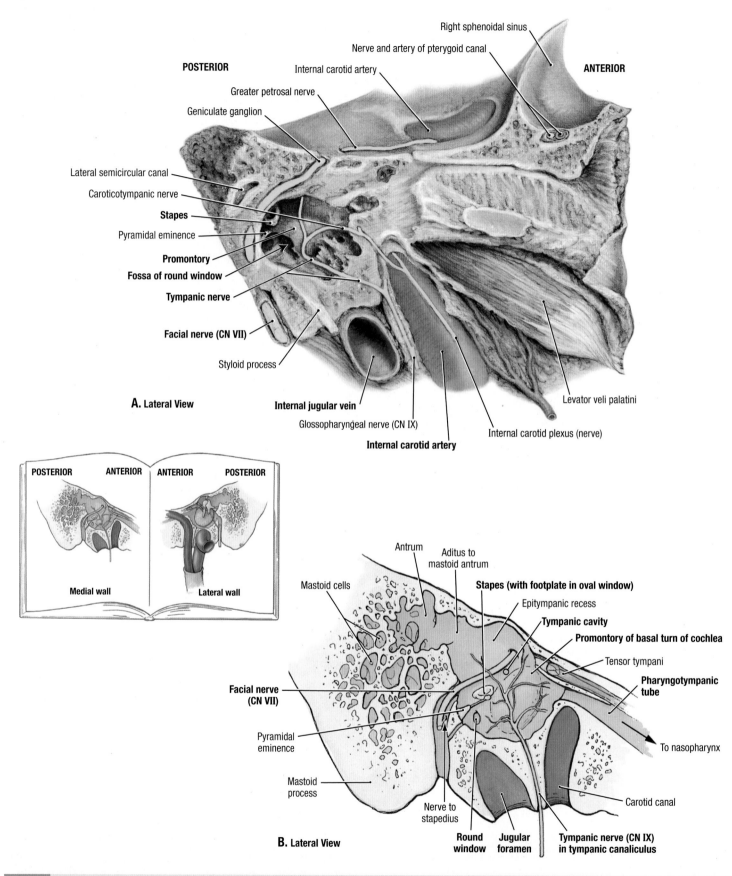

Right sphenoidal sinus

Nerve and artery of pterygoid canal

Internal carotid artery

POSTERIOR

ANTERIOR

Greater petrosal nerve

Geniculate ganglion

Lateral semicircular canal

Caroticotympanic nerve

Stapes

Pyramidal eminence

Promontory

Fossa of round window

Tympanic nerve

Facial nerve (CN VII)

Styloid process

A. Lateral View

Levator veli palatini

Internal jugular vein

Glossopharyngeal nerve (CN IX)

Internal carotid artery

Internal carotid plexus (nerve)

POSTERIOR   ANTERIOR   ANTERIOR   POSTERIOR

Medial wall

Lateral wall

Antrum

Aditus to mastoid antrum

Mastoid cells

Stapes (with footplate in oval window)

Epitympanic recess

Tympanic cavity

Promontory of basal turn of cochlea

Tensor tympani

Facial nerve (CN VII)

Pharyngotympanic tube

Pyramidal eminence

To nasopharynx

Mastoid process

Carotid canal

Nerve to stapedius

Round window   **Jugular foramen**   **Tympanic nerve (CN IX) in tympanic canaliculus**

B. Lateral View

7.88   RIGHT TYMPANIC CAVITY AND PHARYNGOTYMPANIC TUBE

The cut surfaces of this longitudinally sectioned specimen are displayed as pages in a book. **A.** Dissection of medial wall. **B.** Schematic illustration of medial wall.

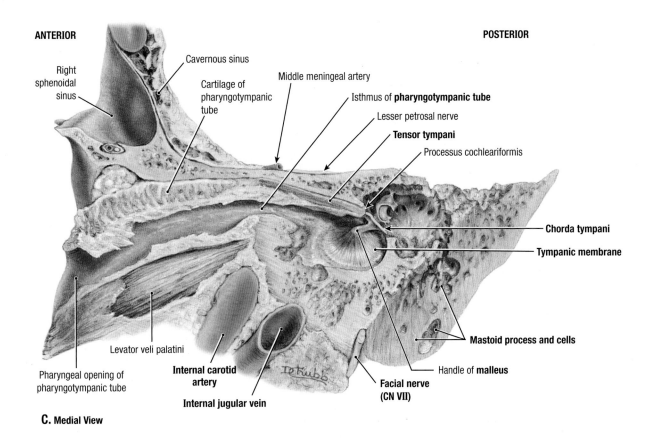

ANTERIOR

POSTERIOR

Right sphenoidal sinus

Cavernous sinus

Cartilage of pharyngotympanic tube

Middle meningeal artery

Isthmus of **pharyngotympanic tube**

Lesser petrosal nerve

Tensor tympani

Processus cochleariformis

Chorda tympani

Tympanic membrane

Mastoid process and cells

Handle of **malleus**

Facial nerve (CN VII)

Levator veli palatini

Internal carotid artery

Internal jugular vein

Pharyngeal opening of pharyngotympanic tube

C. Medial View

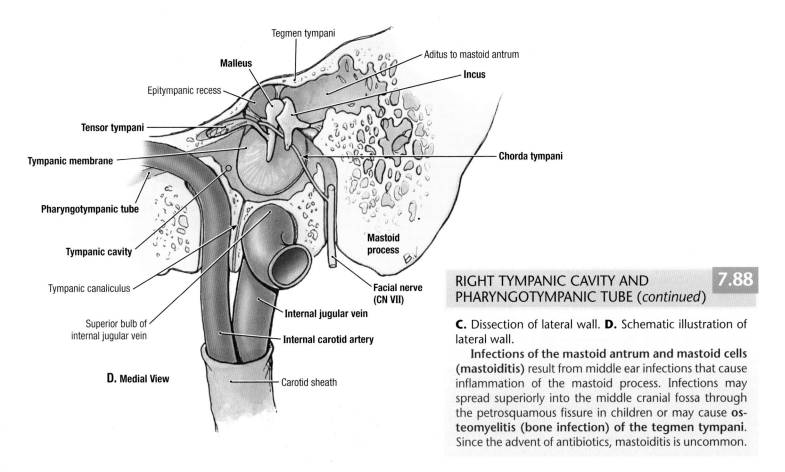

Tegmen tympani

Aditus to mastoid antrum

Malleus

Incus

Epitympanic recess

Tensor tympani

Tympanic membrane

Chorda tympani

Pharyngotympanic tube

Tympanic cavity

Mastoid process

Tympanic canaliculus

Facial nerve (CN VII)

Internal jugular vein

Superior bulb of internal jugular vein

Internal carotid artery

Carotid sheath

D. Medial View

RIGHT TYMPANIC CAVITY AND PHARYNGOTYMPANIC TUBE (*continued*)

7.88

C. Dissection of lateral wall. **D.** Schematic illustration of lateral wall.

Infections of the mastoid antrum and mastoid cells (mastoiditis) result from middle ear infections that cause inflammation of the mastoid process. Infections may spread superiorly into the middle cranial fossa through the petrosquamous fissure in children or may cause **osteomyelitis (bone infection) of the tegmen tympani**. Since the advent of antibiotics, mastoiditis is uncommon.

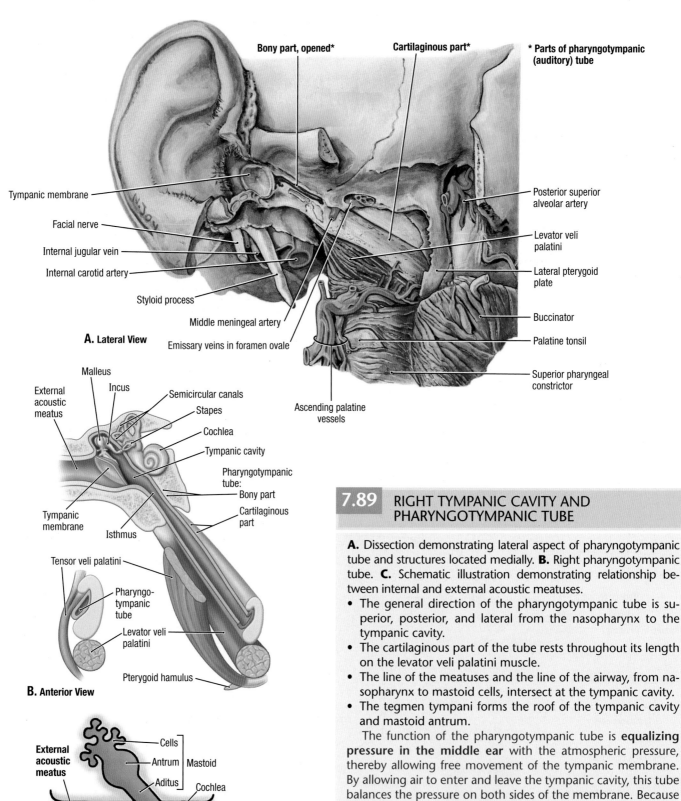

Bony part, opened*

Cartilaginous part*

* Parts of pharyngotympanic (auditory) tube

Tympanic membrane

Facial nerve

Internal jugular vein

Internal carotid artery

Styloid process

Middle meningeal artery

Emissary veins in foramen ovale

A. Lateral View

Posterior superior alveolar artery

Levator veli palatini

Lateral pterygoid plate

Buccinator

Palatine tonsil

Superior pharyngeal constrictor

Ascending palatine vessels

Malleus

Incus

External acoustic meatus

Semicircular canals

Stapes

Cochlea

Tympanic cavity

Pharyngotympanic tube:

Bony part

Cartilaginous part

Tympanic membrane

Isthmus

Tensor veli palatini

Pharyngotympanic tube

Levator veli palatini

Pterygoid hamulus

B. Anterior View

External acoustic meatus

Cells

Antrum | Mastoid

Aditus

Cochlea

Cranial cavity

Internal acoustic meatus

Membrane

Tympanic

Cavity

Pharygotympanic tube

Nasopharynx

C. Schematic Superior View

7.89   RIGHT TYMPANIC CAVITY AND PHARYNGOTYMPANIC TUBE

A. Dissection demonstrating lateral aspect of pharyngotympanic tube and structures located medially. **B.** Right pharyngotympanic tube. **C.** Schematic illustration demonstrating relationship between internal and external acoustic meatuses.

- The general direction of the pharyngotympanic tube is superior, posterior, and lateral from the nasopharynx to the tympanic cavity.
- The cartilaginous part of the tube rests throughout its length on the levator veli palatini muscle.
- The line of the meatuses and the line of the airway, from nasopharynx to mastoid cells, intersect at the tympanic cavity.
- The tegmen tympani forms the roof of the tympanic cavity and mastoid antrum.

The function of the pharyngotympanic tube is **equalizing pressure in the middle ear** with the atmospheric pressure, thereby allowing free movement of the tympanic membrane. By allowing air to enter and leave the tympanic cavity, this tube balances the pressure on both sides of the membrane. Because the walls of the cartilaginous part of the tube are normally in apposition, the tube must be actively opened. The tube is opened by the expanding girth of the belly of the levator veli palatini as it contracts longitudinally, pushing against one wall while the tensor veli palatini pulls on the other. Because these are muscles of the soft palate, equalizing pressure (popping the eardrums) is commonly associated with activities such as yawning and swallowing.

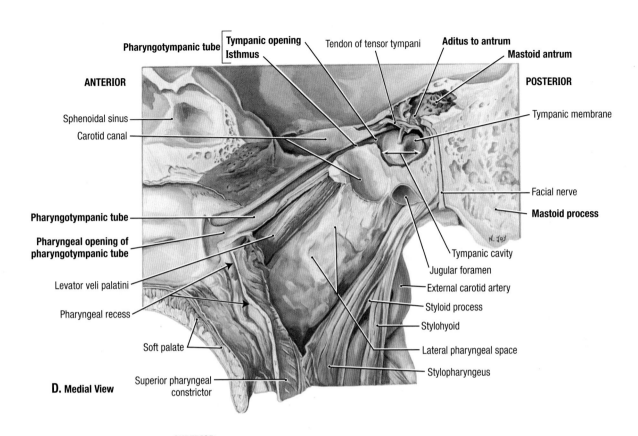

Pharyngotympanic tube **Tympanic opening** Tendon of tensor tympani **Aditus to antrum**
Isthmus **Mastoid antrum**

ANTERIOR POSTERIOR

Sphenoidal sinus
Carotid canal

Tympanic membrane

Facial nerve

Pharyngotympanic tube **Mastoid process**

Pharyngeal opening of
pharyngotympanic tube

Tympanic cavity

Jugular foramen

Levator veli palatini
External carotid artery

Styloid process

Pharyngeal recess
Stylohyoid

Lateral pharyngeal space

Soft palate
Stylopharyngeus

Superior pharyngeal
constrictor

D. Medial View

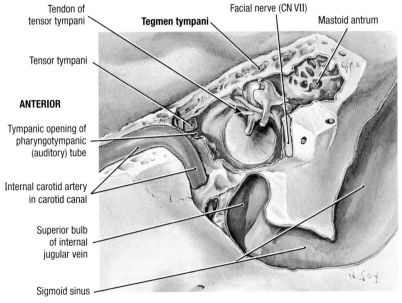

SUPERIOR

Tendon of Facial nerve (CN VII)
tensor tympani **Tegmen tympani** Mastoid antrum

Tensor tympani

ANTERIOR

Tympanic opening of
pharyngotympanic
(auditory) tube

Internal carotid artery
in carotid canal

Superior bulb
of internal
jugular vein

Sigmoid sinus

E. Medial View

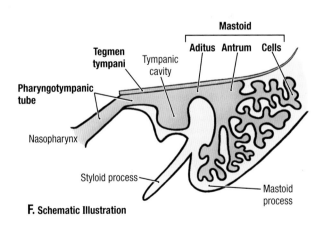

Mastoid

Tegmen Tympanic **Aditus Antrum Cells**
tympani cavity

Pharyngotympanic
tube

Nasopharynx

Styloid process

Mastoid
process

F. Schematic Illustration

RIGHT TYMPANIC CAVITY AND PHARYNGOTYMPANIC TUBE (*continued*)

7.89

D. Spaces of tympanic bone. **E.** Relationship of tympanic cavity to internal carotid artery, sigmoid sinus, and middle cranial fossa. **F.** Diagram of tegmen tympani.

• The internal carotid artery is the primary relationship of the anterior wall, the internal jugular vein is the primary relationship of the floor, and the facial nerve is the primary relationship of the posterior wall.

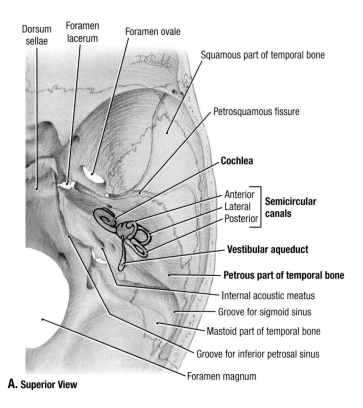

A. Superior View

Dorsum sellae
Foramen lacerum
Foramen ovale
Squamous part of temporal bone
Petrosquamous fissure
Cochlea
Anterior
Lateral
Posterior } **Semicircular canals**
Vestibular aqueduct
Petrous part of temporal bone
Internal acoustic meatus
Groove for sigmoid sinus
Mastoid part of temporal bone
Groove for inferior petrosal sinus
Foramen magnum

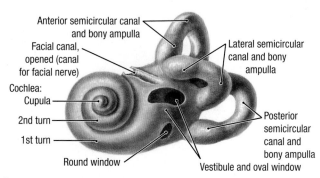

C. Anterolateral View of Left Otic Capsule

Anterior semicircular canal and bony ampulla
Lateral semicircular canal and bony ampulla
Facial canal, opened (canal for facial nerve)
Cochlea:
Cupula
2nd turn
1st turn
Round window
Posterior semicircular canal and bony ampulla
Vestibule and oval window

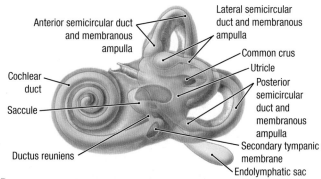

D. Anterolateral View of Left Membranous Labyrinth (through Transparent Otic Capsule)

Anterior semicircular duct and membranous ampulla
Lateral semicircular duct and membranous ampulla
Cochlear duct
Common crus
Utricle
Saccule
Posterior semicircular duct and membranous ampulla
Ductus reuniens
Secondary tympanic membrane
Endolymphatic sac

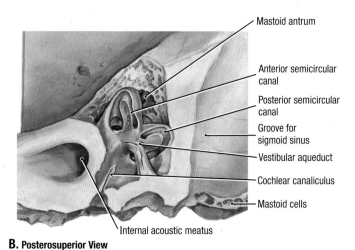

B. Posterosuperior View

Mastoid antrum
Anterior semicircular canal
Posterior semicircular canal
Groove for sigmoid sinus
Vestibular aqueduct
Cochlear canaliculus
Mastoid cells
Internal acoustic meatus

E. Anterolateral View of Left Membranous Labyrinth

Anterior semicircular duct and membranous ampulla
Maculae
Cochlear duct
Saccule
Utriculo-saccular duct
Ductus reuniens
Endolymphatic duct
Lateral semicircular duct
Posterior semicircular duct and ampullary crest
Endolymphatic sac

| 7.90 | BONY AND MEMBRANOUS LABYRINTHS |

A. Location and orientation of bony labyrinth within petrous temporal bone. **B.** Semicircular canals and aqueducts *in situ*. The tegmen tympani has been excised, and the softer bone surrounding the harder bone of the otic capsule has been drilled away. **C.** Walls of left bony labyrinth (otic capsule). The bony labyrinth is the fluid-filled space contained within this formation. **D.** Membranous labyrinth as it lies within the surrounding bony labyrinth. **E.** Isolated left membranous labyrinth.

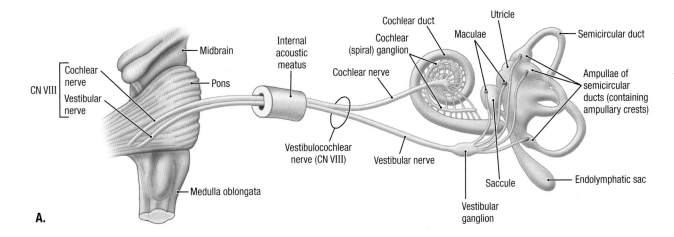

A.

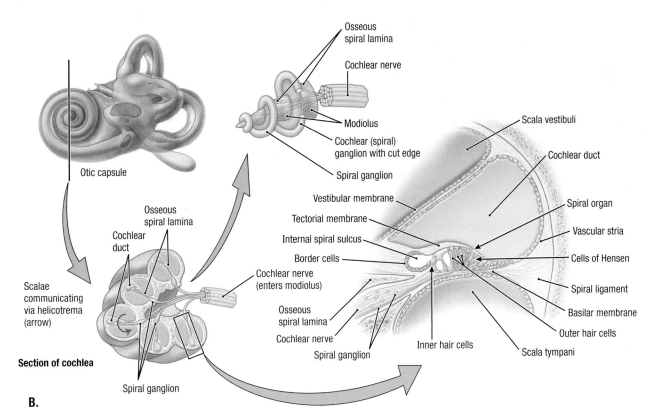

Otic capsule

Section of cochlea

B.

VESTIBULOCOCHLEAR NERVE (CN VIII) AND STRUCTURE OF COCHLEA

7.91

A. Distribution of vestibulocochlear nerve (schematic). **B.** Structure of cochlea. The cochlea has been sectioned along the bony core of the cochlea (modiolus), the axis about which the cochlea winds. An isolated modiolus is shown after the turns of the cochlea are removed, leaving only the spiral lamina winding around it. The large drawing shows the details of the area enclosed in the rectangle, including a cross section of the cochlear duct of the membranous labyrinth.

- The maculae of the membranous labyrinth are primarily static organs, which have small dense particles (otoliths) embedded

among the hair cells. Under the influence of gravity, the otoliths cause bending of the hair cells, which stimulate the vestibular nerve and provide awareness of the position of the head in space; the hairs also respond to quick tilting movements and to linear acceleration and deceleration. **Motion sickness** results mainly from discordance between vestibular and visual stimuli.

- Persistent exposure to excessively loud sound causes degenerative changes in the spiral organ, resulting in **high-tone deafness**. This type of hearing loss commonly occurs in workers who are exposed to loud noises and do not wear protective earmuffs.

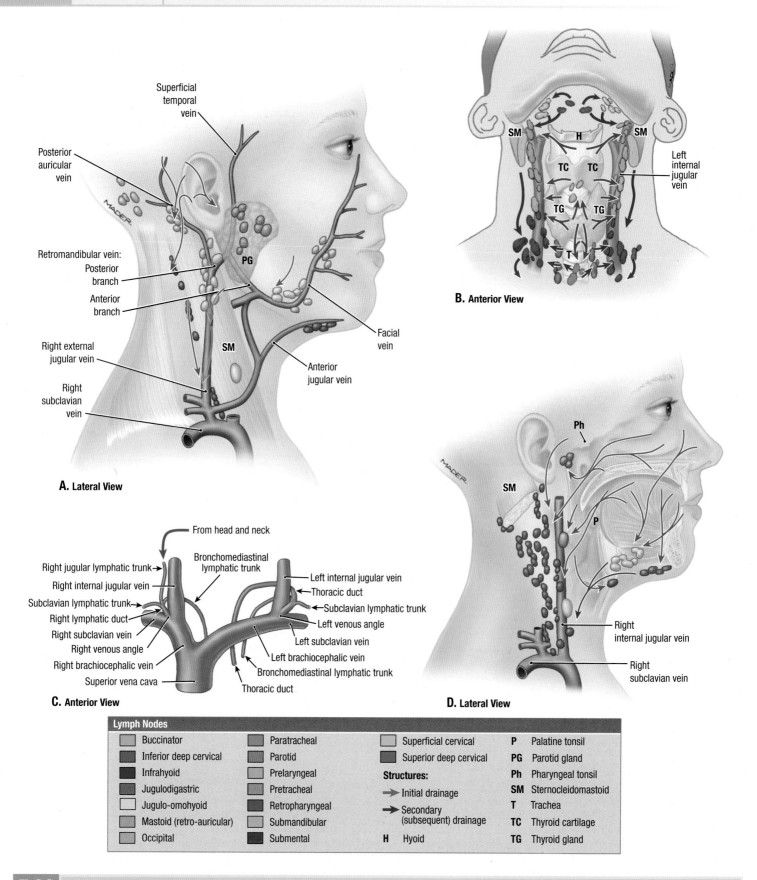

A. Lateral View

B. Anterior View

C. Anterior View

D. Lateral View

Lymph Nodes			
Buccinator	Paratracheal	Superficial cervical	**P** Palatine tonsil
Inferior deep cervical	Parotid	Superior deep cervical	**PG** Parotid gland
Infrahyoid	Prelaryngeal	**Structures:**	**Ph** Pharyngeal tonsil
Jugulodigastric	Pretracheal	Initial drainage	**SM** Sternocleidomastoid
Jugulo-omohyoid	Retropharyngeal	Secondary (subsequent) drainage	**T** Trachea
Mastoid (retro-auricular)	Submandibular		**TC** Thyroid cartilage
Occipital	Submental	**H** Hyoid	**TG** Thyroid gland

7.92 LYMPHATIC AND VENOUS DRAINAGE OF HEAD AND NECK

A. Superficial drainage. **B.** Drainage of the trachea, thyroid gland, larynx, and floor of mouth. **C.** Termination of right and left jugular lymphatic trunks. **D.** Deep drainage.

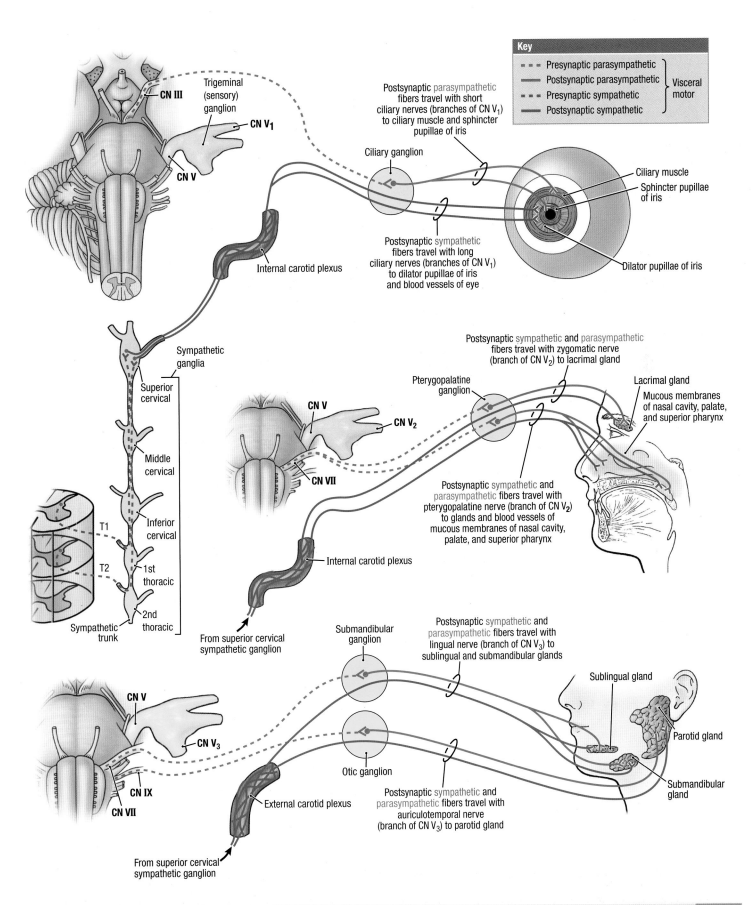

Key

- - - - Presynaptic parasympathetic
———— Postsynaptic parasympathetic ⎤ Visceral
- - - - Presynaptic sympathetic ⎥ motor
———— Postsynaptic sympathetic ⎦

CN III

Trigeminal (sensory) ganglion

CN V₁

CN V

Internal carotid plexus

Postsynaptic parasympathetic fibers travel with short ciliary nerves (branches of CN V₁) to ciliary muscle and sphincter pupillae of iris

Ciliary ganglion

Ciliary muscle

Sphincter pupillae of iris

Dilator pupillae of iris

Postsynaptic sympathetic fibers travel with long ciliary nerves (branches of CN V₁) to dilator pupillae of iris and blood vessels of eye

Sympathetic ganglia

Superior cervical

Middle cervical

Inferior cervical

1st thoracic

2nd thoracic

Sympathetic trunk

T1

T2

From superior cervical sympathetic ganglion

Internal carotid plexus

CN V

CN V₂

CN VII

Pterygopalatine ganglion

Postsynaptic sympathetic and parasympathetic fibers travel with zygomatic nerve (branch of CN V₂) to lacrimal gland

Lacrimal gland

Mucous membranes of nasal cavity, palate, and superior pharynx

Postsynaptic sympathetic and parasympathetic fibers travel with pterygopalatine nerve (branch of CN V₂) to glands and blood vessels of mucous membranes of nasal cavity, palate, and superior pharynx

CN V

CN V₃

CN IX

CN VII

External carotid plexus

From superior cervical sympathetic ganglion

Submandibular ganglion

Otic ganglion

Postsynaptic sympathetic and parasympathetic fibers travel with lingual nerve (branch of CN V₃) to sublingual and submandibular glands

Sublingual gland

Parotid gland

Submandibular gland

Postsynaptic sympathetic and parasympathetic fibers travel with auriculotemporal nerve (branch of CN V₃) to parotid gland

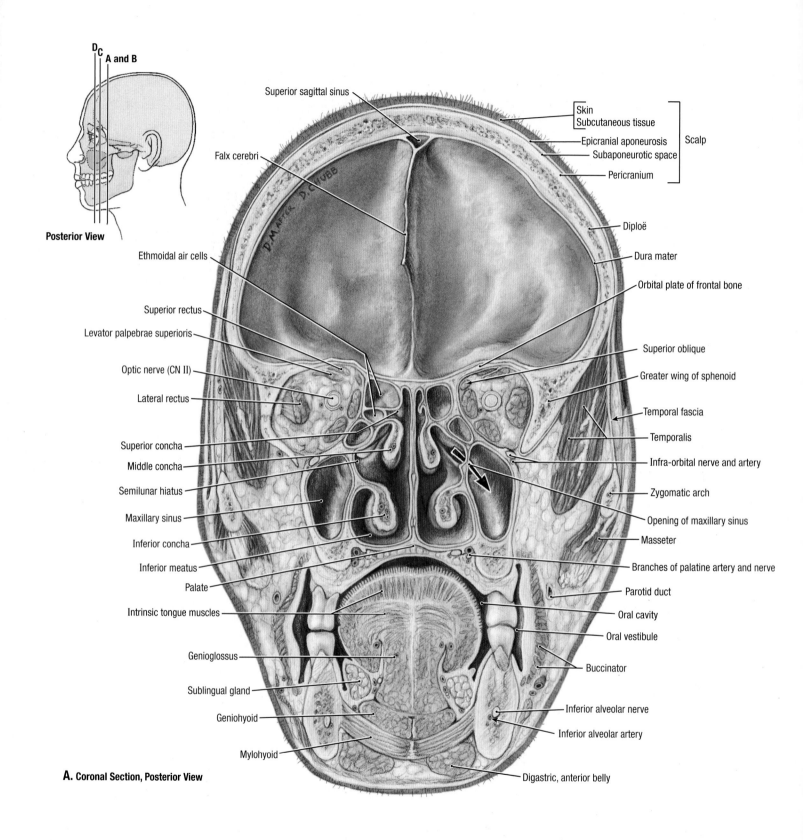

D C **A and B**

Posterior View

Superior sagittal sinus

Skin
Subcutaneous tissue
Epicranial aponeurosis — Scalp
Subaponeurotic space
Pericranium

Falx cerebri

Diploë

Dura mater

Ethmoidal air cells

Orbital plate of frontal bone

Superior rectus

Superior oblique

Levator palpebrae superioris

Greater wing of sphenoid

Optic nerve (CN II)

Temporal fascia

Lateral rectus

Temporalis

Superior concha

Infra-orbital nerve and artery

Middle concha

Zygomatic arch

Semilunar hiatus

Opening of maxillary sinus

Maxillary sinus

Masseter

Inferior concha

Branches of palatine artery and nerve

Inferior meatus

Parotid duct

Palate

Oral cavity

Intrinsic tongue muscles

Oral vestibule

Genioglossus

Buccinator

Sublingual gland

Inferior alveolar nerve

Geniohyoid

Inferior alveolar artery

Mylohyoid

Digastric, anterior belly

A. Coronal Section, Posterior View

7.94 **CORONAL SECTION AND MRIs OF NASOPHARYNX AND ORAL CAVITY**

A. Coronal section.

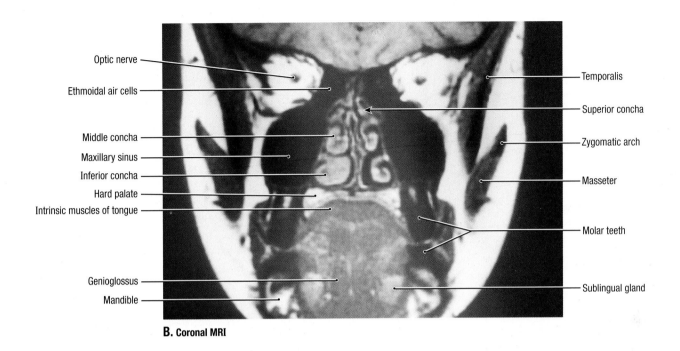

Optic nerve

Ethmoidal air cells

Middle concha

Maxillary sinus

Inferior concha

Hard palate

Intrinsic muscles of tongue

Genioglossus

Mandible

Temporalis

Superior concha

Zygomatic arch

Masseter

Molar teeth

Sublingual gland

B. Coronal MRI

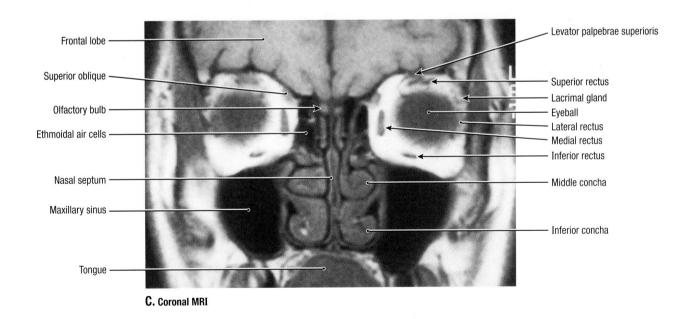

Frontal lobe

Superior oblique

Olfactory bulb

Ethmoidal air cells

Nasal septum

Maxillary sinus

Tongue

Levator palpebrae superioris

Superior rectus

Lacrimal gland

Eyeball

Lateral rectus

Medial rectus

Inferior rectus

Middle concha

Inferior concha

C. Coronal MRI

CORONAL SECTION AND MRIs OF NASOPHARYNX AND ORAL CAVITY (*continued*) | 7.94

B. and **C.** Coronal MRIs.

Deviation of nasal septum. The nasal septum is usually deviated to one side or the other. This could be the result of a birth injury, but more often, the deviation occurs during adolescence and adulthood from trauma. Sometimes, the deviation is so severe that the nasal septum is in contact with the lateral wall of the nasal cavity and often obstructs breathing or exacerbates snoring. The deviation can be corrected surgically.

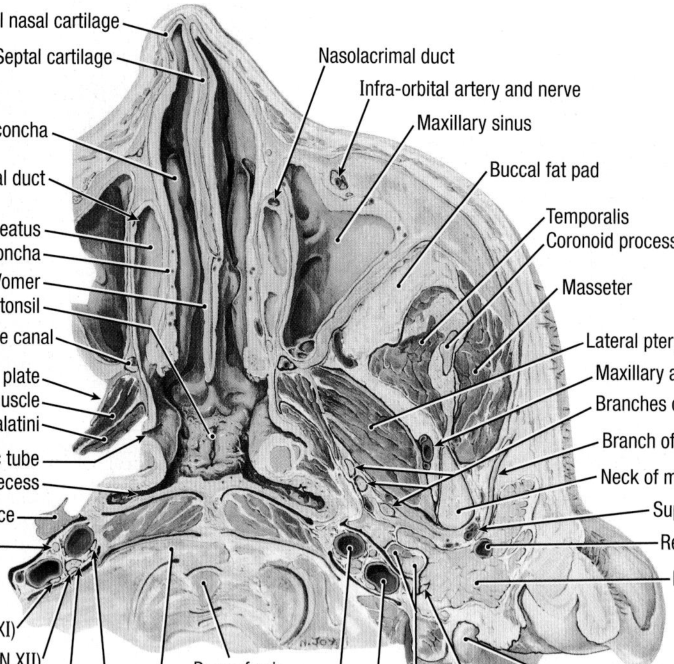

Lateral nasal cartilage
Septal cartilage
Nasolacrimal duct
Infra-orbital artery and nerve
Maxillary sinus
Middle concha
Buccal fat pad
Nasolacrimal duct
Temporalis
Coronoid process
Inferior meatus
Inferior concha
Masseter
Vomer
Pharyngeal tonsil
Lateral pterygoid
Greater palatine canal
Maxillary artery
Lateral pterygoid plate
Branches of mandibular nerve
Medial pterygoid muscle
Tensor veli palatini
Branch of facial nerve
Pharyngotympanic tube
Neck of mandible
Pharyngeal recess
Superficial temporal artery
Lateral pharyngeal space
Retromandibular vein
Glossopharyngeal nerve (CN IX)
Parotid gland
Accessory nerve (CN XI)
Hypoglossal nerve (CN XII)
Tip of mastoid process
Vagus nerve (CN X)
Atlas
Dens of axis
Sympathetic trunk
Internal carotid artery
Facial nerve
Internal jugular vein
Styloid process and stylopharyngeus

A. Inferior View

Nasal septum
Maxillary sinus
Inferior concha
Buccal fat pad
Nasopharynx
Coronoid process of mandible
Temporalis
Medial pterygoid
Lateral pterygoid plate
Masseter
Tensor veli palatini
Lateral pterygoid
Maxillary artery
Branches of mandibular nerve
Pharyngotympanic tube
Neck of mandible
Pharyngeal recess
Internal carotid artery
Superficial temporal vessels
Internal jugular vein
Vertebral artery
Mastoid cells

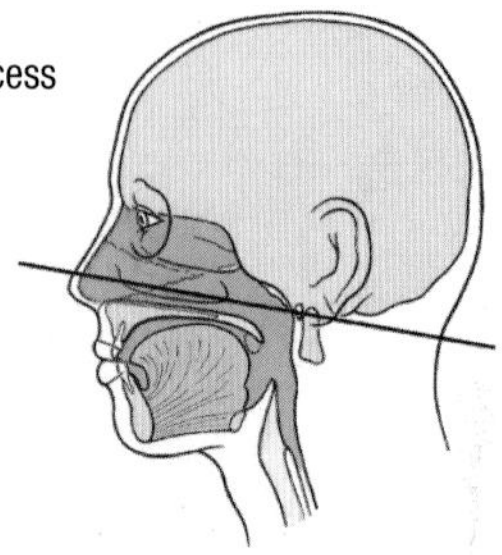

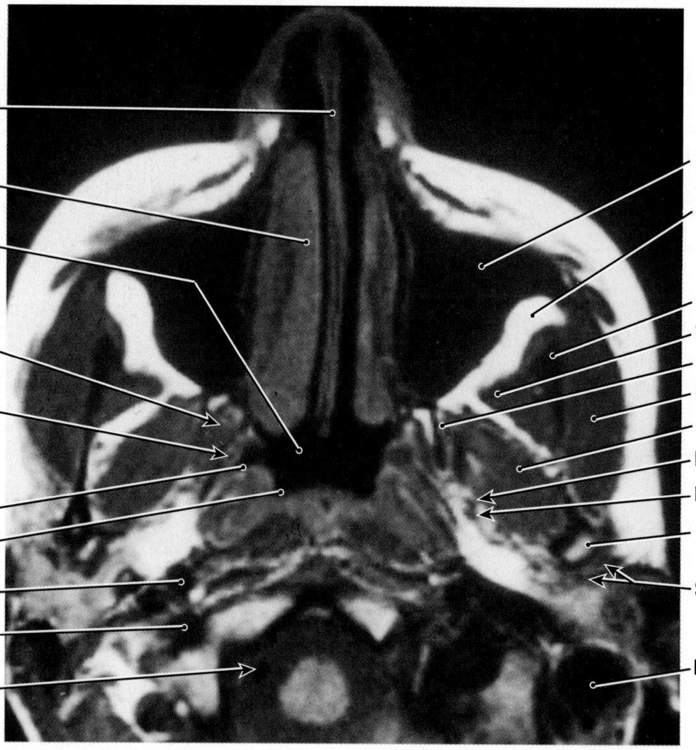

B. Transverse (axial) MRI

7.95 TRANSVERSE SECTION AND MRI OF NASAL CAVITY AND NASOPHARYNX

A. Transverse section of left side of head. **B.** Transverse (axial) MRI.

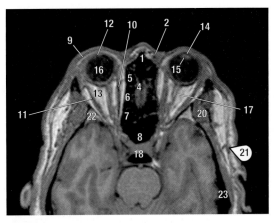

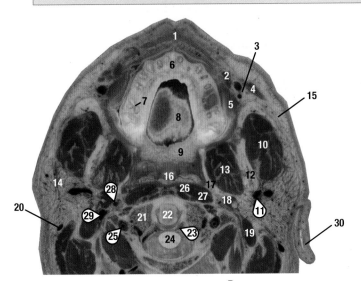

A. Transverse Section and Transverse (Axial) MRI Scan

Key							
1	Nasal bones	7	Posterior ethmoidal air cell	13	Retrobulbar fat	19	Optic tract
2	Angular artery	8	Sphenoid sinus	14	Anterior chamber	20	Temporalis muscle
3	Frontal process of maxilla	9	Orbicularis oculi muscle	15	Lens	21	Superficial temporal vessels
4	Nasal septum	10	Medial rectus muscle	16	Vitreous body	22	Greater wing of sphenoid
5	Anterior ethmoidal cell	11	Lateral rectus muscle	17	Optic nerve	23	Squamous part of temporal bone
6	Middle ethmoidal cell	12	Cornea	18	Optic chiasm		

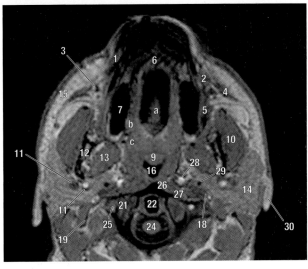

B. Transverse Section and Transverse (Axial) MRI Scan

Key							
1	Orbicularis oris muscle	12	Ramus of mandible	23	Transverse ligament of atlas		
2	Levator anguli oris muscle	13	Lateral pterygoid muscle	24	Spinal cord		
3	Facial artery and vein	14	Parotid gland	25	Vertebral artery in foramina transversaria		
4	Zygomaticus major muscle	15	Subcutaneous tissue	26	Longus colli muscle		
5	Buccinator muscle	16	Region of pharyngeal tubercle	27	Longus capitis muscle		
6	Maxilla	17	Sphenoid bone	28	Internal carotid artery		
7	Alveolar process of maxilla	18	Stylohyoid ligament and muscle	29	Internal jugular vein		
8	Dorsum of tongue	19	Posterior belly of digastric muscle	30	Inferior portion of helix of auricle		
9	Soft palate (uvula apparent in image)	20	Occipital artery	a	Hard palate		
10	Masseter muscle	21	First cervical vertebrae (atlas)	b	Palatoglossus muscle		
11	Retromandibular vein	22	Dens (axis)	c	Palatopharyngeus muscle		

IMAGING OF ORBIT AND ORAL CAVITY/MAXILLARY REGION

A. Transverse section and MRI through in plane of optic nerve. **B.** Transverse section and MRI at level of atlas/dens.

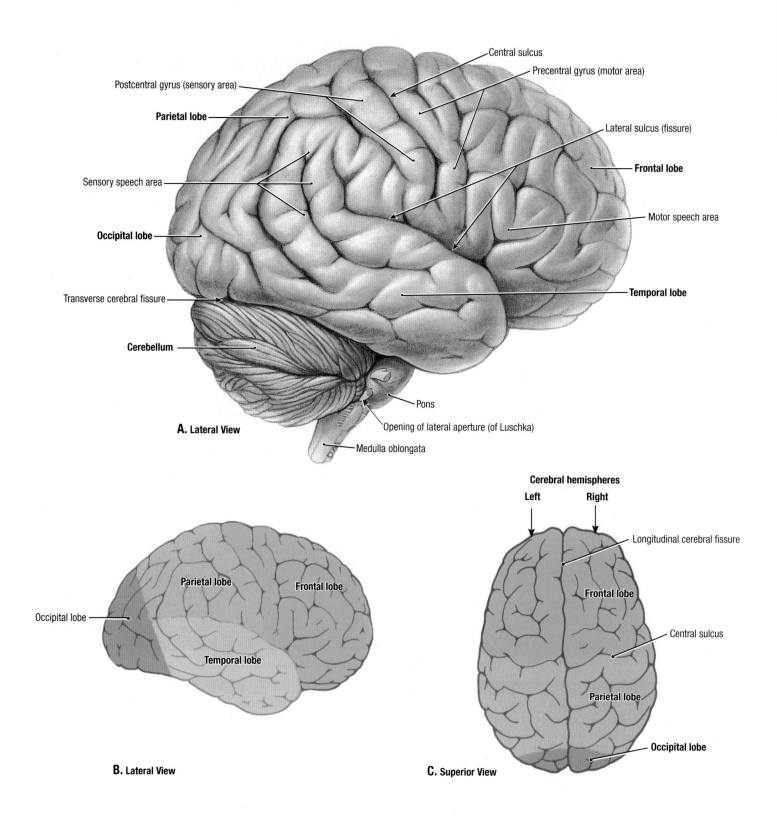

Central sulcus

Precentral gyrus (motor area)

Postcentral gyrus (sensory area)

Parietal lobe

Lateral sulcus (fissure)

Frontal lobe

Sensory speech area

Motor speech area

Occipital lobe

Temporal lobe

Transverse cerebral fissure

Cerebellum

Pons

Opening of lateral aperture (of Luschka)

Medulla oblongata

A. Lateral View

Parietal lobe   **Frontal lobe**

Occipital lobe

Temporal lobe

B. Lateral View

Cerebral hemispheres

Left   **Right**

Longitudinal cerebral fissure

Frontal lobe

Central sulcus

Parietal lobe

Occipital lobe

C. Superior View

7.97	BRAIN

A. Cerebrum, cerebellum, and brainstem, lateral aspect. **B.** Lobes of the cerebral hemispheres, lateral aspect. **C.** Lobes of the cerebral hemispheres, superior aspect.

Cerebral contusion (bruising) results from brain trauma in which the pia is stripped from the injured surface of the brain and may be torn, allowing blood to enter the subarachnoid space. The bruising results from the sudden impact of the moving brain against the stationary cranium or from the suddenly moving cranium against the stationary brain. Cerebral contusion may result in an extended loss of consciousness.

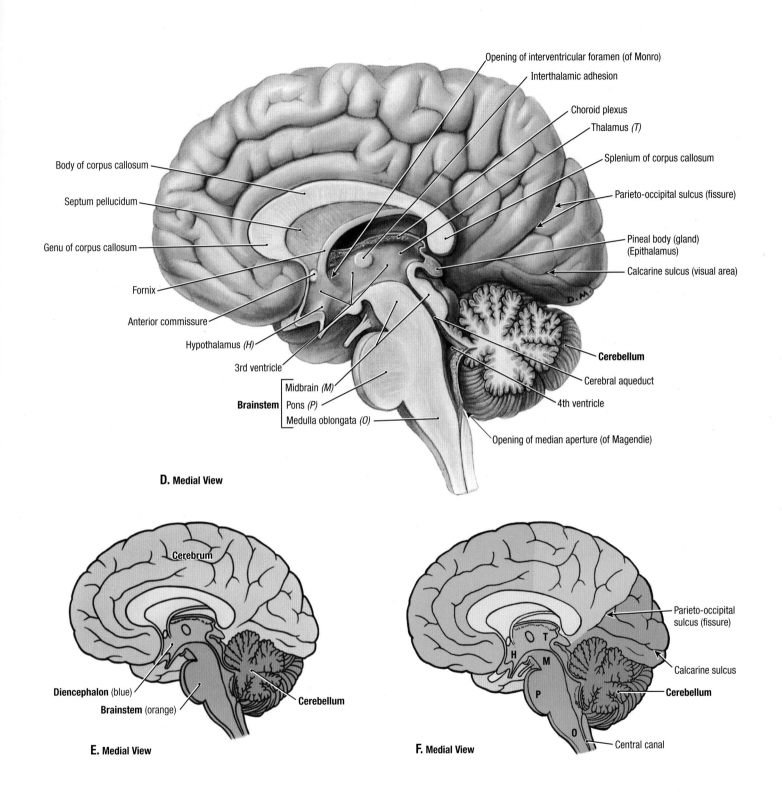

Opening of interventricular foramen (of Monro)

Interthalamic adhesion

Choroid plexus

Thalamus *(T)*

Splenium of corpus callosum

Parieto-occipital sulcus (fissure)

Pineal body (gland) (Epithalamus)

Calcarine sulcus (visual area)

Cerebellum

Cerebral aqueduct

4th ventricle

Opening of median aperture (of Magendie)

Body of corpus callosum

Septum pellucidum

Genu of corpus callosum

Fornix

Anterior commissure

Hypothalamus *(H)*

3rd ventricle

Midbrain *(M)*

Brainstem Pons *(P)*

Medulla oblongata *(O)*

D. Medial View

Cerebrum

Diencephalon (blue)

Brainstem (orange)

Cerebellum

E. Medial View

O T

H M

P

O

Parieto-occipital sulcus (fissure)

Calcarine sulcus

Cerebellum

Central canal

F. Medial View

BRAIN (*continued*)

7.97

D. Cerebrum, cerebellum, and brainstem, median section. **E.** Parts of the brain, median section. **F.** Lobes of the cerebral hemisphere, median section. See **D** for labeling key.

Cerebral compression may be produced by intracranial collections of blood, obstruction of CSF circulation or absorption, intracranial tumors or abscesses, and brain swelling caused by brain edema, an increase in brain volume resulting from an increase in water and sodium content.

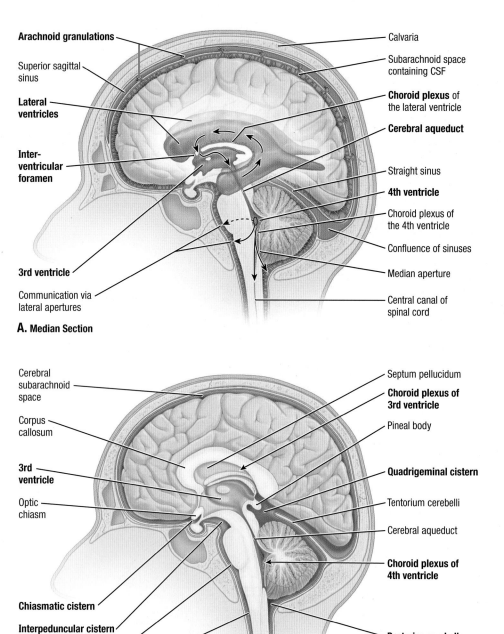

Arachnoid granulations

Superior sagittal sinus

Lateral ventricles

Inter-ventricular foramen

3rd ventricle

Communication via lateral apertures

Calvaria

Subarachnoid space containing CSF

Choroid plexus of the lateral ventricle

Cerebral aqueduct

Straight sinus

4th ventricle

Choroid plexus of the 4th ventricle

Confluence of sinuses

Median aperture

Central canal of spinal cord

A. Median Section

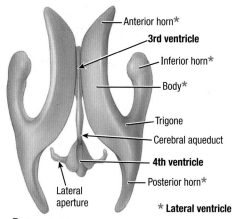

Anterior horn*

3rd ventricle

Inferior horn*

Body*

Trigone

Cerebral aqueduct

4th ventricle

Posterior horn*

Lateral aperture

* **Lateral ventricle**

B. Superior View

Cerebral subarachnoid space

Corpus callosum

3rd ventricle

Optic chiasm

Chiasmatic cistern

Interpeduncular cistern

Pontocerebellar cistern

Spinal subarachnoid space

Septum pellucidum

Choroid plexus of 3rd ventricle

Pineal body

Quadrigeminal cistern

Tentorium cerebelli

Cerebral aqueduct

Choroid plexus of 4th ventricle

Posterior cerebello-medullary cistern

C. Medial Section, Sectioned to Right of Superior Sagittal Sinus

7.98 VENTRICULAR SYSTEM

A. Circulation of cerebrospinal fluid (CSF). **B.** Ventricles: lateral, third, and fourth. **C.** Subarachnoid cisterns.

- The ventricular system consists of two lateral ventricles located in the cerebral hemispheres, a 3rd ventricle located between the right and left halves of the diencephalon, and a 4th ventricle located in the posterior parts of the pons and medulla.
- CSF secreted by choroid plexus in the ventricles drains via the interventricular foramen from the lateral to the 3rd ventricle, via the cerebral aqueduct from the 3rd to the 4th ventricle, and via the

median and lateral apertures into the subarachnoid space. CSF is absorbed by arachnoid granulations into the venous sinuses (especially the superior sagittal sinus).

- **Hydrocephalus.** Overproduction of CSF, obstruction of its flow, or interference with its absorption results in an excess of CSF in the ventricles and enlargement of the head, a condition known as hydrocephalus. Excess CSF dilates the ventricles; thins the brain; and, in infants, separates the bones of the calvaria because the sutures and fontanelles are still open.

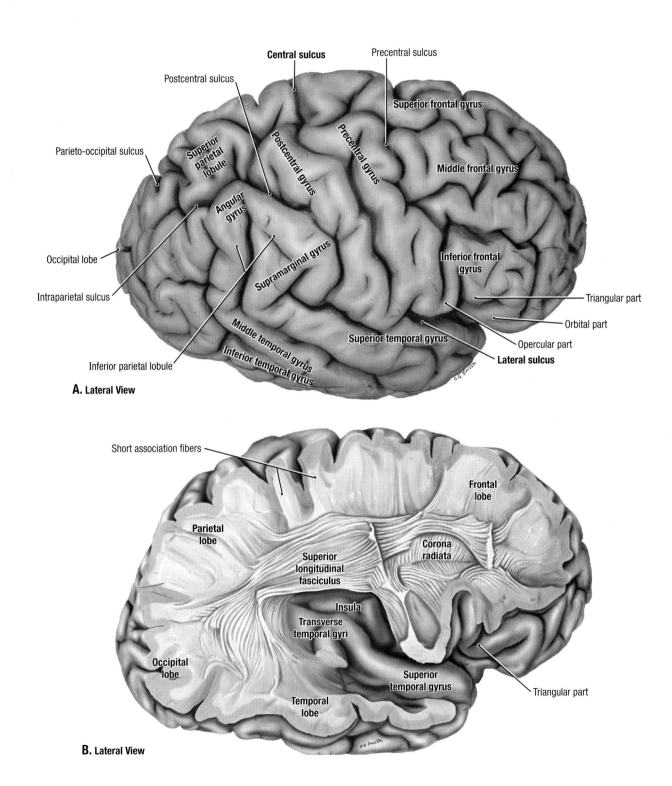

A. Lateral View

Central sulcus
Precentral sulcus
Postcentral sulcus
Superior frontal gyrus
Parieto-occipital sulcus
Superior parietal lobule
Postcentral gyrus
Precentral gyrus
Middle frontal gyrus
Angular gyrus
Inferior frontal gyrus
Occipital lobe
Supramarginal gyrus
Triangular part
Intraparietal sulcus
Orbital part
Superior temporal gyrus
Opercular part
Middle temporal gyrus
Lateral sulcus
Inferior temporal gyrus
Inferior parietal lobule

B. Lateral View

Short association fibers
Frontal lobe
Parietal lobe
Corona radiata
Superior longitudinal fasciculus
Insula
Transverse temporal gyri
Occipital lobe
Superior temporal gyrus
Temporal lobe
Triangular part

SERIAL DISSECTIONS OF LATERAL ASPECT OF CEREBRAL HEMISPHERE

7.99

The dissections begin from the lateral surface of the cerebral hemisphere **(A)** and proceed sequentially medially **(B–F)**.

A. Sulci and gyri of the lateral surface of right cerebral hemisphere. Each gyrus is a fold of cerebral cortex with a core of white matter. The furrows are called sulci. The pattern of sulci and gyri, formed shortly before birth, is recognizable in some adult brains, as shown in this specimen. Usually, the expanding cortex acquires secondary foldings, which make identification of this basic pattern more difficult. **B.** Superior longitudinal fasciculus, transverse temporal gyri, and insula. The cortex and short association fiber bundles around the lateral fissure have been removed.

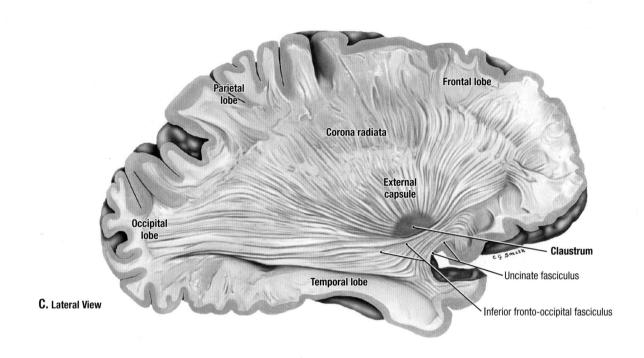

C. Lateral View

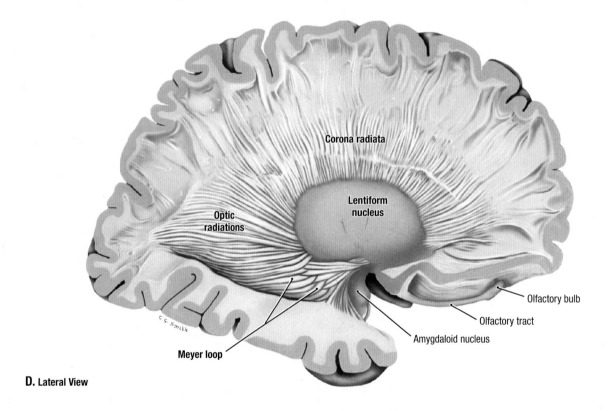

D. Lateral View

| 7.99 | SERIAL DISSECTIONS OF LATERAL ASPECT OF CEREBRAL HEMISPHERE (*continued*) |

C. Uncinate and inferior fronto-occipital fasciculi and external capsule. The external capsule consists of projection fibers that pass between the claustrum laterally and the lentiform nucleus medially. **D.** Lentiform nucleus and corona radiata. The inferior longitudinal and uncinate fasciculi, claustrum, and external capsule have been removed. The fibers of the optic radiations convey impulses from the right half of the retina of each eye; the fibers extending closest to the temporal pole (Meyer's loop) carry impulses from the lower portion of each retina.

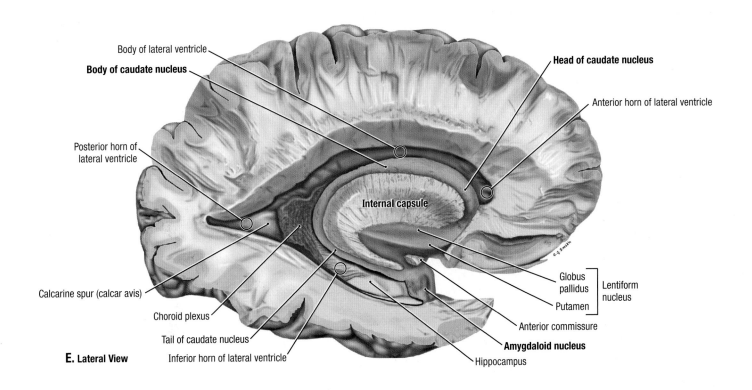

Body of lateral ventricle

Body of caudate nucleus

Head of caudate nucleus

Anterior horn of lateral ventricle

Posterior horn of lateral ventricle

Internal capsule

Calcarine spur (calcar avis)

Globus pallidus

Lentiform nucleus

Putamen

Choroid plexus

Anterior commissure

Tail of caudate nucleus

Amygdaloid nucleus

E. Lateral View   Inferior horn of lateral ventricle

Hippocampus

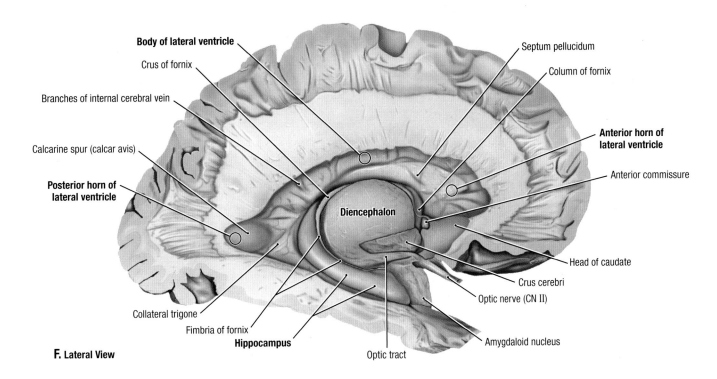

Body of lateral ventricle

Septum pellucidum

Crus of fornix

Column of fornix

Branches of internal cerebral vein

Calcarine spur (calcar avis)

Anterior horn of lateral ventricle

Posterior horn of lateral ventricle

Anterior commissure

Diencephalon

Head of caudate

Crus cerebri

Collateral trigone

Optic nerve (CN II)

Fimbria of fornix

Hippocampus

Amygdaloid nucleus

F. Lateral View   Optic tract

SERIAL DISSECTIONS OF LATERAL ASPECT OF CEREBRAL HEMISPHERE (*continued*)   **7.99**

E. Caudate and amygdaloid nuclei and internal capsule. The lateral wall of the lateral ventricle, the marginal part of the internal capsule, the anterior commissure, and the superior part of the lentiform nucleus have been removed. **F.** Lateral ventricle, hippocampus, and diencephalon. The inferior parts of the lentiform nucleus, internal capsule, and caudate nucleus have been removed.

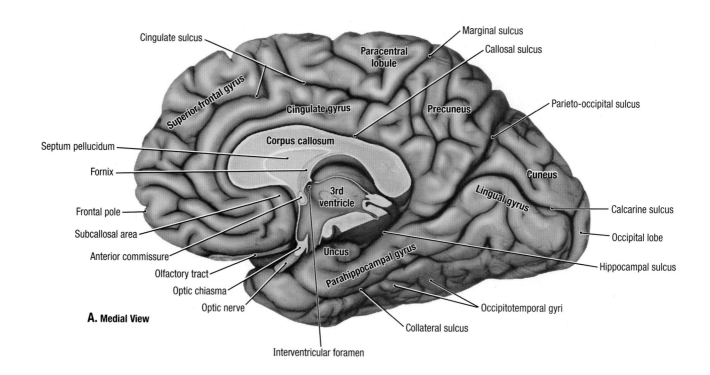

A. Medial View

Cingulate sulcus

Paracentral lobule

Marginal sulcus

Callosal sulcus

Superior frontal gyrus

Cingulate gyrus

Precuneus

Parieto-occipital sulcus

Corpus callosum

Septum pellucidum

Fornix

Cuneus

Lingual gyrus

3rd ventricle

Calcarine sulcus

Frontal pole

Occipital lobe

Subcallosal area

Anterior commissure

Uncus

Hippocampal sulcus

Olfactory tract

Parahippocampal gyrus

Optic chiasma

Optic nerve

Occipitotemporal gyri

Collateral sulcus

Interventricular foramen

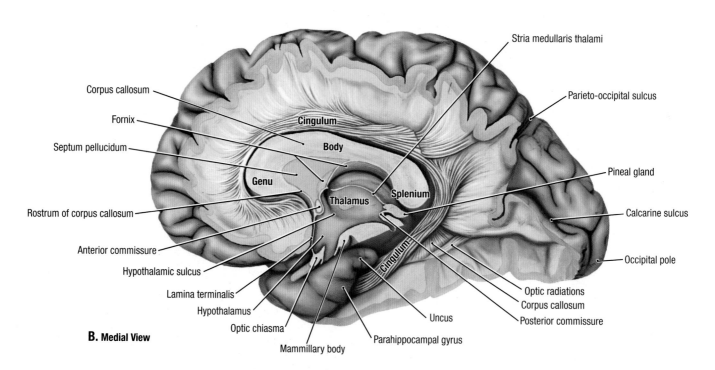

B. Medial View

Stria medullaris thalami

Corpus callosum

Parieto-occipital sulcus

Fornix

Cingulum

Septum pellucidum

Body

Genu

Pineal gland

Thalamus

Splenium

Rostrum of corpus callosum

Calcarine sulcus

Anterior commissure

Cingulum

Occipital pole

Hypothalamic sulcus

Lamina terminalis

Optic radiations

Hypothalamus

Corpus callosum

Optic chiasma

Uncus

Posterior commissure

Mammillary body

Parahippocampal gyrus

7.100 SERIAL DISSECTIONS OF MEDIAL ASPECT OF CEREBRAL HEMISPHERE

The dissections begin from the medial surface of the cerebral hemisphere **(A)** and proceed sequentially laterally **(B–D)**.

A. Sulci and gyri of medial surface of cerebral hemisphere. The corpus callosum consists of the rostrum, genu, body, and splenium; the cingulate and parahippocampal gyri form the limbic lobe. **B.** Cingulum. The cortex and short association fibers were removed from the medial aspect of the hemisphere. The cingulum is a long association fiber bundle that lies in the core of the cingulate and parahippocampal gyri.

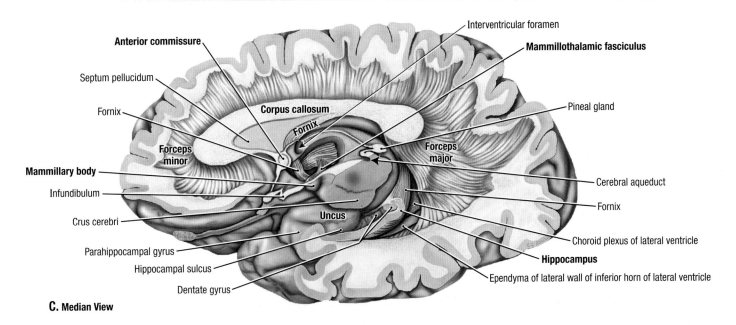

C. Median View

Anterior commissure

Septum pellucidum

Fornix

Forceps minor

Mammillary body

Infundibulum

Crus cerebri

Parahippocampal gyrus

Hippocampal sulcus

Dentate gyrus

Interventricular foramen

Corpus callosum

Fornix

Mammillothalamic fasciculus

Pineal gland

Forceps major

Cerebral aqueduct

Fornix

Choroid plexus of lateral ventricle

Hippocampus

Ependyma of lateral wall of inferior horn of lateral ventricle

Uncus

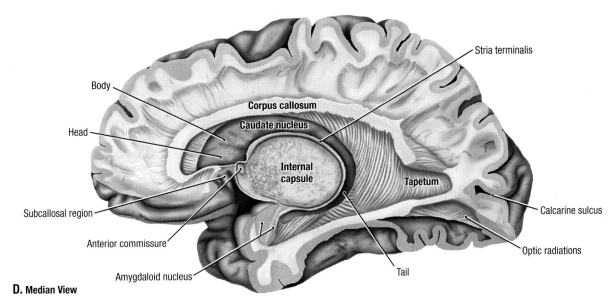

D. Median View

Body

Corpus callosum

Caudate nucleus

Head

Internal capsule

Subcallosal region

Anterior commissure

Amygdaloid nucleus

Stria terminalis

Tapetum

Calcarine sulcus

Optic radiations

Tail

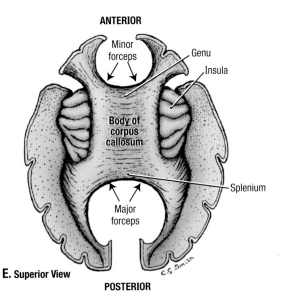

E. Superior View

ANTERIOR

Minor forceps

Body of corpus callosum

Major forceps

POSTERIOR

Genu

Insula

Splenium

SERIAL DISSECTIONS OF MEDIAL ASPECT OF CEREBRAL HEMISPHERE (*continued*) **7.100**

C. Fornix, mammillothalamic fasciculus, and forceps major and minor. The cingulum and a portion of the wall of the 3rd ventricle have been removed. The fornix begins at the hippocampus and terminates in the mammillary body by passing anterior to the interventricular foramen and posterior to the anterior commissure. The mammillothalamic fasciculus emerges from the mammillary body and terminates in the anterior nucleus of the thalamus. **D.** Caudate nucleus and internal capsule. The diencephalon was removed, along with the ependyma of the lateral ventricle, except where it covers the caudate and amygdaloid nuclei. **E.** Corpus callosum. The body of the corpus callosum connects the two cerebral hemispheres; the minor (frontal) forceps (at the genu of corpus callosum) connects the frontal lobes, and the major (occipital) forceps (at splenium) connects the occipital lobes.

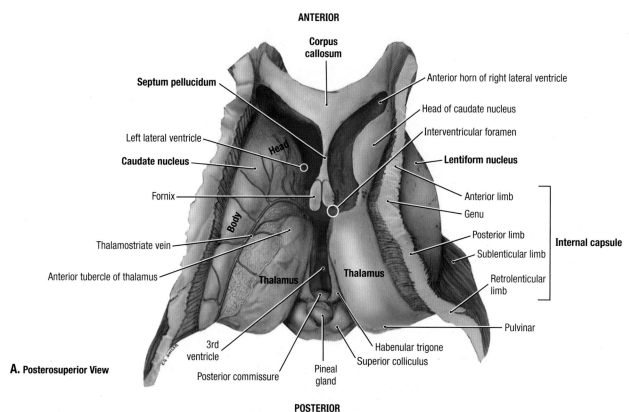

ANTERIOR

Corpus callosum

Septum pellucidum

Anterior horn of right lateral ventricle

Head of caudate nucleus

Interventricular foramen

Left lateral ventricle

Caudate nucleus

Lentiform nucleus

Fornix

Anterior limb

Genu

Thalamostriate vein

Posterior limb

Sublenticular limb

Internal capsule

Anterior tubercle of thalamus

Retrolenticular limb

Thalamus

Thalamus

Pulvinar

3rd ventricle

Habenular trigone

Superior colliculus

Posterior commissure

Pineal gland

A. Posterosuperior View

POSTERIOR

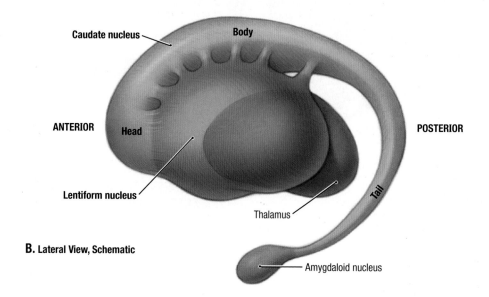

Caudate nucleus

Body

ANTERIOR

Head

POSTERIOR

Lentiform nucleus

Thalamus

Tail

B. Lateral View, Schematic

Amygdaloid nucleus

7.101 CAUDATE AND LENTIFORM NUCLEI

A. Relationship to the lateral ventricles and internal capsule. The dorsal surface of the diencephalon has been exposed by dissecting away the two cerebral hemispheres, except the anterior part of the corpus callosum, the inferior part of the septum pellucidum, the internal capsule, and the caudate and lentiform nuclei. On the right side of the specimen, the thalamus, caudate, and lentiform nuclei have been cut horizontally at the level of the interventricular foramen. The parts of the internal capsule include the anterior, posterior, retrolenticular sublenticular limbs, and genu. **B.** Schematic illustration of nuclei.

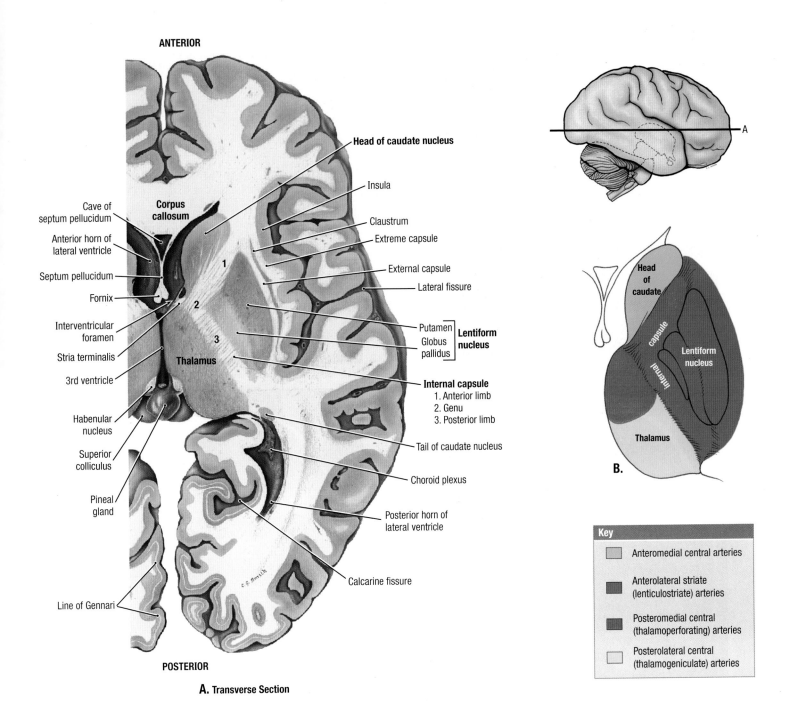

ANTERIOR

Head of caudate nucleus

Insula

Cave of
septum pellucidum

Corpus
callosum

Claustrum

Extreme capsule

Anterior horn of
lateral ventricle

External capsule

Septum pellucidum

1

Lateral fissure

Fornix

2

Putamen

Lentiform
nucleus

Interventricular
foramen

3

Globus
pallidus

Stria terminalis

Thalamus

Internal capsule
1. Anterior limb
2. Genu
3. Posterior limb

3rd ventricle

Habenular
nucleus

Tail of caudate nucleus

Superior
colliculus

Choroid plexus

Pineal
gland

Posterior horn of
lateral ventricle

Calcarine fissure

Line of Gennari

POSTERIOR

A. Transverse Section

A

Head
of
caudate

Internal capsule

Lentiform
nucleus

Thalamus

B.

Key	
	Anteromedial central arteries
	Anterolateral striate (lenticulostriate) arteries
	Posteromedial central (thalamoperforating) arteries
	Posterolateral central (thalamogeniculate) arteries

AXIAL SECTIONS THROUGH THALAMUS, CAUDATE NUCLEUS, AND LENTIFORM NUCLEUS   **7.102**

A. Relationships of the internal capsule. **B.** Blood supply of region.

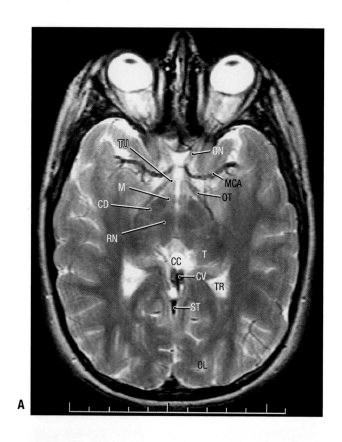

A

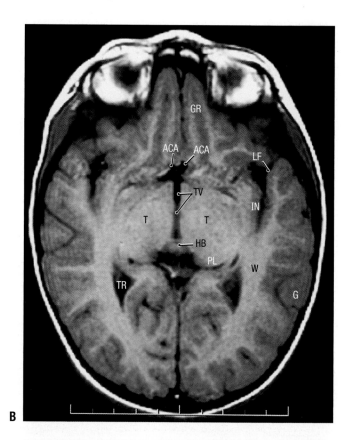

B

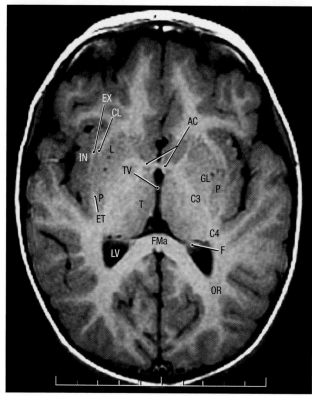

C

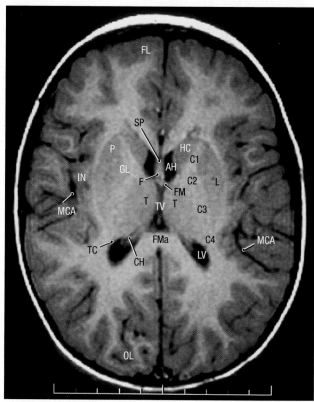

D

7.103 AXIAL (TRANSVERSE) MRIs THROUGH CEREBRAL HEMISPHERES

See orientation drawing for sites of scans **A–F**. **A** is T2-weighted, and **B–F** are T1-weighted.

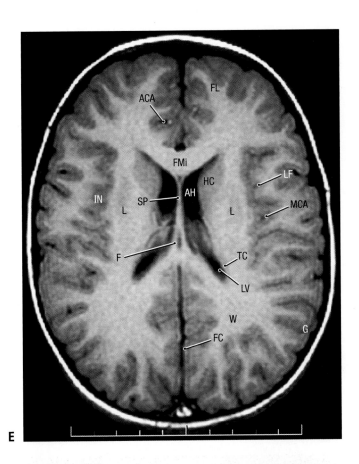

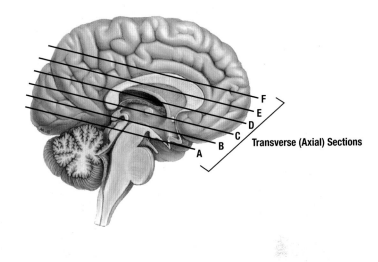

Transverse (Axial) Sections

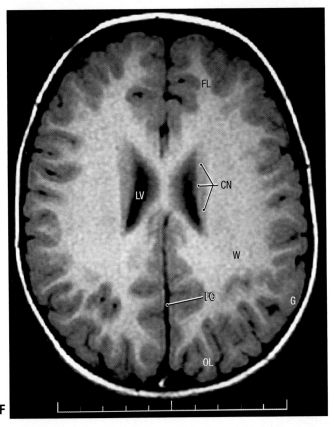

Key			
AC	Anterior commissure	GL	Globus pallidus
ACA	Anterior cerebral artery	GR	Gyrus rectus
AH	Anterior horn of lateral ventricle	HB	Habenular commissure
		HC	Head of caudate nucleus
C1	Anterior limb of internal capsule	IN	Insular cortex
		L	Lentiform nucleus
C2	Genu of internal capsule	LF	Lateral fissure
C3	Posterior limb of internal capsule	LV	Lateral ventricle
		M	Mammillary body
C4	Retrolenticular limb of internal capsule	MCA	Middle cerebral artery
		OL	Occipital lobe
CC	Collicular cistern	ON	Optic nerve
CD	Cerebral peduncle	OR	Optic radiations
CH	Choroid plexus	OT	Optic tract
CL	Claustrum	P	Putamen
CN	Caudate nucleus	PL	Pulvinar
CV	Great cerebral vein	RN	Red nucleus
ET	External capsule	SP	Septum pellucidum
EX	Extreme capsule	ST	Straight sinus
F	Fornix	T	Thalamus
FC	Falx cerebri	TC	Tail of caudate nucleus
FL	Frontal lobe	TR	Trigone of lateral ventricle
FM	Interventricular foramen	TU	Tuber cinereum
FMa	Forceps major	TV	Third ventricle
FMi	Forceps minor	W	White matter
G	Gray matter		

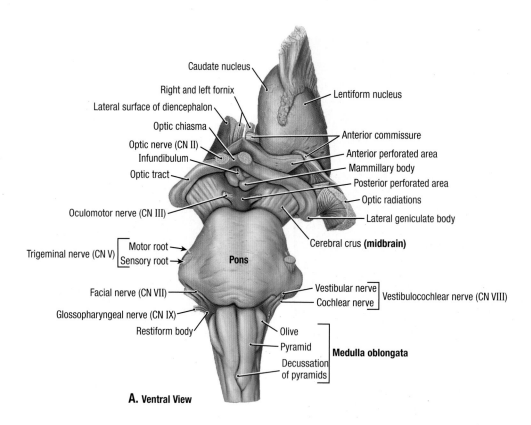

Caudate nucleus

Right and left fornix

Lateral surface of diencephalon

Optic chiasma

Optic nerve (CN II)

Infundibulum

Optic tract

Oculomotor nerve (CN III)

Lentiform nucleus

Anterior commissure

Anterior perforated area

Mammillary body

Posterior perforated area

Optic radiations

Lateral geniculate body

Cerebral crus **(midbrain)**

Trigeminal nerve (CN V) { Motor root →
Sensory root → }

Pons

Facial nerve (CN VII)

Glossopharyngeal nerve (CN IX)

Restiform body

Vestibular nerve
Cochlear nerve } Vestibulocochlear nerve (CN VIII)

Olive

Pyramid **Medulla oblongata**

Decussation
of pyramids

A. Ventral View

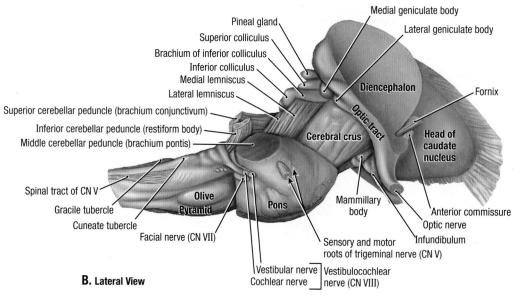

Medial geniculate body

Pineal gland

Superior colliculus

Brachium of inferior colliculus

Inferior colliculus

Medial lemniscus

Lateral lemniscus

Superior cerebellar peduncle (brachium conjunctivum)

Inferior cerebellar peduncle (restiform body)

Middle cerebellar peduncle (brachium pontis)

Spinal tract of CN V

Gracile tubercle

Cuneate tubercle

Facial nerve (CN VII)

Lateral geniculate body

Diencephalon

Fornix

Optic tract

Cerebral crus

**Head of
caudate
nucleus**

Olive

Pyramid

Pons

Mammillary
body

Anterior commissure

Optic nerve

Infundibulum

Sensory and motor
roots of trigeminal nerve (CN V)

Vestibular nerve
Cochlear nerve } Vestibulocochlear
nerve (CN VIII)

B. Lateral View

7.104 BRAINSTEM

The brainstem has been exposed by removing the cerebellum, all of the right cerebral hemisphere, and the major portion of the left hemisphere.

   A. Ventral aspect.
- The brainstem consists of the medulla oblongata, pons, and midbrain.
- The pyramid is on the ventral surface of the medulla; the decussation of the pyramids is formed by the decussating (crossing) lateral corticospinal tract.
- The trigeminal nerve (CN V) emerges as sensory and motor roots.

- The crus cerebri are part of the midbrain.
- The oculomotor nerve emerges from the interpeduncular fossa.
   B. Lateral aspect.
- The vestibulocochlear nerve (CN VIII) consists of two nerves, the vestibular and cochlear nerves.
- The spinal tract of the trigeminal nerve is exposed where it comes to the surface of the medulla to form the tuber cinereum.
- The three are cerebellar peduncles: superior, middle, and inferior.
- The medial and lateral lemnisci on the lateral aspect of the midbrain

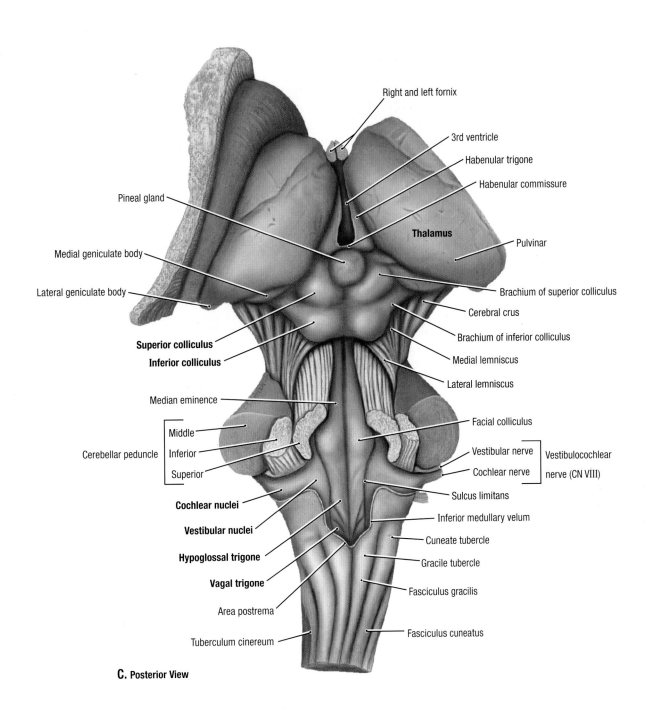

Right and left fornix

3rd ventricle

Habenular trigone

Habenular commissure

Pineal gland

Thalamus

Pulvinar

Medial geniculate body

Brachium of superior colliculus

Lateral geniculate body

Cerebral crus

Brachium of inferior colliculus

Superior colliculus

Medial lemniscus

Inferior colliculus

Lateral lemniscus

Median eminence

Facial colliculus

Middle

Inferior — Cerebellar peduncle

Vestibular nerve

Vestibulocochlear

Superior

Cochlear nerve

nerve (CN VIII)

Cochlear nuclei

Sulcus limitans

Vestibular nuclei

Inferior medullary velum

Hypoglossal trigone

Cuneate tubercle

Gracile tubercle

Vagal trigone

Fasciculus gracilis

Area postrema

Tuberculum cinereum

Fasciculus cuneatus

C. Posterior View

BRAINSTEM (*continued*)

7.104

C. Dorsal aspect.
- Ridges are formed by the fasciculus gracilis and cuneatus.
- The gracile and cuneate tubercles are the sites of the nucleus gracilis and nucleus cuneatus.
- The diamond-shaped floor of the 4th ventricle; lateral to the sulcus limitans are the vestibular and cochlear nuclei and

medially are the hypoglossal and vagal trigones and the facial colliculus.
- The superior and inferior colliculi form the dorsal surface of the midbrain.

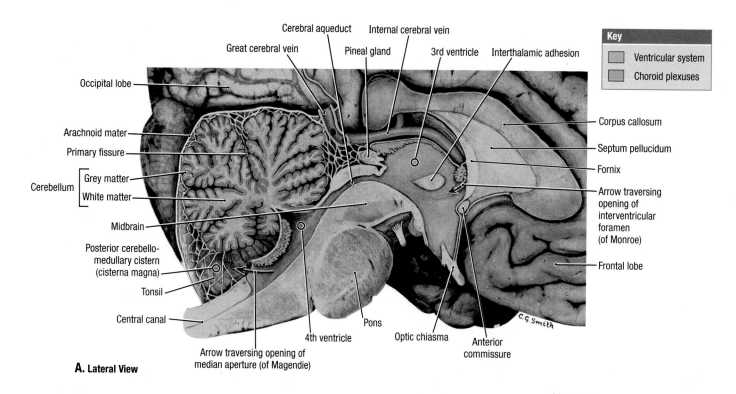

Key
- Ventricular system
- Choroid plexuses

A. Lateral View

Labels (clockwise): Cerebral aqueduct, Internal cerebral vein, Great cerebral vein, Pineal gland, 3rd ventricle, Interthalamic adhesion, Occipital lobe, Corpus callosum, Arachnoid mater, Septum pellucidum, Primary fissure, Fornix, Cerebellum, Grey matter, White matter, Arrow traversing opening of interventricular foramen (of Monroe), Midbrain, Frontal lobe, Posterior cerebello-medullary cistern (cisterna magna), Tonsil, Central canal, Pons, 4th ventricle, Optic chiasma, Anterior commissure, Arrow traversing opening of median aperture (of Magendie)

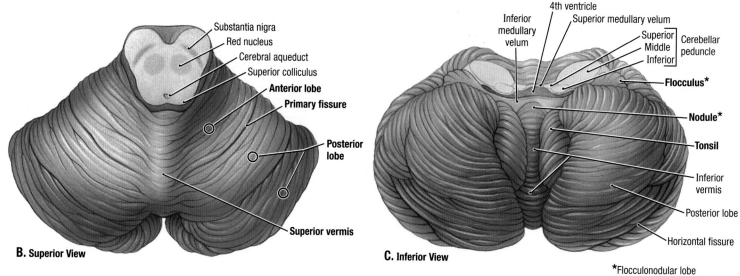

B. Superior View

Labels: Substantia nigra, Red nucleus, Cerebral aqueduct, Superior colliculus, **Anterior lobe**, **Primary fissure**, **Posterior lobe**, **Superior vermis**

C. Inferior View

Labels: 4th ventricle, Inferior medullary velum, Superior medullary velum, Superior / Middle / Inferior Cerebellar peduncle, **Flocculus***, **Nodule***, **Tonsil**, Inferior vermis, Posterior lobe, Horizontal fissure

*Flocculonodular lobe

7.105 CEREBELLUM

A. Median section. The arachnoid mater was removed except where it covered the cerebellum and the occipital lobe. **Cisternal puncture.** CSF may be obtained, for diagnostic purposes, from the posterior cerebellomedullary cistern, using a procedure known as cisternal puncture. The subarachnoid space or the ventricular system may also be entered for measuring or monitoring CSF pressure, injecting antibiotics, or administering contrast media for radiography. **B.** Superior view of the cerebellum. The right and left cerebellar hemispheres are united by the superior vermis; the anterior and posterior lobes are separated by the primary fissure. **C.** Inferior view of cerebellum. The flocculonodular lobe, the oldest part of the cerebellum, consists of the flocculus and nodule; the cerebellar tonsils typically extend into the foramen magnum.

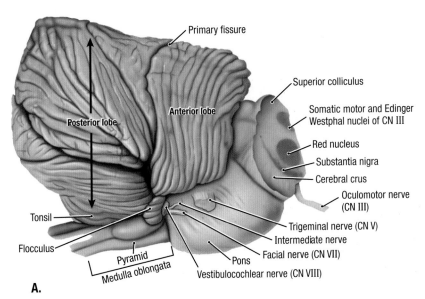

A.

Primary fissure
Superior colliculus
Somatic motor and Edinger Westphal nuclei of CN III
Red nucleus
Substantia nigra
Cerebral crus
Oculomotor nerve (CN III)
Trigeminal nerve (CN V)
Intermediate nerve
Facial nerve (CN VII)
Vestibulocochlear nerve (CN VIII)
Pons
Pyramid
Medulla oblongata
Flocculus
Tonsil
Posterior lobe
Anterior lobe

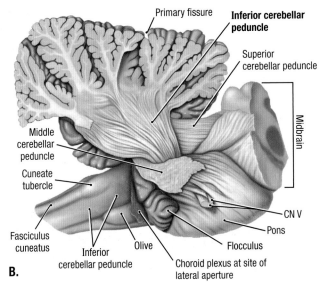

B.

Primary fissure
Inferior cerebellar peduncle
Superior cerebellar peduncle
Midbrain
Middle cerebellar peduncle
Cuneate tubercle
Fasciculus cuneatus
Inferior cerebellar peduncle
Olive
CN V
Pons
Flocculus
Choroid plexus at site of lateral aperture

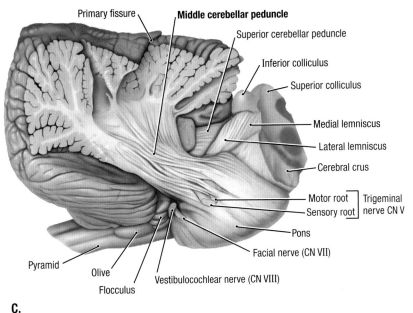

C.

Primary fissure
Middle cerebellar peduncle
Superior cerebellar peduncle
Inferior colliculus
Superior colliculus
Medial lemniscus
Lateral lemniscus
Cerebral crus
Motor root ⎤
Sensory root ⎦ Trigeminal nerve CN V
Pons
Facial nerve (CN VII)
Vestibulocochlear nerve (CN VIII)
Flocculus
Olive
Pyramid

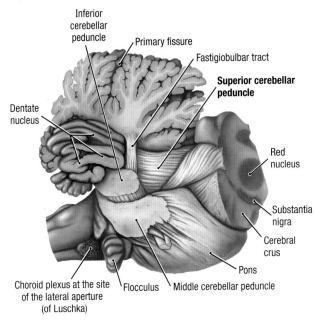

D.

Inferior cerebellar peduncle
Primary fissure
Fastigiobulbar tract
Superior cerebellar peduncle
Dentate nucleus
Red nucleus
Substantia nigra
Cerebral crus
Pons
Middle cerebellar peduncle
Flocculus
Choroid plexus at the site of the lateral aperture (of Luschka)

Lateral Views

SERIAL DISSECTIONS OF THE CEREBELLUM

7.106

The series begins with the lateral surface of the cerebellar hemispheres **(A)** and proceeds medially in sequence **(B–D)**.

A. Cerebellum and brainstem. **B.** Inferior cerebellar peduncle. The fibers of the middle cerebellar peduncle were cut dorsal to the trigeminal nerve and peeled away to expose the fibers of the inferior cerebellar peduncle. **C.** Middle cerebellar peduncle. The fibers of the middle cerebellar peduncle were exposed by peeling away the lateral portion of the lobules of the cerebellar hemisphere. **D.** Superior cerebellar peduncle and dentate nucleus. The fibers of the inferior cerebellar peduncle were cut just dorsal to the previously sectioned middle cerebellar peduncle and peeled away until the gray matter of the dentate nucleus could be seen.

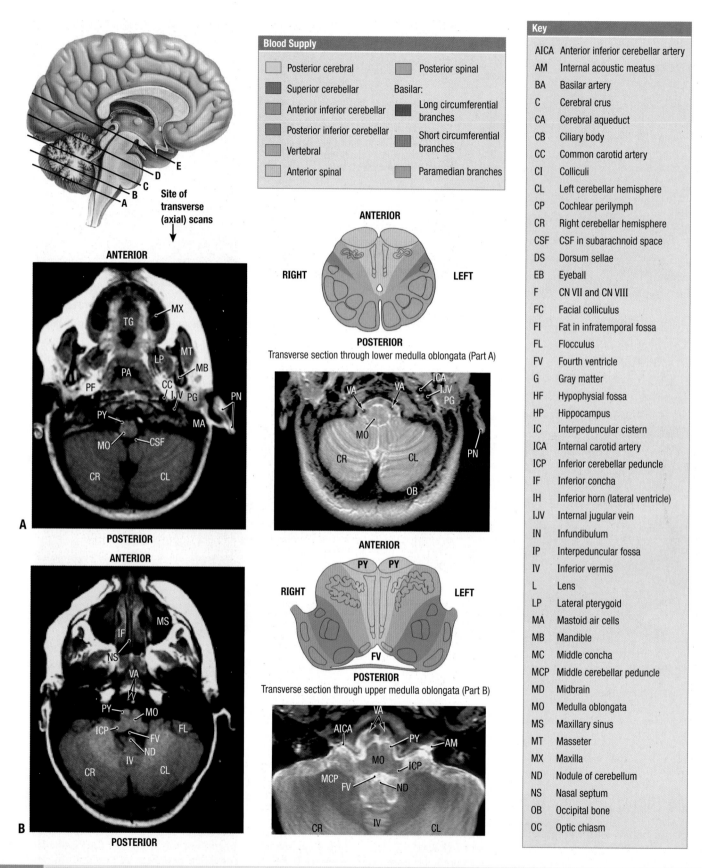

Blood Supply

- ☐ Posterior cerebral
- ☐ Superior cerebellar
- ☐ Anterior inferior cerebellar
- ☐ Posterior inferior cerebellar
- ☐ Vertebral
- ☐ Anterior spinal
- ☐ Posterior spinal
- Basilar:
  - ☐ Long circumferential branches
  - ☐ Short circumferential branches
  - ☐ Paramedian branches

Site of transverse (axial) scans

Transverse section through lower medulla oblongata (Part A)

Transverse section through upper medulla oblongata (Part B)

Key

AICA	Anterior inferior cerebellar artery
AM	Internal acoustic meatus
BA	Basilar artery
C	Cerebral crus
CA	Cerebral aqueduct
CB	Ciliary body
CC	Common carotid artery
CI	Colliculi
CL	Left cerebellar hemisphere
CP	Cochlear perilymph
CR	Right cerebellar hemisphere
CSF	CSF in subarachnoid space
DS	Dorsum sellae
EB	Eyeball
F	CN VII and CN VIII
FC	Facial colliculus
FI	Fat in infratemporal fossa
FL	Flocculus
FV	Fourth ventricle
G	Gray matter
HF	Hypophysial fossa
HP	Hippocampus
IC	Interpeduncular cistern
ICA	Internal carotid artery
ICP	Inferior cerebellar peduncle
IF	Inferior concha
IH	Inferior horn (lateral ventricle)
IJV	Internal jugular vein
IN	Infundibulum
IP	Interpeduncular fossa
IV	Inferior vermis
L	Lens
LP	Lateral pterygoid
MA	Mastoid air cells
MB	Mandible
MC	Middle concha
MCP	Middle cerebellar peduncle
MD	Midbrain
MO	Medulla oblongata
MS	Maxillary sinus
MT	Masseter
MX	Maxilla
ND	Nodule of cerebellum
NS	Nasal septum
OB	Occipital bone
OC	Optic chiasm

7.107 AXIAL (TRANSVERSE) MRIs THROUGH BRAINSTEM, INFERIOR VIEWS

Images on left side of the page are T1-weighted and images on the right side are T2-weighted.

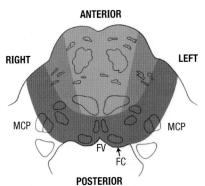

Transverse section through pons (Parts C & D)

Key (continued)

OL	Occipital lobe
ON	Optic nerve (CN II)
P	Pons
PA	Pharynx
PCA	Posterior cerebral artery
PF	Parapharyngeal fat
PG	Parotid gland
PH	Posterior horn (lateral ventricle)
PN	Pinna
PY	Pyramid
RN	Red nucleus
SC	Semicircular canal
SCP	Superior cerebellar peduncle
SE	Suprasellar cistern
SH	Superior concha
SN	Substantia nigra
SS	Superior sagittal sinus
ST	Straight sinus
SV	Superior vermis
TG	Tongue
TL	Temporal lobe
TP	Temporalis
UN	Uncus
VA	Vertebral artery
VP	Vestibular perilymph
VT	Vitreous body
W	White matter

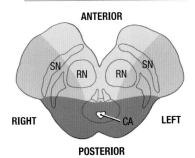

Transverse section through midbrain (Part E)

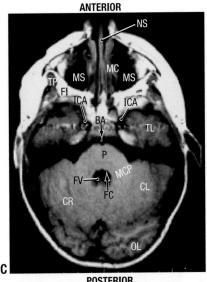

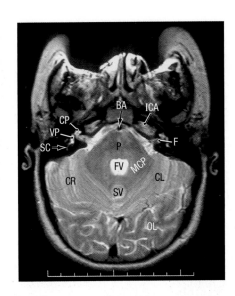

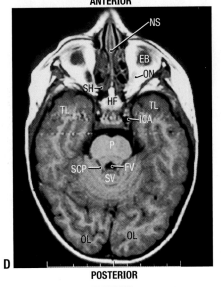

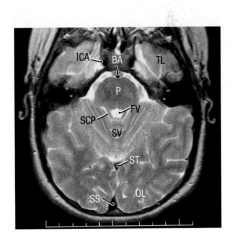

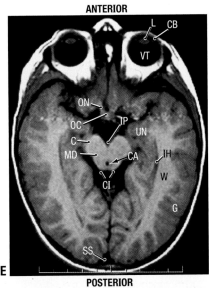

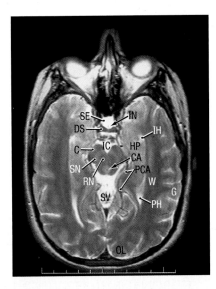

AXIAL (TRANSVERSE) MRIs THROUGH BRAINSTEM, INFERIOR VIEWS (*continued*)

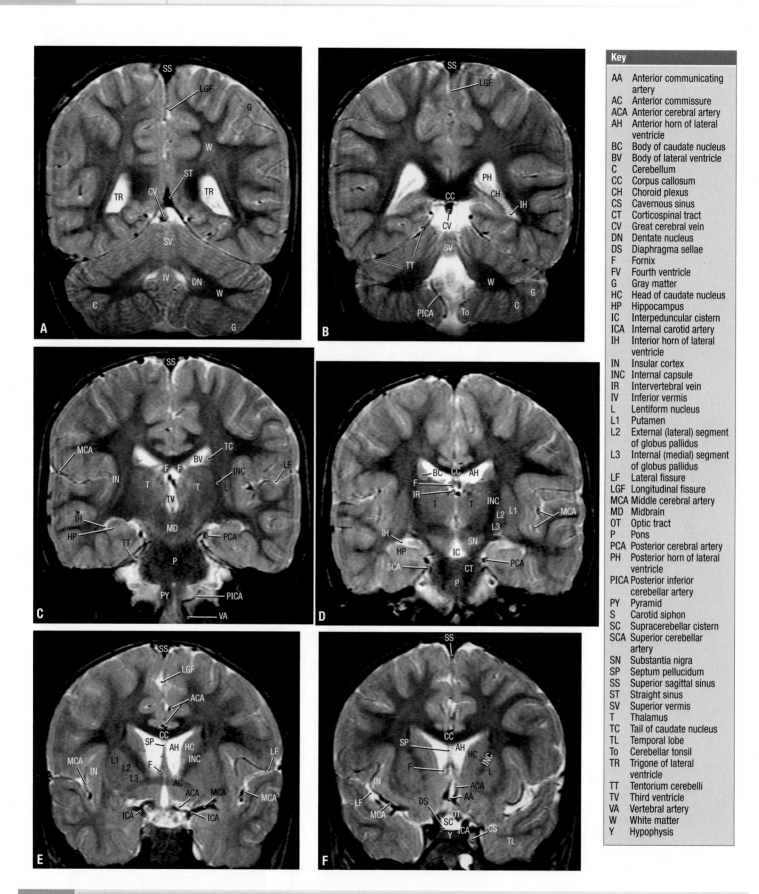

Key

AA	Anterior communicating artery
AC	Anterior commissure
ACA	Anterior cerebral artery
AH	Anterior horn of lateral ventricle
BC	Body of caudate nucleus
BV	Body of lateral ventricle
C	Cerebellum
CC	Corpus callosum
CH	Choroid plexus
CS	Cavernous sinus
CT	Corticospinal tract
CV	Great cerebral vein
DN	Dentate nucleus
DS	Diaphragma sellae
F	Fornix
FV	Fourth ventricle
G	Gray matter
HC	Head of caudate nucleus
HP	Hippocampus
IC	Interpeduncular cistern
ICA	Internal carotid artery
IH	Interior horn of lateral ventricle
IN	Insular cortex
INC	Internal capsule
IR	Intervertebral vein
IV	Inferior vermis
L	Lentiform nucleus
L1	Putamen
L2	External (lateral) segment of globus pallidus
L3	Internal (medial) segment of globus pallidus
LF	Lateral fissure
LGF	Longitudinal fissure
MCA	Middle cerebral artery
MD	Midbrain
OT	Optic tract
P	Pons
PCA	Posterior cerebral artery
PH	Posterior horn of lateral ventricle
PICA	Posterior inferior cerebellar artery
PY	Pyramid
S	Carotid siphon
SC	Supracerebellar cistern
SCA	Superior cerebellar artery
SN	Substantia nigra
SP	Septum pellucidum
SS	Superior sagittal sinus
ST	Straight sinus
SV	Superior vermis
T	Thalamus
TC	Tail of caudate nucleus
TL	Temporal lobe
To	Cerebellar tonsil
TR	Trigone of lateral ventricle
TT	Tentorium cerebelli
TV	Third ventricle
VA	Vertebral artery
W	White matter
Y	Hypophysis

7.108 CORONAL MRIs (T2-WEIGHTED) AND SECTIONS OF BRAIN

A–F. Coronal MRIs. **G–H.** Coronal sections, posterior views.

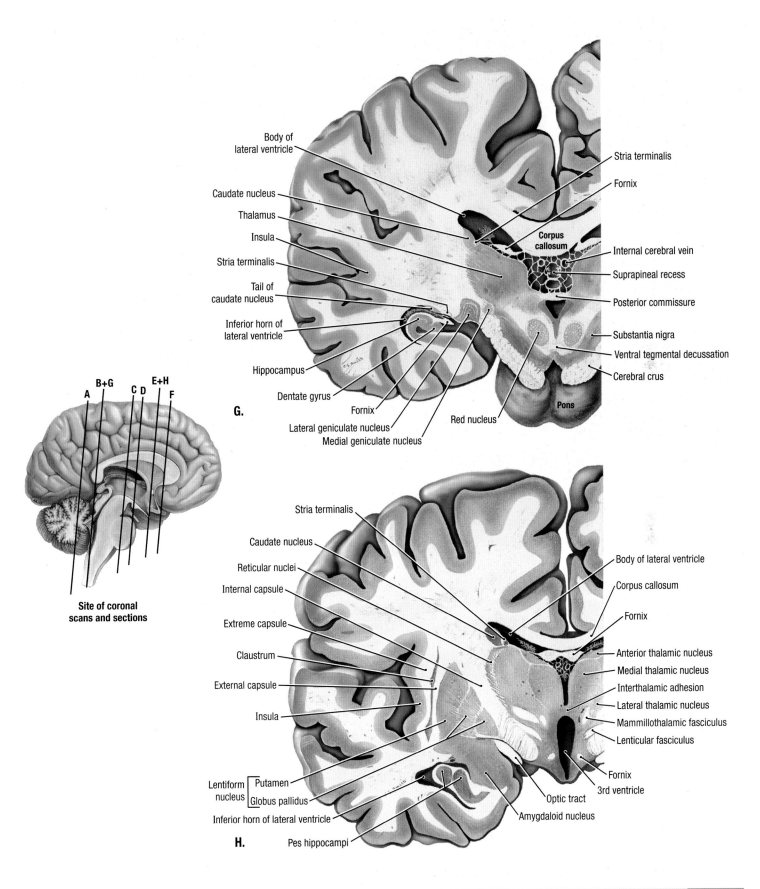

G.

Body of lateral ventricle
Caudate nucleus
Thalamus
Insula
Stria terminalis
Tail of caudate nucleus
Inferior horn of lateral ventricle
Hippocampus
Dentate gyrus
Fornix
Lateral geniculate nucleus
Medial geniculate nucleus

Stria terminalis
Fornix
Corpus callosum
Internal cerebral vein
Suprapineal recess
Posterior commissure
Substantia nigra
Ventral tegmental decussation
Cerebral crus
Pons
Red nucleus

A   B+G   C D   E+H   F

Site of coronal scans and sections

H.

Stria terminalis
Caudate nucleus
Reticular nuclei
Internal capsule
Extreme capsule
Claustrum
External capsule
Insula
Lentiform nucleus { Putamen / Globus pallidus }
Inferior horn of lateral ventricle
Pes hippocampi

Body of lateral ventricle
Corpus callosum
Fornix
Anterior thalamic nucleus
Medial thalamic nucleus
Interthalamic adhesion
Lateral thalamic nucleus
Mammillothalamic fasciculus
Lenticular fasciculus
Fornix
3rd ventricle
Optic tract
Amygdaloid nucleus

CORONAL MRIs (T2-WEIGHTED) AND SECTIONS OF BRAIN (continued)

7.108

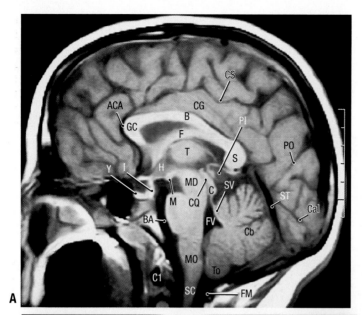

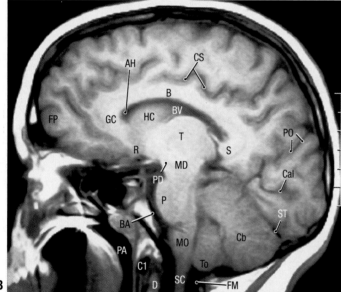

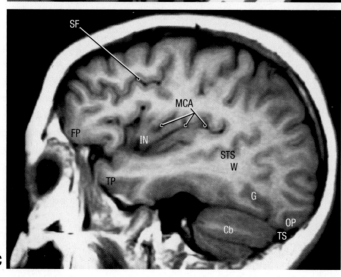

Key

ACA	Anterior cerebral artery
AH	Anterior horn of lateral ventricle
B	Body of corpus callosum
BA	Basilar artery
BV	Body of lateral ventricle
C	Colliculi
C1	Anterior tubercle of atlas
Cal	Calcarine sulcus
Cb	Cerebellum
CG	Cingulate nucleus
CQ	Cerebral aqueduct
CS	Cingulate sulcus
D	Dens (odontoid process)
F	Fornix
FM	Foramen magnum
FP	Frontal pole
FV	Fourth ventricle
G	Cerebral cortex (gray matter)
GC	Genus of corpus callosum
H	Hypothalamus
HC	Head of caudate nucleus
I	Infundibulum
IN	Insular cortex
M	Mammillary body
MCA	Middle cerebral artery
MD	Midbrain
MO	Medulla oblongata
OP	Occipital pole
P	Pons
PA	Pharynx
PD	Cerebral peduncle
PI	Pineal
PO	Parieto-occipital fissure
R	Rostrum of corpus callosum
S	Splenium of corpus callosum
SC	Spinal cord
SF	Superior frontal sulcus
ST	Straight sinus
STS	Superior temporal sulcus
SV	Superior medullary vellum
T	Thalamus
To	Cerebellar tonsil
TP	Temporal pole
TS	Transverse sinus
W	White matter
Y	Hypophysis

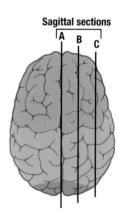

Sagittal sections

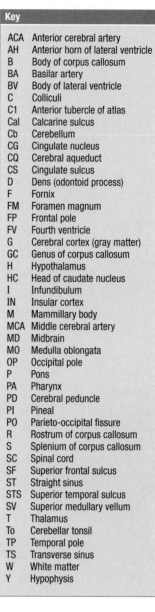

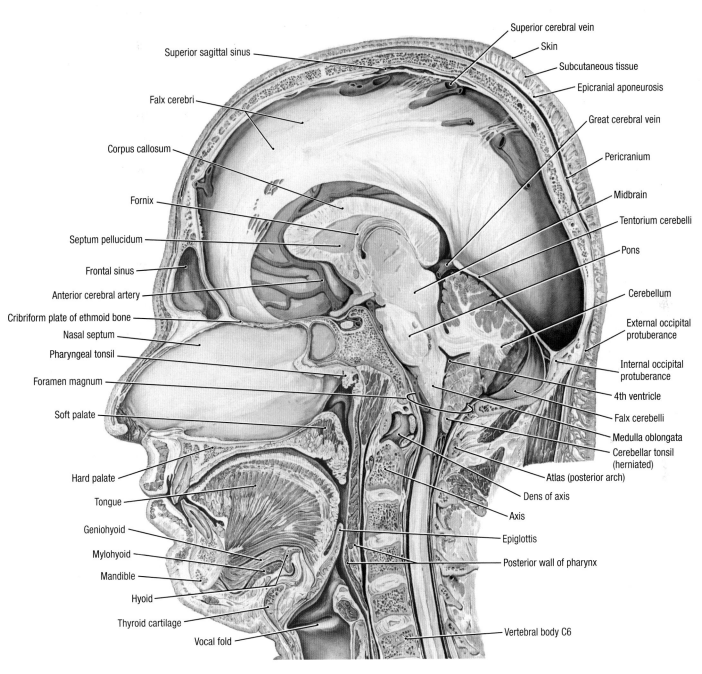

Superior cerebral vein
Skin
Subcutaneous tissue
Epicranial aponeurosis
Great cerebral vein
Pericranium
Midbrain
Tentorium cerebelli
Pons
Cerebellum
External occipital protuberance
Internal occipital protuberance
4th ventricle
Falx cerebelli
Medulla oblongata
Cerebellar tonsil (herniated)
Atlas (posterior arch)
Dens of axis
Axis
Epiglottis
Posterior wall of pharynx
Vertebral body C6

Superior sagittal sinus
Falx cerebri
Corpus callosum
Fornix
Septum pellucidum
Frontal sinus
Anterior cerebral artery
Cribriform plate of ethmoid bone
Nasal septum
Pharyngeal tonsil
Foramen magnum
Soft palate
Hard palate
Tongue
Geniohyoid
Mylohyoid
Mandible
Hyoid
Thyroid cartilage
Vocal fold

D. Median Section

SAGITTAL MRIs (T1-WEIGHTED) AND MEDIAN SECTION OF BRAIN *(continued)*   7.109

See orientation drawing for sites of scans **A–C**.

Increased intracranial pressure (e.g., due to a tumor) may cause displacement of the cerebellar tonsils through the foramen magnum, resulting in a foraminal (tonsillar) herniation. Compression of the brainstem, if severe, may result in respiratory and cardiac arrest.

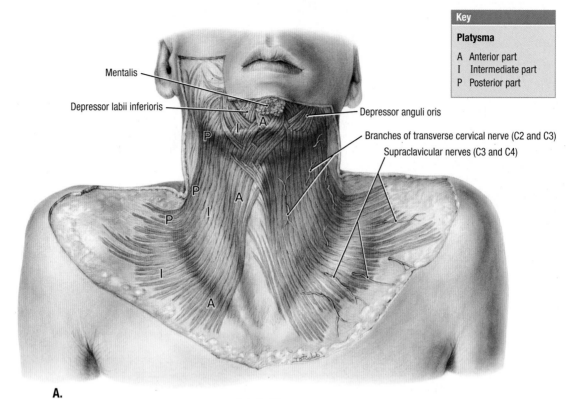

Mentalis

Depressor labii inferioris

Depressor anguli oris

Branches of transverse cervical nerve (C2 and C3)

Supraclavicular nerves (C3 and C4)

Key

Platysma

A Anterior part
I Intermediate part
P Posterior part

A.

Anterior Views

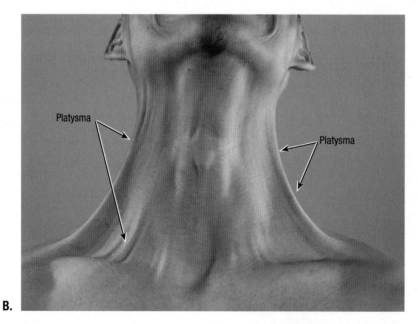

Platysma

Platysma

B.

8.1   PLATYSMA

A. Parts of platysma. **B.** Surface anatomy.

TABLE 8.1	PLATYSMA			
Muscle	**Superior Attachment**	**Inferior Attachment**	**Innervation**	**Main Action**
Platysma	*Anterior part:* Fibers interlace with contralateral muscle *Intermediate part:* Fibers pass deep to depressors anguli oris and labii inferioris to attach to inferior border of mandible *Posterior part:* Skin/subcutaneous tissue of lower face lateral to mouth	Subcutaneous tissue overlying superior parts of pectoralis major and sometimes deltoid muscles	Cervical branch of facial nerve (CN VII)	Draws corner of mouth inferiorly and widens it as in expressions of sadness and fright; draws the skin of the neck superiorly, forming tense vertical and oblique ridges over the anterior neck

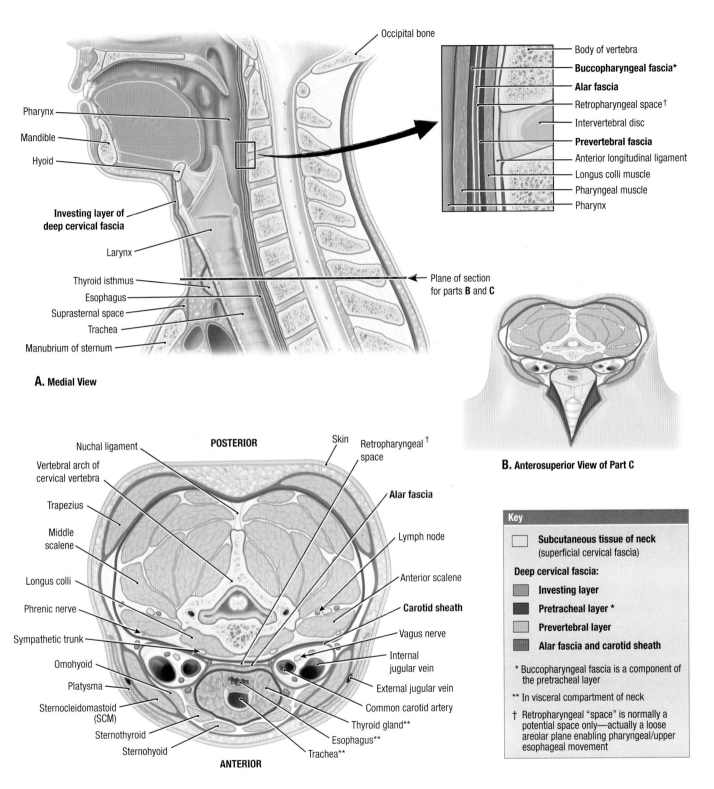

A. Medial View

Labels on Medial View:
- Occipital bone
- Pharynx
- Mandible
- Hyoid
- Investing layer of deep cervical fascia
- Larynx
- Thyroid isthmus
- Esophagus
- Suprasternal space
- Trachea
- Manubrium of sternum
- Plane of section for parts **B** and **C**

Inset labels:
- Body of vertebra
- **Buccopharyngeal fascia***
- **Alar fascia**
- Retropharyngeal space †
- Intervertebral disc
- **Prevertebral fascia**
- Anterior longitudinal ligament
- Longus colli muscle
- Pharyngeal muscle
- Pharynx

B. Anterosuperior View of Part C

Labels on Transverse Section (Part C):
- Nuchal ligament
- Vertebral arch of cervical vertebra
- Trapezius
- Middle scalene
- Longus colli
- Phrenic nerve
- Sympathetic trunk
- Omohyoid
- Platysma
- Sternocleidomastoid (SCM)
- Sternothyroid
- Sternohyoid
- **POSTERIOR**
- Skin
- Retropharyngeal † space
- **Alar fascia**
- Lymph node
- Anterior scalene
- **Carotid sheath**
- Vagus nerve
- Internal jugular vein
- External jugular vein
- Common carotid artery
- Thyroid gland**
- Esophagus**
- Trachea**
- **ANTERIOR**

C. Superior View of Transverse Section (at Level of C7 Vertebra)

Key

Subcutaneous tissue of neck (superficial cervical fascia)

Deep cervical fascia:
- **Investing layer**
- **Pretracheal layer ***
- **Prevertebral layer**
- **Alar fascia and carotid sheath**

* Buccopharyngeal fascia is a component of the pretracheal layer

** In visceral compartment of neck

† Retropharyngeal "space" is normally a potential space only—actually a loose areolar plane enabling pharyngeal/upper esophageal movement

SUBCUTANEOUS TISSUE AND DEEP FASCIA OF NECK

8.2

Sectional demonstrations of the fasciae of the neck. **A.** Fasciae of the neck are continuous inferiorly and superiorly with thoracic and cranial fasciae. The *inset* illustrates the fascia of the retropharyngeal region. **B.** Relationship of the main layers of deep cervical fascia and the carotid sheath. Midline access to the cervical viscera is possible with minimal disruption of tissues. **C.** The concentric layers of fascia are apparent in this transverse section of neck at the level indicated in **A.**

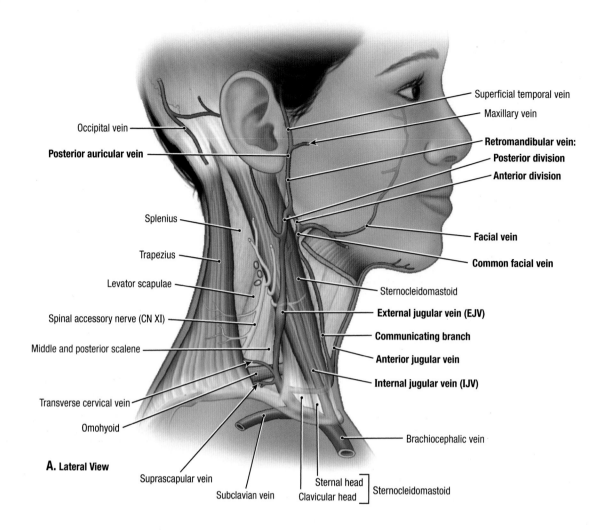

Occipital vein

Posterior auricular vein

Splenius

Trapezius

Levator scapulae

Spinal accessory nerve (CN XI)

Middle and posterior scalene

Transverse cervical vein

Omohyoid

A. Lateral View

Suprascapular vein

Subclavian vein

Superficial temporal vein

Maxillary vein

Retromandibular vein:

Posterior division

Anterior division

Facial vein

Common facial vein

Sternocleidomastoid

External jugular vein (EJV)

Communicating branch

Anterior jugular vein

Internal jugular vein (IJV)

Brachiocephalic vein

Sternal head ⎤
Clavicular head ⎦ Sternocleidomastoid

8.3   SUPERFICIAL VEINS OF NECK

A. Schematic illustration of superficial veins of the neck. The superficial temporal and maxillary veins merge to form the retromandibular vein. The posterior division of the retromandibular vein unites with the posterior auricular vein to form the external jugular vein. The facial vein receives the anterior division of the retromandibular vein, forming the common facial vein that empties into the internal jugular vein. Variations are common. **B.** Surface anatomy of the external jugular vein and the muscles bounding the lateral cervical region (posterior triangle) of the neck.

External jugular vein (EJV). The EJV may serve as an "internal barometer." When venous pressure is in the normal range, the EJV is usually visible superior to the clavicle for only a short distance. However, when venous pressure rises (e.g., as in heart failure), the vein is prominent throughout its course along the side of the neck. Consequently, routine observation for distention of the EJVs during physical examinations may reveal diagnostic signs of heart failure, obstruction of the superior vena cava, enlarged supraclavicular lymph nodes, or increased intrathoracic pressure.

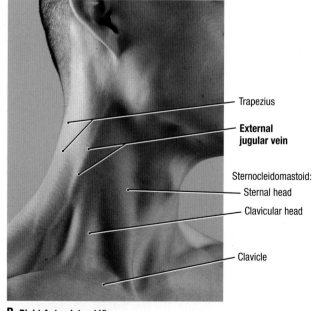

Trapezius

External jugular vein

Sternocleidomastoid:

Sternal head

Clavicular head

Clavicle

B. Right Anterolateral View

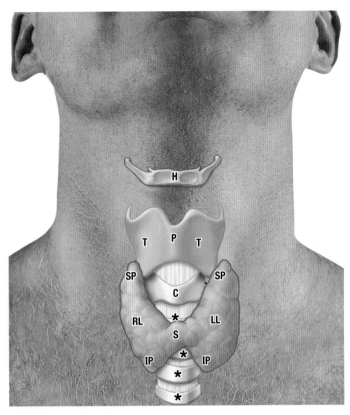

A. Anterior View

Key	
C	Cricoid cartilage
H	Hyoid bone
IP	Inferior pole of thyroid gland
LL	Left lobe of thyroid gland
P	Laryngeal prominence
RL	Right lobe of thyroid gland
S	Isthmus
SP	Superior pole of thyroid gland
T	Thyroid cartilage
★	Tracheal rings

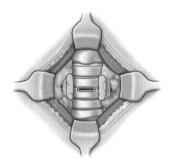

Incision in trachea after
retracting infrahyoid
muscles and incising
isthmus of thyroid gland

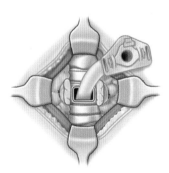

Tracheostomy tube
inserted in tracheal opening

B. Tracheostomy

SURFACE ANATOMY OF HYOID AND CARTILAGES OF ANTERIOR NECK   8.4

A. Surface anatomy. **B.** Tracheostomy. The U-shaped hyoid bone lies superior to the thyroid cartilage at the level of the C4 and C5 vertebrae. The laryngeal prominence is produced by the fused laminae of the thyroid cartilage, which meet in the median plane. The cricoid cartilage can be felt inferior to the laryngeal prominence. It lies at the level of the C6 vertebra. The cartilaginous tracheal rings are palpable in the inferior part of the neck. The 2nd to 4th rings cannot be felt because the isthmus of the thyroid, connecting its right and left lobes, covers them. The 1st tracheal ring is just superior to the isthmus.

Tracheostomy. A transverse incision through the skin of the neck and anterior wall of the trachea (*tracheostomy*) establishes an airway in patients with upper airway obstruction or respiratory failure. The infrahyoid muscles are retracted laterally, and the isthmus of the thyroid gland is either divided or retracted superiorly. An opening is made in the trachea between the 1st and 2nd tracheal rings or through the 2nd through 4th rings. A tracheostomy tube is then inserted into the trachea and secured. To avoid complications during a tracheostomy, the following anatomical relationships are important:

- The inferior thyroid veins arise from a venous plexus on the thyroid gland and descend anterior to the trachea (see Fig. 8.10).
- A small thyroid ima artery is present in approximately 10% of people; it ascends from the brachiocephalic trunk or the arch of the aorta to the isthmus of the thyroid gland (see Fig. 8.21).
- The left brachiocephalic vein, jugular venous arch, and pleurae may be encountered, particularly in infants and children.
- The thymus covers the inferior part of the trachea in infants and children.
- The trachea is small, mobile, and soft in infants, making it easy to cut through its posterior wall and damage the esophagus.

Cricothyrotomy. The incision is made through the cricothyroid membrane, and the tube inserted between the thyroid and cricoid cartilages.

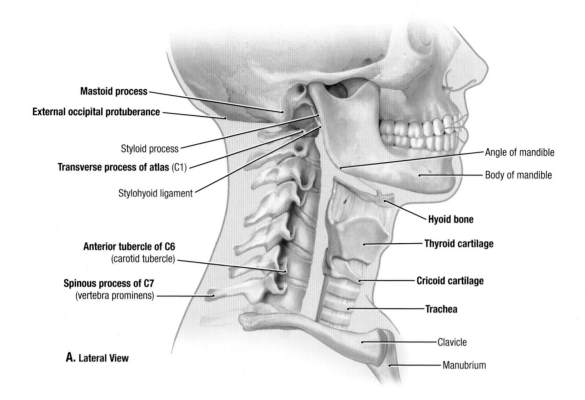

Mastoid process

External occipital protuberance

Styloid process

Transverse process of atlas (C1)

Stylohyoid ligament

Anterior tubercle of C6
(carotid tubercle)

Spinous process of C7
(vertebra prominens)

Angle of mandible

Body of mandible

Hyoid bone

Thyroid cartilage

Cricoid cartilage

Trachea

Clavicle

Manubrium

A. Lateral View

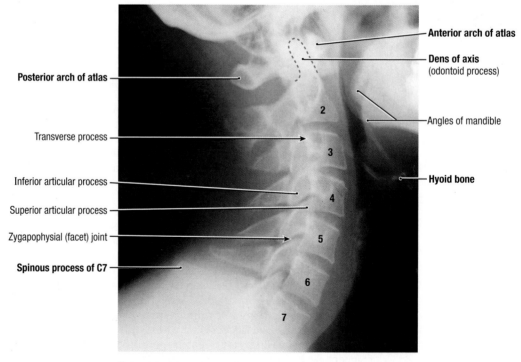

Posterior arch of atlas

Transverse process

Inferior articular process

Superior articular process

Zygapophysial (facet) joint

Spinous process of C7

Anterior arch of atlas

Dens of axis
(odontoid process)

Angles of mandible

Hyoid bone

B. Lateral Radiograph

8.5 BONES AND CARTILAGES OF NECK

A. Bony and cartilaginous landmarks of the neck. **B.** Radiograph of hyoid bone and cervical vertebrae. Because the upper cervical vertebrae lie posterior to the upper and lower jaws and teeth, they are best seen radiographically in lateral views.

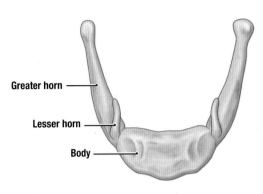

C. and D. views of Hyoid Bone

Greater horn
Lesser horn
Fibrocartilage
Body

C. Right Anterolateral View of Hyoid Bone

Greater horn
Lesser horn
Body

D. Anterosuperior View of Hyoid Bone

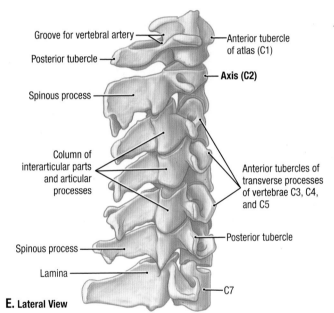

Groove for vertebral artery
Posterior tubercle
Anterior tubercle of atlas (C1)
Axis (C2)
Spinous process
Column of interarticular parts and articular processes
Anterior tubercles of transverse processes of vertebrae C3, C4, and C5
Posterior tubercle
Spinous process
Lamina
C7

E. Lateral View

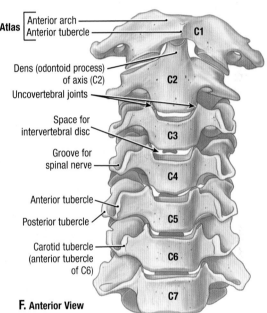

Atlas Anterior arch
Anterior tubercle — C1
Dens (odontoid process) of axis (C2) — C2
Uncovertebral joints
Space for intervertebral disc — C3
Groove for spinal nerve — C4
Anterior tubercle — C5
Posterior tubercle
Carotid tubercle (anterior tubercle of C6) — C6
C7

F. Anterior View

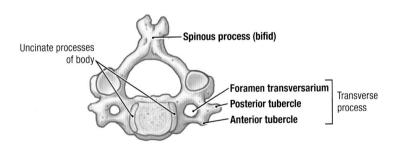

Uncinate processes of body
Spinous process (bifid)
Foramen transversarium
Posterior tubercle
Anterior tubercle
Transverse process

G. Superior View of Typical Cervical Vertebra (e.g., C4)

BONES AND CARTILAGES OF NECK (*continued*)

8.5

C. and **D.** Features of hyoid. **E.** and **F.** Articulated cervical vertebrae. **G.** Features of typical cervical vertebrae.

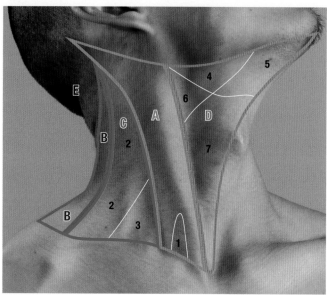

A. Anterolateral View

Key for A and B

A	Sternocleidomastoid region
B	Posterior cervical region
C	Lateral cervical region
D	Anterior cervical region
E	Suboccipital region
SCM	Sternocleidomastoid
	CH Clavicular head
	SH Sternal head
TRAP	Trapezius

B. Lateral View

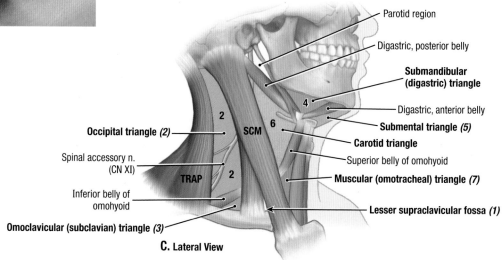

Parotid region
Digastric, posterior belly
Submandibular (digastric) triangle
Digastric, anterior belly
Occipital triangle (2)
Submental triangle (5)
Carotid triangle
Spinal accessory n. (CN XI)
Superior belly of omohyoid
Muscular (omotracheal) triangle (7)
Inferior belly of omohyoid
Lesser supraclavicular fossa (1)
Omoclavicular (subclavian) triangle (3)

C. Lateral View

8.6 CERVICAL REGIONS

A. Surface anatomy. **B.** and **C.** Regions and triangles of neck.

TABLE 8.2	CERVICAL REGIONS AND CONTENTS[a]
Region	**Main Contents and Underlying Structures**
Sternocleidomastoid region (A)	Sternocleidomastoid (SCM) muscle; superior part of the external jugular vein; greater auricular nerve; transverse cervical nerve
Lesser supraclavicular fossa (1)	Inferior part of internal jugular vein
Posterior cervical region (B)	Trapezius muscle; cutaneous branches of posterior rami of cervical spinal nerves; suboccipital region (E) lies deep to superior part of this region
Lateral cervical region (posterior triangle) (C) Occipital triangle (2) Omoclavicular triangle (3)	Part of external jugular vein; posterior branches of cervical plexus of nerves; spinal accessory nerve; trunks of brachial plexus; transverse cervical artery; cervical lymph nodes
	Subclavian artery; part of subclavian vein (variable); suprascapular artery; supraclavicular lymph nodes
Anterior cervical region (anterior triangle) (D) Submandibular (digastric) triangle (4) Submental triangle (5) Carotid triangle (6) Muscular (omotracheal) triangle (7)	Submandibular gland almost fills triangle; submandibular lymph nodes; hypoglossal nerve; mylohyoid nerve; parts of facial artery and vein
	Submental lymph nodes and small veins that unite to form anterior jugular vein
	Common carotid artery and its branches; internal jugular vein and its tributaries; vagus nerve; external carotid artery and some of its branches; hypoglossal nerve and superior root of ansa cervicalis; spinal accessory nerve; thyroid gland, larynx, and pharynx; deep cervical lymph nodes; branches of cervical plexus
	Sternothyroid and sternohyoid muscles; thyroid and parathyroid glands

[a]Letters and numbers in parentheses refer to Figures A, B, and C.

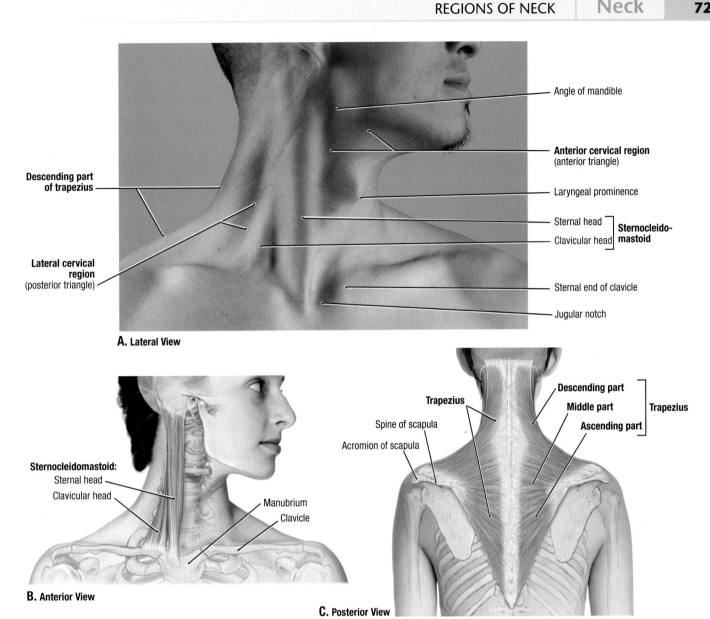

A. Lateral View

Angle of mandible

Anterior cervical region (anterior triangle)

Laryngeal prominence

Sternal head ⎫
Clavicular head ⎬ **Sternocleido-mastoid**

Sternal end of clavicle

Jugular notch

Descending part of trapezius

Lateral cervical region (posterior triangle)

B. Anterior View

Sternocleidomastoid:
Sternal head
Clavicular head

Manubrium
Clavicle

C. Posterior View

Trapezius

Spine of scapula

Acromion of scapula

Descending part
Middle part ⎬ **Trapezius**
Ascending part

STERNOCLEIDOMASTOID AND TRAPEZIUS

8.7

A. Surface anatomy. **B.** Sternocleidomastoid. **C.** Trapezius.

TABLE 8.3	**STERNOCLEIDOMASTOID AND TRAPEZIUS**			
Muscle	*Superior Attachment*	*Inferior Attachment*	*Innervation*	*Main Action*
Sternocleidomastoid	Lateral surface of mastoid process of temporal bone; lateral half of superior nuchal line	*Sternal head:* anterior surface of manubrium of sternum *Clavicular head:* superior surface of medial third of clavicle	Spinal accessory nerve (CN XI) [motor] and C2 and C3 nerves (pain and proprioception)	*Unilateral contraction:* laterally flexes neck; rotates neck so face is turned superiorly toward opposite side *Bilateral contraction:* (1) extends neck at atlanto-occipital joints, (2) flexes cervical vertebrae so that chin approaches manubrium, or (3) extends superior cervical vertebrae while flexing inferior vertebrae, so chin is thrust forward with head kept level; with cervical vertebrae fixed, may elevate manubrium and medial end of clavicles, assisting deep respiration
Trapezius	Medial third of superior nuchal line, external occipital protuberance, nuchal ligament, spinous processes of C7–T12 vertebrae, lumbar and sacral spinous processes	Lateral third of clavicle, acromion, spine of scapula	Spinal accessory nerve (CN XI) [motor] and C2 and C3 nerves (pain and proprioception)	*Descending fibers* elevate pectoral girdle, maintain level of shoulders against gravity or resistance; *middle fibers* retract scapula; and *ascending fibers* depress shoulders; *superior* and *inferior fibers* work together to rotate scapula upward; *when shoulders are fixed,* bilateral contraction extends neck; unilateral contraction produces lateral flexion to same side

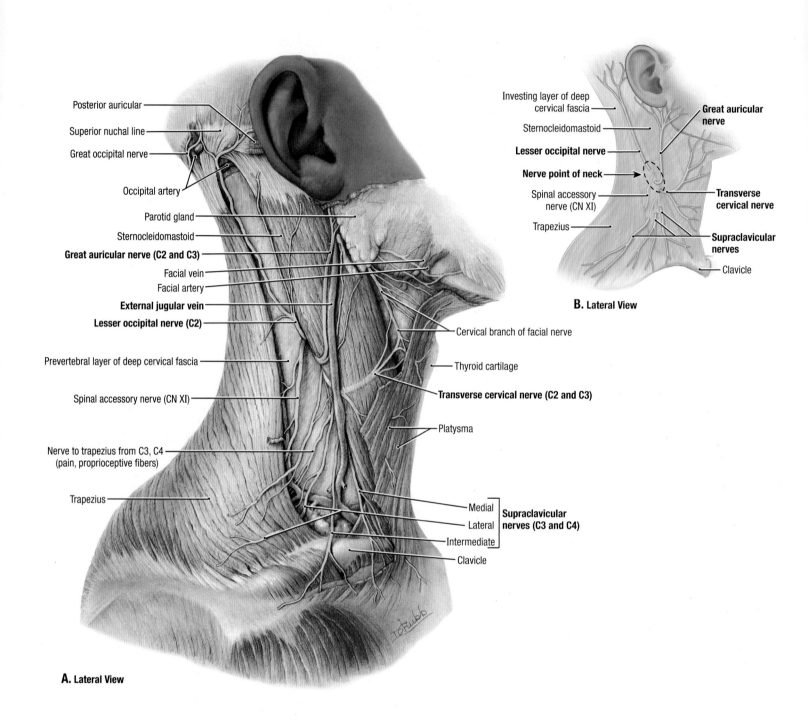

Posterior auricular

Superior nuchal line

Great occipital nerve

Occipital artery

Parotid gland

Sternocleidomastoid

Great auricular nerve (C2 and C3)

Facial vein

Facial artery

External jugular vein

Lesser occipital nerve (C2)

Prevertebral layer of deep cervical fascia

Spinal accessory nerve (CN XI)

Nerve to trapezius from C3, C4 (pain, proprioceptive fibers)

Trapezius

Cervical branch of facial nerve

Thyroid cartilage

Transverse cervical nerve (C2 and C3)

Platysma

Medial

Lateral

Intermediate

Clavicle

Supraclavicular nerves (C3 and C4)

A. Lateral View

Investing layer of deep cervical fascia

Sternocleidomastoid

Lesser occipital nerve

Nerve point of neck

Spinal accessory nerve (CN XI)

Trapezius

Great auricular nerve

Transverse cervical nerve

Supraclavicular nerves

Clavicle

B. Lateral View

8.8 SERIAL DISSECTIONS OF LATERAL CERVICAL REGION (POSTERIOR TRIANGLE OF NECK)

A. External jugular vein and cutaneous branches of cervical plexus. Subcutaneous fat, the part of the plasma overlying the inferior part of the lateral cervical region, and the investing layer of deep cervical fascia have all been removed. The external jugular vein descends vertically across the sternocleidomastoid and pierces the prevertebral layer of deep cervical fascia superior to the clavicle.

• The spinal accessory nerve (CN XI) supplies the sternocleidomastoid (SCM) and trapezius muscles; between them, it courses along the levator scapulae muscle but is separated from it by the prevertebral layer of deep cervical fascia.

B. Nerve point of neck.

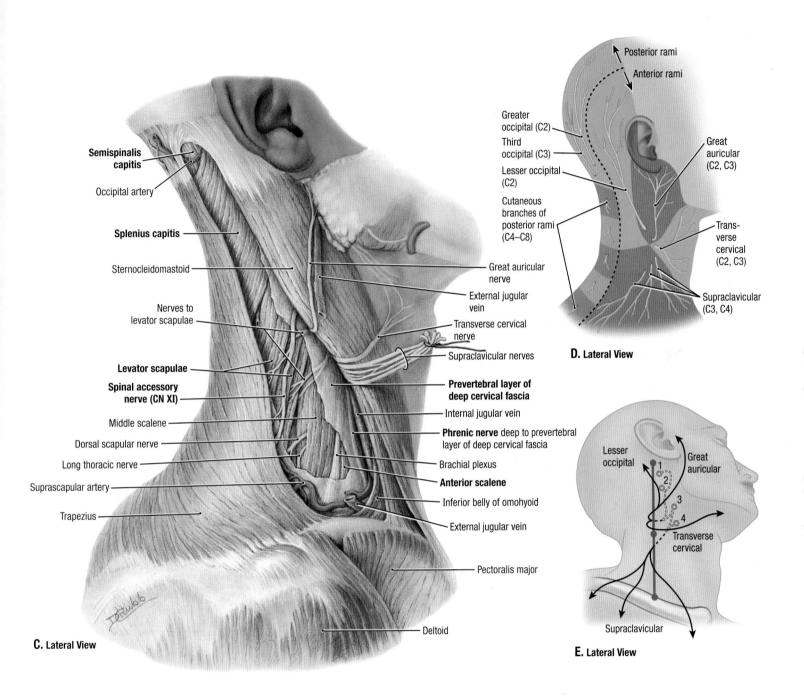

Semispinalis capitis

Occipital artery

Splenius capitis

Sternocleidomastoid

Nerves to levator scapulae

Levator scapulae

Spinal accessory nerve (CN XI)

Middle scalene

Dorsal scapular nerve

Long thoracic nerve

Suprascapular artery

Trapezius

C. Lateral View

Great auricular nerve

External jugular vein

Transverse cervical nerve

Supraclavicular nerves

Prevertebral layer of deep cervical fascia

Internal jugular vein

Phrenic nerve deep to prevertebral layer of deep cervical fascia

Brachial plexus

Anterior scalene

Inferior belly of omohyoid

External jugular vein

Pectoralis major

Deltoid

Posterior rami

Anterior rami

Greater occipital (C2)

Third occipital (C3)

Lesser occipital (C2)

Cutaneous branches of posterior rami (C4–C8)

Great auricular (C2, C3)

Trans-verse cervical (C2, C3)

Supraclavicular (C3, C4)

D. Lateral View

Lesser occipital

Great auricular

1

2

3

4

Transverse cervical

Supraclavicular

E. Lateral View

SERIAL DISSECTIONS OF LATERAL CERVICAL REGION (POSTERIOR TRIANGLE OF NECK) (*continued*)   8.8

C. Muscles forming the floor of the lateral cervical region. The prevertebral layer of deep cervical fascia has been partially removed, and the motor nerves and most of the floor of the region are exposed.

- The phrenic nerve (C3, C4, C5) supplies the diaphragm and is located deep to the prevertebral layer of deep cervical fascia on the anterior surface of the anterior scalene muscle.

Severance of a phrenic nerve results in an ipsilateral paralysis of the diaphragm. A phrenic nerve block produces a short period of paralysis of the diaphragm on one side (e.g., for a lung operation). The anesthetic agent is injected around the nerve where it lies on the anterior surface of the anterior scalene muscle.

D. and **E.** Sensory nerves of cervical plexus. Branches arising from the nerve loop between the anterior rami of C2 and C3 are the lesser occipital, great auricular, and transverse cervical nerves. Branches arising from the loop formed between the anterior rami of C3 and C4 are the supraclavicular nerves, which emerge as a common trunk under cover of the SCM.

Regional anesthesia is often used for surgical procedures in the neck region or upper limb. In a **cervical plexus block,** an anesthetic agent is injected at several points along the posterior border of the SCM, mainly at its midpoint, the nerve point of the neck.

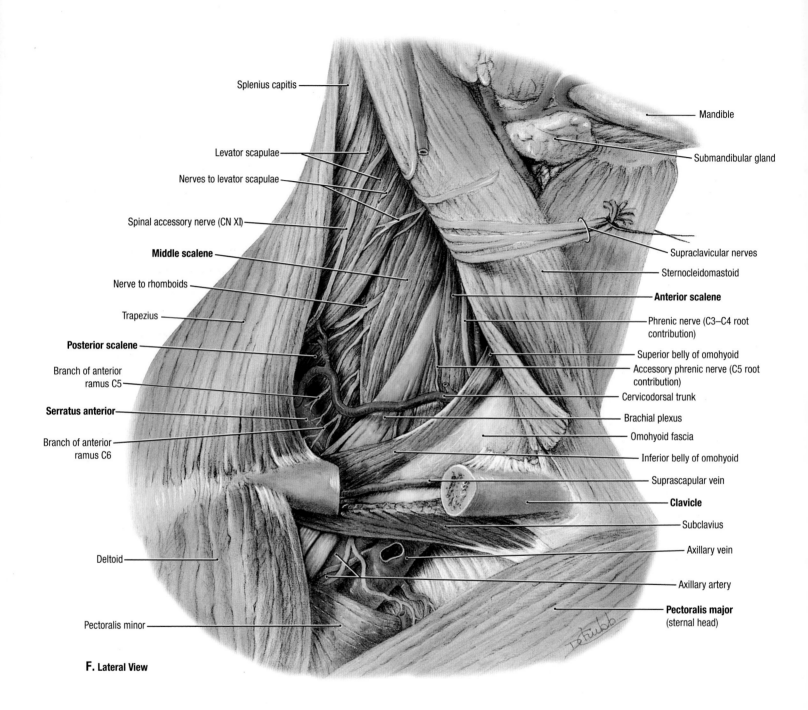

Splenius capitis

Levator scapulae

Nerves to levator scapulae

Spinal accessory nerve (CN XI)

Middle scalene

Nerve to rhomboids

Trapezius

Posterior scalene

Branch of anterior ramus C5

Serratus anterior

Branch of anterior ramus C6

Deltoid

Pectoralis minor

Mandible

Submandibular gland

Supraclavicular nerves

Sternocleidomastoid

Anterior scalene

Phrenic nerve (C3–C4 root contribution)

Superior belly of omohyoid

Accessory phrenic nerve (C5 root contribution)

Cervicodorsal trunk

Brachial plexus

Omohyoid fascia

Inferior belly of omohyoid

Suprascapular vein

Clavicle

Subclavius

Axillary vein

Axillary artery

Pectoralis major (sternal head)

F. Lateral View

8.8   SERIAL DISSECTIONS OF LATERAL CERVICAL REGION (POSTERIOR TRIANGLE OF NECK) *(continued)*

F. Vessels and motor nerves of the lateral cervical region. The clavicular head of the pectoralis major muscle and part of the clavicle have been removed.

The muscles that form the floor of the region are the semispinalis capitis, splenius capitis, and levator scapulae superiorly and the anterior, middle, and posterior scalenes and serratus anterior inferiorly.

A **supraclavicular brachial plexus block** may be utilized for anesthesia of the upper limb. The anesthetic agent is injected around the supraclavicular part of the brachial plexus. The main injection site is superior to the midpoint of the clavicle.

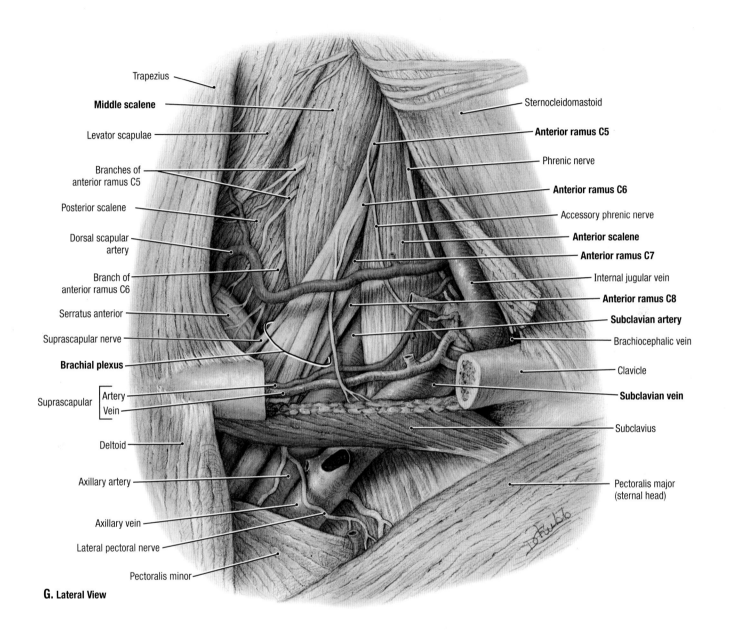

Trapezius

Middle scalene

Levator scapulae

Branches of
anterior ramus C5

Posterior scalene

Dorsal scapular
artery

Branch of
anterior ramus C6

Serratus anterior

Suprascapular nerve

Brachial plexus

Suprascapular — Artery / Vein

Deltoid

Axillary artery

Axillary vein

Lateral pectoral nerve

Pectoralis minor

Sternocleidomastoid

Anterior ramus C5

Phrenic nerve

Anterior ramus C6

Accessory phrenic nerve

Anterior scalene

Anterior ramus C7

Internal jugular vein

Anterior ramus C8

Subclavian artery

Brachiocephalic vein

Clavicle

Subclavian vein

Subclavius

Pectoralis major
(sternal head)

G. Lateral View

G. Structures of the omoclavicular (subclavian) triangle. The omo-hyoid muscle and fascia have been removed, exposing the brachial plexus and subclavian vessels.

- The anterior rami of C5–T1 form the brachial plexus; the anterior ramus of T1 lies posterior to the subclavian artery.
- The brachial plexus and subclavian artery emerge between the middle and anterior scalene muscles.
- The anterior scalene muscle lies between the subclavian artery and vein.

The right or left subclavian vein is often the site of **placement for a central venous catheter**, used to insert intravenous tubes ("central venous lines") for the administration of parenteral nutritional fluids or medications, for testing blood chemistry or central venous pressure, or inserting electrode wires for heart pacemaker devices. The relationships of the subclavian vein to the sternocleidomastoid muscle, clavicle, sternoclavicular joint, and 1st rib are of clinical importance in line placement, and there is danger of puncture of the pleura or subclavian artery if the procedure is not performed correctly.

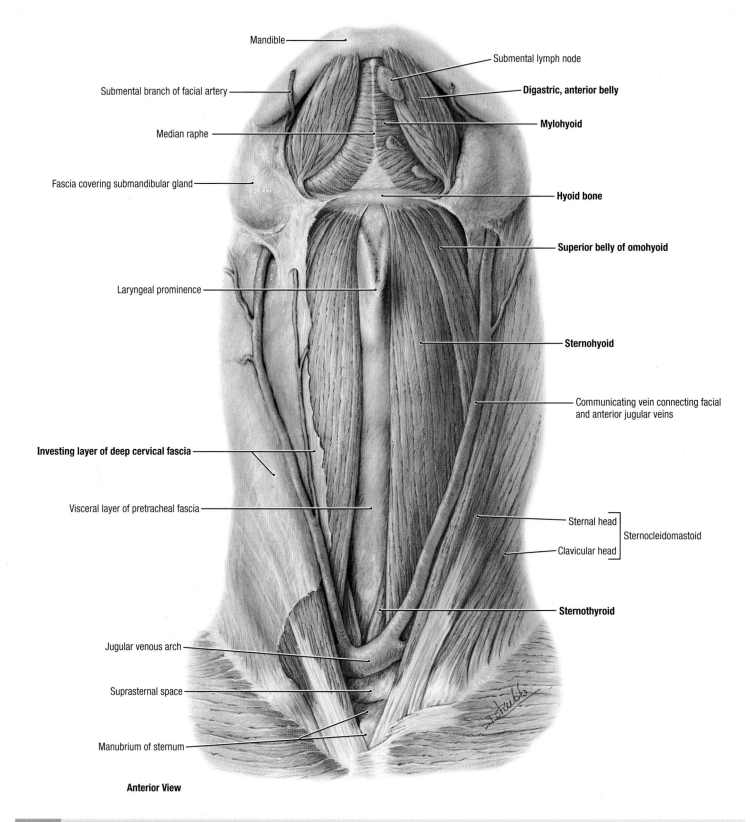

Mandible

Submental branch of facial artery

Median raphe

Fascia covering submandibular gland

Laryngeal prominence

Investing layer of deep cervical fascia

Visceral layer of pretracheal fascia

Jugular venous arch

Suprasternal space

Manubrium of sternum

Submental lymph node

Digastric, anterior belly

Mylohyoid

Hyoid bone

Superior belly of omohyoid

Sternohyoid

Communicating vein connecting facial
and anterior jugular veins

Sternal head

Clavicular head

Sternocleidomastoid

Sternothyroid

Anterior View

8.9   SUPRAHYOID AND INFRAHYOID MUSCLES

Much of the investing layer of deep cervical fascia has been removed.

- The anterior bellies of the digastric muscles form the sides of the suprahyoid part of the anterior cervical region, or submental triangle (floor of mouth). The hyoid bone forms the triangle's base, and the mylohyoid muscles are its floor.

- The infrahyoid part of the anterior cervical region is shaped like an elongated diamond bounded by the sternohyoid muscle superiorly and sternothyroid muscle inferiorly.

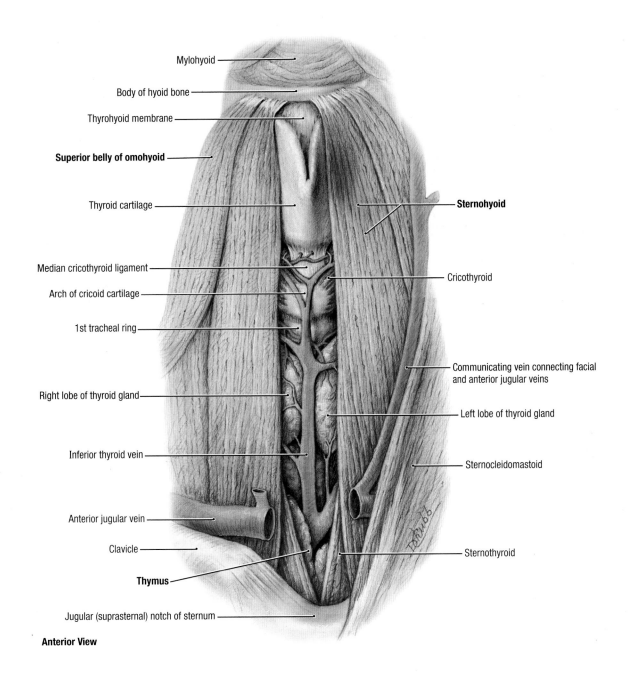

Mylohyoid

Body of hyoid bone

Thyrohyoid membrane

Superior belly of omohyoid

Thyroid cartilage

Median cricothyroid ligament

Arch of cricoid cartilage

1st tracheal ring

Right lobe of thyroid gland

Inferior thyroid vein

Anterior jugular vein

Clavicle

Thymus

Jugular (suprasternal) notch of sternum

Sternohyoid

Cricothyroid

Communicating vein connecting facial and anterior jugular veins

Left lobe of thyroid gland

Sternocleidomastoid

Sternothyroid

Anterior View

INFRAHYOID REGION, SUPERFICIAL MUSCULAR LAYER

8.10

The pretracheal fascia, right anterior jugular vein, and jugular venous arch have been removed.
- A persistent thymus projects superiorly from the thorax.
- The two superficial depressors of the larynx ("strap muscles") are the omohyoid (only the superior belly of which is seen here) and sternohyoid.

Fracture of the hyoid. This results in depression of the body of the hyoid onto the thyroid cartilage. Inability to elevate the hyoid and move it anteriorly beneath the tongue makes swallowing and maintenance of the separation of the alimentary and respiratory tracts difficult and may result in **aspiration pneumonia.**

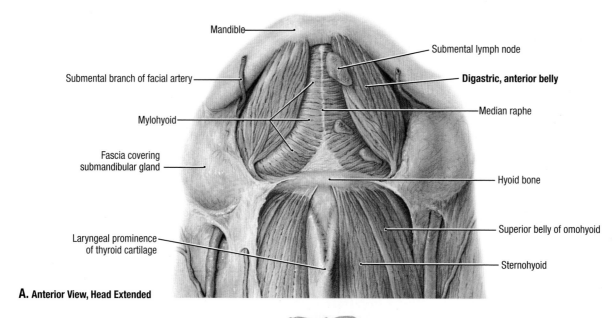

Mandible

Submental lymph node

Submental branch of facial artery

Digastric, anterior belly

Mylohyoid

Median raphe

Fascia covering submandibular gland

Hyoid bone

Superior belly of omohyoid

Laryngeal prominence of thyroid cartilage

Sternohyoid

A. Anterior View, Head Extended

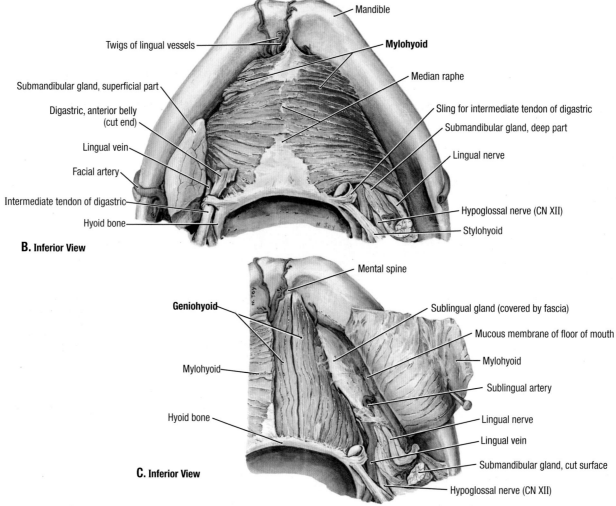

Mandible

Twigs of lingual vessels

Mylohyoid

Submandibular gland, superficial part

Median raphe

Digastric, anterior belly (cut end)

Sling for intermediate tendon of digastric

Submandibular gland, deep part

Lingual vein

Lingual nerve

Facial artery

Intermediate tendon of digastric

Hypoglossal nerve (CN XII)

Hyoid bone

Stylohyoid

B. Inferior View

Mental spine

Geniohyoid

Sublingual gland (covered by fascia)

Mucous membrane of floor of mouth

Mylohyoid

Mylohyoid

Sublingual artery

Hyoid bone

Lingual nerve

Lingual vein

Submandibular gland, cut surface

C. Inferior View

Hypoglossal nerve (CN XII)

8.11 **SUPRAHYOID REGION (SUBMENTAL TRIANGLE)**

A. Superficial layer—anterior belly of digastric. **B.** Intermediate layer—mylohyoid. **C.** Deep layer—geniohyoid.

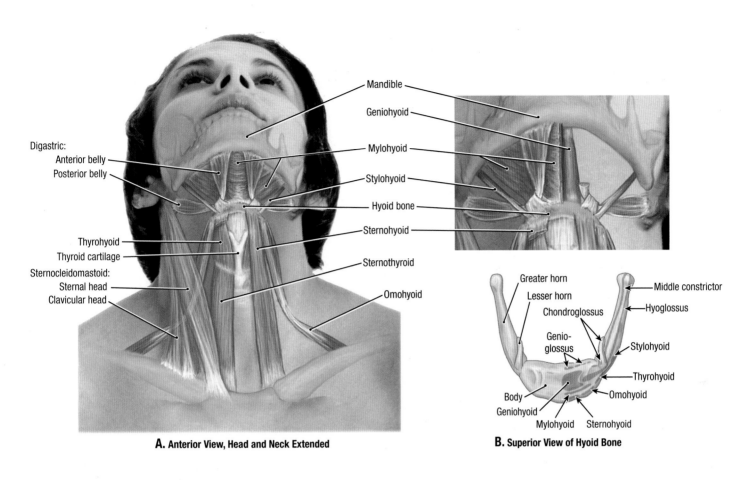

A. Anterior View, Head and Neck Extended

B. Superior View of Hyoid Bone

SUPRAHYOID AND INFRAHYOID MUSCLES

8.12

A. Overview. **B.** Muscular attachments onto the hyoid bone.

TABLE 8.4	**MUSCLES OF ANTERIOR CERVICAL REGION**			
Muscle	*Superior Attachment*	*Inferior Attachment*	*Innervation*	*Main Action*
Suprahyoid muscles				
Mylohyoid	Mylohyoid line of mandible	Raphe and body of hyoid bone	Nerve to mylohyoid, a branch of inferior alveolar nerve (CN V$_3$)	Elevates hyoid bone, floor of mouth and tongue during swallowing and speaking
Digastric	*Anterior belly:* digastric fossa of mandible *Posterior belly:* mastoid notch of temporal bone	Intermediate tendon to body and greater horn of hyoid bone	*Anterior belly:* nerve to mylohyoid, a branch of inferior alveolar nerve (CN V$_3$) *Posterior belly:* facial nerve (CN VII)	Elevates hyoid bone and steadies it during swallowing and speaking; depresses mandible against resistance
Geniohyoid	Inferior mental spine of mandible	Body of hyoid bone	C1 via the hypoglossal nerve (CN XII)	Pulls hyoid bone anterosuperiorly, shortens floor of mouth, and widens pharynx
Stylohyoid	Styloid process of temporal bone		Cervical branch of facial nerve (CN VII)	Elevates and retracts hyoid bone, thereby elongating floor of mouth
Infrahyoid muscles				
Sternohyoid	Body of hyoid bone	Manubrium of sternum and medial end of clavicle	C1–C3 by a branch of ansa cervicalis	Depresses hyoid bone after it has been elevated during swallowing
Omohyoid	Inferior border of hyoid bone	Superior border of scapula near suprascapular notch		Depresses, retracts, and steadies hyoid bone
Sternothyroid	Oblique line of thyroid cartilage	Posterior surface of manubrium of sternum	C2 and C3 by a branch of ansa cervicalis	Depresses hyoid bone and larynx
Thyrohyoid	Inferior border of body and greater horn of hyoid bone	Oblique line of thyroid cartilage	C1 via hypoglossal nerve (CN XII)	Depresses hyoid bone and elevates larynx

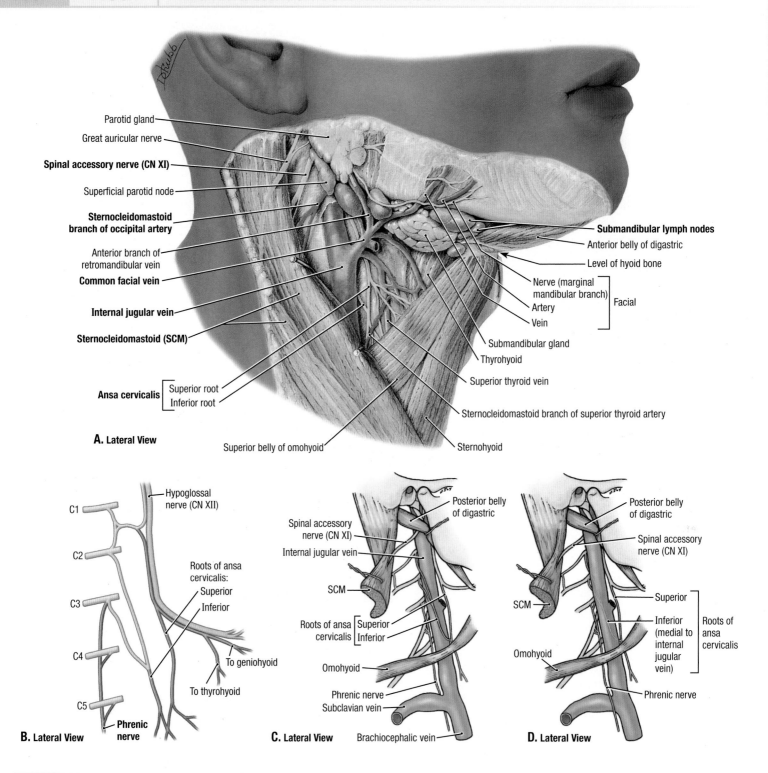

Parotid gland

Great auricular nerve

Spinal accessory nerve (CN XI)

Superficial parotid node

Sternocleidomastoid branch of occipital artery

Anterior branch of retromandibular vein

Common facial vein

Internal jugular vein

Sternocleidomastoid (SCM)

Ansa cervicalis [Superior root / Inferior root]

A. Lateral View

Superior belly of omohyoid

Submandibular lymph nodes

Anterior belly of digastric

Level of hyoid bone

Nerve (marginal mandibular branch)] Facial
Artery
Vein

Submandibular gland

Thyrohyoid

Superior thyroid vein

Sternocleidomastoid branch of superior thyroid artery

Sternohyoid

B. Lateral View

C1
C2
C3
C4
C5

Hypoglossal nerve (CN XII)

Roots of ansa cervicalis:
Superior
Inferior

To geniohyoid

To thyrohyoid

Phrenic nerve

C. Lateral View

Spinal accessory nerve (CN XI)

Internal jugular vein

SCM

Roots of ansa cervicalis [Superior / Inferior]

Omohyoid

Phrenic nerve
Subclavian vein

Posterior belly of digastric

Brachiocephalic vein

D. Lateral View

Posterior belly of digastric

Spinal accessory nerve (CN XI)

SCM

Omohyoid

Superior
Inferior (medial to internal jugular vein)] Roots of ansa cervicalis

Phrenic nerve

8.13 SUPERFICIAL DISSECTION OF CAROTID TRIANGLE

A. The skin, subcutaneous tissue (with platysma), and the investing layer of deep cervical fascia, including the sheaths of the parotid and submandibular glands, have been removed.

- The spinal accessory nerve (CN XI) enters the deep surface of the sternocleidomastoid muscle and is joined along its anterior border by the sternocleidomastoid branch of the occipital artery.
- The (common) facial vein joins the internal jugular vein near the level of the hyoid bone; here, the facial vein is joined by several other veins.

- The submandibular lymph nodes lie deep to the investing layer of deep cervical fascia in the submandibular triangle; some of the nodes lie deep in the submandibular gland.

B. Diagram of the motor branches of cervical plexus. **C.** Typical relationships of ansa cervicalis, spinal accessory nerve (CN XI), and phrenic nerve to the internal jugular and subclavian veins. **D.** Atypical relationships.

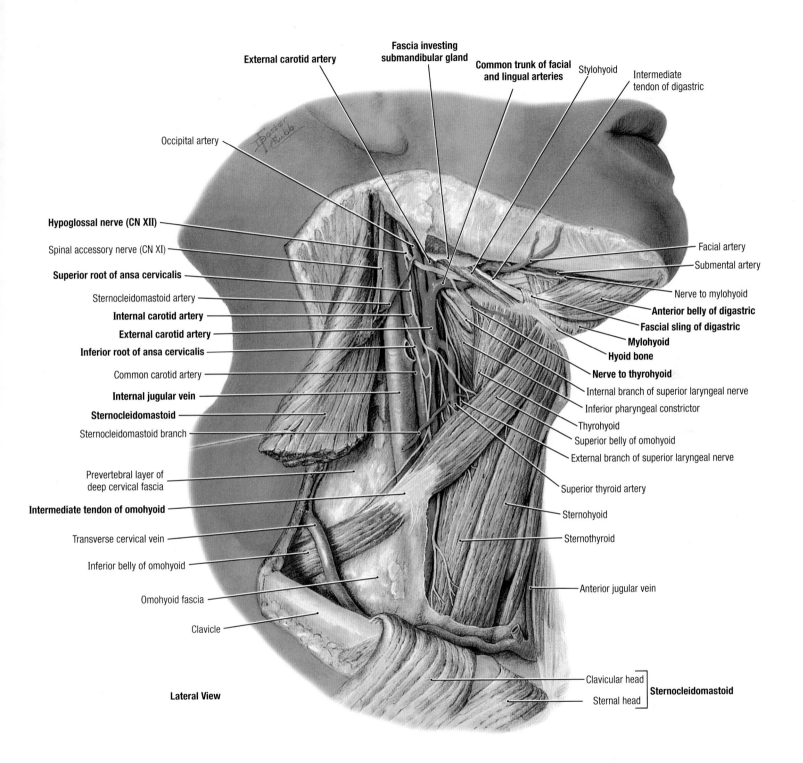

External carotid artery

Fascia investing submandibular gland

Common trunk of facial and lingual arteries

Stylohyoid

Intermediate tendon of digastric

Occipital artery

Hypoglossal nerve (CN XII)

Spinal accessory nerve (CN XI)

Superior root of ansa cervicalis

Sternocleidomastoid artery

Internal carotid artery

External carotid artery

Inferior root of ansa cervicalis

Common carotid artery

Internal jugular vein

Sternocleidomastoid

Sternocleidomastoid branch

Prevertebral layer of deep cervical fascia

Intermediate tendon of omohyoid

Transverse cervical vein

Inferior belly of omohyoid

Omohyoid fascia

Clavicle

Facial artery

Submental artery

Nerve to mylohyoid

Anterior belly of digastric

Fascial sling of digastric

Mylohyoid

Hyoid bone

Nerve to thyrohyoid

Internal branch of superior laryngeal nerve

Inferior pharyngeal constrictor

Thyrohyoid

Superior belly of omohyoid

External branch of superior laryngeal nerve

Superior thyroid artery

Sternohyoid

Sternothyroid

Anterior jugular vein

Clavicular head

Sternal head

Sternocleidomastoid

Lateral View

DEEP DISSECTION OF CAROTID TRIANGLE

8.14

The sternocleidomastoid muscle has been severed; the inferior portion reflected inferiorly and superior portion posteriorly.

- The intermediate tendon of the digastric muscle is connected to the hyoid bone by a fascial sling derived from the muscular part of the pretracheal layer of deep cervical fascia; the tendon of the omohyoid muscle is similarly tethered to the clavicle.
- In this specimen, the facial and lingual arteries arise from a common trunk and pass deep to the stylohyoid and digastric muscles.

- The hypoglossal nerve (CN XII) crosses the internal and external carotid arteries and gives off two branches, the superior root of the ansa cervicalis and the nerve to the thyrohyoid, before passing anteriorly deep to the mylohyoid muscle. In this specimen, the inferior root of the ansa cervicalis lies deep to the internal jugular vein and emerges at its medial aspect.

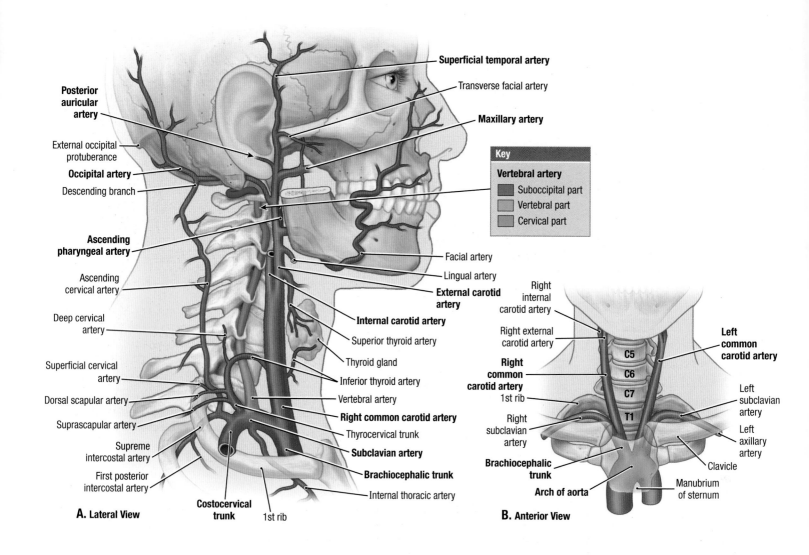

A. Lateral View

B. Anterior View

8.15 ARTERIES OF NECK

A. Overview. **B.** Common carotid and subclavian arteries.

TABLE 8.5	ARTERIES OF NECK	
Artery	*Origin*	*Course and Distribution*
Right common carotid	Bifurcation of brachiocephalic trunk	Ascends in neck within carotid sheath with the internal jugular vein and vagus nerve (CN X). Terminates at superior border of thyroid cartilage (C4 vertebral level) by dividing into internal and external carotid arteries
Left common carotid	Arch of aorta	
Right and left internal carotid	Right and left common carotid	No branches in the neck. Enters cranium via carotid canal to supply brain and orbits. Proximal part location of carotid sinus, a baroreceptor that reacts to change in arterial blood pressure. The carotid body, a chemoreceptor that monitors oxygen level in blood, is located in bifurcation of common carotid
Right and left external carotid		Supplies most structures external to cranium; part of forehead, and scalp are supplied by ophthalmic artery from intracranial internal carotid artery
Ascending pharyngeal	External carotid	Ascends on pharynx to supply pharynx, prevertebral muscles, middle ear, and cranial meninges
Occipital		Passes posteriorly, medial and parallel to the posterior belly of digastric, ending in the posterior scalp
Posterior auricular		Ascends posteriorly between external acoustic meatus and mastoid process to supply adjacent muscles, parotid gland, facial nerve, auricle, and scalp

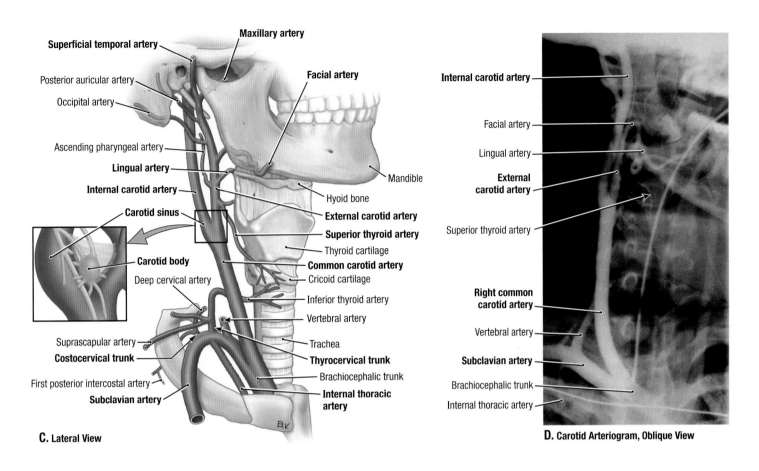

C. Lateral View

D. Carotid Arteriogram, Oblique View

ARTERIES OF NECK (continued)   8.15

C. Branches of external carotid and subclavian arteries. The carotid sinus is a baroreceptor that reacts to changes in arterial blood pressure and is located in the dilatation of the proximal part of the internal carotid artery. The carotid body is an ovoid mass of tissue that lies at the bifurcation of the common carotid artery. It is a chemoreceptor that monitors the level of oxygen in the blood.

TABLE 8.5	**ARTERIES OF NECK (*continued*)**	
Superior thyroid	External carotid	Runs antero-inferiorly deep to infrahyoid muscles to reach thyroid gland. Supplies thyroid gland, infrahyoid muscles, SCM, and larynx via *superior laryngeal artery*
Lingual		Lies on middle constrictor muscle of pharynx; arches supero-anteriorly and passes deep to CN XII, stylohyoid muscle, and posterior belly of digastric then passes deep to hyoglossus, giving branches to the posterior tongue and bifurcating into *deep lingual* and *sublingual arteries*
Facial		After giving rise to *ascending palatine artery* and a tonsillar branch, it passes superiorly under cover of the angle of the mandible. It then loops anteriorly to supply the submandibular gland and give rise to the *submental artery* to the floor of the mouth before entering the face
Maxillary	Terminal branches of external carotid	Passes posterior to neck of mandible, enters infratemporal fossa then pterygopalatine fossa to supply teeth, nose, ear, and face
Superficial temporal		Ascends anterior to auricle to temporal region and ends in scalp
Vertebral	Subclavian	Passes through the foramina transversaria of the transverse processes of vertebrae C1–C6, runs in a groove on the posterior arch of the atlas, and enters the cranial cavity through the foramen magnum
Internal thoracic		No branches in neck; enters thorax
Thyrocervical trunk		Has two branches: the *inferior thyroid artery*, the main visceral artery of the neck; the cervicodorsal trunk sending branches to the lateral cervical region, trapezius, and medial scapular arteries
Costocervical trunk		Trunk passes posterosuperiorly and divides into *superior intercostal* and *deep cervical arteries* to supply the 1st and 2nd intercostal spaces and posterior deep cervical muscles, respectively

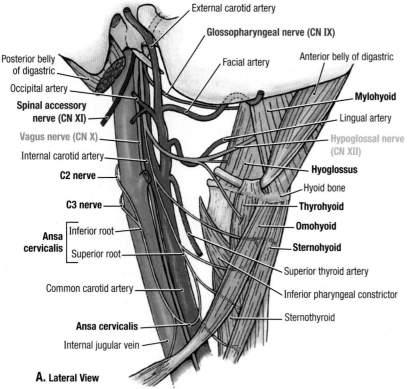

A. Lateral View

Key	
Glossopharyngeal—CN IX	**Vagus—CN X**
Motor: stylopharyngeus, parotid gland **Sensory:** taste: posterior third of tongue; general sensation: pharynx, tonsillar sinus, pharyngotympanic tube, middle ear cavity	**Motor:** palate, pharynx, larynx, trachea, bronchial tree, heart, GI tract to left colic flexure **Sensory:** pharynx, larynx; reflex sensory from tracheo-bronchial tree, lungs, heart, GI tract to left colic flexure
Spinal accessory—CN XI	**Hypoglossal—CN XII**
Motor: sternocleidomastoid and trapezius	**Motor:** all intrinsic and extrinsic muscles of tongue (excluding palatoglossus—a palatine muscle)

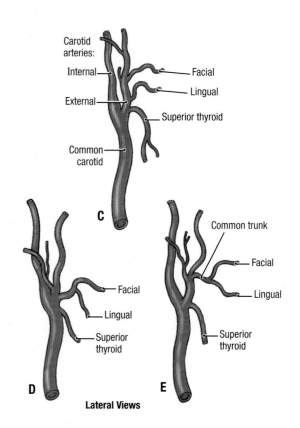

B. Lateral View

Lateral Views

8.16 RELATIONSHIPS OF NERVES AND VESSELS IN CAROTID TRIANGLE OF NECK

A. Ansa cervicalis and the strap muscles. **B.** Hypoglossal nerve (CN XII) and internal and external branches of superior laryngeal nerve (CN X). The palpable tip of the greater horn of the hyoid bone, indicated with a *circle*, is the reference point for many structures. **C–E.** Variation in the origin of the lingual artery as studied by Dr. Grant in 211 specimens. In 80%, the superior thyroid, lingual, and facial arteries arose separately **(C)**; in 20%, the lingual and facial arteries arose from a common stem inferiorly **(D)** or high on the external carotid artery **(E)**. In one specimen, the superior thyroid and lingual arteries arose from a common stem.

Carotid occlusion, causing stenosis (narrowing), can be relieved by opening the artery at its origin and stripping off the atherosclerotic plaque with the artery's lining (intima). This procedure is called **carotid endarterectomy**. Because of the relationships of the internal carotid artery, there is a risk of cranial nerve injury during the procedure involving one or more of the following nerves: CN IX, CN X (or its branch, the superior laryngeal nerve), CN XI, or CN XII.

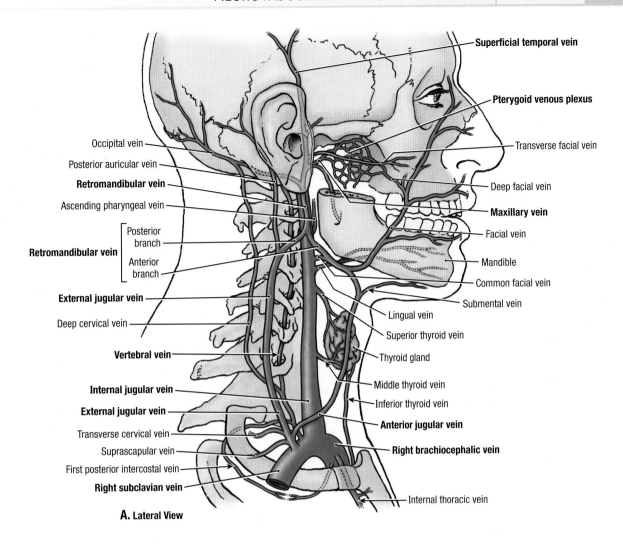

Superficial temporal vein

Pterygoid venous plexus

Occipital vein

Transverse facial vein

Posterior auricular vein

Deep facial vein

Retromandibular vein

Ascending pharyngeal vein

Maxillary vein

Facial vein

Posterior branch

Retromandibular vein

Anterior branch

Mandible

Common facial vein

External jugular vein

Submental vein

Deep cervical vein

Lingual vein

Superior thyroid vein

Vertebral vein

Thyroid gland

Internal jugular vein

Middle thyroid vein

External jugular vein

Inferior thyroid vein

Transverse cervical vein

Anterior jugular vein

Suprascapular vein

Right brachiocephalic vein

First posterior intercostal vein

Right subclavian vein

Internal thoracic vein

A. Lateral View

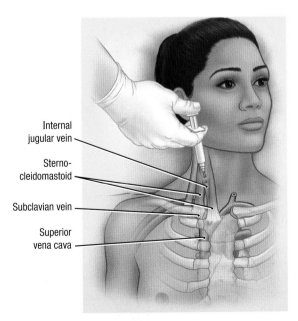

Internal jugular vein

Sterno-cleidomastoid

Subclavian vein

Superior vena cava

B. Internal Jugular Vein Puncture

DEEP VEINS OF NECK   8.17

A. Overview. The internal jugular vein (IJV) begins at the jugular foramen as the continuation of the sigmoid sinus. From a dilated origin, the superior bulb of the IJV, the vein runs inferiorly through the neck in the carotid sheath. Posterior to the sternal end of the clavicle the vein merges perpendicularly with the subclavian vein, forming the "venous angle" that marks the origin of the brachiocephalic vein. The inferior end of the IJV dilates superior to its terminal valve, forming the inferior bulb of the IJV. The valve permits blood to flow toward the heart while preventing backflow into the IJV. The external jugular vein drains blood from the occipital region and posterior neck to the subclavian vein, and the anterior jugular vein the anterior aspect of the neck.

B. Internal jugular vein puncture. A needle and catheter may be inserted into the IJV, using ultrasonic guidance, for diagnostic or therapeutic purposes. The right IJV is preferable to the left because it is usually larger and straighter. The clinician palpates the common carotid artery and inserts the needle into the IJV just lateral to it at a 30-degree angle, aiming at the apex of the triangle between the sternal and clavicular heads of the SCM. The needle is then directed inferolaterally toward the ipsilateral nipple. Venous access can also be achieved by other supra- and infraclavicular approaches.

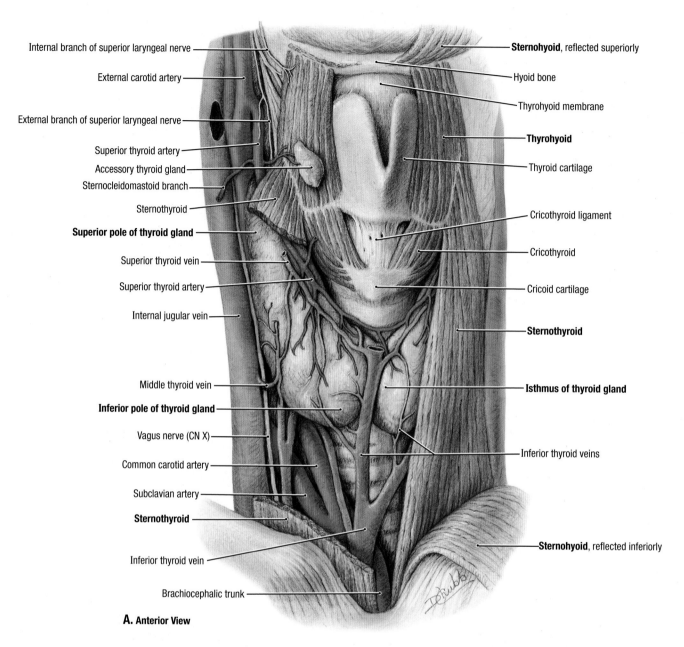

Internal branch of superior laryngeal nerve
External carotid artery
External branch of superior laryngeal nerve
Superior thyroid artery
Accessory thyroid gland
Sternocleidomastoid branch
Sternothyroid
Superior pole of thyroid gland
Superior thyroid vein
Superior thyroid artery
Internal jugular vein
Middle thyroid vein
Inferior pole of thyroid gland
Vagus nerve (CN X)
Common carotid artery
Subclavian artery
Sternothyroid
Inferior thyroid vein
Brachiocephalic trunk

Sternohyoid, reflected superiorly
Hyoid bone
Thyrohyoid membrane
Thyrohyoid
Thyroid cartilage
Cricothyroid ligament
Cricothyroid
Cricoid cartilage
Sternothyroid
Isthmus of thyroid gland
Inferior thyroid veins
Sternohyoid, reflected inferiorly

A. Anterior View

8.18 ENDOCRINE LAYER OF VISCERAL COMPARTMENT I

A. On the left side of the specimen, the sternohyoid and omohyoid muscles are reflected or removed, exposing the sternothyroid and the thyrohyoid muscles; on the right side of the specimen, the sternothyroid muscle is largely excised. **B.** Schematic illustration of the venous drainage of the thyroid gland. Except for the superior thyroid veins, the thyroid veins are not paired with arteries of corresponding names.

The **carotid pulse (neck pulse)** is easily felt by palpating the common carotid artery in the side of the neck, where it lies in a groove between the trachea and the infrahyoid muscles. It is usually easily palpated just deep to the anterior border of the SCM at the level of the superior border of the thyroid cartilage. It is routinely checked during **cardiopulmonary resuscitation (CPR)**. **Absence of a carotid pulse** indicates cardiac arrest.

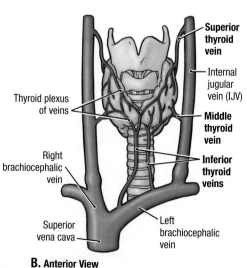

Superior thyroid vein
Internal jugular vein (IJV)
Middle thyroid vein
Inferior thyroid veins
Thyroid plexus of veins
Right brachiocephalic vein
Superior vena cava
Left brachiocephalic vein

B. Anterior View

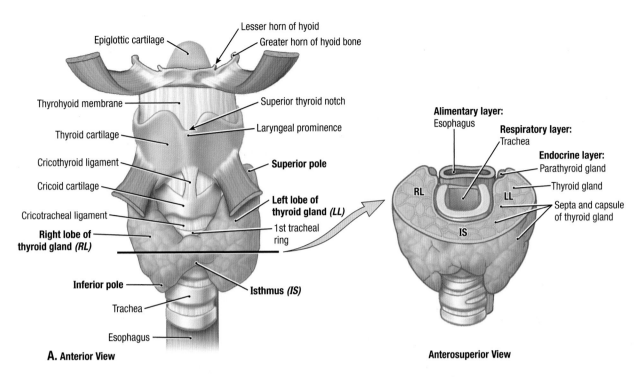

Lesser horn of hyoid

Epiglottic cartilage — Greater horn of hyoid bone

Thyrohyoid membrane — Superior thyroid notch

Thyroid cartilage — Laryngeal prominence

Cricothyroid ligament — **Superior pole**

Cricoid cartilage —

Cricotracheal ligament — **Left lobe of thyroid gland** *(LL)*

Right lobe of thyroid gland *(RL)* — 1st tracheal ring

Inferior pole —

Trachea — **Isthmus** *(IS)*

Esophagus —

A. Anterior View

Alimentary layer: Esophagus

Respiratory layer: Trachea

Endocrine layer: Parathyroid gland — Thyroid gland

RL   LL

IS — Septa and capsule of thyroid gland

Anterosuperior View

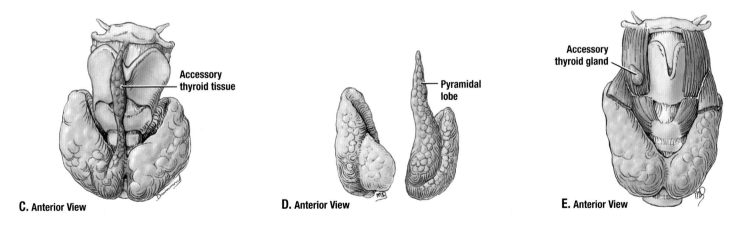

Esophagus   Trachea

Recurrent laryngeal nerve

Visceral layer of pretracheal fascia

Thyroid gland — Common carotid artery

Carotid sheath — Internal jugular vein

Prevertebral fascia — Vagus nerve

Retropharyngeal space — Vertebral body

B. Transverse Section, Inferior View

Accessory thyroid tissue

Accessory thyroid gland

Pyramidal lobe

C. Anterior View   **D. Anterior View**   **E. Anterior View**

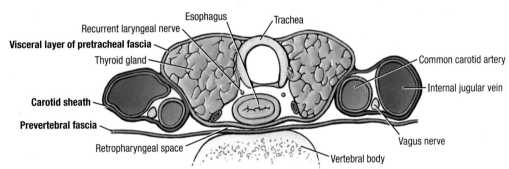

ENDOCRINE LAYER OF VISCERAL COMPARTMENT II

8.19

A. Relations of thyroid gland with transverse section showing alimentary, respiratory, and endocrine layers of visceral compartment. **B.** Fascial relationships. **C.** Accessory thyroid tissue along the course of the thyroglossal duct, which was the path of migration of thyroid tissue from its embryonic site of development.

D. Approximately 50% of glands have a pyramidal lobe that extends from near the isthmus to or toward the hyoid bone; the isthmus is occasionally absent, in which case the gland is in two parts. **E.** An accessory thyroid gland can occur between the suprahyoid region and arch of the aorta (see Fig. 8.18A).

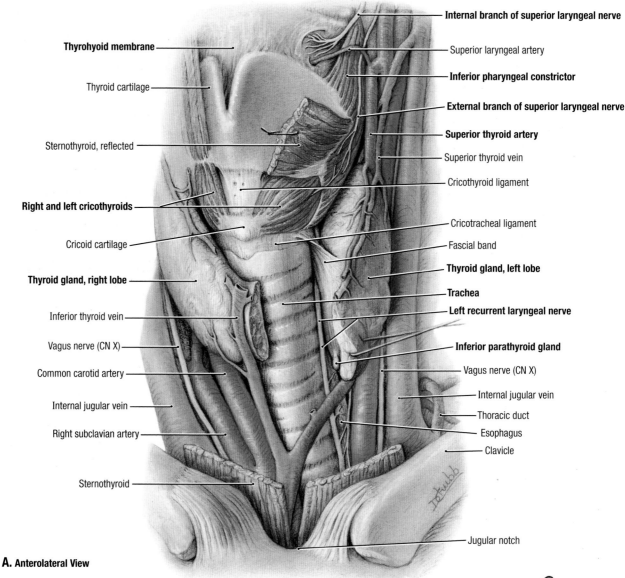

Internal branch of superior laryngeal nerve

Thyrohyoid membrane

Superior laryngeal artery

Thyroid cartilage

Inferior pharyngeal constrictor

External branch of superior laryngeal nerve

Sternothyroid, reflected

Superior thyroid artery

Superior thyroid vein

Cricothyroid ligament

Right and left cricothyroids

Cricotracheal ligament

Cricoid cartilage

Fascial band

Thyroid gland, left lobe

Thyroid gland, right lobe

Trachea

Left recurrent laryngeal nerve

Inferior thyroid vein

Inferior parathyroid gland

Vagus nerve (CN X)

Vagus nerve (CN X)

Common carotid artery

Internal jugular vein

Internal jugular vein

Thoracic duct

Right subclavian artery

Esophagus

Clavicle

Sternothyroid

Jugular notch

A. Anterolateral View

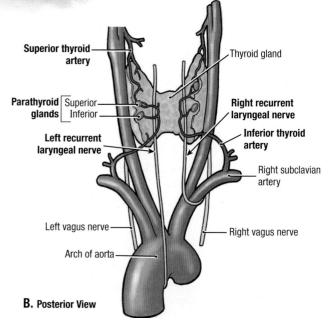

Superior thyroid artery

Thyroid gland

Parathyroid glands — Superior / Inferior

Right recurrent laryngeal nerve

Left recurrent laryngeal nerve

Inferior thyroid artery

Right subclavian artery

Left vagus nerve

Right vagus nerve

Arch of aorta

B. Posterior View

8.20 RESPIRATORY LAYER OF VISCERAL COMPARTMENT

A. The isthmus of the thyroid gland is divided, and the left lobe is retracted. The left recurrent laryngeal nerve ascends on the lateral aspect of the trachea between the trachea and esophagus. The internal branch of the superior laryngeal nerve runs along the superior border of the inferior pharyngeal constrictor muscle and pierces the thyrohyoid membrane. The external branch of the superior laryngeal nerve lies adjacent to the inferior pharyngeal constrictor muscle and supplies its lower portion; it continues to run along the anterior border of the superior thyroid artery, passing deep to the superior attachment of the sternothyroid muscle, and then supplies the cricothyroid muscle. **B.** Blood supply of the parathyroid glands and courses of the left and right recurrent laryngeal nerves.

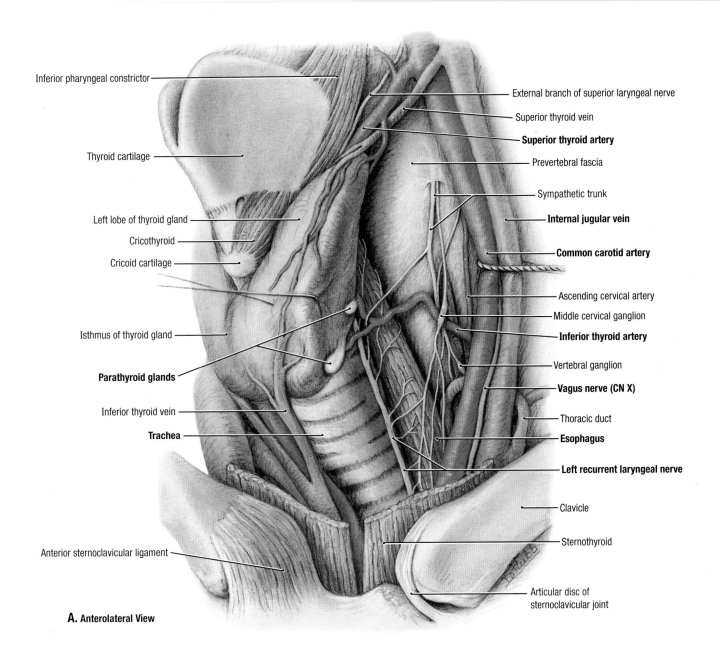

Inferior pharyngeal constrictor

Thyroid cartilage

Left lobe of thyroid gland

Cricothyroid

Cricoid cartilage

Isthmus of thyroid gland

Parathyroid glands

Inferior thyroid vein

Trachea

Anterior sternoclavicular ligament

External branch of superior laryngeal nerve

Superior thyroid vein

Superior thyroid artery

Prevertebral fascia

Sympathetic trunk

Internal jugular vein

Common carotid artery

Ascending cervical artery

Middle cervical ganglion

Inferior thyroid artery

Vertebral ganglion

Vagus nerve (CN X)

Thoracic duct

Esophagus

Left recurrent laryngeal nerve

Clavicle

Sternothyroid

Articular disc of sternoclavicular joint

A. Anterolateral View

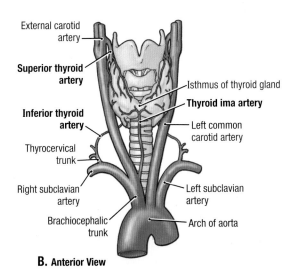

External carotid artery

Superior thyroid artery

Inferior thyroid artery

Thyrocervical trunk

Right subclavian artery

Brachiocephalic trunk

Isthmus of thyroid gland

Thyroid ima artery

Left common carotid artery

Left subclavian artery

Arch of aorta

B. Anterior View

ALIMENTARY LAYER OF VISCERAL COMPARTMENT 8.21

A. Dissection of the left side of the root of the neck. The three structures contained in the carotid sheath (internal jugular vein, common carotid artery, and vagus nerve) are retracted. The left recurrent laryngeal nerve ascends on the lateral aspect of the trachea, just anterior to the recess between the trachea and esophagus. **B.** Arterial supply of thyroid gland. The thyroid ima artery is infrequent (10%) and variable in its origin.

During a **total thyroidectomy** (e.g., excision of a malignant thyroid gland), the parathyroid glands are in danger of being inadvertently damaged or removed. These glands are safe during **subtotal thyroidectomy** because the most posterior part of the thyroid gland usually is preserved. Variability in the position of the parathyroid glands, especially the inferior ones, puts them in danger of being removed during surgery on the thyroid gland. If the parathyroid glands are inadvertently removed during surgery, the patient suffers from **tetany,** a severe convulsive disorder. The generalized convulsive muscle spasms result from a fall in blood calcium levels.

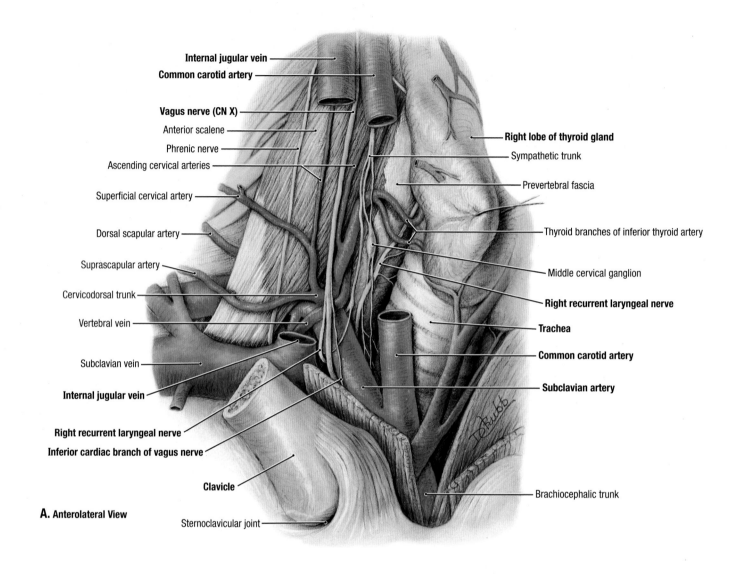

Internal jugular vein

Common carotid artery

Vagus nerve (CN X)

Anterior scalene

Phrenic nerve

Ascending cervical arteries

Superficial cervical artery

Dorsal scapular artery

Suprascapular artery

Cervicodorsal trunk

Vertebral vein

Subclavian vein

Internal jugular vein

Right recurrent laryngeal nerve

Inferior cardiac branch of vagus nerve

Clavicle

A. Anterolateral View

Sternoclavicular joint

Right lobe of thyroid gland

Sympathetic trunk

Prevertebral fascia

Thyroid branches of inferior thyroid artery

Middle cervical ganglion

Right recurrent laryngeal nerve

Trachea

Common carotid artery

Subclavian artery

Brachiocephalic trunk

8.22 ROOT OF NECK

A. Dissection of the right side of the root of the neck. The clavicle is cut, sections of the common carotid artery and internal jugular vein are removed, and the right lobe of the thyroid gland is retracted. The right vagus nerve crosses the first part of the subclavian artery and gives off an inferior cardiac branch and the right recurrent laryngeal nerve. The right recurrent laryngeal nerve loops inferior to the subclavian artery and passes posterior to the common carotid artery on its way to the posterolateral aspect of the trachea.

- **Recurrent laryngeal nerve injury** may occur during thyroidectomy and other surgeries in the anterior cervical region of the neck. Because the terminal branch of this nerve, the inferior laryngeal nerve, innervates the muscles moving the

vocal folds, injury to the nerve results in **paralysis of the vocal folds**.

- A non-neoplastic and noninflammatory enlargement of the thyroid gland, other than the variable enlargement that may occur during menstruation and pregnancy, is called a **goiter**. A goiter results from a lack of iodine. It is common in certain parts of the world where the soil and water are deficient in iodine and iodized salt is unavailable. The enlarged gland causes a swelling in the neck that may compress the trachea, esophagus, and recurrent laryngeal nerves. When the gland enlarges, it may do so anteriorly, posteriorly, inferiorly, or laterally. It cannot move superiorly because of the superior attachments of the sternothyroid and sternohyoid muscles. **Substernal extension of a goiter** is also common.

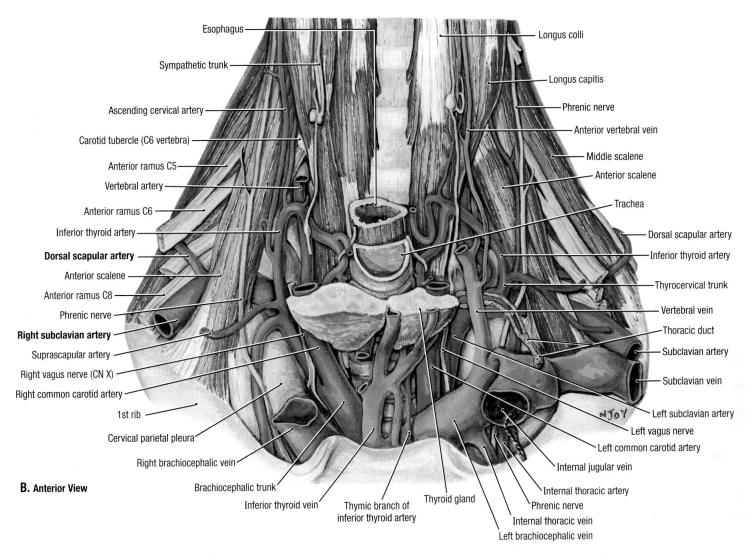

Esophagus

Sympathetic trunk

Ascending cervical artery

Carotid tubercle (C6 vertebra)

Anterior ramus C5

Vertebral artery

Anterior ramus C6

Inferior thyroid artery

Dorsal scapular artery

Anterior scalene

Anterior ramus C8

Phrenic nerve

Right subclavian artery

Suprascapular artery

Right vagus nerve (CN X)

Right common carotid artery

1st rib

Cervical parietal pleura

Right brachiocephalic vein

Longus colli

Longus capitis

Phrenic nerve

Anterior vertebral vein

Middle scalene

Anterior scalene

Trachea

Dorsal scapular artery

Inferior thyroid artery

Thyrocervical trunk

Vertebral vein

Thoracic duct

Subclavian artery

Subclavian vein

Left subclavian artery

Left vagus nerve

Left common carotid artery

Internal jugular vein

Internal thoracic artery

Phrenic nerve

Internal thoracic vein

Left brachiocephalic vein

Thyroid gland

Thymic branch of inferior thyroid artery

Inferior thyroid vein

Brachiocephalic trunk

B. Anterior View

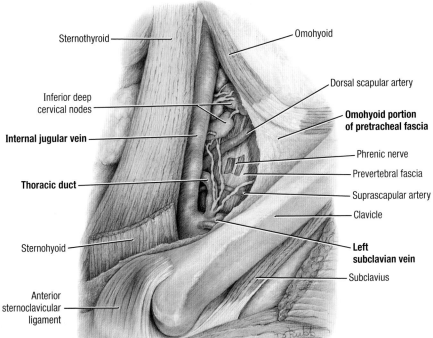

Sternothyroid

Inferior deep cervical nodes

Internal jugular vein

Thoracic duct

Sternohyoid

Anterior sternoclavicular ligament

C. Anterolateral View

Omohyoid

Dorsal scapular artery

Omohyoid portion of pretracheal fascia

Phrenic nerve

Prevertebral fascia

Suprascapular artery

Clavicle

Left subclavian vein

Subclavius

ROOT OF NECK (continued)   8.22

B. Deep anterior dissection. Note that the right dorsal scapular artery arises directly from the subclavian artery, a common variation. **C.** Dissection of termination of the thoracic duct. The sternocleidomastoid muscle is removed, the sternohyoid muscle is resected, and the omohyoid portion of the pretracheal fascia is partially removed. The thoracic duct arches laterally in the neck, passing posterior to the carotid sheath and anterior to the vertebral artery, thyrocervical trunk, and subclavian arteries; it enters the angle formed by the junction of the left subclavian and internal jugular veins to form the left brachiocephalic vein (the left venous angle).

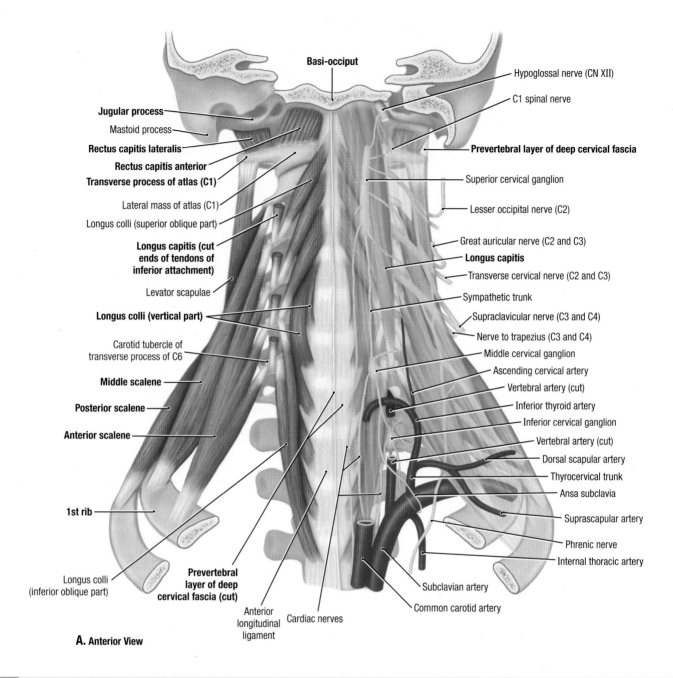

Basi-occiput

Jugular process

Mastoid process

Rectus capitis lateralis

Rectus capitis anterior

Transverse process of atlas (C1)

Lateral mass of atlas (C1)

Longus colli (superior oblique part)

Longus capitis (cut ends of tendons of inferior attachment)

Levator scapulae

Longus colli (vertical part)

Carotid tubercle of transverse process of C6

Middle scalene

Posterior scalene

Anterior scalene

1st rib

Longus colli (inferior oblique part)

Prevertebral layer of deep cervical fascia (cut)

Anterior longitudinal ligament

Cardiac nerves

Hypoglossal nerve (CN XII)

C1 spinal nerve

Prevertebral layer of deep cervical fascia

Superior cervical ganglion

Lesser occipital nerve (C2)

Great auricular nerve (C2 and C3)

Longus capitis

Transverse cervical nerve (C2 and C3)

Sympathetic trunk

Supraclavicular nerve (C3 and C4)

Nerve to trapezius (C3 and C4)

Middle cervical ganglion

Ascending cervical artery

Vertebral artery (cut)

Inferior thyroid artery

Inferior cervical ganglion

Vertebral artery (cut)

Dorsal scapular artery

Thyrocervical trunk

Ansa subclavia

Suprascapular artery

Phrenic nerve

Internal thoracic artery

Subclavian artery

Common carotid artery

A. Anterior View

8.23   PREVERTEBRAL REGION

A. and **B.** Overview of muscles, nerves, and vessels. In **(A)** the prevertebral layer of deep cervical fascia is present on the left side of the specimen but has been removed from the right side.

TABLE 8.6	**PREVERTEBRAL AND SCALENE MUSCLES**			
Muscle	*Superior Attachment*	*Inferior Attachment*	*Innervation*	*Main Action*
Longus colli				
Superior oblique part	Anterior tubercle of atlas (C1)	Anterior tubercles of TVP C3–C5	Anterior rami of C2–C6 spinal nerves (cervical plexus)	Rotation of cervical spine to opposite side (acting unilaterally)
Vertical part	Vertebral bodies of C2–C4	Vertebral bodies C5–T3		
Inferior oblique part	Anterior tubercles of TVP C5–C6	Vertebral bodies T1–T3		Flexion of cervical spine
Longus capitis	Basilar part of occipital bone	Anterior tubercles of TVP C3–C6	Anterior rami of C1–C3 spinal nerves (cervical plexus)	Flexion of head (atlanto-occipital joints)

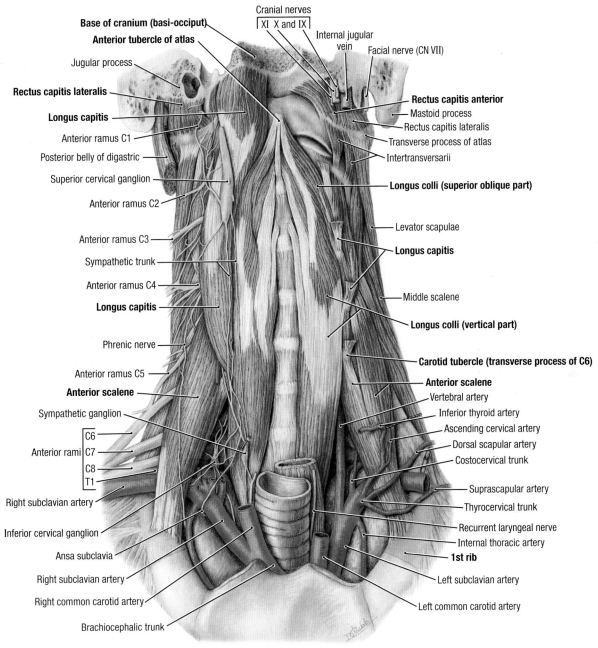

Cranial nerves
XI X and IX

Base of cranium (basi-occiput)
Anterior tubercle of atlas
Jugular process
Rectus capitis lateralis
Longus capitis
Anterior ramus C1
Posterior belly of digastric
Superior cervical ganglion
Anterior ramus C2
Anterior ramus C3
Sympathetic trunk
Anterior ramus C4
Longus capitis
Phrenic nerve
Anterior ramus C5
Anterior scalene
Sympathetic ganglion
C6
Anterior rami C7
C8
T1
Right subclavian artery
Inferior cervical ganglion
Ansa subclavia
Right subclavian artery
Right common carotid artery
Brachiocephalic trunk

Internal jugular vein
Facial nerve (CN VII)
Rectus capitis anterior
Mastoid process
Rectus capitis lateralis
Transverse process of atlas
Intertransversarii
Longus colli (superior oblique part)
Levator scapulae
Longus capitis
Middle scalene
Longus colli (vertical part)
Carotid tubercle (transverse process of C6)
Anterior scalene
Vertebral artery
Inferior thyroid artery
Ascending cervical artery
Dorsal scapular artery
Costocervical trunk
Suprascapular artery
Thyrocervical trunk
Recurrent laryngeal nerve
Internal thoracic artery
1st rib
Left subclavian artery
Left common carotid artery

B. Anterior View

PREVERTEBRAL REGION (*continued*)

8.23

TABLE 8.6	**PREVERTEBRAL AND SCALENE MUSCLES (*continued*)**				
Rectus capitis anterior	Base of cranium, just anterior to occipital condyle	Anterior surface of lateral mass of atlas (C1)	Branches from loop between C1 and C2 spinal nerves	Lateral flexion at atlanto-occipital joints (acting unilaterally)
Rectus capitis lateralis	Base of cranium just lateral to occipital condyle	Transverse process of atlas (C1)		Flexion at atlanto-occipital joints (acting bilaterally)
Anterior scalene	Anterior tubercles of TVP C3–C6	Scalene tubercle of 1st rib	Anterior rami of C3–C8 (cervical and brachial plexus)	Forced inspiration (ribs mobile): elevate superior ribs
Middle scalene	TVP C1–C2	Superior surface of 1st rib; posterior to groove for subclavian artery		
	Posterior tubercles of TVP C3–C7			Ribs fixed: lateral flexion of cervical spine (acting unilaterally)
Posterior scalene	Posterior tubercles of TVP C5–C7	External border of 2nd rib		Flexes neck (acting bilaterally)

TVP, transverse process.

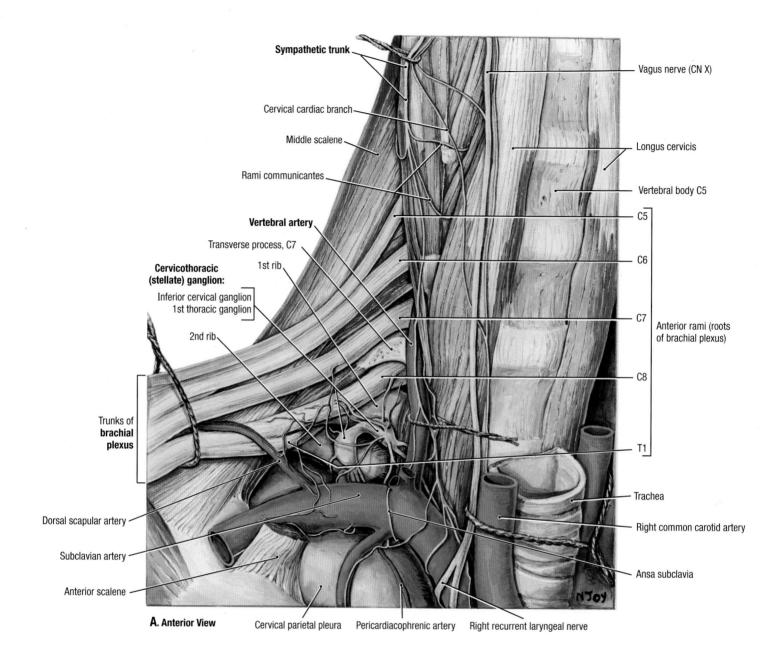

Sympathetic trunk

Cervical cardiac branch

Middle scalene

Rami communicantes

Vertebral artery

Transverse process, C7

1st rib

Cervicothoracic (stellate) ganglion:

Inferior cervical ganglion
1st thoracic ganglion

2nd rib

Trunks of **brachial plexus**

Dorsal scapular artery

Subclavian artery

Anterior scalene

Vagus nerve (CN X)

Longus cervicis

Vertebral body C5

C5

C6

C7

Anterior rami (roots of brachial plexus)

C8

T1

Trachea

Right common carotid artery

Ansa subclavia

A. Anterior View   Cervical parietal pleura   Pericardiacophrenic artery   Right recurrent laryngeal nerve

8.24 BRACHIAL PLEXUS AND SYMPATHETIC TRUNK IN ROOT OF NECK

A. Dissection of right side of specimen. The pleura has been depressed, the vertebral artery retracted medially, and the brachial plexus retracted superiorly to reveal the cervicothoracic (stellate) ganglion (the combined inferior cervical and 1st thoracic ganglia). Anesthetic injected around the cervicothoracic (stellate) ganglion blocks transmission of stimuli through the cervical and superior thoracic ganglia. This **stellate ganglion block** may relieve vascular spasms involving the brain and upper limb. It is also useful when deciding if surgical resection of the ganglion would be beneficial to a person with excess vasoconstriction of the ipsilateral limb. **B.** Relation of brachial plexus and subclavian artery to anterior and middle scalene muscles.

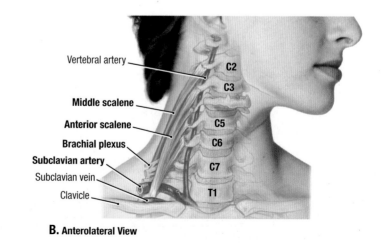

Vertebral artery

C2

C3

Middle scalene

Anterior scalene

C5

C6

Brachial plexus

Subclavian artery

C7

Subclavian vein

T1

Clavicle

B. Anterolateral View

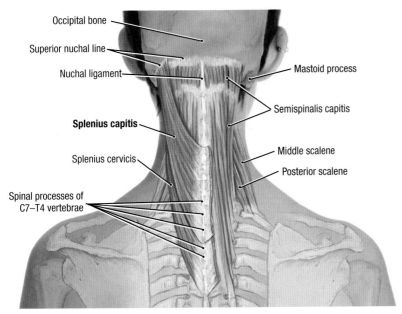

A. Posterior View

Occipital bone
Superior nuchal line
Nuchal ligament
Splenius capitis
Splenius cervicis
Spinal processes of C7–T4 vertebrae
Mastoid process
Semispinalis capitis
Middle scalene
Posterior scalene

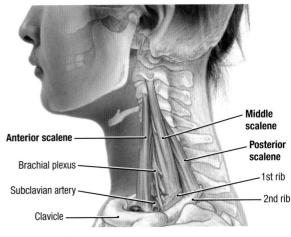

B. Lateral View

Anterior scalene
Brachial plexus
Subclavian artery
Clavicle
Middle scalene
Posterior scalene
1st rib
2nd rib

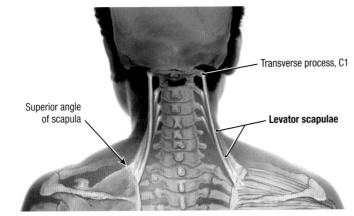

C. Posterior View

Superior angle of scapula
Transverse process, C1
Levator scapulae

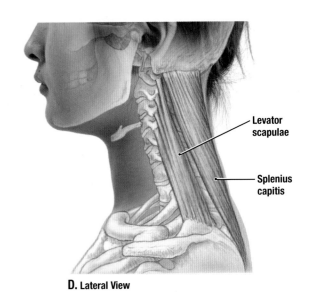

D. Lateral View

Levator scapulae
Splenius capitis

LATERAL VERTEBRAL MUSCLES

8.25

A. Overview. **B.** Scalene muscles. **C.** Levator scapulae. **D.** Levator scapulae and splenius capitis.

TABLE 8.7	**LATERAL VERTEBRAL MUSCLES**[b]			
Muscle	*Superior Attachment*	*Inferior Attachment*	*Innervation*	*Main Action*
Splenius capitis	Inferior half of nuchal ligament and spinous processes of C7 and superior 3–4 thoracic vertebrae	Lateral aspect of mastoid process and lateral third of superior nuchal line	Posterior rami of middle cervical spinal nerves	Laterally flexes and rotates head and neck to same side; acting bilaterally, extends head and neck[a]
Levator scapulae	Posterior tubercles of transverse processes of C1–C4 vertebrae	Superior part of medial border of scapula	Dorsal scapular nerve (C5) and cervical spinal nerves C3 and C4	Elevates scapula and tilts glenoid cavity inferiorly by rotating scapula

[a]Rotation of head occurs at atlanto-axial joints.
[b]Middle and posterior scalene see Table 8.6.

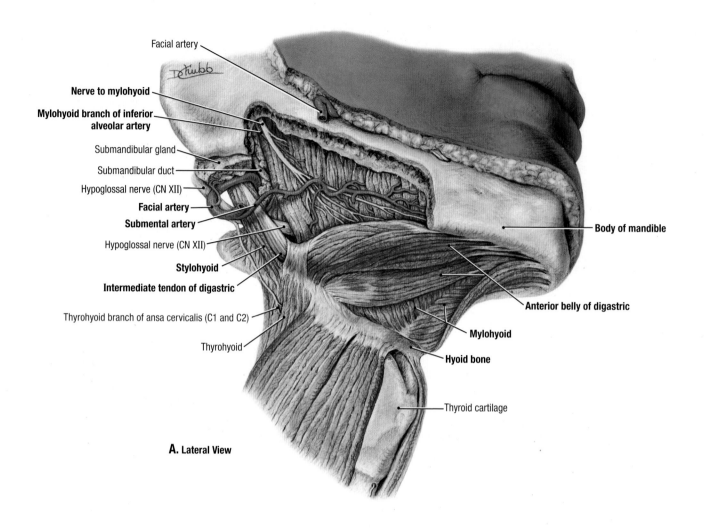

Facial artery

Nerve to mylohyoid

Mylohyoid branch of inferior alveolar artery

Submandibular gland

Submandibular duct

Hypoglossal nerve (CN XII)

Facial artery

Submental artery

Hypoglossal nerve (CN XII)

Stylohyoid

Intermediate tendon of digastric

Thyrohyoid branch of ansa cervicalis (C1 and C2)

Thyrohyoid

Body of mandible

Anterior belly of digastric

Mylohyoid

Hyoid bone

Thyroid cartilage

A. Lateral View

8.26 SERIAL DISSECTION OF SUBMANDIBULAR REGION AND FLOOR OF MOUTH I

Mylohyoid and digastric muscles. **A.** Structures overlying the mandible and a portion of the body of the mandible have been removed.

- The stylohyoid and posterior belly and intermediate tendon of the digastric muscle form the posterior border of the submandibular triangle; the facial artery passes superficial to these muscles.
- The anterior belly of the digastric muscle forms the anterolateral border of the submandibular triangle. In this specimen, the anterior belly has an additional origin from the hyoid. The mylohyoid muscle forms the floor of the triangle and has a thick, free posterior border.
- The nerve to mylohyoid, which supplies the mylohyoid muscle and anterior belly of the digastric muscle, is accompanied by the mylohyoid branch of the inferior alveolar artery posteriorly and the submental artery from the facial artery anteriorly.

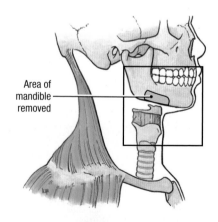

Area of mandible removed

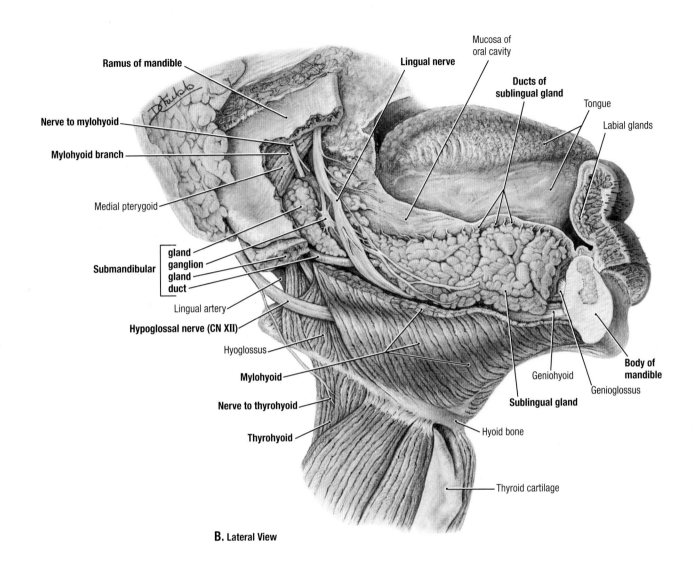

Ramus of mandible

Nerve to mylohyoid

Mylohyoid branch

Medial pterygoid

Submandibular { gland
ganglion
gland
duct }

Lingual artery

Hypoglossal nerve (CN XII)

Hyoglossus

Mylohyoid

Nerve to thyrohyoid

Thyrohyoid

Mucosa of oral cavity

Lingual nerve

Ducts of sublingual gland

Tongue

Labial glands

Body of mandible

Geniohyoid

Genioglossus

Sublingual gland

Hyoid bone

Thyroid cartilage

B. Lateral View

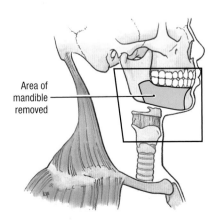

Area of mandible removed

SERIAL DISSECTION OF SUBMANDIBULAR REGION AND FLOOR OF MOUTH II

8.26

B. Sublingual and submandibular glands. The body and adjacent portion of the ramus of the mandible have been removed.

- The sublingual salivary gland lies posterior to the mandible and is in contact with the deep part of the submandibular gland posteriorly.
- Numerous fine ducts pass from the superior border of the sublingual gland to open on the sublingual fold of the overlying mucosa.
- The lingual nerve lies between the sublingual gland and the deep part of the submandibular gland; the submandibular ganglion is suspended from this nerve.
- Spinal nerve C1 fibers, conveyed by the hypoglossal nerve (CN XII), pass to the thyrohyoid muscle before the hypoglossal nerve passes deep to the mylohyoid muscle.

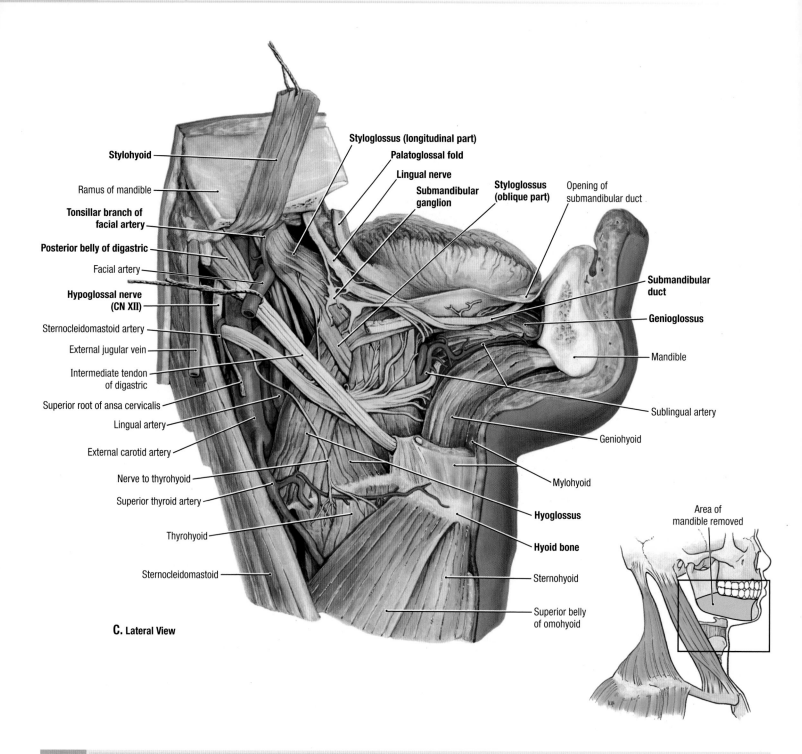

Stylohyoid

Ramus of mandible

Tonsillar branch of facial artery

Posterior belly of digastric

Facial artery

Hypoglossal nerve (CN XII)

Sternocleidomastoid artery

External jugular vein

Intermediate tendon of digastric

Superior root of ansa cervicalis

Lingual artery

External carotid artery

Nerve to thyrohyoid

Superior thyroid artery

Thyrohyoid

Sternocleidomastoid

C. Lateral View

Styloglossus (longitudinal part)

Palatoglossal fold

Lingual nerve

Submandibular ganglion

Styloglossus (oblique part)

Opening of submandibular duct

Submandibular duct

Genioglossus

Mandible

Sublingual artery

Geniohyoid

Mylohyoid

Hyoglossus

Hyoid bone

Sternohyoid

Superior belly of omohyoid

Area of mandible removed

8.26   SERIAL DISSECTION OF SUBMANDIBULAR REGION AND FLOOR OF MOUTH III

C. Hyoglossus muscle, lingual (CN V₃) and hypoglossal (CN XII) nerves. All of the right half of the mandible, except the superior part of the ramus, has been removed. The stylohyoid muscle is reflected superiorly, and the posterior belly of the digastric muscle is left *in situ*.

- The hyoglossus muscle ascends from the greater horn and body of the hyoid bone to the side of the tongue.
- The styloglossus muscle is crossed by the tonsillar branch of the facial artery posterosuperiorly, and its oblique part interdigitates with bundles of the hyoglossus muscle inferiorly.

- The hypoglossal nerve (CN XII) supplies all of the muscles of the tongue, both extrinsic and intrinsic, except the palatoglossus (a palatine muscle, innervated by CN X).
- The submandibular duct runs anteriorly in contact with the hyoglossus and genioglossus muscles to its opening on the side of the frenulum of the tongue.
- The lingual nerve is in contact with the mandible posteriorly, looping inferior to the submandibular duct and ending in the tongue. The submandibular ganglion is suspended from the lingual nerve.

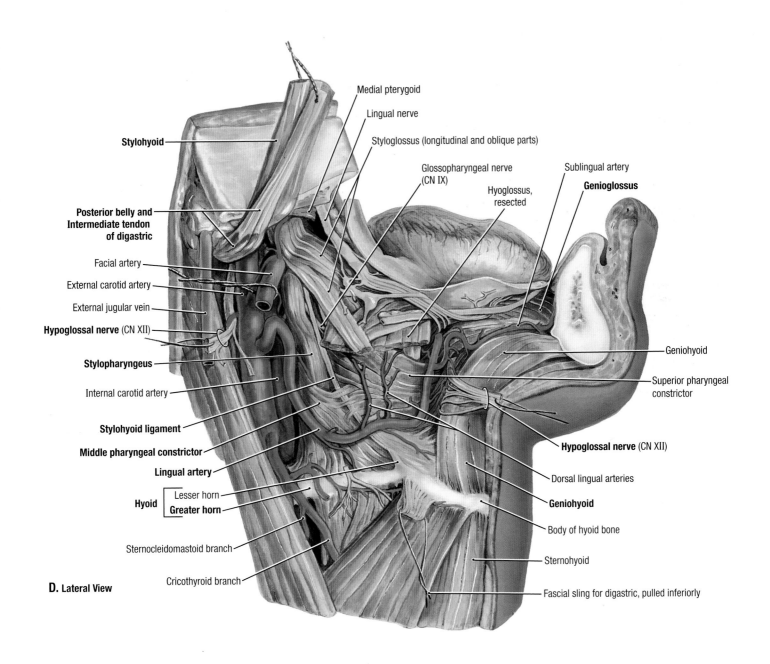

Medial pterygoid

Lingual nerve

Styloglossus (longitudinal and oblique parts)

Glossopharyngeal nerve
(CN IX)

Hyoglossus,
resected

Sublingual artery

Genioglossus

Stylohyoid

**Posterior belly and
Intermediate tendon
of digastric**

Facial artery

External carotid artery

External jugular vein

Hypoglossal nerve (CN XII)

Stylopharyngeus

Internal carotid artery

Stylohyoid ligament

Middle pharyngeal constrictor

Lingual artery

Hyoid ⎰ Lesser horn
        ⎱ **Greater horn**

Sternocleidomastoid branch

Cricothyroid branch

D. Lateral View

Geniohyoid

Superior pharyngeal
constrictor

Hypoglossal nerve (CN XII)

Dorsal lingual arteries

Geniohyoid

Body of hyoid bone

Sternohyoid

Fascial sling for digastric, pulled inferiorly

SERIAL DISSECTION OF SUBMANDIBULAR REGION AND FLOOR OF MOUTH IV

8.26

D. Genioglossus and geniohyoid muscles. The stylohyoid, posterior belly and intermediate tendon of the digastric muscle are reflected superiorly, the hypoglossal nerve (CN XII) is divided, and the hyoglossus muscle is mostly removed.

- The lingual artery passes deep to the hyoglossus muscle (resected here), close to the greater horn of the hyoid, and then passes lateral to the middle pharyngeal constrictor muscle, stylohyoid ligament, and genioglossus muscle and turns into the tongue as the deep lingual arteries.

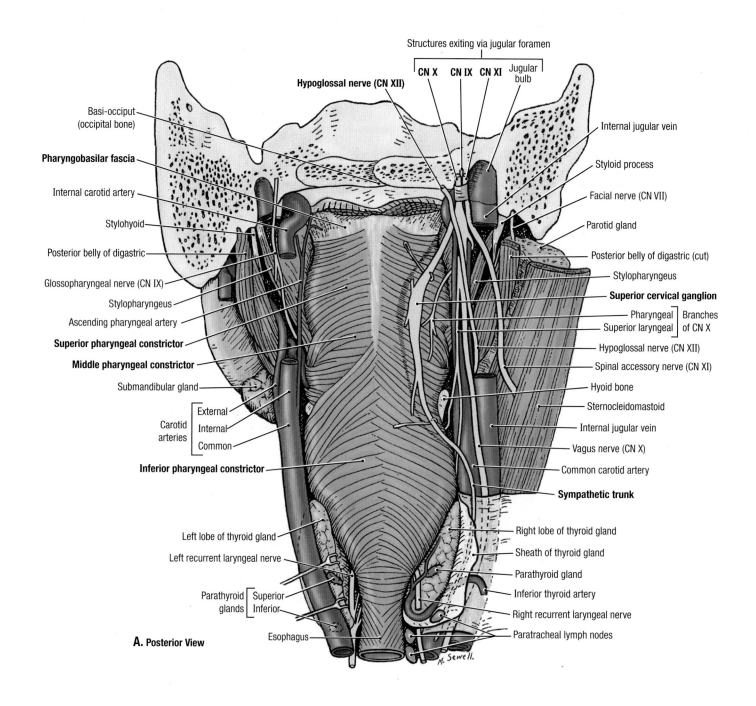

Structures exiting via jugular foramen

CN X   CN IX   CN XI   Jugular bulb

Hypoglossal nerve (CN XII)

Basi-occiput (occipital bone)

Pharyngobasilar fascia

Internal carotid artery

Stylohyoid

Posterior belly of digastric

Glossopharyngeal nerve (CN IX)

Stylopharyngeus

Ascending pharyngeal artery

Superior pharyngeal constrictor

Middle pharyngeal constrictor

Submandibular gland

Carotid arteries { External / Internal / Common }

Inferior pharyngeal constrictor

Left lobe of thyroid gland

Left recurrent laryngeal nerve

Parathyroid glands { Superior / Inferior }

Esophagus

A. Posterior View

Internal jugular vein

Styloid process

Facial nerve (CN VII)

Parotid gland

Posterior belly of digastric (cut)

Stylopharyngeus

Superior cervical ganglion

Pharyngeal ⎤ Branches
Superior laryngeal ⎦ of CN X

Hypoglossal nerve (CN XII)

Spinal accessory nerve (CN XI)

Hyoid bone

Sternocleidomastoid

Internal jugular vein

Vagus nerve (CN X)

Common carotid artery

Sympathetic trunk

Right lobe of thyroid gland

Sheath of thyroid gland

Parathyroid gland

Inferior thyroid artery

Right recurrent laryngeal nerve

Paratracheal lymph nodes

M. Sewell.

8.27 EXTERNAL PHARYNX—POSTERIOR VIEWS

A. Illustration of a dissection similar to B. The sympathetic trunk (including the superior cervical ganglion), which normally lies posterior to the internal carotid artery, has been retracted medially.

- The pharyngobasilar fascia, between the superior pharyngeal constrictor muscle and the base of the skull, attaches the

pharynx to the occipital bone and forms the wall of the pharyngeal recesses.

- As they exit the jugular foramen, CN IX lies anterior to CN X, and CN XI; CN XII, exiting the hypoglossal canal, lies medially.

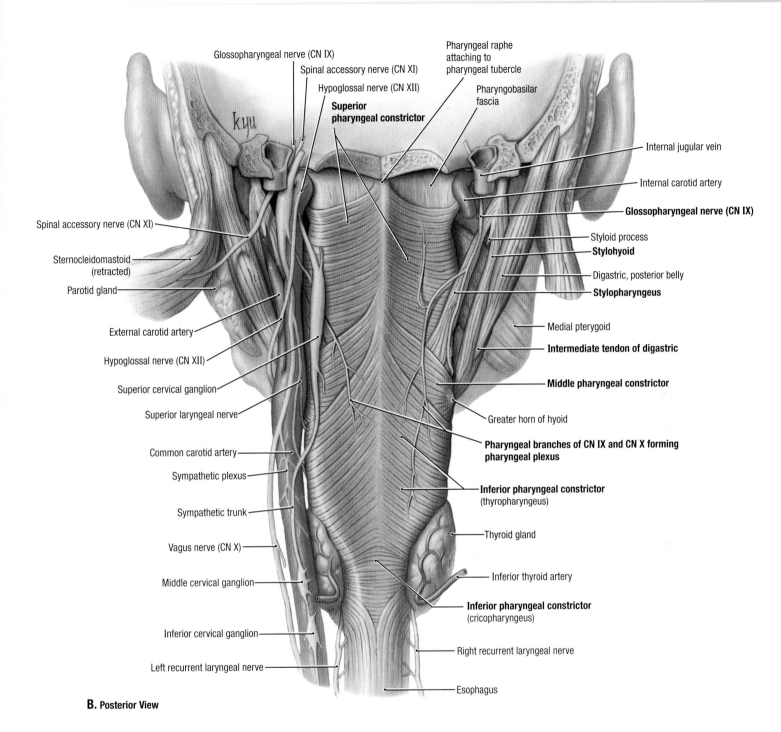

Glossopharyngeal nerve (CN IX)

Spinal accessory nerve (CN XI)

Hypoglossal nerve (CN XII)

Superior pharyngeal constrictor

Pharyngeal raphe attaching to pharyngeal tubercle

Pharyngobasilar fascia

Internal jugular vein

Internal carotid artery

Glossopharyngeal nerve (CN IX)

Styloid process

Stylohyoid

Digastric, posterior belly

Stylopharyngeus

Medial pterygoid

Intermediate tendon of digastric

Middle pharyngeal constrictor

Greater horn of hyoid

Pharyngeal branches of CN IX and CN X forming pharyngeal plexus

Inferior pharyngeal constrictor (thyropharyngeus)

Thyroid gland

Inferior thyroid artery

Inferior pharyngeal constrictor (cricopharyngeus)

Right recurrent laryngeal nerve

Esophagus

Spinal accessory nerve (CN XI)

Sternocleidomastoid (retracted)

Parotid gland

External carotid artery

Hypoglossal nerve (CN XII)

Superior cervical ganglion

Superior laryngeal nerve

Common carotid artery

Sympathetic plexus

Sympathetic trunk

Vagus nerve (CN X)

Middle cervical ganglion

Inferior cervical ganglion

Left recurrent laryngeal nerve

B. Posterior View

EXTERNAL PHARYNX—POSTERIOR VIEWS (*continued*)

B. Dissection. A large wedge of occipital bone (including the foramen magnum) and the articulated cervical vertebrae have been separated from the remainder (anterior portion) of the head and cervical viscera at the retropharyngeal space and removed.

- The pharynx is a unique portion of the alimentary tract, having a circular layer of muscle externally and a longitudinal layer internally.
- The circular layer of the pharynx consists of the three pharyngeal constrictor muscles (superior, middle, and inferior), which overlap one another.
- On the right side of the specimen, the stylopharyngeus muscle and glossopharyngeal nerve (CN IX) pass from the medial side

of the styloid process anteromedially through the interval between the superior and middle pharyngeal constrictor muscles to become part of the internal longitudinal layer. The stylohyoid muscle passes from the lateral side of the styloid process anterolaterally and splits on its way to the hyoid bone to accommodate passage of the intermediate tendon of the digastric.

- Pharyngeal branches of the glossopharyngeal nerve (CN IX) and the vagus nerve (CN X) form the pharyngeal plexus, which provides most of the pharyngeal innervation. The glossopharyngeal nerve supplies the sensory component, while the vagus supplies motor innervation.

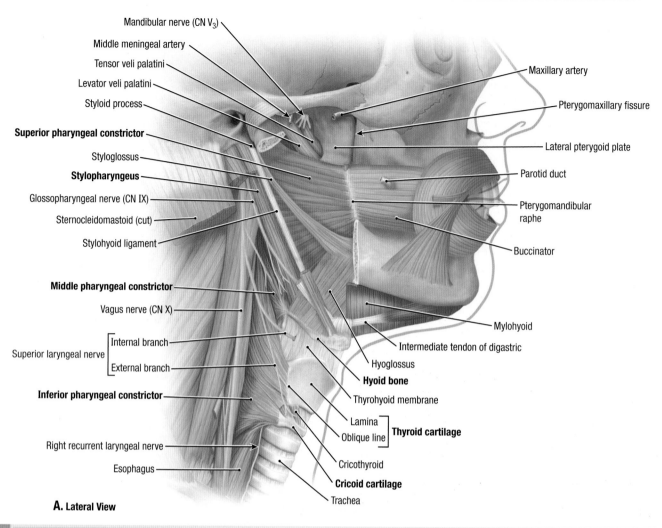

Mandibular nerve (CN V₃)
Middle meningeal artery
Tensor veli palatini
Levator veli palatini
Styloid process
Superior pharyngeal constrictor
Styloglossus
Stylopharyngeus
Glossopharyngeal nerve (CN IX)
Sternocleidomastoid (cut)
Stylohyoid ligament

Middle pharyngeal constrictor
Vagus nerve (CN X)
Superior laryngeal nerve — Internal branch / External branch
Inferior pharyngeal constrictor

Right recurrent laryngeal nerve
Esophagus

A. Lateral View

Maxillary artery
Pterygomaxillary fissure
Lateral pterygoid plate
Parotid duct
Pterygomandibular raphe
Buccinator
Mylohyoid
Intermediate tendon of digastric
Hyoglossus
Hyoid bone
Thyrohyoid membrane
Lamina / Oblique line } **Thyroid cartilage**
Cricothyroid
Cricoid cartilage
Trachea

8.28   EXTERNAL PHARYNX—LATERAL VIEWS

A. Illustration of a dissection similar to **B.**

TABLE 8.8	MUSCLES OF PHARYNX				
Muscle	Origin	Insertion	Innervation	Main Action(s)	
Superior pharyngeal constrictor	Pterygoid hamulus, pterygo-mandibular raphe, posterior end of mylohyoid line of mandible, and side of tongue	Pharyngeal raphe	Pharyngeal and superior laryngeal branches of vagus (CN X) through pharyngeal plexus	Constrict wall of pharynx during swallowing	
Middle pharyngeal constrictor	Stylohyoid ligament and superior (greater) and inferior (lesser) horns of hyoid bone				
Inferior pharyngeal constrictor:					
Thyropharyngeus	Oblique line of thyroid cartilage				
Cricopharyngeus	Side of cricoid cartilage	Contralateral side of cricoid cartilage	Pharyngeal and superior laryngeal branches of vagus (CN X) through pharyngeal plexus + external laryngeal plexus	Serves as superior esophageal sphincter	
Palatopharyngeus (see Fig. 8.29B)	Hard palate and palatine aponeurosis	Posterior border of lamina of thyroid cartilage and side of pharynx and esophagus	Pharyngeal and superior laryngeal branches of vagus (CN X) through pharyngeal plexus	Elevate pharynx and larynx during swallowing and speaking	
Salpingopharyngeus (see Fig. 8.29B)	Cartilaginous part of pharyngotympanic tube	Blends with palatopharyngeus			
Stylopharyngeus	Styloid process of temporal bone	Posterior and superior borders of thyroid cartilage with palatopharyngeus	Glossopharyngeal nerve (CN IX)		

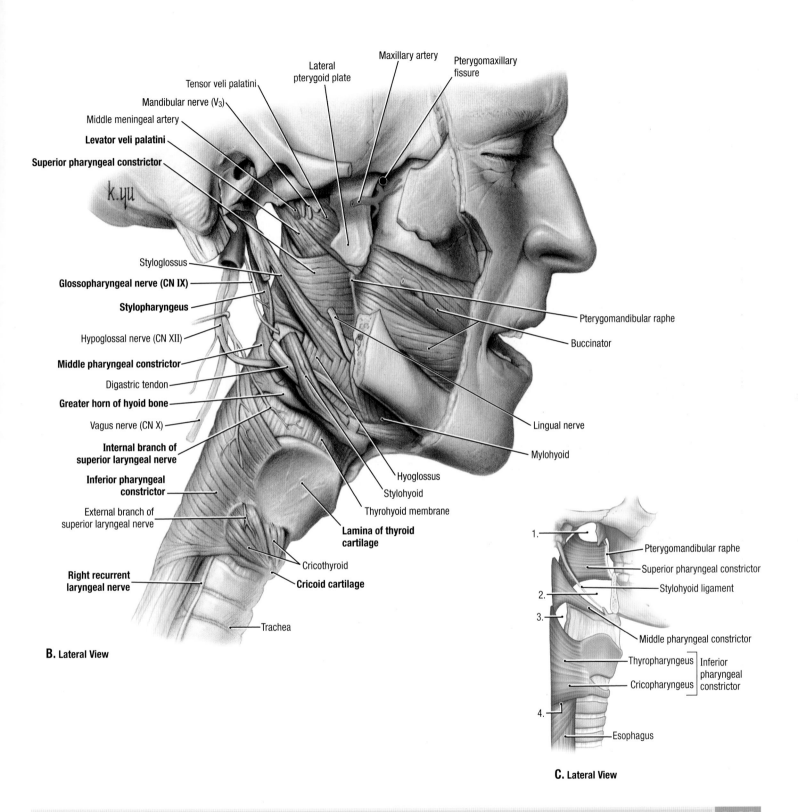

Maxillary artery
Lateral pterygoid plate
Pterygomaxillary fissure
Tensor veli palatini
Mandibular nerve (V₃)
Middle meningeal artery
Levator veli palatini
Superior pharyngeal constrictor
k.yu
Styloglossus
Glossopharyngeal nerve (CN IX)
Stylopharyngeus
Hypoglossal nerve (CN XII)
Middle pharyngeal constrictor
Digastric tendon
Greater horn of hyoid bone
Vagus nerve (CN X)
Internal branch of superior laryngeal nerve
Inferior pharyngeal constrictor
External branch of superior laryngeal nerve
Right recurrent laryngeal nerve
Trachea
Pterygomandibular raphe
Buccinator
Lingual nerve
Mylohyoid
Hyoglossus
Stylohyoid
Thyrohyoid membrane
Lamina of thyroid cartilage
Cricothyroid
Cricoid cartilage

B. Lateral View

1.
2.
3.
4.
Pterygomandibular raphe
Superior pharyngeal constrictor
Stylohyoid ligament
Middle pharyngeal constrictor
Thyropharyngeus ⎤ Inferior pharyngeal constrictor
Cricopharyngeus ⎦
Esophagus

C. Lateral View

EXTERNAL PHARYNX—LATERAL VIEWS (continued)　8.28

B. Dissection. **C.** Observe that there are gaps in the pharyngeal musculature (1 to 4 in **C**) allowing the entry of structures:
1. Superior to the superior constrictor muscle: levator veli palatini muscle and pharyngotympanic (auditory) tube
2. Between the superior and middle constrictors: stylopharyngeus muscle, CN IX, and stylohyoid ligament

3. Between the middle and inferior constrictors: internal branch of superior laryngeal nerve and superior laryngeal artery and nerve
4. Inferior to the inferior constrictor muscle: recurrent laryngeal nerve

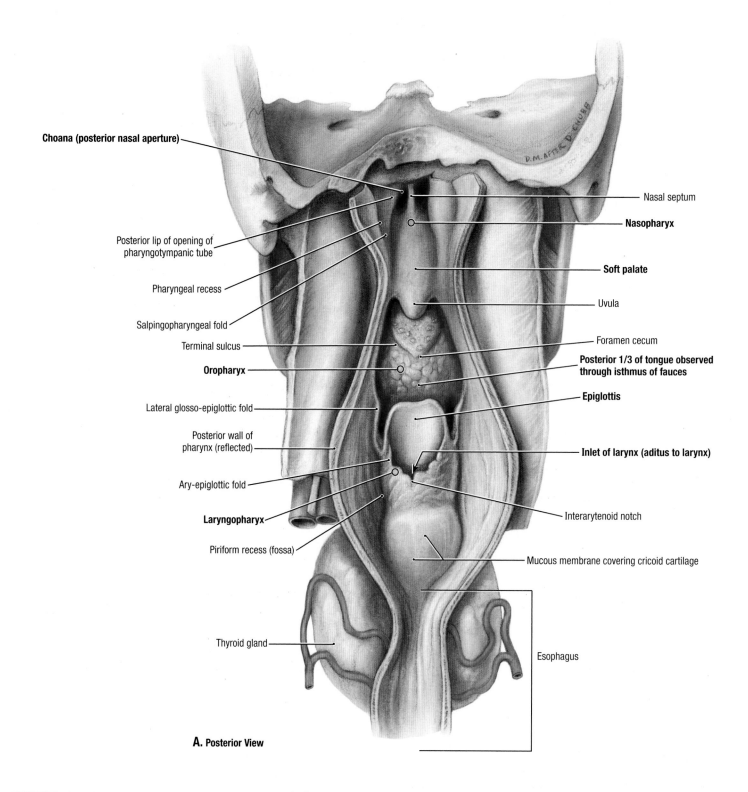

Choana (posterior nasal aperture)

Posterior lip of opening of
pharyngotympanic tube

Pharyngeal recess

Salpingopharyngeal fold

Terminal sulcus

Oropharynx

Lateral glosso-epiglottic fold

Posterior wall of
pharynx (reflected)

Ary-epiglottic fold

Laryngopharynx

Piriform recess (fossa)

Thyroid gland

Nasal septum

Nasopharynx

Soft palate

Uvula

Foramen cecum

**Posterior 1/3 of tongue observed
through isthmus of fauces**

Epiglottis

Inlet of larynx (aditus to larynx)

Interarytenoid notch

Mucous membrane covering cricoid cartilage

Esophagus

A. Posterior View

8.29 INTERNAL PHARYNX

A. Dissection. The posterior wall of the pharynx has been split in the midline and the halves retracted laterally to reveal the internal aspect of the anterior wall of the pharynx. The pharynx consists of three continuous parts: (1) the nasal part (nasopharynx), superior to the level of the soft palate, communicates anteriorly through the choanae with the nasal cavities; (2) the oral part (oropharynx), between the soft palate and the epiglottis, communicates anteriorly through the isthmus of the fauces with the oral cavity; and (3) the laryngeal part (laryngopharynx), posterior to the larynx, communicates with the vestibule of the larynx through the inlet of (aditus to) the larynx. The pharynx extends from the cranial base to the inferior border of the cricoid cartilage.

B. Posterior View

Labels (clockwise from top):
Vagus nerve (CN X) · Nasal septum · Cartilaginous part of pharyngotympanic tube · Internal carotid artery · Pharyngobasilar fascia (wall of pharyngeal recess) · Salpingopharyngeus · **Superior pharyngeal constrictor** · Posterior belly of digastric · **Musculus uvulae** · **Palatopharyngeus** · Uvula · Palatine tonsil · Pharyngo-epiglottic fold · **Ary-epiglottic muscle** · Oblique / Transverse } Arytenoid · Posterior crico-arytenoid · Circular / Longitudinal } Muscle of esophagus · Inferior thyroid artery · Right recurrent laryngeal nerve · Vagus nerve (CN X) · Thyroid gland · Common carotid artery · **Palatopharyngeus** · **Epiglottis** · Root of tongue · Hypoglossal nerve (CN XII) · Vallate papilla · **Levator veli palatini** · Sternocleidomastoid · Spinal accessory nerve (CN XI) · Internal jugular vein

---

INTERNAL PHARYNX (continued)

B. Illustration. The posterior wall of the pharynx has been split in the midline and reflected laterally as in **(A)** then, the mucous membrane was removed to expose the underlying musculature. The muscles of the soft palate, pharynx, and larynx work together during swallowing, elevating the soft palate, narrowing the pharyngeal isthmus (passageway between the nasal and oral parts of the pharynx) and laryngeal inlet, retracting the epiglottis, and closing the glottis, to keep food and drink out of the nasopharynx and larynx as they pass from oral cavity to esophagus. At other times, as when blowing one's nose, the palatopharyngeus muscles, partially encircling the opening to the oral cavity, constrict this opening and depress the soft palate, working with placement and expansion of the posterior tongue to direct expired air through the nasal cavity.

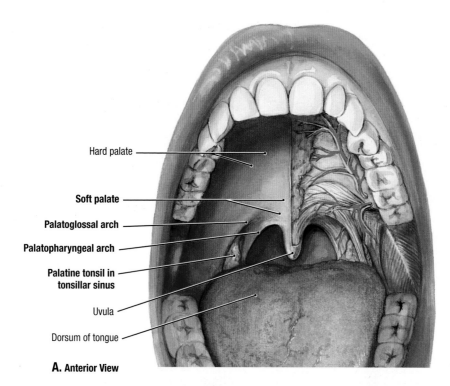

Hard palate

Soft palate

Palatoglossal arch

Palatopharyngeal arch

Palatine tonsil in tonsillar sinus

Uvula

Dorsum of tongue

A. Anterior View

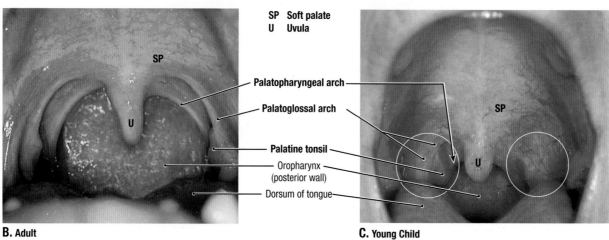

SP   Soft palate
U   Uvula

Palatopharyngeal arch

Palatoglossal arch

Palatine tonsil

Oropharynx (posterior wall)

Dorsum of tongue

B. Adult

C. Young Child

8.30   SURFACE ANATOMY OF ISTHMUS OF THE FAUCES (OROPHARYNGEAL ISTHMUS)

A. Oral cavity and isthmus demonstrating the sinus (bed) of the tonsils. **B.** and **C.** Tonsillar sinuses with palatine tonsils *in situ*, and oropharynx in adult **(B)** and young child **(C)**.

- The fauces (throat), the passage from the mouth to the pharynx, is bounded superiorly by the soft palate, inferiorly by the root (base) of the tongue, and laterally by the palatoglossal and palatopharyngeal arches (folds).
- The palatine tonsils are located between the palatoglossal and palatopharyngeal arches, formed by mucosa overlying the

similarly named muscles; the arches form the boundaries, and the superior pharyngeal constrictor the floor, of the tonsillar sinuses.

- **Normal palatine tonsils.** In the adult, the palatine tonsils are normally involuted, with little glandular tissue in the tonsillar sinuses **(B)**. In contrast in young children, the palatine tonsils are large relative to the adult, since most of the development of the lymphoid system occurs prior to puberty. Despite their large size, as long as the tonsils are not inflamed and not interfering with swallowing/breathing they are considered normal.

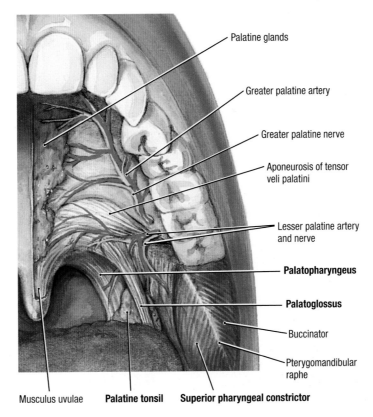

- Palatine glands
- Greater palatine artery
- Greater palatine nerve
- Aponeurosis of tensor veli palatini
- Lesser palatine artery and nerve
- **Palatopharyngeus**
- **Palatoglossus**
- Buccinator
- Pterygomandibular raphe

Musculus uvulae   **Palatine tonsil**   **Superior pharyngeal constrictor**

A. Inferior View

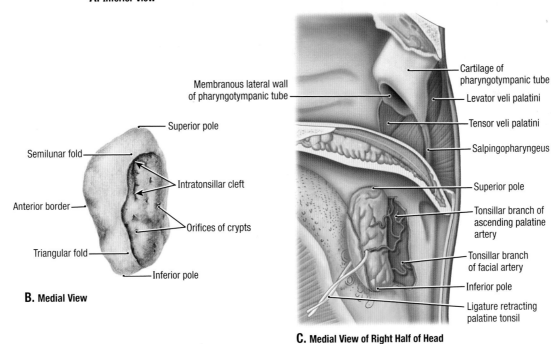

- Membranous lateral wall of pharyngotympanic tube
- Superior pole
- Semilunar fold
- Intratonsillar cleft
- Anterior border
- Orifices of crypts
- Triangular fold
- Inferior pole

B. Medial View

- Cartilage of pharyngotympanic tube
- Levator veli palatini
- Tensor veli palatini
- Salpingopharyngeus
- Superior pole
- Tonsillar branch of ascending palatine artery
- Tonsillar branch of facial artery
- Inferior pole
- Ligature retracting palatine tonsil

C. Medial View of Right Half of Head

PALATINE TONSIL

8.31

A. Palatine tonsil *in situ* and glands of palatine mucosa. **B.** Isolated palatine tonsil. **C. Tonsillectomy**. The procedure involves removal of the tonsil and the fascial sheet covering the tonsillar fossa. Because of the rich blood supply of the tonsil, bleeding commonly arises from the large external palatine vein or less commonly from the tonsillar artery or other arterial twigs. The glossopharyngeal nerve accompanies the tonsillar artery on the lateral wall of the pharynx and is vulnerable to injury because this wall is thin. The internal carotid artery is especially vulnerable when it is tortuous, as it lies directly lateral to the tonsil.

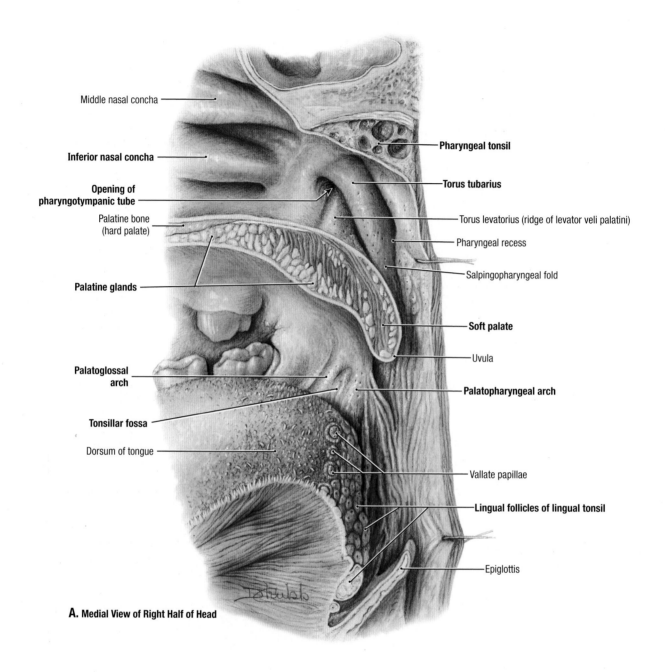

Middle nasal concha

Inferior nasal concha

Opening of pharyngotympanic tube

Palatine bone (hard palate)

Palatine glands

Palatoglossal arch

Tonsillar fossa

Dorsum of tongue

Pharyngeal tonsil

Torus tubarius

Torus levatorius (ridge of levator veli palatini)

Pharyngeal recess

Salpingopharyngeal fold

Soft palate

Uvula

Palatopharyngeal arch

Vallate papillae

Lingual follicles of lingual tonsil

Epiglottis

A. Medial View of Right Half of Head

8.32   SERIAL DISSECTION OF ISTHMUS OF FAUCES AND LATERAL WALL OF NASOPHARYNX I

- The pharyngeal opening of the pharyngotympanic tube is located approximately 1 cm posterior to the inferior concha.
- The pharyngeal tonsil lies in the mucous membrane of the roof and posterior wall of the nasopharynx.
- The palatine glands lie in the soft palate.
- The palatine tonsil lies in the tonsillar fossa between the palatoglossal and palatopharyngeal arches.
- Each lingual follicle has the duct of a mucous gland opening onto its surface; collectively, the follicles are known as the lingual tonsil.

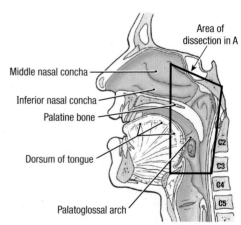

Area of dissection in A

Middle nasal concha

Inferior nasal concha

Palatine bone

Dorsum of tongue

Palatoglossal arch

C2
C3
C4
C5

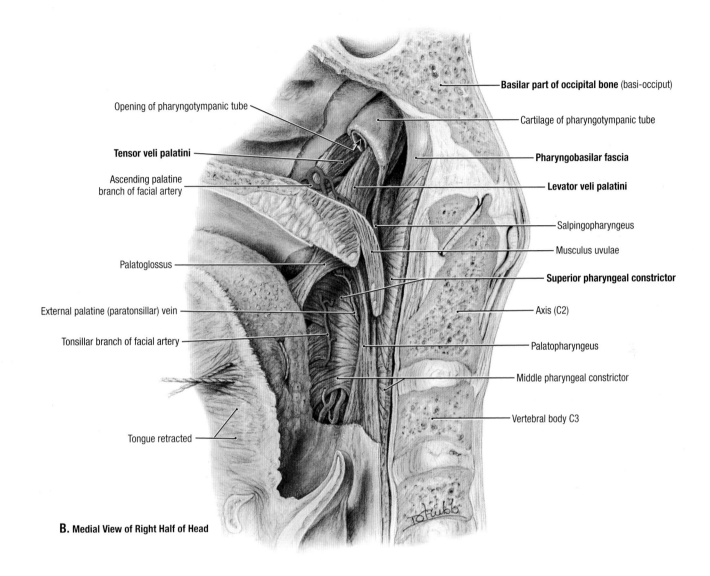

Opening of pharyngotympanic tube

Tensor veli palatini

Ascending palatine
branch of facial artery

Palatoglossus

External palatine (paratonsillar) vein

Tonsillar branch of facial artery

Tongue retracted

Basilar part of occipital bone (basi-occiput)

Cartilage of pharyngotympanic tube

Pharyngobasilar fascia

Levator veli palatini

Salpingopharyngeus

Musculus uvulae

Superior pharyngeal constrictor

Axis (C2)

Palatopharyngeus

Middle pharyngeal constrictor

Vertebral body C3

B. Medial View of Right Half of Head

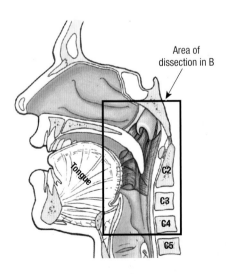

Area of
dissection in B

SERIAL DISSECTION OF ISTHMUS OF FAUCES AND LATERAL WALL OF NASOPHARYNX II **8.32**

Muscles underlying tonsillar fossa and wall of nasopharynx.
The palatine and pharyngeal tonsils and mucous membrane
have been removed. The pharyngobasilar fascia, which attaches
the pharynx to the basilar part of the occipital bone was also
removed, except at the superior, arched border of the superior
pharyngeal constrictor.

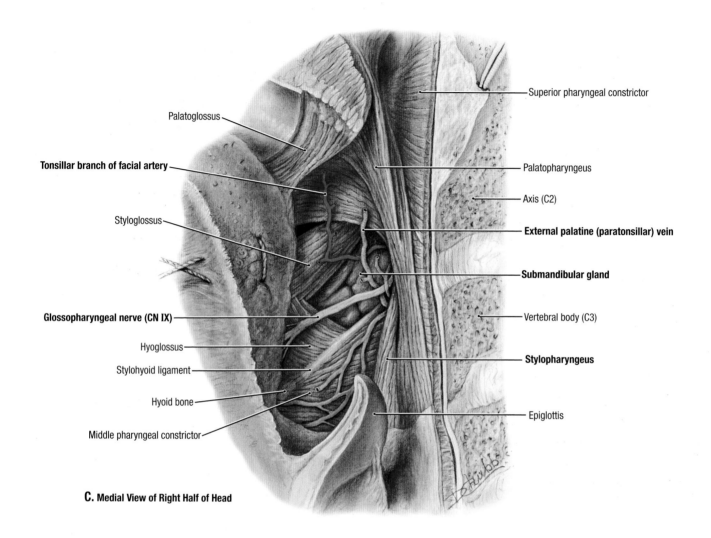

Palatoglossus

Tonsillar branch of facial artery

Styloglossus

Glossopharyngeal nerve (CN IX)

Hyoglossus

Stylohyoid ligament

Hyoid bone

Middle pharyngeal constrictor

Superior pharyngeal constrictor

Palatopharyngeus

Axis (C2)

External palatine (paratonsillar) vein

Submandibular gland

Vertebral body (C3)

Stylopharyngeus

Epiglottis

C. Medial View of Right Half of Head

8.32 SERIAL DISSECTION OF ISTHMUS OF FAUCES AND LATERAL WALL OF NASOPHARYNX III

Neurovascular structures of tonsillar sinus and longitudinal muscles of the pharynx.

- In this deeper dissection, the tongue was pulled anteriorly, and the inferior part of the origin of the superior pharyngeal constrictor muscle was cut away.
- The glossopharyngeal nerve passes to the posterior one third of the tongue and lies anterior to the stylopharyngeus muscle.
- The tonsillar branch of the facial artery sends a branch (cut short here) to accompany the glossopharyngeal nerve to the tongue; the submandibular gland is seen lateral to the artery and external palatine (paratonsillar) vein.

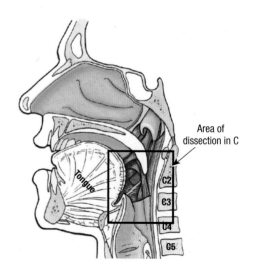

Area of dissection in C

Tongue

C2

C3

C4

C5

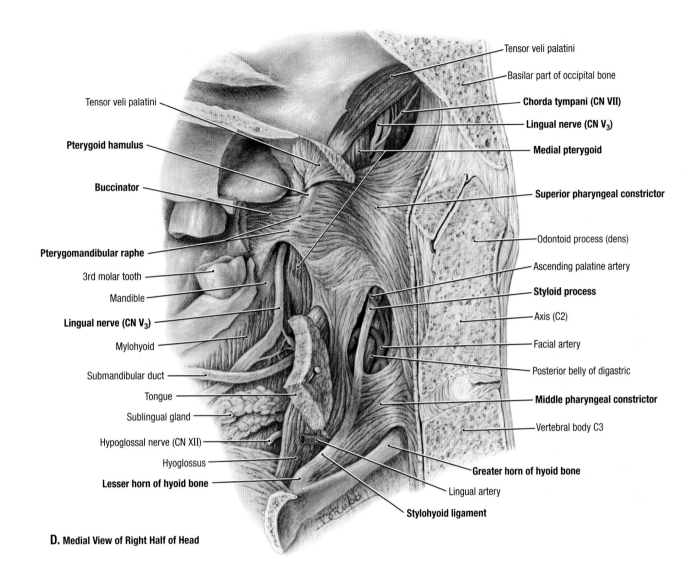

Tensor veli palatini

Basilar part of occipital bone

Chorda tympani (CN VII)

Lingual nerve (CN V₃)

Medial pterygoid

Superior pharyngeal constrictor

Odontoid process (dens)

Ascending palatine artery

Styloid process

Axis (C2)

Facial artery

Posterior belly of digastric

Middle pharyngeal constrictor

Vertebral body C3

Greater horn of hyoid bone

Lingual artery

Stylohyoid ligament

Tensor veli palatini

Pterygoid hamulus

Buccinator

Pterygomandibular raphe

3rd molar tooth

Mandible

Lingual nerve (CN V₃)

Mylohyoid

Submandibular duct

Tongue

Sublingual gland

Hypoglossal nerve (CN XII)

Hyoglossus

Lesser horn of hyoid bone

D. Medial View of Right Half of Head

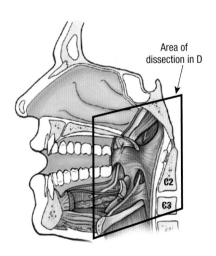

Area of dissection in D

C2

C3

SERIAL DISSECTION OF ISTHMUS OF FAUCES AND LATERAL WALL OF NASOPHARYNX IV — 8.32

- The superior pharyngeal constrictor muscle arises from (1) the pterygomandibular raphe, which unites it to the buccinator muscle; (2) the bones at each end of the raphe, the hamulus of the medial pterygoid plate superiorly and the mandible inferiorly; and (3) the root (posterior part) of the tongue.
- The middle pharyngeal constrictor muscle arises from the angle formed by the greater and lesser horns of the hyoid bone and from the stylohyoid ligament; in this specimen, the styloid process is long and, therefore, a lateral relation of the tonsil.
- The lingual nerve is joined by the chorda tympani, disappears at the posterior border of the medial pterygoid muscle, and reappears at the anterior border to follow the mandible.

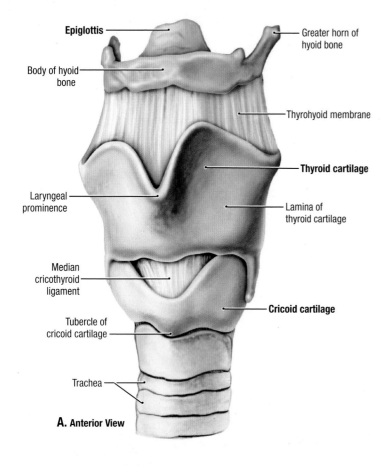

Epiglottis
Body of hyoid bone
Laryngeal prominence
Median cricothyroid ligament
Tubercle of cricoid cartilage
Trachea

Greater horn of hyoid bone
Thyrohyoid membrane
Thyroid cartilage
Lamina of thyroid cartilage
Cricoid cartilage

A. Anterior View

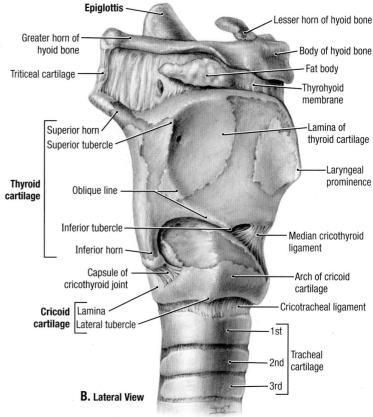

Epiglottis
Greater horn of hyoid bone
Triticeal cartilage
Thyroid cartilage — Superior horn, Superior tubercle
Oblique line
Inferior tubercle
Inferior horn
Capsule of cricothyroid joint
Cricoid cartilage — Lamina, Lateral tubercle

Lesser horn of hyoid bone
Body of hyoid bone
Fat body
Thyrohyoid membrane
Lamina of thyroid cartilage
Laryngeal prominence
Median cricothyroid ligament
Arch of cricoid cartilage
Cricotracheal ligament
1st / 2nd / 3rd Tracheal cartilage

B. Lateral View

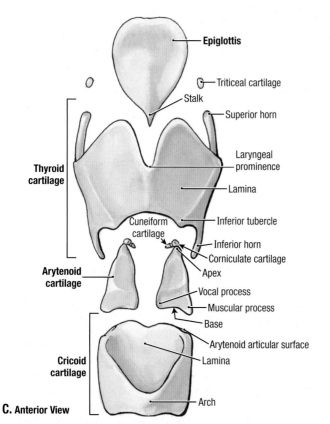

Epiglottis
Triticeal cartilage
Stalk
Superior horn
Laryngeal prominence
Thyroid cartilage
Lamina
Cuneiform cartilage
Inferior tubercle
Inferior horn
Corniculate cartilage
Apex
Arytenoid cartilage
Vocal process
Muscular process
Base
Arytenoid articular surface
Cricoid cartilage
Lamina
Arch

C. Anterior View

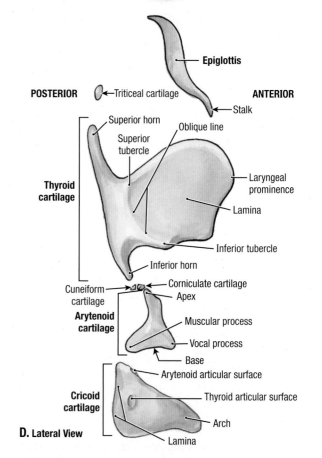

Epiglottis
POSTERIOR — Triticeal cartilage — ANTERIOR
Stalk
Superior horn
Oblique line
Superior tubercle
Thyroid cartilage
Laryngeal prominence
Lamina
Inferior tubercle
Inferior horn
Cuneiform cartilage
Corniculate cartilage
Apex
Arytenoid cartilage
Muscular process
Vocal process
Base
Arytenoid articular surface
Thyroid articular surface
Cricoid cartilage
Arch
Lamina

D. Lateral View

8.33 CARTILAGES OF LARYNGEAL SKELETON

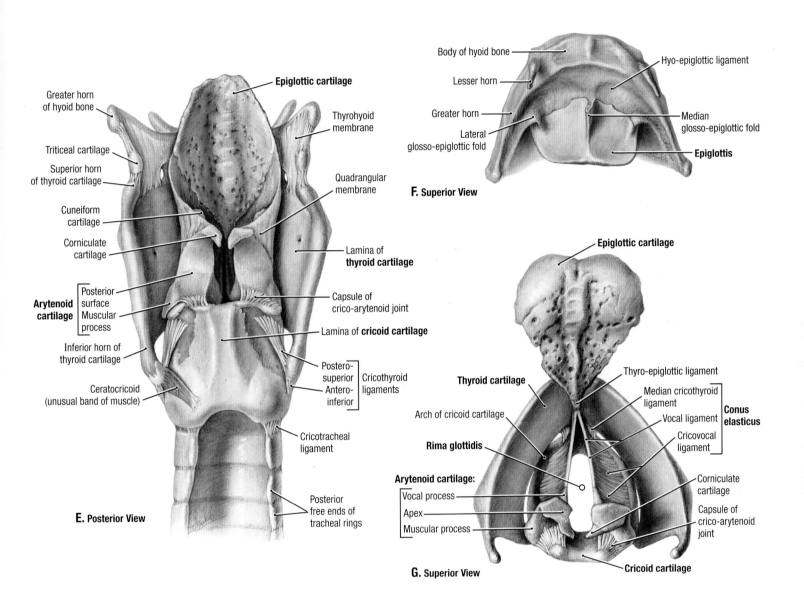

E. Posterior View

- Greater horn of hyoid bone
- Triticeal cartilage
- Superior horn of thyroid cartilage
- Cuneiform cartilage
- Corniculate cartilage
- **Arytenoid cartilage** { Posterior surface / Muscular process
- Inferior horn of thyroid cartilage
- Ceratocricoid (unusual band of muscle)
- **Epiglottic cartilage**
- Thyrohyoid membrane
- Quadrangular membrane
- Lamina of **thyroid cartilage**
- Capsule of crico-arytenoid joint
- Lamina of **cricoid cartilage**
- Postero-superior / Antero-inferior } Cricothyroid ligaments
- Cricotracheal ligament
- Posterior free ends of tracheal rings

F. Superior View

- Body of hyoid bone
- Lesser horn
- Greater horn
- Lateral glosso-epiglottic fold
- Hyo-epiglottic ligament
- Median glosso-epiglottic fold
- **Epiglottis**

G. Superior View

- **Epiglottic cartilage**
- **Thyroid cartilage**
- Arch of cricoid cartilage
- **Rima glottidis**
- **Arytenoid cartilage:** Vocal process / Apex / Muscular process
- Thyro-epiglottic ligament
- Median cricothyroid ligament
- Vocal ligament
- Cricovocal ligament
- Corniculate cartilage
- Capsule of crico-arytenoid joint
- Cricoid cartilage
- **Conus elasticus**

CARTILAGES OF LARYNGEAL SKELETON (*continued*)   8.33

A, B, and **E.** Articulated laryngeal skeleton. **C.** and **D.** Cartilages disarticulated and separated. **F.** Epiglottis and hyo-epiglottic ligament. **G.** Conus elasticus and rima glottidis.

- The larynx extends vertically from the tip of the epiglottis to the inferior border of the cricoid cartilage. The hyoid bone is generally not regarded as part of the larynx.
- The cricoid cartilage is the only cartilage that totally encircles the airway.
- The rima glottidis is the aperture between the vocal folds. During normal respiration, it is narrow and wedge-shaped; during forced respiration, it is wide. Variations in the tension and length of the vocal folds, in the width of the rima glottidis, and in the intensity of the expiratory effort produce changes in the pitch of the voice.
- **Laryngeal fractures** may result from blows received in sports such as kickboxing and hockey or from compression by a shoulder strap during an automobile accident. Laryngeal fractures produce submucous hemorrhage and edema, respiratory obstruction, hoarseness, and sometimes a temporary inability to speak. The thyroid, cricoid, and most of the arytenoid cartilages often ossify as age advances, commencing at approximately 25 years of age in the thyroid cartilage.

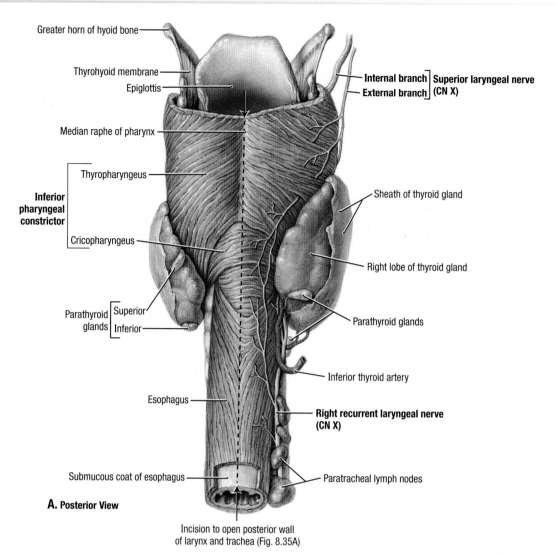

Greater horn of hyoid bone

Thyrohyoid membrane

Epiglottis

Median raphe of pharynx

Thyropharyngeus

Inferior pharyngeal constrictor

Cricopharyngeus

Parathyroid glands — Superior / Inferior

Esophagus

Submucous coat of esophagus

A. Posterior View

Incision to open posterior wall of larynx and trachea (Fig. 8.35A)

Internal branch] **Superior laryngeal nerve**
External branch] **(CN X)**

Sheath of thyroid gland

Right lobe of thyroid gland

Parathyroid glands

Inferior thyroid artery

Right recurrent laryngeal nerve (CN X)

Paratracheal lymph nodes

8.34 EXTERNAL LARYNX AND LARYNGEAL NERVES

A. Posterior aspect.

- The internal branch of the superior laryngeal nerve innervates the mucous membrane superior to the vocal folds, and the external laryngeal branch supplies the inferior pharyngeal constrictor and cricothyroid muscles.
- The recurrent laryngeal nerve supplies the esophagus, trachea, and inferior pharyngeal constrictor muscle. It supplies sensory innervation inferior to the vocal folds and motor innervation to the intrinsic muscles of the larynx, except the cricothyroid.

B. Laryngocele. A laryngocele (enlarged laryngeal saccule) projects through the thyrohyoid membrane and communicates with the larynx through the ventricle. This air sac can form a bulge in the neck, especially on coughing. The inferior laryngeal nerves are vulnerable to injury during operations in the anterior triangles of the neck. **Injury of the inferior laryngeal nerve** results in paralysis of the vocal fold. The voice is initially poor because the paralyzed fold cannot adduct to meet the normal vocal fold. In a bilateral paralysis, the voice is almost absent. **Injury to the external branch of the superior laryngeal nerve** results in a voice that is monotonous in character because the cricothyroid muscle is unable to vary the tension of the vocal fold. Hoarseness is the most common symptom of serious disorders of the larynx.

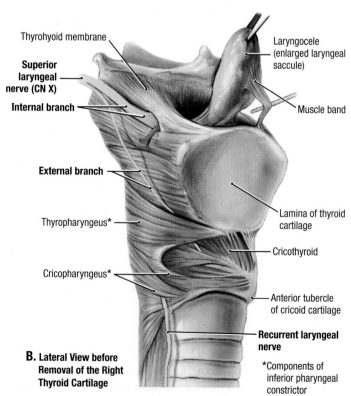

Thyrohyoid membrane

Superior laryngeal nerve (CN X)

Internal branch

External branch

Thyropharyngeus*

Cricopharyngeus*

Laryngocele (enlarged laryngeal saccule)

Muscle band

Lamina of thyroid cartilage

Cricothyroid

Anterior tubercle of cricoid cartilage

Recurrent laryngeal nerve

*Components of inferior pharyngeal constrictor

B. Lateral View before Removal of the Right Thyroid Cartilage

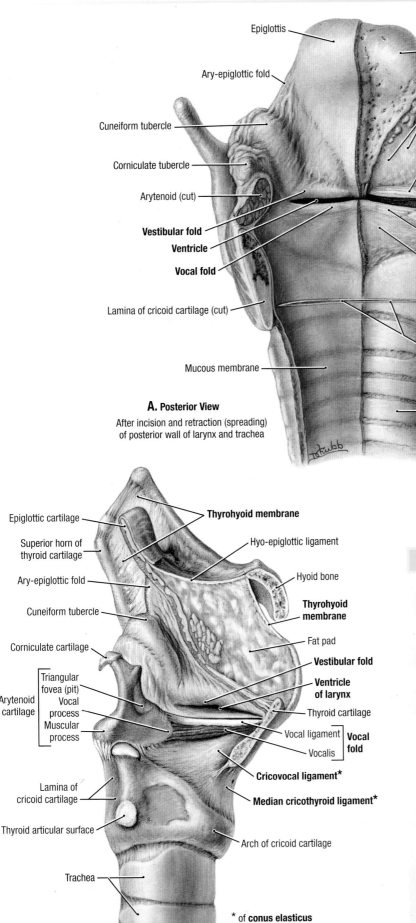

Epiglottis

Epiglottic cartilage

Ary-epiglottic fold

Quadrangular membrane

Vestibular ligament

Cuneiform cartilage

Cuneiform tubercle

Superior horn

Corniculate tubercle

Corniculate cartilage

Arytenoid (cut)

Arytenoid cartilage, medial surface

Thyroid cartilage

Vestibular fold

Posterior crico-arytenoid ligament

Ventricle

Vocal process of arytenoid cartilage

Vocal fold

Vocal ligament

Lamina of cricoid cartilage (cut)

Cricothyroid ligament

Surgical needle spreading cricoid cartilage

Mucous membrane

A. Posterior View

After incision and retraction (spreading)
of posterior wall of larynx and trachea

Tracheal ring

Epiglottic cartilage

Thyrohyoid membrane

Superior horn of
thyroid cartilage

Hyo-epiglottic ligament

Ary-epiglottic fold

Hyoid bone

Cuneiform tubercle

**Thyrohyoid
membrane**

Corniculate cartilage

Fat pad

Triangular
fovea (pit)

Vestibular fold

Arytenoid
cartilage

Vocal
process

**Ventricle
of larynx**

Muscular
process

Thyroid cartilage

Vocal ligament **Vocal
fold**
Vocalis

Lamina of
cricoid cartilage

Cricovocal ligament*

Thyroid articular surface

Median cricothyroid ligament*

Arch of cricoid cartilage

Trachea

* of conus elasticus

B. Lateral View after Removal of the Right Thyroid Cartilage

INTERNAL LARYNX 8.35

A. The posterior wall of the larynx is split in the median plane (see Fig. 8.34A), and the two sides held
apart. On the left side of the specimen, the mucous
membrane is intact; on the right side, the mucous
and submucous coats are peeled off revealing the
cartilages, ligaments, and fibro-elastic membrane. **B.**
Interior of the larynx superior to the vocal folds. The
larynx is sectioned near the median plane to reveal
the interior of its left side. Inferior to this level, the
right side of the intact larynx is dissected.

- The three compartments of the larynx are (1) the
  superior compartment of the vestibule, superior to
  the level of the vestibular folds (false cords); (2) the
  middle, between the levels of the vestibular and
  vocal folds; and (3) the inferior, or infraglottic, cavity, inferior to the level of the vocal folds.
- The quadrangular membrane underlies the ary-
  epiglottic fold superiorly and is thickened inferiorly
  to form the vestibular ligament. The cricothyroid
  ligament (conus elasticus) begins inferiorly as the
  strong median cricothyroid ligament and is thickened superiorly as the vocal ligament. The lateral
  recess between the vocal and vestibular ligaments
  is the ventricle.

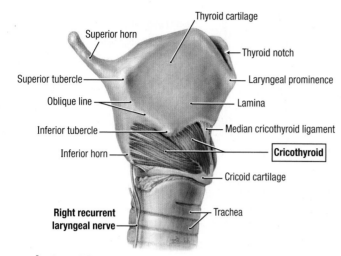

Thyroid cartilage

Superior horn

Thyroid notch

Superior tubercle

Laryngeal prominence

Oblique line

Lamina

Inferior tubercle

Median cricothyroid ligament

Inferior horn

Cricothyroid

Cricoid cartilage

Right recurrent laryngeal nerve

Trachea

A. Lateral View

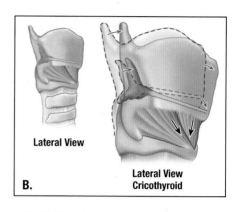

Lateral View

B.

Lateral View Cricothyroid

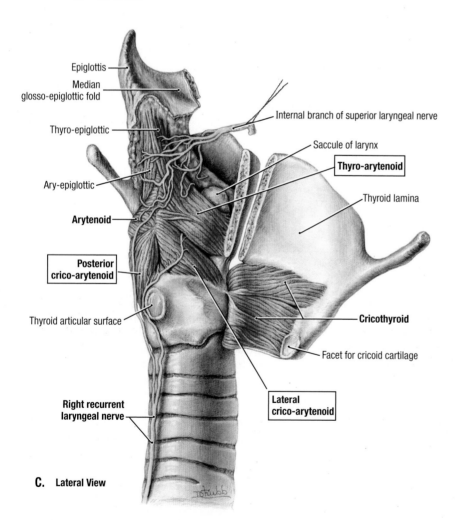

Epiglottis

Median glosso-epiglottic fold

Internal branch of superior laryngeal nerve

Thyro-epiglottic

Saccule of larynx

Thyro-arytenoid

Ary-epiglottic

Thyroid lamina

Arytenoid

Posterior crico-arytenoid

Thyroid articular surface

Cricothyroid

Facet for cricoid cartilage

Right recurrent laryngeal nerve

Lateral crico-arytenoid

C. Lateral View

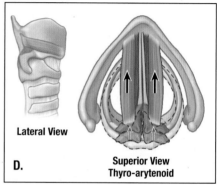

Lateral View

D.

Superior View Thyro-arytenoid

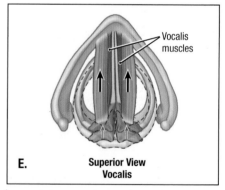

Vocalis muscles

E.

Superior View Vocalis

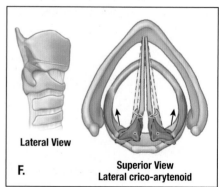

Lateral View

F.

Superior View Lateral crico-arytenoid

8.36 | MUSCLES OF LARYNX

A. and **B.** Cricothyroid. **C.** Muscles of larynx revealed by cutting thyroid cartilage along dashed line **(A)** and reflecting the right thyroid lamina anteriorly. **D.** Thyro-arytenoid. **E.** Vocalis. **F.** Lateral crico-arytenoid.

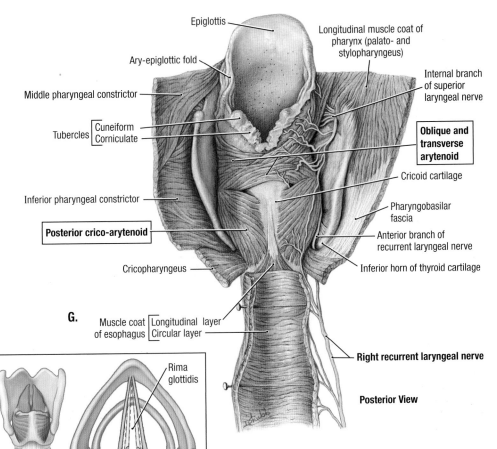

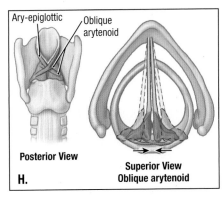

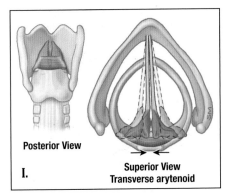

MUSCLES OF LARYNX (continued)

8.36

G. Posterior view of muscles of larynx. **H.** Oblique arytenoid. **I.** Transverse arytenoid. **J.** Posterior crico-arytenoid.

The intrinsic laryngeal muscles move the laryngeal cartilages, making alterations in the length and tension of the vocal folds and in the size and shape of the rima glottidis. All but one of the intrinsic muscles of the larynx is supplied by the recurrent laryngeal nerve (CN X). The cricothyroid muscle (**A** and **B**) is supplied by the external laryngeal nerve, one of the two terminal branches of the superior laryngeal nerve.

TABLE 8.9	**MUSCLES OF LARYNX**			
Muscle	*Origin*	*Insertion*	*Innervation*	*Main Action(s)*
Cricothyroid	Anterolateral part of cricoid cartilage	Inferior margin and inferior horn of thyroid cartilage	External branch of superior laryngeal nerve (CN X)	Tenses vocal fold
Posterior crico-arytenoid	Posterior surface of laminae of cricoid cartilage	Muscular process of arytenoid cartilage	Recurrent laryngeal nerve (CN X)	Abducts vocal fold
Lateral crico-arytenoid	Arch of cricoid cartilage			Adducts vocal fold
Thyro-arytenoid[a]	Posterior surface of thyroid cartilage			Relaxes vocal fold
Transverse and oblique arytenoids[b]	One arytenoid cartilage	Opposite arytenoid cartilage		Close inlet of larynx by approximating arytenoid cartilages
Vocalis[c]	Angle between laminae of thyroid cartilage	Vocal ligament, between origin and vocal process of arytenoid cartilage		Alters vocal fold during phonation

[a]Superior fibers of the thyro-arytenoid muscle pass into the ary-epiglottic fold, and some of them reach the epiglottic cartilage. These fibers constitute the thyro-epiglottic muscle, which widens the inlet of the larynx.
[b]Some fibers of the oblique arytenoid muscle continue as the ary-epiglottic muscle.
[c]This slender muscular slip is derived from inferior deeper fibers of the thyro-arytenoid muscle.

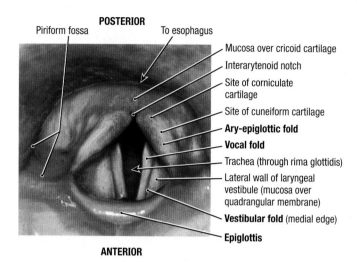

POSTERIOR

Piriform fossa — To esophagus

— Mucosa over cricoid cartilage
— Interarytenoid notch
— Site of corniculate cartilage
— Site of cuneiform cartilage
— **Ary-epiglottic fold**
— **Vocal fold**
— Trachea (through rima glottidis)
— Lateral wall of laryngeal vestibule (mucosa over quadrangular membrane)
— **Vestibular fold** (medial edge)
— **Epiglottis**

ANTERIOR

A. Laryngoscopic Examination

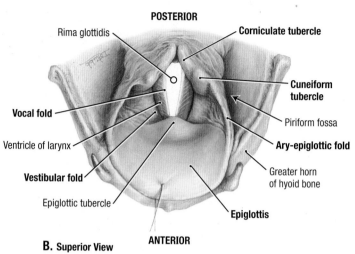

POSTERIOR

Rima glottidis — **Corniculate tubercle**

Vocal fold —
Cuneiform tubercle
Piriform fossa
Ventricle of larynx —
Ary-epiglottic fold
Vestibular fold —
Greater horn of hyoid bone
Epiglottic tubercle —
Epiglottis

ANTERIOR

B. Superior View

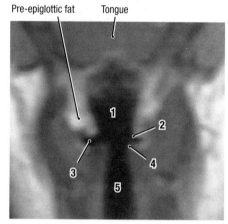

Pre-epiglottic fat — Tongue

1
2
3 4
5

C. Coronal MRI

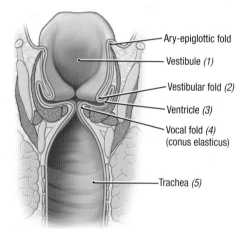

— Ary-epiglottic fold
— Vestibule (1)
— Vestibular fold (2)
— Ventricle (3)
— Vocal fold (4) (conus elasticus)
— Trachea (5)

D. Posterior View

8.37 LARYNGOSCOPIC EXAMINATION AND MRI OF LARYNX

A. Laryngoscopic examination. Laryngoscopy is the procedure used to examine the interior of the larynx. The larynx may be examined visually by indirect laryngoscopy using a laryngeal mirror or it may be viewed by direct laryngoscopy using a tubular and endoscopic instrument, a laryngoscope. The vestibular and vocal folds can be observed. **B.** Vocal folds and rima glottidis. The inlet, or aditus, to the larynx is bounded anteriorly by the epiglottis; posteriorly by the arytenoid cartilages, the corniculate cartilages that cap them, and the interarytenoid fold that unites them; and on each side by the ary-epiglottic fold, which contains the superior end of the cuneiform cartilage. The vocal apparatus of the larynx, the glottis, includes the vocal folds, vocal processes of the arytenoid cartilages, and the rima glottidis, the aperture between the vocal folds. **C.** Coronal MRI. **D.** Coronal section. Numbers in parentheses on diagram refer to numbered structures on MRI.

A **foreign body** such as a piece of steak, may accidentally aspirate through the laryngeal inlet into the vestibule of the larynx, where it becomes trapped superior to the vestibular folds. When a foreign body enters the vestibule, the laryngeal muscles go into spasm, tensing the vocal folds. The rima glottidis closes and no air enters the trachea. **Asphyxiation** occurs, and the person will die in approximately 5 minutes from lack of oxygen if the obstruction is not removed. Emergency therapy must be given to open the airway. The procedure used depends on the condition of the patient, the facilities available, and the experience of the person giving first aid. Because the lungs still contain air, sudden compression of the abdomen (**Heimlich maneuver**) causes the diaphragm to elevate and compress the lungs, expelling air from the trachea into the larynx. This maneuver may dislodge the food or other material from the larynx.

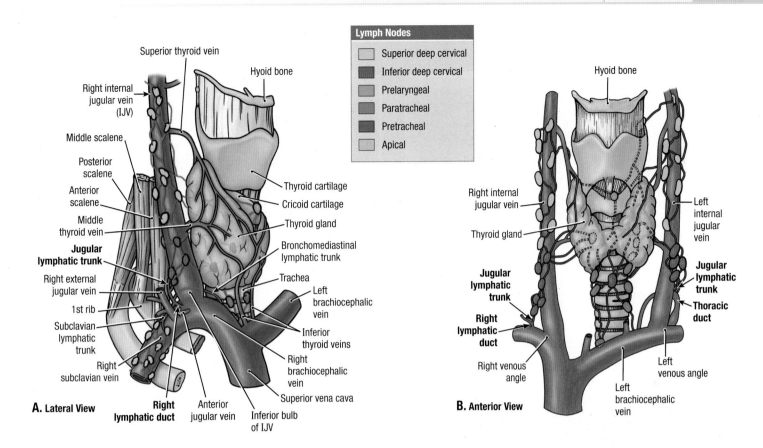

Lymph Nodes

☐	Superior deep cervical
■	Inferior deep cervical
▨	Prelaryngeal
▨	Paratracheal
■	Pretracheal
☐	Apical

A. Lateral View

Superior thyroid vein
Right internal jugular vein (IJV)
Middle scalene
Posterior scalene
Anterior scalene
Middle thyroid vein
Jugular lymphatic trunk
Right external jugular vein
1st rib
Subclavian lymphatic trunk
Right subclavian vein
Right lymphatic duct
Anterior jugular vein
Inferior bulb of IJV

Hyoid bone
Thyroid cartilage
Cricoid cartilage
Thyroid gland
Bronchomediastinal lymphatic trunk
Trachea
Left brachiocephalic vein
Inferior thyroid veins
Right brachiocephalic vein
Superior vena cava

B. Anterior View

Hyoid bone
Right internal jugular vein
Thyroid gland
Jugular lymphatic trunk
Right lymphatic duct
Right venous angle

Left internal jugular vein
Jugular lymphatic trunk
Thoracic duct
Left venous angle
Left brachiocephalic vein

LYMPHATIC DRAINAGE OF THYROID GLAND, LARYNX, AND TRACHEA

8.38

Radical neck dissections are performed when cancer invades the lymphatics. During the procedure, the deep cervical lymph nodes and the tissues around them are removed as completely as possible. Although major arteries, the brachial plexus, CN X, and the phrenic nerve are preserved, most cutaneous branches of the cervical plexus are removed. The aim of the dissection is to remove all tissue that contains lymph nodes in one piece.

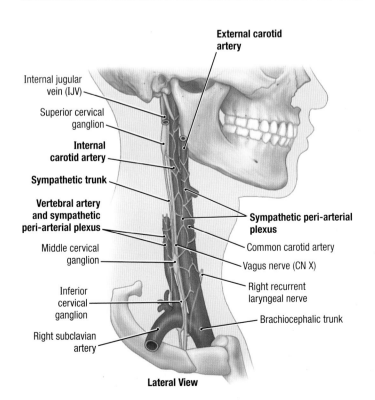

External carotid artery
Internal jugular vein (IJV)
Superior cervical ganglion
Internal carotid artery
Sympathetic trunk
Vertebral artery and sympathetic peri-arterial plexus
Middle cervical ganglion
Inferior cervical ganglion
Right subclavian artery
Sympathetic peri-arterial plexus
Common carotid artery
Vagus nerve (CN X)
Right recurrent laryngeal nerve
Brachiocephalic trunk

Lateral View

SYMPATHETIC TRUNK AND SYMPATHETIC PERI-ARTERIAL PLEXUS

8.39

A **lesion of a sympathetic trunk** in the neck results in a sympathetic disturbance called **Horner syndrome**, which is characterized by the following:

- **Pupillary constriction** resulting from paralysis of the dilator pupillae muscle.
- **Ptosis** (drooping of the superior eyelid), resulting from paralysis of the smooth (tarsal) muscle intermingled with striated muscle of the levator palpebrae superioris.
- Sinking in of the eyeball (**enophthalmos**), possibly caused by paralysis of smooth (orbitalis) muscle in the floor of the orbit.
- Vasodilation and absence of sweating on the face and neck (**anhydrosis**), caused by a lack of sympathetic (vasoconstrictive) nerve supply to the blood vessels and sweat glands.

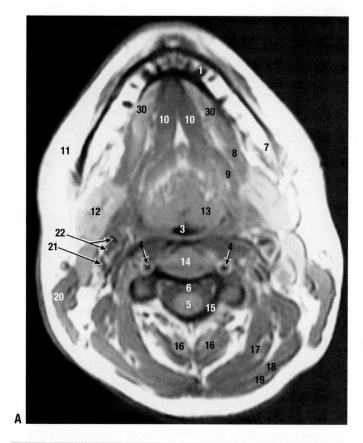

A

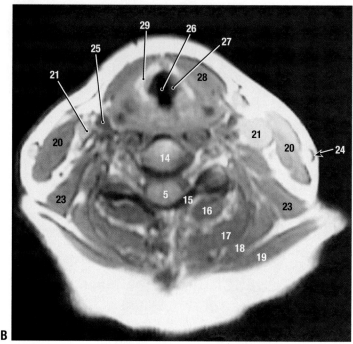

B

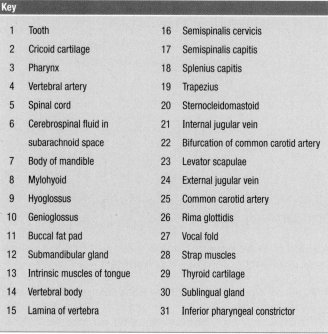

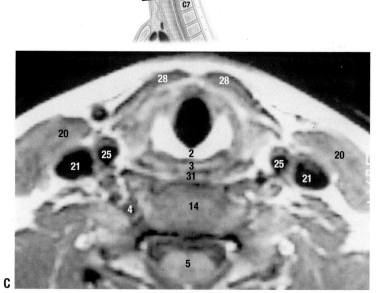

C

Inferior Views

Key			
1	Tooth	16	Semispinalis cervicis
2	Cricoid cartilage	17	Semispinalis capitis
3	Pharynx	18	Splenius capitis
4	Vertebral artery	19	Trapezius
5	Spinal cord	20	Sternocleidomastoid
6	Cerebrospinal fluid in	21	Internal jugular vein
	subarachnoid space	22	Bifurcation of common carotid artery
7	Body of mandible	23	Levator scapulae
8	Mylohyoid	24	External jugular vein
9	Hyoglossus	25	Common carotid artery
10	Genioglossus	26	Rima glottidis
11	Buccal fat pad	27	Vocal fold
12	Submandibular gland	28	Strap muscles
13	Intrinsic muscles of tongue	29	Thyroid cartilage
14	Vertebral body	30	Sublingual gland
15	Lamina of vertebra	31	Inferior pharyngeal constrictor

8.40 TRANSVERSE MRIs OF NECK

The orientation figure indicates the vertebral level of the MRI sections.

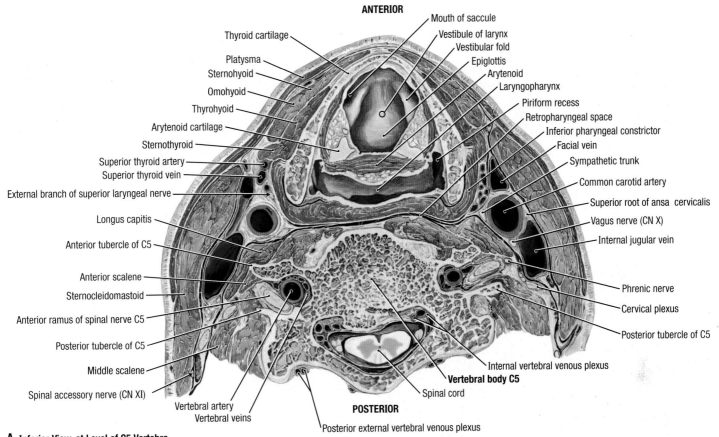

ANTERIOR

Thyroid cartilage
Platysma
Sternohyoid
Omohyoid
Thyrohyoid
Arytenoid cartilage
Sternothyroid
Superior thyroid artery
Superior thyroid vein
External branch of superior laryngeal nerve
Longus capitis
Anterior tubercle of C5
Anterior scalene
Sternocleidomastoid
Anterior ramus of spinal nerve C5
Posterior tubercle of C5
Middle scalene
Spinal accessory nerve (CN XI)

Mouth of saccule
Vestibule of larynx
Vestibular fold
Epiglottis
Arytenoid
Laryngopharynx
Piriform recess
Retropharyngeal space
Inferior pharyngeal constrictor
Facial vein
Sympathetic trunk
Common carotid artery
Superior root of ansa cervicalis
Vagus nerve (CN X)
Internal jugular vein
Phrenic nerve
Cervical plexus
Posterior tubercle of C5

Internal vertebral venous plexus
Vertebral body C5
Spinal cord

Vertebral artery
Vertebral veins

POSTERIOR

Posterior external vertebral venous plexus

A. Inferior View, at Level of C5 Vertebra

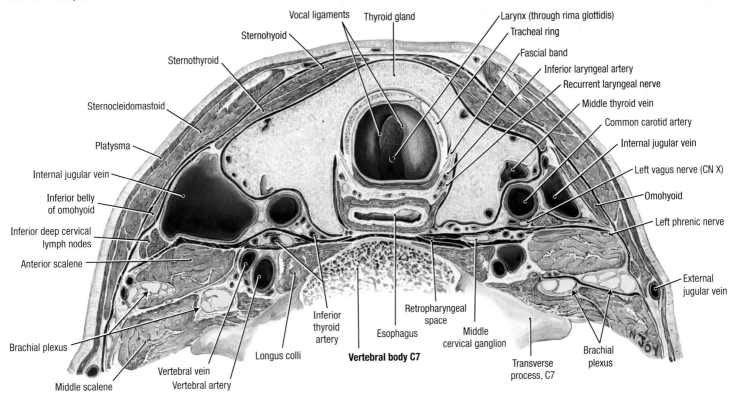

Vocal ligaments
Thyroid gland
Sternohyoid
Sternothyroid
Sternocleidomastoid
Platysma
Internal jugular vein
Inferior belly of omohyoid
Inferior deep cervical lymph nodes
Anterior scalene
Brachial plexus
Middle scalene

Larynx (through rima glottidis)
Tracheal ring
Fascial band
Inferior laryngeal artery
Recurrent laryngeal nerve
Middle thyroid vein
Common carotid artery
Internal jugular vein
Left vagus nerve (CN X)
Omohyoid
Left phrenic nerve
External jugular vein
Brachial plexus
Transverse process, C7

Vertebral vein
Vertebral artery
Longus colli
Inferior thyroid artery
Esophagus
Retropharyngeal space
Vertebral body C7
Middle cervical ganglion

B. Inferior View, at Level of C7 Vertebra

TRANSVERSE ANATOMICAL SECTIONS OF NECK

8.41

A. At level of laryngopharynx. **B.** At level of trachea.

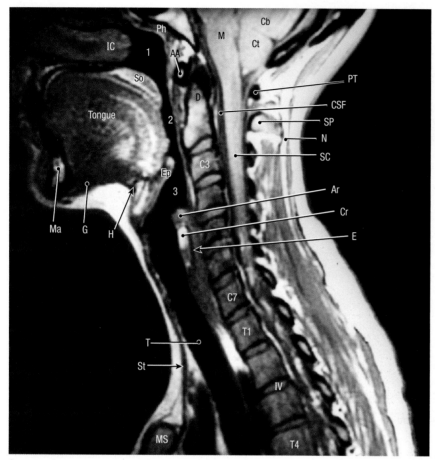

A. Median MRI

Key	
AA	Anterior arch of C1
Ar	Arytenoid cartilage
C3–T4	Vertebral bodies
Cb	Cerebellum
Cr	Cricoid cartilage
CSF	Cerebrospinal fluid in subarachnoid space
Ct	Tonsil of cerebellum
D	Dens
E	Esophagus
Ep	Epiglottis
G	Genioglossus
H	Hyoid bone
IC	Inferior concha
IV	Intervertebral disc
M	Medulla oblongata
Ma	Mandible
MS	Manubrium of sternum
N	Nuchal ligament
Ph	Pharyngeal tonsil (adenoid)
PT	Posterior tubercle of C1
SC	Spinal cord
So	Soft palate
SP	Spinous process
St	Strap muscles
T	Trachea
1	Nasopharynx
2	Oropharynx
3	Laryngopharynx

B. Deglutition (swallowing)

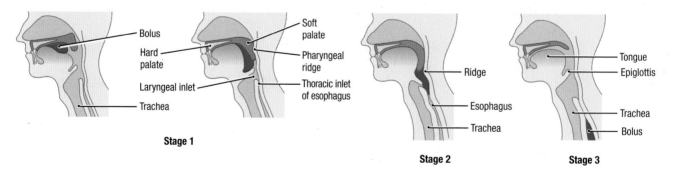

Stage 1   **Stage 2**   **Stage 3**

8.42   **MEDIAN SECTION AND MRI SCAN OF HEAD AND NECK**

A. Median MRI. **B.** Swallowing. There are three main stages of swallowing:

- Stage 1: Voluntary; the bolus is compressed against the palate and pushed from the mouth into the oropharynx, mainly by coordinated movements of the muscles of the tongue and soft palate.
- Stage 2: Involuntary and rapid; the soft palate is elevated, sealing off the nasopharynx from the oropharynx and laryngopharynx.

The pharynx widens and shortens to receive the bolus of food as the suprahyoid and longitudinal pharyngeal muscles contract, elevating the larynx.

- Stage 3: Involuntary; sequential contraction of all three pharyngeal constrictor muscles forces the food bolus inferiorly into the esophagus.

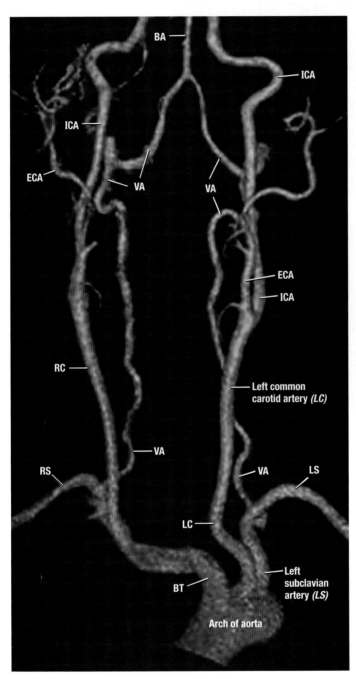

A. Anterior View

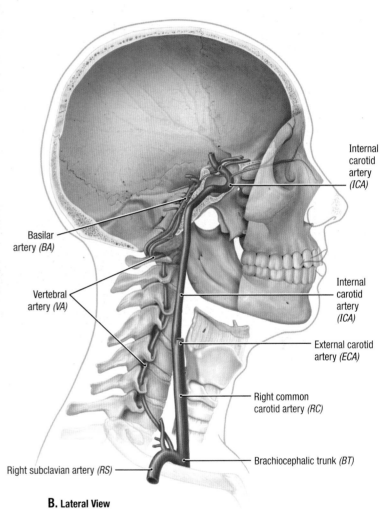

B. Lateral View

IMAGING OF BLOOD SUPPLY OF HEAD AND NECK

8.43

A. CT angiogram. **B.** Schematic illustration.

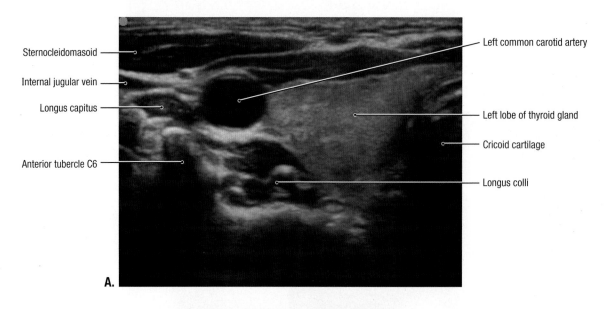

Sternocleidomasoid —
Internal jugular vein —
Longus capitus —
Anterior tubercle C6 —

— Left common carotid artery
— Left lobe of thyroid gland
— Cricoid cartilage
— Longus colli

A.

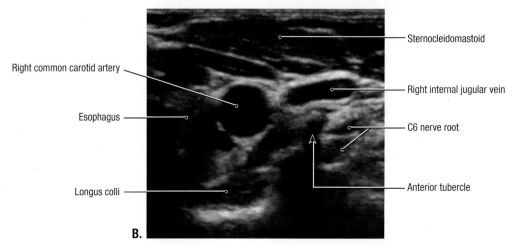

Right common carotid artery —
Esophagus —
Longus colli —

— Sternocleidomastoid
— Right internal jugular vein
— C6 nerve root
— Anterior tubercle

B.

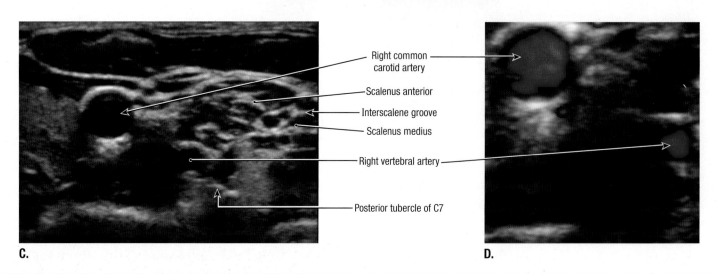

Right common carotid artery —
Scalenus anterior —
Interscalene groove —
Scalenus medius —
Right vertebral artery —
Posterior tubercle of C7 —

C. **D.**

8.44 ULTRASOUND IMAGING OF HEAD AND NECK

Ultrasonography is a useful diagnostic imaging technique for studying soft tissues of the neck. Ultrasound (US) provides images of many abnormal conditions noninvasively, at relatively low cost, and with minimal discomfort. US is useful for distinguishing solid from cystic masses, for example, which may be difficult to deter-mine during physical examination. Vascular imaging of arteries and veins of the neck is possible using intravascular US. The im-ages are produced by placing the transducer over the blood vessel. Doppler ultrasound techniques help evaluate blood flow through a vessel (e.g., for detecting stenosis [narrowing] of a carotid artery).

Cranial Nerves

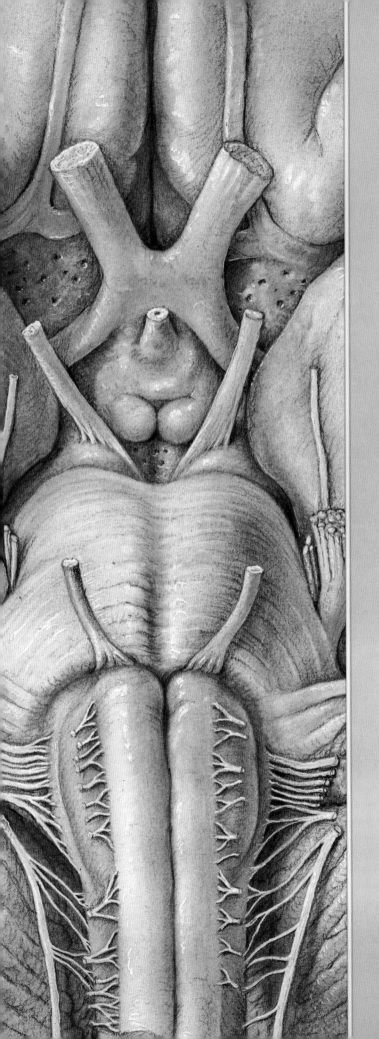

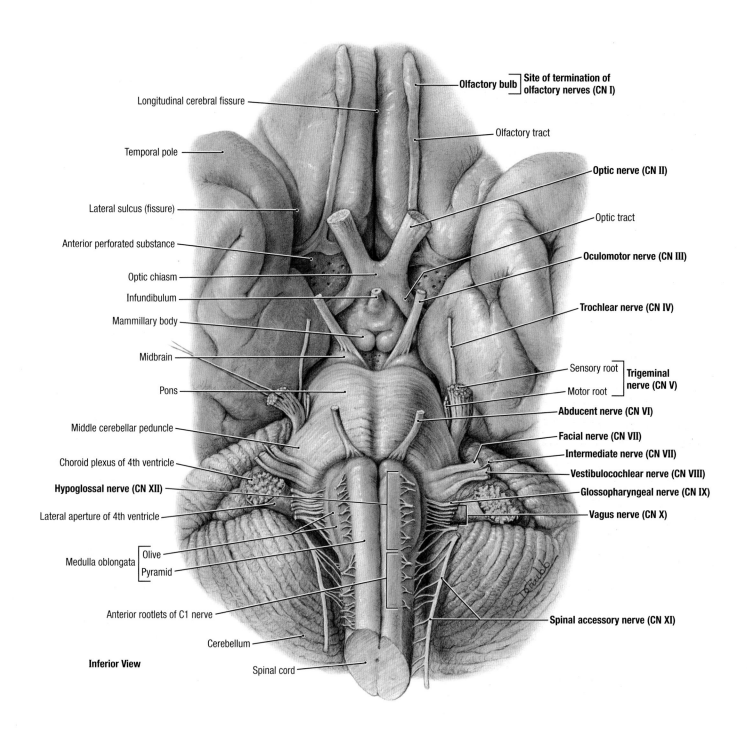

Longitudinal cerebral fissure

Temporal pole

Lateral sulcus (fissure)

Anterior perforated substance

Optic chiasm

Infundibulum

Mammillary body

Midbrain

Pons

Middle cerebellar peduncle

Choroid plexus of 4th ventricle

Hypoglossal nerve (CN XII)

Lateral aperture of 4th ventricle

Medulla oblongata [Olive / Pyramid]

Anterior rootlets of C1 nerve

Cerebellum

Inferior View

Spinal cord

Olfactory bulb] **Site of termination of olfactory nerves (CN I)**

Olfactory tract

Optic nerve (CN II)

Optic tract

Oculomotor nerve (CN III)

Trochlear nerve (CN IV)

Sensory root] **Trigeminal nerve (CN V)**
Motor root

Abducent nerve (CN VI)

Facial nerve (CN VII)

Intermediate nerve (CN VII)

Vestibulocochlear nerve (CN VIII)

Glossopharyngeal nerve (CN IX)

Vagus nerve (CN X)

Spinal accessory nerve (CN XI)

| **9.1** | **CRANIAL NERVES IN RELATION TO THE BASE OF THE BRAIN** |

Cranial nerves are nerves that exit from the cranial cavity through openings in the cranium. There are 12 pairs of cranial nerves that are named and numbered in rostrocaudal sequence of their superficial origins from the brain, brainstem, and superior spinal cord.

The olfactory nerves (CN I, *not shown*) end in the olfactory bulb. The entire origin of the spinal accessory nerve (CN XI) from the spinal cord is not included here; it extends inferiorly as far as the C6 spinal cord segment.

ANTERIOR

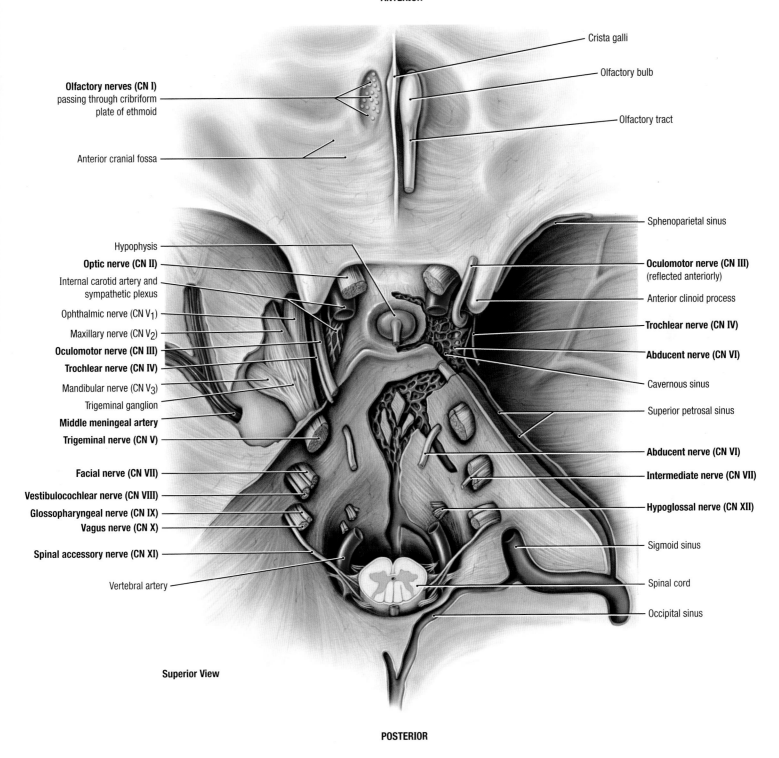

Crista galli

Olfactory bulb

Olfactory nerves (CN I) passing through cribriform plate of ethmoid

Olfactory tract

Anterior cranial fossa

Sphenoparietal sinus

Hypophysis

Oculomotor nerve (CN III) (reflected anteriorly)

Optic nerve (CN II)

Internal carotid artery and sympathetic plexus

Anterior clinoid process

Ophthalmic nerve (CN V$_1$)

Trochlear nerve (CN IV)

Maxillary nerve (CN V$_2$)

Oculomotor nerve (CN III)

Abducent nerve (CN VI)

Trochlear nerve (CN IV)

Cavernous sinus

Mandibular nerve (CN V$_3$)

Trigeminal ganglion

Superior petrosal sinus

Middle meningeal artery

Trigeminal nerve (CN V)

Abducent nerve (CN VI)

Facial nerve (CN VII)

Intermediate nerve (CN VII)

Vestibulocochlear nerve (CN VIII)

Glossopharyngeal nerve (CN IX)

Hypoglossal nerve (CN XII)

Vagus nerve (CN X)

Spinal accessory nerve (CN XI)

Sigmoid sinus

Vertebral artery

Spinal cord

Occipital sinus

Superior View

POSTERIOR

CRANIAL NERVES IN RELATION TO THE INTERNAL ASPECT OF THE CRANIAL BASE

9.2

The venous sinuses have been opened on the right side. The ophthalmic division of the trigeminal nerve (CN V$_1$) and the trochlear (CN IV) and oculomotor (CN III) nerves have been dissected from the lateral wall of the cavernous sinus. Although there are no sympathetic fibers in cranial nerves as they leave the brain, postsynaptic sympathetic nerve fibers "hitch-hike" onto branches of cranial nerves having traveled to the region via major blood vessels.

Trochlear – CN IV

Motor: superior oblique muscle of eye

Abducent – CN VI

Motor: lateral rectus muscle of eye

Oculomotor – CN III

Motor: ciliary muscles, sphincter pupillae, all extrinsic muscles of eye except those listed for CN IV and VI

Optic – CN II

Sensory: vision

Cranial nerve fibers

— Efferent (motor)
— Afferent (sensory)

Facial – CN VII Primary root

Motor: muscles of facial expression

Olfactory – CN I

Sensory: smell

Trigeminal – CN V Sensory root

Sensory: face, sinuses, teeth

Facial – CN VII Intermediate nerve

Motor: submandibular, sublingual, lacrimal glands
Sensory: taste to anterior two thirds of tongue, soft palate

Trigeminal – CN V Motor root

Motor: muscles of mastication

Vestibulocochlear – CN VIII

Vestibular nerve, sensory: orientation, motion
Cochlear nerve, sensory: hearing

Hypoglossal – CN XII

Motor: all intrinsic and extrinsic muscles of tongue (excluding palatoglossus– a palatine muscle)

Spinal accessory – CN XI

Motor: sternocleidomastoid and trapezius

Vagus – CN X

Motor: palate, pharynx, larynx, trachea, bronchial tree, heart, GI tract to left colic flexure
Sensory: pharynx, larynx; reflex sensory from tracheo-bronchial tree, lungs, heart, GI tract to left colic flexure

Glossopharyngeal – CN IX

Motor: stylopharyngeus, parotid gland
Sensory: taste: posterior third of tongue; general sensation: posterior third of tongue, pharynx, tonsillar sinus, pharyngotympanic tube, middle ear cavity

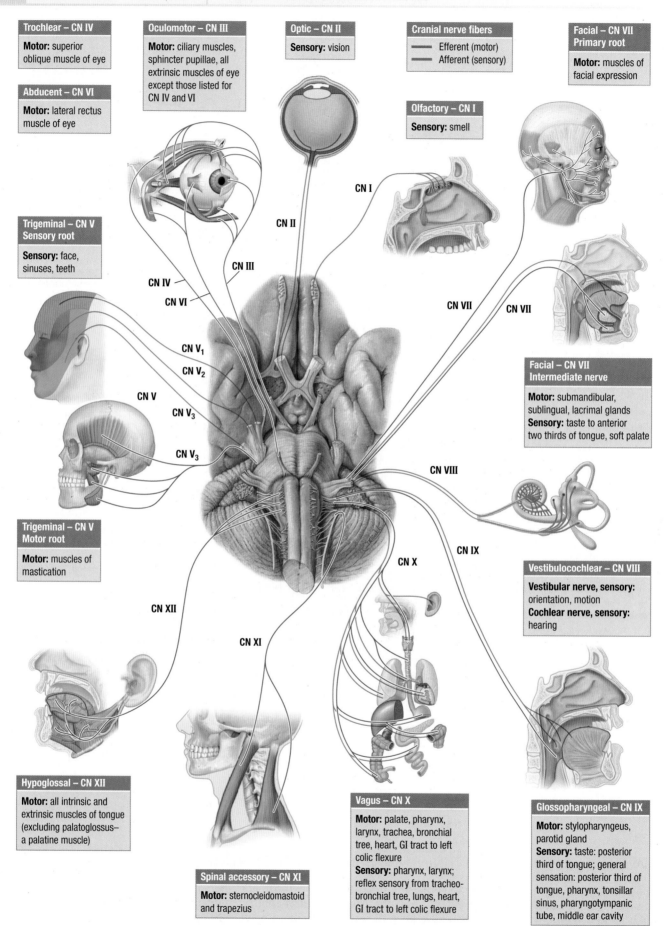

CN I
CN II
CN III
CN IV
CN VI
CN V₁
CN V₂
CN V
CN V₃
CN V₃
CN VII
CN VII
CN VIII
CN IX
CN X
CN XI
CN XII

TABLE 9.1	SUMMARY OF CRANIAL NERVES			
Nerve	Components	Location of Nerve Cell Bodies	Cranial Exit	Function
Olfactory (CN I)	Special sensory	Olfactory epithelium (olfactory cells)	Foramina in cribriform plate of ethmoid bone	Smell from nasal mucosa of roof of each nasal cavity, superior sides of nasal septum and superior concha
Optic (CN II)	Special sensory	Retina (ganglion cells)	Optic canal	Vision from retina
Oculomotor (CN III)	Somatic motor	Midbrain (nucleus of CN III)	Superior orbital fissure	Motor to levator palpebrae superioris, inferior oblique, and superior, inferior and medial rectus muscles that raise upper eyelid and direct gaze superiorly, inferiorly, and medially
	Visceral motor	Presynaptic: midbrain (Edinger-Westphal nucleus) Postsynaptic: ciliary ganglion		Parasympathetic innervation to sphincter pupillae and ciliary muscles that constrict pupil and accommodate lens of eye
Trochlear (CN IV)	Somatic motor	Midbrain (nucleus of CN IV)		Motor to superior oblique that assists in directing gaze inferolaterally
Trigeminal (CN V) Ophthalmic division (CN V$_1$)	Somatic (general) sensory	Trigeminal ganglion Synapse: sensory nucleus of CN V		Sensation from cornea, skin of forehead, scalp, eyelids, nose, and mucosa of nasal cavity and paranasal sinuses
Maxillary division (CN V$_2$)			Foramen rotundum	Sensation from skin of face over maxilla including upper lip, maxillary teeth, mucosa of nose, maxillary sinuses, and palate
Mandibular division (CN V$_3$)			Foramen ovale	Sensation from the skin over mandible, including lower lip and side of head, mandibular teeth, temporomandibular joint, and mucosa of mouth and anterior two thirds of tongue
	Somatic (branchial) motor	Pons (motor nucleus of CN V)		Motor to muscles of mastication, mylohyoid, anterior belly of digastric, tensor veli palatini, and tensor tympani
Abducent (CN VI)	Somatic motor	Pons (nucleus of CN VI)	Superior orbital fissure	Motor to lateral rectus to direct gaze laterally
Facial (CN VII)	Somatic (branchial) motor	Pons (motor nucleus of CN VII)	Internal acoustic meatus, facial canal, and stylomastoid foramen	Motor to muscles of facial expression and scalp; also supplies stapedius of middle ear, stylohyoid, and posterior belly of digastric
	Special sensory	Geniculate ganglion Synapse: nuclei of solitary tract		Taste from anterior two thirds of tongue, and soft palate
	General sensory	Geniculate ganglion Synapse: sensory nucleus of CN V		Sensation from skin of external acoustic meatus
	Visceral motor	Presynaptic: pons (superior salivatory nucleus) Postsynaptic: pterygopalatine ganglion and submandibular ganglion		Parasympathetic innervation to submandibular and sublingual salivary glands, lacrimal gland, and glands of nose and palate
Vestibulocochlear (CN VIII) Vestibular Cochlear	Special sensory	Vestibular ganglion Synapse: vestibular nuclei	Internal acoustic meatus	Vestibular sensation from semicircular ducts, utricle, and saccule related to position and movement of head
	Special sensory	Spiral ganglion Synapse: cochlear nuclei		Hearing from spiral organ
Glossopharyngeal (CN IX)	Somatic (br.) motor	Medulla (nucleus ambiguus)	Jugular foramen	Motor to stylopharyngeus that assists with swallowing
	Visceral motor	Presynaptic: medulla (inferior salivatory nucleus) Postsynaptic: otic ganglion		Parasympathetic innervation to parotid gland
	Visceral sensory	Inferior ganglion		Visceral sensation from parotid gland, carotid body and sinus, pharynx, and middle ear
	Special sensory	Inferior ganglion Synapse: nuclei of solitary tract		Taste from posterior third of tongue
	General sensory	Superior ganglion Synapse: sensory nucleus of CN V		Cutaneous sensation from external ear
Vagus (CN X)	Somatic (branchial) motor	Medulla (nucleus ambiguus)		Motor to constrictor muscles of pharynx, intrinsic muscles of larynx, muscles of palate (except tensor veli palatini), and striated muscle in superior two thirds of esophagus
	Visceral motor	Presynaptic: medulla Postsynaptic: neurons in, on, or near viscera		Smooth muscle of trachea, bronchi, and digestive tract, cardiac muscle
	Visceral sensory	Inferior ganglion Synapse: nuclei of solitary tract		Visceral sensation from base of tongue, pharynx, larynx, trachea, bronchi, heart, esophagus, stomach, and intestine
	Special sensory	Inferior ganglion Synapse: nuclei of solitary tract		Taste from epiglottis and palate
	Somatic (general) sensory	Superior ganglion Synapse: sensory nucleus of trigeminal nerve		Sensation from auricle, external acoustic meatus, and dura mater of posterior cranial fossa
Spinal accessory nerve (CN XI)	Somatic motor	Cervical spinal cord		Motor to sternocleidomastoid and trapezius
Hypoglossal (CN XII)	Somatic motor	Medulla (nucleus of CN XII)	Hypoglossal canal	Motor to muscles of tongue (except palatoglossus)

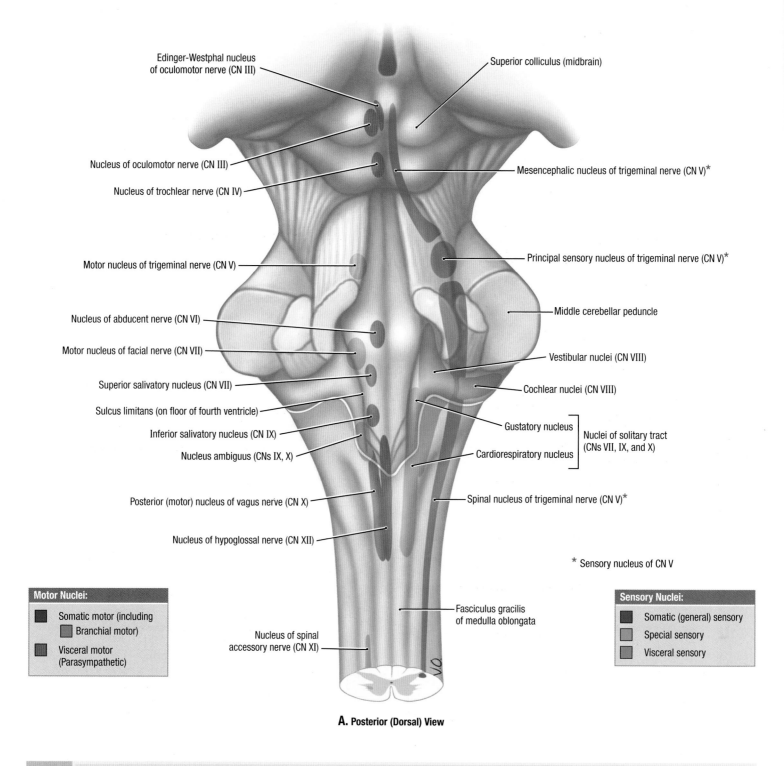

Edinger-Westphal nucleus of oculomotor nerve (CN III)

Superior colliculus (midbrain)

Nucleus of oculomotor nerve (CN III)

Mesencephalic nucleus of trigeminal nerve (CN V)*

Nucleus of trochlear nerve (CN IV)

Motor nucleus of trigeminal nerve (CN V)

Principal sensory nucleus of trigeminal nerve (CN V)*

Nucleus of abducent nerve (CN VI)

Middle cerebellar peduncle

Motor nucleus of facial nerve (CN VII)

Vestibular nuclei (CN VIII)

Superior salivatory nucleus (CN VII)

Cochlear nuclei (CN VIII)

Sulcus limitans (on floor of fourth ventricle)

Gustatory nucleus

Inferior salivatory nucleus (CN IX)

Nuclei of solitary tract (CNs VII, IX, and X)

Nucleus ambiguus (CNs IX, X)

Cardiorespiratory nucleus

Posterior (motor) nucleus of vagus nerve (CN X)

Spinal nucleus of trigeminal nerve (CN V)*

Nucleus of hypoglossal nerve (CN XII)

* Sensory nucleus of CN V

Motor Nuclei:
- ■ Somatic motor (including
  - ■ Branchial motor)
- ▓ Visceral motor (Parasympathetic)

Fasciculus gracilis of medulla oblongata

Nucleus of spinal accessory nerve (CN XI)

Sensory Nuclei:
- ■ Somatic (general) sensory
- ▓ Special sensory
- ▓ Visceral sensory

A. Posterior (Dorsal) View

9.4   CRANIAL NERVE NUCLEI

The fibers of the cranial nerves are connected to nuclei (groups of nerve cell bodies in the central nervous system), in which afferent (sensory) fibers terminate and from which efferent (motor) fibers originate. Nuclei of common functional types (motor, sensory, parasympathetic, and special sensory nuclei) have a generally columnar placement within the brainstem, with the sulcus limitans demarcating motor and sensory columns.

Somatic motor: Motor fibers innervating voluntary (striated) muscle. For the muscles derived from the embryonic pharyngeal arches, their somatic motor innervation can be referred to more specifically as **branchial motor.**

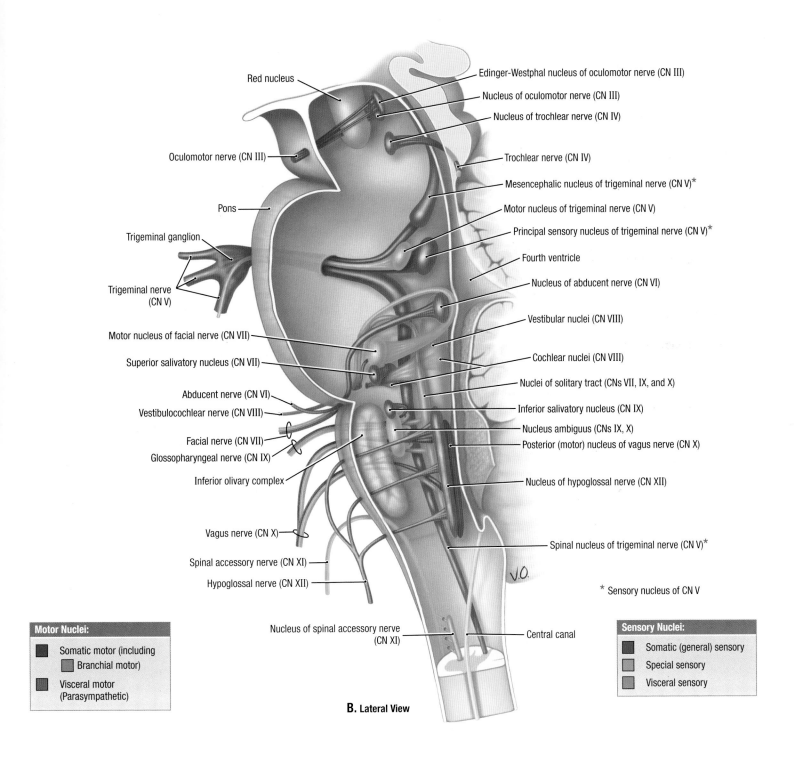

Red nucleus

Edinger-Westphal nucleus of oculomotor nerve (CN III)

Nucleus of oculomotor nerve (CN III)

Nucleus of trochlear nerve (CN IV)

Oculomotor nerve (CN III)

Trochlear nerve (CN IV)

Mesencephalic nucleus of trigeminal nerve (CN V)*

Pons

Motor nucleus of trigeminal nerve (CN V)

Trigeminal ganglion

Principal sensory nucleus of trigeminal nerve (CN V)*

Fourth ventricle

Trigeminal nerve (CN V)

Nucleus of abducent nerve (CN VI)

Motor nucleus of facial nerve (CN VII)

Vestibular nuclei (CN VIII)

Superior salivatory nucleus (CN VII)

Cochlear nuclei (CN VIII)

Nuclei of solitary tract (CNs VII, IX, and X)

Abducent nerve (CN VI)

Inferior salivatory nucleus (CN IX)

Vestibulocochlear nerve (CN VIII)

Nucleus ambiguus (CNs IX, X)

Facial nerve (CN VII)

Posterior (motor) nucleus of vagus nerve (CN X)

Glossopharyngeal nerve (CN IX)

Inferior olivary complex

Nucleus of hypoglossal nerve (CN XII)

Vagus nerve (CN X)

Spinal nucleus of trigeminal nerve (CN V)*

Spinal accessory nerve (CN XI)

Hypoglossal nerve (CN XII)

V.O.

* Sensory nucleus of CN V

Motor Nuclei:
■ Somatic motor (including
■ Branchial motor)
■ Visceral motor (Parasympathetic)

Nucleus of spinal accessory nerve (CN XI)

Central canal

Sensory Nuclei:
■ Somatic (general) sensory
■ Special sensory
■ Visceral sensory

B. Lateral View

CRANIAL NERVE NUCLEI (*continued*)

9.4

Visceral motor: Parasympathetic innervation to glands and involuntary (smooth) muscle.

Somatic (general) sensory: Fibers transmitting general sensation from skin and membranes (e.g., touch, pressure, heat, cold).

Visceral sensory: Fibers conveying sensation from viscera (organs) and mucous membranes.

Special sensory: Taste, smell, vision, hearing, and balance.

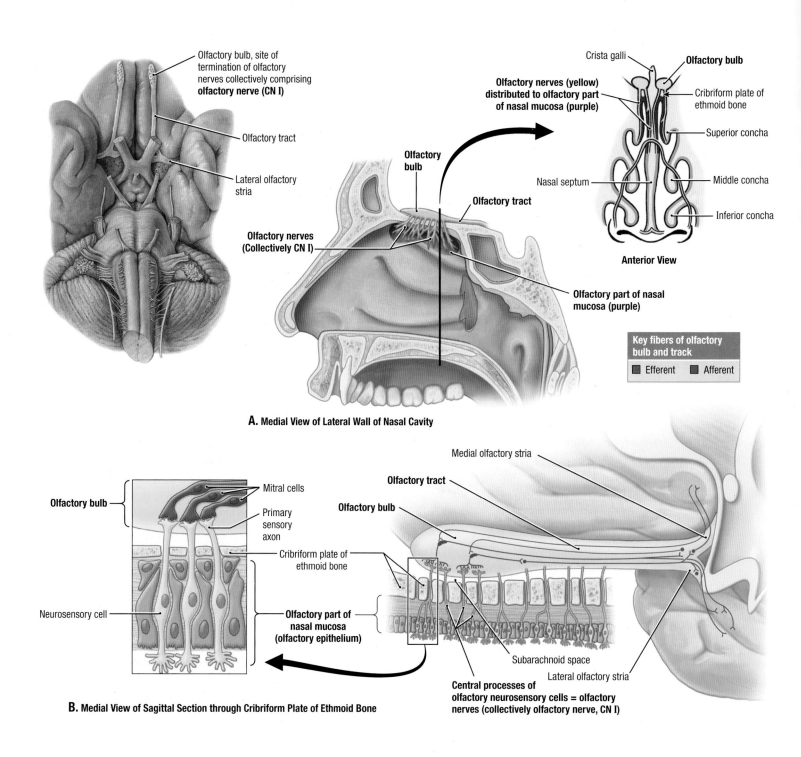

A. Medial View of Lateral Wall of Nasal Cavity

Key fibers of olfactory bulb and track
- Efferent
- Afferent

B. Medial View of Sagittal Section through Cribriform Plate of Ethmoid Bone

9.5 OLFACTORY NERVE (CN I)

A. Relationship of olfactory mucosa to olfactory bulb. **B.** Innervation of olfactory epithelium.

TABLE 9.2	**OLFACTORY NERVE (CN I)**			
Nerve	*Functional Components*	*Cells of Origin/Termination*	*Cranial Exit*	*Distribution and Functions*
Olfactory	Special sensory	Olfactory epithelium (olfactory cells/olfactory bulb)	Foramina of cribriform plate of ethmoid bone	Smell from nasal mucosa of roof and superior sides of nasal septum and superior concha of each nasal cavity

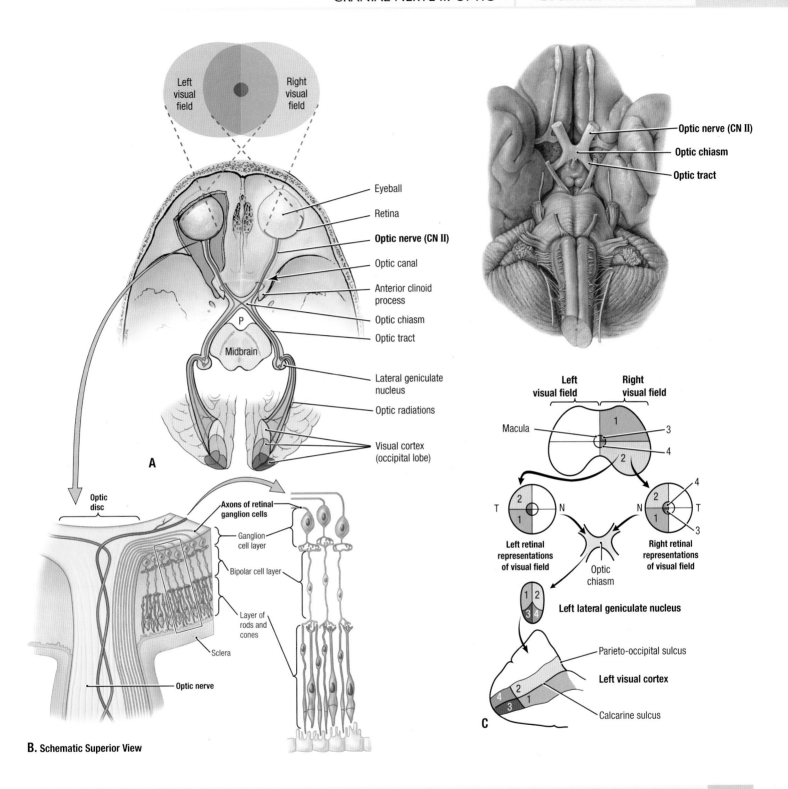

A. Origin and course of visual pathway. **B.** Rods and cones in retina. **C.** Right visual field representation on retinae, left lateral geniculate nucleus, and left visual cortex.

OPTIC NERVE (CN II)   9.6

A. Origin and course of visual pathway. **B.** Rods and cones in retina. **C.** Right visual field representation on retinae, left lateral geniculate nucleus, and left visual cortex.

TABLE 9.3	**OPTIC NERVE (CN II)**			
Nerve	*Functional Components*	*Cells of Origin/Termination*	*Cranial Exit*	*Distribution and Functions*
Optic	Special sensory	Retina (ganglion cells)/lateral geniculate body (nucleus)	Optic canal	Vision from retina

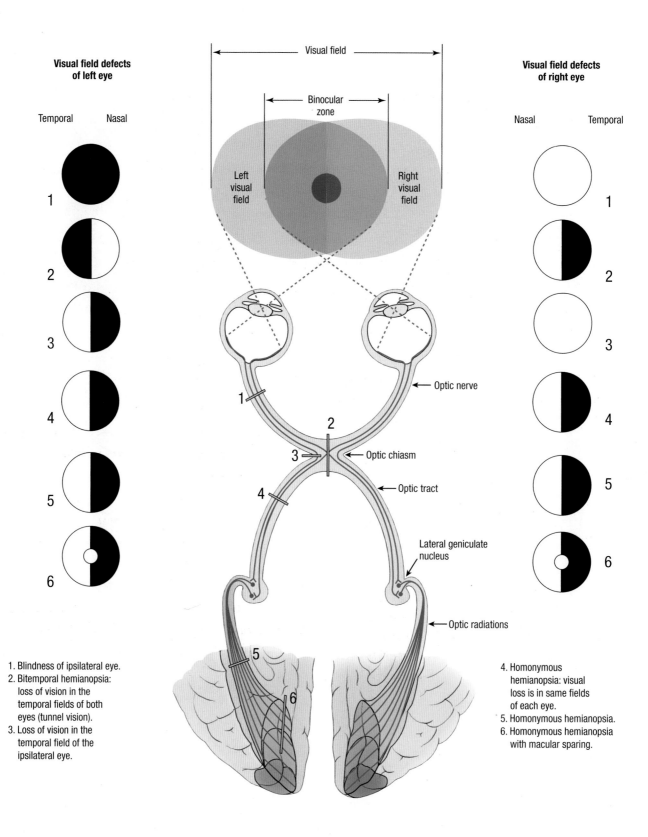

Visual field defects of left eye

Temporal   Nasal

1
2
3
4
5
6

Visual field

Binocular zone

Left visual field

Right visual field

Optic nerve

Optic chiasm

Optic tract

Lateral geniculate nucleus

Optic radiations

Visual field defects of right eye

Nasal   Temporal

1
2
3
4
5
6

1. Blindness of ipsilateral eye.
2. Bitemporal hemianopsia: loss of vision in the temporal fields of both eyes (tunnel vision).
3. Loss of vision in the temporal field of the ipsilateral eye.

4. Homonymous hemianopsia: visual loss is in same fields of each eye.
5. Homonymous hemianopsia.
6. Homonymous hemianopsia with macular sparing.

9.7 VISUAL FIELD DEFECTS (CN II)

Visual field defects may result from a large number of neurologic diseases. It is clinically important to be able to link the defects to a likely location of the lesion.

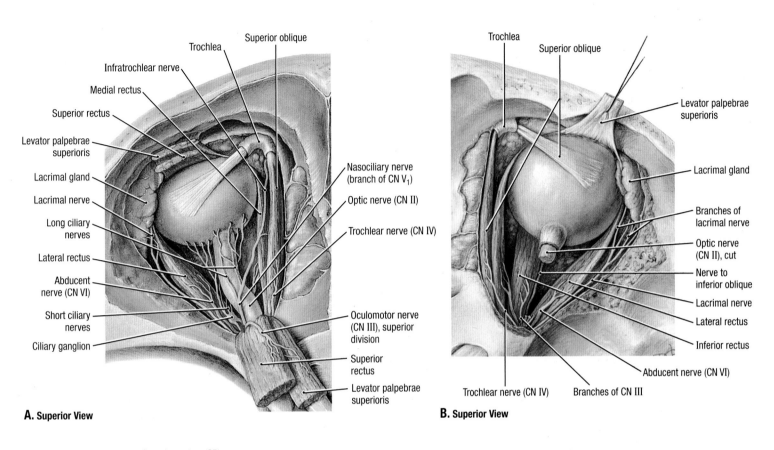

Trochlea

Superior oblique

Infratrochlear nerve

Medial rectus

Superior rectus

Levator palpebrae
superioris

Lacrimal gland

Lacrimal nerve

Long ciliary
nerves

Lateral rectus

Abducent
nerve (CN VI)

Short ciliary
nerves

Ciliary ganglion

Nasociliary nerve
(branch of CN V₁)

Optic nerve (CN II)

Trochlear nerve (CN IV)

Oculomotor nerve
(CN III), superior
division

Superior
rectus

Levator palpebrae
superioris

A. Superior View

Trochlea

Superior oblique

Levator palpebrae
superioris

Lacrimal gland

Branches of
lacrimal nerve

Optic nerve
(CN II), cut

Nerve to
inferior oblique

Lacrimal nerve

Lateral rectus

Inferior rectus

Trochlear nerve (CN IV)   Branches of CN III

Abducent nerve (CN VI)

B. Superior View

Oculomotor nerve (CN III),
superior division

Superior rectus *(SR)*

Levator palpebrae superioris *(LP)*

Trochlea

Superior oblique *(SO)*

Optic nerve (CN II)
fascicles

Ophthalmic artery

Medial rectus *(MR)*

Lacrimal fossa

Oculomotor nerve (CN III),
inferior division

Inferior rectus *(IR)*

Inferior oblique

Trochlear nerve
(CN IV)

Lateral
rectus *(LR)*

Abducent
nerve (CN VI)

Common
tendinous ring

Ciliary ganglion

C. Anterior View

Frontal nerve   Lacrimal nerve   SR   LP

Superior orbital fissure   SO

Superior ophthalmic vein   CN II

CN IV   MR

CN III, superior division   Ophthalmic artery

Nasociliary nerve   Common
tendinous ring

LR   IR

CN VI   CN III, inferior division

Inferior
ophthalmic vein

D. Anterior View

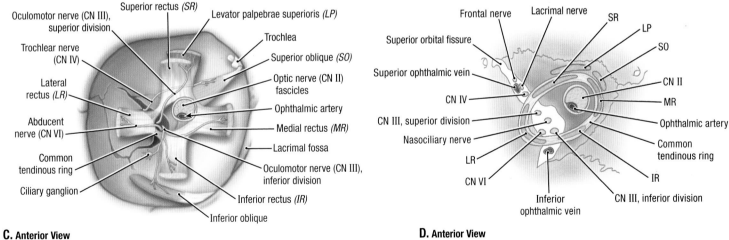

OVERVIEW OF MUSCLES AND NERVES OF ORBIT   9.8

A. and **B.** Orbital cavities, dissected from a superior approach. The optic nerve is intact **(A)** and cut away **(B–D)**. Relationship of muscle attachments and nerves at apex of orbit.

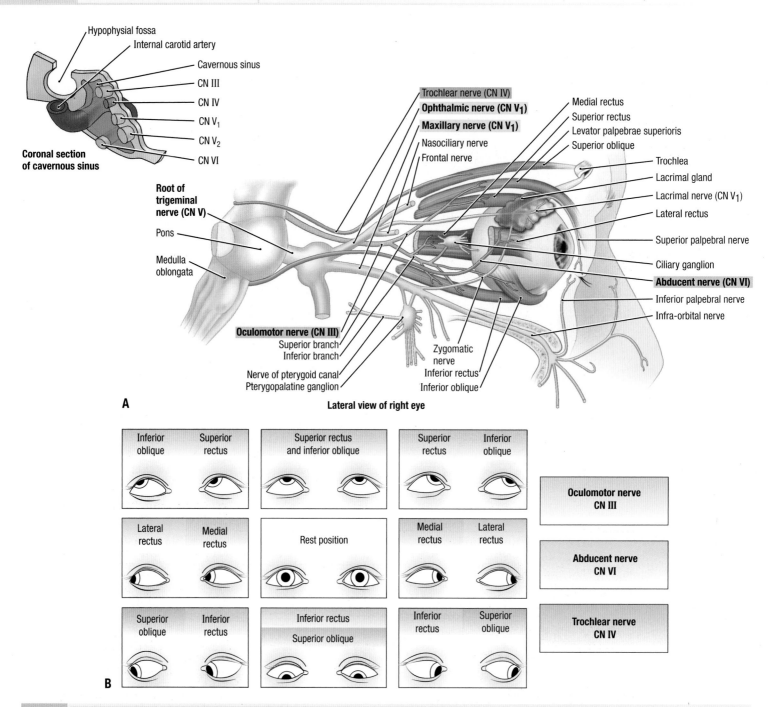

A. Schematic overview. **B.** Binocular movements and muscles producing them. All movements start from the rest (primary) position.

9.9 OCULOMOTOR (CN III), TROCHLEAR (CN IV), AND ABDUCENT (CN VI) NERVES

TABLE 9.4	OCULOMOTOR (CN III), TROCHLEAR (CN IV), AND ABDUCENT (CN VI) NERVES			
Nerve	Functional Components	Cells of Origin/Termination	Cranial Exit	Distribution and Functions
Oculomotor	Somatic motor	Nucleus of CN III	Superior orbital fissure	Motor to superior, inferior, and medial recti, inferior oblique, and levator palpebrae superioris muscles; raises upper eyelid, directing gaze superiorly, inferiorly, and medially
	Visceral motor (parasympathetic)	Presynaptic: midbrain (Edinger-Westphal nucleus) Postsynaptic: ciliary ganglion		Motor to sphincter pupillae and ciliary muscle that constrict pupil and accommodate lens of eyeball
Trochlear	Somatic motor	Nucleus of CN IV		Motor to superior oblique that assists in directing gaze inferolaterally
Abducent	Somatic motor	Nucleus of CN VI		Motor to lateral rectus that directs gaze laterally

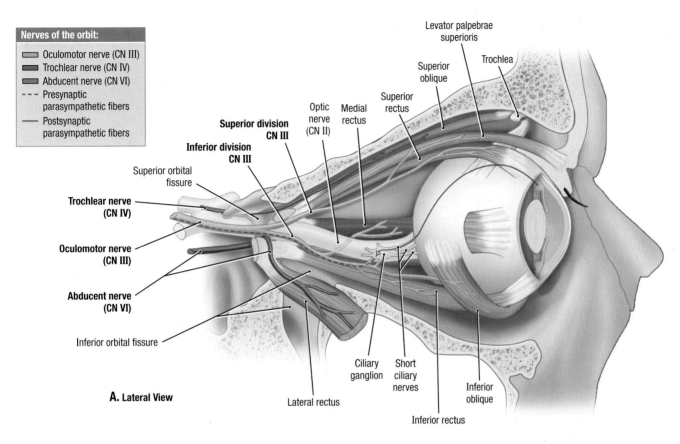

Nerves of the orbit:

◻ Oculomotor nerve (CN III)
▦ Trochlear nerve (CN IV)
▦ Abducent nerve (CN VI)
- - - Presynaptic parasympathetic fibers
—— Postsynaptic parasympathetic fibers

Levator palpebrae superioris
Trochlea
Superior oblique
Superior rectus
Optic nerve (CN II)
Medial rectus
Superior division CN III
Inferior division CN III
Superior orbital fissure
Trochlear nerve (CN IV)
Oculomotor nerve (CN III)
Abducent nerve (CN VI)
Inferior orbital fissure
Ciliary ganglion
Short ciliary nerves
Inferior oblique
A. Lateral View
Lateral rectus
Inferior rectus

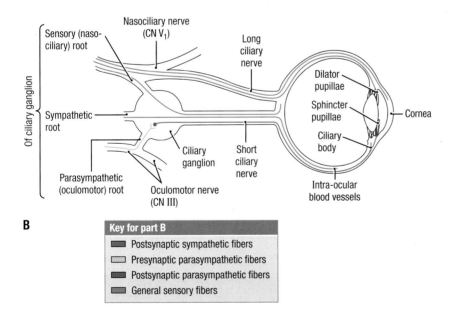

Nasociliary nerve (CN V₁)
Sensory (nasociliary) root
Long ciliary nerve
Of ciliary ganglion
Dilator pupillae
Sphincter pupillae
Cornea
Sympathetic root
Ciliary body
Ciliary ganglion
Short ciliary nerve
Parasympathetic (oculomotor) root
Oculomotor nerve (CN III)
Intra-ocular blood vessels
B

Key for part B

▦ Postsynaptic sympathetic fibers
◻ Presynaptic parasympathetic fibers
▦ Postsynaptic parasympathetic fibers
▦ General sensory fibers

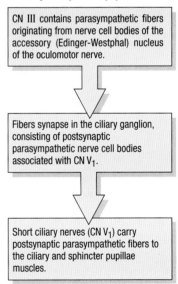

Visceral (parasympathetic) motor innervation of ciliary and sphincter pupillae muscles

CN III contains parasympathetic fibers originating from nerve cell bodies of the accessory (Edinger-Westphal) nucleus of the oculomotor nerve.

⬇

Fibers synapse in the ciliary ganglion, consisting of postsynaptic parasympathetic nerve cell bodies associated with CN V₁.

⬇

Short ciliary nerves (CN V₁) carry postsynaptic parasympathetic fibers to the ciliary and sphincter pupillae muscles.

INNERVATION OF EYEBALL   9.10

A. Nerves of orbit. **B.** Somatic and autonomic innervation of eyeball.

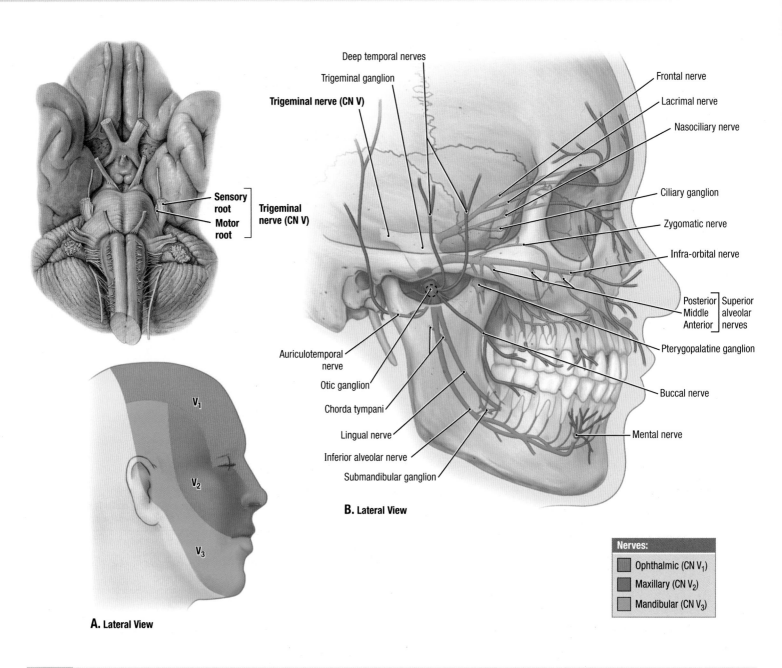

A. Lateral View

B. Lateral View

Nerves:
Ophthalmic (CN V₁)
Maxillary (CN V₂)
Mandibular (CN V₃)

9.11 TRIGEMINAL NERVE (CN V)

A. Cutaneous (somatic sensory) distribution. **B.** Branches of ophthalmic (CN V₁), maxillary (CN V₂), and mandibular (CN V₃) divisions.

TABLE 9.5	**TRIGEMINAL NERVE (CN V)**			
Nerve	*Functional Components*	*Cells of Origin/Termination*	*Cranial Exit*	*Distribution and Functions*
Ophthalmic division (CN V₁)	Somatic (general sensory)	Trigeminal ganglion/ spinal, principal and mesencephalic nucleus of CN V	Superior orbital fissure	Sensation from cornea, skin of forehead, scalp, eyelids, nose, and mucosa of nasal cavity and paranasal sinuses
Maxillary division (CN V₂)			Foramen rotundum	Sensation from skin of face over maxilla including upper lip, maxillary teeth, mucosa of nose, maxillary sinuses, and palate
Mandibular division (CN V₃)			Foramen ovale	Sensation from the skin over mandible, including lower lip and side of head, mandibular teeth, temporomandibular joint, and mucosa of mouth and anterior two thirds of tongue
	Somatic (branchial) motor	Motor nucleus of CN V		Motor to muscles of mastication, mylohyoid, anterior belly of digastric, tensor veli palatini, and tensor tympani

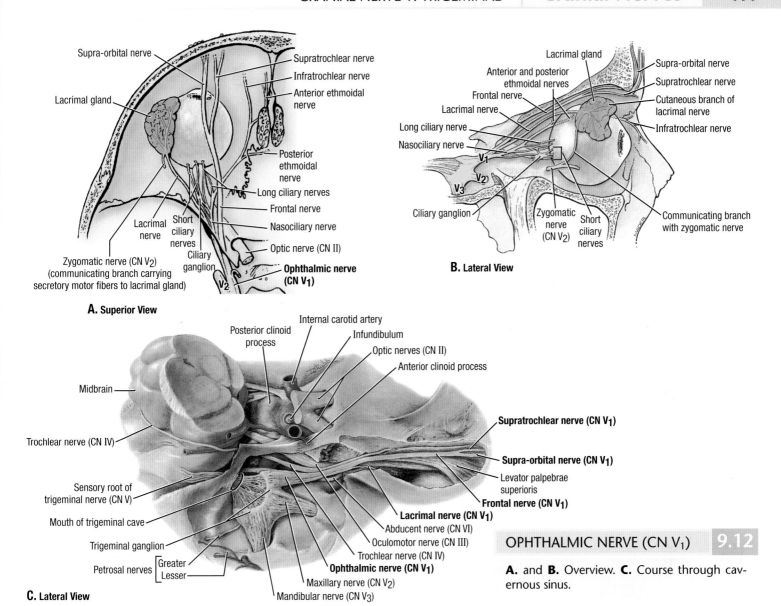

A. Superior View

Supra-orbital nerve
Lacrimal gland
Lacrimal nerve
Zygomatic nerve (CN V₂) (communicating branch carrying secretory motor fibers to lacrimal gland)
Short ciliary nerves
Ciliary ganglion
V₂
Supratrochlear nerve
Infratrochlear nerve
Anterior ethmoidal nerve
Posterior ethmoidal nerve
Long ciliary nerves
Frontal nerve
Nasociliary nerve
Optic nerve (CN II)
Ophthalmic nerve (CN V₁)

B. Lateral View

Lacrimal gland
Anterior and posterior ethmoidal nerves
Frontal nerve
Lacrimal nerve
Long ciliary nerve
Nasociliary nerve
V₁
V₂
V₃
Ciliary ganglion
Zygomatic nerve (CN V₂)
Short ciliary nerves
Supra-orbital nerve
Supratrochlear nerve
Cutaneous branch of lacrimal nerve
Infratrochlear nerve
Communicating branch with zygomatic nerve

C. Lateral View

Midbrain
Trochlear nerve (CN IV)
Sensory root of trigeminal nerve (CN V)
Mouth of trigeminal cave
Trigeminal ganglion
Petrosal nerves — Greater / Lesser
Posterior clinoid process
Internal carotid artery
Infundibulum
Optic nerves (CN II)
Anterior clinoid process
Supratrochlear nerve (CN V₁)
Supra-orbital nerve (CN V₁)
Levator palpebrae superioris
Frontal nerve (CN V₁)
Lacrimal nerve (CN V₁)
Abducent nerve (CN VI)
Oculomotor nerve (CN III)
Trochlear nerve (CN IV)
Ophthalmic nerve (CN V₁)
Maxillary nerve (CN V₂)
Mandibular nerve (CN V₃)

OPHTHALMIC NERVE (CN V₁) 9.12

A. and **B.** Overview. **C.** Course through cavernous sinus.

TABLE 9.6	BRANCHES OF OPHTHALMIC NERVE (CN V₁)
Function	**Branches**

Ophthalmic nerve (CN V₁)
 	
Somatic sensory only at origin from trigeminal ganglion
 	
Visceral motor: extracranially, conveys (1) postsynaptic parasympathetic fibers from ciliary ganglion to ciliary body and sphincter of pupillae; (2) postsynaptic parasympathetic fibers from communicating branch of zygomatic nerve (CN V₂) to lacrimal gland; and (3) postsynaptic sympathetic fibers from internal carotid plexus to dilator pupillae and intra-ocular blood vessels
 	
Passes through superior orbital fissure to enter orbit
 	
Supplies general sensory innervation to cornea; superior bulbar and palpebral conjunctiva; mucosa of anterosuperior nasal cavity; frontal, ethmoidal, and sphenoidal sinuses; anterior and supratentorial dura mater; skin of dorsum of external nose; superior eyelid; forehead; and anterior scalp
 	
Somatic sensory CN V₁
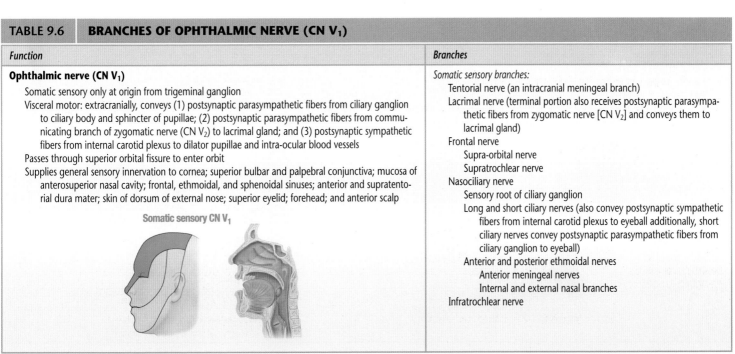
	Somatic sensory branches:
	Tentorial nerve (an intracranial meningeal branch)
	Lacrimal nerve (terminal portion also receives postsynaptic parasympathetic fibers from zygomatic nerve [CN V₂] and conveys them to lacrimal gland)
	Frontal nerve
	   Supra-orbital nerve
	   Supratrochlear nerve
	Nasociliary nerve
	   Sensory root of ciliary ganglion
	   Long and short ciliary nerves (also convey postsynaptic sympathetic fibers from internal carotid plexus to eyeball additionally, short ciliary nerves convey postsynaptic parasympathetic fibers from ciliary ganglion to eyeball)
	   Anterior and posterior ethmoidal nerves
	      Anterior meningeal nerves
	      Internal and external nasal branches
	Infratrochlear nerve

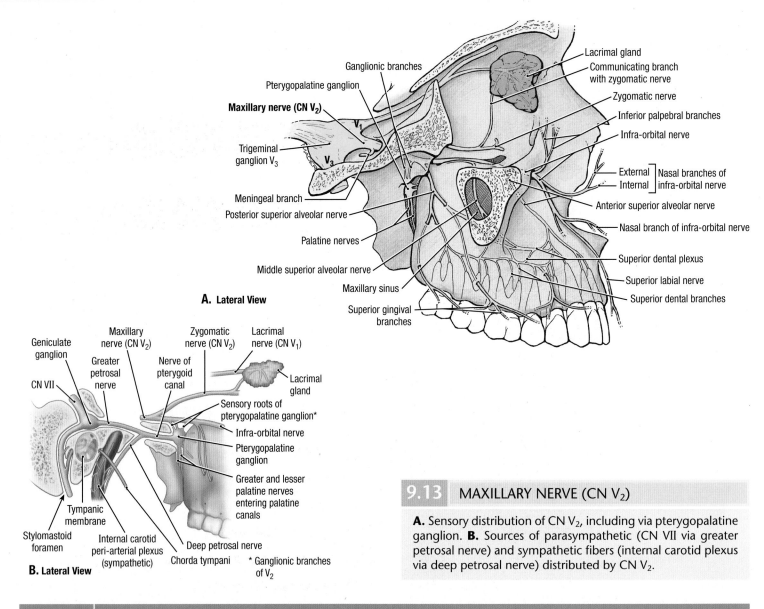

A. Lateral View

Ganglionic branches
Pterygopalatine ganglion
Maxillary nerve (CN V₂)
Trigeminal ganglion V₃
Meningeal branch
Posterior superior alveolar nerve
Palatine nerves
Middle superior alveolar nerve
Maxillary sinus
Superior gingival branches

Lacrimal gland
Communicating branch with zygomatic nerve
Zygomatic nerve
Inferior palpebral branches
Infra-orbital nerve
External / Internal — Nasal branches of infra-orbital nerve
Anterior superior alveolar nerve
Nasal branch of infra-orbital nerve
Superior dental plexus
Superior labial nerve
Superior dental branches

B. Lateral View

Geniculate ganglion
CN VII
Maxillary nerve (CN V₂)
Greater petrosal nerve
Zygomatic nerve (CN V₂)
Nerve of pterygoid canal
Lacrimal nerve (CN V₁)
Lacrimal gland
Sensory roots of pterygopalatine ganglion*
Infra-orbital nerve
Pterygopalatine ganglion
Greater and lesser palatine nerves entering palatine canals
Tympanic membrane
Stylomastoid foramen
Internal carotid peri-arterial plexus (sympathetic)
Deep petrosal nerve
Chorda tympani
* Ganglionic branches of V₂

9.13   MAXILLARY NERVE (CN V₂)

A. Sensory distribution of CN V₂, including via pterygopalatine ganglion. **B.** Sources of parasympathetic (CN VII via greater petrosal nerve) and sympathetic fibers (internal carotid plexus via deep petrosal nerve) distributed by CN V₂.

TABLE 9.7	BRANCHES OF MAXILLARY NERVE (CN V₂)
Function	*Branches*

Maxillary nerve (CN V₂)
Somatic sensory only (proximally, at origin from trigeminal ganglion)
Visceral motor: distally, conveys (1) postsynaptic parasympathetic fibers from pterygopalatine ganglion (presynaptic fibers are from CN VII via greater petrosal nerve and nerve of pterygoid canal); and (2) postsynaptic sympathetic fibers from superior cervical ganglion via internal carotid plexus (presynaptic fibers are from intermediolateral column of gray matter, spinal cord segments T1–T3)
Passes through foramen rotundum to enter pterygopalatine fossa
Supplies dura mater of anterior aspect of lateral part of middle cranial fossa; conjunctiva of inferior eyelid; mucosa of postero-inferior nasal cavity, maxillary sinus, palate, and anterior part of superior oral vestibule; maxillary teeth; and skin of lateral external nose, inferior eyelid, anterior cheek, and upper lip
 	
Somatic sensory CN V₂
	Meningeal branch
	Zygomatic branch
	  Zygomaticofacial branch
	  Zygomaticotemporal branch
	  Communicating branch to lacrimal nerve
	Ganglionic branches to (sensory root of) pterygopalatine ganglion
	Infra-orbital nerve
	  Posterior, middle, and anterior superior alveolar branches
	    Superior dental plexus and branches
	    Superior gingival branches
	  Inferior palpebral branches
	  External and internal nasal branches
	  Superior labial branches
	Greater palatine nerve
	  Posterior inferior lateral nasal nerves
	Lesser palatine nerves
	Posterior superior lateral and medial nasal branches
	Nasopalatine nerve
	Pharyngeal nerve

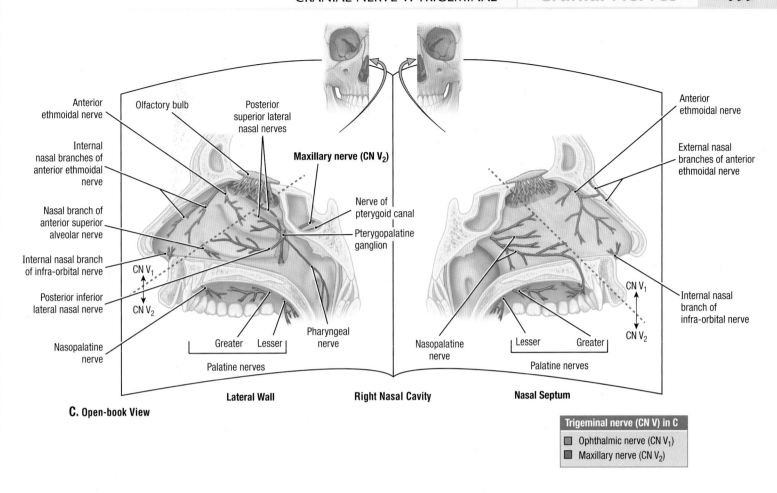

C. Open-book View

Anterior ethmoidal nerve

Olfactory bulb

Posterior superior lateral nasal nerves

Maxillary nerve (CN V₂)

Internal nasal branches of anterior ethmoidal nerve

Nasal branch of anterior superior alveolar nerve

Internal nasal branch of infra-orbital nerve

CN V₁

CN V₂

Posterior inferior lateral nasal nerve

Nasopalatine nerve

Nerve of pterygoid canal

Pterygopalatine ganglion

Greater   Lesser

Palatine nerves

Pharyngeal nerve

Lateral Wall

Right Nasal Cavity

Nasal Septum

Anterior ethmoidal nerve

External nasal branches of anterior ethmoidal nerve

CN V₁

CN V₂

Internal nasal branch of infra-orbital nerve

Nasopalatine nerve

Lesser   Greater

Palatine nerves

Trigeminal nerve (CN V) in C
☐ Ophthalmic nerve (CN V₁)
☐ Maxillary nerve (CN V₂)

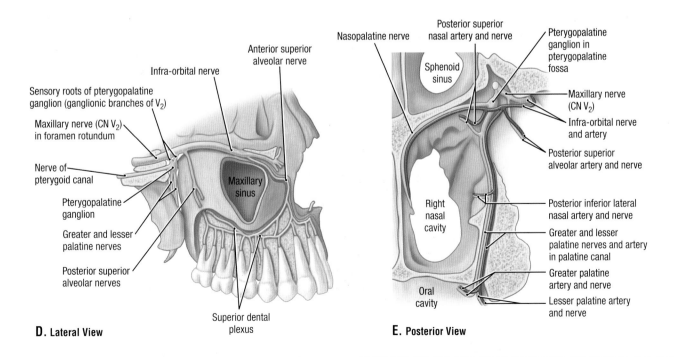

D. Lateral View

Anterior superior alveolar nerve

Infra-orbital nerve

Sensory roots of pterygopalatine ganglion (ganglionic branches of V₂)

Maxillary nerve (CN V₂) in foramen rotundum

Nerve of pterygoid canal

Pterygopalatine ganglion

Greater and lesser palatine nerves

Posterior superior alveolar nerves

Superior dental plexus

Maxillary sinus

E. Posterior View

Nasopalatine nerve

Posterior superior nasal artery and nerve

Pterygopalatine ganglion in pterygopalatine fossa

Sphenoid sinus

Maxillary nerve (CN V₂)

Infra-orbital nerve and artery

Posterior superior alveolar artery and nerve

Right nasal cavity

Posterior inferior lateral nasal artery and nerve

Greater and lesser palatine nerves and artery in palatine canal

Greater palatine artery and nerve

Oral cavity

Lesser palatine artery and nerve

MAXILLARY NERVE (CN V₂) *(continued)*

9.13

C. Innervation of lateral wall and septum of right side of nasal cavity and palate. **D.** Relationship of nerves to maxillary sinus. **E.** Coronal section showing course of the nasopalatine and greater and lesser palatine nerves.

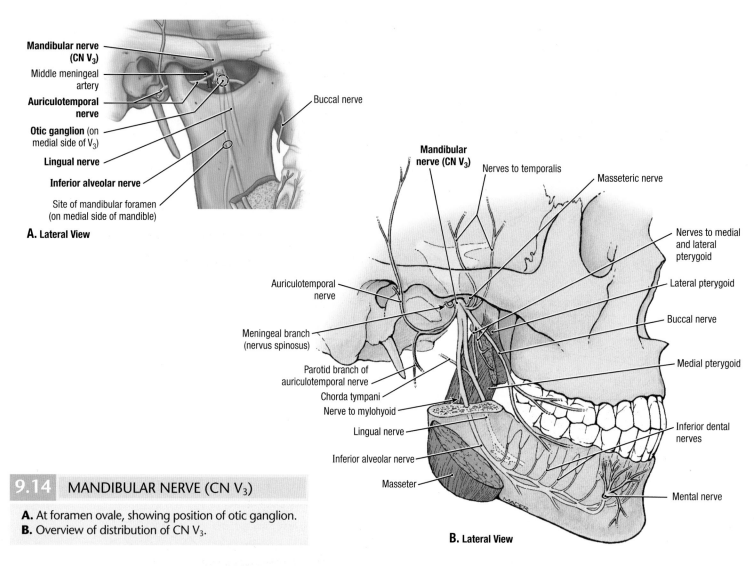

A. Lateral View

Mandibular nerve (CN V₃)
Middle meningeal artery
Auriculotemporal nerve
Otic ganglion (on medial side of V₃)
Lingual nerve
Inferior alveolar nerve
Site of mandibular foramen (on medial side of mandible)
Buccal nerve

Mandibular nerve (CN V₃)
Nerves to temporalis
Masseteric nerve
Auriculotemporal nerve
Nerves to medial and lateral pterygoid
Lateral pterygoid
Meningeal branch (nervus spinosus)
Buccal nerve
Parotid branch of auriculotemporal nerve
Medial pterygoid
Chorda tympani
Nerve to mylohyoid
Lingual nerve
Inferior dental nerves
Inferior alveolar nerve
Masseter
Mental nerve

B. Lateral View

9.14 MANDIBULAR NERVE (CN V₃)

A. At foramen ovale, showing position of otic ganglion.
B. Overview of distribution of CN V₃.

TABLE 9.8	BRANCHES OF MANDIBULAR NERVE (CN V₃)
Function	**Branches**
Maxillary nerve (CN V₃) Somatic sensory and somatic (branchial) motor Special sensory: extracranially, conveys taste fibers (from CN VII via chorda tympani nerve) to anterior two thirds of tongue Visceral motor: extracranially, conveys (1) presynaptic parasympathetic fibers to submandibular ganglion (presynaptic fibers are from CN VII via chorda tympani nerve); (2) postsynaptic parasympathetic fibers from submandibular ganglion to submandibular and sublingual glands; and (3) postsynaptic parasympathetic fibers from otic ganglion to parotid gland Passes through foramen ovale to enter infratemporal fossa Supplies general sensory innervation to mucosa of anterior two thirds of tongue, floor of mouth, and posterior and anterior inferior oral vestibule; mandibular teeth; and skin of lower lip, buccal and temporal regions of face, and external ear (anterior superior auricle, upper external auditory meatus, and tympanic membrane) Supplies motor innervation to all four muscles of mastication, mylohyoid, anterior belly of digastric, tensor tympani and tensor veli palatin	*Somatic sensory branches:* Meningeal branch (nervus spinosum) Buccal nerve Auriculotemporal nerve (also conveys *visceral motor fibers*) Superficial temporal branches Parotid branches Lingual nerve (also conveys *visceral motor* and *special sensory fibers*) Inferior alveolar nerve Nerve to mylohyoid Inferior dental plexus Inferior dental branches Inferior gingival branches Mental nerve *Somatic (branchial) motor branches:* Masseteric nerve Medial and lateral pterygoid branches Deep temporal nerves Nerve to mylohyoid Nerve to tensor tympani Nerve to tensor veli palatini

Somatic sensory CN V₃

Somatic motor CN V₃

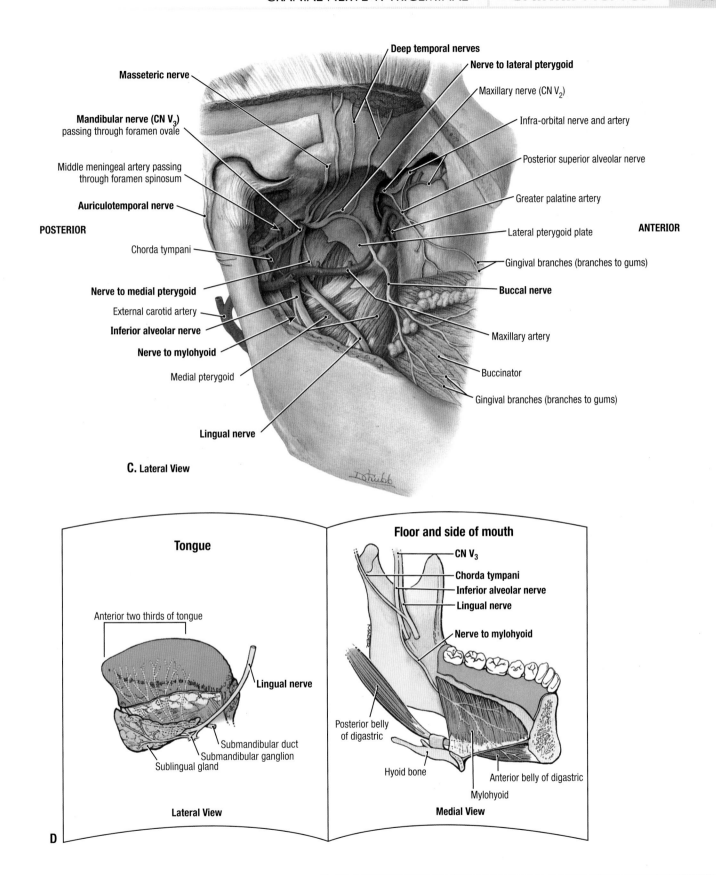

Deep temporal nerves

Nerve to lateral pterygoid

Masseteric nerve

Maxillary nerve (CN V₂)

Mandibular nerve (CN V₃)
passing through foramen ovale

Infra-orbital nerve and artery

Posterior superior alveolar nerve

Middle meningeal artery passing
through foramen spinosum

Greater palatine artery

Auriculotemporal nerve

Lateral pterygoid plate

POSTERIOR

ANTERIOR

Chorda tympani

Gingival branches (branches to gums)

Nerve to medial pterygoid

Buccal nerve

External carotid artery

Inferior alveolar nerve

Maxillary artery

Nerve to mylohyoid

Buccinator

Medial pterygoid

Gingival branches (branches to gums)

Lingual nerve

C. Lateral View

Tongue

Floor and side of mouth

CN V₃

Chorda tympani
Inferior alveolar nerve
Lingual nerve

Anterior two thirds of tongue

Nerve to mylohyoid

Lingual nerve

Posterior belly
of digastric

Submandibular duct
Submandibular ganglion

Sublingual gland

Hyoid bone

Anterior belly of digastric

Mylohyoid

Lateral View

Medial View

D

MANDIBULAR NERVE (CN V₃) *(continued)*

9.14

C. Deep dissection of CN V₃ and branches at foramen ovale. **D.** Lateral aspect of tongue and medial aspect of mandible displayed as pages in an open book that is, the tongue has been reflected from the mandible.

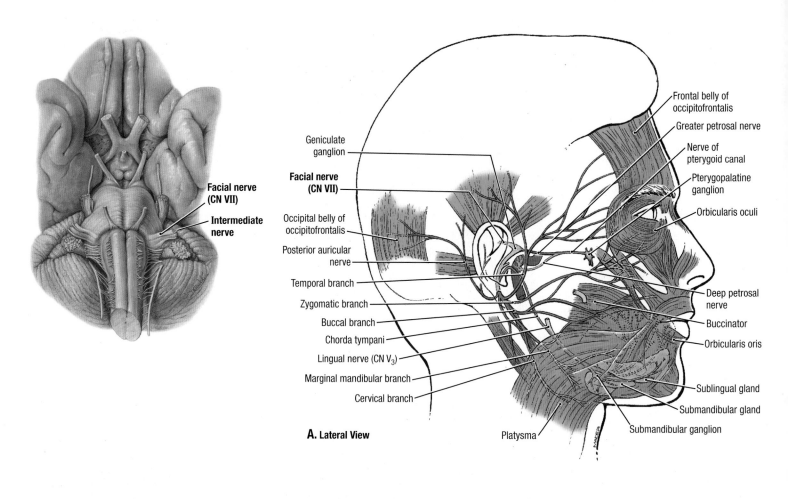

A. Lateral View

9.15   FACIAL NERVE (CN VII)

A. Overview. **B.** Parasympathetic motor innervation of lacrimal, submandibular, and sublingual glands. **C.** Nerve of pterygoid canal.

TABLE 9.9	FACIAL NERVE (CN VII), INCLUDING MOTOR ROOT AND INTERMEDIATE NERVE[a]			
Nerve	*Functional Components*	*Cells of Origin/Termination*	*Cranial Exit*	*Distribution and Functions*
Temporal, zygomatic, buccal, mandibular, cervical, and posterior auricular nerves, nerve to posterior belly of digastric, nerve to stylohyoid, nerve to stapedius	Somatic (branchial) motor	Motor nucleus of CN VII	Stylomastoid foramen	Motor to muscles of facial expression and scalp, also supplies stapedius of middle ear, stylohyoid, and posterior belly of digastric
Intermediate nerve through chorda tympani	Special sensory	Geniculate ganglion/solitary nucleus	Internal acoustic meatus/facial canal/petrotympanic fissure	Taste from anterior two thirds of tongue, through chorda tympani floor of mouth, and palate
Intermediate nerve	Somatic (general) sensory	Geniculate ganglion/spinal trigeminal nucleus	Internal acoustic meatus	Sensation from skin of external acoustic meatus
Intermediate nerve through greater petrosal nerve	Visceral sensory	Nuclei of solitary tract	Internal acoustic meatus/facial canal/foramen for greater petrosal nerve	Visceral sensation from mucous membranes of nasopharynx and palate
Greater petrosal nerve Chorda tympani	Visceral motor	Presynaptic: superior salivatory nucleus Postsynaptic: pterygopalatine ganglion (greater petrosal nerve) and submandibular ganglion (chorda tympani)	Internal acoustic meatus/facial canal/foramen for greater petrosal nerve (greater petrosal nerve) petrotympanic fissure (chorda tympani)	Parasympathetic innervation to lacrimal gland and glands of the nose and palate (greater petrosal nerve); submandibular and sublingual salivary glands (chorda tympani)

[a]See Table 9.15.

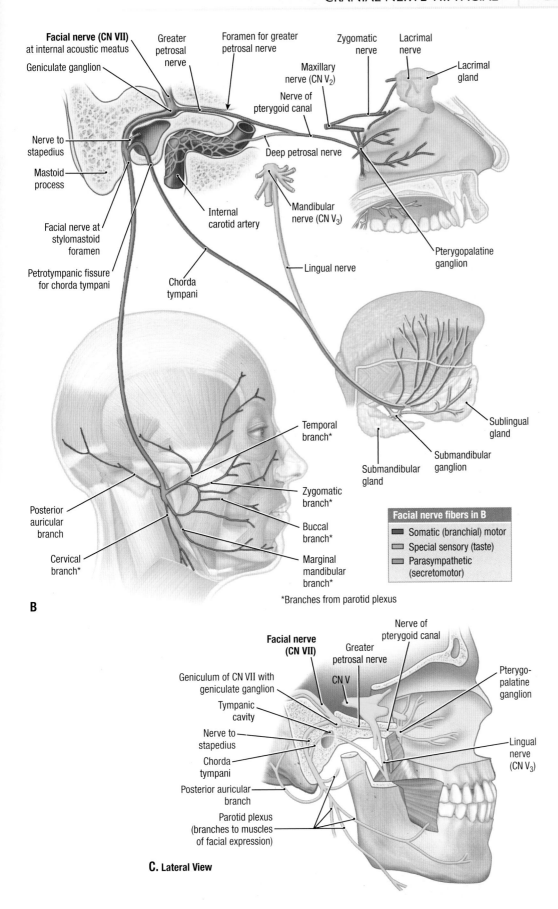

Facial nerve (CN VII) at internal acoustic meatus

Geniculate ganglion

Nerve to stapedius

Mastoid process

Facial nerve at stylomastoid foramen

Petrotympanic fissure for chorda tympani

Chorda tympani

Greater petrosal nerve

Foramen for greater petrosal nerve

Nerve of pterygoid canal

Maxillary nerve (CN V₂)

Zygomatic nerve

Lacrimal nerve

Lacrimal gland

Deep petrosal nerve

Internal carotid artery

Mandibular nerve (CN V₃)

Lingual nerve

Pterygopalatine ganglion

Temporal branch*

Zygomatic branch*

Buccal branch*

Marginal mandibular branch*

Posterior auricular branch

Cervical branch*

Sublingual gland

Submandibular ganglion

Submandibular gland

*Branches from parotid plexus

B

Facial nerve fibers in B
- ■ Somatic (branchial) motor
- □ Special sensory (taste)
- ■ Parasympathetic (secretomotor)

Facial nerve (CN VII)

Geniculum of CN VII with geniculate ganglion

Tympanic cavity

Nerve to stapedius

Chorda tympani

Posterior auricular branch

Parotid plexus (branches to muscles of facial expression)

Greater petrosal nerve

CN V

Nerve of pterygoid canal

Pterygo-palatine ganglion

Lingual nerve (CN V₃)

C. Lateral View

Visceral motor (parasympathetic) to lacrimal gland

Greater petrosal nerve arises from CN VII at the geniculate ganglion and emerges from the superior surface of the petrous part of the temporal bone to enter the middle cranial fossa.

Greater petrosal nerve joins the *deep petrosal nerve* (sympathetic) at the foramen lacerum to form the nerve of the pterygoid canal.

Nerve of the pterygoid canal travels through the pterygoid canal and enters the pterygopalatine fossa.

Parasympathetic fibers from the nerve of pterygoid canal in pterygopalatine fossa synapse in the *pterygopalatine ganglion*.

Postsynaptic parasympathetic fibers from this ganglion innervate the *lacrimal gland* via the zygomatic branch of CN V₂ and the lacrimal nerve CN V₁.

Visceral motor (parasympathetic) to submandibular and sublingual glands

The *chorda tympani* branch arises from CN VII superior to stylomastoid foramen.

The chorda tympani crosses tympanic cavity medial to handle of malleus.

The chorda tympani passes through the petrotympanic fissure between the tympanic and petrous parts of the temporal bone to join the lingual nerve (CN V₃) in infratemporal fossa; parasympathetic fibers of the chorda tympani synapse in the submandibular ganglion; postsynaptic fibers follow arteries to glands.

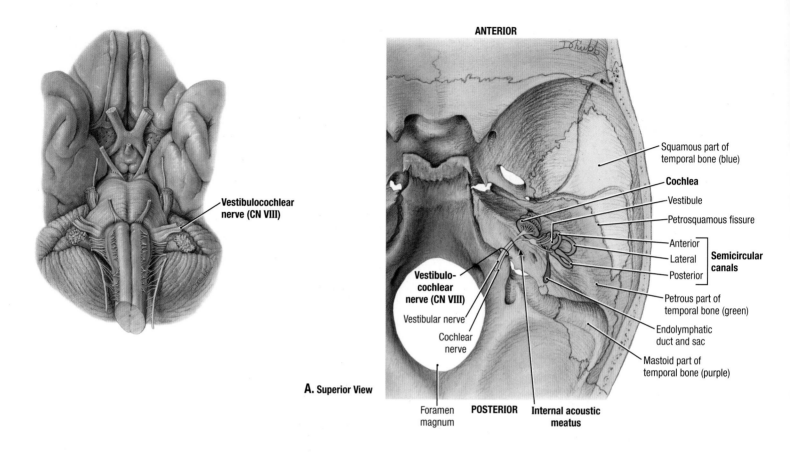

A. Superior View

A. Cochlea and semicircular canals in the cranium. **B.** Schematic overview of distribution.

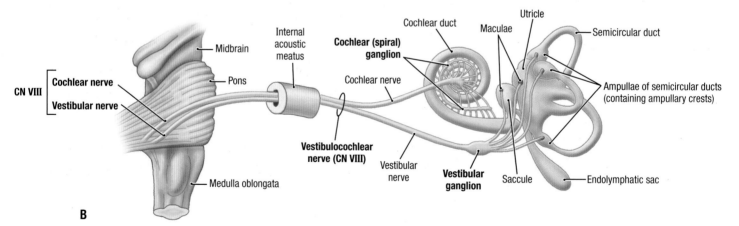

B

9.16 VESTIBULOCOCHLEAR NERVE (CN VIII)

A. Cochlea and semicircular canals in the cranium. **B.** Schematic overview of distribution.

TABLE 9.10	VESTIBULOCOCHLEAR NERVE (CN VIII)			
Part of Vestibulocochlear Nerve	*Functional Components*	*Cells of Origin/Termination*	*Cranial Exit*	*Distribution and Functions*
Vestibular nerve	Special sensory	Vestibular ganglion/vestibular nuclei	Internal acoustic meatus	Vestibular sensation from semicircular ducts, utricle, and saccule related to head position and movement.
Cochlear nerve		Spiral ganglion/cochlear nuclei		Hearing from spiral organ

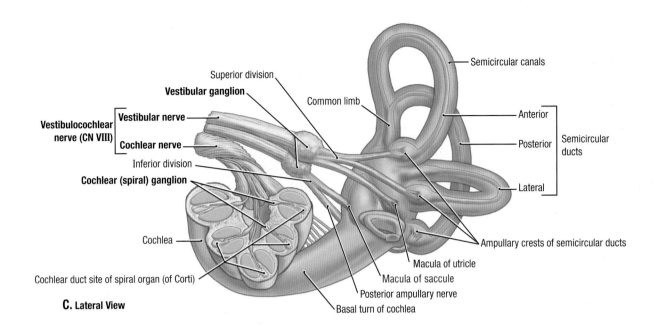

C. Lateral View

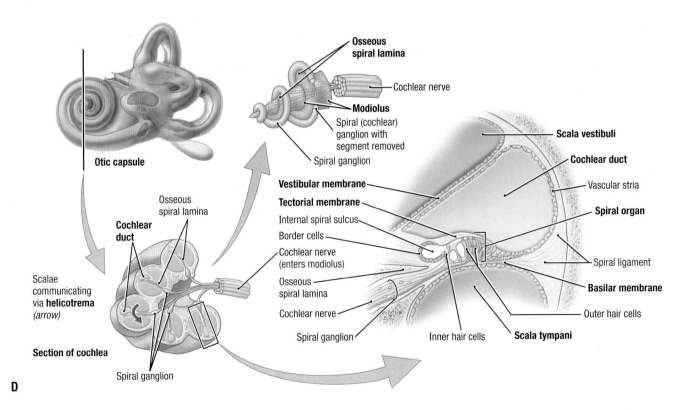

D

VESTIBULOCOCHLEAR NERVE (CN VIII) (*continued*)

9.16

C. Labyrinthine and cochlear apparatus, nerves and ganglia. **D.** Structure of cochlea. Observe:

- The triangular cochlear duct is a spiral tube between the osseous spiral lamina and the external wall of the cochlear canal (spiral ligament).
- The roof of the cochlear duct is formed by the vestibular membrane and the floor by the basilar membrane and osseous spiral lamina.

- The receptor of auditory stimuli is the spiral organ (of Corti), situated on the basilar membrane; it is overlaid by the gelatinous tectorial membrane.
- The spiral organ contains hair cells that respond to vibrations induced in the perilymph by sound waves.
- The fibers of the cochlear nerve are axons of neurons of the spiral ganglion; the peripheral processes enter the spiral organ (of Corti).

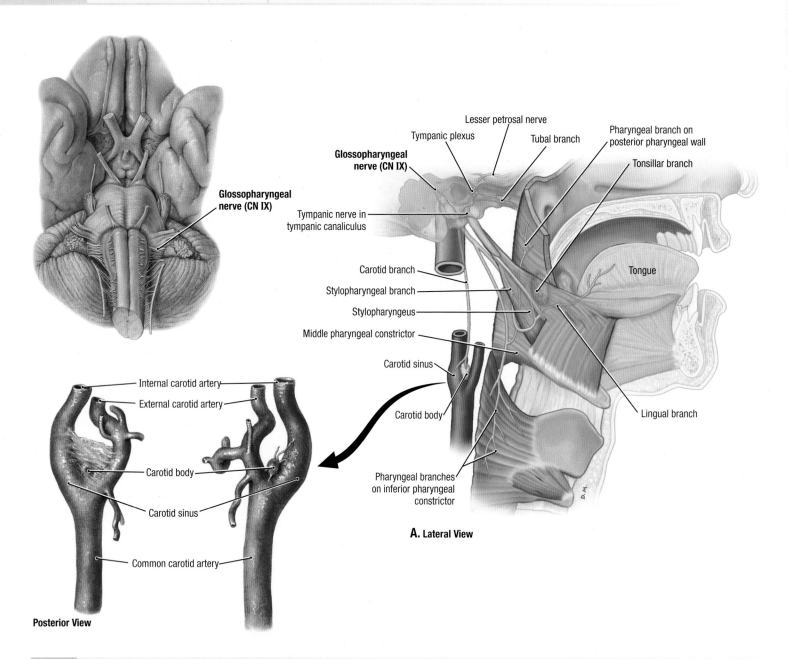

Glossopharyngeal nerve (CN IX)

Glossopharyngeal nerve (CN IX)

Tympanic nerve in tympanic canaliculus

Lesser petrosal nerve

Tympanic plexus

Tubal branch

Pharyngeal branch on posterior pharyngeal wall

Tonsillar branch

Tongue

Carotid branch

Stylopharyngeal branch

Stylopharyngeus

Middle pharyngeal constrictor

Carotid sinus

Carotid body

Lingual branch

Pharyngeal branches on inferior pharyngeal constrictor

A. Lateral View

Internal carotid artery

External carotid artery

Carotid body

Carotid sinus

Common carotid artery

Posterior View

9.17   GLOSSOPHARYNGEAL NERVE (CN IX)

A. and **B.** Overview of distribution.

TABLE 9.11	**GLOSSOPHARYNGEAL NERVE (CN IX)**[a]			
Nerve	*Functional Components*	*Cells of Origin/Termination*	*Cranial Exit*	*Distribution and Functions*
Glossopharyngeal	Somatic (branchial) motor	Nucleus ambiguus	Jugular foramen	Motor to stylopharyngeus that assists with swallowing
	Visceral motor	Presynaptic: inferior salivatory nucleus Postsynaptic: otic ganglion		Parasympathetic innervation to parotid gland
	Visceral sensory	Nuclei of solitary tract, spinal trigeminal nucleus/inferior ganglion		Visceral sensation from parotid gland, carotid body, carotid sinus, pharynx, and middle ear
	Special sensory	Nuclei of solitary tract/inferior ganglion		Taste from posterior third of tongue
	General sensory	Spinal trigeminal nucleus/superior ganglion		External ear, posterior third of tongue, tympanic membrane, isthmus of fauces, and pharyngotympanic tube

[a]See Table 9.15.

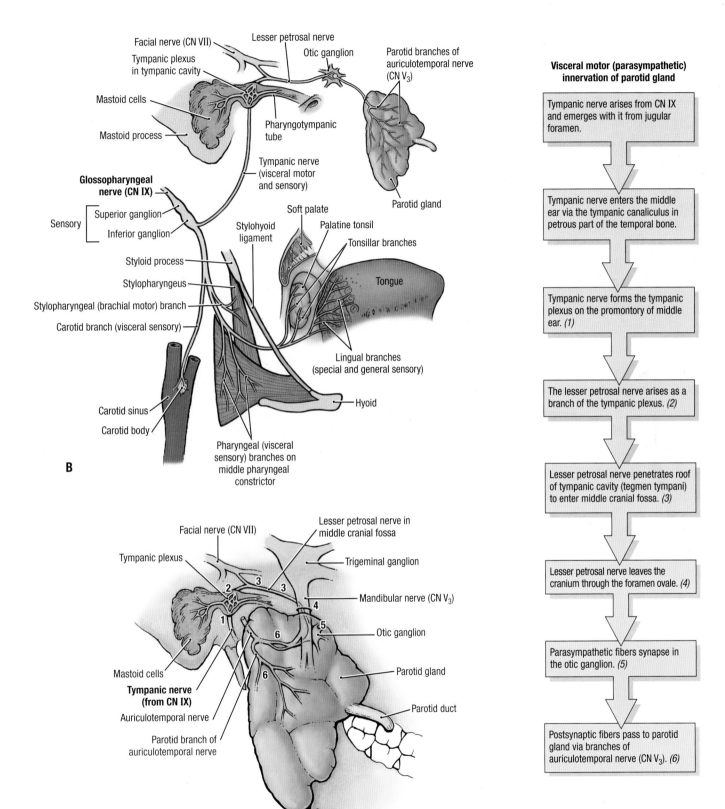

Visceral motor (parasympathetic) innervation of parotid gland

Tympanic nerve arises from CN IX and emerges with it from jugular foramen.

Tympanic nerve enters the middle ear via the tympanic canaliculus in petrous part of the temporal bone.

Tympanic nerve forms the tympanic plexus on the promontory of middle ear. *(1)*

The lesser petrosal nerve arises as a branch of the tympanic plexus. *(2)*

Lesser petrosal nerve penetrates roof of tympanic cavity (tegmen tympani) to enter middle cranial fossa. *(3)*

Lesser petrosal nerve leaves the cranium through the foramen ovale. *(4)*

Parasympathetic fibers synapse in the otic ganglion. *(5)*

Postsynaptic fibers pass to parotid gland via branches of auriculotemporal nerve (CN V₃). *(6)*

B — diagram labels:
- Facial nerve (CN VII)
- Lesser petrosal nerve
- Otic ganglion
- Parotid branches of auriculotemporal nerve (CN V₃)
- Tympanic plexus in tympanic cavity
- Mastoid cells
- Mastoid process
- Pharyngotympanic tube
- Tympanic nerve (visceral motor and sensory)
- Parotid gland
- **Glossopharyngeal nerve (CN IX)**
- Sensory: Superior ganglion, Inferior ganglion
- Soft palate
- Stylohyoid ligament
- Palatine tonsil
- Tonsillar branches
- Styloid process
- Stylopharyngeus
- Tongue
- Stylopharyngeal (brachial motor) branch
- Carotid branch (visceral sensory)
- Lingual branches (special and general sensory)
- Hyoid
- Carotid sinus
- Carotid body
- Pharyngeal (visceral sensory) branches on middle pharyngeal constrictor

C — Lateral View — diagram labels:
- Facial nerve (CN VII)
- Lesser petrosal nerve in middle cranial fossa
- Tympanic plexus
- Trigeminal ganglion
- Mandibular nerve (CN V₃)
- Otic ganglion
- Mastoid cells
- Parotid gland
- **Tympanic nerve (from CN IX)**
- Auriculotemporal nerve
- Parotid duct
- Parotid branch of auriculotemporal nerve

GLOSSOPHARYNGEAL NERVE (CN IX) *(continued)* | 9.17

C. Parasympathetic distribution to parotid gland.

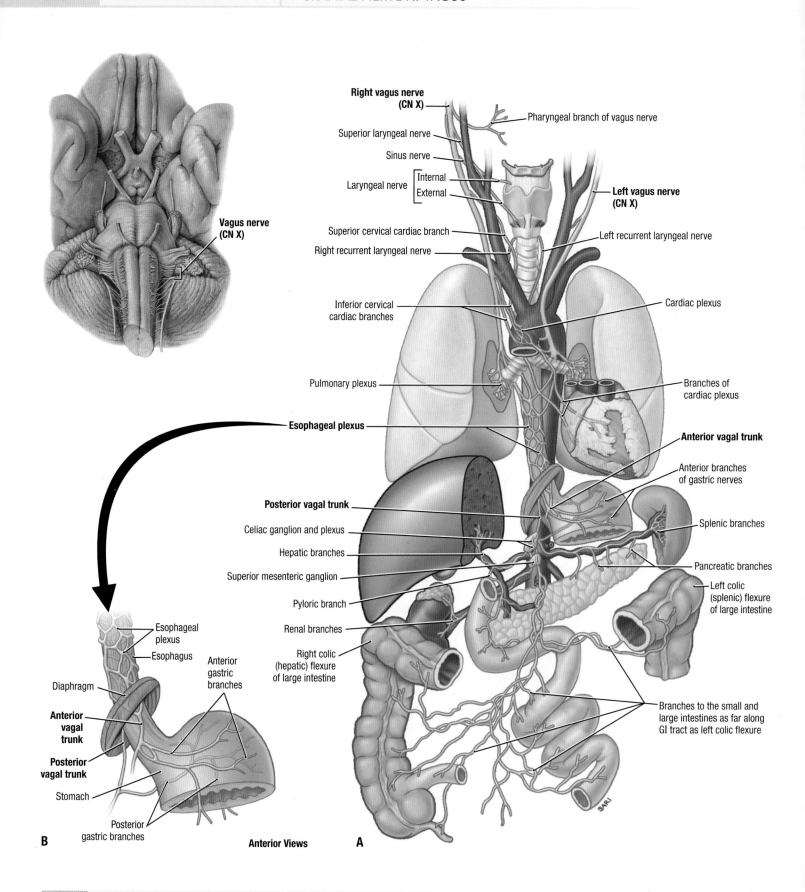

Right vagus nerve (CN X)

Pharyngeal branch of vagus nerve

Superior laryngeal nerve

Sinus nerve

Laryngeal nerve [Internal / External]

Left vagus nerve (CN X)

Superior cervical cardiac branch

Right recurrent laryngeal nerve

Left recurrent laryngeal nerve

Inferior cervical cardiac branches

Cardiac plexus

Pulmonary plexus

Branches of cardiac plexus

Esophageal plexus

Anterior vagal trunk

Anterior branches of gastric nerves

Posterior vagal trunk

Splenic branches

Celiac ganglion and plexus

Hepatic branches

Pancreatic branches

Superior mesenteric ganglion

Left colic (splenic) flexure of large intestine

Pyloric branch

Renal branches

Right colic (hepatic) flexure of large intestine

Branches to the small and large intestines as far along GI tract as left colic flexure

Vagus nerve (CN X)

Esophageal plexus

Esophagus

Anterior gastric branches

Diaphragm

Anterior vagal trunk

Posterior vagal trunk

Stomach

Posterior gastric branches

B

Anterior Views

A

9.18 VAGUS NERVE (CN X)

A. Course in neck, thorax, and abdomen. **B.** Anterior and posterior vagal trunks.

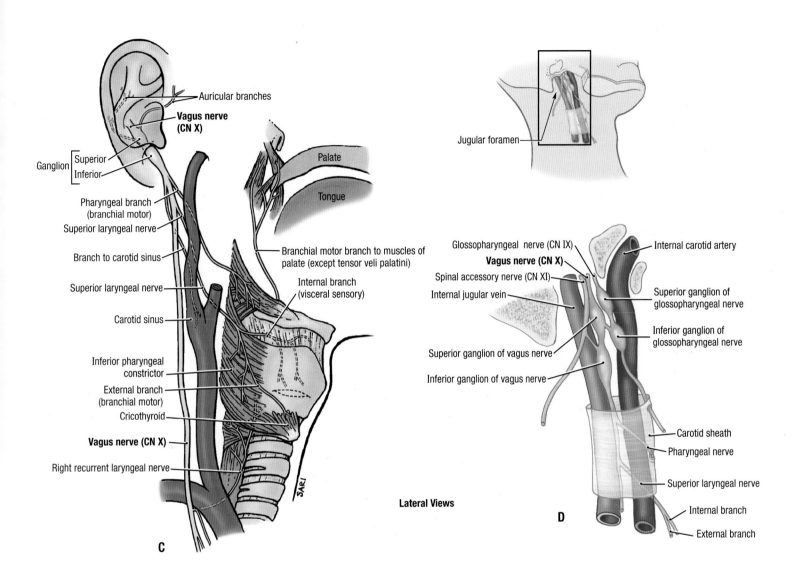

Lateral Views

C

D

VAGUS NERVE (CN X) (*continued*)

C. Branches in neck. **D.** Superior and inferior sensory ganglia of vagus and glossopharyngeal nerves.

TABLE 9.12	**VAGUS NERVE (CN X)**			
Nerve	*Functional Components*	*Cells of Origin/Termination*	*Cranial Exit*	*Distribution and Functions*
Vagus	Branchial motor	Nucleus ambiguus	Jugular foramen	Motor to constrictor muscles of pharynx, intrinsic muscles of larynx, muscles of palate (except tensor veli palatini), and striated muscle in superior two thirds of esophagus
	Visceral motor	Presynaptic: posterior (dorsal) nucleus of CN X Postsynaptic: neurons in, on, or near viscera		Parasympathetic innervation to smooth muscle of trachea, bronchi, and digestive tract, cardiac muscle
	Visceral sensory	Nuclei of solitary tract, spinal trigeminal nucleus/inferior ganglion		Visceral sensation from base of tongue, pharynx, larynx, trachea, bronchi, heart, esophagus, stomach, and intestine
	Special sensory	Nuclei of solitary tract/inferior ganglion		Taste from epiglottis and palate
	General sensory	Spinal trigeminal nucleus/superior ganglion		Sensation from auricle, external acoustic meatus, and dura mater of posterior cranial fossa

Labels on figure C:
- Auricular branches
- Vagus nerve (CN X)
- Ganglion — Superior / Inferior
- Palate
- Tongue
- Pharyngeal branch (branchial motor)
- Superior laryngeal nerve
- Branch to carotid sinus
- Branchial motor branch to muscles of palate (except tensor veli palatini)
- Internal branch (visceral sensory)
- Superior laryngeal nerve
- Carotid sinus
- Inferior pharyngeal constrictor
- External branch (branchial motor)
- Cricothyroid
- Vagus nerve (CN X)
- Right recurrent laryngeal nerve

Labels on figure D:
- Jugular foramen
- Glossopharyngeal nerve (CN IX)
- Vagus nerve (CN X)
- Spinal accessory nerve (CN XI)
- Internal jugular vein
- Superior ganglion of vagus nerve
- Inferior ganglion of vagus nerve
- Internal carotid artery
- Superior ganglion of glossopharyngeal nerve
- Inferior ganglion of glossopharyngeal nerve
- Carotid sheath
- Pharyngeal nerve
- Superior laryngeal nerve
- Internal branch
- External branch

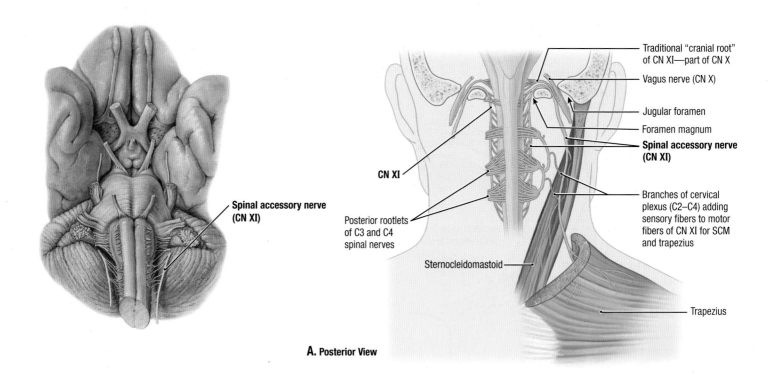

Spinal accessory nerve
(CN XI)

Traditional "cranial root"
of CN XI—part of CN X

Vagus nerve (CN X)

Jugular foramen

Foramen magnum

**Spinal accessory nerve
(CN XI)**

Branches of cervical
plexus (C2–C4) adding
sensory fibers to motor
fibers of CN XI for SCM
and trapezius

CN XI

Posterior rootlets
of C3 and C4
spinal nerves

Sternocleidomastoid

Trapezius

A. Posterior View

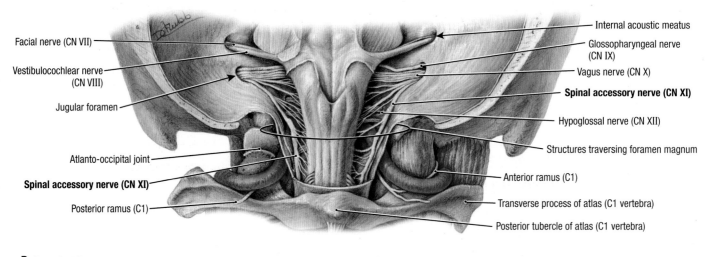

Facial nerve (CN VII)

Vestibulocochlear nerve
(CN VIII)

Jugular foramen

Atlanto-occipital joint

Spinal accessory nerve (CN XI)

Posterior ramus (C1)

Internal acoustic meatus

Glossopharyngeal nerve
(CN IX)

Vagus nerve (CN X)

Spinal accessory nerve (CN XI)

Hypoglossal nerve (CN XII)

Structures traversing foramen magnum

Anterior ramus (C1)

Transverse process of atlas (C1 vertebra)

Posterior tubercle of atlas (C1 vertebra)

B. Posterior View

9.19   SPINAL ACCESSORY NERVE (CN XI)

A. Schematic illustration of distribution. **B.** Intracranial course.

TABLE 9.13	**SPINAL ACCESSORY NERVE (CN XI)**			
Nerve	*Functional Components*	*Cells of Origin/Termination*	*Cranial Exit*	*Distribution and Functions*
Spinal accessory	Somatic motor	Accessory nucleus of spinal cord	Jugular foramen	Motor to sternocleidomastoid and trapezius

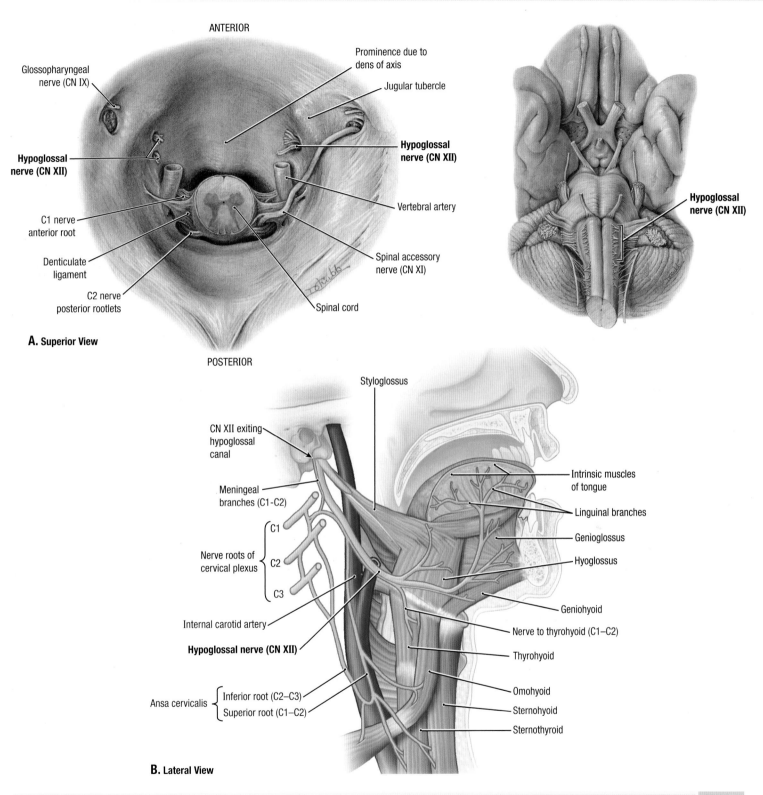

ANTERIOR

Glossopharyngeal nerve (CN IX)

Prominence due to dens of axis

Jugular tubercle

Hypoglossal nerve (CN XII)

Hypoglossal nerve (CN XII)

Vertebral artery

C1 nerve anterior root

Denticulate ligament

C2 nerve posterior rootlets

Spinal accessory nerve (CN XI)

Spinal cord

A. Superior View

POSTERIOR

Hypoglossal nerve (CN XII)

Styloglossus

CN XII exiting hypoglossal canal

Meningeal branches (C1-C2)

Nerve roots of cervical plexus

C1
C2
C3

Internal carotid artery

Hypoglossal nerve (CN XII)

Ansa cervicalis { Inferior root (C2–C3)
Superior root (C1–C2)

Intrinsic muscles of tongue

Linguinal branches

Genioglossus

Hyoglossus

Geniohyoid

Nerve to thyrohyoid (C1–C2)

Thyrohyoid

Omohyoid

Sternohyoid

Sternothyroid

B. Lateral View

HYPOGLOSSAL NERVE (CN XII)

9.20

A. Intracranial exit from cranium into hypoglossal canal. **B.** Schematic illustration of distribution.

TABLE 9.14	HYPOGLOSSAL NERVE (CN XII)			
Nerve	*Functional Components*	*Cells of Origin/Termination*	*Cranial Exit*	*Distribution and Functions*
Hypoglossal	Somatic motor	Nucleus of CN XII	Hypoglossal canal	Motor to muscles of tongue (except palatoglossus)

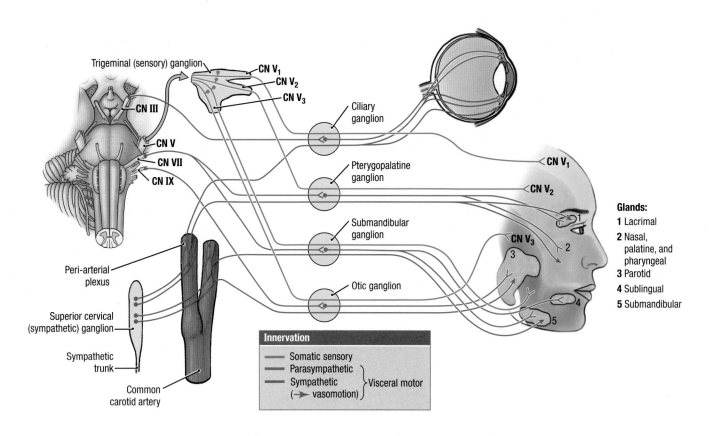

Trigeminal (sensory) ganglion — CN V₁ / CN V₂ / CN V₃

CN III
CN V
CN VII
CN IX

Ciliary ganglion
Pterygopalatine ganglion
Submandibular ganglion
Otic ganglion

CN V₁
CN V₂
CN V₃

Glands:
1 Lacrimal
2 Nasal, palatine, and pharyngeal
3 Parotid
4 Sublingual
5 Submandibular

Peri-arterial plexus
Superior cervical (sympathetic) ganglion
Sympathetic trunk
Common carotid artery

Innervation
— Somatic sensory
— Parasympathetic
— Sympathetic (→ vasomotion) } Visceral motor

9.21 SUMMARY OF AUTONOMIC INNERVATION OF HEAD

Both sympathetic and parasympathetic innervation is outlined.

TABLE 9.15	AUTONOMIC GANGLIA OF HEAD			
Ganglion	Location	Parasympathetic Root (Nucleus of Origin)[a]	Sympathetic Root	Main Distribution
Ciliary	Between optic nerve and lateral rectus, close to apex of orbit	Inferior branch of oculomotor nerve (CN III) (Edinger-Westphal nucleus)	Branch from internal carotid plexus in cavernous sinus	Parasympathetic postsynaptic fibers from ciliary ganglion pass to ciliary muscle and sphincter, pupillae of iris; sympathetic postsynaptic fibers from superior cervical ganglion pass to dilator pupillae and blood vessels of eye
Pterygopalatine	In pterygopalatine fossa, where it is attached by pterygopalatine branches of maxillary nerve; located immediately anterior to opening of pterygoid canal and inferior to CN V₂	Greater petrosal nerve from facial nerve (CN VII) (superior salivatory nucleus)	Deep petrosal nerve, a branch of internal carotid plexus that is continuation of postsynaptic fibers of cervical sympathetic trunk; fibers from superior cervical ganglion pass through pterygopalatine ganglion and enter branches of CN V₂	Parasympathetic postsynaptic fibers from pterygopalatine ganglion innervate lacrimal gland through zygomatic branch of CN V₂; sympathetic postsynaptic fibers from superior cervical ganglion accompany branches of pterygopalatine nerve that are distributed to the nasal cavity, palate, and superior parts of the pharynx
Otic	Between tensor veli palatini and mandibular nerve; lies inferior to foramen ovale	Tympanic nerve from glossopharyngeal nerve (CN IX); tympanic nerve continues from tympanic plexus as lesser petrosal nerve (inferior salivatory nucleus)	Fibers from superior cervical ganglion travel via plexus on middle meningeal artery	Parasympathetic postsynaptic fibers from otic ganglion are distributed to parotid gland through auriculotemporal nerve (branch of CN V₃); sympathetic postsynaptic fibers from superior cervical ganglion pass to parotid gland and supply its blood vessels
Submandibular	Suspended from lingual nerve by two short roots; lies on surface of hyoglossus muscle inferior to submandibular duct	Parasympathetic fibers join facial nerve (CN VII) and leave it in its chorda tympani branch, which unites with lingual nerve (superior salivatory nucleus)	Sympathetic fibers from superior cervical ganglion travel via the plexus on facial artery	Postsynaptic parasympathetic fibers from submandibular ganglion are distributed to the sublingual and submandibular glands; sympathetic fibers supply sublingual and submandibular glands and appear to be secretomotor

[a]For location of nuclei, see Figure 9.3.

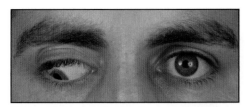

Right eye: Downward and outward gaze, dilated pupil, ptosis of eyelid   Left

A. Right Oculomotor (CN III) Nerve Palsy

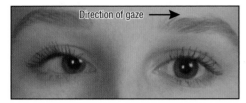

Direction of gaze ⟶

Right   Left eye: Does not abduct

B. Left Abducent (CN VI) Nerve Palsy

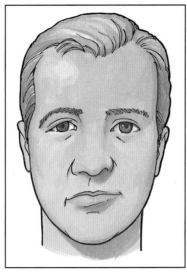

C. Right Facial (CN VII) Palsy (Bell Palsy)

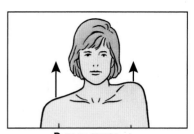

D. Right CN XI Lesion

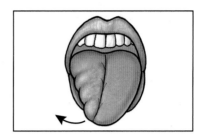

E. Right CN XII Lesion

CRANIAL NERVE LESIONS

9.22

TABLE 9.16	**SUMMARY OF CRANIAL NERVE LESIONS**	
Nerve	**Lesion Type and/or Site**	**Abnormal Findings**
CN I	Fracture of cribriform plate	Anosmia (loss of smell); cerebrospinal fluid (CSF) rhinorrhea (leakage of CSF through nose)
CN II	Direct trauma to orbit or eyeball; fracture involving optic canal	Loss of pupillary constriction
	Pressure on optic pathway; laceration or intracerebral clot in temporal, parietal, or occipital lobes of brain	Visual field defects
	Increased CSF pressure	Swelling of optic disc (papilledema)
CN III	Pressure from herniating uncus on nerve; fracture involving cavernous sinus; aneurysms	Dilated pupil, ptosis, eye rotates inferiorly and laterally (down and out), pupillary reflex on the side of the lesion will be lost **(A)**
CN IV	Stretching of nerve during its course around brainstem; fracture of orbit	Inability to rotate adducted eye inferiorly
CN V	Injury to terminal branches (particularly CN V$_2$) in roof of maxillary sinus; pathologic processes (tumors, aneurysms, infections) affecting trigeminal nerve	Loss of pain and touch sensations/paresthesia on face; loss of corneal reflex (blinking when cornea touched); paralysis of muscles of mastication; deviation of mandible to side of lesion when mouth is opened
CN VI	Base of brain or fracture involving cavernous sinus or orbit	Inability to rotate eye laterally; diplopia on lateral gaze **(B)**
CN VII	Laceration or contusion in parotid region	Paralysis of facial muscles; eye remains open; angle of mouth droops; forehead does not wrinkle **(C)**
	Fracture of temporal bone	As above, plus associated involvement of cochlear nerve and chorda tympani; dry cornea and loss of taste on anterior two thirds of tongue
	Intracranial hematoma ("stroke")	Weakness (paralysis) of lower facial muscles contralateral to the lesion, upper facial muscles are not affected because they are bilaterally innervated
CN VIII	Tumor of nerve	Progressive unilateral hearing loss; tinnitus (noises in ear); vertigo (loss of balance)
CN IX^a	Brainstem lesion or deep laceration of neck	Loss of taste on posterior third of tongue; loss of sensation on affected side of soft palate; loss of gag reflex on affected side
CN X	Brainstem lesion or deep laceration of neck	Sagging of soft palate; deviation of uvula to unaffected side; hoarseness owing to paralysis of vocal fold; difficulty in swallowing and speaking
CN XI	Laceration of neck	Paralysis of sternocleidomastoid and superior fibers of trapezius; drooping of shoulder **(D)**
CN XII	Neck laceration; basal skull fractures	Protruded tongue deviates toward affected side; moderate dysarthria, disturbance of articulation **(E)**

^{a}Isolated lesions of CN IX are uncommon; usually, CNs IX, X, and XI are involved together as they pass through the jugular foramen.

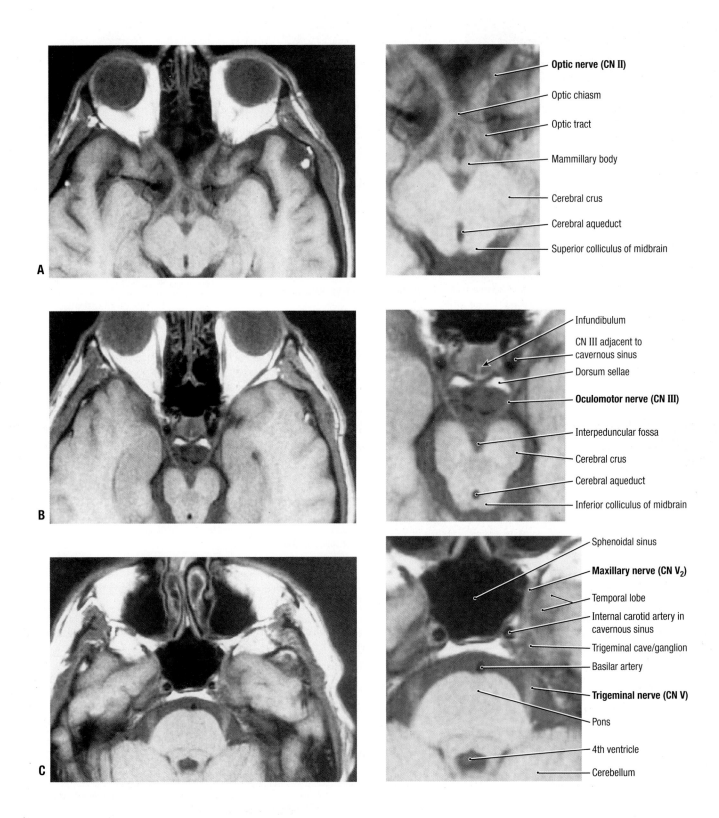

A. Optic nerve (CN II)
- Optic chiasm
- Optic tract
- Mammillary body
- Cerebral crus
- Cerebral aqueduct
- Superior colliculus of midbrain

B. Infundibulum
- CN III adjacent to cavernous sinus
- Dorsum sellae
- **Oculomotor nerve (CN III)**
- Interpeduncular fossa
- Cerebral crus
- Cerebral aqueduct
- Inferior colliculus of midbrain

C. Sphenoidal sinus
- **Maxillary nerve (CN V₂)**
- Temporal lobe
- Internal carotid artery in cavernous sinus
- Trigeminal cave/ganglion
- Basilar artery
- **Trigeminal nerve (CN V)**
- Pons
- 4th ventricle
- Cerebellum

9.23 | TRANSVERSE MRIs THROUGH HEAD, SHOWING CRANIAL NERVES

A. Optic nerve (CN II). **B.** Oculomotor nerve (CN III). **C.** Trigeminal nerve (CN V).

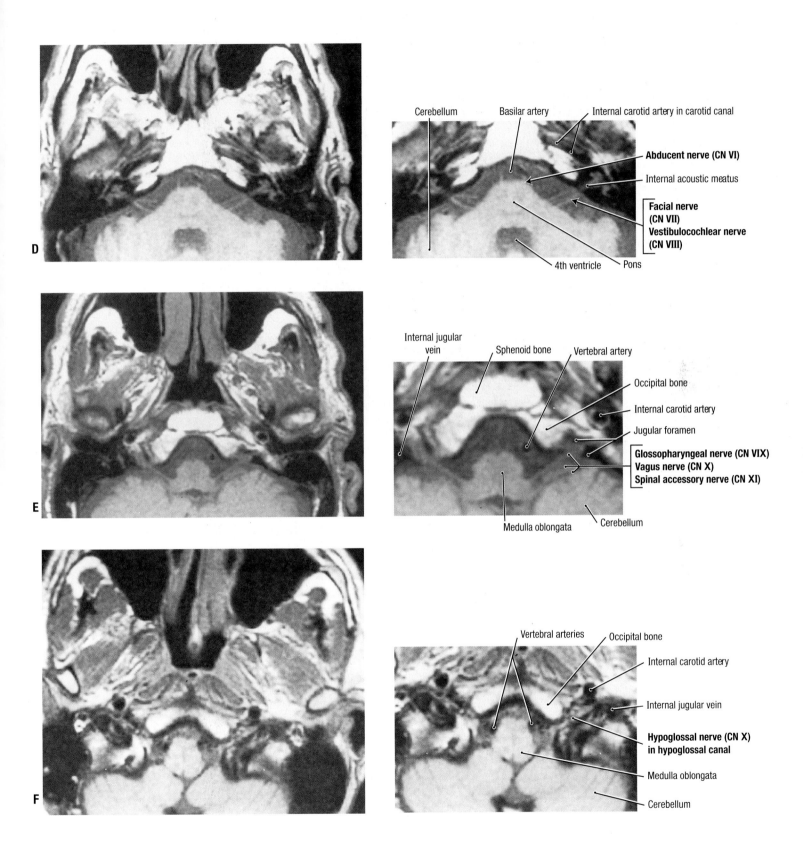

Cerebellum   Basilar artery   Internal carotid artery in carotid canal

Abducent nerve (CN VI)

Internal acoustic meatus

Facial nerve (CN VII)
Vestibulocochlear nerve (CN VIII)

4th ventricle   Pons

Internal jugular vein   Sphenoid bone   Vertebral artery

Occipital bone

Internal carotid artery

Jugular foramen

Glossopharyngeal nerve (CN VIX)
Vagus nerve (CN X)
Spinal accessory nerve (CN XI)

Medulla oblongata   Cerebellum

Vertebral arteries   Occipital bone

Internal carotid artery

Internal jugular vein

Hypoglossal nerve (CN X) in hypoglossal canal

Medulla oblongata

Cerebellum

TRANSVERSE MRIs THROUGH HEAD, SHOWING CRANIAL NERVES (*continued*)   9.23

D. Abducent (CN VI), facial (CN VII), and vestibulocochlear (CN VIII) nerves. **E.** Glossopharyngeal (CN IX), vagus (CN X), and spinal accessory (CN XI) nerves. **F.** Hypoglossal nerve (CN XII).

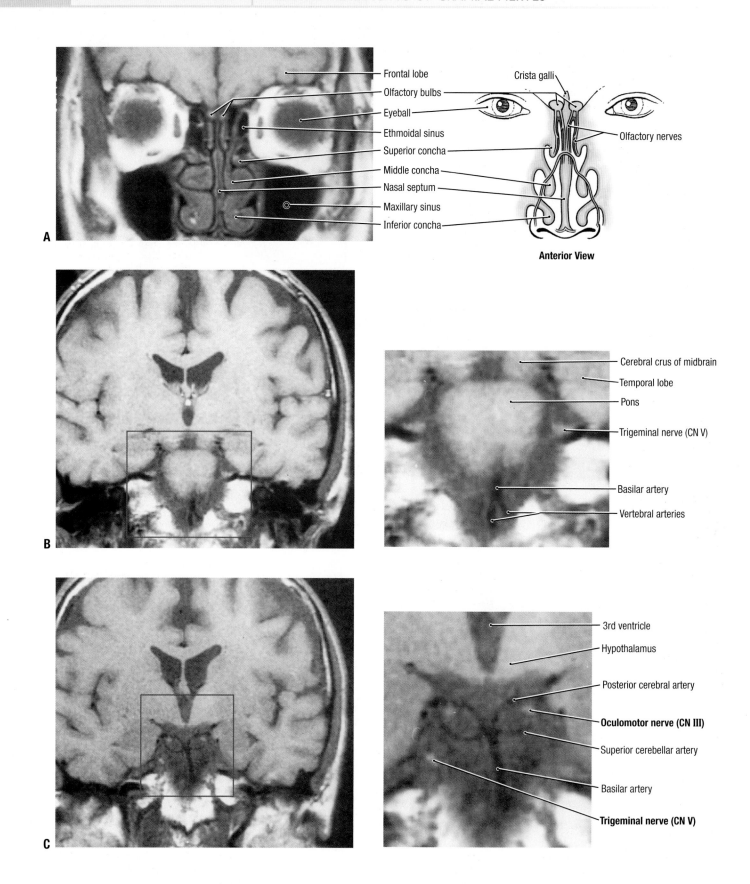

Frontal lobe
Crista galli
Olfactory bulbs
Eyeball
Olfactory nerves
Ethmoidal sinus
Superior concha
Middle concha
Nasal septum
Maxillary sinus
Inferior concha

Anterior View

Cerebral crus of midbrain
Temporal lobe
Pons
Trigeminal nerve (CN V)
Basilar artery
Vertebral arteries

3rd ventricle
Hypothalamus
Posterior cerebral artery
Oculomotor nerve (CN III)
Superior cerebellar artery
Basilar artery
Trigeminal nerve (CN V)

9.24 CORONAL MRIs THROUGH HEAD, SHOWING CRANIAL NERVES

A. Olfactory bulb. **B.** Trigeminal (CN V) nerve. **C.** Oculomotor (CN III) and trigeminal (CN V) nerves.

Index

Page numbers followed by "t" denote tables.